The Mesozoic Pacific: Geology, Tectonics, and Volcanism

Geophysical Monograph Series

Including

IUGG Volumes

Maurice Ewing Volumes

Mineral Physics Volumes

GEOPHYSICAL MONOGRAPH SERIES

Geophysical Monograph Volumes

1 **Antarctica in the International Geophysical Year** *A. P. Crary, L. M. Gould, E. O. Hulburt, Hugh Odishaw, and Waldo E. Smith (Eds.)*
2 **Geophysics and the IGY** *Hugh Odishaw and Stanley Ruttenberg (Eds.)*
3 **Atmospheric Chemistry of Chlorine and Sulfur Compounds** *James P. Lodge, Jr. (Ed.)*
4 **Contemporary Geodesy** *Charles A. Whitten and Kenneth H. Drummond (Eds.)*
5 **Physics of Precipitation** *Helmut Weickmann (Ed.)*
6 **The Crust of the Pacific Basin** *Gordon A. Macdonald and Hisashi Kuno (Eds.)*
7 **Antarctica Research: The Matthew Fontaine Maury Memorial Symposium** *H. Wexler, M. J. Rubin, and J. E. Caskey, Jr. (Eds.)*
8 **Terrestrial Heat Flow** *William H. K. Lee (Ed.)*
9 **Gravity Anomalies: Unsurveyed Areas** *Hyman Orlin (Ed.)*
10 **The Earth Beneath the Continents: A Volume of Geophysical Studies in Honor of Merle A. Tuve** *John S. Steinhart and T. Jefferson Smith (Eds.)*
11 **Isotope Techniques in the Hydrologic Cycle** *Glenn E. Stout (Ed.)*
12 **The Crust and Upper Mantle of the Pacific Area** *Leon Knopoff, Charles L. Drake, and Pembroke J. Hart (Eds.)*
13 **The Earth's Crust and Upper Mantle** *Pembroke J. Hart (Ed.)*
14 **The Structure and Physical Properties of the Earth's Crust** *John G. Heacock (Ed.)*
15 **The Use of Artificial Satellites for Geodesy** *Soren W. Henricksen, Armando Mancini, and Bernard H. Chovitz (Eds.)*
16 **Flow and Fracture of Rocks** *H. C. Heard, I. Y. Borg, N. L. Carter, and C. B. Raleigh (Eds.)*
17 **Man-Made Lakes: Their Problems and Environmental Effects** *William C. Ackermann, Gilbert F. White, and E. B. Worthington (Eds.)*
18 **The Upper Atmosphere in Motion: A Selection of Papers With Annotation** *C. O. Hines and Colleagues*
19 **The Geophysics of the Pacific Ocean Basin and Its Margin: A Volume in Honor of George P. Woollard** *George H. Sutton, Murli H. Manghnani, and Ralph Moberly (Eds.)*
20 **The Earth's Crust: Its Nature and Physical Properties** *John C. Heacock (Ed.)*
21 **Quantitative Modeling of Magnetospheric Processes** *W. P. Olson (Ed.)*
22 **Derivation, Meaning, and Use of Geomagnetic Indices** *P. N. Mayaud*
23 **The Tectonic and Geologic Evolution of Southeast Asian Seas and Islands** *Dennis E. Hayes (Ed.)*
24 **Mechanical Behavior of Crustal Rocks: The Handin Volume** *N. L. Carter, M. Friedman, J. M. Logan, and D. W. Stearns (Eds.)*
25 **Physics of Auroral Arc Formation** *S.-I. Akasofu and J. R. Kan (Eds.)*
26 **Heterogeneous Atmospheric Chemistry** *David R. Schryer (Ed.)*
27 **The Tectonic and Geologic Evolution of Southeast Asian Seas and Islands: Part 2** *Dennis E. Hayes (Ed.)*
28 **Magnetospheric Currents** *Thomas A. Potemra (Ed.)*
29 **Climate Processes and Climate Sensitivity (Maurice Ewing Volume 5)** *James E. Hansen and Taro Takahashi (Eds.)*
30 **Magnetic Reconnection in Space and Laboratory Plasmas** *Edward W. Hones, Jr. (Ed.)*
31 **Point Defects in Minerals (Mineral Physics Volume 1)** *Robert N. Schock (Ed.)*
32 **The Carbon Cycle and Atmospheric CO_2: Natural Variations Archean to Present** *E. T. Sundquist and W. S. Broecker (Eds.)*
33 **Greenland Ice Core: Geophysics, Geochemistry, and the Environment** *C. C. Langway, Jr., H. Oeschger, and W. Dansgaard (Eds.)*
34 **Collisionless Shocks in the Heliosphere: A Tutorial Review** *Robert G. Stone and Bruce T. Tsurutani (Eds.)*
35 **Collisionless Shocks in the Heliosphere: Reviews of Current Research** *Bruce T. Tsurutani and Robert G. Stone (Eds.)*
36 **Mineral and Rock Deformation: Laboratory Studies—The Paterson Volume** *B. E. Hobbs and H. C. Heard (Eds.)*
37 **Earthquake Source Mechanics (Maurice Ewing Volume 6)** *Shamita Das, John Boatwright, and Christopher H. Scholz (Eds.)*
38 **Ion Acceleration in the Magnetosphere and Ionosphere** *Tom Chang (Ed.)*
39 **High Pressure Research in Mineral Physics (Mineral Physics Volume 2)** *Murli H. Manghnani and Yasuhiko Syono (Eds.)*
40 **Gondwana Six: Structure Tectonics, and Geophysics** *Gary D. McKenzie (Ed.)*

41 **Gondwana Six: Stratigraphy, Sedimentology, and Paleontology** *Garry D. McKenzie (Ed.)*

42 **Flow and Transport Through Unsaturated Fractured Rock** *Daniel D. Evans and Thomas J. Nicholson (Eds.)*

43 **Seamounts, Islands, and Atolls** *Barbara H. Keating, Patricia Fryer, Rodey Batiza, and George W. Boehlert (Eds.)*

44 **Modeling Magnetospheric Plasma** *T. E. Moore and J. H. Waite, Jr. (Eds.)*

45 **Perovskite: A Structure of Great Interest to Geophysics and Materials Science** *Alexandra Navrotsky and Donald J. Weidner (Eds.)*

46 **Structure and Dynamics of Earth's Deep Interior (IUGG Volume 1)** *D. E. Smylie and Raymond Hide (Eds.)*

47 **Hydrological Regimes and Their Subsurface Thermal Effects (IUGG Volume 2)** *Alan E. Beck, Grant Garven, and Lajos Stegena (Eds.)*

48 **Origin and Evolution of Sedimentary Basins and Their Energy and Mineral Resources (IUGG Volume 3)** *Raymond A. Price (Ed.)*

49 **Slow Deformation and Transmission of Stress in the Earth (IUGG Volume 4)** *Steven C. Cohen and Petr Vaníček (Eds.)*

50 **Deep Structure and Past Kinematics of Accreted Terranes (IUGG Volume 5)** *John W. Hillhouse (Ed.)*

51 **Properties and Processes of Earth's Lower Crust (IUGG Volume 6)** *Robert F. Mereu, Stephan Mueller, and David M. Fountain (Eds.)*

52 **Understanding Climate Change (IUGG Volume 7)** *Andre L. Berger, Robert E. Dickinson, and J. Kidson (Eds.)*

53 **Plasma Waves and Istabilities at Comets and in Magnetospheres** *Bruce T. Tsurutani and Hiroshi Oya (Eds.)*

54 **Solar System Plasma Physics** *J. H. Waite, Jr., J. L. Burch, and R. L. Moore (Eds.)*

55 **Aspects of Climate Variability in the Pacific and Western Americas** *David H. Peterson (Ed.)*

56 **The Brittle-Ductile Transition in Rocks** *A. G. Duba, W. B. Durham, J. W. Handin, and H. F. Wang (Eds.)*

57 **Evolution of Mid Ocean Ridges (IUGG Volume 8)** *John M. Sinton (Ed.)*

58 **Physics of Magnetic Flux Ropes** *C. T. Russell, E. R. Priest, and L. C. Lee (Eds.)*

59 **Variations in Earth Rotation (IUGG Volume 9)** *Dennis D. McCarthy and Williams E. Carter (Eds.)*

60 **Quo Vadimus *Geophysics for the Next Generation* (IUGG Volume 10)** *George D. Garland and John R. Apel (Eds.)*

61 **Cometary Plasma Processes** *Alan D. Johnstone (Ed.)*

62 **Modeling Magnetospheric Plasma Processes** *Gordon K. Wilson (Ed.)*

63 **Marine Particles Analysis and Characterization** *David C. Hurd and Derek W. Spencer (Eds.)*

64 **Magnetospheric Substorms** *Joseph R. Kan, Thomas A. Potemra, Susumu Kokubun, and Takesi Iijima (Eds.)*

65 **Explosion Source Phenomenology** *Steven R. Taylor, Howard J. Patton, and Paul G. Richards (Eds.)*

66 **Venus and Mars: Atmospheres, Ionospheres, and Solar Wind Interactions** *Janet G. Luhmann, Mariella Tatrallyay, and Robert O. Pepin (Eds.)*

67 **High-Pressure Research: Application to Earth and Planetary Sciences (Mineral Physics Volume 3)** *Yasuhiko Syono and Murli H. Manghnani (Eds.)*

68 **Microwave Remote Sensing of Sea Ice** *Frank Carsey, Roger Barry, Josefino Comiso, D. Andrew Rothrock, Robert Shuchman, W. Terry Tucker, Wilford Weeks, and Dale Winebrenner*

69 **Sea Level Changes: Determination and Effects (IUGG Volume 11)** *P. L. Woodworth, D. T. Pugh, J. G. DeRonde, R. G. Warrick, and J. Hannah*

70 **Synthesis of Results from Scientific Drilling in the Indian Ocean** *Robert A. Duncan, David K. Rea, Robert B. Kidd, Ulrich von Rad, and Jeffrey K. Weissel (Eds.)*

71 **Mantle Flow and Melt Generation at Mid-Ocean Ridges** *Jason Phipps Morgan, Donna K. Blackman, and John M. Sinton (Eds.)*

72 **Dynamics of Earth's Deep Interior and Earth Rotation (IUGG Volume 12)** *Jean-Louis Le Mouël, D.E. Smylie, and Thomas Herring (Eds.)*

73 **Environmental Effects on Spacecraft Positioning and Trajectories (IUGG Volume 13)** *A. Vallance Jones (Ed.)*

74 **Evolution of the Earth and Planets (IUGG Volume 14)** *E. Takahashi, Raymond Jeanloz, and David Rubie (Eds.)*

75 **Interactions Between Global Climate Subsystems: The Legacy of Hann (IUGG Volume 15)** *G. A. McBean and M. Hantel (Eds.)*

76 **Relating Geophysical Structures and Processes: The Jeffreys Volume (IUGG Volume 16)** *K. Aki and R. Dmowska (Eds.)*

Maurice Ewing Volumes

1 **Island Arcs, Deep Sea Trenches, and Back-Arc Basins** *Manik Talwani and Walter C. Pitman III (Eds.)*

2 **Deep Drilling Results in the Atlantic Ocean: Ocean Crust** *Manik Talwani, Christopher G. Harrison, and Dennis E. Hayes (Eds.)*

3 **Deep Drilling Results in the Atlantic Ocean: Continental Margins and Paleoenvironment** *Manik Talwani, William Hay, and William B. F. Ryan (Eds.)*

4 **Earthquake Prediction—An International Review** *David W. Simpson and Paul G. Richards (Eds.)*

5 **Climate Processes and Climate Sensitivity** *James E. Hansen and Taro Takahashi (Eds.)*

6 **Earthquake Source Mechanics** *Shamita Das, John Boatwright, and Christopher H. Scholz (Eds.)*

IUGG Volumes

1 **Structure and Dynamics of Earth's Deep Interior** *D. E. Smylie and Raymond Hide (Eds.)*

2 **Hydrological Regimes and Their Subsurface Thermal Effects** *Alan E. Beck, Grant Garven, and Lajos Stegena (Eds.)*

3 **Origin and Evolution of Sedimentary Basins and Their Energy and Mineral Resources** *Raymond A. Price (Ed.)*

4 **Slow Deformation and Transmission of Stress in the Earth** *Steven C. Cohen and Petr Vaníček (Eds.)*

5 **Deep Structure and Past Kinematics of Accreted Terrances** *John W. Hillhouse (Ed.)*

6 **Properties and Processes of Earth's Lower Crust** *Robert F. Mereu, Stephan Mueller, and David M. Fountain (Eds.)*

7 **Understanding Climate Change** *Andre L. Berger, Robert E. Dickinson, and J. Kidson (Eds.)*

8 **Evolution of Mid Ocean Ridges** *John M. Sinton (Ed.)*

9 **Variations in Earth Rotation** *Dennis D. McCarthy and William E. Carter (Eds.)*

10 **Quo Vadimus Geophysics for the Next Generation** *George D. Garland and John R. Apel (Eds.)*

11 **Sea Level Changes: Determinations and Effects** *Philip L. Woodworth, David T. Pugh, John G. DeRonde, Richard G. Warrick, and John Hannah (Eds.)*

12 **Dynamics of Earth's Deep Interior and Earth Rotation** *Jean-Louis Le Mouël, D.E. Smylie, and Thomas Herring (Eds.)*

13 **Environmental Effects on Spacecraft Positioning and Trajectories** *A. Vallance Jones (Ed.)*

14 **Evolution of the Earth and Planets** *E. Takahashi, Raymond Jeanloz, and David Rubie (Eds.)*

15 **Interactions Between Global Climate Subsystems: The Legacy of Hann** *G. A. McBean and M. Hantel (Eds.)*

16 **Relating Geophysical Structures and Processes: The Jeffreys Volume** *K. Aki and R. Dmowska (Eds.)*

Mineral Physics Volumes

1 **Point Defects in Minerals** *Robert N. Schock (Ed.)*

2 **High Pressure Research in Mineral Physics** *Murli H. Manghnani and Yasuhiko Syona (Eds.)*

3 **High Pressure Research: Application to Earth and Planetary Sciences** *Yasuhiko Syono and Murli H. Manghnani (Eds.)*

Geophysical Monograph 77

The Mesozoic Pacific: Geology, Tectonics, and Volcanism

A Volume in Memory of Sy Schlanger

Malcolm S. Pringle
William W. Sager
William V. Sliter
Seth Stein
Editors

American Geophysical Union

Published under the aegis of the AGU Books Board.

Library of Congress Cataloging-in-Publication Data

The Mesozoic Pacific : geology, tectonics, and volcanism / Malcolm S.
Pringle . . . [et al.] editors.
p. cm. — (Geophysical monograph, ISSN 0065-8448 ; 77)
Includes bibliographical references.
ISBN 0-87590-036-4 (alk. paper)
1. Geology, Stratigraphic—Mesozoic. 2. Geology—Pacific Ocean.
I. Pringle, Malcolm S. II. Series.
QE675.M38 1993
551.7'6'01964—dc20 93-33017
CIP

ISSN: 0065-8448

ISBN: 0-87590-036-4

This book is printed on acid-free paper. ♾

Printed in the United States of America.

CONTENTS

CONTENTS

PREFACE

This volume provides an interdisciplinary summary of current research on the history and evolution of the Mesozoic Pacific Ocean. The importance of the work lies in not only describing the early history of the largest modern tectonic plate, but also in studying the geologic record and possible causes of the Cretaceous Greenhouse. Thus, it offers important lessons as we try to understand modern man's impact on the interaction between the atmosphere, oceans, and solid earth.

The idea for this summary originated nearly five years ago, during a discussion between Sy Schlanger and Malcolm Pringle on the need to bring together various groups working on Cretaceous Pacific volcanism. After the untimely death of Sy Schlanger in the summer of 1990, it was decided to dedicate this work to him. Because of the breadth of Sy Schlanger's research and interests, the scope of the volume was broadened from Cretaceous Pacific volcanism in particular to the history of the Mesozoic Pacific Ocean as a whole. It is a testament to Sy's impact on Pacific research that he made a significant contribution to the scientific problems addressed by almost every paper included herein.

The volume begins with a summary of Sy's contributions to the geology of the Pacific Ocean, followed by sections on the Mesozoic Ocean, the Pacific Ocean seafloor, and Cretaceous Pacific seamounts and plateaus.

The Mesozoic Ocean

In the lone paper on Pacific paleoceanography, Bralower et al. use high-resolution foraminiferal and nannofossil biostratigraphy to conclude that the long Albian-Aptian Oceanic Anoxic Event consisted of at least three distinct, relatively brief episodes of widespread dysoxia/anoxia within generally oxygenated oceanographic conditions. These events seem to correlate with global sea-level transgressions and volcanism. The best developed of these episodes occurred in the lower Aptian, apparently contemporaneous with the eruption of the Ontong Java Plateau flood basalts. The other episodes occurred in the early Albian and early late Albian, and may have been more sensitive to regional and/or local factors.

The Pacific Ocean Seafloor

Bathymetry, Heat Flow, and Seismic Stratigraphy

The first two papers in this section examine the relationship between intraplate volcanism and seafloor depth anomalies. Davies and Pribac use a half-space model to test for depth anomalies with causes other than simple conductive cooling. They find a superswell 9000 km long in the central and western Pacific, including both the Cretaceous Darwin Rise and the modern volcanoes of French Polynesia. They argue that some depth anomalies can persist for hundreds of millions of years, in regions where shallow, sub-lithospheric plume-derived material and deep mantle material are relatively warm due to the absence of subduction for more than 200 m.y.

Stein and Stein use a plate model with a hotter and thinner lithosphere than previous reference models. They find that mid-plate swells such as Hawaii, the area including French Polynesia, and the Cretaceous Darwin Rise have heat flow values consistent with (or only slightly higher than) unperturbed lithosphere of the same age, in spite of the current (Hawaii and French Polynesia) and original (Darwin Rise) depth anomalies. They favor a primarily dynamic rather than thermal origin for these swells, and suggest that the anomalously thin elastic thickness of the lithosphere found under many of the modern volcanoes of French Polynesia and the Cretaceous seamounts of the Darwin Rise is the result of mechanical, rather than thermal, weakening.

The extent of Cretaceous volcanism covering the Jurassic ocean crust of the western Pacific has been examined with multichannel seismic reflection techniques calibrated with deep sea drilling results. In the East Mariana and southeast Pigafetta Basins, Abrahms et al. document 500,000 km^2 of middle Cretaceous basaltic flows and sills that overlie Jurassic/Early Cretaceous sediments and ocean crust. In the northern Nauru Basin, Shipley et al. find similarly extensive Cretaceous volcanism, but are unable to find definitive indications of the Jurassic crust that should lie beneath. They do find Jurassic age crust in the southern Nauru Basin, but no evidence for extensive Cretaceous volcanism.

Nakanishi describes the location and features of five fracture zones in the northwestern Pacific that previously were inferred only from offsets between magnetic anomaly lineations. Possible deviations from the classic transform fault/fracture zone model include volcanism associated with plate reorganization, Cretaceous intraplate sources, and rift propagation.

Tectonic History of the Late Cretaceous Pacific

The tectonic history of the Late Cretaceous Pacific plate is difficult to decipher because of (1) the lack of seafloor magnetic lineations in crust formed during the Cretaceous Long Normal Superchron, (2) apparent mismatches in the geometries of the boundaries of the Quiet Zone crust, and (3) an apparent excess in the amount of ocean crust formed during the Superchron. Three papers examine this problem by looking at fracture zone morphology. Atwater et al. examine the north and central Pacific, and suggest a significant northern relocation of the Mendocino Fracture Zone formed during the Superchron. Searle et al. report a detailed analysis of the Molokai Fracture Zone near Hawaii. Joseph et al. look at the central and southern Pacific, and conclude that the Nova-Canton Trough is the middle Cretaceous extension of the Clipperton Fracture Zone.

All three papers emphasize a Late Cretaceous reorganization of spreading at the Pacific/Farallon ridge. Searle et al. develop a detailed tectonic model for the growth of 'leaky' transforms following transtensive changes in spreading direction. Atwater et al. require one or more major spreading-ridge jumps to explain geometric differences in the boundaries of the Quiet Zone crust and argue that the jumps occurred early in the Superchron. Joseph et al. concur with previous suggestions that a southward jump of the Pacific-Phoenix ridge, to the location of initial volcanism on the Manihiki Plateau, trapped portions of the Phoenix plate on the Pacific plate. They also suggest that the eastern edges of the Manihiki Plateau are rifted margins of the Pacific-Farallon spreading system.

Cretaceous Pacific Seamounts and Plateaus

Geochronology and Petrology

The hot spot hypothesis describing the origin of ocean island and seamount chains has been one of the most successful predictive tools of plate tectonic theory. Pringle uses the rate of age progressive volcanism along a NW-SE trending chain of Musicians Seamounts to estimate the velocity of the Pacific plate from 96 to 82 Ma. Garcia et al. examine the geochemistry of lavas from the Line Islands chain, and suggest that much of the chain could have originated from multiple hot spot sources originally located in the northeastern portion of the South Pacific Isotopic and Thermal Anomaly (SOPITA). Similarly, Lincoln et al. document several episodes of Early and Late Cretaceous volcanism and reef-building in the Marshall Islands that could be explained by the passage of the region over several hot spots, some of which may still be active today in the central and western SOPITA.

Flood basalts are thought to represent the initial outpouring of lavas associated with the beginning of hot spot volcanic chains. Mahoney et al. examine lavas from the largest of these, the Ontong Java Plateau, and conclude that the bulk of the plateau was constructed rapidly above a surfacing mantle plume head. However, at least some volcanism continued on the plateau for approximately 30 m.y. Also, the isotopic signature of lavas from the presumed hot spot trail (the Louisville Ridge) is distinct from the plateau basalts and may signal a change in mantle source between plume head and plume tail. Pringle and Dalrymple find that ages of lavas from the Hess Rise and the Wentworth Seamount Chain are consistent with a plume head/plume tail origin for the pair, but that the hot spot track only lasted 14 m.y. Also, active volcanism continued on Hess Rise for at least 12 m.y. Thus, continuing volcanism may be characteristic of oceanic flood basalts, even if they form by a mechanism consistent with the plume initiation model.

Geology and Geophysics of Western Pacific Guyots

Winterer et al. and van Waasbergen and Winterer conclude that pre-Albian seamounts in the northwestern Pacific evolved into atolls with up to 800 m of limestone cap. Albian edifices only developed fringing- and barrier-reef deposits before a regional emergence led to the cessation of shallow-water-limestone accumulation. Post-Albian guyots formed mainly by sea-level erosion, without significant reef development. Modeling of gravity data shows that many of these Cretaceous guyots formed on crust with anomalously thin elastic plate thickness, similar to the volcanoes of French Polynesia.

Bergersen shows that multiple episodes of volcanism, uplift, platform growth, block faulting, and mass wasting can create a complicated picture of guyot formation and evolution. Bryan et al. use seamount paleomagnetic models to estimate a Late Cretaceous age for Loen Guyot in the Marshall Islands. Sager et al. use similar techniques to show that most of the Japanese and Wake seamounts formed during the Cretaceous Long Normal Superchron, with some older volcanism but little younger volcanism north of the Marshall Islands. The current data can not distinguish between models of multiple hot spot and/or regional, episodic volcanism for the origin of seamounts and guyots in the western Pacific.

Suggestions for Future Work

An interdisciplinary summary of current research sometimes leads to an integrated synthesis that will stand the test of time. More often than not, however, even more questions present themselves for study. The Mesozoic Pacific will continue as a fruitful focus in the study of the dynamic earth, and the studies in this volume suggest some of the next steps in that pursuit:

-- Just how closely is mid-Cretaceous volcanism of the western Pacific linked to oceanographic changes such as the Aptian-Albian dysoxia/anoxia episodes, global sea-level change, and, ultimately, the development of the Cretaceous Greenhouse? Are the ocean-wide Albian dysoxia/anoxia episodes related to the basin-wide late Albian emergence of guyots?

-- Can we reconcile different models for the origin and duration of mantle thermal anomalies responsible for mid-plate swells?

-- Do tectonic models of 'leaky' transform faults associated with spreading direction change describe the origin of volcanic ridges approximately parallel to spreading direction, such as Horizon, Necker, and those in the Musicians Seamounts?

-- Does the incorporation of pieces of Farallon and Phoenix plate into Cretaceous Pacific Quiet Zone crust significantly reduce estimates of excess ocean crust formed during that time?

-- Are rifted margins common on the edges of oceanic plateaus, and what are the implications of such margins for the plume-initiation model of the formation of oceanic flood basalts?

-- Will we be able to document age progressive seamount chains and Pacific plate velocity for periods older than 96 Ma, and between 82 and 70 Ma?

Many of the studies presented here began as site surveys for the Atolls and Guyots Project of the Ocean Drilling Program, which successfully completed drilling during Legs 143 and 144 last year. Some of the scientists aboard those legs have added "Notes added in proof" to their manuscripts in this volume; undoubtedly some of the final reports from those legs will modify some of the conclusions presented here.

However, answering some of these questions will take much more extensive work. Most will require studies using techniques involving all of the geological disciplines; answers will have equally wide-ranging impact. Rather than being content with the completion of this volume, we can easily imagine that Sy Schlanger would have been busy designing new experiments to test the hypotheses proposed here.

Malcolm S. Pringle
William W. Sager
William V. Sliter
Seth Stein
Editors

S.O. Schlanger and his Role in Western Pacific Ocean Geology

EDWARD L. WINTERER

Scripps Institution of Oceanography, La Jolla, CA 92093

Papers in this monograph honor the late Seymour O. (Sy) Schlanger, who for a period of more than 35 years, from 1955 to his death in 1990, was at the forefront of research on the geology of the Pacific Ocean. His work touched almost every strand of Pacific research, from biogeography through sedimentology and stratigraphy to volcanology and tectonics, and his influence is felt in all these branches of earth science. He saw the strands as all woven into a seamless whole, and this unity of vision expresses itself repeatedly in the dozens of papers he contributed to the literature.

Introduction to Pacific Geology

At the very beginning of his professional career, Sy was introduced to Pacific Geology at its most complex: he was recruited by the U.S. Geological Survey from graduate school (Rutgers) in 1952 to join the World War II stellar team of military geologists then working on the geology of Guam, under the leadership of Harry Ladd and Josh Tracey. Here was a whole array of special rocks: Eocene-Miocene island arc pillow basalts intercalated with pelagic sediments, and encrusted with a Pleistocene reef, in the form of a barrier reef with all its requisite pieces of anatomy — lagoon, reef rim, outer slope — all uplifted and exposed to study. Sy plunged in and became the master of most of the major oceanic sedimentary facies (Schlanger, 1957; Forman and Schlanger, 1957; Schlanger, 1964; Tracey and Schlanger, 1964; Hathaway and Schlanger, 1965; Garrison et al., 1975).

Shallow-water carbonate diagenesis and sea level

It was on Guam and on a nearby atoll where he was sent to study reefal hydrology, that Sy came to realize the importance of meteoric-water diagenesis, which left imprints in the petrology and geochemistry of the reefal limestone (Schlanger and Brookhart, 1955; Schlanger, 1964). During emersion, fresh water alters the limestone in the phreatic zone, yielding a decipherable record of the history of relative sea level fluctuations. On returning to Washington, D.C., Sy was assigned the task of describing and interpreting the drill cores from the deep holes drilled at Enewetak Atoll, in the western part of the Marshall Islands, in connection with the nuclear bomb tests. His descriptions and interpretations of the cores are models of scholarship, reflecting painstaking observation and tightly controlled logic (Ladd and Schlanger, 1961; Schlanger, 1963; this work also formed part of his doctoral theses, under Francis Pettijohn, at the Johns Hopkins). He recognized a number of "solution unconformities" in the Eocene - Recent sequence of shallow-water limestones, each showing evidence of a relative fall of sea level that exposed the sediments to meteoric-water dissolution, recrystallization, and cementation. Sy then went on the correlate these emergence events from Enewetak to Bikini Atoll, a few hundred km east in the Marshalls, and then to Midway Atoll, in the Hawaiian Islands chain. Sy concluded that these widely separated reefs preserved a record of eustatic sea level fluctuations, though each atoll had its own background rate of tectonic subsidence.

Confirmation of Schlanger's model has come in recent years by extensive drilling on Enewetak Atoll, in which it was demonstrated first that numerous correlatable solution unconformities punctuate the Pleistocene strata of the reef plate and the adjacent lagoon, and then that the seismic reflectors traceable across the lagoon in Miocene-Recent sediments are associated with acoustic impedance contrasts between poorly cemented limestones and better-cemented limestones, which were diagenetically altered at times of atoll emergence [*Henry and Wardlaw*, 1986; *Grow et al.*, 1986]. This is a powerful tool in stratigraphy, and is being exploited in the JOIDES drilling of Cretaceous sunken atolls in the western Pacific. Schlanger was one of the principal architects of the experimental plan for this drilling.

Schlanger's interests in the geology and petrology of reefs extended to southeast Asia [*Fulthorpe and Schlanger*, 1989], The Ryukyu Islands [*Konishi et al.*, 1970; *Schlanger and Konishi*, 1975], and the Hawaiian chain [*Rotondo et al.*, 1981; *Lincoln and Schlanger*, 1986; *Lincoln and Schlanger*, in press]. In the Emperor chain, drilling during Leg 55 [*McKenzie et al.*, 1980] showed that the drowned Paleocene carbonate bank that caps the Suiko seamount is made of coralline algae and bryozoans, rather than hermatypic corals that *Schlanger and Konishi* [1975] had

The Mesozoic Pacific: Geology, Tectonics, and Volcanism
Geophysical Monograph 77

shown should be expected at the latitude where the volcanoes were believed to have formed, i.e., at the 19°N latitude of Kilauea volcano, the modern hotspot at the end of the Hawaiian-Emperor chain. The implication of this finding is to this day generally unappreciated by Pacific geologists and geophysicists: either the hotspot has migrated latitudinally since Paleocene times, as suggested also by the 27°N paleolatitude indicated by the inclination of remanent magnetism in cored basalt samples from Suiko [*Kono*, 1980], or the biogeographic boundary has moved significantly.

Deep-water carbonate diagenesis and paleoceanography : DSDP

Schlanger's first experience in deep-water ocean drilling was on Leg 17, in the Central Pacific, and he participated later, as co-chief scientist, on Leg 33 [*Jackson and Schlanger*, 1976; *Schlanger and Winterer,* 1976] Leg 61, and Leg 89. From study of the cores from the thick pelagic limestone recovered on Magellan Plateau during Leg 17, he and Bob Douglas helped lay the foundations for understanding the progressive burial diagenesis of pelagic carbonate sediments, and introduced the concept of diagenetic potential, that is the length of the diagenetic path still potentially to be realized by a layer of newly deposited sediment [*Coplen and Schlanger*, 1973; *Schlanger et al*., 1973; *Schlanger and Douglas*, 1974]. Some layers are so dissolved or mixed with clay at the sea floor that they have little remaining potential; others may become buried with unstable components, such as aragonite, still unaltered. They then applied this concept to explaining the origin of seismic reflectors, arguing that the high-potential layers become more indurated during burial than the interbedded low-potential layers, leading to acoustic impedance contrasts. They then dated biostratigraphically the most prominent reflectors in the central Pacific Cenozoic pelagic carbonates and saw that they correlated over long distances. They speculated that simultaneous oceanographic changes, such as changes in the depth of the calcium carbonate compensation depth, were responsible for the changing diagenetic potentials. Schlanger followed up on these early studies with a series of papers relating the physical and petrographic properties of pelagic carbonates to their acoustic response [*Milholland et al*., 1980; *Manghnani* et al., 1981; *Kim et al*., 1983; *Kim et al*., 1985; *Fulthorpe et al*., 1989], and on the effects on the global budget of strontium by the deposition and dissolution of marine carbonates during sea level rise and fall [*Schlanger*, 1988].

His predictions have been amply borne out by subsequent drilling, where the effects of original bulk density (a proxy for carbonate content) are seen to dominate impedance in the upper part of the column, while sound velocity comes into play in more deeply buried parts [*Mayer et al*., 1986]. The most astonishing result has come from comparison of the detailed sea-level history worked out on Enewetak Atoll with the record of oceanographic changes in the pelagic carbonates: there is apparent correspondence [*Wardlaw*, 1989].

Black shale

At an earlier stage in his career, while working in the Cretaceous strata of the Reconcavo Basin of Brazil [*Murphy and Schlanger*, 1965], Sy encountered the black shale problem, a problem he turned almost all his energies to for nearly a decade, beginning in 1979 [*Arthur and Schlanger*, 1979; *Schlanger and Cita*, 1982; *Arthur et al*., 1985; *Arthur et al*., 1987; *Schlanger et al*., 1987]. Black shales show up in the mid-Cretaceous on Pacific oceanic plateaus (Manihiki, Ontong Java), while correlative sediments in deep-water basins are red. This was regarded as indicating that anoxia was confined to an intensified oxygen-minimum zone, whereas in the Atlantic oxygen-poor conditions extended in some places to the deep sea floor [*Schlanger and Jenkyns*, 1976; *Winterer*, 1977].

Volcanic seamount chains

His focus on atolls led Sy naturally to their volcanic foundations, and especially to the plate tectonic significance of the many seamount chains in the Pacific. He was one of the guiding spirits in organizing an early test of the hotspot hypothesis, during Leg 33 of the Deep Sea Drilling Project, when several holes were drilled close to the Line Islands Chain. Sy reasoned that a nearly complete record of volcanism, reef-building, and reef drowning could be obtained by drilling into the flanking aprons of interbedded pelagic and turbidite sediments. The result was ambiguous [*Jackson and Schlanger*, 1976]: the oldest reefal sediments were of sensibly the same Late Cretaceous age (*ca.* 85 Ma) at holes some 400 km apart along the chain. Later radiometric dating of dredged and drilled basalts along the chain were interpreted [*Schlanger et al*., 1984] as showing progressive volcanism, in spite of great scatter in the data.

The use of oceanic carbonate banks such as atolls and guyots as monitors of relative sea level was one of the main themes of Schlanger's research, for the rest of his life [*Schlanger*, 1986], but he progressively broadened this theme to include the more general question of the history of vertical movements of the sea floor, combining the effects of eustatism and local and regional tectonism. The work in the Line Islands had opened up the possibility of multiple episodes of volcanism along the same chain, at very widely separated times, in contrast to the modest lag times in modern chains. Radiometric data from the Line Islands pointed to Eocene volcanism (*ca.* 40 Ma) overprinting the Late Cretaceous edifice building, and some of the dredged rocks have peperite-like textures involving Eocene pelagic sediments and basalt fragments [*Haggerty et al*., 1982]. Repeated volcanism also raised the possibility of creation of a thermal bulge, and hence of tectonically induced changes of relative sea level recorded along with eustatic changes in the reefal caps [*Hsü and Schlanger*, 1968; *Schlanger*, 1981; *Schlanger and Premoli Silva*, 1982; *Schlanger et al*., 1981; *Schlanger and Premoli Silva*, 1986; *Schlanger et al*., 1987]. Sy's work in the Marshall Islands chain has demonstrated that as many as three different ages of reefs may be stacked on a single seamount, with resultant complexity in the vertical trajectories of such seamounts. Simple age-*vs*.-depth equations are simply inapplicable.

Mid-plate flood basalts

Drilling along the flanks of the Marshall Islands and Line Islands chains during DSDP Leg 17 and Leg 33 confirmed the importance of emplacement of large volumes of flood basalts on the deep ocean floor, especially in the vicinity of seamount chains, where the basalts form a major component of the archipelagic apron. Seismic reflection profiling in the western Pacific, especially in the Marianas and Nauru Basins, had shown [*Houtz and Ludwig*, 1979] that acoustic basement was abnormally flattish over very large regions, and the preliminary drill results suggested this might be mid-plate flood basalt, separated from underlying rough oceanic basement by an interval of sediments.

Sy and Roger Larson attacked this problem head-on in Leg 61, by drilling a deep hole in the Nauru Basin, where a seemingly bottomless pile of mid-plate, Aptian (possibly Albian) basalt was encountered beneath the regional flattish acoustic basement reflector [*Batiza et al.*, 1980; *Jenkyns and Schlanger*, 1981; *Larson and Schlanger*, 1981a; *Larson and Schlanger*, 1981b; *Schlanger and Premoli Silva*, 1982]. Sy and Ralph Moberly reconfirmed these results during Leg 89 in the Marianas basin, where thick volcanoclastic sediments occur beneath the reflector [*Moberly et al.*, 1983; *Moberly et al.*, 1986; *Schlanger and Moberly*, 1986]. Only during Leg 129, after very assiduous pre-cruise reflection seismic surveying, was it possible to drill through a window in these volcanic rocks and sample a normal section of pelagic sediments to Jurassic basement (Shipboard Scientific Party, 1990). It was owing in large measure to Sy's careful definition of the problem and his willingness to spend enormous effort in trying to drill *through* the mid-plate basalts, that ultimate success came to those who followed his lead.

At the time of his death Sy and coworkers had just completed a major synthesis, for JOIDES, of all the data on Western Pacific mid-plate volcanic rocks, and this great compendium serves as a capstone on the monumental contributions he made to Pacific geology.

I cannot end without a few personal reflections on the man himself, for he was my friend. He was an intellectual omnivore, devouring and relishing everything that came into his orbit. His reading ranged insatiably over an astonishing spread of material, from Mad Comics to Wittgenstein. While the evidence of his scientific curiosity and his logical powers is preserved in his writings and the accomplishments of students he trained, the evidence of his powerful intellect is in the personal memories of the host of students, colleagues, and friends that shared his company. Warmly convivial, he delighted in conversation on political and moral philosophy (he was no political partisan), history, literature, the arts, and, of course, geology. He was a sensitive seismometer for detecting sham and pretence, but reacted more with gentle amusement than with censure. Finally, he was a wise man, the kind that students and younger colleagues seek out for career advice, that scientific peers trust to chair important planning groups, and that senior colleagues and Deans come to for counsel.

REFERENCES

Arthur, M. and S.O. Schlanger, Cretaceous "Oceanic Anoxic Events" as causal factors in development of reef-reservoired giant oil fields, *Am. Assoc. Pet. Geol. Bull.*, *63*, 870-885, 1979.

Arthur, M. A., W.E. Dean and S.O. Schlanger, Variations in the global carbon cycle during the Cretaceous related to climate, volcanism, and changes in atmospheric CO_2, *in*: *The Carbon Cycle and Atmospheric CO_2: Natural Variations Archean to Present* (Sundquist, E.T. and W.S. Broeker, eds.), Geophys. Mon. 32, Amer. Geophys. Union, Washington, D.C., 504-530, 1985.

Arthur, M.A., S.O. Schlanger, H.C. Jenkyns, The Cenomanian-Turonian Anoxic Event, II. Paleoceanographic controls on organic matter production and preservation, in: *Marine Petroleum Source Rocks* (J. Brooks and A. Fleet, eds.), Geol. Soc. London Spec. Pub. No. 26, 401-420, 1987.

Batiza, R., R.L. Larson, S.O. Schlanger, S.A. Sheka and H.Tokuyama, Trace element abundances in basalts of Nauru Basin (DSDP Leg 61): Late Cretaceous off-ridge volcanism in the Western Pacific, *Nature*, *286*, 476-478, 1980.

Coplen, T.C. and S.O. Schlanger, Oxygen and carbon isotope studies of carbonate sediments from site 167, Magellan Rise, Leg 17, *in* Winterer, E.L., et al., *Initial Reports Deep Sea Drilling Project*, *17*, 505-509, Wash. D.C., 1973.

Forman, M.J. and S.O. Schlanger, Tertiary reef and associated limestone facies from Louisiana and Guam, *Jour. Geol.*, *65*, 611-627, 1957.

Fulthorpe, C.S., S.O. Schlanger and R.D. Jarrard, In situ acoustic properties of pelagic carbonate sediments on the Ontong-Java Plateau, *J. Geophys. Res.*, *94*, 4025-4032, 1989.

Fulthorpe, C.S., and S.O. Schlanger, Paleoceanographic and tectonic settings of Early Miocene reefs and associated carbonates of offshore southeast Asia, *Bull. Amer. Assoc. Petrol. Geol.*, *73*, 729-756, 1989.

Garrison, R.E., S.O. Schlanger and D. Wachs, Petrology and paleogeographic significance of Tertiary nannoplankton-foraminiferal limestones, Guam, *Palaeoclim. Palaeoecol. Palaeogeog.*, *17*, 49-64, 1975.

Grow, J.A., M.W. Lee, J.J. Miller, W.F. Agena, J.C. Hampson, D.S. Foster, and R.A. Woellner, Multichannel seismic-reflection survey of KOA and OOK craters, in: *Sea-Floor Observations and Subbottom Seismic Characteristics of OAK and KOA Craters, Eniwetak Atoll, Marshall Islands*, F.G. Folger (ed.), U.S. Geol. Surv. Bull. 1978, D1-D46, 1986.

Haggerty, J.A., S.O. Schlanger, and I. Premoli-Silva, Late Cretaceous and Eocene volcanism in the southern Line Islands and implications for hot spot theory, *Geology*, *10*, 433-438, 1982.

Hathaway, J. and S. O. Schlanger, Nordstandite from Guam, *Am. Mineralogist*, *50*, 1029-1037, 1965.

Henry, T.W., and B.R. Wardlaw (eds.), *Stratigraphic analysis and other geologic and geophysical studies in vicinity of OAK and KOA craters*, U.S. Geol. Surv. Open File Rept. 86-555, Part 3, 1986.

Houtz, R., and W.J. Ludwig, Distribution of reverberant sub-bottom layers in the southwest Pacific basin, *J. Geophys. Res.*, *84*, 3497-3504, 1979.

Hsü, K.J. and S.O. Schlanger, Thermal history of the upper mantle and its relation to crustal history in the Pacific Basin, *Proc. Int. Geol. Congr. 23rd Sess.*, Czechoslovakia, *1*, 91-105, 1968.

Jackson, E.D., and S.O. Schlanger, Regional syntheses, Line Island Chain, Tuamotu Island Chain, and Manihiki Plateau, Central Pacific Ocean, *in*: Schlanger, S.O., E.D. Jackson, et al., *Initial Reports Deep Sea Drilling Project*, *33*, 915-927, 1976.

Jenkyns, H.C. and S.O. Schlanger, Significance of plant remains in redeposited Aptian sediments, Hole 462A, Nauru Basin to Cretaceous oceanic oxygenation models, *in* Larson, R.L. and S.O. Schlanger, et al., *Initial Reports Deep Sea Drilling Project*, *61*, 557-562, 1981.

Kim, D-C, K.W. Katahara, M. H. Manghnani, and S.O. Schlanger, Velocity and attenuation anisotropy in deep-sea carbonate sediments, *J. Geophys. Res.*, *88*, 2337-2343, 1983.

Kim, D-C., M. Manghnani and S.O. Schlanger, The role of diagenesis in the development of physical properties of deep-sea carbonate sediments, *Marine Geol.*, *69*, 69-91, 1985.

Konishi, K, S.O. Schlanger and A. Omura, Neotectonic rates in the

Central Ryukyu Islands derived from ^{230}Th coral ages, *Marine Geol.*, *9*, 225-240, 1970.

Kono, M., Paleomagnetism of DSDP Leg 55 basalts and implications for the tectonics of the Pacific plate, in: Jackson, E.D., I. Koisumi, et al., *Initial Reports Deep Sea Drilling Project*, *55*, 737-752, 1980.

Ladd, H.S. and S.O. Schlanger, Drilling operations on Eniwetok Atoll, *U.S. Geol. Surv. Prof. Paper 260-Y*, 42 pp., 1961.

Larson, R.L. and S.O. Schlanger, Geological evolution of the Nauru Basin and regional implications, *in* Larson, R.L. and S.O. Schlanger, et al., *Initial Reports Deep Sea Drilling Project*, *61*, 841-862, 1981.

Larson, R.L. and S.O. Schlanger, Cretaceous volcanism and Jurassic magnetic anomalies in the Nauru Basin, western Pacific Ocean, *Geology*, *9*, 480- 484, 1981.

Lincoln, J. and S.O. Schlanger, Miocene sea level falls related to the geological history of Midway Atoll, *Geology*, *15*, 458-351, 1986.

Lincoln, J. and S.O. Schlanger, Atoll stratigraphy as a record of sea-level changes, Amer. Geophys. Union Monogr., in press.

Manghnani, M, S.O. Schlanger, and P.D. Milholland, Elastic properties related to depth of burial, strontium content and age, and diagenetic stage in pelagic carbonate sediments, *in*: *Bottom Interacting Ocean Acoustics*, Plenum Press, New York, 41-51, 1981.

McKenzie, J.D., D. Bernoulli and S.O. Schlanger, Shallow-water carbonate sediments from the Emperor Seamounts: Their diagenesis and paleogeographic significance, *in* Jackson, E.D. and I. Ozumi, et al., *Initial Reports Deep Sea Drilling Project*, *55*, 415-455, 1980.

Mayer, L.A., T.H. Shipley, and E.L. Winterer, Equatorial Pacific seismic reflectors as indicators of global oceanographic events, *Science*, *233*, 761-764, 1986.

Milholland, P., M. Manghnani, S.O. Schlanger and G. Sutton, Geoacoustic modeling of deep-sea sediment sequences, *Jour. Acoust. Soc. Amer.*, *68*, 1351-1360, 1980.

Moberly, R.M., S.O. Schlanger, et al., The Mesozoic superocean, *Nature*, *302*, 381, 1983.

Moberly, R.M., S.O. Schlanger, et al., *Initial Reports Deep Sea Drilling Project*, *89*, U.S. Gov't. Printing Office, Washington D.C., 678 pp., 1986.

Murphy, M.A. and S.O. Schlanger, Sedimentary structures in Ilhas and Sao Sebastiao formations, Reconcavo Basin, Brazil, *Bull. Amer. Assoc. Pet. Geol.*, *46*, 457-477, 1962.

Rotondo, G.M., V.C. Springer, G.A. J. Scott and S.O. Schlanger, Plate movements and island integration - a possible mechanism in the formation of endemic biotas, with special reference to the Hawaiian Islands, *Systematic Zoology*, *30*, 12-21, 1981.

Schlanger, S.O., Dolomite growth in coralline algae, *Jour. Sed. Petrology*, *27*, 181-186, 1957.

Schlanger, S.O., Subsurface geology of Eniwetok Atolls, *U.S. Geol. Surv. Prof. Paper 260-BB*, 74 pp., 1963.

Schlanger, S.O., Petrology of the limestones of Guam, *U.S. Geol. Surv. Prof. Paper 403-D*, 51 pp., 1964.

Schlanger, S.O., Dolomite-evaporite relations on Pacific islands, *Sci. Rep. of the Tokohu Univ.*, Japan, *37*, 15-29, 1965.

Schlanger, S.O, Shallow-water limestones in oceanic basins as tectonic and paleogeographic indicators, *in*: *Deep Sea Drilling - A Decade of Progress* (Warme, J., E.L. Winterer, and R.G. Douglas, eds.), Soc. Econ. Paleont. Mineral. Spec. Pub., *32*, 209-225, 1981.

Schlanger, S.O., High frequency sea-level fluctuations in Cretaceous time: an emerging geophysical problem, in: *History of Mesozoic and Cenozoic Oceans* (Hsü, K.J., ed.), Amer. Geophys. Union Geodynam. Ser. *15*, 61-74, 1986.

Schlanger, S.O., Strontium storage and release during deposition and diagenesis of marine carbonates related to sea-level variations, in: Lerman, A., and M. Meybeck (eds.), *Physical and Chemical Weathering in Geochemical Cycles*, Kluwer, Dordrecht, 323-339, 1988.

Schlanger, S. O., and J.W. Brookhart, Geology and water resources of Falalop Island, Western Caroline Islands, *Am. Jour. Sci.*, *253*, 553-573, 1955.

Schlanger, S.O., M.A. Arthur, H.C. Jenkyns and P.A. Scholle, The Cenomanian- Turonian Oceanic Anoxic Event, I. Stratigraphy of organic carbon-rich beds and the marine $\sigma^{13}C$ excursion, in: *Marine Petroleum Source Rocks* (J. Brooks and A. Fleet, eds.), Geol. Soc. London Spec. Pub. No. 26, 371-399, 1987.

Schlanger, S.O., F. Cambell, M. Jackson, Post-Eocene subsidence of the Marshall Islands recorded by drowned atolls on Harrie and Sylvania Guyots, in: Keating, B., P. Fryer, R. Batiza and G. Boehlert (eds.), *Seamounts, Islands and Atolls, A Memorial to Henry William Menard*, Amer. Geophys. Union. Monogr., *43*, pp. 165-174, 1987.

Schlanger, S.O. and M.B. Cita, (eds.), *Nature and Origin of Cretaceous Organic Carbon-Rich Facies*, Academic Press, London, 420p., 1982.

Schlanger, S.O., R.G. Douglas, Y. Lancelot, T.C. Moore, and P.H. Roth, Fossil preservation and diagenesis of pelagic carbonates from the Magellan Rise, *in* Winterer, E.L., et al., *Initial Reports Deep Sea Drilling Project*, *17*, 407-427, Washington, D.C., 1973.

Schlanger, S.O. and R.G. Douglas, Diagenesis of pelagic carbonates and its implications for marine stratigraphy, *in: Pelagic Sediments: on Land and Under the Sea* (K.J. Hsü and H. Jenkyns, eds), Spec. Pub. 1 of the Int. Ass. Sedimentologists, Blackwells, London, 403-434, 1974.

Schlanger, S.O., M. Garcia, B. Keating, J. Naughton, J. Haggerty, and W. Sager, Geology and geochronology of the Line Islands, *J. Geophys. Res.*, *89*, 11261-11272, 1984.

Schlanger, S.O., and H.C. Jenkyns, Cretaceous oceanic anoxic events: causes and consequences, *Geologie en Mijnbouw*, *55*, 179-184, 1976.

Schlanger, S.O., H.C. Jenkyns and I. Premoli-Silva, Volcanism and vertical tectonics in the Pacific Basin related to the global Cretaceous transgression, *Earth Planet. Sci. Lett.*, *52*, 435-449, 1981.

Schlanger, S.O., and K. Konishi, The geographic boundary between the Coral-Algal and the Bryozoan-Algal facies: a paleolatitude indicator, *Proc. 9th Int. Congr. Sedimentol.*, 187-191, 1975.

Schlanger, S.O. and R.M. Moberly, Sedimentary and volcanic history: East Marianas Basin and Nauru Basin, in: Moberly, R.M., Schlanger, S.O., et al., *Initial Reports of the Deep Sea Drilling Project*, *89*, U.S. Gov't. Printing Office, Washington, D.C., 653-678, 1986.

Schlanger, S.O. and I. Premoli-Silva, Tectonic, volcanic and paleogeographic implications of redeposited reef faunas of Late Cretaceous and Tertiary age from the Nauru Basin and the Line Islands, *in:* Larson, R.L. and S.O. Schlanger, et al., *Initial Reports Deep Sea Drilling Project*, *61*, 817-828, 1981.

Schlanger, S.O. and I. Premoli-Silva, Oligocene sea-level falls recorded in mid- Pacific atoll and archipelagic settings, *Geology*, *14*, 392-396, 1986.

Schlanger, S.O., and E.L. Winterer, Underway geophysical data: Navigation, bathymetry, magnetic and seismic profiles, *in*: Schlanger, S.O., E.D. Jackson et al., *Initial Reports Deep Sea Drilling Project*, *33*, 655-694, 1976.

Shipboard Scientific Party, Site 801, *In:* Lancelot, Y, R. Larson, et al., *Proc. ODP, Init. Repts.*, *129*, College Station, TX (Ocean Drilling Program), 91-170, 1990.

Tracey, J.I. and S.O. Schlanger, et al., General geology of Guam, *U.S. Geol. Surv. Prof. Paper 403-A*, 103 pp., 1964.

Wardlaw, B.R., Strontium-isotope stratigraphy of Enewetak atoll, *Geology*, *19*, 190-191, 1989.

Winterer, E.L., Carbonate sediments on the flanks of seamount chains, *in*: Andersen, N.R., and A. Malahoff (eds.), *The Fate of Fossil Fuel CO_2 in the Oceans*, Plenum, New York, 693-715, 1977.

E.L. Winterer, Scripps Institute of Oceanography, La Jolla, CA 92093.

Dysoxic/Anoxic Episodes in the Aptian-Albian (Early Cretaceous)

TIMOTHY J. BRALOWER

Department of Geology, University of North Carolina, Chapel Hill, NC

WILLIAM V. SLITER

Branch of Paleontology and Stratigraphy, USGS, Menlo Park, CA

MICHAEL A. ARTHUR

Department of Geosciences, Pennsylvania State University, University Park, PA

R. MARK LECKIE

Department of Geology, University of Massachusetts, Amherst, MA

DAVID ALLARD

Graduate School of Oceanography, University of Rhode Island, Narragansett, RI

SEYMOUR O. SCHLANGER

Department of Geological Sciences, Northwestern University, Evanston, IL

New biostratigraphic data suggest, for the first time, that the long Aptian-Albian Oceanic Anoxic Event was marked by at least three distinct, relatively brief episodes of widespread dysoxia/anoxia which interrupted generally oxygenated conditions. The high-resolution, integrated foraminiferal and nannofossil biostratigraphy derived here allows recognition of an apparently ocean-wide dysoxic/anoxic episode in the early Aptian (*Globigerinelloides blowi* foraminiferal Zone, *Chiastozygus litterarius* nannofossil Zone, *Conusphaera rothii* nannofossil Subzone, shortly after magnetic Chron CM0). Equally widespread, but distributionally patchier dysoxic/anoxic episodes occurred in the early Albian (*Hedbergella planispira* foraminiferal Zone, *Prediscosphaera columnata* nannofossil Zone, subzone NC8B) and early late Albian (*Biticinella breggiensis* foraminiferal Zone, *Axopodorhabdus albianus* nannofossil Zone, subzone NC9B). These episodes can be best recognized in highly carbonaceous sediments deposited in epicontinental basins, now exposed on land, including the lower Aptian "Livello Selli" of the Italian Apennines, the lower Albian "Niveau Paquier" of the Fosse Vocontienne, France, and the lower upper Albian Toolebuc Formation of the Eromanga Basin, Queensland, Australia. Our data indicate that these horizons correlate at the nannofossil subzonal level to carbonaceous intervals in DSDP/ODP sites which were deposited in pelagic and hemipelagic oceanic settings.

Although none of these episodes is associated with major biotic extinctions, they are characterized by changes, of variable magnitude, in the community structure of planktonic foraminifera, which commonly consist of a low-diversity assemblage of opportunistic taxa or are entirely absent. Nannofossil taxa in these horizons do not change as radically, but in some sites show marked nearshore affinities or, in other sequences, are possibly replaced by other phytoplankton. Dissolution events cannot be entirely ruled out as a cause of some of the changes in calcareous biota. The proposed dysoxic/anoxic episodes appear to correlate with sea level transgressions or highstands. Rising sea level and climatic consequences of volcanism conditioned the oceans to be prone to dysoxia/anoxia. The spatial and sedimentological variability of organic carbon-rich intervals, and the fact that at least two of them have a patchy geographic distribution, indicate that regional or local climatic, tectonic or oceanographic factors induced or triggered dysoxia/anoxia.

INTRODUCTION

Sy Schlanger's interest in the enigma of Cretaceous black shales stemmed from his life-long fascination with the geology of the Pacific Ocean. Some of the first truly pelagic organic-rich sediments were recovered in the Pacific during Deep Sea Drilling Project Legs 17 and 33, of which Schlanger was co-chief scientist. His classic paper coauthored with Hugh Jenkyns in 1976, based partly on the results of these legs, recognized that these organic-rich sediments were not deposited randomly through time, but tend to be concentrated in particular time intervals with suitable oceanographic and/or climatic conditions. Schlanger and Jenkyns [1976] termed these Oceanic Anoxic Events (OAEs), and this theory and term is now cemented in the literature. The original OAEs proposed were

The Mesozoic Pacific: Geology, Tectonics, and Volcanism
Geophysical Monograph 77

localized within the Aptian-Albian (Early Cretaceous) and Cenomanian/Turonian boundary (Late Cretaceous).

Subsequent work by Schlanger contributed to various aspects of black shale research. His work on Cretaceous volcanism in the Pacific, which culminated in a paper with Jenkyns and Isabella Premoli Silva published in 1981, documented the major episode of mid-plate volcanism in the Pacific from the Aptian through the Santonian (125-80 Ma). This paper correlated this volcanic episode to "the great Late Cretaceous transgression" [e.g., Hancock and Kauffman, 1979], which involved the highest sea levels of the Phanerozoic Era [Vail et al., 1977; Haq et al., 1987]. High sea level caused by intraplate volcanism and thermal rejuvenation of the sea floor are cited as critical factors by most authors who have discussed the origin of Cretaceous black shales [e.g., Jenkyns, 1980; Waples, 1983; Weissert, 1989]. The hypsographic and atmospheric effects of this volcanic activity both indirectly led to black shale deposition. High sea level, higher-than-present atmospheric carbon-dioxide concentrations [e.g., Berner et al., 1983] and vastly different paleogeography [e.g., Barron and Washington, 1982] led to a Cretaceous climate that is considered to have been warm and equable, with continental ice sheets generally absent. As the result of relatively low thermal gradients and an active hydrologic cycle, ocean circulation may have been influenced more by haline-dominated convection than at present [e.g., Brass et al., 1982; Barron and Peterson, 1990]. High mean temperatures and reduced oxygen solubility, as well as generally slow deep circulation, created an ocean highly susceptible to dysoxic/anoxic conditions [e.g., Thiede and van Andel, 1977; Roth, 1978; Thierstein and Berger, 1978; Arthur and Natland, 1979; Demaison and Moore, 1980; Pedersen and Calvert, 1990]. The profundity of Schlanger's work on Pacific volcanism is now just being realized with the discovery that individual volcanic episodes such as the formation of the Ongtong Java Plateau, coincide with OAEs [Larson, 1991a; Tarduno et al., 1991].

One of Schlanger's favorite projects concerned the detailed investigation of the origin and effects of the Cenomanian/Turonian boundary OAE on the marine environment. This OAE, which was clearly the most dramatic and widespread of these episodes, is marked by turnover of macrofossil groups [e.g., Raup and Sepkoski, 1982; Elder, 1989] and corresponds to a widespread excursion of carbon isotopes [e.g., Scholle and Arthur, 1980], apparently reflecting the rapid removal of large quantities of organic material from the oceanic system. In collaboration with Michael Arthur, Jenkyns and Peter Scholle, Schlanger carried out a detailed investigation of the sedimentology, distribution and carbon isotopic stratigraphy of organic-rich horizons in a number of Cenomanian/Turonian boundary sequences. The culmination of this project was the publication in 1987 of an oceanographic model that elegantly accounted for the global distribution of black shale horizons in this time interval. Through the flooding of shallow epicontinental seaways and the production of saline bottom waters, Schlanger and colleagues proposed widespread displacement and upwelling of nutrient-rich deep waters. This upwelling led to the expansion of oxygen minimum zones in shelf settings which was conducive to black shale deposition.

Schlanger was involved in the inception, funding and early stages of this present study which concerns the origin, spatial and temporal distribution of organic-rich sediments during OAE 1 in the Aptian-Albian. The term OAE is not strictly appropriate for this long (20 m.y.) interval. Although sediments enriched in TOC (Total Organic Carbon) are common, their spatial and temporal distribution has not been well documented. A salient question remains as to whether there are discrete intervals in which carbonaceous sediments are concentrated within the overall Aptian-Albian time window, or whether sedimentation of TOC-rich units occurred at different times in different places. Sliter [1989a] correlated organic-rich horizons from Pacific DSDP Sites 167, 305 and 463 to the early Aptian *Globigerinelloides blowi* foraminiferal Zone and proposed that they were the same age as the "Livello Selli" of Italy and the "Niveau Goguel" of southern France based on the work of Coccioni et al. [1987; 1989], Erba [1988] and Bréhéret [1988]. The regional nature of an early Albian organic-rich episode also was demonstrated in European land sections by Bréhéret [1985]. By correlating calcareous nannofossil and planktonic foraminiferal biochronologies, and by surveying a host of additional land and DSDP/ODP sections, we have taken these previous studies one step further by defining exact temporal and spatial limits for these intervals of oxygen deficiency. Following the terminology introduced by Arthur et al. [1990], we term these episodes "oceanic anoxic subevents".

This paper is dedicated to the memory of Sy Schlanger. Although he did not see this project come to fruition, his guidance and energy during its early stages greatly helped us address the problems that we faced. While Sy might not have been entirely convinced by our conclusions, we certainly attribute much of the scientific groundwork that led to this project to his previous studies of Cretaceous black shales.

TECHNIQUES

We have conducted a thorough literature review of 39 Aptian-Albian sections recovered by DSDP/ODP (Table 1). Certain sections were too poorly recovered (e.g., DSDP Site 418) or too poorly fossiliferous (e.g., DSDP Site 317) to be studied further. Other sections (e.g. DSDP Site 547) were almost devoid of organic material but were still an integral part of our analysis of the spatial distribution of anoxia/dysoxia. This study is based upon combined lithostratigraphic and biostratigraphic investigations of 6 land and 15 DSDP/ODP sections (Fig. 1). Certain sequences have been studied largely for biostratigraphic purposes. These sections are expanded, mostly TOC-poor, nearly continuously recovered and contain well-preserved microfossil assemblages. Other sequences, including all of the land sections, have been studied largely because they contain horizon(s) enriched in TOC. The latter sections generally contain less well-preserved microfossil assemblages characterized by moderate to extreme amounts of overgrowth and etching which systematically remove solution-susceptible taxa and reduce diversity [e.g., Thierstein, 1980]. There is some evidence that the TOC-rich units of the land sections are stratigraphically condensed relative to correlative deposits from certain deep-sea sites [Bralower, et al., submitted]. Several of the DSDP/ODP sections

TABLE 1. DSDP/ODP Sections Considered for this Investigation

Site	Location	Stage	Range TOC	Reason Unselected
105	W. N. Atlantic	Apt.-Alb.	0-4%	Poor recovery, preservation
144	Demerara Rise	U.Apt.	Unknown	Poor recovery
167	**Magellan Rise**	**U.Barr.-U.Apt.**	**Unknown**	
258	**Naturaliste Plat.**	**M.Alb.-U.Alb.**	**0-3%**	
288	Ontong Java Plat.	U.Apt., U.Alb.	~0	Poor recovery
289	Ontong Java Plat.	U.Apt.	~0	Poor recovery
305	Shatsky Rise	L.Apt.-U.Alb.	0-9.3%	Poor recovery, preservation
306	Shatsky Rise	Aptian, U.Alb.	~0	Poor recovery, preservation
317	Manihiki Plateau	L.Apt.-U.Alb.	0-29%	Poor preservation
327	Falkland Plateau	L.Apt-M.Alb	0-3.5%	Same section as Site 511
330	Falkland Plateau	L.Apt-M.Alb	0-4%	Same section as Site 511
356	Rio Grande Rise	U.Alb.	0-2%	Poor preservation
361	Cape Basin	L.Apt.-M.Alb.	0-14%	Poor preservation
363	Walvis Ridge	U.Apt.-U.Alb.	<1%	Poor preservation, low TOC
364	**Angola Basin**	**L. Apt**	**0-40%**	
367	Cape Verde Basin	L.Apt-U.Alb.	0-35%	Poor preservation
370	**Morocco Basin**	**L.Apt.-U.Alb.**	**0-7%**	
386	**Bermuda Rise**	**L.Alb-U.Alb.**	**0-15%**	
387	Bermuda Rise	L.Apt.-U.Alb.	0-4%	Poor preservation
391	Blake-Bahama Basin	L.Apt.-M.Alb.	0-3%	Poor recovery
398	**Vigo Seamount**	**U.Barr.-U.Apt.**	**0-3%**	
400	**Bay of Biscay**	**U.Apt.-U.Alb.**	**0-3%**	
402	**Bay of Biscay**	**L.Apt.-U.Apt.**	**0-12%**	
417	**Bermuda Rise**	**L.Apt.-U.Apt.**	**0-21%**	
418	Bermuda Rise	L.Apt.-U.Alb.	0-10%	Poor recovery
463	**Mid. Pac. Mtns.**	**L.Apt.**	**0-8%**	
464	Hess Rise	U.Apt.-U.Alb.	0-4%	Poor recovery
465	Hess Rise	U.Alb.	0-9%	Poor recovery, preservation
466	Hess Rise	U.Alb.	0-5%	Poor recovery
511	**Falkland Plateau**	**U.Barr.-U.Alb.**	**0-10%**	
530	Walvis Bay	U.Alb.	0-9%	Poor preservation
534	Blake-Bahama Bas.	L.Apt.-M.Alb.	0-3%	Poor preservation
535	Gulf of Mexico	L.Apt-M.Alb.	0-6%	Poor preservation, low TOC
540	Gulf of Mexico	L.Alb-U.Alb.	0-4%	Poor preservation, low TOC
545	**Mazagan Esc.**	**U.Apt.-U.Alb.**	**0-5%**	
547	**Mazagan Esc.**	**U.Alb-L.Cen.**	**0-1.5%**	
641	**Galicia Bank**	**U.Barr.-M.Alb.**	**0-26%**	
693	Weddell Sea	U.Apt.-L.Alb.	0-3%	No zonal markers
763	**Exmouth Plat.**	**U.Barr.-Cen.**	**<1%**	

Selected sites shown in bold. Stratigraphic intervals shown for these are those studied here.

investigated, including Sites 258, 364, 402 and 417, are poorly recovered (20-40%) but still reveal interesting trends.

Calcareous nannofossil biostratigraphy was carried out by Bralower on all of the sequences investigated. Thickly-concentrated smear slides were prepared so that even the rarest of occurrences could be noted. Preservation of nannofossils ranges from poor to good in the investigated sections. Special attention was paid to samples in critical stratigraphic intervals, especially towards the ends of species ranges, where slides were examined for several hours. Foraminiferal micropaleontology was carried out by Leckie on the Atlantic DSDP/ODP Sites and by Sliter on the Italian land sections and Pacific DSDP Sites. Leckie concentrated mostly on marlstones and claystones which could be washed using standard techniques, and Sliter worked predominantly on limestones using thin sections. The two workers collaborated closely and exchanged samples in critical intervals. Detailed foraminiferal taxonomy and stratigraphic distributions will be presented elsewhere [Leckie and Sliter, in prep.] A detailed analysis of the early Aptian interval has been offered in Bralower et al. [submitted]. That publication includes comprehensive nannofossil range charts of all of the sections investigated. In this publication, we display abbreviated range charts of the Albian sections investigated which include only the stratigraphic distributions of important biomarker taxa. Complete sample-by-sample range charts of all of the taxa identified in the other sections are illustrated in Sliter and Bralower [in press]. A list of all

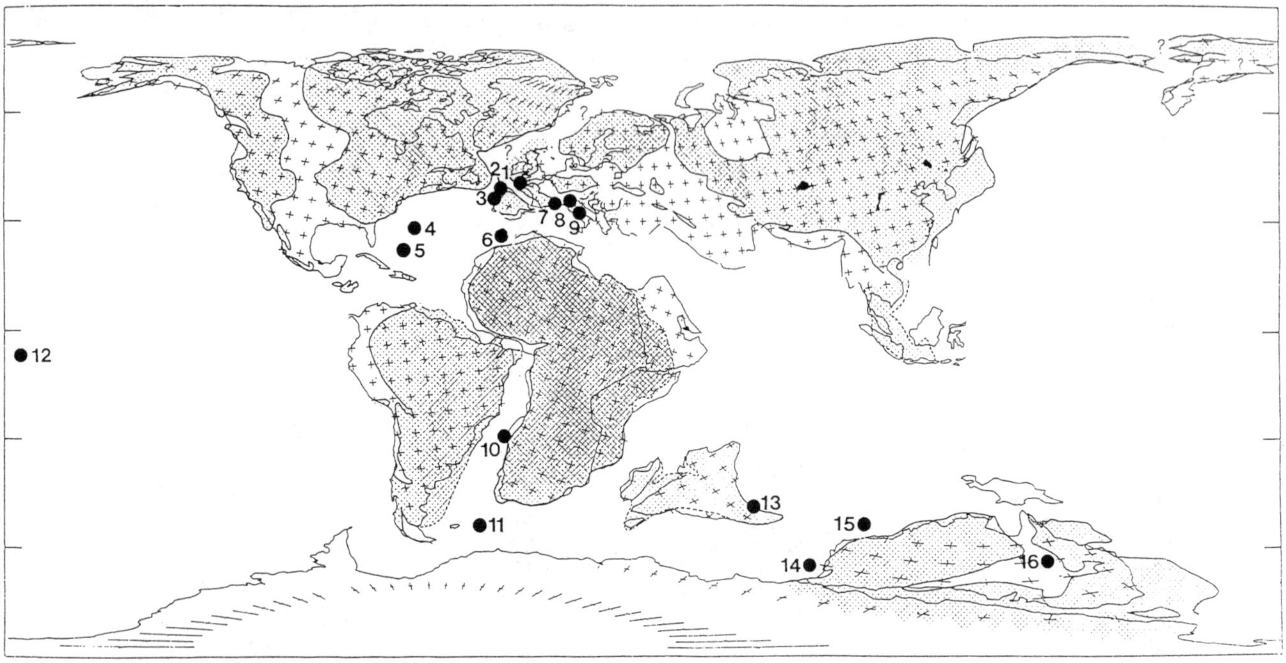

Fig. 1. Paleogeographic location of DSDP/ODP sites and land sections studied in this investigation. 1. Bay of Biscay (DSDP 400, 402), 2. ODP 641, 3. DSDP 398, 4. DSDP 386, 5. DSDP 417, 6. Morocco Basin, DSDP Sites 370, 545, 547, 7. Fosse Vocontienne (Col de Palluel, Vergons), 8. Cismon, 9. Apennines (Gorgo A Cerbara), 10. DSDP 364, 11. DSDP 511, 12. DSDP 463, 13. Cauvery Basin, 14. DSDP 258, 15. ODP 763, 16. Eromanga Basin. Paleogeographic reconstruction of Barron et al. [1981].

nannofossil and planktonic foraminifer taxa discussed here is provided in Table 2.

TOC and carbonate contents of samples from DSDP/ODP sites were taken from the literature and analyzed as part of this investigation (Table 3). Analyses were carried out either at GSO/URI or at UNC-CH using Coulometrics coulometers. A large number of samples of different lithologies at individual sites have been analyzed in order to avoid the problem of representative sampling and single TOC values for long intervals. The absolute concentration of organic carbon in any sample is a result of the carbon flux to the sea floor and decomposition rates during burial, as well as the supply of dilutants such as carbonate, silica and clastic material. Higher TOC contents or peaks in TOC profiles in DSDP/ODP sections have been interpreted to indicate enhanced organic matter preservation and accumulation regardless of the cause. Because supply rates of carbon and dilutants vary considerably from site to site, we have not erected criteria as to what constitutes "TOC-rich" or "TOC-poor" and instead we use relative terms such as "higher" and "lower" to describe trends in TOC in individual locations. We would prefer to interpret accumulation rates of TOC, a factor which takes into account differential compaction and negates the effect of dilutants on TOC contents. However, given the current uncertain status of radiometric control in the middle Cretaceous, it is not possible to calculate precisely the component accumulation rates.

INTEGRATED APTIAN-ALBIAN MICROFOSSIL BIOCHRONOLOGY

Original zonations for Aptian-Albian planktonic foraminifera and calcareous nannofossils were developed up to 20 years ago [e.g., Bolli, 1966; Moullade, 1966; Thierstein, 1971; Longoria, 1974] in sections from the Tethyan province. These zonations have been modified somewhat since in part by the addition of subzones [e.g., Manivit et al., 1977; Sigal, 1977; Roth, 1978]; however, the resolution remains low, averaging 4-6 m.y. per zone and 2-3 m.y. per subzone. Recently, there have been advances both in techniques and taxonomy. For both microfossil groups, techniques have evolved allowing the study of indurated sedimentary rocks such as limestones [e.g., Monechi and Thierstein, 1985; Sliter, 1989b]. Taxonomic advances which include the clarification of existing taxa, and the discovery of new taxa also lead to increased potential resolution as there are more species available for biostratigraphy. Continued operation of the Deep Sea Drilling Project and Ocean Drilling Program has led to recovery of new Aptian-Albian sections in the last twenty years. These include sites from high latitudes and from all of the different ocean basins. Many of these sections are expanded relative to those in which original zonations were developed. In this paper, we propose an upper Barremian-lower Cenomanian calcareous nannofossil biostratigraphic scheme with higher resolution than those presently available. In addition, we propose a preliminary integration of

TABLE 2. Microfossil Taxa Discussed

Calcareous Nannofossils
Parhabdolithus achlyostaurion Hill, 1976
Corollithion achylosum [Stover, 1966] Thierstein, 1971
Corollithion acutum Thierstein in Roth and Thierstein, 1972
Lithraphidites acutum Verbeek and Manivit in Manivit et al. [1977]
Cretarhabdus angustiforatus [Black, 1971] Bukry, 1973
Parhabdolithus angustus [Stradner, 1963] Stradner, Adimaker & Maresch, 1968
Prediscosphaera columnata [Stover, 1966] Perch-Nielsen, 1984
Grantarhabdus coronadventis [Reinhardt, 1966] Grün in Grün and Allemann, 1975
Axopodorhabdus decorus [Deflandre, 1954] Wind and Wise in Wise and Wind, 1977
Broinsonia enormis [Shumenko, 1968] Manivit, 1971
Eiffellithus cf. *E. eximius* [e.g., Hill and Bralower, 1987]
Sollasites falklandensis Filewicz et al., in Wise and Wind, 1977
Eprolithus floralis [Stradner, 1962] Stover, 1966
Percivalia hauxtonensis Black, 1973
Micrantholithus hoschulzii [Reinhardt, 1966] Thierstein, 1971
Parhabdolithus infinitus [Worsley, 1971] Thierstein, 1972
Rucinolithus irregularis Thierstein in Roth and Thierstein, 1972
Corollithion kennedyi Crux, 1981
Chiastozygus litterarius [Gorka, 1957] Manivit, 1971
Octocyclus magnus Black, 1972
Flabellites oblongus [Thierstein, 1973] Crux, 1982
Conusphaera rothii [Thierstein, 1971] Jakubowski, 1986
Corollithion signum Stradner, 1963
Prediscosphaera spinosa [Bramlette and Martini, 1964] Gartner, 1968
Nannoconus steinmannii [Kamptner, 1931] subsp. *steinmannii*
Cruciplacolithus striatum subsp. *constansii* Wise and Parker, in Wise, 1983
Eiffellithus trabeculatus Gorka, 1957
Eiffellithus turriseiffelii [Deflandre in Deflandre and Fert, 1954] Reinhardt, 1965

Planktonic Foraminifera
Biticinella breggiensis [Gandolfi, 1942]
Globigerinelloides algerianus Cushman and ten Dam, 1948
Globigerinelloides blowi [Bolli, 1959]
Globigerinelloides duboisi [Chevalier, 1961]
Globigerinelloides ferreolensis [Moullade, 1961]
Hedbergella gorbachikae Longoria, 1974
Hedbergella planispira [Tappan, 1940]
Hedbergella similis Longoria, 1974
Hedbergella trochoidea [Gandolfi, 1942]
Leupoldina cabri [Sigal, 1952]
Planomalina buxtorfi Gandolfi, 1942
Planomalina cheniourensis Sigal, 1952
Rotalipora appenninica Renz, 1936
Rotalipora brotzeni Sigal, 1948
Rotalipora subticinensis [Gandolfi, 1957]
Rotalipora ticinensis [Gandolfi, 1942]
Schackoina cenomana [Schacko, 1897]
Ticinella bejaouaensis Sigal, 1966
Ticinella praeticinensis Sigal, 1966
Ticinella primula Luterbacher, 1963
Schackoina spp.

biostratigraphies based on nannofossils and planktonic foraminifera.

The nannofossil zonation scheme developed (Fig. 2) is based largely on studies of continuous DSDP/ODP sites with good microfossil preservation. The sections and intervals applied include DSDP Sites 398 (upper Barremian to upper Aptian), 511 (upper Barremian to upper Albian), 545 (upper Aptian to upper Albian), 547 (upper Albian to lower Cenomanian), and ODP Sites 641 (upper Barremian to lower Albian) and 763 (upper Barremian to lower Cenomanian). Detailed range charts for these sections are illustrated in Bralower et al. [submitted; late Barremian-early Aptian interval of all sections], Bralower and Siesser [1992; ODP Site 763] and Sliter and Bralower [in press; upper Aptian to Cenomanian portions of Sites 511, 545, 547 and 641].

In order to be fully compatible with previously established zonations, we propose a series of informal subzonal divisions for the widely-used Roth [1978] nannofossil zonation which was modified after the zonation of Thierstein [1971, 1973]. The two subzones of the *Chiastozygus litterarius* Zone (NC6) have been formally defined by Bralower et al. [submitted]. Subzonal units are recognizable in all sections except those with the most poorly preserved microfossils. We propose a series of additional events, termed biohorizons, the relative order of which is not as well established partly because they can only be precisely determined in expanded sequences with good microfossil preservation (e.g., Sites 398, 545, 547, 641, 763). A full discussion of these biohorizons and a detailed correlation of zonal and subzonal units applicable in low latitude sites with those in high latitude settings is given in Bralower [1992]. We stress that the order proposed for these biohorizons, and their correlation with zonal and subzonal units, is tentative and will continue to be revised as new sequences are found and as taxonomic concepts are clarified.

The definitions of the zonal and new subzonal units, which are also classified using the "NC" prefix of Roth [1978] are as follows:

Chiastozygus litterarius Zone (NC6)
Interval from the first occurrence (FO) of *Rucinolithus irregularis* to the FO of *Eprolithus floralis* [Thierstein, 1971; emended by Wiegand, 1984; Erba and Quadrio, 1987; Erba, 1988].

Subzone NC6A: *Conusphaera rothii* Subzone
Interval from the FO of *Rucinolithus irregularis* to the last occurrence (LO) of *Conusphaera rothii*.
Subzone NC6B: *Grantarhabdus coronadventis* Subzone
Interval from the LO of *C. rothii* to the FO of *E. floralis*.

Comments: The last occurrence of *C. rothii* can be determined precisely in deep-sea sections, but this event is more difficult to establish in the land sections due to poorer preservation and the rarity of this taxon.

Parhabdolithus angustus Zone (NC7)
Interval from the FO of *E. floralis* to the FO of *Prediscosphaera columnata* [Thierstein, 1971, 1973].

TABLE 3. Sources of Lithological, Biostratigraphical and Magnetostratigraphical Data in Sections Investigated

Section	Lithostratigraphy	Biostratigraphy*	Magnetostratigraphy	Carbon/Carbonate
DSDP 167	Winterer et al., 1973	Roth, 1973 (n); Douglas, 1973 (f); Tarduno et al., 1989 (n,f)	Tarduno et al., 1989	----
DSDP 258	Davies et al., 1974	Thierstein, 1974 (n)	----	Davies et al., 1974; this study
DSDP 364	Bolli et al., 1978	Proto Decima et al., 1978 (n); Caron, 1978	----	Foresman, 1978; this study
DSDP 370	Lancelot et al., 1977	Cepek, 1977 (n); Pflaumann and Krasheninnikov, 1977 (f)	----	Erdman and Schorno, 1977 this study
DSDP 386	Tucholke et al., 1979	Okada and Thierstein, 1979 (n)	----	Cameron, 1979; Bralower, 1984
DSDP 398	Arthur, 1979; Sibuet et al., 1979	Blechschmidt, 1979 (n); Sigal, 1979 (f)	----	Sibuet et al., 1979; Erdman and Schorno, 1979; Johnson et al., 1979
DSDP 400	Montadert, Roberts et al., 1979	Dupeuble, 1979 (f); Müller, 1979 (n)	----	Deroo et al., 1979; this study
DSDP 402	Montadert, Roberts et al., 1979	Müller, 1979 (n)	----	Deroo et al., 1979; this study
DSDP 417	Donnelly et al., 1980	Gartner, 1980 (n); Siesser, 1980 (n)	----	Deroo et al.,1980; this study
DSDP 463	Dean et al., 1981	Cepek, 1981 (n); Roth, 1981 (n); Tarduno et al., 1989 (n,f)	Tarduno et al., 1989	Dean, 1981; this study
DSDP 511	Ludwig et al., 1983	Krashenninikov and Basov, 1983 (f); Wise, 1983 (n)	Salloway, 1983	Deroo et al., 1983; Bode, 1983; Parker et al., 1983; this study
DSDP 545	Hinz et al., 1984	Leckie, 1984 (f) Wiegand, 1984 (n)	----	Hinz et al., 1984; Deroo et al., 1984
ODP 641	Boillot et al., 1988	Applegate and Bergen, 1988 (n)	Ogg, 1988	Boillot et al., 1988; Dunham et al., 1988; this study
ODP 763	Haq et al., 1990	Bralower and Siesser, 1992 (n)	----	this study
Palluel	----	----	----	this study
Vergons	----	----	----	this study
Cismon	----	Channell et al., 1979 (n,f); Bralower, 1987 (n)	Channell et al., 1979	this study
Piobbico	----	----	----	this study
Cerbara	Coccioni et al., 1989	Lowrie et al., 1980 (n,f); Bralower, 1987 (n)	Lowrie and Alvarez, 1984	this study
Augathella	----	Shafik, 1985 (n)	----	----
Cauvery	----	Kale and Phansalkar, 1992 (n)	----	----

* n - nannofossils; f - foraminifera. The nannofossil biostratigraphy of all sections was investigated as part of this study. Foraminifera were investigated in Sites 370, 400, 511, 545 and 641 and in the Palluel, Vergons, Cismon and Cerbara Sections.

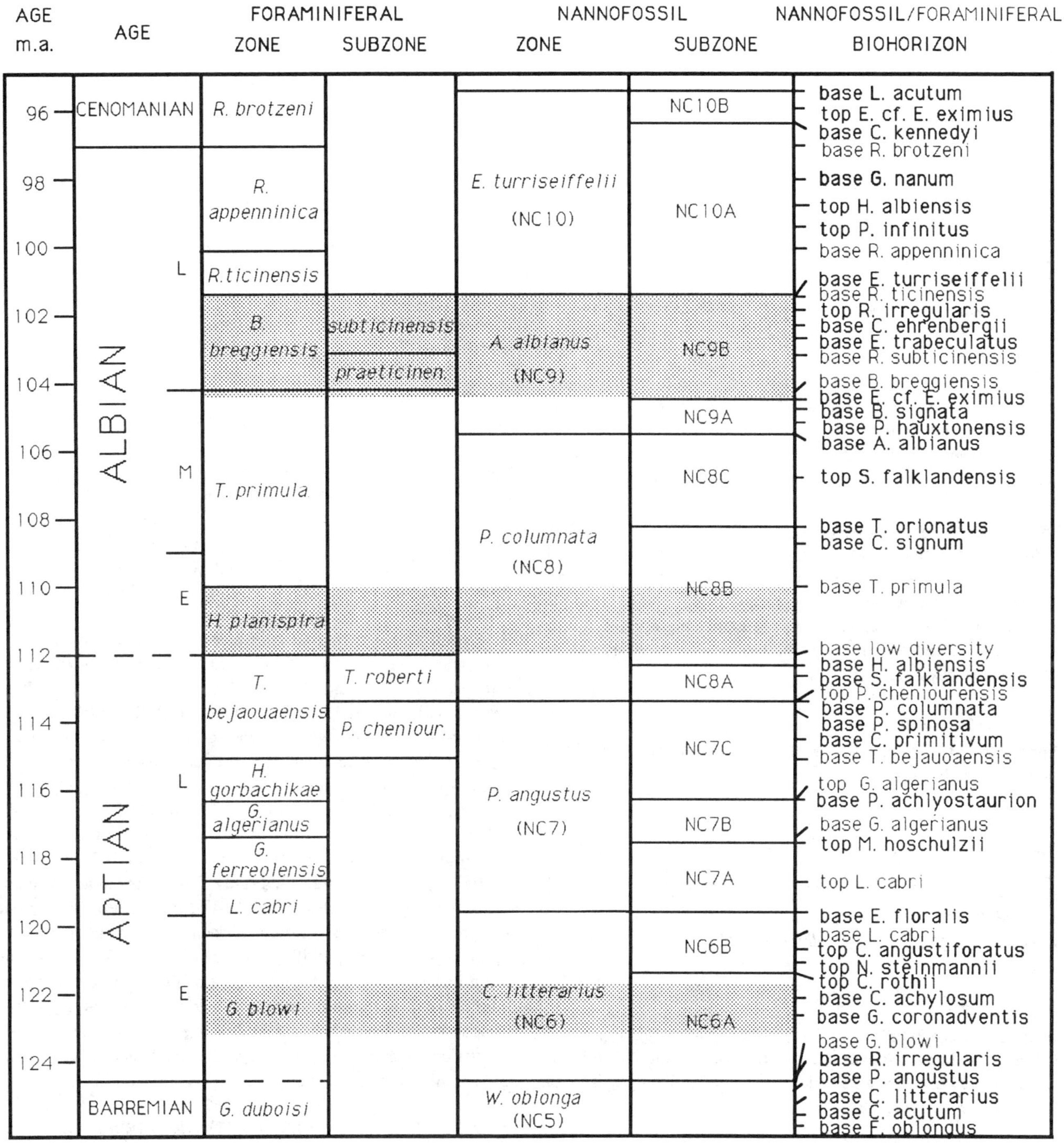

Fig. 2. Aptian-Albian planktonic foraminiferal-calcareous nannofossil biostratigraphy and the timing of the three proposed dysoxic/anoxic episodes (shown in grey). Planktonic foraminifera biostratigraphy is modified after Sigal [1977] and Sliter [1989b], calcareous nannofossil biostratigraphy and correlation with planktonic foraminifera biostratigraphy is described in the text. Nannofossil biohorizons in bold face. The durations of dysoxic/anoxic episodes illustrated are based on zonal definitions and are longer than those calculated in text which are based on thicknesses of TOC-rich horizons (see text for details).

Subzone NC7A
Interval from the FO of *E. floralis* to the LO of *Micrantholithus hoschulzii.*

Subzone NC7B
Interval from the LO of *M. hoschulzii* to the FO of *Parhabdolithus achlyostaurion.*

Subzone NC7C
Interval from the last occurrence of *P. achlyostaurion* to the first occurrence of *Prediscosphaera columnata.*

Comments: The original definition of Thierstein [1971, 1973] suggested that the first occurrence of *E. floralis* could substitute for the first occurrence of *P. angustus* in marking the base of the *P. angustus* Zone. Since we consistently find *P. angustus* to evolve close to or before the first occurrence of *R. irregularis*, this no longer serves as a suitable marker for the base of this zone. We therefore use the first occurrence of *E. floralis* as suggested previously by Erba [1988] and Applegate and Bergen [1988]. The long *P. angustus* Zone is not an easy interval to subdivide as there is little turnover of nannofossil taxa. Both of the events used to define subzones in this interval have problems. The last occurrence of *M. hoschulzii* (we include *M. obtusus* in this taxon) may have been affected by reworking at Sites 398 and 641 where rare specimens are observed up into the Albian. The first occurrence of *P. achlyostaurion* has been utilized by Erba [1988] as a zonal marker in the middle Albian. This species is extremely rare in the land sections, but we believe that its occurrence in the upper Aptian of Sites 364 and 545 is a distinctive event which should be recognizable elsewhere.

Prediscosphaera columnata Zone (NC8)
Interval from the FO of *P. columnata* to the FO of *Axopodorhabdus albianus* [Roth, 1978].

Subzone NC8A
Interval from the FO of *P. columnata* to the FO of *Hayesites albiensis.*

Subzone NC8B
Interval from the FO of *H. albiensis* to the FO of *Tranolithus orionatus.*

Subzone NC8C
Interval from the FO of *T. orionatus* to the FO of *A. albianus.*
Comments: The first occurrences of *H. albianus* and *T. orionatus* are extremely useful events which can be determined precisely in land and deep-sea sections.

Axopodorhabdus albianus Zone (NC9)
Interval from the FO of *A. albianus* to the FO of *Eiffellithus turriseiffelii* [Roth, 1978].

Subzone NC9A
Interval from the FO of *A. albianus* to the FO of *Eiffellithus* cf. *E. eximius.*

Subzone NC9B
Interval from the FO of *E.* cf. *E. eximius* to the FO of *E. turriseiffelii.*

Comments: The first occurrence of *E.* cf. *E. eximius*, a taxon which has been discussed in some length by Hill and Bralower [1987], can be determined precisely in sections with well-preserved nannofossil assemblages. This taxon is not usually observed in land sections.

Eiffellithus turriseiffelii Zone (NC10)
Interval from the FO of *E. turriseiffelii* to the FO of *Lithraphidites acutum* [Thierstein, 1971; emended by Manivit et al., 1977].

Subzone NC10A
Interval from the FO of *E. turriseiffelii* to the FO of *Corollithion kennedyi.*

Subzone NC10B
Interval from the FO of *C. kennedyi* to the FO of *L. acutum.*

Comments: The first occurrence of *C. kennedyi* has been used to define the Albian/Cenomanian boundary [see discussion in Bralower and Siesser [1992]), however, recent investigations by E. Erba [Pers. Comm., 1992] suggest that this event lies well within the lower Cenomanian. *C. kennedyi* has a cosmopolitan distribution, but is more abundant in high-latitude sites.

The resolution provided by this biostratigraphy varies through time from 1 to 4 m.y. based on nannofossil subzones, 0.5 to 3 m.y. when nannofossil subzones and foraminiferal zones are utilized together, and 0.25 to 2 m.y. based on subsidiary nannofossil biohorizons (Fig. 2). Correlation of zonal and subzonal units with stage and substage divisions is illustrated in Figure 2. These correlations are well determined for nannofossils in the Aptian and for both groups in the Albian, although the exact placement of the Barremian/Aptian and Aptian/Albian boundaries is still in question. Integration of calcareous nannoplankton and planktonic foraminiferal zonations is not without problems and will be the subject of further detailed study. The correlations illustrated are based largely on Sites 545 and 547 for the late Aptian to late Albian interval as well as results of detailed biostratigraphy of the Piobbico Core [Erba, 1988; Tornaghi et al., 1989]. Planktonic foraminiferal zonation schemes in the Barremian-early Aptian interval is in a state of flux (see Leckie and Bralower [1991], Coccioni et al. [in press], and Bralower et al. [submitted] for full discussion). Hence the correlation with nannofossil biostratigraphy and chronostratigraphy for this interval is tentative.

STRATIGRAPHIC AND GEOGRAPHIC DISTRIBUTION OF ORGANIC CARBON-RICH INTERVALS

The sites have been divided into two categories in terms of their TOC contents. In the first, major TOC-rich horizons are interbedded with sediments almost devoid of organic matter. At other sites, however, variable TOC enrichments occur throughout the Aptian-Albian interval. Correlation of the former category can

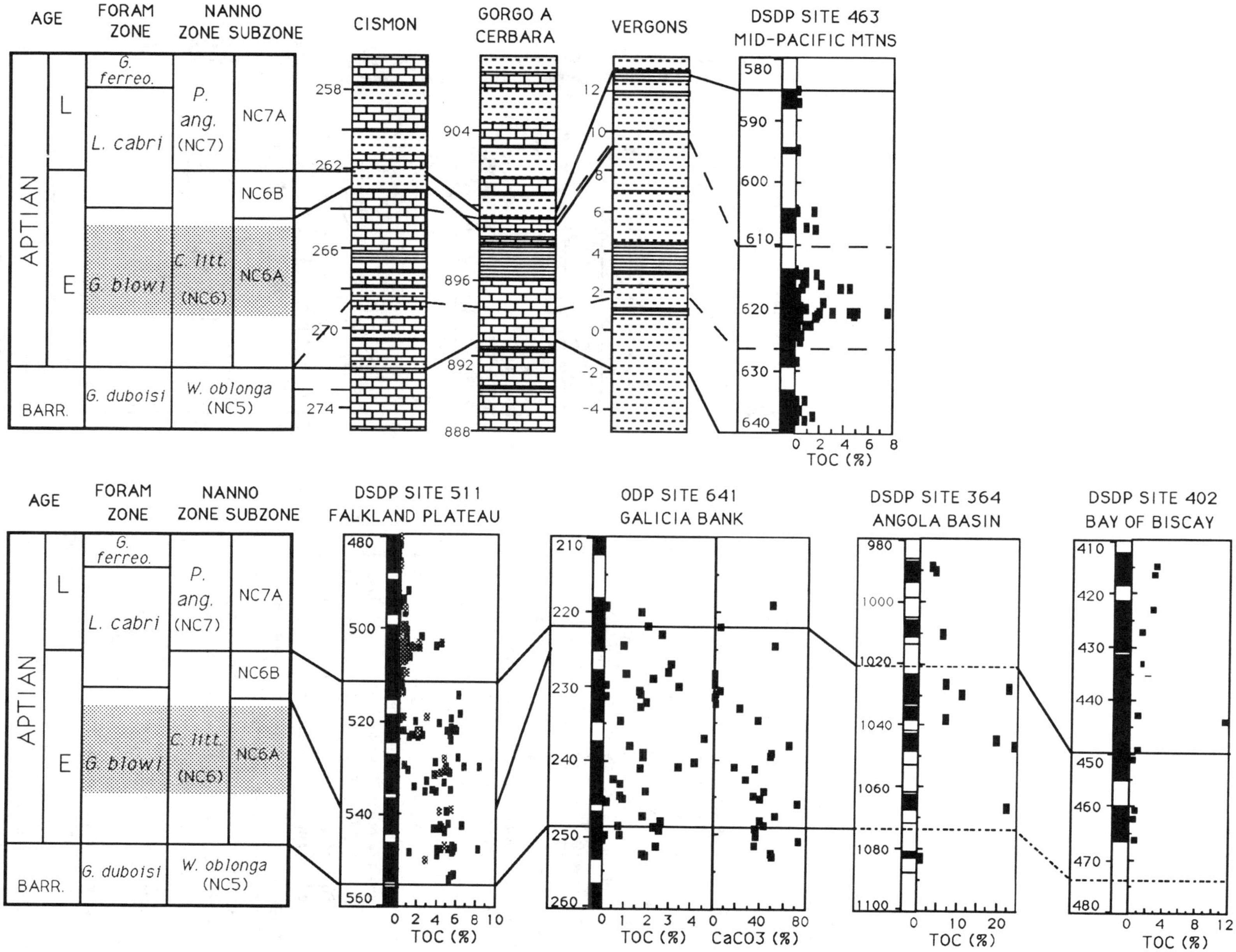

Fig. 3. The early Aptian dysoxic/anoxic episode. Planktonic foraminiferal zonation is from Sliter [1989b], nannofossil zonation is after Bralower et al. [submitted]. Solid thin lines: nannofossil events, dashed lines: foraminiferal events, solid thick lines: unconformities, dotted lines: uncertain event positions. Columns for DSDP/ODP sites indicate recovered (black) and unrecovered (white) intervals. See Bralower et al. [submitted] for detailed lithologic and biostratigraphic range data.

usually be established; this might be quite complicated for the latter type. We presume that enhanced preservation of organic matter in the horizons that we have studied resulted from some combination of oxygen deficiency, high fluxes of organic matter from surface waters, and reduced dilution of organic matter by other sedimentary components. However, there is currently no foolproof technique that allows detection of water column dysoxia/anoxia from the properties of ancient sedimentary rocks. Dysoxic (ie, <0.2 ml/l O_2) or anoxic conditions at or above the sediment/water interface are commonly indicated by the presence of sedimentary laminations [e.g., Rhoads and Morse, 1971; Savrda et al., 1984]. However, because seasonality of sedimentation might be suppressed or dysoxic/anoxic conditions need only extend slightly above the sediment/water interface, presence or absence of lamination is not an absolute indicator of the vertical extent or intensity of dysoxic/anoxic conditions. In several sites (e.g., DSDP Site 370 and ODP Site 641), laminated sediments occur intermittently through the Aptian-Albian interval. In such locations, TOC preservational events are inferred from a combination of discrete peaks in TOC and/or changes in planktonic microfossil assemblages (e.g., assemblage composition, diversity, barren intervals). A more complete discussion of the significance of these biotic indicators is given below.

Our biostratigraphic results indicate that laminated units most enriched in TOC, and sediments characterized by changes in the planktonic assemblages cluster in three narrow biostratigraphic intervals (Fig. 2): (1) the early Aptian *Globigerinelloides blowi* foraminiferal Zone, *Chiastozygus litterarius* nannofossil Zone, *Conusphaera rothii* Subzone (NC6A; Fig. 3); (2) the early Albian *Hedbergella planispira* foraminiferal Zone, *Prediscosphaera columnata* nannofossil Zone, subzone NC8B (Fig. 4), and (3) the early late Albian *Biticinella breggiensis* foraminiferal Zone, *Axopodorhabdus albianus* nannofossil Zone, subzone NC9B (Fig. 5). Using the terminology of Arthur et al. [1990], we term these subevents OAE 1a, 1b and 1c, respectively. It should be noted that Arthur et al. [1990] assigned a late late Albian age for OAE 1c. The proposed subevents correspond to lithologically-distinctive and TOC-rich intervals in pelagic/hemipelagic sections exposed on land including the lower Aptian "Livello Selli" of the Umbrian Apennines and Southern Alps of Italy, the lower Albian "Niveau Paquier" of the Fosse Vocontienne of France and the lower upper Albian Toolebuc Formation of the Eromanga basin, Queensland, Australia (Figs. 3-5). These horizons correlate well with carbonaceous units in the deep sea (DSDP/ODP sites), strongly increasing the significance of the three proposed subevents.

The lower Aptian "Livello Selli" at Cismon (Southern Alps) and several sections in the Italian Apennines correlates with the "Niveau Goguel" in the Fosse Vocontienne. These horizons lie within the *G. blowi* foraminiferal Zone which is defined based on the presence of the nominate species, but absence of *Leupoldina cabri*, the *C. litterarius* nannofossil Zone which is here defined as the interval between the first occurrences of *Rucinolithus irregularis* and *Eprolithus floralis*, and the *Conusphaera rothii* nannofossil Subzone (NC6A) defined as the interval between the first occurrence of *R. irregularis* and the last occurrence of *C. rothii* [Bralower et al., submitted; Figs. 2,3]. These horizons also correlate at the

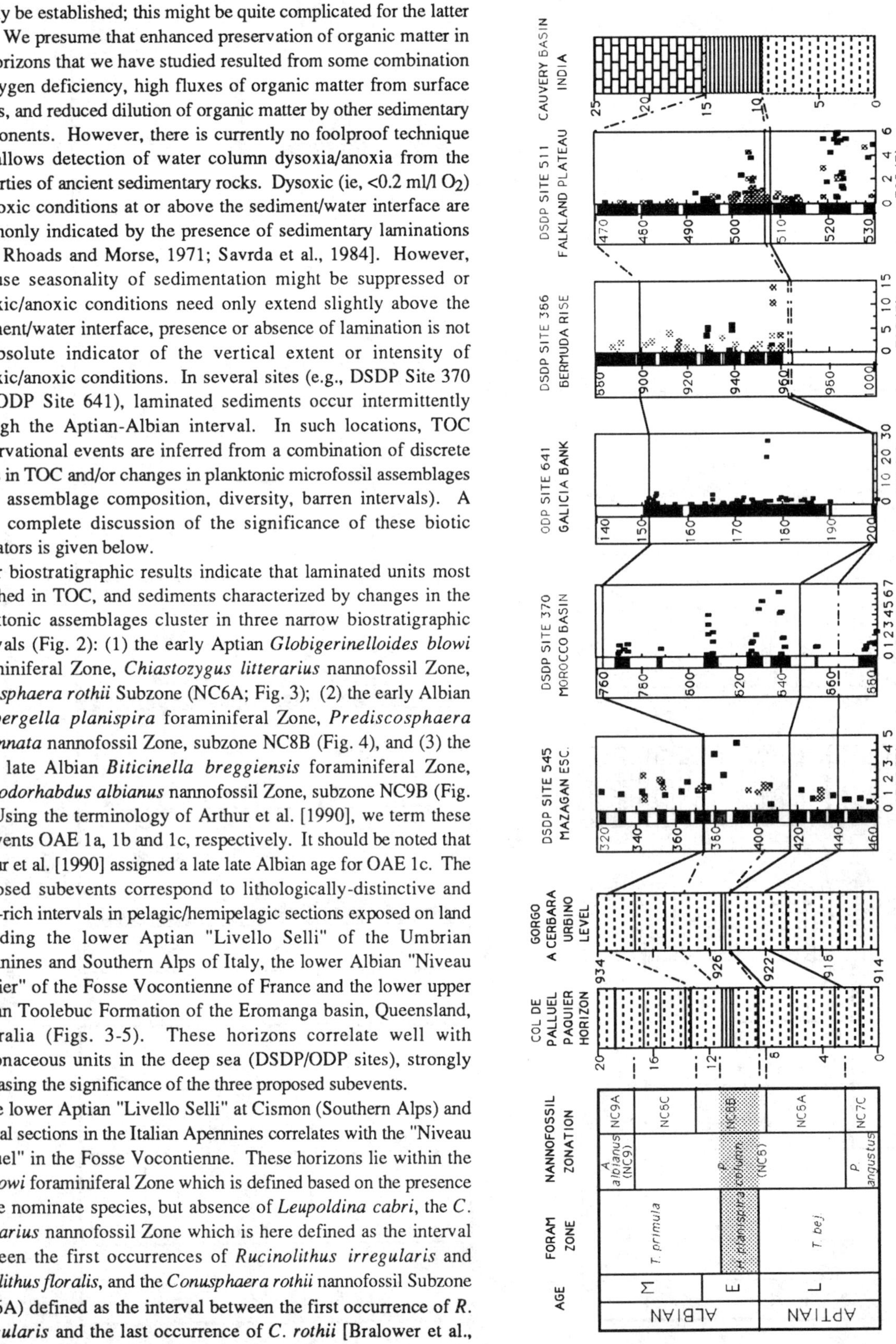

Fig. 4. The early Albian dysoxic/anoxic episode. Planktonic foraminiferal zonation is from Sliter [1989b], nannofossil zonation is described in this paper. Lines/columns as in Fig. 3. See Figs. 6-13 for more detailed biostratigraphic/lithostratigraphic data for each section.

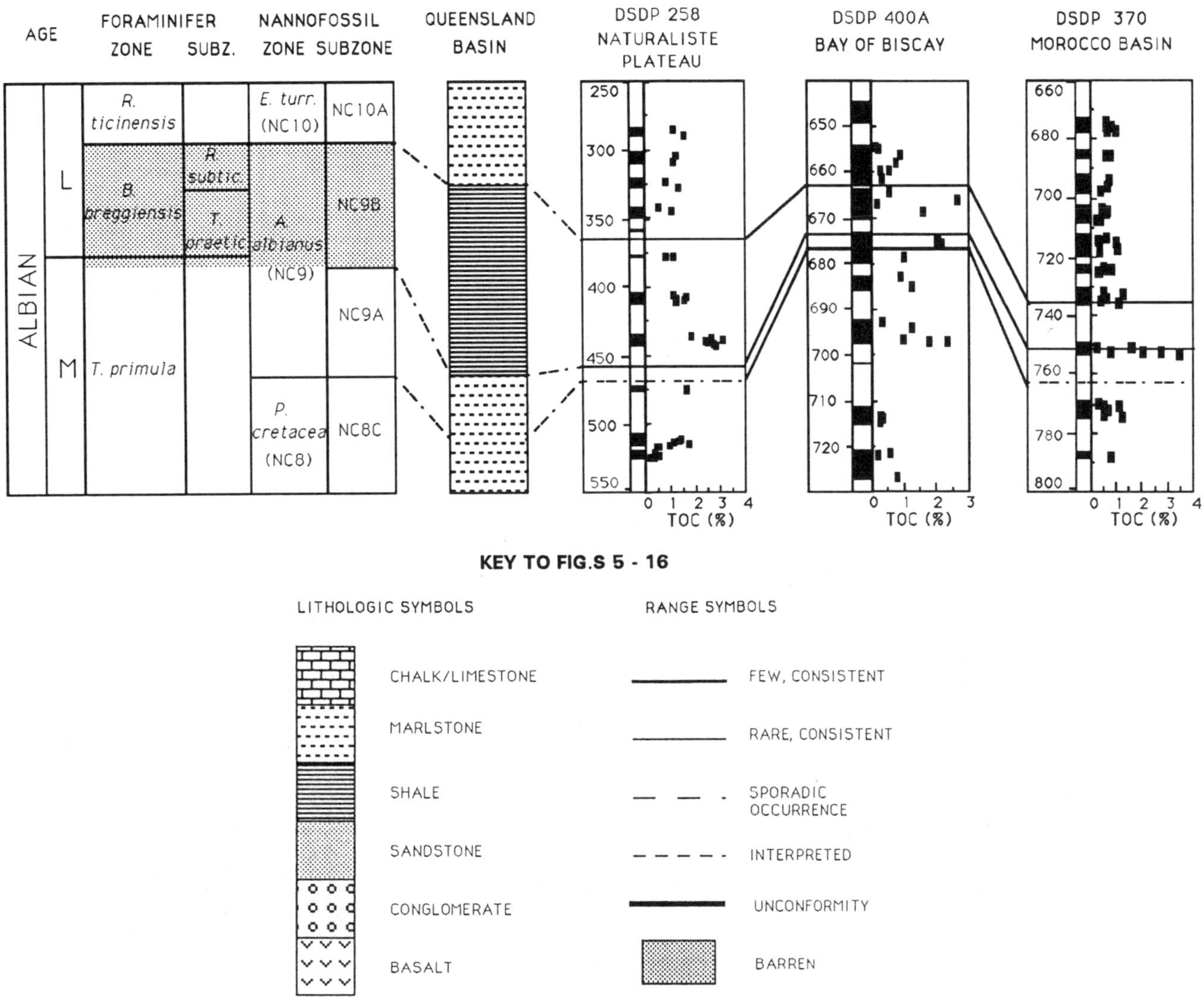

Fig. 5. The late middle Albian dysoxic/anoxic episode. Planktonic foraminiferal zonation is from Sliter [1989b], nannofossil zonation is described in this paper. Lines/columns as in Fig. 3. See Figs. 14-17 for more detailed biostratigraphic/lithostratigraphic data for each section.

nannofossil subzonal level with carbonaceous limestones (up to 8% TOC) in Cores 70 and 71 of DSDP Site 463 in the Mid-Pacific Mountains (Fig. 3), and at the zonal level with black shales in Cores 42 to 45 of DSDP Site 364 in the Angola Basin, South Atlantic (up to 24% TOC). TOC contents are variable throughout the *C. litterarius* Zone at ODP Site 398 drilled on Vigo Seamount, but the correlative subevent (interval barren of microfossils in Cores 398D-118 to -120) occurs at the top of an interval of higher TOC and lies just above the *C. rothii* Subzone (see discussion in Bralower et al. [submitted]). The age of TOC-rich sediments in Cores 60-62 of DSDP Site 511 on the Falkland Plateau is not easily established as the low diversity nannoflora lacks zonal marker species such as *R. irregularis* and planktonic foraminifera are absent. However, we place these sediments within and slightly above the *C. rothii* nannofossil Subzone based on the last occurrence of *C. rothii* in Section 61-2 and the absence of *E. floralis* throughout this interval (Fig. 3). TOC contents are also highly variable throughout the *C. litterarius* Zone at ODP Site 641 drilled on Galicia Bank with no identifiable peak associated with the interval barren of microfossils (Sections 641C-9R-1 and -2, 228-230 meters below sea floor; Fig. 3) that is thought to represent OAE 1a. This interval occurs within the *C. rothii* nannofossil Subzone and some 10 m above magnetic polarity zone M0 [Ogg, 1988] in a similar position to organic carbon-rich horizons at DSDP Site 463 and Italian land sections.

The "Niveau Paquier" of the Col de Palluel Section, Fosse Vocontienne, France, lies within the early Albian *H. planispira* foraminiferal Zone based on the presence of a low-diversity

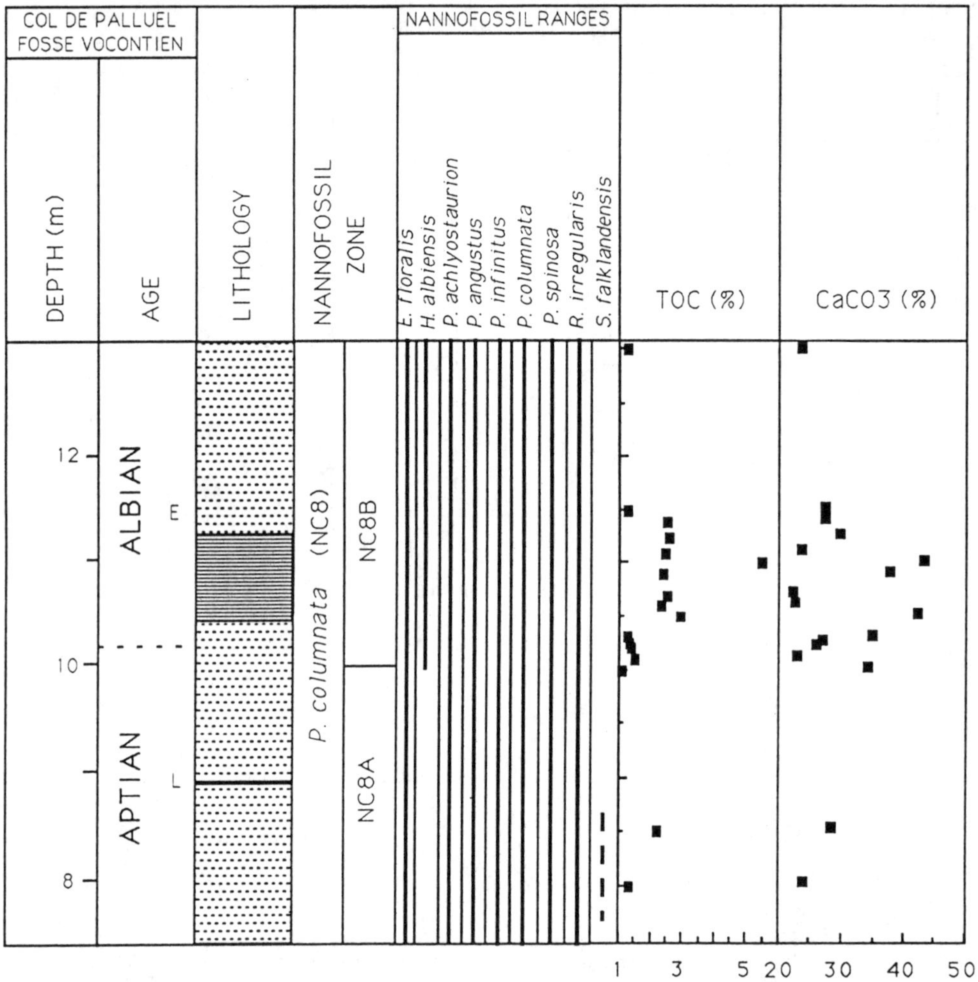

KEY TO FIG.S 5 - 16

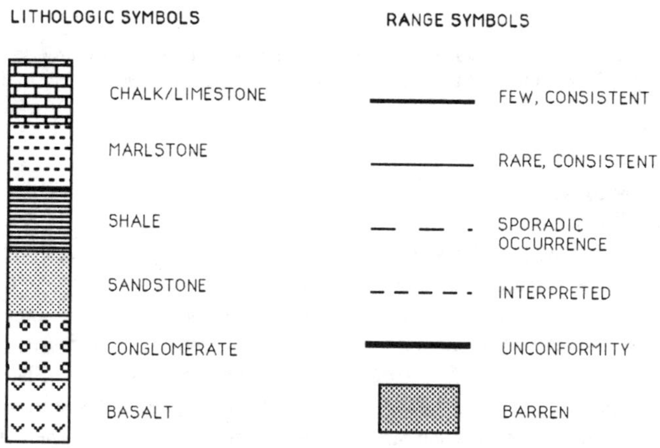

Fig. 6. Upper Aptian-lower Albian stratigraphy of the Col de Palluel Section, Fosse Vocontienne, France. Depths are from an arbitrary sampling scheme. Lithologic column shows position of the organic-rich "Niveau Paquier". Nannofossil biostratigraphy is part of this investigation. See key for explanation of lithologic and range symbols. Position of stage boundary is from nannofossil biostratigraphy. Organic carbon and carbonate contents are from this investigation.

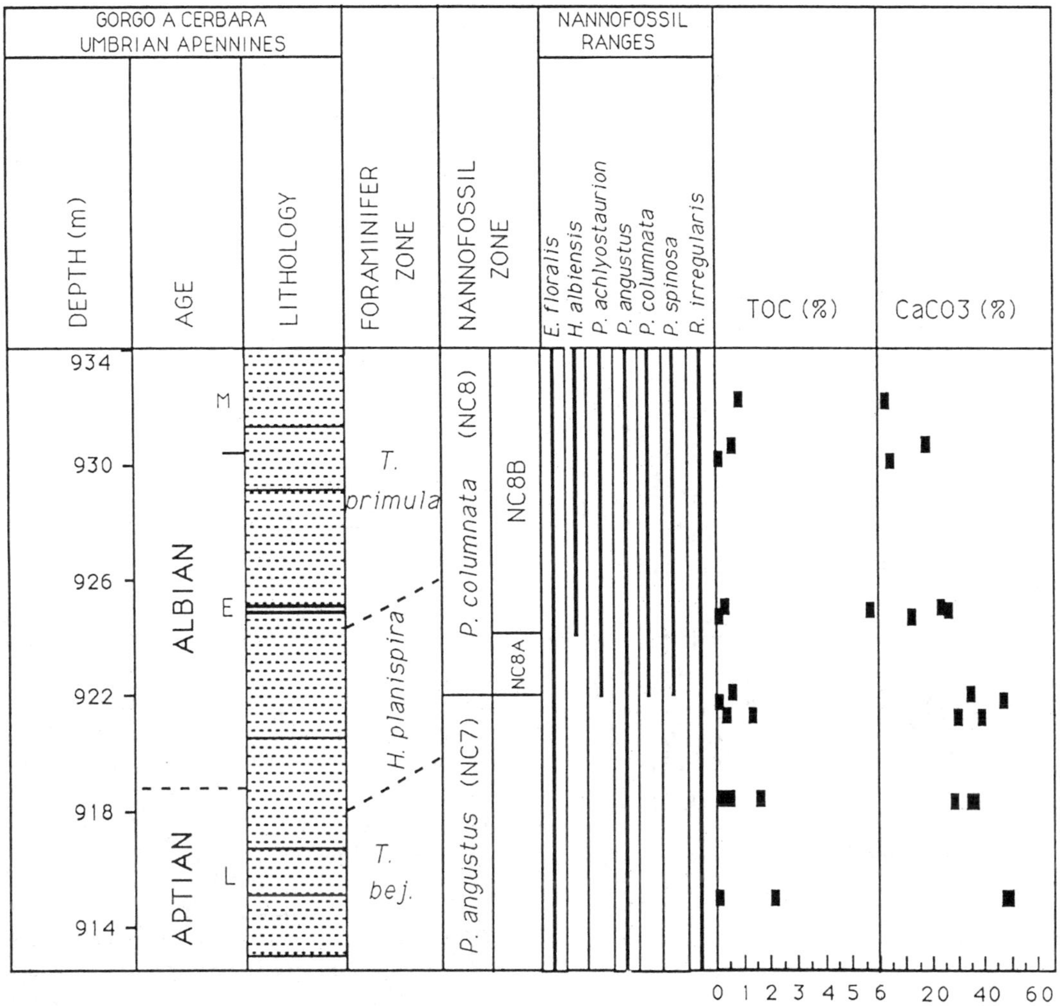

KEY TO FIG.S 5 - 16

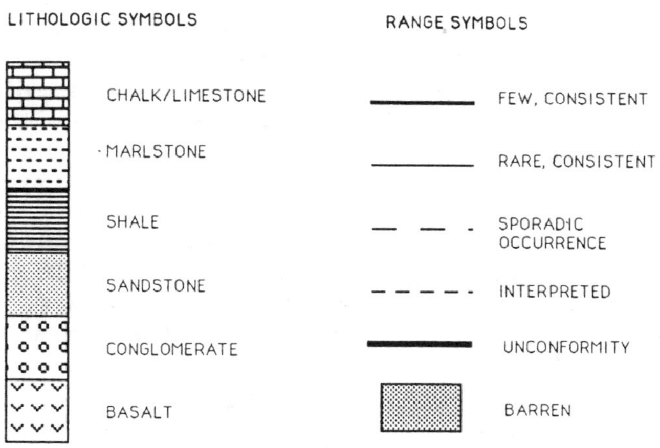

Fig. 7. Upper Aptian-middle Albian stratigraphy of the Gorgo a Cerbara Section, Umbrian Apennines, Italy. Depths are from sampling scheme which is an extension of that of Lowrie and Alvarez [1984]. Lithologic column shows position of the organic-rich "Livello Urbino" at 924.5m. Foraminifera and nannofossil biostratigraphy are part of this investigation. See key for explanation of lithologic and range symbols. Positions of stage and substage boundaries are from nannofossil biostratigraphy. Organic carbon and carbonate contents are from this investigation.

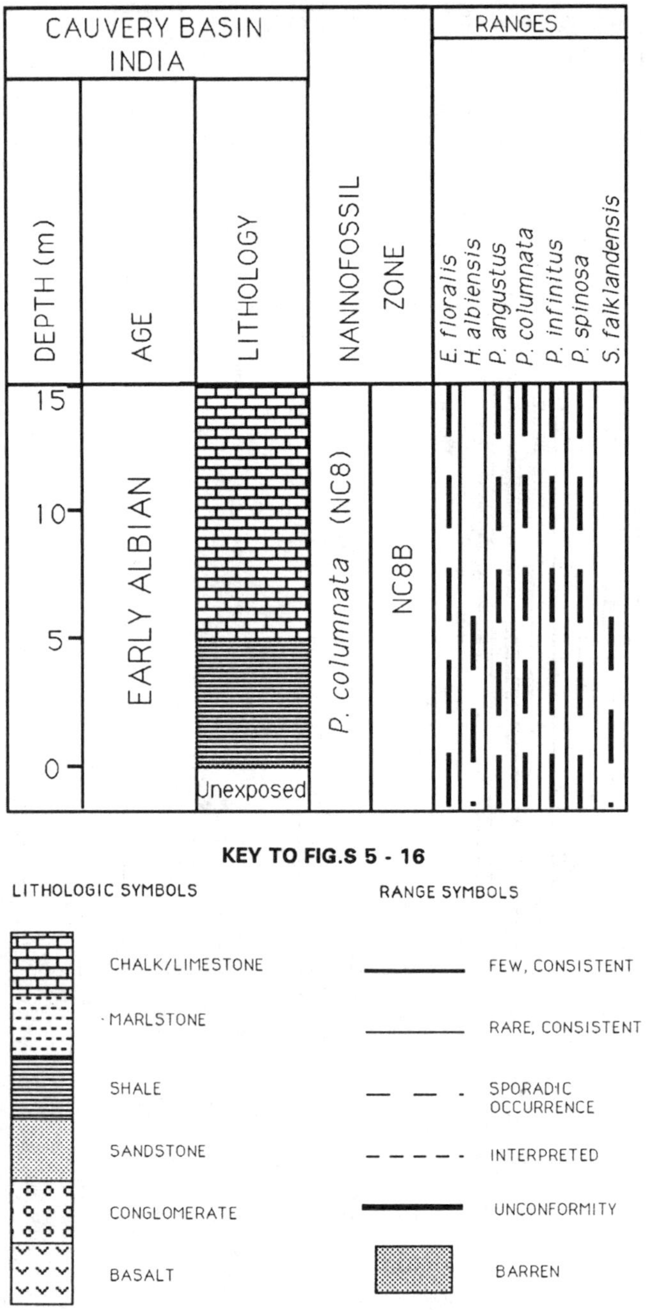

Fig. 8. Lower Albian stratigraphy of the Dalmiapuram Grey Shale, Cauvery Basin, India. Depths are an arbitrary scheme used in this investigation. Illustrated is generalized lithology. Nannofossil biostratigraphy was obtained as part of this investigation. See key for explanation of lithologic and range symbols.

assemblage of hedbergellids, and the absence of *Ticinella primula*, and nannofossil subzone NC8B based on the occurrence of *P. columnata* and *Hayesites albiensis*, and the absence of *Tranolithus orionatus* (Figs. 2, 4, 6). Based on identical nannofossil biostratigraphic criteria, the "Niveau Paquier" correlates with the "Livello Urbino" in the Gorgo a Cerbara (Fig. 7) and other sections from the Italian Apennines. Our planktonic foraminifera biostratigraphy does not disagree with this correlation, although because of the lack of limestones and the corresponding meagre number of samples analyzed in this interval, we cannot differentiate between the *H. planispira* and the overlying *T. primula* Zones (Fig. 7). An unresolved problem exists in that Tornaghi et al. [1989] have correlated the "Livello Urbino" not with the *H. planispira* foraminiferal Zone, but with the underlying *T. bejaouaensis* Zone. The nannofossil biostratigraphy carried out here would correlate the TOC-rich Dalmiapuram shales of the Cauvery Basin, India [Govinden, 1982] with the "Niveau Paquier" based on the presence of *H. albiensis*, but absences of *T. orionatus* and *A. albianus* in our samples (Fig. 8). However, Kale and Phansalkar [1992] have reported the latter two species from this unit and would thus correlate it with the early late Albian episode (based on the absences of *E.* cf. *E. eximius* and *E. turriseiffelii*).

Based on nannofossil biostratigraphy, these units exposed on land correlate with intervals which are not entirely TOC-rich, but which do possess individual TOC-rich horizons, at DSDP Site 545 off the Mazagan Escarpment (Fig. 9; up to 5% TOC), ODP Site 641 on Galicia Bank (Fig. 10; up to 27% TOC), DSDP Site 511 on the Falkland Plateau (Fig. 11; up to 4% TOC) and DSDP Site 386 on Bermuda Rise (Fig. 12; up to 14% TOC). This correlation is very clear at Sites 545 and 641 where the ranges of *P. columnata*, *H. albiensis* and *T. orionatus* have been established precisely, but slightly more difficult at Site 386, where long intervals are barren of calcareous microfossils. However, our data combined with those of Okada and Thierstein [1979] allow limits to be placed for the boundaries of subzone NC8B (Fig. 12). At Site 511, there is a general paucity of traditional marker species, however, a peak in TOC contents, and a microfossil-poor interval in Core 57 lies in nannofossil subzone NC8B based on the combined occurrence of *Prediscosphaera spinosa* (but not *P. columnata* [see Bralower, 1992]) and *H. albiensis* in a sample from Section 57-6 (Fig. 11). Poor sediment recovery precludes an in-depth study of DSDP Site 370; however, higher TOC contents (up to 7%) are observed in subzone NC8B than in NC8A (Fig. 13).

Foraminiferal biostratigraphy provides age constraints in some of the deep-sea sites. The peak in TOC contents at Site 545 lies within the *H. planispira* Zone (Fig. 9). This zone cannot be differentiated from the *T. bejaouaensis* Zone at Site 511 and the boundary between these combined units and the underlying *H. gorbachikae* Zone cannot be defined exactly based on the paucity of low latitude marker species. The peak in TOC in Core 511-57 lies within this transitional interval (Fig. 11). Although absence of markers inhibits zonal assignment, the interval proposed to correlate to OAE1b at Site 370 based on nannofossil biostratigraphy also appears to be early Albian in age based on the foraminifera *G. blowi* and "*G.*" *gyroidinaeformis* in Section 30-4 (Fig. 13). Cores 27-29 and 31 are largely barren of planktonic foraminifera.

The lower upper Albian carbonaceous Toolebuc Formation, Eromanga Basin, Queensland, Australia recovered in the Augathella Borehole lies in nannofossil subzone NC9B based on the combined occurrence of *Axopodorhabdus albianus* and *Eiffellithus* cf. *E.*

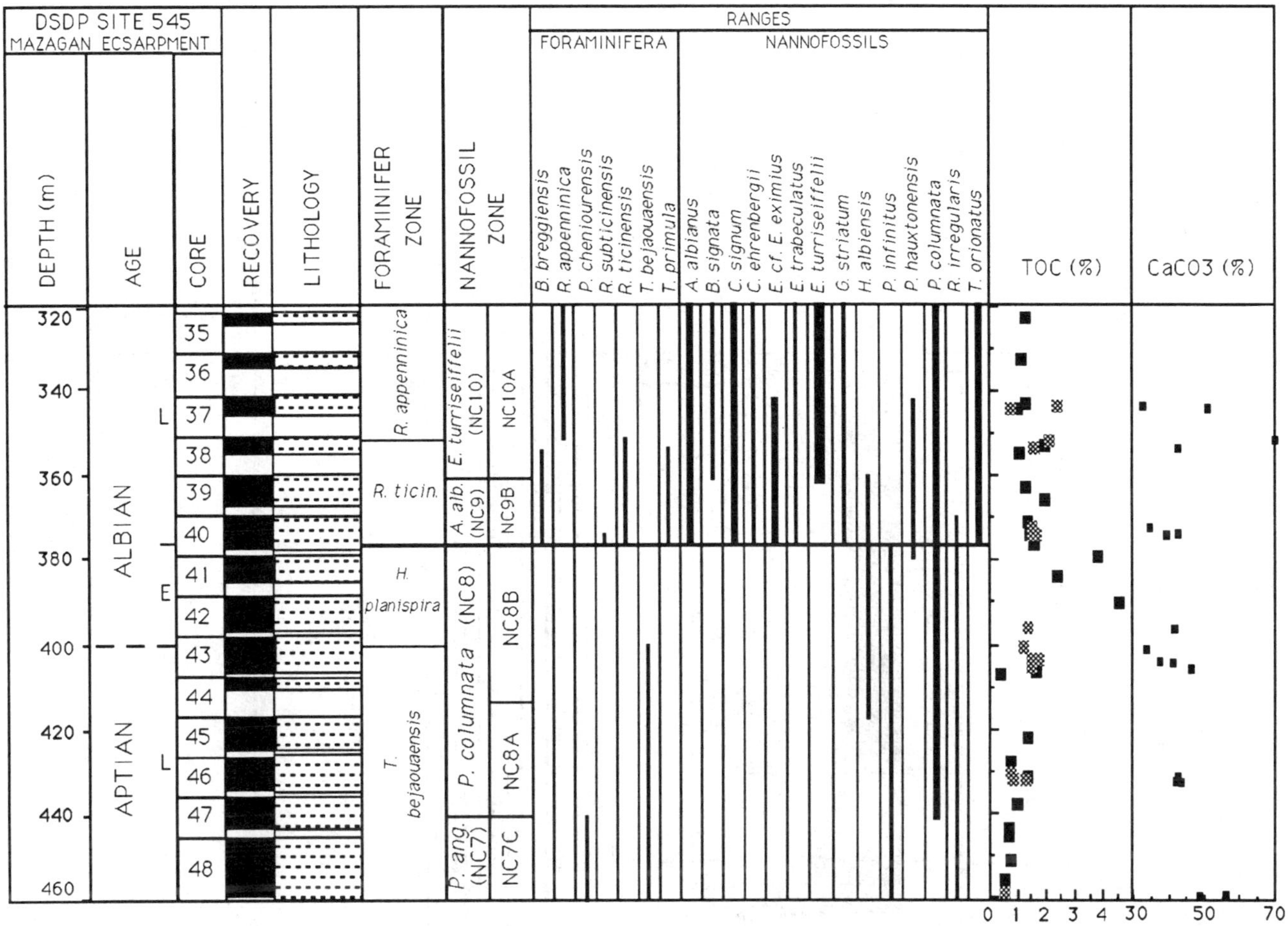

KEY TO FIG.S 5 - 16

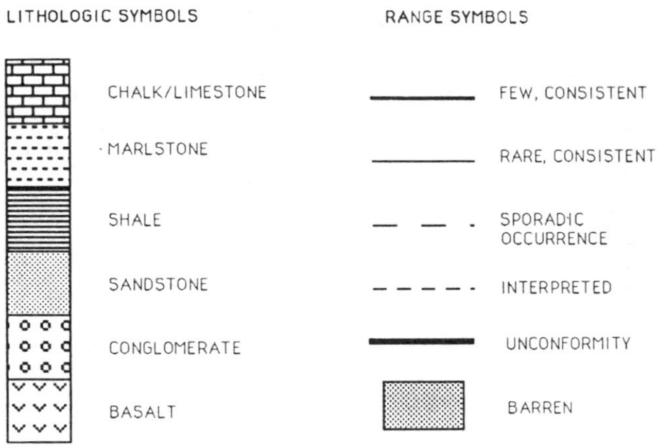

Fig. 9. Upper Aptian-upper Albian stratigraphy of DSDP Site 545, Mazagan Escarpment. Depths are in meters below sea floor. Illustrated are core recovery and generalized lithology. Foraminifera biostratigraphy is modified after Leckie [1984]. Recovery column shows unrecovered intervals in white. Thick horizontal line indicates position of unconformity. Nannofossil biostratigraphy was obtained as part of this investigation. See key for explanation of lithologic and range symbols. Positions of stage and substage boundaries are from nannofossil biostratigraphy. Organic carbon and carbonate contents are from Hinz et al. [1984; grey symbols] and Deroo et al. [1984; black symbols].

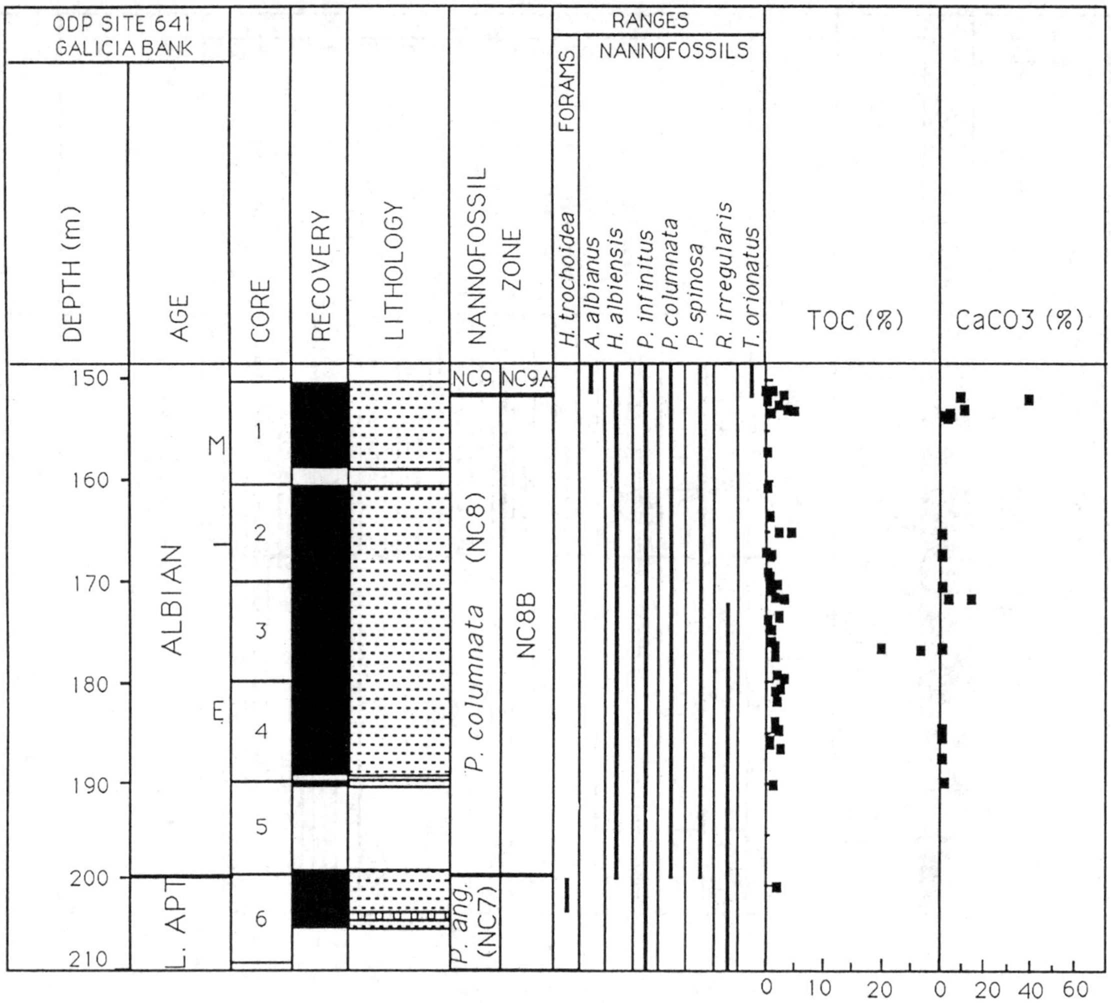

KEY TO FIG.S 5 - 16

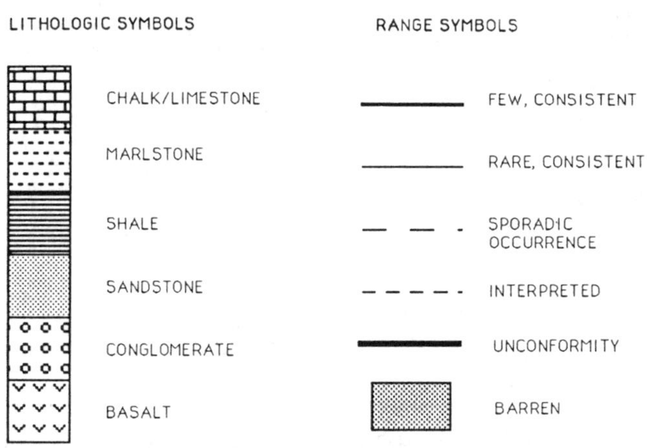

Fig. 10. Upper Aptian-upper Albian stratigraphy of ODP Site 641, Galicia Bank. Depths are in meters below sea floor. Illustrated are core recovery and generalized lithology. Foraminifera and nannofossil biostratigraphy were obtained as part of this investigation. See key for explanation of lithologic and range symbols. Positions of stage and substage boundaries are from nannofossil biostratigraphy. Organic carbon and carbonate contents are from Boillot et al. [1988] and this study.

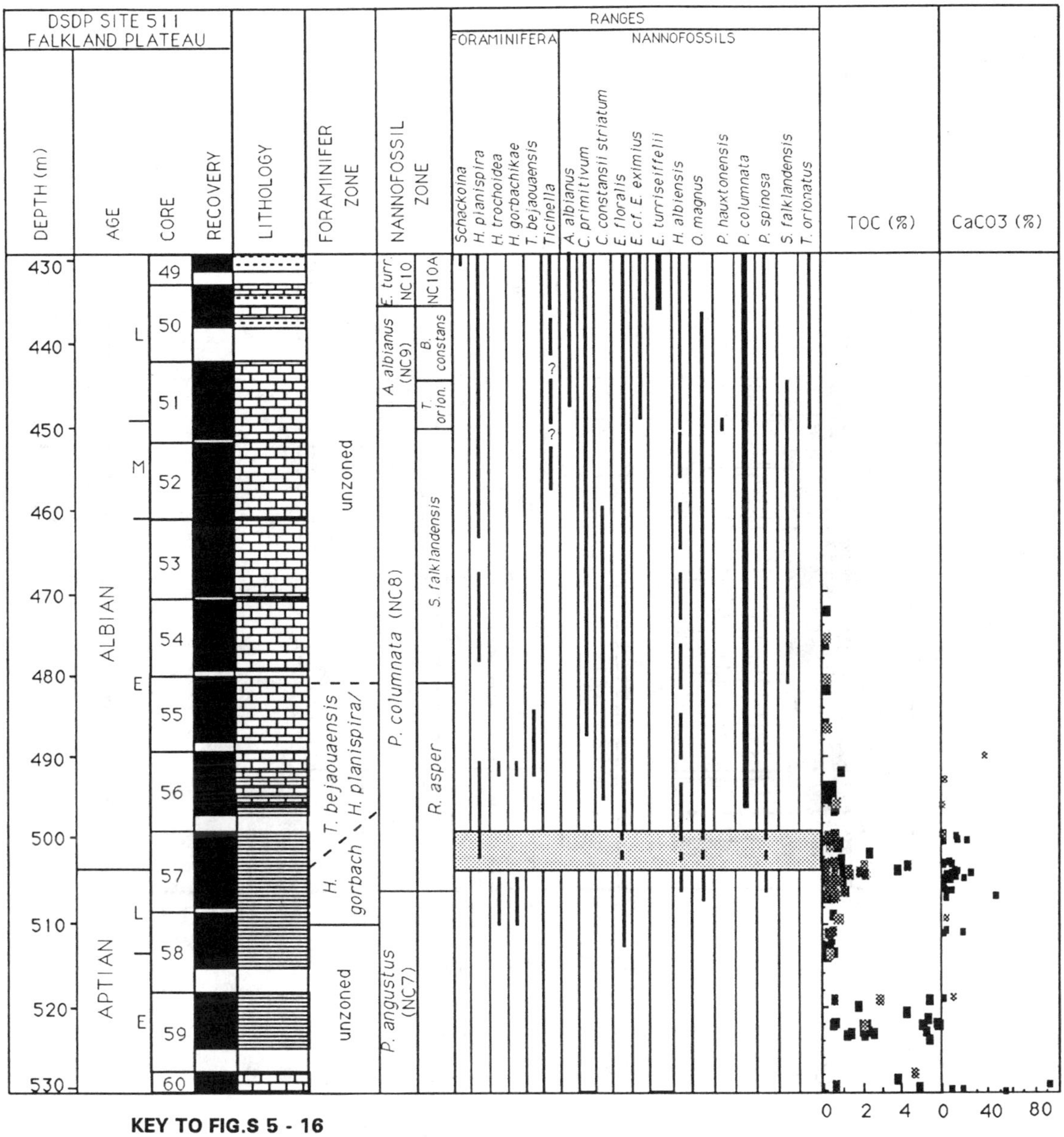

KEY TO FIG.S 5 - 16

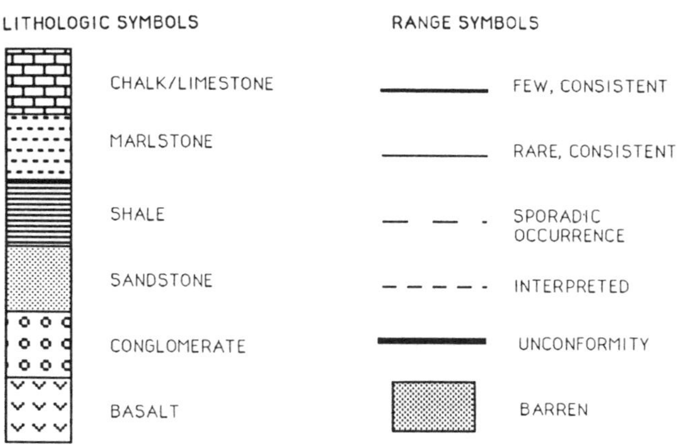

Fig. 11. Aptian-Albian stratigraphy of DSDP Site 511, Falkland Plateau. Depths are in meters below sea floor. Illustrated are core recovery and generalized lithology. Foraminifera biostratigraphy is modified after Krashenninikov and Basov [1983]. Nannofossil biostratigraphy was combined from Wise [1983] and this investigation. Subzones shown are from the high latitude scheme of Wise [1983]. See key for explanation of lithologic and range symbols. Dotted range area indicates depths of unfossiliferous section. Positions of stage and substage boundaries are from nannofossil biostratigraphy. Organic carbon and carbonate contents are from Deroo et al. [1983], Bode [1983], Parker et al. [1983] (all grey symbols)] and this investigation (black symbols).

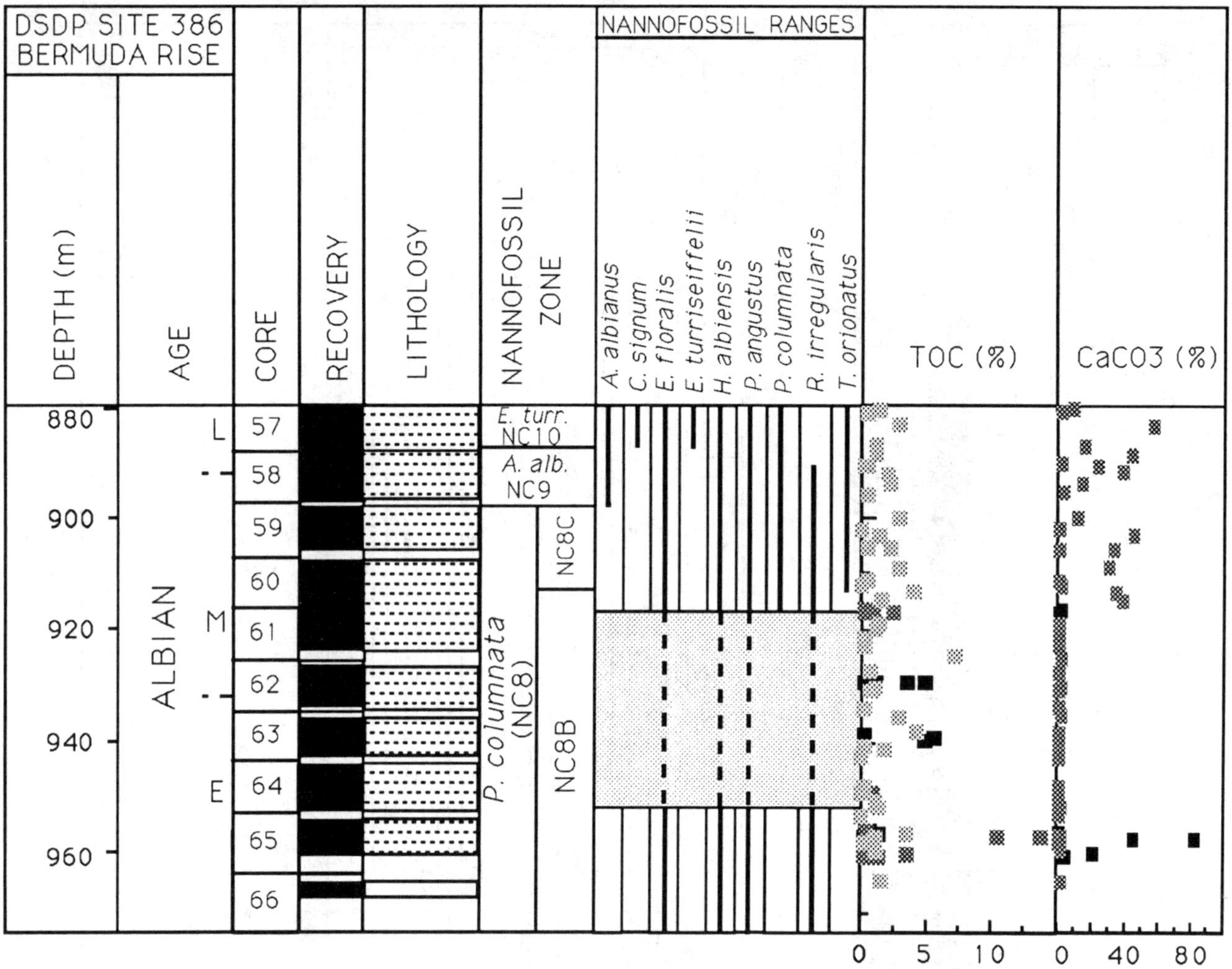

KEY TO FIG.S 5 - 16

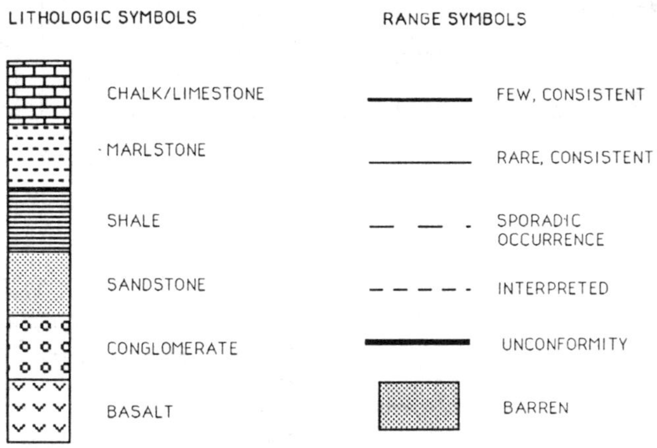

Fig. 12. Albian stratigraphy of DSDP Site 386, Bermuda Rise. Depths are in meters below sea floor. Illustrated are core recovery and generalized lithology. Nannofossil biostratigraphy was combined from Okada and Thierstein [1979] and the results of this investigation. See key for explanation of lithologic and range symbols. Dotted range area indicates depths of unfossiliferous section. Positions of substage boundaries are from nannofossil biostratigraphy. Organic carbon and carbonate contents are from Cameron [1979; light grey symbols] and Bralower [1984; dark grey and black symbols].

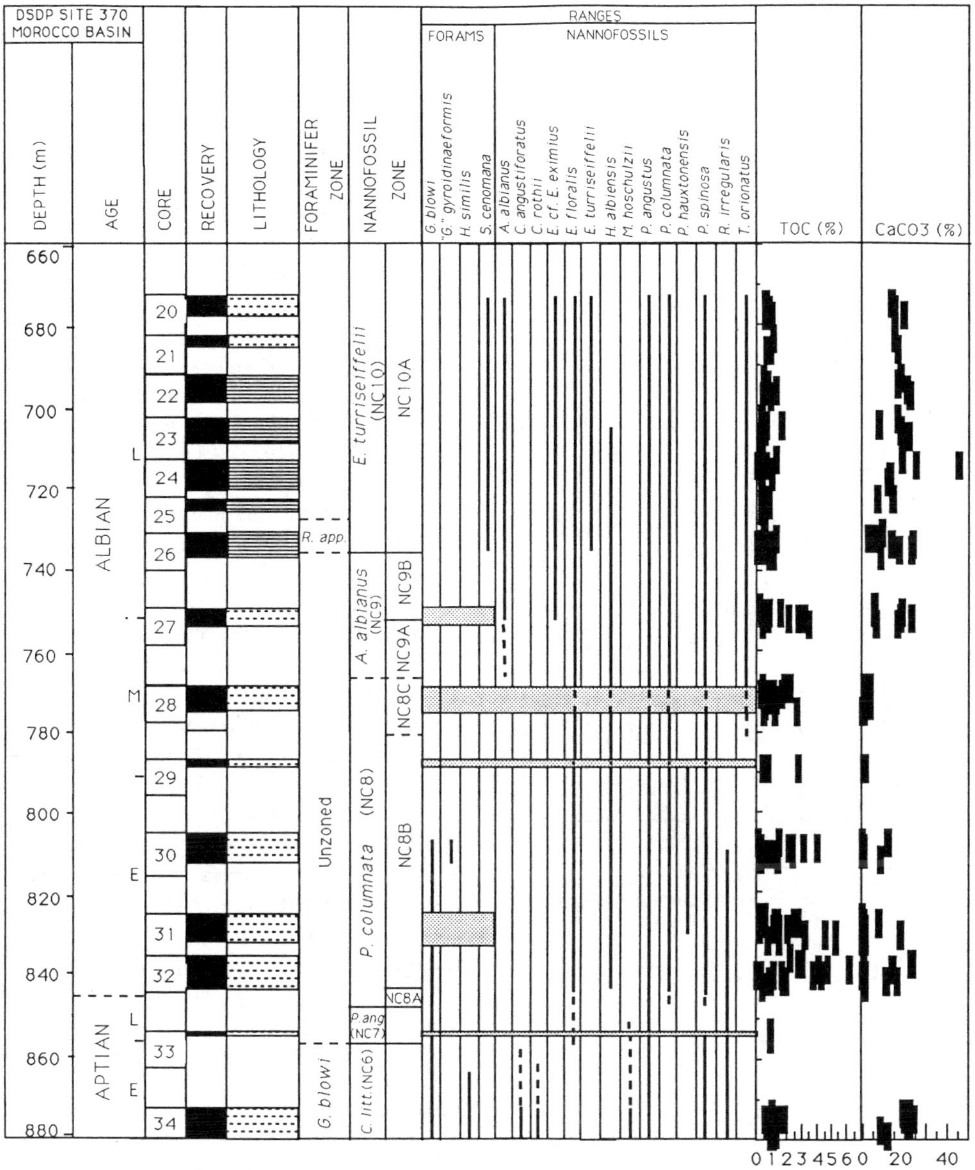

KEY TO FIG.S 5 - 16

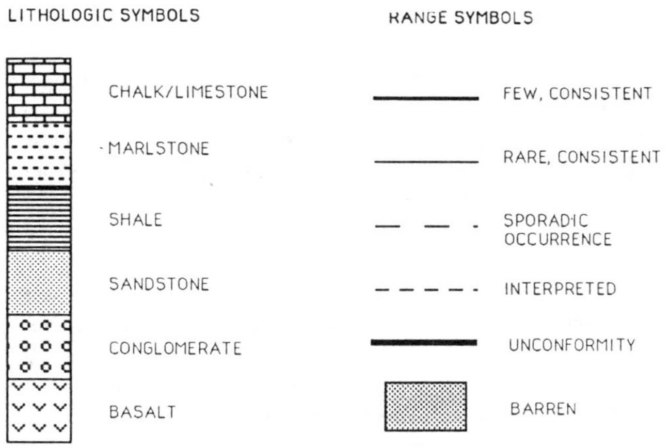

Fig. 13. Lower Aptian-upper Albian stratigraphy of DSDP Site 370, Morocco Basin. Depths are in meters below sea floor. Illustrated are core recovery and generalized lithology. Foraminifera biostratigraphy is modified after Pflaumann and Krashenninikov [1977]. Nannofossil biostratigraphy was obtained as part of this investigation. See key for explanation of lithologic and range symbols. Positions of stage and substage boundaries are from nannofossil biostratigraphy. Organic carbon and carbonate contents are from Erdman and Schomo [1977] and this study.

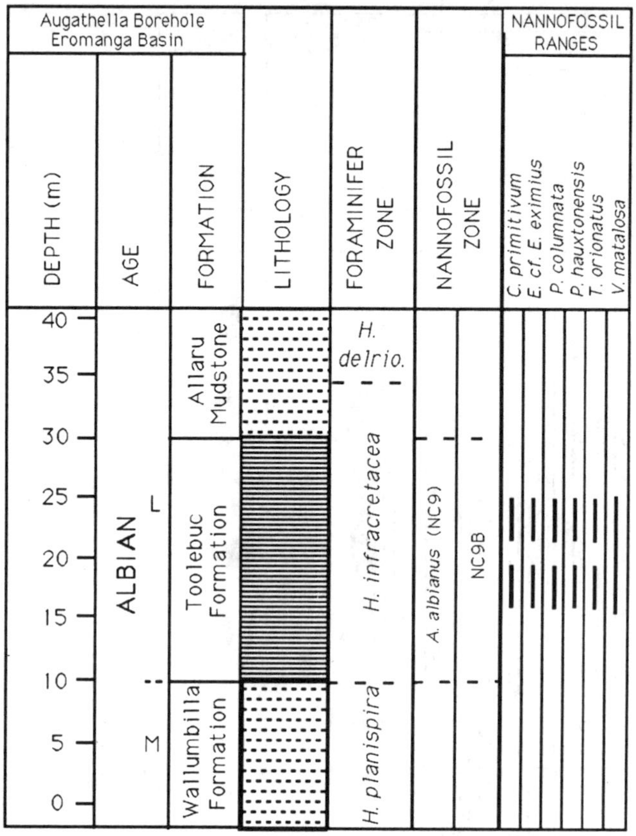

KEY TO FIG.S 5 - 16

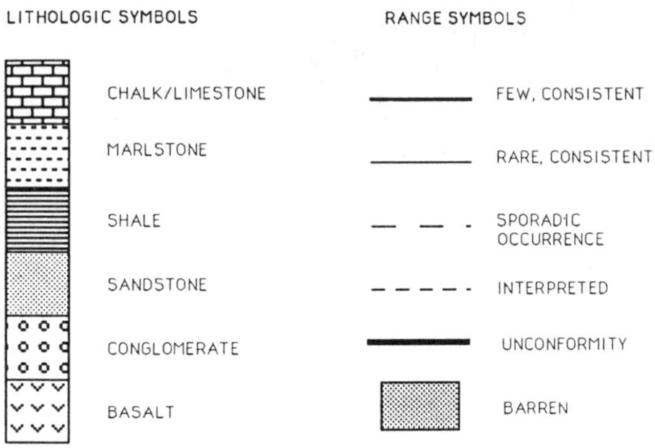

Fig. 14. Middle and upper Albian stratigraphy of Augathella Borehole, Eromanga Basin, Australia. Depths are an arbitrary scheme used in this investigation. Generalized lithology is illustrated. Foraminifera biostratigraphy is modified after Haig [1979]. Nannofossil biostratigraphy was combined from Shafik [1985] and this investigation. See key for explanation of lithologic and range symbols. Positions of substage boundaries are from nannofossil biostratigraphy.

eximius, but absence of *E. turriseiffelii* (Figs. 2, 14; see also Shafik, 1985). It is not possible to define exact zonal and subzonal boundaries in this section as the five fossiliferous samples observed all lie within the same subzone. Correlation of the Toolebuc with peaks in TOC contents in deep-sea sites is less certain than for the other subevents owing to poor recovery in most sections studied (Fig. 5). However, it is clear that this interval corresponds with peaks in TOC contents at DSDP Site 370 from the Morocco Basin (Fig. 13), Site 258 from the Naturaliste Plateau, Indian Ocean (Fig. 15) and Site 400 from the Bay of Biscay (Fig. 16). The TOC peak at Site 258 lies within Core 22, an interval lithologically identical to adjacent cores (Fig. 15). More significant than the absolute magnitude of the TOC peaks at Sites 370 and 400 (Figs. 13 and 16) is the similar decrease in TOC at the base of the overlying *E. turriseiffelii* Zone.

Foraminiferal biostratigraphy provides little information on the timing of proposed OAE1c as the sites investigated are either barren of foraminifera in the late Albian interval (Site 258) or contain a low diversity fauna which lacks diagnostic marker species (Toolebuc Formation; Haig [1979]). The late Albian peak of TOC at Site 400 lies within an interval in the transition between the *T. primula/H. planispira* and the basal *B. breggiensis* Zones. The relevant sediments at Site 370 are largely barren of planktonic foraminifera, however the assemblage in Section 26-4 appears to be latest Albian in age based on the occurrence of *Schackoina cenomana* and *Guembelitria* spp. indicating a possible unconformity between Cores 26 and 27. This assignment in not inconsistent with the correlation of OAE1c at this site.

Results of other studies confirm the identification of these three dysoxic/anoxic episodes. The lower Aptian TOC-rich Fischschiefer of northern Germany is a likely equivalent of OAE 1a [Gaida et al., 1981]. TOC contents are high throughout the Aptian interval of DSDP Site 361 in the Cape Basin, South Atlantic, but reach a peak of 25% in the *Chiastozygus litterarius* nannofossil Zone [Proto Decima et al. [1978]; Herbin et al. [1986, 1987]. Poor nannofossil preservation does not allow subzonal subdivision of this section. Koutsoukos et al. [1991] identified a peak in TOC contents in the lower Albian of several wells from the Sergipe Basin, offshore Brazil. Although no precise biostratigraphic control was cited, these sediments are possible equivalents of OAE 1b. The lower Albian bituminous Pariatambo Formation of Peru (*Douvilleiceras mammilatum* and *Hoplites dentatus* ammonite Zones; Benavides-Caceres [1956]; Jaillard [1987]) also is a likely equivalent. TOC-rich black shales of the Skull Creek Formation of the U.S. Western Interior (*Inoceramus bellvuensis* and *I. comanchaeanis* Zones of the early late Albian; Kauffman et al., in press) are likely equivalents of OAE 1c.

The occurrence of TOC-rich horizons in the three proposed dysoxic/anoxic intervals is by no means ubiquitous at all of the sites investigated. Figures 17a-c illustrate the geographic distribution of all sites investigated for the three intervals of interest. These figures differentiate, for each proposed subevent, sites at which TOC-rich sediments exist, or which show diagnostic changes in microfossil assemblages, and those which possess no evidence for dysoxia/anoxia. For these same sites, Figure 17d shows occurrences of TOC-rich sediments which lie in other calcareous nannofossil subzones. Certain localities are characterized by high TOC at times other than the proposed episodes. For example, the Aptian-Albian section in the Umbrian Apennines contains numerous other black shale horizons beside the "Livelli Selli" and "Urbino",

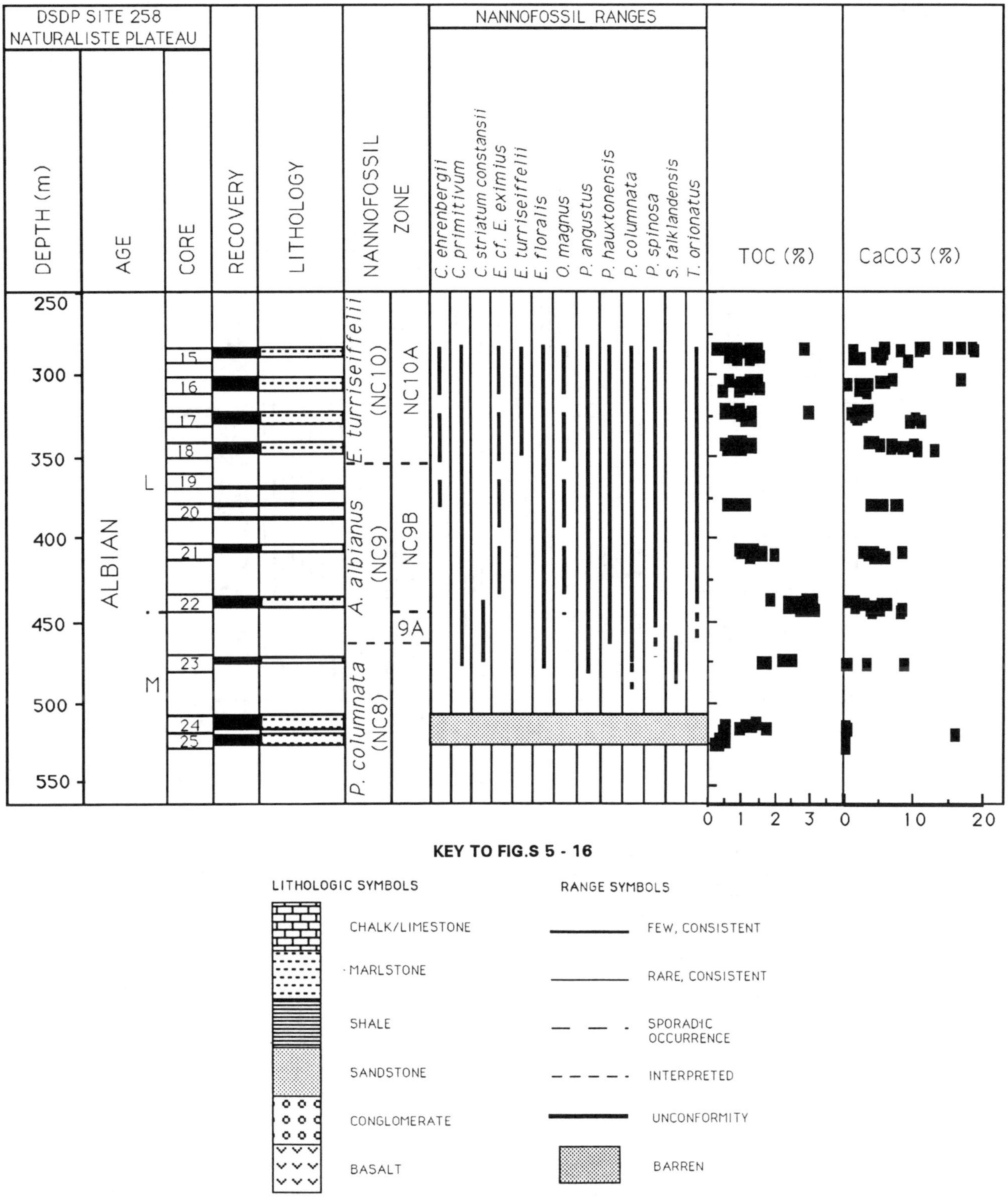

Fig. 15. Middle and upper Albian stratigraphy of DSDP Site 258, Naturaliste Plateau. Depths are in meters below sea floor. Illustrated are core recovery and generalized lithology. Nannofossil biostratigraphy was combined from Thierstein [1974] and this investigation. See key for explanation of lithologic and range symbols. Positions of substage boundaries are from nannofossil biostratigraphy. Organic carbon and carbonate contents are from Davies et al. [1974] and this investigation..

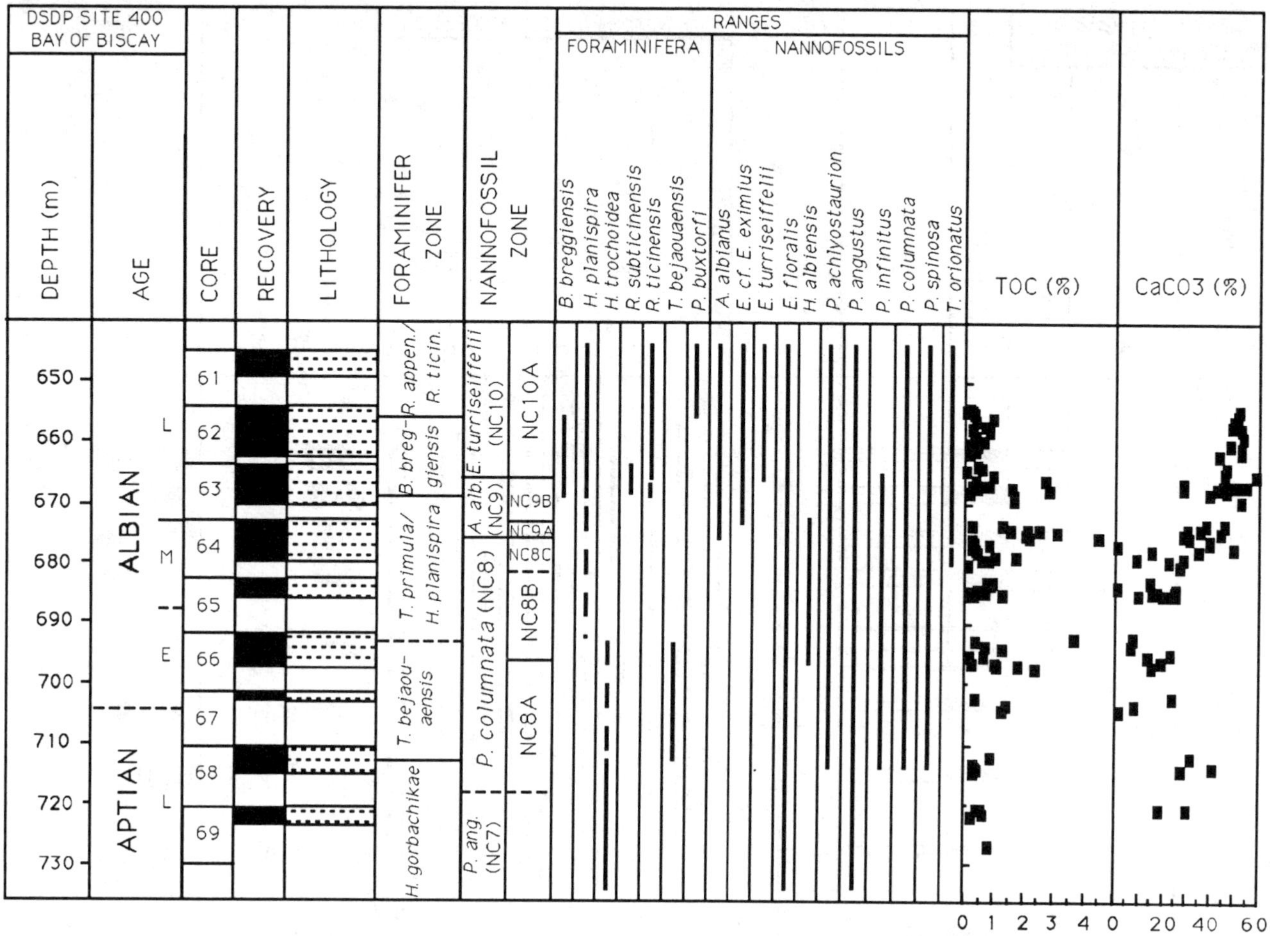

KEY TO FIG.S 5 - 16

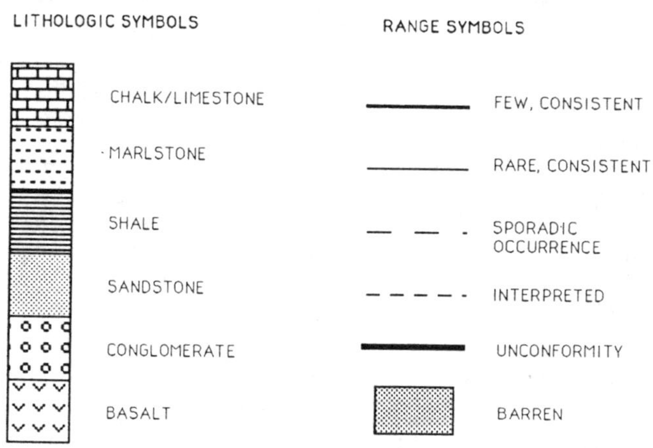

Fig. 16. Middle and upper Albian stratigraphy of DSDP Site 400, Bay of Biscay. Depths are in meters below sea floor. Illustrated are core recovery and generalized lithology. Foraminifera biostratigraphy is modified after Depeuble [1979]. Nannofossil biostratigraphy was obtained as part of this investigation. See key for explanation of lithologic and range symbols. Positions of substage boundaries are from nannofossil biostratigraphy. Organic carbon and carbonate contents are from Deroo et al. [1979] and this investigation.

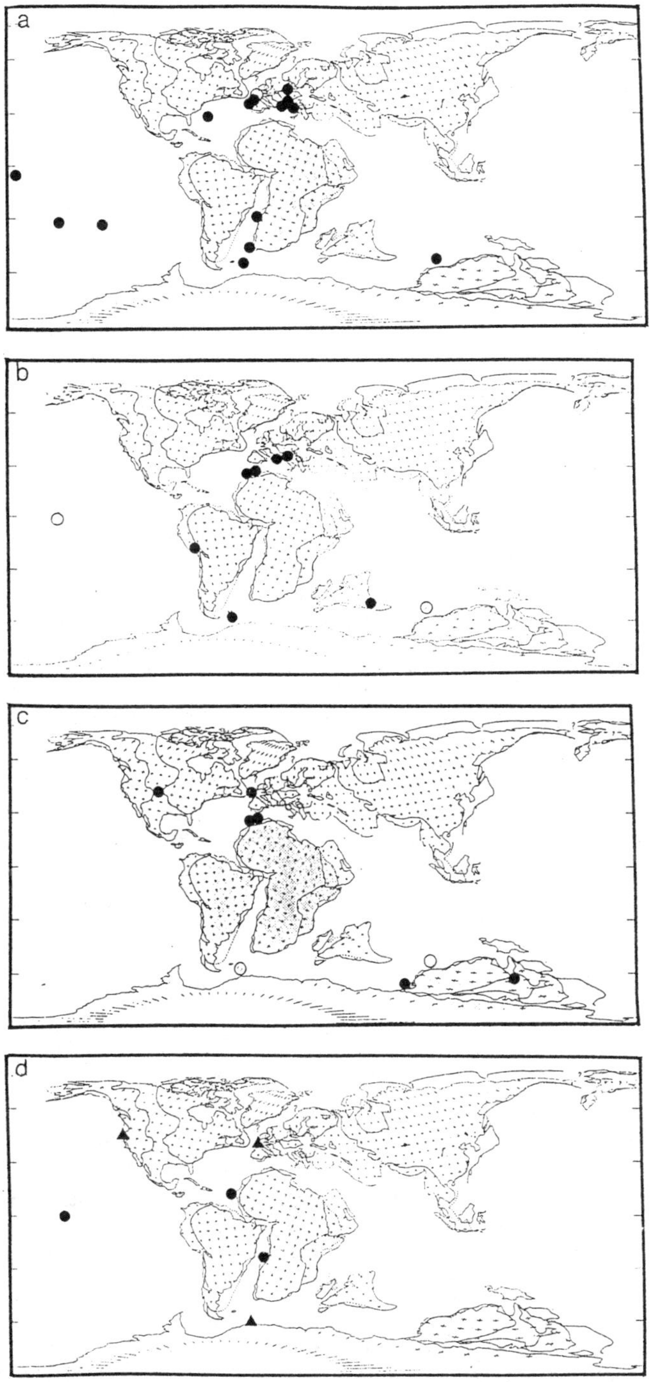

Fig. 17a-c Location of sites which show evidence of anoxia/dysoxia in the three proposed dysoxic/anoxic episodes. a. Early Aptian OAE1a, b. early Albian OAE1b and c. early late Albian OAE1c. These sites are shown by filled circles. Open circles indicate locations of sites which show no evidence of dysoxia/anoxia in these episodes. d. Location of sites which show evidence of anoxia/dysoxia in other intervals of the Aptian-Albian: triangles indicate late Aptian (*P. angustus* nannofossil Zone and subzone NC8A), squares indicate late early Albian-middle Albian (nannofossil subzones NC8C and NC9A) and circles indicate late late Albian (*E. turriseiffelii* nannofossil Zone). Paleogeographic reconstruction of Barron et al. [1981].

but these other horizons usually have lower organic carbon contents and can be differentiated by other geochemical and microfossil parameters [e.g., Pratt and King, 1986; Premoli Silva et al., 1989]. DSDP Site 465 on the Hess Rise in the paleo-equatorial Pacific includes a thick section of laminated, TOC-rich limestone in the upper Albian *E. turriseiffelii* nannofossil Zone. DSDP Site 402 in the Bay of Biscay recovered sediments with low TOC in the lower Aptian *C. litterarius* nannofossil Zone, and sediments with higher TOC contents in the overlying *P. angustus* Zone (Fig. 3). Certain sites (e.g., ODP Site 763, Exmouth Plateau) contain no evidence of high TOC or laminated sediments in the intervals of the proposed Albian subevents. South Atlantic DSDP sequences (e.g. Sites 361 and 364) contain TOC-rich Albian sediments, but with insufficient data and biostratigraphic control [Arthur and Natland, 1979; Graciansky et al., 1982; Müller et al., 1983] to be interpreted at this point.

At this time there is an insufficient data set to carry out a statistical analysis of the significance of correlation of the three subevents defined. In other words, it is not possible to quantify how closely all available TOC-rich samples correlate to the three subevents or, alternatively, how closely they approach a random distribution. In a purely qualitative sense there is a strong stratigraphic clustering of TOC-rich samples to the early Aptian subevent. However, as mentioned, in other sequences, this subevent is recognized by criteria other than TOC. Partly because of the smaller number of sites, there appears to be a weaker temporal clustering of TOC-rich horizons to the two Albian subevents. However, the validity of these episodes is strengthened considerably by the stratigraphic correlation of carbonaceous horizons deposited in deep-sea sites with prominent TOC-rich units which were laid down in epicontinental settings. Obviously, a final quantitative analysis of the objective basis of all of the subevents will have to await the recovery of more continuous Aptian-Albian sections and the refinement of zonations and chronologies.

Because of the various uncertainties described, the biostratigraphic intervals defined (Fig. 2) are considered as time "envelopes" for these subevents, the duration of which, at any given locality, may have been considerably shorter. Moreover, it is not possible at present, given current biostratigraphic resolution, to distinguish whether the subevents were synchronous or somewhat diachronous in different locations. This question is extremely difficult to answer in this interval in which magnetostratigraphy is inapplicable. The duration of any Mesozoic event is difficult to estimate because of the paucity of reliable radiometric ages, and uncertainties in the correlation of available radiometric ages with biostratigraphy. However, we can make some preliminary calculations. The durations of nannofossil subzones NC8B (4.2 m.y.) and NC9B (2.7 m.y.) were estimated assuming constant sedimentation rates at Site 763 utilizing the timescale of Harland et al. [1990]. We assume a duration of 0.56 m.y. for magnetic chron M0 [Harland et al., 1990] to estimate sedimentation rates at Site 641. Based on sedimentation rates and thicknesses of the correlative units at Site 641 (OAE 1a), Col de Palluel (OAE 1b) and Site 258 (OAE 1c), the durations of the subevents are as follows; OAE 1a and 1b: approximately 0.6 m.y., and OAE 1c: approximately 0.8 m.y.

A significant geographic trend emerges from the admittedly small data set analyzed. The early Aptian subevent appears to have affected a much larger number of sites than the later two subevents. These include locations in a variety of different paleogeographic settings such as open ocean plateaus, continental margins and epicontinental seaways (Fig. 17a). The other two subevents affected an equally widespread set of locations and oceanographic settings (Figs. 17b,c). However, it is clear that dysoxic/anoxic watermasses had a much patchier extent during these subevents as there are a large number of locations which show no evidence of dysoxia/anoxia. Site 763 provides a particularly striking comparison in this regards. The nearly complete upper Barremian-middle Turonian section recovered at this site includes TOC-rich sediments in the early Aptian (OAE 1a; Fig. 17a) and at the Cenomanian/Turonian boundary (OAE 2; e.g., Bralower and Siesser, 1992), however, the apparently complete Albian section contains no evidence of significant TOC enrichment resulting from OAEs 1b and 1c (Figs. 17b,c).

CHANGES OF PLANKTON ASSEMBLAGES

The effect of oxygen depletion on establishment of benthic communities had led to the association of such environments with intervals in sedimentary sections in which micro and macrobenthos are sparse or entirely absent [e.g., Eicher and Worstell, 1970; Savrda et al., 1984; Sageman et al., 1991]. In several sequences investigated, the intervals of the proposed dysoxic/anoxic episodes correspond to the absence of calcareous nannofossils and/or planktonic foraminifera. Most of these intervals are extremely enriched in calcified and pyritized radiolarians [e.g., Sigal, 1979; Thurow, 1988]. A complete lack of carbonate is observed in some of the most TOC-rich units such as the "Livello Selli".

In all of these cases, early diagenetic dissolution of relatively low amounts of carbonate may have occurred accompanying the oxidation of large quantities of organic matter [e.g., Emerson and Bender, 1981]. At other sites, such as DSDP Sites 398 and 511, and ODP Site 641, there are intervals in which all calcareous plankton have disappeared, and others in which planktonic foraminifera are absent but nannofossils are abundant, and in some cases, well preserved. Nannofossils are, in general, less susceptible to dissolution during diagenesis [Schlanger and Douglas, 1974] and this may explain their diversity and abundance in samples without planktonic foraminifera. However, the absence of obvious dissolution of the nannofossils in several samples from these sites within critical intervals indicates that the lack of microfossil carbonate in other samples within critical intervals may be, at least partially, of primary origin. Additionally, there is no apparent correlation between TOC and microfossil preservation in the sections studied, i.e. in some cases higher TOC values correspond to better microfossil preservation. However, as it is not possible to reconstruct the history of oxidation of organic matter in an individual horizon, a diagenetic interpretation cannot be totally ruled out.

It is possible that there was a rise in the CCD coincident with the onset of the overall OAE1 episode [e.g., Thierstein, 1979] and that the paleodepth of many of the sites studied was below the CCD for a short period of time. We view this as an unlikely explanation of the entire phenomenon due to the large range of paleodepths of the investigated sites, some of which are at least as shallow as several hundred meters (e.g., Vergons Section). Theoretically, nannoplankton and planktonic foraminifera should respond differently to dysoxic/anoxic conditions and associated changes in water column properties because of significant differences in life modes, habitats and ecological tolerances. Local disappearance of planktonic foraminifera may have been induced by changes in oxygenation or salinity. These organisms have fairly narrow salinity tolerances which do not differ by more than 5-6 ppt from open ocean values [e.g., Bé, 1977; Lipps, 1979; Hemleben et al., 1989]. Nannoplankton in general tend to be more tolerant to changes in water column conditions. Rare modern coccolithophores can endure salinities up to 250 ppt and below 5 ppt. However, the salinity tolerance of common species, such as *Emiliania huxleyi*, lies between approximately 16 and 40 ppt (see summary in Tappan, 1980). High latitude DSDP Site 511 on the Falkland Plateau is a location in which lowered salinity might have led to a low diversity, moderately well-preserved nannoflora and to the exclusion of planktonic foraminifera (Cores 511-60 to -62). The development of an oxygen-minimum zone also may have had a profound effect on planktonic foraminiferal assemblages, leading to a low diversity fauna of very shallow-dwelling opportunistic species [e.g., Caron and Homewood, 1983; Bréhéret et al., 1986; Arthur et al., 1987; Leckie, 1987].

A feasible explanation for the absence of nannoplankton in many of these TOC-rich intervals is related to productivity. These sediments are often enriched in dinoflagellates [Habib, 1979], and it is therefore possible that nannofossils were uncompetitive during a brief interval of high nutrient supply to surface waters. The production of modern coccolithophores, such as *E. huxleyi*, decreases in highly turbulent/highly fertile water masses [Margalef, 1978]. This is not a result of intolerance of coccolithophores to any ecological variable, but rather that diatoms and dinoflagellates have higher potential growth rates [Raymont, 1980], utilize nutrients more efficiently, and therefore outcompete coccolithophores under these conditions. Coccolithophores also become less competitive than dinoflagellates in highly stratified/low turbulent/low fertility water masses because they do not have resting cysts or functional flagella which allow them to leave the uppermost part of the water column in search of nutrients [Margalef, 1978]. Modern planktonic foraminifera seem to be more competitive than nannoplankton in eutrophic regimes [e.g., Hemleben et al., 1989], producing higher total abundances but lower species diversities than in oligotrophic settings. As with oxygenation, unstable, eutrophic environments led to the proliferation of opportunistic foraminifer taxa which can effectively compete with radiolarians and other zooplankters for food. There is no record of diatom productivity in these sections that might support an increase in nutrient supply, either because they were not present or abundant in the Early Cretaceous or because of dissolution and silica diagenesis. Our working hypothesis, therefore, is that these units which are barren of calcareous microfossils have a primary origin and were produced during brief eutrophic intervals.

One of the most pervasive questions concerning mid-Cretaceous organic carbon-rich sediments is whether they formed as a result of high oceanic productivity or effective preservation of organic carbon that was produced [e.g., Schlanger and Jenkyns, 1976; Waples, 1983]. Much of the debate is centered on the suitability of modern environments, such as the anoxic Black Sea, as modern analogs for ancient black-shale forming environments [e.g., Demaison and Moore, 1980; Pedersen and Calvert, 1990]. The largest source of error in these studies lies not with the calculation of organic carbon accumulation rates [e.g., Calvert et al., 1991], but rather with estimation of primary production rates which are known to be variable on short time-scales and difficult to measure accurately. Our paleoecological interpretation of these calcareous planktonic-free intervals follows some earlier ideas expressed by Arthur et al. [1987] and Leckie [1989] who attributed biotas of organic-rich horizons to higher productivity. We point out that such high productivity is intimately coupled with effective preservation of organic material on the sea floor in an oxygen-minimum zone setting in modern environments in which upwelling is perennial. The suggestion that individual TOC-rich units (which represent fractional zonal durations of less than 1 m.y.) were produced as a result of higher fertility conditions does not negate the conclusions that Bralower and Thierstein [1984], Dean et al. [1986] and Arthur et al. [1988] reached independently using different techniques, that average long-term mid-Cretaceous productivity (estimated at zonal durations of several millions of years) was a fraction of that at the present day. We also stress that this conclusion does not imply that other black shale horizons could not have been deposited under dramatically different conditions [e.g., Herbert and Fischer, 1986; Erba et al., 1989], and in particular, lower fertility.

PLANKTON EVOLUTION AND DYSOXIA/ANOXIA

Marine plankton experienced considerable diversification and turnover during the mid-Cretaceous [e.g., Tappan and Loeblich, 1973; Roth, 1987, 1989; Leckie, 1989]. Barremian-Aptian time was characterized by the first major radiation of planktonic foraminifera in terms of numbers of taxa and morphologic diversity [Leckie, 1989]. This same interval witnessed the demise of the dominant low-latitude assemblage of calcareous nannofossils, including *Nannoconus*, and the radiation of a more cosmopolitan coccolithophorid assemblage [Bergen, 1992]. Roth [1987] suggested that black shale deposition during the mid Cretaceous was associated with rising sea level, an increase in oceanic stratification, a decrease in oceanic fertility, and an increase in calcareous nannofossil diversification rates. Needless to say, there are numerous errors involved in the calculation of diversification and turnover rates and problems with the analysis of measures of species diversity but we show general trends in Figure 18. The problems with synthesizing such data are discussed in detail in the publications from which the data are taken. The relationship between calcareous plankton evolution and the development of widespread dysoxia/anoxia is complex as evidenced by the different diversity patterns between the three Oceanic Anoxic Subevents (Fig. 18).

Early Aptian OAE1a, probably the most widespread of the three Oceanic Anoxic Subevents, developed during a time of radiation in both the nannofossils and planktonic foraminifera (Fig. 18). According to Roth [1987], rates of nannofossil diversification peaked during early Aptian time. The planktonic foraminifera were in the midst of their first major radiation. The first presumed deeper dwelling morphotypes (e.g., *Ticinella* and *Planomalina*) did not appear until late Aptian time, and therefore, if OAE1a was associated with the development or expansion of an oxygen minimum zone, it would probably not have had a deleterious effect on a planktonic foraminiferal population composed exclusively of near-surface dwellers. Roth [1987] suggested that this time of overall nannofossil diversification represents the filling of niches to attain equilibrium diversity. This interpretation may be particularly relevant for the wide open niche-space being exploited by the planktonic foraminifera. In a study of the evolutionary history of Neogene planktonic foraminifera, Wei and Kennett [1986] suggested that equilibrium diversity is determined by the availability of resources or ecospace in a given physical environment. The late Barremian-early Aptian, somewhat similar to the early Miocene, was a time of overall diversification, where rates of speciation exceeded rates of extinction, despite the tectonic, climatic or oceanographic changes responsible for widespread dysoxia/anoxia during OAE1a.

In certain hemipelagic sites, blooms of nannoconids occur slightly above OAE 1a (e.g. at Site 641 and Gorgo a Cerbara), but species richness does not appear to change significantly. Nannoconid abundance increases in continental margin and epicontinental settings [e.g., Roth and Krumbach, 1986], but it is not possible to attribute this increase to an individual paleoecologic variable. It has been postulated [Busson and Noël, 1991] that this group are resting cysts produced by calcareous dinoflagellates in response to adverse conditions, in an analogous fashion to modern *Thoracosphaera*. While there are abundance peaks in certain samples, the late Barremian-early Aptian interval is accompanied by a significant overall decrease in the abundance of nannoconids (e.g., Coccioni et al., in press). The extinction of the dominant taxon, *N. steinmannii* occurred shortly after OAE1a.

Early Albian OAE1b, on the other hand, developed during a time of reduced simple diversities in the calcareous plankton and dinoflagellates, and reduced rates of nannofossil turnover and diversification (Fig. 18). Nearly all presumed deeper dwelling species of planktonic foraminifera disappeared, including the keeled morphotype [e.g., Leckie, 1989; Premoli Silva et al., 1989]. *H. planispira* dominates faunas in its nominate zone. Leckie [1989] suggested that this loss of planktonic foraminiferal diversity and morphologic variety was in response to the decrease of density gradients in the upper water column and proposed that increased rates of formation of warm, saline intermediate and/or deep water masses characterized latest Aptian and early Albian time. An alternative hypothesis, although physically improbable, is that an expanded oxygen-minimum zone occasionally reached to within tens of meters of the sea surface and thereby impinged on the niche of the deeper dwelling species of planktonic foraminifera. Increased oceanic fertility at this time may also have favored opportunists amongst the plankton and thus tended to lower overall

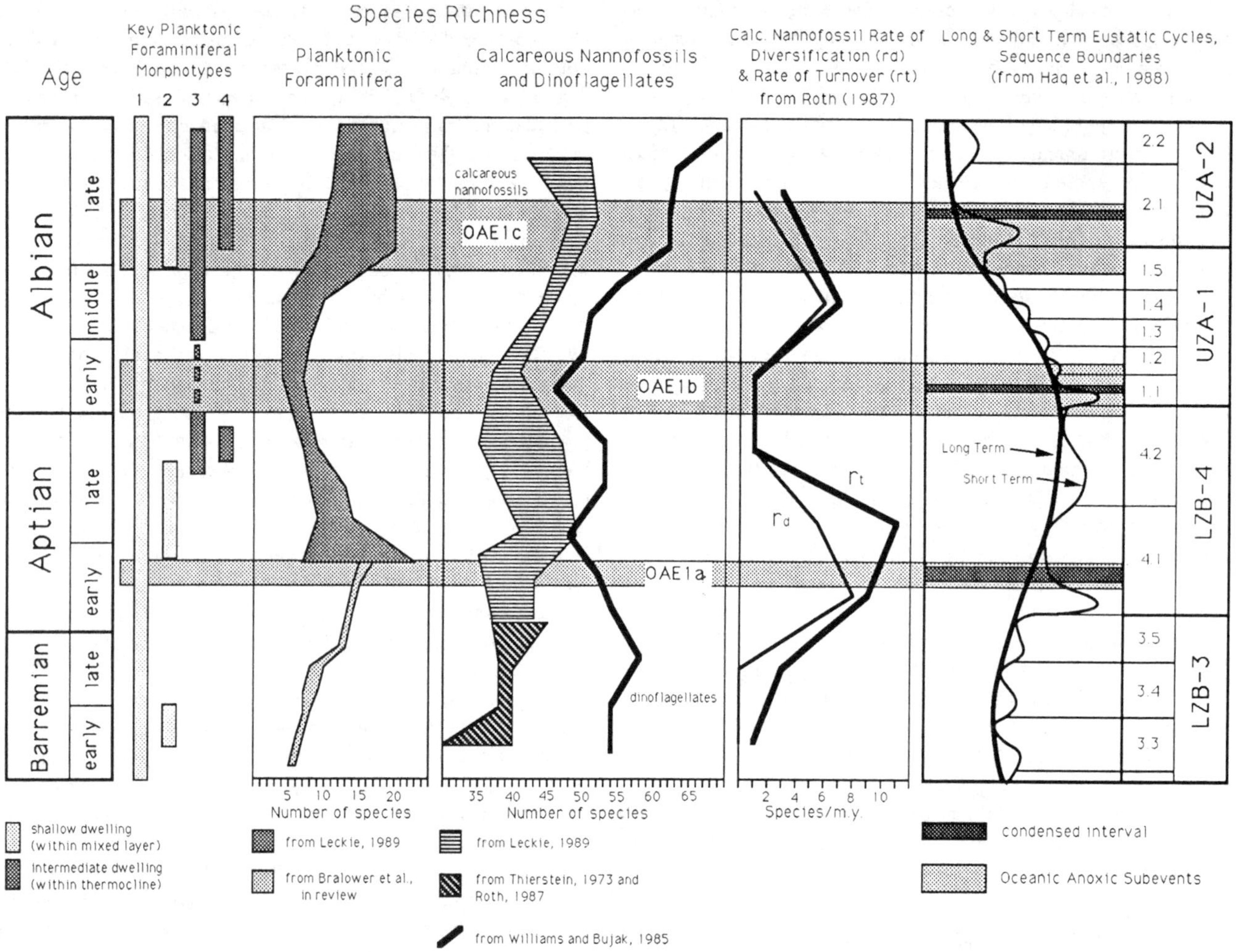

Fig. 18. Microfossil evolution during the late Barremian to Albian and relation to eustatic change. Illustrated from left are: A. Planktonic foraminiferal morphotypes: 1. Trochospiral and planispiral taxa (e.g., *Hedbergella, Globigerinelloides*); 2. Taxa with radially elongate chambers (e.g., *Schackoina, Leupoldina, Clavihedbergella*); 3. Taxa with complex apertural systems (e.g., *Ticinella*); 4. Keeled taxa (e.g., *Planomalina, Rotalipora, Praeglobotruncana*). B. Species Richness (number of species observed in individual samples; data sources indicated at base of figure) of planktonic foraminifera, calcareous nannofossils and dinoflagellates. C. Calcareous nannofossil rate of diversification and rate of turnover in number of species per million years [from Roth, 1987]. D. 2nd and 3rd order eustatic cycles [from Haq et al., 1988]. Also illustrated are the stratigraphic distribution of OASEs and condensed intervals. See text for full discussion.

diversity [Roth, 1987; Leckie, 1989]. These hypotheses are not mutually exclusive. Where observed, ammonite communities may indicate the effects of dysoxia/anoxia or related factors; the "Niveau Paquier" is dominated by Boreal and juvenile specimens of Tethyan taxa [Bréhéret et al., 1986] suggesting cooler, more fertile waters or a stressed environment. Calcareous nannofossil community structure does not appear to change nearly as dramatically. Abundance peaks of *Nannoconus* occur in coincidence with OAE 1b in some sites (e.g. in the "Niveau Paquier"; Bréhéret [1983]), also suggesting stressed or fertile surface waters.

Calcareous nannofossils experienced a marked increase in diversification rates during the mid-Albian, which began, perhaps coincidentally, in the same general time interval as OAE1b. Planktonic foraminifera, on the other hand, had an explosion in diversity and turnover rates during the late Albian, roughly coincident with the development of OAE1c (Fig. 18). The diversification of planktonic foraminifera during the late Albian signalled the development of strong density gradients close to or within the photic zone. New high-latitude water masses may have developed as sea level rose, epicontinental seas grew, and warm, saline water mass sources in the low latitudes were shut down, thereby permitting the establishment of enhanced upper water column density gradients [Leckie, 1989].

The evolutionary history of the calcareous plankton was very different between the time of OAE1a (early Aptian), OAE1b (early Albian), and OAE1c (late Albian) perhaps suggesting that the oceanographic processes responsible for the development of dysoxia/anoxia differed between these times, or alternatively, that factors controlling the preservation and accumulation of organic matter in the deep sea were decoupled from the uppermost water column where plankton evolution occurred. In addition, the evolutionary trends considered here are much longer-term phenomena than the relatively brief dysoxic/anoxic subevents. Even though nannoplankton and planktonic foraminiferal communities may have been affected by dysoxic/anoxic episodes in the Aptian-Albian, it is clear from the high diversity in samples overlying these intervals that the flora and fauna quickly reestablished themselves. The early Albian episode may have had a profound effect on the evolution of the calcareous plankton as this interval appears to correlate with major faunal and floral turnover and significantly reduced simple diversities. Both the early Aptian and early late Albian episodes come in the middle of an interval of diversification of nannofossil and planktonic foraminiferal assemblages. However, this diversification started earlier than the onset of dysoxic/anoxic conditions and there is no exact correlation with the few extinctions which occur in these intervals [Leckie, 1989; Bralower et al., submitted; Fig. 18]. The Aptian-Albian OAE has previously been evoked as having a major effect on the evolution of calcareous plankton [Roth, 1987, 1989; Leckie, 1989]; however this was clearly an interval of rapid paleoceanographic change and it is difficult to associate this evolution with an individual parameter.

The relative stasis of nannoplankton assemblages during the Aptian-Albian subevents compared to those of planktonic foraminifera and ammonites is similar to the situation at the Cenomanian/Turonian boundary OAE. In this latter interval, foraminifera and ammonites underwent dramatic changes in community structure, including decreased diversity and extinction of deeper-dwelling taxa [e.g., Hart, 1980; Caron and Homewood, 1983; Leckie, 1989; Elder, 1989], whereas nannoplankton community structure and diversity did not change significantly [Watkins, 1986; Bralower, 1988].

CAUSES OF "OCEANIC ANOXIC SUBEVENTS"

The correlation of many Aptian-Albian TOC-rich horizons to these three "oceanic anoxic subevents" indicates operation of a forcing mechanism over a widespread set of sites and oceanographic locations. Our results indicate that two of these events (OAE 1b and OAE 1c) had patchier, although apparently as widespread distributions as the other subevent (OAE 1a). Also, when compared in detail, it is clear that each episode left a variable stratigraphic record in each site. The stratigraphic and geographic variability displayed by these subevents leads us to distinguish between ocean-wide forcing mechanisms which caused water masses to be prone to oxygen-deficiency during the three time "envelopes" and factors which operate on a regional or even local scale, which actually induced or triggered dysoxia/anoxia in different water masses. In the following we briefly explore some of these global and regional forcing mechanisms [see Allard et al., in prep. for a more detailed analysis].

The concatenation of a number of global forcing mechanisms caused the oceans to be susceptible to oxygen deficiency during the three "ocean anoxic subevents". Evidence exists in lower Aptian sections in the Pacific Ocean basin that there may have been a relationship between volcanism and dysoxia/anoxia; for example, Vogt [1989] proposed that increased heat flow associated with midplate volcanic activity would lead to warming of bottom waters and would induce rapid turnover of nutrients, stimulating productivity followed by oxygen depletion in deep-water masses. Evidence for this suggestion includes lower Aptian carbonaceous limestones at DSDP Site 463 which are highly tuffaceous [e.g., Dean et al., 1981]. Recently, Tarduno et al. [1991] proposed that tuff-bearing sediments overlying basement on the Ontong Java Plateau correlate with the *G. blowi* foraminiferal Zone and that the production of this massive oceanic plateau occurred in a 1-3 m.y. episode in the latest Barremian to earliest Aptian. The interval of OAE 1a may, therefore, correlate with widespread midplate volcanic activity and the possible (but, as yet, undated) initiation of rapid seafloor spreading at the beginning of the Cretaceous long normal polarity zone [e.g., Larson, 1991a] (Fig. 19). Active volcanism may have led to "black shale" deposition as a result of increased pCO_2 (increased CO_2 outgassing rates), global warming and consequent decreases in oxygen solubility in deep-water masses [Arthur et al., 1991]. However, a more indirect mechanism may also be inferred.

Because of the absence of direct correlation, it is difficult to interpret fully the relationship between the proposed subevents and sea level. As a first approximation, however, the three OASE's appear to correspond in time with second order seismic sequence boundaries (super-cycles) as recognized by Haq et al. [1988] who defined three super-cycle boundaries in the Barremian through Albian interval (Fig. 19) [Arthur et al., 1990; Leckie et al., 1992]. All three represent Type 1 boundaries based on the recognition of low-stand fan deposits. The Haq et al. [1988] sequence stratigraphy and eustatic sea level curve represent a model. The stratigraphic expression of eustatic sea level varies from basin to basin. While the supercycles may represent second-order changes of global sea level, their boundaries may not everywhere be expressed as a Type 1 sequence boundary. The sequence stratigraphic interpretation of these three supercycles is that each is bounded by a relatively rapid rate of sea level fall, exceeding the rate of subsidence, followed by a relatively rapid rate of sea level rise. The subsequent sea level rise may be associated with the development of a condensed interval in offshore marine sections with the potential to accumulate greater concentrations of organic matter [Loutit et al., 1988]. The development of condensed intervals in such areas may be related to reduced sedimentary supply during relatively rapid transgressive sea level episodes, to a rising CCD as the locus of carbonate deposition shifts landward, or to increased geostrophic circulation and erosion or non-deposition near continental margins associated with increased deepwater renewal rates. However, it is not clear that OC concentration would increase simply as a result of reduced dilution (note that TOC concentration and accumulation rates in modern environments increase with increasing sedimentation rate [e.g., Müller and Suess, 1979; Sarnthein et al. 1988]). It appears more likely to us that

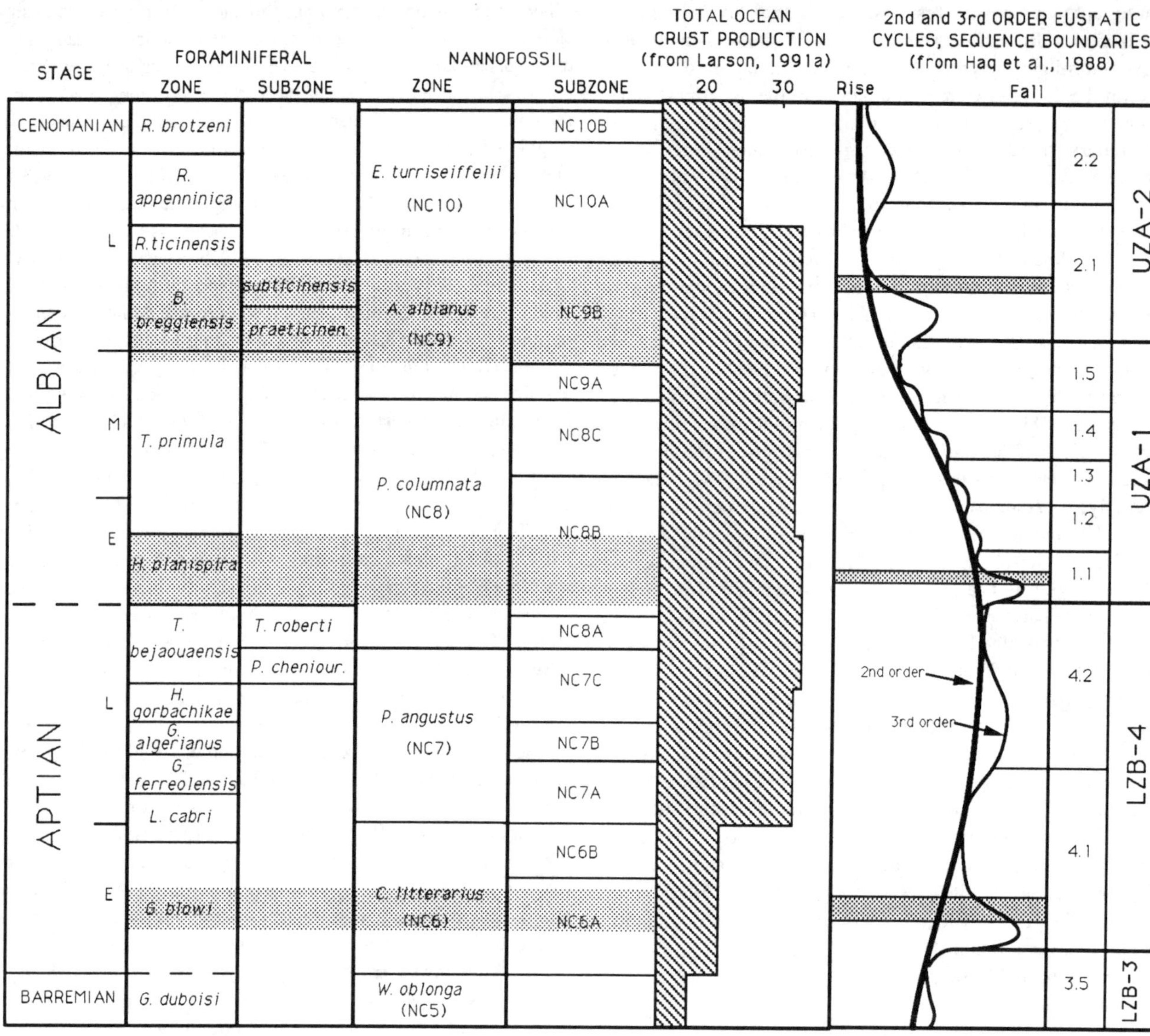

Fig. 19. Correlation of dysoxic/anoxic episodes with 2nd and 3rd order eustatic cycles and sequence boundaries of Haq et al. [1988] and total ocean crust production curve of Larson [1991a]. Terminology as in Haq et al. [1988]. Ages of eustatic cycles have been modified slightly based on apparent disparities in nannofossil zonations between Haq et al. [1988] and this study. Light shading refers to stratigraphic record of dysoxia/anoxia; dark shading refers to position of condensed section as predicted by sequence stratigraphy. Dark lines refer to Type 1 Sequence Boundaries. Crustal production rates are in 10^6 km^3/m.y. See text for full discussion.

some other mechanism is operating to enhance TOC flux or preservation at the same time.

Although this mechanism is not entirely clear, in some cases the intervals of the hypothesized oceanic dysoxic/anoxic subevents correspond to unconformities and condensed section. For example, Bralower et al. [submitted] showed that the interval encompassing or directly overlying early Aptian OAE 1a is condensed in numerous sequences. The *H. planispira* Zone (OAE 1b) is absent from a fairly continuous pelagic section in the Calera Limestone of California [Sliter, 1991]. The upper part of the *Rotalipora subticinensis* Subzone of the *B. breggiensis* Zone (OAE 1c) is also missing in the Calera Limestone. A pronounced unconformity correlates with OAE 1c at DSDP Site 545 off the Mazagan Escarpment (Fig. 9). Therefore the rock record, although extremely indirect, does support a correlation of Aptian/Albian organic carbon enrichments with transgressive phases in sea level.

A more direct relationship has been established for the interval of the Cenomanian/Turonian OAE which can be directly tied to a late transgressive phase of sea level in outcrops of the Western Interior Cretaceous Seaway [e.g., Kauffman, 1977] or in northwest

Europe [e.g., Hancock and Kauffman, 1979]. Based on this correlation, Arthur et al. [1987] proposed that flooding of low-latitude evaporative marginal seas led to increased rates of "warm, saline" deep-water formation, which consequently caused upwelling of nutrients over large areas of the oceans, particularly at ocean margins. Because increased seafloor spreading rates and mid-plate volcanism are likely driving mechanisms for sea level changes [e.g., Larson, 1991a, b], they may have indirectly led to the correlation of intervals of elevated sea level and Aptian-Albian dysoxic/anoxic episodes but were not implicated in the Cenomanian/Turonian event [Schlanger, 1986].

While the previously discussed factors might have caused the ocean to be more susceptible to oxygen deficiency, none of them could have directly induced a water mass to become fully oxygen depleted. A different set of factors, which operate on more regional or local scales, directly affect the supply or demand of oxygen in a water mass. These factors might include upwelling events and higher OC fluxes from surface waters which can stress the supply of oxygen in an individual water mass, but which are unlikely to have affected large areas of the ocean simultaneously, or tectonic factors which lead to restriction of a basin. Regional climatic fluctuations such as increased precipitation and ensuing runoff might create stable stratification on the longer term [e.g., Pratt, 1984] and lead to less replenishment of oxygen by mixing down of surface waters during storms or other events such as winter cooling and convection. Other potential factors depleting oxygen include: 1) advection of oxygen minima into epicontinental seaways or onto shelves upon sea level rise [e.g., Frush and Eicher, 1975; Heckel, 1977; Fischer and Arthur, 1977], and 2) rising temperature and consequent decrease of oxygen solubility of deep waters.

In conclusion, it appears that the oceans were more prone to dysoxia/anoxia during the early Aptian and Cenomanian/Turonian events than during the two Albian subevents. It is likely that similar global forcing mechanisms, such as high sea level and an overall warm, equable climate, operated in all of these time intervals. However the patchier extent of dysoxic/anoxic water masses in the Albian subevents indicates either that the oceans were less susceptible to dysoxia/anoxia at these times or that local/regional triggering factors had a stronger influence.

CONCLUSIONS

The Aptian-Albian Oceanic Anoxic Event is characterized by at least three individual dysoxic/anoxic episodes: OAE 1a, early Aptian *G. blowi* foraminiferal Zone, *C. litterarius* nannofossil Zone, *C. rothii* nannofossil Subzone; OAE 1b, early Albian *H. planispira* foraminiferal Zone, *P. columnata* nannofossil Zone (subzone NC8B); and OAE 1c, early late Albian (*B. breggiensis* foraminiferal Zone, *A. albianus* nannofossil Zone (subzone NC9B). These subevents are observed in sections deposited in pelagic environments and along continental margins in all ocean basins, and in several epicontinental seaways. All three episodes appear to have had a nearly ocean-wide extent, but the distribution of dysoxic/anoxic watermasses in the Albian intervals appears to have been much patchier than during the early Aptian episode.

Although none of the subevents is associated with major extinctions, they appear to be marked by changes in the community structure of planktonic foraminifera, but the pattern differs for each subevent. The absence of planktonic foraminifera in some intervals may be related to dissolution and/or adverse ecological factors such as low salinity and/or oxygen deficiency. Nannofossil taxa in these horizons may show marked epicontinental affinities or may have been excluded competitively by other phytoplankton.

The correlation of the subevents with sea level is only indirect at this time. However, rising sea level and climatic consequences of volcanism are likely factors which conditioned the oceans to be prone to dysoxia/anoxia through a variety of feedbacks. Regional or even local climatic or tectonic factors or high biologic productivity actually induced or triggered dysoxia/anoxia at mid- to deep-water depths.

Acknowledgments. We gratefully acknowledge discussions with W. Balch, W. Dean, E. Erba and J. Mutterlose. We thank D. Bottjer, D. Eicher, W. Elder, E. Erba, T. Herbert, H. Jenkyns, S. Kidwell, K. McDougall, J. Mutterlose, C. Paull and B. Sageman for critical reviews of a previous manuscript. We are grateful to C. Botelho, J.T. Ellis, C. Mellor, M. Parrow and L. Sierra for laboratory assistance. We thank S. Shafik and V. Phansalkar for generously donating material from Australia and India, respectively, and the Ocean Drilling Program for numerous samples. Research supported by NSF (EAR-8721350).
RML would like to thank Richard Cashman for laboratory assistance.
RML's contributions to this project have been partially funded by the Petroleum Research Fund, administered by the American Chemical Society.

REFERENCES

Applegate, J.L., and J.A. Bergen, Cretaceous calcareous nannofossil biostratigraphy of sediments recovered from the Galicia Margin, ODP Leg 103, *Proc. Ocean Drilling Program, Sci. Results*, 103, 293-319, 1988.

Arthur, M.A., North Atlantic Cretaceous black shales: the record at Site 398 and a brief comparison with other occurrences, *Initial Rep. Deep Sea Drilling Project*, 47B, 719-738, 1979.

Arthur, M.A., and J.H. Natland, Carbonaceous sediments in the North and South Atlantic: the role of salinity in stable stratification of early Cretaceous basins, in M. Talwani, W. Hay, and W.B.F. Ryan (eds.), *Deep Drilling in the Atlantic Ocean: Continental Margins and Paleoenvironment*, Maurice Ewing Series, 3, Am. Geophys. Union, Washington, D.C., pp. 375-401, 1979.

Arthur, M.A., S.O. Schlanger, and H.C.Jenkyns, The Cenomanian-Turonian Oceanic Anoxic Event, II, Paleoceanographic controls on organic matter production and preservation, in *Marine Petroleum Source Rocks, Spec. Publ.*, 24, J. Brooks and A. Fleet (eds.), Geological Society of London, pp. 399-418, 1987.

Arthur, M.A., W.E. Dean, and G.E. Claypool, Anomalous ^{13}C enrichment in modern marine organic carbon, *Nature*, 315, 216-218, 1988.

Arthur, M.A., H.-J. Brumsack, H.C. Jenkyns, and S.O. Schlanger, Stratigraphy, geochemistry, and paleoceanography of organic carbon-rich Cretaceous sequences, in: R.N. Ginsburg, and B. Beaudoin (eds.), *Cretaceous Resources, Events and Rhythms*, Kluwer Academic Publishers, pp. 75-119, 1990.

Arthur, M.A., L.R. Kump, W.E. Dean, and R.L. Larson, Superplume?, Supergreenhouse? *EOS*, 72, 301, 1991.

Barron, E.J., and W.M. Washington, Cretaceous climate: a comparison of atmospheric simulations with the geologic record, *Palaeogeogr., Palaeoclimatol., Palaeoecol.*, 40, 103-133, 1982.

Barron, E.J., and W.H. Peterson, W.H., Mid-Cretaceous ocean circulation: results from model sensitivity studies, *Paleoceanography*, 5, 319-338, 1990.

Barron, E.J., C.G. Harrison, J.L. Sloan, II, and W.W. Hay,

Paleogeography, 180 million years ago to the present: *Ecolog. Geol. Helv.*, 74, 443-470, 1981.

Bé, A.W.H., An ecological, zoogeographic and taxonomic review of Recent planktonic foraminifera: in A.T.S. Ramsay (ed.), *Oceanic Micropaleontology*, Academic Press, New York, 1, 1-100, 1977.

Benavides-Caceres, V.E., Cretaceous System in Northern Peru, *Bull. Am. Mus. Nat. Hist.*, 108, 357-493, 1956.

Bergen, J.A., Assemblage turnover in Mesozoic calcareous nannofossils: periodicity and distinction from the terminal Cretaceous event, *5th North American Paleontological Congress, Abstracts with Programs*, 26, 1992.

Berner, R.A., A.C. Lasaga, and R.M. Garrels, The carbonate-silicate geochemical cycle and its effect on atmospheric carbon dioxide over the past 100 million years: *Amer. Jour. Sci.*, 283, 641-683, 1983.

Blechschmidt, G., Biostratigraphy of calcareous nannofossils: leg 47B, Deep Sea Drilling Project: *Initial Rep. Deep Sea Drilling Project*, 47B, 327-360, 1979.

Bode, G.W., Appendix II. Carbon and carbonate analyses, *Initial Rep. Deep Sea Drilling Project*, 73, 1185-1187, 1983.

Boillot, G., E.L. Winterer, A.W. Meyer, et al., *Proc. Ocean Drilling Program, Sci. Results*, 103, 663pp, 1988.

Bolli, H.M., Zonation of Cretaceous to Pliocene marine sediments based on planktonic foraminifera, *Bol. Inform. Asoc. Venezolana de Geol., Min y Petrol.*, 9, 3-32, 1966.

Bolli, H.M., W.B.F. Ryan, et al., *Initial Rep. Deep Sea Drilling Project*, 40, 1079 pp, 1978.

Bralower, T.J., Organic carbon and transition metal accumulation rates in Holocene and mid-Cretaceous sediments: data and techniques, *Scripps Institution of Oceanography Data Series*, 84-24, 55pp, 1984.

Bralower, T.J., Valanginian to Aptian calcareous nannofossil stratigraphy and correlation with the upper M-sequence magnetic anomalies, *Mar. Micropaleontol.*, 11, 293-310, 1987.

Bralower, T.J., Calcareous nannofossil biostratigraphy and assemblages of the Cenomanian-Turonian boundary interval: implications for the origin and timing of oceanic anoxia, *Paleoceanography*, 3, 275-316, 1988.

Bralower, T.J., Aptian-Albian calcareous nannofossil biostratigraphy of ODP Site 763, and the correlation between high and low latitude zonations, in: Duncan, R.A., and D.K. Rea, (eds.), *Am. Geophys. Union Monograph*, 70, 1992.

Bralower, T.J., and H.R. Thierstein, Low productivity and slow deep-water circulation in mid-Cretaceous oceans: *Geology*, 12, 614-618, 1984.

Bralower, T.J., and W.G. Siesser, Cretaceous calcareous nannofossil biostratigraphy of ODP Leg 122 Sites 761, 762 and 763, Exmouth and Wombat Plateaus, N.W. Australia, *Proc. Ocean Drilling Program, Sci. Results*, 122, 529-556, 1992.

Bralower, T.J., M.A. Arthur, R.M. Leckie, W.V. Sliter, D.J. Allard, and S.O. Schlanger, Oceanic anoxia/dysoxia in the early Aptian: lithostratigraphy, biostratigraphy and paleoecological consequences, *Paleoceanography*, submitted.

Brass, G.W., J.R. Southam, and W.H. Peterson, Warm saline bottom water in the ancient ocean, *Nature*, 296, 620-623, 1982.

Bréhéret, J.-G., Sur des niveaux de black shales dans l'Albien inférieur et moyen du domaine vocontien (sud-est de la France), étude de nannofacies et signification des paléoenvironments, *Bull. Mus. natn. Hist. nat., Paris*, 5, 113-159, 1983.

Bréhéret, J.-G., Indices d'un événement anoxique étendu à la Téthys alpine, à l'Albien inférieur (événement Paquier), *C. R. Acad. Sc. Paris*, 300, 355-358, 1985.

Bréhéret, J.-G., Épisodes de sédimentation riche en matière organique dans les marnes bleues d'âge aptien et albien de la partie pélagique du bassin vocontien, *Bull Soc. géol. France*, 8, 349-356, 1988.

Bréhéret, J.-G., M. Caron, and M. Delamette, Niveaux riches en matière organique dans Albien vocontien, *Docum. Bur. Rech. géol. min.*, 110, 141-191, 1986.

Busson, G., and D. Noël, Les nannoconidés, indicateurs environmentaux des océans et mers épicontinentales du Jurassique terminal et du Crétacé inférieur, *Oceanologica Acta*, 14, 333-356, 1991.

Calvert, S.E., R.E. Karlin, L.J. Toolin, D.J. Donahue, J.R. Southon, and J.S. Vogel, Low organic carbon accumulation rates in Black Sea sediments, *Nature*, 350, 692-695, 1991.

Cameron, D.H., Grain size and carbon/carbonate analyses, Leg 43, *Initial Rep. Deep Sea Drilling Project*, 43, 1043-1048, 1979.

Caron, M., Cretaceous planktonic foraminifers from DSDP Leg 40, Southeastern Atlantic Ocean, *Initial Rep. Deep Sea Drilling Project*, 40, 651-678, 1978.

Caron, M. and Homewood, P., Evolution of early planktonic foraminifers, *Mar. Micropaleontol.*, 7, 453-462, 1983.

Cepek, P., Mesozoic calcareous nannoplankton of the eastern North Atlantic, Leg 41, *Initial Rep. Deep Sea Drilling Project*, 41, 667-688, 1977.

Cepek, P., Mesozoic calcareous nannoplankton stratigraphy of the Central North Pacific (Mid-Pacific Mountains and Hess Rise), Deep Sea Drilling Project, Leg 62, *Initial Rep. Deep Sea Drilling Project*, 62, 397-418, 1981.

Channell, J.E.T., W. Lowrie, and F. Medizza, Middle and Early Cretaceous magnetic stratigraphy from the Cismon section, Northern Italy, *Earth Planet. Sci. Lett.*, 42, 153-166, 1979.

Coccioni, R., O. Nesci, M. Tramontana, F.-C. Wezel, and E. Moretti, Descrizione di un livello-guida "radiolaritico-bituminoso-ittiolitico" alla base delle Marne a Fucoidi nell Appennino Umbro-Marchigiano, *Boll. Soc. Geol. It.*, 106, 183-192, 1987.

Coccioni, R., R. Franchi, O. Nesci, F.-C. Wezel, F. Battistini, and P. Pallecchi, Stratigraphy and mineralogy of the Selli Level (Early Aptian) at the base of the Marne a Fucoidi in the Umbro-Marchean Apennines (Italy), in J. Wiedmann (ed.) Cretaceous of the Western Tethys, *Proc. 3rd Int. Cretaceous symposium, Tübingen*, 1989.

Coccioni, R., E. Erba, and I. Premoli Silva, Barremian-Aptian calcareous plankton biostratigraphy from the Gorgo a Cerbara section (Marche, central Italy) and implications for plankton evolution, *Cret. Res.*, in press.

Davies, T.A., B.P. Luyendyk, et al., *Initial Rep. Deep Sea Drilling Project*, 26, 1129pp, 1974.

Dean, W.E., Calcium carbonate and organic carbon in samples from Deep Sea Drilling Project Sites 463, 464, 465 and 466, *Initial Rep. Deep Sea Drilling Proj.*, 62, 869-876, 1981.

Dean, W.E., G.E. Claypool, and J. Thiede, Origin of organic carbon-rich mid-Cretaceous limestones, mid-Pacific mountains and southern Hess Rise, *Initial Rep. Deep Sea Drilling Project*, 61, 877-890, 1981.

Dean, W.E., M.A. Arthur, and G.E. Claypool, Depletion of ^{13}C in Cretaceous marine organic matter: Source, diagenetic or environmental signal?, *Mar. Geol.*, 70, 119-158, 1986.

Demaison, G.J., and G.T. Moore, Anoxic environments and oil source bed genesis, *Organic Geochem.*, 2, 9-31, 1980.

Deroo, G., J.P. Herbin, J. Roucaché, and B. Tissot, Organic geochemistry of Cretaceous shales from DSDP Site 398, Leg 47B, Eastern North Atlantic, *Initial Rep. Deep Sea Drill. Project*, 47, pt. B, 513-522, 1979.

Deroo, G., J.P. Herbin, J. Roucaché, and B. Tissot, Organic geochemistry of Cretaceous sediments from DSDP Holes 417D (Leg 51), 418A (Leg 52) and 418B (Leg 53), Western North Atlantic, *Initial Rep. Deep Sea Drilling Project*, vs. 51-53, 737-746, 1980.

Deroo, G., J.P. Herbin, and J. Roucaché, Organic geochemistry of Upper Jurassic-Cretaceous sediments from Site 511, Leg 71, Western South Atlantic, *Initial Rep. Deep Sea Drilling Project*, 71, 1001-1013, 1983.

Deroo, G., J.P. Herbin, and J. Roucaché, Organic geochemistry of Cenozoic and Mesozoic sediments from Deep Sea Drilling Project Sites 544 to 547, Leg 79, Eastern North Atlantic, *Initial Rep. Deep Sea Drilling Project*, 79, 721-742, 1984.

Donnelly, T., J. Francheteau, et al., *Initial Rep. Deep Sea Drilling Project*, 51, 52 and 53, 1613pp, 1980.

Douglas, R.G., Planktonic foraminiferal biostratigraphy in the Central North Pacific Ocean, *Initial Rep. Deep Sea Drilling Project*, 17, 607-694, 1973.

Dunham, K.W., P.A. Meyers, and E.S. Ho, Organic geochemistry of Cretaceous black shales and adjacent strata from the Galicia Margin, North

Atlantic Ocean, *Proc. Ocean Drilling Program, Sci. Results*, 103, 557-565, 1988.

Dupeuble, P.A., Mesozoic foraminifers and microfacies from Holes 400A, 401 and 402A of the DSDP Leg 48, *Initial Rep. Deep Sea Drill. Project*, 48, 451-474, 1979.

Eicher D.L., and P. Worstell, Cenomanian and Turonian foraminifera from the Great Plains, United States, *Micropaleontology*, 16, 269-324, 1970.

Elder, W.P., Molluscan extinction patterns across the Cenomanian-Turonian stage boundary in the Western Interior of the United States, *Paleobiology*, 15, 299-320, 1989.

Emerson, S., and M. Bender, Carbon fluxes at the sediment-water interface of the deep-sea: calcium carbonate preservation, *J. Mar. Res.*, 39, 139-162, 1981.

Erba, E., Aptian-Albian calcareous nannofossil biostratigraphy of the Scisti a Fucoidi cored at Piobbico (Central Italy), *Riv. It. Paleont. Strat.*, 94, 249-284, 1988.

Erba, E., and B. Quadrio, Biostratigrafia a Nannofossili calcarei, Calpionellidi e Foraminiferi planctonici della Maiolica (Titoniano superiore-Aptiano) nelle Prealpi Bresciane (Italia settentrionale), *Riv. It. Paleont. Strat.*, 93, 3-108, 1987.

Erba, E., I. Premoli Silva, and M.E. Tornaghi, The Aptian-Albian scisti a fucoidi, Italy, record changes in fertility and carbonate productivity, *Terra Abstracts*, 1, 42, 1989.

Erdman, J.G., and K.S. Schorno, Geochemistry of carbon, Deep Sea Drilling Project, Leg 41, *Initial Rep. Deep Sea Drilling Project*, 41, 849-854, 1977.

Erdman, J.G., and K.S. Schorno, Geochemistry of carbon: Deep Sea Drilling Project, Legs 47A and 47B, *Initial Rep. Deep Sea Drilling Project*, 47, pt. B, 553-560, 1979.

Fischer, A.G., and M.A. Arthur, Secular variations in the pelagic realm, *Soc. Econ. Paleont. Min., Spec. Publ.*, 25, 19-50, 1977.

Foresman, J.B., Organic geochemistry DSDP Leg 40, continental rise of southwest Africa, *Initial Rep. Deep Sea Drilling Project*, 40, 557-568, 1978.

Frush, M.P., and D.L. Eicher, Cenomanian and Turonian foraminifera and paleoenvironments in the Big Bend region of Texas and Mexico, in W.G.E. Caldwell (ed.), *The Cretaceous system in the Western Interior of North America, Geological Assoc. of Canada, Spec. Pap.*, 13, 277-301, 1975.

Gaida, K.-H., R. Gedenik, E. Kemper, W. Michaelis, R. Scheuch, H.-H. Schmitz, and W. Zimmerle, litholosche, mineralogische und organisch-geochemische Untersuchungen an Tonsteinen und Tonmergelsteinen der Unterkreide Nordwest-deutschlands (unter besonder Berücksichtigung der Schwarzschiefer), *Geol. Jahrb.*, A, 58, 15-47, 1981.

Gartner, S. Jr., Calcareous nannofossils, Deep Sea Drilling Project Holes 418A and 418B, *Initial Rep. Deep Sea Drilling Project*, 51, 52, 53, 815-821, 1980.

Govinden, A., Imprint of global "Cretaceous Anoxic Events" in east coast basins of India and their implications, *Bull. ONGC*, 19, 257-270, 1982.

Graciansky, P.C. de, E. Brosse, G. Deroo, J.P. Herbin, L. Montadert, C. Müller, J. Sigal, and A. Schaaf, Les formations d'Age Crétacé de l'Atlantique Nord et leur matière organique paléogéographie et milieux de dépôt, *Revue de l'Inst. France Petrole*, 37, 276-336, 1982.

Habib, D., Sedimentary origin of North Atlantic Cretaceous palynofacies, in M. Talwani, W. Hay and W.B.F. Ryan (eds.), *Deep Drilling Results in the Atlantic Ocean; Continental Margins and Paleoenvironment*, Maurice Ewing Series, v, 3, Am. Geophys. Union, Washington, D.C., 420-437, 1979.

Haig, D.W., Cretaceous foraminiferal biostratigraphy of Queensland, *Alcheringa*, 3, 171-187, 1979.

Hancock, J.M., and E.G. Kauffman, The great transgressions of the Late Cretaceous. *J. Geol. Soc. Lond.*, 136, 175-186, 1979.

Haq, B.U., J. Hardenbol, and P.R. Vail, Chronology of fluctuating sea levels since the Triassic (250 million years to Present), *Science*, 235, 1156-1167, 1987.

Haq, B.U., J. Hardenbol, and P.R. Vail, Mesozoic and Cenozoic chronostratigraphy and cycles of sea-level change, *Soc. Econ. Paleont. Min., Spec. Publ.*, 42, 71-108, 1988.

Haq, B.U., U. von Rad, S. O'Connell, et al., *Proc. Ocean Drilling Program, Sci. Results*, 122, 826pp, 1990.

Harland, W.B., R.L. Armstrong, A.V. Cox, L.E. Craig, A.G. Smith, and D.G. Smith, *A geologic time scale*, Cambridge University Press, 263pp, 1990.

Hart, M.B., A water depth model for the evolution of the planktonic Foraminiferida, *Nature*, 286, 252-254, 1980.

Heckel, P.H., Origin of phosphatic black shale facies in Pennsylvanian cyclothems of mid-continent North America, *Am. Assoc. Petrol. Geol. Bull.*, 61, 1045-1068, 1977.

Hemleben, C., M. Spindler, and O.R. Anderson, *Modern Planktonic Foraminifera*, 363pp, Springer Verlag, New York, 1989.

Herbert, T.D., and A.G. Fischer, Milankovitch climate origin of mid-Cretaceous black shale rhythms, central Italy, *Nature*, 321, 739-743, 1986.

Herbin, J.P., F. Magniez-Jannin, and C. Müller, Mesozoic organic-rich sediments in the South Atlantic: distribution in time and space, in Biogeochemistry of Black Shales, E.T. Degens, P.A. Meyers, and S.C. Brassell,(eds.), *Mitteilingen aus dem Geologisch-Paläontologischen Institut der Universität Hamburg*, 71-98, 1986.

Herbin, J.P., C. Müller, P.C. de Graciansky, T. Jaquin, F. Magniez-Jannin, and P. Unternehr, Cretaceous anoxic events in the South Atlantic, *Revista Brasieleira de Geociencias*, 17, 92-99, 1987.

Hill, M.E., and T.J. Bralower, Early evolution, stratigraphy and taxonomy of *Eiffellithus eximius* and related species, *Abh. Geol. Bundesanst.*, 39, 98-97, 1987.

Hinz, K., E.L. Winterer, et al., *Initial Rep. Deep Sea Drilling Project*, 79, 934pp, 1984.

Jaillard, E., Sedimentary evolution of an active margin during middle and Upper Cretaceous times: the north Peruvian margin from late Aptian up to Senonian, *Geol. Rund.*, 76, 677-697, 1987.

Jenkyns, H.C., Cretaceous anoxic events: From continents to oceans, *J. Geol. Soc. London*, 137, 171-188, 1980.

Johnson, D.L., R.D. McIver, and M.A. Rogers, Insoluble organic matter bitumens in Leg 47 samples, *Initial Rep. Deep Sea Drilling Project*, 47, pt. B, 543-546, 1979.

Kale, A.S., and V.G. Phansalkar, Nannofossil biostratigraphy of the Utatur Group, Trichinopoly District, South India, *Mem. di Scienze Geol., Univ. di Padova*, 63, 89-107, 1992.

Kauffman, E.G., Geological and biological overview, Western Interior Cretaceous basin, in Cretaceous facies, faunas and paleoenvironments across the Western Interior Basin: edited by E.G. Kauffman, *Mountain Geologist*, 14, 75-99, 1977.

Kauffman, E.G., B.B. Sageman, J.I. Kirkland, W.P. Elder, T.J. Harries, and T. Villamil, Molluscan biostratigraphy of the Western Interior Cretaceous Basin, North America, in W.G.E. Caldwell and E.G. Kauffman (eds.), *Cretaceous Evolution of the Western Interior Basin of North America, Geological Assoc. of Canada, Spec. Pap.*, in press.

Koutsoukos, E.A.M., M.R. Mello, N.C. de Azambuja Filho, M.B. Hart, and J.R. Maxwell, The Upper Aptian-Albian succession of the Sergipe Basin, Brazil: an integrated paleoenvironmental assessment, *Am. Assoc. Petrol. Geol. Bull.*, 75, 479-498, 1991.

Krasheninnikov, V.A., and I.A. Basov, Stratigraphy of Cretaceous sediments of the Falkland Plateau based on planktonic foraminifers, Deep Sea Drilling Project, Leg 71, *Initial Rep. Deep Sea Drilling Project*, 71, 789-820, 1983.

Lancelot, Y., E. Seibold, et al., *Initial Rep. Deep Sea Drilling Project*, 41, 1259pp, 1977.

Larson, R.L., Latest pulse of the Earth: evidence for a mid-Crataceous super plume, *Geology*, 19, 547-550, 1991a.

Larson, R.L., Geological consequences of superplumes, *Geology*, 19, 963-966, 1991b.

Leckie, R.M., Mid-Cretaceous planktonic foraminiferal biostratigraphy off

Central Morocco, Deep Sea Drilling Project Leg 79, Sites 545 and 547. *Initial Rep. Deep Sea Drilling Project*, 79, 579-620, 1984.

Leckie, R.M., Paleoecology of mid-Cretaceous planktonic foraminifera: a comparison of open ocean and epicontinental sea assemblages, *Micropaleontology*, 33, 164-176, 1987.

Leckie, R.M., An oceanographic model for the early evolutionary history of planktonic foraminifera, *Paleogeogr. Paleoclimatol. Paleoecol.*, 73, 107-138, 1989.

Leckie, R.M., and T.J. Bralower, New ages for planktonic foraminiferal datums in the Barremian-Lower Aptian, *Geol. Soc. America, Abs. with Prog.*, 23, 167, 1991.

Leckie, R.M., R. Scott, R., T.J. Bralower, and W.V. Sliter, Relationships between sequence boundaries and the evolutionary history of planktonic foraminifera, calcareous nannofossils and reef communities in the mid Cretaceous (Barremian-Cenomanian), *5th North American Paleontology Abstracts with Programs*, 179, 1992.

Lipps, J.H., The ecology and paleoecology of planktic foraminifera, in *Foraminiferal Ecology and Paleoecology*, SEPM Short Course, 6, 62-104, 1979.

Longoria, J.F., Stratigraphic, morphologic and taxonomic studies of Aptian planktonic foraminifera, *Riv. Esp. Micropal. Num. Extra*, 107pp, 1974.

Loutit, T.S., J. Hardenbol, P.R. Vail, and G.R. Baum, Condensed sections: the key to age determination and correlation of continental margin sequences, *Soc. Econ. Paleont. Min., Spec. Publ.*, 42, 183-213, 1988.

Lowrie, W., and W. Alvarez, Lower Cretaceous magnetic stratigraphy in Umbrian pelagic limestone sections, *Earth Planet. Sci. Lett.*, 71, 315-328, 1984.

Lowrie, W., W. Alvarez, I. Premoli Silva, and S. Monechi, Lower Cretaceous magnetic stratigraphy in Umbrian pelagic carbonate rocks, *Geophys. J.R. Astron. Soc.*, 60, 263-281, 1980.

Ludwig, W.J., V. Krasheninnikov, et al., Site 511 Summary, *Init. Rep. Deep Sea Drilling Project*, 71, pt. 1, 477pp, 1983.

Manivit, H., K. Perch-Nielsen, B. Prins, and J.W. Verbeek, Mid Cretaceous calcareous nannofossil biostratigraphy, *Proc. K. Ned. Akad. Wet.* Ser. B, 80, 169-181, 1977.

Margalef, R., Life-forms of phytoplankton as survival alternatives in an unstable environment, *Oceanologica Acta*, 1, 493-509, 1978.

Monechi, S., and H.R. Thierstein, Late Cretaceous-Eocene nannofossil and magnetostratigraphic correlations near Gubbio, Italy, *Mar. Micropaleontol.*, 9, 419-440, 1985.

Montadert, L., D.G. Roberts, et al., *Initial Rep. Deep Sea Drilling Project*, 48, 1183pp, 1979.

Moullade, M., Etude stratigraphique and micropaléontologique du Crétacé inférieur de la "Fosse vocontienne". *Doc. Lab. Fac. Sci. Lyon*, 15, 1-369, 1966.

Müller, C., Calcareous nannofossils from the North Atlantic (Leg 48), *Initial Rep. Deep Sea Drilling Project*, 48, 589-639, 1979.

Müller, C., A. Schaaf, and J. Sigal, Biochronostratigraphy des formations d'âge Crétacé dans les forages du DSDP dans l'ocean Atlantique Nord, *Rev. Inst. Fr. Pet.*, 38, 683-708, 1983.

Müller, P.J. and Suess, E., Productivity, sedimentation rate and sedimentary organic carbon in the oceans. 1. Organic carbon preservation, *Deep Sea Res.*, 26A, 1347-1362, 1979.

Ogg, J.G., Early Cretaceous and Tithonian magnetostratigraphy of the Galicia Margin (Ocean Drilling Program Leg 103), *Proc. Ocean Drill. Prog. Sci. Res.*, 103, 659-682, 1988.

Okada, H., and H.R. Thierstein, Calcareous nannoplankton-Leg 43, Deep Sea Drilling Project, *Initial Rep. Deep Sea Drilling Project*, 43, 507-574, 1979.

Parker, M.E., M.A. Arthur, S.W. Wise, Jr., and C.R. Wenkham, Carbonate and organic carbon cycles in Aptian-Albian black shales at Deep Sea Drilling Project Site 511, Falkland Plateau, *Initial Rep. Deep Sea Drilling Project*, 71, 481-550, 1983.

Pedersen, T.F., and S.E. Calvert, Anoxia vs. productivity: what controls the formation of organic-carbon-rich sediments and sedimentary rocks?, *Am. Assoc. Petrol. Geol. Bull.*, 74, 454-466, 1990.

Pflaumann, U., and V.A. Krasheninnikov, Early Cretaceous planktonic foraminifers from eastern North Atlantic, DSDP Leg 41, *Initial Rep. Deep Sea Drilling Project*, 41, 539-564, 1977.

Pratt, L.M., Influence of paleoenvironmental factors on preservation of organic matter in middle Cretaceous Greenhorn Formation, Pueblo, Colorado, *Am. Assoc. Petrol. Geol. Bull.*, 68, 1146-1159, 1984.

Pratt, L.M., and J.D. King, Variable marine productivity and high eolian input recorded by rhythmic black shales in mid-Cretaceous pelagic deposits from Central Italy, *Paleoceanography*, 1, 507-522, 1986.

Premoli Silva, I., E. Erba, and M.E. Tornaghi, Paleoenvironmental signals and changes in surface fertility in mid-Cretaceous Corg-rich pelagic facies of the fucoid marls (central Italy), *Geobios, mém. spéc.*, 11, 225-236, 1989.

Proto-Decima, F., F. Medizza, and L. Todesco, Southeastern Atlantic Leg 40 calcareous nannofossils, *Initial Rep. Deep Sea Drilling Project*, 40, 571-634, 1978.

Raup, D.M., and J.J Sepkoski, Mass extinction in the marine fossil record, *Science*, 215, 1510-1503, 1982.

Raymont, J.E.G., *Plankton and Productivity in the Oceans, 1, Phytoplankton*, Pergamon, Oxford, 489 pp, 1980.

Rhoads, D.C., and J.W. Morse, Evolutionary and ecologic significance of oxygen-deficient basins, *Lethaia*, 4, 413-428, 1971.

Roth, P.H., Calcareous nannofossils, Leg 17, Deep Sea Drilling Project, *Initial Rep. Deep Sea Drilling Project*, 17, 695-795, 1973.

Roth, P.H., Cretaceous nannoplankton biostratigraphy and oceanography of the northwestern Atlantic Ocean, *Initial Rep. Deep Sea Drilling Project*, 44, 731-759, 1978.

Roth, P.H., Mid-Cretaceous calcareous nannoplankton from the central Pacific: implications for paleoceanography, *Initial Rep. Deep Sea Drilling Project*, 62, 471-489, 1981.

Roth, P.H., Mesozoic calcareous nannofossil evolution: relation to paleoceanographic events, *Paleoceanography*, 2, 601-612, 1987.

Roth, P.H., Ocean Circulation and calcareous nannoplankton evolution during the Jurassic and Cretaceous, *Paleogeog. Paleoclimatol., Paleoecol.*, 74, 111-126, 1989.

Roth, P.H., and K.P. Krumbach, Middle Cretaceous calcareous nannofossil biogeography and preservation in the Atlantic and Indian Oceans: Implications for paleoceanography, *Mar. Micropaleontol.*, 10, 235-266, 1986.

Sageman, B.B., P.B. Wignall, and E.G. Kauffman, Biofacies models for organic-rich facies: tool for paleoenvironmental analysis, in *Cycles and Events in Stratigraphy*, edited by G. Einsele, W. Ricken, and A. Seilacher, Springer Verlag, pp. 542-564, 1991.

Salloway, J.C., Paleomagnetism of sediments from Deep Sea Drilling Project Leg 71, *Initial Rep. Deep Sea Drilling Project*, 71, 1073-1091, 1983.

Sarnthein, M., K. Winn, J.-C. Duplessy, and M.R. Fontugne, Global variations of surface ocean productivity in low and mid latitudes: influence on CO_2 reservoirs of the deep ocean and atmosphere during the last 21,000 years, *Paleoceanography*, 3, 361-399, 1988.

Savrda, C.E., D.J. Bottjer, and D.S. Gorsline, Development of a comprehensive oxygen-deficient marine biofacies model: evidence from Santa Monica, San Pedro, and Santa Barbara Basins, California Continental Borderland, *Am. Assoc. Petrol. Geol. Bull.*, 68, 1179-1192, 1984.

Schlanger, S.O., High frequency sea-level fluctuations in Cretaceous time: an emerging geophysical problem, in K.J. Hsu (ed.) *Mesozoic and Cenozoic Oceans, AGU Geodynamics Series*, vol. 15, pp. 61-74, 1986

Schlanger, S.O., and R.G. Douglas, The pelagic ooze-chalk-limestone transition and its implications for marine stratigraphy, *Spec. Publ. Int. Assoc. Sediment.*, 1, 117-148, 1974.

Schlanger, S.O., and H.C. Jenkyns, Cretaceous oceanic anoxic events: causes and consequences, *Geol. en Mijnbouw*, 55, 179-184, 1976.

Schlanger, S.O., H.C. Jenkyns, and I. Premoli Silva, Volcanism and vertical tectonics in the Pacific basin related to global Cretaceous transgression, *Earth Planet. Sci. Lett.*, 52, 435-449, 1981.

Schlanger, S.O., M.A. Arthur, H.C. Jenkyns, and P.A. Scholle, The Cenomanian-Turonian oceanic anoxic event, I. Stratigraphy and distribution of organic carbon-rich beds and the marine $\partial^{13}C$ excursion, in *Marine Petroleum Source Rocks, Spec. Publ.*, vol. 24, edited by J. Brooks and A. Fleet, Geological Society of London, pp. 347-375, 1987.

Scholle, P.A., and M.A. Arthur, Carbon isotope fluctuations in Cretaceous pelagic limestones: Potential stratigraphic and petroleum exploration tool, *Am. Assoc. Petrol. Geol. Bull.*, 64, 67-87, 1980.

Shafik, S., Calcareous nannofossils from the Toolebuc, Eromanga Basin, Australia, *BMR Jour. of Australian Geology and Geophysics*, 9, 171-181, 1985.

Sibuet, J.-C., W.B.F. Ryan, et al., *Initial Rep. Deep Sea Drilling Project*, 47B, 787pp, 1979.

Siesser, W., Calcareous nannofossils: Legs 51 and 52 of Deep Sea Drilling Project, *Initial Rep. Deep Sea Drilling Project*, 51, pt. B, 823-845, 1980.

Sigal, J., Essai de zonation du Crétacé méditerranéen à l'aide des foraminifères planctoniques, *Geol. Medit.*, 4, 99-108, 1977.

Sigal, J., Chronostratigraphy and ecostratigraphy of Cretaceous formations recovered on DSDP Leg 47B, Site 398, *Initial Rep. Deep Sea Drilling Project*, 47, pt. B, 287-327, 1979.

Sliter, W.V., Aptian anoxia in the Pacific Basin, *Geology*, 17, 909-91, 1989a.

Sliter, W.V., Biostratigraphic zonation for Cretaceous planktonic foraminifers examined in thin section, *Jour. Foram. Res.*, 19, 1-19, 1989b.

Sliter, W.V., Cretaceous hiatuses, black shales and Pacific paleoceanography, *Geol. Soc. Am. Abs. with Prog.*, 23, A341, 1991.

Sliter, W.V., and T.J. Bralower, USGS Open File report, in press.

Tappan, H., *The Paleobiology of Plant Protists*, Freeman,1028pp, 1980.

Tappan, H., and A.R. Loeblich, Jr., Evolution of the Oceanic Plankton, *Earth Sci. Rev.*, 9, 207-240, 1973.

Tarduno, J.A., W.V. Sliter, T.J. Bralower, M. McWilliams, I. Premoli Silva, and J.G. Ogg, M-sequence reversals recorded in DSDP Sediment Cores from the Western Mid Pacific Mountains and Magellan Rise, *Geol. Soc. Am. Bull.*, 101, 1306-1316, 1989.

Tarduno, J.A., W.V. Sliter, L. Kroenke, R.M. Leckie, H. Mayer, J.J. Mahoney, R. Musgrave, M. Storey, and E.L. Winterer, Rapid formation of Ontong Java Plateau by Aptian mantle volcanism, *Science*, 254, 399-403, 1991.

Thiede, J., and T.H. van Andel, The paleoenvironment of anaerobic sediments in the late Mesozoic South Atlantic, *Earth Planet. Sci. Lett.*, 33, 301-309, 1977.

Thierstein, H.R., Tentative Lower Cretaceous calcareous nannoplankton zonation, *Eclog. Geol. Helv.*, 64, 459-488, 1971.

Thierstein, H.R., Lower Cretaceous calcareous nannoplankton biostratigraphy, *Abhandlung. Geolog. Bund.*, 29, 3-53, 1973.

Thierstein, H.R., Calcareous nannoplankton-Leg 26, Deep Sea Drilling Project, *Initial Rep. Deep Sea Drilling Project*, 26, 619-688, 1974.

Thierstein, H.R., Paleoceanographic implications of organic carbon and carbonate distribution in Mesozoic deep-sea sediments, in M. Talwani, W. Hay, and W.B.F. Ryan (eds.), *Deep Drilling in the Atlantic Ocean; Continental Margins and Paleoenvironment*, Maurice Ewing Series 3, Am. Geophys. Union, Washington, D.C., pp. 249-274, 1979.

Thierstein, H.R., Selective dissolution of Late Cretaceous and Earliest Tertiary calcareous nannofossils, experimental evidence, *Cretaceous Res.*, 2,165-176, 1980.

Thierstein, H.R., and W.H. Berger, Injection events in ocean history, *Nature*, 276, 461-466, 1978.

Thurow, J., Diagenetic history of Cretaceous radiolarians, North Atlantic Ocean (ODP Leg 103 and DSDP Holes 398D and 603B), *Proc. Ocean Drill. Prog. Sci. Res.*, 103, 531-555, 1988.

Tornaghi, M.E., I. Premoli Silva, and M. Ripepe, Lithostratigraphy and planktonic foraminiferal biostratigraphy of the Aptian-Albian "Scisti a Fucoid" in the Piobbico core, Marche, Italy: background for cyclostratigraphy, *Riv. It. Paleont. Strat.*, 95, 223-264, 1989.

Tucholke, B.E., P.R. Vogt, et al., *Initial Rep. Deep Sea Drilling Project*, 43, 1115pp, 1979.

Vail, P.R., R.M. Mitchum, Jr., R.G. Todd, J.M. Widmier, S. Thompson, III, J.B. Sangree, J.N. Bubb, and W.G. Hatfield, Seismic stratigraphy and global changes in sea level, in, C.E. Payton, (ed.), *Seismic Stratigraphy-Applications to Hydrocarbon Exploration, Am. Assoc. Petrol. Geol. Mem.*, 26, 49-212, 1977.

Vogt, P.R., Volcanogenic upwelling of anoxic, nutrient-rich water: a possible factor in carbonate-bank/reef demise and benthic faunal extinctions?, *Geol. Soc. Am. Bull.*, 101, 1225-1245, 1989.

Waples, D.W., A reappraisal of anoxia and richness of organic material, with emphasis on the Cretaceous North Atlantic, *Initial Rep. Deep Sea Drilling Project*, 80, 999-1016, 1983.

Watkins, D.K., Calcareous nannofossil paleoceanography of the Cretaceous Greenhorn Sea, *Geol. Soc. Am. Bull.*, 97, 1239-1249, 1986.

Wei, K.-Y., and J.P. Kennett, Taxonomic evolution of Neogene planktonic foraminifera and paleoceanographic relations, *Paleoceanography*, 1, 67-84, 1986.

Weissert, H., C-isotope stratigraphy, a monitor of paleoenvironmental change: a case study from the early Cretaceous, *Surveys in Geophysics*, 10,1-61, 1989.

Wiegand, G.E., Cretaceous nannofossils from the Northwest African margin, Deep Sea Drilling Project Leg 79, *Initial Rep. Deep Sea Drilling Project*, 79, 563-578, 1984.

Williams, G.L., and J.P. Bujak, Mesozoic and Cenozoic dinoflagellates, in, *Plankton Stratigraphy*, H.M. Bolli, J.B. Saunders, and K. Perch-Nielsen, (eds.), Cambridge University Press, 847-964, 1985.

Winterer, E.L., J.I. Ewing, et al., *Initial Rep. Deep Sea Drilling Project*, 17, 930pp, 1973.

Wise, S.W., Jr., Mesozoic and Cenozoic calcareous nannofossils recovered by Deep Sea Drilling Project Leg 71 in the Falkland Plateau Region, Southwest Atlantic Ocean, *Initial Rep. Deep Sea Drilling Project*, 71, 481-550, 1983.

Timothy J. Bralower, Department of Geology, University of North Carolina, Chapel Hill, NC 27599-3315.

William V. Sliter, Branch of Paleontology and Stratigraphy, USGS, Menlo Park, CA 94025.

Michael A. Arthur, Department of Geosciences, Pennsylvania State University, University Park, PA 16802.

R. Mark Leckie, Department of Geology, University of Massachusetts, Amherst, MA 01003.

David Allard, Graduate School of Oceanography, University of Rhode Island, Narragansett, RI 02882.

Seymour O. Schlanger, Department of Geological Sciences, Northwestern University, Evanston, IL 60201.

Mesozoic Seafloor Subsidence and the Darwin Rise, Past and Present

GEOFFREY F. DAVIES AND FRIDERIK PRIBAC

Research School of Earth Sciences, Australian National University

The expected subsidence behavior of old (Mesozoic) seafloor is re-examined in terms of the behavior expected to arise from a convecting mantle. Empirically, old seafloor does not "flatten" (that is, asymptotically approach a constant depth), and there is no good reason to expect that it should. It is argued that a square-root-of-age subsidence curve provides a more useful physical reference for revealing depth anomalies with causes other than conductive cooling from the earth's surface. Even without this reference, there is clear evidence in seafloor topography of the influence of sublithospheric thermal anomalies of the kinds expected in a convecting fluid. The Pacific seafloor is re-analysed from the new perspective, and a superswell is revealed that extends for about 9000 km through the central and western Pacific. Independent evidence indicates that the superswell is due to the presence of both shallow, sublithospheric, plume-derived material and deep mantle that is relatively warm because of the absence of subduction in the area for more than 200 m.y. It is argued that such uplifts can be expected to persist for hundreds of millions of years, and on this basis it is proposed that this superswell is the modern continuation of the Cretaceous Darwin Rise.

INTRODUCTION

Two factors have motivated us to reassess the long-wavelength topography of the sea floor. One is the existence of direct evidence, in the broad-scale undulation of mid-ocean rise crests away from hotspots, for the influence of the sub-lithospheric mantle on long-wavelength seafloor topography. The other is the inadequacy of the conventional view of the subsidence of the old sea floor. In this view, the old sea floor "flattens" (i.e., asymptotically approaches a constant depth) for ages greater than about 70 Ma [Parsons and Sclater, 1977], but this is not an adequate description of its empirical behavior, nor is the conventional explanation for this so-called "flattening" adequate.

The approach adopted here is to proceed from well-established principles of thermal convection. Davies [1988b] noted that it is useful to think of three potential sources of topography at the surface of a convecting fluid: the two thermal boundary layers at the top and bottom of the fluid, and the small-amplitude, large-scale temperature variations away from the boundary layers that are always present in a convecting fluid because the interior fluid has been incompletely homogenized. Davies' results indicated that the latter could be a significant contributor to the topography of the earth's surface.

It is possible to identify topography of the sea floor that plausibly corresponds to each of these three contributions. Thus the dominant topography, the mid-ocean rise system, is clearly due to the top thermal boundary layer, of which the cooling and thickening oceanic lithosphere comprises the main part [Davies, 1988a]; this topography is due simply to thermal contraction of the cooled material in the boundary layer. The hotspot swells [Crough, 1983] are very plausibly due to upwelling buoyant mantle that most likely comes from a lower, hot boundary layer at the base of the convecting layer [Davies, 1988b, 1992]. The broad-scale undulation of mid-ocean rise crests, referred to above, is unlikely to be due to lithospheric effects, as will be discussed below, so it is likely to be due to large-scale, small amplitude variations in mantle temperature of the type expected in a convecting fluid.

If a long-wavelength component of topography is visible at ridge crests, then presumably it may exist elsewhere. Since the subsidence of cooling plates also contributes to long-wavelength topography, these two contributions will need to be separated. To do this, we first discuss the inadequacies of the flattening seafloor subsidence curve, which is still widely used as a reference for attempting to define depth anomalies. We then argue that the "square-root-of-age" subsidence curve is a

The Mesozoic Pacific: Geology, Tectonics, and Volcanism
Geophysical Monograph 77

more useful reference since it is based on well-defined physics that is clearly a good approximation for younger sea floor. Deviations from this reference can be attributed unequivocally to other physical processes.

Having defined the methodology, we exemplify its application by discussing the largest anomalous feature to emerge from the analysis: a very large swell in the Pacific Ocean. We propose that this swell is the modern continuation of the Darwin Rise, which was proposed by Menard [1964, 1984] to have existed in the Cretaceous. In other words, we propose that the Darwin Rise is still extant. We then discuss possible mantle origins of the Darwin Rise. Larson [1991] has stimulated ineterest in events in the Pacific region during the Cretaceous era, when a great outburst of volcanism seems to correlate with a number of significant global phenomena, including high sea level and greenhouse climatic conditions. These events may be related to the origin of the Darwin Rise.

The approach presented here raises questions about the interpretation of all the older areas of seafloor. Detailed discussion of these areas will be deferred because their analysis is a substantial task, but a brief indication of some possibilities will be given. This paper is logically the third in a series on ocean bathymetry and mantle convection [Davies, 1988b, c].

THE OLD SEA FLOOR DOES NOT "FLATTEN"

The mid-ocean rise system stands high because the lithosphere cools by thermal conduction and subsides by thermal contraction as it moves away from the spreading centers, and this gives rise to the observed increase of depth of the oceans in proportion to the square-root of the age of the sea floor [Davis and Lister, 1974]. The remarkable success of this simple theory for sea floor less than about 70 Ma old has been well-remarked upon. Less well noted is the implicit assumption in this theory that the influence of the underlying convecting mantle can be ignored, to first order. Davies [1988b] exploited this observation to argue against the existence of strong, buouyant mantle upwellings (other than plumes, discussed below), which in turn weighs against the existence of a barrier to flow in the mantle transition zone at 670 km depth. Davies [1988b] also noted the implication that upwelling under normal spreading centers is passive.

Despite the first-order success of the simple conductive cooling model, there are clear deviations from its predictions that are of three types. One type is the hotspot swells [Crough, 1983] which we attribute to the buoyancy of the mantle plumes inferred to underlie volcanic hotspots [Davies, 1988b, 1992]. A second type, which has received little attention to date, is the variation in depth of the crests of mid-ocean rises. If the mantle were passive with uniform temperature, rise crests should all be at the same depth. Some variations, most notably near Iceland, are clearly associated with mantle plumes. Others are not, the clearest example being south of Australia, where the rise crest reaches about 1 km deeper than average in a low region extending over thousands of kilometers east-west. Such variations cannot be attributed to plumes, seafloor age (it is all zero age) or the strength of the lithosphere (it can support loads only over a few tens of kilometers distance near rise crests [Bodine et al., 1981]). They must be an expression of the dynamic mantle.

The third type of deviation from the square-root-of-age subsidence has received a great deal of attention: in many areas, especially of older sea floor, the sea floor subsides at less than the predicted rate. Although this deviation has been characterized as "flattening", implying an asymptotic approach to a constant level [Parsons and Sclater, 1977], the data presented by Marty and Cazenave [1989] show that "flattening" is not an accurate description of the sea floor topography. They divided the sea floor into 32 regions and plotted depth versus square-root of age for each. A selection from their plots is shown in Figure 1, where the reader's attention is directed just to the data, rather than to the regression lines. In some regions the sea floor continues to subside to the greatest ages (160 Ma), while in others it turns and rises again, and in others it rises and falls again as a function of age. In only one or two cases, the most favorable being region 3 (Figure 1), could it reasonably be said that the sea floor asymptotically approaches a constant depth.

In fairness it should be noted that the early analyses were based on a much more restricted data set [Parsons and Sclater, 1977], and that the interpretation was influenced by the apparent flattening of heat flow, both data sets being apparently consistent with the asymptotic approach of the lithosphere to a constant thickness [Sclater and Francheteau, 1970]. However, all of the reasons for holding this interpretation have been in question for some time. Thus the flattening interpretation of the heat flow data is no longer held [Sclater et al., 1980]. Heestand and Crough [1981] proposed that much of the deviation from root-age subsidence could be attributed to mantle plumes (a view not necessarily supported here). They and Schroeder [1984] showed that much of the seafloor away from hotspot tracks followed root-age subsidence more closely, although the uncertainties were large. The efficacy of the mechanism supposed to account for the approach of the lithsophere to constant thickness (small-scale convection) was also criticised [O'Connell and Hager, 1981].

Davies [1988c] examined the whole question of small-scale convection and its supposed relation to flattening bathymetry. He showed with numerical models what O'Connell and Hager [1981] had already argued, that small-scale convection modes, that are supposed to pervade the upper mantle and to deliver extra heat to the base of the lithosphere, cause faster subsidence, not slower subsidence. The opposite, incorrect, conclusion was reached [Parsons and McKenzie, 1978] because no account was taken of the cooler, descending part of the hypothetical convection. Davies [1988c] also argued that such small-scale convection should produce pervasive bathymetric and gravity anomalies over the old sea floor, but none have been convincingly distinguished from hotspot tracks and oceanic plateaus [Sandwell and Renkin, 1988]. More recently Zhang and Tanimoto [1991] have presented Love-wave seismic data that indicates that the oceanic lithosphere continues to thicken to the greatest ages.

Despite these severe problems, the flattening bathymetry and

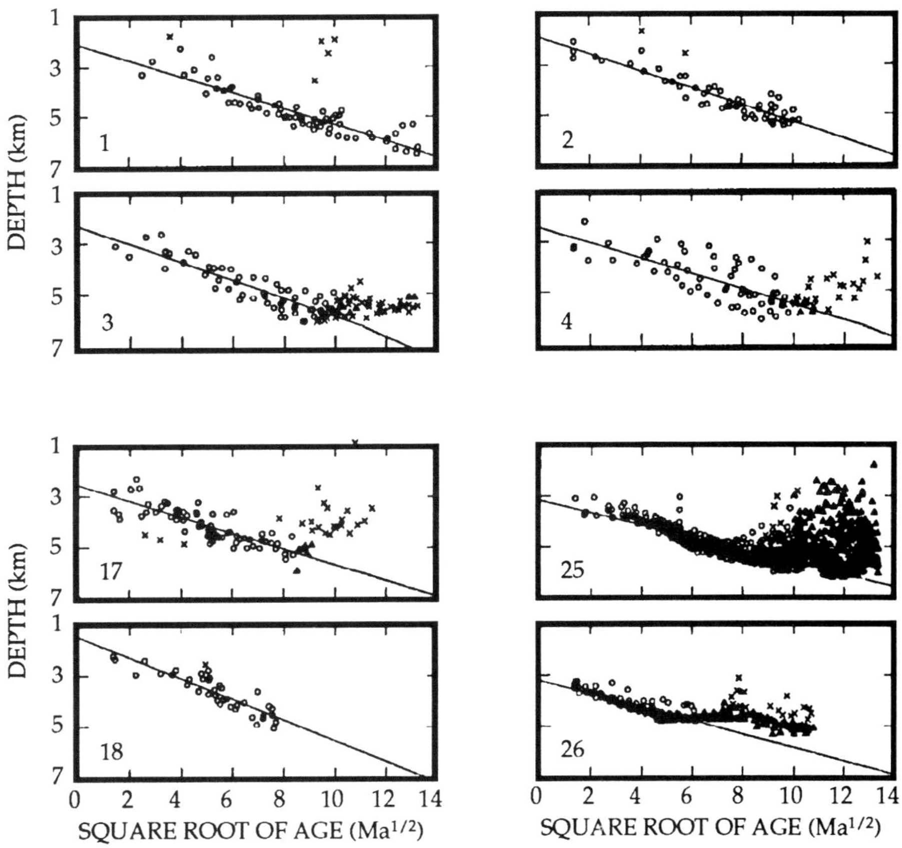

Fig. 1. A selection of seafloor depth versus age plots for some of the regions analysed by Marty and Cazenave[1989]. Attention is directed here to the data, not to their regression lines. The data show a variety of behaviors: subsidence in proportion to the square-root of age to 160 Ma; subsidence followed by uplift; subsidence, uplift, subsidence; and (region 3) subsidence followed by constant depth. Only two or three of the 32 regions shows behavior that can reasonably be characterized as subsidence followed by an asymptotic approach to a constant depth, region 3 being the best candidate. It is concluded that the latter is not an accurate description of the behavior of the old sea floor. After Marty and Cazenave [1989].

small-scale convection interpretations have still frequently been taken as the default interpretations [e.g., Watts et al., 1985; Craig and McKenzie, 1986; Colin and Fleitout, 1990; McNutt et al., 1990; McNutt and Judge, 1990].

THE COOLING HALFSPACE REFERENCE

Our approach is to assume that the physics of plate subsidence is well-known to first order: it is simply one-dimensional conductive cooling [Davis and Lister, 1974]. Since we do not expect the thermal conductivity of the oceanic lithosphere to vary significantly with location, any deviations from the subsidence predicted from this cooling must be due to other physical causes, including the influence of the sublithospheric mantle. If the topography due to this boundary layer cooling is adopted as a reference and subtracted out, we can then attempt to identify the physical causes of any residual topography.

The age-depth relationship resulting from the conductive cooling of the oceanic lithosphere, modelled as an infinite halfspace, has the form [Davis and Lister, 1974]

$$d = a + bt^{1/2} \qquad (1)$$

where d is ocean depth, t is age and a and b are constants. The value of a is not important here, since we are interested in relative depths of the sea floor. We have tried a range of values of the coefficient b.

Schroeder's [1984] analysis of Pacific bathymetry reveals values of b consistently in the range 300 to 315 m/Myr$^{1/2}$. Hayes [1988] recommends a global average of 300 m/Myr$^{1/2}$, but finds a range of 270 - 473 m/Myr$^{1/2}$ in regions of the South Atlantic Ocean and the Southern Ocean. Marty and Cazenave [1989] find a global average of 317 m/Myr$^{1/2}$ with a range of 214 - 389 m/Myr$^{1/2}$ for various regions in cases where the ridge depth was separately fixed to the local average value. Parsons and Sclater [1977] obtained 350 m/Myr$^{1/2}$ for seafloor ages less than 70 Myr using a more limited data set.

The intrinsic subsidence rate due to the cooling of the lithosphere is difficult to separate empirically from the long-wavelength anomalies being sought. The only reasonable approach to removing other long-wavelength components seems to be to use global averages. The apparent net subsidence may also be affected by shorter-wavelength effects such as hotspots and seamounts. It is plausible that younger seafloor will have less probability of having been affected by the short-wavelength phenomena, so the value b = 350 m/Myr$^{1/2}$ of Parsons and Sclater for younger seafloor may be a reasonable estimate. However Schroeder [1984] obtained 305 m/Myr$^{1/2}$ and 337 m/Myr$^{1/2}$ for the north and south Pacific, respectively, for seafloor younger than 80 Myr.

To cover these uncertainties, we have used b = 320 m/Myr$^{1/2}$ as a base value, and 280 m/Myr$^{1/2}$ and 360 m/Myr$^{1/2}$ to represent a plausible range.

DEPTH AND AGE DATA AND ANALYSIS

We demonstrate the application of this approach by presenting a preliminary analysis of the depth and age of the Pacific sea floor. The sea floor bathymetry was obtained from the ETOP05 Topographic Data Set, sampled on a grid of 5 by 5 minutes, available from the National Oceanographic and Atmospheric Administration in Boulder, Colorado. In conjunction we used a data set of age of sea floor, digitized from magnetic lineation maps of Larson et al. [1985] and gridded at 30 by 30 minutes by Marty and Cazenave [1989]. This data set provides an age coverage for most of the oceanic lithosphere.

While the seafloor age data set used here has uncertainties in the magnetic quiet zones and has been superseded in a number of areas, most of its uncertainties and deficiencies are in the older parts of the sea floor, where subsidence rates are slow and where uncertainties in depth corrections are therefore smaller. Also, we are interested here in the largest-scale features, and so some localised uncertainties will be less significant. Thus it is expected that the data set is quite sufficient for the present purpose.

The seafloor depth also needs to be corrected for the presence of sediment. For our purposes, this correction is not crucial beacuse sediment thicknesses are generally less than 100 m, though it is significant in some areas, as will be seen. We obtained gridded sediment thicknesses also from Marty and Cazenave [1989]. For the Pacific region considered here, these are based on the map of Ludwig and Houtz [1979]. A correction was then calculated according to Schroeder's [1984] formula

$$s = 0.224\ S + 900[1 - \exp(-S/1800)] \tag{2}$$

where s is the correction, S is the sediment thickness, and both are in meters.

There are in the Pacific many regions of thickened crust, mostly in the form of volcanic chains or oceanic plateaus, such as the Ontong-Java Plateau. Since they are smaller than the scale of the features to be discussed here, their locations are well known and their internal density structure is not well known, no attempt has been made here to remove their contribution to seafloor topography.

There are also regions in the western Pacific where it is believed that there are extensive basalt flows embedded in the sedimentary sequence [Schlanger et al.,1981]. Renkin and Sclater [1988] argue that these contribute only a small correction because of the higher densities of basalts. No correction for these has been attempted here, although it may be important to test this effect in some limited regions.

RESIDUAL DEPTH ANOMALIES IN THE PACIFIC

A map of depth anomalies in the Pacific region remaining after subtraction of a subsidence curve, given by equation (1) with b = 320 m/Myr$^{1/2}$, and the sediment correction is shown in Figure 2. Although the crest of the East Pacific Rise is still recognizable, it is greatly reduced in amplitude, indicating that the chosen subsidence rate is reasonably appropriate.

Many oceanic plateaus and volcanic seamount chains are visible in Figure 2 (mostly in red and orange), as well as some hotspot swells (yellow-buff), with the Hawaiian Swell being particularly prominent. The Ontong-Java Plateau is the most prominent oceanic plateau. Most of these features are quite distinctly visible and we have not attempted to remove them for the reasons stated in the last section; their relative significance can be judged by the reader.

In addition to these smaller-scale features, there is a geographically very extensive region of anomalously shallow sea floor in the central and west Pacific, coloured in light blue, greens and yellows in Figure 2. This region extends from the Japan seamounts, in the northwest, about 9,500 km to the French Polynesian region in the southeast. It is about 3,500 km in width and is flanked to the north by the Hawaiian seamounts and to the south by New Guinea. The "background" sea floor (neglecting smaller-scale features) in this region is about 600 m shallower than seafloor in the northern and eastern Pacific.

The effect of using different subsidence rates for the reference curves is illustrated in Figures 3(a,b), for which b values of 360 and 280 were used, respectively. In Figure 3(a), the rise crest is still visible, but as a local feature. Away from the crest, the residual depth is nearly constant in the southeast Pacific, suggesting that b = 360 is within the range of reasonable estimates of the intrinsic subsidence rate. The main swell is very prominent and has an amplitude of 900-1200 m, with a slightly larger extent than in Figure 2.

The swell in Figure 3(b) might be viewed as a series of smaller swells, rather than a single large one, but the low value of b = 280 has no real empirical justification, as is reflected by the failure to completely remove the East Pacific Rise over a broad region. This case is included mainly to illustrate the robustness of the main swell.

The effect of the sediment correction is illustrated in Figure 3(c), which is similar to Figure 2 except that the sediment correction is omitted. The main changes are in the basin southeast of the Shatsky Rise and in the Mariana Basin, and the

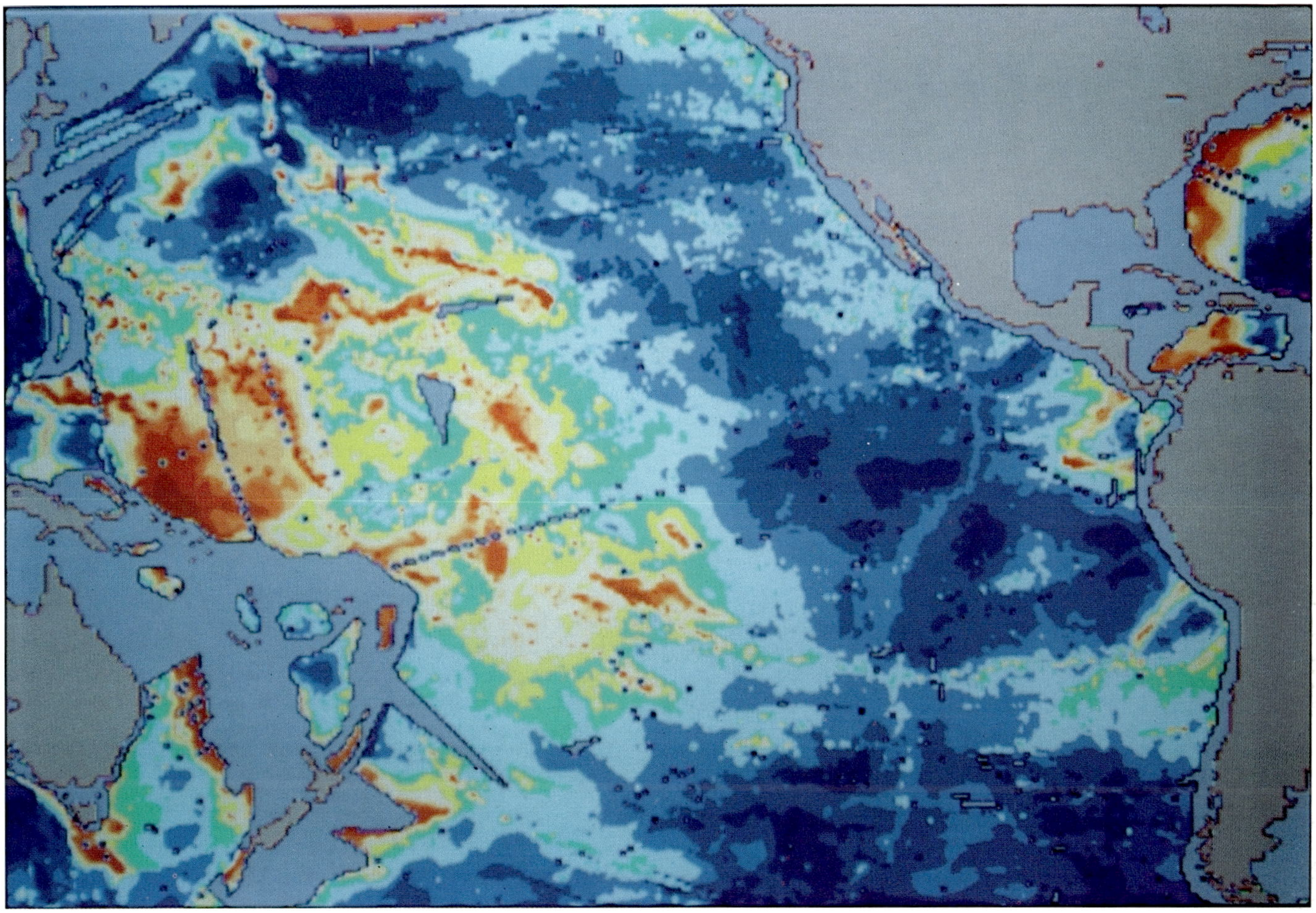

Fig. 2. The residual depth map for the north Pacific basin after the removal of the sediment correction and a subsidence curve given by equation (1) with b = 320 m/Myr$^{1/2}$. Colours represent residual depths and are incremented in 300 m contours from dark blue (deepest) to red (shallowest). The neutral grey signifies gaps in the data set of age of the sea floor: some of the diagonal streaks are artifacts of these gaps.

main effect is to extend the swell northwest towards the Kurile Islands and Japan. It is possible, as noted in the last section, that there should be other locally large corrections, for example in the Nauru Basin, northeast of the Ontong-Java Plateau [Renkin and Sclater, 1988].

THE DARWIN RISE?

The huge swell visible in Figure 2 is consistently present for all reasonable parameters of the reference subsidence curve. It conforms closely in its amplitude and its location on the Pacific Plate with the Mesozoic rise proposed by Hess [1962] and named the Darwin Rise by Menard [1964, 1984]. They originally assumed (not unreasonably for the 1960s) that it was analogous to modern mid-ocean ridges, and concluded that it had been ephemeral, since they could not see any evidence for it at present. However, there is a clear suggestion in Schlanger et al.'s [1981] discussion of the Cretaceous volcanic episode that a depth anomaly might have persisted for long afterwards. This point will be made explicit and quantitative below. We propose that the Cretaceous Darwin Rise has in fact persisted to the present, and we identify it with the large superswell revealed in Figure 2. The cause of the present superswell and the nature of its relationship with the Cretaceous Darwin Rise will be discussed below.

RELATED PROPOSALS

Several recent papers have discussed aspects of Pacific superswells and their possible relationship with the Darwin Rise. McNutt and Fischer [1987] argued that there is a region of elevated sea floor and thinned lithosphere around French Polynesia that cannot be explained solely by the hotspots in the region. This "South Pacific superswell" is only the southeast edge of the larger superswell proposed here. Pribac and Davies [1987] independently proposed an anomalous

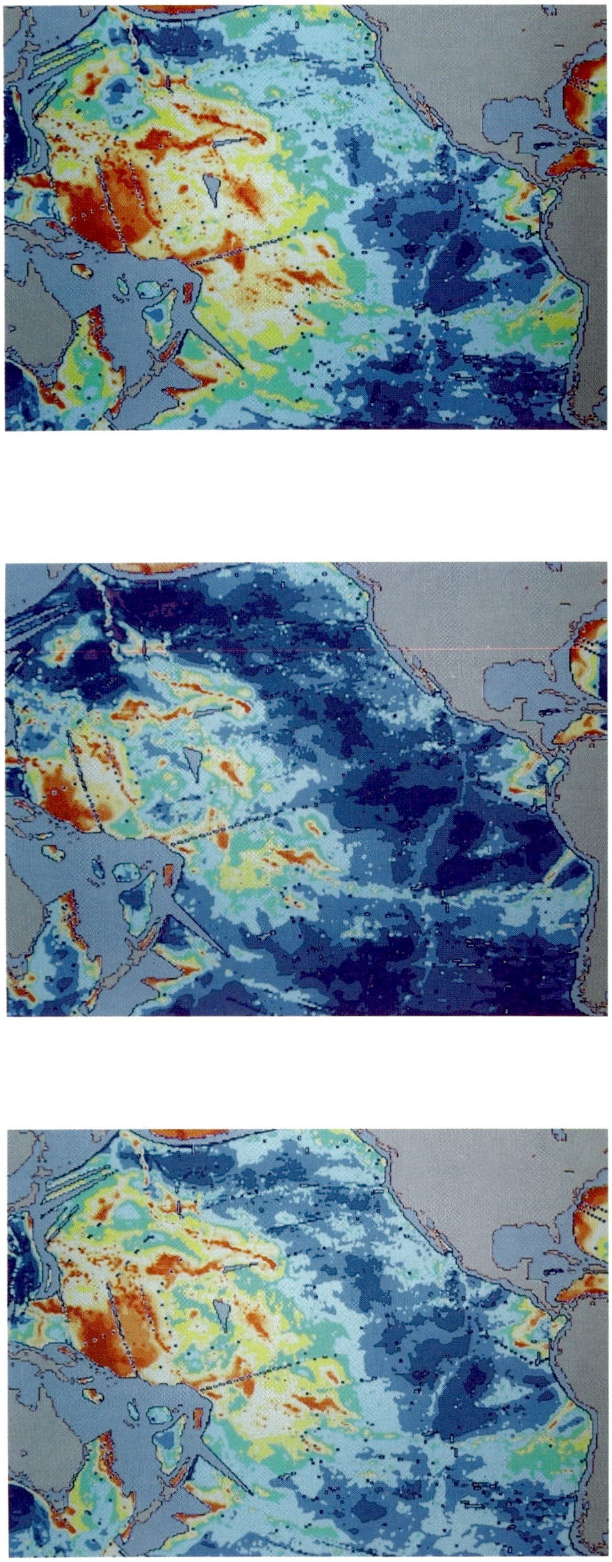

(a)

(b)

(c)

elevation of the central and western Pacific, with a theoretical rationale but with no detailed analysis of topography. Both papers proposed the term "superswell".

McNutt et al. [1990] and McNutt and Judge [1990] have argued that the putative South Pacific superswell may be a modern expression of the phenomenon that produced the Cretaceous Darwin Rise. A suggestion of a western Pacific superswell is visible in the map of anomalous bathymetry of Cazenave et al. [1988], although they used an asymptotic flattening curve as a reference. Our proposed Darwin Rise is clearly revealed by Cazenave and Lago [1991], who removed a square-root of age subsidence very similar to ours, though they said little about its cause.

The present paper extends the previous work by proposing the essential identity of the present central- and western-Pacific superswell with the Cretaceous Darwin Rise, and by discussing its mantle origin more explicitly and quantitatively.

HYPOTHESES ON THE MANTLE ORIGIN OF THE DARWIN RISE

Thinned Lithosphere.

McNutt and Fischer [1987] and McNutt and Judge [1990], following the hypothesis advanced by Detrick and Crough [1978], attributed their putative superswell to thinning of the lithosphere to less than 75 km, citing small values of the flexural thickness of the lithosphere and low velocities of seismic surface waves. However, there are reasons to doubt this conclusion.

The seismic velocity anomaly was calculated relative to a regional velocity-age trend attributed solely to variations in the geotherm, but the required temperature derivative of velocity is twice that measured in the laboratory [McNutt and Judge, 1990], suggesting that other effects may also be involved regionally. The increase of the local Love wave velocity anomaly with period was then used to argue that the source must be within about 100 km of the surface, but the scatter in these data is about as large as the claimed trend, which casts doubt on its reality.

Stein and Abbott [1991] noted the absence of a heat flow anomaly that would be expected to accompany such thin lithosphere, and suggested that the thinning deduced by McNutt and Fischer [1987] and McNutt and Judge [1991] from the response to the loads of volcanoes may be only local or may be mechanical rather than thermal weakening. Devey et al. [1990] concluded from geochemical evidence that large degrees of melting did not occur under Polynesian hotspots, and this is not compatible with the presence of hot, upwelling mantle under a thin lithosphere [McKenzie and Bickle, 1988], despite the claim by McNutt and Judge [1990] to the contrary. The

Fig. 3. Alternative residual depth maps. *(a)* With sediment correction and b = 360 m/Myr$^{1/2}$. *(b)* With sediment correction and b = 280 m/Myr$^{1/2}$. *(c)* Without sediment correction and with b = 320 m/Myr$^{1/2}$, as in Figure 2. See Figure 2 for further explanation.

seismic results of Zhang and Tanimoto [1991] indicate that the lithosphere continues to thicken with age throughout the Darwin Rise region.

Other Possibilities

The analysis of bathymetry presented above and the preceeding discussion make the case that the Darwin Rise is a feature not arising in the upper thermal boundary layer of the mantle (the cooling lithosphere). Accordingly, it should then be attributed either to effects from a lower thermal boundary layer or to small-amplitude temperature variations in the interior of the mantle. Three such possibilities will be considered here: plume "tails" [Morgan, 1972; Davies, 1988b; Sleep, 1990], plume "heads" [Morgan, 1981; Richards et al., 1989; Griffiths and Campbell, 1990], and a large warm region of the mantle interior [Davies, 1988b].

Plume Heads and Tails

The term "plume head" refers to the large spherical blob by which a new plume is expected to force a path through the mantle, while a "plume tail" is the narrower conduit through which additional buoyant material can ascend once a path is established [Griffiths and Campbell, 1990]. A plume tail corresponds to Morgan's [1972] original concept of a plume. In either case, hotter, buoyant mantle material would be expected to ascend to the base of the lithosphere and uplift it. The analysis by Davies [1992] has shown that in the case of the Hawaiian swell the uplift can be satisfactorily explained by this simple mechanism, without invoking thinning of the lithosphere [Detrick and Crough, 1978], and there is seismic evidence against such thinning [Woods et al., 1991].

The possibility of explaining the deviation of older seafloor from square-root of age subsidence as due to the effects of plume tails has been examined by Heestand and Crough [1981] and Schroeder [1984], who concluded that it was possible. However their studies were permissive rather than conclusive, since they showed that when putatively hotspot-affected areas were excluded the remaining areas were not inconsistent with square-root-of-age subsidence. McNutt and Fischer [1987] argued that the "South Pacific superswell" could not be explained entirely as a superposition of hotspot swells and that there is an additional background elevation of several hundred meters in the area.

It is possible that several plume heads have been emplaced under the Pacific lithosphere. These would be expected to be about 2000 km in diameter and to cause several hundred meters of uplift [Griffiths and Campbell, 1990]. Their locations should correspond with the beginnings of hotspot tracks, but the age data from Pacific islands and seamounts are so sparse that it is difficult to clearly define many of the tracks, and their beginnings are unclear [Duncan and Clague, 1985]. This uncertainty makes it difficult to test in any detail for the presence of plume heads. Therefore, plume heads will be regarded here as just another means by which hot plume material could have been emplaced under the lithosphere in the Darwin Rise region.

Deep Warm Mantle

The "plate-scale" mode of mantle convection accounts for most of the heat removed from the earth's interior, with plumes accounting for only about 10 per cent of the heat transported through the mantle [Davies, 1988b; Davies and Richards, 1992]. This implies that the temperature of the mantle will be most affected by the cycle of plate formation and subduction. Richards and Engebretson [1992] have shown that the temperature variations implied by both lateral seismic velocity variations and the geoid (see below) correlate with regions of mantle that have been cooled by subduction in the past 180 m.y. We thus hypothesize, following also Davies [1984], that there will be regions of mantle that are relatively warm because they have been slowly warming by radioactive heating but have not undergone any recent cooling by the injection of cold subducted lithosphere. Note that the important factor is the absence of subduction, rather than, for example, the presence of a spreading ridge, which is only a near-surface structure that has no short-term effect on the mantle deeper than about 100 km [Davies, 1988a; Davies and Richards, 1992].

This tendency for warm regions of internally heated convecting fluid to develop away from descending flows is enhanced if the deeper fluid has higher viscosity [Gurnis and Davies, 1986; Pribac, 1991]. There is evidence for high viscosity in the deep mantle, as will be noted below.

OTHER GEOPHYSICAL EVIDENCE

Three dimensional seismic imaging of the sub-Pacific mantle reveals that there is a large volume of material with anomalously slow seismic P-wave velocities in the lower mantle beneath the region of the Darwin Rise [Dziewonski and Woodhouse, 1987]. There is also an extensive positive geoid anomaly in this region with an amplitude of about 60 m [Hager et al., 1985]. Hager et al. have shown that these observations are consistent with the presence of mantle material with temperature elevated above the ambient by about 20∞C. They predicted that this warm mantle should cause anomalous elevations, and their predicted topography corresponds well in location and amplitude with the anomalous topography revealed in Figure 2. Richards and Engebretson [1992] have shown that the geoid, seismic structure and hotspot distribution all correlate strongly with the time-integrated locations of subduction during the last 180 m.y., suggesting that the deep mantle thermal structure is controlled to first order by the occurrence or absence of subduction. The Darwin Rise region has been devoid of subduction during this period. These observations and results favor deep-seated, small-amplitude variations in mantle temperature as a contributing cause of the Darwin Rise.

The evidence for the shallower mantle is less clear. McNutt and Judge [1990] report low velicities of Love waves in French Polynesia relative to the Pacific region. The Love wave velocities of Zhang and Tanimoto [1991] tend to be slightly low in the central Pacific relative to the north Pacific, but a clear association with the Darwin Rise is not evident. Nevertheless, the prevalence of strong hotspots in the region, the likelihood that some of them started relatively recently [Duncan and Clague, 1985], and the existence of visible local swells around some hotspots in Figure 2 indicates that a significant amount of plume material, both heads and tails, has been emplaced under the lithosphere of this region, though it may not yet have been well resolved seismically.

On the basis of this evidence, it will be assumed here that the Darwin Rise is caused by a combination of deep and shallow material in the sub-lithospheric mantle.

TOPOGRAPHY FROM DEEP, WARM MANTLE

We now present a numerical model of mantle convection designed to illustrate how a convecting mantle can give rise to the inferred large volume of slightly warmer mantle, and to give some preliminary quantification to the resulting surface uplift. This model is intended only to illustrate the general phenomenon, and not to match the particular circumstances of the Darwin Rise.

Realistic models of mantle convection must include the effects of rigid but mobile lithospheric plates [Davies, 1988a]. In this study this is achieved by combining prescribed velocity boundary conditions with temperature-dependent viscosity, so as to produce a stiff upper boundary layer that nevertheless moves like a plate. The efficacy and accuracy of this approach has been demonstrated by Davies [1988a, 1989], where the numerical methods have been described.

There are several observational indications that the lower mantle viscosity is a factor of 10 to 100 times that of the upper mantle. Hager et al. [1985] deduced a viscosity increase of this magnitude through the transition zone from geoid anomalies over subduction zones. Nakada and Lambeck [1989] inferred from relative sea level changes that the lower mantle viscosity might be about a factor of 10 higher than that of the upper mantle. Gurnis and Hager [1988] deduced an increase by a factor of about 30 from the change in dip (or kink) in subducted lithosphere as it passes through the transition zone, as indicated by inferred aseismic extensions of deep seismic zones [Creager and Jordan, 1986]. Gurnis and Davies [1986] have shown that higher viscosities at depth would lead to higher temperatures because of the longer residence time of material there.

To include this effect, a depth-dependence of viscosity is combined multiplicatively with the temperature dependence. In addition to the overall increase with depth just discussed, a sub-lithospheric low viscosity zone is often postulated, with viscosity reduced by a factor between 10 and 100 [e.g., Craig and McKenzie, 1986; Ceuleneer et al., 1988]. In the model presented here we have chosen a viscosity profile with an exponential increase with depth by a factor of 30 and a low viscosity layer where the viscosity is reduced by a factor of 20. The superimposed temperature dependence of the viscosity has a variation by a factor of 500 between warmest and coldest mantle material.

Spreading centers commonly migrate relative to other tectonic features such as subduction zones and hotspots. This is a consequence of plates evolving according to some simple empirical rules that seem to reflect the mechanics of the lithosphere [Davies, 1988a]. In the present context, this has the important implication that a spreading center may migrate relative to deep mantle structure, and this should be reflected in numerical models. However, the basic phenomenon to be described here is present regardless of ridge migration, which merely modifies its surface expression.

Figure 4 shows the temperature, viscosity and streamlines in a convection model with internal heating and no heat input from below [Davies, 1988b]. The model was initially run to

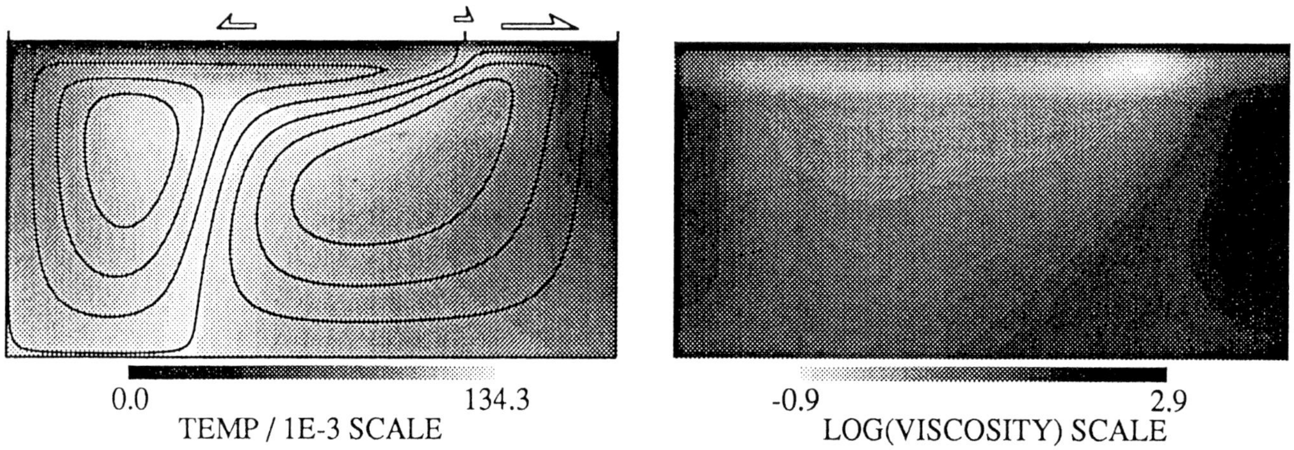

Fig. 4. Temperature, viscosity and streamlines for a convection model with normalized viscosity exponentially increasing with depth (maximum contrast of 30), a superimposed temperature-dependent viscosity (maximum contrast of 500) and a low viscosity layer with viscosity reduced by a factor of 20. Normalized temperature and viscosity are shown side by side in grey scales, as indicated, and the flow streamlines are shown superimposed on the temperature map. The spreading centre near the right hand side of the mantle has migrated from the left hand side because the left-hand plate speed is half that of the right-hand plate. Rayleigh number = 10^6, Peclet number = 75.

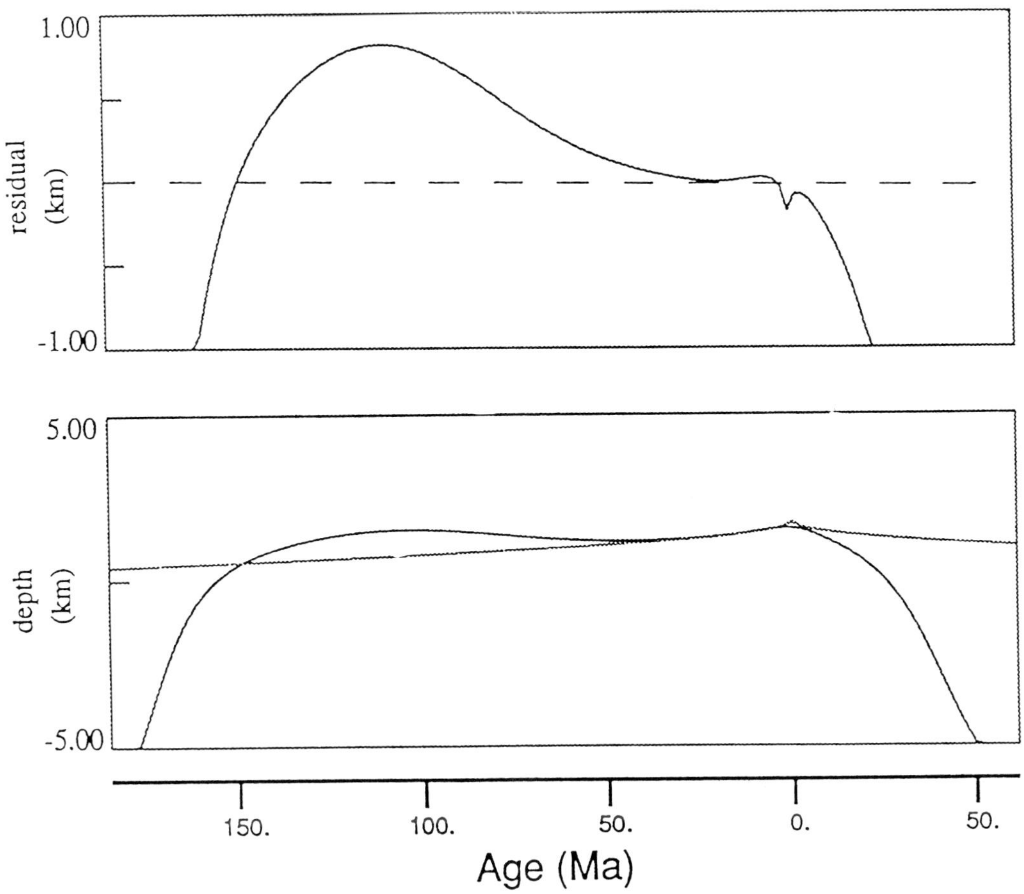

Fig. 5. Topography resulting from the convection model shown in Figure 4. The lower box shows calculated topography (heavy curve), along with the contribution to the topography just from conductive cooling of the lithosphere. The difference is shown in the upper box: this is the topography due to the temperature differences in the sublithospheric fluid. The model has not been designed to represent accurately the topography at downwelling regions (the ends of the box).

near steady state with the spreading center stationary at the left margin, during which time a region of warm mantle developed under the spreading center, due to the fact that cool, descending material from the surface did not penetrate directly into the region. Then the plate velocities were modified so that the spreading center migrated to the right. The region of warm mantle lagged behind the spreading center, and subsequently began rising (very slowly) under the influence of its own buoyancy.

Figure 5 shows the topography calculated from this model. As well as the topography resulting from the cooling, subsiding "lithosphere", there is an additional broad uplift where the warm fluid is rising from depth. For this model, this anomalous topography has an amplitude of up to 700 m, in good agreement with the observed amplitude of the Darwin Rise, given the preliminary nature of the model.

The effects of other viscosity structures have been investigated [Pribac, 1991]. A model without any depth dependence of viscosity yielded much smaller anomalous topography. Preliminary indications suggest that the inclusion of a low viscosity zone has negligible effect on the amplitude of such long-wavelength bathymetry and that, when normalized appropriately, viscosity structures stepped at the 670km discontinuity give results within a factor of 2 of those from exponential variations. A more detailed account of such models will be presented at a later time.

PERSISTENCE OF SWELLS AND SUPERSWELLS

Topography due to temperature differences that are below the lithosphere can be expected to persist for 100 Myr or more, because it takes that long for heat to conduct through the lithosphere. We illustrate this in two ways.

The anomalous topography illustrated in Figure 5 is due to the deep thermal anomaly visible in the model shown in Figure 4. The model time scales to the equivalent of 450 Myr for this model, and the thermal structure is clearly persistent on this timescale.

Even for much shallower thermal anomalies, such as those

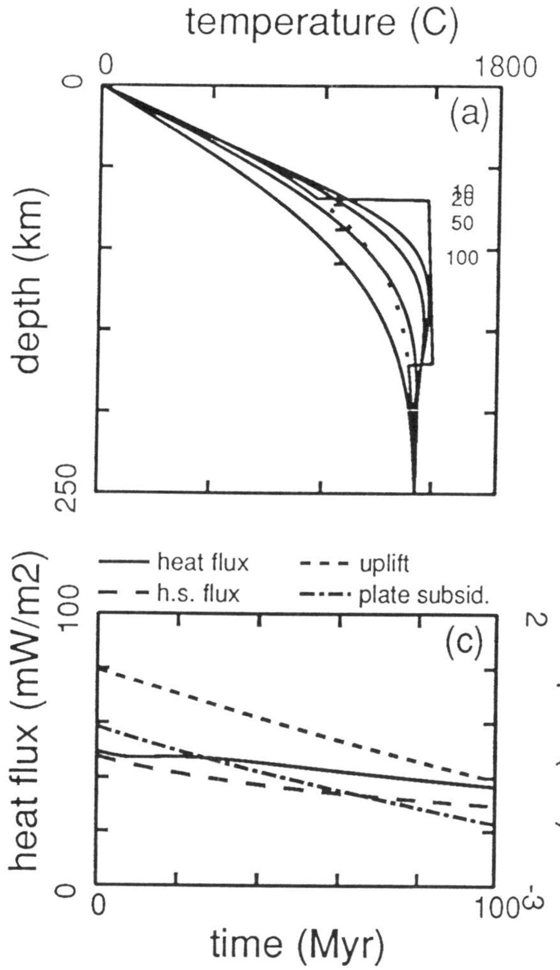

Fig. 6. Geotherms, topography and heat flux due to the emplacement of warmer (plume) material between depths of 70 km and 170 km under oceanic lithosphere of 70 Ma age. *(a)* The unperturbed geotherm (dotted) represents cooling from a mantle adiabat with a surface temperature of 1350°C. The warm material has a decompressed (potential) temperature of 1450∞∞C. Conductive geotherms are labelled with the time after emplacement (in Myr). *(b)* Surface heat flux and elevation (*uplift*) with and without the effects of the emplaced material. The unperturbed curves (*h. s. flux* and *plate subsid.*) are calculated from the cooling halfspace model. The emplaced material causes a modest increase in heat flux and an anomalous elevation of about 1 km that persists for more than 100 Myr.

expected from the emplacement of buoyant plume material below the lithosphere, the resulting elevation, relative to seafloor of the same age, is very persistent. This is illustrated in Figure 6. Figure 6a shows a sequence of conductive geotherms resulting from the emplacement of hotter material between depths of 70 km and 170 km under lithosphere of age 70 Ma. Figure 6b shows the resulting surface elevation and surface heat flux, compared with the predictions for undisturbed cooling lithosphere. The emplaced material causes an initial uplift of about 1010 m. After 100 Myr (corresponding to a total plate age of 170 Ma), this has declined only to 760 m relative to the undisturbed subsidence. The surface heat flux is not perturbed very much, being maintained between 40 and 50 mW/m2 for most of the period. This is consistent with the absence of a heat flow anomaly through the Darwin Rise region [Stein and Abbott, 1991].

The essential point is that unless hot material is emplaced at or near the earth's surface, it is insulated by the overlying lithosphere and takes a long time to lose its heat.

ORIGIN AND PERSISTENCE OF THE DARWIN RISE: SUMMARY

In the interpretation developed here, the present Darwin Rise is due to a combination of hot plume material immediately under the lithosphere and deeper mantle material that is a few tens of degrees warmer than normal. The depth range and lateral distribution of the shallower material is not well constrained. The presence of the deep material is indicated by seismic tomography and by the geoid. The presence of the shallow material is inferred from the presence of several hotspots in the region, from the presence of smaller swells localized around some hotspot tracks, from the likelihood that some of the hotspots are young and will have plume heads located at the start of their tracks, and from some indications that seismic surface wave velocities are low in parts of the region.

The Rise is inferred to have existed at least since the mid-Cretaceous, about 100 Ma ago. This inference comes from the coincidence in location on the Pacific plate, and from the above arguments that its amplitude, relative to normal seafloor of the same age, should not have decreased substantially in that time, since most of the thermal anomaly is sublithospheric, and thus largely insulated from surface cooling.

If the Rise has moved with the Pacific plate, then some of the buoyant material must also have moved with the plate. This suggests that a significant part of the anomalous elevation is due to the material immediately under the lithosphere, which is the most likely to have moved with the plate. However a clear discrimination between shallow and deep sources on this basis is unlikely because the plate will in any case take the order of 100 m.y. to pass over the deep thermal anomaly, and some of the deep thermal anomaly may also move in the direction of the plate.

Figure 7 shows a sketch of the sequence of mantle events suggested by these inferences. The sequence begins with a region of deep, warm mantle left by a prolonged absence of subduction. This warm mantle first rises slowly and then begins to move in the direction of the overlying large plate, although more slowly than the plate. The warm mantle uplifts the surface by hundreds of meters over a broad region throughout this period. At the same time, a series of new, hot plumes rises from the thermal boundary layer at the base of the mantle, which is thickened and destabilized by the converging large-scale flow associated with subduction in the surrounding

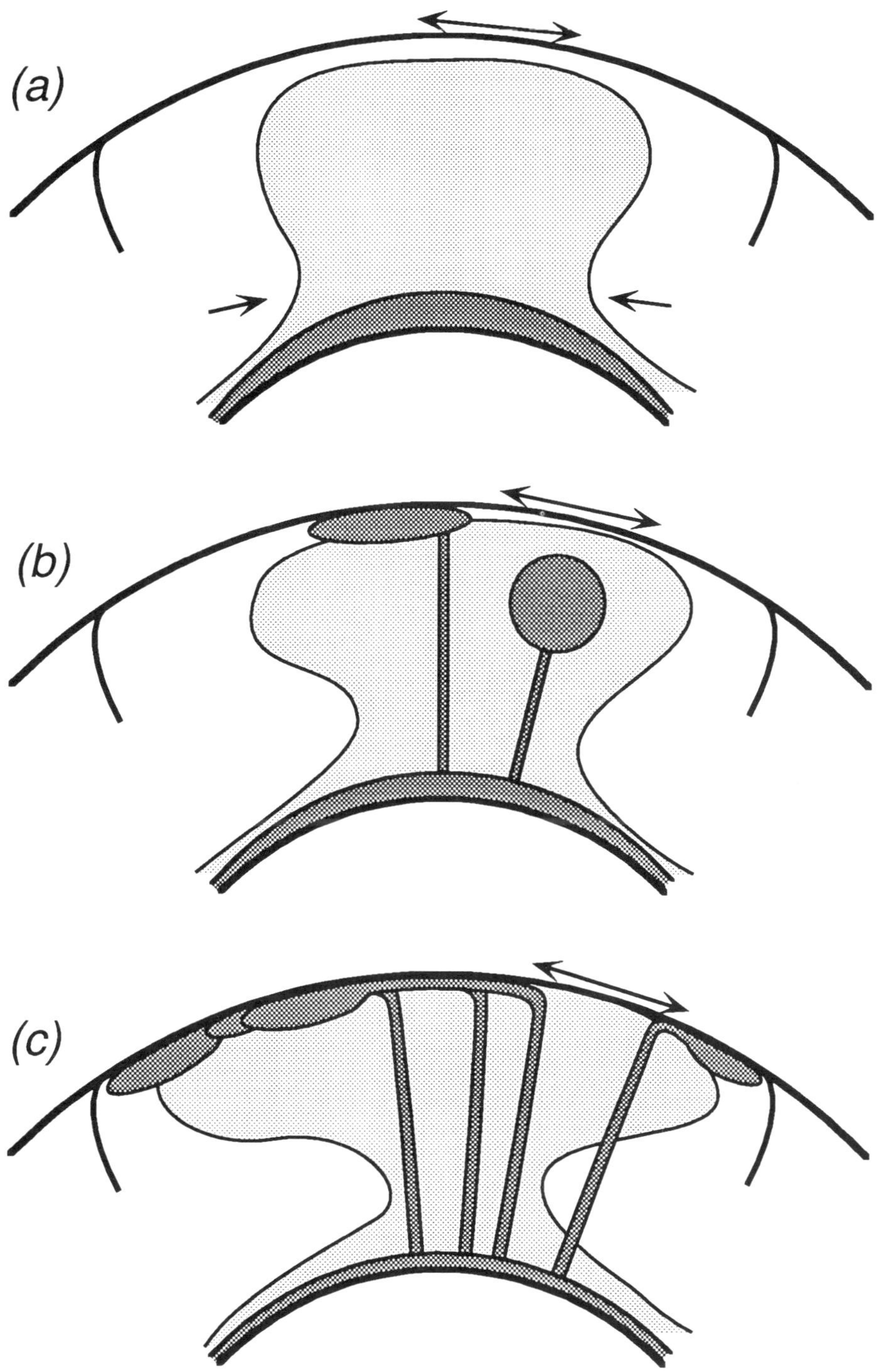

Fig. 7. Sketch of the sequence of events that may have caused the Darwin Rise and its modern continuation. *(a)* Prolonged subduction in surrounding regions leaves a large region of warmer-than-average deep mantle that causes an elevation of a few hundred meters over a large area. The thermal boundary layer at the base of the mantle is thickened and destabilized by the slowly converging deep flow. *(b)* The warm mantle slowly rises and begins to move with the overlying plates. New plumes form from the lower thermal boundary layer, causing additional local uplift and outbursts of flood basalt and hotspot volcanism. *(c)* The process has continued, leaving several active plume tails feeding from the thermal boundary layer, while the warm mantle and plume head material have moved away with the plate.

regions. These plumes punch through the warm mantle and emplace hot material at the base of the lithosphere, causing volcanism and more localized uplift. This plume material is carried with the overlying plate.

Although the Rise is inferred to have retained much of its Cretaceous amplitude, it will nevertheless have subsided relative to sea level since that time, as is evident in the results presented in Figure 6: both the swell and normal seafloor would have subsided by a comparable amount. This is consistent with evidence that the Cretaceous Darwin Rise has subsided by about a kilometer relative to sea level [Menard, 1984]. The Rise is currently not obvious because it is surrounded by younger sea floor.

DISCUSSION

Subsidence of the Old Sea Floor

The inadequacy of the asymptotic flattening picture of the subsidence of the old sea floor means that the behavior of the old sea floor must be re-evaluated. Empirically, some sea floor follows a square-root-of-age subsidence to ages up to 160 m.y., and the slope of the subsidence (the coefficient b in equation 1) varies from region to region [Hayes, 1988; Marty and Cazenave, 1989]. However there are many deviations from this trend, usually positive [Marty and Cazenave, 1989], implying that there are influences from the sub-lithospheric mantle that tend to uplift the lithosphere. On the other hand, when a single square-root-of-age reference curve is subtracted from all sea floor, some large-scale negative deviations are evident, most notably at the Australian-Antarctic discordance south of Australia [Cazenave and Lago, 1991; Pribac, 1991].

These features are, in general terms, what is to be expected for a convecting mantle [Davies, 1988b]. The cooling, thickening and subsidence of the top thermal boundary layer will continue at a steady rate unless interrupted by a strong upwelling (though even at the strongest plume, Hawaii, it is not clear that significant thinning of the lithosphere has occurred [Woods et al., 1991; Davies, 1992]). Simultaneously, the lithosphere may be displaced vertically by temperature variations in the sublithospheric mantle. These may be localized, as in the hotspot swells, or very large in scale, causing apparent regional variations in the rate of subsidence. The localized uplifts are likely to be due to plumes upwelling from the lower thermal boundary layer. The larger-scale variations may be due to the small-amplitude temperature variations expected in the interior of any convecting fluid, because of incomplete mixing of thermal boundary layers, and they would be expected to be both positive and negative relative to a global mean.

Turning to specific features, the Hawaiian swell has been interpreted in terms of sublithospheric mantle plume material by Davies [1992]. The Darwin Rise has been interpreted here as due to a combination of plume uplifts and large-scale internal temperature variations. The Darwin Rise correlates with a region where there has been no subduction for more than 200 m.y., but that has been fringed by subduction for most of that time [Richards and Engebretson, 1992]. The large-scale low south of Australia correlates with the location of Cretaceous subduction [Richards and Engebretson, 1992]. More speculatively, the positive deviations of older Atlantic seafloor may be related to the presence under the adjacent continents of old plume heads associated with the opening of the Altlantic [Morgan, 1981; Hill et al., 1992]. These possibilities will need to be investigated, but this task is well beyond the scope of the present paper.

Plumes, the Darwin Rise and the Cretaceous Outburst

Despite the evidence for a contribution from deep, warm mantle to the origin of the superswell, there is a clear association of the Darwin Rise with unusual plume activity, as evidenced by the large number of hotpot tracks through it and the presence of several oceanic plateaus that are arguably flood basalts [Richards et al., 1989, 1991; Pringle, 1991]. This association can plausibly be related to the presence of a large volume of warm ascending mantle because the instabilities in the thermal boundary layer at the bottom of the mantle that are thought to give rise to plumes are more likely under a region of ascending flow, since the boundary layer will be thickened by the horizontally converging flow there [Davies, 1990].

Recently Larson [1991] has pointed out that a number of regional and global phenomena correlate with the early-mid-Cretaceous period which is plausibly the time of origin of the Darwin Rise. These phenomena include increased rates of seafloor spreading, the eruption of several major flood basalt provinces, high relative sea levels, the cessation of magnetic reversals [Larson and Olson, 1991], deposition of major hydrocarbon deposits, a period of global greenhouse conditions [Caldeira and Rampino, 1991; Arthur et al., 1991], and important extinctions of marine nannofossils [Erba and Larson, 1991].

Larson [1991] attributes these events to the ascent through the mantle of a "superplume" whose nature is not specified in any detail. The impression of a superplume may be due to the combination of several strong plumes occurring in an area where a broad mantle thermal anomaly already existed.

Other Effects of Superswells

Superswells of this size and magnitude should produce significant contributions to long-term global sea level changes, and indeed the correlation between the Cretaceous uplift of the Darwin Rise and a sea level high at the time has already been noted [Schlanger et al., 1981; Larson, 1991]. Such a long-lived superswell occurring beneath a continent would certainly produce a sea level regression, a plausible present example being Africa. A superswell might also contribute to continental rifting due to uplift and gravity sliding [Houseman and England, 1986]. Whether a superswell alone could lead to continental breakup is not yet clear.

Acknowledgements. We thank Dr. J. Braun for developing and lending access to his mapping graphics package "2DMap", Dr A. Cazenave at the Groupe de Recherche de Geodesie Spatiale,

Centre National d'Etudes Spatiales for kindly providing the data sets of age of the oceans and sediment thickness, and Walter Smith and Norm Sleep for constructive reviews.

REFERENCES

Arthur, M. A., L. R. Kump, W. E. Dean, and R. L. Larson, Superplume, supergreenhouse? (abstract), Eos, Trans. Amer. Geophys. Union, 72, 301, 1991.

Bodine, J. H., M. S. Steckler, and A. B. Watts, Observations of flexure and the rheology of the oceanic lithosphere, *J. Geophys. Res.*, *86*, 3695-3707, 1981.

Caldeira, K., and M. R. Rampino, The mid-Cretaceous super plume, carbon dioxide, and global warming, *Geophys. Res. Lett.*, *18*, 987-990, 1991.

Cazenave, A., K. Dominh, M. Rabinowicz, and G. Ceuleneer, Geoid and depth anomalies over ocean depths and troughs, *J. Geophys. Res.*, *93*, 8064-8077, 1988.

Cazenave, A., and B. Lago, Long wavelength topography, seafloor subsidence and flattening, *Geophys. Res. Lett.*, *18*, 1257-1260, 1991.

Ceuleneer, G., M. Rabinowicz, M. Monnereau, A. Cazenave, and C. Rosemberg, Viscosity and thickness of the sub-lithospheric low-viscosity zone: constraints from geoid and depth over oceanic swells, *Earth Planet. Sci. Lett.*, *89*, 84-102, 1988.

Colin, P. and L. Fleitout, Topography of the ocean floor: Thermal evolution of the lithosphere and interaction of deep mantle heterogeneities with the lithosphere, *Geophys. Res. Lett.*, *17*, 1961-1964, 1990.

Craig, C. H., and D. McKenzie, The existence of a thin, low viscosity layer beneath the lithosphere, *Earth Planet. Sci. Lett.*, *78*, 420-426, 1986.

Creager, K. C., and T. H. Jordan, Slab penetration into the lower mantle beneath the Mariana and other island arcs of the northwest Pacific, *J. Geophys. Res.*, *91*, 3573-3589, 1986.

Crough, S. T., Hotspot swells, *Ann. Rev. Earth Planet. Sci.*, *11*, 165-193, 1983.

Davies, G. F., Lagging mantle convection, the geoid and mantle structure, *Earth Planet. Sci. Lett.*, *69*, 197-194, 1984.

Davies, G. F., Role of the lithosphere in mantle convection, *J. Geophys. Res.*, *93*, 10,451-10,466, 1988a.

Davies, G. F., Ocean bathymetry and mantle convection, 1. Large-scale flow and hotspots, *J. Geophys. Res.*, *93*, 10.467-10,480, 1988b.

Davies, G. F., Ocean bathymetry and mantle convection, 2. Small-scale flow, *J. Geophys. Res.*, *93*, 10,481-10,488, 1988c.

Davies, G. F., Mantle convection with a dynamic plate: topography, heat flow and gravity anomalies, *Geophys. J. Int.*, *98*, 461-464, 1989.

Davies, G. F., Mantle plumes, mantle stirring, and hotspot chemistry, *Earth Planet. Sci. Lett.*, *99*, 94-109, 1990.

Davies, G. F., Temporal vatiation of the Hawaiian plume flux, *Earth Planet. Sci. Lett.*, *113*, 277-286, 1992.

Davies, G. F., and M. A. Richards, Mantle convection, *J. Geol.*, *100*, 151-206, 1992.

Davis, E. E., and C. R. B. Lister, Fundamentals of ridge crest topography, *Earth Planet. Sci. Lett.*, *21*, 405-413, 1974.

Detrick, R. S., and S. T. Crough, Island subsidence, hot spots, and lithospheric thinning, *J. Geophys. Res.*, *83*, 1236-1244, 1978.

Devey, C. W., F. Albarede, J.-L. Cheminee, A. Michard, R. Muhe, and P. Stoffers, Active submarine volcanism on the Society hotspot swell (West Pacific): a geochemical study, *J. Geophys. Res.*, *95*, 5049-5066, 1990.

Duncan, R. A., and D. A. Clague, Pacific plate motion recorded by linear volcanic chains, in *The Ocean Basins and Margins*, edited by A. E. M. Nairn, F. G. Stehli, and S. Uyeda, 89-119, Plenum Publishing Corp., 1985.

Dziewonski, A. M. and J. H. Woodhouse, Global images of the earth's interior, *Science*, *236*, 37-48 , 1987.

Erba, E., and R. L. Larson, Nannofossils and superplumes (abstract), *Eos, Trans. Amer. Geophys. Union*, *72*, 301, 1991.

Griffiths, R. W., and I. H. Campbell, Stirring and structure in mantle starting plumes, *Earth Planet. Sci. Lett.*, *99*, 66-78, 1990.

Gurnis, M., and G. F. Davies, Numerical study of high Rayleigh number convection in a medium with depth-dependent viscosity, *Geophys. J. R. Astr. Soc.*, *85*, 523-541, 1986.

Gurnis, M., and B. H. Hager, Controls on the structure of subducted slabs and the viscosity of the lower mantle, *Nature*, *335*, 317-321, 1988.

Hager, B. H., R. W. Clayton, M. A. Richards, R. P. Comer, and A. M. Dziewonski, Lower mantle heterogeneity, dynamic topography, and the geoid, *Nature*, *313*, 541-545, 1985.

Hayes, D. E., Age-depth relationships and depth anomalies in the southeast Indian Ocean and South Atlantic Ocean, *J. Geophys. Res.*, *93*, 2937-2954, 1988.

Heestand, R. L., and S. T. Crough, The effect of hotspots on the oceanic age-depth relation, *J. Geophys. Res.*, *86*, 6107-6114, 1981.

Hess, H. H., History of ocean basins, in *Petrological Studies: a Volume in Honor of A.F. Buddington*, edited by A. E. J. Engel et al., 599-620, *Geol. Soc. Am.*, Boulder, Co., 1962.

Houseman, G., and P. England, A dynamical model of lithosphere extension and sedimentary basin formation, *J. Geophys. Res.*, *91*, 719-729, 1986.

Larson, R. L., Latest pulse of the earth: evidence for a mid-Cretaceous superplume, *Geology*, *19*, 547-550, 1991.

Larson, R. L., W. C. Pitman, III, X. Golovchenko, S. C. Cande, J. F. Dewey, W. F. Haxby, and K. L LaBrecque, *The bedrock geology of the world, Freeman and Company, New York, N.Y.*, 1985.

Larson, R. L., and P. Olson, Mantle plumes control magnetic reversal frequency, *Earth Planet. Sci. Lett.*, *107*, 437-447, 1991.

Ludwig, W. J., and R. E. Houtz, Isopach map of sediments in the Pacific Ocean basins and marginal sea basins, *Amer. Assoc. Petr. Geol. Rep.*, 1979.

Marty, J. C. and A. Cazenave, Regional variations in subsidence rate of oceanic plates: a global analysis, *Earth Planet. Sci. Lett.*, *94*, 301-315, 1989.

McKenzie, D., and M. J. Bickle, The volume and composition of melt generated by the extension of the lithosphere, *J. Petrol.*, *29*, 625-679, 1988.

McNutt, M. K., and K. M. Fischer, The South Pacific Superswell, in Seamounts, Islands and Atolls, B.H. Keating, P. Fryer, R. Batiza & G.W. Boehlert, Eds., *Geophys. Monograph 43, Amer. Geophys. Union, Washington D.C.*, 1987.

McNutt, M. K., and Judge, A. V., The Superswell and mantle

dynamics beneath the South Pacific, *Science, 248*, 969-975, 1990.

McNutt, M. K., E. L. Winterer, W. W. Sager, J. H. Natland and G. Ito, The Darwin Rise: a Cretaceous superswell? *Geophys. Res. Lett., 17*, 1101-1104, 1990.

Menard, H. W., *Marine Geology of the Pacific*, 271 pp., McGraw-Hill, New York, 1964.

Menard, H. W., Darwin reprise, *J. Geophys. Res., 89*, 9960-9968, 1984.

Morgan, W. J., Plate motions and deep mantle convection, *Geol. Soc. Amer. Mem., 132*, 7-22, 1972.

Morgan, W. J., Hotspot tracks and the opening of the Atlantic and Pacific Oceans, in *The Sea, Vol. 7*, edited by C. Emiliani, Wiley Interscience, New York, 443-487, 1981.

Nakada, M., and K. Lambeck, Late Pleistocene and Holocene sea-level change in the Australian region and mantle rheology, *Geophys. J. R. Astr. Soc., 96* , 497-517, 1989.

O'Connell, R. J., and B. H. Hager, On the thermal state of the earth, in *Physics of the Earth's Interior, Proc. Enrico Fermi Int. Sch. Phys.*, edited by A. Dziewonski and E. Boschi, *78*, 270-317, 1980.

Parsons, B., and D. McKenzie, Mantle convection and the thermal structure of the plates, *J. Geophys. Res., 83*, 4485-4496, 1978.

Parsons, B., and J. G. Sclater, An analysis of the variation of ocean floor bathymetry and heat flow with age, *J. Geophys. Res., 82*, 803-827, 1977.

Pribac, F., *Superswells due to Mantle Convection*, Ph. D. Thesis, Australian National University, 174 pp., 1991.

Pribac, F., and G. F. Davies, Mantle superswells: regressions and rifts? (abstract), *Eos, Trans. Amer. Geophys. Union, 68*, 1451, 1987.

Pringle, M. S., Rolling thunder of the early Cretaceous: plateau basalts, EM1 from the lower mantle, and the origin of the Dupal anomaly (abstract), *Eos, Trans. Amer. Geophys. Union, 72*, 300, 1991.

Renkin, M. L., and J. G. Sclater, Depth and age in the North Pacific, *J. Geophys. Res., 93*, 2919-2935, 1988.

Richards, M. A., R. A. Duncan, and V. E. Courtillot, Flood basalts and hotspot tracks: plume heads and tails, *Science, 246*, 103-107, 1989.

Richards, M. A., D. L. Jones, R. A. Duncan, and D. J. DePaolo, A mantle plume initiation model for the Wrangellia flood basalt and other oceanic plateaus, *Science, 254*, 263-267, 1991.

Richards, M. A., and D. C. Engebretson, Large-scale mantle convection and the history of subduction, *Nature, 355*, 437-440, 1992.

Sandwell, D. T., and M. L. Renkin, Compensation of swells and plateaus in the North Pacific: no direct evidence for mantle convection, *J. Geophys. Res., 93*, 2775-2783, 1988.

Schlanger, S. E., H. C. Jenkyns & I. Premoli-Silva., *Earth Planet. Sci. Lett., 52*, 435-449, 1981.

Schroeder, W., The empirical age-depth relationand depth anomalies in the Pacific Ocean basin, *J. Geophys. Res., 89*, 9873-9884, 1984.

Sclater, J. G., and J. Francheteau, The implications of terrestrial heat flow observations on current tectonic and geochemical models of the crust and upper mantle of the earth, *Geophys. J. Roy. Astr. Soc., 20*, 509-542, 1970.

Sclater, J. G., C. Jaupart, and D. Galson, The heat flow through the oceanic and continental crust and the heat loss of the earth, *Rev. Geophys., 18*, 269-312, 1980.

Sleep, N. H., Hotspots and mantle plumes: some phenomenology, *J. Geophys. Res., 95*, 6715-6736, 1990.

Stein, C. A., and D. H. Abbott, Heat flow constraints on the South Pacific superswell, *J. Geophys. Res., 96*, 16083, 1991.

Watts, A. B., D. P. McKenzie, B. E. Parsons, and M. Roufosse, The relationship between gravity and bathymetry in the Pacific Ocean, *Geophys. J. Roy. Astr. Soc., 83*, 263-298, 1985.

Woods, M. T., J.-J. Leveque, and E. A. Okal, Two-station measurements of Rayleigh wave group velocity along the Hawaiian swell, *Geophys. Res. Lett., 18*, 105-108, 1991.

Zhang, Y.-S., and T. Tanimoto, Global Love wave phase velocity variation and its significance to plate tectonics, *Phys. Earth Planetary Interiors, 66*, 160-202, 1991.

G. F. Davies and F. Pribac, Research School of Earth Sciences, Australian National University, GPO Box 4, Canberra ACT 2601, Australia.

Constraints on Pacific Midplate Swells From Global Depth-Age and Heat Flow-Age Models

CAROL A. STEIN

Department of Geological Sciences, University of Illinois at Chicago, Chicago, IL 60680

SETH STEIN

Department of Geological Sciences, Northwestern University, Evanston, IL 60208

Oceanic midplate swells are identified by shallow seafloor depths. In turn, models of the processes giving rise to these regions rely on assessments of how their depths, surface heat flow, and flexural properties differ from those for lithosphere which is presumed not to have been affected by these processes. Such comparisons have been inhibited because reference thermal models, which are assumed to describe unperturbed lithosphere, predict deeper depths and lower heat flow than typically observed for lithosphere older than 70 Ma. As a result, both depth and heat flow anomalies can be overestimated. To address this difficulty, we have derived model GDH1 (Global Depth and Heat flow) by joint fitting of heat flow and bathymetry. GDH1, which has a hotter and thinner lithosphere than previous models, fits the depth and heat flow data significantly better, including the data from older lithosphere previously treated as anomalous. It also provides an improved fit to depth-to-basement data for ocean drilling sites, and to geoid offsets across fracture zones. The improved fit occurs for depth-age data from both the DBDB-5 digital bathymetry, and from regional medians from ship tracks, which yield comparable depth-age curves. We use GDH1 to study three classes of midplate swells: the Hawaiian and other hot spot swells, the Darwin Rise area of widespread Cretaceous volcanism, and the Superswell, considered a present analogue to the Darwin Rise. Heat flow on the Hawaiian swell, though anomalously high with respect to previous reference models, is at most slightly high relative to GDH1. The situation is similar for the Bermuda, Cape Verde, and Crozet hot spots. The absence of a significant heat flow anomaly favors a primarily dynamic, rather than thermal, origin for these swells. Similarly, the present depths and heat flow for the Darwin Rise are consistent with GDH1, although they were anomalous with respect to previous reference models. The depth and heat flow data thus provide no evidence for the Darwin Rise being thermally different at present from lithosphere of the same age elsewhere. The situation for the Superswell is similar to that for Hawaii, in that the heat flow data are consistent with those for unperturbed lithosphere, despite the shallow bathymetry. Flexural data for the Superswell indicate that the lithosphere was anomalously weak at the time of seamount loading. This also appears to have been the case for the Darwin Rise, but not for the Hawaiian or other swells. Although the weakness has been interpreted as due to elevated temperatures in the lithosphere, the required high temperatures should give rise to heat flow much higher than observed. Hence the flexural strength anomaly with respect to Hawaii and other swells suggests that the lithosphere for the Superswell and Darwin Rise has been mechanically weakened by the cumulative action of multiple hot spots.

Attention is once again focusing on the timing and mechanisms of midplate volcanism as a major factor in the vertical tectonic history of the Central Pacific.

Schlanger and Premoli-Silva, 1981

The Mesozoic Pacific: Geology, Tectonics, and Volcanism
Geophysical Monograph 77

INTRODUCTION

Understanding the nature and origin of significant vertical crustal motion in the Pacific Basin has been a challenge since its identification by Darwin [1845]. Menard [1964] proposed that a large region of shallow bathymetry in the Western Pacific, which he termed the Darwin Rise, had undergone major volcanism and uplift during the Cretaceous (Figure 1). This region (Figure 2) includes the Mid-Pacific Mountains, the Marshall Islands, the Magellan Seamounts, and the Wake Guyots. With the advent of plate tectonics, it became clear that the formation

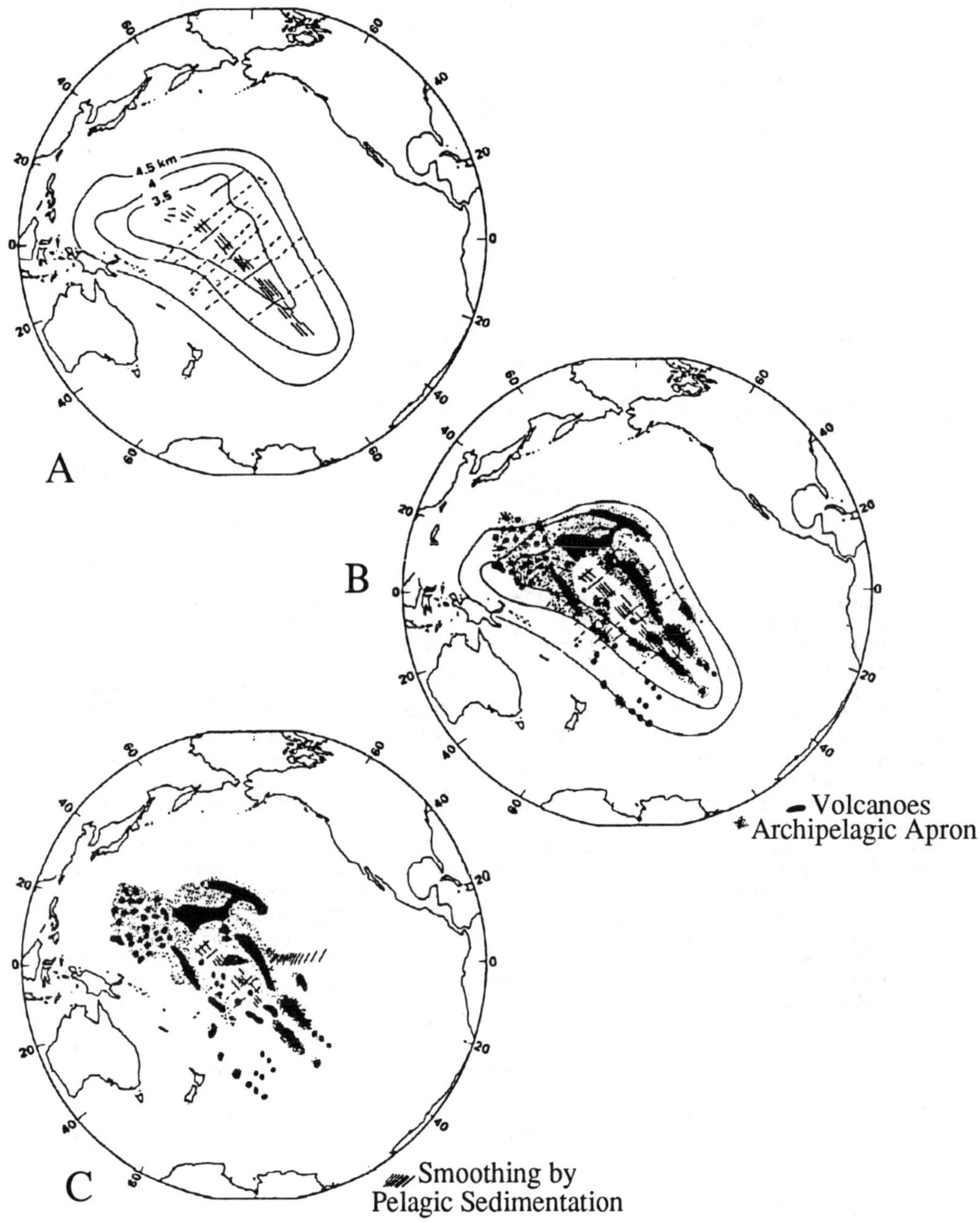

Fig. 1. Menard's [1964] model for the evolution of the Darwin Rise by elevation of a broad region (A), widespread Late Mesozoic volcanism (B), and Cenozoic subsidence (C).

of the Darwin Rise was a striking example of midplate volcanism. Subsequently, considerable effort has been devoted to investigating the processes responsible for the Darwin Rise and other midplate volcanism and uplifts.

At present, the origin of the Darwin Rise is generally thought to be related to the effects of mantle plumes [Morgan, 1972]. Crough [1979] noted that the present Easter and Macdonald hot spots would have been under the Darwin Rise during the period of uplift and volcanism, and proposed that the region represented a series of seamounts formed along these tracks, rather a broad area simultaneously uplifted. By analogy to the present Hawaiian hot spot, Crough proposed that the hot spots were also associated with broad swells [Crough, 1978; Detrick and Crough, 1978] which gave rise to regional uplifts. Deep sea drilling, however, provided evidence for widespread volcanism in the area in addition to the seamounts, leading to the

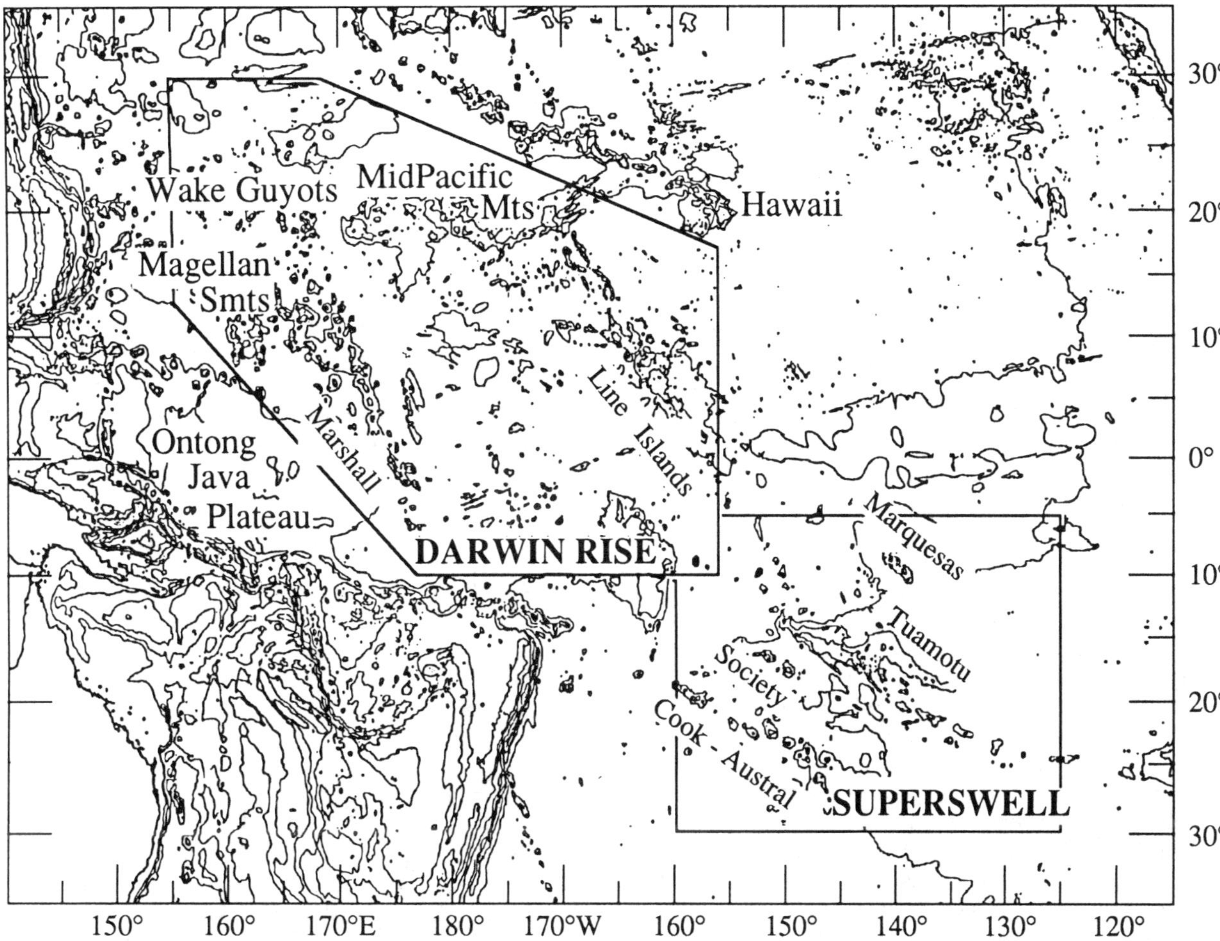

Fig. 2. Bathymetry of the Pacific, showing the regions we used for Darwin Rise and Superwell depth and heat flow analyses. Because the extent of the Superwell and Darwin Rise are ill-defined, the boundaries of the regions we used are somewhat arbitrary.

suggestion that the areal extent and magnitude of the volcanism and uplift exceeded that which might be expected from multiple hot spot tracks [Winterer, 1973; 1976; Jackson and Schlanger, 1976; Schlanger and Premoli-Silva, 1981; Larson and Schlanger, 1981; Schlanger et al., 1981]. Menard [1984] therefore proposed that the hot spot model could be reconciled with the regional uplift and volcanism by assuming that the Cretaceous Darwin Rise was similar to the present area extending from the East Pacific Rise to the Society Islands. He considered this area, termed the Polynesian Plume Province (PPP) [Vogt, 1981], to contain both a broad regional uplift and a number of hot spot swells, including the Cook-Austral, Marquesas, Pitcairn, and Society seamount chains. These hot spot tracks have formed volcanic edifices in the past 18 Ma [Duncan and Clague, 1985], giving rise to a complex pattern of volcanic ages and types. Holocene volcanism occurs at several sites on the chains, and the NW-SE trend of the plateaus and island chains is similar to that of the Hawaiian hot spot track. Because many of these hot spot tracks can be traced back to the Darwin Rise, a similar situation might have formed the Darwin Rise.

McNutt and Fisher [1987] proposed the term "Superswell" for the PPP, and further developed the concept of this area as a present-day analog to the Darwin Rise. They noted that the Superswell area, from west of the East Pacific Rise to approximately 160°W, 9-30°S is substantially shallower than expected for its age, which ranges from 20 to 90 Ma. In addition, the effective elastic thicknesses of the lithosphere calculated from the loading of volcanoes and seamounts are substantially less than expected for the age of loading [McNutt and Menard, 1978; Calmant and Cazenave, 1987; Calmant et al., 1990]. McNutt and Fisher suggested that the shallow bathymetry resulted from the lithosphere in the area having the temperature structure of an anomalously thin 75-km thick thermal plate. McNutt and Judge [1990] further suggested that the weak flexural strengths, low surface wave velocities [Nishimura and Forsyth, 1985], and geochemical anomalies [Hart, 1984, 1988;

Castillo, 1988] were consequences of the combined effects of a thin thermal plate and a deeper low-density plume. In this model, the lithosphere is thinned by enhanced heat flux from the mantle and low viscosity beneath the plate, such that the weak plate is easily penetrated by hot spot volcanism. The net volume flux of the Superswell plume has been estimated as approximately 30% of the total global plume flux, a flux exceeding that of the Hawaiian plume, thought to be the single largest [Sleep, 1990].

Extending this analysis, McNutt and Fisher [1987] interpreted the Darwin Rise as also anomalously shallow for its age. They noted other similarities to the Superswell, including similar petrologic anomalies [Hart, 1984; Natland and Wright, 1984]. In addition, the effective elastic thicknesses are lower than expected for the age of loading [Calmant, 1987; Calmant and Cazenave, 1987; Smith et al., 1989; Wolfe and McNutt, 1991]. McNutt et al. [1990] suggested a schematic history in which the Darwin Rise was dynamically uplifted during the Cretaceous, and was then similar to the present Superswell until about 70–80 Ma. The present depth was thus interpreted as anomalously shallow, indicating a residual effect of the transient Cretaceous reheating. Larson [1991] termed the Cretaceous event a "Superplume," which produced both the Darwin Rise and very large amounts of lithosphere at midocean ridges. In this model, the present Superswell reflects the Superplume's waning phase. Davies and Pribac [this volume] propose that the Darwin Rise and Superswell reflect the effect of both plume material and anomalously warm deep mantle material.

Models for the Darwin Rise and Superswell thus combine aspects of the two basic models that have been proposed for the origin of hot spot swells. In one model, the swell is a thermal effect due to the hot spot thinning and heating the lithosphere at depth [Detrick and Crough, 1978; Von Herzen et al., 1982]. In the second model, the uplift is primarily due to the dynamic effects of the upwelling plume [Parsons and Daly, 1983; Courtney and White, 1986; Robinson and Parsons, 1988; Sleep, 1992], which may largely reflect thermal buoyancy forces within the upwelling mantle [Liu and Chase, 1989; 1991]. Thus the Darwin Rise and Superswell may reflect, at least in part, perturbations to the thermal structure of the lithosphere, as evidenced by depth and flexural strength anomalies relative to unperturbed lithosphere of the same age. Similarly, the possible thermal perturbations associated with smaller "ordinary" swells are modeled using as constraints depth, heat flow, and flexural strength anomalies. As a result, tectonic interpretations can depend significantly on the reference model which is assumed to characterize the thermal evolution of unperturbed ("normal") lithosphere, and its corresponding predictions for the variation in depth and heat flow as functions of age. In this paper, we discuss the selection of a reference model, and explore its implications for tectonic interpretations of hot spot swell regions.

REFERENCE MODEL GDH1

Motivation

Studies of lithospheric processes, such as the origin of the Darwin Rise and Superswell, make use of a reference model in two complementary ways. First, the predicted variations in depth and heat flow as functions of age are used to identify and characterize anomalous regions. Second, the thermal structure model is assumed to characterize unperturbed lithosphere, and thus provide a basis by which the inferred depth and heat flow perturbations are used to estimate thermal perturbations and corresponding perturbations in temperature-dependent physical properties such as material strength or seismic velocity.

Reference models are developed using as constraints the decrease in heat flow and increase in seafloor depth with age as lithosphere formed at high temperature cools as it spreads away from the ridge (Figure 3). These data jointly reflect the evolution with age of the lithospheric geotherm, because the bathymetry depends on the temperature integrated over depth and the heat flow depends on the temperature gradient at the sea floor.

The simplest reference model is one in which the lithosphere behaves as the cold upper boundary layer of a cooling halfspace, such that depth and heat flow vary as $age^{1/2}$ and

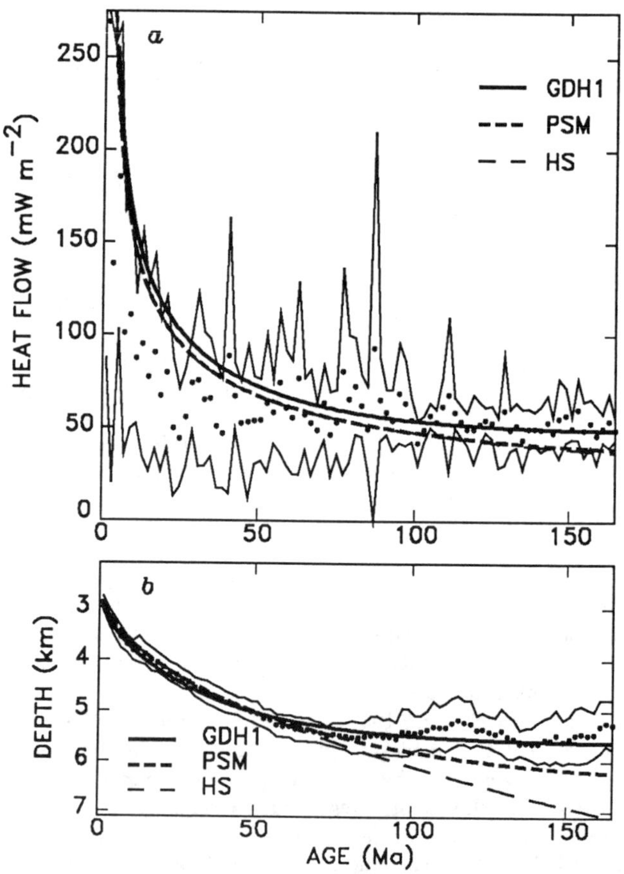

Fig. 3. Data and models for heat flow and ocean depth as a function of age. Data are averaged in two-m.y. bins, and one standard deviation about the mean value for each is shown by the envelope. Also shown are the predicted values for the plate model of Parsons and Sclater [1977] (PSM), a cooling halfspace model with the same thermal parameters (HS), and the GDH1 plate model. Heat flow for the HS and PSM curves overlap for ages younger than ~120 Ma. [Stein and Stein, 1992].

age$^{-1/2}$, respectively [Davis and Lister, 1974]. This model, with reasonable values for the physical parameters, describes the primary features of the depth and heat flow variations. Subsequently, however, it was recognized that the depth and heat flow data "flatten", varying more slowly with age for older lithosphere (Figure 3). Most recent studies thus use a reference model in which the lithosphere is treated as a cooling plate with an isothermal lower boundary [Langseth et al., 1966; McKenzie, 1967; Parsons and Sclater, 1977]. For young ages, the plate behaves like a cooling halfspace, but for older ages, the effects of the lower boundary cause the depth and heat flow curves to flatten.

The lower isothermal boundary condition models the additional heat input from below, which is assumed to prevent the halfspace from continuing to cool for older ages. Various mechanisms including radiogenic heat [Crough, 1977], shear heating [Schubert et al., 1976], small-scale convection [Parsons and McKenzie, 1978; Parsons and Richter, 1981], and mantle plumes [Heestand and Crough, 1981] have been proposed as sources of this additional heat. In most of these formulations, the asymptotic plate thickness to which the lithosphere evolves corresponds to the depth at which the additional heat is supplied, and above which temperature changes cause bathymetric variations. Because the thermal structure within the lithosphere does not depend critically on how this heat is supplied [Parsons and McKenzie, 1978], the plate model has become a standard for comparing the predicted temperatures within the lithosphere to other data including seismic surface wave velocities [Forsyth, 1977; Sato et al., 1989], flexure due to applied loads [Bodine et al., 1981] and the depths of intraplate earthquakes [Wiens and Stein, 1983; Chen and Molnar, 1983].

Parsons and Sclater [1977] found that such a model, with a 125 km thick plate with 1350°C basal temperature, provided good fits to the data then available. It described the general shape of the depth curve, including the flattening for ages > 70 Ma, and the heat flow for ages > 50 Ma. Its major limitation, the underprediction of heat flow for ages < 50 Ma, is thought to result from the transport of significant amounts of heat by water circulation [Lister, 1972; Anderson and Hobart, 1976; Wolery and Sleep, 1976]. The Parsons and Sclater model (a term often applied to a plate model with the parameters they estimated from the available data), has become a standard model for the lithosphere.

For studies of processes that may have formed hot spot swells, however, the Parsons and Sclater model (hereafter, PSM) has serious limitations. It predicts deeper depths and lower heat flow than generally observed for lithosphere older than 70-100 Ma (Figure 3). As a result, depth data [Kaula and Philips, 1981; Sclater and Wixon, 1986; Renkin and Sclater, 1988; Marty and Cazenave, 1989; Colin and Fleitout, 1990; Smith, 1990; Stein and Abbott, 1991] and heat flow data [Detrick et al., 1986; Louden et al., 1987; Von Herzen et al., 1989; Lister et al., 1990; Langseth et al., 1990; Stein and Abbott, 1991] for older lithosphere typically appear anomalous. Hence an area with depth or heat flow anomalous with respect to PSM need not differ from the average lithosphere of that age.

Derivation

To alleviate this difficulty, we have developed a new reference model by a combined analysis of recent depth and heat flow data [Stein and Stein, 1992]. We used a global heat flow data set derived primarily from a published compilation [Louden, 1989], with additional data some of which have been reported elsewhere [Stein and Abbott, 1991]. Because the heat flow and depth for old lithosphere are crucial joint constraints, the model was derived with only data from regions where the published depth data extend to at least 166 Ma and have been corrected for sediment thicknesses. The combined depth data for these areas, the Northwestern Atlantic [Sclater and Wixon, 1986] and North Pacific [Renkin and Sclater, 1988], and the corresponding heat flow data are shown in Figure 3, and listed in Table 1.

We used the depth data for all ages and the heat flow data for ages greater than 50 Ma to constrain the primary parameters of a plate model: lithospheric thickness (a), basal temperature (T_m), and coefficient of thermal expansion (α). Figure 4a-c shows contours of the misfit to the data as a function of basal temperature and thickness, for the best fitting coefficient of thermal expansion, 3.1×10^{-5} °C^{-1}. These plots show a slice of the three dimensional model space with the coefficient of thermal expansion fixed. Figure 4d shows the misfit as a function of the coefficient of thermal expansion and basal temperature, for the best fitting thickness of 95 km, and Figure 4e shows the misfit as a function of the coefficient of thermal expansion and thickness, for the best fitting basal temperature, 1450°C.

The misfit contours for the depth data show a diagonal pattern, because the asymptotic depth for infinitely old age is proportional to the product of plate thickness and basal temperature. The best fit is for a thinner plate and somewhat higher basal temperature than in PSM, consistent with the observed depths (Figure 3) being shallower than predicted by PSM. The heat flow data for lithosphere older than 50 Ma provide a separate constraint. The misfit contours are elongated in the opposite sense as for the depths, because the asymptotic heat flow for infinitely old age depends on the ratio of basal temperature to thickness. As for the depths, the best fit occurs for a thinner plate and hence higher T_m/a than for PSM, because the observed heat flow exceeds that predicted by PSM. The coefficient of thermal expansion plots show the best fitting value, and the partial tradeoffs due to the fact that the depths for young and old ages depend on the products αT_m and $\alpha T_m a$, respectively. Joint fitting of the combined depth and heat flow data gives a best fit for a 95 km thick plate, with a basal temperature of 1450°C, and a coefficient of thermal expansion of 3.1×10^{-5} °C^{-1}. The model parameters are listed in Table 2.

The depth and heat flow predictions for this model, which we call GDH1 (for Global Depth and Heat flow), are accurately approximated using a halfspace model with the same parameters for young lithosphere, and the first term of the series solution for old lithosphere [Stein and Stein, 1992]. The depth (m) is related to the age (Ma) by

$$d(t) = 2600 + 365\, t^{1/2} \qquad t<20 \text{ Ma}$$

$$= 5651 - 2473 \exp(-0.0278\, t) \qquad t\geq 20 \text{ Ma}$$

TABLE 1: Data for GDH1 Model

Age range (Ma)	Depth (m)	Heat flow (mW m^{-2})	Age range (Ma)	Depth (m)	Heat flow (mW m^{-2})
0 - 2	2896 ± 139	269 ± 182	82 - 84	5519 ± 304	62 ± 17
2 - 4	3176 ± 162	138 ± 118	84 - 86	5516 ± 298	50 ± 23
4 - 6	3378 ± 196	185 ± 82	86 - 88	5482 ± 380	93 ± 117
6 - 8	3568 ± 208	101 ± 66	88 - 90	5477 ± 385	65 ± 22
8 - 10	3700 ± 162	110 ± 61	90 - 92	5514 ± 338	59 ± 12
10 - 12	3832 ± 187	87 ± 35	92 - 94	5507 ± 363	54 ± 16
12 - 14	3811 ± 264	95 ± 61	94 - 96	5542 ± 321	69 ± 27
14 - 16	3898 ± 207	77 ± 51	96 - 98	5415 ± 476	63 ± 23
16 - 18	4001 ± 227	90 ± 52	98 - 100	5417 ± 460	55 ± 12
18 - 20	4065 ± 198	67 ± 41	100 - 102	5435 ± 357	42 ± 12
20 - 22	4142 ± 189	79 ± 41	102 - 104	5416 ± 333	49 ± 7
22 - 24	4244 ± 194	50 ± 37	104 - 106	5377 ± 389	57 ± 7
24 - 26	4303 ± 197	44 ± 26	106 - 108	5305 ± 496	54 ± 7
26 - 28	4388 ± 188	56 ± 27	108 - 110	5319 ± 410	62 ± 15
28 - 30	4430 ± 173	74 ± 26	110 - 112	5320 ± 479	68 ± 42
30 - 32	4547 ± 183	75 ± 47	112 - 114	5232 ± 488	59 ± 8
32 - 34	4605 ± 203	65 ± 36	114 - 116	5177 ± 487	54 ± 8
34 - 36	4652 ± 226	66 ± 32	116 - 118	5213 ± 440	50 ± 9
36 - 38	4715 ± 217	51 ± 33	118 - 120	5249 ± 466	50 ± 19
38 - 40	4769 ± 224	47 ± 29	120 - 122	5372 ± 294	53 ± 12
40 - 42	4836 ± 209	89 ± 75	122 - 124	5355 ± 330	55 ± 7
42 - 44	4871 ± 193	67 ± 18	124 - 126	5402 ± 310	53 ± 9
44 - 46	4915 ± 175	53 ± 22	126 - 128	5446 ± 331	42 ± 14
46 - 48	4967 ± 208	53 ± 41	128 - 130	5388 ± 444	60 ± 30
48 - 50	5015 ± 198	54 ± 24	130 - 132	5495 ± 436	49 ± 11
50 - 52	5056 ± 225	54 ± 22	132 - 134	5545 ± 394	52 ± 12
52 - 54	5151 ± 187	63 ± 37	134 - 136	5628 ± 252	44 ± 16
54 - 56	5181 ± 212	58 ± 17	136 - 138	5625 ± 296	50 ± 11
56 - 58	5287 ± 224	74 ± 38	138 - 140	5639 ± 319	50 ± 9
58 - 60	5306 ± 221	61 ± 31	140 - 142	5628 ± 405	50 ± 12
60 - 62	5303 ± 217	55 ± 30	142 - 144	5558 ± 427	54 ± 14
62 - 64	5358 ± 237	77 ± 52	144 - 146	5567 ± 472	48 ± 8
64 - 66	5367 ± 233	57 ± 20	146 - 148	5470 ± 505	55 ± 21
66 - 68	5374 ± 236	53 ± 29	148 - 150	5602 ± 437	57 ± 15
68 - 70	5400 ± 210	49 ± 15	150 - 152	5555 ± 455	50 ± 6
70 - 72	5436 ± 204	64 ± 20	152 - 154	5474 ± 541	60 ± 16
72 - 74	5489 ± 185	47 ± 23	154 - 156	5449 ± 565	49 ± 15
74 - 76	5466 ± 236	53 ± 17	156 - 158	5476 ± 518	53 ± 13
76 - 78	5484 ± 252	81 ± 56	158 - 160	5456 ± 450	46 ± 16
78 - 80	5496 ± 262	66 ± 34	160 - 162	5299 ± 533	50 ± 8
80 - 82	5519 ± 322	72 ± 26	162 - 164	5229 ± 465	54 ± 14
			164 - 166	5270 ± 489	49 ± 5

and the heat flow (mW m^{-2}) is

$$q(t) = 510\, t^{-1/2} \qquad t \le 55 \text{ Ma}$$

$$= 48 + 96 \exp(-0.0278\, t) \qquad t > 55 \text{ Ma} .$$

(The t^2 term in the expression for young depths in Stein and Stein [1992] is a typographical error.)

GDH1 fits the data significantly better than PSM, especially for old lithosphere (Figure 3). For the dataset we use, the mean and standard deviation of the depth and heat flow for lithosphere older than 140 Ma are 5464 ± 484 m and 52 ± 13 mW m^{-2}. GDH1 predicts an asymptotic depth for infinitely old lithosphere of 5651 m, considerably shallower than the 6400 m for PSM. The asymptotic heat flow for GDH1 is 48 mW m^{-2}, approximately 40% higher than the 34 mW m^{-2} value for PSM. An F-ratio test indicates that the improved fit to the depth and heat flow data is significant above the 99.9% level. Figure 4f shows that the improved fit of GDH1 relative to PSM is comparable to

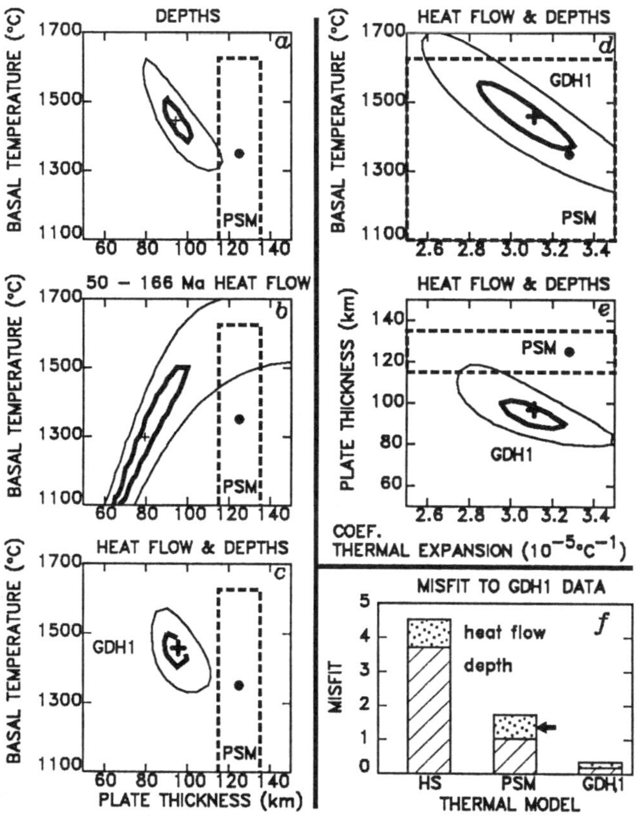

Fig. 4. (a-e): Contour plots of the misfit to subsets of the data as a function of combinations of the three parameters. In each panel, the best fit is denoted by a cross, and the contours show values 1.25 and 2.5 times the minimum misfit. Parsons and Sclater's values (PSM) and their estimated uncertainties are also indicated. The best fitting solution for the combined depth and heat flow (c-e), a 95 km thick plate with a basal temperature of 1450°C and coefficient of thermal expansion, 3.1×10^{-5} °C^{-1} is model GDH1. (f): Misfit to the data for GDH1, PSM, the halfspace model, and a version of the PSM model including radioactivity (arrow), which predicts the same depths but higher heat flow than PSM. [Stein and Stein, 1992].

that for PSM relative to the halfspace model. An F-ratio test shows that GDH1 also provides an improved fit above the 99.9% level relative to a version of the PSM model in which in addition to the heat flow due to plate cooling, 4 mW m^{-2} is assumed to be generated by radioactivity [Parsons and Sclater, 1977], giving the same depths as PSM but a better fit to the heat flow.

GDH1 was developed by joint fitting of heat flow and bathymetry. The heat flow data alone [Stein and Stein, 1992] are best fit by a model with an 80 km thick lithosphere and a 1300°C basal temperature, corresponding to a 6% higher (51 mW m^{-2}) asymptotic heat flow than for GDH1. Such a heat flow-only model is poorly constrained due to the lack of depth data. It illustrates, however, that GDH1 slightly underpredicts the heat flow data.

TABLE 2: GDH1 model parameters

a	plate thickness	95 km
T_m	basal temperature	1450 °C
α	thermal expansion coefficient	3.1×10^{-5} °C^{-1}
k	thermal conductivity	3.138 W m^{-1} °C^{-1}
C_p	specific heat	1.171 kJ kg^{-1} °C^{-1}
ρ_m	mantle density	3330 kg m^{-3}
ρ_w	water density	1000 kg m^{-3}
d_r	ridge depth	2600 m

GDH1 is representative of the plate models that fit the data well. Quantification of the true uncertainties in model parameters is difficult, given the limitations of the data and model. From examination of fits to the data, we estimate the plate thickness as 95 ± 15 km, basal temperature as 1450 ± 250°C, and α as $3.1 \pm 0.8 \times10^{-5}$ °C^{-1}, where the uncertainties are one standard deviation, and do not include the uncertainty in the model or other parameters (e.g. conductivity). Alternatively, if we regard α as known a priori, we estimate the thickness and T_m uncertainties as 10 km and 100°C, respectively. Thus our solutions have plate thicknesses generally less than Parsons and Sclater's estimate (125 ± 10 km), basal temperatures comparable to their estimate (1350 ± 275°C), and α comparable to their estimate ($3.2 \pm 1.1 \times10^{-5}$ °C^{-1}).

Properties Of Reference Models

Thermal models of the lithosphere are used in several ways for tectonic studies. First, they provide reference models to characterize the average variation in depth and heat flow with age for assumed-"normal" lithosphere. It is thus possible to identify regions which are anomalous with respect to a reference model, and to estimate how anomalous the depth and heat flow are. Second, a model predicts a temperature structure for "normal" lithosphere, and hence can be used to draw inferences about the processes giving rise to the temperature structure of both "normal" and "anomalous" lithosphere. These two applications are somewhat decoupled, in that it is useful to characterize "normal" lithosphere and identify anomalies, although the assumed thermal structure presumably approximates a more complex situation.

The reference model is thus used to distinguish "normal" lithosphere, that characterized by normal heat flow and depth, from "anomalous" lithosphere. Moreover, lithosphere can be characterized as either thermally anomalous, in that both its depth and heat flow are anomalous, presumably due to elevated temperatures in the lithosphere, or topographically anomalous, presumably due primarily to mantle flow. Table 3 summarizes these end member cases. As discussed shortly, hot spot swells appear to be closer to topographic anomalies than thermal ones.

This classification relies on the fact that we, like Parsons and Sclater [1977], find a consistent thermal model by fitting the depth and heat flow jointly. Although empirical curves could be derived for either data type alone, they are generally less useful in investigating thermal structure. First, there is no requirement that the other data type be well fit. Second, an empirical depth-

TABLE 3: End Member Lithospheric Types

Normal Lithosphere	Topographically Anomalous Lithosphere	Thermally Anomalous Lithosphere
Heat flow and depth fit reasonably well by GDH1	Depth significantly shallow with respect to GDH1: Heat flow normal or slightly elevated	Depth significantly shallow and heat flow significantly high with respect to GDH1
Most >70 Ma lithosphere; Darwin Rise (at present)	Hotspot swells (Hawaii, Bermuda, Cape Verde, Crozet) and Superswell	?
Normal lithospheric thermal structure	Primarily dynamic uplift	Anomalous lithospheric thermal structure

age curve often achieves a better fit to data because it treats the young and old depths separately [Smith, 1990] and hence contains a greater number of parameters than a plate model. The improved fit is thus meaningful only if it exceeds that expected purely by chance due to the additional free parameters. Third, because constants in such models which correspond to combinations of the thermal model parameters are estimated independently, they can yield inconsistent values of the thermal model parameters, and thus be of little value in predicting the geotherm. In contrast, joint fitting of depth and heat flow assumes that both variations reflect the cooling of the lithosphere.

GDH1, like most thermal models for the lithosphere, is phenomenological in that it was derived to be the simplest model that describes the observed variation in depth and heat flow. These variations may in part also reflect other possible effects, some of potential significance, which are not included in the model [Stein and Stein, 1992]. Of special importance for our discussion here is that the variation in depth with age is assumed to result entirely from the temperature structure of the lithosphere [Parsons and McKenzie, 1978], whereas depth effects can also reflect sublithospheric processes [Hager and O'Connell, 1980; Buck and Parmentier, 1986; Davies, 1988; Jarvis and Peltier, 1989; Davies and Pribac, this volume]. In addition, the plate model used here does not incorporate time dependent effects beyond cooling, such as might be associated with the proposed Cretaceous superplume [Larson, 1991]. Determination of the magnitude of these possible effects not included in the plate model has long been challenging precisely because the simple plate model fits the two primary surface observables, depth and heat flow, reasonably well. Although the depth-age and heat flow-age curves are derived assuming a thermal model, their utility is not directly tied to the appropriateness of the thermal model.

The choice of the data used to derive the model is crucial. Some data must be excluded because of biases, but as more data are excluded, the resulting model is increasingly influenced by a priori assumptions about the processes being described. For example, we have chosen not to fit the heat flow data in young lithosphere, because of the assumed bias from hydrothermal flow. We have not attempted to remove the effect of hot spot swells, because removing swells requires using an a priori reference model to decide which lithosphere is anomalously shallow, which in turn influences the resulting view of what depths are "normal." Our approach differs from Parsons and Sclater's largely in the choice of which depths for old lithosphere to fit. Their approach, fitting a model to the deepest old lithosphere, caused the remaining old lithosphere to appear shallow. We instead fit a model to all the depth data. By doing so, we assume that whether the temperature structure causing flattening of the depth curve reflects small scale convection, hot spot reheating, or other processes, an appropriate global reference model should reflect their average effects.

An alternative view is to adopt the cooling halfspace model as the reference [Davies and Pribac, this volume]. With this reference almost all old lithosphere is anomalous, and the degree to which it is anomalous indicates the perturbation to the cooling halfspace. Thus lithosphere which would be called "normal" relative to GDH1 would be perturbed by the "usual" amount for its age relative to a halfspace. A disadvantage with the halfspace reference model is that it is harder to identify regions of old lithosphere which are significantly more perturbed than most old lithosphere.

Implicit in our approach, in which the model is fit to the data, is that the model depends on the data used to derive it. In particular, given that old lithosphere is restricted to a few regions, the choice of data has an important effect. Successive reference models are improvements over previous models, in that they better fit more of the data that the previous ones could not. Nonetheless, the differences in subsidence rates between plates and within plates are substantial enough [Davis and Lister, 1974; Cochran, 1986; Hayes, 1988; Marty and Cazenave, 1989] that no single model can perfectly represent the data. Similarly, our fitting of mean depths, which are convenient for statistical analysis, or the alternative of fitting modal depths, which are less affected by processes which produce shallow depths, affects the result somewhat, as discussed shortly. As a result, any global reference model is an approximation to an unattainable ideal.

TESTS OF THE GDH1 MODEL

Before applying the GDH1 model to study midplate swells, we examined how well the model fit several other data sets.

Drill Site Data

First, we considered a global data set of the basement depths (corrected for sediment) versus age compiled for all DSDP and ODP drill sites which recovered normal tholeiitic extrusive basalts [Johnson and Carlson, 1992]. This test has several advantages. First, these data were not used in deriving GDH1. Second, the basement depths and ages are well constrained for the drill site data. Third, these data include sites in the Indian Ocean, whereas no Indian Ocean data were used to derive GDH1. Figure 5 shows that GDH1 offers a significantly better fit than PSM (a 56% misfit reduction). An independent analysis of these depth data [Johnson and Carlson, 1992] yielded a best-fitting plate model with a thicknesses of 105 ± 10 km and a basal temperature of 1400 ± 140°C, values quite consistent with GDH1.

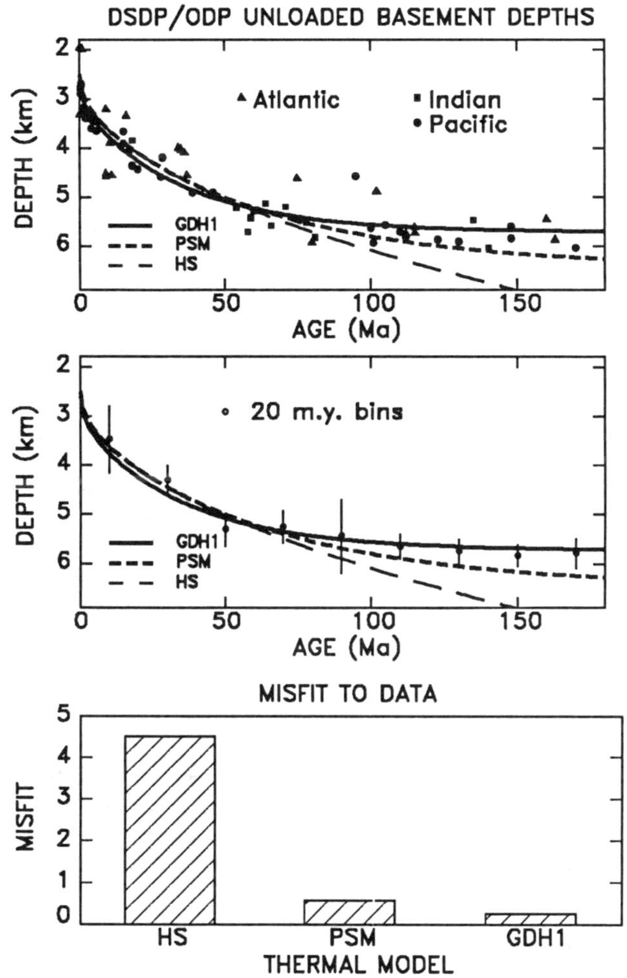

Fig. 5. Top: Fits to a global data set of the basement depths (corrected for sediment) versus age for DSDP and ODP drill sites [Johnson and Carlson, 1992]. The curves are for GDH1, the plate model of Parsons and Sclater [1977] (PSM), and a cooling halfspace model with the same thermal parameters (HS). These data were not used to derive GDH1. Center: Fits to this data set combined into 20 My bins. Bottom: Net misfit to the binned data. GDH1 offers a significantly better fit than PSM (a 56% misfit reduction).

Ocean-specific Data

Next, we tested the possibility that GDH1's good fits to the old lithosphere in the Pacific, which are especially important in this paper, might merely reflect the use of these depths in deriving it. To address this issue, we found best fitting plate model parameters separately for the North Pacific and Northwest Atlantic data, using the same procedure as for GDH1. The resulting models, NPC1 and NWA1, both have 90 km thick plates with a basal temperature of 1375°C. The coefficients of thermal expansion differ, being 3.3×10^{-5} °C^{-1} and 3.5×10^{-5} °C^{-1}, respectively. Alternatively, if we require that the coefficient of thermal expansion be that used in the GDH1 model, 3.1×10^{-5} °C^{-1}, we find models NPC2, a 95 km thick plate with a 1425°C basal temperature, and NWA2, a 95 km thick plate with a 1400°C basal temperature. For convenience, we refer to these models by name, with those ending in "2" being those derived with a fixed coefficient of thermal expansion. Chosing between the two members of the pairs of models is largely a matter of taste; keeping the coefficient of thermal expansion the same is attractive philosophically, but the improved fits resulting from allowing the additional free parameter are statistically significant as measured by an F-ratio test. As discussed later, the apparent variation in the coefficient of thermal expansion probably reflects limitations of the simple plate model.

Comparison of ocean-specific models NPC1 and NWA1 to the data (Figure 6) illustrates both similarities and differences. The average heat flow as a function of age is similar for the two plates, although as previously noted the Atlantic data have larger scatter to older ages [Anderson and Skilbeck, 1981]. The predicted heat flow, which does not depend on the coefficient of thermal expansion, is the same for these two models. The depths, however, are different between the two oceans. For ages less than about 70 Ma, the depths are similar. For older ages, however, the Pacific is generally shallower. For the dataset we use (Table 1), the mean depth and standard deviation for ages older than 100 Ma is 5423 ± 432 m, an average of the Atlantic and Pacific values of 5649 ± 308 and 5197 ± 807 m, respectively. NWA1, with a larger coefficient of thermal expansion, fits the deeper Atlantic depths for old lithosphere reasonably well. NPC1, however, does not fit the shallower Pacific depths for old lithosphere as well. The misfit presumably reflects the fact that much of the shallow topography in the Pacific, such as

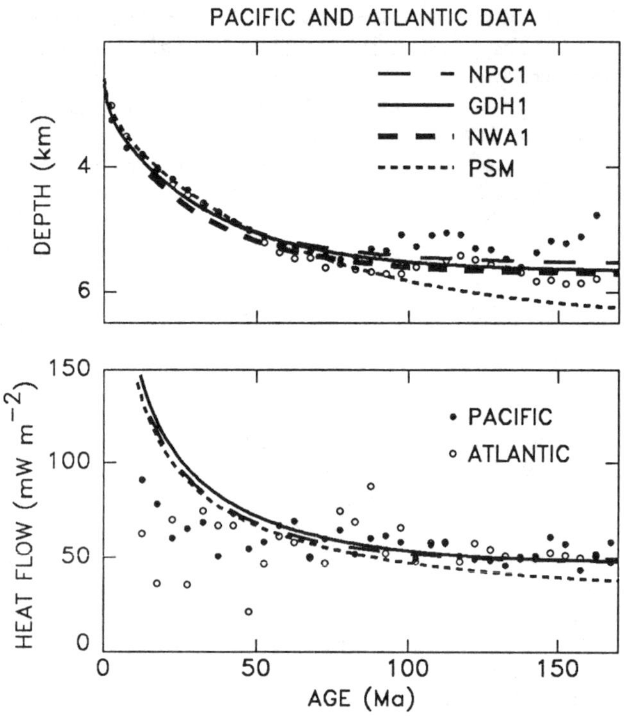

Fig. 6. Northwest Atlantic and North Pacific heat flow and depths as a function of age. Data are averaged in five-m.y. bins. Also shown are the predicted values for global models GDH1 and PSM, and models NWA1 and NPC1, each derived using data for a single ocean basin. NWA1 and NPC1 have the same predicted heat flow. The predicted depths and heat flow for the ocean-specific models closely resemble those for GDH1, indicating that GDH1 is a suitable average model for both ocean basins.

hot spot swells, plateaus, or flood basalts, is produced by processes not represented by the plate model.

The ocean-specific model parameters, and the predicted depths and heat flow, are quite similar to each other and GDH1. In particular, the predicted heat flow for GDH1 is essentially the same as for either ocean-specific model. Thus GDH1 is not unduly biased by data from one plate with respect to the other. Given the scatter in the data and the range of model parameters which give similar fits, we consider GDH1 a good compromise model for both ocean basins and data types. Hence although no reference model can fully characterize oceanic lithosphere, due to the intrinsic differences between and among ocean basins [Marty and Cazenave, 1989], GDH1 is a good approximation.

Digital Bathymetry versus Regional Median Depths

The depth data we used are from studies using the mean depths from DBDB-5 (or ETOPO-5) digital bathymetric dataset, which have some biases. To assess the effect of such biases, we compared the Pacific data to a recent set of depth data for the Pacific (Figure 7), which are computed as regional median depths taken directly from ship tracks [Smith, 1990]. These data, which span the Pacific Plate, vary with age in essentially the same way as the mean depths for the North Pacific Ocean from the digital database [Renkin and Sclater, 1988], which we used in deriving GDH1. Given that the two data sets only partially overlap in space, the agreement is gratifying. Some of the differences, such as the North Pacific being slightly deeper for young ages, may reflect the different areas sampled as much as the different data types. For older ages, where the two data sets largely sample the same area, the median depths smooth the curve, by reducing the effects of seamounts, plateaus, and other structures which bias the mean to shallower depth. As the figure illustrates, the two datasets are quite comparable, both before and after we applied an approximate sediment correction (a linear increase of sediment thickness with age rising to 300 m at 180 Ma) to Smith's data. The median absolute difference between the two datasets is approximately 130 m. The apparent difference in the uncertainties reflect Smiths' use of quartiles rather than standard deviations. Thus although the regional medians can offer better local detail and some smoothing, their overall depth-age behavior is quite comparable to the digital bathymetry, especially given the intrinsic scatter in the data.

Given that the differences between the regional medians and digital database are small compared to the differences in the thermal model predictions, we expect little difference between the lithospheric thermal structure inferred from the two. This is in fact the case. We inverted the corrected median depths by assuming a coefficient of thermal expansion of $3.1 \pm 0.8 \times 10^{-5}$ °C^{-1}, because α cannot be found independently from the basal temperature with depth data alone. The resulting model, SCD, is a 105 km thick plate with a basal temperature of 1300°C. Inversion of these depths together with the Pacific heat flow data yields model SCDH1, a 105 km thick plate with a basal temperature of 1550°C, and coefficient of thermal expansion 2.6×10^{-5} °C^{-1}. Alternatively, fixing the coefficient of thermal expansion at 3.1×10^{-5} °C^{-1} yields a poorer-fitting model, SCDH2, a 90 km thick plate with a 1375°C basal temperature. SCDH1 is almost identical from the depth-only result in its depth predictions, but also fits the heat flow. The depth-age curves for SCDH1 are similar to those for NPC1 (Figure 7), differing largely in their predictions for the young depths, and are thus similar to those for GDH1. As discussed shortly, the predicted temperature structures are also quite similar. Thus for the Pacific, the mean depths from digital bathymetry are essentially as good for depth-age studies as regional median depths.

Geoid Data

Another test of GDH1 is to examine how well it fits geoid data, which reflect a depth-weighted integral of the density distribution, and hence provide a different constraint on the geotherm from those offered by bathymetry or heat flow [Turcotte, 1986]. Lithospheric thickness as a function of age can be inferred [Crough, 1979; Sandwell and Schubert, 1982; Cazenave, 1984] from the geoid offset resulting from the age difference across fracture zones. This technique has the advantage that it suppresses the long wavelength geoid variations, and concentrates on the offset which should reflect lithospheric thickness. Figure 8 shows the fit to geoid data across a set of

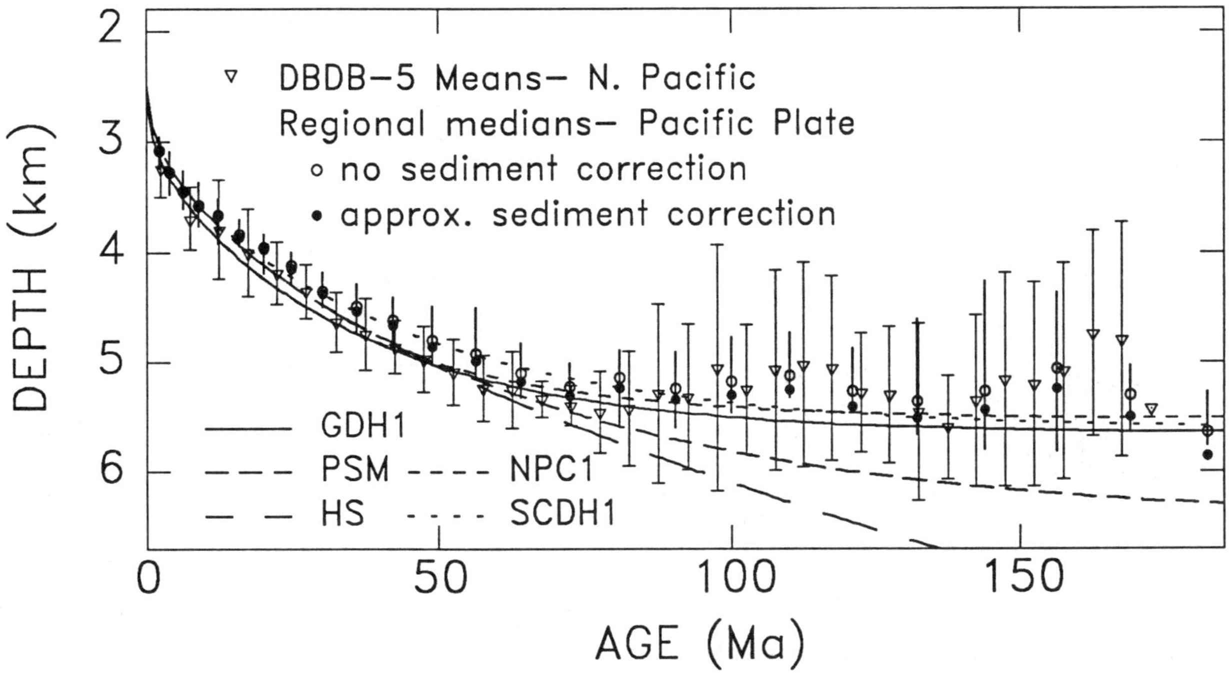

Fig. 7. Top: Comparison of different depth-age data for the Pacific. The North Pacific data [Renkin and Sclater, 1988] are mean depths, whereas the Pacific-wide data [Smith, 1990] are regional median depths from ship tracks. The data sets are quite comparable, both before and after we applied an approximate sediment correction to the median data. The apparent difference in uncertainties is because the mean depths are shown with the standard deviation, and the median depth with the quartile range. Both data sets are fit better by GDH1 than PSM. The depth-age curves for SCDH1, derived from the median depths, and NPC1, derived from the North Pacific data, are similar to each other and GDH1. The mean depths from digital bathymetry and regional median depths from ship tracks are thus essentially equivalent for our purposes.

Pacific fracture zones [Cazenave, 1984]. As noted in that paper, the data are much better fit by lithosphere thinner than the 125 km thickness for PSM. GDH1 provides a significantly better fit, but still overpredicts the geoid step, especially for young (<30 Ma) lithosphere. This misfit at the young ages may reflect the mechanics of fracture zones being more complex than a simple thermal age offset [Parmentier and Haxby, 1986; Marty et al., 1988; Driscoll and Parsons, 1988; Robinson et al., 1988; Wessel and Haxby, 1990].

COMPARISON OF TEMPERATURE MODELS

The discussion of the thermal models resulting from inverting different data sets brings out the question of how to compare the models. Each model is specified by four primary parameters; the plate thickness a, basal temperature T_m, coefficient of thermal expansion α, and thermal conductivity k. The other parameters, densities, specific heat, and ridge depth are generally assumed, although the latter can also be estimated directly. For simplicity, we and others generally treat the conductivity as the same for all models. For GDH1, the improved fit from estimating the conductivity from the data is not meaningfully better than using an a priori value [Stein and Stein, 1992].

We could compare the other three parameters for different models and see whether they agree to within their difficult-to-estimate confidence limits. We instead consider it more useful to compare models in terms of those properties that are of greatest interest. As discussed earlier, such models are used primarily either to characterize depth-age and heat flow-age data, or to predict temperatures in the lithosphere. Although we fit the depth data for all ages, and the heat flow for all ages above a certain value (50 Ma), it is useful to consider three limiting features easily observable from the data. The predicted values of these features correspond to combinations of the model parameters. One such combination is the asymptotic heat flow for old lithosphere, $k\ T_m/a$, which is proportional to the asymptotic linear geotherm. A second combination is the asymptotic depth, which is proportional to $\alpha T_m a$, the heat lost as the plate cools. A third combination is the slope of the depths in young lithosphere versus the square root of age, which is proportional to $k^{1/2}\alpha T_m$. Because these depths can be equally well fit assuming a cooling halfspace, they are insensitive to plate thickness. Thus we can think of the models as having two parameters (plate thickness and basal temperatures) which reflect the thermal structure, and two parameters (conductivity and coefficient of thermal expansion) which reflect the average physical properties of the lithosphere. The latter two act as scale factors which

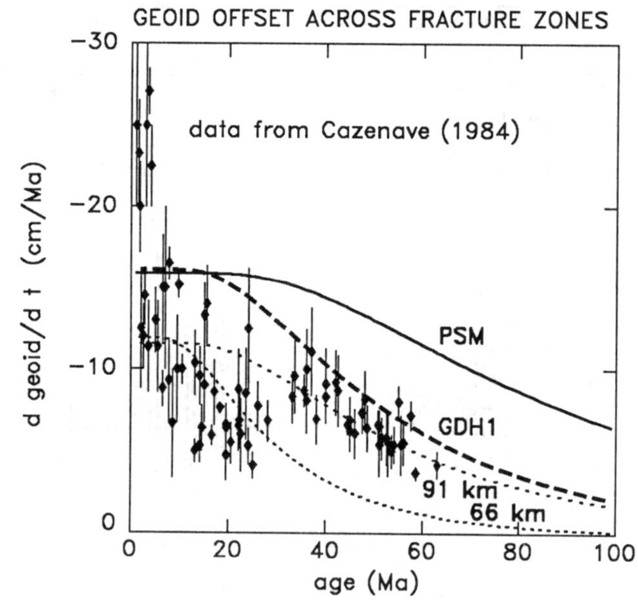

Fig. 8. Change in geoid height across Pacific fracture zones [Cazenave, 1984] and the predictions of models with different lithospheric thicknesses. The data are better fit by GDH1 than by the 125 km thickness for PSM. The remaining misfit for young (<30 Ma) lithosphere may reflect the mechanics of fracture zones being more complex than a simple thermal age offset.

map the thermal parameters into the primary observable features (young depths, old depths, and old heat flow) of the data.

Consideration of the three observable features shows why GDH1 differs from PSM in the way it does. Because GDH1 and PSM have similar predictions for the depths of young lithosphere, the product of their αT_m values must be similar. The two models have α and T_m values differing by less than 8%, and a product differing by only 4%. Thus for old lithosphere, GDH1 predicts shallower depths and higher heat flow primarily via a 24% thinner lithosphere, although the slightly higher (7%) basal temperature also contributes.

Asymptotic Depth and Heat Flow

We often compare models using the asymptotic depth and heat flow, two of the three observables used in deriving the model from the data, because both reflect the plate thickness and basal temperature. Although the asymptotic values are for infinitely old lithosphere, they are good approximations for old lithosphere. For example, the depth and heat flow predicted by GDH1 for 100 Ma lithosphere are only 154 m less and 6 mW m^{-2} higher, respectively, than their asymptotic values.

Figure 9 (top) compares the asymptotic depth and heat flow for models GDH1 and PSM to those for the ocean-specific models NWA1, NPC1 and SCDH1. PSM clearly differs significantly from these other models, which are all relatively similar to GDH1. The asymptotic depths for NWA1, NPC1 and SCDH1 are within 150 m of the 5651 m value for GDH1, whereas PSM has a noticeably deeper (6400 m) asymptotic depth. The asymptotic heat flow for these ocean-specific models is within 2 mW m^{-2} of GDH1's 48 mW m^{-2}, compared to the 34 mW m^{-2} for PSM. Some of this similarity is expected, given that the ocean-specific models were found by fitting subsets of the GDH1 heat flow data. The comparison illustrates the advantage of comparing models in terms of the asymptotic depth and heat flow. Models that are nominally different when quoted in terms of plate thickness, basal temperature, and coefficient of thermal expansion, can be quite similar in their predictions for the two observables.

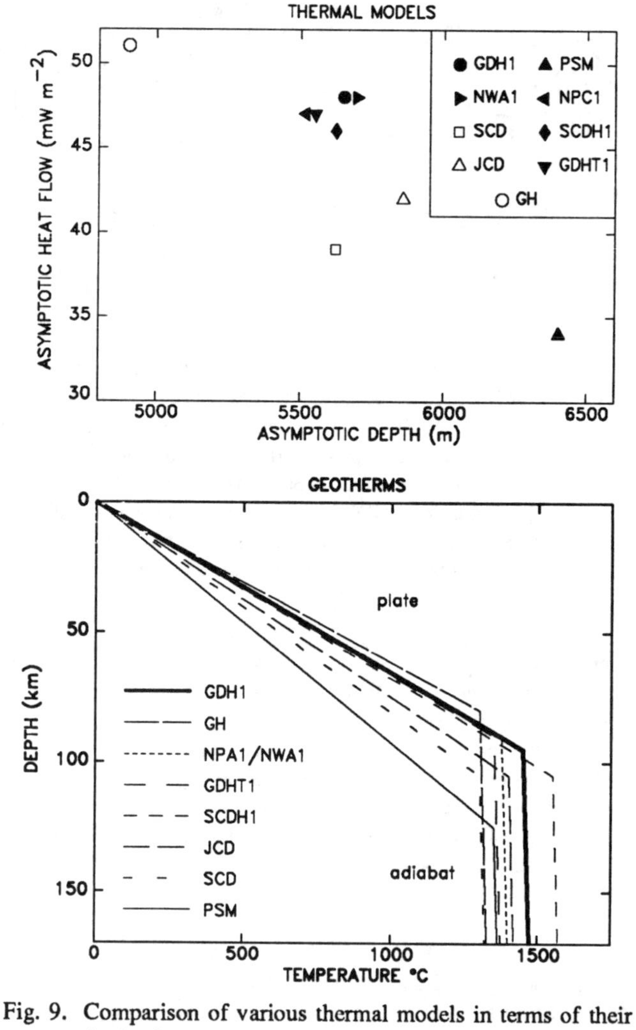

Fig. 9. Comparison of various thermal models in terms of their asymptotic depth and heat flow (top) and geotherms (bottom). In the top panel, open symbols denote models derived using either depth or heat flow data alone; closed symbols are models derived using both. Ocean-specific models derived using depth and heat flow data (NWA1, NPC1 and SCDH1) and a model derived for a 1350°C basal temperature (GDHT1) are similar to GDH1. The two models (JCD, SCD) derived using only depth data and the model derived using only global heat flow (GH) are closer to GDH1 than PSM. All models predict higher temperatures than PSM at depth within the lithosphere.

In particular, this comparison of models bears out that the differing coefficients of thermal expansion found in various inversions are features of models with similar asymptotic depth and/or heat flow. The coefficient of thermal expansion is not tightly constrained in these models, as indicated by the uncertainties estimated for GDH1 ($3.1 \pm 0.8 \times 10^{-5}$ °C^{-1}) and by Parsons and Sclater ($3.2 \pm 1.1 \times 10^{-5}$ °C^{-1}). As a result, we do not place much credence on the apparent variation in the coefficient of thermal expansion between the results of various inversions. Comparison of models SCD and SCDH1 resulting from inverting the Smith [1990] depth data with and without heat flow data illustrate that the basal temperature changed to fit the heat flow, and α changed to compensate and keep the depth prediction the same. Some of the apparent variation in α probably reflects the thermal structure of the lithosphere, and the resulting depth-age and heat flow-age variation, being more complicated than the simple plate model. The difference in the apparent α values for the Pacific models (NPC1 vs SCDH1) supports this idea.

The asymptotic depth and heat flow are useful in comparing models derived from different datasets. Johnson and Carlson's [1992] plate model was derived using only depth data, which are independent of those used in deriving GDH1. Nonetheless, as they noted, the depths predicted by their model (denoted by "JCD") are much closer to GDH1 than PSM. Similarly, inverting the Smith [1990] depth data gives model SCD, whose asymptotic depth is closer to GDH1 than PSM. If heat flow data is added to this inversion, the resulting model SCDH1 is quite similar to GDH1 in its depth and heat flow predictions. The effect of adding the heat flow data can be visualized from the position of the model (denoted by "GH") which was derived from the heat flow data alone.

We can also use this approach to see how much a model would be changed by inverting the data under different assumptions. For example, although the differences between GDH1 and PSM primarily reflect GDH1's thinner lithosphere, GDH1 also has a higher basal temperature. This 1450°C basal temperature is slightly (7%) higher than the value of approximately 1350°C often inferred for the temperature of midocean ridges from the thickness of oceanic crust [e.g. Sleep and Windley, 1982; McKenzie and Bickle, 1988]. Given the uncertainties in the GDH1 temperature (± 250°C) and in the inferred ridge temperature, it is hard to say if the difference is meaningful. The difference may be real, and reflect the basal temperature representing the heat addition to old lithosphere via mantle plumes, which are thought to be several hundred degrees hotter than the ridge temperature [Sleep, 1992]. The alternative is to ask what model results from inverting the GDH1 data, with the basal temperature fixed at 1350°C. This inversion yields model GDHT1, a 90 km thick plate with coefficient of thermal expansion 3.4×10^{-5} °C^{-1}. Also fixing α, to 3.1×10^{-5} °C^{-1}, gives a poorer-fitting model GDHT2, a 100 km thick plate. Thus, as expected, the required shallower asymptotic depths and higher asymptotic heat flow can be predicted by models with a basal temperature the same as in PSM, so long as the plate is thinner. GDHT1 and GDHT2 are quite similar to GDH1, in terms of their asymptotic depth (5552 and 5291 m) and heat flow (47 mW m^{-2}). However, the improved fit of GDH1 over these models, from also inverting for the basal temperature, is significant as measured by F-ratio test. Thus the higher basal temperature still is a better fit, at least for the data set we used. Our sense is that the thinner lithosphere will be a robust result for different data sets, but the higher basal temperature may be less so. As discussed next, we consider the hotter geotherm, corresponding to the T_m/a ratio, to be the primary feature of GDH1 and similar models.

Geotherms

Comparing models via their asymptotic depth and heat flow is a way of comparing how they fit depth and heat flow data. The asymptotic heat flow has the further advantage that it is proportional to the geotherm for old lithosphere. Given that one of our primary interests is the geotherm, we can compare the geotherms between models. The comparison is especially easy when the conductivity is assumed to be the same for different models. Figure 9 (bottom) compares the geotherms for GDH1 and PSM to those for other models. Temperatures below the plate base, which are not predicted by the models, are shown as rising along a shallow adiabatic gradient of 0.3°C km^{-1}. Although GDH1, NPC1, NWA1, SCDH1, and GDHT1 have different combinations of plate thickness and basal temperature, the geotherms are essentially the same within the lithosphere, because the models were found by fitting the heat flow data. In contrast, because PSM was found assuming lower heat flow in old lithosphere, it has a lower gradient and thus lower temperatures at any depth. Both the models derived from depth and heat flow data, and models derived from depth data alone (JCD, SCD), predict temperatures at depth higher than for PSM.

From these comparisons, we regard GDH1 as representative of the models that fit the data well. Different sets of data, and different modeling assumptions, yield different models. It would be hard to argue that one can select a "best" set of data or a "best" model. Successful models have to predict shallower asymptotic depths and higher asymptotic heat flow than PSM, and thus require higher T_m/a and lower $\alpha T_m a$, both of which can be achieved by assuming a thinner plate. We thus consider the hotter geotherm and thinner plate more important than the specific values of the basal temperature and plate thickness. Both due to the range of solutions and the simplicity of the model, we do not ascribe great significance to the details of the predicted temperature structures. This is especially the case for the thermal structure in the lower portions of the plate, beyond the general property that the models are hotter at depth than PSM.

HOT SPOT SWELLS

The GDH1 model has major advantages for investigation of midplate volcanism and swells. Because PSM overpredicts the depth and underpredicts the heat flow for most older lithosphere, its use as a reference model has the unsatisfying implication that most of the data for older lithosphere are anomalous. This creates a situation in which apparent perturbations can result from comparing data to a model which does not appropriately represent the unperturbed lithosphere.

To illustrate this effect, we first consider the Hawaiian swell,

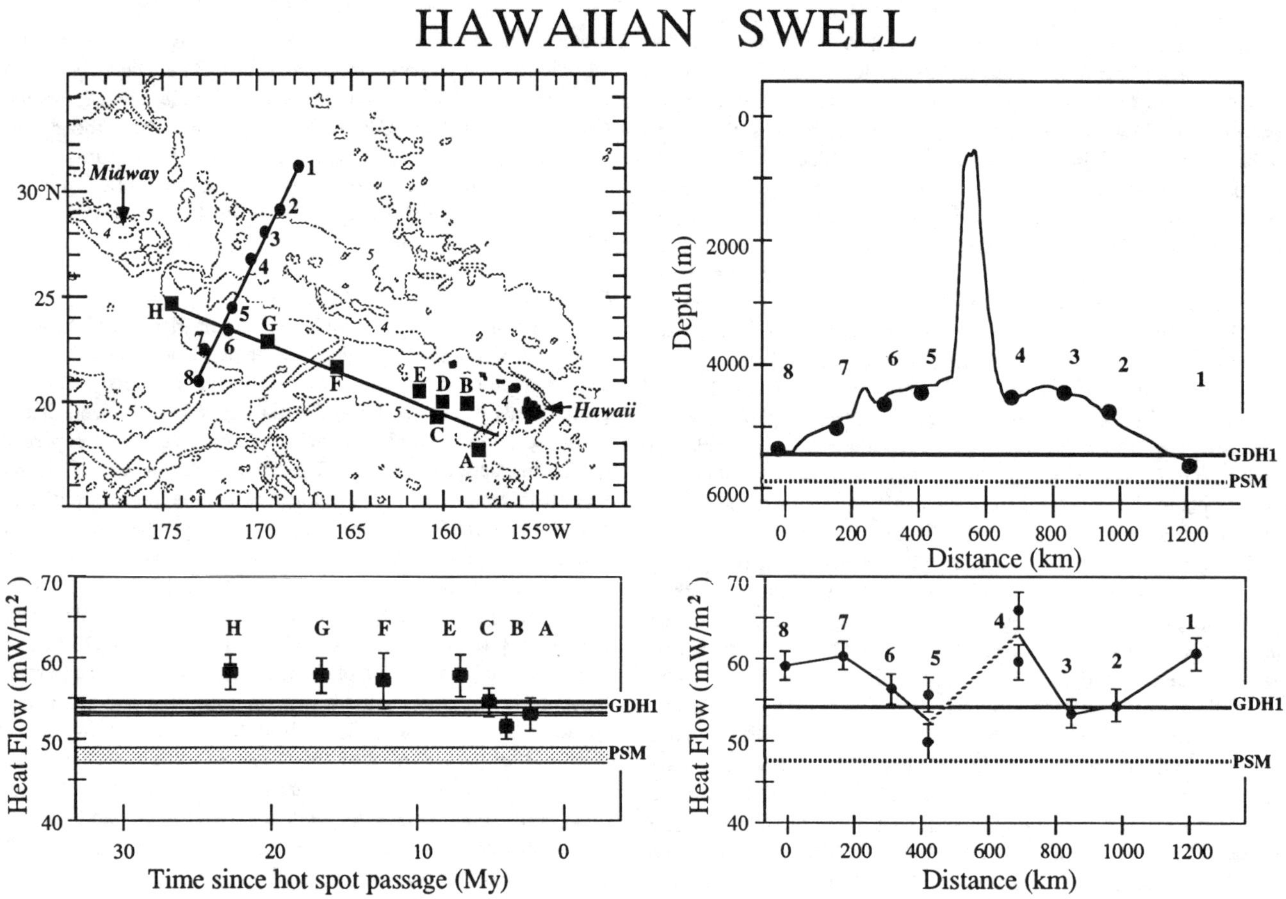

Fig. 10. Heat flow data for transects along (lower left) and across (lower right) the Hawaiian Swell at the locations shown (upper left). The heat flow, though anomalously high with respect to the Parsons and Sclater (PSM) model, is at most slightly above that expected for GDH1. The predicted heat flow values are for 100 Ma (lower right) and 95-110 Ma (lower left). Figure modified from Von Herzen et al. [1982; 1989].

the largest and best studied hot spot swell. Hawaii is the type example for hot spot studies, because of its size and isolation from other perturbing processes (including ridges and other hot spots). The observation that heat flow on the Hawaiian swell was higher than that predicted for PSM (Figure 10) was initially treated as consistent with the elevated heat flow expected for a reheating model [Von Herzen et al., 1982]. A subsequent transect across the swell showed that the heat flow differs at most slightly from that for lithosphere of comparable ages [Von Herzen et al., 1989]. Thus much of the apparent anomaly resulted from comparing the heat flow to PSM, which systematically underpredicts the heat flow for old lithosphere. In contrast, the swell heat flow is only slightly above that expected for GDH1 and hence at most slightly higher than for older lithosphere elsewhere. Even the heat flow anomaly relative to GDH1 may be an overestimate, because GDH1 slightly underpredicts the heat flow data. The swell is, of course, identified by its being shallow with respect to lithosphere of the same age elsewhere. It is thus shallow with respect to GDH1 or PSM.

The tectonic inferences that would be drawn from these data (Table 3) depend on the reference model used. Anomalies relative to a reference model are assumed to represent processes, for example plumes, not included in the reference model assumptions. Because both the depths and the heat flow are significantly anomalous with respect to PSM, one would infer that the hot spot gives rise to additional heat flow and shallow topography. In contrast, using GDH1 implies that the uplift has little heat flow signature, suggesting that the topography results primarily from dynamic effects. Hence the plume is providing uplift beyond that observed for most lithosphere of this age. These data can thus be used as modeling constraints [e.g. Von Herzen et al., 1982; Liu and Chase, 1989]. The interpretation that the Hawaiian swell heat flow data favor a dynamic model is consistent with seismological data, which find no evidence for a low velocity zone under the swell [Woods et al., 1991].

A similar situation applies for the Bermuda, Cape Verde and Crozet hot spot swells (Figures 11, 12, and 13). The heat flow data [Detrick et al., 1986; Courtney and White, 1986; Courtney

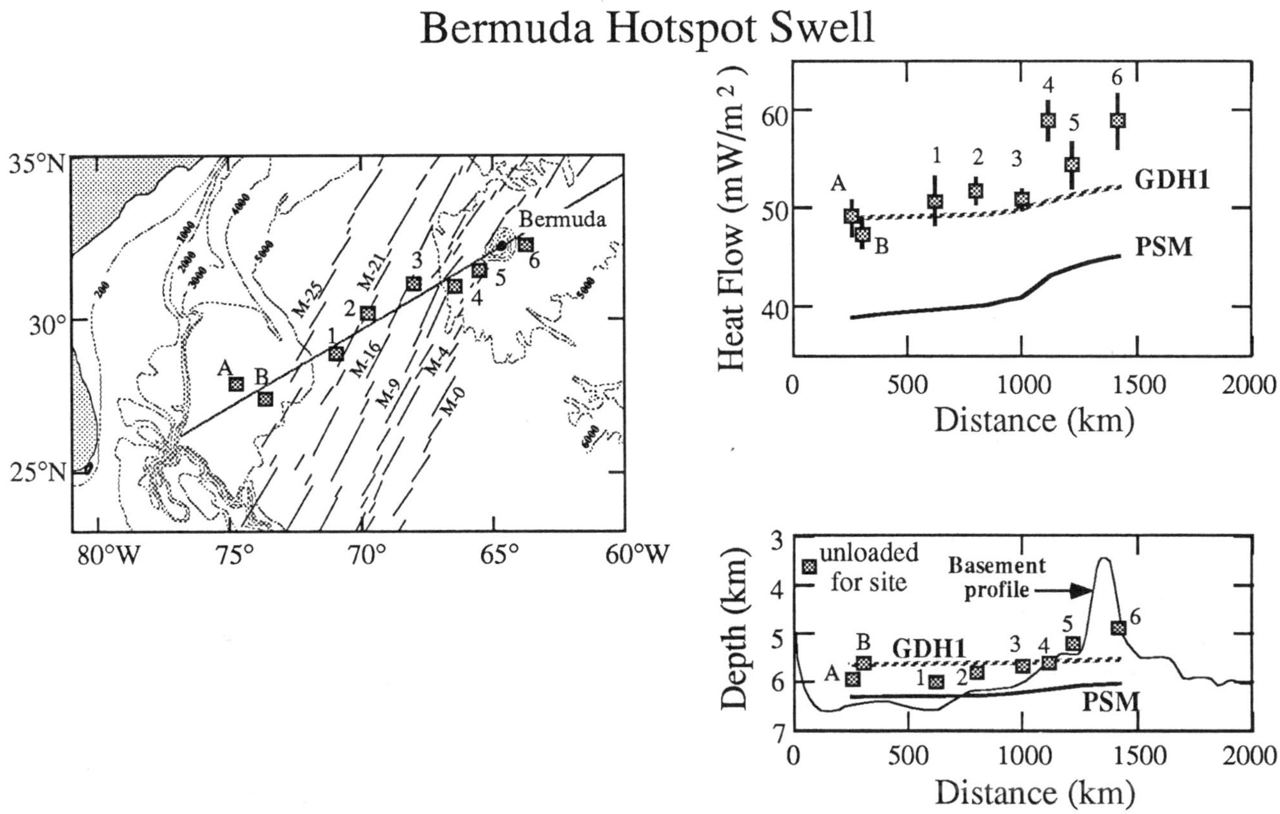

Fig. 11. Heat flow data and depth data for the Bermuda Swell. The heat flow data [Detrick et al., 1986] show a large anomaly with respect to PSM, but a much smaller one with respect to GDH1.

and Recq, 1986] show a large anomaly with respect to PSM, but a much smaller one with respect to GDH1.

The fractions of thermal and dynamic uplift can be estimated crudely [Sleep, 1990] by using the observed heat flow to determine the thermal age, the age at which a reference model predicts the observed heat flow. We then find the thermal depth anomaly, the difference between the predicted depth for the thermal age and that observed, and compare it to the total depth anomaly, the difference between the predicted depth for the lithospheric age and that observed. Table 4 shows these estimates, illustrating that GDH1 predicts higher heat flow and hence less thermal uplift than PSM.

We consider the estimates of the thermal uplift fraction crude for several reasons. The thermal uplift fraction is an lower bound, in that some of the perturbed heat flow may not yet be detectable at the surface, so the thermal age is an upper bound assuming the data and reference model were known exactly. In fact, both the data and model have uncertainties. Thus the thermal uplift fraction estimates depend on the assumed values of the swell depth (taken here from Detrick et al. [1989]) and heat flow. In addition, the estimates depend on the assumed depth and heat flow for "normal" lithosphere, which although predicted better by GDH1 than PSM, still have uncertainties. Nonetheless, it appears that the data favor a primarily dynamic origin for the swells. The dynamic component is 2-3 times larger than would be inferred using the PSM reference model. We thus regard these estimates as illustrating how improved reference models like GDH1 can be useful in studies using more sophisticated models of hot spots.

Darwin Rise Depth and Heat Flow

To determine whether the Darwin Rise retains a thermal anomaly at present from the process which caused the widespread Cretaceous volcanism and uplift, we examined the depth and heat flow data [Stein and Abbott, 1991]. For this purpose, the Darwin Rise was assumed to be the region (Figure 2) surrounding the continuation of hot spot tracks from the present-day Superswell, an area roughly south and west of Hawaii and north of the Samoan hot spot track. The heat flow data for the sites, averaged in 20 m.y. bins, are shown in Figure 14. The Darwin Rise heat flow values are anomalously high with respect to PSM, but consistent with GDH1.

In addition, the Darwin Rise data can be compared directly with data from the rest of the Pacific plate. In each of the age bins the difference of the means between the Darwin Rise data and that elsewhere in the Pacific is less than 10 mW m^{-2} and the maximum difference in median value is 7 mW m^{-2}. The Darwin Rise data thus show no evidence for a heat flow ano-

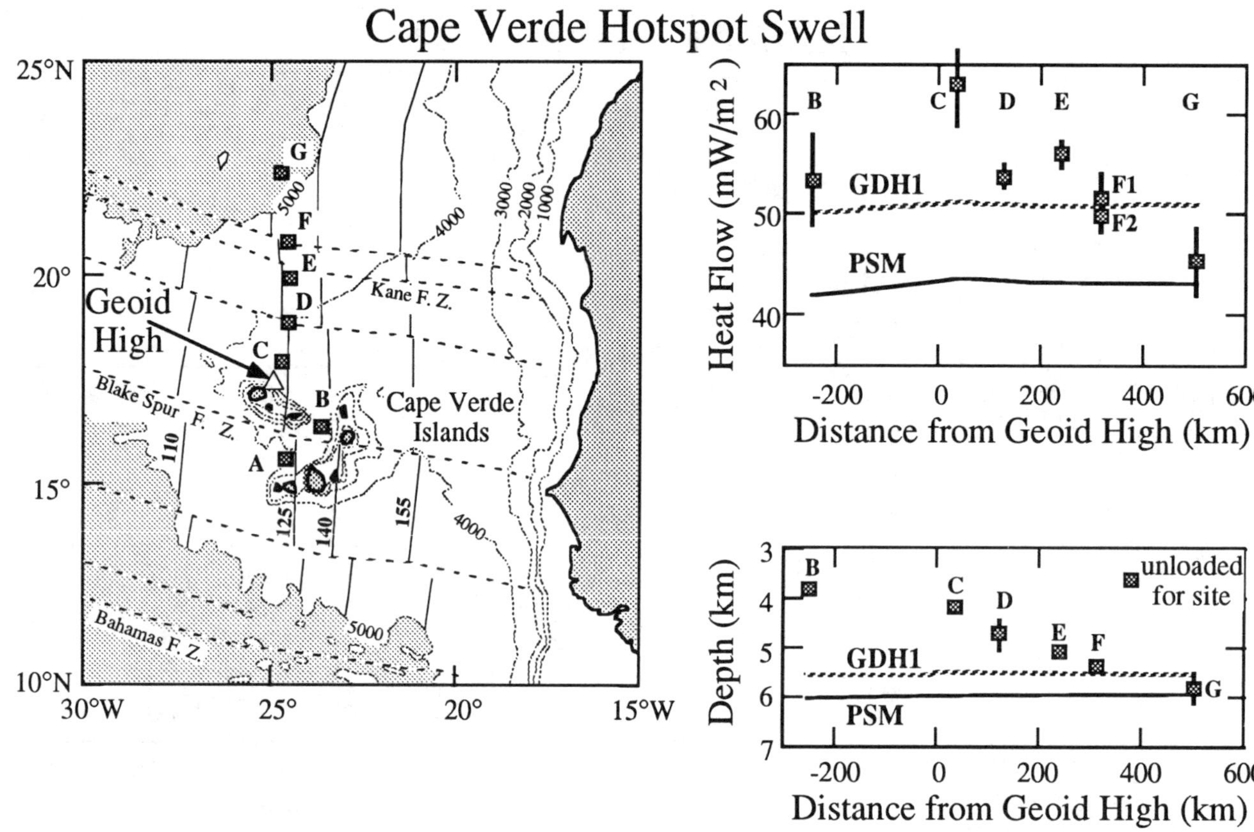

Fig. 12. Heat flow data and depth data for the Cape Verde Swell. The heat flow data [Courtney and White, 1986] show a large anomaly with respect to PSM, but a much smaller one with respect to GDH1.

maly either relative to lithosphere of these ages elsewhere in the Pacific, or the global dataset. Moreover, the lack of an anomaly is not a result of some condition specific to the Pacific Plate, because the predicted heat flow for the model derived from the Atlantic data alone (Figure 6) is essentially the same as for GDH1. Direct comparison with the Atlantic data also shows no anomaly [Stein and Abbott, 1991].

The Darwin Rise has been considered anomalously shallow, based on comparisons with the PSM model [Schlanger and Premoli-Silva, 1981; McNutt et al., 1990]. This interpretation implies that the area is characterized not only by many seamounts, but by a regional bathymetric anomaly. To test this idea, we compared the depths of the Darwin Rise heat flow sites to those elsewhere in Pacific, and to the two reference models. The depths in Figure 14 differ from those shown by Stein and Abbott [1991] because we corrected for sediment thickness, using seismic reflection records for 128 sites and isopachs [Ludwig and Houtz, 1979] for the remaining 135.

The depths are anomalously shallow with respect to PSM, but are consistent with GDH1. Similarly, the depths are comparable to those of lithosphere for these ages elsewhere in the Pacific. (The shallow average depth for the 135-155 Ma data elsewhere in the Pacific is due to sites on the Hess and Shatsky Rises and the Ontong-Java Plateau). We thus find no evidence for the Darwin Rise being anomalously shallow, either with respect to the rest of the Pacific or GDH1.

The lack of depth and heat flow anomalies has interesting implications for the history of the Darwin Rise. McNutt et al. [1990] (Figure 15) inferred that the Darwin Rise was similar to the present Superswell, which they assume to be characterized by an anomalously thin plate, and thus subsided at a slow rate from its formation (113 ± 8 Ma) until about 70–80 Ma, after which it subsided at the rate of normal seafloor. The present depth, which was interpreted as anomalously shallow with respect to PSM, would thus indicate that the thermal structure differs from that of comparable age lithosphere elsewhere. In contrast, we find no depth anomaly with respect to GDH1, and thus no evidence for the Rise lithosphere presently retaining a significant thermal signature of the Cretaceous events.

SUPERSWELL DEPTH AND HEAT FLOW

We conducted a similar depth and heat flow analysis for the Superswell. The heat flow sites in the area [Stein and Abbott, 1991] provide good regional coverage of the Superswell and, for comparison, the region surrounding it. The sites sample the region of shallower bathymetry described by McNutt and Fisher [1987].

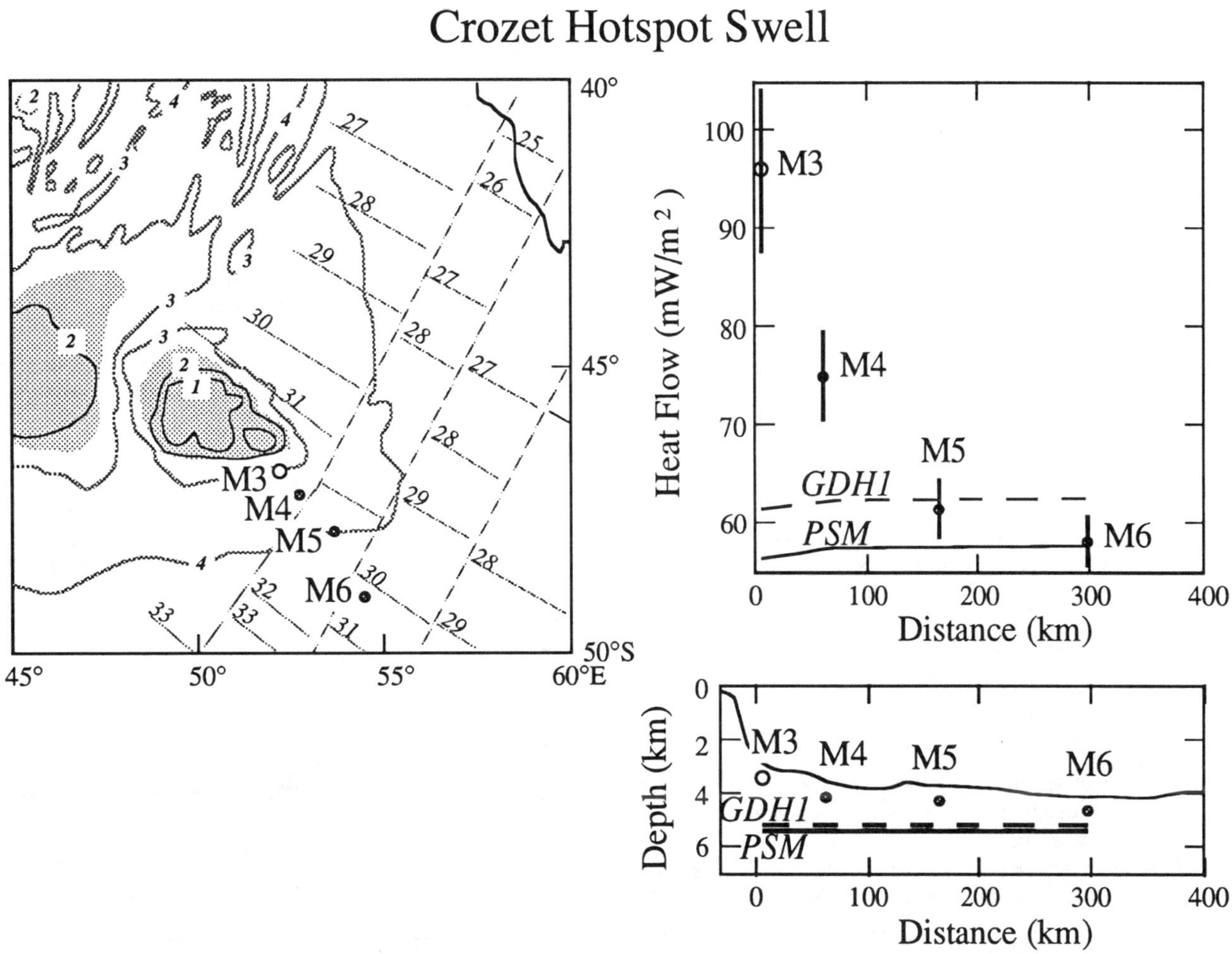

Fig. 13. Heat flow data and depth data for the Crozet Swell. The heat flow data [Courtney and Recq, 1986] show a smaller anomaly with respect to GDH1 than PSM. The magnitude of the anomaly estimated in Table 4 depends on whether the swell heat flow is assumed to be that for site M3 or M4; the M3 value is from one measurement and was thus not considered very reliable by Courtney and Recq.

Figure 16 shows the heat flow data versus age for the Superswell region and elsewhere on the Pacific plate. The data are divided into four age bins, each containing an approximately equal number of data for the Superswell. Comparison with GDH1 shows the characteristically lower than predicted heat flow for young (<50 Ma) lithosphere, that is observed elsewhere (Figure 3) and presumably reflects hydrothermal circulation. For the older ages, the data show no anomaly with respect to GDH1. Similarly, the heat flow values for the Superswell are no higher than for the corresponding ages elsewhere on the Pacific plate. The heat flow data thus neither require nor exclude the possibility of a thinner thermal lithosphere, as shown by predictions for the 75 km-thick plate with a 1385°C basal temperature suggested by McNutt and Fisher [1987]. The poor fit of the heat flow for ages > 60 Ma predicted by a 60 km-thick plate argues against further thinning.

The depth data provide another constraint. As shown, the depths are shallower than expected for GDH1. The depths in Figure 16 differ from those shown by Stein and Abbott [1991] because we corrected for sediment thickness, using seismic reflection records for 112 sites and isopachs [Ludwig and Houtz, 1979] for the remaining 245. The Superswell heat flow sites are shallower than sites of the same age elsewhere on the Pacific plate. The depths shown here for a model with a 75 km plate are somewhat deeper than shown by McNutt and Fisher [1987] and Stein and Abbott [1991] because we use a formulation in which the ridge depth is fixed to the observed value [Stein and Stein, 1992]. In the other formulation, the ridge depth varies depending on the temperature structure, and is thus shallower for hotter plates [McNutt and Fisher, 1987]. The subsidence with respect to the ridge, which reflects the cooling, is the same in both formulations. GDH1, which corresponds to a thicker plate, better fits the depth of the other Pacific sites.

The depth and heat flow situation for the Superswell is thus

TABLE 4: Hotspot Thermal Anomaly Estimates

	Thermal age	Thermal depth anomaly	Thermal fraction of total depth anomaly
Hawaii (lithospheric age 100 Ma; heat flow 58 mW m^{-2})			
PSM	67 Ma	449 m	37 %
GDH1	81 Ma	106 m	12 %
Bermuda (lithospheric age 116 Ma; heat flow 58 mW m^{-2})			
PSM	67 Ma	595 m	69 %
GDH1	81 Ma	161 m	36 %
Cape Verde (lithospheric age 125 Ma; heat flow 62 mW m^{-2})			
PSM	59 Ma	819 m	43 %
GDH1	69 Ma	285 m	19 %
Crozet (lithospheric age 67 Ma; heat flow 75-96 mW m^{-2}*)			
PSM	40-24 Ma	593-1081 m	21-39 %
GDH1	46-28 Ma	299-745 m	11-28 %

*Values given assuming that the swell heat flow is that for either site M3 or M4; the higher (M3) value was considered less reliable.

similar in general to that for Hawaii (Figure 10), in that the shallow depth anomaly has no corresponding high heat flow anomaly. The Superswell has drawn such interest, however, because of the large area over which the depth anomaly extends. Much of this depth anomaly might be expected simply from the fact that the Superswell region contains a number of hot spot swells. McNutt and Fisher [1987] and McNutt and Judge [1990], however, find that a depth anomaly remains even after removal of the estimated effects of the swells, and interpret the residual anomaly as evidence for dynamic uplift by a large scale plume.

The depth and heat flow anomaly data for the Hawaiian swell, the Superswell, and the Darwin Rise thus are generally consistent. The present swells, Hawaii and the Superswell, show significant depth but at most minimal heat flow anomalies, implying that the uplift is primarily dynamic. The Darwin Rise shows neither a heat flow anomaly nor a depth anomaly, suggesting that the effects of the Cretaceous volcanism have dissipated.

FLEXURAL DATA

The flexural data for these three regions, however, are inconsistent and more challenging to interpret. The strength of the lithosphere can be inferred from the shape of the deflection induced by the loads of seamounts and islands. The effective elastic thickness, a representation of the vertically integrated strength when the load was applied, generally increases approximately as the square root of the loading age [Watts et al., 1980], and corresponds to depths where the predicted temperatures are approximately 300-600°C (Figure 17) [Bodine et al., 1981]. These observations are consistent with ductile flow laws in which strength decreases exponentially with temperature [Kirby, 1977, 1980; Goetze and Evans, 1979], because cooling of the lithosphere would give increased strength at depth.

The effective elastic thicknesses beneath islands and seamounts in the Superswell and Darwin Rise are anomalously low for their age of loading (Figure 17) compared with sites along the Hawaiian swell, elsewhere in the Pacific, or in the Atlantic and Indian Oceans. It has been suggested that the low effective elastic thicknesses for the Superswell and Darwin Rise reflect weakening due to elevated temperatures in the lithosphere [Menard and McNutt, 1982; Calmant and Cazenave, 1987; McNutt and Fisher, 1987; Smith et al., 1989; McNutt and Judge, 1990]. This idea can be tested via the simple assumption

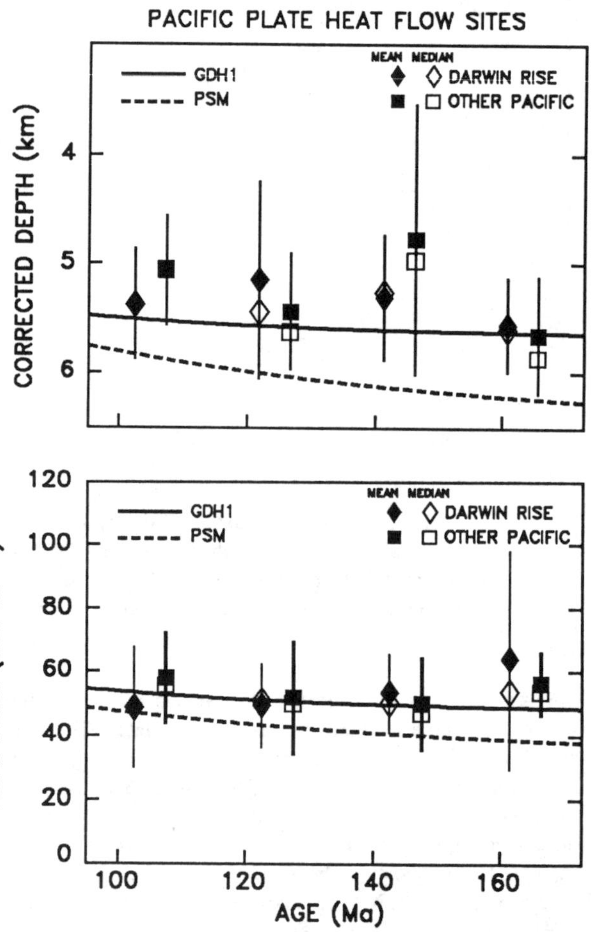

Fig. 14. Depth (top), and heat flow (bottom) as a function of age for sites on the Darwin Rise and on lithosphere of the same age elsewhere in the Pacific. Data are averaged in 20-m.y. bins. The depths and heat flow, though anomalously shallow and high, respectively, relative to PSM, are consistent with GDH1. Because the Darwin Rise data also do not differ significantly from those for the remainder of the Pacific, the Rise appears not to be anomalously shallow or hot at present.

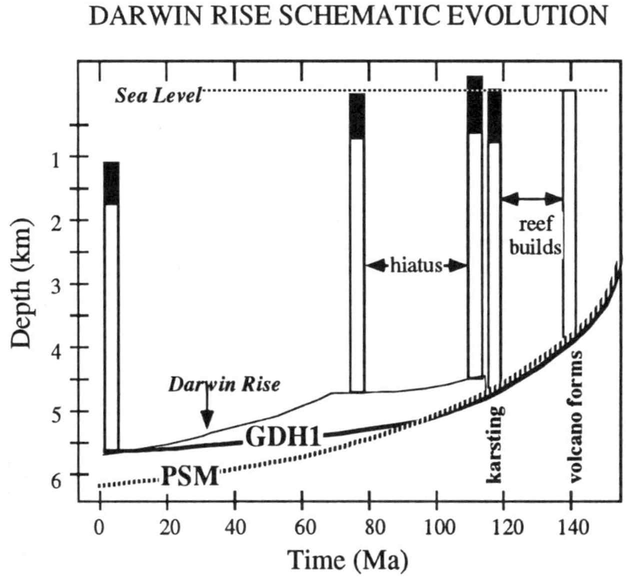

Fig. 15. Schematic history for the Darwin Rise [McNutt et al., 1990], for the case of a 140-Ma volcano on 155-Ma lithosphere. The Darwin Rise is presently anomalously shallow with respect to PSM, but not with respect to GDH1, suggesting that the Rise presently does not retain a significant thermal signature of the Cretaceous events.

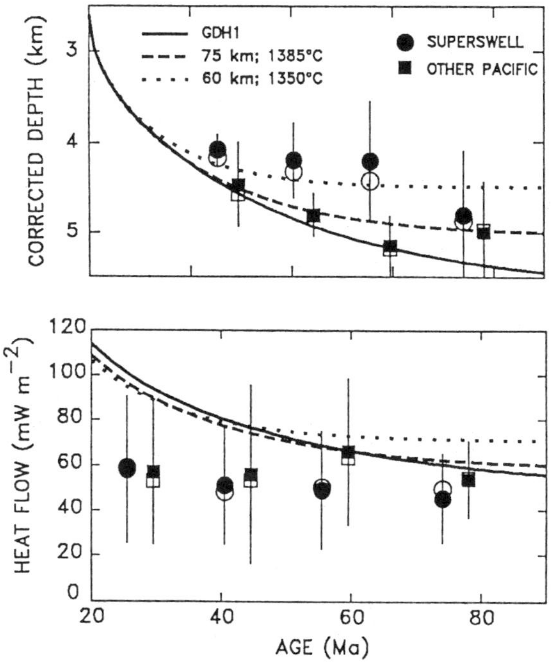

Fig. 16. Corrected depth (top), and heat flow (bottom) as a function of age for sites in the Superswell and on lithosphere of the same age elsewhere in the Pacific. Data are averaged in 20-m.y. bins. Closed and open symbols indicate means and medians, respectively. For ages >50 Ma the Superswell heat flow data do not differ significantly from those for the remainder of the Pacific or from those expected for GDH1. The Superswell depths are shallower than for the remainder of the Pacific, or than expected for GDH1.

that if the thinner effective elastic thicknesses result from thermal perturbations, the perturbed 300° and 600°C isotherms should rise to corresponding depths.

Figure 17 shows the predicted depths of the isotherms for GDH1 and the 75-km-thick plate proposed by McNutt and Fisher [1987]. The depths to the isotherms in the upper lithosphere for the 75-km-thick plate are only slightly shallower, because of the relatively young ages at the time of loading for the Superswell. This elevation of the isotherms should produce little change in effective elastic thickness.

Alternatively, much more dramatic heating could reduce the effective elastic thickness to that observed. The required high temperatures, however, appear to be excluded by the heat flow and depth data. Figure 18 (top) shows that a 40-km-thick plate with a 1350°C basal temperature or a 75-km-thick plate with a 2000°C basal temperature would have 400°C isotherms shallow enough to be consistent with the Superswell and Darwin Rise effective elastic thicknesses. Such models, however, predict heat flow substantially greater than observed (Figure 18, bottom) and seafloor depths much shallower than observed [Stein and Abbott, 1991].

These calculations are for steady state models, in which the boundary conditions do not change. An alternative approach is to consider models in which the thermal structure of the lithosphere is perturbed and then evolves. Figure 19 shows the temperature structure and heat flow predicted by a halfspace reheating model in which the base of the lithosphere is reset to asthenospheric temperatures and then cools [Von Herzen et al., 1982] The reheating depth determines the anomalous bathymetry

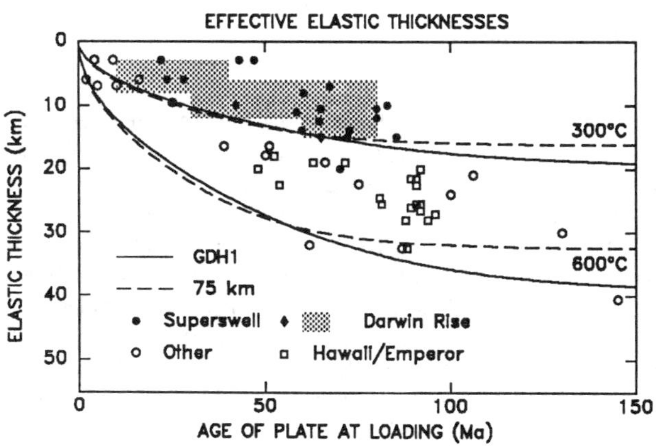

Fig. 17. Effective elastic thicknesses for locations on the Superswell and Darwin Rise compared to sites elsewhere. Data compiled from Smith et al. [1989], Calmant et al. [1990], and Wolfe and McNutt [1991] (shaded area). The thicknesses for the Darwin Rise and Superswell are less than for the other regions, including the Hawaiian Swell. Isotherms predicted for model GDH1 and for a model with a 75 km thick plate are also shown.

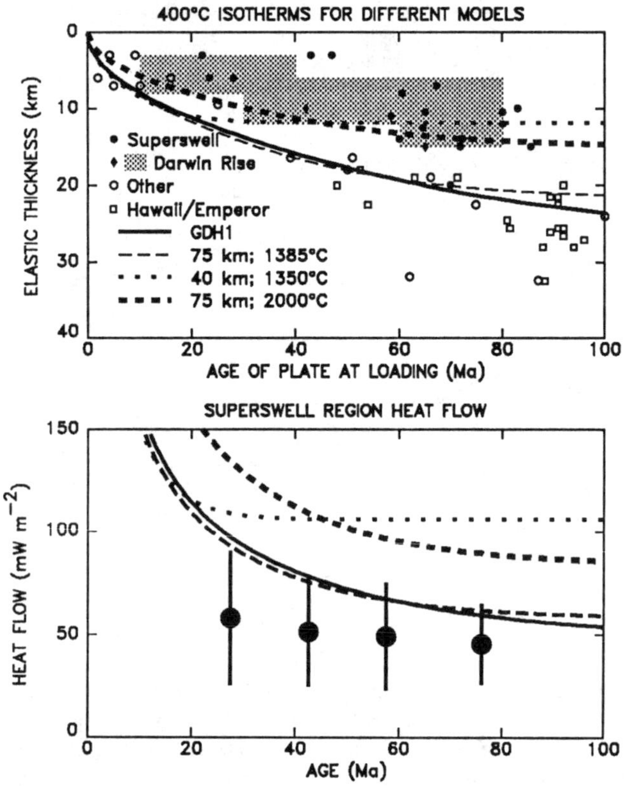

Fig. 18. Comparison of observed effective elastic thicknesses and predicted isotherms for various models. Effective elastic thicknesses for areas other than the Superswell and Darwin Rise correspond approximately to the GDH1 400°C isotherm. Models which raise the temperatures enough to make the Superswell thicknesses correspond to the 400°C isotherm (top) would yield heat flow much higher than observed for the Superswell (bottom).

and heat flow with time: the shallower the reheating depth, the greater the depth anomaly, the sooner the maximum surface heat flow anomaly occurs, and the greater its magnitude. Reheating 45-Ma lithosphere to the depth of the 900°C isotherm, as proposed by McNutt [1987], results in too little shallowing of the 400°C isotherm to explain the flexural data. Much shallower reheating is required to elevate the temperatures sufficiently to fit the effective elastic thicknesses data. In this example, thinning to 15 km depth would be required to place the effective elastic thicknesses in the expected temperature range. Such thinning would cause much higher heat flow than observed.

These results are interesting for the comparison with the Hawaiian swell. For Hawaii, the lack of a heat flow anomaly precludes major reheating of the lithosphere, an inference which is also consistent with the normal effective elastic thickness values. For the Superswell, although the effective elastic thicknesses indicate anomalously weak lithosphere, the heat flow excludes significant reheating. Thus the difference between the small effective elastic thicknesses for the Superswell and Darwin Rise and the larger values for Hawaii and the other regions appears not to result from general thermal weakening of the entire lithosphere.

There are several obvious other possibilities for the mechanism causing the Superswell and Darwin Rise lithosphere to have thinner effective elastic thicknesses than the other swells [Stein and Abbott, 1991]. First, shallow reheating might be localized under the seamounts, and thus not reflected in the heat flow data. This seems unlikely, given that the Superswell appears to be a widespread phenomenon. Second, the weakness might be only apparent, due to the action of intraplate stresses. Third, the lithosphere might be weakened mechanically.

In the second case, the low effective elastic thicknesses could reflect intraplate stresses. The flexural response of the lithosphere to an applied load depends on both the mechanical properties and the intraplate stress. Figure 20 (top) shows this effect schematically for a line seamount load on a 20 km thick elastic plate, computed using the formulation of Hetenyi [1974]. 500 MPa (5 Kb) of compressional stress reduces the effective elastic thickness to 16 km, and a 1 GPa (10 Kb) stress reduces it to 14 km. The high stresses in this elastic example would be less for a realistic depth dependent rheology [McAdoo and Sandwell, 1985]. McNutt et al [1991], for example, propose that approximately 50 MPa of in-plane tension could explain the Superswell effective elastic thicknesses.

This hypothesis is difficult to test, in that the relevant intraplate stress is that at the time of loading. There is, however, no apparent reason to expect the effect of regional intraplate stresses to have been dramatically greater for the Superswell than for other swells. The present intraplate seismicity is not particularly intense [Wysession et al., 1991], especially in terms of large events, and the present intraplate stresses predicted from driving force models are not unusually high [Wortel et al., 1991]. This situation is quite different from the Indian Ocean, where large earthquakes occur [Stein and Okal, 1978; Bergman and Solomon, 1985] and the predicted high stresses [Cloetingh and Wortel, 1985, 1986; Stein et al., 1987] are consistent with the observed large scale deformation [Stein et al., 1989]. In principle, the Superswell-forming process might instead somehow have provided the necessary stresses. It is unclear, however, how this might occur. Although hot spots might give rise to significant thermal stresses [Zhu and Wiens, 1991], these stresses should occur for all hot spots, not just those in the Superswell. We thus have difficulty envisioning how the apparent differences in lithospheric strength between the Superswell and other swells could be a consequence of intraplate stress.

An alternative possibility is that the lithosphere beneath the Superswell and Darwin Rise have been weakened mechanically, giving rise to a lower effective elastic thickness. Figure 20 (bottom) illustrates this schematically for the limiting case of a fractured lithosphere: a broken 20 km thick plate looks quite similar to an unbroken 11 km thick plate with the same physical properties. Hence an analysis which assumed that the lithosphere was not broken would yield an 11 km effective elastic thickness. In such a scenario, the weakening might occur in several ways [Stein and Abbott, 1991]. The Superswell-forming

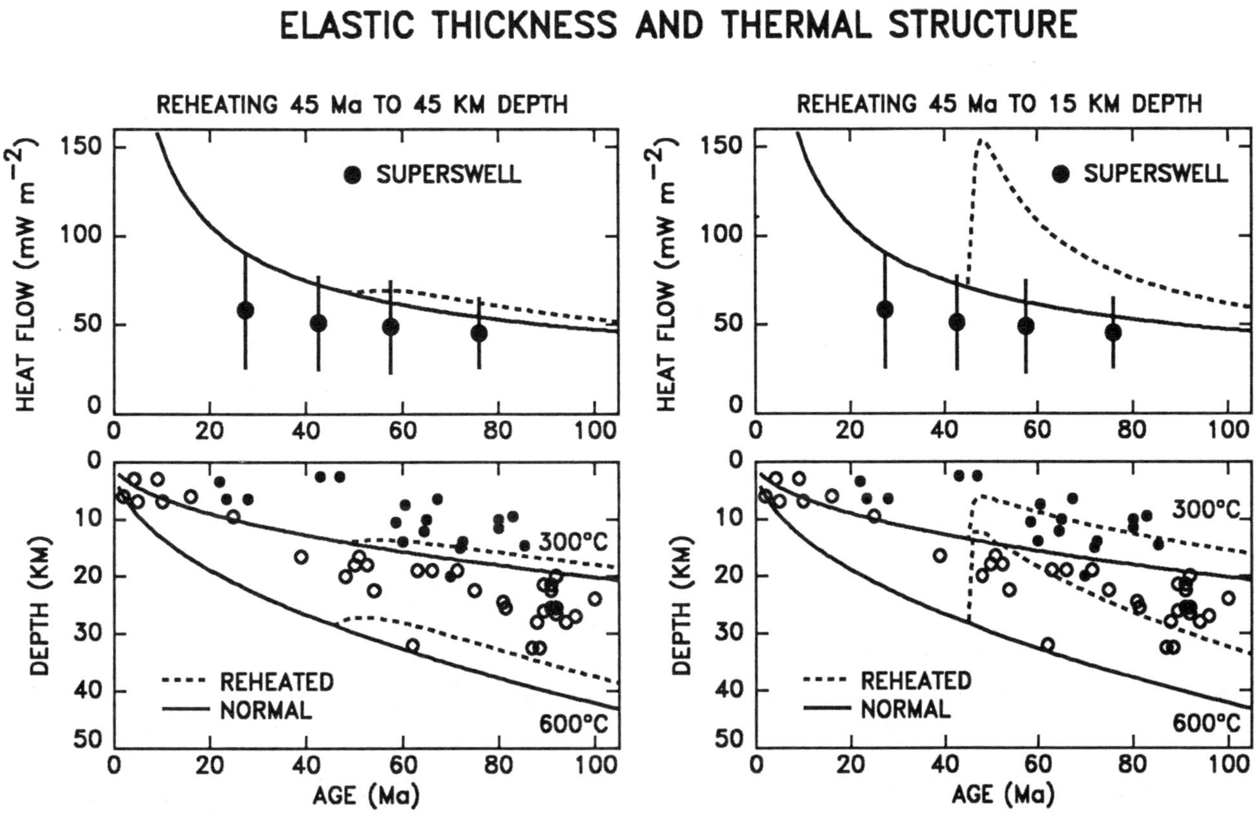

Fig. 19. Application of reheating models to the Superswell heat flow and effective elastic thickness data. Superswell seamount chains are indicated by closed symbols; other chains are shown as open symbols. Right: Model for reheating 45-Ma lithosphere to a depth of 45 km. Although the calculated heat flow anomaly (top) is plausible given the observed heat flow, such a model does not give sufficient reheating of the upper lithosphere (bottom) to account for the low effective elastic thicknesses. Left: Model for reheating 45-Ma lithosphere to a depth of 15 km. Although this model produces sufficient reheating in the upper lithosphere to account for the effective elastic thicknesses of the superswell sites, it predicts a heat flow anomaly significantly larger than that observed. [Stein and Abbott, 1991].

process might introduce significant amounts of volatiles such as water, giving rise to the weaker rheology expected from laboratory experiments [Kirby, 1983]. Alternatively, weakening might be due to successive intrusion events cracking and flexing the lithosphere [Diament and Baudry, 1987].

This hypothesis has the advantage that the unusual weakness in the Superswell area is related to the area's history. Given the Superswell's complex history of volcanism due at least in part to multiple hot spots, it is conceivable that the hot spots and possible Superplume have affected the lithosphere in the way required. The difficulty with this idea, however, is that it is hard to test. We can exclude general thermal weakening from the absence of its observable consequences for heat flow and depth and the absence of evidence from mantle xenoliths for high lithospheric temperatures [Tracy, 1980]. In contrast, a mechanically weakened lithosphere would be difficult to observe except in the flexural data that lead to the suggestion of this property. Although in principle earthquake depths might constrain the weakening [Wiens and Stein, 1983; Liu and Chase, 1989], the low level of seismicity makes such analysis difficult.

CONCLUSIONS

Because much of the data on the processes giving rise to midplate swells and volcanism are derived from depth and heat flow anomalies, the choice of the reference model with respect to which these anomalies are computed is significant. GDH1, a reference model which describes the depth and heat flow in older lithosphere reasonably well, thus is helpful in assessing anomalies. In particular, it indicates that heat flow in swell regions, which previously appeared anomalously high, is at best slightly higher than for lithosphere of the same age elsewhere. This observation tends to favor largely dynamic, rather than thermal, models for swell uplifts. It similarly excludes the possibility that the weak lithosphere inferred for the Superswell results from elevated temperatures.

More generally, these results bear out the desirability of

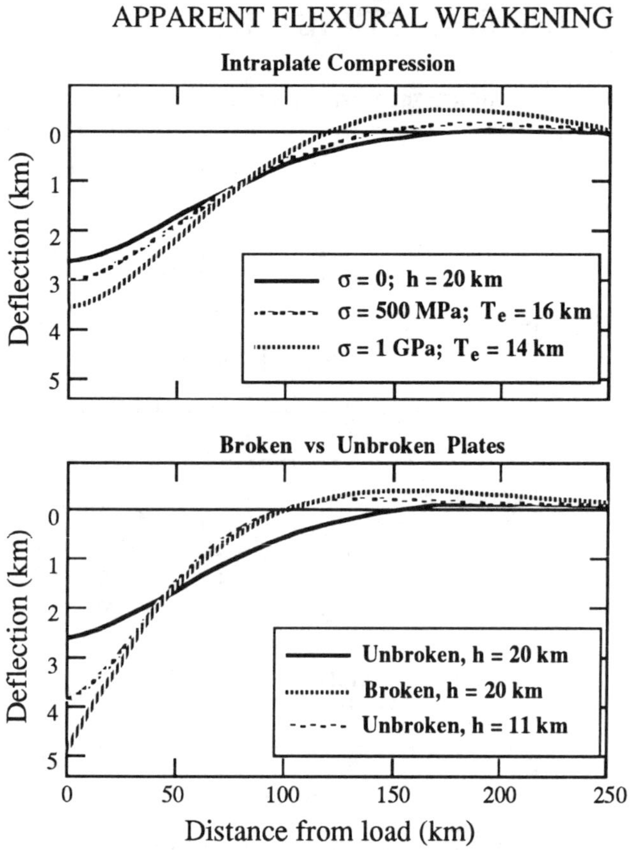

Fig. 20. Schematic illustration of two possible models for apparent weakening of the lithosphere, as inferred from flexural data. Top: Deflection of a 20 km thick lithosphere assuming different values of intraplate compression. Larger stresses reduce the effective elastic thickness, T_e. Bottom: Deflection of 20 km thick lithosphere, for unbroken and broken cases. The effective elastic thickness inferred from the flexural shape for the broken lithosphere would be 11 km.

developing improved reference models, both for comparison with observable data and for interpretation of the corresponding anomalies in temperature and physical properties in the lithosphere. It is worth noting that this situation is a common one in the earth sciences. For example, corresponding reference model artifacts occur in seismology, where an inappropriate velocity reference model can produce apparent anomalies [van der Hilst and Spakman, 1989].

Acknowledgments. Our thinking about midplate swells was influenced greatly by many pleasant discussions with Sy Schlanger. We learned much from his deep familiarity with a wide range of data, his interest in treating the data as constraints on large scale tectonic processes, and his geological intuition. For example, when we told him why we favored mechanical rather than thermal weakening of the Superswell lithosphere, Sy said he had long believed that seamounts weakened the plate as they erupted, and that their loads should not be treated "as though they were dropped from balloons." Sy was a fine scientist, a warm person, and a good friend.

We thank Aristeo Pelayo for assistance in preparing the figures, Richard Carlson, Paul Johnson, Norm Sleep, Tom Shoberg, and Mark Woods for helpful discussions, and Clem Chase and Cecily Wolfe for useful reviews. This research was supported by NSF grant OCE-9019572 and NASA grant NAG-5-1944.

REFERENCES

Anderson, R. N., and M. A. Hobart, The relation between heat flow, sediment thickness, and age in the eastern Pacific, *J. Geophys. Res.*, *81*, 2968–2989, 1976.

Anderson, R. N., and J. N. Skilbeck, Oceanic heat flow, in *Oceanic Lithosphere, The Sea, 7*, edited by C. Emiliani, pp. 489–523, Wiley-Interscience, New York, 1981.

Bergman, E. A., and S. C. Solomon, Earthquake source mechanisms from body waveform inversion and intraplate tectonics in the northern Indian Ocean, *Phys. Earth Planet. Inter.*, *40*, 1–23, 1985.

Bodine, J. H., M. S. Steckler, and A. B. Watts, Observations of flexure and the rheology of the oceanic lithosphere, *J. Geophys. Res.*, *86*, 3695–3707, 1981.

Buck, W. R., and E. M. Parmentier, Convection beneath young oceanic lithosphere: implications for thermal structure and gravity, *J. Geophys. Res.*, *91*, 1961–1974, 1986.

Calmant, S., The elastic thickness of the lithosphere in the Pacific Ocean, *Earth Planet. Sci. Lett.*, *85*, 277–288, 1987.

Calmant, S., and A. Cazenave, Anomalous elastic thickness of the oceanic lithosphere in the south-central Pacific, *Nature*, *328*, 236–238, 1987.

Calmant, S., J. Francheteau, and A. Cazenave, Elastic layer thickening with age of the oceanic lithosphere a tool for prediction of age of volcanoes or oceanic crust, *Geophys. J. Int.*, *100*, 59–67, 1990.

Castillo, P., The Dupal anomaly as a trace of the upwelling lower mantle, *Nature*, *336*, 667–670, 1988.

Cazenave, A., Thermal cooling of the lithosphere: new constraints from geoid height data, *Earth Planet. Sci. Lett.*, *70*, 395–406, 1984.

Chen, W.-P., and P. Molnar, Focal depths of intracontinental and intraplate earthquakes and their implications for the thermal and mechanical properties of the lithosphere, *J. Geophys. Res.*, *88*, 4183–4214, 1983.

Cloetingh, S., and R. Wortel, Regional stress field of the Indian plate, *Geophys. Res. Lett.*, *12*, 77–80, 1985.

Cloetingh, S., and R. Wortel, Stress in the Indo-Australian plate, *Tectonophysics*, *132*, 49–67, 1986.

Cochran, J. R., Variations in subsidence rates along intermediate and fast spreading mid-ocean ridges, *Geophys. J. R. astron. Soc.*, *87*, 421–454, 1986.

Colin, P., and L. Fleitout, Topography of the ocean floor: Thermal evolution of the lithosphere and interaction of deep mantle heterogeneities with the lithosphere, *Geophys. Res. Lett.*, *17*, 1961–1964, 1990.

Courtney, R. C., and M. Recq, Anomalous heat flow near the Crozet Plateau and mantle convection, *Earth Planet. Sci. Lett.*, *79*, 373–384, 1986.

Courtney, R. C., and R. S. White, Anomalous heat flow and geoid across the Cape Verde Rise: evidence for dynamic support from a thermal plume in the mantle, *Geophys. J. R. astron. Soc.*, *87*, 815–867, 1986.

Crough, S. T., Approximate solutions for the formation of the lithosphere, *Phys. Earth Planet. Inter.*, *14*, 365–377, 1977.

Crough, S. T., Thermal origin of mid-plate hot-spot swells, *Geophys. J. R. astron. Soc.*, *55*, 451–469, 1978.

Crough, S. T., Geoid anomalies across fracture zones and the thickness of the lithosphere, *Earth Planet. Sci. Lett.*, *44*, 224–230, 1979.

Darwin, C., *The Voyage of the Beagle (1975 reissue)*, J. M. Dent, London, 1845.

Davies, G. F., Ocean bathymetry and mantle convection, 1. Large-scale flow and hot spots, *J. Geophys. Res.*, *93*, 10,467–10,480, 1988.

Davies, G. F., and F. Pribac, Mesozoic seafloor subsidence and the Darwin Rise past and present, this volume.

Davis, E. E., and C. R. B. Lister, Fundamentals of ridge crest topography, *Earth Planet. Sci. Lett.*, *21*, 405–413, 1974.

Detrick, R. S., and S. T. Crough, Island subsidence, hot spots, and lithospheric thinning, *J. Geophys. Res., 83,* 1236–1244, 1978.

Detrick, R. S., R. P. Von Herzen, B. Parsons, D. Sandwell, and M. Dougherty, Heat flow observations on the Bermuda Rise and thermal models of midplate swells, *J. Geophys. Res., 91,* 3701–3723, 1986.

Detrick, R. S., R. S. White, R. C. Courtney, and R. P. Von Herzen, Heat flow on midplate swells, in *Handbook of seafloor heat flow,* edited by J. A. Wright and K. E. Louden, pp. 325–485, CRC Press, Inc., Boca Raton, Florida, 1989.

Diament, D., and N. Baudry, Structural trends in the Southern Cook and Austral archipelagoes (south central Pacific) based on an analysis of Seasat data: Geodynamic implications, *Earth Planet Sci. Lett., 85,* 427–438, 1987.

Driscoll, M. L., and B. Parsons, Cooling of the oceanic lithosphere - evidence from geoid anomalies across the Udintsev and Eltanin fracture zones, *Earth Planet. Sci. Lett., 88,* 289–307, 1988.

Duncan, R. A., and D. A. Clague, Pacific plate motion recorded by linear volcanic chains, in *The Ocean Basins and Margins, 7A: The Pacific Ocean,* edited by A. E. Nairn, F. G. Stehli and S. Uyeda, pp. 89–121, Plenum Press, New York, 1985.

Forsyth, D. W., The evolution of the upper mantle beneath mid-ocean ridges, *Tectonophysics, 38,* 89–118, 1977.

Goetze, C., and B. Evans, Stress and temperature in the bending lithosphere as constrained by experimental rock mechanics, *Geophys. J. R. astron. Soc., 59,* 463–478, 1979.

Hager, B. H., and R. J. O'Connell, Lithospheric thickening and subduction, plate motions and mantle convection, in *Physics of the Earth's Interior, Proc. Int. Sch. Phys. "Enrico Fermi"* edited by A. M. Dziewonski and E. Boschi, pp. 464–492, North Holland, Amsterdam, 1980.

Hart, S. R., A large-scale isotope anomaly in the Southern Hemisphere mantle, *Nature, 309,* 753–757, 1984.

Hart, S. R., Heterogeneous mantle domains: signatures, genesis and mixing chronologies, *Earth Planet. Sci. Lett., 90,* 273–296, 1988.

Hayes, D. E., Age-depth relationships and depth anomalies in the southeast Indian Ocean and south Atlantic Ocean, *J. Geophys. Res., 93,* 2937–2954, 1988.

Heestand, R. L., and S. T. Crough, The effect of hot spots on the oceanic age-depth relation, *J. Geophys. Res., 86,* 6107–6114, 1981.

Hetenyi, M., *Beams on Elastic Foundation,* The University of Michigan Press, Ann Arbor, 1974.

Jackson, E. D., and S. O. Schlanger, Regional syntheses, Line island chain, Tuamotu island chain, and Manihiki Plateau, Central Pacific Ocean, *Init. Repts. DSDP, 33,* 915–927, 1976.

Jarvis, G. T., and W. R. Peltier, Convection models and geophysical observations, in *Mantle Convection,* edited by W. R. Peltier, pp. 479–594, Gordon and Breach, New York, 1989.

Johnson, H. P., and R. L. Carlson, The variation of sea floor depth with age: a test of existing models based on drilling results, *Geophys. Res. Lett., 19,* 1971-1974, 1992.

Kaula, W. M., and R. J. Philips, Quantitative tests for plate tectonics on Venus, *Geophys. Res. Lett., 8,* 1187–1190, 1981.

Kirby, S. H., State of stress in the lithosphere: Inferences from the flow laws of olivine, *Pure and Applied Geophys., 115,* 245–258, 1977.

Kirby, S. H., Tectonic stresses in the lithosphere: Constraints provided by the experimental deformation of rocks, *J. Geophys. Res., 85,* 6353–6363, 1980.

Kirby, S. H., Rheology of the lithosphere, *Rev. Geophys. Space Phys., 21,* 1458–1487, 1983.

Langseth, M. G., X. Le Pichon, and M. Ewing, Crustal structure of the mid-ocean ridges, 5, Heat flow through the Atlantic Ocean floor and convection currents, *J. Geophys. Res., 71,* 5321–5355, 1966.

Langseth, M. G., A. H. Lachenbruch, and B. V. Marshall, Geothermal observations in the Arctic region, in *The Geology of North America: The Arctic Ocean Region, L,* edited by A. Grantz, L. Johnson and J. F. Sweeney, pp. 188–152, GSA, Boulder, Colorado, 1990.

Larson, R. L., Latest pulse of Earth: Evidence for a mid-Cretaceous superplume, *Geology, 19,* 547–550, 1991.

Larson, R. L., and S. O. Schlanger, Cretaceous volcanism and Jurassic magnetic anomalies in the Nauru Basin, western Pacific Ocean, *Geology, 9,* 480–484, 1981.

Lister, C. R. B., On the thermal balance of a mid-ocean ridge, *Geophys. J. R. astron. Soc., 26,* 515–535, 1972.

Lister, C. R. B., J. G. Sclater, E. E. Davis, H. Villinger, and S. Nagihara, Heat flow maintained in oceanic basins of great age: investigations in the north-equatorial west Pacific, *Geophys. J. Int., 102,* 603–630, 1990.

Liu, M., and C. G. Chase, Evolution of midplate hot spot swells: numerical solutions, *J. Geophys. Res., 94,* 5571–5584, 1989.

Liu, M., and C. G. Chase, Boundary layer model of mantle plumes with thermal and chemical diffusion and buoyancy, *Geophys. J. Int., 104,* 433–440, 1991.

Louden, K. E., D. O. Wallace, and R. C. Courtney, Heat flow and depth versus age for the Mesozoic northwest Atlantic Ocean: results from the Sohm abyssal plain and implications for the Bermuda Rise, *Earth Planet. Sci. Lett., 83,* 109–122, 1987.

Louden, K. E., Marine heat flow data listing, Appendix B, in *Handbook of seafloor heat flow,* edited by J. A. Wright and K. E. Louden, pp. 325–485, CRC Press, Inc., Boca Raton, Florida, 1989.

Ludwig, W. J., and R. E. Houtz, *Isopach map of sediments in the Pacific Ocean Basin and Margina Sea Basins,* American Association of Petroleum Geologists, 1979.

Marty, J. C., A. Cazenave, and B. Lago, Geoid anomalies across Pacific fracture zones, *GJ, 93,* 1–23, 1988.

Marty, J. C., and A. Cazenave, Regional variations in subsidence rate of oceanic plates: a global analysis, *Earth Planet. Sci. Lett., 94,* 301–315, 1989.

McAdoo, D. C., and D. T. Sandwell, Folding of oceanic lithosphere, *J. Geophys. Res., 90,* 8563–8569, 1985.

McKenzie, D. P., Some remarks on heat flow and gravity anomalies, *J. Geophys. Res., 72,* 6261–6273, 1967.

McKenzie, D. P., and M. J. Bickle, The volume and composition of melt generated by extension of the lithosphere, *J. Petrol., 29,* 625–679, 1988.

McNutt, M. K., and H. W. Menard, Lithospheric flexure and uplifted atolls, *J. Geophys. Res., 83,* 1206–1212, 1978.

McNutt, M. K., Temperature beneath midplate swells: the inverse problem, in *Seamounts, Islands, and Atolls, Geophysical Monograph #43* edited by B. H. Keating, P. Fryer, R. Batiza and G. W. Boehlert, pp. 123–132, American Geophysical Union, Washington, D.C., 1987.

McNutt, M. K., and K. M. Fisher, The South Pacific Superswell, in *Seamounts, Islands, and Atolls, Geophysical Monograph #43* edited by B.H. Keating, P. Fryer, R. Batiza and G. W. Boehlert, pp. 25–34, American Geophysical Union, Washington, D.C., 1987.

McNutt, M. K., and A. V. Judge, The superswell and mantle dynamics beneath the South Pacific, *Science, 248,* 969–975, 1990.

McNutt, M. K., E. L. Winterer, W. W. Sager, J. H. Natland, and G. Ito, The Darwin Rise: A Cretaceous Superswell?, *Geophys. Res. Lett., 17,* 1101–1104, 1990.

McNutt, M., D. Caress, J. Mutter, and R. Detrick, Thickness of the elastic plate beneath the Marquesas islands: why is it so much greater than values elsewhere in French Polynesia?, *Eos Trans. AGU, 72 (supplement),* 437, 1991.

Menard, H. W., *Marine Geology of the Pacific,* McGraw-Hill, New York, 1964.

Menard, H. W., Darwin reprise, *J. Geophys. Res., 89,* 9960–9968, 1984.

Menard, H. W., and M. McNutt, Evidence for and consequences of thermal rejuvenation, *J. Geophys. Res., 87,* 8570–8580, 1982.

Morgan, W. J., Deep mantle convection plumes and plate motions, *Am. Assoc. Petrol. Geol. Bull., 56,* 203–213, 1972.

Natland, J. H., and E. Wright, Magmatic lineages and mantle sources of Cretaceous seamounts in the Central Pacific, *Eos Trans. AGU, 65,* 1075–1076, 1984.

Nishimura, C. E., and D. W. Forsyth, Anomalous Love-wave phase velocities in the Pacific: sequential pure-path and spherical harmonic inversion, *Geophys. J. R. astron. Soc., 81,* 389–407, 1985.

Parmentier, E. M., and W. F. Haxby, Thermal stresses in the oceanic lithosphere: evidence from geoid anomalies at fracture zones, *J. Geophys. Res., 91,* 7193–7204, 1986.

Parsons, B., and J. G. Sclater, An analysis of the variation of ocean floor bathymetry and heat flow with age, *J. Geophys. Res., 82,* 803–827, 1977.

Parsons, B., and D. P. McKenzie, Mantle convection and the thermal structure of the plates, *J. Geophys. Res., 83,* 4485–4496, 1978.

Parsons, B., and F. M. Richter, Mantle convection and the oceanic lithosphere, in *Oceanic Lithosphere, (The Sea, vol. 7),* edited by C. Emiliani, pp. 73–117, Wiley-Interscience, New York, 1981.

Parsons, B., and S. Daly, The relationship between surface topography, gravity anomalies, and temperature structure of convection, *J. Geophys. Res., 88,* 1129–1144, 1983.

Renkin, M. L., and J. G. Sclater, Depth and age in the North Pacific, *J. Geophys. Res., 93,* 2919–2935, 1988.

Robinson, E. M., B. Parsons, and M. Driscoll, The effect of a shallow low-viscosity zone on the mantle flow, the geoid anomalies and geoid and depth-age relationships at fracture zones, *Geophys. J., 93,* 25–43, 1988.

Robinson, E. M., and B. Parsons, Effect of a shallow low-viscosity zone on the formation of midplate swells, *J. Geophys. Res., 93,* 3144–3156, 1988.

Sandwell, D. T., and G. Schubert, Geoid height-age relation from SEASAT altimeter profiles across the Mendocino fracture zone, *J. Geophys. Res., 87,* 3949–3958, 1982.

Sato, H., I. S. Sacks, and T. Murase, Use of laboratory velocity data for estimating temperature and partial melt fraction in the low velocity zone: comparison with heat flow and electrical conductivity studies, *J. Geophys. Res., 94,* 5689–5704, 1989.

Schlanger, S. O., H. C. Jenkyns, and I. Premoli-Silva, Volcanism and vertical tectonics in the Pacific Basin related to global Cretaceous transgressions, *Earth Planet. Sci. Lett., 52,* 435–449, 1981.

Schlanger, S. O., and I. Premoli-Silva, Tectonic, volcanic, and paleogeographic implications of redeposited reef faunas of Late Cretaceous and Tertiary age from the Nauru Basin and the Line Islands, *Initial Rep. Deep Sea Drill. Proj., 61,* 817–828, 1981.

Schubert, G., C. Froidevaux, and D. A. Yuen, Oceanic lithosphere and asthenosphere: Thermal and mechanical structure, *J. Geophys. Res., 81,* 3525–3540, 1976.

Sclater, J. G., and L. Wixon, The relation between depth and age and heat flow and age in the western North Atlantic, in *The Geology of North America: The Western North Atlantic Region, M,* edited by P. R. Vogt and B. E. Tucholke, pp. 257–270, The Geological Society of America, 1986.

Sleep, N. H., Hot spots and mantle plumes: some phenomenology, *J. Geophys. Res., 95,* 6715–6736, 1990.

Sleep, N. H., Hot spots and mantle plumes, *Ann. Rev. Earth Planet. Sci., 20,* 19–43, 1992.

Sleep, N. H., and B. F. Windley, Archean plate tectonics: constraints and inferences, *J. Geol., 90,* 363–379, 1982.

Smith, W. H. F., H. Staudigel, A. B. Watts, and M. S. Pringle, The Magellan Seamounts: Early Cretaceous record of the South Pacific isotopic and thermal anomaly, *J. Geophys. Res., 94,* 10,501–10,523, 1989.

Smith, W. H. F., Marine geophysical studies of seamounts in the Pacific ocean basin, Ph.D. dissertation, Columbia University, New York, 1990.

Stein, C., and D. Abbott, Heat flow constraints on the South Pacific Superswell, *J. Geophys. Res., 96,* 16,083–16,100, 1991.

Stein, C. A., S. A. P. L. Cloetingh, and M. J. R. Wortel, Seasat-derived gravity constraints on stress and deformation in the northeastern Indian Ocean, *Geophys. Res. Lett., 16,* 823–826, 1989.

Stein, C. A., and S. Stein, A model for the global variation in oceanic depth and heat flow with lithospheric age, *Nature, 359,* 123–129, 1992.

Stein, S., S. Cloetingh, D. Wiens, and R. Wortel, Why does near-ridge extensional seismicity occur preferentially in the Indian Ocean?, *Earth Planet. Sci. Lett., 82,* 107–113, 1987.

Stein, S., and E. A. Okal, Seismicity and tectonics of the Ninetyeast Ridge area: Evidence for internal deformation of the Indian plate, *J. Geophys. Res., 83,* 2233–2245, 1978.

Tracy, R. J., Petrology and genetic significance of an ultramafic xenolith suite from Tahiti, *Earth Planet. Sci. Lett., 48,* 80–96, 1980.

Turcotte, D. L., Properties of the lithosphere and asthenosphere deduced from geoid observations, in *Space Geodesy and Geodynamics,* edited by A. J. Anderson and A. Cazenave, pp. 451–475, 1986.

van der Hilst, R. D., and W. Spakman, Importance of the reference model in linearized tomography and images of subduction below the Caribbean plate, *Geophys. Res. Lett., 16,* 1093–1096, 1989.

Vogt, P. R., On the applicability of thermal conduction models to midplate volcanism: comments on a paper by Gass et al., *J. Geophys. Res., 86,* 950–960, 1981.

Von Herzen, R. P., R. S. Detrick, S. T. Crough, D. Epp, and U. Fehn, Thermal origin of the Hawaiian swell: Heat flow evidence and thermal models, *J. Geophys. Res., 87,* 6711–6723, 1982.

Von Herzen, R. P., M. J. Cordery, R. S. Detrick, and C. Fang, Heat flow and the thermal origin of hot spot swells: the Hawaiian swell revisited, *J. Geophys. Res., 94,* 13,783–13,799, 1989.

Watts, A. B., J. H. Bodine, and M. S. Steckler, Observations of flexure and the state of stress in the oceanic lithosphere, *J. Geophys. Res., 85,* 6369–6376, 1980.

Wessel, P., and W. Haxby, Thermal stress, differential subsidence, and flexure at oceanic fracture zones, *J. Geophys. Res., 95,* 375–391, 1990.

Wiens, D. A., and S. Stein, Age dependence of oceanic intraplate seismicity and implications for lithospheric evolution, *J. Geophys. Res., 88,* 6455–6468, 1983.

Winterer, E. L., Sedimentary facies and plate tectonics of the equatorial Pacific, *AAPG, 57,* 265–282, 1973.

Winterer, E. L., Anomalies in the tectonic evolution of the Pacific, in *The Geophysics of the Pacific Ocean basin and its Margin, 19,* edited by G. H. Sutton et al., pp. 269–280, AGU, Washington, D. C., 1976.

Wolery, T. J., and N. H. Sleep, Hydrothermal circulation and geochemical flux at mid-ocean ridges, *J. Geol., 84,* 249–275, 1976.

Wolfe, C. J., and M. K. McNutt, Compensation of Cretaceous seamounts of the Darwin Rise, Northwest Pacific Ocean, *J. Geophys. Res., 96,* 2363–2374, 1991.

Woods, M. T., J.-J. Leveque, and E. A. Okal, Two-station measurements of Rayleigh wave group velocity along the Hawai'ian swell, *Geophys. Res. Lett., 18,* 105–108, 1991.

Wortel, M. J. R., M. J. N. Remkes, R. Govers, S. A. P. L. Cloetingh, and P. Th. Meijer, Dynamics of the lithosphere and the intraplate stress field, *Phil. Trans. R. Soc. Lond., Ser. A, 337,* 111–126, 1991.

Wysession, M. E., E. A. Okal, and K. L. Miller, Intraplate seismicity of the Pacific basin, *Pure Appl. Geophys., 135,* 261–359, 1991.

Zhu, A., and D. A. Wiens, Thermoelastic stress in oceanic lithosphere due to hot spot reheating, *J. Geophys. Res., 96,* 18,323–18,334, 1991.

C.A. Stein, Department of Geological Sciences, University of Illinois at Chicago, Chicago, IL 60680.

S. Stein, Department of Geological Sciences, Northwestern University, Evanston, IL 60208.

Cretaceous Volcanic Sequences And Jurassic Oceanic Crust In The East Mariana And Pigafetta Basins Of The Western Pacific

L. J. ABRAMS AND R. L. LARSON

Graduate School of Oceanography, University of Rhode Island
Narragansett, RI

T. H. SHIPLEY

Institute for Geophysics, University of Texas
Austin, TX

Y. LANCELOT

Laboaratoire de Geologie du Quaternaire, CNRS-Luminy
Marseille, France

We report the results of the first regional multichannel seismic studies of both the East Mariana and Pigafetta basins of the western Pacific which imaged oceanic crust, mapped the overlying strata and identified a site where Ocean Drilling Program (ODP) Leg 129 recovered Jurassic sediments and oceanic crust for the first time in the Pacific. The calibration of our seismic surveys and the extrapolation of drilling results throughout these oldest Pacific basins indicate that the flat-lying, high-amplitude reflection known as Horizon B corresponds to the top of middle Cretaceous flows/sills that overlie Jurassic/ Early Cretaceous sediments and oceanic crust throughout an area ~ 500,000 km^2 in the East Mariana Basin and the southeast Pigafetta Basin. Seismic refraction and reflection data allow an estimate of 400 m for the maximum thickness of these flows/sills. A distinct change in the reflection character of Horizon B from flat-lying and "smooth" to higher relief and "rough" observed in the southeast Pigafetta Basin marks the change from middle Cretaceous sills/flows overlying Jurassic oceanic crust to oceanic crust without a massive igneous overburden. Jurassic oceanic crust and overlying Upper Jurassic - Lower Cretaceous sediments unquestionably exist at ODP Site 801 in the Pigafetta Basin and are intermittently covered by sills/flows over the 550 km between Site 801 and Site 800. The Ogasawara Fracture Zone, which separates the East Mariana Basin and Pigafetta Basin, acted as a locus for middle Cretaceous volcanic activity as indicated by the coincident alignment and age of seamounts. Intrabasin morphotectonic features other than seamounts which could represent a possible source for the extensive cover of deep sea sills and flows are not obvious on seismic or bathymetric records. The ubiquitous presence of middle-to-Late Cretaceous volcanogenic turbidites and widespread occurrences of middle Cretaceous deep sea sills/flows throughout these basins indicate a pulse in voluminous off-ridge volcanic activity beginning in late Barremian/early Aptian.

INTRODUCTION

Magnetic anomaly sequences indicate that the world's oldest in situ ocean crust lies centered in the far western Pacific [Larson and Chase, 1972; Hilde et al., 1976; Larson, 1976]. An area, approximately the size of the contiguous United States, enclosed by the M17 isochron is inferred to contain oceanic crust and sediment of Jurassic age (Figure 1). Prior to ODP Leg 129 (Nov-Jan 1989-90), no Jurassic material had been sampled within this region by eight scientific drilling legs since 1969 and numerous dredging expeditions beginning in the 1950's. Instead the oldest material recovered has almost invariably been of mid- to Late Cretaceous age and often volcanic in origin. The results from the earliest drilling expeditions led Winterer [1976] to conclude that the recovered basalts probably buried older Early Cretaceous-Jurassic sediments and oceanic crust in the deep portions of western Pacific basins. This hypothesis represented one aspect of Menard's [1964] "Darwin Rise" which calls for the emplacement of abyssal flood basalts in Cretaceous time in order to account for the burial of abyssal hills surrounding Cretaceous seamounts.

The Mesozoic Pacific: Geology, Tectonics, and Volcanism
Geophysical Monograph 77

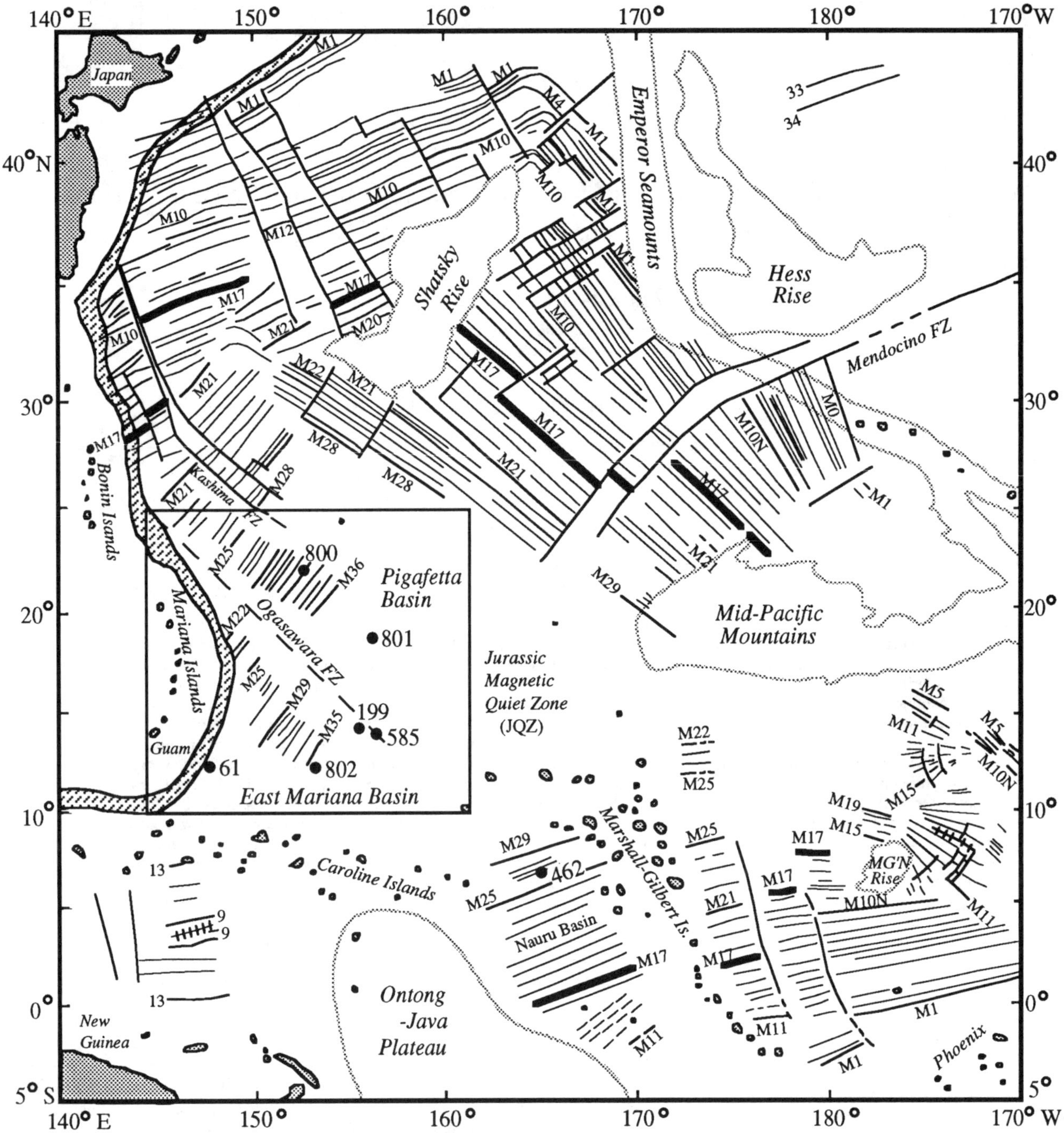

Fig. 1. Magnetic lineations of the western Pacific compiled by R. L. Larson from Hilde et al. [1976], Larson [1976], Weissel and Anderson [1978], Mammerickx and Sharman [1988], Tamaki and Larson [1988], Handschumacher et al. [1988], and Sager et al. [1988], Nakanishi et al. [1989]. Magnetic isochron M17 is highlighted and approximates the Cretaceous - Jurassic boundary. Dots locate selected DSDP Sites 199, 585, 462 and all of ODP Leg 129 Sites 800, 801 and 802. Boxed region is shown in Figures 2, 4 and, 7.

The middle Cretaceous volcanism affecting these Jurassic basins manifested itself in several ways. Volcanism is associated with chains of seamounts and islands such as the Magellan and Marcus-Wake seamounts, which were near or above sea level by early Aptian (~124 Ma) [Moberly and Schlanger et al., 1986; Lancelot and Larson et al., 1990; Winterer et al., this volume], the Mid-Pacific Mountains of Barremian to Campanian age (~132-74 Ma) [Hamilton, 1956;

Winterer and Ewing et al., 1973; Thiede and Vallier et al., 1981], and the Marshall/Gilbert Seamounts of at least Albian to Paleocene age (112-56.5 Ma) [Lincoln et al., this volume; Haggerty and Primoli Silva, et al., in press]. There is also excess mid-plate volcanism in the form of huge oceanic plateaus such as the Ontong-Java Plateau of earliest Aptian age (~124 Ma) [Kroenke and Berger et al., 1991; Tarduno et al., 1991]. Finally, the most poorly mapped and dated mid-plate volcanism in the form of deep-sea sills and flows (Nauru Basin - Larson and Schlanger et al., 1981; Central Pacific Basin - Winterer and Ewing et al., 1973; East Mariana and Pigafetta basins - Lancelot and Larson et al., 1990). The extent of this mid - to Late Cretaceous deep sea volcanism, its relationship to numerous seamounts, guyots, underlying sediments, and oceanic crust are of primary concern in this study.

Seismic reflection and refraction studies of the East Mariana Basin (EMB) and Pigafetta Basin (PB) presented in this paper show images of oceanic crust, define its seismic structure, map the overlying strata and identify a site where the D/VJOIDES Resolution was able to recover Jurassic oceanic crust during ODP Leg 129 [Abrams et al., 1988; Lancelot and Larson et al., 1990]. The successful completion of Leg 129, resulting in the first and only holes to penetrate igneous basement (middle Cretaceous basalt or Jurassic oceanic crust) in the EMB and PB, allowed the calibration of this extensive MCS and SCS data set, the extrapolation of drilling results throughout these basins and comparisons with other western Pacific basins of Jurassic age (e.g. Nauru Basin see Shipley et al., this volume). This basinwide correlation permits a physical and age determination of the flat-lying, high-amplitude reflection(s) known as "Horizon B" or "reverberant layer", observed in extensive areas of the western Pacific including the EMB and PB. These results enable us to place geologic age and volumetric bounds on deep sea volcanism in the EMB and PB and to discuss its relationship with the worldwide occurrence of voluminous middle Cretaceous volcanism.

TECTONIC SETTING

The bathymetric map of this region (Figure 2) reveals a huge area (~ 1 x 10^6 km^2) at depths greater than 5500 m, which we consider the EMB and PB. The EMB is bounded to the west-northwest by the Mariana Trench, to the southwest by the Caroline Ridge, and to the northeast by the northwest-trending Ogasawara Fracture Zone and Magellan Seamounts [Tamaki et al., 1987; Handschumacher et al., 1988] (Figures 1 and 2). The PB is bounded to the northeast by the Kashima Fracture Zone and Marcus-Wake Seamounts and to the southeast by the Ogasawara Fracture Zone and Magellan Seamounts and extends to the northwest, where it terminates at the Bonin Trench.

In this paper age is assigned according to the Harland et al. [1990] geologic time scale. The younger parts of the EMB toward the northwest are believed to be Kimmeridgian to Oxfordian (anomalies M22 to M25, ~150 - 155 Ma) and just seaward of the Bonin Trench the PB may be as young as Valanginian-Tithonian (M13-M17, ~139-145 Ma). Within our survey area the central parts of these basins are extrapolated to be Oxfordian to Callovian (anomalies M29 to M36, ~157-161 Ma) and toward the southeast, within the so-called Jurassic magnetic quiet zone (JQZ), the age of the crust may be as old as Bathonian to Bajocian (Middle Jurassic, ~161-174 Ma) [Tamaki et al.,1987; Handschumacher et al., 1988; Nakanishi et al., 1989; and Nakanishi et al., 1992]. These inferences were confirmed by the 166.8 +/- 4.5 m.y. age and geochemical attributes of the tholeiitic basalts recovered at Site 801 in the JQZ that indicate they represent Jurassic oceanic crust [Floyd et al., 1991; Pringle, 1992a; Castillo et al., 1992]. The exact nature of the low field uncorrelated magnetic anomalies in the JQZ remains unknown, but the equivalent time interval in land sections was found to be a period of extremely frequent magnetic reversals by Steiner et al. [1987]. A crustal age of ~167 Ma at Site 801 results in pre-M25 half-spreading rates of approximately 80 mm/yr assuming constant rates of spreading and an M25 date of 155 Ma [Harland et al., 1990]. This spreading rate is comparable to the fastest half-spreading rates presently documented on the East Pacific Rise (80 - 90 mm/yr, Macdonald, 1982).

PREVIOUS STUDIES

Western Pacific Seismic Stratigraphy

The seismic stratigraphy of large portions of the western Pacific including the PB and EMB consists of four units originally defined by Ewing et al. [1968]: (1) an upper transparent layer, (2) an upper opaque layer, (3) a lower transparent layer, and (4) Horizon B. This paper will concentrate primarily on the Horizon B reflection and sub-Horizon B velocity structure throughout the EMB and PB. The stratigraphy overlying Horizon B is discussed in Abrams et al. [1992].

Horizon B has been referred to as the "deep opaque layer" [Heezen and MacGregor et al., 1973], as the "reverberant layer" by Houtz et al. [1970] and Houtz and Ludwig [1979], and as "acoustic basement". Horizon B has been characterized from observations of SCS analog airgun profiles by the presence of a prominent zone of flat-lying ("smooth"), high-amplitude, and closely spaced reflections. Houtz and Ludwig [1979] completed an extensive review of the areal extent, thickness and velocity of the reverberant layer/Horizon B seismic facies throughout the western Pacific and speculated (as did Ewing et al., 1968) that this layer is composed of highly stratified calcareous or redeposited volcaniclastic sediments that overlie oceanic crust. Drilling in the western Pacific prior to Leg 129 has demonstrated that Horizon B can be produced by the impedance contrast between sediments and volcanic sills or flows (e.g. Nauru Basin-DSDP Site 462) and, in some cases, the sediment/oceanic crust interface in areas interpreted to contain particularly smooth crust. It is now recognized that the reverberant nature of Horizon B in airgun profiles is usually the result of a trailing bubble pulse oscillation reflecting from a high impedance boundary implying that the thickness or stratification inferred for this layer is an artifact of the airgun source and gain setting in the analog recording [Shipley et al., 1983]. This characteristic of

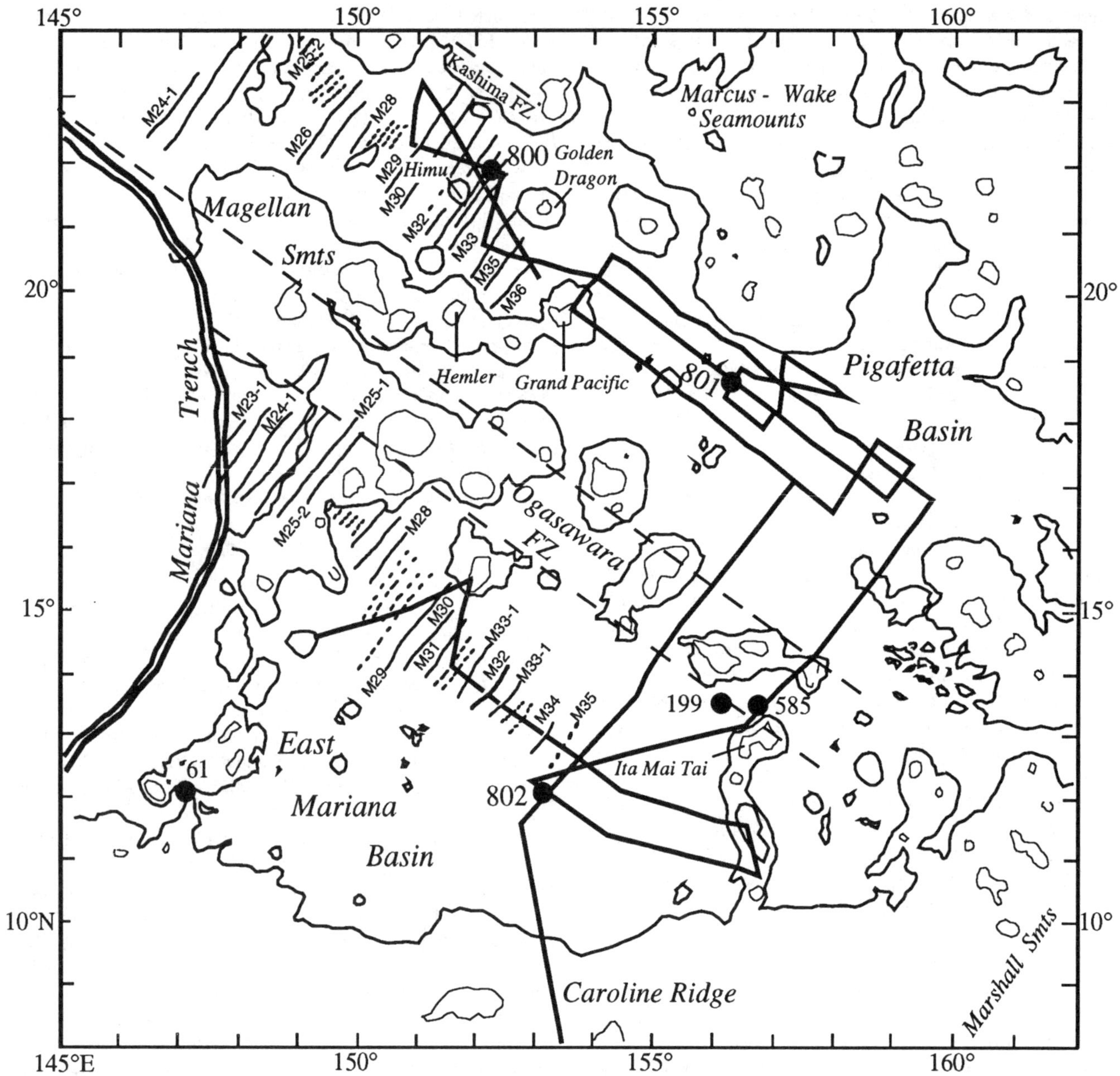

Fig. 2. Bathymetry of the central western Pacific modified from Brenner and Angell [1992] with location of selected DSDP sites and all of ODP Leg 129 sites. The 5500 m contour is highlighted and the 3000 m contour is shown as a lighter line. Magnetic anomalies modified from Tamaki et al. [1987] and Handschumacher et al. [1988]. The bold solid line indicates coverage of the MESOPAC II and FM35-12 seismic surveys. The broad area inferred for the location of the Ogasawara Fracture Zone is bounded by the dashed line.

the airgun source is now routinely compensated for through the use of tuned airgun arrays and/or deconvolution techniques and the use of waterguns, which provide a more implosive and bubble pulse-free source signature resulting in a high-resolution record with less processing effort [Hutchinson and Detrick, 1984].

Leg 129 Site Review

Sites drilled during ODP Leg 129 penetrated and recovered material correlated to all four general seismic units of Ewing et al. [1968], including Horizon B. The seismic stratigraphy at ODP Leg 129 Sites 800, 801, and 802 is summarized in Figure 3 (from Abrams et al. 1992).

Horizon B at all three Leg 129 Sites corresponds to the top of intrusive and/or extrusive dolerite/basalt at approximately 500 mbsf. In the northwest PB, at Site 800, dolerite sills radiometrically dated at 126.1 +/- 0.6 Ma intrude Berriasian (~140-145 Ma) pelagic sediment [Lancelot and Larson et al., 1990; Pringle, 1992]. Oceanic crust at Site 800, predicted to

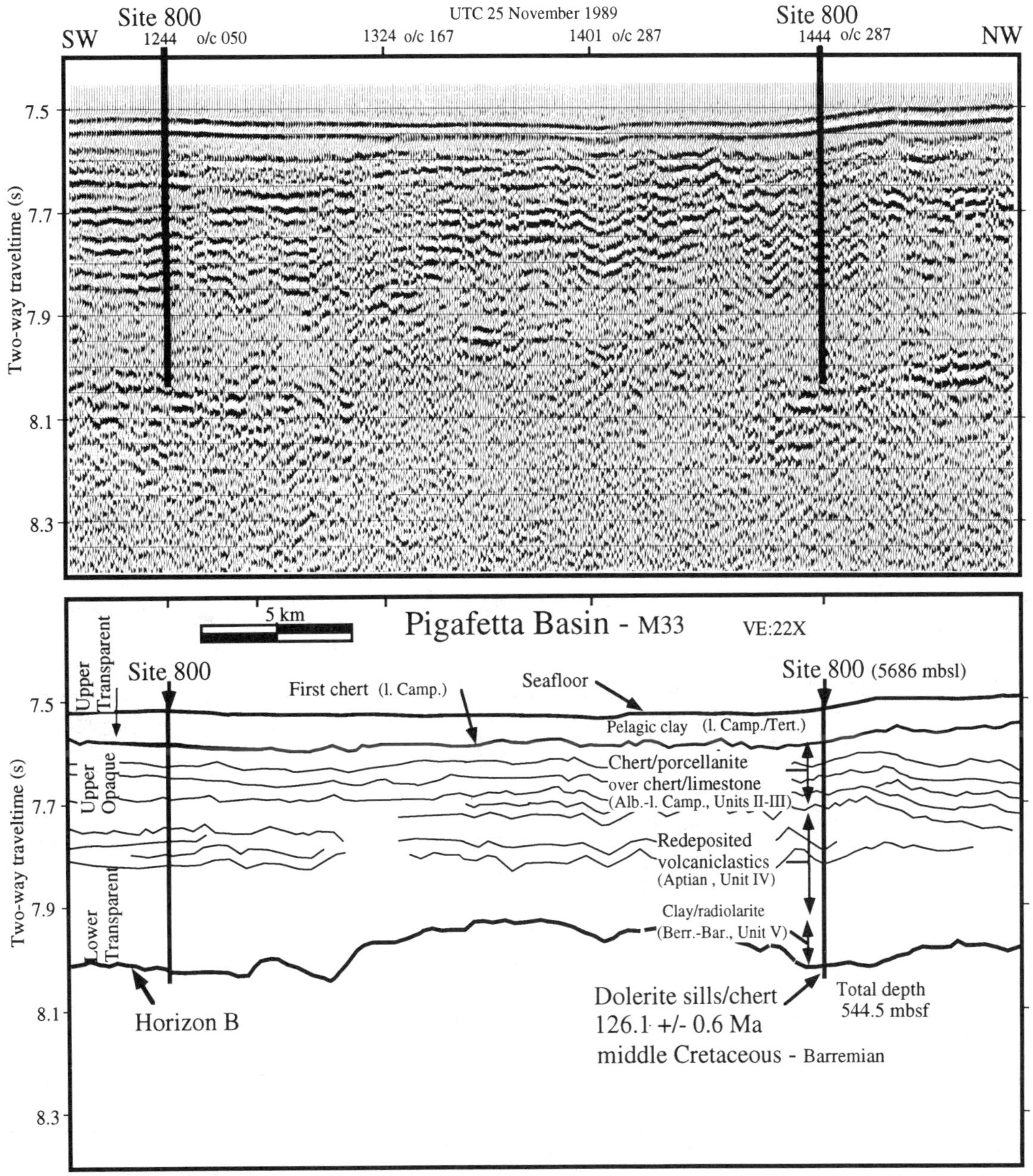

Fig. 3a.

Fig. 3a-c. Summary of seismic stratigraphy at ODP Leg 129 Sites 800 (a), 801 (b), 802 (c) showing generalized correlations with lithologic units, ages and depths, with emphasis on key reflective horizons such as igneous basement and shallowest chert ("first" chert) [Lancelot and Larson et al.,1990, Pringle, 1992]. Site 800 imaged with SCS waterguns (two 80 in^3) obtained on approach to Site 800 during Leg 129; seismic profiles for closest point of approach (CPA) to Sites 801 and 802 during MESOPAC II. These single channel watergun records were processed and displayed with the following parameters: water-velocity F/K migration, mute, 2-trace mix, band-pass filter 25-100 Hz, 500 msec AGC, and vertical exaggeration ~22X at 1.5 km/s. O/C = on course, UTC = Universal Time Code.

be ~ 160 Ma (M33), was not recovered. Horizon B at Site 802, in the central EMB, also correlates to a middle Cretaceous volcanic surface. At this location pillow and flow basalts dated at 114.6 +/- 3.2 Ma were recovered in an area where oceanic crust is predicted to be >161 Ma (M35). Site 801, in the JQZ of the PB is the only location where oceanic crust was recovered. The radiometric age (166.8 +/- 4.5 Ma, Middle Jurassic) of the tholeiitic basalt flows is almost exactly that predicted from linear extrapolation of post M25 magnetic lineations. Finally, thick sequences of middle to Late

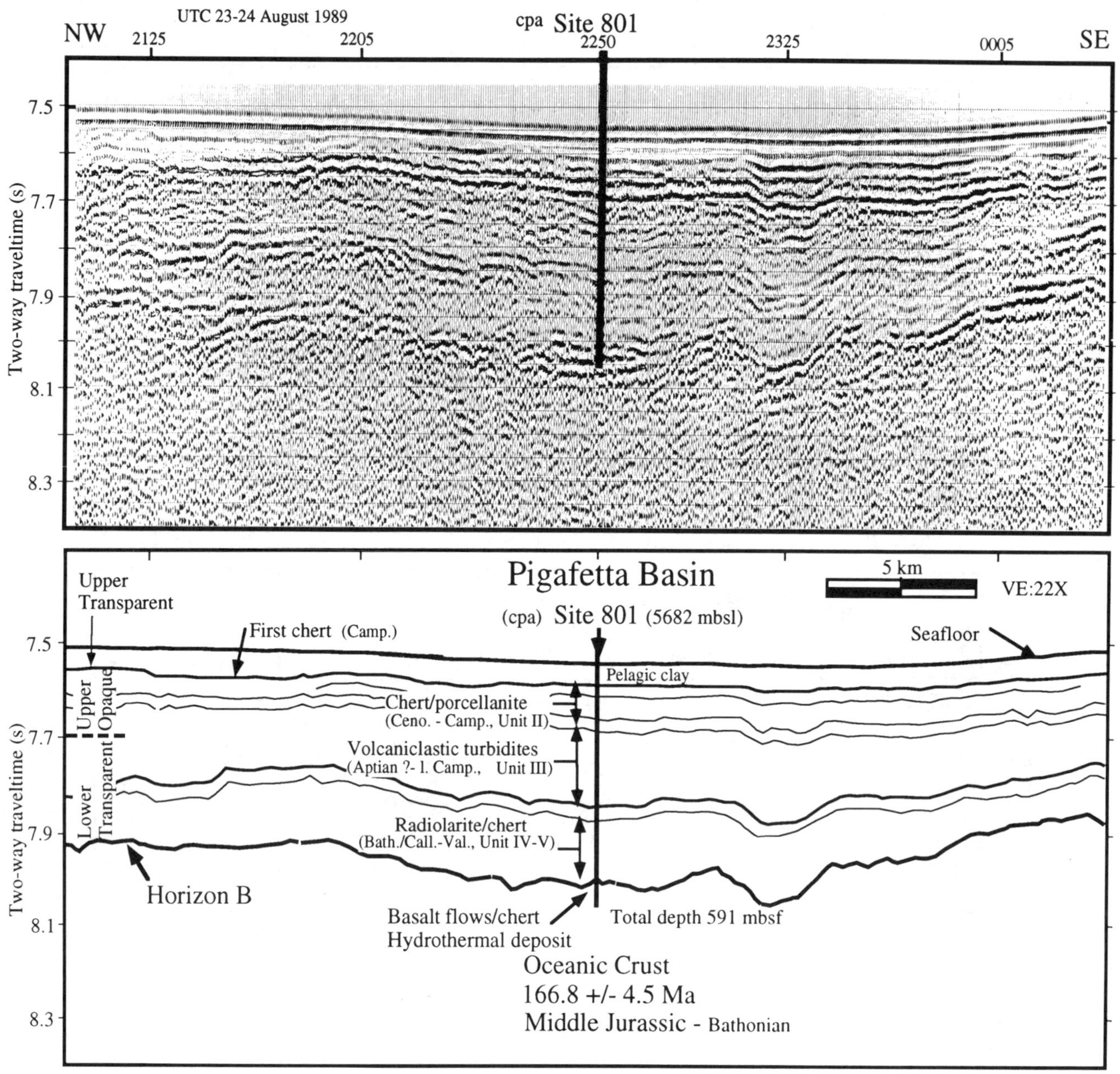

Fig. 3b.

Cretaceous volcanogenic turbidites were penetrated at all three Leg 129 sites as well as at DSDP Site 585, indicating that edifice building and emplacement of deep sea sills/flows (e.g., Sites 800, 802) were essentially synchronous [Lancelot and Larson et al., 1990]

DATA ACQUISITION AND PROCESSING

In November - December 1987 the R/V Fred H. Moore expedition FM35-12 collected over 3900 km of multichannel seismic (MCS) data using a 3200-m-long receiving array consisting of 96 hydrophone groups each 33.33 m in length (Figures 2 and 4). The seismic data were sampled at 4 ms, demultiplexed and recorded on a GUS 4200 Marine System. The Moore's seismic source consisted of various combinations of airguns fired at approximately 13.8 MPa (2000 psi) every 20 s. The maximum airgun volume achieved was 84.15 liters (5135 in^3) but the mean airgun volume during the 17 days of MCS acquisition was 40.65 liters (2481 in^3). Refraction data collected by seven long-range sonobuoys were also digitally recorded during this regional survey (sonobuoys 7, 8, 18, 20, 21, 22, 23). The MESOPAC II MCS survey conducted with the N/O Le Suroit during August - September 1989 focused on

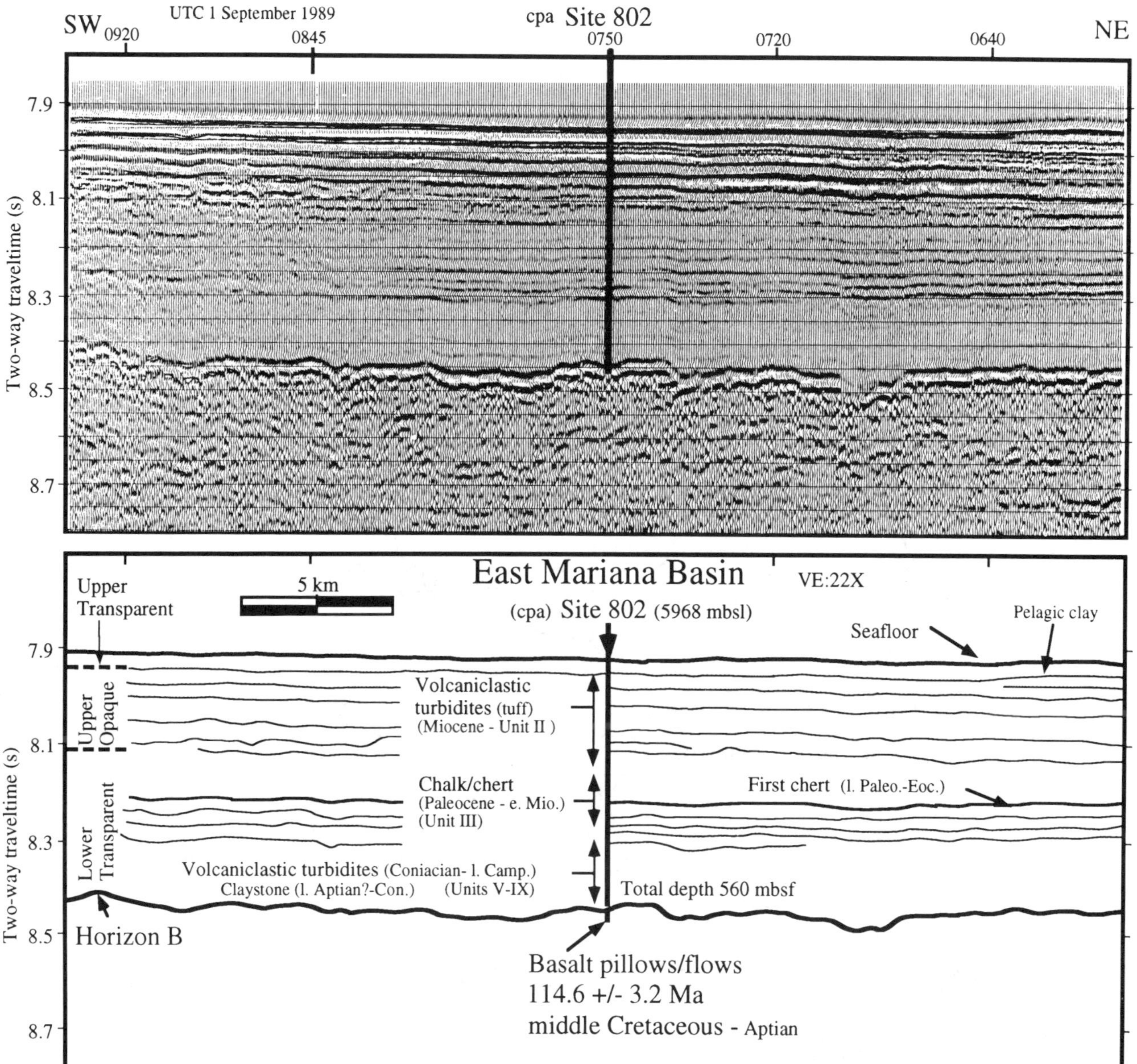

Fig. 3c.

potential drilling targets first identified on FM35-12 as well as investigating the northern PB within the oldest portion of the M-sequence lineation pattern. Only processed single channel data from MESOPAC II are used in this study. The MESOPAC II expedition also recorded approximately 3900 km of MCS data using an array of four to six 1.31 liter (80 in^3) waterguns fired every 13 s during vertical incident profiling and two 16.39 liter (1000 in^3) airguns fired every 31 s while digitally recording refraction data from six long-range sonobuoys (sonobuoys 1m, 2m, 3m, 4m, 6m, 8m).

During both cruises the ships maintained a speed of approximately 5 to 6 knots and navigated primarily by transit satellite but also by NAVSTAR global positioning system (GPS) with coverage of approximately 5 hours a day. Shot spacing was calculated for the FM35-12 data using final navigation along individual lines, and for MESOPAC II SCS watergun data was estimated at 40 m per shot. FM35-12 CDP gathers were created with 66.66-m bin size, 33.33-m hydrophone group spacing with variable shot spacing (~50 m/shot) resulting in 100 to 120 fold gathers. Stacking velocities from the seafloor to Horizon B were derived from semblance analysis of four adjacent CDP gathers at selected areas along each line and sub-Horizon stacking velocities were derived from sonobuoy results. Shot spacing for FM3512 and

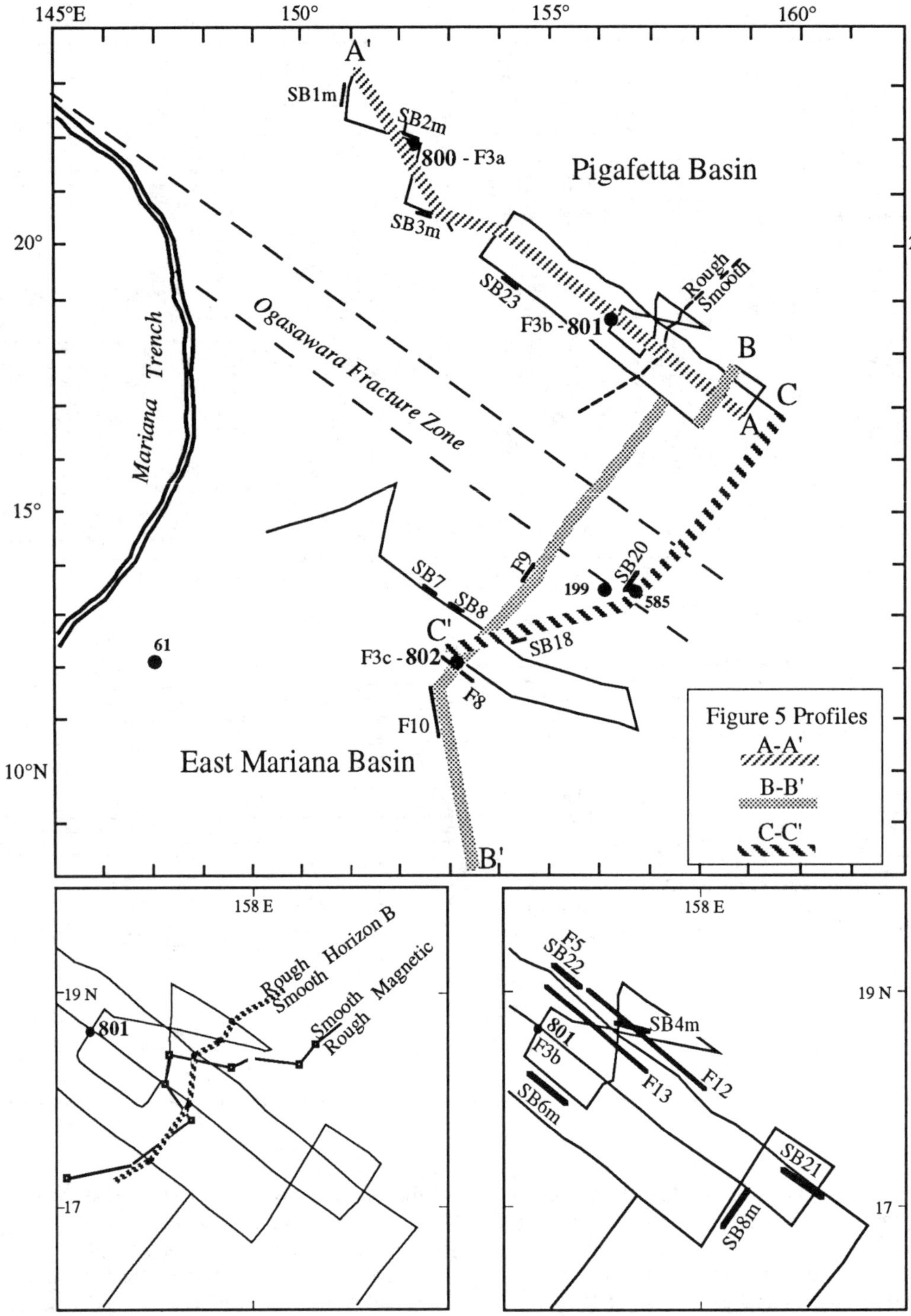

Fig. 4. Solid line indicates coverage of the FM35-12 and MESOPAC II regional seismic surveys with location of sonobuoy stations and transects in subsequent figures (heavy solid lines marked SB# and F# respectively). The endpoints of composite profiles shown in Figure 6 are marked A-A', B-B' and C-C' and highlighted as shown in the legend. The boxed region is enlarged and shown in two inset figures. The spatial relationship of rough/smooth Horizon B and the magnetic boundary as identified by Handschumacher et al. [1988] is shown in the left inset and the right inset more clearly displays track geometry and figure/sonobuoy locations. Dots show DSDP/ODP site locations.

MESOPAC II sonobuoys was 50-70 m and 80-90 m, respectively.

Sonobuoy Methods and Analyses

Travel time-range (T/X) data are used to determine the velocity beneath Horizon B, establish its relationship to the top of oceanic crust and define the general crustal structure throughout the EMB and PB. The sonobuoy T/X data from both FM35-12 and MESOPAC II were inverted to a layered velocity-versus-depth model by simply measuring the slopes and intercept times of first-arrival refractions and employing the Tau-p method of Diebold and Stoffa [1981]. This velocity model was used as a starting point for iterative forward modeling of T/X data using structures consisting of laterally homogeneous, constant velocity layers and/or constant velocity gradient. Observed T/X data were usually modeled within +/- 0.02 s, except for longer range mantle arrivals which have uncertainties on the order of 0.1 s. Amplitudes were not modeled, but relative amplitudes were used to qualitatively assess velocity gradients and to infer the location of critical points on mantle reflections in order to estimate mantle velocities and place further constraints on overlying velocities.

The primary limitations of this modeling technique are: (1) the assumption of one-dimensionality and, (2) the nonuniqueness of any "best fit" model, particularly the inherent ambiguity between variations in layer thickness and velocity in modeling travel time alone. The assumption of one-dimensionality with respect to topography appears justified in this area where the seafloor is essentially flat (+/- 75m maximum) over the 30-40 km range of these sonobuoys. The highest amplitude sub-bottom reflections are most often flat-lying but at some locations Horizon B displays up to 150 m of relief. There is also evidence on some sonobuoys for out-of-plane disturbances and discontinuities in basement structure. Direct wave arrivals were used to calculate ranges assuming a water velocity of 1.54 km/s estimated from CTD measurements. Sediment velocities, based on analysis of four contiguous CDP gathers from the 3.3 km-long 96-channel streamer, ranged from 1.6 to 3.0 km/sec, usually increasing with depth. This range in velocities is also consistent with modeled reflections within the sediment layer. Only one or two of the highest amplitude reflectors within the sediment were matched by velocity contrasts in the model, and sediments were divided into isovelocity layers even though it is recognized that velocity gradients may also occur.

In a few cases (e.g. sonobuoys 1m, 4m, 6m, 22) refracted second arrivals were observed within the reflection hyperbola of the seafloor and sediment and traced to their tangency to the Horizon B sub-bottom reflection (e.g., Figure 5). These continuously curved refractions are indicative of relatively steep velocity gradients and provide constraints on the velocity immediately below Horizon B. These refracted arrivals are continuous with high-amplitude first arrivals beyond 7 km range. In the remaining instances, where first arrival refractions could not be followed to shorter ranges within the seafloor reflection, a single velocity gradient was assumed, resulting in continuously curved modeled refraction arrivals tangent to the Horizon B reflection. The resulting velocities and velocity gradients are characteristic of oceanic layer 2. It was with this evidence that high-velocity igneous material was successfully predicted to occur at the depth of the Horizon B reflection at all three ODP Leg 129 Sites.

Oceanic layer 3 is well defined on many sonobuoys by the weak first arrivals beyond 18-20 km range (Figure 5). The relatively low amplitude of these refractions along with the high-amplitude reflection from the Moho indicate a low velocity gradient typical of oceanic layer 3 [White, 1979; Purdy, 1983]. The velocity at the bottom of layer 2 and top of layer 3 is constrained by arrivals between 10-20 km that are modeled as a layer with a small positive velocity gradient overlying a zero gradient layer 3. The velocity within layer 3 was usually assumed to be continuous with the overlying velocity structure.

Wide-angle mantle arrivals were modeled as reflection events from a single discontinuity in velocity at the layer 3/mantle interface. The distinct travel time separation between refracted arrivals from layer 3 (P3) and Moho reflections (PmP) places constraints on the thickness of layer 3, and therefore, total crustal thickness. The velocity and thickness of layer 3 were chosen to best fit the curvature and travel time of the Moho reflection. Reflections from the Moho also appear on FM35-12 CDP data at 1.8 to 2.2 stwt below Horizon B (10.1 - 10.5 stwt) and are used as further constraints on crustal thickness. The underlying mantle velocity is not well constrained by travel time data primarily because most sonobuoys were unable to record arrivals at ranges greater than 30 km, however upper mantle velocities of 8.2 km/s were recorded on ocean bottom seismometers (OBSs) from explosive charges at ranges from 50 to 500 km in the EMB [Asada et al., 1985]. In addition, the observed amplitude build up at 20-22 km range was used to estimate critical distance which required upper mantle velocities between 8.0-8.3 km/s for all of the models. Although the top of the Moho is well placed in these models, mantle arrivals over their entire range cannot be matched by a single discontinuity; the travel times in some cases are better fit by models which include a transition zone with a velocity gradient at the base of the crust.

SEISMIC OBSERVATIONS AND RESULTS

The results of our regional seismic profiles are illustrated in three cross sections (Figure 6) accompanied by several key seismic reflection examples. The cross sections were created by digitizing distinctive reflection surfaces which seemed to maintain unique characteristics (i.e. seismic facies) over basinwide distances. The selected horizons include the seafloor, the shallowest abundant chert/porcellanite/clay or chert/chalk, middle Cretaceous basaltic sills/flows and/or the top of Jurassic oceanic crust. The lithofacies significance of these distinct horizons and the intervals which they bound are based on the lithologies recovered during ODP Leg 129 as well as DSDP Sites 61, 199 and 585 [Abrams et al., 1992].

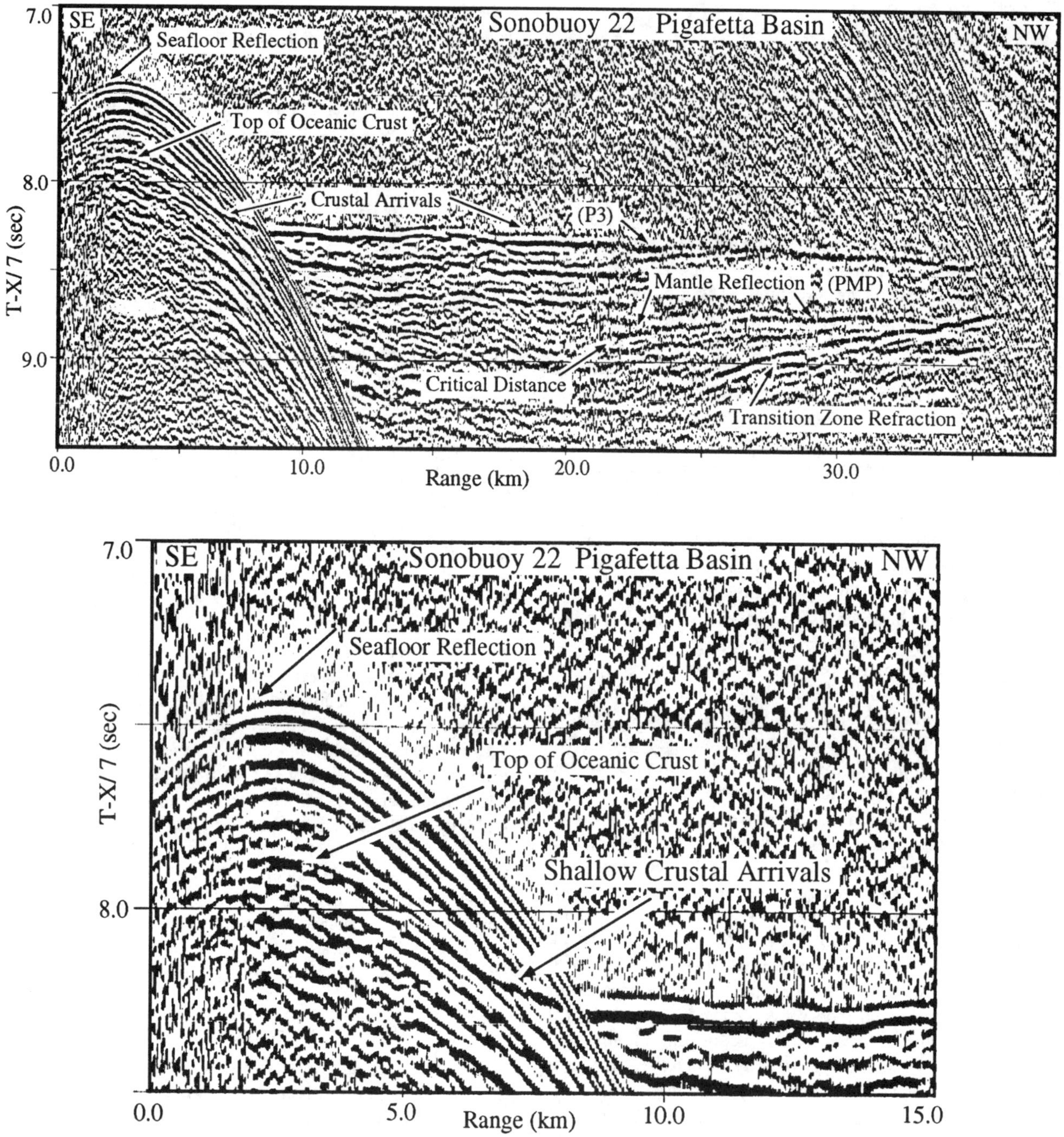

Fig. 5a-b. Traveltime - range data from sonobuoy 22 in the JQZ of the PB located 55 km Northwest of ODP Site 801 is shown at reduced travel time (7km/s), 500 msec AGC, band-pass filter 8-25 Hz. Figure 5a (top) displays entire record section out to 40 km range. Figure 5b (bottom) is an enlargement of this data from 0-15 km. Refracted/reflected arrivals of note are labeled. Refracted second arrivals (labeled "shallow crustal arrivals") that are tangent to Horizon B (labeled "top of oceanic crust") are clearly visible between 5-8 km and continuous with refracted first arrivals observed beyond 30 km range. Shallower high-velocity first arrivals are not observed in contrast to those revealed in sonobuoys 8 and 18 in the EMB (Figure 14).

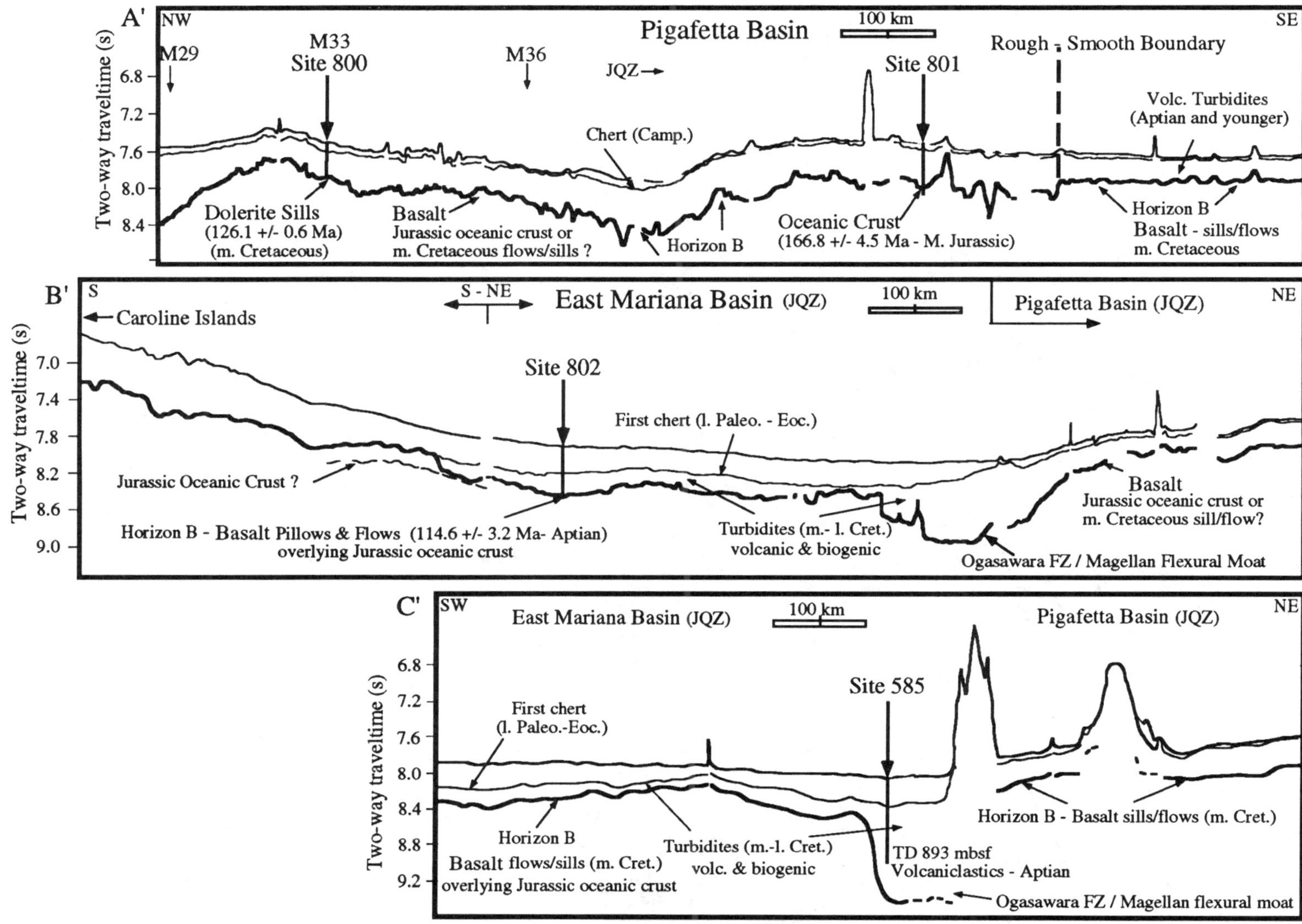

Fig. 6a-c . Three 900 - 1300-km long profiles showing selected reflections digitized from MCS airgun and near-channel watergun data (see Figure 4 track chart for locations (a) A-A', (b) B-B' and (c) C-C'). The deepest horizon (heavy line) represents the top of high-velocity igneous material. This horizon is characterized and interpreted from direct sampling at three widely separated ODP sites, reflection character and from velocity structure. The middle horizon represents the top of the shallowest chert - porcellanite - silicified limestone (First chert). The top horizon represents the seafloor. Note the depth difference between the PB and EMB, the deepening of Horizon B associated with the Ogasawara Fracture Zone and flexural moat of the Magellan Seamounts and the uniformly flat Horizon B surface (even at VE=133X) in the EMB and southeast PB compared to the high-relief basement of the PB northwest of the rough-smooth boundary. Radiometric ages are from Pringle [1992a].

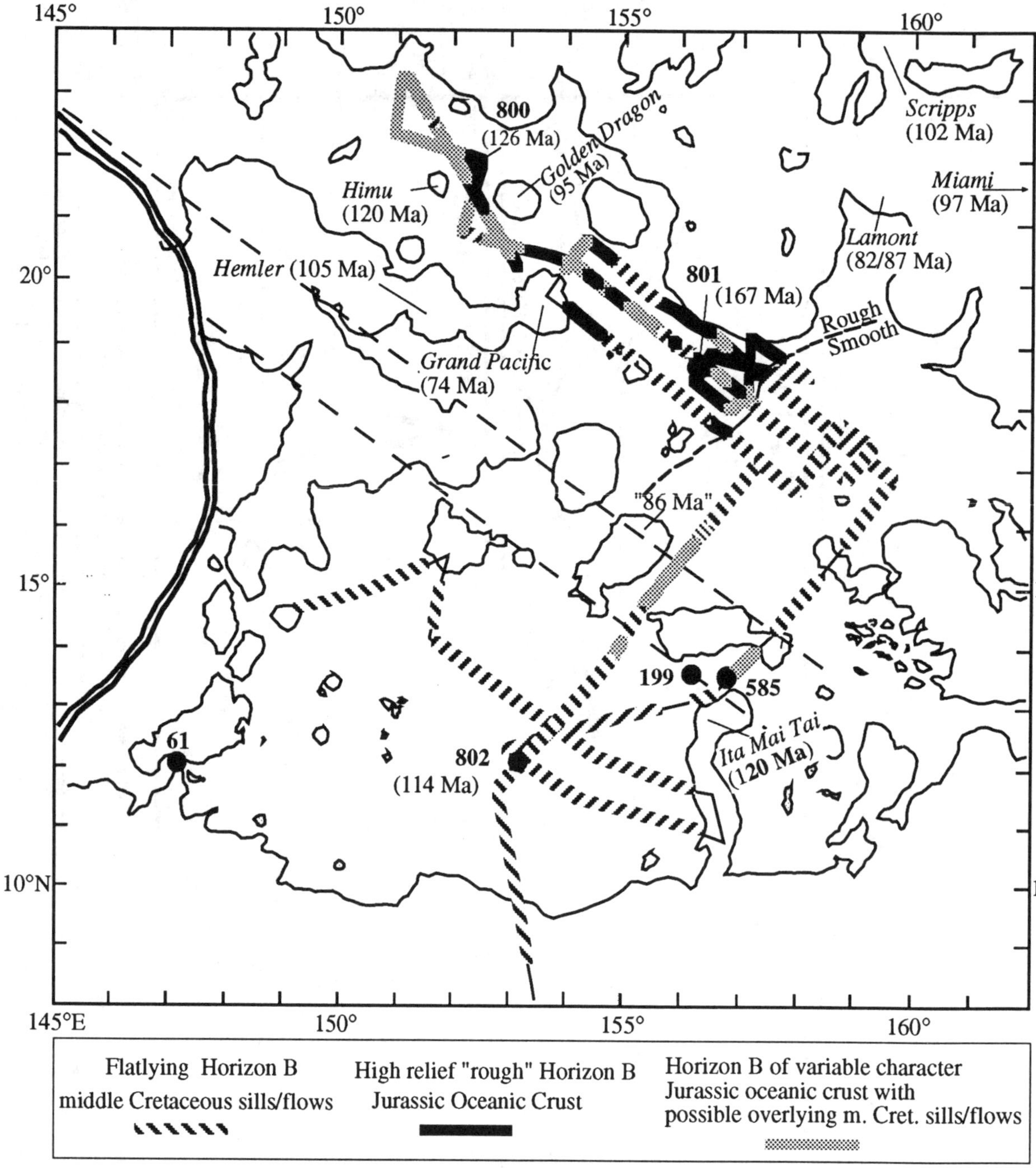

Fig. 7. Simplified bathymetry (5500 m contour is shown) of the central western Pacific modified from Brenner and Angell, [1992]. The reflection character of Horizon B and its interpretation are indicated along track from Abrams et al. [1992]. The oceanic crust in the entire EMB and southeast PB (~ 500,000 km^2) is overlain by mid-Cretaceous sills/flows (l. Barremian or younger). Only a few restricted areas in the PB are confidently interpreted to contain Jurassic oceanic crust with no massive igneous overburden. Horizon B in extended areas (primarily in the PB) is more tentatively interpreted as Jurassic oceanic crust with mid-Cretaceous sill/flow overburden possible. Radiometric ages of basalt are given in Ma for Leg 129 Sites from Pringle [1992a], and for seamounts from Ozima et al. [1983], Pringle [1992b], and Winterer et al., [this volume].

General Characterization of Horizon B

Horizon B is associated with the onset of high velocities (>3.6 km/s), and we believe it is often comparable in acoustic character to the unusually smooth basement observed in parts of the western North Atlantic and Caribbean Basin that was designated as Horizon B" [Ewing et al., 1967; Edgar and Saunders et al., 1973; Houtz and Ludwig, 1977]. Its reflection character and stratigraphic position determines whether

Horizon B is interpreted as the top of sills/flows overprinting true oceanic crust, or as Jurassic oceanic crust with no such massive igneous overburden.

In general, Horizon B can be characterized by two end-members: 1) a relatively high-relief, (50-100 m over tens of kilometers), lower reflection amplitude, diffractive ("rough") surface which always appears as acoustic basement and is interpreted as the top of oceanic crust, and 2) a flat-lying, high reflection amplitude, semicontinuous surface which often appears nondiffractive ("smooth"), does not necessarily appear as acoustic basement and is interpreted as sills/flows that overlie oceanic crust. A map view of the acoustic character of Horizon B and its interpretation is presented along seismic track lines in Figure 7. Figure 7 indicates that there are extensive areas in the northwest PB where the acoustic character does not clearly fall within the above-defined modes and/or these end-member characteristics alternate over distances of only tens of kilometers and as a result these areas have been more tentatively interpreted as the top of oceanic crust with possible sill/flow overburden.

Horizon B in The East Mariana Basin

In the majority of the EMB, Horizon B (heavy line in Figure 6) is high-amplitude, flat-lying, continuous and "smooth" (e.g. end-member 2). There are, however, several significant variations in its character. In restricted areas of the EMB

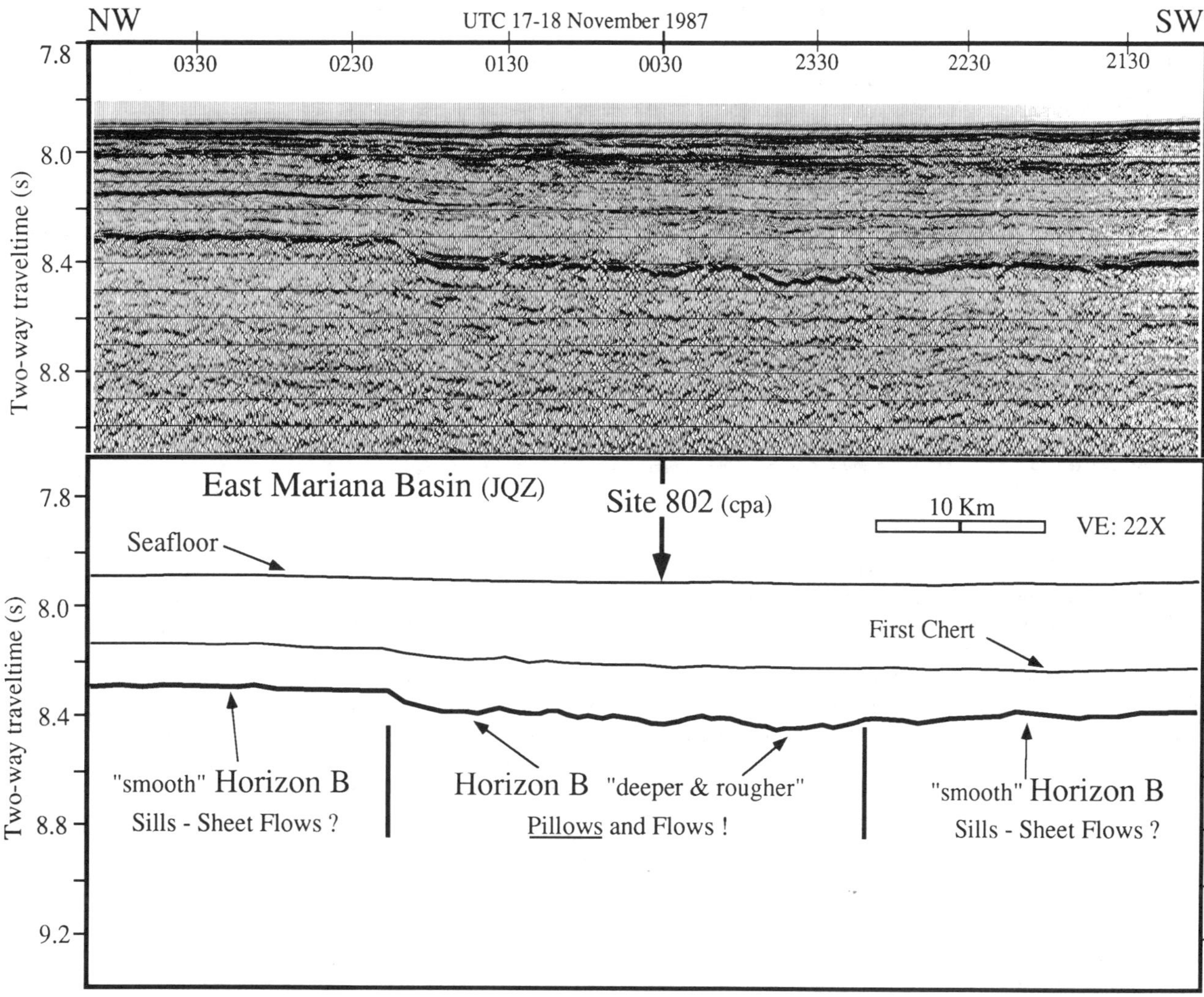

Fig. 8. FM35-12 MCS airgun profile at its closest point of approach (cpa) to Site 802 (approximately 4 km to the east). The depth and character of the basement reflector changes from a flat-lying, "smooth" event at 8.3 stwt to a "rougher" reflection at 8.4 stwt (marked by vertical lines). The "rough" (diffractive) Horizon B reflection correlates to the top of extrusive basalt (pillow units) and the shallow, "smooth" (non-diffractive) Horizon B reflection is interpreted as igneous intrusives or sheet flows. The relatively rough surface, however, remains flat-lying and shows much less overall relief than does the Horizon B in the PB northwest of the rough/smooth boundary. The processing and display parameters are as follows: pre-stack spike deconvolution, 120-fold stack, band-pass 20-60 filter Hz, F/K migration, 500 msec AGC and vertical exaggeration of ~22X at 1.5 km/s.

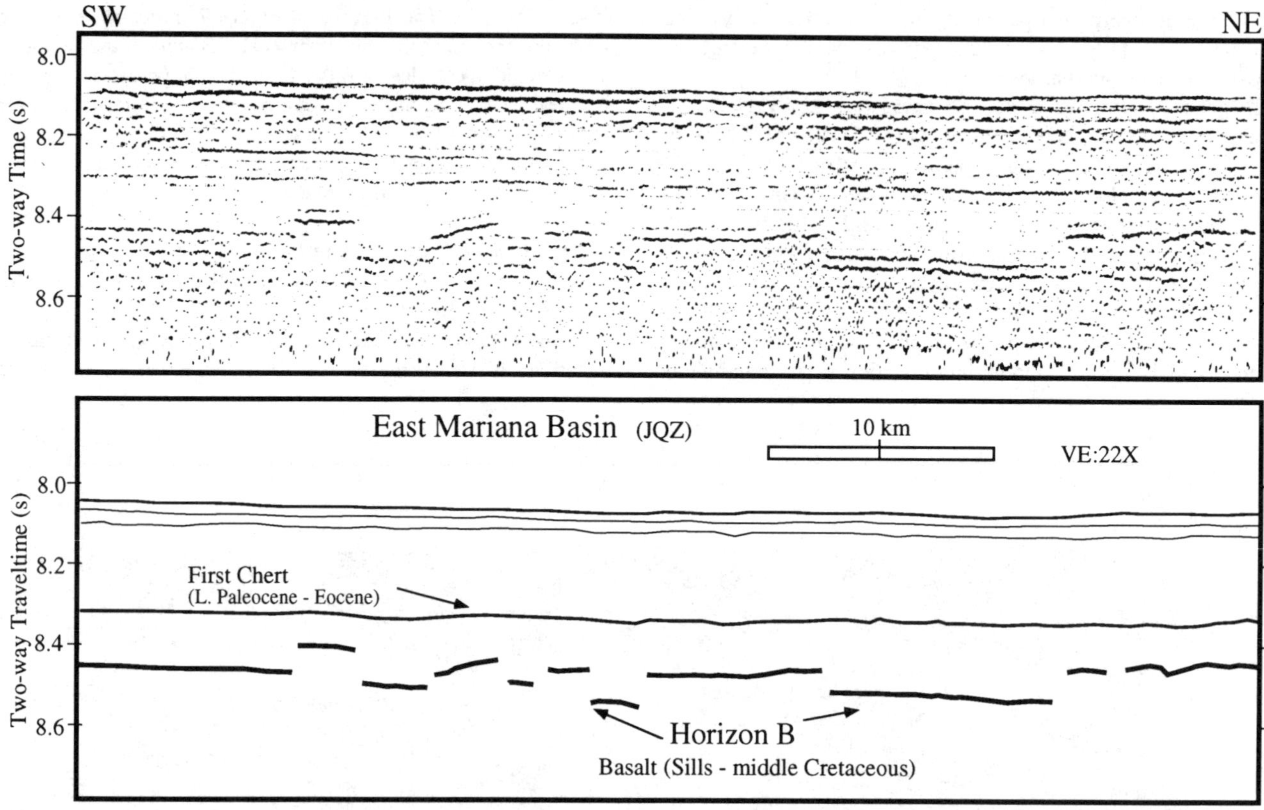

Fig. 9. MESOPAC II line 18 near-channel watergun profile in the EMB (location Figure 4) showing Horizon B where it appears flat-lying, "smooth" and laterally discontinuous "slablike". This type of reflection geometry is not characteristic of the top of oceanic crust and is interpreted as sills emplaced at different stratigraphic levels overlying an unimaged oceanic crust. This seismic record is processed and displayed as indicated in Figure 3.

Horizon B deepens slightly and becomes noticeably more diffractive (but still flat-lying), for example, at Site 802 where Horizon B corresponds to pillow and flow basalt (Figure 3c and 8). In other instances Horizon B appears to be divided by and contiguous with zones of poor reflectivity giving the appearance of "windows" through this ubiquitous horizon. A similar geometry is evident along the eastern EMB margin where Horizon B appears as a series of slabs which are offset laterally as well as vertically with no corresponding offset of the overlying sediment (Figure 9). To the south of Site 802, Horizon B appears continuous as it rises gradually from 6000 m to the 5000 m contour which marks the base of the Caroline seamounts, atolls and guyots (Figures 2, 10). The continuity of the gently rising Horizon B is only disrupted near the base of the Caroline Ridge by an offset (150 msec) which is not matched by offsets in the overlying reflections.

Two seismic profiles crossing the eastern margin of the EMB (Figures 6b and 6c) reveal the deepening of Horizon B from approximately 300-500 mbsf (8.3-8.5 stwt) to the west within the EMB to over 1150 mbsf (> 9.2 stwt) where the southern profile crosses DSDP Site 585. Correlations of seismic stratigraphy with drilling results from Site 585 indicate that the total depth reached at this site corresponds to a zone of relatively high-amplitude conformable reflections at approximately 8.9 to 9.0 stwt [Moberly and Schlanger et al., 1986, Abrams et al., 1992] and may correlate with the well-cemented, Aptian volcaniclastic turbidites and debris flows lying approximately 260 m above the high-velocity basement at 9.2 stwt as determined by sonobuoy 20 (Figure 11a). In this area there is no well-developed flat-lying, high-amplitude Horizon B observed. Finally, there are also sub-Horizon B reflection events observed in certain areas of the EMB. The suppression of bubble pulse multiples from Horizon B on processed FM35-12 MCS data revealed weaker, discontinuous reflections approximately 100 - 150 ms below Horizon B. The identification of these weak reflections, which are often identical in shape to the overlying Horizon B, as primary reflection interfaces rather than bubble pulse or interbed multiples is strengthened by isolated instances where these weak reflections display significant relief independent of flat-lying Horizon B and overlying sediments. A clear example of this type of reflection geometry is displayed in Figure 10.

The flat-lying, typically smooth character of Horizon B in the EMB and southeastern PB does not appear to change from profiles shot perpendicular to those shot parallel to magnetic lineations. This indicates that any abyssal hill morphology is

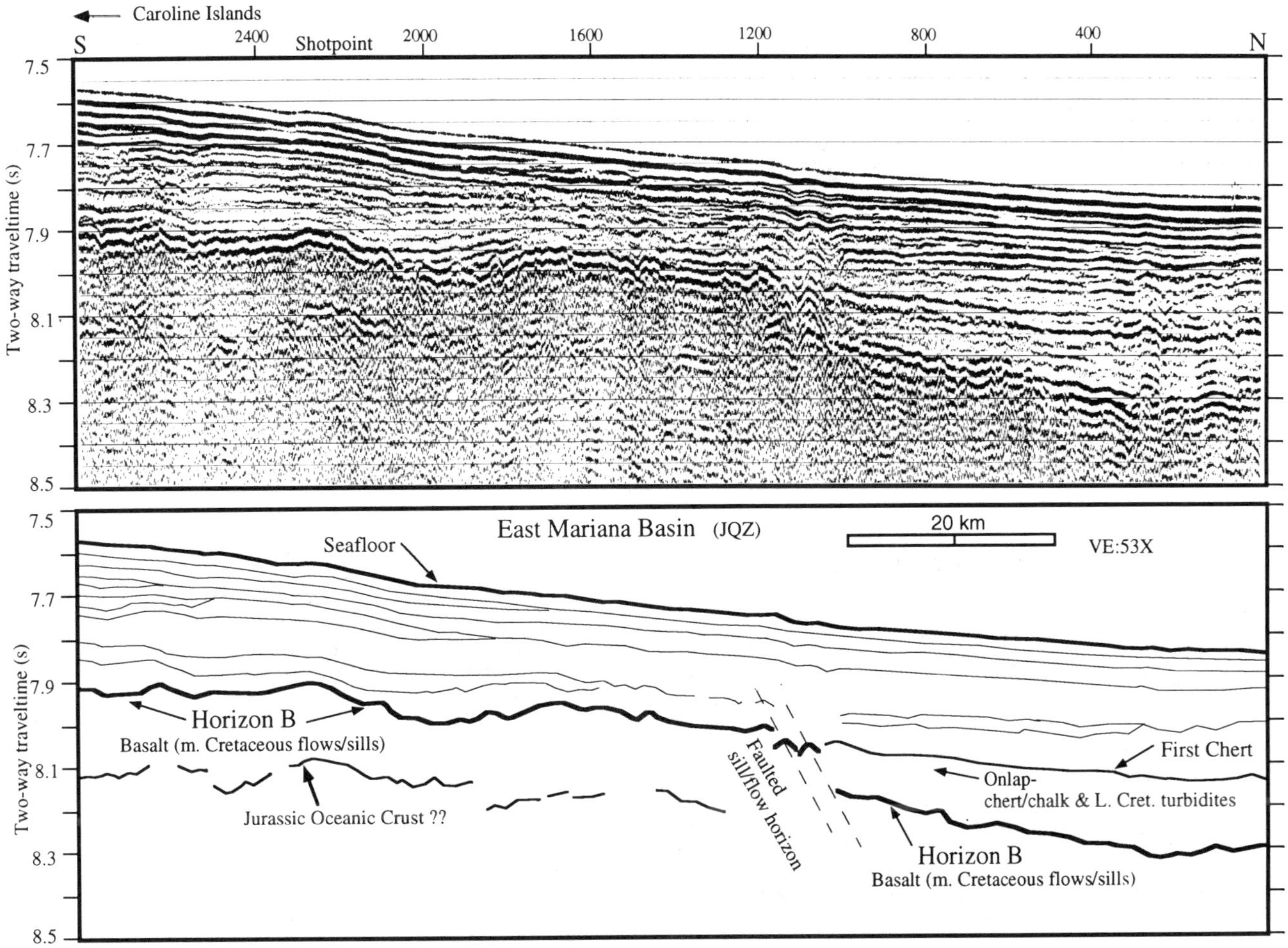

Fig. 10. MESOPAC II SCS watergun profile in the southern EMB (location Figure 4). Site 802 penetrated extrusive basalt dated at 114 +/- 3.2 m.y. beneath Aptian sediment, this horizon (Horizon B) appears faulted and is onlapped by late Paleocene chert/chalks and underlying Late Cretaceous turbidites. The basalt surface penetrated at Site 802 continues up onto the Caroline Ridge. A weak semicontinuous, relatively high-relief horizon imaged beneath the rising flow/sill surface is interpreted as the top of oceanic crust. A similar reflection geometry is observed on both airgun and watergun records in other restricted portions of the EMB. This MESOPAC II watergun record is processed and displayed as indicated in Figure 3.

muted, absent or buried and suggests that this is not the upper surface of oceanic crust. In addition, the Aptian age (124.5-112 Ma) of sediment overlying ~114 m.y. old extrusive basalt is approximately 50 m.y. younger than predicted from magnetic lineation correlations which had so successfully predicted the basement age at Site 801 (e.g. ~167 Ma) [Tamaki et al., 1987; Lancelot and Larson et al., 1990; Pringle, 1992a]. Thus, throughout the EMB, including the Caroline Ridge, Horizon B is interpreted as the top of middle Cretaceous flows and/or sills (probably late Barremian - Aptian) which overlie Jurassic oceanic crust. The weak sub-Horizon B reflections, though not penetrated at ODP Site 802, are interpreted as the top of oceanic crust.

Horizon B in The Pigafetta Basin

In the southeastern PB, Horizon B is semicontinuous, flat-lying and shallow (250-300 mbsf, 7.9-8.0 stwt). It is observed to monotonically deepen (400-480 mbsf) to the northwest becoming distinctly more diffractive (rougher), undulating, and lower in reflection amplitude beginning approximately 130 km southeast of, and including ODP Site 801 (Figures 6a and 12). This change in acoustic character from flat-lying to high-relief appears as a distinct boundary rather than an extended transition zone on our southeast to northwest seismic profiles and is designated as the "rough/smooth boundary" (Figures 4, 7 and 12). This boundary was located along profiles running southeast to northwest where flat-lying, semicontinuous Horizon B is first disrupted by a distinctly higher relief (perhaps tectonized) surface. To the northwest of the rough/smooth boundary, Horizon B displays a variety of reflection characteristics, which are superimposed upon a subtle, longer wavelength Horizon B surface (Figure 6a). Horizon B at Site 801 appears as a rough, undulating horizon that results primarily from a

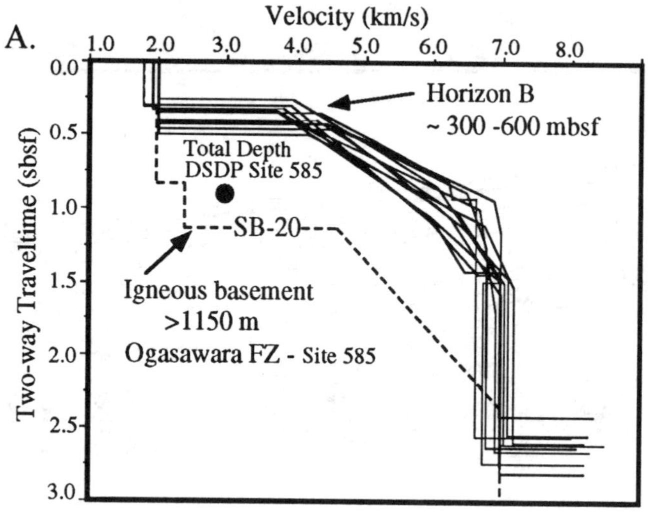

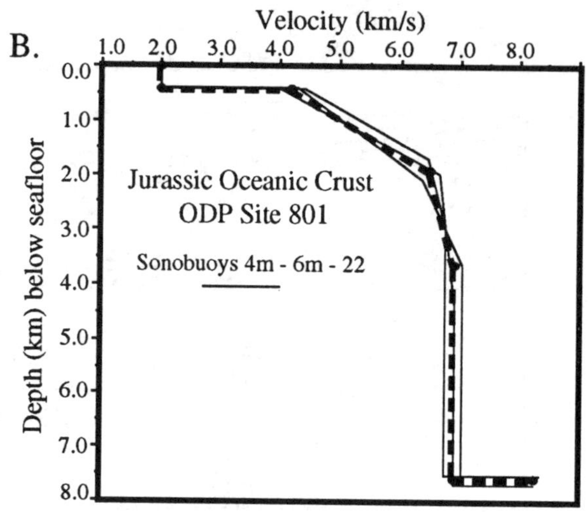

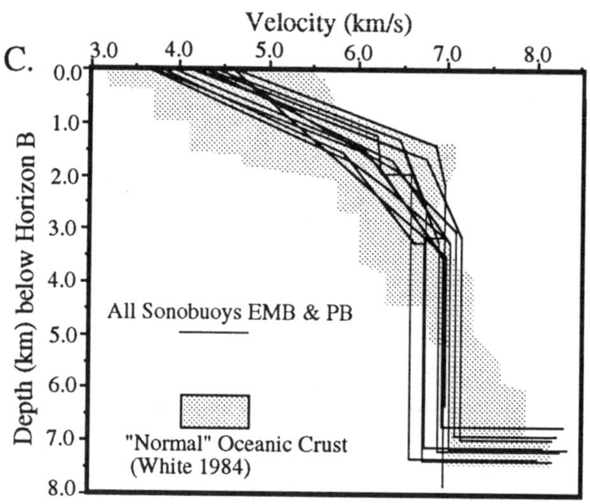

Fig. 11a-c. a) Velocity verses two-way travel time in seconds below seafloor (sbsf) from sonobuoy data collected during FM35-12 and MESOPAC II in the EMB and PB. Crustal velocities begin from 0.3 to 0.6 sbsf (approximately 300 to 600 mbsf), with the exception of sonobuoy 20. Sonobuoy 20 (dashed line) was shot over DSDP 585 and indicates that approximately 1150 m of sediment overlie the relatively deep basement imaged at the far eastern edge of the EMB. The location of deep basement lies along the inferred trend of the Ogasawara Fracture Zone but may also be associated with flexural depression of the lithosphere due to the loading effects of surrounding seamounts. The onset of velocities and velocity gradients characteristic of oceanic crust correspond to the Horizon B reflection imaged throughout both basins. b) Velocity versus depth below seafloor of individual profiles for sonobuoys 4m, 6m and 22 all located within 45-90 km of Site 801. The nearly identical velocity-depth profiles derived for these sonobuoys provide the best estimate of seismic structure for Jurassic oceanic crust penetrated at Site 801 in the PB. c) Velocity versus depth below the top of Horizon B in the PB and EMB compared to the bounds of "normal" oceanic crust given by White [1984].

large impedance contrast between the Callovian/Bathonian radiolarite clays and the top of interbedded chert/basalt and hydrothermal deposits which unquestionably represent the top of Jurassic oceanic crust (Figure 3b) [Lancelot and Larson et al., 1990]. In the areas to the southeast of Site 801 where Horizon B becomes flat-lying, "smooth" and shallow, and the overlying seismically transparent layer thins, sills and/or flows that are significantly younger (probably late Barremian - Aptian) than those recovered at Site 801 are expected to overlie oceanic crust (Figures 7 and 12 right side). Further to the northwest, Horizon B appears tectonized, sometimes exhibiting smooth, block faulted slabs having an apparent dip to the northwest.

Within the area of correlated magnetic lineations M29-M36 (Figures 1 and 2), Horizon B rarely appears similar to the rough, undulating basement horizon observed in the vicinity of ODP Site 801 (Figures 3b and 12 left side), however, it is not characterized by the extended areas of flat-lying, high-amplitude and "smooth" reflections observed in the southeast PB and the majority of the EMB. Specifically, there are no examples where Horizon B appears discontinuous and/or vertically offset as in the EMB (e.g. Figure 9). Horizon B at Site 800 correlates to the top of dolerite sills intruding Berriasian age [140-145 Ma] radiolarite clays. Although drilling at Site 800 penetrated through a thick middle Cretaceous volcaniclastic sediment unit and reached earliest Cretaceous pelagic sediments, Horizon B here does not represent oceanic crust because of the age (~126 Ma, Pringle, 1992a), thickness and obvious intrusive character of the dolerite, and because the oldest sediments would be expected to be Oxfordian in age (~160 Ma, M33).

The Horizon B reflections can be followed over the 550 km separating Sites 800 and 801. Results from these sites suggest that this horizon corresponds to the top of Jurassic oceanic crust where it appears particularly high-relief, undulating, lower in reflection amplitude, and diffractive. It corresponds to a thin, (approximately 100-400 m) middle Cretaceous igneous unit overlying oceanic crust where it appears flat,

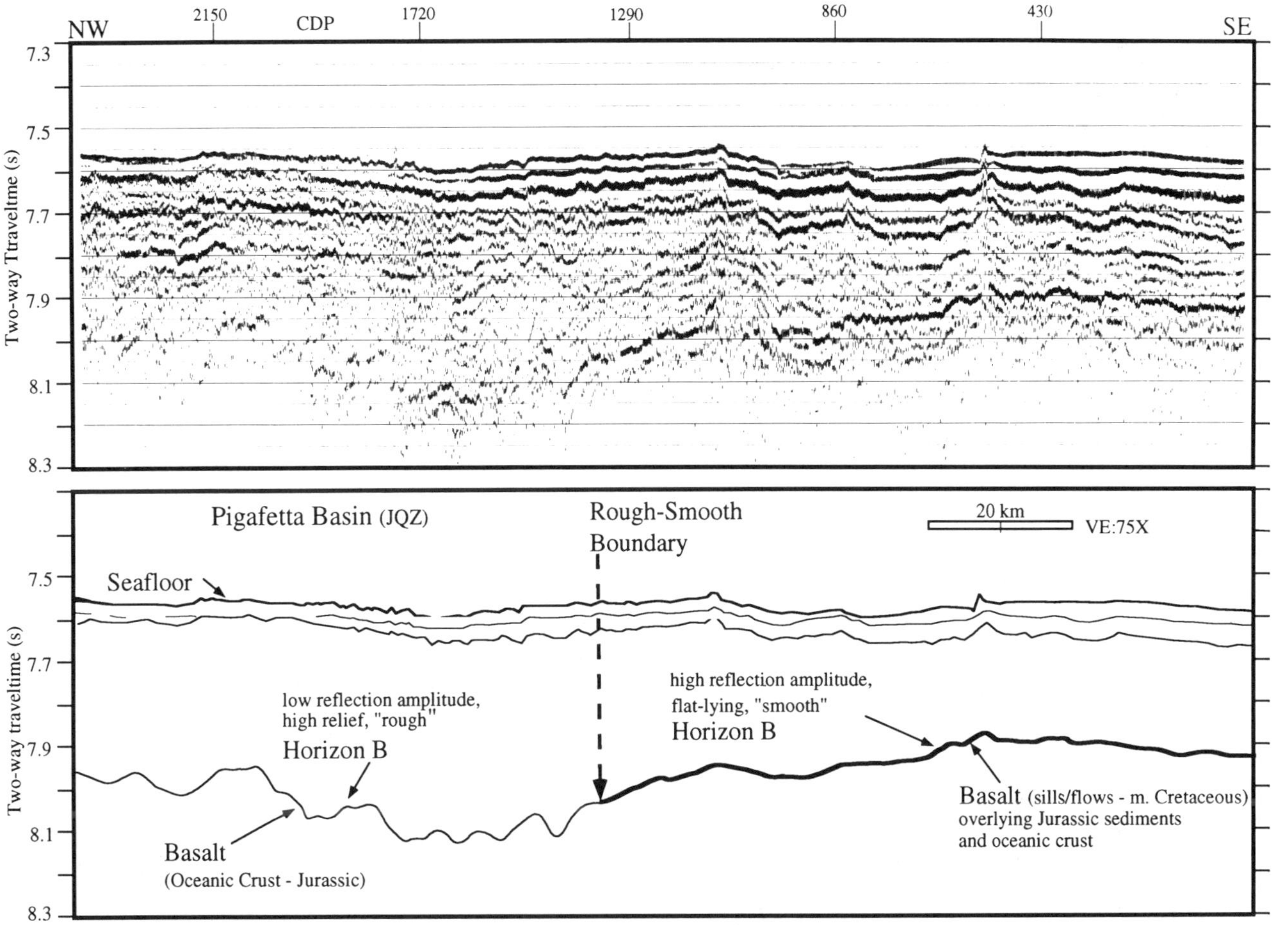

Fig. 12. FM35-12 MCS airgun profile in the JQZ of the PB along an inferred flowline course and displayed at true relative amplitude. The continuous, flat-lying (at VE:75x) and high-amplitude Horizon B reflection southeast of the rough/smooth boundary deepens and appears to be contiguous with an undulating, diffractive and lower reflection amplitude "rough" Horizon B. The "rough" Horizon B correlates to the top of middle Jurassic oceanic crust (e.g. Site 801) and the "smooth" horizon is interpreted as mid-Cretaceous sills and flows overlying the unimaged top of oceanic crust. The rough/smooth boundary appears on all five of our tracks crossing this area and a comparable boundary is not apparent in the EMB. We suggest that the boundary marks the limit of semicontinuous mid-Cretaceous (Aptian?) flows/sills. The processing and display parameters are the same as in Figure 8 except displayed with true relative amplitude, band-pass filter 8-25 Hz and vertical exaggeration of ~75X at 1.5 km/s.

high-amplitude, "smooth" and slablike. The actual thickness of the middle Cretaceous sills and flows, however, is uncertain, and aspects of this determination are discussed in the following section.

Sill/Flow/Sediment Thickness

To establish the relationship between Horizon B and the top of oceanic crust by focusing on evidence for sills/flows overlying oceanic layer 2, we first describe the area in the PB where Horizon B correlates to the top of Jurassic oceanic crust. We then compare the velocity-depth profiles and record sections from this area to areas in the PB and EMB where Horizon B correlates to the top of middle Cretaceous sills and/or flows.

In Figure 11 and Table 1 we show Vp versus depth functions obtained from 13 sonobuoys. The velocity-depth functions below the seafloor are given in terms of a series of constant velocity gradient layers. Figure 11a indicates that velocities and/or velocity gradients characteristic of oceanic crust begin at or just below the dominant Horizon B reflection, imaged regionally at 300 to 600 mbsf. Sonobuoy 20, shot over DSDP Site 585, is the one exception to this relationship and indicates that approximately 1150 m of sediment overlie high-velocity basement in the area that we identify as the Ogasawara Fracture Zone (Figures 2, 6).

Sonobuoys 4m, 6m and 22 (Figures 5 and 11b) within 45-95 km of ODP Site 801 provide the best constraints on crustal structure, not only because a complete suite of arrivals are

Table 1. Velocity - Depth functions below seafloor given in terms of a combination of constant velocity gradient layers with top and bottom velocities for each layer given in column 3. Dashes indicate constant velocity layer. The thickness of each layer given in seconds (s) and kilometers (km) (column 4). Velocity gradient (s-1) for each layer is given in column 5.

SB#	Layer	Water Depth (m)	Vp (km/s)	Thickness (s)	- (km)	Gradient (s -1)
4m	1	5730	2.0 -	.42	.42	0.0
	2		4.41-6.47	.47	1.28	1.61
	3		6.47-6.76	.38	1.24	0.23
	4		6.76 -	1.36	4.59	0.0
6m	1	5660	2.0 -	.40	.40	0.0
	2		4.15-6.38	.64	1.69	1.32
	3		6.38-7.03	.46	1.54	0.42
	4		7.03 -	1.10	3.89	0.0
22	1	5680	2.0 -	.41	.41	0.0
	2		3.98-6.67	.59	1.58	1.71
	3		6.67-6.92	.70	2.38	0.11
	4		6.92 -	.96	3.33	0.0
1m	1	5585	2.0 -	.50	.50	0.0
	2		4.13-6.75	.61	1.65	1.58
	3		6.75-7.17	.43	1.48	0.28
	4		7.17 -	1.06	3.82	0.0
2m	1	5680	2.0 -	.46	.46	0.0
	2		4.63-6.87	.49	1.41	1.59
	3		6.87-6.98	.22	0.75	0.15
	4		6.98 -	1.44	5.03	0.0
3m	1	5755	2.0 -	.35	.35	0.0
	2		4.32-6.21	.47	1.23	1.54
	3		6.21-6.26	.12	0.73	0.07
	4		6.6 -	1.62	5.36	0.0
23	1	5660	2.0 -	.43	.43	0.0
	2		4.45-6.43	.69	1.87	1.06
	3		6.43-7.09	.35	1.17	0.57
	4		7.09 -	1.08	3.83	0.0
8m	1	5745	2.0 -	.34	.34	0.0
	2		3.70-6.03	.63	1.52	1.53
	3		6.03-6.24	.09	0.29	0.73
	4		6.24-6.98	.53	1.74	0.42
	5		6.98-	1.16	4.27	0.0
21	1	5685	2.0 -	.26	.26	0.0
	2		3.92-6.06	.59	1.46	1.46
	3		6.06-6.96	.62	2.01	0.45
	4		6.96 -	.95	3.25	0.0
20	1	6100	1.8 - 1.8	.30	.27	0.0
	2		2.0 - 2.0	.52	0.52	0.0
	3		2.4 - 2.4	.30	0.36	0.0
	4		4.60-7.00	1.23	3.50	0.68
	5		7.00-	.80	2.80	0.0
7	1	5925	1.9 -	.32	.30	0.0
	2		3.81-5.83	.70	1.69	1.2
	3		5.83-6.63	.42	1.31	0.61
	4		6.77 -	.70	2.37	0.0
8	1	5940	1.9 -	.33	.32	0.0
	2		3.66-5.94	.72	1.73	1.32
	3		5.94-6.33	.16	0.49	0.8
	4		6.33-6.98	.28	0.93	0.7
	5		6.73-	1.25	4.19	0.0
18	1	5930	1.78 -	.30	.27	0.0
	2		3.90-6.04	.82	2.04	1.05
	3		6.04-6.45	.3	0.92	0.44
	4		6.98 -	1.39	4.86	0.0

Table 2. Layer 2 thicknesses as defined by velocity and velocity gradient, and total crustal thicknesses are given. Mantle arrivals were not observed for sonobuoy 7 (SB7).

SB#	Layer 2 Thickness (km) (<6.4 km/s)	- (1-2 s-1)	Total Crustal Thickness (km)
4m	1.3	1.3	7.12
6m	1.7	1.7	7.12
22	1.42	1.58	7.28
1m	1.44	1.65	6.95
2m	1.11	1.41	7.18
3m	1.96	1.23	7.31
23	1.84	1.87	6.86
8m	1.90	1.52	7.83
21	2.22	1.46	6.76
20	2.61	--	6.35
7	2.6	1.7	--
8	2.32	1.73	7.34
18	2.85	2.04	7.82

observed, but also because Site 801 is the single location where Jurassic oceanic crust undoubtedly exists without any younger igneous overburden. Velocity-depth functions of these three sonobuoys are shown separately in Figure 11b. The normal-incidence two-way reflection time through this structure agrees to better than 0.1 s with the location of the Moho reflection observed on FM35-12 MCS profiles collected during the acquisition of sonobuoy 22 (Figure 13). The total crustal thickness is 7.2 km (2.22 stwt), overlain by an average 410 m of sediment. The remaining sonobuoys are divided into groups representative of distinct areas of the EMB and PB (Figures 2 and 4 location). Layer thicknesses derived from these sonobuoys are defined and listed in Table 2.

Analysis of T/X data along with the drilling results from ODP Sites 800, 801 and 802 clearly establish that Horizon B represents the top of high-velocity (>3.6 km/s), massive igneous material throughout these basins. The greatest limitation of this travel time-range study is that in those cases where short range refracted second arrivals in the critical range (5-8 km) are obscured by seafloor and Horizon B reflections it is impossible to distinguish between a variety of probable velocity structures (resolvable at seismic wavelengths) between the Horizon B reflection and the structure revealed by longer range (deeper) refracted arrivals. The necessity of assuming or calculating a velocity for the uppermost crust (layer 2a, b or in this case the layer beneath Horizon B) because of the lack of observable refracted arrivals is well documented [Ewing and Purdy 1982]. The problem here is that other evidence indicates that sills/flows and sediment actually overlie Jurassic oceanic crust (layer 2) in the majority of the EMB, and in the PB southeast of the rough-smooth boundary. Modeling the structure immediately below Horizon B as a single velocity gradient in these areas of the EMB and PB in effect includes middle Cretaceous sills/flows and sediments as part of layer 2.

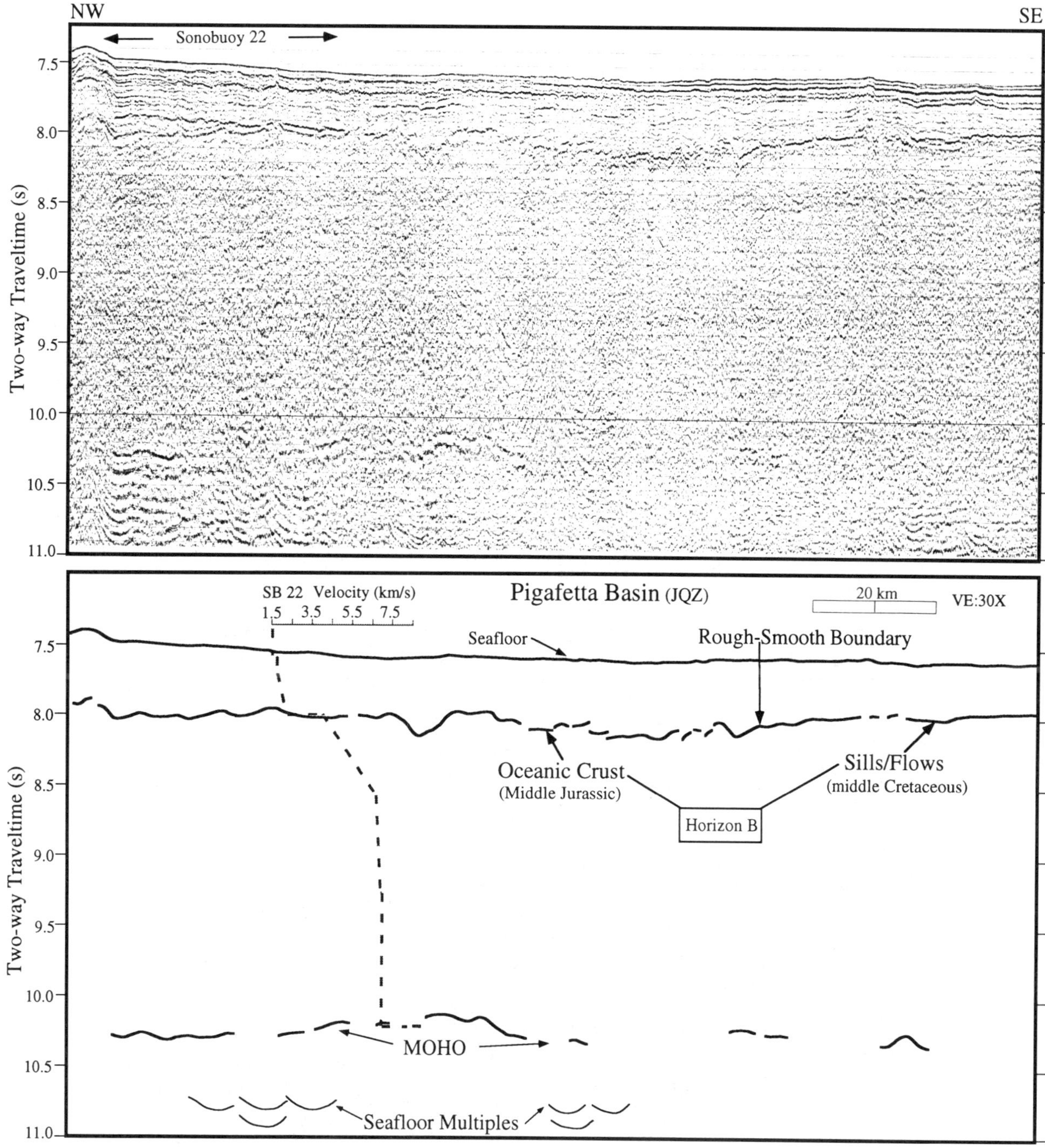

Fig. 13. FM35-12 MCS airgun profile in the JQZ of the PB with the velocity-depth profile derived from sonobuoy 22 superimposed. The onset of velocities and velocity gradients characteristic of oceanic crust correspond to the Horizon B reflection. Sonobuoy 22, 6m and 4m all located within 45-90 km of Site 801 provide nearly identical velocity-depth profiles including wide-angle mantle reflections with intercept times matching those observed in the MCS profile labeled MOHO at ~10.2 stwt. Processing and display parameters are the same as in Figure 8 except band-pass filter 8-25 Hz and vertical exaggeration of ~30X at 1.5 km/s.

Velocity-depth structures calculated in this way fall within the wide bounds defining normal oceanic crust (Figure 11c; Houtz and Ewing, 1976; Spudich and Orcutt, 1980; White, 1984) and total crustal thicknesses do not appear significantly different between the EMB and PB (Table 2). Assuming layer 2 thicknesses determined from sonobuoys 22, 4m and 6m are representative of Jurassic oceanic crust in this area implies that the middle Cretaceous sill/flow/sediment complex in the

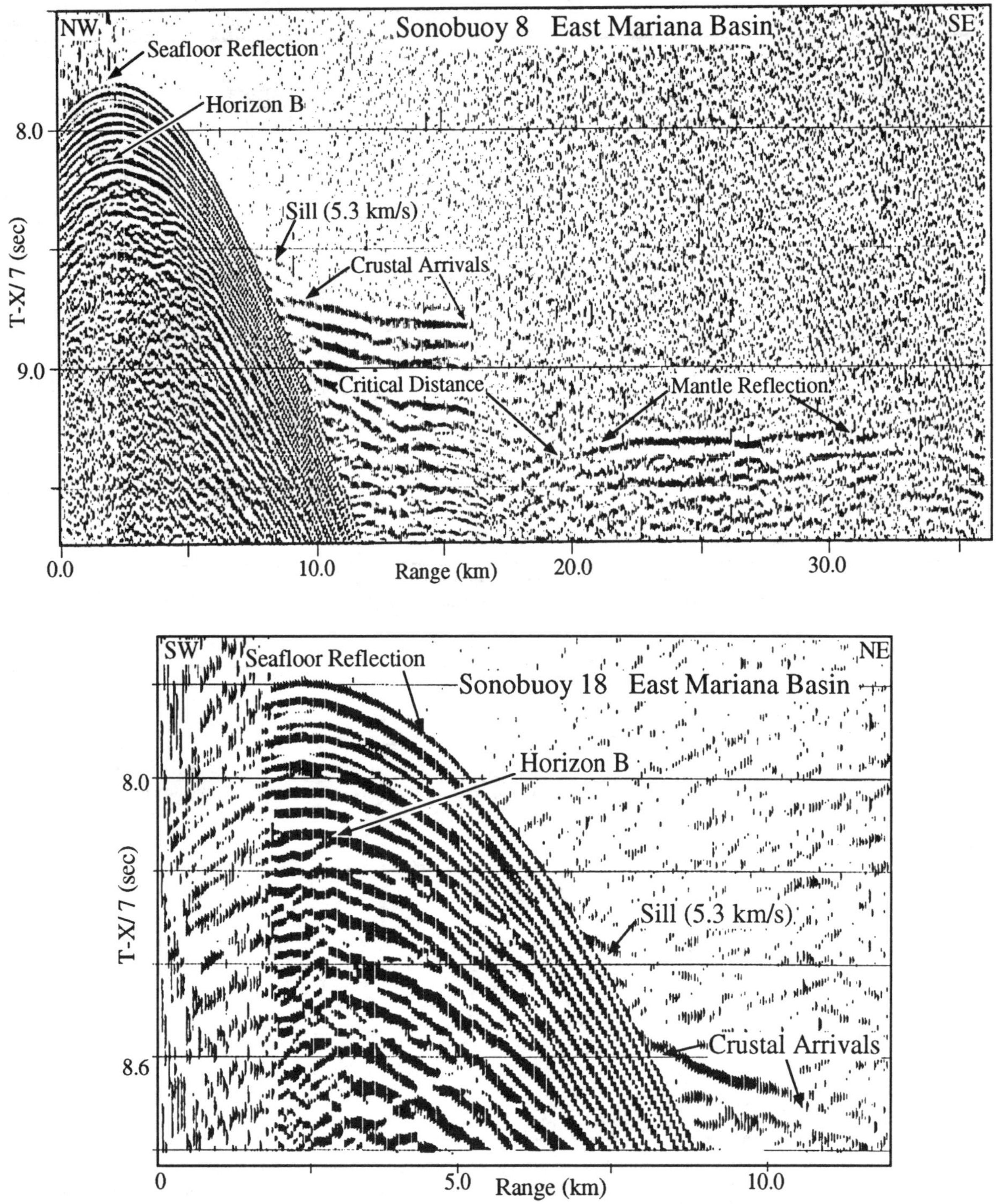

Fig. 14a-b. a) Sonobuoy 8 in the JQZ of the EMB located 120 km North of Site 802 is shown over the entire range of the sonobuoy (>35 km). b) Sonobuoy 18 in the JQZ of the EMB located 130 km northeast of Site 802 is shown over 0-15 km range. Note the lower amplitude, shallow first refracted arrival with a phase velocity of 5.3 km/s (labeled "sill") which appears tangent to the Horizon B reflection (SB18) or to reflections just below Horizon B (SB8). This short range arrival is indicative of a high-velocity layer (sill/flow) overlying lower velocity sediment and oceanic crust. Similar arrivals are not observed on sonobuoys northwest of the rough/smooth boundary (compare to Figure 5). Sonobuoy data displayed at reduced travel time (7km/s), band-pass filter 8-25 Hz.

EMB may range in thickness from 0 to approximately 1 km depending on how layer 2 is defined (e.g. Table 2, 1.7 to 2.85 km vs 1.3 to 1.7 km). Thus, there is no seismically resolvable, systematic differences in crustal velocities and thicknesses between areas where middle Cretaceous sills/flows are present and where they are absent. This observation is significant when compared to anomalously thick crust observed in portions of other basins where a smooth basement reflection (Horizon B or B") corresponds to the surface of younger (middle to Late Cretaceous) sills/flows throughout an area predicted to contain Jurassic sediment and oceanic crust such as the Nauru Basin (Figure 1) and basins of the Caribbean Sea [Edgar and Saunders et al., 1973; Larson and Schlanger et al., 1981; Wipperman et al., 1981; Diebold et al., 1981].

The results from sonobuoys in the EMB, however, provide further support for the presence of sills/flows overlying sediment and oceanic crust. In two instances (sonobuoys 8 and 18) low-amplitude refracted arrivals of limited range appear tangent to or slightly below the Horizon B reflection (Figure 14a,b, arrival labeled "sill"). These weak but distinct events extend as first arrivals that overlie the higher amplitude, continuous refractions that characterize normal oceanic crust. These arrivals have a phase velocity of approximately 5.3 km/s and show a rapid decay in amplitude with range. Such observations are consistent with the existence of a high-velocity igneous layer capping lower velocity sediments which overlie oceanic crust. Shallow "sill" refractions are not observed on any sonobuoy records from the PB (e.g., Sites 800 and 801; compare Figures 5b and 14b). We note that the velocity-depth solutions given in Table 1 were derived by matching the stronger, more coherent and longer range arrivals that occur approximately 0.15 sec below the weaker first arrivals labeled "sill". This was done primarily because ray theory cannot simultaneously match both sets of arrivals over the same range when the velocity of the shallow first arrival is greater than that of the deeper second arrival.

In order to be imaged as a separate unit, the sill/flow sediment complex must create an impedance contrast with layer 2 at least 1/4 of a seismic wavelength in thickness (> 50 - 125 m). If one is looking for a velocity discontinuity representing the base of sill/flow/sediment and top of layer 2 then the most appropriate factor to examine is the presence or absence of reflections. In the EMB, low-amplitude, discontinuous reflections observed 0.1-0.15 stwt below Horizon B may be due to the impedance contrast between Jurassic oceanic crust and pre-Aptian (>124.5 Ma) pelagic sediment with Aptian sills/ flows. The velocity at the top of this interval is approximately 5.3 km/s as revealed by weak, short-range refractions observed on sonobuoys in the EMB (Figure 14a,b) and approximately 5.8 km/s as determined by the average velocity of basalt samples from Site 802 [Lancelot and Larson et al. 1990]. Assuming an average velocity of 5.3 km/s for the interval between the top of Horizon B and top of the weak sub-Horizon B reflections (top of oceanic crust) results in a maximum thickness of 265-398 m for the Aptian flows/sills and interbedded sediment.

DISCUSSION

The thickness of middle Cretaceous igneous overburden and pre-middle Cretaceous sediments in the EMB and portions of the PB is estimated at 100 - 400 m based on the following corroborative evidence.

1) The similarities of upper crustal velocity structure and total crustal thicknesses derived from T/X sonobuoy data throughout the PB and EMB indicate a maximum thickness of approximately 1 km. However, based on weak sub-Horizon B reflections, the maximum thickness for this unit is better estimated at 400 m, as discussed above.

2) Synthetic seismograms created from a simplified impedance structure at Site 802, which employ the source waveform from FM35-12 and Mesopac II, indicate that a minimum sub-Horizon B layer thickness of 200 m can be resolved, that is the top of Horizon B and the top of oceanic crust appear as two separate reflection events. We interpret the discontinuous and low amplitude nature of reflections from the top of oceanic crust as due to the vertical resolution of the seismic signal and, to a lesser degree, the lack of seismic energy penetration resulting from the high reflection coefficient of Horizon B. In the majority of the EMB and PB the thickness of the sill/flow/sediment complex above oceanic crust is on the order of our ability to easily resolve seismically (e.g. ~200 m). When this layer thins, the already weak reflection from the top of oceanic crust merges with and becomes part of an interference pattern created by closely spaced impedance contrasts that begin with Horizon B. Furthermore, we expect that only 100 to 150 m of sediments are intruded and or capped by sills/flows at Sites 800 and 802, respectively, assuming that a sedimentary section comparable to Site 801 is representative of the PB and EMB for sediment deposited prior to the onset of middle Cretaceous volcanism.

3) The relatively high-relief, diffractive reflection associated with the top of oceanic crust at Site 801 is observed to extend semicontinuously to the northwest where the basement horizon alternates between rough, high-relief (e.g. Site 801) and relatively smooth and flat-lying (e.g. sills/flows) while remaining at the same depth. This type of reflection geometry is inconsistent with a widespread, thick overburden of igneous material.

In the EMB and PB middle Cretaceous sills/flows are identified along more than 4000 km of trackline throughout an area of approximately 500,000 km^2 (Figure 7). Assuming a maximum thickness of 500 m allows a volume estimate for deep sea middle Cretaceous sills/flows in the EMB/PB of ~0.25 x 10^6 km^3. Although areally extensive, the estimated volume is small compared to volume estimates for well-known flood basalts such as the Deccan Traps (~2 x 10^6 km^3 - Courtillot et al., 1987) and to large oceanic plateaus such as Ontong-Java (~ 50 x 10^6 km^3 - Schubert and Sandwell, 1989).

Tectonic Setting and Source of Deep Sea Volcanism

The MCS coverage of the EMB and PB along with bathymetry provide additional constraints on the morphotectonic expression of possible volcanic sources. The geometry of the "rough/smooth boundary" in the PB and the continuation of the middle Cretaceous igneous surface up onto the Caroline Ridge south of the EMB indicate that sources for these semicontinuous sills/flows extend to the south. Basalts recovered in the EMB are similar in age and in major, trace and isotopic element composition to basalts recovered in the Nauru Basin and the Ontong-Java Plateau (Figure 1), all showing strong affinities to MORB with a variably enriched mantle I (EM I) deep mantle source component [Castillo and Carlson, 1990; Mahoney and Tarduno, 1990; Castillo and Pringle, 1991; Tarduno et al., 1991]. These authors have speculated that the Nauru Basin and EMB were both sites of deep sea volcanism associated with the rapid formation of the Ontong-Java Plateau. Our seismic observations are consistent with this notion and indicate that the southeast PB was similarly affected. Northwest of the "rough/smooth boundary" (e.g., Sites 800 and 801) middle Cretaceous deep sea volcanism is much more restricted, being associated with individual seamounts and or seamount provinces as indicated by the comparable ages and distinct geochemistries of both the sills penetrated at Site 800 and samples dredged from Himu and Golden Dragon seamounts (Figures 2 and 7) [Smith et al., 1989; Castillo and Pringle, 1991, Pringle, 1992b; Castillo et al., 1992]. Continuity between these types of volcanic structures (seamounts, plateau, and deep sea flood basalts) is not observed seismically, so any inference of genetic relationships requires geochemical comparisons. However, the syncronicity in emplacement of these volcanic constructs with increased world-wide spreading rates (increased total crustal production) and the onset of the Cretaceous Normal Polarity Superchron (~124 Ma) is clear and may be the consequence of sudden heat and material transfer from the core/mantle boundary in the form of a rising plume resulting in excess mantle melt generation [Larson, 1991a,b].

The local volcanic sources and actual mode of emplacement of laterally extensive deep sea igneous provinces is still unknown. The most probable tectonically-controlled source location for middle Cretaceous volcanism is the Ogasawara Fracture Zone because of its association with large middle Cretaceous seamounts. Magnetic anomalies M22-M36 are offset approximately 500-600 km left-laterally from the EMB to the PB and are separated by a region approximately 300 km wide (Figure 2). The Ogasawara Fracture Zone that offsets these lineations lies somewhere within this broad zone, the two seismic profiles across this region indicate that the fracture zone lies along the northeastern margin of the EMB (e.g. southwestern dashed line in Figures 2, 4, and 7) where the depth to seafloor and to Horizon B is greatest (Figure 6). Seamounts of the Magellan Group are scattered within this region and appear to be aligned along the trend inferred for a fracture zone offset. Intra-plate volcanism in the form of volcanoes and elongate ridges that are localized along and within fracture zones is well documented and has been used as an argument for a weakened lithosphere along fracture zones [Vogt, 1974; Batiza, 1981; Lowrie et al., 1986]. The apparent alignment of the southern Magellan Seamounts along this fracture zone in addition to age constraints from dredge samples and cored volcanogenic material indicate that it became one of many focal points for mid-plate volcanism beginning in the Aptian and extending into the Santonian. The observation of the middle Cretaceous flow/sill unit (Horizon B) possibly onlapping Ita Mai Tai and another unnamed large guyot in the EMB, as well as the distinct isotopic signatures of both Himu seamount and the sills recovered at Site 800, indicate that these Aptian age seamounts may have acted as a local source for deep sea volcanism. The recovery of Aptian pillow basalt at Site 802, located more than 300 km from the most proximal seamount, and the huge areal extent of these basalts, indicate that individual seamounts are not the direct volcanic sources in the majority of the EMB and southeast PB.

Large seamounts of the Magellan Group, however, may provide another, more indirect, mechanism for deep sea lava flow emplacement. Lipman et al. [1989] and Clague et al. [1990] report the presence of widespread, young lava flows at several sites along the Hawaiian arch approximately 200 km from the island chain. These remarkably flat-lying areas of up to 25,000 km^2, which are characterized by high-acoustic backscatter and hyperbolic reflections on GLORIA and seismic reflection images respectively, correspond to young lava flows. The spatial association of sheet flows with the Hawaiian Flexural Arch, as well as with clusters of Cretaceous seamounts, led Clague et al. [1990] to speculate that lavas are derived from magma concentrated beneath the upbowed lithosphere of the flexural arch and delivered to the surface where the lithosphere locally responds to the seamount load by fracturing.

Individual flows may extend for 100 km, however, the vents for these extensive Hawaiian Arch sheetflows have not been identified, despite the lack of sediment cover [Clague et al., 1990]. Similarly, the morphotectonic expression of such vents, fissures or rifts (spreading or not) is not obvious in the available bathymetric maps or limited seismic profiles in the PB/EMB. In fact, it is difficult to detect tectonic fabric of any kind in the EMB or southeastern PB, most likely due to the masking effects of the pervasive flows and sills. The example of Hawaiian Arch flows also indicates that the seismic expression of potential vents and fissures can be quite subdued. However, the subtle, localized deepening and associated increase in roughness of Horizon B, which we have interpreted as pillow/flow units in the EMB, may represent such fissures and vents (Figure 8).

Magnetic Rough-Smooth Boundary

The distinctive change in the seismic character of Horizon B in the PB from "smooth" to "rough" can also be correlated to a change in magnetic anomaly amplitudes and field strength [Handschumacher et al., 1988]. The "rough/smooth" Horizon B boundary (Figure 4) is nearly coincident with the magnetic boundary interpretation of Handschumacher et al. [1988] which marks a change in magnetic anomalies from higher field/lower-amplitude anomalies to lower field/higher-amplitude anomalies. A comparable seismic-magnetic boundary within the EMB, which is characterized by the nearly ubiquitous presence of the high-amplitude flat-lying Horizon B, is not apparent. Handschumacher et al. [1988] suggested that the distinct change in magnetic character in the PB may represent a change in spreading rate and/or direction, or a fossil plate boundary. We suggest an additional possibility, that both seismic and magnetic rough-smooth boundaries mark the geographic extent of continuous, middle Cretaceous, igneous overburden in the PB. If this is the case then the change in magnetic anomaly character at the "rough/smooth boundary" is due to the edge of normally magnetized Cretaceous sills/flows overlying and adjacent to mixed polarity Jurassic oceanic crust.

A similar seismic-magnetic "rough/smooth" boundary associated with the central Venezuelan Basin Fault Zone (CVF) in the Caribbean Sea has also been recognized. The CVF has been interpreted as marking the boundary between an anomalously thick Late Cretaceous igneous province built upon pre-existing oceanic crust and normal thickness oceanic crust [Talwani et al. 1977; Diebold et al., 1981]. While the crustal thickness variation across the "rough/smooth boundary" in the PB is not as large, there is also the possibility that this "boundary" may mark the location of a fracture zone (Mendocino?) offsetting the proto-Pacific/Farallon ridge [Handschumacher et al., 1988], which in middle Cretaceous time acted as a structural boundary to sills and flows originating from sources to the south.

CONCLUSIONS

1) Analysis of seismic travel time-range data along with drilling results from ODP Leg 129 clearly establish that Horizon B represents the top of high-velocity, massive igneous material throughout the EMB and PB. Horizon B displays a range of reflection characteristics between two end-member modes. It corresponds to the top of Jurassic oceanic crust where it has a high-relief, undulating, lower amplitude and diffractive character. It corresponds to middle Cretaceous sills/flows overlying oceanic crust where it appears flat-lying, high-amplitude, discontinuous and often "smooth". These Horizon B reflection characteristics calibrated by Leg 129 drill sites indicate that a semicontinuous sequence of middle Cretaceous deep sea flows/sills overlies Jurassic sediments and oceanic crust throughout an area of approximately 500,000 km^2 in the EMB and southeast PB.

2) Sonobuoy refraction data indicate no significant (> 1 km) difference in crustal thicknesses between areas overlain by documented middle Cretaceous sills/flows and areas with Jurassic oceanic crust free of such younger igneous overburden. Based on sub-Horizon B reflections and velocity information from sonobuoys, we estimate the maximum thickness of the middle Cretaceous sill/flow/sediment complex at 100 - 400 m. The volume of deep sea sills and flows in the EMB/PB is estimated at 0.25×10^6 km^3, approximately 200 times less than the volume estimated for the Ontong-Java Plateau.

3) A distinct smooth-to-rough Horizon B boundary is found in the southeast PB where the continuous overburden of middle Cretaceous flows/sills pinch out and expose Jurassic oceanic crust. Northwest of this boundary, occurrences of late Barremian - early Aptian sills/flows are documented, but are believed to be localized and associated with individual seamounts of similar age and isotopic character.

4) The change in magnetic anomaly character in the so called Jurassic Magnetic Quiet Zone of the PB from rough to smooth is associated with the change in reflection character of Horizon B from smooth to rough. This kind of magnetic and seismic boundary is not observed in the EMB, and it is believed both boundaries mark the edge of pervasive deep sea sill and flow overburden. It is suggested that this "rough/smooth boundary" was created by a fracture zone offset of the proto-Pacific/Farallon plate boundary.

5) The Ogasawara Fracture Zone acted as a focal point for Aptian-age and younger extrusive activity as indicated by the alignment and age of seamounts. There appears to be morphotectonic and geophysical evidence for both fracture zone and flexural effects within the broad region separating the PB and EMB. This feature has separated the two basins since at least 160 Ma.

6) Intrabasin morphotectonic structures other than seamounts, which could suggest a local (<100 km) source location for the extensive cover of deep sea sills and flows, are not evident on seismic or bathymetric records. However, the continuity of the sill/flow surface up the Caroline Ridge, the orientation of the "rough/smooth" Horizon B boundary, age and geochemical similarities between MORBs recovered in the Nauru Basin, EMB and the Ontong-Java Plateau suggest that the volcanism in these areas are closely related. We speculate that the deep sea sills/flows of varying thicknesses found in the EMB, southeast PB were locally emplaced through seamounts such as Ita Mai Tai with associated arch volcanism, and regionally emplaced along fissures/failed rifts which are manifested as subtle changes in depth and reflection character of volcanic basement.

Acknowledgement. This work was part of a joint U.S. - French effort to investigate the western Pacific basins. We wish to thank the ships crew and scientific party of the R/V Fred Moore (RIP) cruise FM35-12 and N/O Le Suroit. Perry Crampton and fellow airgun techs worked beyond the call of duty. Steffen Saustrup, Mark Wiederspahn, Patty Ganey at UTIG and Marc Schaming, Jacques Renault, Cecille Pourcellot and Roland Schlich at IPG, Strasbourg provided invaluable assistance in the processing of FM35-12 and MESOPAC II seismic data. John Mutter kindly donated the use of the L-DGO processing facilities during sonobuoy analysis. John Diebold contributed interactive software (JDSEIS), and along with Emilio Vera, provided helpful advice and discussion concerning sonobuoy data analysis/interpretation. Discussions with Nancy Grindlay, and Malcolm Pringle and reviews from Beth Ambos, Tom Brocher and Will Sager improved this paper. This project was funded by NSF grants OCE 86-13867 (R.Larson), OCE 86-13641 (T. Shipley) and from JOI/USSAC Ocean Drilling Fellowship (JSG-CY5-4) and a USSAC grant (20431) to the first author.

REFERENCES

Abrams, L.J., R.L. Larson, T. Shipley and Y. Lancelot, Cretaceous volcanic sequences and Jurassic(?) crust in the western Pacific, *Eos ,Trans. AGU, 69*, 1442, 1988.

Abrams, L.J., R. L. Larson, T. Shipley and Y. Lancelot, The seismic stratigraphy of the East Mariana and Pigafetta Basins of the western Pacific, *Proc. ODP Sci. Res., 129*, 1992.

Asada, T., H. Shimamura, S. Asano, K. Kobayashi, and Y. Tomoda, Explosion seismological experiments on long-range profiles in the Northwestern Pacific and the Marianas Sea, in *The Ocean Basins and Margins*, Vol. 7A, edited by A. E. M. Nairn, Stehli, and S. Uyeda, 1985.

Batiza, R., Lithospheric age dependence of off-ridge volcano production in the North Pacific, *Geophys. Res. Lett., 8*, 853-856, 1981.

Brenner, C. and M. Angell, Bathymetric map of the East Mariana and Pigafetta Basins in the western Pacific, *Proc. ODP, Sci. Res., 129*, 1992.

Castillo, P. R., and R. W. Carlson, Possible origin of widespread Cretaceous volcanism in the Pacific: constraints from Sr, Nd, and Pb isotope geochemistry of Nauru Basin thoeliites, *Eos, Trans. AGU, 71*, 1667, 1990.

Castillo, P. R., and M. S. Pringle, Cretaceous volcanism in the western Pacific sampled at Sites 800 and 802, ODP Leg 129, *Eos, Trans. AGU, 72*, 300, 1991.

Castillo, P. R., P. A., Floyd, and C, France-Lanord, Isotope geochemistry of ODP Leg 129 basalts: Implications for the origin of a widespread volcanic event in the Pacific, *Proc. ODP Sci. Res., 129,* 1992.

Clague, D. A., R. T. Holcomb, J. M. Sinton, R. S. Detrick, and M. E. Torresan, Pliocene and Pleistocene alkalic flood basalts on the seafloor north of the Hawaiian islands, *Earth Planet. Sci. Lett., 98*, 175-191, 1990.

Courtillot, V. E., J. Besse, D. Vandamme, R. Montigny, J. J. Jeager, and H. Cappetta, Deccan flood basalts at the Cretaceous/Tertiary boundary?, *Earth Planet. Sci. Lett., 80*, 361-374, 1987.

Diebold, J. B., and P. L. Stoffa, The traveltime equation, tau-p mapping and inversion of common midpoint data, *Geophysics, 46*, 238-254, 1981.

Diebold, J. B., P. L. Stoffa, P. Buhl, and M. Truchan, Venezuela Basin Crustal Structure, *J. Geophys. Res., 86*, 7901-7923, 1981.

Edgar, N. T., and J. B. Saunders et al., *Init. Repts. DSDP, 15*: pp. 1137, 1973.

Ewing, J., M. Talwani, M. Ewing, and T. Edgar, Sediments of the Caribbean, in *International Conference on Tropical Oceanography Proc., 5*, pp. 88-102, Univ. of Miami, Miami, Fla., 1967.

Ewing, J., M. Ewing, T. Aitken, and W. J. Ludwig, North Pacific sediment layers measured by seismic profiling, in *The Crust and Upper Mantle of the Pacific Area*, edited by C. L. Drake, and P. J. Hart, pp. 147-173, AGU Monogr. Ser. v. 12, AGU, Washington, D. C., 1968.

Ewing, J. I., and G. M. Purdy, Upper crustal velocity structure in the ROSE area of the East Pacific Rise, *J. Geophys. Res., 87*, 8397-8402, 1982.

Floyd, P. A., P. R. Castillo, and M. S. Pringle, Tholeiitic and alkalic basalts of the oldest Pacific Ocean crust, *Terra Nova, 3*, 257-265, 1991.

Haggerty, J. A., and I. Primoli Silva et al., *Proc. ODP Init. Repts., 144:* College Station, TX (Ocean Drilling Program), in press, 1993.

Hamilton, E. L., Sunken islands of the Mid-Pacific Mountains, *Geol. Soc. Am. Mem., 64,* pp. 97, 1956.

Handschumacher, D. W., W. W. Sager, T. W. C. Hilde, and D. R. Bracey, Pre-Cretaceous tectonic evolution of the Pacific plate and extension of the geomagnetic polarity reversal time scale with implications for the origin of the Jurassic "Quiet Zone", *Tectonophysics, 155*, 365-380, 1988.

Harland, W. B., R. L. Armstrong, A. V. Cox, L. E. Craig, A. G. Smith, D. G. Smith, *A geologic time scale 1989*, 263 p., Cambridge Univ. Press, Cambridge, UK, 1990.

Heezen, B. C. and, I. D. MacGregor, et al., Mesozoic chalks beneath the Caroline abyssal plain: DSDP Site 199, *Init. Repts. DSDP, 20*, 65-85, 1973.

Hilde, T. W. C., Isezaki, N., and Wageman, J. M., Mesozoic seafloor spreading in the North Pacific, in *The Geophysics of the Pacific Ocean Basin and its Margins*, edited by G. H. Sutton, M. H. Manghnani, and R. Moberly, pp. 205-226, AGU Geophys. Monogr. Ser., vol. 19, 1976.

Houtz, R. E., J. Ewing, and P. Buhl, Seismic data from sonobuoy stations in the northern and equatorial Pacific, *J. Geophys. Res., 75*, 5093-5111, 1970.

Houtz, R., and J. Ewing, Upper crustal structure as a function of plate age, *J. Geophys. Res., 81*, 2490-2498, 1976.

Houtz, R. E., and W. J. Ludwig, Structure of Columbia Basin, Caribbean Sea, from profiler-sonobuoy measurements, *J. Geophys. Res., 82*, 4861-4868, 1977.

Houtz, R. E., and W. J. Ludwig, Distribution of reverberant subbottom layers in the southwest Pacific basin, *J. Geophys. Res., 84*, 3497-3505, 1979.

Hutchinson, D. R., and R. S. Detrick, Water gun vs air gun: a comparison, *Mar. Geophys. Res., 6*, 295-310, 1984.

Kroenke, L. W., and W. H. Berger et al., *Proc. ODP Init. Repts., 130*: College Station, TX (Ocean Drilling Program), pp. 1240, 1991.

Lancelot, Y., and R. L. Larson et al., *Proc. ODP Init. Repts., 129:* College Station, TX (Ocean Drilling Program), pp. 488, 1990.

Larson, R. L. and C. G. Chase, Late Mesozoic evolution of the western Pacific Ocean, *Geol. Soc. Am. Bull., 83*, 3627-3644, 1972.

Larson, R. L., Late Jurassic and Early Cretaceous evolution of the western central Pacific Ocean, *J. Geomag. Geoelectric., 28*, 219-236, 1976.

Larson, R. L. and, S. O. Schlanger, et al., *Init. Repts. DSDP, 61*: pp. 885, 1981.

Larson, R. L., Latest pulse of the Earth: Evidence for a mid-Cretaceous super plume, *Geology, 19*, 547-550, 1991a.

Larson, R. L., Geologic consequences of super plumes, *Geology, 19*, 963-966, 1991b.

Lincoln, J. M., M. S. Pringle, and I. Primoli Silva, Early and Late Cretaceous volcanism and reef-building in the Marshall Islands: New fossil evidence and Ar^{40}/Ar^{39} ages, this volume.

Lipman, P. W., D. A. Clague, and R. T. Holcomb, South arch volcanic field - newly identified young lava flows on the sea floor south of the Hawaiian Ridge, *Geology, 17*, 611-614, 1989.

Lowrie, A., C. N. Smoot, and R. Batiza, Are oceanic fracture zones strong or weak? New evidence for volcanic activity and weakness, *Geology, 14*, 242-245, 1986.

Macdonald, K. C., Mid-ocean ridges: Fine scale tectonic, volcanic and hydrothermal processes within the plate boundary zone, *Ann. Rev. Earth Planet. Sci., 10*, 155-190, 1982.

Mahoney, J. J., and J. A. Tarduno, Cretaceous volcanism and the formation of the Ontong-Java plateau: New insights from ODP Leg 130, *Eos, Trans. AGU, 71*, 1668, 1990.

Mammerickx, J., and G. F. Sharman, Tectonic evolution of the North Pacific during the Cretaceous quiet period, *J. Geophys. Res., 93*, 3009-3040, 1988.

Menard, H. W., *Marine Geology of the Pacific*, pp. 271, McGraw Hill Book Co., New York, NY, 1964.

Moberly, R., and S. O. Schlanger et al., *Init. Repts. DSDP, 89*, pp. 678, 1986.

Nakanishi, M., K. Tamaki, and K. Kobayashi, Mesozoic magnetic anomaly lineations and seafloor spreading history of the northwestern Pacific, *J. Geophys. Res., 94*, 15,437-15,462, 1989.

Nakanishi, M., K. Tamaki, and K. Kobayashi, Magnetic anomaly lineations from Late Jurassic to Early Cretaceous in the west-central Pacific Ocean, *Geophys. J. Int., 109*, 701-719, 1992.

Ozima, M., I. Kaneoka, K. Saito, M. Honda, M. Yanagisawa, and Y. Takigami, Summary of geochronological studies of submarine rocks from the western Pacific Ocean, in *Geodynamics of the western Pacific-indonesian region, v.11*, edited by T .W .C. Hilde, and S. Uyeda, 137-142, 1983.

Pringle, M. S., Radiometric ages of basaltic basement recovered by ODP Leg 129, *Proc. ODP Sci. Res., 129*, 1992a.

Pringle, M. S., Geochronology and petrology of the Musicians Seamounts and the search for hot spot volcanism in the Cretaceous Pacific, (Ph.D. dissertation), University of Hawaii, Honolulu, 1992b.

Purdy, G. M., The seismic structure of 140 my old crust in the Central Atlantic Ocean, *Geophys. J. R. Astron. Soc., 72*, 115-138, 1983.

Sager, W. W., D. W. Handschumacher, T. W. C. Hilde, and D. R. Bracey, Tectonic evolution of the northern Pacific plate and Pacific - Farallon - Izanagi triple junction in the Late Jurassic and Early Cretaceous (M21-M10), *Tectonophysics, 155*, 345-364, 1988.

Schubert, G., and D. Sandwell, Crustal volumes of the continents and of oceanic plateaus and continental submarine plateaus, *Earth Planet. Sci. Lett., 92*, 234-246, 1989.

Shipley, T. H., J. M. Whitman, F. K. Duennebier, and L. D. Peterson, Seismic stratigraphy of the East Mariana Basin, western Pacific, *Earth Planet. Sci. Lett., 64*, 257-275, 1983.

Shipley, T. H., L. J. Abrams, R. L. Larson, and Y. Lancelot, Extent of Cretaceous volcanic sequences in the Jurassic Nauru Basin, western Pacific, this volume.

Smith, W. H. F., H. Staudigel, A. B. Watts, and M. S. Pringle, The Magellan Seamounts: Early Cretaceous record of the South Pacific isotopic and thermal anomaly, *J. Geophys. Res., 94*, 10,501-10,523, 1989.

Spudich, P., and J. A. Orcutt, A new look at the seismic velocity structure of the oceanic crust, *Rev. Geophys. Space Phys., 18*, 627-645, 1980.

Steiner, M., J. Ogg, and J. Sandoval, Jurassic magnetostratigraphy, 3. Bathonian-Bajocian of Carcabuey, Sierra Harana and Campillo de Arenas (Subbetic Cordillera southern Spain), *Earth Planet. Sci. Lett., 82*, 357-372, 1987.

Talwani, M., C. C. Windisch, P. L. Stoffa, P. Buhl, and R. E. Houtz, Multi-channel seismic study in the Venezuelan Basin and the Curacao ridge, in *Island Arcs, Deep Sea Trenches and Back-Arc Basins*, edited by M. Talwani, and W. C. Pitman, pp. 83-98, AGU Monogr. M. Ewing Ser. v. 1, 1977.

Tamaki, K. and R. L. Larson, The Mesozoic tectonic history of the Magellan microplate in the Western Central Pacific, *J. Geophys. Res., 93*, 2857-2874, 1988.

Tamaki, K., M. Nakanishi, K. Sayanagi and K. Kobayashi, Jurassic magnetic anomaly lineations of the Western Pacific and the origin of the Pacific plate, *Eos, Trans. AGU, 68*, 1493, 1987.

Tarduno, J. A., W. V. Sliter, L. Kroenke, M. Leckie, H. Mayer, J. J. Mahoney, R. Musgrave, M. Storey, and E. L. Winterer, Rapid formation of the Ontong-Java Plateau by Aptian mantle plume volcanism, *Science, 254*, 399-403, 1991.

Thiede, J., and T. L. Vallier et al., *Init. Repts. DSDP, 62*, pp. 1120, 1981.

Vogt, P. R., Volcano spacing, fractures and thickness of the lithosphere, *Earth Planet. Sci. Lett., 21*, 235-252, 1974.

Weissel, J. K. and R. N. Anderson, Is there a Caroline plate?, *Earth Planet. Sci. Lett., 41*, 143-158, 1978.

Wipperman, L. K., R. L. Larson, and D. M. Hussong, The geological and geophysical setting near site 462, *Init. Repts. DSDP, 61*, 763-770, 1981.

Winterer, E. L., and J. Ewing et al., *Init. Repts. DSDP, 17:*, pp. 930, 1973.

Winterer, E. L., Anomalies in the tectonic evolution of the Pacific, in *The Geophysics of the Pacific Ocean Basin and its Margin*, edited by G. H. Sutton, M. H. Manghnani, and R. Moberly, pp. 269-278, AGU Monogr. Ser. v. 19, AGU, Washington, D.C., 1976.

Winterer, E. L., Duncan, R. A., McNutt, M. K., Natland, J. H., Premoli Silva, Isabella, Sager, W. W., Sliter, W. V., Van Waasbergen, R., and Wolfe, C. J., Cretaceous guyots in the northwest Pacific: An overview of their geology and geophysics, this volume.

White, R. S., Oceanic upper crustal structure from variable angle seismic reflection - refraction profiles, *Geophys. J. R. astr. Soc., 57*, 683-726, 1979.

White, R. S., Atlantic oceanic crust: seismic structure of a slow-spreading ridge, in *Ophiolites and Oceanic Lithosphere*, edited by I. G. Gass, S. J. Lippard and A. W. Shelton, pp. 101-111 Geol. Soc. London, 1984.

L. J. Abrams, now at Department of Geology, University of Puerto Rico, P.O. Box 5000, Mayaguez, PR 00709.

R. L. Larson, Graduate School of Oceanography, University of Rhode Island, Narragansett, RI 02882.

T. H. Shipley, Institute for Geophysics, University of Texas, Austin, TX 78759.

Y. Lancelot, Laboratoire de Geologie du Quaternaire, CNRS-Luminy, Case 907, 13288 Marseille, Cedex 9, France.

Late Jurassic-Early Cretaceous Oceanic Crust and Early Cretaceous Volcanic Sequences of the Nauru Basin, Western Pacific

Thomas H. Shipley

University of Texas, Institute for Geophysics, Austin, Texas 78759

Lewis J. Abrams

Graduate School of Oceanography, University of Rhode Island 02882

Yves Lancelot

Laboratoire de Geologie du Quaternaire, CNRS, Case 907, 13288 Marseille Cedex 9

Roger L. Larson

Graduate School of Oceanography, University of Rhode Island 02882

Seismic reflection data totaling 2000 km in the Nauru Basin collected in 1986-1989 reveal widespread Early Cretaceous volcanic sequences and the division of the Nauru Basin into northern and southern sub-basins. Oceanic crust is observed in the southern Nauru Basin. Acoustic basement in the northern Nauru Basin is a fairly smooth surface that gently rises towards the northeast. It corresponds to the top of drilled Cretaceous volcanics. Near Site 462 discontinuous reflections correlate with the upper portion of the drilled section of intrusives, but the top of the oceanic crust is not observed in the seismic data. The nature of the northern Nauru Basin remains enigmatic considering the thick Cretaceous volcanics which overlie what is thought to be oceanic crust characterized by Jurassic magnetic anomalies. One 1300 km long N-S seismic profile extends from 10° N to 2° S along 165° E longitude. This track is perpendicular to magnetic anomalies M29 to M15. Along this profile, south of about 2.5° N, the acoustic basement is twice as deep and we refer to this area as the southern Nauru Basin. The additional section beneath a thin (~ 100 m) Cretaceous volcanic interval (Kv) is 400 to 700 m thick and has a velocity of about 2.0 km/s. This section overlies acoustic basement with relief of 150 m to 300 m and a velocity of 5.9 km/s interpreted as the top of M21 to M15 oceanic crust. Observed intermittently along a 300 km long portion of the profile are a number of high amplitude reflection segments within the thick sedimentary section, some of which are probably sills, and others discrete volcanic flows. We still have little understanding of the emplacement mechanism for widespread Cretaceous volcanic flows and sills, though we speculate that several broad low-relief bathymetric highs that separate the Nauru Basin into northern and southern sub-basins may be relict volcanic centers or rift sources.

Introduction

Early Cretaceous volcanism in the central and western Pacific is manifested as linear seamount chains (e.g. Marshall-Gilbert, Wake), thick plateaus (e.g., Hess, Shatsky, Ontong-Java), and basin filling flood basalts. Recent work has established that the origins of all three types of volcanics are related in time and space [Castillo, et al., 1991; Schlanger and Premoli Silva, 1981; Tarduno et al., 1991]. One recent speculation for the cause of this voluminous, widespread volcanism is that a Cretaceous super plume in the central Pacific produced a regional Early Cretaceous peak in volcanic production [Larson, 1991]. Many seamount chains have time and magmatic progressions most easily related to "fixed" mantle hot spots and are often associated with plateau or flood basalt provinces [Wilson, 1965; Morgan, 1972; Duncan and Richards, 1991; Coffin and Eldholm, 1991]. Little attention has been given to accurately estimating the volume of basin

The Mesozoic Pacific: Geology, Tectonics, and Volcanism
Geophysical Monograph 77

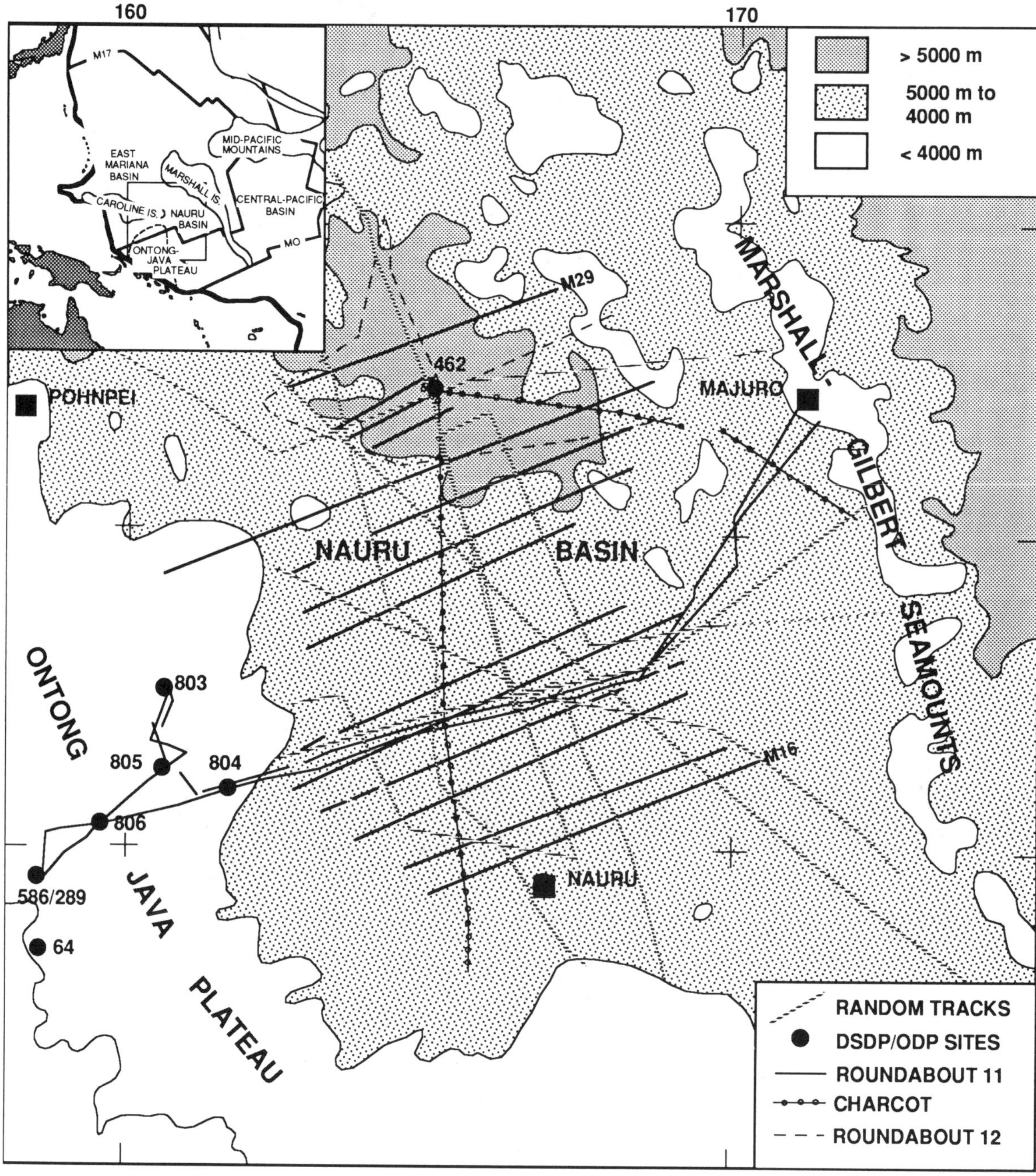

Fig. 1. A simplified bathymetric map of the west central Pacific after Iwabuchi [1979] and Monahan et al. [1982]. Tracks show location of seismic data used in this study. Magnetic anomalies (M29 to M16) are from Cande et al. [1989]. Inset shows regional location.

volcanics, though it is thought to be dwarfed by both plateau and seamount production. For example, recent estimates of the Ontong Java Plateau volume range from about 50 to 60 million km^3 [Coffin and Eldholm, 1991; Larson, 1991; Schubert and Sandwell, 1989]. Even if basin volcanics are a relatively small volume compared to the plateaus, they are areally extensive and their underlying volcanic and tectonic mechanisms of emplacement are not understood.

The purpose of this paper is to investigate the Cretaceous volcanic sections in the deep Nauru Basin in the western Pacific using seismic reflection data and the drilling results at DSDP Site 462 (Figure 1).

The Nauru Basin extends from about 12° N to 3° S along longitude 165° E and lies at depths greater than 4000 m (Figure 1). It also extends east-west from the Marshall-Gilbert Seamount chain to the Ontong-Java Plateau, both Cretaceous volcanic structures. Thus, we define the Nauru Basin as somewhat larger than the typical description, covering more than 1.7 million km^2. On the basis of shallow structure, we divide the basin into north and south sub-basins, separated by a 4200 m high ridge trending WNW at 2 to 4° N. The sub-basins are referred to throughout the following discussion (Figure 2) [Mammerickx and Smith, 1985]. Another paper in this special volume examines Cretaceous volcanism in two basins to the north, the East Mariana and Pigafetta Basins [Abrams et al., this volume]. Emplacement mechanisms for basin wide volcanism usually assume multiple point or line sources that create flows that are unable to advance more than a few hundred kilometers at most, although shallow injection of sills might lead to greater lateral extent [Clague et al., 1990; Lipman et al., 1989]. This tectonics allows multi-centered rifts that do not disrupt the pre-existing Jurassic and Early Cretaceous magnetic anomalies. Alternatively, Castillo et al. [1991] have speculated that the northern Nauru Basin may have been a Cretaceous ridge crest developed by rifting pre-existing Jurassic crust. However, this theory does not explain the preservation of the M-series anomalies. The mode of emplacement of the volcanic sequence and the tectonic setting and age of the Nauru Basin continue to be problematic.

Larson [1976] and Cande et al. [1978] identified Late Jurassic-Early Cretaceous magnetic anomalies in the Nauru Basin. However, depth to 'basement' in this region is 1000 m shallower than its Jurassic age predicts [Renkin and Sclater, 1988]. In contrast to the Oxfordian age predicted by the magnetic anomalies, DSDP Site 462 demonstrated that either the Cretaceous volcanism in the deep sea is much more extensive than previously recognized, or that the magnetic anomaly identifications were incorrect [Larson et al., 1976 and Moberly et al., 1986]. Over 649 m of mostly Albian (~110 Ma) basalts were recovered beneath Cenomanian to Recent sedimentary rocks [Pringle, 1992; Ozima et al., 1981]. Following earlier discoveries of unusually shallow and young basement at other sites in the region, Winterer [1976] concluded that younger basalts overlie older Cretaceous and Jurassic sediments and oceanic crust in significant parts of the western Pacific basins. Fourteen years after discovery of thick Cretaceous volcanics at Site 462, we still do not know if the magnetic anomaly age is correct or if the thick sequence of volcanics is true younger oceanic crust [Castillo et al., 1991].

Random ship tracks with seismic data exist in the basin but all were collected with small sound sources (Figure 1). All correlations of seismic reflection data are tied to DSDP Site 462 site. It should be noted that Site 462 is located in a 10 km wide subsurface depression in which the top of Cretaceous volcanics is about 150 ms (~ 250 m) deeper than the surrounding area (Figure 3). This offset must be considered when correlating DSDP Site 462 results to regional seismic reflection and velocity data.

The basic stratigraphy at Site 462 consists of:

1. 649 m of volcanic flows and sills with few intercalated sediments of mainly Albian age [Tarduno et al., 1991; Pringle, 1992]. The section is divided into two parts: the lower 489 m consists of basaltic sheet flows, minor pillow lavas and minor red brown silt units and the upper 160 m consists of basalt sills and small amounts of interbedded volcanogenic sediments; and [Larson et al., 1981; Moberly et al., 1986; Castillo et al., 1991]. The top of the volcanics is identified as K in the figures.
2. 112 m of volcanogenic and zeolitic sandstone, mudstone and limestones of Albian to Maestrichtian age (this section may be thicker at Hole 462A than elsewhere because of the siting within a localized depression).
3. 150 m of cherts, chalks and limestones, Maestrichtian to Eocene age. The top of this unit is identified as mE in the figures.
4. 297 m of calcareous and radiolarian oozes and chalks, mainly turbidite deposits of late Eocene and younger age identified as mM (middle Miocene) and eM (early Miocene) in the figures.

SEISMIC DATA

Our investigation relies on various seismic reflection data, limited velocity data, and Site 462 results. The seismic data presented here were collected between 1986 and 1989 on a cruise of the N/O *Charcot* (Mesopac 1) and two on the R/V *Thomas Washington* (Roundabout 11 and 12). The *Charcot* used a high resolution, implosive source sound (flexichoc) and a 24-channel receiving array. Most of the Roundabout 11 data were collected as single channel data using a 1.3 l water gun and did not penetrate the Eocene cherts and Cretaceous volcanics. On one Roundabout 11 line crossing of the *Charcot* line where deep reflections were observed near 1.9° N, we deployed a commercial low-frequency sonobuoy and used a 32.8 l air gun to obtain a useful reflection line and Sonobuoy 2 (Figure 2). Roundabout 12 collected 8-channel, 33.3 m group, 32.8 l air gun data and commercial sonobuoys in the northern Nauru Basin. The reflection data were deconvolved and stacked into 133 m bins of about 14-fold. Our goal was to see through the thick Cretaceous volcanic section and to identify true oceanic basement. No previous seismic work in the vicinity of Site 462 is relevant to these basement studies because of the reverberant nature of the section and low energy of the sources used [Houtz, 1976; Wipperman et al., 1981; Whitman, 1985]. The data we present have a high vertical exaggeration for basin wide correlations.

Our results appear in three cross sections of the entire basin and in some key seismic reflection examples (Figures 4, 5 and 6). Because of sparse seismic coverage, we were restricted in the choice of reflections that maintain unique characteristics over long distances (e.g., seismic facies). These four are labeled mM (middle Miocene), eM (early Miocene), mE (middle Eocene) which is approximately the top of the zone containing variable amounts of chert, and K (Albian (?)), which is the top of the volcanics or their time-equivalent. The object of this paper is the volcanism and not the post-Late Cretaceous stratigraphy. Therefore the younger stratigraphy will only be briefly discussed. We will begin with the seismic correlation with the drilled volcanics at Site 462 in the northern Nauru Basin (Figure 3).

BASIN STRATIGRAPHY

Basement to Cretaceous Volcanics

Northern Basin. The top of the Cretaceous volcanics, K, is easily identified in relative true-amplitude plots by its very high amplitude.

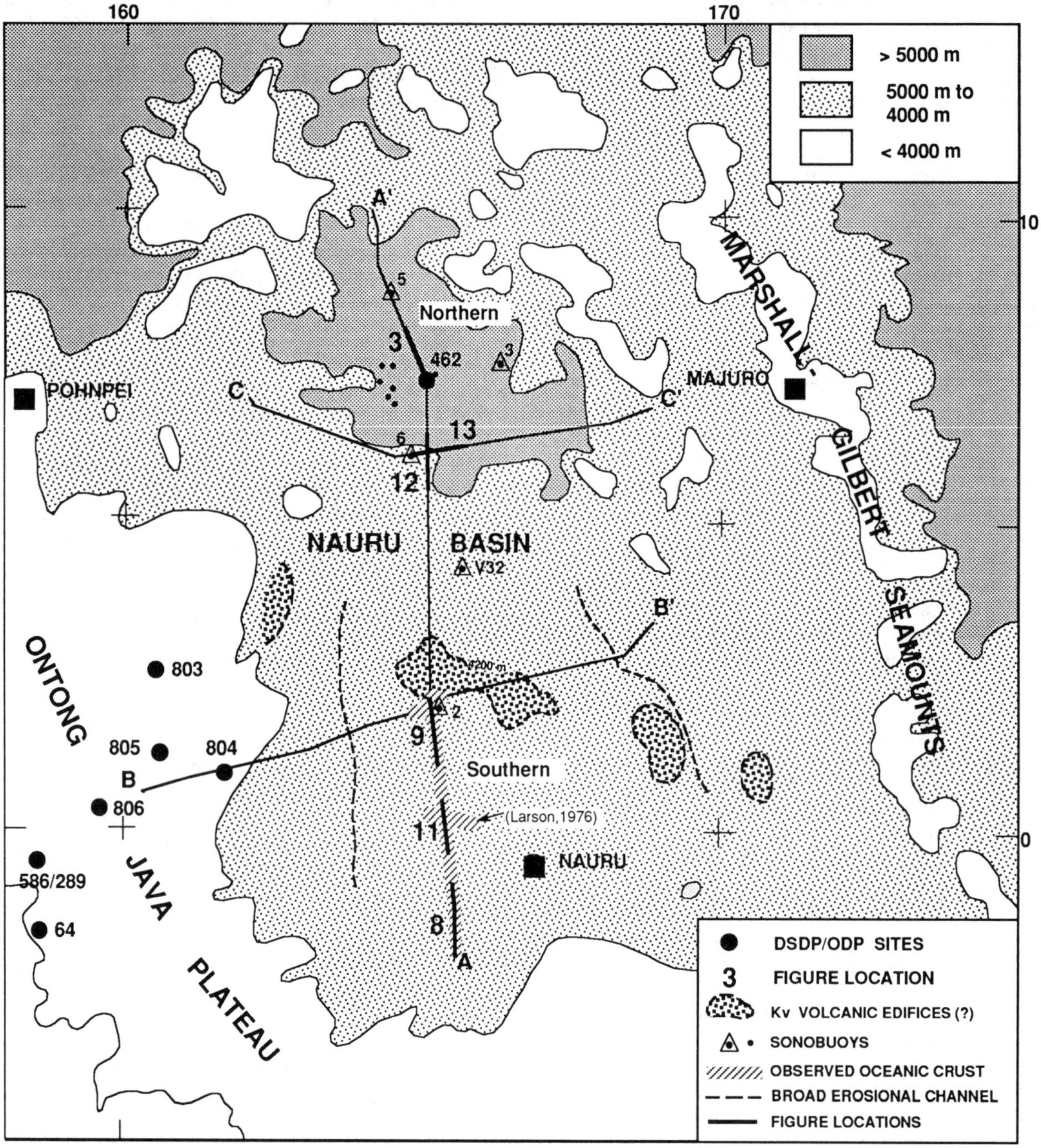

Fig. 2. Locations of cross sections, seismic sections and sonobuoy data shown in subsequent figures and referred to in the text. Hash marks along track outlines the known limit of the normal oceanic crust observed in the seismic data (contours after Mammerickx and Smith [1985]). Bold stipples refer to speculation in text that these low rises may represent Cretaceous volcanic centers in the Nauru Basin. Triangles with dots are sonobuoys collected on Roundabout Legs 11 and 12, except V32 from Houtz [1976]. Seven dots without triangles are locations of Wipperman et al. [1981] sonobuoys near Site 462. Because of scale, OBS 1 location is not shown but is about 20 km northwest of Site 462. Short line segment crossing A-A' at 0.5° N is from Larson [1976].

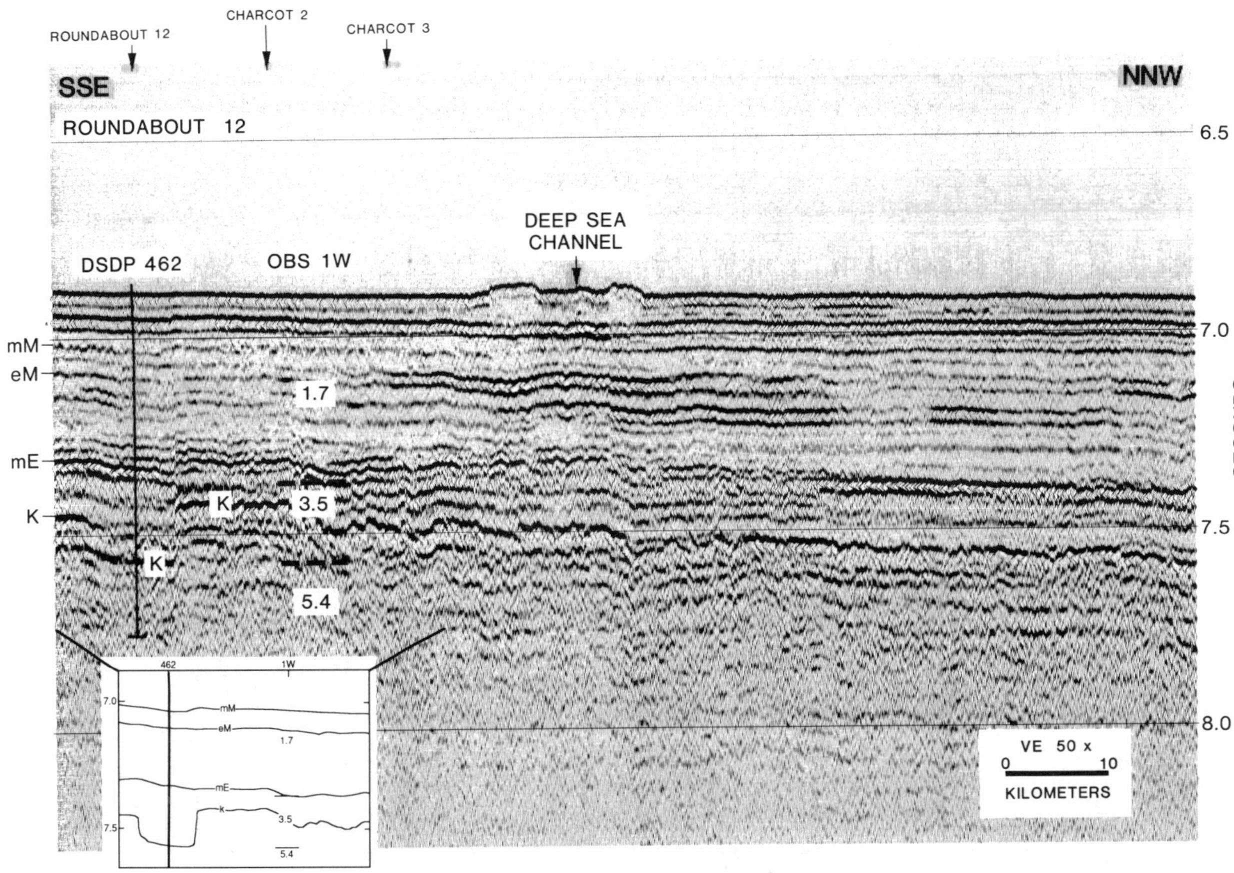

Fig. 3. Seismic section from near DSDP Site 462. The drill site is actually in a local low as defined by the top of the volcanics (K). Indistinct seismic layering in Kv is common in the basin north of 6° N and may correlate with the upper portion of the drilled section of intrusives as Site 462. OBS 1W is projected onto this line [Wipperman et al., 1981]. The drill site is in a local depression offset from the OBS cross section. This observation makes the refraction data fit better with the velocities derived from the drilling program.

It is usually a simple, single reflection which is the highest amplitude in the section, even exceeding that of the seafloor. It is often highly diffractive suggesting small scale roughness (Figure 3). It is usually overlain by closely spaced high amplitude reflections that define a unit of variable thickness, and correlates to the chert-rich section. Identification of the top of the volcanics at Site 462 in previous seismic data was difficult [Wipperman et al., 1981]. Interpretation of the stratigraphy and particularly basement is possible because of modern seismic systems and processing [e.g., Shipley et al., 1983; Kim et al., 1987]. Seismic layering is observed within the Kv unit only in the northern basin (Figure 3). By using true amplitude sections to predict reflections likely to produce multiples, and by estimating their reflection times, we were able to identify several sub-basement reflections as being interbed multiples, particularly where the chert or top of Cretaceous volcanics is highly reflective and/or horizontal. In addition, the *Charcot* data suffer from reverberations associated previous shots, related to the rapid repetition rate of the source. These could usually be identified by the relatively higher frequencies, lower amplitudes and similarities in structure to overlying sections, but some ambiguities remain.

In the northern Nauru Basin, the high amplitude K reflection correlates well with the onset of relatively high velocities, at least in Sonobuoy 3 (Figure 7). This disagrees with the sonobuoy data of Wipperman et al. [1981]. On the basis of the averages of seven sonobuoys (all sonobuoy wide angle reflection estimates) they correlate the top of a 3.5 km/s layer (with a mean thickness of 452 m) to the top of the chert sequence, whereas we correlate their 3.5 km/s layer with the top of the volcanics (Figure 3). As discussed by these authors, their correlation created a discrepancy with the drilling results that suggest the chert unit above the volcanics should have a velocity of only 2.5 km/s, and be 175 m thick. Our newer data with a broad bandwidth and simple source function, demonstrated that the problem was related to difficulties in identifying the top of the volcanics consistently in seismic data and the location of Site 462 in a local depression (Figure 3). Wipperman et al. [1981] reported that the top of the igneous complex "appear as a sequence of reflectors which are generally difficult to correlate from line to line and even along the same record." The combination of a complex source function and closely spaced reflections in the overlying chert section may have degraded efforts at interpreting the analog records. We did

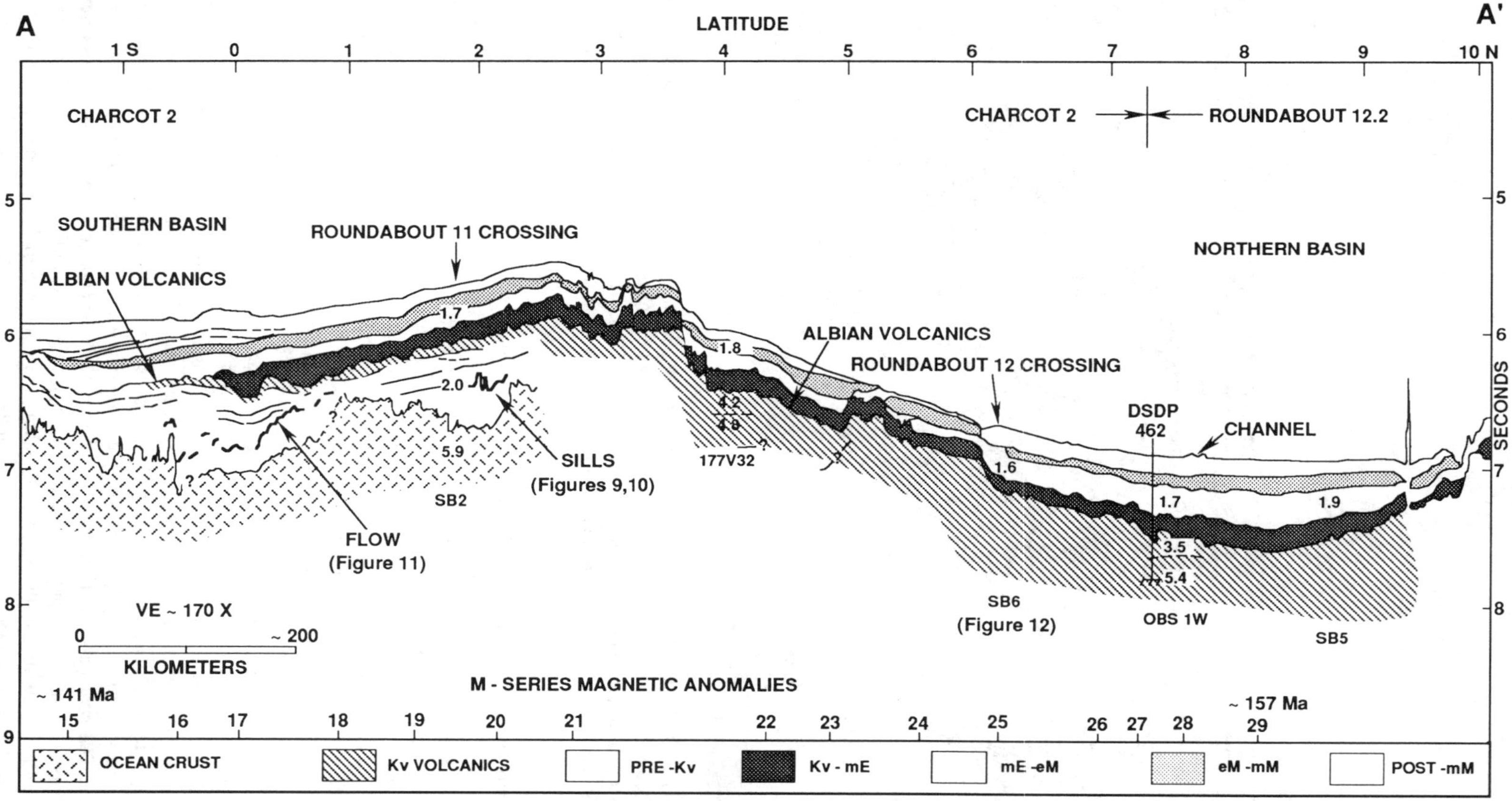

Fig. 4. A 1300 km cross section along 165° E. This section is a composite of two long seismic lines collected by the N/O *Charcot* and R/V *Washington*. The Nauru Basin consists of two sub-basins (northern and southern) separated by a gentle WNW trending high between 2° and 4° N. All correlations were made from DSDP Site 462. See Figure 2 for location of this and following sections. Note high vertical exaggeration of 170 x. M-series magnetic anomalies are from Cande et al. [1989] shown with the 1989 Harland et al. [1990] time scale. V32 sonobuoy is from Houtz [1976].

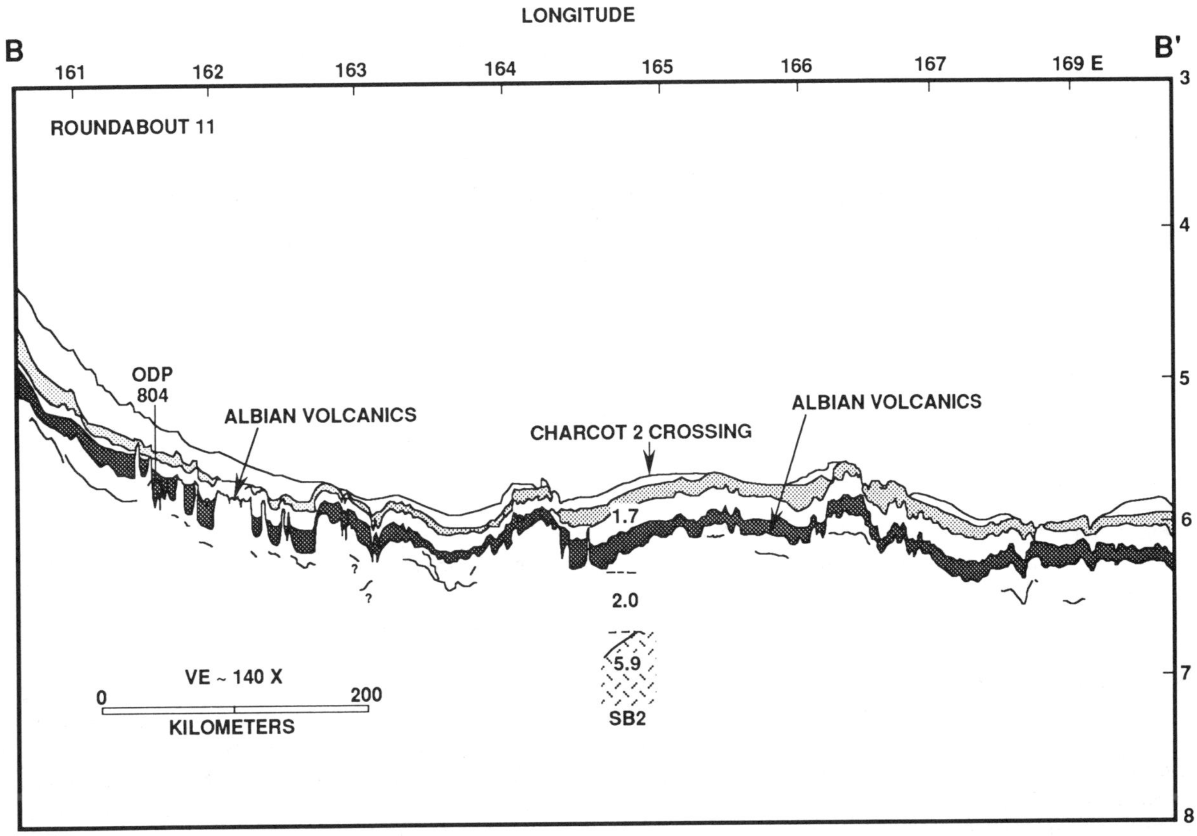

Fig. 5. East-west trending cross section at about 2° N. Collected with a single 1.3 l water gun this section does not penetrate beneath the younger volcanics. An exception was near the crossing with the *Charcot* line where we used a 32.8 l air gun and sonobuoy (SB 2). The deep west dipping reflections correspond to the top of the 5.9 km/sec unit. Cross section on the flank of the Ontong-Java Plateau is after Mayer et al. [1991]. ODP Site 804 is not suitable for correlation purposes since it penetrated a slump. Results from ODP Site 805 generally confirmed the correlations made down from DSDP Site 462 [Mosher et al., in press]. See Figure 2 for location of profile and sonobuoy. Legend is as in Figure 4.

not observe high velocity first arrivals in most of our sonobuoy data. To reproduce absence of first arrivals, we modeled gradient layers which resulted in velocity models similar to those reported by Wipperman et al. [1981] (Figure 7).

Houtz [1976] also reported an unusually low velocity for the shallow structure of the Nauru Basin in characterizing shallow crustal structure in the Pacific. Of six sonobuoys in the Nauru Basin, three had layer 2A-like velocities of 3.0 to 4.2 km/sec, and thicknesses of 590 to 1400 m. In almost every case high velocity (post-critical) reflections or refractions were not observed in the Nauru Basin, and Houtz [1976] believed that this may "imply pronounced inhomogeneities, such as alternating bands of volcanics and sediments in layer 2A (now known to be the Cretaceous volcanics) which could make the layer reflective but not necessarily coherent enough to propagate head waves." Even with our much larger sound source and better low frequency sonobuoys, we only detected high velocity first arrivals in Sonobuoy 3. Modern digital recording and analysis techniques should have constrained the shallow structure better, but we detected no higher velocity post-critical first arrivals. Where high velocity basement is not detected in the northern Nauru Basin is also where weak discontinuous reflections are present below K. One of Houtz's [1976] interpretations based on a high velocity first arrival is projected onto line A-A' where we have no other data (south of 6° N, Figure 4, 177V32).

Transition between sub-basins. Along section A-A' the K reflection is rather smooth in the northern Nauru Basin (Figure 4). An abrupt vertical offset at about 6° N seems to mark the southern limit of the particularly smooth K and of the underlying weak and discontinuous reflections (Figure 4). This geometry is also observed on random profiles and seems to define a northwesterly trend at least west of 165° E. Farther south a gentle bathymetric rise near 3° N is mimicked by K in the subsurface. This corresponds to the gentle seafloor high which trends WNW (Figure 2). Even further south, the section beneath K is more complicated, commonly containing several closely spaced, high amplitude reflections.

Southern Nauru Basin. In the south a low velocity sedimentary

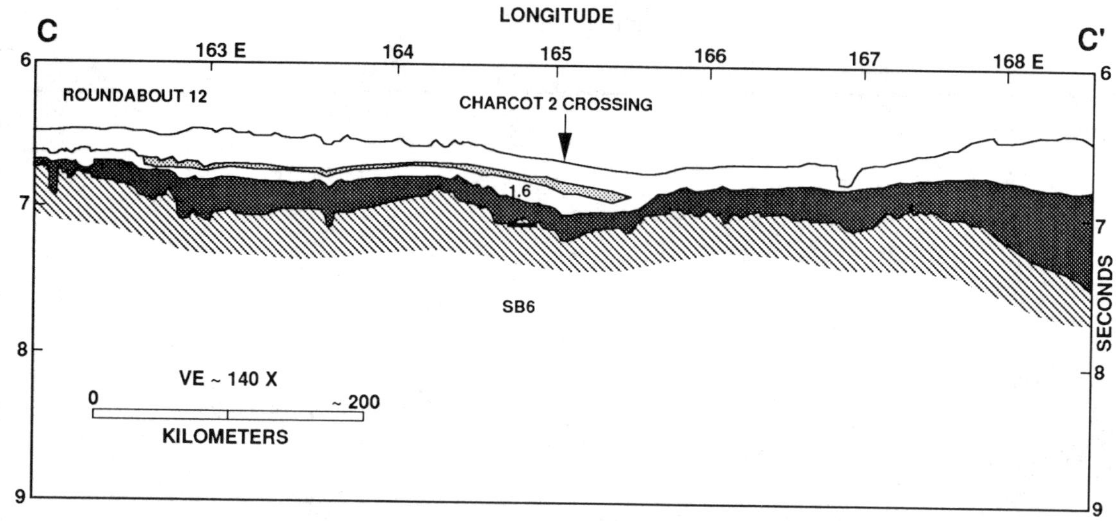

Fig. 6. East-west cross section in the northern Nauru basin at about 6° N. Much of the younger section pinches out by downlap near the center of the basin. Location of profile and sonobuoy is in Figure 2. A portion of the seismic data is shown in Figure 13. Legend is as in Figure 4.

section was discovered beneath Kv and is of similar thickness to the entire post Kv sedimentary section. The lack of high velocities associated with Kv at 1.9° N indicates a much thinner Kv sequence than in the north (Sonobuoy 2, Figures 4 and 9). However, the high reflection amplitude and diffractive nature suggest that it persists as a thin volcanogenic unit at least to 1.2° S. Near 1.3° S, the top of Kv loses its relatively high amplitude and diffractive nature entirely (Figure 8). Where structure beneath K is masked north of 2.5° N, this may be attributed to thicker sequences of heterogeneous volcanics that scatter energy producing few laterally extensive seismic reflections.

For the first time, oceanic crust is identified as a rough surface with 150 m to 300 m of relief south of 2.5° N (section A-A'). It correlates with the high velocities (5.9 km/sec, Sonobuoy 2). At 2.5° N there is an abrupt tectonic-volcanic transition (Figures 2 and 4). If true oceanic crust to the north is masked by a thicker Kv section, then an abrupt southward thinning at 2.5° N is implied. No seismic reflection or refraction evidence indicates that the 400 to 700 m thick sedimentary section beneath the Kv section extends north of 2.5° N. The seismic line coincident with Sonobuoy 2 shows west dipping oceanic crust over a 53 km section where the large air gun was used (B-B'). One other seismic line exhibiting an unequivocal deep reflection in the southwestern portion of the Nauru Basin shows an east dip (Figure 2, Larson, 1976).

The structure within the upper Cretaceous, pre-Kv sediment section, is generally characterized by basin filling sequences which onlap a rough surface with 150 to 300 m of relief, which is interpreted as the top of normal oceanic crust (Figure 8). If the age of the seafloor in the southern region as defined by magnetic anomalies is correct, then the rate of sediment accumulation in this interval is much higher than post-Kv. Some extremely high amplitude reflections have the characteristics of sills which abruptly change in stratigraphic level of emplacement (Figures 9 and 10). These are the seismic characteristics of sills that have been documented elsewhere (e.g., Juan de Fuca sedimented ridge area [Davis, 1982]). The ages of the sills and flows are not known. They could be nearly the same age as Kv. Evidence for sills and flows is extensive along a 300 km section from 2.2° N to 0.8° S (Figure 4).

Post Late Cretaceous Sediments

Sediments overlying the Kv section extend throughout the Nauru Basin. Three units are identified which extend over the entire area. For completeness they are briefly described here. The Cretaceous (~Albian) to middle Eocene sequence (K-mE) is characterized by varying, moderate to high amplitude reflections, probably associated with interbedded chert. This section is widespread, of nearly constant thickness and is missing only in the most southerly part of the basin (Figures 4 and 11). The top is usually recognized by a series of diffractions as illustrated in the south side of Figure 12, possibly a result of chert nodules and lenses.

The middle Eocene to early Miocene (mE to eM) is a seismic unit within the turbidite section. Reflections are fairly continuous but of low amplitude relative to overlying sections. It is thickest in the northern Nauru Basin. Clear evidence of downlap and onlap are shown in Figures 8, 11 and 13. The early Miocene to middle Miocene (eM to mM) seismic unit of turbidites is manifested as several strong and continuous reflections. Downlapping is best illustrated in Figure 8 but thinning is evident elsewhere such as in Figure 11. The thickness of these units varies by 50% but does not follow a clear pattern related to the bathymetry. The youngest part of the turbidite section (mM to Recent) has fairly continuous reflections and is much thicker in the deeper portions of the Nauru Basin (Figure 4). It is thinnest in some areas where the seafloor has been eroded (e.g., Figure 13). A deep channel is evident along the eastern side of the basin in section C-C' and is probably associated with an unmapped submarine canyon originating on the Marshall-Gilbert pedestal. A

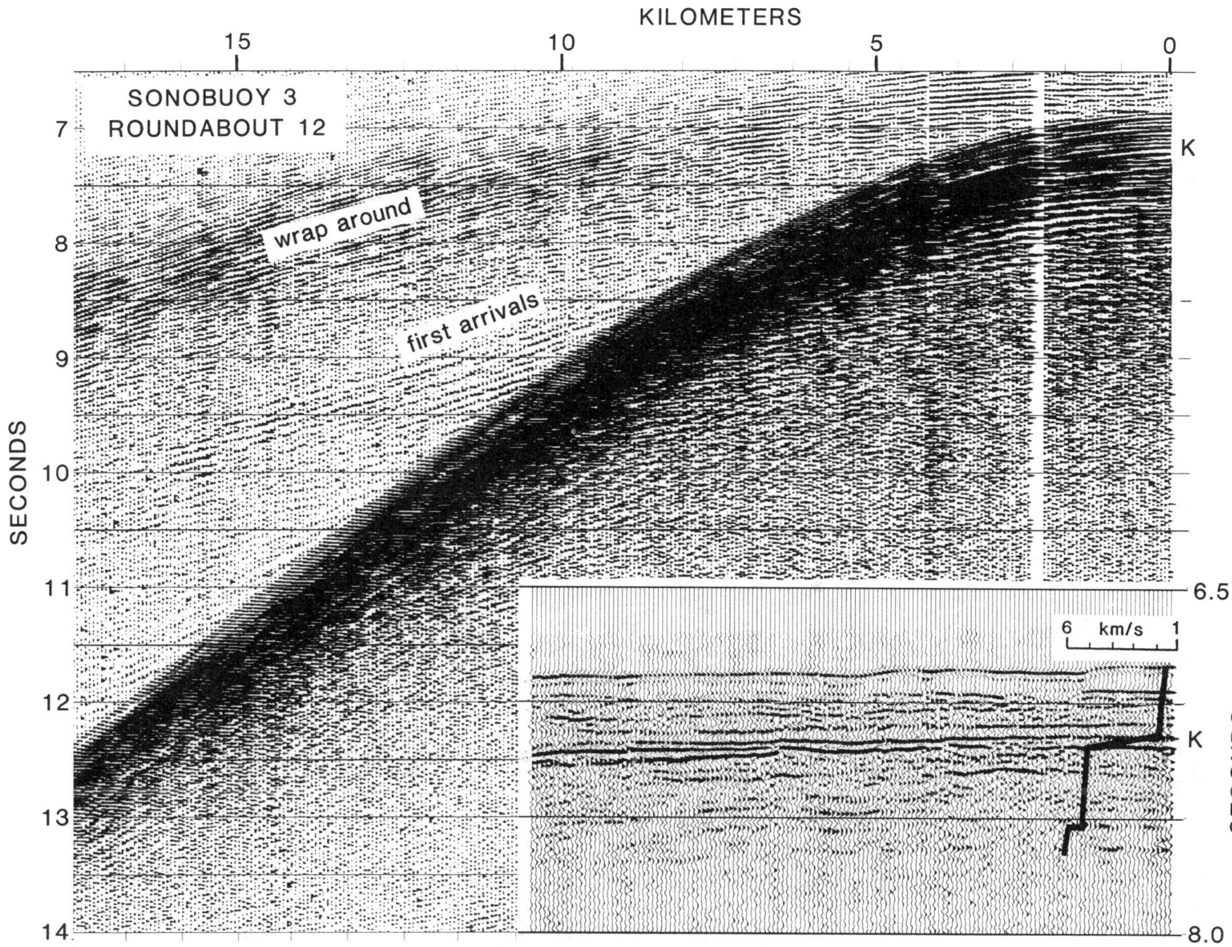

Fig. 7. Roundabout Sonobuoy 3 from the northern Nauru Basin. Weak first arrivals were observed from this sonobuoy using a large 32.8 l sound source, much larger than used in previous work. Inset is seismic line coincident with the sonobuoy showing velocity model. Location is shown in Figure 2.

deep sea channel with levee deposits wanders through the northern Nauru Basin (Figure 3).

The distribution of post-middle Eocene sediment in the Nauru Basin is controlled by processes typical of deep ocean settings. For instance, two broad gaps occur on the sides of the broad high (see B-B'). Both are erosional in nature, and are presumably caused by bottom currents moving through the region. This is also apparent on section C-C' (Figure 13) suggesting an intensification between the middle Eocene and early Miocene time based on downlapping and missing section. Regardless of whether all of the sedimentary sections are turbidites or some units on the broad high are pelagic, the regional thickening into the two basins suggests a major turbidite component (even if mainly composed of pelagic sediments). Also, the northern basin is surrounded on three sides by seamount chains, providing a likely turbidite source. Significant downlap in the early Miocene to Recent section indicates sources from the west and north (Figures 6, 11 and 13), although the northeast side of the Ontong-Java Plateau was also a source [Wipperman et al., 1981; Schlanger and Premoli Silva, 1981]. The southern basin also contains major post-Kv sediment accumulations with southerly downlap, suggesting a northern source. The absence of crossing lines prevents calculating the true dip, but the likely source is the Ontong Java Plateau. There is ample evidence of Neogene slumping on the eastern slope [Berger and Mayer, 1976; Berger et al., 1991; Mosher et al., in press.; Hagen et al., in press].

In the northern basin, K dips towards the Marshall-Gilbert Seamounts (C-C' Figure 6). The overlying thick onlapping section near the Marshall-Gilbert Seamounts indicates until perhaps the middle Eocene that much of the first sediment from the seamounts was trapped nearby in a flexural moat close by the seamount, though some undoubtedly traveled farther.

None of the younger sedimentary history bears significantly on Kv or older sections, but does suggest that the present setting of seamounts, plateaus and basins existed by the end of the Cretaceous, with later only isolated younger volcanic islands of Holocene age in the Caroline Islands.

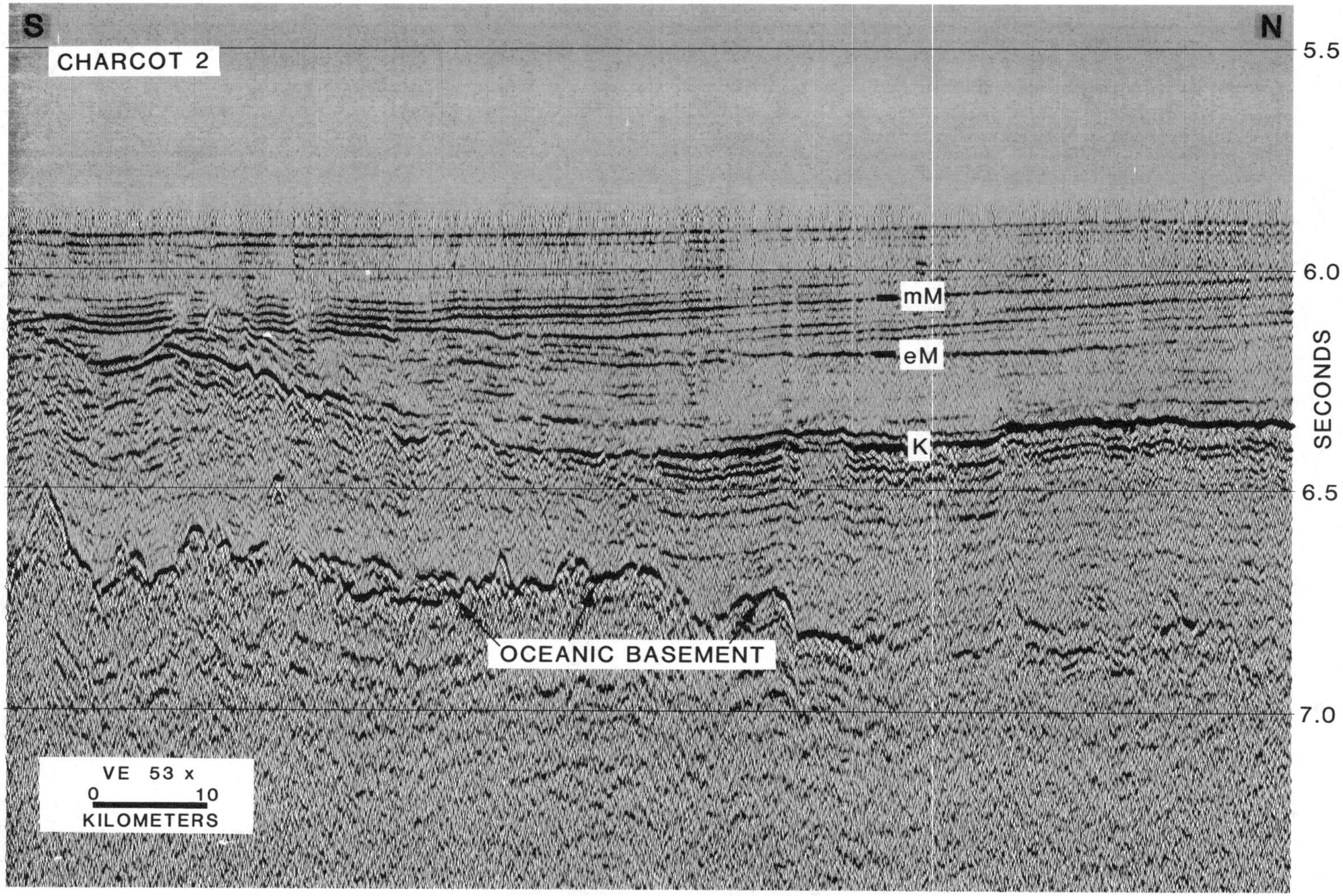

Fig. 8. Section at southernmost end of the *Charcot* line showing rough oceanic basement. Well-defined stratigraphy is evident between basement and the surface correlated with Cretaceous volcanics. The reduced reflectivity of K suggests that is not volcanic rock but a time-equivalent surface, perhaps volcanoclastic (?). Note downlap occurs onto the early Miocene surface. Processing information and the initial reports on this data are presented in Froger [1989; Froger and Lancelot, 1990]. The data were stacked 24-fold and subsequently deconvolved and filtered preserving original relative amplitude information that is important in identifying volcanics and cherts. Location is shown in Figure 2.

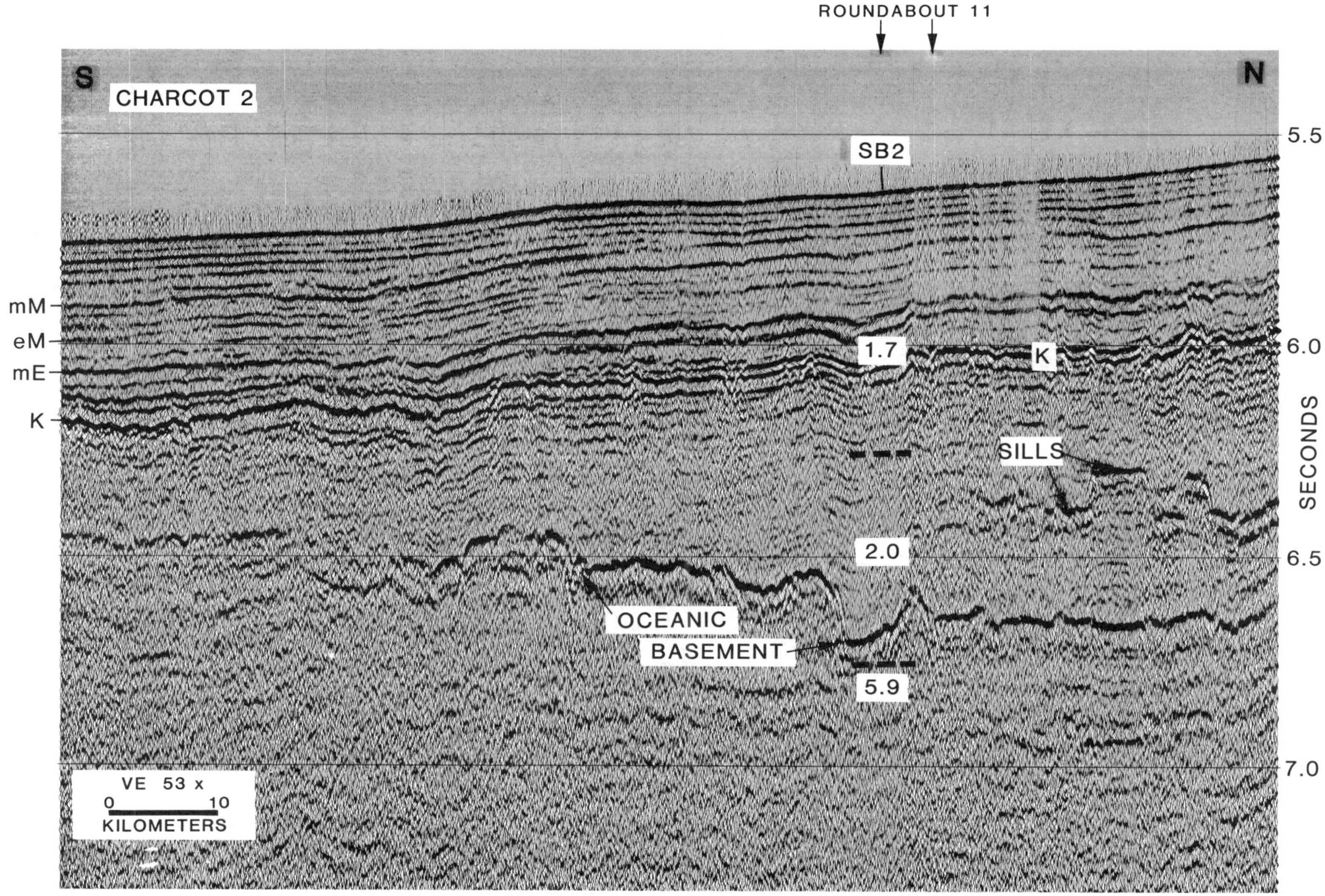

Fig. 9. *Charcot* section illustrating the rough surface thought to be the top of oceanic crust and thick section beneath it and K. Note well-developed high amplitude "bright spots" on right hand side at about 6.4 seconds that are most likely volcanic sills (Fig. 10). Note low velocity indicating a sedimentary section about 900 m thick. Location is shown in Figure 2.

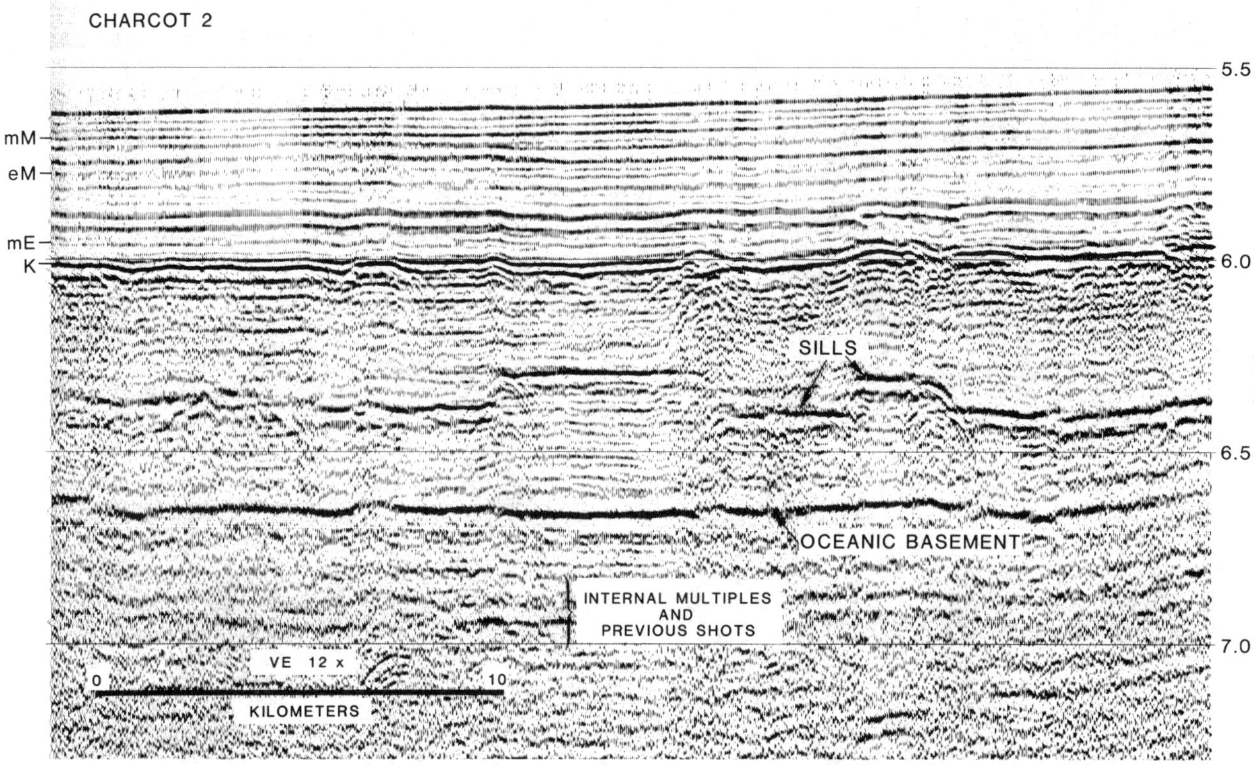

Fig. 10. Blowup of part of the previous figure, illustrating bright spots and abrupt shifts in stratigraphic level associated with them. These are known characteristic of sills from other settings. Coherent energy beneath the reflection labeled ocean basement is from both internal multiples and from previous shots still reverberating in the water column. The *Charcot* data presented problems with amplitude recovery in the first 0.5 s subbottom, which were never fully understood, and preclude quantitative treatment, but they do not unduly affect interpretation.

DISCUSSION

The distribution of the sedimentary section since the Late Cretaceous episode of volcanism demonstrates that the approximate geomorphic shape of the Nauru region has been maintained as two basins separated by a gentle ~ 500 m high, WNW trending arch (Figure 2). There is strong evidence that we have identified Latest Jurassic (Tithonian) to Early Cretaceous (Berriasian) oceanic crust in the southern area from (1) correlation of K from DSDP Site 462, (2) the thick pre-K sedimentary section, and (3) from the high relief associated with the basement-sediment contact. Sediments onlapping the oceanic crust within the overlying Tithonian to Albian (?) section (Figure 4, A-A') indicate rapid basin filling. Also, the high amplitude reflections within the section indicate coeval (?) volcanic flows (Figure 11) and post-depositional sills (Figure 10) within a 300 km long profile.

Sub-Kv structure in the northern Nauru Basin consists of common though indistinct reflections but puts little constraint on whether true oceanic crust is slightly or much deeper. Sub-Kv reflections probably originate from volcanic sills and intercalated sediments, with homogeneous sheet flows underneath. Castillo et al. [1991] demonstrate that the Early Cretaceous volcanics at Site 462 have geochemical affinities to ocean ridge systems and they speculate that Site 462 may have drilled through sills into oceanic crust of Albian (?) or earlier age. Only the well identified M-series anomalies upset this hypothesis. The northern Nauru Basin region also roughly corresponds to that identified by Houtz [1976] as having "abnormal" layer 2A velocities, and from which few high-velocity first arrivals are detected. Random seismic profiles and sonobuoy data suggest the sill sequence probably extends southward to only about 6° N. Fewer sub-Kv reflections occur south of the boundary, perhaps indicative of fewer sills and sediments and more extrusives than sampled at Site 462. In any event, the observed structures in the northern Nauru Basin strongly resemble those in the Venezuela Basin another possible location of Jurassic oceanic crust [Stoffa et al., 1981; Diebold et al., 1981].

After correction for sediment loading [after Renkin and Sclater, 1988], the depth to the top of the oceanic crust in the southern basin is about 5100 m. The depth to K in the northern basin is about 5500 m. Both depths are about 1000 m shallower than the plate model predicts for crust of Jurassic or Early Cretaceous age, in fact the depth is more appropriate for about 80 Ma old crust [Parsons and McKenzie, 1978]. This is probably not an important local discrepancy, as there is evidence that old oceanic crustal regions in the western Pacific are often as much as 800 m shallower than the plate model would predict [Stein and Abbott, 1991]. Our southern basin depth coincides with

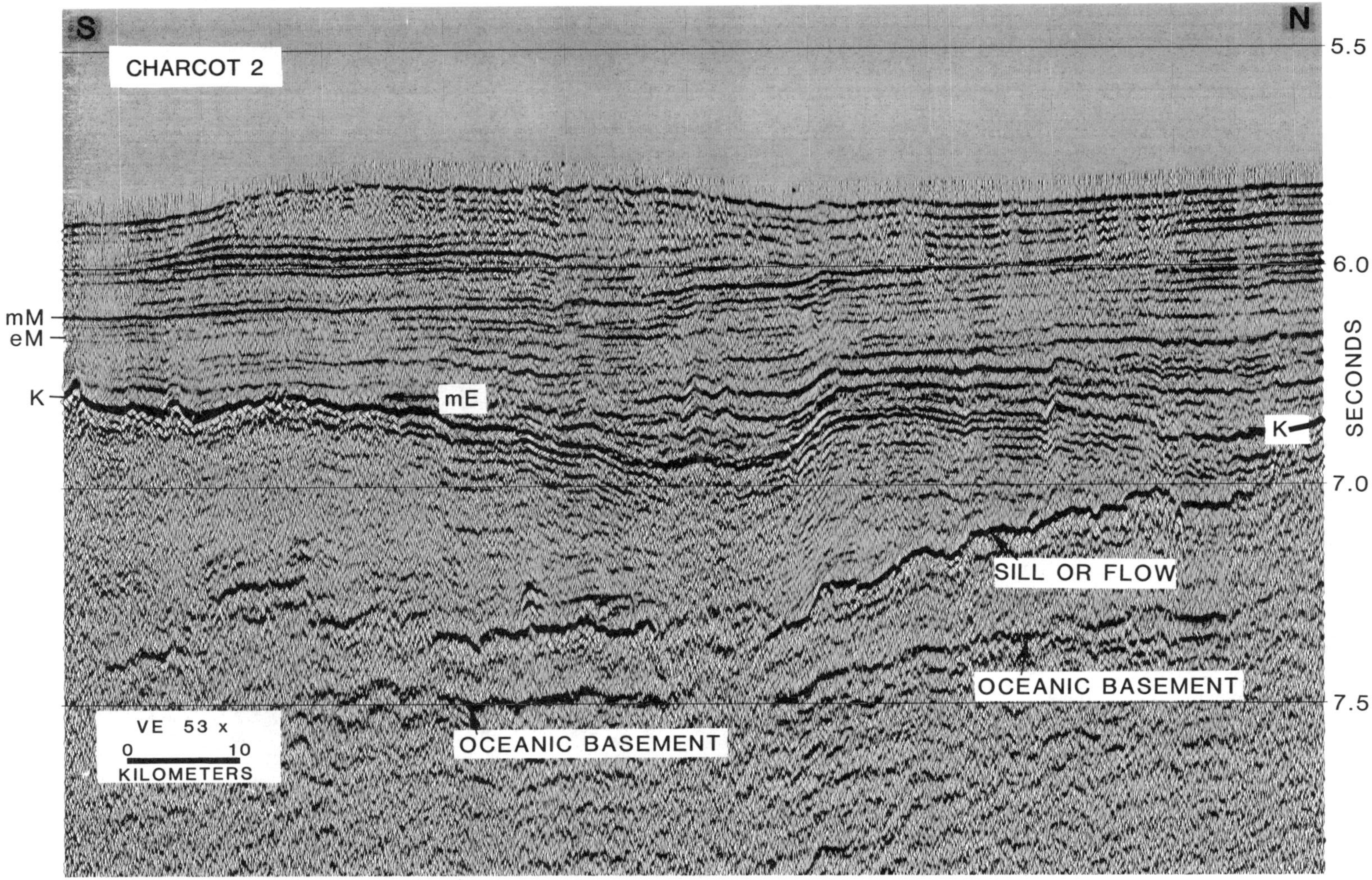

Fig. 11. Oceanic basement is visible in the center of this section. Overlying high amplitude event on the right side is interpreted as a flow (or perhaps sill). The section beneath the flow (?) contains few coherent reflections, probably due to scattering associated with the surface of the unit. The overlying section shows onlap onto the dipping event. Higher in the section the middle Eocene cherts and sediments pinch out against K. Location is shown in Figure 2.

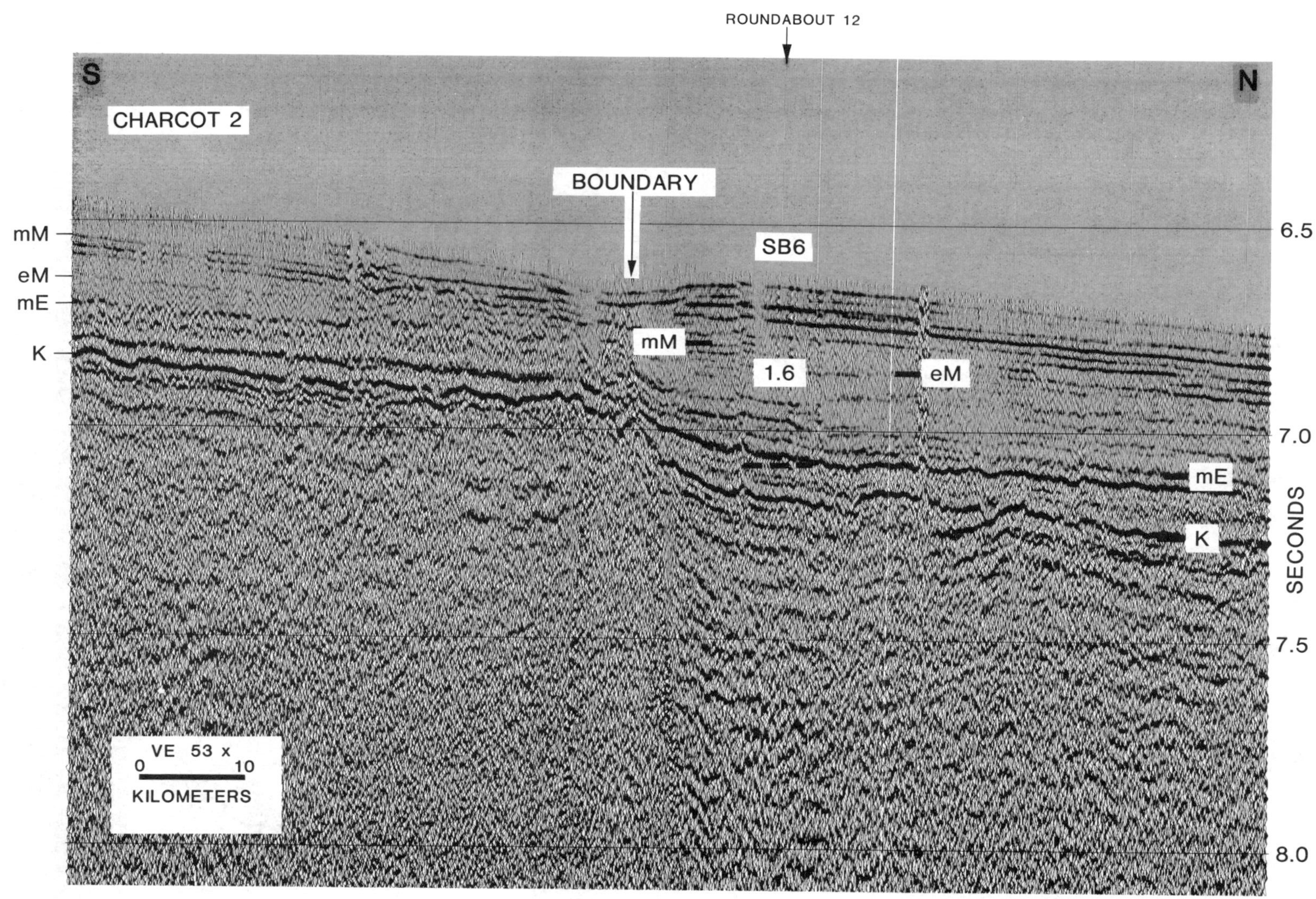

Fig. 12. Correlation of the section across boundary in the northern Nauru Basin. The results of Sonobuoy 6 from the crossing Roundabout line are projected onto the section. North of this boundary, the sub-K section contains more reflections than elsewhere; probably representing layers in the shallow part of the volcanic section associated with sills and intercalated sediments drilled at DSDP 462. Stoffa et al. [1981] and Diebold et al. [1981] observe similar discontinuous layering in far higher-quality data from the Venezuela Basin that may contain Jurassic ocean crust as in the Nauru Basin. Stoffa et al. [1981] also had problems distinguishing internal multiples from true seismic layering in the Venezuela Basin. Location is shown in Figure 2.

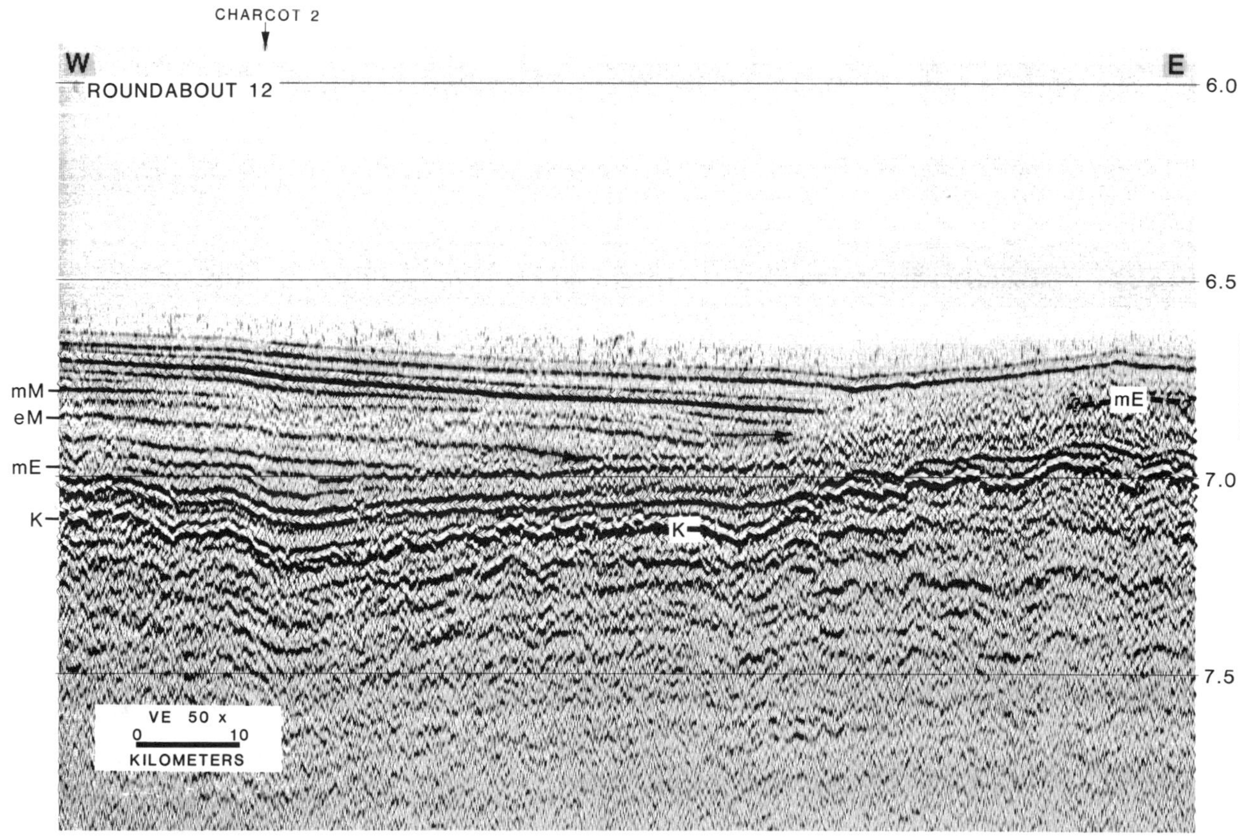

Fig. 13. East-west section in the northern Nauru Basin showing strong downlap of the middle Eocene to Recent section. The east side of this figure includes part of the region where the sediment thickness is a minimum, and represents modern and ancient scouring of the seafloor. Location is shown in Figure 2.

the median of Renkin and Sclater [1988]. The discrepancy between the plate model predictions and observed depth to basement is not restricted to the Nauru region and thus is not a particularly helpful indicator of local crustal structure.

Cretaceous volcanism in the Nauru Basin can be separated into three areas with varying degrees of uncertainty to estimate maximum volumes. These geographic divisions are based on seismic reflection, velocity and morphology data. The northern basin, defined by the absence of high velocity first arrivals in numerous sonobuoys and on indistinct layering beneath K, includes an area roughly enclosed by the 5000 m isobath totaling 150,000 km^2. If the entire depth anomaly, after adjusting for the known sediment thickness, is attributed to a volcanic sequence, the section could be 1700 m thick (and with no adjustment made for the empirical observation that basement of Late Jurassic age is about 800 m shallower than the plate model predicts) [Froger, 1989; Froger and Lancelot, 1990]. This translates to a volume of 255,000 km^3. The southern basin volcanics south of 2.5° N may cover another 700,000 km^2, but their thickness must be on the order of 100 m since we are able to image sediments underneath. The resulting volume is another 70,000 km^3. Finally, all of the remaining area between the 4000 and 5000 m isobaths totals 900,000 km^2 with an unconstrained thickness. If we use the Houtz sonobuoy results, Kv might be at least 590 m thick, yielding a volume of 531,000 km^3. Thus, the likely extreme volume of Kv in the Nauru Basin is less than one million km^3 which is a relatively small amount compared to a single plateau such as the Ontong Java Plateau (<2%) [Coffin and Eldholm, 1991; Schubert and Sandwell, 1989].

We had hoped to find seismic evidence for multiple sources, to explain the widespread areal extent of the basin wide volcanic rocks. These discrete sources would be compatible with preservation of Jurassic magnetic anomalies. Without separated discrete sources, widespread rifting would raise the temperature of the underlying basement and destroy the old magnetic anomalies. We had also hoped to find some evidence relevant to testing the Castillo et al. [1991] model of Cretaceous rifting of the northern Nauru Basin. However, we observed little evidence for fissures or long-lived erupting sites associated with the K surface and few structures were apparent within the volcanic section. Similarly, Clague et al. [1990], in their study of a large, relatively young lava field on the seafloor off Hawaii, were unable to locate fissures responsible for the largest known recent seafloor eruption (25,000 km^2) even with side-scan sonar data. We speculate that the broad, WNW trending, low relief highs across the central part of the Naru Basin may be Cretaceous volcanic centers in the pre-existing ocean basin, and may have been a source for some of the volcanics in the northern basin. Whether this represents a plate boundary is unclear. Satellite derived short wavelength gravity anomaly charts do not reveal any obvious plate tectonic structures in this region [Haxby, 1981; Sandwell and Renkin, 1988]. Yet, the abrupt increase in sediment thickness and abrupt thinning of the Cretaceous volcanics to the south of the WNW tending highs cannot be easily explained. With the very general understanding of the spatial relationships revealed in this work, more subsurface mapping of the transition into the southern sub-basin and along the eastern flank of the Ontong Java Plateau might reveal more about the potential volcanic sources. Investigations need to be concentrated where the volcanics are thin enough to acoustically penetrate and perhaps sample in wells, rather than in the northern Nauru Basin. Evaluation of gross crustal structure in the northern Nauru Basin will be of help. However, as in the Venezuela Basin the crustal structure is not enough to explain the tectonic origin of the basin or the volcanism. Exploration of the relationship between the Ontong Java Plateau volcanism and basin volcanism seems quite possible in the southern Nauru Basin.

Conclusions

Our work summarizes the current state of knowledge of the shallow structure of the igneous crust and sediments within the Nauru Basin. We still have little direct evidence for "normal" oceanic crust in the northern Nauru Basin besides previously identified Jurassic magnetic anomalies. The northern Nauru Basin is characterized by indistinct seismic layering in the shallow crustal section that correlate to interbedded sills and sediments sampled in the upper 160 m of "basement" at Site 462. M-series magnetic anomalies require that underlying sheet flows are not the top of normal oceanic crust of Cretaceous age. Intrusives overlying sheet flows seem to characterize the structure of the northern Nauru Basin, south to about 6° N, where seismic layering beneath K is less common. The zone between 6° N and 2.5° N may be the source for some of the basin wide volcanism. K does correlate to the WNW trending bathymetric high that could possibly be Aptian/Albian volcanic centers (Figure 2). The basic enigma of how Jurassic magnetic anomalies are preserved, given the areally extensive volcanics and hence expected broadly distributed feeders through pre-existing Jurassic crust, remains unanswered.

Unequivocal oceanic crust exists in the Nauru Basin south of 2.5° N and is directly overlain by low velocity rocks with recognizable sedimentary structures. This sediment section also contains strong evidence for volcanic flows and sill intrusion. Whether this volcanic activity is associated with the Cretaceous volcanic event of the northern Nauru Basin or significantly pre-dates it is not known, but some flows appear to be nearly coeval with turbidite sedimentation. We speculate that the thickness of Cretaceous volcanics decreases abruptly at about 2.5° N, coincident with an increase in sediment thickness of about 400 to 700 m. We do not believe that the thick Tithonian to Albian (?) sediments exist north of this position since the velocity contrast, even with overlying volcanics, should not completely mask such a strong reflection.

Acknowledgments. This work was part of a joint U.S.-French effort to investigate the western Pacific basins. We thank Veronique Froger who was responsible for processing of the *Charcot* data under the direction of Marc Schaming at the Strasbourg lab of Roland Schlich. The authors also benefited from discussions about the western Pacific with S. Schlanger (a participant on the *Charcot* cruise), Pat Castillo and E.L. Winterer; and wide-angle reflection analysis with John Diebold. Dan Barker re-interpreted the sonobuoy data using an interactive software program of John Diebold (JDSEIS). We thank Mike Coffin, Malcom Pringle, Beth Ambos and Uri Ten Brink for their comments on this manuscript. Roundabout 11 work was supported by an NSF OCE grant to E.L. Winterer, and we thank him for the use of part of this data. Our work was supported by NSF OCE 86-13641 (Shipley) and NSF OCE 86-13867 (Larson). This paper is contribution 896 of the Institute for Geophysics, University of Texas at Austin

References

Abrams, L., and R. Larson, T. Shipley and Y. Lancelot, Cretaceous volcanic sequences and Jurassic oceanic crust in the East Mariana and Pigafetta Basins of the western Pacific, this volume.

Berger, W., L. Kroneke, et al., *Proceedings of the Ocean Drilling Program, Part A, Leg 130*, 1240 pp., U.S. Govt. Printing Office, Washington, D.C., 1991.

Berger, W.H., and L.A. Mayer, Deep-sea carbonates: dissolution and mass wasting on the Ontong Java Plateau, *Science, 192*, 785-787, 1976.

Cande, S.C., R.L. Larson, and J.L. LaBrecque, Magnetic lineations in the Pacific Jurassic Quiet Zone, *Earth Planet. Sci. Lett., 41*, 434-440, 1978.

Cande, S.C., J.L. LaBrecque, R.L. Larson, W.C. Pitman III, X. Golovchenko, and W.F. Haxby, *Magnetic lineations of the World's ocean basins (chart)*, Am. Assoc. Petrol. Geol., Tulsa, Oklahoma, 1989.

Castillo, P.R., R.W. Carlson, and R. Batiza, Origin of Nauru Basin igneous complex: Sr, Nd and Pb isotope and REE constraints, *Earth Planet. Sci. Lett., 103*, 200-213, 1991.

Clague, D.A., R.T. Holcomb, J.M. Sinton, R.S. Detrick, and M.E. Torresan, Pliocene and Pleistocene alkalic flood basalts on the seafloor north of the Hawaiian Islands, *Earth Planet. Sci. Lett., 98*, 175-191, 1990.

Coffin, M.F., and O. Eldholm, *Large Igneous Provinces, a workshop to develop scientific drilling initiatives on volcanic margins and oceanic plateaus*, 79 pp., Joint Oceanographic Institutions, Washington, D.C., 1991.

Davis, E.E., Evidence for extensive basalt flows on the seafloor, *Geol. Soc. Am. Bull., 93*, 1023-1029, 1982.

Diebold, J.B., P.L. Stoffa, P. Buhl, and M. Truchan, Venezuela Basin crustal structure, *J. Geophys. Res., 86*, 7901-7923, 1981.

Duncan, R.A., and M.A. Richards, Hotspots, mantle plumes, flood basalts, and true polar wander, *Reviews of Geophysics, 29*, 31-50, 1991.

Froger, V., Recherches sur l'origine et l'evolution des bassins Mesozoiques du Pacifique Centre Quest, These de Doctorat, Universite Pierre et Marie Curie, Paris, 1989.

Froger, V. and Y. Lancelot, Evenements volcaniques cretaces dans le Pacifique ouest et origine du bassin de Nauru, (campagne Mesopac), *Oceanologica Acta, Special Volume 10*, 241-257, 1990.

Hagen, R.A., L. Mayer, D. Mosher, L. Kroenke, T. Shipley, and E.L. Winterer, Basement structure of the northern Ontong Java Plateau, *Proceedings of Ocean Drilling Program, Part B, 130*, in press.

Harland, W.B., R.L. Amstong, A.V. Cox, L.E. Craig, A.G. Smith, and D.G. Smith, *A geologic time scale 1989*, 263 pp., Cambridge Univ. Press, Cambridge, 1990.

Haxby, W.F., Gravity field of the world's ocean, (chart) National Geophysical Data Center, NOAA, Boulder, Colorado, 1981.

Houtz, R.E., Seismic properties of Layer 2A in the Pacific, *J. Geophys Res., 81*, 6321-6331, 1976.

Iwabuchi, Y., *General bathymetric chart of the oceans 5-06*, Canadian Hydrographic Survey, scale 1:10,000,000, Ottawa, Canada, 1979.

Kim, I.I., D.K. Smith, H.W. Menard, J.A. Orcutt, and T.H. Jordan, Seismic reflection site survey: correlation with physical properties Leg 91, Deep Sea Drilling Project, *Initial Reports Deep Sea Drilling Project, 91*, 271-305, U.S. Govt. Printing Office, Washington, D.C., 1987.

Larson, R.L., Late Jurassic and Early Cretaceous evolution of the western central Pacific Ocean, *J. Geomagnetism and Geoelect.*, *28*, 219-236, 1976.

Larson, R.L., 1991, Latest pulse of the Earth: Evidence for a mid-Cretaceous super plume, *Geology*, *19*, 547-550, 1991.

Larson, R.L., S.O. Schlanger, et al., *Initial Reports of the Deep Sea Drilling Project*, *61*, 885 pp., U.S. Govt. Printing Office, Washington, D.C., 1981.

Lipman, P.W., D.A. Clague, J.G. Moore, and R.T. Holcomb, South Arch volcanic field—Newly identified young lava flows on the seafloor south of the Hawaiian Ridge, *Geology*, *17*, 611-614, 1989.

Mammerickx, J., and S.M. Smith, Bathymetry of the North Central Pacific, *Geol. Soc. Am. Map and Chart Ser.*, *No. MC-52*, 1985.

Mayer, L., T. Shipley, E.L. Winterer, D. Mosher, and R.A. Hagen, Seabeam and seismic reflection surveys on the Ontong-Java Plateau, *Proceedings of the Ocean Drilling Program, Part A, 130*, 45-97, U.S. Govt. Printing Office, Washington, D.C., 1991.

Moberly, R., S.O. Schlanger, et al., *Initial Reports of the Deep Sea Drilling Project*, *89*, 678 pp., U.S. Govt. Printing Office, Washington, D.C., 1986.

Morgan, W.J., Deep mantle convection plumes and plate motions, *Am. Assoc. Petrol. Geol. Bull.* *56*, 203-213, 1972.

Mosher, D., L. Mayer, T. Shipley, E.L. Winterer, R. Hagen, J. Marsters, F. Bassinot, R. Wilkens, and M. Lyle, Seismic stratigraphy of the Ontong Java Plateau, *Proceedings of Ocean Drilling Program Part B, 130*, in press.

Monahan, D., R.H.K. Falconer, and M. Tharp, *General bathymetric chart of the oceans 5-10*, Ottawa, Canada, Canadian Hydrographic Survey, scale 1:10,000,000, 1982.

Ozima, M., K. Saito, and Y. Takigami, 40Ar-39Ar Geochronological studies on rocks drilled at holes 462 and 462A, Deep Sea Drilling Project Leg 61, *Initial Reports of Deep Sea Drilling*, *61*, 701-703, 1981.

Parsons, B. and D. McKenzie, Mantle convection and the thermal structure of the plate, *J. Geophys. Res.* *83*, 4485-4496, 1978.

Renkin, M., and J.G. Sclater, Depth and age in the North Pacific, *J. Geophys. Res.*, *93*, 2919-2935, 1988.

Pringle, M., Geochronology and petrology of the Musicians Seamounts, and the search for hot spot volcanism in the Cretaceous Pacific, Ph. D. dissertation, University of Hawaii, Manoa, Hawaii, 1992.

Sandwell, D.T., and M.L. Renkin, 1988, Compensation of swells and plateaus in the North Pacific: No direct evidence for mantle convection, *J. Geophys. Res.*, *93*, 2775-2783, 1988.

Schlanger, S.O., and I. Premoli Silva, Tectonic, volcanic and paleogeographic implications of redeposited reef faunas of Late Cretaceous and Tertiary age, from the Nauru Basin and the Line Islands, *Initial Reports Deep Sea Drilling Project*, *61*, pp. 817-827, 1981.

Schubert, G., and D. Sandwell, Crustal volumes of the continents and of oceanic and continental submarine plateaus, *Earth Planet. Sci. Lett.*, *92*, 234-246, 1989.

Shipley, T.H., J.M. Whitman, F.K. Duennebier, and L.D. Petersen, Seismic stratigraphy and sedimentation history of the East Mariana Basin, western Pacific, *Earth Planet. Sci. Lett.*, *64*, 257-275, 1983.

Stein, C.A., and D.H. Abbott, Heat flow constraints on the south Pacific superswell, *J. Geophys. Res.*, *96*, 16083-16099, 1991.

Stoffa, P.L., A. Mauffret, M. Truchan, and P. Buhl, Intrabasement layering in the Southern Caribbean: The Aruba gap and Venezuela basin, *Earth Planet. Sci. Lett.*, *53*, 131-146, 1981.

Tarduno, J.A., W.V. Sliter, L. Kroenke, M. Leckie, H. Mayer, J.J. Mahoney, R. Musgrave, M. Storey, and E.L. Winterer, Rapid formation of Ontong-Java Plateau by Aptian Mantle Plume Volcanism, *Science*, *254*, 399-403, 1991.

Whitman, J.M., Seismic stratigraphic sections: East Mariana Basin (Site 585), Nauru Basin (Site 462), and Ontong-Java Plateau (Site 586): *Initial Reports of the Deep Sea Drilling Project*, *61*, 589-596, U.S. Govt. Printing Office, Washington, D.C., 1985.

Wilson, J.T., Evidence from oceanic islands suggesting movement in the Earth, *Philos. Trans. R. Soc. London, Ser. A*, *258*, 145-165, 1965.

Winterer, E.L., Anomalies in the tectonic evolution of the Pacific, *Geophysical Monograph 19*, Am. Geoph. Un., p. 269-278, 1976.

Wipperman, L.K., R.L. Larson, and D.M. Hussong, The geological and geophysical setting near site 462, *Initial Reports Deep Sea Drilling Project*, *61*, 763-770, U.S. Govt. Printing Office, Washington, D.C., 1981.

T.H. Shipley, Institute for Geophysics, University of Texas at Austin, 8701 North Mopac Blvd., Austin, TX 78759-8397.

L.J. Abrams and R.L. Larson, Graduate School of Oceanography, University of Rhode Island, Narragansett Bay Campus, Narragansett, RI 02882-1197.

Y. Lancelot, Laboratoire de Geologie du Quaternaire, CNRS, Case 905, 13288 Marseille Cedex 9, France.

Topographic Expression of Five Fracture Zones in the Northwestern Pacific Ocean

M. NAKANISHI

Ocean Research Institute, University of Tokyo, Tokyo, Japan

Bathymetric and seismic reflection profiles in the northwestern Pacific Ocean were examined to fix precise positions and characteristic features of fracture zones previously inferred only from offsets of magnetic anomaly lineations. The topographic expression of the Kashima Fracture Zone is characterized by a complex trough with a ridge on its northern segment and by a chain of seamounts on the south of the Ogasawara Plateau. The transition from complex trough to seamount chain is concordant with a change in strike of the fracture zone caused by reorganization of spreading on the Pacific-Izanagi ridge. It is postulated that a leaky transform fault has created the complex trough after the plate reorganization. The Nosappu Fracture Zone is manifested by more complicated topographic features. A deep-sea channel is developed along its northern segment. A wide trough associated with the South Shatsky Fracture Zone, striking obliquely to the surrounding magnetic lineations, suggests a rift-propagation origin. The Phoenix Fracture Zone is the most remarkable fracture zone in the Phoenix lineation set. Its great vertical relief may be due to Cretaceous volcanism. The Central Pacific Fracture Zone, previously located only by offsets of magnetic anomaly lineations, is now confirmed by its topographic expression. Its topographic expression is either simply a ridge or a scarp with a ridge. The vertical displacement across the Cretaceous Kashima and Nosappu Fracture Zones indicates that their morphology did not originate from subsidence of the lithosphere following a simple conductive-cooling model.

INTRODUCTION

A fracture zone was originally defined as a long and narrow "band of grossly irregular topography" characterized by chains of seamounts, linear ridges and scarps [Menard and Chase, 1970]. It is, in a typical case, recognized by a distinct difference in water depths from one side to the other. Magnetic lineations are largely offset by the fracture zone, which is most easily detectable in magnetic anomaly maps. Orientation of the fracture zone provides an important clue in deciphering previous plate motions, because it is formed as a remnant of the transform fault that is usually parallel to the direction of spreading.

The topographic expression of fracture zones and transform faults in the eastern Pacific and Atlantic Oceans have been investigated using satellites, side scan sonar and multi-beam echo sounder [e.g., Embley and Wilson, 1992; Tucholke and Schouten, 1988; Wessel and Haxby, 1990]. One of the most remarkable topographic characteristics of fracture zones in the north Atlantic Ocean is asymmetry with valleys on its younger side and high walls or scarps on its older side [Collette, 1986; Roest and Collette, 1986]. The topographic expression of the Mendocino Fracture Zone in the northeastern Pacific Ocean is represented by a steep escarpment, whereas the topography of the Clarion and Clipperton Fracture Zones in the central Pacific Ocean are varied and irregular. In the northwestern Pacific Ocean, however, many fracture zones have been identified only by offsets of magnetic anomaly lineations. All the spreading ridges with active transform faults have long been subducted under trenches, so that old inactive fracture zones alone are observed there. Surface topographic expression of such ancient fracture zones is generally buried by sediment cover thicker than 500 m [Ewing et al., 1968; Ludwig and Houtz, 1979]. To obtain more precise constraints on their locations and basement topography, I have examined bathymetric and seismic reflection data obtained by a number of geophysical cruises.

Nakanishi et al. [1989, 1992] compiled Mesozoic magnetic anomaly lineations in the northwestern Pacific Ocean (Figure 1, Plate 1) to reconstruct four Mesozoic plates: Pacific, Izanagi, Farallon and Phoenix. The Magellan microplate [Tamaki and Larson, 1988] was situated near the junction among the Pacific, Farallon and Phoenix plates from chron M15 to chron M9 (140-129 Ma). Consequently the anomalies in this region consist of five lineation sets, the Japanese, Hawaiian, Phoenix, Magellan and Mid-Pacific Mountains lineation sets. The Japanese lineation set was formed at the Pacific-Izanagi ridge in two strikes; NE-SW and ENE-WSW. The Hawaiian lineation set was formed at the Pacific-Farallon ridge and tends from N55°W at lineation M29 to N20°W at lineation Ml. The Phoenix lineation set

The Mesozoic Pacific: Geology, Tectonics, and Volcanism
Geophysical Monograph 77

Fig. 1. Mesozoic magnetic anomaly lineations and fracture zones in the northwestern Pacific Ocean [Nakanishi et al., 1992]. Solid lines are the lineations identified with M numbers. Positions of fracture zones are indicated with heavy lines. Major topographic features are shown for reference.

was formed at the Pacific-Phoenix ridge and has an E-W trend. The Magellan lineation set was formed at the Pacific-Magellan ridge and exhibits a fan-shape pattern. The Mid-Pacific Mountains lineation set was formed by the Magellan-Farallon ridge, the trend of which is NW-SE.

Sixty-one fracture zones in the northwestern Pacific Ocean were identified by Nakanishi et al. [1989, 1992]. New names were proposed for the five most prominent ones. "Kashima", "Nosappu" and "South Shatsky" Fracture Zones were named after names of towns on land or prominent submarine topography close to the fracture zones [Nakanishi et al., 1989]. That of "Phoenix" Fracture Zone was proposed because the fracture zone is most remarkable in the Phoenix lineation set and that of "Central Pacific" Fracture Zone was proposed because the fracture zone is situated in the Central Pacific Ocean [Nakanishi et al., 1992]. Mammerickx and Smith [1985] named a part of the Phoenix Fracture Zone "Victoria Fracture Zone". I extended it southward and further clarified its location, so that I proposed a new name Phoenix Fracture Zone. A part of the Central Pacific Fracture Zone was previously located only by offsets of magnetic anomaly lineations [Larson et al., 1972; Larson, 1976], but I confirmed its location by its topographic expression as well as offsets of magnetic anomaly lineations.

The Kashima Fracture Zone strikes roughly N20°W to N45°W. The strike of the Nosappu Fracture Zone is about N30°W. The South Shatsky Fracture Zone has a strike of N20°E. The Phoenix and Central Pacific Fracture Zones strike about N-S. Details of topographic expression and tectonic features of these fracture zones are described in this article with some discussion of their origins.

Data and Method of Analysis

Data used by the present investigation to identify fracture zones consist of bathymetric data from 357 geophysical cruises and seismic data from 28 cruises (see Nakanishi et al. 1989; 1992 for a detailed cruise list). All cruise data were compiled in the MGD77 format [Hittelman et al., 1977]. Bathymetric data in the format are written in two forms; one in two-way travel time (in seconds) and the other in corrected water depth (in meters). To compute depths, the two-way travel time data were converted to water depths according to the revised Matthews table [Carter, 1980].

Topographic expressions along each offset of the magnetic anomaly lineations were examined. If an irregularity in bathymetry was found, we concluded that it is a fracture zone. Sixty-one fracture zones were confirmed at 290 locations by bathymetric and seismic profiling data [Nakanishi et al., 1989, 1992]. The geomagnetic reversal time scale proposed by Kent and Gradstein [1985] was used in this study.

Topographic Expression of the Fracture Zones

Kashima Fracture Zone

The Kashima Fracture Zone is located along the western margin of the northwestern Pacific Ocean south of the Japan Trench (Figure 2). Its strike is nearly parallel to the axes of the Izu-Ogasawara and northern Mariana Trenches. The northern end of this fracture zone seems to have been subducted at the Japan Trench near Daiichi Kashima Seamount, although topographic expression is not clear at its northern portion. The horizontal offset of lineation M15 (139 Ma) along this fracture zone is nearly 500 km, which is about half the offset of the Mendocino Fracture Zone in the northeastern Pacific Ocean (Table 1). Ages of the ocean floor west of the fracture zone are roughly 10 m.y. younger than those east of it at the same latitudes.

The half spreading rate around the Kashima Fracture Zone at 32°N is 5.85 cm/yr on its east, whereas it is 4.62 cm/yr on its west [Nakanishi et al., 1989]. Because of this difference, the amount of horizontal offset decreases with age (Figure 3). The rate of increase in ridge offset is nearly 1.0 cm/yr after chron M20 (148 Ma) and about 0.6 cm/yr before. Our previous study of magnetic lineations [Nakanishi et al., 1989] indicated that this transition in a rate and amount of offset took place during a plate reorganization between chrons M21 and M20 (149.5-148 Ma). A similar increase in the offset across the Marquesas Fracture Zone was reported by Kruse [1988], who postulated its ridge jump origin. However, as no evidence of ridge jump was found near the Kashima Fracture Zone, we concluded that the origin of variation in the horizontal offset across the Kashima Fracture Zone is a difference of spreading rates.

The Kashima Fracture Zone north of 28°N is characterized by a complex trough morphology. The width of the trough ranges from 30 to 100 km. The trough is bounded by ridges with maximum height of 2,000 m above seafloor. This feature becomes more pronounced northward as seen in Figure 4.

Figure 5 shows a seismic reflection profile obtained by R/V *Vema* (V2106) across the Kashima Fracture Zone near 30°N. Sediment thickness is approximately 100 m on both sides of the fracture zone. Vertical displacement of both seafloor and the underlying acoustic basement is about 300 m. The younger (western) side of the Kashima Fracture Zone is shallower than that of the older (eastern) side.

Several seamounts exist along this fracture zone in the region south of the Ogasawara Plateau (Figure 2). The transition from complex trough to chain of seamounts is concordant with the change in strike of the fracture zone from N20°W in the north to N45°W in the south as well as the rate increase of the offset at 28°N. The timing of this change is consistent with the reorganization of the Pacific-Izanagi ridge between chrons M21 and M20 [Nakanishi et al., 1989]. A similar change in topographic expression due to the change in a strike of a fracture zone was observed at the Mendocino Fracture Zone at 152°20'W [Caress et al., 1988]. The topographic expression of the Mendocino Fracture Zone changes from a trough with a linear ridge to a scarp.

The trend of the Kashima Fracture Zone at a point where its trend changes is not orthogonal to the lineations M21 and M20 in its older (eastern) side but orthogonal to those in its younger (western) side. This is in accord with the Menard and Atwater's observation [1969] that major fracture zones are orthogonal to the magnetic anomaly lineations on their younger side.

Nosappu Fracture Zone

The Nosappu Fracture Zone extends more than 1,000 km from the northwestern edge of the Shatsky Rise to the Kuril Trench (Figure 1). It is apparently discontinuous at least at 3 points; 40°N, 38°N and 35°N (Figure 6). The amount of horizontal offset of lineation M5 by the Nosappu Fracture Zone is about 350 km corresponding to an age-difference of slightly greater than 6 m.y., which is one-third of that

Kashima Fracture Zone

Fig. 2. Location and geometry of the Kashima Fracture Zone. Magnetic anomaly lineations are shown with M numbers. Small dots on the lineations are the fixed points. Small open triangles show points of fracture zones fixed on bathymetric profiles and seismic profiles. Shaded areas represent seamounts along the Kashima Fracture Zone. The track line (V2106) for Figure 5 is shown by a heavy dashed line (AA'). The 5000 m and 7000 m GEBCO bathymetric contours are shown for reference.

TABLE 1. Offset of Major Fracture Zones in the North Pacific Ocean

Fracture Zone	Offset (km)	Lateral Direction	Reference
Kashima	500	right	this study
Nosappu	350	left	this study
South Shatsky	90	right	this study
Phoenix	600	right	this study
Central Pacific	150	right	this study
Mendocino	1170	left	Vacquier, et al. [1961]
Pioneer	250	left	Vacquier [1959]
Murray (eastern)	150	right	Mason [1958]
Murray (western)	640	right	Raff [1962]

of the Mendocino Fracture Zone (Table 1). The horizontal offset along the Nosappu Fracture Zone is variable and has no linear correlation with age in contrast to the Kashima Fracture Zone. The Nosappu Fracture Zone is a long and narrow belt of grossly irregular topography characterized by seamounts, linear ridges, and scarps (Figure 7). It can be divided into several segments with poor continuity. The acoustic basement is higher on the younger (eastern) side of the Nosappu Fracture Zone than on the older (western) side in a similar manner to usual fracture zones (Figure 8).

The Nakwe channel [Mammerickx, 1980] follows the northern part of the Nosappu Fracture Zone. The two most northern segments, between 38°N and 42°N, are situated in the Nakwe channel. However the channel changes its course to a southerly direction apart from the southeastern continuation of the fracture zone.

A seismic reflection profile recorded by R/V *Hakurei-maru* (cruise GH76-2) at the northern end of this fracture zone shows that vertical displacement of the acoustic basement across the fracture zone is about 800 m (Figure 8a), whereas that in seafloor topography is only 150 m. The vertical displacement of the acoustic basement is comparable with that of the Mendocino Fracture Zone (800 m)

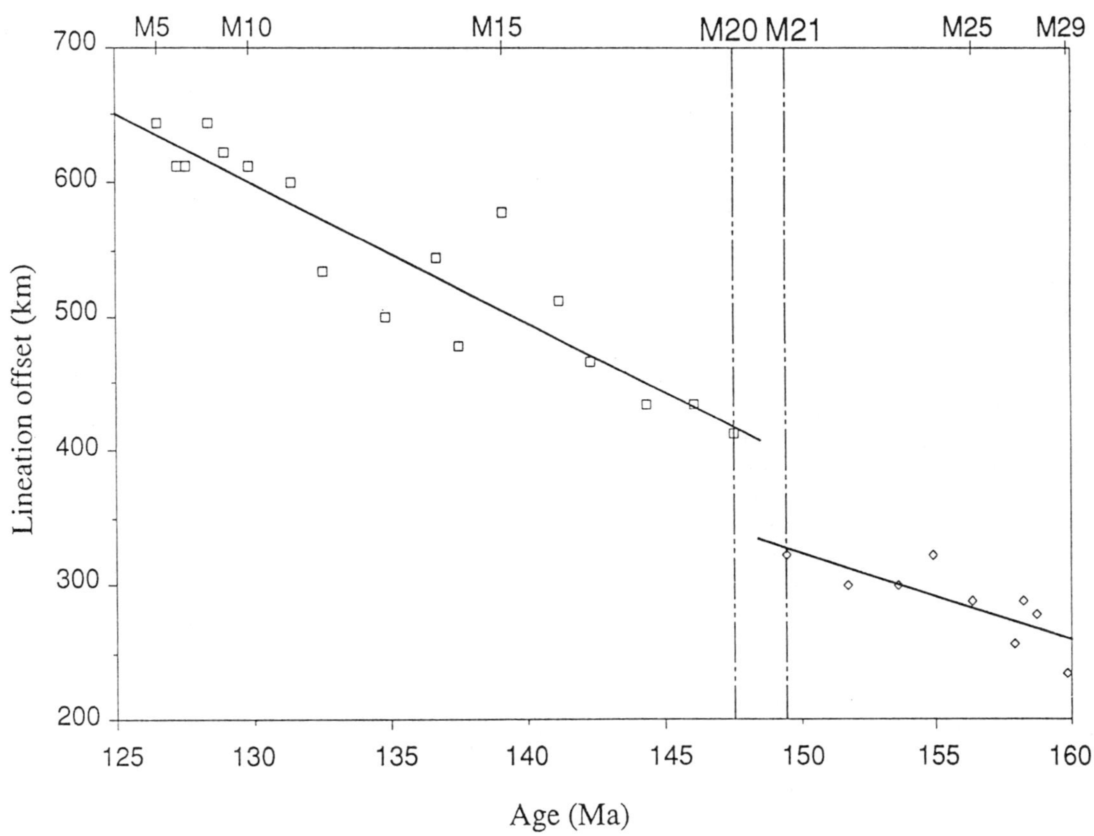

Fig. 3. Offset of magnetic lineations (in km) versus age (in Ma) along the Kashima Fracture Zone. M number of anomalies are also indicated on the upper abcissa. A heavy line represents the best fitted lines. Note the gap at about 149 Ma between chron M21 and M20 when the plate boundaries were reorganized. Ages are assigned according to the geomagnetic time scale by Kent and Gradstein [1985].

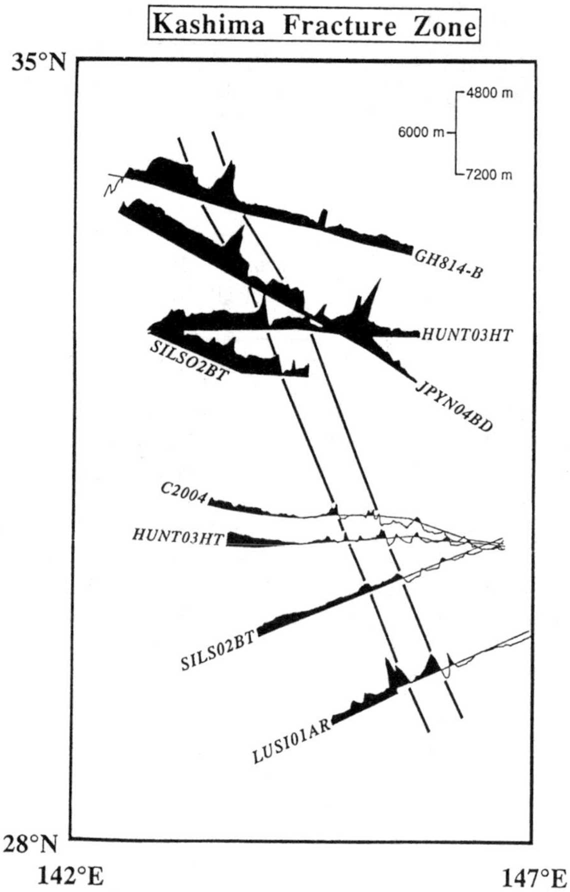

Fig. 4. Selected bathymetric profiles across the Kashima Fracture Zone. Cruise identifications are given at an end of each profile. Bathymetric expression of the fracture zone is a series of complex troughs and ridges. The heavy lines represent correlation of the fracture zones shown in Figure 2. The topography shallower than 6000 m is colored black.

[Menard and Dietz, 1952]. Figure 8b shows a seismic reflection profile recorded by D/V *Glomar Challenger* (DSDP Leg 32) across the Nosappu Fracture Zone near 41°N. This record was also used by Mammerickx [1980] to identify the Nakwe channel. About 500 m of sediment masks the vertical displacement of the acoustic basement that amounts to 300 m. Figure 8c is a seismic reflection profile by R/V *Conrad* (cruise C1405) near 38°N. Here, the fracture zone is a ridge with a height of about 1,800 m above. No vertical displacement of the acoustic basement exists here. A depression owing to lithospheric flexure by a ridge is recognized on the eastern (young) side of the ridge, although a depression along a fracture zone is normally observed at the older side [e.g., Sandwell and Schubert, 1982]. Figure 8d represents a seismic reflection profile by D/V *Glomar Challenger* (DSDP Leg 86) near 36°N. At this location, the fracture zone is characterized by a gentle trough. The depth of the trough is about 100 m and its width is about 10 km. Both the bottom and scarp of the trough are covered by a sediment layer with a thickness of a few hundred meters, which appears to be continuous from the surrounding seafloor. The vertical displacement of the acoustic basement across this trough is about 400 m.

South Shatsky Fracture Zone

The South Shatsky Fracture Zone is approximately 600 km long and obliquely cuts the Hawaiian lineation set (Figure 9). Its general bathymetric expression is a wide trough (about 60 km wide) bounded by ridges (Figure 10). The maximum height of the ridges above seafloor is about 1,000 m. The horizontal offset is 90 km, which is much smaller than those of the Kashima and Nosappu Fracture Zones. The difference in age across the fracture zone is 2 m.y. Its northern extremity reaches the southern edge of the Shatsky Rise, where lineation M21 is identified. I could not recognize an extension of the fracture zone within the Shatsky Rise.

Phoenix Fracture Zone

The Phoenix Fracture Zone is located east of the Marshall Islands and its bathymetry is the most remarkable of all the fracture zones in the Phoenix lineation set (Figure 11). The topographic expression is a trough bounded by large ridges, similar to the Kashima Fracture Zone (Figures 12 and 13). The continuity of the Phoenix Fracture Zone is much better than that of the Kashima Fracture Zone and is depicted in the GEBCO bathymetric map [GEBCO, 1984]. The maximum height of ridges above seafloor is 3000 m. The horizontal offset of lineations older than chron M20 is nearly 600 km, which is about half of that in the Mendocino Fracture Zone (Table 1). That of lineations younger than chron M14 is about 400 km. The horizontal offset between lineations M19 and M15 is obscured because these lineations on the eastern side were not identified. The age-difference across the Phoenix Fracture Zone is greater than 5 m.y. The vertical displacement of the seafloor is about 500 m and that of the underlying acoustic basement is about 700 m (Figure 13), which is comparable to that of the Nosappu Fracture Zone.

Central Pacific Fracture Zone

The Central Pacific Fracture Zone is situated in the central Pacific Basin east of the Phoenix Fracture Zone (Figure 11). This fracture zone had previously been identified only from the offset of magnetic anomaly lineations by Larson et al. [1972] and Larson [1976], until I confirmed its existence with bathymetric and seismic reflection data. Its topographic expressions are a scarp with a ridge and a simple ridge (Figure 14). The maximum height of ridges above seafloor is 1,000 m. The horizontal offset is nearly 150 km. The age-difference across the Central Pacific Fracture Zone is less than 5 m.y. The

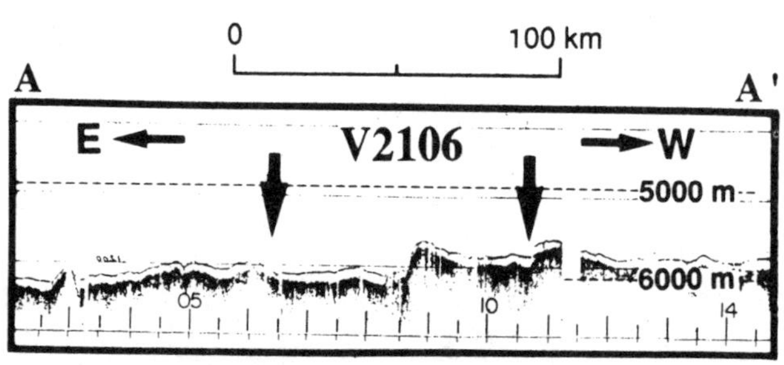

Fig. 5. A seismic reflection profile crossing the Kashima Fracture Zone obtained by R/V *Vema*, cruise V2106. Region between two arrows represents the fracture zone. The location of the profile is shown in Figure 2 by a heavy dashed line.

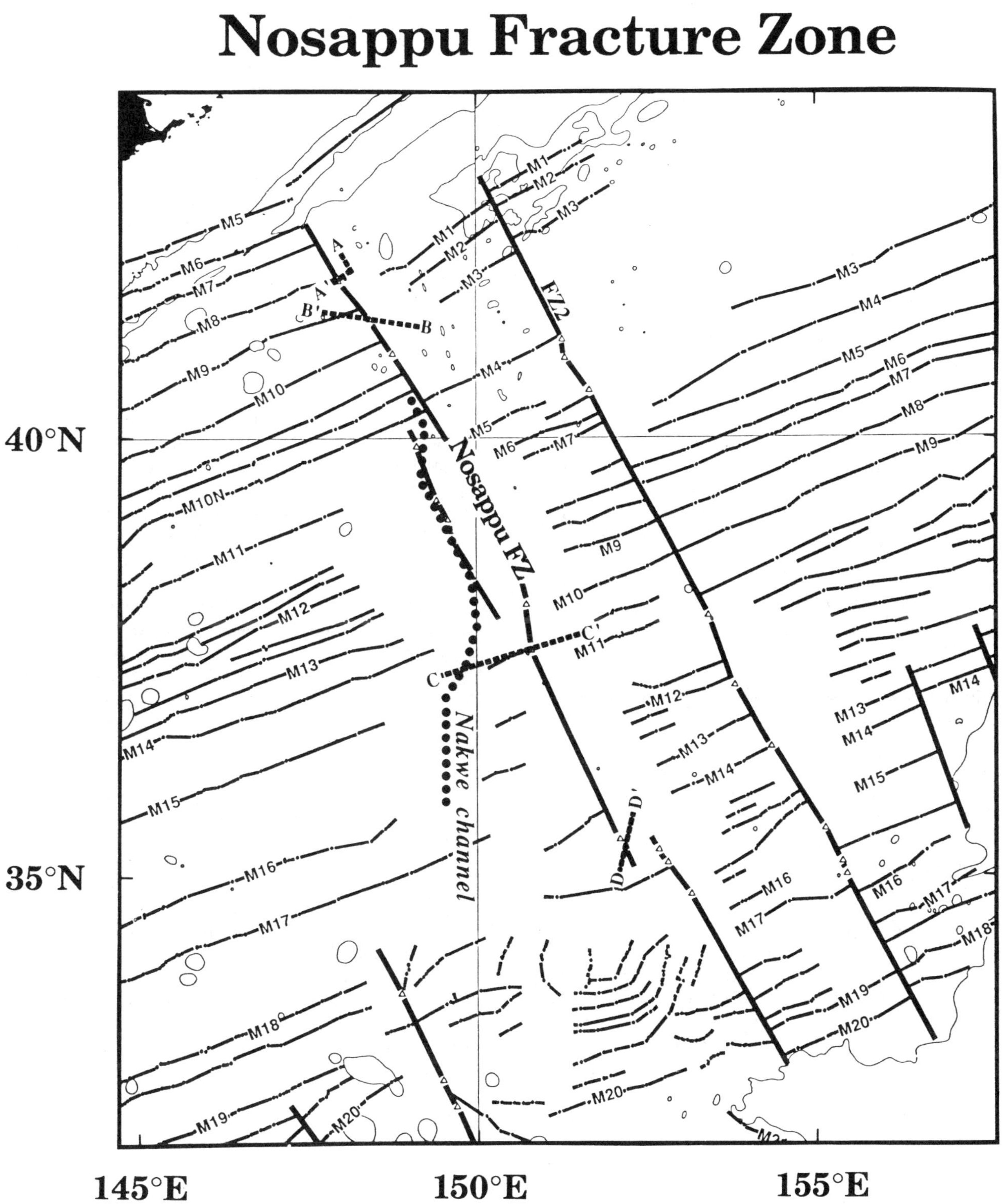

Fig. 6. Location and geometry of the Nosappu Fracture Zone. Conventions are identical to Figure 2. A dotted line indicates the Nakwe channel. Heavy dashed lines are tracks of seismic profiles of Figure 8.

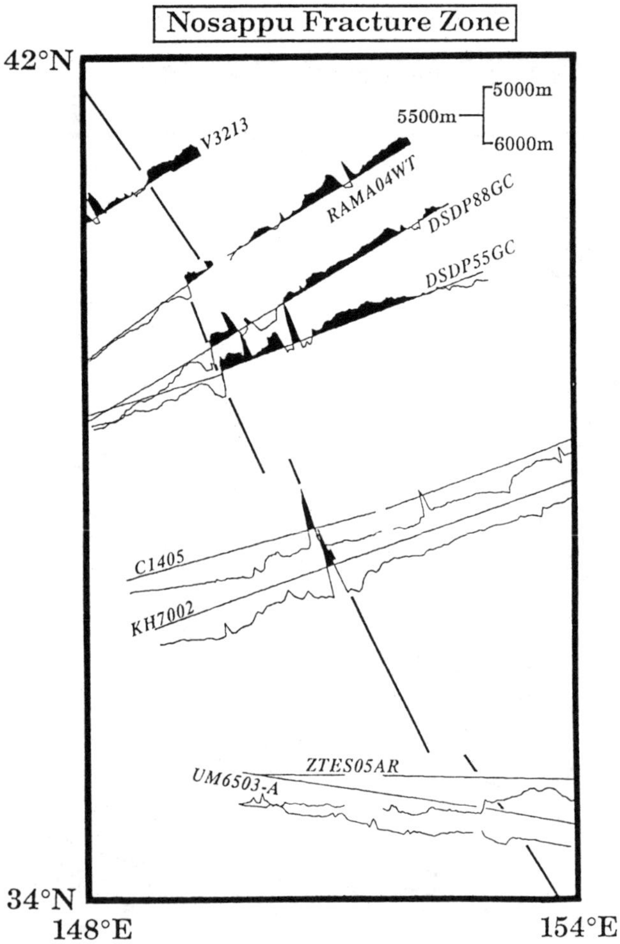

Fig. 7. Selected bathymetric profiles across the Nosappu Fracture Zone. Cruise identifications are given at each profile. The heavy lines represent correlation of the fracture zones shown in Figure 6. Topography shallower than 5500 m is colored black.

vertical displacement of the seafloor is about 200 m. Figure 15 shows that the height of the scarp is about 800 m.

Other Fracture Zones

Existence of the fracture zone south of the Ogasawara Plateau and further west of the Kashima Fracture Zone has been confirmed by the detailed bathymetric survey of Hydrographic Department, Marine Safety Agency of Japan [Hydrographic Department, Maritime Safety Agency, 1989]. The fracture zone is remarked as FZ1 as Figure 2. The survey revealed that the principal topographic expression is a scarp with a ridge, striking parallel to the southern portion of the Kashima Fracture Zone.

A fracture zone (FZ2, Figure 6) east of the Nosappu Fracture Zone identified as a displacement of magnetic anomaly lineations has nearly the same length as the Nosappu Fracture Zone, but its topographic expression is not as pronounced. Several small fracture zones have been recognized in a region north and east of the Shatsky Rise from bathymetric data as well as magnetic data. In the area southeast of the Shatsky Rise, a few fracture zones have been identified by bathymetric data as well as magnetic offsets. Mammerickx [1983] proposed the Mendocino North and South Fracture Zones in the Hawaiian lineation set as a westward extension of the Mendocino Fracture Zone (Figure 1). Later she found that these fracture zones offset the Pacific plate before its major reorganization at 84 Ma, independently of the growth of the Mendocino Fracture Zone in the northeastern Pacific Ocean [Mammerickx personal communication, 1989]. Mammerickx and Smith [1985] proposed a new name, "Waghenaer Fracture Zone", in place of Mendocino North Fracture Zone. Nakanishi et al. [1989] proposed a provisional name "Waghenaer South" to the previous designated Mendocino South Fracture Zone. The strike of the Waghenaer and Waghenaer South Fracture Zones changes according to the change in trends of lineations on both sides which are orthogonal to them.

ORIGIN OF TOPOGRAPHIC EXPRESSION OF MAJOR FRACTURE ZONES

Kashima Fracture Zone

Complex troughs, similar to those observed in the Kashima Fracture Zone, were reported in the western portion of the Murray Fracture Zone [Menard and Chase, 1970] and associated with some fracture zones in the central Pacific Ocean [Menard, 1967]. Menard and Atwater [1969] proposed that such a lineated topographic feature along a fracture zone was formed by a leaky transform fault associated with a reorganization of the spreading system (e.g., a change of spreading direction). The part of the Kashima Fracture Zone associated with a complex trough was formed after the reorganization of the Pacific-Izanagi-Farallon triple junction at a period between chrons M21 and M20 [Nakanishi et al., 1989]. Fox and Gallo [1984] suggested that multi-transform fracture zones are common adjacent to fast-spreading ridges. Searle [1983] concluded that multi-transform fracture zones are a physical embodiment of leaky transform faults in fast-spreading ridges. Although half spreading rates around the Kashima Fracture Zone are 4 to 7 cm/yr, medium to fast, its characteristic features imply a leaky transform origin created by the reorganization of the Pacific-Izanagi-Farallon triple junction at a period between chrons M21 and M20.

The origin of seamounts along the southern segment of the Kashima Fracture Zone older than chron M21 may be either post-spreading volcanism or diapiric intrusion of serpentinites along a fracture zone. These seamounts belong to the Marcus-Wake chain and geological and paleomagnetic studies by Winterer et al. [this volume] and Sager et al. [this volume] suggest that they originated from Mid-Cretaceous volcanism. As a part of the western Pacific Basin was rejuvenated by basaltic volcanism in the Cretaceous [e.g., Larson and Schlanger, 1981; Lancelot, Larson, et al., 1990], it seems most plausible that an old fracture zone with a peculiar crustal structure was especially affected by such off-ridge magmatic activities. The other possibility of serpentine diapirs cannot be excluded. Francis [1981] suggested that serpentinization along the margin of a transform fault occurs preferentially on its older, colder side, causing uplift of the seafloor on this portion. In the Kashima Fracture Zone several seamounts exist on its west (younger side), although majority is seen on its east (older side). These hypotheses may be tested by petrographic studies and age determination of these seamounts with basaltic samples collected either by dredge hauls or ocean drillings.

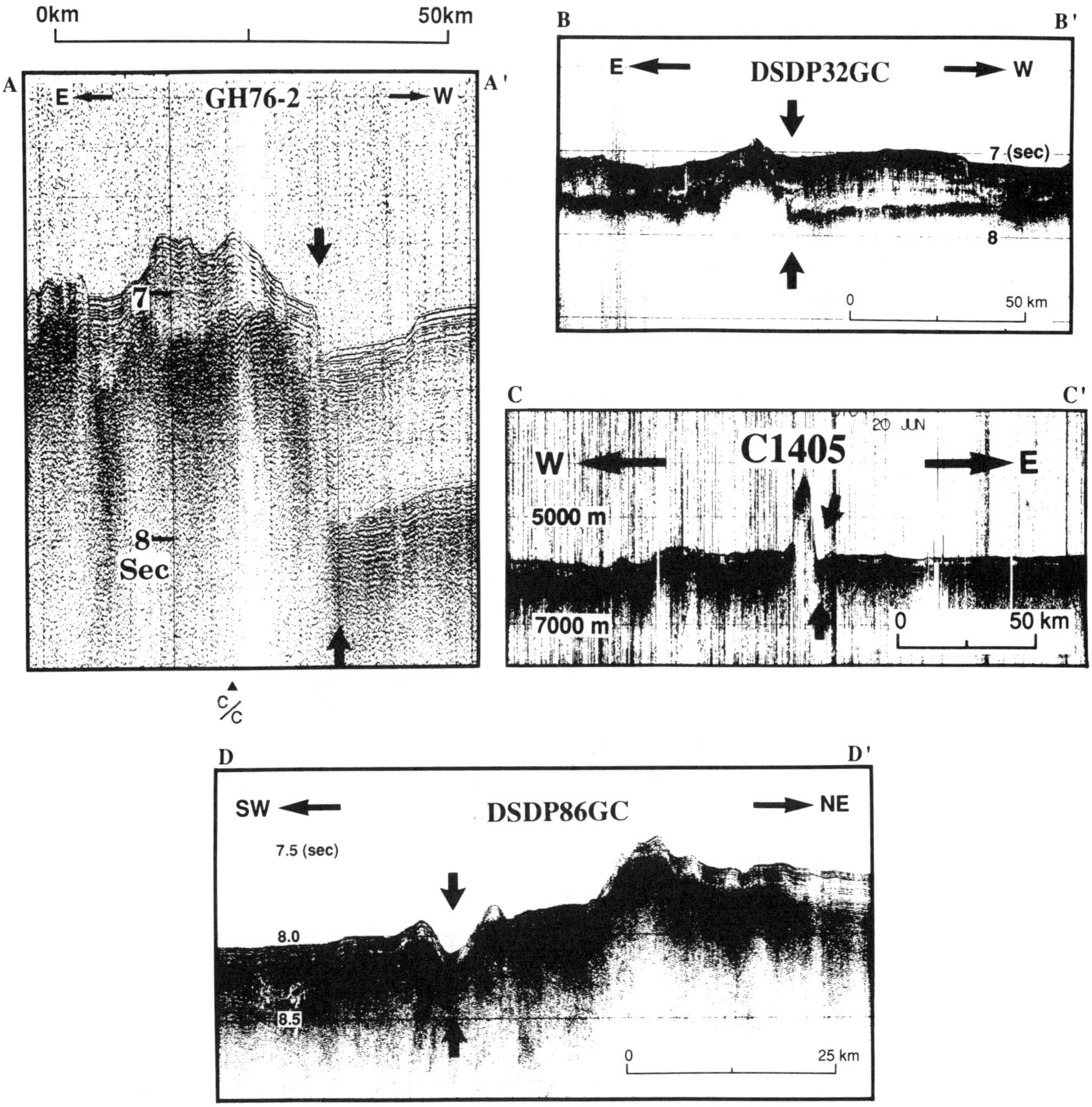

Fig. 8. Four seismic reflection profiles across the Nosappu Fracture Zone The locations of profiles are shown in Figure 6. Solid arrows indicate the fracture zone. (a) Profile AA' at 42°N obtained by R/V *Hakurei-maru* (Geological Survey of Japan, cruise GH76-2). (b) Profile BB' crossing the Nosappu Fracture Zone near 41°N obtained by D/V *Glomar Challenger*, cruise DSDP32GC. (c) Profile CC' crossing the Nosappu Fracture Zone near 38°N obtained by R/V *Conrad*, cruise C1405. (d) Profile DD' crossing the Nosappu Fracture Zone near 36°N obtained by D/V *Glomar Challenger*, cruise DSDP86GC.

Nosappu Fracture Zone

The Nosappu Fracture Zone is more complicated than the Kashima Fracture Zone and manifests various topographic features such as a ridge, a trough and a scarp. A similar variety in topographic features is observed along the Clipperton and Clarion Fracture Zones in the eastern Pacific Ocean [Menard and Chase, 1970]. Bonatti [1978] attributed such a topographic variation in a fracture zone to the effect of differential cooling and subsidence.

Magnetic anomaly lineations in the vicinity of the Nosappu Fracture Zone are less easily recognized than those in the vicinity of the

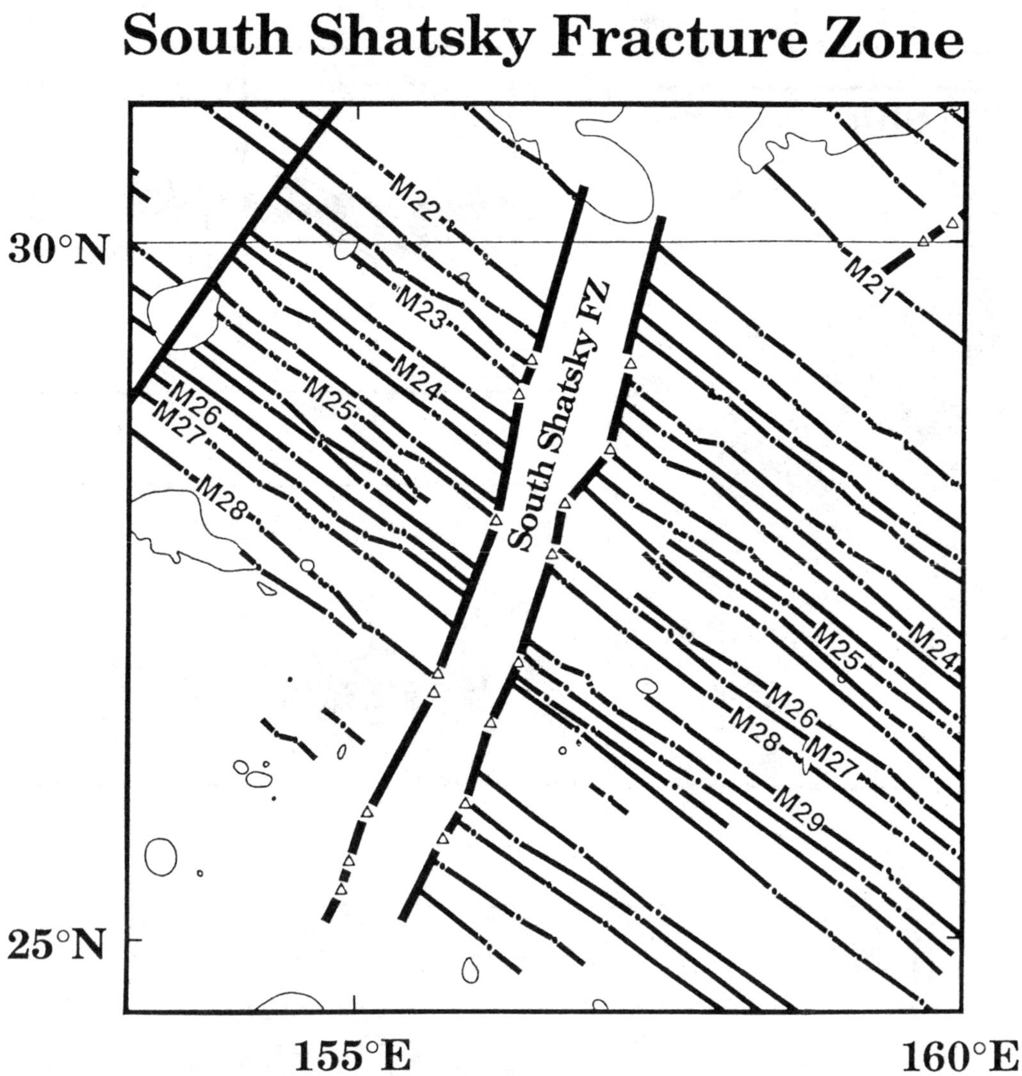

Fig. 9. Location and geometry of the South Shatsky Fracture Zone. Conventions are identical to Figure 2.

Kashima Fracture Zone, perhaps implying that the activity of the spreading ridges near the Nosappu Fracture Zone had been less stable. This implies that the irregular topographic features of the Nosappu Fracture Zone are due to instability of the activity of the spreading ridges.

The proximity of positions of the northern portion of the Nosappu Fracture Zone with the Nakwe Channel implies that an erosional channel was formed in post-Miocene time in connection with topographic troughs of the fracture zone. It has not been completely understood why the southern channel departs southwestward from the fracture zone. The channel of a gravity flow may have simply followed the topographic gradient from high to low places. The topographic expression of the Nosappu Fracture Zone changes from a trough to a ridge, so the channel followed the local gradient.

South Shatsky Fracture Zone

The South Shatsky Fracture Zone is characterized by a wide trough bounded by ridges that obliquely cut the surrounding lineations. Obliqueness suggests that the fracture zone may be a pseudofault formed by rift propagation along the Pacific-Farallon Ridge. The rift propagation occurred from chron M29 (165 Ma) to chron M21 at a rate of about 1.6 cm/yr. The rift propagation ceased at the reorganization of the Pacific-Izanagi-Farallon triple junction associated with the birth of the Shatsky hotspot, which created the Shatsky Rise [Nakanishi et al., 1989].

Similar rift propagation along the East Pacific Rise near Easter Island has occurred at an average rate of about 15 cm/yr in the past 1.2 m.y., which is 1.3 times faster than its whole spreading rate (12 cm/yr) [Naar and Hey, 1986]. The propagation rate on the Galapagos

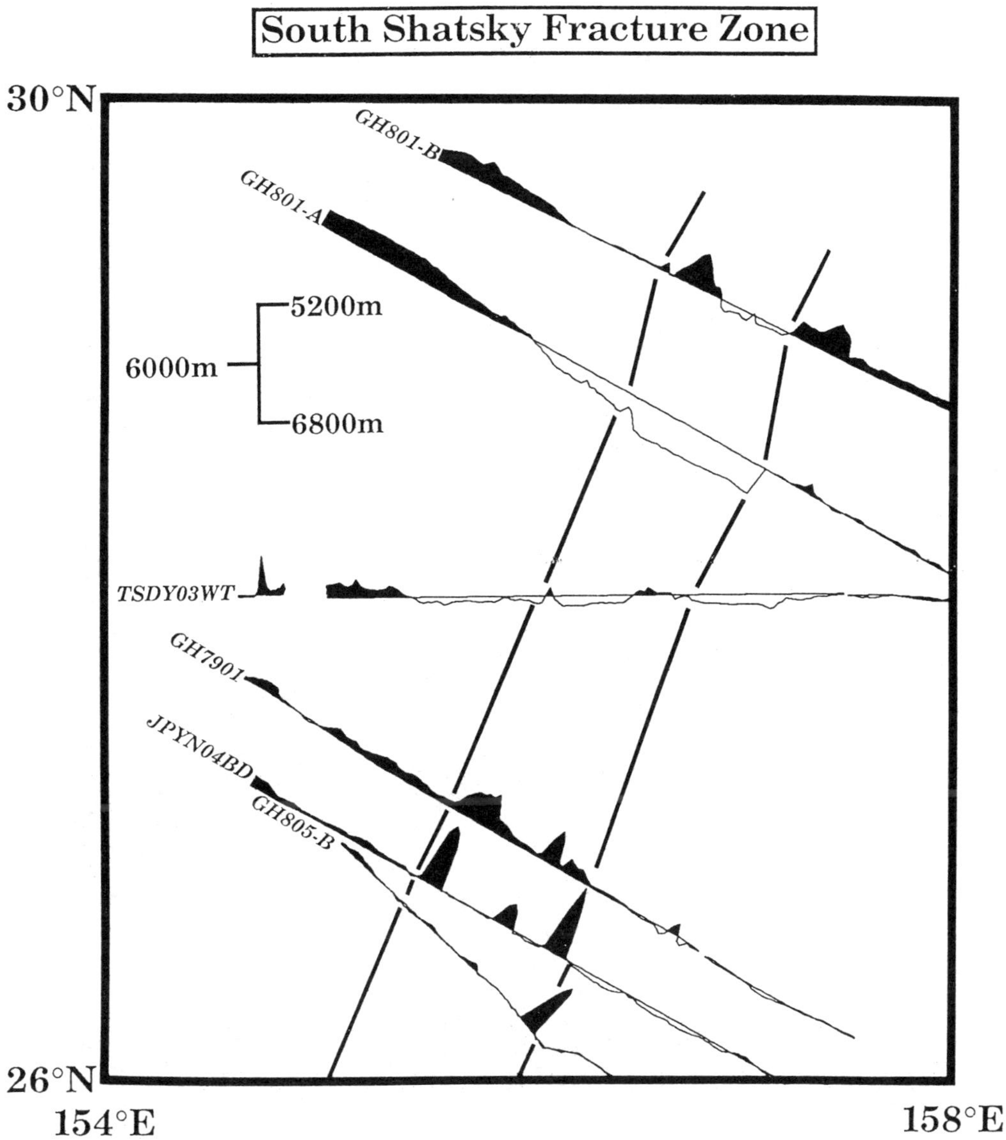

Fig. 10. Selected bathymetric profiles across the South Shatsky Fracture Zone. Cruise identifications are given at the left of each profile. The heavy lines represent correlation of the fracture zones shown in Figure 9. The topography shallower than 6000 m is colored black.

ridge at 95.5°W has been about 5.2 cm/yr, which is 0.85 times as fast as the whole spreading rate (5.8 cm/yr) [Hey et al., 1986]. Propagation rate of the Juan de Fuca Ridge varies between 2.6 and 5.4 time the whole spreading rates [Wilson, 1988]. The propagation rate associated with the South Shatsky Fracture Zone is comparable with these spreading centers, although the ratio of propagation/spreading is 0.16 which is much smaller than the others. Tamaki and Larson [1988] and Sager et al. [1988] indicated that propagation without changes in spreading direction along the Pacific-Magellan Ridge and Pacific-Izanagi Ridge, respectively, occurred in the Mesozoic. The rift propagation of the Pacific-Farallon Ridge that made the South Shatsky Fracture Zone was not accompanied with changes in spreading direction either.

If rift propagation created the South Shatsky Fracture Zone, it may be the oldest pseudofault among those so far discovered in the world oceans [e.g., Fernandez and Hey, 1991; Fullerton et al., 1989; Mammerickx, 1984; Okal and Cazenave, 1985; Mammerickx and Sandwell, 1986; Mammerickx and Sharman, 1988; Tamaki and Larson, 1988].

Phoenix Fracture Zone

In contrast to the sense of offset usually observed across a fracture zone, the younger side of the Phoenix Fracture Zone is deeper than the older side. Mammerickx [1983] indicated that the depth of the seafloor west of the Phoenix Fracture Zone is shallower than that predicted by theoretical age-depth curve [e.g., Parsons and Sclater,

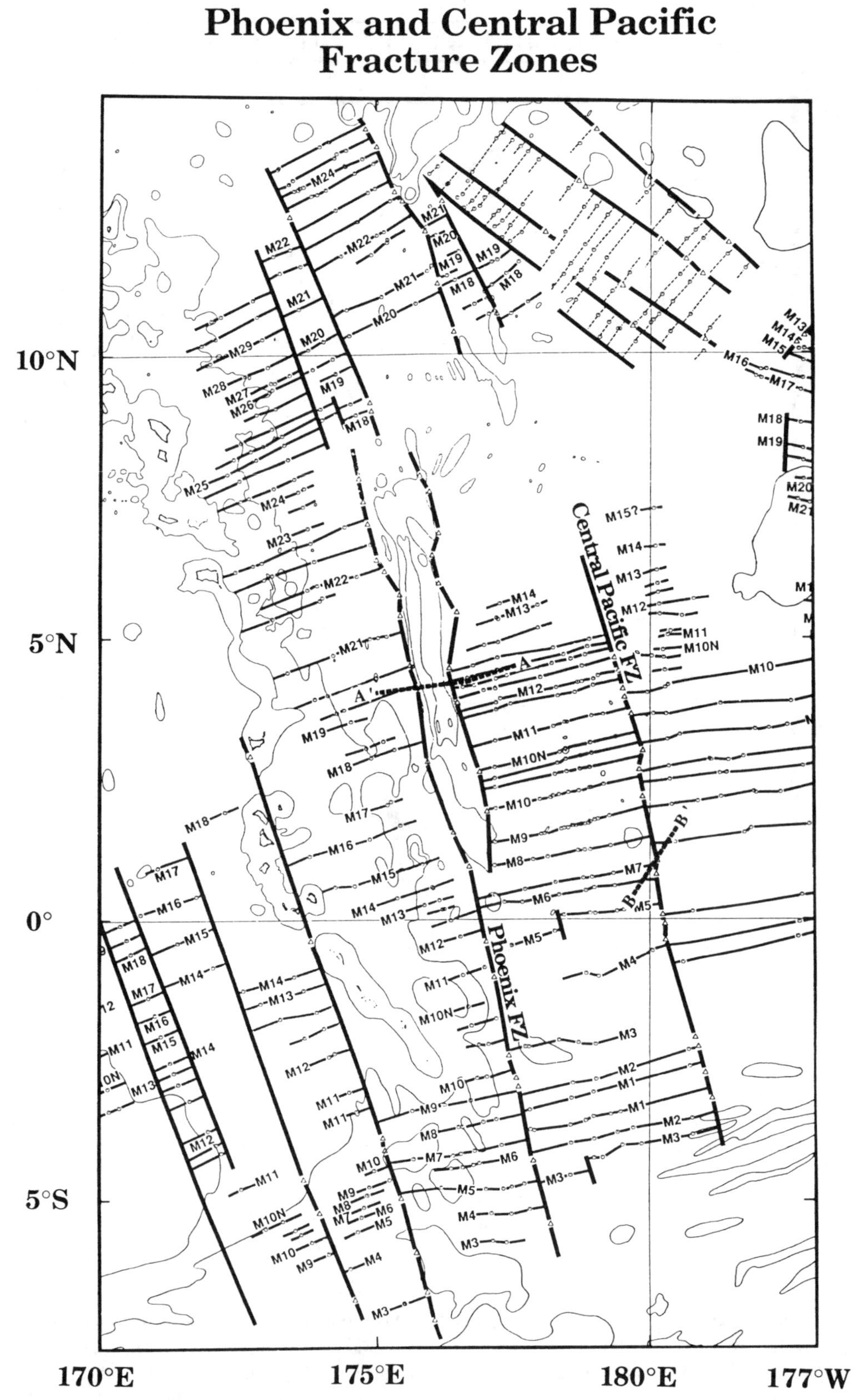

Fig. 11. Location and geometry of the Phoenix and the Central Pacific Fracture Zones. Conventions are identical to Figure 2. Heavy dashed lines are tracks of seismic profiles of Figures 13 and 15.

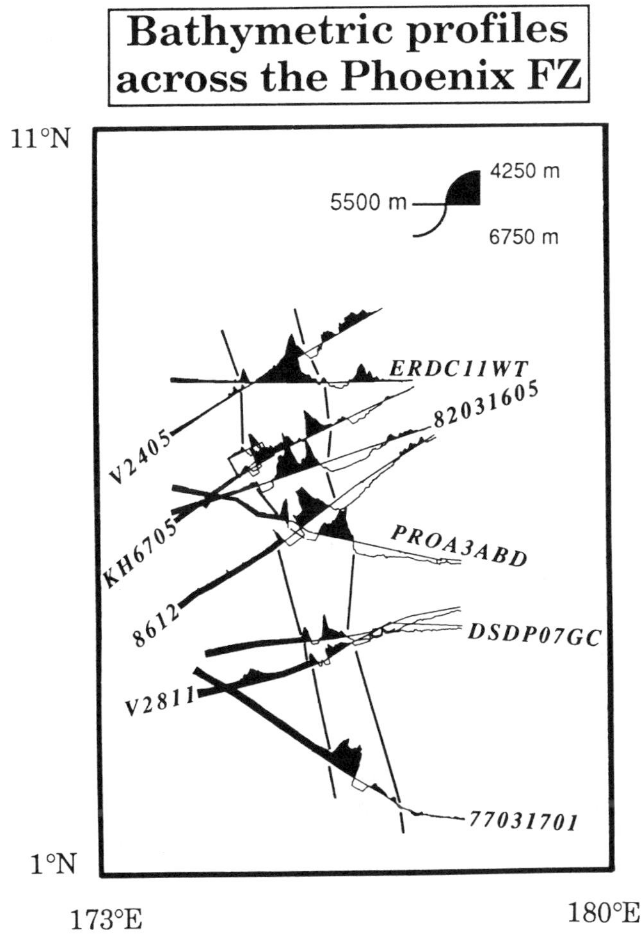

Fig. 12. Selected bathymetric profiles across the Phoenix Fracture Zone. The heavy lines represent correlation of the fracture zones shown in Figure 11. Cruise identifications are given at each profile. The topography shallower than 5500 m is colored black.

1977]. Larson and Schlanger [1981] concluded that the depth anomaly in the Nauru Basin west of the Phoenix Fracture Zone was produced by Cretaceous mid-plate volcanism accompanied by thermally induced uplift. The volcanism covered a large area of the central Pacific Ocean, but DSDP Sites 166 and 167, which are situated northeast of the Phoenix Fracture Zone, did not encounter Mid-Cretaceous volcanic materials [Schlanger and Premoli Silva, 1981]. This implies that the seafloor east of the Phoenix Fracture Zone did not undergo the thermally induced uplift by the volcanism. Thus the origin of the uplift of the younger side of the Phoenix Fracture Zone is probably related to the Mid-Cretaceous volcanism.

The Vertical Displacement along a Fracture Zone

It is generally accepted that the younger side of a fracture zone is normally shallower than the older portion. In young ocean floor the offset in water depths across a fracture zone can be attributed to differential subsidence because of the age offset. Both empirical and theoretical analyses indicate that depth increases with the square root of crustal age [Parsons and Sclater, 1977]. The vertical displacement of the fracture zones is also observed in the Cretaceous oceans. For example, that of the Kashima Fracture Zone is about 500 m and that of the Nosappu Fracture Zone changes from 400 to 800 m. The age difference of the Kashima and Nosappu Fracture Zones is nearly 10 m.y., although the vertical displacement of the Phoenix Fracture Zones is reverse of those of the Kashima and Nosappu Fracture Zone as mentioned before.

Although an offset of 500 m is predicted by an age difference of 10 m.y. if the ocean floor is younger than 100 Ma, this relation is not valid in the northwestern Pacific Ocean where the seafloor is older than 120 Ma. If an oceanic lithosphere subsided according to a simply conductive-cooling model [e.g., Parsons and Sclater, 1977], there should be no significant vertical displacement across the three fracture zones. Sandwell and Schubert [1982] and Sandwell [1984] postulated that evolution of the fracture zone topography is extremely sensitive to the initial thermal structure and age offset near the ridge-transform intersection because of differential subsidence. Lowrie et

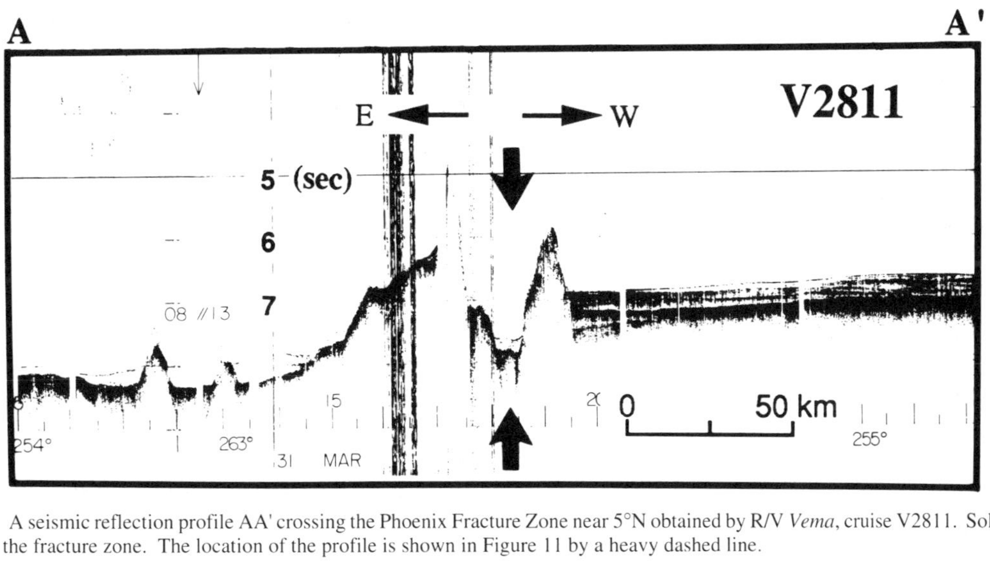

Fig. 13. A seismic reflection profile AA' crossing the Phoenix Fracture Zone near 5°N obtained by R/V *Vema*, cruise V2811. Solid arrows indicate the fracture zone. The location of the profile is shown in Figure 11 by a heavy dashed line.

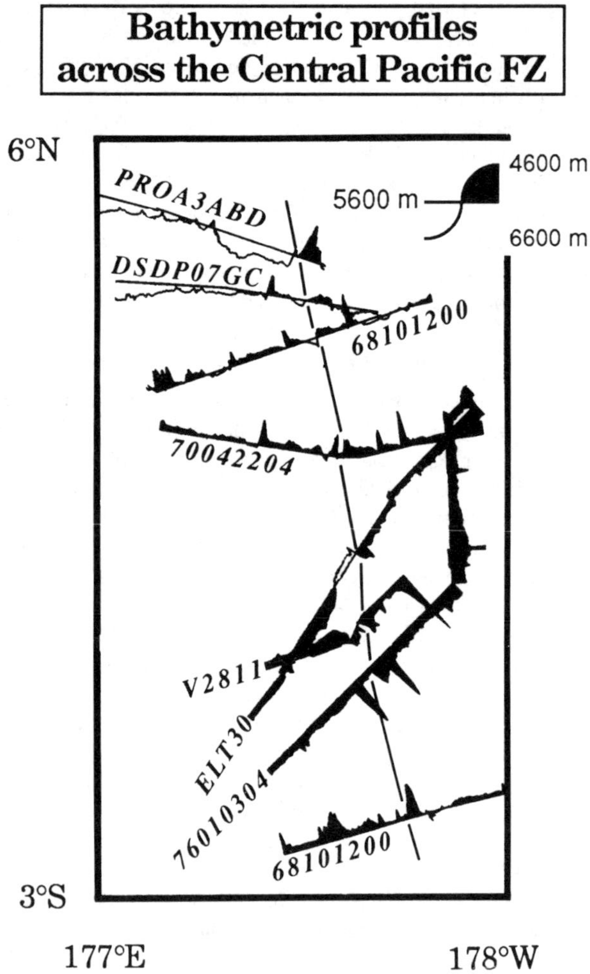

Fig. 14. Selected bathymetric profiles across the Central Pacific Fracture Zone. Cruise identifications are given at each profile. The heavy lines represent correlation of the fracture zones shown in Figure 11. The topography shallower than 5600 m is colored black.

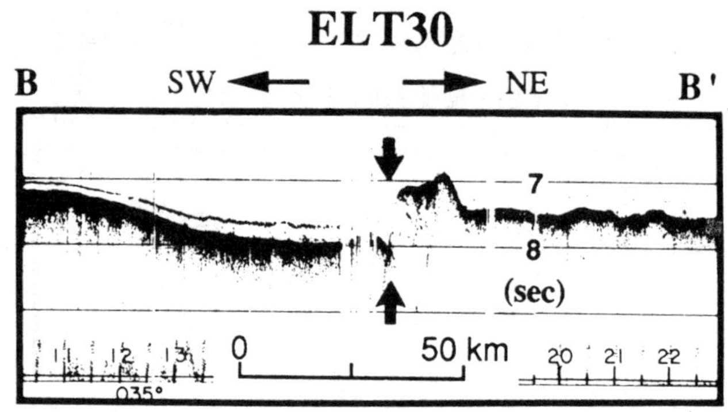

Fig. 15. A seismic reflection profile BB' crossing the Central Pacific Fracture Zone near 1°N obtained by R/V *Eltanin*, cruise ELT30. Solid arrows indicate the fracture zone. The location of the profile is shown in Figure 11 by a heavy dashed line.

al. [1986] proposed that the vertical displacement along some parts of fracture zones is preserved shortly after moving away from ridge-transform intersections. However, Figure 3 indicates that the Kashima Fracture Zone was not frozen until long time after spreading. On the other hand, Parmentier and Haxby [1986] and Wessel and Haxby [1990] suggested that the vertical displacement occurred in response to thermal stresses in the oceanic lithosphere. More detailed data of the Kashima and Nosappu Fracture Zones, such as the regional crustal structure and petrography of the walls, are needed to elucidate the vertical displacement of these fracture zones.

Fracture Zones in a Subduction Zone

The Kashima and Nosappu Fracture Zones are now being subducted in the Japan and Kuril Trenches, respectively. The strike of the Japan Trench changes from N-S to NNE-SSW near 36°N, where the Kashima Fracture Zone intersects it. In the latitudes south of the intersection its strike returns from NNE-SSW to N-S. It might be suspected that there is correlation between the change in the strike of the Japan Trench and the past subduction of the Kashima Fracture Zone.

In the case of the Nosappu Fracture Zone, no appreciable change in the strike of the Kuril Trench is observed at its intersection. However, the seafloor of the western part of the intersection between Nosappu Fracture Zone and Kuril Trench is generally shallower than that of the eastern part. There is a fault (Tuscarora) on the landward side of the trench where the fracture zone is subducted [Gnibidenko, 1985]. It is an active fault which seems to be correlatable to subduction of the Nosappu Fracture Zone. The variation in the trench depth may be caused by subduction of the Nosappu Fracture Zone.

Summary and Conclusions

Analysis of bathymetric and seismic data together with magnetic anomalies have allowed us clear identification of 61 fracture zones in the northwestern Pacific Ocean. The Kashima Fracture Zone is divided into two portions according to its topographic expression. The south portion is characterized by several seamounts aligned along the fracture zone and the north is composed of a complex trough, the western edge of which is bounded by asymmetric ridges. Transition between these two features was caused by change in strike of the transform motion owing to reorganization of the Pacific-Izanagi ridges. The complex trough observed in the Kashima Fracture Zone was formed by a leaky transform caused by the change in strike of the Pacific-Izanagi ridge from N45°E to N70°E.

The Nosappu Fracture Zone can be divided into several discontinuous segments. This fracture zone manifests various topographic features such as a ridge, trough and scarp. The Nakwe channel [Mammerickx, 1980] is developed along its northern segment.

The South Shatsky Fracture Zone is characterized by a wide trough bounded by ridges that obliquely cut the surrounding lineations formed by the Pacific-Farallon Ridge. I propose that the origin of this fracture zone is a pseudofault due to rift propagation of the Pacific-Farallon Ridge. It would be the oldest example of rift propagation so far discovered.

The Phoenix Fracture Zone has the most remarkable topographic expression in the Phoenix lineation set. Its topographic expression is a trough bounded by large ridges, possibly due to Cretaceous volcanism.

The Central Pacific Fracture Zone was confirmed in this study by its topographic expression, although it was mapped only by offsets of magnetic anomaly lineations in previous works. It is characterized by a scarp with a ridge.

The vertical displacements across the Kashima and Nosappu Fracture Zones are greater than expected for their ages, although no regional tectonic evidence has been yet obtained. Clearly, the oceanic lithosphere at a fracture zone does not subside according to a simple conductive-cooling model.

Acknowledgements. The author acknowledges National Oceanic and Atmospheric Administration/National Geophysical and Solar-Terrestrial Data Center (NOAA/NGDC) for providing us with 342 cruise data and 28 seismic data. He also thanks Japan Ocean Data Center (JODC) for their thoughtful arrangement regarding the seismic data. He also thanks E. Honza for providing a seismic profile across the Nosappu Fracture Zone.
The author is grateful to K. Kobayashi and K. Tamaki for their illuminating discussions and comments. He also acknowledges reviews and comments from W. W. Sager, D. F. Naar, and D. Joseph.

REFERENCES

Bonatti, E., Vertical tectonism on ocean fracture zones, *Earth Planet. Sci. Lett.*, *37*, 369-379, 1978.

Caress, D.W., H.W. Menard, and R.N. Hey, Eocene reorganization of the Pacific-Farallon spreading center north of the Mendocino Fracture Zone, *J. Geophys. Res.*, *93*, 2813-2838, 1988.

Carter, D.J.T., Echo-Sounding Correction Tables, Third edition, Hydrographic Department Ministry of Defense, Taunton, 150 pp., 1980.

Collette, B.J., Fracture zones in the north Atlantic: morphology and a model, *J. Geol. Soc., London*, *143*, 763-774, 1986.

Embley, R., and D.S. Wilson, Morphology of the Blanco Transform Fault Zone-NE Pacific: implication for its tectonic evolution, *Mar. Geophys. Res.*, *14*, 25-45, 1992.

Ewing, J., M. Ewing, T. Aitken, and W.J. Ludwig, North Pacific sediment layers measured by seismic profiling, in *The Crust and Upper Mantle of the Pacific Area, Geophys. Monogr.*, vol. 12, edited by L. Knopoff, C.L. Drake and P.J. Hart, AGU, Washington D. C., pp. 147-173, 1968.

Fernandez, L.S., and R.N. Hey, Late Tertiary tectonic evolution of the seafloor spreading system off the coast of California between the Mendocino and Murray Fracture Zones, *J. Geophys. Res.*, *96*, 17,955-17,979, 1991.

Fox, P.J., and D.G. Gallo, A tectonic model for ridge-transform-ridge plate boundaries: implications for the structure of oceanic lithosphere, *Tectonophysics*, *104*, 205-242, 1984.

Francis, T.J.G., Serpentinization faults and their role in the tectonics of slow spreading ridges, *J. Geophys. Res.*, *12*, 11,616-11,622, 1981.

Fullerton, L.G., W.W. Sager, and D.W. Handschumacher, Late Jurassic-Early Cretaceous evolution of the eastern Indian Ocean adjacent to northwest Australia, *J. Geophys. Res.*, *94*, 2937-2953, 1989.

GEBCO, General Bathymetric chart of the oceans Sheet 5-06, Canadian Hydrographic Chart Service, Ottawa, 1984.

Gnibidenko, H.S., The sea of Okhotsk-Kuril Islands Ridge and Kuril-Kamchatka Trench, in *The Ocean Basins and Margins,* vol. 7A, *The Pacific Ocean*, edited by A.E.M. Nairn, F.G. Stehli and S. Uyeda, Plenum Press, New York, pp. 377-418, 1985.

Hey, R.N., M.C. Kleinrock, S.P. Miller, T.M. Atwater, and R.C. Searle, Sea Beam/Deep-Tow investigation of an active propagating rift system Galapagos 95.5°W, *J. Geophys. Res.*, *91*, 3369-3393, 1986.

Hittelman, A.M., R.C. Groman, R.T. Haworth, T.L. Holcombe, G. McHendrie, and S.M. Smith, The Marine Geophysical Data Exchange Format-'MGD77' (bathymetry, magnetics, and gravity), *KGRD No. 10*, Nat. Geophys. and Solar-Terr. Data Center, NOAA, Boulder, Colo., 18 pp., 1977.

Hydrographic Department Maritime Safety Agency, Data report of hydrographic observation, Series of continental shelf survey, No. 5, Hydrographic Department Maritime Safety Agency, Tokyo, 274 pp., 1989.

Kent, D.V., and F.M. Gradstein, A Cretaceous and Jurassic geochronology, *Geol. Soc. Am. Bull.*, *96*, 1419-1427, 1985.

Kruse, S.E., Magnetic lineations on the flanks of the Marquesas swell: implications for the age of the seafloor, *Geophys. Res. Lett.*, *15*, 573-576, 1988.

Lancelot, Y., R.L. Larson, et al., *Proc. ODP, Init. Repts.*, *129*, pp. 488, 1990.

Larson, R.L., Late Jurassic and Early Cretaceous evolution of the western Central Pacific Ocean, *J. Geomagn. Geoelectr.*, *28*, 219-236, 1976.

Larson, R.L., and S.O. Schlanger, Geological evolution of the Nauru Basin, and regional implications, *Init. Repts. Deep Sea Drill. Proj.*, *61*, 841-862, 1981.

Larson, R.L., S.M. Smith, and C.G. Chase, Magnetic lineations of Early Cretaceous age in the western equatorial Pacific Ocean, *Earth Planet. Sci. Lett.*, *15*, 315-319, 1972.

Lowrie, A., N.C. Smoot, and R. Batiza, Arc oceanic fracture zones locked and strong or weak?: new evidence for volcanic activity and weakness, *Geology*, *14*, 242-245, 1986.

Ludwig, W.J., and R.E. Houtz, Isopach map of sediments in the Pacific Ocean Basin and marginal sea basins, scale 1:13,999,369, AAPG Map series, 1979.

Mammerickx, J., A deep-sea channel in the Northwest Pacific Basin, *Marine Geol.*, *34*, 207-218, 1980.

Mammerickx, J., Depth anomalies over Mesozoic crust in the western Pacific, in *Geodynamics of the western Pacific-Indonesian region, Geodynamics Series,* vol. 11, edited by T.W.C. Hilde, and S. Uyeda, AGU, Washington, D. C., pp. 63-73, 1983.

Mammerickx, J., The morphology of propagating spreading centers: new and old, *J. Geophys. Res.*, *89*, 1817-1828, 1984.

Mammerickx, J., and D. Sandwell, Rifting of old oceanic lithosphere, *J. Geophys. Res.*, *91*, 1975-1988, 1986.

Mammerickx, J., and G.F. Sharman, Tectonic evolution of the North Pacific during the Cretaceous quiet period, *J. Geophys. Res.*, *93*, 3009-3024, 1988.

Mammerickx, J., and S.M. Smith, Bathymetry of the northcentral Pacific, Geol. Soc. Am., Map Chart Ser., scale 1:6,442,194, *MC-52*, 1985.

Mason, R.G., A magnetic survey off the west cost of the United States between latitudes 32° and 36°N, longitudes 121° and 128°W, *Geophys. J.*, *1*, 320-329, 1958.

Menard, H.W., Extension of northeastern Pacific fracture zones, *Science*, *155*, 72-74, 1967.

Menard, H.W., and T. Atwater, Origin of fracture zone topography, *Nature*, *222*, 1037-1040, 1969.

Menard, H.W., and T.E. Chase, Fracture zones, in *The Sea*, vol. 4, edited by A.E. Maxwell, Wiley-Interscience, New York, pp. 421-443, 1970.

Menard, H.W., and R.S. Dietz, Mendocino submarine escarpment, *J. Geol.*, *60*, 266-278, 1952.

Naar, D.F., and R.N. Hey, Fast rift propagation along the East Pacific Rise near Easter Island, *J. Geophys. Res.*, *91*, 3425-3438, 1986.

Nakanishi, M., K. Tamaki, and K. Kobayashi, Mesozoic magnetic anomaly lineations and seafloor spreading history of the northwestern Pacific, *J. Geophys. Res.*, *94*, 15,437-15,462, 1989.

Nakanishi, M., K. Tamaki, and K. Kobayashi, Magnetic anomaly lineations from Late Jurassic to Early Cretaceous in the west-central Pacific Ocean, *Geophys. J. Int.*, *109*, 701-719, 1992.

Okal, E.A., and A. Cazenave, A model for the plate tectonic evolution of the east-central Pacific based on SEASAT investigations, *Earth Planet. Sci. Lett.*, *72*, 99-116, 1985.

Parmentier, E.M., and W.F. Haxby, Thermal stresses in the oceanic lithosphere: evidence from geoid anomalies at fracture zones., *J. Geophys. Res.*, *91*, 7193-7204, 1986.

Parsons, B., and J.G. Sclater, An analysis of the variation of ocean floor bathymetry and heat flow with age, *J. Geophys. Res.*, *82*, 803-827, 1977.

Raff, A.D., Further magnetic measurements along the Murray fault, *J. Geophys. Res.*, *67*, 417-418, 1962.

Roest, W.R., and B.J. Collette, The fifteen twenty fracture zones and the north American-south American plate boundary, *J. Geol. Soc., London*, *143*, 833-843, 1986.

Sager, W.W., R.A. Duncan, and D.W. Handschumacher, Paleomagnetism of the Japanese and Marcus-Wake Seamounts, Western Pacific Ocean, in *The Mesozoic Pacific*, Geophys. Monogr., edited by M.S. Pringle, W.W. Sager, W.V. Sliter, and S. Stein, AGU, Washington, D. C., in press.

Sager, W.W., D.W. Handschumacher, T.W.C. Hilde, and D.R. Bracey, Tectonic evolution of the northern Pacific plate and Pacific-Farallon-Izanagi triple junction in the Late Jurassic and Early Cretaceous (M21-M10), *Tectonophysics*, *155*, 345-364, 1988.

Sandwell, D.T., Thermomechanical evolution of oceanic fracture zones, *J. Geophys. Res.*, *89*, 11,401-11,413, 1984.

Sandwell, D.T., and G. Schubert, Lithospheric flexure at fracture zones, *J. Geophys. Res.*, *87*, 4657-4667, 1982.

Schlanger, S.O., and I. Premoli Silva, Tectonic, volcanic, and Paleogeographic implications of redeposited reef faunas of Late Cretaceous and Tertiary age from the Nauru Basin and the Line Islands, *Init. Repts. Deep Sea Drill. Proj.*, *61*, 817-827, 1981.

Searle, R.C., Multiple, closely spaced transform faults in fast-slipping fracture zones, *Geology*, *11*, 607-610, 1983.

Tamaki, K., and R.L. Larson, The Mesozoic tectonic history of the Magellan microplate in the western central Pacific, *J. Geophys. Res.*, *93*, 2857-2874, 1988.

Tucholke, B.E., and H. Schouten, Kane Fracture Zone, *Mar. Geophys. Res.*, *10*, 1-39, 1988.

Vacquier, V., Measurement of horizontal displacement along faults in the ocean floor, *Nature*, *183*, 452-453, 1959.

Vacquier, V., A.D. Raff, and R.E. Warren, Horizontal displacements in the floor of the northeastern Pacific Ocean, *Bull. Geol. Soc. Am.*, *72*, 1251-1258, 1961.

Wessel, P., and W.F. Haxby, Thermal stresses, differential subsidence, and flexure at oceanic fracture zones, *J. Geophys. Res.*, *95*, 375-391, 1990.

Wilson, D.S., Tectonic history of the Juan de Fuca Ridge over the Last 40 million years, *J. Geophys. Res.*, *93*, 11,863-11,876, 1988.

Winterer, E.L., R.A. Duncan, M.K. McNutt, J.H. Natland, I. Premoli Silva, W.W. Sager, W.V. Sliter, R. van Waasbergen, and C.J. Wolfe, Cretaceous guyots in the northwest Pacific: an overview of their geology and geophysics, in *The Mesozoic Pacific*, Geophys. Monogr., edited by M.S. Pringle, W.W. Sager, W.V. Sliter, and S. Stein, AGU, Washington, D. C., in press.

M. Nakanishi, Ocean Research Institute, University of Tokyo, 1-15-1 Minamidai Nakano, Tokyo 164, Japan.

Fracture Zone Traces across the North Pacific Cretaceous Quiet Zone and their Tectonic Implications

TANYA ATWATER

Department of Geological Sciences, University of California, Santa Barbara, California

JOHN SCLATER[1] AND DAVID SANDWELL

Geological Research Division, Scripps Institution of Oceanography, University of California at San Diego, La Jolla, California

JEFF SEVERINGHAUS

Lamont-Doherty Geological Observatory, Columbia University, Palisades, New York

MICHAEL S. MARLOW

U. S. Geological Survey, Menlo Park, California

We present a new, more complete mapping of the fracture zones as they cross the Cretaceous Quiet Zone in the central north Pacific. We compile and combine observations of lineations from three distinct data bases: deflection-of-the-vertical profiles from GEOSAT altimetry measurements, magnetic and topographic profiles collected on closely spaced north-south ship tracks by the Pioneer survey, and magnetic and Gloria side-scan records from the concentrated Exclusive Economic Zone surveys around the Hawaiian chain and Johnston Island. We adopt and emphasize a northern relocation of the Quiet Zone portion of the Mendocino fracture zone. We find that all the fracture zones from the Surveyor to the Clarion reflect the same plate history; they were all formed on the Pacific-Farallon spreading system by relatively smooth spreading. One or more major ridge jumps are required to explain geometric differences in the Quiet Zone edges; we argue that these occurred early in the Superchron. Most of the fracture zones are seen to consist of groups of multiple strands that widen and narrow appropriately following changing plate motion directions. We contend that the shape of the Mendocino contra-indicates the existence of a Chinook plate in the late Superchron. We explored the fitting of Euler stage poles to our fracture zone shapes and to other Quiet Zone features. We conclude that the youngest Superchron stage pole must lie very near the spin axis. Stage poles for earlier stages of the Superchron probably lie at lower latitudes in the south Pacific but their locations are poorly constrained by the data presently in hand.

INTRODUCTION

The formation of the deep sea floor of the North Pacific basin is summarized, in broad outline, by Atwater [1989]. The seafloor that presently fills the western North Pacific was formed during the Jurassic and early Cretaceous by seafloor spreading between the Pacific plate and at least three others: the Izanagi plate to the north, the Farallon plate to the east, and the Phoenix plate to the south. This history is clearly documented by seafloor magnetic anomaly isochrons of the M-series reversals [Larson and Pitman, 1972; Hilde et al., 1976; Sager et al., 1988; Tamaki and Larson, 1988; Nakanishi et al., 1989, 1992]. Similarly, magnetic isochrons in the northern and eastern North Pacific document its formation during the latest Cretaceous and the Cenozoic by seafloor spreading between the Pacific plate and the Kula and Farallon plates [Grim and Erickson, 1969; Atwater and Menard, 1970; Wilson, 1988; Atwater and Severinghaus, 1989]. In the central North Pacific, between these two sets of isochrons, lies an area that has no

[1]Also at the Institute for Geophysics, The University of Texas at Austin, Austin, Texas.

The Mesozoic Pacific: Geology, Tectonics, and Volcanism
Geophysical Monograph 77

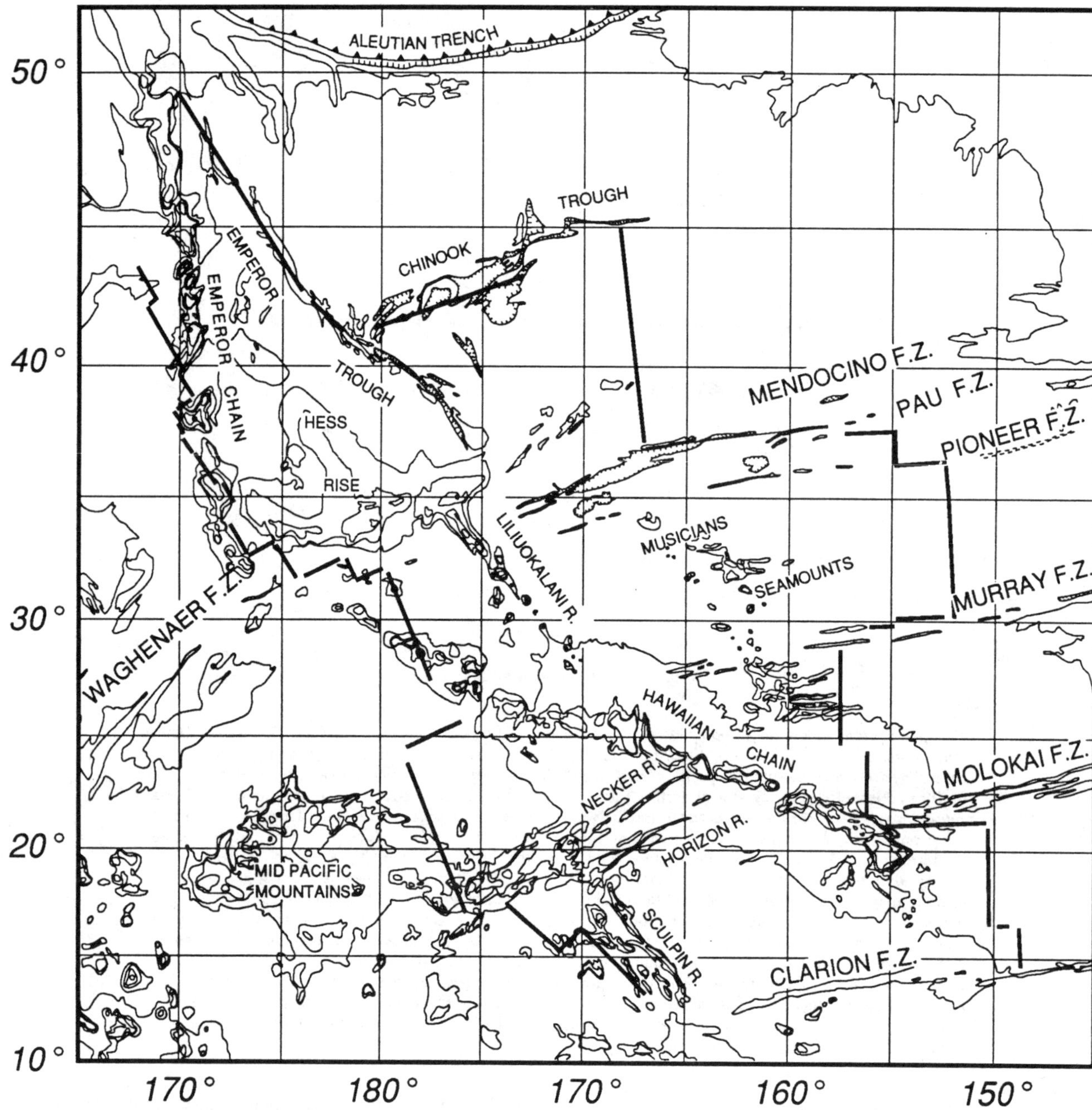

Fig. 1. Bathymetric and tectonic features of the central North Pacific. Isobaths are from Mammerickx [1989] given in 1 km intervals with 5.5 km isobaths added as needed to delineate the fracture zones. Heavy lines mark the edges of the Cretaceous Quiet Zone, seafloor presumed to have been formed during the late Cretaceous Superchron.

known magnetic isochrons, a magnetic "quiet zone", Figure 1. This region was formed during the Cretaceous Normal Superchron when the earth's magnetic field had a stable normal polarity for nearly 40 million years. (In this paper we shall refer to this reversal-free time period as the "Superchron" and to the area of seafloor formed during this time as the "Quiet Zone".) The lack of magnetic isochrons in the Quiet Zone leaves us with relatively little information about the Late Cretaceous tectonic history of the central North Pacific. However, the area is crossed by a set of fracture zones and these contain some information concerning the seafloor spreading system that formed them. In this paper we refine the locations of these fracture zones and examine their tectonic implications.

It is clear that a number of important and interesting events occurred in the North Pacific during the Cretaceous. A number of very voluminous volcanic eruptions occurred (Manihiki Plateau, Ontong-Java Plateau, Hess rise, etc.) leading Larson [1991] to postulate the establishment of a "superplume" in the area. A number of plate tectonic reorganizations also occurred during the Superchron. The Kula plate was formed, probably by the breaking off of the northern Pacific plate near the end of this time period, and other small, short-lived plates may have existed for a time [Mammerickx and Sharman, 1988]. The plate boundary between the Pacific and Farallon plates continued spreading through the Superchron, but offsets across its transform faults were greatly altered by some spreading center jumps of unknown location and timing, and the spreading system was lengthened greatly to the south [Joseph et al., 1992 and this volume]. A major change in Pacific-Farallon relative motion, from nearly east-west to east-northeast-west-southwest, occurred during Chron 33R (the reversed magnetic period preceding Chron 33: about 79-83 Ma according to Cande and Kent, 1992] leaving a distinct bend in all the northeast Pacific fracture zones [Sager and Pringle, 1987b; Joseph et al., 1992 and this volume; Searle et al., this volume]. In the text we shall refer to this feature as the "Chron 33R bend". By tracking the fracture zones from this bend, backward across the North Pacific Quiet Zone and checking their consistency with rigid plate models, we believe we can constrain the timing of the ridge jumps and establish the northern extent of the Pacific-Farallon system.

Fracture zones are the fossil traces of transform faults, and from their transform fault origins they inherit regional and local topographic disturbances, distinctive gravity signatures, and magnetic anomalies. Their topographic traces usually include a regional step, reflecting the original transform fault offset and the resulting juxtaposition of seafloor of different ages on the two sides of the zone [Menard and Atwater, 1969]. This topographic step can be quite large for large offsets in young crust, but the step tends to decay with age since the age-depth curve nearly levels off for seafloor older than about 100 million years age [Parsons and Sclater, 1977; Renkin and Sclater, 1988]. If the fracture zone remains locked during this decline of the regional step, a local step may be preserved, accompanied by a large flexural gravity anomaly over a characteristic scale of about 50 km on either side of the fracture zone [Sandwell, 1984a]. An additional topographic signature common at fracture zones is a narrow (10-20 km wide) strip of ground that was sheared and disrupted while the transform fault was an active plate boundary. The relief associated with this shear zone is relatively small, but it is permanently embedded in the seafloor so that it forms a long-term signal. When a fracture zone is offsetting magnetic isochrons, it can be the site of large block-end magnetic anomalies. Furthermore, transform fault shear zones often contain crushed and metamorphosed rock and serpentine and may have anomalously shallow mantle, creating additional permanent perturbations in the surrounding magnetic and gravity fields.

Many Pacific fracture zones consist of groups of nearly parallel, linear strands and the detailed geometric relationships among these strands provide clues for the plate motion history during the time they were formed. The multi-strand configuration results from a plate boundary configuration where a ridge-ridge transform system consists of two or more transform fault segments linked by one or more short spreading center segments. This configuration is common in presently active Pacific transform systems [Searle, 1983; Embley and Wilson, 1992; Fornari et al., 1989] and appears to have been the rule during the Cretaceous Superchron. Nearly all of the fracture zones reported in this paper are multi-stranded. We shall use the observed strand relationships to help distinguish among various plate motion models.

The topographic signatures of the fracture zones that cross the Quiet Zone are subtle. Farther east, where they cross the relatively young sea floor of the northeastern Pacific, the major fracture zones dominate the topographic chart [Menard, 1964; Mammerickx, 1989]. The traces of those fracture zones extend westward into the Quiet Zone but they become steadily more subdued as they cross increasingly older crust. The seafloor in the Quiet Zone is believed to be about 83 to 119 million years old [Kent and Gradstein, 1985; Cande and Kent, 1992], so that regional step offsets across the fracture zones have all but disappeared and the local shear topography has been partially smoothed by sediment accumulation. Additional topographic "noise" has been added by other Late Cretaceous features such as the Chinook and Emperor troughs to the north, the Hess rise and Liliuokalani ridge in the center and the Mid-Pacific Mountains and Sculpin ridge to the south. Some of these may be traces of ridge jumps and other non-steady seafloor spreading events, while others are surely the results of off-axis volcanic events. Finally, the Cenozoic Hawaiian/Emperor chain and its associated swells, moats, volcanism and sedimentation have been superimposed upon the older sea floor structures. It is no wonder that traditional attempts to use topography to map the fracture zones through this region have led to a number of ambiguities and controversies.

The fracture zones in the central North Pacific can best be mapped by combining shipboard and satellite data bases. The shipboard geophysical data coverage of the area is patchy. A few regions have been mapped in great detail: by the north-south surveys of the *R/V Pioneer* in the 1960's and by the U. S. Geological Survey Gloria side-scan sonar surveys of the Exclusive Economic Zone (EEZ) around the Hawaiian chain in the 1980's and 1990's. Elsewhere, there are only single, sparsely scattered ship tracks. Fortunately, the fracture zones create a clear gravity signal which is easily discerned on satellite altimetry profiles [Sandwell, 1984b; Haxby, 1985]. These traces tend to be longer wavelength and thus are less precise locators for the fracture zones, but they have the great advantage that the areal coverage is relatively uniform. They give an independent estimate of the locations and offsets of the various fracture zone strands and they allow us to chart them across otherwise uncharted territory.

In this paper we present topographic and magnetic profiles from the Pioneer survey and information from the EEZ magnetic and side-scan sonar surveys. We present deflection-of-the-vertical profiles from the GEOSAT altimeter across the same region. In each of these data sets we identify and trace likely fracture zone lineations. We combine these traces with recent compilations of the surrounding magnetic isochron patterns [Atwater and Severinghaus, 1989; Nakanishi et al., 1992] to provide a revised tectonic chart of the Cretaceous North Pacific. Finally, we use this chart to address some unresolved questions about the tectonic evolution of the Pacific plate.

OBSERVATIONS

Magnetic and topographic profiles from the Pioneer systematic surveys

The marine geophysical profiles that have been collected in the central North Pacific are, for the most part, sparse and unsystematically located. However, a major detailed survey of portions of the area was carried out as part of the U. S. Coast and Geodetic systematic survey by the *R/V Pioneer* in 1961 through 1963 [Elvers et al., 1967]. Many of the lines of this survey run north-south between 175° and 155°W and 40° and 25°N and have a spacing of about one degree of longitude. We present as Figures 2a and 2b the bathymetric and residual magnetic anomaly profiles plotted at right angles to track along these north-south lines.

The major fracture zones can be seen quite clearly crossing the survey lines in Figure 2. On the topographic profiles, Figure 2a, they appear as linear groupings of subtle ridges, troughs and steps. They are small and somewhat discontinuous, presumably due to their age and sediment cover. Ironically, these fracture zones, so named because of their topographic expression, show up much more clearly in the shipboard magnetic profiles in Figure 2b, where the individual strands are marked by high amplitude, short wavelength magnetic anomalies. We traced a number of linear features across the survey, using the magnetic data and cross-checking with the topographic data for "ground truth".

Magnetic profiles and side-scan sonar maps from the Hawaiian/Johnston Island EEZ surveys

In the 1980's and 1990's, the U. S. Geological Survey conducted detailed geophysical surveys of most seabeds within 200 nautical miles of the shores of the United States and its territories (the "Exclusive Economic Zone" or EEZ). This mapping included magnetic profiling and the complete insonification of the seabed with the Gloria side-scan sonar system. The resulting maps present the fabric and texture of the deep seafloor in spectacular detail. For the present study, this data set provides excellent control on the location and character of the multi-stranded Molokai and Murray fracture zones as they approach the Hawaiian ridge and Johnston Island (Figure 3). It provides our only information for location of the Murray fracture zone strands south of the Hawaiian ridge (near 175°W, 25°N).

We used the EEZ data set in several ways. The side-scan sonar technique is excellent for mapping linear seafloor features. We traced the fracture zone strands directly off the photo-montages of the shipboard records (presented in Normark et al. [1987, 1989]; Clague et al. [1988, 1989]; Torresan et al. [1989]; Searle et al. [this volume] and Holcomb [in preparation]). These traces appear as lines and dashes on Figure 3b. In addition, the side-scan technique brings out the ordinary abyssal hill fabric that was formed during seafloor spreading. This fabric is generally aligned perpendicular to the fracture zones, commonly curving as it approaches them. While this fabric is not included on our map, we used it to help distinguish and validate some of the fracture zone traces. The EEZ magnetic profiles, Figure 3a, record the fracture zone locations as sharp linear anomalies. We used these to extend the side-scan fracture zone traces toward the Hawaiian ridge, across regions where their seafloor traces had become buried in debris from the chain. On Figure 3b we present fracture zone traces located by side-scan data as lines and dashes and those located by magnetic profiles as connected dots.

GEOSAT Altimetry

In recent years, fracture zones have been mapped in many parts of the world by tracing their distinct signatures in the records collected by the altimetry satellites SEASAT and GEOSAT. Many authors [e.g., Sandwell, 1984a; Haxby, 1985; Shaw, 1987; Cande et al., 1988, 1989] have used SEASAT altimetry data to trace the positions of fracture zones, and Mayes et al. [1990] averaged GEOSAT repeat orbits to construct detailed deflection-of-the-vertical charts for tectonic mapping of the South Pacific. The validity and accuracy of altimetric fracture zone mapping was tested by Muller et al. [1991] when they compared stacked deflection-of-the-vertical profiles from GEOSAT with detailed topographic mapping of the Kane fracture zone presented by Tucholke and Schouten [1988]. Muller et al. [1991] showed that the main peak (or trough) in the geoid or the deflection-of-the-vertical signature traces the main topographic expression to within $\pm$ 5 km over the entire length of a fracture zone.

GEOSAT data for the central North Pacific are presented in Figure 4. Each profile in the figure represents the average of 44 repeat GEOSAT altimetry profiles [Sandwell and McAdoo, 1990], presented in Figures 4a and 4b as the ascending and descending profiles of the deflection-of-the-vertical (horizontal gravity) along track. To remove random noise and the long wavelength gravity field, we used a Gaussian filter to remove wavelengths shorter than 18 km and the spherical harmonic coefficients of the PGS3337 gravity model of Marsh et al. [1990] to remove wavelengths greater than 4000 km.

The two charts in Figure 4 show signals from all the major topographic features of the area. The Hawaii-Emperor chain runs across the region from the northwest to the southeast; the Mid-Pacific Mountains and the Necker ridge show up in the lower center of the chart; the Emperor trough and Chinook trough appear in the upper center. In addition, a whole suite of less prominent, but strikingly linear features trend east by northeast across the chart. These obviously coincide with the

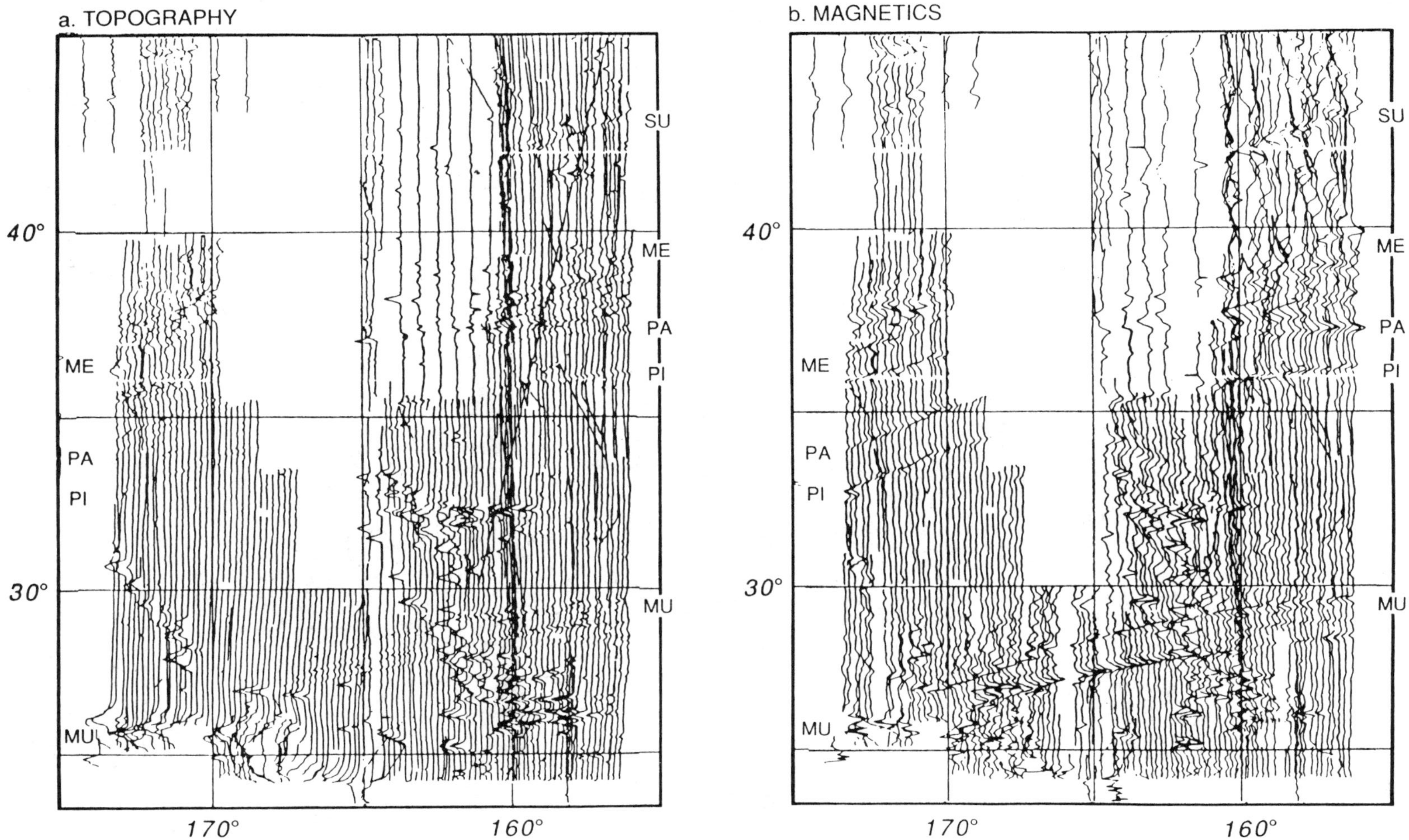

Fig. 2. Profiles of shipboard data from the *R/V Pioneer* systematic survey [Elvers et al., 1967]. Labeled fracture zone traces are SU: Surveyor, ME: Mendocino, PA: Pau, PI: Pioneer, and MU: Murray fracture zones. (a) Topographic profiles; positive elevations are plotted to the west. (b) Magnetic anomaly profiles; positive magnetic anomalies are plotted to the west.

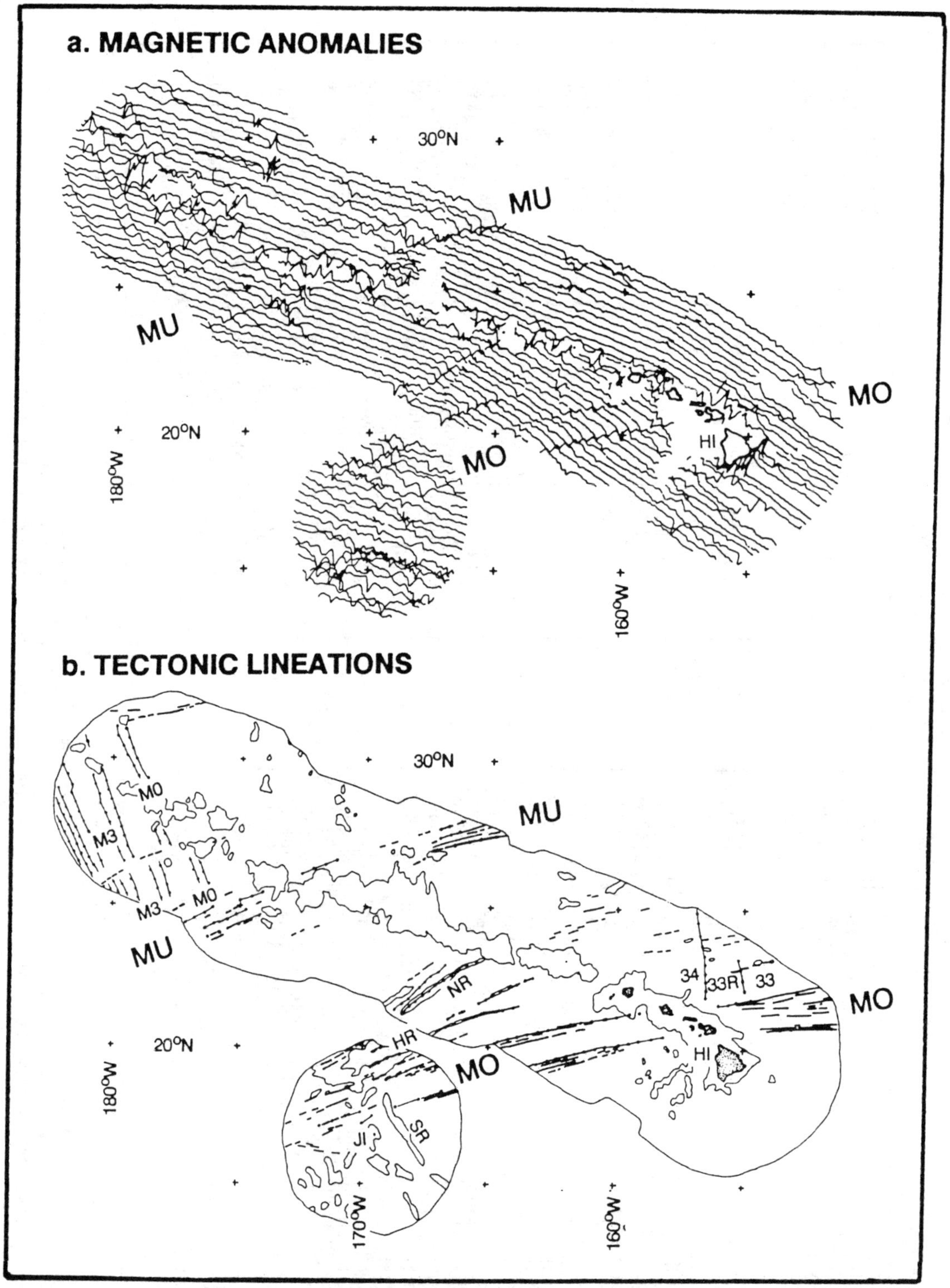

Fig. 3. Data from the Hawaiian and Johnston island Exclusive Economic Zone surveys (EEZ SCAN) from Normark et al. [1987, 1989], Clague et al. [1988, 1989], Torresan et al. [1989], and Searle et al. [this volume]. Labeled features are MU: Murray fracture zone, MO: Molokai fracture zone, NR: Necker ridge, HR: Horizon ridge, SR: Sculpin ridge, HI: Hawaii and JI: Johnston Island. (a) Magnetic anomalies collected along survey ship tracks. (b) Tectonic lineations. Solid lines are lineations traced from EEZ SCAN side-scan records; lines with dots are lineated features crossing the magnetic profiles, with the dots at ship track crossings. 4 km isobath from Mammerickx [1989].

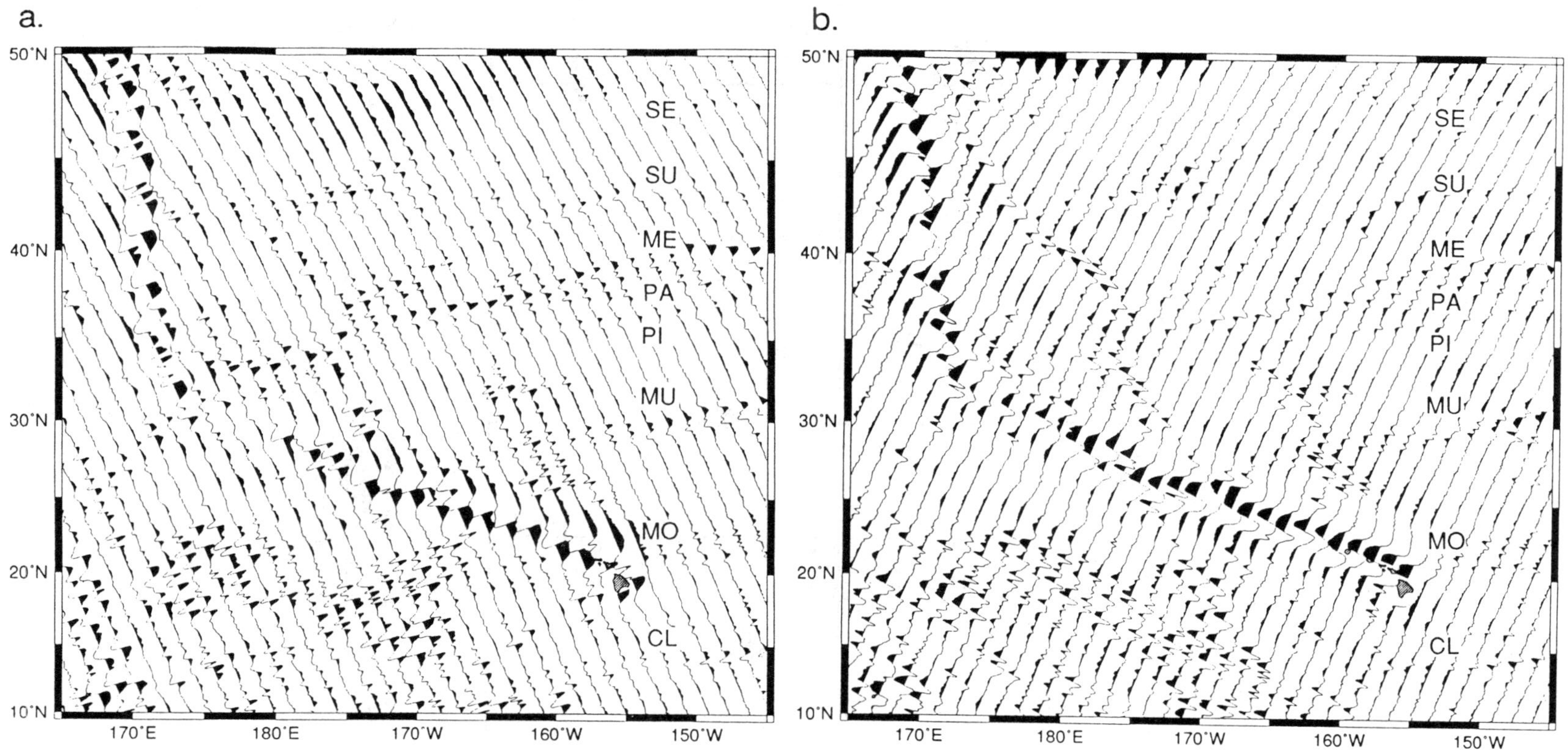

Fig. 4. Profiles of the deflection-of-the-vertical at right angles to track from the GEOSAT altimeter. Labeled fracture zone traces are SE: Sedna, SU: Surveyor, ME: Mendocino, PA: Pau, PI: Pioneer, MU: Murray, MO: Molokai, and CL: Clarion fracture zones. (a) Ascending profiles. These tracks ascend to the northwest. (b) Descending profiles. These tracks descend to the southwest.

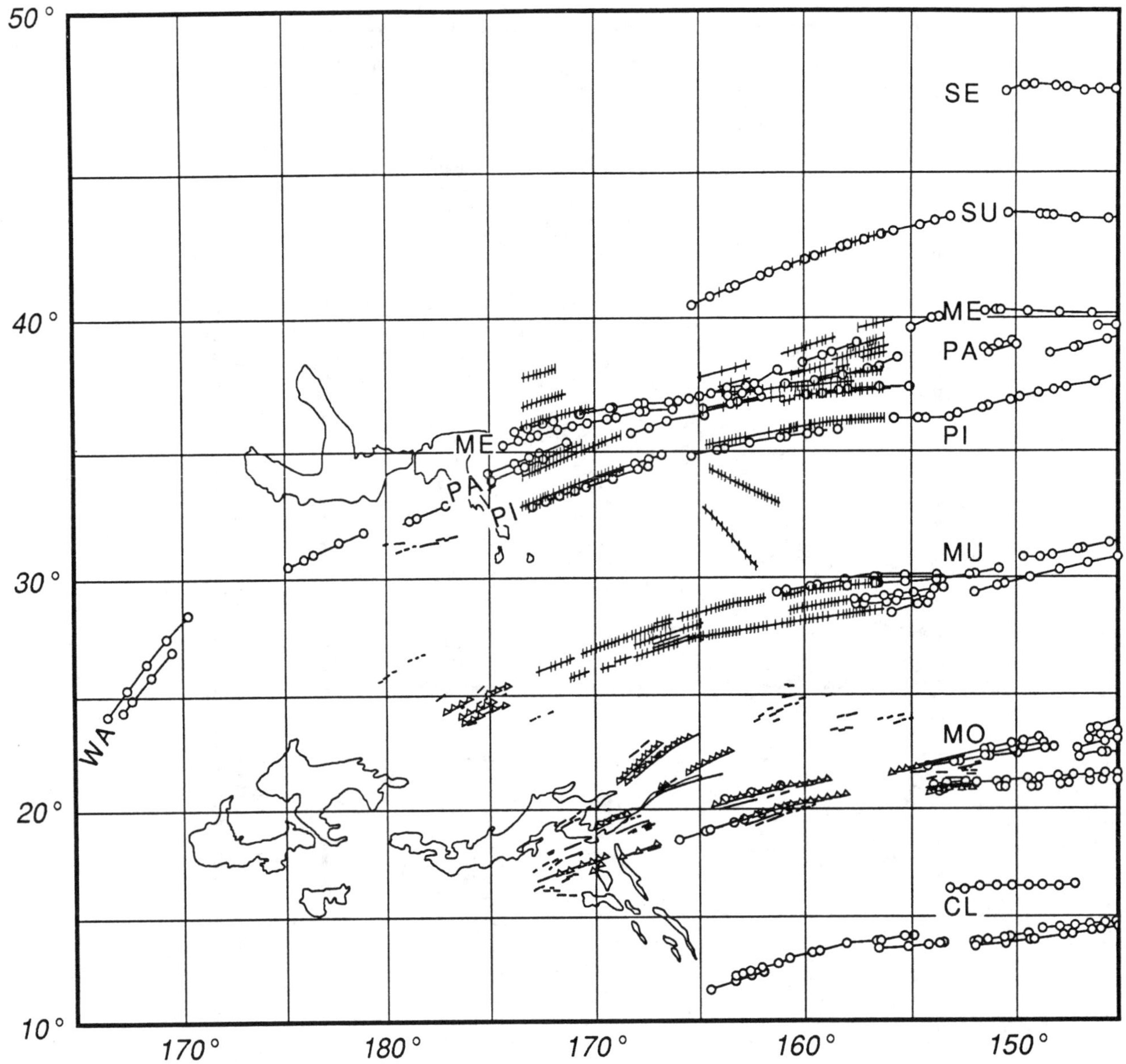

Fig. 5. Compilation of observed fracture zone traces in the central North Pacific. Hatched lines are fracture zone locations from the Pioneer survey magnetic and topographic profiles in Figure 2; each hatchure represents a ship track crossing point. Plain lines and dashes are side-scan traces and lines with open triangles are magnetic lineations, both from the EEZ SCAN data in Figure 3; each triangle represents a track crossing point. Lines with open circles are fracture zone locations from the GEOSAT altimetry data in Figure 4; each cicle represents the crossing of an averaged line. Labeled fracture zone traces are SE: Sedna, SU: Surveyor, ME: Mendocino, PA: Pau, PI: Pioneer, MU: Murray, MO: Molokai, and CL: Clarion fracture zones.

major fracture zones where they are known from shipboard data and they provide the means to continue the fracture zones across areas where they were not previously surveyed. For our compilation, we picked both peaks and troughs in the deflection-of-the-vertical profiles, mapping several sub-parallel strands for each fracture zone. The traces can be followed across most of the Quiet Zone, although they disappear near the Hawaiian ridge, presumably because they have been overprinted.

Fracture zone compilation

Our various determinations of the locations of the fracture zones, described above, are compiled in Figure 5 with a different symbol for each different data set. The lineations

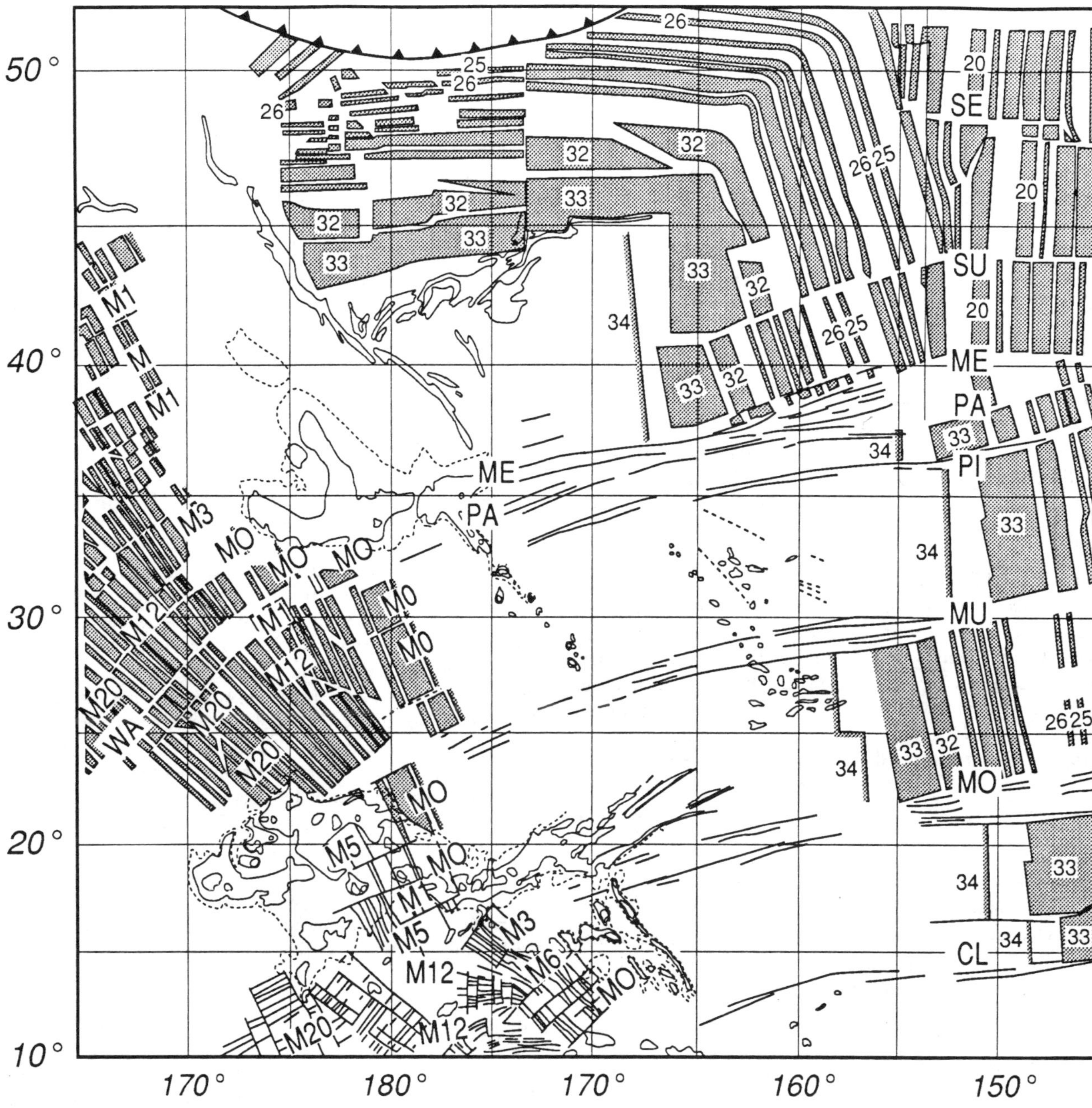

Fig. 6. Tectonic chart of the Cretaceous North Pacific. The numbered, shaded areas represent identified magnetic isochrons from Atwater and Severinghaus [1989] modified using EEZ SCAN magnetic data from Figure 3. Line pattern in the lower left are M-series isochron locations from Nakanishi et al. [1992]. Locations of fracture zone strands are simplified from Figure 5. The light continuous lines represent the 2 and 4 km contours, the light dashed line portions of 5 km contour. The contours of the Hawaiian-Emperor chain have been omitted to emphasize that this feature was superimposed upon the Mesozoic seafloor during the Cenozoic.

from these very different sources are in close agreement wherever they overlap, demonstrating the validity of the individual methods and the power of their combination. By combining the data sets we were able to map the fracture zones and their separate strands across much of the central North Pacific.

Revised Mesozoic Tectonic Chart

We have created a revised Mesozoic tectonic chart of the central North Pacific (Figure 6) by combining our new fracture zone traces with the other major topographic features of the area from Mammerickx [1989] and with the magnetic isochrons compiled by Nakanishi et al. [1992] and by Atwater and Severinghaus [1989], modified using EEZ isochrons from Figure 3a and Searle et al. [this volume]. We have deliberately excluded the Hawaiian-Emperor chain from this map in order to concentrate on the possible Mesozoic sea floor spreading elements. The primary contribution of this chart is a more complete and consistent mapping of the fracture zones than has previously been presented. In the next sections we present the details of each fracture zone, working from north to south.

Surveyor Fracture Zone

The Surveyor fracture zone is recorded very clearly in both shipboard and satellite data bases. In the Pioneer topographic profiles it appears as a linear ridge and trough; in the magnetic data it creates a linear high-amplitude disturbance and an offset of identified isochrons; in the deflection-of-the-vertical profiles it is a linear anomaly. This fracture zone appears to have been created as part of the accommodation of the Farallon-Pacific spreading system to the change in relative plate motion direction, described above, that occurred during Chron 33R. None of the data bases shows any evidence of its existence before that time.

Mendocino, Pau and Pioneer Fracture Zones

Our most important departure from traditional interpretations concerns the identification of the Mendocino fracture zone and its relationship to the Pau and Pioneer fracture zones. Most previous workers [e.g., Menard, 1964; Engebretson et al., 1984; Mammerickx, 1989] did not recognize the Chron 33R bend in the Mendocino fracture zone but, rather, traced it approximately straight to the west-southwest. We adopt the new location for the Mendocino fracture zone proposed by Atwater and Severinghaus [1989], following a more northerly strand westward across the eastern Quiet Zone and west-southwestward near the southern edge of the Hess rise. To emphasize this important difference, Atwater and Severinghaus renamed the more southerly strand that was formerly labelled as the Mendocino. They named it the Pau fracture zone. While this reinterpretation may seem to be a minor alteration, it has significant implications for the kinematic history of the area, described below.

The Atwater and Severinghaus [1989] reinterpretation of the Mendocino and Pau fracture zones is driven primarily by differences in the signatures of these features in the satellite altimetry data in the eastern Quiet Zone. The Mendocino, Pau, and Pioneer fracture zones and several additional intermittent strands can all be mapped across the shipboard magnetic and topographic profiles around 170°W, Figures 2a and 2b. Their traces in this traditional data set all look about the same, so that they do not help us to distinguish the huge-offset Mendocino from its relatively minor-offset relatives. However, there is no ambiguity in the deflection-of-the-vertical profiles crossing these features, Figures 4a and 4b. The Pioneer and Pau traces are barely visible in this data, whereas the northern strand, the newly labelled Mendocino, has a very large anomaly, presumably reflecting its much greater offset.

These new fracture zone identifications are also supported by some tectonic considerations. This new interpretation brings the shape of the Mendocino fracture zone into agreement with the shapes of the other fracture zones that were formed during the same period of time and that might have been reacting to the same plate motion changes. The fracture zones that we identify as the Pioneer and Pau fracture zones can be seen in both the satellite and Pioneer data sets. They trace out very similar shapes and are nearly parallel on the map. A similar shape is seen on the Mendocino but the entire pattern is shifted more than 1000 km to the west so that adjacent segments of the Pau and Mendocino are rarely parallel. Indeed, they approach and depart from one another with each change in direction, until the Pau meets its demise when the Mendocino runs over it (near 142°W) following a particularly abrupt change in plate motion direction.

Murray Fracture Zone

Five hundred kilometers to the south, the Murray fracture zone shows up as several distinct strands crossing the deflection-of-the-vertical profiles (Figure 4) and the Pioneer survey magnetic profiles (Figure 2b). It can also be detected on the Pioneer topographic profiles (Figure 2a), but it is more difficult to follow the individual strands as they are obscured by topographic noise and sedimentary cover. The combined data permit the tracing of this fracture zone all the way from the easternmost North Pacific to just north of the Hawaiian ridge. On the south side of the Hawaiian ridge it is not detectable in the deflection-of-the-vertical profiles but can be clearly seen in the EEZ magnetic records, Figure 3a. The Gloria side-scan records of this region are quite smooth, but they do contain an intermittent group of linear features in the vicinity of the traces in the magnetic records, Figure 3b. They also contain a widely scattered set of lineations perpendicular to the first, not shown, suggesting occasional glimpses of ordinary sea floor spreading fabric. On this basis, we trace the Murray fracture zone across the Hawaiian ridge to this location and to the major offset in the M-series isochrons, farther to the west.

Molokai Fracture Zone

The next fracture zone to the south is the Molokai. It is clearly traced in the deflection-of-the-vertical profiles and spectacularly laid out in the Gloria side-scan records [see Searle et al., this volume]. It has a complex multi-strand nature which increases in width and complexity to the east, in response to the Chron 33R change. To the west it encounters the Hawaiian Islands at the latitude of Molokai, hence its name. When it emerges from the south side of the chain, its location and multi-strand nature is again discernible in both the satellite and side-scan data sets. It becomes increasingly obscure and intermittent as it crosses the Sculpin ridge and approaches the M-series isochrons, but it is not clear if this is due to hot-spot overprinting or to original tectonic reorganizations. The Horizon ridge lies along the northern edge of the fracture zone and is sub-parallel to its trend, so that it may be part of it. Farther to the north, the Necker ridge lies oblique to all the other structures and seems to require a separate explanation.

Clarion Fracture Zone

The most southerly of the fracture zones in the region, the Clarion, shows up clearly in the deflection-of-the-vertical profiles. It appears to have a simpler structure than the Quiet Zone fracture zones to the north, consisting of only one or two strands. It forms a continuous feature from the easternmost North Pacific to the southern end of the Sculpin ridge.

IMPLICATIONS FOR TECTONIC HISTORY

The Pacific/Farallon Plate Boundary and its Reorganizations

Perhaps the most important result of our new tectonic map is the fact that the shapes of the various fracture zones are all quite similar. From west to east, i.e. from old to young, all are concave south, gradually changing trend from east-northeast to nearly east, then all abruptly shift back to east-northeast at the Chron 33R bend. The similarity suggests that all were formed along transform faults of a single plate pair (the Farallon and Pacific plates) and that those two plates experienced changes in relative plate motion direction that were recorded simultaneously by all the transform faults.

Unfortunately, although it seems clear that most of the central Pacific Quiet Zone was formed by Pacific-Farallon spreading, it is equally clear that the spreading was not symmetrical and continuous but, rather, that the system suffered major ridge crest jumps sometime during the Superchron. This is required by the fact that the fracture zone displacements offsetting magnetic isochron M0 are distinctly different from those offsetting isochron 34 (Figures 1 and 6). The timings of these jumps is not established, but several lines of evidence suggest that they occurred early in the Superchron. The similarities in the fracture zone shapes, described above, are particularly distinctive in the eastern part of the Quiet Zone. Since a major spreading center jump would be likely to change the shape of its adjoining fracture zone segments, this shape-similarity suggests that the jumps occurred in the western (older) part of the Quiet Zone. Furthermore, there are a number of topographic anomalies of unknown origin that could be the results of spreading center jumps. These occur primarily in the western part of the Quiet Zone. Finally, the proposed timing fits with the larger observation that the western Pacific region was the site of numerous, extremely voluminous volcanic outpourings that began at the beginning of the Superchron [Larson, 1991]. Such an event might facilitate plate boundary reorganizations by weakening the lithosphere on the flank of a spreading center. Engebretson et al. [1991] and Joseph et al. [this volume] postulate such a relationship between the Manihiki plateau volcanism and a southward jump of the Pacific-Farallon-Phoenix triple junction, for example.

In general, spreading center "jumps" are accomplished when a new propagating rift tears off a piece of one plate, creating a microplate between it and the old, established ridge. If activity on the old ridge then dies, the microplate has been transferred to the other plate and the spreading center has "jumped". In some well-documented cases of such transfers, the extinct ridge is marked by a topographic high while the rift propagation line (i.e. the initiation site) is a trough. Such features are seen around the Mathematician plate [Mammerickx et al., 1988] and around the Murray-Molokai "disturbed zone" [Malahoff and Handschumacher, 1971, as modified by Atwater and Severinghaus, 1989].

There are several anomalous topographic features in the western Quiet Zone, Figure 1, that might have been formed by ridge jumps. The Hess, Liliuokalani, and Sculpin ridges might all be extinct spreading centers, although each could, rather, have been superimposed on the crust later by off-axis volcanic events. The southern Emperor trough could be an initiation site, though it also may have been superimposed later, during the breakoff of the Kula plate, as described next. Initiation troughs are not observed in the other segments, but these may have been diminished by the flattening of the age-depth curve. Given the ambiguous origins of all these features, we are left concluding only that if they were formed by ridge jumps, the jumps occurred early in the superchron.

One group of topographic features that does not correspond to any common explanation is the Necker ridge and adjacent, parallel ridges near 22°N, 165°W on Figure 3b. These features are extremely linear, narrow and high and they lie oblique to the nearby Molokai fracture zone and to the seafloor topographic grain. Thus, they resemble neither common seafloor spreading features nor common superimposed volcanic features. In some ways they resemble a series of ridges reported by Winterer and Sandwell [1987] that run east-southeastward from Christmas Island. Winterer and Sandwell propose that those features formed along leaky tension cracks. If the Necker ridge group formed in some similar way, it is of great interest for plate stress considerations at its origin time, but it was probably superimposed on aged lithosphere so does not impact our Cretaceous seafloor spreading scenarios.

Development of the Kula Plate (and Chinook Plate?)

The patterns and relationships shown in Figure 6 and discussed above have implications for the history of the Kula plate. We agree with both Rea and Dixon [1983] and Mammerickx and Sharman [1988] that spreading on the Kula-Pacific boundary initiated in the Chinook trough. Magnetic isochrons north of the trough mapped by Mammerickx and Sharman [1988] and by Atwater and Severinghaus [1989] are shown in Figure 6. They demonstrate that Kula-Pacific spreading initiated along a break that propagated from west to east during Chron 33R (i.e., after the Superchron), breaking off a northern piece of the Pacific plate. If the northern Emperor trough and the Chinook trough are assumed to have been formed as the breaks along the western and southern edges of the new Kula plate, it is possible that the southern Emperor trough may have been formed during this same rifting event, as a third arm or "failed arm" similar to those formed during continental breakups [Burke, 1976].

The breakoff of the northern part of the Pacific plate has important ramifications for plate driving forces. The break would have dramatically reduced the northward pull on the Pacific plate from north Pacific subduction zones. Given its timing, the breakoff is the likely cause for the major change of spreading direction observed in Pacific/Farallon motion corresponding to the Chron 33R bend.

It has been suggested by various authors that the region between the Chinook trough and the Mendocino fracture zone was occupied by another plate, the Chinook plate [Rea and Dixon, 1983; Mammerickx and Sharman, 1988]. However, our contention that the Mendocino was a fracture zone in the Pacific-Farallon system implies that the region to the north of it was part of the Pacific plate, not some other. We are firmly convinced of this interpretation for the eastern part of the Quiet Zone because of the similarity of fracture zone shapes approaching the Chron 33R bend. Furthermore, the sea floor between the Mendocino fracture zone and the Chinook trough lacks any indication of an abandoned plate boundary, such as a north-south ridge, and Chron 34 is extremely straight, lacking the usual offsets and mismatches that indicate a recent reorganization. Thus, we conclude that, if the Chinook plate existed at all, it was deactivated before late Superchron time.

Variations in fracture zone widths

One of the more interesting aspects of transform systems is the process by which they accommodate changes in relative plate motion directions. The geometric problem is illustrated for a flat earth case in Figure 7. During steady spreading periods (Figure 7: steps 1-3) transform systems generate fracture zones with strands following parallel straight lines (or concentric small circle paths on a sphere). When there is a change in spreading direction (steps 3-4), the transform faults must take on the new direction and the resulting fracture zones are widened and narrowed, depending upon their senses of offset (step 7).

The fracture zones in our area have large offsets in both senses and they change widths in a manner similar to that illustrated in Figure 7. These width changes are particularly clear at the Chron 33R bend, where the Molokai and Mendocino-Pau-Pioneer systems become significantly wider from west to east while the Murray and Clarion systems become narrower. Some subtle width changes that are in the opposite senses from those just described can be detected going from west to east (from old to young) across the Quiet Zone. Particularly clear is a widening of the Murray system, but width changes can be seen or inferred on the others as well. These changes in width provide an important constraint for plate kinematic models, as will be described next.

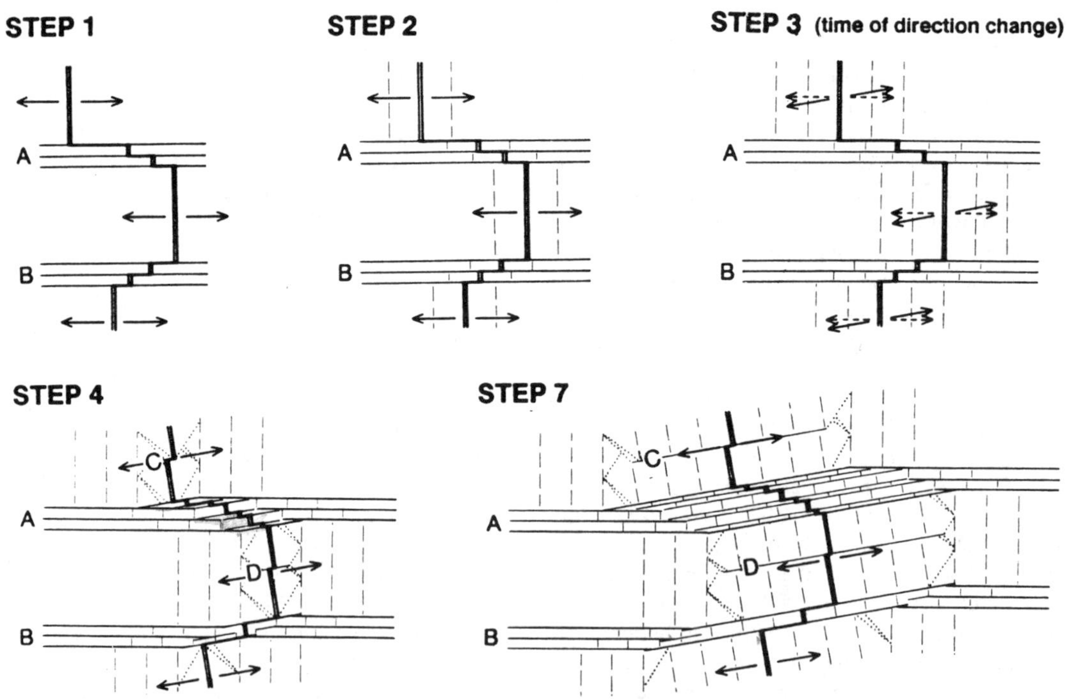

Fig. 7. Theoretical sea floor patterns formed when a spreading system containing two multi-stranded fracture zones (A and B) suffers an abrupt change in direction of relative plate motion. Heavy double lines are active spreading centers; heavy single lines are active transform faults; light lines are fracture zone traces; vertical dashed lines are seafloor isochrons. Numbers represent equal time steps. From step 1 to step 3 spreading was steady E-W. At the time of step 3, the relative plate motion changed by 10° to east-northeast-west-southwest. From step 3 to step 4 the the spreading centers and transform faults adjusted to the new direction, as follows: transform A narrowed and consolidated its strands; transform B widened and created additional strands; propagating rifts (traced by oblique dotted lines) gradually established spreading centers perpendicular to the new direction, creating new transform faults when they stalled (fracture zones C, D, and E). Steady spreading continued in the new direction through steps 5 and 6 (not shown) to step 7.

Stage pole descriptions of Pacific-Farallon relative motions

The relative spreading directions between plates during the formation of oceanic lithosphere can be described by rotations about Euler "stage poles". Stage poles have been fit to the Pacific-Farallon isochron patterns that were formed during the M-Series reversals by Engebretson et al. [1984] and to the patterns formed during the early Heirtzler scale reversals by Francheteau et al. [1970], Engebretson et al. [1984] and Rosa and Molnar [1988]. Some of these poles are listed in Table 1 and are plotted on Figure 8. For the Quiet Zone, some poles have been published that describe the entire Zone as one single stage [Francheteau et al., 1970; Engebretson et al., 1984; Sager and Pringle, 1987a; Joseph et al., 1992 and this volume] but none have been published for multiple stages within the Quiet Zone.

This lack of poles for stages during the Superchron is a direct result of the lack of datable magnetic reversals during this time period. The procedures employed for fitting stage poles use fracture zones to measure spreading directions and magnetic reversal isochrons to choose the appropriate segments of the fracture zones and to measure half-spreading distances. Without isochrons the process is severely hampered. However, if there were significant Euler pole shifts, the fracture zone shapes alone could contain age information since fracture zone segments formed during each given stage should all follow concentric small circles about the pole for that stage. We explored this possibility for the Quiet Zone fracture zones with mixed results. In the next sections we describe our attempt to fit the fracture zones with a series of seven high latitude stage poles, an attempt by Joseph et al. [1992] to fit them with a single low latitude pole and a possible compromise between these approaches.

TABLE 1.
Stage poles for Cretaceous Pacific-Farallon relative plate motions.

Time Interval			Stage Pole Parameters		
Pole	Chrons	Ma.*	Lat,°S	Lon,°W	Rot,°c
Final M-Series Stage [Engebretson et al., 1984]					
M_f	M11-M0	131-119	9	164	15
Superchron Alternative 1 [this paper]					
S_1	M0 - ?	119 - ?	64	114	4.0
S_2	?	?	70	114	4.0
S_3	?	?	74	114	3.0
S_4	?	?	77	114	3.0
S_5	?	?	80	114	2.0
S_6	?	?	84	114	2.0
S_f	? - 33R	? - 81	87	114	2.1
Superchron Alternative 2 [Joseph et al., 1992]					
S_J	M0 - 33R	119-81	30	152	25
Superchron Alternative 3 [this paper]					
S_J	M0 - ?	119 - ?	30	152	15
S_6	?	?	84	114	2.0
S_f	? - 33R	? - 81	87	114	2.1
First Heirtzler Series Stage [Engebretson et al., 1984]					
H_1	34 - 25	83 - 56	66	116	21

*absolute ages from time scales of Kent and Gradstein (1985) and Cande and Kent (1992)

A crucial decision one must make before searching for Quiet Zone stage poles is whether to include the new configuration of the Mendocino fracture zone in the analysis. The special importance of the Mendocino for the pole fitting process arises from the fact that it is significantly offset in longitude from the other fracture zone traces. When the Mendocino is included, the stage pole for the youngest part of the Superchron (just preceding the Chron 33R bend) must lie near the spin axis in order to fit the nearly east-west trends of all the fracture zones over such a wide longitudinal spread. For reasons described in an earlier section, we assume that it must be included for the youngest part of the Superchron. Thus, we conclude that the pole for the final Superchron stage is relatively well constrained to lie near the spin axis: pole S_f in Table 1 and filled circle on Figure 8.

We found that we could fit flow lines to the Quiet Zone fracture zones using a series of high latitude stage poles. We fit poles to the fracture zones working westward (backwards in time) from our late Superchron stage pole S_f. The fracture zones curve quite smoothly and we fit them using a series of six stage poles, S_6-S_1, that drift smoothly from S_f to S_1. This series is listed under Superchron Alternative 1 in Table 1 and is plotted as open and filled circles on Figure 8. However, this is not the only approach for fitting Quiet Zone poles.

Joseph et al. [1992 and this volume] consider features in the equatorial and southern central Pacific and propose a southern Quiet Zone history and kinematic model. They elaborate on the model of Engebretson et al. [1991] to suggest that the Pacific-Farallon-Phoenix triple junction, shown to lie near 2°N, 167°W by the M1 isochrons of Nakanishi et al. [1992], jumped south to about 20°S. They suggest that this occurred near the beginning of the Superchron when the Manihiki Plateau (M on Figure 8) erupted and broke the seafloor along its eastern edge. This transferred a microplate (the region outlined by the dotted line on Figure 8) to the Pacific plate and created a new southern segment of the Pacific-Farallon Quiet Zone spreading system. Joseph et al. [1992] propose an elegantly simple kinematic model for the relative motion that created all of this greatly extended Quiet Zone: a rotation about a single, low latitude pole, S_J, at 30°S, 152°W (Superchron Alternative 2 in Table 1 and circle with dot in Figure 8).

One obvious test of these alternative kinematic models is to construct theoretical flow lines about the poles and compare

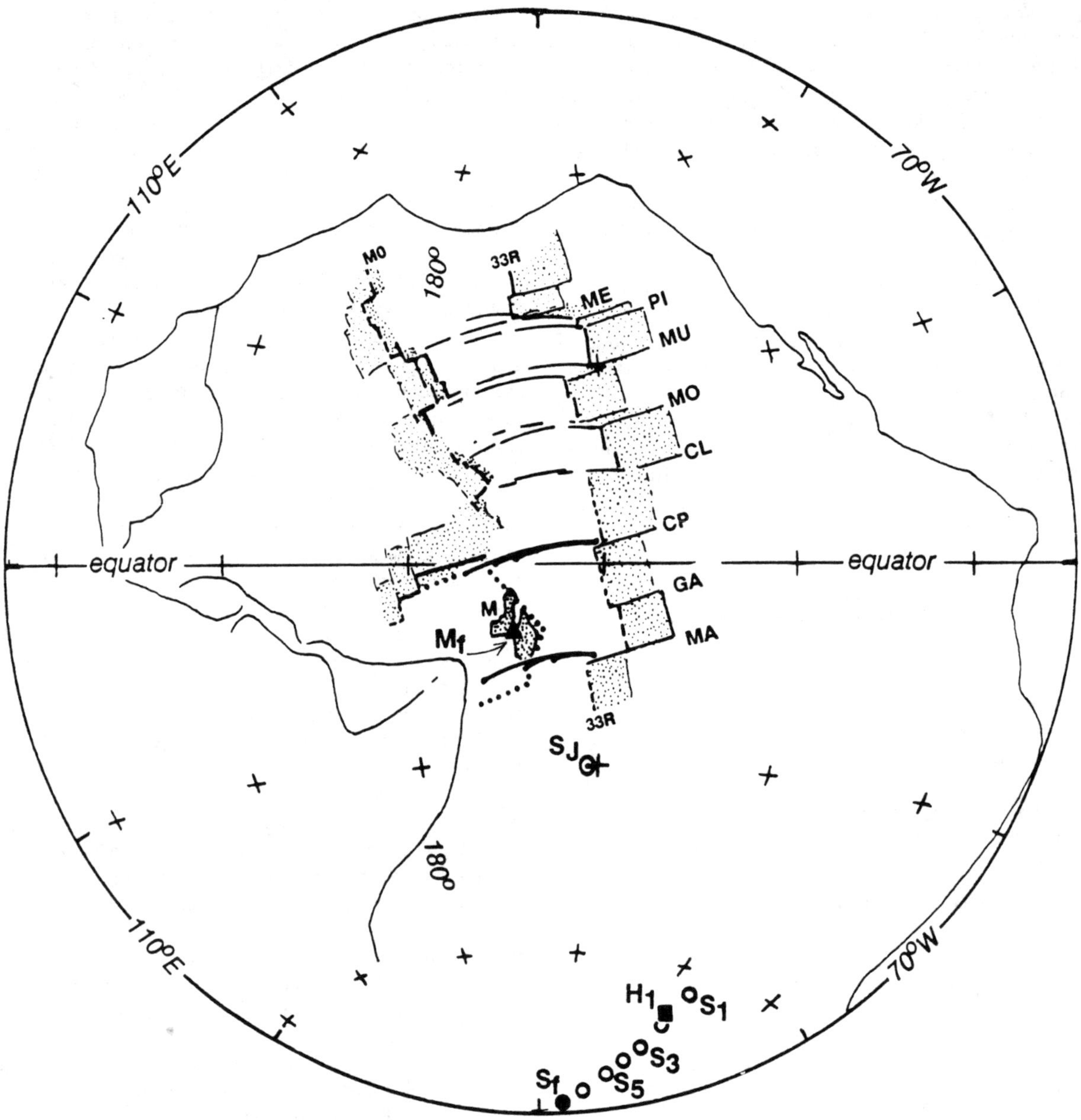

Fig. 8. Configuration of the central Pacific Quiet Zone and possible Euler pole descriptions of the Pacific-Farallon relative plate motions during the Cretaceous. Western and eastern stippled regions are approximate areas of seafloor formed preceding the Superchron (chrons M11-M0) and following the Superchron (chrons 33R-25), respectively. Heavy dotted line shows approximate outline of Phoenix mini-plate, postulated by Engebretson et al. [1991] and Joseph et al. [1992] to have been transferred to the Pacific plate about M0 time. Euler pole locations are M_f (filled triangle) describing the final M-series motions and H_1 (filled square) describing early Heirtzler-scale motions, both from Engebretson et al. [1984], and Superchron alternatives: S_J (circle with dot) - single low latitude pole of Joseph et al. [1992] and S_1-S_6, S_f (open and filled circles) - high latitude series. Two sets of nested curves each include Superchron flowlines predicted by the three alternative plate kinematic models listed in Table 1: longest curves are predicted by Alternative 1; shortest curves are predicted by Alternative 2; intermediate length curves are predicted by Alternative 3. Labelled features are ME: Mendocino, PI: Pioneer, MU: Murray, MO: Molokai, CL: Clarion, CP: Clipperton, GA: Galapagos, and MA: Marquesas fracture zones and M: Manihiki Plateau. Western stippled region is bordered by Chron M11 and M1 or M0 where known after Nakanishi et al. [1992], Atwater and Severinghaus [1989] and Figure 3 (details of Magellan plate patterns have been omitted). Eastern stippled regions are bordered by approximate locations of Chrons 33 and 25 (dashed where extrapolated) and fracture zone traces after Mayes et al. [1990], Eittreim et al. [submitted], Atwater and Severinghaus [1989], Candy et al. [1989], Haxby [1987], and Figures 3 and 6. Light lines outline features of the present Pacific rim, for orientation. Stereo projection is centered on 160°W and the equator.

them to our north Pacific fracture zones. We have done this in Figure 9. Since we believe that most of the ridge jumps in the Quiet Zone occurred early in the Superchron, we constructed flow lines working backward in time (from east to west) starting at the Chron 33R bend. We note that, since the bend locations are not perfectly known and, in any case, were probably shifted somewhat by propagating rifts at the time of the bend, we made minor adjustments in the bend location to maximize the fit for each kinematic model that was tried. Predicted flow lines from the two models are presented in Figures 9a and 9b, as theoretical lines on the left and as grey curves overlain on the observed fracture zone lineations on the right. The predicted patterns are quite similar to each other except for two aspects, as follows.

One important difference between the two kinematic models concerns the fitting of the Mendocino fracture zone. Flow lines constructed for pole S_J, Figure 9b, miss it entirely. Joseph et al. do not claim to fit this fracture zone; they extend their analysis only as far north as the Pau fracture zone. If we accept S_J as the single pole to describe the entire Quiet Zone creation, then we must postulate a non-Farallon-Pacific origin for the Mendocino fracture zone right up to Chron 33R time or else we must abandon the Atwater and Severinghaus [1989] mapping of it. As described in earlier sections, we find both of these alternatives to be unlikely.

A second, more subtle test of the Figure 9 flow line fits is to note the extent of widening and narrowing of fracture zones as they cross the Quiet Zone. Our drifting, high latitude poles, S_1-S_f, predict distinct width changes, as the individual strands converge or diverge, Figure 9a. The single pole, S_J, predicts constant-width fracture zones with concentric strands, Figure 9b. While the observational data is fragmentary, it is clear that the observed fracture zones do widen or narrow, depending on their offsets, in concert with one another. However, the width changes do not seem to be as pronounced as those predicted by the flow lines in Figure 9a.

We constructed a compromise solution that incorporates the more successful attributes of both the models just discussed. It adopts S_J for the older two-thirds to three-quarters of the Quiet Zone and S_6 and S_f for the youngest stages (Superchron Alternative 3 in Table 1). Flow lines for this compromise model are shown in Figure 9c. They predict modest width changes and they fit the Mendocino trend.

We also compared predictions from the three alternative kinematic models for Quiet Zone features in the equatorial and south Pacific. Predicted flow lines are illustrated by the two sets of nested curves crossing the Quiet zone on Figure 8. Each set includes three curves, one for each of the three alternative models. A possible test would be to compare the observed fracture zone curvatures to the curves predicted by the models. Joseph et al. [1992 and this volume] present a test similar to this for pole S_J, showing a fair match to the traces of the fracture zones in the SEASAT data. Unfortunately, as shown by the nested curves, the other kinematic models predict quite similar curvatures. Furthermore, the SEASAT data for the southern Quiet Zone fracture zones are somewhat indistinct and ambiguous, so that the predicted subtle differences are not likely to be distinguishable.

A simpler test of the kinematic models concerns the breadth of the southern Quiet Zone. If the history proposed by Engebretson et al [1991] and Joseph et al. [this volume] is accepted, the eastern edge of the Manihiki plateau (eastern dotted lines on Figure 8) should approximately correspond to the M0 isochron. Approximate locations for the Chron 33R bend can be found from SEASAT traces and from isochron mappings and extrapolations. Thus the breadth of the Quiet Zone is approximately known, as shown on Figure 8. The theoretical breadths predicted by the three kinematic models for the southern Quiet Zone are dramatically different from one another. In each set of nested curves on Figure 8, the shortest curve corresponds to the low latitude pole S_J, and the longest to the high latitude poles S_1-S_f. The former clearly underestimates the breadth of the southern Quiet Zone while the latter clearly overestimates it. Several major additional ridge jumps would be required to fit either of these models. The compromise model (Superchron Alternative 3 in Table 1) predicts the intermediate length curves and is more compatible with the observed breadth.

The most obvious conclusion from all these pole fitting exercises is that the kinematic description of the Quiet Zone is still very poorly constrained. We were hard-pressed to distinguish among poles that were nearly sixty degrees apart on the globe! A final kinematic description, if we ever are able to find one at all, will probably have a high latitude pole for the late Superchron (to fit the Mendocino), shifting poles (to fit the variable fracture zone widths) and lower latitude poles for earlier stages (to fit the southern Quiet Zone breadth). While the compromise model, Superchron Alternative 3, has these characteristics, it is by no means unique.

CONCLUSIONS

We have combined shipboard topographic and magnetic profiles, Gloria side-scan data, and satellite altimetry data to refine the chart of fracture zone traces in the central North Pacific. Adding the known magnetic isochrons to these fracture zone traces and leaving out the Hawaiian/Emperor chain, we have produced a new Cretaceous tectonic chart.

The mapped fracture zone and isochron patterns shed new light on various aspects of the spreading history that created the sea floor in the central Pacific Quiet Zone. They support the idea that all of this area was formed by spreading between the Pacific and Farallon plates and they cast doubt on the existence of the Chinook plate, at least in the late Superchron. They show that one or more major ridge jumps occurred, probably early in the Superchron. They document smooth spreading with relative direction gradually changing from east-northeast to east, then a sudden change back to east-northeast during Chron 33R. They show patterns of fracture zone widening and narrowing, intricately following the geometric constraints imposed by plate direction changes.

When combined with data and ideas from Joseph et al. [1992 and this volume] for the Quiet Zone in the equatorial and

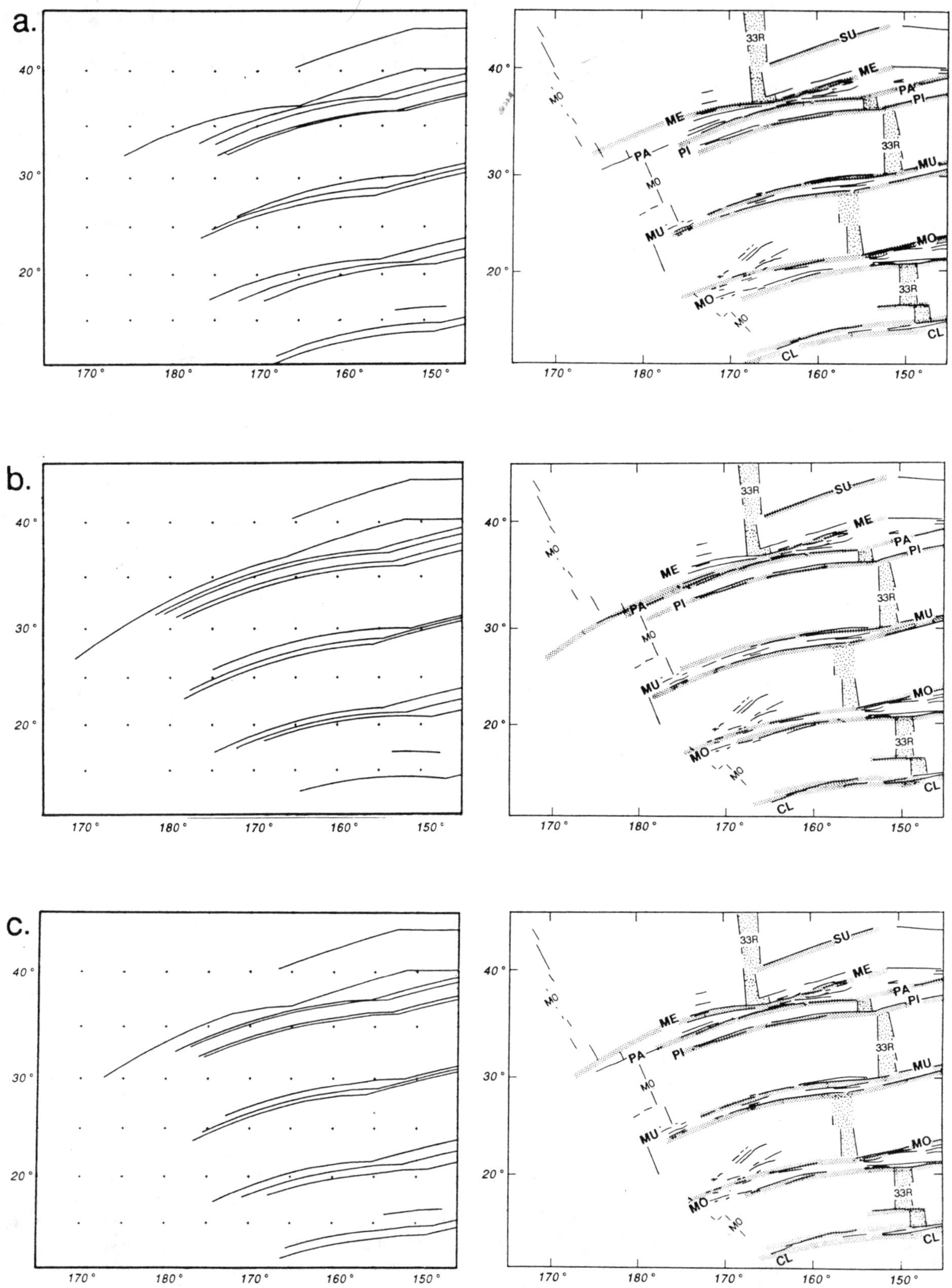
a.
40°
30°
20°
170°
180°
170°
160°
150°
33R
SU
ME
ME
PA
PI
MO
PA
PI
33R
MU
MO
MU
MO
MO
33R
MO
CL
CL
b.
c.

Fig. 9. Comparison between fracture zone traces mapped in the central north Pacific and synthetic flow lines predicted by alternative kinematic models. Black lines and dashes are observed linear features; M0 lines show observed or extrapolated locations of Chron M0; coarsely stippled areas are Chron 33R regions (polarity reversal between between Chrons 34 and 33), all from Figure 6. Fat grey lines show flowlines constructed starting at points near the Chron 33R bends; as described in the text, these points were slightly adjusted to maximize each fit. The flow lines were constructed eastward using pole H_1 and westward using the three alternative Superchron kinematic models described in Table 1 and the text. Labelled fracture zones are SU: Surveyor, ME: Mendocino, PA: Pau, PI: Pioneer, MU: Murray, MO: Molokai, and CL: Clarion fracture zones. (a) Flowlines predicted by Superchron Alternative 1: high latitude series of poles, S_f-S_6-S_5-S_4-S_3-S_2-S_1. (b) Flowlines predicted by Superchron Alternative 2: single low latitude pole of Joseph et al. [1992], S_J. (c) Flowlines predicted by Superchron Alternative 3 (compromise model): S_f-S_6-S_J.

south Pacific, our north Pacific tectonic patterns supply some constraints for Superchron plate kinematic models. The changing widths of our fracture zones imply that there were several distinct stage poles (as opposed to a single pole for the whole Superchron). The fracture zone trends for the late Superchron (immediately pre-Chron 33R), when the Mendocino is included, require a stage pole very near the spin axis. The narrowing breadth of the Quiet Zone in the south Pacific implies lower latitude, south Pacific poles for the earlier stages of the Superchron. Locations for the various stage poles (except, perhaps, for the youngest stage) are poorly constrained by the present data sets. Future shipboard mapping of the Quiet Zone traces of the Marquesas and Galapagos fracture zones will be particularly important for this problem, since these features lie quite near the likely locations of the early Superchron stage poles.

Acknowledgments. This work was supported by the Paleoceanographic Mapping Project of The University of Texas at Austin. We thank Dietmar Muller, Doug Wilson and John LaBrecque for help with various steps in the fitting of poles and creation of flowlines for Figures 8 and 9. Roger Larson, an unknown reviewer and, especially, Will Sager gave extremely helpful reviews.

We thank Dave Clague and the U.S.G.S. marine geophysics group for allowing us to trace lineations from their unpublished Gloria side-scan charts.

REFERENCES

Atwater, T., Plate tectonic history of the northeast Pacific and western North America, in *The Eastern Pacific Ocean and Hawaii, Geology of North America, vol. N*, edited by E.L. Winterer, D.M. Hussong, and R.W. Decker, Geol. Soc. of America, Boulder, CO, 21-72, 1989.

Atwater, T. and H.W. Menard, Magnetic lineations in the northeast Pacific, *Earth and Plan. Sci. Lett., 7*, 445-450, 1970.

Atwater, T. and J. Severinghaus, Tectonic maps of the northeastern Pacific, in *The Eastern Pacific Ocean and Hawaii, Geology of North America, vol. N*, edited by E.L. Winterer, D.M. Hussong, and R.W. Decker, Geol. Soc. of America, Boulder, CO, 15-20 and 3 maps, 1989.

Burke, K., Development of graben associated with the initial ruptures of the Atlantic Ocean, *Tectonophysics, 36*, 93-112, 1976.

Cande, S.C., J.L. LaBrecque, and W.F. Haxby, Plate kinematics of the south Atlantic: Chron 34 to present, *J. Geophys. Res., 93*, 13479-13492, 1988.

Cande, S.C., J.L. LaBrecque, R.L. Larson, W.C. Pitman, X. Golovchenko, and W.F. Haxby, Magnetic lineations of the world's ocean basins, 1:27,400,000, Am. Assoc. Petrol. Geol., Tulsa, OK, 1989.

Cande, S.C., and D.V. Kent, A new geomagnetic polarity time scale for the late Cretaceous and Cenozoic, *J. Geophys. Res., 97*, 13917-13951, 1992.

Clague, D.A., Holcomb, R.T., Torresan, M.E., and Ross, S.L., Shipboard report for Hawaii ground-truth cruise F11-88-HW, 25 October - 7 November 1988, *U.S. Geological Survey Open-File Report 89-109*, 37 p., 1989.

Clague, D.A., Moore, J.G., Torresan, M.E., Holcomb, R.T., and Lipman, P.W., Shipboard report for Hawaii GLORIA ground-truth cruise F2-88-HW, 25 February - 9 March 1988, *U.S. Geological Survey Open-File Report 88-292*, 54 p., 1988.

EEZ SCAN 84 Scientific Staff, Atlas of the Exclusive Economic Zone, Western Coterminous United States, *U.S. Geol. Surv. Misc. Invest. Series 1-1792*, 152 p., 1986.

EEZ SCAN 85 Scientific Staff, Atlas of the U.S. Exclusive Economic Zone, Gulf of Mexico and Eastern Caribbean Areas, *U.S. Geol. Surv. Misc. Invest. Series 1-1864-A*, 162 p., 1987.

Eittreim, S.L., N. Ragozin, H.S. Gnibidenko and C.E. Helsley, Crustal age in the Clipperton-Clarion Zone, *Geophys. Res. Lett.*, submitted.

Elvers, D.J., C.C. Mathewson, R.E. Kohler and R.L. Moses, Systematic ocean surveys by the U.S.C.& G.S.S. Pioneer 1961-1963, *Coast and Geod. Surv. Op. Data Report C.&G.S.D.R.-1*, 19 p., 1967.

Embley, R.W., and D.G. Wilson, Morphology of the Blanco transform fault zone - NE Pacific: implications for its tectonic evolution, *Mar. Geophys. Res., 14*, 25-45, 1992.

Engebretson, D.C., A. Cox, and R.G. Gordon, Relative motions between oceanic plates of the Pacific basin, *J. Geophys. Res., 89*, 10291-10310, 1984.

Engebretson, D.C., J. Mammerickx, and C.A. Raymond, Tonga lineations - the Phoenix plate has arisen? (abstract), *EOS, Trans. Am. Geophys. Union, 72*, 444, 1991.

Fornari, D.J., D.G. Gallo, M.H. Edwards, J.A. Madsen, M.R. Perfit, and A.N. Shor, Structure and topography of the Siqueiros transform fault system: evidence for the development of intra-transform spreading centers, *Mar. Geophys. Res., 11*, 263-299, 1989.

Grim, P.J., and B.H. Erickson, Fracture zones and magnetic anomalies south of the Aleutian trench, *J. Geophys. Res., 74*, 1488-1494, 1969.

Francheteau, J., J.G. Sclater, and H.W. Menard, Pattern of relative motion from fracture zone and spreading rate data in the north-eastern Pacific, *Nature, 226*, 746-748, 1970.

Haxby, W.F., Gravity Field of the World's Oceans, 1:40,000,000, Nat. Geophys. Data Center, NOAA, Boulder, CO, 1987.

Hilde, T.W.C., N. Isezaki ad J.M. Wageman, Mesozoic sea-floor spreading in the north Pacific, in *The Geophysics of the Pacific Ocean Basin and its Margin - A Volume in Honor of George P. Woollard, Geophys. Monograph 19*, edited by G.H. Sutton, M.H. Manghnani and R. Moberly, Am. Geophys. Union, Wash. D.C., 205-226, 1976.

Joseph, D., B. Taylor, and A.N. Shor, New sidescan sonar and gravity evidence that the Nova-Canton Trough is a fracture zone, *Geology, 20*, 435-438, 1992.

Joseph, D., B. Taylor, A.N. Shor, and T. Yamazaki, The Nova-Canton trough and the late Cretaceous Evolution of the central Pacific, in *The Mesozoic Pacific*, Geophys. Mon. Ser., edited by M.S. Pringle, W.W. Sager, W.V. Sliter, and S. Stein, Am. Geophys. Union, Washington, D. D., this volume.

Kent, D.V. and F.M. Gradstein, A Cretaceous and Jurassic geochronology, *Geol. Soc. Am. Bull., 96*, 1419-1427, 1985.

Larson, R.L., Latest pulse of the Earth: Evidence for a mid-Cretaceous superplume, *Geology, 19*, 574, 1991.

Larson, R.L., and W.C. Pitman, World-wide correlation of Mesozoic magnetic anomalies and its implications, *Bull. Geol. Soc. Am., 83*, 3645-3662, 1972.

Malahoff, A., and D.W. Handschumacher, Magnetic anomalies south of the Murray fracture zone - new evidence for a secondary seafloor spreading center and strike-slip movement, *J. Geophys. Res., 76*, 6265-6275, 1971.

Mammerickx, J., Large scale undersea features of the northeast Pacific, in *The Eastern Pacific Ocean and Hawaii, The Geology of North America, vol. N*, edited by E.L. Winterer, D.M. Hussong, and R.W. Decker, Geol. Soc. of America, Boulder, CO, 5-13 and 3 maps, 1989.

Mammerickx, J., D.F. Naar and R.L. Tyce, The Mathematician paleoplate, *J. Geophys. Res., 93*, 3025-3040, 1988.

Mammerickx, J. and G.F. Sharman, Tectonic evolution of the north Pacific during the Cretaceous Quiet Period, *J. Geophys. Res., 93*, 3009-3024, 1988.

Marsh, J.G., C.J. Koblinsky, F. Lerch, S.M. Klosko, J.W. Robbins, R.G. Williamson and G.B. Patel, Dynamic sea surface topography, gravity and improved orbit accuracies from the direct evaluation of Seasat altimeter data, *J. Geophys. Res. 95*, 13129-13150, 1990.

Mayes, C.L., L.A. Lawver and D.T. Sandwell, Tectonic history and new isochron chart of the South Pacific, *J. Geophys. Res., 95*, 8543-8567, 1990.

Menard, H.W., *Marine Geology of the Pacific*, McGraw-Hill, New York, 1964.

Menard, H.W., and T. Atwater, Origin of fracture zone topography, *Nature, 219*, 463-467, 1969.

Muller, R.D., D.T. Sandwell, B.E. Tucholke, J.G. Sclater and P.R. Shaw, Depth to basement and geoid expression of the Kane Fracture Zone: A comparison, *Mar. Geophys. Res., 13*, 105-1129, 1991.

Nakanishi, M., K. Tamaki, and K. Kobayashi, Mesozoic magnetic anomaly lineations and seafloor spreading history of the northwestern Pacific, J. Geophys. Res., 94, 15437-15462, 1989.

Nakanishi, M., K. Tamaki, and K. Kobayashi, Magnetic anomaly lineations from the Late Jurassic to Early Cretaceous in the west-central Pacific Ocean, *Geophys. J. Int., 109*, 701-719, 1992.

Normark, W.R., P.W. Lipman, J.B. Wilson, C.L. Jacobs, D.P. Johnson and C.E. Gutmacher, Preliminary Cruise Report, Hawaiian GLORIA Leg 2, F6-86-HW, November 1986, *U.S. Geol. Surv. Open-File Report 87-298*, 42 p., 1987.

Normark, W.R., R.T. Holcomb, R.C. Searle, and M.L. Somers, Preliminary Cruise Report, Hawaiian GLORIA Legs 3 and 4, F3-88-HW and F4-88-HW, *U.S. Geol. Surv. Open-File Report*, 1989.

Parsons, B., and J.G. Sclater, An analysis of the variation of ocean floor bathymetry and heat flow with age, *J. Geophys. Res., 82*, 803-827, 1977.

Rae, D.K. and J.M. Dixon, Late Cretaceous and Paleogene tectonic evolution of the North Pacific Ocean, *Earth Planet. Sci. Lett., 65*, 145-166, 1983.

Renkin, M.L., and J.G. Sclater, Depth and age in the north Pacific, *J. Geophys. Res., 93*, 2919-2935, 1988.

Rosa, J.W.C., and P. Molnar, Uncertainties in reconstructions of the Pacific, Farallon, Vancouver, and Kula plates and constraints on the rigidity of the Pacific and Farallon (and Vancouver) plates between 72 and 34 Ma, *J. Geophys. Res., 93*, 2997-3008, 1988.

Sager, W.W., D.W. Handschumacher, T.W.C. Hilde, and D.R. Bracey, Tectonic evolution of the Northern Pacific plate and Pacific-Farallon-Izanagi triple junction in the late Jurassic and early Cretaceous (M21-M10), *Tectonophysics, 155*, 345-364, 1988.

Sager, W.W., and M.S. Pringle, Tectonic evolution of the central Pacific during the Cretaceous Quiet Period (abstract), *EOS, Trans. Am. Geophys. Union, 68*, 1451, 1987a.

Sager, W.W., and M.S. Pringle, Paleomagnetic constraints on the origin and evolution of the Musicians and south Hawaiian seamounts, central Pacific Ocean, in *Seamounts, Islands, and Atolls, Geophysical Monograph 43*, edited by B.H. Keating, P. Fryer, R. Batiza, and G.W. Boehlert, Am. Geophys. Union, Washington D.C., 133-162, 1987b.

Sandwell, D.T., Thermomechanical evolution of oceanic fracture zones, *J. Geophys. Res., 89*, 11401-11413, 1984a.

Sandwell, D.T., Along-track deflection of the vertical from Seasat: GEBCO overlays, *NOAA Tech. Memo. NOS NGS-40*, 8 p., 1984b.

Sandwell, D.T. and D.C. McAdoo, High-accuracy, high-resolution gravity profiles from 2 years of the Geosat exact repeat mission, *J. Geophys. Res., 95*, 3049-3060, 1990.

Searle, R.C., Multiple, closely spaced transform faults in fast-slipping fracture zones, *Geology, 11*, 607-610, 1983.

Searle, R., R.T. Holcomb, J.B. Wilson, R.J. Whittington, M.L. Holmes, B.A. McGregor, and A.N. Shor, The Molokai fracture zone near Hawaii and the late Cretaceous change in Pacific/Farallon spreading direction, in *The Mesozoic Pacific*, Geophys. Mon. Ser., edited by M.S. Pringle, W.W. Sager, W.V. Sliter, and S. Stein, Am. Geophys. Union, Washington, D. D., *this volume*.

Shaw, P.R., Investigations of relative plate motions in the South Atlantic using SEASAT altimeter data, *J. Geophys. Res., 92*, 9363-9375, 1987.

Tamaki, K. and R.L. Larson, The Mesozoic tectonic history of the Magellan microplate in the western central Pacific, *J. Geophys. Res., 93*, 2857-2874, 1988.

Tucholke, B.E., and H. Schouten, Kane fracture zone, *Marine Geophys. Res., 10*, 1-39, 1988.

Torresan, M.E., A.N. Shor, J.B. Wilson and J. Campbell, Cruise Report, Hawaiian GLORIA Leg 5, F5-88-HW, *U. S. Geol. Surv. Open-File Report 89-198*, 56 p, 1989.

Wilson, D.S., Tectonic history of the Juan de Fuca Ridge over the last 40 million years, *J. Geophys. Res., 93*, 11863-11876, 1988.

Winterer, E.L., and D.T. Sandwell, Evidence from en-echelon cross-grain ridges for tensional cracks in the Pacific plate, *Nature, 329*, 534-537, 1987.

T. Atwater, Department of Geological Sciences, University of California, Santa Barbara, CA 93106.

J. Sclater and D. Sandwell, Geological Research Division, Scripps Institution of Oceanography, La Jolla, CA 92093-0215.

J. Severinghaus, Lamont-Doherty Geological Observatory, P.O. Box 718, Palisades, NY 10964.

M. Marlow, U. S. Geological Survey, Mail Stop 999, 345 Middlefield Road, Menlo Park, CA 94025.

The Molokai Fracture Zone near Hawaii, and the Late Cretaceous Change in Pacific/Farallon Spreading Direction

Roger C. Searle[1,2], Robin T. Holcomb[3], John B. Wilson[4], Mark L. Holmes[3], Robert J. Whittington[5], Ellen S. Kappel[6], Bonnie A. McGregor[7], Alexander N. Shor[8]

[1]*Department of Geological Sciences, University of Durham DH1 3LE, United Kingdom*
[2]*Formerly at Institute of Oceanographic Sciences Deacon Laboratory, Wormley GU8 5UB, United Kingdom*
[3]*U.S. Geological Survey, University of Washington WB-10, Seattle, Washington 98195*
[4]*Institute of Oceanographic Sciences Deacon Laboratory, Wormley GU8 5UB, United Kingdom*
[5]*Institute of Earth Studies, University College of Wales, Aberystwyth SY23 3DB, United Kingdom*
[6]*Joint Oceanographic Institutions, Inc., 1755 Massachusetts Ave., NW, Washington, D.C. 20036*
[7]*U.S. Geological Survey MS 915, 12201 Sunrise Valley Drive, Reston. Virginia 22092*
[8]*Hawaii Institute of Geophysics, University of Hawaii, Honolulu, Hawaii 96822*

A survey of the U.S. Exclusive Economic Zone around the Hawaiian Islands using the GLORIA sidescan sonar has revealed the structure of the Molokai Fracture Zone in considerable detail, in particular the structures associated with a 10° counterclockwise change of spreading direction in the Late Cretaceous. The fracture zone consists of several strong ENE-WSW tectonic lineaments in two separate bands, which we propose to call the Oahu and Maui strands, crossing beneath the Hawaiian Island chain along arcuate paths that are concave to the south, trending 076° in the west and 087° in the east. Each band consists of four or five component lineaments up to 12 km apart and 300 km long, all separated by sinistral offsets. Inferred offsets on the Oahu and Maui strands are about 530 and 70 km, respectively. Two additional strands diverging from these east of Hawaii are explained as the traces of the realigned fracture zone following a change of spreading direction. These results have enabled us to develop detailed tectonic models for the development of 'leaky' transforms following a transtensive change in spreading direction. The change in spreading direction has been recognized as a Pacific-wide event, which we date as very near isochron 34 (the Campanian/Santonian boundary). Its closeness to the end of the Cretaceous Normal Superchron suggests the possibility of an intimate relationship between plate motions and behavior of the earth's core.

Introduction

The Molokai Fracture Zone (MFZ) is one of the great fracture zones of the northeast Pacific [Atwater, 1989]. The existence of the MFZ was predicted by Menard [1955], who subsequently named the fracture zone from its apparent intersection with the Hawaiian island chain in the vicinity of Molokai [Menard, 1962]. Only a part of the MFZ passes near Molokai, however; Smith and Menard [1965] later mapped the MFZ west of 135°W as a band more than 300 km wide whose intersection with the Hawaiian chain would extend from southern Oahu to northern Hawaii.

The MFZ is buried by the younger Hawaiian Ridge but reappears farther west. Southwest of Oahu, Eckhoff and Wilkes [1963] recognized a magnetic lineation and fracture-zone topography (consisting of one strongly defined band and a separate graben about 100 km farther north), which Smith and Menard [1965] took to be an extension of the MFZ west of the Hawaiian chain. That inference was confirmed by Malahoff et al. [1966], who traced the magnetic signature of the MFZ from east to west across the Hawaiian chain, finding two bands of magnetic anomalies about 100 km apart.

The trend of the MFZ changes in the vicinity of the Hawaiian chain. Malahoff et al. [1966] noted that the trend changed from E east of the island chain to ENE west of the chain, similar to central-Pacific changes in trend of the Murray and Mendocino fracture zones. Menard [1967] suggested that these changes in trend reflect westward splaying of the great fracture zones where they extend into the central Pacific, with each fracture zone beginning to splay where it crosses a 'branching line' that extends NNW past Oahu.

When the great Pacific fracture zones were recognized as the traces of former transform faults, changes in trend of the fracture zones were inferred to represent changes in direction of seafloor spreading. Menard and Atwater [1968] recognized five successive

The Mesozoic Pacific: Geology, Tectonics, and Volcanism
Geophysical Monograph 77

directions of spreading along major fracture zones of the northeast Pacific, three occurring east of the 'branching line' and two to the west. Rea [1970] argued that a major change of trend occurred without branching, and he renamed the 'branching line' the 'bending line'. Later Sager and Pringle [1987a, 1987b] recognized from bathymetry that west of the bending line all of the great fracture zones are concave to the south, whereas to the east the fracture zones have a slightly more northerly trend and are essentially linear. The changes in trend and the curvature of the fracture zones are even more obvious in the Seasat gravity field [Haxby, 1987; Figure 1 herein] and reflect a major change in seafloor spreading that occurred in the late Cretaceous near the end of the Cretaceous Normal Superchron [Atwater, 1989; Atwater et al., this volume].

We have used the GLORIA long-range sidescan sonar to survey a 1500-km-long section of the Molokai Fracture Zone that is crossed by the Hawaiian Ridge, the section of the MFZ that records the Late Cretaceous change in Pacific-Farallon spreading direction. In this paper we describe the detailed structure of the fracture zone, which we have confirmed to consist of two principal strands separated by up to 100 km of normal seafloor. Each principal strand is composed of multiple component strands. The response of the MFZ to the change in spreading direction appears to follow closely predictions made from simple plate-tectonic theory. The timing of the change is found to be very near the end of the Cretaceous Normal Surperchron.

GLORIA SURVEY

Our work represents a phase of the program by the United States Geological Survey and the British Institute of Oceanographic Sciences Deacon Laboratory to survey the United States Exclusive Economic Zone (EEZ), principally using GLORIA [McGregor and Lockwood, 1985]. We surveyed a segment of the EEZ extending 200 nm (370 km) from the Hawaiian Islands (Figure 2). Other results of this survey have been published by Holcomb et al. [1988], Lipman et al. [1989], and Moore et al. [1989].

Methods

The GLORIA survey tracks shown in Figure 3 represent seven cruises of the research vessel Farnella between October 1986 and October 1988. Five of these cruises [F5-86-HW, F3-88-HW, F4-88-HW, F5-88-HW and F10-88-HW: Holmes et al., 1986; Normark et al., 1989; Torresan et al., 1989; McGregor et al., 1991] produced data that covered the Molokai Fracture Zone. Survey tracks were planned to give complete GLORIA coverage within the EEZ. Navigation was by a combination of LORAN C, Transit Satellite and GPS; accuracy of navigated ship tracks is generally better than 200 m.

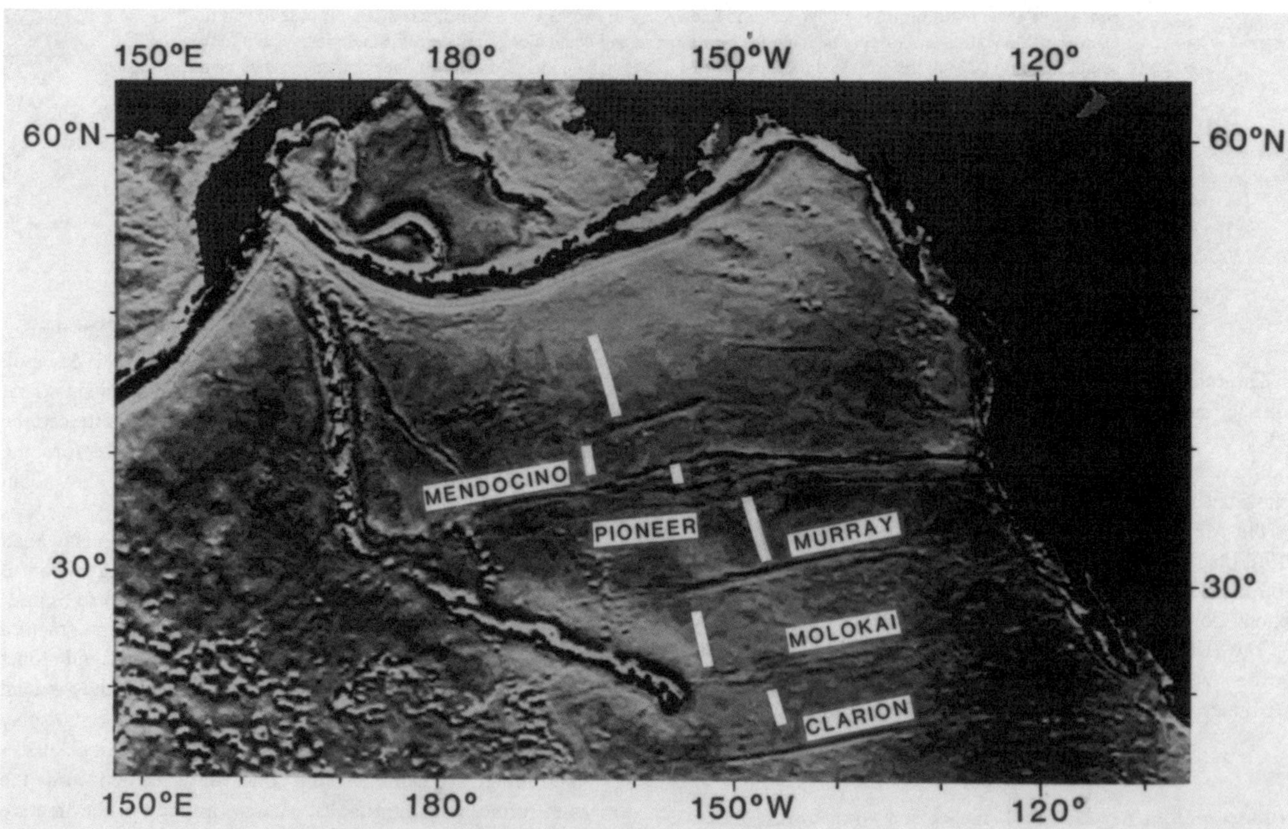

Fig. 1. Gravity field of the northeast Pacific, after Haxby [1987]. Note the changes in trend and radius of curvature of major fracture zones near the longitude of Hawaii. Magnetic anomaly 32 (white lines) is indicated for reference.

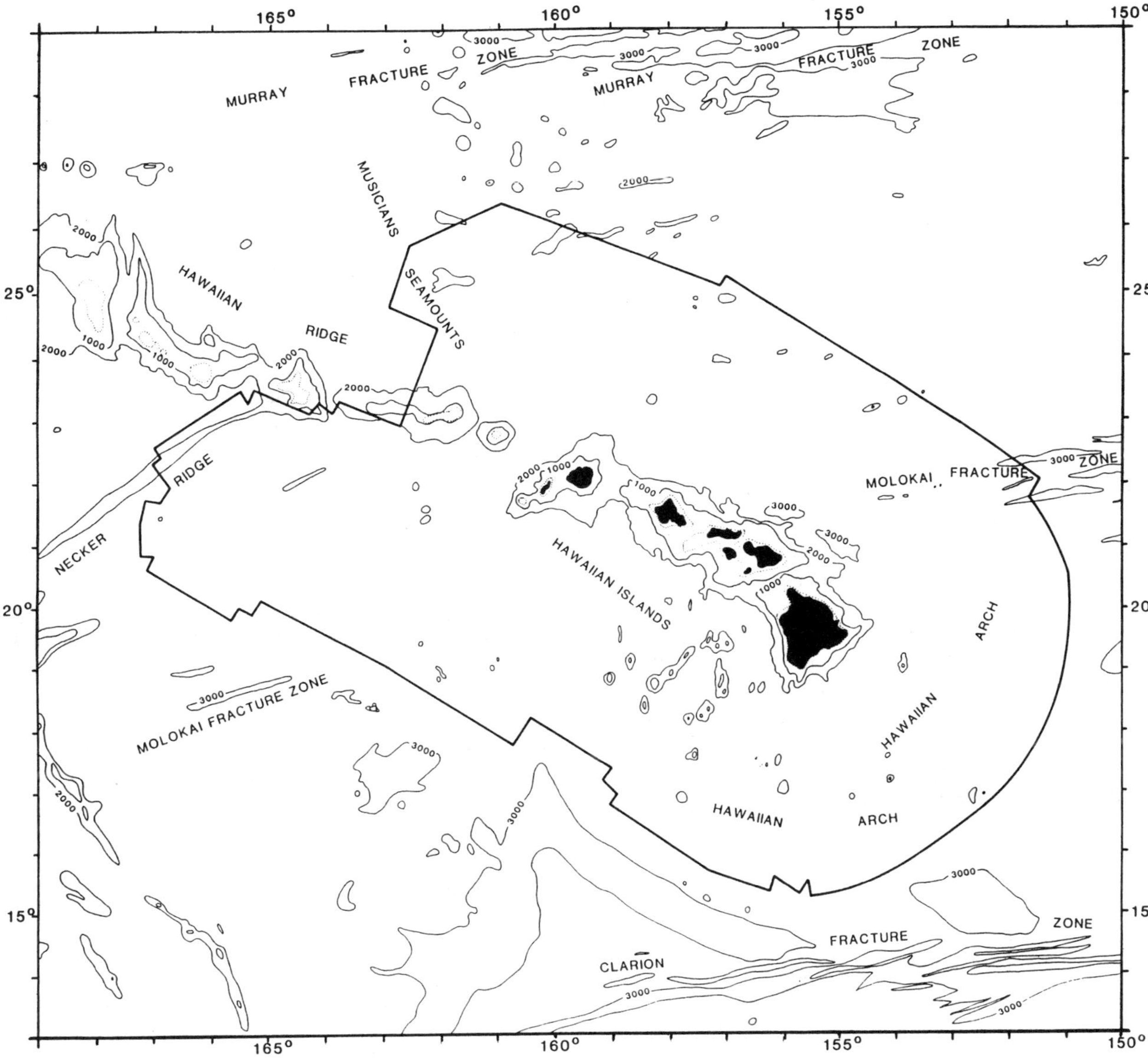

Fig. 2. Simplified bathymetry of Hawaiian region [after U.S. Naval Oceanographic Office, 1973], showing outline of GLORIA coverage used in this study. Contour interval 1000 fathoms (1830 m).

In addition to obtaining the GLORIA data, standard cruise operations included single-channel seismic reflection profiling using a 2630 cm^3 (160 cubic-inch) air-gun sound-source, 3.5-kHz subbottom profiling, 10-kHz echo-sounding, magnetic and gravity potential-field measurements, and upper water-column temperature profiles using expendable bathythermographs. Although we concentrate herein mainly on a structural interpretation using the GLORIA sonograph data, we do refer to the other data where appropriate; magnetic data are especially critical to some parts of the interpretation.

Recent descriptions of the GLORIA system and image processing are given by Searle et al. [1990] and Chavez [1986]. The GLORIA data are being published by the USGS as an atlas of digitally processed sonograph mosaics. The interpretations in this paper are based on study of unprocessed shipboard images that were corrected only for slant-range distortion and assembled into preliminary mosaics at a scale of 1:375,000.

General Results

The GLORIA survey revealed seafloor structures around the Hawaiian Islands in great detail. Figure 4a is a sample of the processed GLORIA mosaic showing a part of the MFZ northeast of Hawaii, and Figure 4b shows an interpretation of that sample. Broad gray areas on the mosaic are inferred to represent expanses of

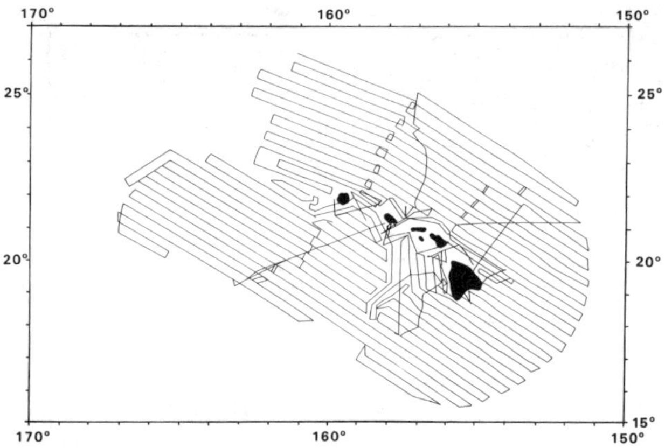

Fig. 3. Track chart for GLORIA cruises around Hawaii. Mercator projection.

thick sediment having medium to low backscatter, whereas small dark areas are interpreted as acoustic shadows. Sharp-edged bright lineaments are interpreted as high-backscatter outcrops or talus along fault scarps. Some escarpments are revealed only by dark shadows, whereas other linear features that appear narrow or sharp-edged but only weakly backscattering are interpreted as traces of tectonic structures buried by sediment. A final class of lineaments, all trending E to ENE, can be inferred along terminations of close-spaced faults, trending N or NNW, which either end abruptly or make sharp sinistral bends as they approach the E-trending lineaments. Other features have irregular or circular shapes; especially common are circular hills, commonly having summit depressions, that are interpreted as small central volcanoes.

Results of the survey are summarized in Figure 5, a volcano-tectonic map derived from analysis of the preliminary GLORIA mosaics. Most features on the map can be placed into one of four categories: 1) closely-spaced faults having N to NNW trends, interpreted as 'seafloor-spreading fabric' formed at mid-ocean ridges; 2) bands of parallel lineaments trending E to ENE, and short lineaments subparallel and oblique to them, interpreted as fracture zones; 3) linear volcanic ridges trending ENE parallel to the fracture zones; and 4) other features of volcanic origin (circular volcanoes and irregularly-shaped lava fields and clusters of volcanoes). Near the Hawaiian island chain the original features of the oceanic crust have been buried by the Hawaiian Ridge and its archipelagic apron, young features that are well imaged by GLORIA but are omitted from Figure 5 for simplicity.

Structural Pattern of the Molokai Fracture Zone

The GLORIA mosaics show that the MFZ consists of strong tectonic lineaments striking E to ENE. East of the Hawaiian Ridge these lineaments cluster into four distinct bands, each 15-30 km wide, which will be referred to herein as bands A, B, C, and D, going from south to north (Figure 5). West of the Hawaiian Ridge are two bands, up to 50 km wide, referred to herein as E and F.

The trend of bands C and D (076°) differs by about 10.5° from the trend of bands A and B (087°), so that the two sets of traces diverge eastward. Bands C and D appear to emerge from the north side of band B, the north edge of band C diverging from band B near 153°W, and the north edge of band D diverging from band B near 156°W.

Normal ridge-parallel (N to NNW) spreading fabric occurs between the bands of the MFZ and at some places within the bands (Figure 4). This fabric is less prominent south of the MFZ, apparently because of greater sediment cover on that side of the fracture zone. Almost everywhere the spreading fabric is orthogonal to the nearest band, striking 353°-356° south of A and B and 345°-348° north of C and D. (Spreading fabric between intersecting bands B and D is normal to band D, but not enough fabric is visible between bands B and C to distinguish a clear trend.) The spreading fabric departs from these dominant trends chiefly within 5-10 km of the MFZ, where structures of the spreading fabric curve sharply in a sinistral sense to approach some MFZ structures at acute angles. Along other lineaments of the MFZ, the spreading fabric is truncated sharply without bending.

Our soundings (Figure 6) and published charts [Chase et al., 1971; U.S. Naval Oceanographic Office, 1973] show that on average the seafloor shoals northward from 4900-5200 m south of band A to 5000-5200 m between bands A and B, 4800-5200 m between C and D, and 4600-5000 m north of D. A similar decrease in seafloor depth also occurs northward across bands E and F. We covered insufficient area to estimate reliably the depth between bands B and C.

Each band of the MFZ contains along most stretches at least two and as many as four or five individual lineaments 5-12 km apart. Lineaments within the bands are composed of fault scarps or colinear terminations of curved faults in the spreading fabric (Figure 4). Continuous segments of lineaments range in length from 10 to 80 km in band A and up to 150 km in band D; if small offsets or gaps a few kilometers long between segments are ignored, some individual lineaments can be traced for up to 300 km.

Bathymetric profiles across the MFZ (Figure 6), together with directions of GLORIA backscatter (Figure 4a), show that most ENE-WSW lineaments are associated with north-facing escarpments several hundreds of meters high. High escarpments face south at some places, especially along outer boundaries of the MFZ such as the south side of band A and north side of band D.

High scarps form both flanks of a major ridge about 10 km wide by 150 km long along the intersection between bands B and D near 21.7°N between 153.5-154.5°W. The relief across the MFZ is greatest along this ridge, which is more than 500 m high over much of its length. Our soundings (Figure 6) and published charts [Chase et al., 1971; U.S. Naval Oceanographic Office, 1973] suggest that the ridge shoals to 3090 m a short distance west of 154°W. A similar ridge does not occur along the intersection of bands B and C, where instead the fracture-zone structure is obscured by thick sediment atop a structural low. A similar ridge does occur farther east along band C in the vicinity of 152°W.

The relief across fracture-zone escarpments increases progressively from band A through bands B and C to band D. Whereas seafloor relief is 400-500 m across band A, it is about 700 m across bands B and C, and is 1200-1600 m across D. (In each

Fig. 4a. Digitally processed GLORIA sonograph mosaic showing part of Molokai Fracture Zone, as indicated on Fig. 5. Light tones represent strong backscatter. NNW-SSE seafloor-spreading fabric is cut by ENE-WSW lineaments inferred to be fossil transform fault traces. The latter lineaments normally consist of north-facing scarps (revealed as narrow, bright reflectors on south-looking sonographs or linear shadows on north-looking ones) or colinear terminations of curved spreading fabric.

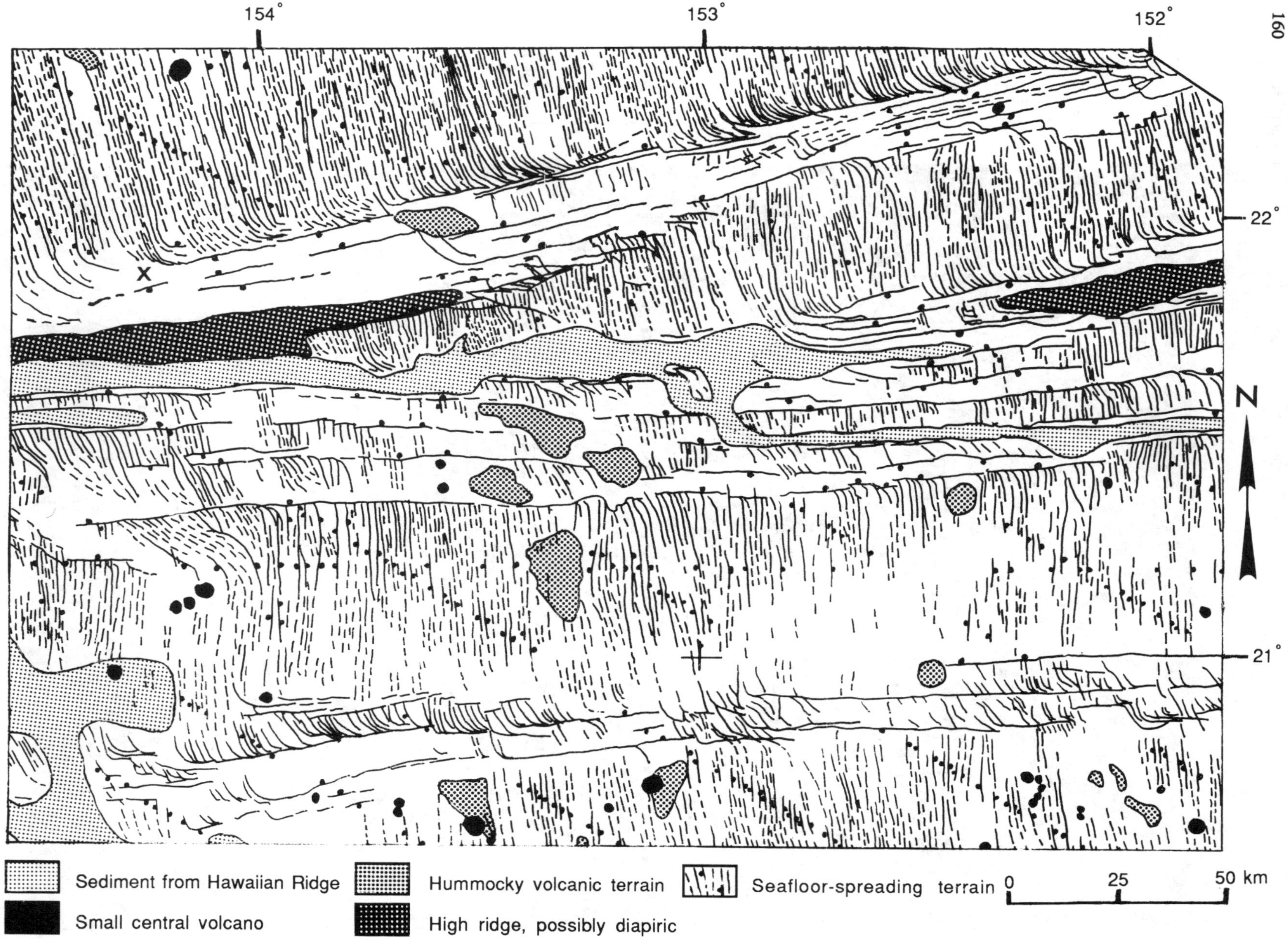

Fig. 4b. Geologic map of the area shown in Fig. 4a. Faults are shown by continuous lines where sharply defined and by dashed lines where poorly defined beneath thick sediment. Balls indicate downthrown side of faults where displacement is known, usually from NW-SE subbottom profiles along ship tracks.

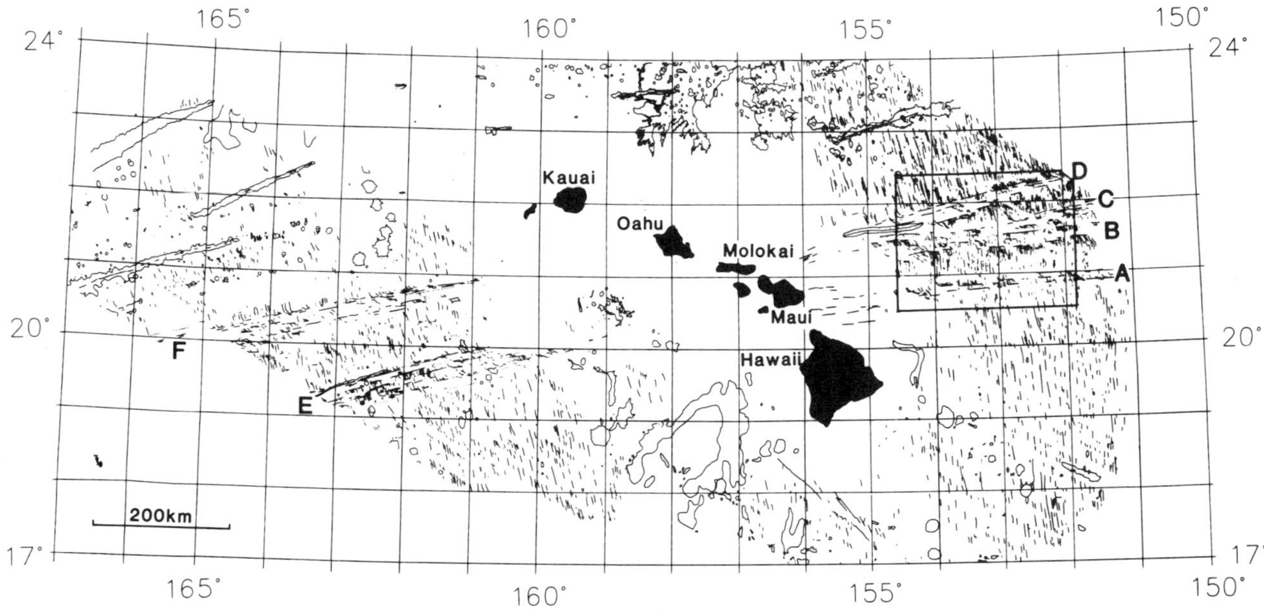

Fig. 5. Volcano-tectonic features interpreted from the GLORIA survey. Lines indicate interpreted tectonic lineaments; closed, unshaded outlines are constructional volcanic features. Original structures of the oceanic crust along the island chain are largely buried by the constructional Hawaiian Ridge. Letters A - F indicate separate fracture-zone bands discussed in the text. Box shows area of Fig. 4. Universal Transverse Mercator projection, central meridian 159°W.

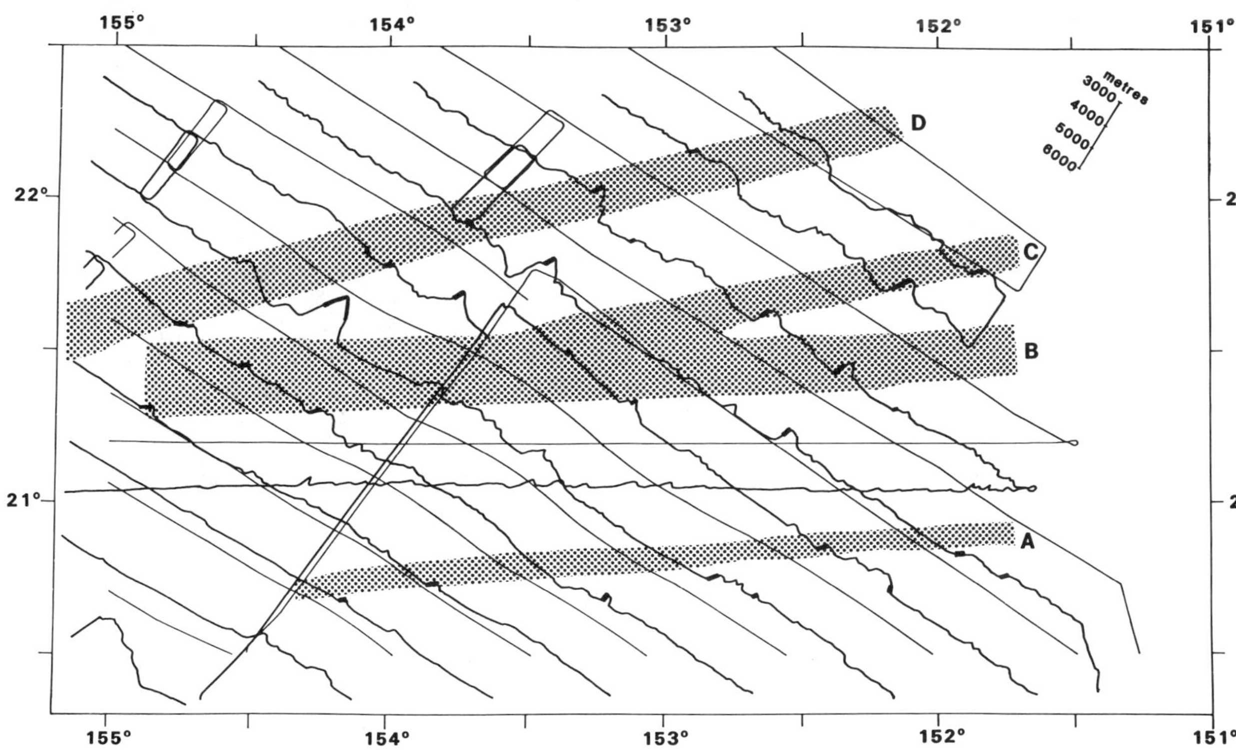

Fig. 6. Bathymetric profiles across the Molokai Fracture Zone, showing the dominant structure of steep north-facing scarps and gentler south-dipping slopes. Thickened lines on profiles indicate areas giving rise to prominent ENE-WSW reflections on GLORIA. Stippled pattern shows positions of MFZ bands A to D. Ship track drawn at 3000 m depth, and bathymetry projected in direction of scale bar. Mercator projection.

case corresponding basement relief is actually greater, because sediment has ponded at the base of each scarp. Whereas the thickness of draped sediment is less than 200 m on dip slopes, the thickness of ponded sediment is typically about 600 m along the bases of principal fracture-zone escarpments.) The north-facing escarpment along the south side of band D shoals to a depth of 3500 m near (21.9°N, 153-153.5°W), ENE of the high ridge along the intersection of bands B and D.

INTERPRETATION

The structural pattern within each band of the MFZ is similar to the pattern of presently-active multi-transform fault systems such as the fast-slipping Quebrada-Gofar [Searle, 1983], Garrett [Gallo et al., 1983], and Siqueiros [Fornari et al., 1989] transforms on the East Pacific Rise and the medium-slipping Cox Transform on the southern Mid-Atlantic Ridge [Grindlay et al., 1991]. Because of this similarity, we suggest that each band of the MFZ is itself a fracture zone representing a separate transform system. Spreading fabric and more numerous lineaments within some bands of the MFZ, suggest that certain stretches of those bands (e.g. band B west of 153°W) may themselves represent more than one transform fault.

Also by analogy with modern transform systems [e.g. Searle, 1986], we infer from the sense of curvature of the spreading fabric that all ridge offsets along the MFZ were sinistral, so that a section across the MFZ should encounter progressively younger seafloor going from south to north. This is consistent with the observed variations in regional sediment cover and depth of the seafloor, described above. An overall sinistral offset across the MFZ is consistent also with offsets of magnetic polarity boundaries within our survey area and farther east [Atwater and Menard, 1970; Atwater, 1989; Atwater and Severinghaus, 1989; Atwater et al., this volume]. For example, anomaly 34 [young end of the Quiet Zone representing the Cretaceous Normal Superchron, near the Santonian/Campanian boundary: 84.0 Ma in time scale of Berggren et al., 1985] occurs in the vicinity of 156.5°W on the north side of the MFZ but is near 150.5°W on the south side of the MFZ.

Anomaly 34 south of the MFZ occurs nearly opposite anomaly 31 (approx. 69 Ma) on the north side of the MFZ [Atwater and Severinghaus, 1989], suggesting an age contrast of about 15 Ma across the entire fracture zone. Age offsets across the MFZ are not constant, however, indicating differences in average half-spreading rate on opposite sides of the fracture zone. Such differences might be brought about through ridge propagation in response to the change in spreading direction [see, for example, Caress et al., 1988; Atwater, 1989; and Atwater and Severinghaus, 1989]. From the maps of Atwater and Severinghaus [1989], we estimate that the average half-spreading rate from isochron 34 to 32 was about 35 mm/yr north of the MFZ and 50 mm/yr south of it. We cannot judge spreading rates between different strands of the MFZ during this interval because we do not know the locations of magnetic anomalies 32 and 34 between the strands.

Correlation of Fracture-zone Strands across the Hawaiian Ridge

Although band E appears to be approximately colinear with C or D, and F with B, such a correlation would imply the impossibility of crossing fracture zones. We therefore prefer to correlate E with

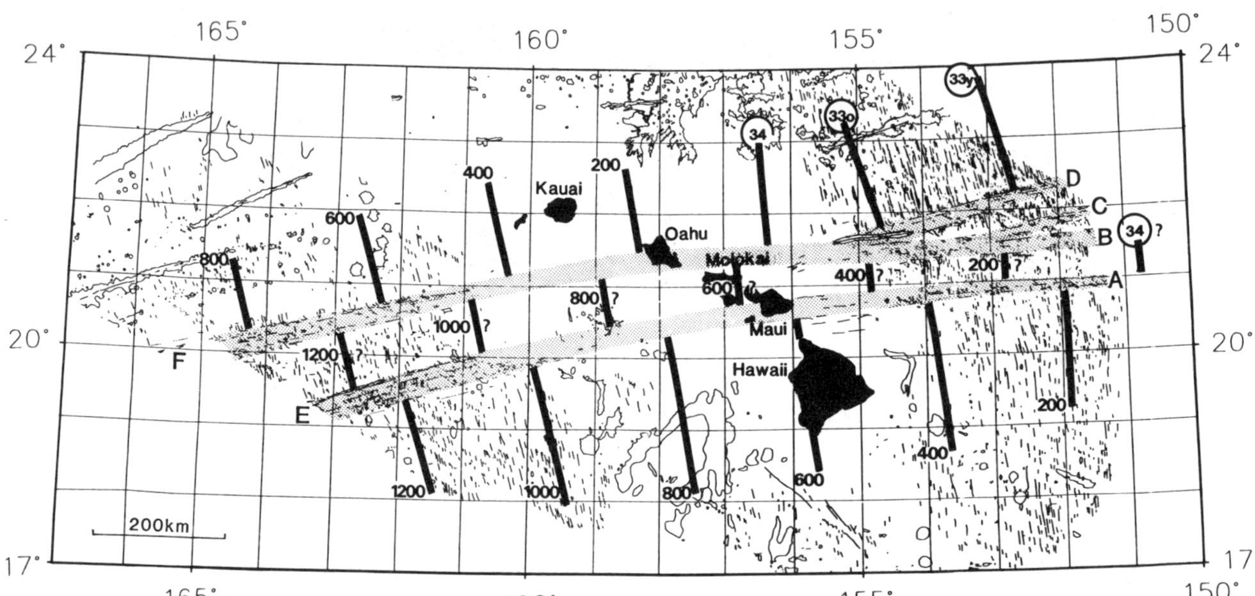

Fig. 7. Observed tectonic pattern of the Molokai fracture zones, with proposed correlations across the Hawaiian Ridge (wide gray bands). Thick black lines labeled 33y, 33o, and 34, are young and old edges of magnetic normal anomaly 33 and anomaly 34, as located by Atwater and Severinghaus [1989] and our survey. Other black lines drawn at intervals of 200 km west of anomaly 34; lines between bands A-E and B-F are queried because relevant location of anomaly 34 is uncertain. Universal Transverse Mercator Projection, central median 159°W.

A and F with B to form a pair of curved fracture zones that are concave to the south and converge gradually from a spacing of 100 km at 163°W to 50 km at 152°W (Figure 7). Bands C and D can then be explained as younger traces of the fracture zone following an approximately 10° counterclockwise change of spreading direction at or shortly after isochron 34.

Several observations support our proposed correlation. First, the Seasat-derived gravity field [Haxby, 1987] shows that the Mendocino, Pioneer, Murray, and probably also Molokai and Clarion fracture zones all follow curved traces (concave to the south) through most of the Quiet Zone, striking ENE in the western part of the Quiet Zone and curving to nearly due E near the eastern edge of the zone (Figure 1). The concave-south curvature is confirmed by studies of the regional bathymetry [Sager and Pringle, 1987a, 1987b] and magnetic anomalies [Atwater, 1989]. Moreover, the gravity, bathymetry, and magnetic anomalies all support Rea's [1970] inference of a counterclockwise change in spreading direction, from E-W to ENE-WSW, near or shortly after isochron 34. The concave-south curvature of fracture zones within the Quiet Zone and the change in spreading direction near anomaly 34 are consistent with our proposed correlation of MFZ bands A-E and B-F and divergence of bands C and D from band B near isochron 34.

We conclude that, during much of the late Cretaceous prior to the end of the Superchron, the MFZ consisted of two systems of transform faults — forming bands A-E and B-F — about 100 km apart, each having sinistral offset. Because A-E passes beneath southern Maui and B-F passes beneath southern Oahu, we propose to call them the Maui and Oahu strands, respectively, of the MFZ. Traces C and D are younger strands of the MFZ that developed in response to a counterclockwise change in spreading direction that occurred near isochron 34, but we refrain from naming these.

Time of the Change in Spreading Drection

A Late Cretaceous change in spreading direction is recorded widely by fracture zones and magnetic anomalies in the northeast Pacific. There are, however, several inconsistencies in the reported timing of the change (which may have been diachronous), and its timing near Hawaii has not been precisely resolved from data previously available.

Using regional bathymetric and magnetic data, Atwater [1989, p. 30] judged the change in spreading direction to occur abruptly during the middle of anomaly 33 time, which she gave parenthetically as 79 Ma but is nearer 77 Ma in the chronologies of Kent and Gradstein [1985] and Berggren et al. [1985]. An even later time of about 75 Ma for the spreading change was also given by Atwater [1989; caption of her Figure 5, p. 28]. Some of the regional evidence is quite inconsistent however; as shown on Atwater's Figure 5, sharp bends in the Pau and Pioneer fracture zones suggest that the change in spreading direction occurred near isochron 34, or at about 84 Ma., and Atwater et al. (this volume) now judge the change to have occurred sometime during Chron 33R (between anomalies 34 and 33 of Figure 8, prior to 79 Ma).

We infer that the change in spreading direction occurred along the MFZ at or very near the time of anomaly 34, and certainly before the onset of anomaly 33. We infer this timing from 1) seafloor-spreading fabric west of anomaly 33 having post-rotation trends, 2) the projected intersection of band D with band B near anomaly 34, 3) a well-defined difference in trend between anomaly 34 and the old side of anomaly 33, and 4) irregularities in seafloor-spreading fabric associated with a dogleg in the trend of anomaly 34.

Much of the seafloor-spreading fabric south of 24°N and west of anomaly 33 (Figure 8) is covered by the Hawaiian archipelagic apron [Menard, 1956], but some structures of this fabric can be seen as far west as about 155.8°W, halfway between anomaly 34 and the old side of anomaly 33 (33o of Figure 7). Offsets in anomaly 33o have been suggested to indicate that adjustments were still taking place then by ridge propagation (T. Atwater, personal communication, 1992), but we see no sign of such propagation (e.g. pseudofault traces or magnetic offsets) in our data. All structures east of 156°W trend approximately 346° — parallel to post-rotation fabric and magnetic anomalies farther east — indicating that the change in spreading direction must have been complete by about 82 Ma.

If the northern edges of MFZ bands B and D are projected westward, they are found to intersect along the base of the Hawaiian Ridge, slightly east of anomaly 34 (point S of Figure 8). Because anomaly 34 should bend sharply eastward with the seafloor-spreading fabric as it approaches the MFZ, the change in spreading direction must have occurred very near the time of anomaly 34.

Anomaly 34 follows a different trend than the anomalies farther east. We used the regional compilation of Atwater and Severinghaus [1989] to approximately locate each anomaly, which we then adjusted using our more closely spaced magnetic profiles (Figure 8). From two profiles having highly variable navigational control, Atwater and Severinghaus [1989] tentatively mapped anomaly 34 near 156.4°W, a clue that we used to identify the anomaly in ten profiles of our own data set. The southern six profiles between 22-23.5°N define an anomaly striking about 355°, parallel to older seafloor-spreading fabric (south of the MFZ) that predates the change in spreading direction. This segment of anomaly 34 appears to have been unaffected by the change in spreading, and must therefore predate that change.

The magnetic profiles seem to require, however, a slight dogleg in the trace of anomaly 34, such that it bends near 23.5°N to trend about 350° farther north (Figure 8). This trend is intermediate between those predating and postdating the change in spreading direction, and if projected southward it would pass only slightly west of the projected intersection of bands B and D (point S of Figure 8). Moreover, an area of poorly imaged seafloor-spreading fabric near anomaly 34, between 24-25°N (Figure 9), appears to be complicated by unusually oblique structures. Whereas most fabric west of 157°W has pre-rotation trends and most fabric east of anomaly 34 has post-rotation trends, much of the fabric between is relatively complex. Some of the oblique structures, particularly those trending ESE and ENE, could represent later faulting or additional fracture zones subparallel to the MFZ; but we suspect that much of the complexity arose from southward rift propagation related to the change in spreading direction. If so, the change in direction must have begun prior to anomaly 34. The intermediate

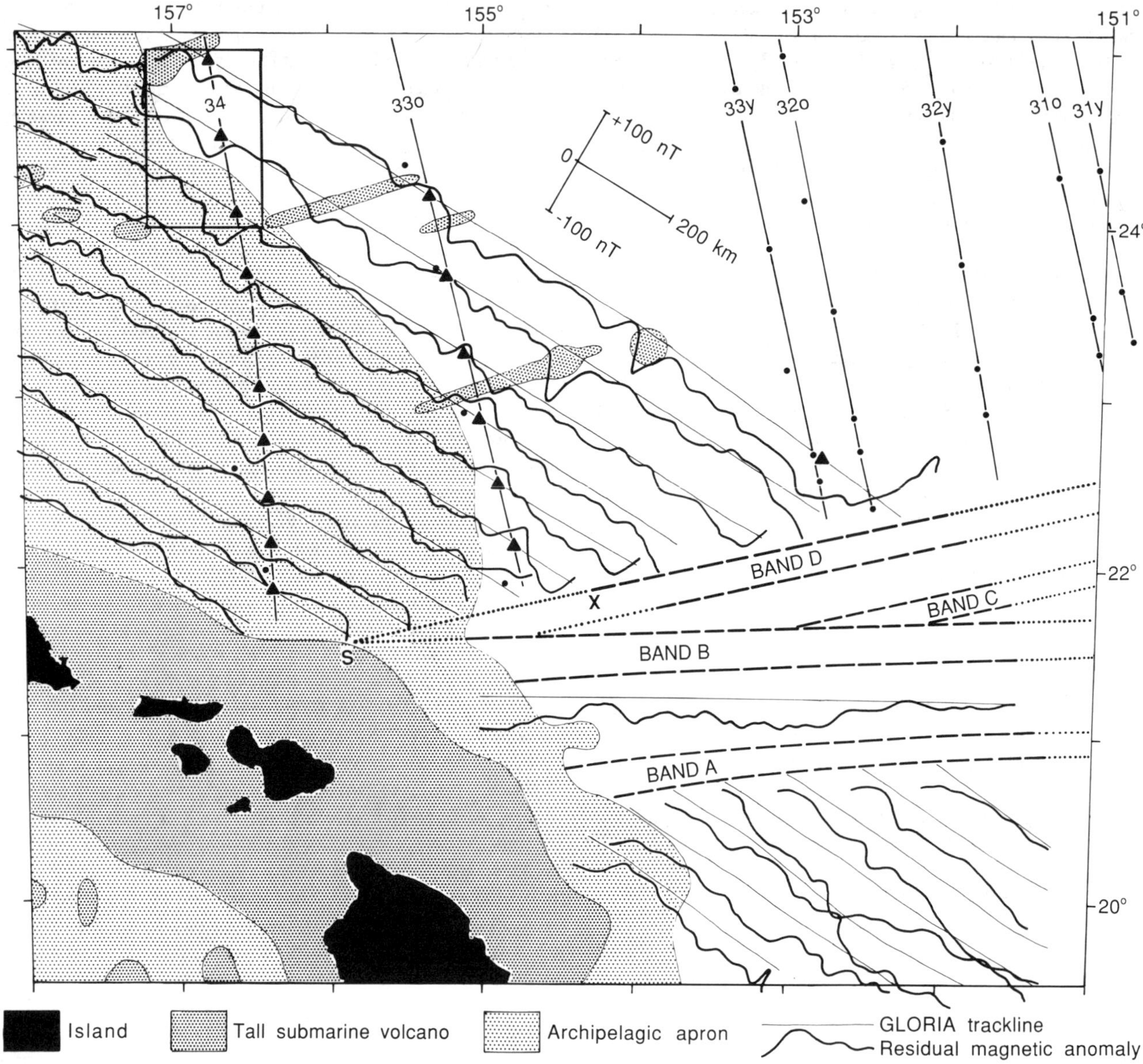

Fig. 8. Residual magnetic-anomaly profiles obtained during GLORIA survey, plotted on a simplified geologic map. Small filled circles are isochron crossings as taken from Atwater and Severinghaus [1989], whereas larger filled triangles are isochron crossings as inferred from our profiles. Approximate margins of MFZ bands are dashed where observed and dotted where projected; X is at the end of the most sharply defined segment of band D, as shown on Fig. 4b, and S is the projected intersection of northern edges of bands B and D. Rectangle at upper left indicates area of Fig. 9.

trend of anomaly 34 north of the dogleg (Figure 8) may represent an average across a series of offsets of the anomaly, now mostly buried.

In summary, the change in spreading direction here must have occurred very near the end of the Cretaceous Magnetic Superchron, and it may have already been underway at the end of the Superchron. Detailed surveys across and along anomaly 34 may prove valuable for defining the exact time and duration of the change in spreading direction, as well as the pattern of rift propagation by which the change was effected.

Lateral Offsets along the Molokai Fracture Zone

The nature of the tectonic pattern produced at the MFZ by the spreading change does not depend critically on the lengths of the fracture-zone offsets, but it is helpful to estimate these approximately. Atwater and Severinghaus [1989] mapped anomaly

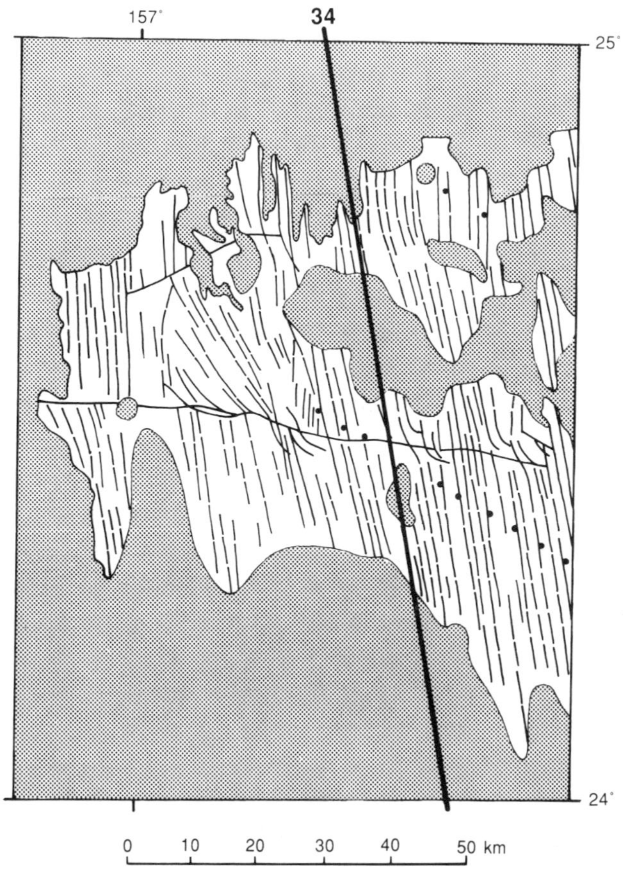

Fig. 9. Structures inferred from GLORIA mosaic in the vicinity of anomaly 34 between 24-25°N. Balls indicate downthrown side of faults where displacement is known from NW-SE subbottom profiles along ship track. Shading indicates areas where structures are completely obscured by younger materials.

34 at about 150.5°W on the south side of the MFZ and at about 156.4°W on the north side of the MFZ, for a total sinistral offset of about 600 km along the double fracture zone. In order to draw cartoon models of MFZ response to the change in spreading direction, it is necessary to partition the total offset into separate components along bands A and B of the fracture zone.

Atwater and Severinghaus [1989] reported no recognized anomalies between bands A and B, and our own profiles trending WNW oblique to the MFZ are dominated by large anomalies (omitted from Figure 8 for simplicity) associated with structures of the fracture zone. We ran a single E-W profile along latitude 21.2°N near the axis of the 50-km-wide zone between these bands, but it shows no sign of anomaly 34 (Figure 8). Nor is there a significant change in trend of the spreading fabric in that zone (Figure 4). All of the surveyed strip between bands A and B must therefore predate anomaly 34 and the associated change in spreading direction. This sets a lower limit on the offset across the Oahu strand (band B) of 480 km, leaving no more than 120 km to be taken up on the Maui strand (band A).

On the other hand, the offset along the Maui strand cannot have been much less than 30 km, which is about the lower limit at which transform faults are thought to develop on slowly spreading seafloor [Tamsett and Searle, 1988]. Somewhat shorter transforms (about 20 km) can form in fast-slipping, multi-strand transform systems, but total offset along such systems still seems to be at least 35 km [Searle, 1983; Fornari et al., 1989]. We therefore judge the offset along the Maui strand to be between 30 and 120 km. In drawing Figures 7 and 10 we have assumed offsets of 70 km along the Maui strand and 530 km along the Oahu strand.

Response of the MFZ to the Change in Spreading Direction

Whereas two main strands comprised the MFZ prior to the change in spreading direction, more than two distinct strands evidently comprised it following the change: we observe two (C and D), and at least one additional strand must extend ENE from the unsurveyed ends of bands A and B. Figures 10 and 11 show how additional strands may have developed as the MFZ adjusted to the change in spreading direction.

Because the Oahu and Maui strands had sinistral offsets, the net response to a counterclockwise change in spreading direction would have been to open rhomboid-shaped gaps along both strands [Menard and Atwater, 1968; Figure 10A, B herein]. Initially, the old spreading segments probably broke up into shorter en echelon segments normal to the new spreading direction (Figure 10C), and the old transforms probably 'leaked', perhaps producing intrusive ridges such as the one at the intersection between bands B and D. Leakage is likely to have quickly become localized at multiple discrete sites along the old transforms, forming additional very short spreading segments within the incipient rhomboids. [Such a process was probably responsible for the development of the short spreading segments in the Quebrada-Gofar transform system, if that became leaky following a transtensive change in spreading direction as proposed by Searle, 1983.] As spreading continued in the new direction, widening the rhomboids, some of the new spreading centers probably propagated at the expense of others [cf. Caress et al., 1988], resulting in fewer centers, larger separations between adjacent centers, and new transforms between them. (Such propagation may have produced some of the oblique structures shown in Figure 9) Continued rift propagation would tend to increase the length of spreading segments and lessen the number of spreading segments and transforms within the opening rhomboids (Figure 10D). However, spreading segments probably would not have propagated across edges of rhomboids because the opposite flanks of the rhomboids would have consisted of old, strong lithosphere along the far side of the fracture zone. As a result, a new fracture zone (band C) would bisect the rhomboid gap between the separating sides of band B — the Oahu strand (Figure 10D).

The development of additional strands like band C might be explained in other ways also. For example, closely-spaced component strands within the Oahu strand might have 'unraveled' during the change in spreading direction to become distinctly separate bands (Figure 11). In Figure 10 we assumed (for simplicity of drafting) that, although band B may have been multistranded, it possessed only one transform long enough to be

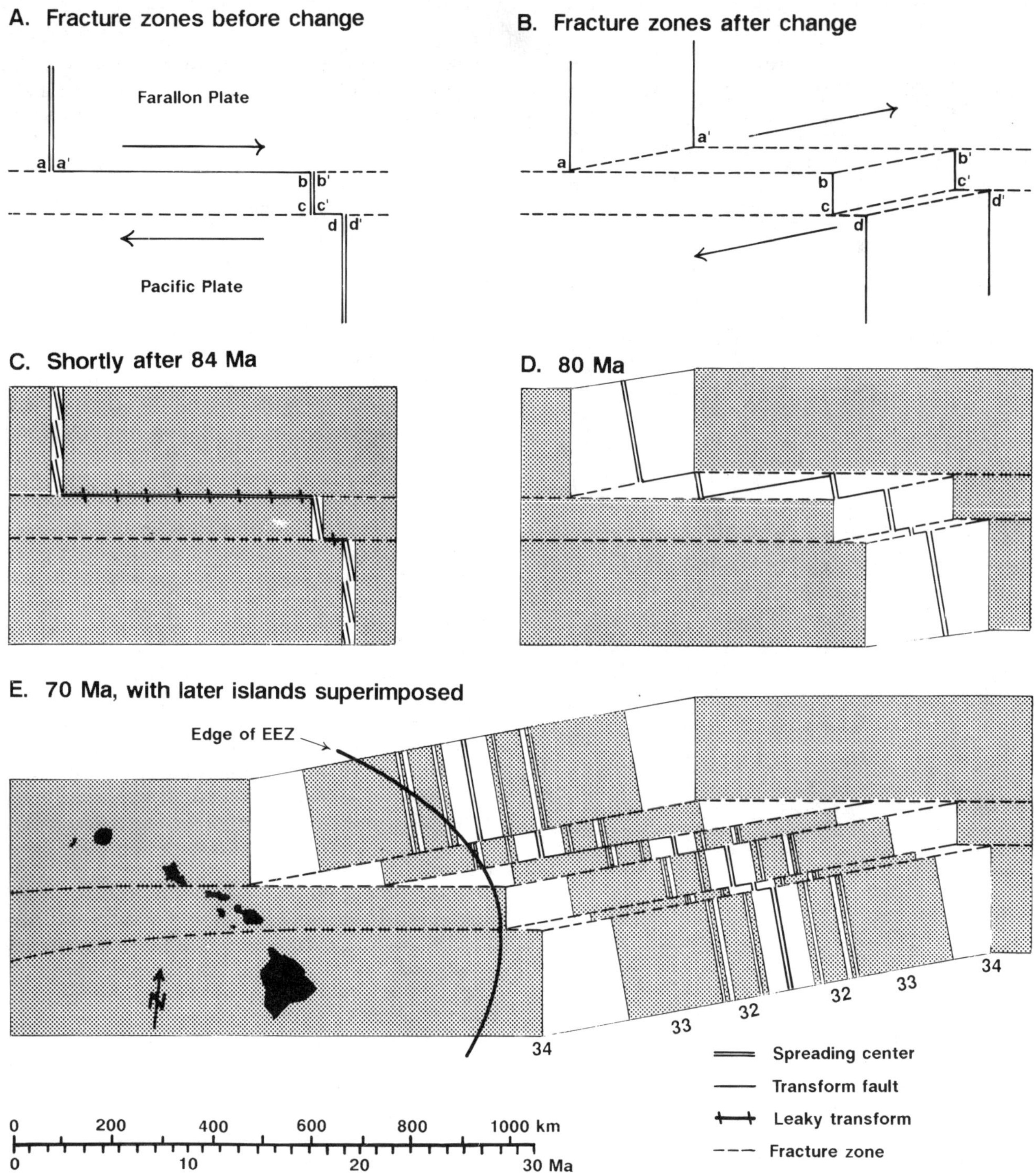

Fig. 10. Cartoon showing general geometry of MFZ (A, B) and its possible adjustments to change in spreading direction (C, D, E). For simplicity, the change is shown to occur after anomaly 34, the Oahu and Maui strands are shown as single transforms, and fracture zones younger than isochron 34 are drawn as straight lines. Arrows show relative plate motion. Shading shows seafloor of normal magnetic polarity. A: Spreading segments offset by major transforms a'b and c'd. B: Rhomboids aa'b'b and cc'd'd have opened up along old transforms. C: Long spreading segments have broken up into short en echelon segments, and transforms have become leaky. D: Long ridge segments have propagated perpendicular to new spreading direction, and leaky transforms have been replaced by chains of short new transforms and spreading segments. Details within the rhomboids are uncertain; the southern one is shown with a single spreading center, but the northern one is shown with two segments separated by an intermediate transform. The latter transform and one spreading segment might disappear with time. E: Pattern of seafloor magnetization well after the change of direction. Superimposed for reference are the modern Hawaiian Islands. Heavy curved line is eastern limit of GLORIA survey.

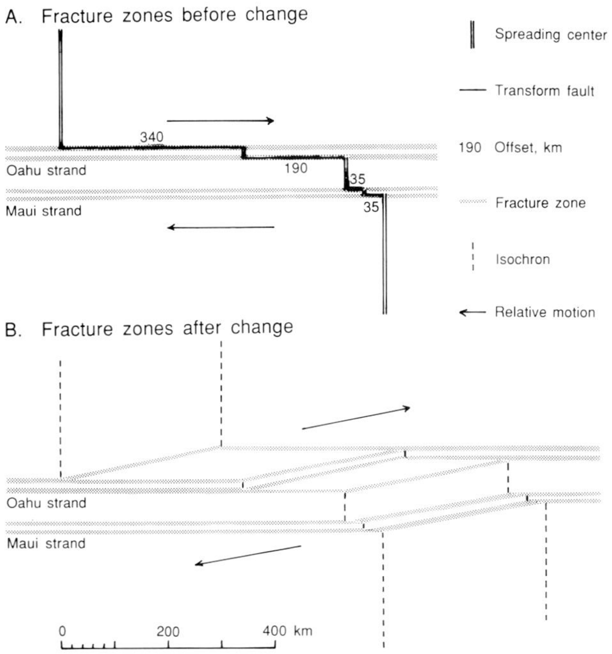

Fig. 11. Modification of Fig. 10A, B to account for probable multi-transform complexity of Oahu and Maui strands, showing how change in spreading direction might cause an isolated short spreading center to diverge from the Oahu strand.

treated as a separate fault. But if band B possessed a second long transform (as is apparent from the GLORIA data, Figure 4), the short spreading center between the two transforms could have responded by itself to the change in spreading direction, diverging from band B to form a separate band of spreading fabric. Consistent with this alternative is an abrupt decrease in width of band B from about 30 km to about 15 km near 153°W. According to this interpretation, the Oahu strand would have consisted of a northern strand having an offset of not more than 340 km and a southern strand having an offset of at least 190 km. The Maui strand also seems to have had two (in some places, three) component strands (Figure 4) whose transforms must have been connected by very short spreading centers similar to those now present within the Blanco Transform Zone [Embley et al., 1986]. If both the Oahu and Maui strands were multistranded, their unraveling could have produced several additional strands beyond the eastern limit of our survey.

DISCUSSION

The Molokai Fracture Zone just east of Hawaii represents one of the best surveyed examples of the effect of a transtensional change in spreading direction on the tectonic pattern and morphology of a fracture zone. This example poses several questions for further study. Especially pertinent are those regarding the mechanisms by which transforms respond to abrupt large changes in spreading direction, and the possible mechanisms by which events in the Earth's outer core are related to lithospheric spreading.

Comparison with Changes along the Kane Fracture Zone

We have suggested ways in which the Molokai Fracture Zone might have responded to a moderate change in spreading direction. Similar changes (but at different times) have been observed along Atlantic fracture zones. Those on the Kane Fracture Zone have been mapped and discussed in detail by Tucholke and Schouten [1988/89], who observed both transtensional (like Molokai) and transpressional changes in spreading direction. The two responses are quite different, and we shall consider only the transtensional one.

Transtensional changes on the Kane Fracture Zone are accompanied by the formation of transverse ridges: 'median ridges' within the fracture valley, and what we will call 'flanking ridges' at their edges. Median ridges are observed by Tucholke and Schouten at most sites of transtensional change, and are inferred by them to represent serpentinite diapirs. Their formation is facilitated by transtension, which first allows seawater to reach deeply into the lithosphere, promoting serpentinization, and then allows the serpentinite and associated rocks to rise. Such ridges were considered by Tucholke and Schouten to be a 'moderate' response to a change in spreading direction, where the angle of change is small enough that the transform fault has room to become reoriented within its existing fracture valley and no new transforms need be created.

Tucholke and Schouten also identify what they call a 'severe response'. This occurs when the angular change is so great that a single existing transform cannot reorient without breaking out of the fracture valley into older, stronger lithosphere. Because such breakouts cannot easily occur, the preexisting transform breaks up into new, shorter transforms in the way that we have suggested for the MFZ, with creation and propagation of short spreading centers inside rhomboidal gaps. Tucholke and Schouten propose that when a spreading center propagates to the edge of a rhomboid it reheats the old, bounding lithosphere, which then rises — probably along the plane of weakness already established by the serpentinite intrusion below the median ridge. This creates the second type of transverse ridge, the 'flanking ridge'.

The high, symmetrical ridge along the junction of MFZ bands B and D (Figure 6) is of similar size, shape and orientation to the Kane ridges, and it is associated with large magnetic anomalies (omitted from Figure 8 for simplicity), as would be expected for a serpentinite ridge. So it too may reflect diapirism in response to N-S tension associated with early stages of the change of spreading direction. It thus represents the physical effects of transform leakage. However, the topography of the Molokai Fracture Zone is complex (Figure 6), and available bathymetric information does not yet allow us to distinguish between the median or flanking types. Because Tucholke and Schouten's model predicts the occurrence of both types, this is an important question for future work.

Tucholke and Schouten found that all but one of the new transforms that developed along the Kane Fracture Zone immediately after a transtensive change were subsequently eliminated by ridge propagation during the next several million years. Our survey did not extend far enough east to test whether such readjustment occurred along the MFZ. This question represents another opportunity for future work.

Tucholke and Schouten also found that spreading fabric bent more strongly toward the direction of transform offset during transtensive episodes, producing a set of 'J-shaped' ridges. Tucholke and Schouten attributed this enhanced bending to an increase in the minimum compressive stress, σ_3, which is oriented at 45° to the transform slip direction. We certainly observe strong curvature of the spreading fabric along all of the imaged stretch of the MFZ, but without additional bathymetric data we are unable to say whether enhanced J-shaped topographic ridges occur at the change in spreading direction.

Finally, Tucholke and Schouten suggested that the reorientation of the spreading centers flanking a transform causes the centers to break into segments separated by short, non-transform offsets, which can then migrate along the ridge to form chevron patterns. (The fast-spreading equivalents would be propagating ridges and their pseudofaults.) We see no such effect in the spreading fabric following the change of direction along the MFZ, except possibly for rapid propagation across anomaly 34. Perhaps this difference is an effect of spreading rate. The seafloor was spreading somewhat faster along the MFZ (35-50 mm/yr) than along the Kane Fracture Zone (15-25 mm/yr), and fast-spreading ridges can reorient and propagate much more readily because of their thinner, weaker lithosphere.

Timing and Cause of the Change in Spreading Direction

As nearly as we can tell from our combination of magnetic-anomaly and structural data, the ocean-wide Late Cretaceous change of plate motions occurred at the same time that the earth's magnetic field began reversing again at the end of the Cretaceous Normal Superchron. It is therefore tempting to consider whether there could be some fairly direct connection between plate motions of the lithosphere and the state of the Earth's magnetic field or, ultimately, the core.

We note that Larson and Olson [1991] have suggested a mechanism whereby enhanced mantle plume activity may have precipitated the Cretaceous magnetic quiet period by increasing heat flow across the core/mantle boundary and thus enhancing convection, poloidal field components, and magnetic stability of the core. They show that, within the ±3 Ma resolution of available dating, the onset of the Cretaceous Normal Superchron coincided with a sudden twelve-fold increase in plume productivity. We are intrigued that the end of the Superchron, like its beginning, coincides with such a major lithospheric event as the late Cretaceous change in Pacific/Farallon spreading direction, although we are unable to suggest a mechanism to account for it.

Acknowledgments. We thank the shipboard scientific parties of cruises F5-86-HW, F3-88-HW, F4-88-HW, F5-88-HW, and F10-88-HW for assistance with data collection, and Carolyn Degnan for data processing. We are indebted to Tanya Atwater for discussions of the magnetic anomalies and central Pacific tectonics, for making her magnetic maps available prior to publication, and for her valuable review of our manuscript. We also appreciate reviews by Alan Cooper, Kim Klitgord and Brian Taylor.

REFERENCES

Atwater, T., Plate tectonic history of the northeast Pacific and western North America, *in* Winterer, E.L., D.M. Hussong, and R.W. Decker, eds., The Eastern Pacific Ocean and Hawaii, *The Geology of North America, N*, Geological Society of America, 21-72, 1989.

Atwater, T.M., and H.W. Menard, Magnetic lineations in the northeast Pacific, *Earth Planet. Sci. Lett.*, *7*, 445-450, 1970.

Atwater, T., and J. Severinghaus, Tectonic maps of the northeast Pacific, *in* Winterer, E.L., D.M. Hussong, and R.W. Decker, eds., The Eastern Pacific Ocean and Hawaii, *The Geology of North America, N*, Geological Society of America, 15-19, 1989

Atwater, T., J. Sclater, J. Severinghaus, D. Sandwell, and M. Marlow, Fracture zone traces across the north Pacific Cretaceous Quiet Zone and their tectonic implications, this volume.

Berggren, W.A., D.V. Kent, J.J. Flynn, and J.A. Van Couvering, Cenozoic geochronology, *Geol. Soc. Am. Bull.*, *96*, 1407-1418, 1985.

Caress, D.W., H.W. Menard, and R.N. Hey, Eocene reorganization of the Pacific-Farallon spreading center north of the Mendocino Fracture Zone, *J. Geophys. Res.*, *93*, 2813-2838, 1988.

Chase, T.E., H.W. Menard, and J. Mammerickx, *Topography of the North Pacific* (map), Scripps Institution of Oceanography, Institute of Marine Resources, 1971.

Chavez, P.S., Processing techniques for digital sonar images from GLORIA, *Photogramm. Eng. Remote Sens.*, *52*, 1133-1145, 1986.

Dalrymple, G.B., M.A. Lanphere, and D.A. Clague, Conventional and $^{40}Ar/^{39}Ar$ K-Ar ages of volcanic rocks from Ojin (Site 430), Nintoku (Site 432), and Suiko (Site 433) seamounts and the chronology of volcanic propagation along the Hawaiian-Emperor chain, *Initial Rep. Deep Sea Drill. Proj.*, *55*, 659-676.

Duncan, R.A., and D.A. Clague, Pacific plate motion recorded by linear volcanic chains, *in* Nairn, A.E.M., F.G. Stehli, and S. Uyeda, eds., *The Ocean Basins and Margins*, *7A*, Plenum, New York, 89-121, 1985.

Eckhoff, O.B., and G.E. Wilkes, Submarine topography southwest of the Hawaiian Islands, *Trans. AGU*, *44*, 67, 1963.

Embley, R.W., L.D. Kulm, G. Massoth, D. Abbott, and M. Holmes, Morphology, structure, and resource potential of the Blanco Transform Fault Zone, *in* Scholl, D.W., A. Grantz, and J.G. Vedder, eds., *Geology and Resource Potential of the Continental Margin of Western North America and Adjacent Ocean Basins – Beaufort Sea to Baja California, 6*, Circum-Pacific Council for Energy and Mineral Resources, Earth Science Series, 549-561, 1986.

Fornari, D.J., D.G. Gallo, M.H. Edwards, J.A. Madsen, M.R. Perfit, and A.N. Shor, Structure and topography of the Siqueiros transform fault system: evidence for the development of intra-transform spreading centers, *Mar. Geophys. Res.*, *11*, 263-299, 1989.

Gallo, D.G., P.J. Fox, P. Lonsdale, and J. Madsen, The morphotectonic expression of the world's fastest-slipping plate boundary: the Garrett transform, *EOS Trans. AGU*, *64*, 855, 1983.

Grindlay, N.R., P.J. Fox, and K.C. Macdonald, Second-order ridge axis discontinuities in the south Atlantic: morphology, structure and evolution, *Mar. Geophys. Res.*, *13*, 21-49, 1991.

Haxby, W.F., *Gravity Field of the World's Oceans* (map), National Geophysical Data Center, Boulder, Colo., 1987.

Holcomb, R.T., J.G. Moore, P.W. Lipman, and R.H. Belderson, Voluminous submarine lava flows from Hawaiian volcanoes, *Geology*, *16*, 400-404, 1988.

Holmes, M.L., J.G. Moore, R.T., Holcomb, and R.H. Belderson, Preliminary cruise report, research vessel Farnella, windward Hawaii, F5-86-HW, U.S. Geol. Surv. admin. rept., 17 p., 1986.

Kent, D.V. and F.M. Gradstein, A Jurassic to recent chronology, *in* Vogt, P.R., and B.E. Tucholke, eds., The Western North Atlantic Region, *The Geology of North America, M*, Geological Society of America, 45-50, 1986.

Larson, R.E., and P. Olson, Mantle plumes control magnetic reversal frequency, *Earth Planet. Sci. Lett.*, *107*, 437-447,1991.

Lipman, P.W., D.A. Clague, J.G. Moore, and R.T. Holcomb, South Arch volcanic field -- newly identified young lava flows on the sea floor south of the Hawaiian Ridge: *Geology*, *17*, 6112-6114, 1989.

Malahoff, A., W.E. Strange, and G.P. Woolard, Molokai Fracture Zone: continuation west of the Hawaiian Ridge, *Science*, *153*, 521-522, 1966.

Mammerickx, J., and G.F. Sharman, Tectonic evolution of the north Pacific during the Cretaceous quiet period, *J. Geophys. Res.*, 93, 3009-3024.

McGregor, B.A., E.S. Kappel, J.B. Wilson, F. Trincardi, R.J. Whittington, W. Mayer, and J. Campbell, Cruise Report, Hawaiian GLORIA Leg 7, F10-88-HW, *U.S. Geol. Surv. Open-File Rept. 91-333*, 52 p, 1991.

McGregor, B.A., and Lockwood, M. *Mapping and Research in the Exclusive Economic Zone*, USGS/NOAA Special Publication, 1985.

Menard, H.W., Deformation of the northeastern Pacific basin and the west coast of North America, *Geol. Soc. Am. Bull.*, *66*, 1149-1198, 1955.

Menard, H.W., Archipelagic aprons, *Am. Assoc. Pet. Geol. Bull.*, *40*, 2195-2210, 1956.

Menard, H.W., Correlation between length and offset on very large wrench faults, *J. Geophys. Res.*, *67*, 4096-4098, 1962.

Menard, H.W., Extension of northeastern Pacific fracture zones, *Science*, *155*, 72-74, 1967.

Menard, H.W., and T. Atwater, Changes in direction of sea floor spreading, *Nature*, *219*, 463-467, 1968.

Menard, H.W., and T. Atwater, Origin of fracture zone topography, *Nature*, *222*, 1037-1040, 1969.

Moore, J.G., D.A. Clague, R.T. Holcomb, P.W. Lipman, W.R. Normark, and M.E. Torresan, Prodigious submarine landslides on the Hawaiian Ridge, *J. Geophys. Res.*, *94*, 17465-17484, 1989.

Normark, W.R., R.T. Holcomb, R.C. Searle, M.L. Somers, and C.E. Gutmacher, Cruise report, Hawaiian GLORIA legs 3 and 4, F3-88-HW and F4-88-HW, *U.S. Geol. Surv. Open-File Rept. 89-213*, 56 p., 1989.

Normark, W.R., P.W. Lipman, J.B. Wilson, C.L. Jacobs, D.P. Johnson, and C.E. Gutmacher, Preliminary cruise report, Hawaiian GLORIA leg 2, F6-86-HW, November 1986, *U.S. Geol. Surv. Open-File Rept. 87-298*, 42 p., 1987

Rea, D.K., Changes in structure and trend of fracture zones north of the Hawaiian Ridge and relation to sea-floor spreading, *J. Geophys. Res.*, *75*, 1421-1430, 1970.

Sager, W.W., and M.S. Pringle, Tectonic evolution of the central Pacific during the Cretaceous quiet period, *EOS Trans. AGU*, *68*, 1451, 1987a.

Sager, W.W., and M.S. Pringle, Paleomagnetic constraints on the origin and evolution of the Musicians and South Hawaiian seamounts, central Pacific ocean, *in* Keating, B., P. Fryer, and G. Boehlert, eds., Seamounts, Islands and Atolls, *AGU Geophys. Mon. 43*, p. 133-162, 1987b.

Sager, W.W., and M.S. Pringle, Mid-Cretaceous to early Tertiary apparent polar wander path of the Pacific plate, *J. Geophys. Res.*, 93, 11753-11771, 1988.

Searle, R.C., Multiple, closely-spaced transform faults in fast slipping fracture zones, *Geology*, *11*, 607-610, 1983.

Searle, R.C., GLORIA investigations of oceanic fracture zones: comparative study of the transform fault zone, *J. Geol. Soc. London*, *143*, 743-756, 1986.

Searle, R.C., T.P. LeBas, N.C. Mitchell, M.L. Somers, L.M. Parson, and P. Pariat, GLORIA image processing: the state of the art, *Mar. Geophys. Res.*, 12, 21-39, 1990.

Smith, S.M., and H.W. Menard, The Molokai Fracture Zone, *Progress in Oceanography*, *3*, 333-345, 1965.

Torresan, M.E., A.N. Shor, J.B. Wilson, and J. Campbell, Cruise report, Hawaiian GLORIA leg 5 F5-88-HW, *U.S. Geol. Surv. Open-File Rept. 89-198*, 56 p., 1989.

Tucholke, B.E., and H. Schouten, Kane Fracture Zone, *Mar. Geophys. Res.*, *10*, 1-39, 1988/89.

U.S. Naval Oceanographic Office, *Bathymetric Atlas of the North Pacific Ocean*, U.S. Nav. Oceanog. Off. Publ. 1301-2-3, 172p, 1973.

The Nova-Canton Trough and the Late Cretaceous Evolution of the Central Pacific

DEVORAH JOSEPH, BRIAN TAYLOR, AND ALEXANDER N. SHOR

School of Ocean and Earth Science and Technology, University of Hawaii at Manoa, Honolulu, Hawaii, 96822

TOSHITSUGU YAMAZAKI

Geological Survey of Japan, Ibaraki, 305 Japan

Free-air gravity anomalies derived from satellite altimetry data show that the major Pacific fracture zones, from the Pau to Marquesas, are co-polar about an Euler pole located at 150.5°W, 34.6°S for the period preceding chron 33 and including a large portion of the Cretaceous Normal Superchron. They also show continuity of the Clipperton Fracture Zone through the Line Islands to the Nova-Canton ridge and trough; this Canton-Clipperton trend is co-polar to the same pole. Sidescan-sonar and bathymetry data in the Nova-Canton Trough region reveal N140°E-striking abyssal hill topography south of the N70°E-striking structures of the Nova-Canton Trough and crustal fabric striking normal to the trough (N160°E) to the north. We conclude that the Nova-Canton Trough is the Middle Cretaceous extension of the Clipperton Fracture Zone. We propose that the anomalous depths (7000-8400 m) of the trough between 167°30'-168°30'W are the result of a complex plate reorganization. Conjugate magnetic anomaly lineations M1-M3 in the Phoenix lineations between the Central Pacific Fracture Zone and the Phoenix Fracture Zone and the absence of lineations younger than anomaly M3 west of the Phoenix Fracture Zone suggest that spreading may have gradually ceased along the Pacific-Phoenix system from west to east. We infer that the remaining active segment of the Pacific-Phoenix spreading system after anomaly M1 time was the easternmost section of the Phoenix lineations. At ~M0 time, the Pacific-Phoenix spreading axis stretched from lineated bathymetric depressions lying between 180°W and the Phoenix Islands to ~168°W and included the western deep of the Nova-Canton Trough. We hypothesize that accretion terminated on the Pacific-Phoenix spreading axis shortly after M0 time and that the absence of an M0 isochron in the region between the eastern Phoenix lineations and the Nova-Canton Trough, or along the Nova-Canton Trough itself, may be due to a decrease in spreading rate prior to termination. We concur with previous hypotheses that portions of the Phoenix plate were trapped on the Pacific plate by a ridge jump south to the nascent Manihiki Plateau; some portions were overprinted by the Aptian volcanism that formed the Manihiki Plateau and Robbie Ridge. Pacific-Farallon spreading south of the Nova-Canton Trough jumped westwards, initiating transcurrent motion along the easternmost section of the failed ~M0 spreading axis (the western deep of the Nova-Canton Trough) which subsequently became the western end of the Clipperton (Pacific-Farallon) transform. In our reconstruction, the northeast and southeast margins of the Manihiki Plateau are rifted margins that form the western limit of Pacific-Farallon spreading between the Clipperton, Galapagos, and Marquesas fracture zones.

INTRODUCTION

Although great progress has been made over the past twenty years in reconstructing the tectonic evolution of the Pacific basin, plate movements during the Cretaceous Normal Superchron remain uncertain. The absence of magnetic reversals during this period in the evolution of the Central Pacific Basin has hampered attempts to explain the change in spreading configurations from those inferred from the N80°E-trending Mesozoic Phoenix magnetic anomaly lineations to spreading trends evidenced by the long, linear scars of the eastern Pacific fracture zones (Figures 1, 2). The magnetic bight between the Phoenix and Hawaiian lineations indicates a three-plate system in effect in the Central Pacific Basin prior to M0 time. Following the Cretaceous Normal Superchron, however, a two-plate configuration is observed in the region east of the Hawaiian, Line, and Marquesas seamount chains. The subsequent passage of the Central Pacific Basin over the south Pacific hotspots compounds the problem by obscuring original crustal fabric with a profusion of seamount chains and volcanic islands.

We corroborate this fracture zone interpretation with data from recent sidescan sonar [Joseph et al., 1992] and bathymetry [Tanahashi, 1992; Yamazaki, 1992] surveys of the Nova-Canton

The Mesozoic Pacific: Geology, Tectonics, and Volcanism
Geophysical Monograph 77

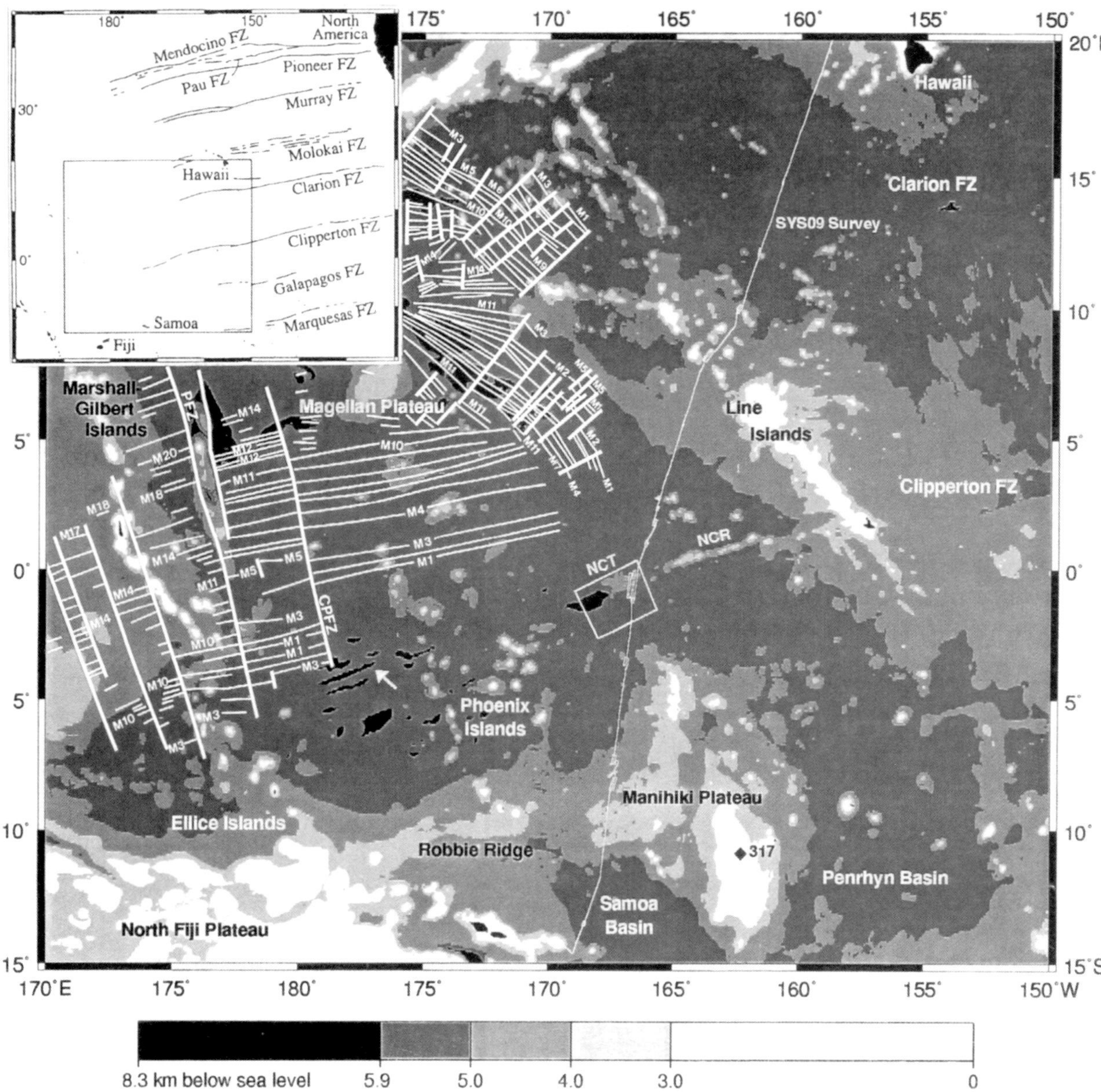

Fig. 1. Bathymetric map of the Central Pacific Basin showing the location of the major seafloor features. The diamond locates DSDP Site 317 [Scientific Staff, 1974]. The thick white lines show the Phoenix magnetic anomaly lineations [Nakanishi et al., 1992]; the thin white line indicates the route of the SYS09 survey during the first leg of the PacRim cruise. The white box surrounding the Nova-Canton Trough shows the region enlarged in the oblique projection in Fig. 3. NCT: Nova-Canton Trough; NCR: Nova-Canton Ridge; CPFZ: Central Pacific Fracture Zone; PFZ: Phoenix Fracture Zone. The white arrow shows the location of lineated bathymetric depressions discussed in the text. This gray-tone map was produced from ETOPO-5 data [National Geophysical Data Center, 1988] using the GMT System, Version 2.0 [Wessel and Smith, 1991]. The inset location map shows trends of major eastern Pacific fracture zones and their central Pacific extensions.

As in previous attempts to unravel the genesis of plate geometries in the central Pacific during the Cretaceous Normal Superchron, we reexamine the two major structures in that region, the Nova-Canton Trough and Manihiki Plateau (Figure 1), and their relationship to the Phoenix lineations. We use GEOSAT altimetry data to show continuity of the Clipperton Fracture Zone through the Line Islands volcanic chain and sediment apron to the N75°E-striking Nova-Canton Ridge and the N70°E-striking Nova-Canton Trough. We also show that the Nova-Canton Trough-Clipperton Fracture Zone trend is co-polar about a pole located at 150.5°W, 34.6°S with other coeval Pacific-Farallon fracture zone segments, from the Pau to Marquesas fracture zones.

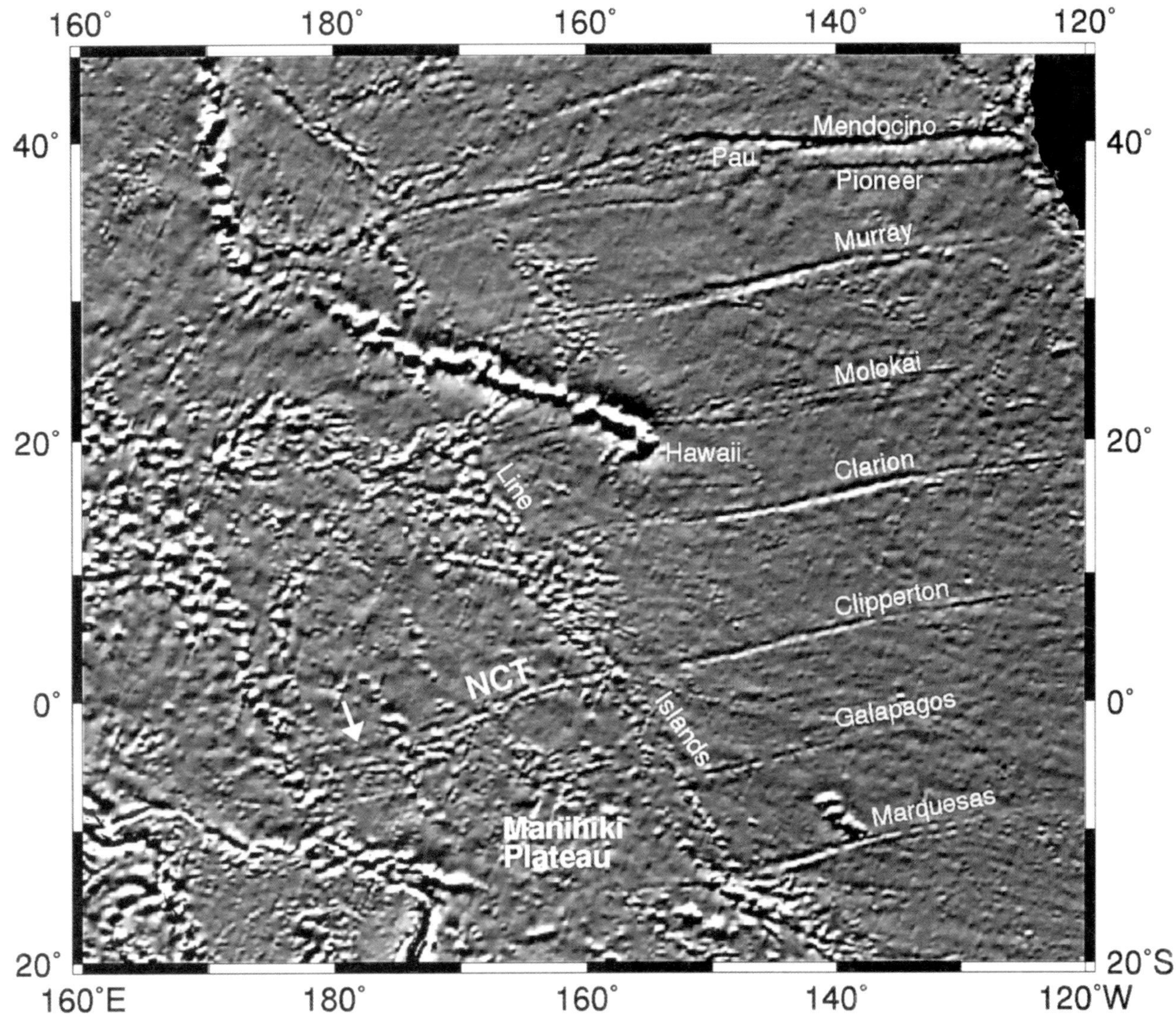

Fig. 2. Mercator projection of the slope of the free-air gravity field. The major features in the Pacific Ocean are located for ease in interpreting Figure 3. Note that at the change in trend of the Murray, Clarion, and Clipperton fracture zones, there are small "Y's" in the fracture zone trace. The white arrow indicates the location of the N80°E-striking lineated bathymetric depressions which we do not consider to have been formed by the same mechanism responsible for the formation of the Nova-Canton Trough. Compare the strike of these features in this Mercator projection to their strike when projected relative to our estimated pole of opening for the Middle Cretaceous. This image was produced from GEOSAT free-air gravity data.

Trough region. Both surveys reveal pronounced crustal fabric paralleling the trend of the northeast margin of the Manihiki Plateau (N140°E) and at a high angle to trough structures. The combined evidence supports the interpretation that the Nova-Canton Trough is the Middle Cretaceous extension of the Clipperton Fracture Zone and that the northeast edge of the Manihiki Plateau is a rifted margin. On this basis we propose a new model for the transition from the Mesozoic three-plate geometry to the two-plate configuration that spans the central and northeastern Pacific.

It is necessary at this point to address the confusion regarding the name and precise location of the feature we refer to as the Nova-Canton Trough to differentiate it from other structures in the general region to be examined in this paper. Engel and Engel [1970] first applied the name "Argo Trench" to a trough located at 3°S, 178°W with depths of 8400 m and slopes averaging 32°; current sea-floor charts do not show any such extreme relief at the location described by these authors. The designation "Canton Trough", which first appeared in Halunen et al. [1970], was applied to "a deep trough in excess of 8000 m located northeast of Canton Island at approximately 1°S, 168°W (striking) N68°E and (extending) for more than 200 km." This designation (also "Nova Trough") appears in the Gazetteer of Undersea Features [Defense Mapping Agency, 1981]; however, in this publication the feature

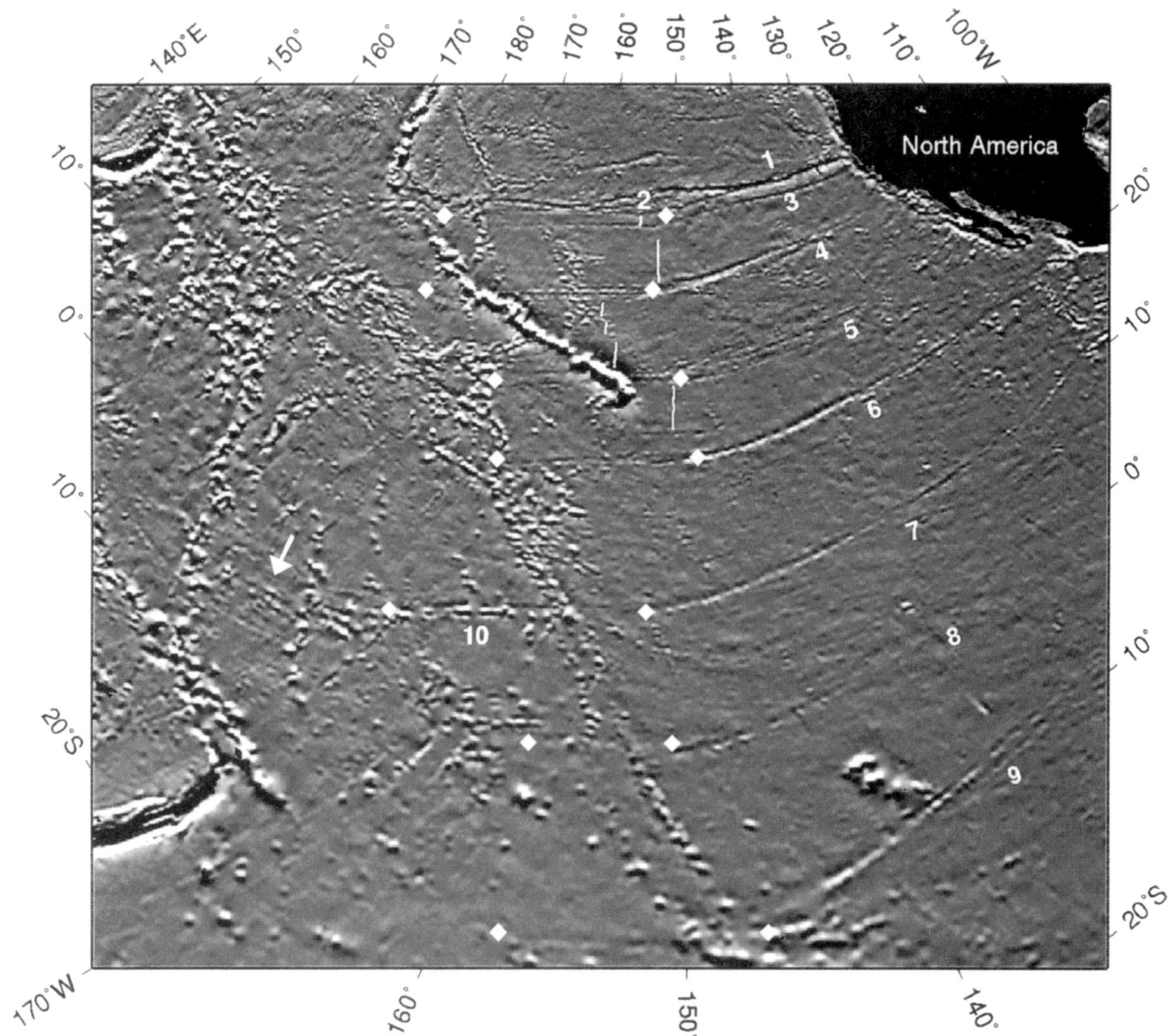

Fig. 3. Oblique Mercator projection of the slope of the free-air gravity field from GEOSAT altimetry data, using our best-fit Middle Cretaceous Pacific-Farallon pole (150.5°W, 34.6°S). The western fracture zone extensions (between the white diamonds) are generally parallel and horizontal, forming small circles around this pole. In this projection, the pole change following chron 34 (white vertical lines) is especially apparent. 1: Mendocino FZ; 2: Pau FZ; 3: Pioneer FZ; 4: Murray FZ; 5: Molokai FZ; 6: Clarion FZ; 7: Clipperton FZ; 8: Galapagos FZ; 9: Marquesas FZ; 10: Nova-Canton trough and ridge. The white arrow shows the location of the lineated bathymetric depressions inferred to have formed at the abandoned Pacific-Phoenix spreading axis at ~M0 time. Note the difference in the strike of these lineations, as compared to the strike of the Nova-Canton trough and ridge (between the white diamonds at "10") further to the east.

is located at 2°30'S, 173°W, a site in the middle of the Phoenix Island group. The Scripps maps of the Pacific [Mammerickx et al., 1971; Mammerickx and Smith, 1982] attach the name "Nova-Canton Trough" to a series of bathymetric lineaments approximately 5800 m deep, spanning over 5° north to south that lie in the region between the Phoenix Islands and 180°. These charts designate Halunen et al.'s "Canton Trough" the "Nova Trough" [Mammerickx et al., 1973]. However, in the 1985 edition of Bathymetry of the Northcentral Pacific [Mammerickx and Smith, 1985] the "Nova-Canton Trough" consists of the "Canton Trough," as defined by Halunen et al. [1970] and the region between the Phoenix Islands and 180°. We prefer to retain the designation "Nova-Canton Trough" when referring to the deep trough system extending east from approximately 168°30'W because this usage has become common. We do not consider, however, the lineated depressions lying in the region between the Phoenix Islands and 180°W (Figure 1) to have been formed by the same mechanism responsible for the extreme relief found in the region east of 168°30' W, and will refer to these bathymetric features by their location. These lineaments are designated on Figures 1, 2, and 3 by a white arrow.

PREVIOUS WORK

Menard [1967] first reported the ridge-trough topography present in the generally flat equatorial Central Pacific Basin and postulated that these structures are the western extension of the Clipperton Fracture Zone. He further noted that several of the major Pacific fracture zones, (from the Mendocino Fracture Zone in the north Pacific to the Marquesas Fracture Zone near 10°S) extend into the central Pacific beyond the mid-plate swells created by the Hawaiian, Line, and Marquesas seamount chains (Figures 1, 2) and branch west of these island groups. Rea [1970] suggested that rather than branching, the fracture zones change trend by 10° to 15° throughout a 125 km to 150 km wide strip extending in a great circle from ~156°W, 21°N to ~174°W, 43.5°N, which he named the "bending line." Noting the concave southward expression of the Pioneer, Murray, Molokai, and Clarion fracture zones (Figure 2), Sager and Pringle [1987a] proposed an Euler pole located at 155°W, 11°S to describe motion on the Pacific-Farallon paleo-spreading axis responsible for these fracture zone traces.

In the 1960's, dredges of the scarps of Engel and Engel's [1970] "Argo Trench" returned relatively fresh tholeiitic and alkali basalts, as well as gabbros altered in part to greenschist facies. The 1970 R/V *Mahi* survey of the Canton Trough [Halunen et al., 1970; Rosendahl and Halunen, 1971] revealed ridge-trough topography with depths locally exceeding 8 km, which are the greatest depths in the Pacific not associated with subduction zones (Plate 1). Asymmetric ridges flanking the main trough rise to 2.8 km below sea level in some locations, but ridge-to-trough relief is generally ~3 km, with slopes averaging 20°. Rosendahl et al. [1975] and Rosendahl [1972] explained the extreme relief, thin (3-4 km) crust beneath the axial trough, and elongate magnetic anomaly highs as indicating the presence of ENE-striking linear intrusive bodies in an extensional environment. He concluded that the ridges and troughs were created by slow spreading within the Nova-Canton transform fault during a slight reorientation of the Pacific-Farallon spreading axis, or volcanism and faulting during a later period of rejuvenation. He argued that the Nova-Canton Trough mainly exhibits structures suggestive of fracture zone morphology, but the "occurrence of relatively fresh intrusive rocks and some instances of faulted sediments" [Rosendahl, 1972] suggest that recent rejuvenation (volcanism and faulting) may have occurred.

The entire central Pacific, from the Nauru Basin in the west to the Line Islands in the east, is hypothesized by many researchers [e. g., Menard, 1964; Crough, 1978; Schlanger et al., 1981] to have experienced a thermo-volcanic event in the Late Jurassic which resulted in thermal rejuvenation and uplift of the entire region (Menard's "Darwin Rise"). More recently, Larson [1991] proposed a "superplume" episode beginning at ~125 Ma in the mid-Cretaceous south Pacific that so perturbed the thermal structure at the core-mantle boundary that the earth's magnetic field ceased reversing for over 40 m.y. The Central Pacific Basin formed ~4000 km southeast of where it presently lies. The gradual northwest movement of the Pacific plate during the last 125 m.y. would have necessitated movement of the Central Pacific Basin over the Darwin Rise (or "superplume") at some time in its history, which may lend credence to Rosendahl's [1972] theory of thermal rejuvenation of the Nova-Canton Trough region. Indeed, ocean floor depths in the Central Pacific Basin are less than depths predicted by the Parsons-Sclater [1977] seafloor subsidence curve.

Northwest of the Nova-Canton Trough, Larson et al. [1972] identified the Phoenix magnetic lineations (Figure 1), which record Early Cretaceous spreading between the Pacific plate to the north and Phoenix plate to the south. Larson and Chase's [1972] tectonic reconstruction of the Central Pacific Basin interpreted the Nova-Canton Trough as a Pacific-Farallon transform fault at ~100 Ma, north of a Pacific-Phoenix-Farallon triple junction.

The identification of the Phoenix lineations by Larson et al. [1972] led Winterer et al. [1974] to hypothesize that the Nova-Canton Trough originated as "a deep slot associated with a spreading direction change," and later Winterer [1976a, b] postulated a spreading center origin for the Nova-Canton Trough. In Winterer's reconstruction of Early Cretaceous Pacific basin tectonics, the Nova-Canton Trough forms the western axis of a ridge-ridge-ridge triple junction between the Pacific, Phoenix, and Farallon plates at about anomaly M0 time. Winterer et al. [1974] considered the trough to be an extension of ~N80°E-striking linear bathymetric depressions lying between 180°W and the Phoenix Islands (Figures 1 and 2, arrow). They proposed that an abrupt change in spreading geometries preserved the extreme relief found in the trough when the triple junction jumped south to the vicinity of the Manihiki Plateau.

Magnetic anomaly mapping by Nakanishi et al. [1992] of Early Cretaceous lineations in the western Pacific identified conjugate anomalies M1 through M3 in the Phoenix lineation set between the Central Pacific and Phoenix fracture zones (Figure 1). These conjugate isochrons imply a cessation of spreading on the western portion of the Pacific-Phoenix spreading center just after M1 time. No conjugate anomaly lineations have been identified in the region between the eastern Phoenix lineations and the Nova-Canton Trough [Cande et al., 1989; Nakanishi et al., 1992; Yamazaki and Tanahashi, 1992]. In addition, Nakanishi et al. [1992] have identified NNW-trending M-series anomalies in the region between 170°W and the Line Islands (Figure 1). The presence of this magnetic bight confirms the existence of a triple junction between the Pacific, Phoenix, and Farallon plates prior to M0 time.

Southeast of the Nova-Canton Trough lies the Manihiki Plateau, a large, heavily sedimented region of anomalously shallow, thick oceanic crust [Winterer et al., 1974; Hussong et al., 1979] (Figure 1). Deep troughs, hypothesized to be of fault origin, divide the plateau into three distinct morphological provinces: the high, western, and north plateaus [Winterer et al., 1974]. The trends defined by these structural boundaries led Winterer et al. [1974] to posit a triple junction in the northern Manihiki Plateau near the junction between the high, western, and north plateaus. The high and western plateaus are composed of narrow ridges and

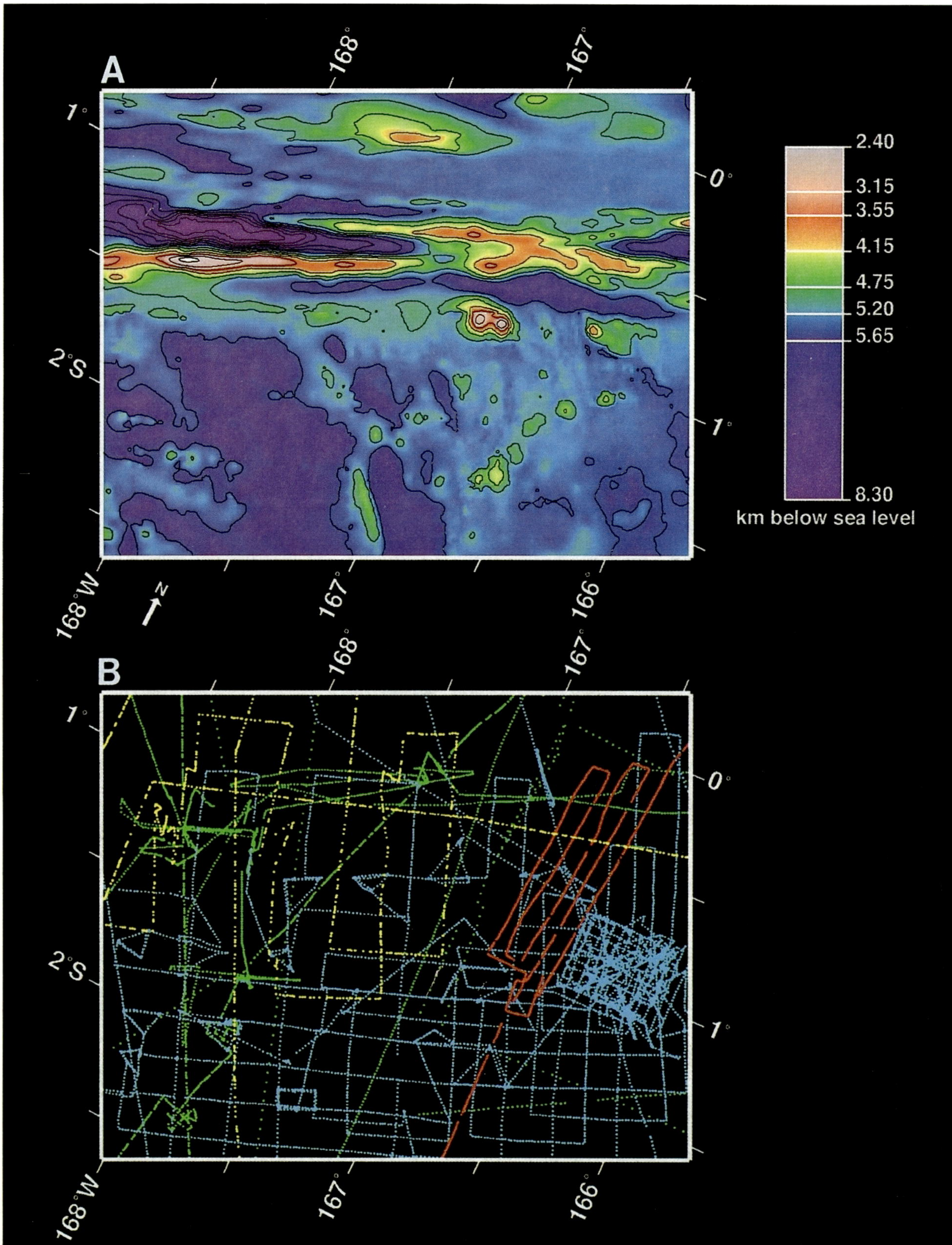
A
168°
167°
1°
0°
2°S
1°
168°W
167°
166°
N
2.40
3.15
3.55
4.15
4.75
5.20
5.65
8.30
km below sea level
B
168°
167°
1°
0°
2°S
1°
168°W
167°
166°

sediment-filled troughs forming a sub-orthogonal pattern along NNE- and NW-trends, interpreted by Winterer et al. [1974] as offset ridge-axis segments. The northeast margin of the north plateau is bounded by a region of N140°E-trending tilted fault blocks, similar to the topography bounding the eastern edge of the high plateau [Sharman and Mammerickx, 1990; Hill et al., 1991].

SATELLITE ALTIMETRY DATA

Due to the paucity of high density bathymetric information, we examined satellite altimetry data to determine if a genetic relationship exists between the Nova-Canton Trough and the Clipperton Fracture Zone. The slope of the free-air gravity field derived from the altimetry data (Figure 2) shows a lineament stretching from the Clipperton Fracture Zone through the Line Islands volcanic edifice and sediment apron to the Nova-Canton Ridge, then west to join with the Nova-Canton Trough. We interpret this lineament to indicate structural continuity between the Clipperton Fracture Zone and Nova-Canton Trough. There is a slight break in the trace immediately east of the Line Islands. We interpret this discontinuity as a response to the change in the pole of rotation of the Pacific-Farallon spreading axis, which occurred shortly after chron 34 time [Sager and Pringle, 1987a, b; Atwater, 1989; Searle et al., this volume]. The Clarion and Murray fracture zones also exhibit discontinuities of this type, in the form of small overlaps or "Y's" at the point where the pole of rotation changed and the concave southward fracture zone traces change trend (Figure 2).

The Nova-Canton Trough is not atypical in extending to almost 169°W. As Menard [1967] and Rea [1970] pointed out, several Pacific fracture zones extend into the central Pacific, well beyond the mid-plate swells created by the Hawaiian, Line, and Marquesas island chains. GLORIA data [Searle et al., this volume] record two strands of the Molokai Fracture Zone extending west of the Hawaiian Islands. Free-air gravity data show the Clarion Fracture Zone extending to 165°W, where it is overprinted by the Line and Crosstrend seamounts (Figure 2). The Murray Fracture Zone extends westward beyond the Hawaiian-Emperor chain, and the Marquesas and Galapagos fracture zones extend west of the Marquesas Islands (Figure 2). By using the change in relative plate motion after chron 34 time [Sager and Pringle, 1987a, b; Atwater, 1989] and the accompanying change in strike of the fracture zone traces as a reference point (Figure 3), it is evident that the fracture zones extend a similar angular distance west of the bend. In addition, these fracture zones trace out the same broad, concave-southward arc [Sager and Pringle, 1987a, b] as the lineament connecting the Nova-Canton Trough to the Clipperton Fracture Zone (Figures 1, 2). Given this similarity in shape and extent, it is unlikely that the Nova-Canton Trough, lying between these fracture zones, should have formed by a different mechanism.

If the Nova-Canton Trough and Nova-Canton Ridge are indeed the western extension of the Clipperton Fracture Zone, these structures must be co-polar about the relevant stage pole of Pacific-Farallon motion with other coeval Pacific fracture zones. By analyzing the trends of western fracture zone extensions from the Pau to the Marquesas fracture zones (Figures 1, 2) and determining the intersection of great circles perpendicular to these fracture zones, we have derived a congruent pole at 150.5°W, 34.6°S of Pacific-Farallon opening for the period prior to and including chron 34. This pole of opening is slightly better than our original estimate (150°W, 32°S) [Joseph et al., 1992], around which we searched for the minimum error in the least squares sense using the estimates of fracture zone azimuths given in Table 1. The method of Gordon and Cox [1980] determines a 95% confidence ellipse for this pole by the linear propagation of the standard error of the fracture zone azimuths, which we assume to be 2°. The fracture zone extensions used in the analysis lie on crust created during the Cretaceous Normal Superchron. Therefore, the constraining influence of spreading velocities cannot be applied to the analysis and a precise pole of opening is difficult to determine [Francheteau et al., 1970]; however, by projecting the free-air gravity field about this pole (Figure 3) we show that the western fracture zone extensions do, indeed, form small circles such that in the oblique projection the fracture zone traces are horizontal and parallel. The 95% confidence ellipse is shown in Figure 4. As expected, the ellipse is elongated in a direction orthogonal to the fracture zone trends. Several shorter duration stage poles may better fit these fracture zone traces, but this pole works very well and serves to corroborate Menard's hypothesis regarding the relationship between the Nova-Canton Trough and Clipperton Fracture Zone.

SIDESCAN SONAR AND BATHYMETRY DATA

The Seafloor Surveys International/OTC Limited PacRim cable route cruise in January 1990 mapped 6700 km^2 in the eastern

Plate 1. (a) Bathymetry of the Nova-Canton Trough region projected about a pole located at 150.5°W, 34.6°S produced from 3.5 and 12 kHz echo sounder data, and SYS09 swath bathymetry (see Figure 1 for location). Note the N140°E-striking spreading fabric extending across the region south of the Nova-Canton Trough. Data sources include the National Geophysical Data Center as well as the Geological Survey of Japan (cruises GH80-1, GH81-4, and GH82-4). The data was surfaced and gridded using the GMT System, Version 2.0 [Wessel and Smith, 1991]. Contour lines are drawn at 500 meter depth intervals. (b) The cruises included in the gridding process used to produce the bathymetric map in (a), viewed in the same projection. The red points indicate PacRim ship tracks (R/V *Moana Wave*); blue lines are Geological Survey of Japan cruises conducted aboard the R/V *Hakurei-maru*; yellow lines are R/V *Mahi* cruises; green lines are R/V *Conrad*, the Scripps' Eurydice Expedition, R/V *Kana Keoki*, and R/V *Dmitri Mendeleyev* cruises. The circles indicate individual data points. In the gridding process, the PacRim cruise (swath bathymetry) was weighted at 1, the R/V *Hakurei-maru* (GSJ) cruises at 2, the R/V *KanaKeoki* at 3, the R/V *Dmitri Mendeleyev* and Eurydice Expedition at 4, and the R/V *Conrad* and R/V *Mahi* cruises at 5.

TABLE 1. Observation Site Locations and Model Results Used in Euler Pole (150.5°W, 34.6°S) Calculations for Pacific-Farallon Opening During the Cretaceous Normal Superchron.

[1]Site	Site Latitude (°N)	Site Longitude (°E)	Obs. Trend (°)	[2]Calc.. Trend (°)	Difference (°)	Weight
Galapagos	-5.5	-160.0	75	74.4	0.6	0.16
Clipperton	1.50	-161.3	71	75.3	-4.3	0.10
Clipperton	2.1	-158.8	75	78.8	-3.8	0.07
Clipperton	3.0	-154.4	84	84.8	-0.8	0.10
Clipperton	-1.2	-167.2	69	66.8	2.2	0.36
Clipperton	0.7	-163.7	72.5	72.0	0.5	0.17
Clipperton	3.1	-151.2	90	89.1	0.9	0.09
Clarion	14.5	-149.5	93	91.1	1.9	0.17
Clarion	14.4	-152.7	85	87.6	-2.6	0.10
Clarion	13.2	-159.7	75	79.9	-4.9	0.04
Molokai	20.6	-163.0	77	77.7	-0.7	0.04
Molokai	19.5	-162.5	79	78.0	1.0	0.04
Molokai	21.1	-152.0	90	88.5	1.5	0.10
Molokai	21.5	-152.8	87	87.7	-0.7	0.09
Murray	28.6	-163.9	83	77.8	5.2	0.03
Murray	28.0	-164.4	79	77.3	1.7	0.03
Murray	27.3	-165.3	79	76.4	2.6	0.03
Murray	28.4	-158.0	86	83.1	2.9	0.04
Murray	29.3	-159.0	84	82.2	1.8	0.04
Pioneer	35.2	-164.2	75	78.1	-3.1	0.03
Pioneer	35.7	-161.2	80	80.7	-0.7	0.03
Pioneer	36.1	-157.0	85	84.3	0.7	0.05
Pau	36.4	-165.5	77	77.1	-0.1	0.03
Pau	36.6	-163.7	79	78.6	0.4	0.03
Pau	37.6	-158.4	84	83.2	0.8	0.04
Pau	36.7	-165.4	78	77.2	0.8	0.03

[1] Sources of fracture zone trend observations as follows: Galapagos FZ, [Mammerickx, unpublished map]; Clipperton and Clarion FZs, [Mammerickx and Smith, 1982; Mammerickx and Smith, 1985]; Molokai FZ, [Searle et al., this volume]; Murray, Pioneer, and Pau FZs [Mammerickx, 1989a, b; Atwater and Severinghaus, 1989a, b].

[2] Standard deviation of 2° used in calculations.

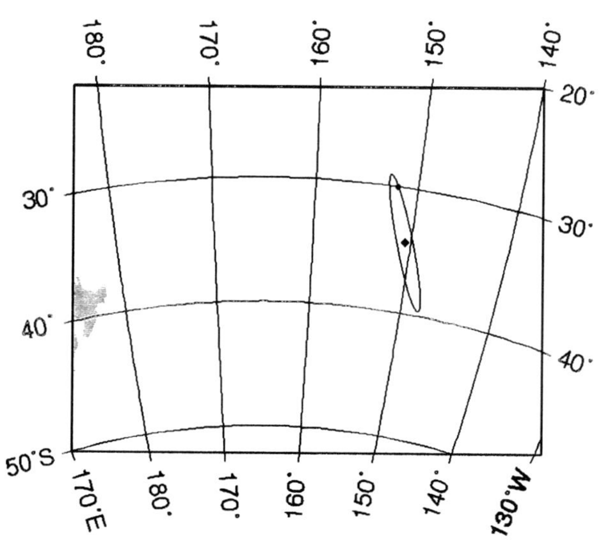

Fig. 4. The 95% confidence region for an Euler pole located at 150.5°W, 34.6°S (diamond) for Pacific-Farallon opening during the Cretaceous Normal Superchron. The long axis of the ellipse is oriented NNW, being unconstrained by spreading velocity information. The circle locates our initial Euler pole position [Joseph et al., 1992] at 152°W, 30°S. This confidence ellipse was calculated by the method of Gordon and Cox [1980] using the 26 observations given in Table 1, with a standard deviation of 2°.

region of the Nova-Canton Trough between latitude 0° and 1°15'S, longitude 166°22' to 166°54'W. Data collected during the survey include five parallel lines of SYS09 sidescan sonar imagery and bathymetry [Blackinton et al., 1991] (Figure 5, Plate 2), single-channel seismic reflection profiles (60-200 kHz, 40 in^3 airgun), 3.5 kHz echo-sounder, and gravity data.

The backscatter data show thick accumulations of sediments, mainly turbidites [Kroenke, 1970; Orwig, 1981], limited to the northern plain (Figure 5). The bathymetric map (Plate 2) shows a trough more than 6 km deep on the eastern edge of the survey area, similar to the deep trough surveyed by the R/V *Mahi* farther to the west (Plate 1) [Rosendahl, 1972; Rosendahl et al., 1975]. A series of N60°E-N80°E-striking ridges separate this trough from a fault-bounded basin to the south. Seismic reflection profiles show turbidites covering a chain of small seamounts in the northern part of the survey area that parallel the strike of the ridges. Thin accumulations of pelagic sediments drape lower-lying parts of the region south of the fault-bounded basin. Despite the different character of the sediments north and south of the ridge-trough structures, analysis of bathymetric data from the PacRim survey did not show a clear offset of basement depths across the Nova-Canton Trough even when sediment loading was taken into account.

The strike of abyssal hill fabric may be used in the absence of magnetic anomaly lineations to define the strike of a paleo-

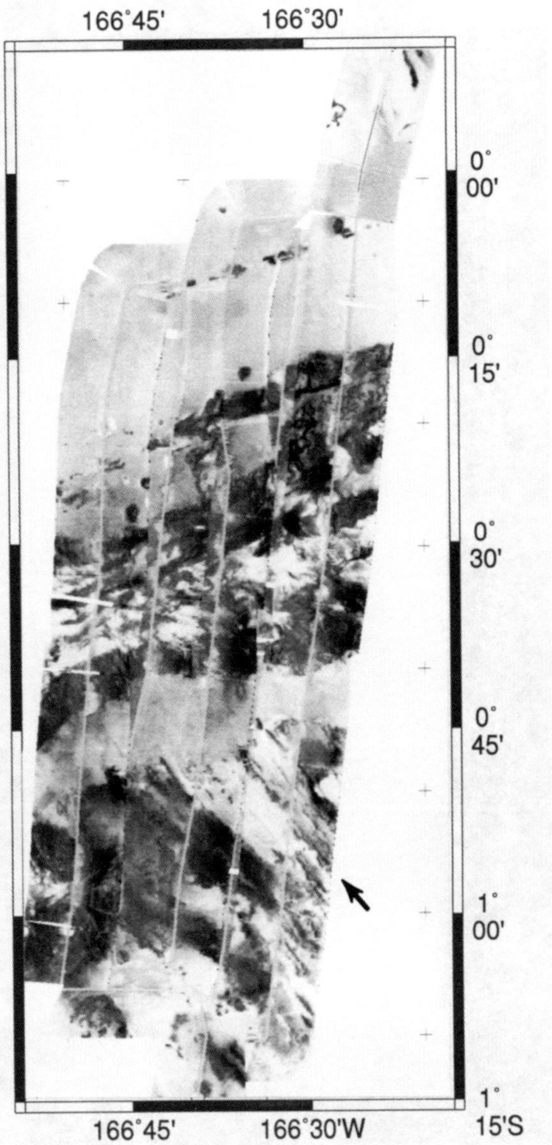

Fig. 5. SYS09 sidescan-sonar mosaic of the Seafloor Surveys International/OTC Limited PacRim Nova-Canton Trough survey (swath width ~14 km). The SYS09 sidescan data were reprocessed at the Hawaii Institute of Geophysics using SeaMARC II sidescan enhancement programs to refine the fabric information contained in the reflectivity data [Sender et al., 1989]. Gains are first normalized to a common level, then a global contrast map is generated to ensure that gray scales are uniform throughout the survey. The arrow indicates the location of the N140°E-striking abyssal hill fabric.

spreading axis. Abyssal hill topography striking N140°E ± 5° lies on a broad arch in the southern part of the surveyed area. Seismic data show that this crustal fabric is composed of normal-faulted blocks rising ~100-250 m above the surrounding sea-floor. SYS09 sidescan data indicate that the N140°E-striking crustal fabric extends to at least 2°S (Figure 6). The combined bathymetry from cruise data available through the National Geophysical Data Center and a survey conducted by the Geological Survey of Japan during a 1982 cruise of the R/V *Hakurei-maru* [Tanahashi, 1992; Yamazaki, 1992; Yamazaki and Tanahashi, 1992] (Plate 1) indicate a pervasive N140°E trend south of the Nova-Canton Trough as far as 167°30'W. This bathymetric fabric strikes at a high angle to the overall N70°E strike of the Nova-Canton Trough and parallels the fault structures of the northeast margin of the Manihiki Plateau (Figure 1). West of an elongate ridge at 167°10'W, 2°S (Plate 1), a wide turbidite-filled basin [Tanahashi, 1992] spans the remainder of the region south of the Nova-Canton Trough.

North of the Nova-Canton Trough, thick accumulations of turbidites cover the sea-floor. Crustal fabric striking ~N160°E, orthogonal to the N70°E-striking Nova-Canton Trough, can be detected on sidescan sonar images, though muted by sedimentary cover (Figure 7). At the extreme northern edge of the SYS09 survey area (Figure 5, Plate 2) a normal fault striking N162°E steps into the northern plain. This suggests that immediately north of the ridge-trough system, spreading was oriented orthogonal to the Nova-Canton transform fault.

TECTONIC RECONSTRUCTION OF THE CENTRAL PACIFIC BASIN

The Pacific-Phoenix-Farallon magnetic bight (Figure 1) indicates that the plate configuration in the Central Pacific Basin at M1 time was characterized by a ridge-ridge-ridge triple junction between the Pacific, Phoenix and Farallon plates (Figure 8a) [Larson and Chase, 1972; Winterer et al., 1974; Winterer, 1976a, b]. The southernmost identified magnetic anomaly lineations young from the Nauru Basin (M11) [Larson, 1976; Nakanishi et al., 1992] eastward to the spreading segment between the Central Pacific and Phoenix fracture zones, where spreading ceased just after M1 time, as evidenced by conjugate anomalies M1 through M3 [Nakanishi et al., 1992] (Figure 1). It is possible that spreading ceased on all of these western Phoenix-Pacific spreading axes by M1 time, although the absence of isochrons as young as M1 on the most westerly spreading segments (Figure 1) suggests to us that spreading ceased progressively eastwards. We concur with Engebretson et al. [1991] that accretion continued on the easternmost segment of the Pacific-Phoenix spreading axis until about M0 time. A spreading center either abandoned shortly before M0 time or decreasing in spreading rate to zero just after M0 time would account for the absence of an M0 magnetic anomaly in this region [Cande et al., 1989; Nakanishi et al., 1992; Yamazaki and Tanahashi, 1992]. We postulate that as the triple junction moved southeastward, the Pacific-Phoenix spreading axis lengthened to include the westernmost region of the Nova-Canton Trough (the region of >7 km depths). The ~N80°E-striking linear bathymetric depressions lying between the Phoenix Islands and 180° (Figure 1, arrow), which also have some expression in the gravity data (Figures 2 and 3, arrow), may be associated with the abandoned Pacific-Phoenix spreading axis along this segment further west. When projected about our best-fit Euler pole at 150.5°W, 34.6°S (compare Figure 2 to Figure 3), it is obvious that these depressions, if they are indeed related to spreading, were not formed about the same pole as the Nova-Canton Trough or the western fracture zone extensions of the entire central Pacific. A

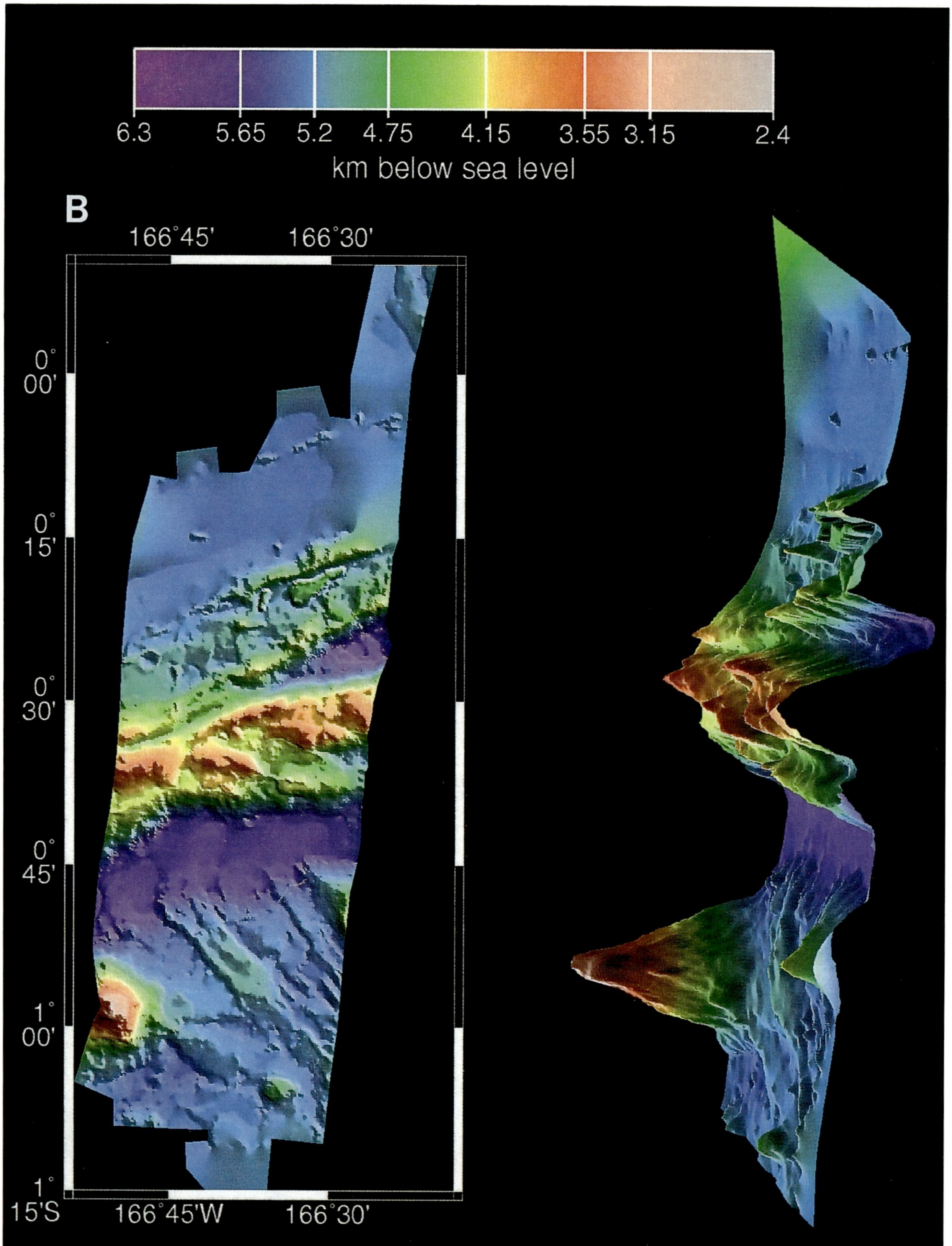

Plate 2. SYS09 bathymetry of the Seafloor Surveys International/OTC Limited PacRim Nova-Canton Trough survey. The original shipboard bathymetry mosaic was hand-contoured by D. M. Hussong at 50 m intervals and digitized by the Seafloor Surveys International staff. The swath width is 3.4 times tow-fish altitude; bathymetry is accurate to 2% of tow-fish altitude. The image shows a plan view of SYS09 bathymetry on the left, and a 3-D perspective image on the right, viewed from an azimuth of N70°E and elevation of 20°.

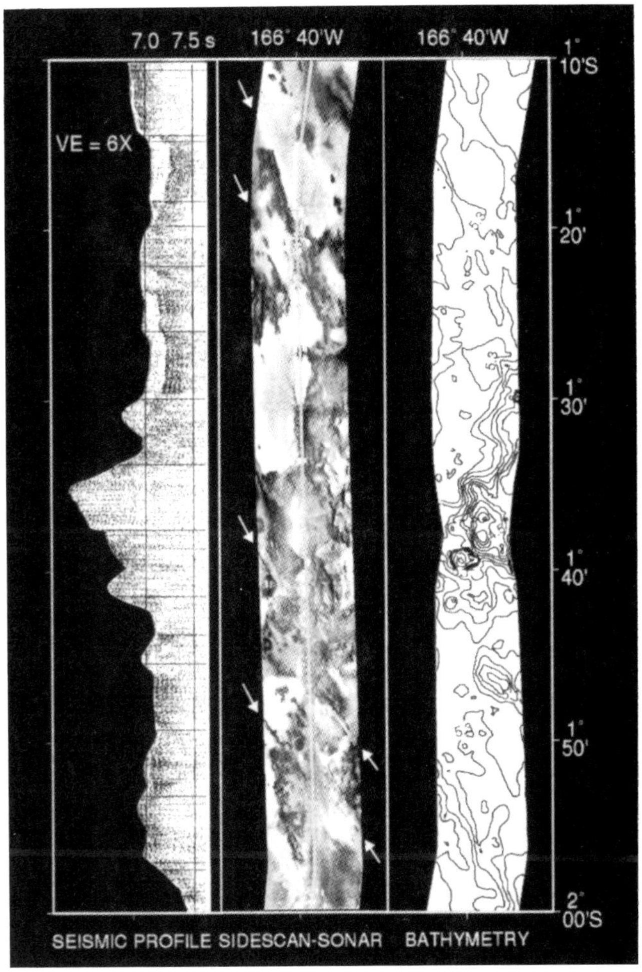

Fig. 6. Composite image of single-channel seismic reflection profiles, SYS09 sidescan-sonar and bathymetry of the region immediately south of the SYS09 Nova-Canton Trough survey. The bathymetry is contoured at 100 m intervals, white arrows indicate the location of the N140°E-striking spreading fabric, and the seismic profiles show this fabric in profile. The bathymetry is surfaced and gridded from SSI digitized contours using the GMT System, Version 2.0 [Wessel and Smith, 1991].

chain of small seamounts stretches symmetrically away from the inferred remnant spreading axis (Figure 8a).

We propose that formation of the Manihiki Plateau coincided with or immediately preceded cessation of spreading on the easternmost Pacific-Phoenix spreading axis. This event may even have initiated the shutdown of the M-sequence spreading system in the Central Pacific Basin. An ^{40}Ar-^{39}Ar incremental heating plateau age of basalt recovered at Site 317 (Figure 1) on the Manihiki Plateau indicates that it formed at 123 ± 1.5 Ma [Mahoney et al., this volume, citing R. A. Duncan], immediately prior to M0 time. On the basis of Ontong Java Plateau basalts and overlying microfossils [Kroenke, Berger et al., 1991], Mahoney et al. [this volume] date M0 at 122.4 ± 0.8 Ma. The sharp relief bordering the southeast and northeast margins of the Manihiki Plateau and the orthogonal trends defined by ridge-trough structures within the Plateau suggest a more complex origin and subsequent evolution than that generated by a surfacing mantle plume head. This hypothesis has been applied to other oceanic plateaus and continental flood basalt provinces [Richards et al., 1989; Duncan and Richards, 1990; Duncan and Richards, 1991; Mahoney and Spencer, 1991; Tarduno et al., 1991]. We suggest that upon the shutdown of the ~M0 Pacific-Phoenix spreading ridge, the triple junction jumped south to the nascent Manihiki Plateau. The tilted fault-block morphology of the northeast margin of the Manihiki Plateau [e. g., Hill et al., 1991] strikes N140°E. The similarity in strike of this boundary to the abyssal hill fabric south of the Nova-Canton Trough suggests a related origin [Tanahashi, 1992]. We postulate that the Pacific-Farallon spreading axis south of the Clipperton (Nova-Canton) transform jumped westwards, rifting the northeast margin of the Plateau and forming the oceanic crust south of the Nova-Canton Trough (Figure 8b). This rift extended into the abandoned Pacific-Phoenix spreading axis, converting the eastern end into a transform fault, and creating the anomalous depths observed in the western deep of the Nova-Canton Trough. We have insufficient bathymetric data from the PacRim and R/V *Hakurei-maru* cruises to determine the eastern extent of the oblique spreading fabric but assume that the ridge reoriented later to produce spreading fabric orthogonal to the Clipperton Fracture Zone.

The recent identification of NNE-trending conjugate M22 through M29 magnetic anomaly lineations in the Samoan Basin south of the Manihiki Plateau [Engebretson et al., 1991] suggests that a large portion of the Phoenix plate was trapped on the Pacific plate when the spreading center jumped south (Figure 8b). The formation of the Manihiki Plateau and Robbie Ridge probably overprinted many of the younger conjugate Phoenix magnetic lineations.

The mode of formation of the Manihiki Plateau is a problem that remains unsolved, but is presently receiving a great deal of scrutiny together with other oceanic plateaus. Mahoney [1987] and Mahoney and Spencer [1991] find that although Pb, Nd, and Sr isotopic ratios are within the range of oceanic island basalts, samples from DSDP Site 317 (Figure 1) [Jackson et al., 1976] also possess mid-ocean ridge basalt affinities. This suggests that the basaltic flood event responsible for the formation of the Manihiki Plateau may have been generated by the initiation of a mantle plume in a near-ridge crest setting [Mahoney and Spencer, 1991]. Various theories advanced to explain the origin of the Shatsky Rise in the northwest Pacific also invoke the interaction of triple junction movements [Sager et al., 1988; Sharman and Risch, 1988] and hotspots [Nakanishi et al., 1989]. Indeed, the confluence of mantle plume initiation, ridge-crest processes, and concomitant rifting seems necessary to explain both the thickened oceanic crust and internal, as well as external, structural expression of the Manihiki Plateau. We cannot confidently suggest any tectonic configurations for the intervening period between the shutdown of the Pacific-Phoenix spreading system and the initiation of the Pacific-Antarctic-Farallon triple junction to the south [Winterer et al., 1974]. However, the Pacific-Phoenix-Farallon triple junction may have jumped south in discrete steps through the Manihiki Plateau, trapping more

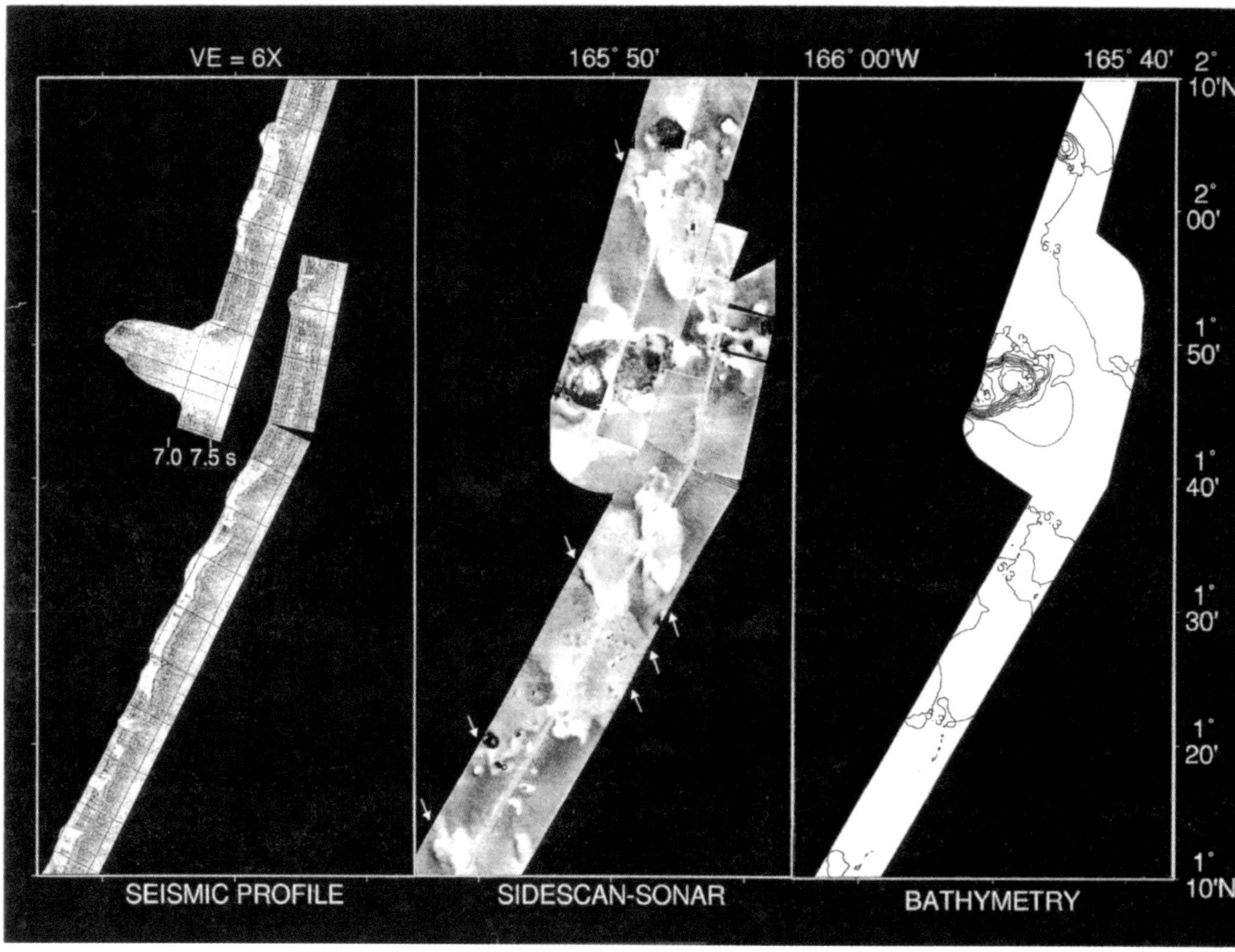

Fig. 7. Composite image of single-channel seismic reflection profiles, SYS09 sidescan-sonar and bathymetry of the region between 1°10'N to 2°10'N. The bathymetry is contoured at 100 m intervals. The original crustal fabric is heavily obscured by turbidites, however, abyssal hill fabric striking ~N160E° (white arrows) can be detected, as shown in the seismic profile and reflectivity image. The bathymetry is surfaced and gridded from SSI digitized contours using the GMT System, Version 2.0 [Wessel and Smith, 1991].

portions of the Phoenix plate [Winterer et al., 1974, Figures 13 and 14], while the Galapagos and Marquesas fracture zones organized in the wake of the southward movement of the Pacific-Farallon spreading ridge. A "Phoenix microplate" between the Pacific and Antarctic plates may be a viable alternative [see also Engebretson et al., 1991], but there is not enough information to constrain its motion.

We considered the possibility that the N140°E-striking crustal fabric south of the Nova-Canton Trough was the result of Phoenix-Farallon spreading south of a ridge-ridge-transform (R-R-F) triple junction (Figure 8a). This could explain its sub-normal trend with respect to the Nova-Canton transform, while keeping the Nova-Canton trend co-polar to the Pacific-Farallon pole. A R-R-F triple junction with the transform between the Pacific and Farallon plates is not stable, given the known spreading rates. Furthermore, the Pacific-Farallon fabric trend is not compatible with the ESE Pacific-Farallon spreading direction predicted by the vector diagram of such a triple junction (Figure 8a). It is oblique to both Pacific-Farallon and Phoenix-Farallon expected spreading fabric. We prefer the R-R-R triple junction configuration, given the similar strike of the seafloor fabric and the rifted northeast Manihiki Plateau margin and that the Plateau formed about when Phoenix-Farallon spreading ceased. In addition, we cannot discount the possibility that the Nova-Canton Trough formed on the Phoenix plate and was later transferred to the Pacific plate by a ridge jump. However, this interpretation would not explain the co-polarity of the Nova-Canton trough and ridge to the entire Cretaceous Normal Superchron system of fracture zone trends formed by Pacific-Farallon spreading (Figure 3).

North of the Nova-Canton Trough, many small ridge segments offset by fracture zones define the Early Cretaceous Pacific-Farallon plate boundary (Figure 1). The profusion of small spreading segments probably resulted from the reorganization of this plate boundary following the transfer of the Magellan microplate [Tamaki and Larson, 1988] to the Pacific plate. We suggest that by the Late Cretaceous these had been transformed

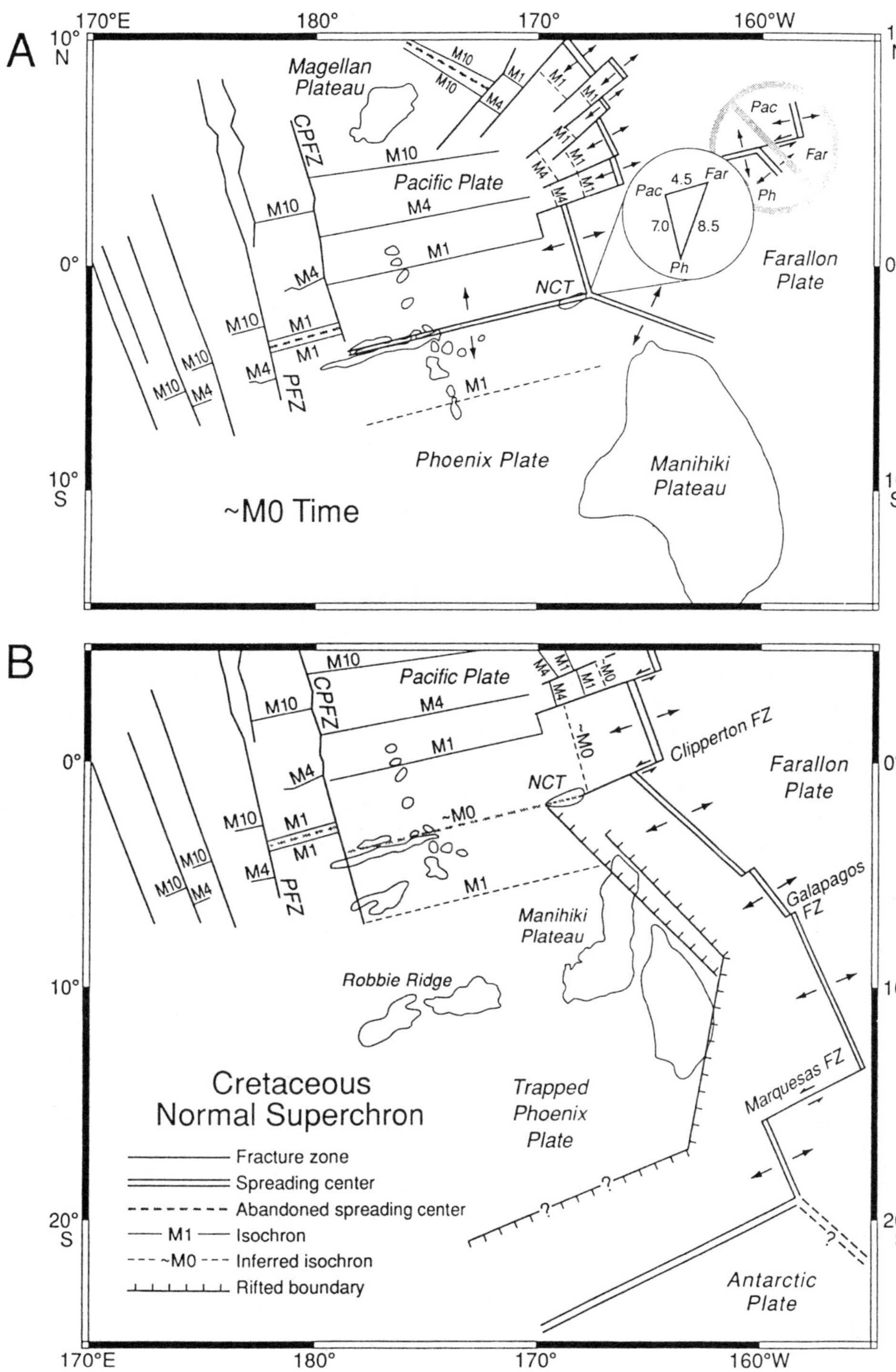

Fig. 8. (a) Tectonic reconstruction of the Central Pacific Basin at ~M0 time showing the ridge-ridge-ridge triple junction active at that time. The ~M0 spreading axis between the Pacific and Phoenix plates lies in the region between the lineated bathymetric depressions near 180°W and the westernmost deep of the Nova-Canton Trough. Further west, spreading ceased between M1 and M4 time. The Manihiki Plateau is in the process of forming , but has not yet undergone any structural modification. (b) Tectonic reconstruction of the Central Pacific Basin during the early part of the Cretaceous Normal Superchron. The triple junction moved through the Manihiki Plateau [see Winterer et al., 1974, Figs. 13 and 14], and jumped south of the Plateau trapping Phoenix plate on the Pacific plate. A rift bordering the northeastern boundary of the Manihiki Plateau extended into the western deep of the Nova-Canton Trough and initiated spreading between the nascent Nova-Canton (Clipperton) and Galapagos transform faults. Similarly, spreading between the Galapagos and Marquesas transform faults was initiated by a rift propagating along the southeastern boundary of the Manihiki Plateau [Sharman and Mammerickx, 1990; Hill et al., 1991]

into one or two long spreading segments between the Clarion and Clipperton fracture zones.

SUMMARY

By analyzing free-air gravity anomalies derived from GEOSAT altimetry data, we have shown co-polarity of the eastern Pacific fracture zones, from the Pau to Marquesas, about a pole located at 150.5°W, 34.6°S, through most of the Cretaceous Normal Superchron. This pole also fits the Nova-Canton trough and ridge, which we conclude are the western extension of the Clipperton Fracture Zone, as Menard [1967] originally proposed. This interpretation is supported by sidescan sonar and bathymetry data which show abyssal hill topography at a high angle to the N70°E-striking Nova-Canton Trough. The crustal fabric strikes N140°E over a large area to the south of the trough and limited data suggest a N160°E trend to the north.

Recently identified magnetic lineations better define the magnetic bight north of the Nova-Canton Trough, indicate (a probably eastward progressive) cessation of Pacific-Phoenix spreading at M3 to M0 time [Nakanishi et al., 1992], and also show that much of the Phoenix plate was trapped west of the Manihiki Plateau when spreading initiated on the Pacific-Antarctic Ridge south of the Samoan Basin [Winterer et al., 1974; Engebretson et al., 1991]. Aptian volcanism along the Manihiki Plateau and Robbie Ridge overprinted the trapped crust during the southward jump to the Pacific-Antarctic-Farallon triple junction. This transition may have involved a series of jumps, as Winterer et al. [1974] suggested, and/or the short existence of a "Phoenix microplate" between the Pacific and Antarctic plates.

The block-faulted northeast and southeast margins of the Manihiki Plateau are rifted margins that form the western limit of Pacific-Farallon spreading between the Canton-Clipperton, Galapagos, and Marquesas fracture zones. We suggest that the 7000-8400 m deep western part of the Nova-Canton Trough initially formed the eastern limit of the failed Pacific-Phoenix spreading axis, but later became the western limit of the Canton-Clipperton transform fault.

Acknowledgments. We thank OTC Limited of Sydney, Australia for allowing us to publish the PacRim sidescan data, especially Gregory Sachs and Michael Malikides for their support; Donald M. Hussong, Diane Hills, Debbie Peat and the staff of Seafloor Surveys International for invaluable assistance; Brooks Bays for the tectonic reconstruction illustrations; May Izumi and Diane Henderson for copyediting assistance; Dave Engebretson and Bruce Rosendahl for helpful and constructive reviews; Will Sager for his guidance, advice, and Euler pole error analysis; Paul Wessel and W. H. F. Smith for access to GMT programs used to produce included bathymetric and free-air gravity figures; and W. F. Haxby for use of the GEOSAT data base used in Figures 2 and 3. D. Joseph wishes to thank the ARCS Foundation for their generous support. School of Ocean and Earth Science and Technology Contribution No. 2948.

REFERENCES

Atwater, T., Plate tectonic history of the northeast Pacific and western North America, in *The Eastern Pacific Ocean and Hawaii, The Geology of North America, vol. N*, edited by E. L. Winterer, D. M. Hussong, and R. W. Decker, pp. 21-72, Geological Society of America, Boulder, Colorado, 1989.

Atwater, T., and J. Severinghaus, Tectonic map of the north central Pacific Ocean, scale 1:6,442,194, Plate 3A, in *The Eastern Pacific Ocean and Hawaii, The Geology of North America, vol. N*, edited by E. L. Winterer, D. M. Hussong, and R. W. Decker, Geological Society of America, Boulder, Colorado, 1989a.

Atwater, T., and J. Severinghaus, Tectonic map of the north central Pacific Ocean, scale 1:6,442,194, Plate 3B, in *The Eastern Pacific Ocean and Hawaii, The Geology of North America, vol. N*, edited by E. L. Winterer, D. M. Hussong, and R. W. Decker, Geological Society of America, Boulder, Colorado, 1989b

Blackinton, J. G., D. M. Hussong, and J. Steenstrup, Seafloor cable surveys: Swath bathymetry from a "fish," *Sea Tech., 32*, 33-39, 1991.

Cande, S. C., J. L. LaBrecque, R. L. Larson, W. C. Pitman, X. Golovchenko, and W. F. Haxby, *Magnetic Lineations of the World's Ocean Basins*, scale 1:27,400,000, AAPG, Tulsa, 1989.

Crough, S. T., Thermal origin of mid-plate hot-spot swells, *Geophys. J., 55*, 451–469, 1978.

Defense Mapping Agency, *Gazetteer of Undersea Features*, 3rd edition, Washington, D. C., 250 pp., 1981.

Duncan, R. A., and M. A. Richards, Early Cretaceous global flood basalt volcanism (abstract), *EOS, Trans. AGU, 71*, 1668, 1990.

Duncan, R. A., and M. A. Richards, Hotspots, mantle plumes, flood basalts, and true polar wander, *Rev. Geophys., 29*, 31-50, 1991.

Engebretson, D. C., J. Mammerickx, and C. A. Raymond, Tonga lineations - the Phoenix plate has arisen? (abstract), *EOS, Trans. AGU, 72*, 444, 1991.

Engel, A. E. J., and C. G. Engel, Mafic and Ultramafic Rocks, in *The Sea*, vol. 4, edited by M. N. Hill, ,pp. 465-519, Wiley-Interscience, New York, 1970.

Francheteau, J., J. G. Sclater, and H. W. Menard, Pattern of relative motion from fracture zone and spreading rate data in the North-eastern Pacific, *Nature, 226*, 746-748, 1970.

Gordon, R. G., and A. Cox, Calculating paleomagnetic poles for oceanic plates, *Geophys. J. R. astr. Soc., 69*, 619-640, 1980.

Halunen, A. J., L. W. Kroenke, and B. Rosendahl, Geophysical investigation of a deep trough in the central Pacific (abstract), *EOS, Trans. AGU, 51*, 762, 1970.

Hill, P. J., W. T. Coulbourn, and D. D. Bergerson, Manihiki Plateau, structured eastern margin, SeaMARC II sidescan sonar imagery, bathymetry, and interpretation, 1:250,000, in *Pacific Seafloor Atlas*, Sheet No. 18, edited by D. J. Henderson, Hawaii Institute of Geophysics, Honolulu, 1991.

Hussong, D. M., L. K. Wippeman, and L. W. Kroenke, The crustal structure of the Ontong Java and Manihiki oceanic plateaus, *J. Geophys. Res., 84*, 6003-6010, 1979.

Jackson, E. D., K. E. Bargar, B. P. Fabbi, and C. Heropoulos, Petrology of the basaltic rocks drilled on Leg 33 of the Deep Sea Drilling Project, in *Initial Reports of the Deep Sea Drilling Project, vol. 33*, edited by S. O. Schlanger and E. D Jackson et al., pp. 571-630, U. S. Govt. Printing Office, Washington, D. C., 1976.

Joseph, D., B. Taylor, and A. N. Shor, New sidescan sonar and gravity evidence that the Nova-Canton Trough is a fracture zone, *Geology, 20*, 435-438, 1992.

Kroenke, L. W., Line Island area sediment types, in *Annual Progress Report, Contract N00014-70-A-0016-0001, January 1, 1970-July 31, 1970*, pp. 40-41, edited by K. Wyrtki, G. H. Sutton, and K. Chave, Hawaii Inst. of Geophysics Report 70-27, 1970.

Kroenke, L. W., W. H. Berger, T. R. Janacek, et al., *Proc. ODP, Init. Repts., 130*: College Station, TX (Ocean Drilling Program), 1991.

Larson, R. L. , Late Jurassic and Early Cretaceous evolution of the western central Pacific Ocean, *J. Geomag. Geoelec., 28*, 219-236, 1976.

Larson, R. L., Latest pulse of Earth: evidence for a Mid-Cretaceous superplume, *Geology, 19*, 547-550, 1991.

Larson, R. L., and C. G. Chase, Late Mesozoic evolution of the western Pacific Ocean, *Geol. Soc. Amer. Bull., 83*, 3627-3644, 1972.

Larson, R. L., S. M. Smith, and C. G. Chase, Magnetic lineations of Early Cretaceous age in the western equatorial Pacific Ocean, *Earth Planet. Sci. Lett., 15*, 315-319, 1972.

Mahoney, J. J., An isotopic survey of Pacific oceanic plateaus: implications for their nature and role, in *Seamounts, Islands, and Atolls, Geophys. Monogr. 43*, edited by B. H. Keating, P. Fryer, R. Batiza, and

G. W. Boehlert, pp. 207-220, American Geophysical Union, Washington, D. C., 1987.

Mahoney, J. J., and K. J. Spencer, Isotopic evidence for the origin of the Manihiki and Ontong Java oceanic plateaus, *Earth Planet. Sci. Lett.*, *104*, 196-210, 1991.

Mahoney, J. J., M. Storey, R. A. Duncan, K. J. Spencer, M. Pringle, Geochemistry and age of the Ontong-Java Plateau, in *The Mesozoic Pacific , Geophys. Monogr. Ser.*, edited by M. Pringle, W. Sager, W. Sliter, and S. Stein, this volume.

Mammerickx, J. Bathymetry of the North Pacific Ocean, scale 1:6,442,194, Plate 1A, in *The Eastern Pacific Ocean and Hawaii, The Geology of North America, vol. N*, edited by E. L. Winterer, D. M. Hussong, and R. W. Decker, Geological Society of America, Boulder, Colorado, 1989a.

Mammerickx, J. Bathymetry of the North Pacific Ocean, scale 1:6,442,194, Plate 1B, in *The Eastern Pacific Ocean and Hawaii, The Geology of North America, vol. N*, edited by E. L. Winterer, D. M. Hussong, and R. W. Decker, Geological Society of America, Boulder, Colorado, 1989b.

Mammerickx, J., T. E. Chase, S. M. Smith, and I. L. Taylor, *Bathymetry of the South Pacific*, Chart 12, Scripps Institution of Oceanography, La Jolla, California, 1971.

Mammerickx, J., T. E. Chase, S. M. Smith, and I. L. Taylor, *Bathymetry of the South Pacific*, Chart 13, Scripps Institution of Oceanography, La Jolla, California, 1973.

Mammerickx, J., and S. M. Smith, *General bathymetric chart of the oceans*, scale 1:10,000,000, panel 5.10, 5th ed., International Hydrographic Organization/Intergovernmental Oceanic Commission/Canadian Hydrographic Service, Ottawa, Ontario, 1982.

Mammerickx, J., and S. M. Smith, *Bathymetry of the Northcentral Pacific*, scale 1:6,442,194, Map and Chart Series MC-52, Geological Society of America, Boulder, Colorado, 1985.

Menard, H. W., *Marine Geology of the Pacific*, McGraw-Hill, New York, New York, 255 pp. 1964.

Menard, H. W., Extension of northeastern-Pacific fracture zones, *Science*, *155*, 72-74, 1967.

Nakanishi, M., K. Tamaki, and K. Kobayashi, Mesozoic magnetic anomaly lineations and seafloor spreading history of the Northwestern Pacific, *J. Geophys. Res.*, *94*, 15437-15462, 1989.

Nakanishi, M., K. Tamaki, and K. Kobayashi, A new Mesozoic isochron chart of the northwestern Pacific Ocean: paleomagnetic and tectonic implications, *Geophys. Res. Lett.*, *19*, 693-696, 1992.

National Geophysical Data Center, ETOPO-5 bathymetry/topography data: National Oceanic and Atmospheric Administration, Data Announcement 88-M6-02, 1988.

Orwig, T. L., Channeled turbidites in the eastern Central Pacific Basin, *Mar. Geol.*, *39*, 33-57, 1981.

Parsons, B., and J. G. Sclater, An analysis of the variation of ocean floor bathymetry and heat flow with age, *J. Geophys. Res.*, *82*, 803–827, 1977.

Rea, D. K., Changes in structure and trend of fracture zones north of the Hawaiian Ridge and relation to sea-floor spreading, *J. Geophys. Res.*, *75*, 1421-1430, 1970.

Richards, M. A., R. A. Duncan, and V. E. Courtillot, Flood basalts and hot-spot tracks: plume heads and tails, *Science*, *246*, 103-107, 1989.

Rosendahl, B. R., *Geological and geophysical studies of the Canton Trough*, M. S. thesis, 120 pp., University of Hawaii at Manoa, Honolulu, 1972.

Rosendahl, B. R., and J. A. John Halunen, The petrology and structure of the Canton Trough (abstract), *GSA, abs. with progs.*, *3*, 689, 1971.

Rosendahl, B. R., R. Moberly, A. J. Halunen, J. C. Rose, and L. W. Kroenke, Geological and geophysical studies of the Canton Trough region, *J. Geophys. Res.*, *80*, 2565-2574, 1975.

Sager, W. W., and M. S. Pringle, Tectonic evolution of the Central Pacific during the Cretaceous Quiet Period (abstract), *EOS, Trans. AGU*, *68*, 1451, 1987a.

Sager, W. W., and M. S. Pringle, Paleomagnetic constraints on the origin and evolution of the Musicians and South Hawaiian Seamounts, central Pacific Ocean, in *Seamounts, Islands, and Atolls, Geophys. Monogr. 43*, edited by B. H. Keating, P. Fryer, R. Batiza, and G. W. Boehlert, pp. 133-162, Amer. Geophys. Union, Washington, D. C., 1987b.

Sager, W. W., D. W. Handschumacher, T. W. C. Hilde, and D. R. Bracey, Tectonic evolution of the northern Pacific plate and Pacific-Farallon-Izanagi triple junction in the Late Jurassic and Early Cretaceous (M21-M10), *Tectonophysics*, *155*, 345-364, 1988.

Schlanger, S. O., H. C. Jenkyns, and I. Premoli-Silva, Volcanism and vertical tectonics in the Pacific related to global Cretaceous transgressions, *Earth Planet. Sci. Lett.*, *52*, 435-449, 1981.

Scientific Staff, Deep Sea Drilling Project: Leg 33, Testing a hot spot theory, *Geotimes*, *19*, no. 3, 16-20, 1974.

Searle, R. C., R. T. Holcomb, J. B. Wilson, R. J. Whittington, M. L. Holmes, E. S. Kappel, B. A. McGregor, and A. N. Shor, The Molokai Fracture Zone near Hawaii and the Late Cretaceous Pacific/Farallon spreading direction, in *The Mesozoic Pacific, Geophys. Monogr. Ser.*, edited by M. Pringle, W. Sager, W. Sliter, and S. Stein, this volume.

Sender, K. L., A. Shor, and R. Hagen, SeaMARC II side-scan processing techniques (abstract), *EOS, Trans. AGU*, *70*, 1304, 1989.

Sharman, G. F., and J. Mammerickx, Eastern boundary of the Manihiki Plateau: a propagating rift site (abstract), *EOS, Trans. AGU*, *71*, 1668, 1990.

Sharman, G. F., and D. L. Risch, Northwest Pacific tectonic evolution in the Middle Mesozoic, *Tectonophysics*, *155*, 331-344, 1988.

Tamaki, K., and R. L. Larson, The Mesozoic history of the Magellan microplate in the western central Pacific, *J. Geophys. Res.*, *93*, 2857-2874, 1988.

Tanahashi, M., Seismic reflection survey in the southern part of the Central Pacific Basin (GH82-4 area), in *Marine geology, geophysics, and manganese nodule deposits in the southern part of the Central Pacific Basin, August-October 1982 (Hakurei-Maru cruise GH82-4), Cruise report No. 22*, edited by A. Usui, pp. 39-69, Geological Survey of Japan, Ibaraki, Japan, 1992.

Tarduno, J. A., W. V. Sliter, L. Kroenke, M. Leckie, H. Mayer, J. J. Mahoney, R. Musgrave, M. Storey, and E. L. Winterer, Rapid formation of the Ontong Java Plateau by Aptian mantle plume volcanism, *Science*, *254*, 399-403, 1991.

Wessel, P., and W. H. F. Smith, GMT: A public-domain UNIX software package for data manipulation and presentation using the PostScript page description language, *EOS, Trans. AGU* , *72*, 441, 445-446, 1991.

Winterer, E. L., Bathymetry and regional tectonic setting of the Line Islands chain, in *Initial Reports of the Deep Sea Drilling Program, vol. 33*, edited by S. O. Schlanger, and E. D. Jackson et al., pp. 731-747, U. S. Government Printing Office, Washington, D. C, 1976a.

Winterer, E. L., Anomalies in the tectonic evolution of the Pacific, in *The Geophysics of the Pacific Ocean Basin and Its Margins, Geophys. Monogr. 19*, edited by G. H. Sutton, M. H. Manghnani, and R. Moberly, pp. 269-278, American Geophysical Union, Washington, D. C., 1976b.

Winterer, E. L., P. F. Lonsdale, J. L. Matthews, and B. R. Rosendahl, Structure and acoustic stratigraphy of the Manihiki Plateau, *Deep-Sea Res.*, *21*, 793-814, 1974.

Yamazaki, T., Heat flow in the south of the Nova-Canton Trough, Central Equatorial Pacific (GH82-4 area), in *Marine geology, geophysics, and manganese nodule deposits in the southern part of the Central Pacific Basin, August-October 1982 (Hakurei-Maru cruise GH82-4), Cruise report No. 22*, edited by A. Usui, pp. 71-83, Geological Survey of Japan, Ibaraki, Japan, 1992.

Yamazaki, T., and T. Tanahashi, Magnetic and gravity anomalies in the Nova-Canton trough region, central equatorial Pacific (GH82-4 area), in *Marine geology, geophysics, and manganese nodule deposits in the southern part of the Central Pacific Basin, August-October 1982 (Hakurei-Maru cruise GH82-4), Cruise report No. 22*, edited by A. Usui, pp. 19-28, Geological Survey of Japan, Ibaraki, Japan, 1992.

D. Joseph, B. Taylor, and A. N. Shor, School of Ocean and Earth Science and Technology, University of Hawaii at Manoa, Honolulu, HI 96822.

T. Yamazaki, Geological Survey of Japan, Higashi 1-1-3, Tsukuba-shi, Ibaraki-ken, 305, Japan.

Age Progressive Volcanism in the Musicians Seamounts: A Test of the Hot Spot Hypothesis for the Late Cretaceous Pacific

MALCOLM S. PRINGLE

U.S. Geological Survey, Menlo Park, CA

$^{40}Ar/^{39}Ar$ ages of 12 of the Musicians Seamounts, located north of the Hawaiian Islands in the central Pacific Ocean, describe generally age progressive volcanism from 96 Ma in the NW corner of the province to 75 Ma in the SE. In particular, from 95.7 to 82.5 Ma, volcanism along a NW-SE trending chain of seamounts progressed at a linear rate of 52.0 ±5.8 km/m.y., or 0.503 ±0.055 °/m.y. about a rotation pole at 36°N, 76°W. Several tests are consistent with, but not conclusive proof of, a fixed hot spot hypothesis for this age progressive trend. Assuming the fixed hot spot model and combining these results with previously published results from the northern Line Islands, the best estimate for the rate of Pacific plate/hot spot motion for the period 96 to 82 Ma is 0.508 ± 0.053 °/m.y. No age progression could be found along a N-S lineament of seamounts within the Musicians Seamounts province, and there is no evidence for volcanism younger than 75 Ma, coeval with volcanism in the Emperor Seamounts. If the hot spot model is correct, then the Pacific apparent polar wander path describes a unique period of rapid true polar wander at the end of the Cretaceous normal polarity superchron, about 85 to 82 Ma.

The $^{40}Ar/^{39}Ar$ age data presented here highlight the difficulty in determining accurate ages of even relatively fresh whole rock and phenocryst minerals that have been subjected to submarine alteration since Cretaceous time. Feldspar conventional K-Ar ages in general, individual feldspar $^{40}Ar/^{39}Ar$ total-fusion ages, and even groups of whole rock $^{40}Ar/^{39}Ar$ total-fusion ages all proved unable to provide the high quality age estimates necessary to decipher the history of the Late Cretaceous seamounts examined here. The results of this study emphasize the necessity of using incremental-heating and total-fusion isochron analysis, with its inherent internal concordance check, in choosing which ages from partially altered samples may be considered reliable.

INTRODUCTION

According to the hot spot hypothesis, linear island and seamount chains form as the lithosphere moves over deep, fixed melting anomalies. An important step in our understanding the origin and evolution of island and seamount chains, the concept of hot spots consists of at least two distinct hypotheses: the *kinematic* hot spot hypothesis, where age progressive chains of volcanoes form as the lithosphere moves relative to a melting anomaly [Wilson, 1963a, b; Christofferson, 1968], and the *fixed* hot spot hypothesis, where these melting anomalies are fixed with respect to each other in some absolute motion reference frame [Morgan, 1971, 1972a, 1972b]. Morgan [1971, 1972a, 1972b] also proposed a mechanism of hot spot formation, where hot spots are the surface expression of thermal plumes rising from deep within the mantle.

The importance of the hot spot hypothesis for plate tectonic theory cannot be overestimated. The absolute pole of rotation *and* the angular rotation rate about that pole for a particular lithospheric plate can be determined from a single hot spot track if both the kinematic and fixed hot spot hypotheses can be shown to be true for that volcanic lineament. Multiple hot spot tracks on a single plate can confirm the fixed hot spot hypothesis and may allow the plate velocity to be defined with greater precision. Almost all our knowledge of plate velocity comes from either magnetic lineations on the sea floor (relative velocity constraints) or well-defined hot spot age progressions (absolute velocity constraints). For areas without mappable magnetic lineations, such as seafloor formed during the Cretaceous normal polarity superchron or continental plates without any attached sea floor, hot spot tracks may be the only way to determine plate velocities.

The Pacific Basin has been the focus of much of the research on linear island and seamount chains. Morgan [1972a, 1972b] suggested a hot spot origin for four Pacific seamount chains, although few radiometric age data were available to actually test the hypothesis. Since 1972, at least some radiometric age data have indicated age progressions for 10 or so Pacific seamount and island chains (Figure 1). Most of the age data are for hot spot tracks younger than 42 Ma, and most of those ages are less

The Mesozoic Pacific: Geology, Tectonics, and Volcanism
Geophysical Monograph 77

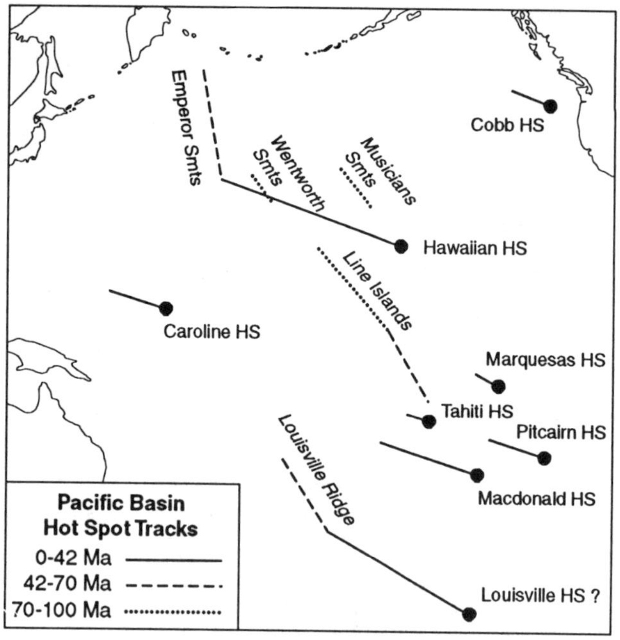

Figure 1. Documented age progressive hot spot tracks in the Pacific Basin.

than 20 Ma. Between 42 and 65 Ma, a well-documented age progression has been found only for the Emperor Seamounts. Less well-documented age progressions have been found in the Line Islands and along the Louisville ridge. Older than about 70 Ma, age progressive volcanism has been documented for only a few seamounts in the northern Line Islands.

The purpose of this work was to test the hot spot hypothesis for the Musicians Seamounts, a Late Cretaceous seamount province in the north-central Pacific, and to compare these results with those previously published for the northern Line Islands [Schlanger et al., 1984]. A primary goal of this work was to estimate the Late Cretaceous velocity of the Pacific plate with respect to the hot spot reference frame.

Geologic Setting

The Musicians Seamounts form a complex province of isolated seamounts and east-west trending ridges in the north-central Pacific Basin (Figure 2). Two distinct seamount lineaments were examined in this study. One lineament trends NW-SE, parallel to the northern Line Islands. It starts within a cluster of unnamed seamounts in the NW corner of the province, and continues southeast until it crosses the Murray Fracture Zone and merges with the western edge of the east-west trending ridges. This trend is not distinct within the east-west trending ridges south of the Murray Fracture Zone, and Campbell et al. [1980] found no evidence of its extension further to the southeast. The other lineament trends N-S, parallel to the Emperor Seamounts. It starts in the northern part of the province, within the small plateau of seamounts called the Musicians Horst by Rea and Naugler [1971], and continues south to Mendelssohn, the southernmost Musicians Seamount.

The Musicians Seamounts sit atop seafloor that apparently formed during the Cretaceous normal polarity superchron. The Murray Fracture Zone, consisting of a band of numerous linear troughs and ridges approximately 100 km wide, trends east-northeast through the middle of the province. It causes no perceptible offset of the seamounts north and south of the fracture zone, suggesting that many, if not all, formed outside the active fracture zone. Based on a westward extrapolation of the Cretaceous-Tertiary-Quaternary magnetic lineations, Sager and Pringle [1987] concluded that the seafloor beneath the Musicians north of the Murray Fracture Zone is approximately 95-100 Ma, while the age of the seafloor beneath the Musicians south of the fracture zone is approximately 80-85 Ma.

The Musicians Seamounts were apparently formed within a relatively short period after the formation of the seafloor upon which they rest. Several studies of compensation depth and lithospheric flexure have concluded that most of the Musicians Seamounts were formed near the spreading ridge [Watts et al., 1980; Schwank and Lazarewicz, 1982; Freedman and Parsons, 1986]. However, some variation in the amplitudes of the gravity anomalies derived from Seasat altimeter data and modeled compensation depths led Schwank and Lazarewicz [1982] and Freedman and Parsons [1986] to suggest that some of the Musicians Seamounts could be as much as 20 m.y. younger than the seafloor. Further 2-dimensional modeling, based on the limited available ship-based gravity data, found no evidence for seamount volcanism more than 15 m.y. older than the seafloor [W.H.F. Smith and M.S. Pringle, unpublished data]. These conclusions agree with the conclusions of Sager and Pringle [1987], who found no evidence for volcanism younger than about 70 Ma in the paleomagnetic data.

Only two radiometric ages were available for the Musicians Seamounts before this study, both from a reconnaissance study of Cretaceous seamounts in the northern Pacific [Clague and Dalrymple, 1975]. Both were determined on plagioclase separates from basaltic rocks but were determined by different techniques: an age of 89 Ma for Rachmaninoff Seamount was determined by the $^{40}Ar/^{39}Ar$ total-fusion technique, whereas an age of 67 Ma age for Khatchaturian was determined by conventional K-Ar analysis (all ages in this study have been calculated using the decay constants recommended by Steiger and Jäger, 1977).

Clague and Jarrard [1973] suggested that the Musicians and Line Islands seamount chains might be hot spot tracks tracing a pre-Emperor, i.e. older than 70 Ma, pole of Pacific plate motion. Epp [1978] suggested either a one or two hot spot model for the two seamount lineaments in the Musicians Seamounts, and detemined a rotation pole of 36°N, 76°W for the older trend. The two hot spot model predicts an older age progression along the NW-SE lineament in the Late Cretaceous, followed by a N-S trending age progression coeval with the Emperor Seamounts. The single hot spot model [W. J. Morgan, personal communication, 1978] predicts a hot spot trail initially trending SE and then changing direction around 70 Ma as the Pacific rotation direction changed. Both these models could form the Musicians Horst in the north and the east-west trending ridges in the south of the

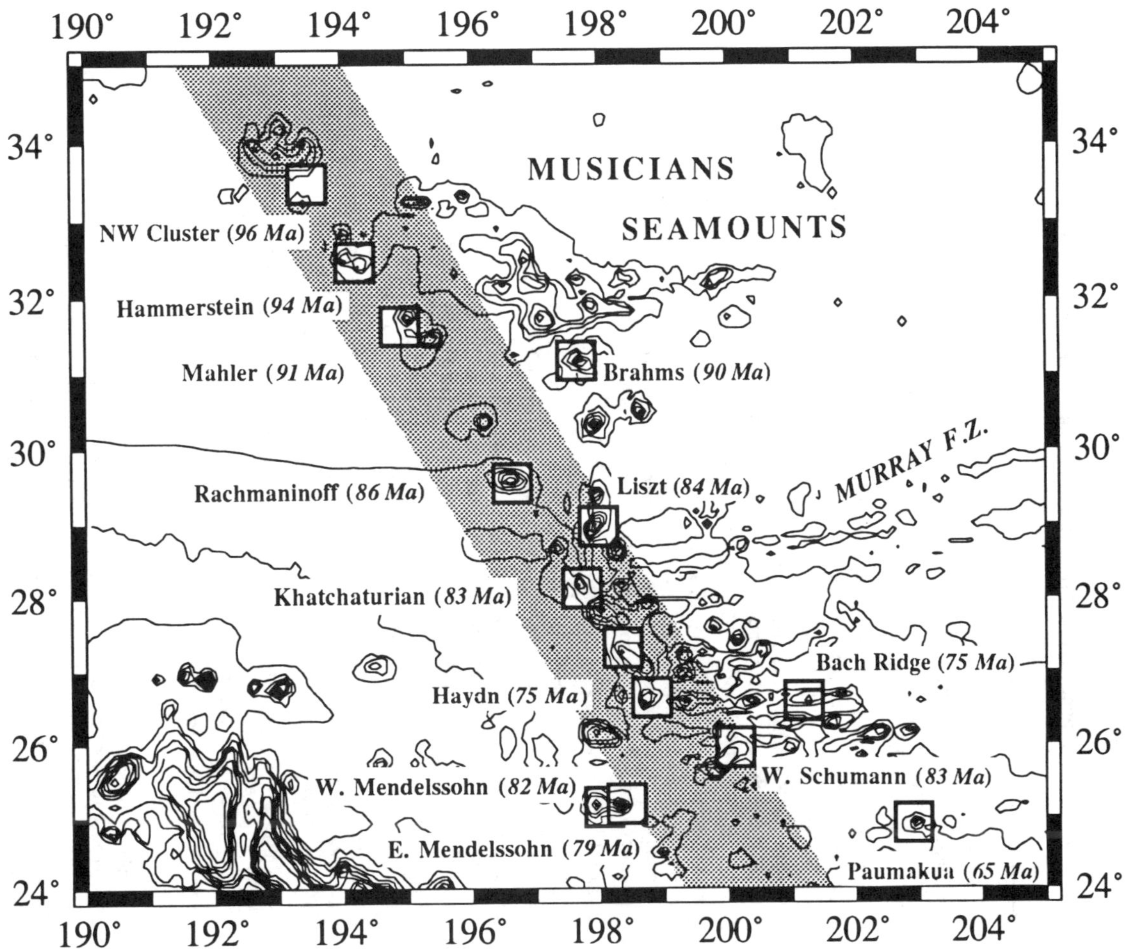

Figure 2. The Musicians Seamounts, with seamount ages determined here and in Sager and Pringle [1987] for Paumakua. Age for unlabeled seamount marked between Khatchaturian and Haydn Seamounts (Ravel) has not been determined yet. A 200 km wide, model hot spot track centered on the NW-SE chain and drawn about the Late Cretaceous rotation pole of 36°N, 76°W is shown in gray. Figure drawn using GMT utilities of Wessel and Smith [1991].

province through interaction between the hot spot (or hot spots) and the nearby spreading ridge.

Samples Studied

Samples were collected in 35 successful dredges from 27 seamounts in the Musicians Seamounts. Most were collected during two Hawaii Institute of Geophysics (HIG) cruises in 1980, KK8040414 (Leg 2) and KK800715, whose purpose was to gather geophysical data and samples to test various models for the origin of the Musicians Seamounts. Mendelssohn and Schumann Seamounts were resampled in 1984 during an HIG study assessing the Mn-crust resources of seamounts within the Hawaiian Exclusive Economic Zone (EEZ), cruise KK84082402. Mussorgski, Rachmaninoff, and Khatchaturian Seamounts were dredged during the Aries VII cruise of the Scripps Institute of Oceanography in 1971; some of these samples were studied by Clague and Dalrymple [1975].

Samples suitable for testing the two hot spot models using $^{40}Ar/^{39}Ar$ dating techniques were recovered from 12 of these seamounts (Figure 1). Selection criteria are summarized in Pringle [1992]. Table 1 includes a description of the selected samples and a list of their dredge locations and depths; Pringle [1992] contains a complete list of all Musicians Seamounts dredges. Even the freshest of the collected samples show some signs of alteration, and none of the samples met the usual selection criteria for conventional K-Ar dating of whole rock samples [Dalrymple and Lanphere, 1969; Mankinen and Dalrymple, 1972].

Although the potential for excess ^{40}Ar in glassy submarine lavas presents a potentially serious problem in attempting to date such rocks, Dalrymple and Moore [1968] and Noble and Naughton [1968] found that samples dredged from water depths less than 1400 m contained little or no excess ^{40}Ar. Also,

TABLE 1. Dredge Locations and Descriptions of Samples Suitable for $^{40}Ar/^{39}Ar$ Dating Recovered from the Musicians Seamounts

Dredge No.	Feature	Dredge Location Lat. (°N)	Long. (°W)	Depth (m)	Sample Number	Description (including suitability for dating)
Cruise: KK80041402 (abbreviated: KK804)						
1	Uncharted Smt	33°29′	166°32′	3220-3800	1-9	microvesicular (2-3%), feldspar and sparsely aegerine-augite-phyric trachyte; fresh, essentially holocrystalline, fine-grained groundmass with no vesicle filling or other signs of alteration. Excellent whole-rock IH candidate.
					1-10	microvesicular (1-2%), sparsely plagioclase-phyric mugearite with a fine-grained, trachytic groundmass containing 15-25% clay and glass. Fair to marginal whole-rock IH candidate.
					1-19	vesicular (5-10%), plagioclase and sparsely clinopyroxene-phyric hawaiite with a feldspar-rich groundmass including 10-15% pale brown glass and clay. Fair to marginal whole-rock IH candidate.
5	Hammerstein Smt	32°28′	165°48′	3150-2900	5-16	massive, banded, sparsely plagioclase-microphyric mugearite; alteration confined to streaks and bands in a very fine-grained, nearly holocyrstalline trachytic groundmass with abundant highly oxidized olivine and rare micro-phenocrysts of orthopyroxene. Fair whole-rock IH candidate.
					5-27	same lithology as 5-16.
7	Mahler Smt	31°38′	165°06′	2740-2700	7-1	vesicular (2-3%), aphyric hawaiite with a feldspar-rich groundmass containing 20% yellow clay after glass. Marginal whole-rock IH candidate.
					7-17	vesicular (5-10%), altered porphyritic basalt with 20% plagioclase phenocrysts in a nearly holocrystaline matrix of plagioclase and 25% pale yellow clay after pyroxene. For plagioclase separate only.
21	Brahms Smt	31°07′	162°22′	2270-2530	21-6	highly altered, trachytic(?) scoria with sparse but relatively fresh plagioclase and potassium feldspar phenocrysts. For K-spar separate only.
22	Brahms Smt	31°13′	162°20′	2160-3660	22-1	massive, plagioclase and olivine-phyric basalt; olivine phenocrysts, most groundmass pyroxene and even some plagioclase completely replaced by pale to yellow clay. For plagioclase and pyroxene separate only.
23	Liszt Smt	28°58′	162°01′	2820-3270	23-8	somewhat vesicular (1-2%), plagioclase-phyric, altered ferrobasalt; groundmass 20% relatively fresh skeletal plagioclase and 40% yellow to brown clay after glass and mafics. For plagioclase separate only.
					23-10	same lithology as 23-10.
Cruise: KK80071500 (abbreviated: KK807)						
27	Bach Ridge	26°35′	158°50′	3430-3530	27-2	massive, plagioclase and clinopyroxene-microphyric basalt; groundmass of plagioclase and clinopyroxne contains 5% yellow clays and 10% clouded brown glass. Marginal whole-rock IH candidate (tholeiitic composition).
					27-3	somewhat vesicular (2-3%), plagioclase and sparsely clinopyroxene-phyric basalt with a nearly holocrystalline groundmass containing plagioclase, clinopyroxne, and 10-15% pale yellow clays. For plagioclase separate only.
					27-7	same lithology as 27-3.

TABLE 1 (continued). Dredge Locations and Descriptions of Samples Suitable for $^{40}Ar/^{39}Ar$ Dating Recovered from the Musicians Seamounts

Dredge No.	Feature	Dredge Location Lat. (°N)	Long. (°W)	Depth (m)	Sample Number	Description (including suitability for dating)
33	Haydn Smt	26°38′	161°12′	2000-3000	33-6	vesicular (2-5%), plagioclase-phyric altered basalt with 25% plagioclase phenocrysts in a matrix of plagioclase and yellow to brown clay with no groundmass mafics left. For plagioclase separate only.
					33-7	same lithology as 33-6.
37	E Mendelssohn Smt	25°11′	161°36′	1800-2600	37-2	microvesicular (5-10%), plagioclase-microphyric mugearite; microvesicles filled with streaks of calcite ± zeolite, otherwise trachytic groundmass nearly unaltered with sparse patches of clay. Good whole-rock IH candidate.
38	W Mendelssohn Smt	25°08′	161°57′	2700-3600	38-1	massive, plagioclase-phyric basalt with 10% plagioclase and rare clinopyroxene phenocrysts in a groundmass containing plagioclase, clinopyroxene, 2% brown glass, and 5-20% clear to yellow clay. Pagioclase separate; fair whole-rock IH candidate.
					38-2	massive, sparsely plagioclase-phyric olivine ferrobasalt with 10% brown clay replacing holocrystalline groundmass and olivine oxidized to throughly replaced by iddingsite. Fair whole-rock IH candidate.
Cruise: KK84082402 (abbreviated: KK848)						
52	W. Schumann Smt	25°57′	159°54′	2700-3000	52-51	somewhat vesicular (1-2%), porphyritic basalt with 25% plagioclase and rare clinopyroxene phenocrysts in an intersertal matrix including 20% yellow clay after glass and pyroxene. For plagioclase separate only.
Cruise: ARIES7 (abbreviated: A7)						
58	Rachmaninoff Smt	29°33′	163°22′	1560-2500	58-0	vesicular (ca. 10%), plagioclase and sparsely pyroxene-phyric altered basalt; carbonate, zeolite, and phosphate filling some of the vesicles; groundmass includes 30% clay. For plagioclase separate only.
					58-25	sparsely vesicular (ca. 1%), plagioclase-phyric altered basalt; groundmass contains 15% clays after glass; plagioclase phenocrysts are partially resorbed and strongly zoned. For plagioclase separate only.
					58-26	same lithology as 58-25.
					58-33*	same lithology as 58-25.
59	Khatchaturian Smt	28°08′	162°17′	2270-2850	59-15	vesicular (5-10%), plagioclase and pyroxene-phyric, altered basalt; groundmass contains 15% clay after glass. For plagioclase separate only.
					59-31	same lithology as 59-15.

* includes sample originally labeled 58-34 but later found to be a matching half to 58-33

Dalrymple and Moore [1968] showed a significant decrease in the excess ^{40}Ar inward from the rim of a pillow basalt. Because (1) Pringle [1992] concluded that all of the samples studied here were erupted in waters less than 1300 m deep (and many could have erupted subaerially), (2) whole rock samples were taken from the interior of pillow fragments when neccessary, and (3) whole rock samples with significant amounts of glass were avoided, excess ^{40}Ar should not be a significant problem for this study.

K-AR GEOCHRONOLOGY

The K-Ar Clock

The K-Ar clock has been widely used to date oceanic lavas, but several fundamental assumptions of the technique may be violated by samples taken from the submarine environment. During normal submarine weathering, low-temperature alteration phases (clays, zeolites, phosphates, etc.) replace the original magmatic phases, and both K-addition and radiogenic ^{40}Ar loss will lower the apparent K-Ar age. Clague et al. [1975] and Dalrymple and Clague [1976] found that $^{40}Ar/^{39}Ar$ total-fusion ages of altered Hawaiian basalts were a better estimate of the crystallization age than the conventional K-Ar ages of these same rocks. They attributed this effect to the proportional loss of radiogenic ^{40}Ar and K-derived ^{39}Ar from the low-temperature phases which lowered the conventional K-Ar ages. Apparently, ^{39}Ar derived from potassium in alteration phases is lost during the irradiation and extraction process from precisely the same sites which lost ^{40}Ar during the alteration process. This approach seems to work for the less visibly altered Hawaiian-Emperor seamount samples, but no independent criteria are available to indicate when a sample is too altered for the total-fusion technique to yield an accurate crystallization age.

In the more general case, the different argon isotopes may be lost from a given sample by different mechanisms. Radiogenic ^{40}Ar escapes when alteration products replace the original crystallization phases over geological time, and from non-retentive glass and alteration products over geological time and during the extraction process [Dalrymple and Lanphere, 1969; Baksi, 1974]. If the extraction line bakeout is incomplete, the lighter argon isotopes may be lost at a rate greater than the heavier isotopes because of differential diffusion out of low-retention phases such as clays and zeolites [Baksi, 1974].

The irradiation process itself can induce the loss of some of the reactor-derived argon isotopes. K-derived ^{39}Ar, because of the high energy of the $^{39}K(n,p)^{39}Ar$ reaction (>2Mev), is subject to recoil loss during and subsequent to the irradiation process, especially in fine-grained phases [Turner and Cadogan, 1974; Huneke and Smith, 1976]. Yanase et al. [1975], Dalrymple and Clague [1976], Duncan [1978], Halliday [1978], Seidemann [1978], Walker and McDougall [1982], and Foland et al. [1984] all documented ^{39}Ar loss from fine-grained geological materials. Moreover, Walker and McDougall [1982] found that some alteration phases that do not retain ^{39}Ar may still retain some radiogenic ^{40}Ar, causing anomalously old apparent ages. Ca-derived ^{37}Ar, because of the even higher recoil distance of the $^{41}Ca(\alpha)^{37}Ar$ reaction in the reactor (3 times that of ^{39}Ar), also is subject to recoil loss during and subsequent to the irradiation process [Hess and Lippolt, 1986].

The different argon isotopes may also be redistributed *within* a given sample. $^{40}Ar/^{39}Ar$ incremental-heating experiments of altered samples have shown that ^{40}Ar, ^{39}Ar, and ^{37}Ar are all subject to redistribution. Fleck et al. [1977] suggested that progressive devitrification of glassy tholeiitic basalts leads to redistribution of radiogenic ^{40}Ar from sites that release their argon at high temperatures to sites which release their argon at low temperatures. Turner and Cadogan [1974] and Huneke and Smith [1976] first described the recoil of K-derived ^{39}Ar out of high-K, low-temperature steps into low-K, high-temperature steps. Baksi [1990] documented that Ca-derived ^{37}Ar recoil out of high-Ca, high temperature steps produces too small a correction for Ca-derived ^{36}Ar and results in calculated ages which are too young. The general effect of all three redistribution mechanisms is to produce anomalously old ages for increments of gas released at low temperatures and anomalously young ages for increments of gas released at high temperatures. The resulting age spectrum of a sample so affected would have a decreasing step-like pattern, a pattern observed almost ubiquitously for relatively fine-grained, altered whole rock samples.

It is extremely difficult, if not impossible, to identify and separately evaluate all of these effects. Rather, the approach of Lanphere and Dalrymple [1978] and Dalrymple et al. [1980] is employed here. $^{40}Ar/^{39}Ar$ incremental-heating experiments are used to test the degree to which K-Ar crystallization ages have been disturbed and to conclude whether or not the sample is too disturbed or altered to yield reliable, independent crystallization ages. The final conclusions of this study rely almost exclusively on the $^{40}Ar/^{39}Ar$ incremental-heating experiments. A substantial amount of conventional K-Ar and $^{40}Ar/^{39}Ar$ total-fusion data will also be presented, in an attempt to better understand the limitations of those two techniques in deciphering the crystallization age of altered submarine lavas.

Sample Preparation

For whole rock K-Ar analysis, slabs 1-2 cm in thickness were taken from the freshest part of a dredged sample with a water-cooled trim saw. For $^{40}Ar/^{39}Ar$ measurements, small cores 6 mm in diameter weighing 0.5 to 0.8 g were taken from the centers of these slabs. The remainder of the slabs was broken into small blocks less than 1 cm thick with a small hydraulic splitter and cleaned in an ultrasonic bath with distilled water. Splits for conventional K-Ar argon measurements were either the 6 mm cores or one of the small blocks. The remainder of the small blocks was pulverized to a fine powder for chemical analysis, including K_2O by flame photometry [Ingamells, 1970] or XRF [Pringle, 1992].

Plagioclase feldspar is often the only phase remaining in altered submarine lavas still useful for K-Ar geochronology. While standard density, magnetic, and HF-etching techniques can be used to produce relatively pure plagioclase separates [e.g. Dalrymple and Lanphere, 1969], further acid-cleaning with warm 3 to 6N HCl can be quite important in dissolving altered plagioclase, carbonates, zeolites, and phosphates. A grain size range of 125-250μ, usually much smaller than the original phenocrysts, has proven to be a good compromise between being fine enough to ensure that most of the alteration was found on the surface of cleavage fragments (accessible to the cleaning action of the acids), but coarse enough to avoid sample handling problems and the ^{39}Ar and ^{37}Ar recoil problems described above. The HCl cleaning also seems to rid the plagioclase separate of material which is usually degassed in the low-temperature, cleaning steps of incremental-heating experiments.

Analytical Techniques

Samples for $^{40}Ar/^{39}Ar$ analysis were sealed in air in quartz vials and irradiated in the core of the U.S. Geological Survey TRIGA reactor (GSTR). Whole rock samples were irradiated for 24-30 hours, where they received a neutron dose of approximately 2.4×10^{18} to 3×10^{18} nvt. Plagioclase separates were irradiated for shorter periods, 12-24 hours, to minimize the effect of corrections from the Ca-derived argon isotopes. However, the decrease in the precision of the measurements because of the smaller K-derived ^{39}Ar quantities sometimes offset the increase in analytical precision resulting from the smaller Ca-derived corrections. The neutron flux monitor, J, was measured with the intralaboratory standard biotites SB-2 and SB-3, which are from the same rock and have ages of 161.4 Ma and 162.9 Ma, respectively. Some of the results reported here differ slightly from those reported in Sager and Pringle [1987] because of a monitor recalibration, which resulted in an age 0.9% older than previously used for SB-3. Further details of the GSTR flux characteristics, sample encapsulation and geometry, and the corrections for interfering K- and Ca-derived Ar isotopes are given in Dalrymple and Lanphere [1971] and Dalrymple et al. [1981].

Conventional K-Ar and $^{40}Ar/^{39}Ar$ argon extractions were done in a glass argon extraction line, following a 12 hour bakeout at 280°C, using radio-frequency (RF) induction heating and cleanup with CuO and Ti furnaces [Dalrymple and Lanphere, 1969]. Temperatures for incremental-heating experiments were estimated using an optical pyrometer and RF power settings previously calibrated using the two thermocouple crucible described in Lanphere and Dalrymple [1971]; these temperatures are accurate to no better than 50°C. Feldspar total-fusion experiments and incremental-heating steps were held at temperature for 30 minutes each; whole rock total-fusion experiments were held at temperature for 15-20 minutes.

Argon mass ratios for the conventional K-Ar and most of the $^{40}Ar/^{39}Ar$ measurements were measured with the computerized multiple-collector mass spectrometer described by Stacey et al. [1981]. A few $^{40}Ar/^{39}Ar$ measurements were done in analog mode on a single collector Nier-type mass spectrometer [Dalrymple and Lanphere, 1969].

All errors are reported as one standard deviation, or at the 67% confidence level. Errors reported for the calculated ages of individual age measurements are estimates of the standard deviation of analytical precision. The conventional K-Ar errors were derived by differentiation of the isotope dilution and relevant age equations [Cox and Dalrymple, 1967]. The $^{40}Ar/^{39}Ar$ errors were derived by differentiation of the relevant age equations, similar to the derivation in Dalrymple et al. [1981] except that the uncertainties in the mass discrimination, K- and Ca- correction factors, and the half-life of ^{37}Ar were also taken into account. The pooled standard deviation of over 70 flux monitor pairs analyzed during the course of this study was 0.46%; all age errors reported here include an estimate in the error in J of 0.5%.

Mean ages were calculated as weighted means, where each age is weighted by the inverse of its variance. This weighting allows data with different analytical errors to be combined without the poorer quality data having a disproportionate effect on the result. The variance of a weighted age is calculated as $1/\Sigma(1/\sigma_i^2)$, where σ_i^2 is the estimated variance of the individual increment age (Taylor, 1982).

Both age spectra and isochron ages were calculated from $^{40}Ar/^{39}Ar$ incremental-heating experiments. Plateau ages were calculated as weighted means, as described above. Isochron ages are calculated for both the $^{40}Ar/^{36}Ar$ versus $^{39}Ar/^{39}Ar$ and $^{36}Ar/^{40}Ar$ versus $^{39}Ar/^{40}Ar$ correlation diagrams using the York2 least-squares cubic fit with correlated errors [York, 1969]. We have found [Dalrymple et al., 1988] that the difference between the two correlations is negligible when the error correlation coefficients are determined using the method recommended by York [as quoted in Ozima et al., 1977].

Interpretation of Incremental-heating Experiments

The real power in using $^{40}Ar/^{39}Ar$ incremental-heating experiments to decipher the crystallization age of altered submarine rocks is that crystallization ages can be differentiated from unreliable ages using a set of *internal* tests for *each* experiment. It may not be possible to differentiate *why* a particular experiment is unreliable, i.e. because of geologic error or experimental artifact or both. However, it is critical to use these internal tests to determine *whether* the K-Ar system of a particular sample is too disturbed to reveal a meaningful geologic age. The need for such a consistency check is becoming increasingly important as an increasing number of ages from altered rocks are being reported.

The simple thermal history of volcanic rocks allows one to avoid many of the pitfalls in the interpretation of $^{40}Ar/^{39}Ar$ incremental-heating experiments encountered when studying metamorphic rocks [for example, Harrison, 1983]. Here, the conservative criteria of Lanphere and Dalrymple [1978] and Dalrymple et al. [1980] are modified so that *each* of these criteria involves the use of a rigorous statistical test. We accept an age as an accurate estimate of the crystallization age only if:

(1) A well-defined, high temperature age spectrum plateau is formed by three or more concordant, contiguous steps representing at least 50% of the ^{39}Ar released, i.e. no age difference can be detected between *any* two plateau steps at the 95% confidence level (the critical value test discussed below),

TABLE 2. F-Distibution Critical Values for Detecting Excess Scatter in Regression Analysis

Number of Data Regressed	Degrees of Freedom	F Variate Critical Value [a]		
		95% CL	99% CL	99.5% CL
3	1	3.84	6.63	7.88
4	2	3.00	4.61	5.30
5	3	2.60	3.78	4.28
6	4	2.37	3.32	3.72
7	5	2.21	3.02	3.35
8	6	2.10	2.80	3.09
9	7	2.01	2.64	2.90
10	8	1.94	2.51	2.74
11	9	1.88	2.41	2.62
12	10	1.83	2.32	2.52
14	12	1.75	2.18	2.36
17	15	1.67	2.04	2.19
22	20	1.57	1.88	2.00
26	24	1.52	1.79	1.90
32	30	1.46	1.70	1.79
42	40	1.39	1.59	1.67
62	60	1.32	1.47	1.53
122	120	1.22	1.32	1.36
-- infinite --		1.00	1.00	1.00

(a) F variate statistic with N-2 degrees of freedom (N is the number of data regressed) in the demominator and an infinite degree of freedom in the numerator. This is the proper degrees of freedom in the numerator for techniques where analytical errors are an estimate of the population variance, as is the case for K-Ar data. For techniques which base their estimates of analytical precision on the sample variance, such as a finite number of duplicate analyses, a slightly higher F statisitic should be used (for example, see Brooks, et al. 1972). For numbers of data regressed not tabulated, use the reciprocals of the degrees of freedom to interpolate linearly between tabulated values.

(2) A well-defined isochron exists for the plateau points, i.e. the F-distribution statistic for the York2 fit, SUMS/(N-2), is sufficiently small at the 95% confidence level (also discussed below),

(3) The plateau and isochron ages are concordant at the 95% confidence level, and

(4) The $^{40}Ar/^{36}Ar$ intercept on the isochron diagram is not significantly different from the atmospheric value of 295.5 at the 95% confidence level.

The Critical Value test used for criteria (1) and (3) simply states that no difference between ages can be detected if the difference in the two ages is less than $1.960(\sigma_1^2 + \sigma_2^2)^{1/2}$, where σ_1 and σ_2 are the standard deviations of the two ages [Dalrymple and Lanphere, 1969].

For the test of a well-defined isochron in criterion (2), most authors follow Brooks et al. [1972] and McIntyre et al. [1966], who suggest a general cutoff value of 2.5 for the ratio between the scatter about the isochron and how much of that scatter can be explained within the limits of analytical error alone. This departs from standard statistical practice somewhat in not using some level of significance to reject the model age, and does not take into account that different cutoff values should be used for isochrons with different numbers of regressed data points. The more rigorous F-distribution test is preferred; Table 2 tabulates the critical values for different numbers of data regressed. For example, the proper cutoff value at the 95% confidence level for isochrons with 3, 4, or 5 data is 3.84, 3.00 and 2.60, respectively. If this ratio is exceeded, then there is more scatter about the isochron than can be explained by analytical error alone, i.e. some additional geologic or experimental disturbance is significant. Further justification of an age derived from such an isochron is necessary before that age can be accepted as having geological meaning.

General Results

Table 3 lists the best age estimate for each of the twelve seamounts studied here, based on the weighted mean of the isochron ages for each of the experiments which meets the criteria for a concordant $^{40}Ar/^{39}Ar$ age. The isochron ages are prefered over the age specta plateau ages because they (1) combine both an estimate of analytical error and an estimate of the degree of internal disturbance -- the scatter about the isochron line -- in the final error estimate, and (2) make no assumption about the composition of the initial non-radiogenic, or trapped, argon component. Table 3 also presents a summary of the age spectra and isochron analysis of the 19 incremental-heating and 4 plagioclase total-fusion experiments upon which these best ages are based. Individual data for the incremental-heating experiments are presented in Table 4; data from the total-fusion experiments are presented in Table 5.

When this study was initiated, it was thought that reliable ages with errors on the order of 0.5-1.5% could be achieved through carefully selected feldspar conventional K-Ar and whole rock $^{40}Ar/^{39}Ar$ total-fusion experiments, with some verification from $^{40}Ar/^{39}Ar$ total-fusion experiments on the feldspar separates and incremental-heating experiments on the whole rock samples. For example, from data consisting almost exclusively of whole rock $^{40}Ar/^{39}Ar$ total-fusion ages, Schlanger et al. [1984] and Watts et al. [1987] were able to conclude that age progressive volcanism could explain at least parts of the Line Islands and Louisville Ridge, respectively.

However, most of the conventional K-Ar and $^{40}Ar/^{39}Ar$ total-fusion ages determined for 33 samples from the Musicians Seamounts (Table 5) are significantly different from the most reliable sample and seamount ages (Table 3). After a discussion on the determination of each of the individual seamount ages, we shall explore how the conventional and $^{40}Ar/^{39}Ar$ total-fusion ages differ from these ages, and conclude that accurate ages can-

TABLE 3. Age Spectra and Isochron Analysis of $^{40}Ar/^{39}Ar$ Experiments on Samples from the Musicians Seamounts
Best age for each seamount, as discussed in text, indicated in brackets

SAMPLE	MATERIAL	RECOMBINED TF AGE (Ma)	AGE SPECTRA: INCREMENTS USED (°C)	^{39}Ar (%)	WEIGHTED AGE (Ma)	ISOCHRON ANALYSIS: AGE (Ma)	$^{40}Ar/^{36}Ar$ INTERCEPT	SUMS/N-2	N
KHATCHATURIAN SEAMOUNT		[82.5 ±2.5 Ma]							
A7-59D-15	Plagioclase	74.2 ±7.0	800 - FUSE	59.4	80.4 ±1.3	81.5 ±5.0	294.5 ±4.5	0.17	5
A7-59D-31	Plagioclase	78.1 ±3.5	680 - 900	69.7	82.1 ±1.1	82.5 ±2.9	294.4 ±7.1	0.94	4
NW CLUSTER		[95.7 ±0.7 Ma]							
KK804 1-9	Trachyte	96.1 ±0.7	500 - FUSE	100.0	96.0 ±0.4	95.7 ±0.9	296.1 ±2.0	1.95	9
			with TF	----	95.9 ±0.4	95.7 ±0.7	296.2 ±1.5	1.71	10
KK804 1-10	Mugearite	91.1 ±0.8	700 - 850	53.9	94.0 ±0.8	97.0 ±3.4	220.2 ±71.3	4.48	3
			with TF	----	93.7 ±0.7	97.0 ±2.4	218.9 ±45.9	2.24	4
HAMMERSTEIN SEAMOUNT		[94.0 ±0.8 Ma]							
KK804 5-16	Mugearite	105.1 ±0.7	700 - 800	66.0	94.1 ±0.5	93.4 ±1.1	301.2 ±6.3	----	2
KK804 5-27	Mugearite	104.1 ±0.7	620 - 800	65.0	94.7 ±0.5	94.6 ±1.0	297.8 ±15.0	0.10	4
MAHLER SEAMOUNT		[91.0 ±0.7 Ma; 83.9 ±0.9 Ma]							
KK804 7-1	Mugearite	88.5 ±0.5	600 - 800	42.9	91.2 ±0.4	91.6 ±0.7	290.2 ±5.6	1.90	3
			with TF	----	91.0 ±0.6	91.4 ±0.7	290.8 ±5.2	1.95	4
KK804 7-17	Plagioclase	83.9 ±2.8	550 - 1160	90.8	83.2 ±1.4	84.0 ±2.4	293.0 ±4.2	1.96	8
	50-125μ	75.2 ±3.9	700 - FUSE	95.6	80.5 ±0.7	81.4 ±2.5	293.5 ±2.8	7.26	7
BRAHMS SEAMOUNT		[89.7 ±0.6 Ma]							
KK804 21-6	K-feldspar	90.0 ±0.6	800 - 1150	94.0	90.2 ±0.4	89.7 ±0.6	317.7 ±20.3	0.12	4
EAST MENDELSSOHN		[78.5 ±0.9 Ma]							
KK807 37-2	Mugearite	70.0 ±0.5	600 - 800	56.0	75.8 ±0.4	76.6 ±0.6	287.8 ±3.0	0.52	3
	HCl leach	74.1 ±0.6	600 - 720	55.7	77.9 ±0.5	79.7 ±0.9	285.3 ±4.2	1.22	3
	HF leach	74.2 ±0.7	600 - 750	53.7	78.8 ±0.6	78.5 ±0.9	303.9 ±20.8	0.50	3
WEST MENDELSSOHN		[82.4 ±1.3 Ma]							
KK807 38-1	Plag-Basalt	85.5 ±1.2	650 - 1000	79.5	80.8 ±0.9	84.0 ±2.9	270.0 ±22.0	0.33	4
	Plagioclase	83.9 ±1.2	550 - 1020	95.1	83.1 ±0.6	82.0 ±1.4	298.2 ±3.0	0.98	7
CENTRAL BACH RIDGE		[74.6 ±1.7 Ma]							
KK807 27-2	Basalt	74.8 ±2.4	600 - 850	68.5	74.3 ±1.9	74.4 ±3.3	295.3 ±2.4	0.54	6
			with TF	----	74.4 ±1.3	74.6 ±1.7	295.3 ±1.7	0.46	7
KK807 27-3	Plagioclase	62.3 ±6.2	780 - 1250	71.9	62.6 ±2.6	73.1 ±8.1	290.7 ±3.2	2.02	5
WEST SCHUMANN SEAMOUNT		[83.0 ±1.0 Ma]							
KK848 52-51	Plagioclase	83.8 ±0.7	720 - 1040	83.1	83.7 ±0.6	83.0 ±1.0	298.7 ±3.8	0.17	5
PLAGIOCLASE TOTAL FUSION ISOCHRONS									
Mahler Seamount					[83.9 ±0.9]	82.0 ±1.4	308.5 ±7.5	0.46	3
Rachmaninoff Seamount					[86.4 ±1.2]	84.5 ±1.9	301.5 ±5.1	0.39	6
Liszt Seamount					[84.4 ±1.5]	79.8 ±5.2	301.2 ±6.3	1.38	3
Haydn Seamount					[75.1 ±1.2]	74.4 ±2.6	296.4 ±3.1	4.49	4

"Age Spectrum" age for plagioclase total-fusion data is weighted average of the total-fusion ages.

TABLE 4. $^{40}Ar/^{39}Ar$ Incremental-Heating Data of Samples from the Musicians Seamounts

TEMP (°C)	$^{40}Ar/^{39}Ar$	$^{37}Ar/^{39}Ar^a$	$^{36}Ar/^{39}Ar$	$^{36}Ar_{Ca}{}^b$ (%)	$^{40}Ar_K{}^b$ (%)	$^{40}Ar_R{}^b$ (%)	^{39}Ar (%)	Apparent Age [c] (Ma)	
A7-59D-15		Plagioclase J=0.002474			Recombined TF age:			74.18	± 6.97
550	1785.5	44.83	6.055	0.2	0.0	0.0	6.4	0.002	± 90.8
650	1063.2	111.9	3.561	0.9	0.0	1.9	2.3	95.60	± 159.
700	412.5	135.3	1.384	2.7	0.0	3.6	4.4	71.08	± 11.0
750	345.8	130.8	1.159	3.2	0.0	4.1	3.0	67.69	± 12.0
800	148.6	97.04	0.4732	5.8	0.0	11.3	4.2	78.82	± 5.34
850	79.73	68.77	0.2296	8.4	0.0	22.1	19.8	80.60	± 1.83
950	87.29	53.96	0.2506	6.0	0.0	20.3	31.0	80.21	± 1.85
LOST	74.00	27.00	0.2000	3.8	0.0	23.2	24.5	76.27	± 5.64
FUSE	131.4	87.35	0.4050	6.1	0.0	14.4	4.4	87.72	± 16.7
A7-59D-31	Plagioclase 125-250μ		J=0.002260			Recombined TF age:		78.09	± 3.49
500	2125.3	42.14	7.139	0.2	0.0	0.9	1.8	79.49	± 178.
600	399.4	139.9	1.339	2.9	0.0	3.8	5.4	67.75	± 14.1
680	54.36	116.4	0.1514	21.6	0.0	35.5	15.3	83.43	± 2.41
750	32.34	110.1	0.07589	40.7	0.0	58.9	19.6	82.04	± 1.85
820	49.11	102.8	0.1292	22.3	0.0	39.6	17.3	83.39	± 2.13
900	55.27	95.97	0.1506	17.9	0.0	33.9	17.5	79.89	± 2.16
1100	99.03	85.76	0.3041	7.9	0.0	16.4	18.8	69.19	± 2.57
FUSE	114.3	52.98	0.3499	4.2	0.0	13.4	4.3	63.41	± 8.62
KK804 1-9	Trachyte Core	J=0.005489				Recombined TF age:		96.11	± 0.69
500	34.68	0.02073	0.08242	0.0	0.0	29.8	12.9	99.42	± 1.46
600	117.9	0.000000	0.3666	0.0	0.0	8.1	7.0	92.71	± 5.38
680	24.91	0.01244	0.05009	0.0	0.0	40.6	19.7	97.38	± 1.02
750	17.65	0.01219	0.02595	0.0	0.0	56.5	17.9	96.16	± 0.77
840	22.68	0.02695	0.04384	0.0	0.0	42.9	14.9	93.79	± 0.95
900	19.92	0.01827	0.03395	0.0	0.0	49.6	13.2	95.28	± 0.87
980	19.87	0.01463	0.03342	0.0	0.0	50.3	6.5	96.27	± 1.06
1050	24.92	0.01185	0.05057	0.0	0.0	40.0	3.6	96.16	± 1.59
FUSE	15.86	0.01349	0.01999	0.0	0.0	62.7	4.2	95.93	± 1.29
TF	10.24	0.01599	0.001051	0.4	0.1	96.9	0.0	95.68	± 0.78
KK804 1-10	WR CORE	J=0.005500				Recombined TF age:		91.11	± 0.79
600	17.62	0.6203	0.02426	0.7	0.0	59.6	18.6	101.3	± 1.90
700	11.07	0.4623	0.003826	3.4	0.1	90.1	23.1	96.34	± 1.23
800	10.75	0.5370	0.003782	4.0	0.1	90.0	11.7	93.51	± 1.29
850	11.75	0.7080	0.008095	2.5	0.1	80.1	19.1	91.04	± 1.50
950	12.57	0.9600	0.01370	2.0	0.0	68.4	9.9	83.42	± 2.47
1000	9.547	2.010	0.005711	9.9	0.1	84.0	3.6	77.98	± 3.28
1100	9.611	3.612	0.006406	15.8	0.1	83.4	13.2	77.98	± 1.59
FUSE	20.65	39.52	0.06717	16.5	0.0	19.7	0.8	41.05	± 20.1
TF	11.56	1.582	0.007441	6.0	0.1	82.1	0.0	91.87	± 1.90
KK804 5-16	WR CORE	J=0.006528				Recombined TF age:		105.1	± 0.70
600	33.78	0.2198	0.06451	0.1	0.0	43.6	17.3	165.6	± 1.60
700	14.61	0.3548	0.02159	0.4	0.0	56.5	31.6	94.74	± 0.83
800	9.752	0.5179	0.005491	2.6	0.1	83.7	34.5	93.71	± 0.71
850	11.61	0.6600	0.01338	1.3	0.1	66.3	9.8	88.51	± 1.69
920	23.26	0.9615	0.05961	0.4	0.0	24.6	2.7	66.10	± 6.09
980	42.03	1.550	0.1197	0.4	0.0	16.1	2.3	78.14	± 7.17
1050	37.02	1.353	0.09983	0.4	0.0	20.6	1.1	87.68	± 14.8
FUSE	50.26	0.9286	0.1402	0.2	0.0	17.7	0.8	102.0	± 19.8

TABLE 4 (continued). $^{40}Ar/^{39}Ar$ Incremental-Heating Data of Samples from the Musicians Seamounts

TEMP (°C)	$^{40}Ar/^{39}Ar$	$^{37}Ar/^{39}Ar^{a}$	$^{36}Ar/^{39}Ar$	$^{36}Ar_{Ca}{}^{b}$ (%)	$^{40}Ar_{K}{}^{b}$ (%)	$^{40}Ar_{R}{}^{b}$ (%)	^{39}Ar (%)	Apparent Age [c] (Ma)	
KK804 5-27	WR CORE	J=0.005489				Recombined TF age:		104.1	± 0.70
450	68.09	0.3184	0.1575	0.1	0.0	31.7	7.0	201.8	± 3.20
550	28.94	0.3284	0.05522	0.2	0.0	43.7	9.8	121.1	± 1.90
620	11.03	0.3490	0.004165	2.4	0.1	89.0	16.4	94.71	± 1.06
680	11.32	0.4269	0.005308	2.3	0.1	86.4	18.4	94.39	± 0.99
740	10.53	0.5177	0.002499	5.8	0.1	93.3	16.9	94.86	± 1.05
800	13.67	0.6115	0.01309	1.3	0.0	72.0	13.3	94.99	± 1.30
900	11.20	1.118	0.006037	5.2	0.1	84.8	10.1	91.81	± 1.62
1100	39.37	5.076	0.1045	1.4	0.0	22.6	8.1	86.44	± 2.50
KK804 7-1	WR CORE	J=0.004612				Recombined TF age:		88.52	± 0.50
500	41.79	0.9898	0.08956	0.3	0.0	36.9	6.7	123.9	± 1.70
600	13.98	0.9047	0.009939	2.6	0.0	79.5	11.1	90.25	± 0.82
680	13.22	1.058	0.006700	4.4	0.0	85.6	15.4	91.87	± 0.70
740	13.84	0.7632	0.008759	2.4	0.0	81.7	10.5	91.80	± 0.85
800	25.93	0.6686	0.05048	0.4	0.0	42.7	5.9	89.81	± 1.43
870	13.21	1.224	0.01043	3.3	0.0	77.4	7.4	83.16	± 1.06
950	13.21	1.119	0.009526	3.3	0.0	79.3	11.9	85.19	± 0.78
1030	12.12	0.9701	0.005967	4.6	0.0	86.1	22.0	84.84	± 0.60
FUSE	18.04	2.941	0.03395	2.4	0.0	45.7	9.2	67.45	± 0.98
KK804 7-17	Plagioclase 125-250μ		J=0.002260			Recombined TF age:		83.91	± 2.82
460	1024.9	12.49	3.401	0.1	0.0	2.0	2.4	83.72	± 93.0
550	282.1	48.65	0.9092	1.5	0.0	6.2	4.8	72.30	± 16.1
650	72.06	62.80	0.1874	9.4	0.0	30.4	11.7	90.87	± 6.24
720	39.38	66.89	0.08275	22.7	0.0	52.0	15.0	85.39	± 3.15
780	36.11	59.12	0.06737	24.6	0.0	58.4	17.2	87.51	± 2.76
870	33.77	53.63	0.06348	23.7	0.0	57.6	17.8	80.50	± 2.69
940	52.10	40.12	0.1211	9.3	0.0	37.7	8.9	80.52	± 5.37
1040	51.70	23.13	0.1245	5.2	0.0	32.5	6.3	68.35	± 7.71
1160	35.25	29.98	0.06323	13.3	0.0	54.0	9.1	77.58	± 5.23
FUSE	97.55	58.34	0.2612	6.3	0.0	25.8	6.7	103.9	± 7.60
KK804 7-17	Plagioclase 80-125μ		J=0.002474			Recombined TF age:		75.23	± 3.92
550	1790.7	27.10	6.067	0.1	0.0	0.0	4.1	0.049	± 94.0
700	231.2	61.07	0.7429	2.3	0.0	7.2	7.2	76.35	± 5.28
780	251.7	65.71	0.8163	2.3	0.0	6.3	13.6	72.91	± 5.63
850	51.66	67.30	0.1300	14.5	0.0	36.4	24.5	85.98	± 1.29
950	155.7	55.49	0.4882	3.2	0.0	10.3	13.4	72.91	± 5.75
1050	29.51	46.69	0.05372	24.4	0.0	59.3	17.3	78.92	± 1.18
1200	74.16	38.19	0.2056	5.2	0.0	22.3	13.9	74.35	± 1.89
FUSE	108.1	48.49	0.3181	4.3	0.0	16.8	6.1	81.81	± 3.50
KK848 52-51A	Plagioclase 125-250μ		J=0.005049			Recombined TF age:		83.80	± 0.74
550	237.6	49.61	0.7755	1.8	0.0	5.3	2.3	115.0	± 11.1
650	38.32	64.21	0.1186	15.2	0.0	22.4	3.7	80.06	± 3.80
720	32.17	71.31	0.09794	20.4	0.0	28.4	7.4	85.41	± 2.33
780	18.43	75.80	0.05360	39.7	0.0	48.1	9.2	83.30	± 1.81
850	14.96	75.79	0.04150	51.3	0.0	60.0	15.1	84.27	± 1.39
950	12.55	73.61	0.03296	62.7	0.0	71.0	32.8	83.47	± 1.11
1040	12.96	72.54	0.03419	59.6	0.0	68.4	18.6	83.06	± 1.24
1150	19.97	55.54	0.05468	28.5	0.0	42.1	8.5	77.96	± 1.83
FUSE	48.60	51.27	0.1479	9.7	0.0	18.8	2.3	84.22	± 5.88

TABLE 4 (continued). $^{40}Ar/^{39}Ar$ Incremental-Heating Data of Samples from the Musicians Seamounts

TEMP (°C)	$^{40}Ar/^{39}Ar$	$^{37}Ar/^{39}Ar$ [a]	$^{36}Ar/^{39}Ar$	$^{36}Ar_{Ca}$ [b] (%)	$^{40}Ar_K$ [b] (%)	$^{40}Ar_R$ [b] (%)	^{39}Ar (%)	Apparent Age [c] (Ma)	
KK807 37-2	WR CORE	J=0.006677				Recombined TF age:		70.02	± 0.47
500	26.20	0.6207	0.06451	0.3	0.0	27.4	7.5	84.56	± 1.76
600	7.931	0.5633	0.005221	3.0	0.1	81.1	16.9	75.85	± 0.72
700	7.612	0.8644	0.004035	6.0	0.1	85.2	25.2	76.52	± 0.63
800	16.82	1.068	0.03615	0.8	0.0	37.0	13.9	73.46	± 1.06
900	6.545	1.130	0.003670	8.6	0.1	84.8	11.9	65.67	± 0.89
950	7.725	1.120	0.007220	4.4	0.1	73.5	3.9	67.18	± 2.47
1030	5.636	1.460	0.003501	11.7	0.1	83.7	11.6	55.99	± 0.90
FUSE	6.107	3.745	0.007958	13.2	0.1	66.5	9.1	48.38	± 1.13
KK807 37-2	WR 250-500µ HCl leached		J=0.004612			Recombined TF age:		74.14	± 0.63
500	69.05	0.4607	0.1962	0.1	0.0	16.1	11.5	90.10	± 2.62
600	14.90	0.4230	0.01764	0.7	0.0	65.2	16.3	79.08	± 0.96
660	12.83	0.5054	0.01094	1.3	0.0	75.1	20.6	78.47	± 0.80
720	23.82	1.188	0.04957	0.7	0.0	38.9	18.8	75.53	± 1.07
780	14.12	1.619	0.01865	2.4	0.0	61.9	10.0	71.34	± 1.40
840	13.38	1.502	0.01769	2.4	0.0	61.8	6.7	67.58	± 2.03
910	12.28	1.448	0.01801	2.3	0.0	57.6	7.7	57.99	± 1.78
1000	18.23	3.009	0.04337	1.9	0.0	31.0	4.5	46.57	± 3.06
FUSE	30.39	5.134	0.08020	1.8	0.0	23.4	3.9	58.39	± 3.62
KK807 37-2	WR 105-250µ HF leached		J=0.004266			Recombined TF age:		74.27	± 0.69
500	39.78	0.6743	0.09499	0.2	0.0	29.6	8.0	88.35	± 2.65
600	12.68	0.6998	0.008162	2.4	0.0	81.4	8.9	77.73	± 2.17
700	14.15	0.6014	0.01222	1.4	0.0	74.8	12.6	79.67	± 1.56
750	11.03	1.153	0.002290	14.1	0.1	94.7	32.1	78.68	± 0.74
800	10.73	2.209	0.003295	18.8	0.1	92.6	11.6	74.95	± 1.66
850	10.37	2.312	0.005297	12.2	0.1	86.7	8.6	67.97	± 2.20
930	9.587	1.716	0.004983	9.7	0.1	86.1	10.7	62.48	± 1.77
1050	7.995	1.488	0.004920	8.5	0.1	83.3	7.3	50.58	± 2.52
FUSE	131.2	3.075	0.4381	0.2	0.0	1.5	0.2	15.23	± 143.3
TF	26.54	1.479	0.05647	0.7	0.0	37.6	0.0	75.18	± 0.88
KK807 38-1	PLAGIOCLASE 125-250µ		J=0.001929			Recombined TF age:		83.93	± 1.23
450	756.1	42.63	2.523	0.5	0.0	1.9	0.9	49.62	± 81.5
550	81.06	56.27	0.2103	7.5	0.0	29.1	7.0	83.36	± 3.07
600	63.20	58.83	0.1483	11.1	0.0	38.4	12.7	85.86	± 1.80
670	52.48	57.65	0.1140	14.2	0.0	44.9	20.7	83.41	± 1.25
750	63.49	55.59	0.1508	10.4	0.0	37.1	28.1	83.24	± 1.19
850	43.24	53.35	0.08264	18.1	0.0	53.7	16.4	82.04	± 1.37
960	40.06	50.59	0.07373	19.3	0.0	56.1	7.3	79.20	± 2.71
1020	224.9	48.66	0.6946	2.0	0.0	10.5	2.9	83.08	± 9.78
1120	635.1	52.20	2.017	0.7	0.0	6.8	1.6	150.5	± 37.1
FUSE	131.3	49.21	0.3771	3.7	0.0	18.2	2.5	84.21	± 9.13
KK807 38-1	PLAG BASALT CORE	J=0.006677				Recombined TF age:		85.23	± 1.17
500	53.02	1.179	0.1518	0.2	0.0	15.6	10.2	96.78	± 4.20
650	11.06	2.141	0.01524	3.9	0.1	60.8	25.3	79.39	± 1.46
800	9.352	4.924	0.009665	14.3	0.1	73.8	30.9	81.51	± 1.23
900	10.95	5.714	0.01505	10.7	0.1	63.6	10.5	82.34	± 3.34
1000	9.374	5.595	0.009496	16.5	0.1	75.0	12.8	83.02	± 4.89
1050	15.83	8.320	0.02817	8.3	0.0	51.7	4.0	96.58	± 8.67
FUSE	10.56	28.12	0.01238	63.8	0.1	87.4	6.3	109.9	± 5.50

TABLE 4 (continued). $^{40}Ar/^{39}Ar$ Incremental-Heating Data of Samples from the Musicians Seamounts

TEMP (°C)	$^{40}Ar/^{39}Ar$	$^{37}Ar/^{39}Ar^{a}$	$^{36}Ar/^{39}Ar$	$^{36}Ar_{Ca}{}^{b}$ (%)	$^{40}Ar_{K}{}^{b}$ (%)	$^{40}Ar_{R}{}^{b}$ (%)	^{39}Ar (%)	Apparent Age [c] (Ma)	
KK807 38-2	FERROBASALT CORE		J=0.006677			Recombined TF age:		107.6	± 1.00
500	21.71	0.7536	0.02130	1.0	0.0	71.3	35.7	177.4	± 1.8
600	9.194	2.006	0.008373	6.7	0.1	74.8	14.9	81.13	± 2.27
700	8.154	3.664	0.006317	16.3	0.1	80.8	23.2	77.82	± 1.16
800	9.919	4.220	0.01742	6.8	0.1	51.6	9.9	60.77	± 2.55
900	6.080	4.639	0.007971	16.3	0.1	67.5	5.8	48.91	± 4.36
1000	6.558	25.16	0.01784	39.6	0.1	51.3	9.7	40.78	± 2.68
FUSE	25.50	12.86	0.07390	4.9	0.0	18.5	0.8	56.56	± 30.3
KK807 27-2	BASALT CORE	J=0.005380				Recombined TF age:		74.84	± 2.44
400	595.1	2.303	1.956	0.0	0.0	2.9	5.2	159.9	± 32.5
500	125.7	2.347	0.3844	0.2	0.0	9.8	4.2	115.9	± 13.4
600	95.75	3.088	0.2993	0.3	0.0	7.9	10.4	72.06	± 6.14
650	36.40	5.912	0.09901	1.7	0.0	21.0	18.1	72.85	± 3.04
700	38.16	9.339	0.1035	2.5	0.0	21.8	12.4	79.63	± 4.19
750	113.6	10.05	0.3606	0.8	0.0	6.9	7.3	75.25	± 8.46
800	28.54	7.815	0.07302	3.0	0.0	26.6	12.5	72.73	± 4.10
850	37.34	4.344	0.1014	1.2	0.0	20.7	7.8	73.75	± 6.50
950	39.34	3.459	0.1130	0.9	0.0	15.9	7.2	59.67	± 7.14
FUSE	19.28	45.12	0.06327	20.0	0.0	22.4	15.0	42.73	± 3.53
TF	13.34	12.18	0.02221	15.4	0.0	58.3	0.0	74.54	± 1.74
KK807 27-3	PLAGIOCLASE 105-250μ		J=0.002474			Recombined TF age:		62.31	± 6.15
550	1661.6	79.70	5.616	0.4	0.0	0.5	9.6	39.96	± 55.3
650	422.5	221.9	1.453	4.3	0.0	2.7	3.6	59.30	± 35.0
720	653.2	239.1	2.249	3.0	0.0	1.3	6.7	45.19	± 29.9
780	185.2	265.7	0.6585	11.3	0.0	6.8	15.5	67.51	± 6.30
850	215.2	257.5	0.7640	9.5	0.0	5.0	12.5	57.57	± 7.91
930	64.56	234.3	0.2377	27.7	0.0	21.3	13.9	71.66	± 4.61
1080	148.1	127.5	0.4980	7.2	0.0	7.8	15.7	55.47	± 5.10
1250	221.8	64.04	0.7291	2.5	0.0	5.3	14.2	53.66	± 6.72
FUSE	183.0	131.1	0.5763	6.4	0.0	12.9	8.2	112.2	± 9.10
KK804 21-6	K-SPAR 125-250μ	J=0.004612				Recombined TF age:		90.01	± 0.57
500	588.8	2.111	1.962	0.0	0.0	1.5	0.1	74.19	± 159.
560	234.0	0.6205	0.7543	0.0	0.0	4.8	0.3	90.44	± 31.4
620	36.54	0.2885	0.08862	0.1	0.0	28.4	0.7	84.31	± 9.96
680	18.51	0.1712	0.02498	0.2	0.0	60.2	1.5	90.34	± 4.69
740	13.92	0.1014	0.01016	0.3	0.0	78.4	3.1	88.64	± 2.33
800	13.22	0.07741	0.006945	0.3	0.0	84.5	6.7	90.64	± 1.37
900	11.87	0.05982	0.002635	0.6	0.1	93.4	28.8	90.02	± 0.60
1040	11.47	0.04534	0.001297	1.0	0.1	96.6	44.1	89.96	± 0.55
1150	12.73	0.04002	0.005177	0.2	0.0	88.0	14.3	90.81	± 0.81
FUSE	73.75	0.1991	0.2219	0.0	0.0	11.1	0.2	66.99	± 48.5

a Corrected for ^{37}Ar and ^{39}Ar decay, 35.1 days and 259 years, respectively.
b Subscripts indicate radiogenic (R), calcium-derived (Ca), and potassium-derived (K) argon (%).
c Lamda E = $0.581 \times 10^{-10} yr^{-1}$, lamda B = $4.962 \times 10^{-10} yr^{-1}$. Errors are s.d. of analytical precision.

TABLE 5. Analytical Data for Conventional K-Ar and $^{40}Ar/^{39}Ar$ Total-Fusion Ages of Samples from the Musician Seamounts

Sample No.	Material	CONVENTIONAL K-Ar: K2O Weight %	Argon: $^{40}Ar_R$ (mol/gm)	Argon: $^{40}Ar_R$ [b] (%)	AGE [a] (Ma)	$^{40}Ar/^{39}Ar$ TOTAL FUSION: J	$^{40}Ar/^{39}Ar$	$^{37}Ar/^{39}Ar$ [c]	$^{36}Ar/^{39}Ar$	$^{36}Ar_{Ca}$ (%)	$^{39}Ar_{Ca}$ (%)	$^{40}Ar_K$ (%)	$^{40}Ar_R$ [b] (%)	AGE [a] (Ma)
UNNAMED SEAMOUNT / NW CLUSTER														
1-9	Trachyte Core	6.603	8.859E-10	93.6	91.1 ±0.9	0.005489	10.24	0.016	0.00105	0.4	0.0	0.06	96.9	95.7 ±0.8
			8.890E-10	84.1		0.005489	9.97	*****	*****	**	**	**	**	96.1 ±0.7
1-10	Mugearite Core	2.616	3.181E-10	87.9	84.2 ±0.8	0.005500	11.56	1.582	0.00744	6.0	0.1	0.05	82.1	91.9 ±1.9
			3.308E-10	93.2		0.005500	9.42	*****	*****	**	**	**	**	91.1 ±0.8
1-19	Hawaiite Core	1.450	7.100E-11	41.6	43.5 ± 14	------	------	------	------	------	------	------	------	------
			1.821E-10	35.8										
HAMMERSTEIN SEAMOUNT														
5-16	Mugearite Core	2.448	2.468E-10	80.3	71.5 ±4.0	0.005868	14.57	0.460	0.01360	0.9	0.0	0.04	72.6	108.8 ±0.7
			2.676E-10	63.5		0.006528	9.19	*****	*****	**	**	**	**	105.1 ±0.7
5-27	Mugearite Core	2.398	2.642E-10	71.4	74.4 ±0.8	0.005489	18.18	0.875	0.02411	1.0	0.1	0.03	61.1	106.9 ±0.8
			2.596E-10	60.0		0.005489	10.82	*****	*****	**	**	**	**	104.1 ±0.7
MAHLER SEAMOUNT														
7-1	Hawaiite Core	3.230	3.950E-10	85.7	83.3 ±0.8	0.005489	11.58	1.471	0.00790	5.2	0.1	0.05	80.8	90.5 ±0.6
			3.965E-10	78.4		0.004612	10.90	*****	*****	**	**	**	**	88.5 ±0.5
7-17	Plag 75-125µ	0.123	1.583E-11	12.0	87.4 ±4.3	0.003054	35.79	55.919	0.08323	18.8	3.8	0.02	44.2	88.3 ±2.9
						0.002474	17.21	*****	*****	**	**	**	**	75.2 ±3.9
7-17	Plag 125-250µ	0.147	1.659E-11	59.0	76.1 ±1.3	0.003054	44.93	52.149	0.11372	12.9	3.6	0.01	34.8	87.3 ±3.9
			1.610E-11	43.0		0.003054	20.18	52.126	0.03249	45.0	3.6	0.03	73.8 (H)	83.2 ±1.0
			1.662E-11	18.1		0.002260	21.07	*****	*****	**	**	**	**	83.9 ±2.8
BRAHMS SEAMOUNT														
21-6	Kspar 125-250µ	5.558	6.646E-10	86.6	81.2 ±0.7	0.004612	11.09	*****	*****	**	**	**	**	90.0 ±0.6
22-1	Plag 74-125µ	------	------	------	------	0.003054	37.14	40.424	0.08232	13.0	2.7	0.02	43.0 (H)	88.2 ±1.8
RACHMANINOFF SEAMOUNT														
58D-0	Plag 125-250µ	------	------	------	------	0.003240	20.54	30.600	0.02904	29.6	2.1	0.03	70.6 (H)	84.5 ±1.8
						0.003240	33.34	31.803	0.07177	12.4	2.2	0.02	44.3 (L)	86.1 ±2.5
58D-25	Plag 100-250µ	------	------	------	------	0.003240	45.84	39.144	0.11557	9.5	2.7	0.01	32.6	87.5 ±8.3
58D-26	Plag 100-250µ	------	------	------	------	0.003240	115.80	37.166	0.35063	3.0	2.5	0.01	13.2	89.3 ±15.
58D-33	Plag 125-250µ	------	------	------	------	0.003240	31.71	35.247	0.06542	15.1	2.4	0.02	48.2 (H)	89.4 ±3.0
						0.003240	43.23	38.172	0.10586	10.1	2.6	0.01	35.0 (L)	88.5 ±2.9
LISZT SEAMOUNT														
23-8	Plag 125-250µ	0.107	1.279E-11	16.0	81.2 ±2.0	0.003240	76.42	59.394	0.22589	7.4	4.1	0.01	19.1	86.8 ±3.0
	70-125µ	0.110	1.235E-11	5.3	76.4 ±4.0	0.003240	46.20	58.932	0.12609	13.1	4.0	0.01	29.9	82.3 ±2.1
23-10	Plag 125-250µ	0.112	1.172E-11	11.8	74.9 ±3.6	0.003240	45.89	55.686	0.12166	12.8	3.8	0.01	31.7	86.3 ±2.9
			1.292E-11	11.5										
	100-125µ	0.110	1.215E-11	3.4	75.4 ±5.0	------	------	------	------	------	------	------	------	------

TABLE 5 (continued). Analytical Data for Conventional K-Ar and $^{40}Ar/^{39}Ar$ Total-Fusion Ages of Samples from the Musician Seamounts

Sample No.	Material	CONVENTIONAL K-Ar: K2O Weight %	Argon: $^{40}Ar_R$ (mol/gm)	Argon: $^{40}Ar_R$ [b] (%)	AGE [a] (Ma)	$^{40}Ar/^{39}Ar$ TOTAL FUSION: J	$^{40}Ar/^{39}Ar$	$^{37}Ar/^{39}Ar$ [c]	$^{36}Ar/^{39}Ar$	$^{36}Ar_{Ca}$ (%)	$^{39}Ar_{Ca}$ (%)	$^{40}Ar_K$ (%)	$^{40}Ar_R$ [b] (%)	AGE [a] (Ma)
KHATCHATURIAN SEAMOUNT														
59D-15	Plag 125-250μ	0.151	1.507E-11	14.0	66.9 ±2.7	0.002558	86.94	36.646	0.24999	4.5	2.7	0.01	18.8	76.0 ±2.4
			1.441E-11	18.5										
59D-31	Plag 125-250μ	0.071	8.482E-12	16.0	80.8 ±3.0	0.002558	75.37	93.396	0.23047	11.4	6.4	0.01	19.9	72.5 ±1.6
						0.002260	19.57	*****	*****	**	**	**	**	78.1 ±3.5
	74-125μ	0.071	8.596E-12	20.5	81.9 ±1.9									
HAYDN SEAMOUNT														
33-6	Plag 125-250μ	0.081	7.597E-12	31.8	62.6 ±1.4	0.003069	208.55	85.821	0.68604	3.5	5.9	0.00	6.2	74.5 ±5.9
			7.264E-12	37.5		0.003069	22.32	81.123	0.05675	40.1	5.6	0.03	55.0 (H)	70.6 ±2.1
33-7	Plag 125-250μ	0.096	8.436E-12	42.0	62.7 ±2.7	0.003069	135.00	74.322	0.42913	4.9	5.1	0.00	10.6	81.9 ±6.2
			9.207E-12	13.7		0.003069	23.27	112.115	0.06577	47.8	7.7	0.03	56.4 (H)	77.0 ±1.5
	74-125μ	0.099	8.049E-12	3.5	55.7 ±3.2									
CENTRAL BACH RIDGE														
27-2	Basalt Core	0.540	3.745E-11	40.3	50.9 ±3.3	0.005380	13.34	12.183	0.02221	15.4	0.8	0.04	58.3	74.5 ±1.7
			4.265E-11	57.9		0.005380	7.87	*****	*****	**	**	**	**	74.8 ±2.4
27-3	Plag 74-150μ					0.003240	54.46	196.800	0.20330	27.2	13.5	0.01	19.7 (H)	71.0 ±3.0
27-7	Plag-basalt	0.710	5.332E-11	45.0	51.4 ±1.5	0.005380	46.50	13.974	0.12917	2.9	0.9	0.01	20.3	90.2 ±2.3
W. MENDELSSOHN SEAMOUNT														
38-1	Plag-basalt	0.808	6.610E-11	46.7	57.2 ±2.9	0.006204	16.37	6.552	0.03451	5.0	0.4	0.03	40.8	73.5 ±0.8
			6.415E-11	49.8		0.006677	7.25	*****	*****	**	**	**	**	85.2 ±1.2
	Plag 125-250μ	0.121	1.385E-11	42.7	77.8 ±0.8	0.005868	34.13	55.339	0.10320	15.1	3.8	0.02	24.1	88.2 ±4.0
			1.390E-11	21.9		0.001929	24.69	*****	*****	**	**	**	**	83.9 ±1.2
						0.003069	33.35	57.254	0.07952	20.2	3.9	0.02	43.8 (H)	82.2 ±2.3
						0.003069	23.54	54.572	0.04757	32.2	3.7	0.02	59.5 (L)	78.8 ±1.3
38-2	Ferrobasalt	0.998	7.232E-11	56.5	49.7 ±0.8	0.006204	18.37	5.182	0.03671	3.7	0.3	0.03	43.1	86.9 ±0.8
						0.006677	9.20	*****	*****	**	**	**	**	107.6 ±1.0
E. MENDELSSOHN SEAMOUNT														
37-2	Mugearite Core	3.125	2.575E-10	57.7	59.7 ±3.3	0.006204	8.07	1.231	0.00582	5.6	0.1	0.07	79.8	70.7 ±0.5
			2.885E-10	47.7		0.006677	5.93	*****	*****	**	**	**	**	70.0 ±0.5
	HCl lch	2.750	2.608E-10	61.7	64.5 ±1.1	0.004612	9.10	*****	*****	**	**	**	**	74.1 ±0.6
			2.591E-10	38.7										
	HF lch	2.405	2.331E-10	79.3	65.5 ±0.8	0.004266	9.85	*****	*****	**	**	**	**	74.3 ±0.7

**: Recombined total-fusion ages, $^{40}Ar/^{39}Ar$ reported is $^{40}Ar_R/^{39}Ar_K$. (H),(L): High or low temperature step (c.f. Table 6).

a Calculated using Lamda E = $0.581 \times 10^{-10} yr^{-1}$, lamda B = $4.962 \times 10^{-10} yr^{-1}$. Errors are standard deviation of analytical precision.

b Subscripts indicate radiogenic (R), calcium-derived (Ca), and potassium-derived (K) argon (%).

c Corrected for ^{37}Ar decay, half-life = 35.1 days.

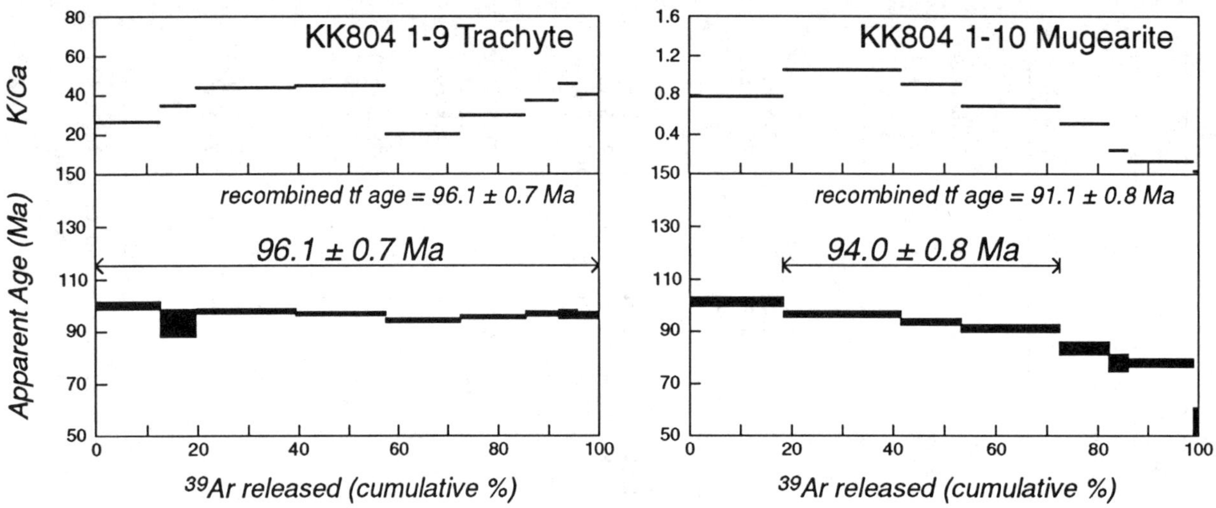

Figure 3. $^{40}Ar/^{39}Ar$ age spectra and K/Ca plot for two whole rock samples from an unnamed seamount in the NW seamount cluster of the Musicians Seamounts. Age spectra steps drawn 2 s.d. tall.

not be determined from (1) weighted averages of plagioclase conventional K-Ar ages, (2) individual plagioclase $^{40}Ar/^{39}Ar$ total-fusion ages, and (3) even some groups of whole rock $^{40}Ar/^{39}Ar$ total-fusion ages.

Individual Seamount Ages

NW Cluster. A previously uncharted seamount on the southeast flank of the cluster of seamounts at 34°N, 167°W yielded several differentiated alkalic rocks suitable for incremental-heating analysis. A potassic tracyhte, KK804 1-9, yielded a completely undisturbed age spectrum (Figure 3, left; Table 3), with 100% of the gas released. Since the sample proved to be undisturbed, a second isochron analysis (Table 3) included a separate total-fusion age in an attempt to improve the precision of the age estimate. This yielded the best age estimate for this seamount, with an isochron age of 95.7 ±0.7 Ma, SUMS/N-2 of 1.71, and a $^{40}Ar/^{36}Ar$ intercept of 296.2 ±1.5. The concordance of the whole gas analysis with each of the incremental-heating steps lends further support to the validity of this age.

A second sample, KK804 1-10, did not yield a reliable age according to the criteria used here. The age spectrum (Figure 3, right) shows the series of decreasing ages usually attributed to reactor-induced ^{39}Ar recoil. The middle temperature argon release of this experiment yielded a discordant pseudo-plateau, with an isochron age of 97.0 ±3.4 Ma and SUMS/N-2 value of 4.48 (Table 3), indicating a significant source of non-analytical error (Table 2). Including a total-fusion age with the pseudo-plateau data improved the isochron fit to an acceptable SUMS/N-2 of 2.24 and age of 97.0 ±2.4 Ma. The concordance between the isochron analysis with and without the total-fusion datum, and the fact that the whole gas datum significantly improves the scatter about the isochron, suggests that recoil experienced by the sample was more an internal redistribution of the argon isotopes rather than a recoil loss out of the sample. In any case, the age from this disturbed sample is consistent with the age determination from the trachyte described above.

A third sample, hawaiite KK804 1-19, seemed suitable for $^{40}Ar/^{39}Ar$ dating according to the general criteria described in Pringle [1992]. However, two conventional K-Ar whole rock analyses at least 50% younger than the fresher samples from this seamount (Table 5), suggested that this sample had been significantly affected by alteration. The poor reproducibility of the two analyses also suggested alteration effects, and the sample was not examined further.

Hammerstein Seamount. Two samples of mugearite from Hammerstein Seamount both yielded concordant incremental-heating experiments, despite evidence of significant ^{39}Ar redistribution (Figure 4). The microcrystalline trachytic groundmass of these samples, partially altered to fine-grained clays, proved susceptible to significant ^{39}Ar loss. The total-fusion and recombined total-fusion ages of these samples are 6-15% too old. The low temperature steps of the age spectra are even older (Figure 4), with apparent ages approximately twice the crystallization age. The low temperature steps also had the highest K/Ca ratios, indicating the presence of high potassium clays which released their argon at these temperatures. The incremental-heating experiment on KK804 5-16 yielded two concordant steps with 66.0% of the ^{39}Ar released (Figure 4, left). Dividing the lower temperature argon release into more steps, sample KK804 5-27 gave four concordant mid-temperature steps with 65.0% of the ^{39}Ar released (Figure 4, right). The best age estimate for Hammerstein Seamount is the weighted average of the isochron ages for these two experiments, 94.0 ±0.8 Ma (Table 3).

Mahler Seamount. Neither of the two samples from Mahler Seamount yielded completely concordant results (Figure 5; Table 3). The incremental-heating experiment on sample KK804 7-1, a fine-grained mugearite with 10% clays in the groundmass,

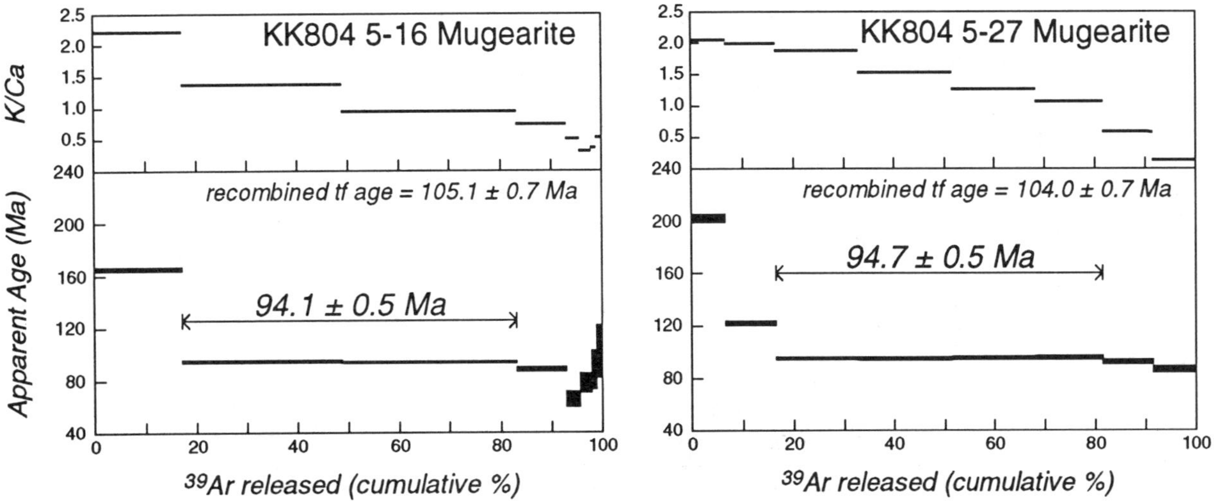

Figure 4. $^{40}Ar/^{39}Ar$ age spectra and K/Ca plot for two samples from Hammerstein Seamount. Age spectra steps drawn 2 s.d. tall.

revealed a disturbed age spectrum suggestive of ^{39}Ar redistribution (Figure 5). The three temperature steps from 600 to 800°C yielded a concordant age spectrum and isochron analysis, however, they account for only 42.9% of the ^{39}Ar released. Like the two samples for the NW Cluster described above, the recombined age from the incremental-heating experiment and the separate total-fusion age were concordant with the plateau described above, implying ^{39}Ar redistribution within rather than ^{39}Ar loss out of the sample. Including the total-fusion age in the isochron calculation resulted in an isochron age of 91.4 ±0.7 Ma with a SUMS/N-2 of 1.95 and a $^{40}Ar/^{36}Ar$ intercept of 290.8 ±5.2, almost identical to the results without the whole gas datum (Table 3). This age is the best age estimate for this sample, but it did not completely meet the criteria used for this study since less than 50% of the ^{39}Ar released was on the plateau.

Several different size fractions of plagioclase separated from a relatively altered, vesicular plagioclase-phyric basalt, KK804 7-17, yielded a different age estimate for Mahler Seamount. An incremental-heating experiment on the coarser (125-250μ) size fraction revealed a concordant but relatively imprecise age of 84.0 ±2.4 Ma (Table 3). The inverse correlation between apparent age and K/Ca ratio (Figure 5) suggests that at least some of the disturbance in the sample is the result of ^{39}Ar and/or ^{37}Ar redistribution. A incremental-heating experiment on the finer (88-125μ) size fraction (Table 4) did not reveal a concordant age spectrum or isochron analysis (age spectrum not shown). However, an isochron analysis of the three $^{40}Ar/^{39}Ar$ total-fusion experiments on different splits of this plagioclase proved concordant, with an isochron age of 82.0 ±1.4 Ma (Figure 6). As with the other plagioclase total-fusion analyses, the

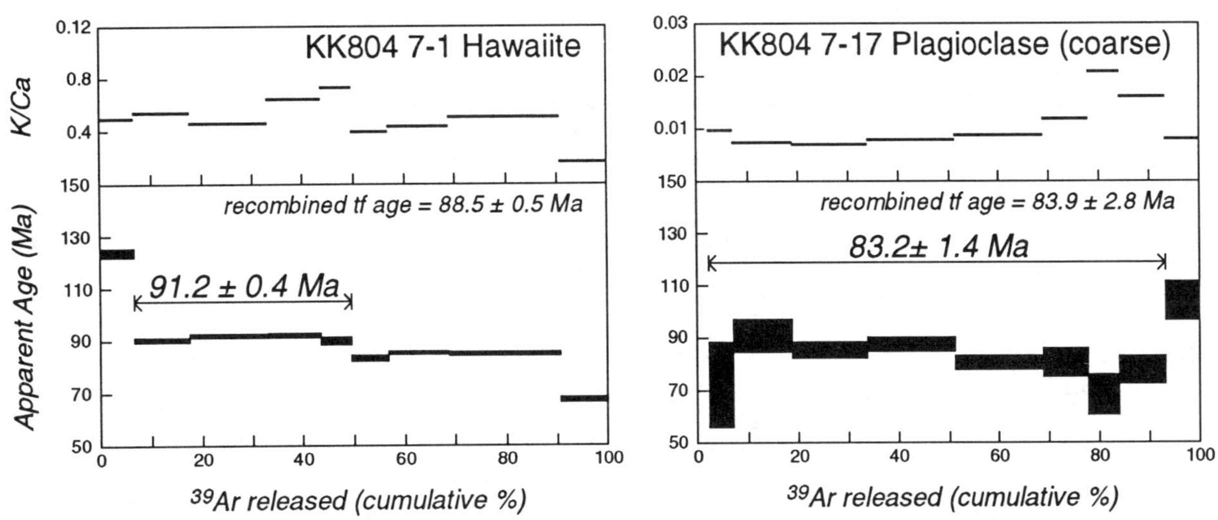

Figure 5. $^{40}Ar/^{39}Ar$ age spectra and K/Ca plot for two samples from Mahler Seamount. Age spectra steps drawn 2 s.d. tall.

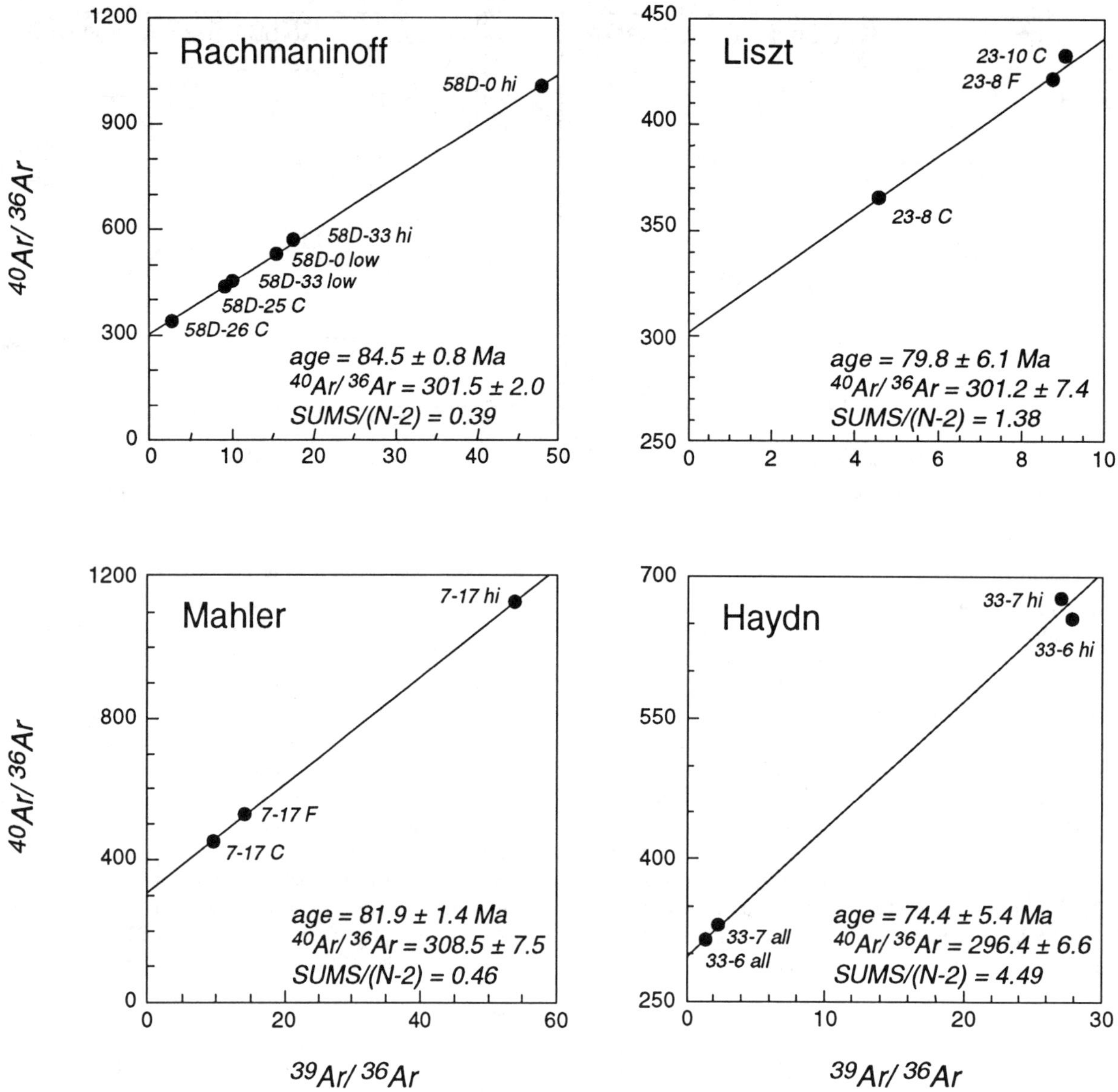

Figure 6. Plagioclase $^{40}Ar/^{39}Ar$ total-fusion isochrons for samples from Rachmaninoff, Liszt, Mahler, and Haydn Seamounts. Sample points are labeled by last digits of sample indentification number and *C* for coarse grained split, *F* for fine-grained split, *all* for whole gas total-fusion age, and *hi* and *low* for high and low temperature steps of two-step experiments.

weighted mean of the total-fusion ages is considered the best estimate of the crystallization age of this sample, 83.9 ±0.9 Ma.

The two age estimates for the two samples from this seamount are different at the 95% confidence level. One age estimate for the whole seamount is the simple arithmetic mean of the two samples, 87.7 ±5.3 Ma. Alternatively, the ages may indicate inception of volcanism at 91.4 ±0.7 Ma, continuing at least until 83.9 ±0.9 Ma. The latter explanation is preferred because it is more consistent with the paleomagnetic modeling for Mahler Seamount [Sager and Pringle, 1987] and the ages for other nearby seamounts (Figure 2). Also, the plagioclase separated from KK804 7-17 is more altered than any of the other plagioclase separates studied here and may only represent a minimum age, in spite of the $^{40}Ar/^{39}Ar$ analysis used here.

Rachmaninoff Seamount. Six $^{40}Ar/^{39}Ar$ total-fusion ages on four different plagioclase separates provided the most well-constrained total-fusion isochron analysis in this study (Table 3; Figure 6). The ages for A7-58D-25 and A7-58D-26 were published previously by Clague and Dalrymple [1975], and are recalculated here using the decay corrections and isotopic abundances recommended by Steiger and Jäger [1977]. The age from the finer-grained aliquot (nominally 74-100μ) from sample A7-

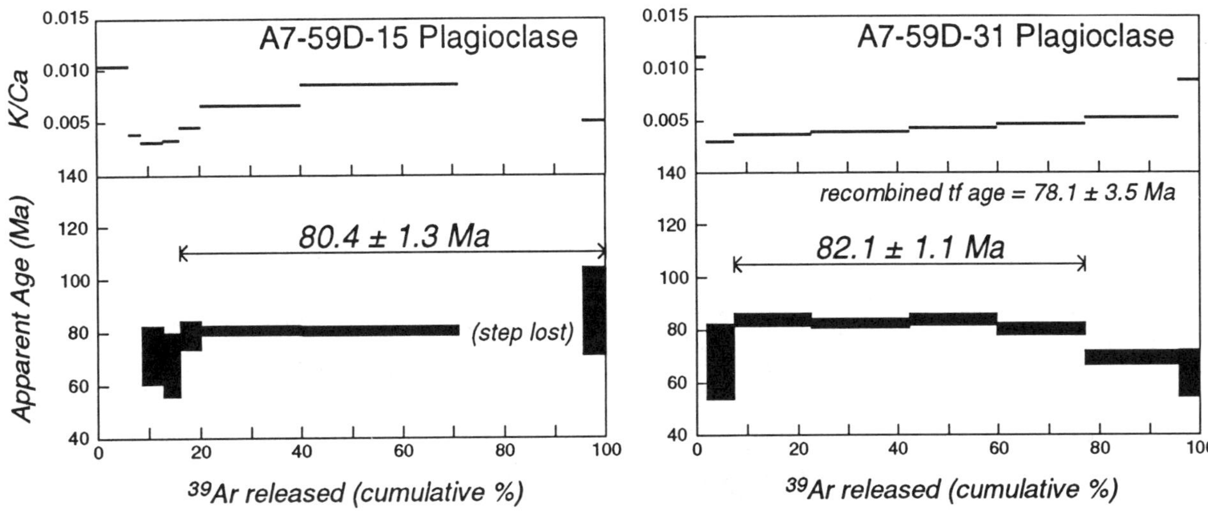

Figure 7. $^{40}Ar/^{39}Ar$ age spectra and K/Ca plot for two samples from Khatchaturian Seamount. Age spectra steps drawn 2 s.d. tall.

58D-25 was not included in this analysis because a re-examination of the original data discovered a computational mistake in the reported age, which should have been over 102 Ma. This was significantly older than the age from the coarser aliquot (100-250μ), possibly because of greater ^{39}Ar recoil from the finer-grained material, and this age was not considered further. The preferred age estimate from the other five samples is the weighted total-fusion age, 86.4 ±1.2 Ma. The isochron analysis confirms the internal consistency of the age, with an isochron age of 84.5 ±1.9 Ma, $^{40}Ar/^{36}Ar$ intercept of 301.5 ±5.1, and SUMS/N-2 of 0.39.

Liszt Seamount. Three $^{40}Ar/^{39}Ar$ total-fusion ages on two plagioclase separates from two samples of the same lithology yielded a weighted age of 84.4 ±1.5 Ma (Table 3; Figure 6), which is the preferred age estimate for Liszt Seamount. The total-fusion isochron analysis is concordant with this weighted age but much less precise, with an isochron age of 79.8 ±5.2, an $^{40}Ar/^{36}Ar$ intercept of 301.2 ±6.3, and a SUMS/N-2 of 1.38.

Haydn Seamount. Four $^{40}Ar/^{39}Ar$ total-fusion ages on plagioclase separates from two samples of the same lithology yielded a weighted mean age of 75.1 ±1.2 Ma (Table 3; Figure 6), which is the preferred age estimate for Haydn Seamount. Two of these total-fusion ages are whole gas analyses, with an estimated analytical precision of 7.5 and 8.0%. The other two ages are the higher temperature gas samples from two different two-step total-fusion analyses, with an estimated analytical precision of 1.9 and 3.0%, respectively. The two-step total-fusion analyses clearly increased the precision of the best age estimate for this

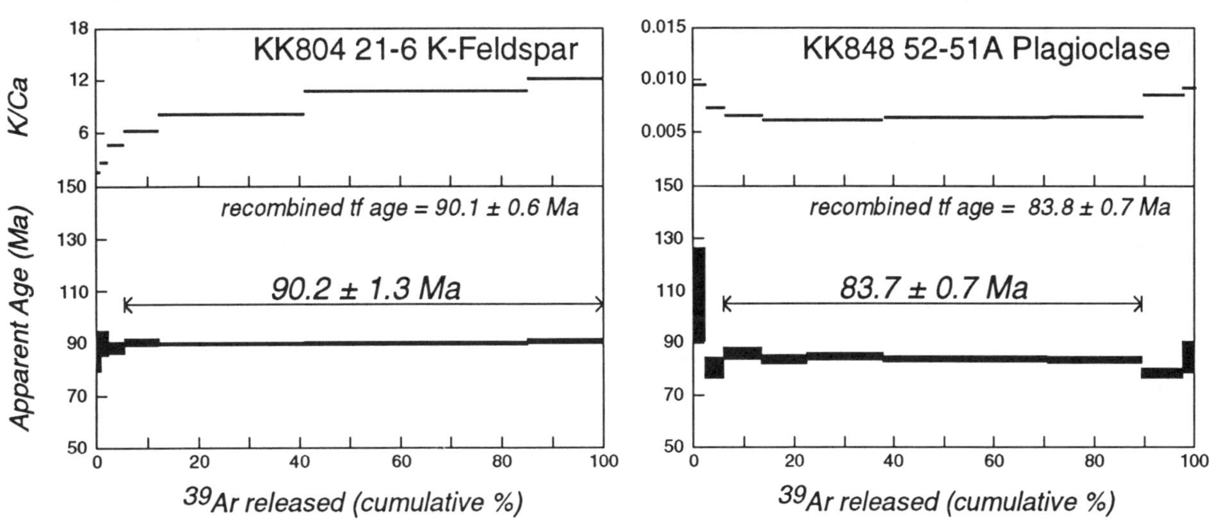

Figure 8. $^{40}Ar/^{39}Ar$ age spectra and K/Ca plot for samples from Brahms Seamount (left) and Schumann Seamount (right). Age spectra steps drawn 2 s.d. tall.

TABLE 6. Summary of $^{40}Ar/^{39}Ar$ Two-Step-Heating Experiments on Mineral Separates from the Musicians Seamounts

SAMPLE	FUSION STEP				600° STEP				PERCENT ARGON IN 600° STEP	
	AGE (Ma)	±1 sd	$^{40}Ar_R$ (%)	K/Ca	AGE (Ma)	±1 sd	$^{40}Ar_R$ (%)	K/Ca	$^{40}Ar_{atm}$	$^{39}Ar_K$
HF treatment only: Plagioclase										
KK804 23-8	82.3	±2.0	29.9	0.0080	-27.7	±30.4	-0.5	0.0095	78.6	15.2
KK807 33-6	70.6	±2.1	55.0	0.0057	4.8	± 5.0	1.0	0.0034	42.3	16.4
KK807 27-3	71.0	±3.0	19.7	0.0022	44.5	±14.3	3.4	0.0020	47.0	19.6
A7-59D-15	76.0	±2.4	18.8	0.0120	93.5	±22.8	3.4	0.0046	61.8	19.1
A7-59D-31	72.5	±1.6	19.9	0.0049	7.1	±17.0	0.2	0.0042	65.0	16.8
HF and warm HCl treatment: Plagioclase										
KK804 7-17	83.2	±1.0	73.8	0.0091	59.0	±17.6	6.1	0.0209	21.8	3.2
KK804 22-1	88.2	±1.8	43.0	0.0118	82.5	± 5.0	15.8	0.0117	24.5	11.3
KK807 33-7	77.0	±1.5	56.4	0.0044	77.2	±10.8	22.4	0.0045	19.9	8.9
KK807 38-1	78.8	±1.3	59.5	0.0086	82.2	± 2.3	43.8	0.0082	24.1	18.3
A7-58D-0	84.5	±1.8	70.6	0.0157	86.1	± 2.5	44.3	0.0151	24.0	16.3
A7-58D-33/34	89.4	±3.0	48.2	0.0136	88.5	± 2.9	35.0	0.0136	59.6	51.9
HF and warm HCl treatment: Pyroxene										
KK804 22-1	-69.1	±36.6	-2.5	0.0003	54.0	±78.6	3.1	0.0800	22.7	30.9

seamount. Similar to Liszt Seamount, the total-fusion isochron analysis is concordant with, but much less precise than, the weighted total-fusion age, with an isochron age of 74.4 ±2.6 Ma and $^{40}Ar/^{36}Ar$ intercept of 296.4 ±3.1. However, the SUMS/N-2 of 4.49 indicates more scatter about the isochron than could be explained by experimental error alone, and some caution should be used in the interpretation of this age.

Khatchaturian Seamount. Two incremental-heating experiments on plagioclase separates from two samples of the same lithology both yielded concordant ages for Khatchaturian Seamount (Figure 7; Table 3). Sample A7-59D-31 revealed the more precise age estimate, with an isochron age of 82.5 ±2.9 Ma representing 69.7% of the ^{39}Ar released. Unfortunately, the 1100° step from the incremental-heating experiment on sample A7-59D-15 was lost during the extraction process, but the remaining steps yielded an isochron age of 81.5 ±5.0 Ma representing 59.4% of the ^{39}Ar released. The best age estimate for Khatchaturian Seamount is the weighted average of these two isochron ages, 82.2 ±2.5 Ma.

Clague and Dalrymple [1975] reported a significantly younger conventional K-Ar age of 66.9 ±2.6 Ma on the same plagioclase separate used here for A7-59D-15, which highlights the problem of dating even fresh-looking plagioclase separates with the conventional K-Ar technique.

Brahms Seamount. Incremental-heating analysis of a potassium feldspar separate from a highly altered trachyte, KK804 21-6, yielded a concordant age of 89.7 ±0.6 Ma (Table 3). K/Ca increased with temperature for this sample (Figure 8, left), suggesting that potassium feldspar is not the principal potassium-bearing phase in samples that have expeirienced significant amounts of recoil-induced argon redistribution, where K/Ca tends to decrease with temperature.

A two-step total-fusion experiment on a plagioclase separate from the plagioclase-phyric basalt KK804 22-1 yielded an age of 88.2 ±1.8 for the high temperature step, with 43.0% radiogenic argon and 88.7% of the total K-derived ^{39}Ar released (Tables 5 and 6). The age of the lower temperature step is not significantly different, 82.9 ±5.0 Ma, and does have a higher atmospheric component, with only 16.1% radiogenic ^{40}Ar. An attempt was made to date a pyroxene separate from this same sample with another $^{40}Ar/^{39}Ar$ two step experiment, but the results from both the fusion and low temperature step could not be distinguished from atmospheric argon (Table 5). These results from the plagioclase separate confirm the incremental-heating age of the potassium feldspar from KK804 21-6, which is the preferred age for Brahms Seamount.

Schumann Seamount. The best age estimate for Schumann Seamount is from a single incremental-heating experiment on a plagioclase separate from KK848 52-51A (Figure 8, right; Table 3). The preferred age is the isochron age of 83.0 ±1.0 Ma. This is the only seamount age that is based on the results from a single sample. Some caution should be exercised with the interpretation of this age until additional samples from this seamount can be analyzed.

West Mendelssohn Seamount. Incremental-heating experiments revealed concordant ages on both a whole rock core and a plagioclase separate from KK807 38-1 (Figure 9, upper and middle left; Table 3). The plagioclase separate yielded isochron

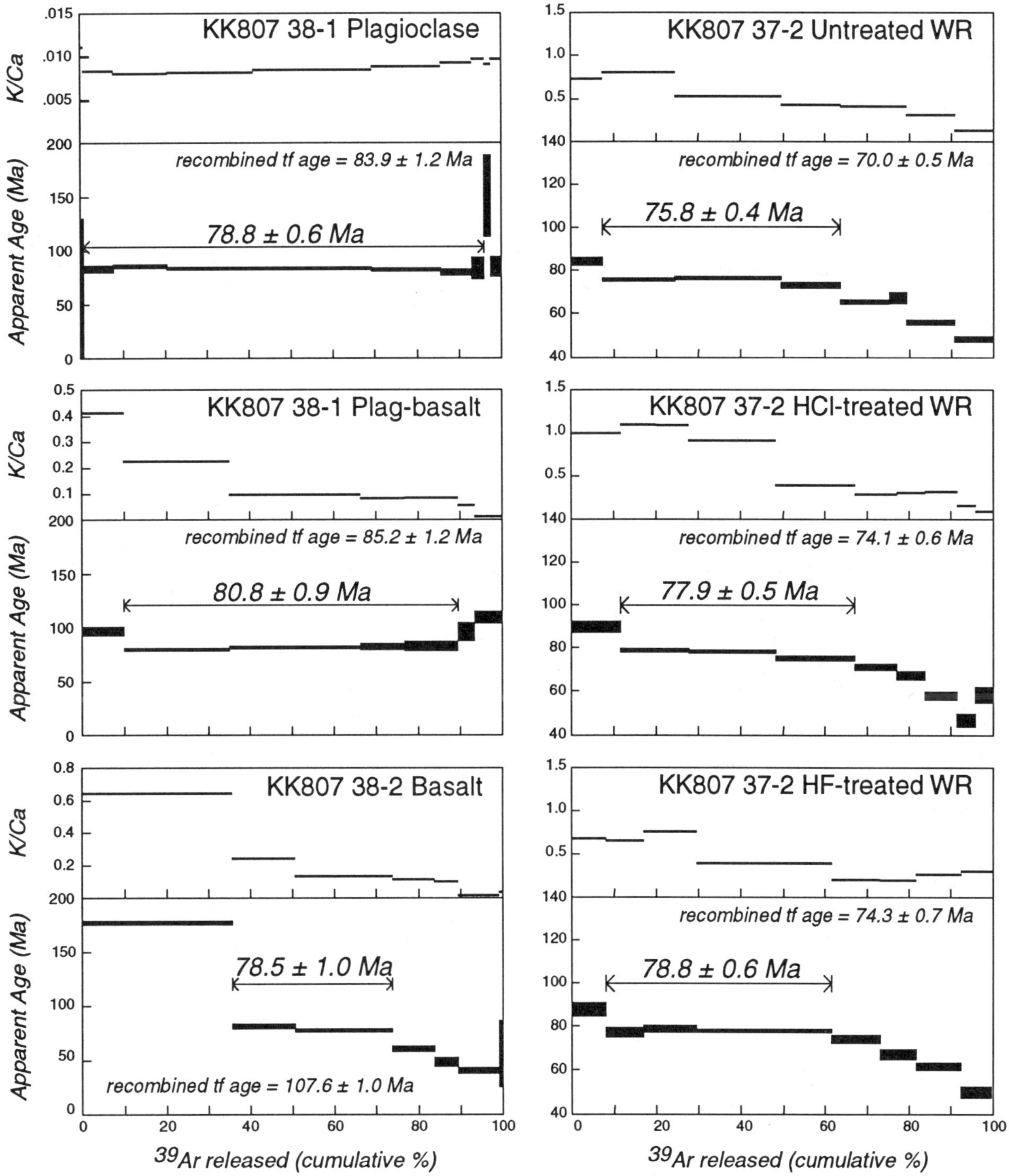

Figure 9. $^{40}Ar/^{39}Ar$ age spectra and K/Ca plot for three experiments each on samples from West Mendelssohn Seamount (left) and East Mendelssohn (right). Age spectra steps drawn 2 s.d. tall.

age of 82.0 ±1.4 Ma. The whole rock analysis was less precise, with an isochron age of 84.0 ±2.9. The best age estimate for West Mendelssohn Seamount is the weighted mean of the two isochron ages, 82.4 ±1.3 Ma.

The incremental-heating experiment on ferrobasalt KK807, 38-2 (Figure 9, lower left) shows strong evidence of ^{39}Ar loss and redistribution, gives a recombined age 32% higher than the concordant experiments described above, and has no age plateau. It provides no useful information on the age of the seamount.

East Mendelssohn Seamount. Sample KK807 37-2, a fine-

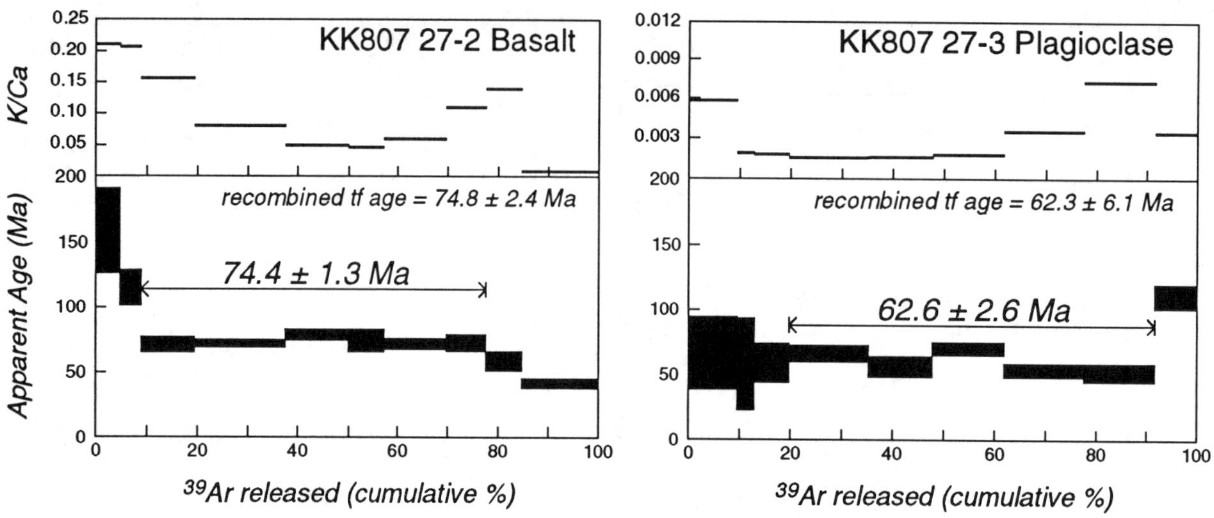

Figure 10. $^{40}Ar/^{39}Ar$ age spectra and K/Ca plot for two samples from central Bach Ridge. Age spectra steps drawn 2 s.d. tall.

grained, holocrystalline mugearite, was the only sample recovered from East Mendelssohn Seamount that was a suitable candidate for $^{40}Ar/^{39}Ar$ analysis. The incremental-heating experiment on an untreated whole rock core showed strong evidence of ^{39}Ar redistribution (Figure 9, upper right). The isochron age of 76.6 ±0.6 is almost completely concordant, with only the slight but significant difference between the $^{40}Ar/^{36}Ar$ intercept of 287.8 ±3.0 (Table 3) and the air ratio of 295.5 violating the reliability criteria used for this study.

In an attempt to minimize the effects of ^{39}Ar recoil and redistribution, two different acid treatments were tried on this sample. First, a mild 2.5N HCl treatment was tried on a 250-500μ size fraction. The incremental-heating experiment revealed an age distribution very similar to the untreated core sample (Figure 9, middle right). As with the untreated sample, only the small but significant difference between the $^{40}Ar/^{36}Ar$ intercept of 285.3 ±4.2 and the atmospheric value suggests some problems with the isochron age of 79.7 ±0.9 Ma (Table 3). Second, the HF treatment used routinely for plagioclase separates was tried on a 150-250μ size fraction. The incremental-heating experiment on this aliquot also yielded a very similar age distribution: with 53.7% of the ^{39}Ar released, a $^{40}Ar/^{36}Ar$ intercept of 303.9 ±20.8, a SUMS/N-2 of 0.50, and an isochron age of 78.5 ±0.9 Ma (Figure 9, bottom right; Table 3). Because only this last sample resulted in a $^{40}Ar/^{36}Ar$ intercept indistinguishable from the atmospheric ratio of 295.5 at the 95% confidence level, and because the predicted age is between the isochron ages of the other two splits, the isochron age of 78.5 ±0.9 Ma from the HF treated split is preferred as the best age estimate for East Mendelssohn. Using a weighted average of all three experiments would not result in a significantly different age determination nor alter the conclusions of this study.

Central Bach Ridge. The incremental-heating experiment on an aphyric tholeiitic basalt from central Bach Ridge was concordant, with an isochron age of 74.4 ±3.3 Ma (Figure 10; Table 3). Since both the recombined total-fusion age of 74.8 ±2.4 Ma (Table 4) and the whole gas total-fusion age of 74.5 ±1.7 Ma (Table 5) are concordant with this age, the age spectrum suggests an internal redistribution of ^{39}Ar rather than a loss or gain of ^{40}Ar or ^{39}Ar in or out of the sample. Like the samples from the NW Cluster, the precision of the age estimate can be improved by including the whole gas total-fusion analysis (Table 5) with the incremental-heating data into one isochron analysis (Table 3). This results in the best age estimate for the central Bach Ridge, 74.6 ±1.7 Ma. A concordant incremental-heating experiment on a plagioclase separated from a different lithology agreed with the age estimate from the aphyric tholeiite, but was so imprecise (73.1 ±8.1 Ma) that it has no effect on the best age estimate for Bach Ridge.

Conventional and $^{40}Ar/^{39}Ar$ Total-Fusion ages

As expected for altered samples, the K-Ar and $^{40}Ar/^{39}Ar$ apparent ages from the same sample commonly do not agree (Table 5). The age discordance, defined as the difference between the conventional K-Ar and $^{40}Ar/^{39}Ar$ total-fusion ages for each sample and the best age estimate for that seamount, are shown in Figure 11 as a percentage of the best age estimate. All of the feldspar separates shown in Figure 11 yielded concordant incremental-heating ages. The group of whole rock samples plotted in the center of Figure 11 had internally concordant incremental-heating ages; those on the right were too disturbed to yield concordant ages themselves.

Also as expected, conventional K-Ar whole rock ages of altered samples are younger than the most reliable estimates of crytallization age. What is surprising in Figure 11 is that almost all of feldspar data are also younger than the best estimates of crystallization age. Apparently, even plagioclase separates without much visible signs of alteration may not give accurate conventional K-Ar ages. However, conventional feldspar K-Ar ages are still better minimum age estimates than conventional whole

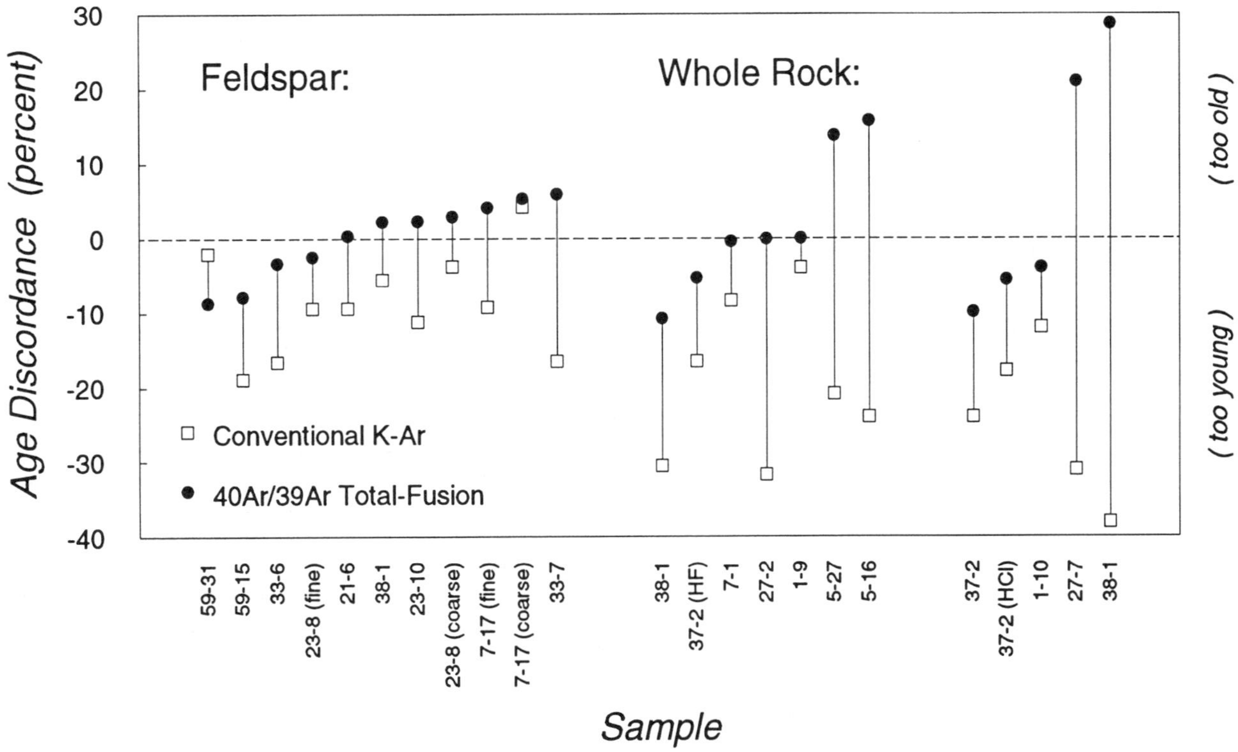

Figure 11. Age discordance of conventional K-Ar and $^{40}Ar/^{39}Ar$ total-fusion ages shown as percentage of best estimate of the appropriate seamount age. Feldspar apparent ages on left, whole rock apparent ages on right. Whole rock samples in center had concordant $^{40}Ar/^{39}Ar$ age spectra, those on right were more disturbed. Note that conventional K-Ar ages are almost always less than best age estimate, whereas $^{40}Ar/^{39}Ar$ total-fusion ages can either be higher or lower.

rock ages. In this study, conventional feldspar ages ranged from 0-18% too young, while conventional whole rock ages ranged from 5-38% too young.

Feldspar $^{40}Ar/^{39}Ar$ total-fusion ages are much more accurate. Most are within analytical error of the concordant incremental-heating experiment ages. However, some feldspar total-fusion ages are up to 8% too young. These anomalously young apparent ages may be the result of radiogenic ^{40}Ar loss from sites which have retained some of their K-derived ^{39}Ar, from ^{37}Ar recoil loss (which would underestimate the Ca-derived ^{36}Ar and overestimate the amount of atmospheric ^{36}Ar and ^{40}Ar), or from the presence of younger alteration phases which were not completely separated from the original feldspar. There are no independent criteria for evaluating the reliability of individual total-fusion feldspar ages, but a set of total-fusion ages can be tested with the same criteria used for incremental-heating experiments. In the worst case, feldspar K-Ar ages may be considered reliable minimum age estimates.

Many of the plagioclase separates dated by the conventional K-Ar technique had a higher atmospheric component than normally would be expected from fresh plagioclase separates of this age (Table 5). About one third of the feldspar total-fusion experiments were conducted as two-step incremental-heating experiments, using an initial low temperature (≤600°C) step in an attempt to remove a significant fraction of this atmospheric contaminant. As much as 80% of the trapped atmospheric component, but less than 20% of the radiogenic ^{40}Ar and K-derived ^{39}Ar, is released during these low temperature steps (Table 6, Figure 12). The difference between atmospheric argon and the composition of these low temperature steps cannot be detected, the small loss of radiogenic ^{40}Ar and K-derived ^{39}Ar does not effect the $^{40}Ar/^{39}Ar$ age of the fusion step, and the precision of the high temperature age determined for the plagioclase is greatly improved. Note that this cleaning step is more effective for those plagioclase separates which had been treated only with HF. Apparently, the warm HCl leaching step accomplishes some, but not all, of the same cleaning as the low temperature degassing step (Figure 12).

The age discordance for the whole rock total-fusion data is greater than for the feldspar data. Most samples can be constrained to only ±20-30%. Note that it is possible for $^{40}Ar/^{39}Ar$ whole rock total-fusion ages to be significantly older than the "true" crystallization age, here up to 32% too old (Figure 11, right hand side). This is most likely because of K-derived ^{39}Ar loss from alteration phases that have retained at least some of their radiogenic ^{40}Ar, as discussed above. Even whole rock samples exhibiting concordant incremental-heating age spectra, i.e. those with relatively undisturbed $^{40}Ar/^{39}Ar$ signatures, may

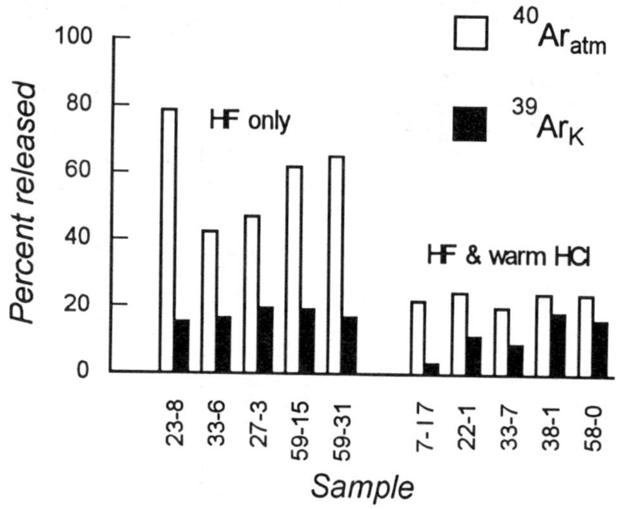

Figure 12. Atmospheric ^{40}Ar and K-derived ^{39}Ar released in low temperature steps of $^{40}Ar/^{39}Ar$ two-step experiments on plagioclase. Separates on left were treated with 10% cold HF only, separates on right were also treated with warm 6N HCl. Note that the additional HCl leach cleans plagioclase of most, but not all, of the low temperature, high atmospheric argon component.

have total-fusion ages up to 10% too old or too young (Figure 11, center group). Thus, unlike the results from conventional K-Ar analysis of altered samples, whole rock total-fusion ages may not be used as minimum age estimates. For the samples examined in this study, the range in age discordance for the whole rock total-fusion experiments, -10 to +32%, is actually greater than the range observed for conventional whole rock experiments, -5 to -38%. This conclusion emphasizes the fact that there are no independent criteria with which to evaluate the reliability of individual $^{40}Ar/^{39}Ar$ whole rock total-fusion ages, and such ages should be interpreted cautiously.

DISCUSSION

Test of the Kinematic Hot Spot Hypothesis

Six seamounts along the NW-SE chain show a well-defined age progression from 95.7 Ma for the seamount cluster in the NW to 82.5 Ma for Khatchaturian Seamount in the SE (Figure 2). Thus, the kinematic hot spot mechanism of formation can be accepted for this seamount chain. The older of the two ages for Mahler Seamount is part of this trend, i.e. inception of volcanism on Mahler Seamount defines part of the hot spot trace. The 75 Ma age for Haydn Seamount is also consistent with this age progression. However, the trace of this lineament is not clear after it passes south of the Murray fracture zone. Also, a change in Pacific plate spreading direction around 80 Ma [Sager and Pringle, 1987] may correspond to rotation about a different stage pole for the Pacific plate between about 70 and 80 Ma. This 70-80 Ma pole could be distinct from both older motion described by the Musicians NW-SE chain and younger motion described by the Emperor Seamount chain. Therefore, Haydn is not included in the following discussion of volcanic

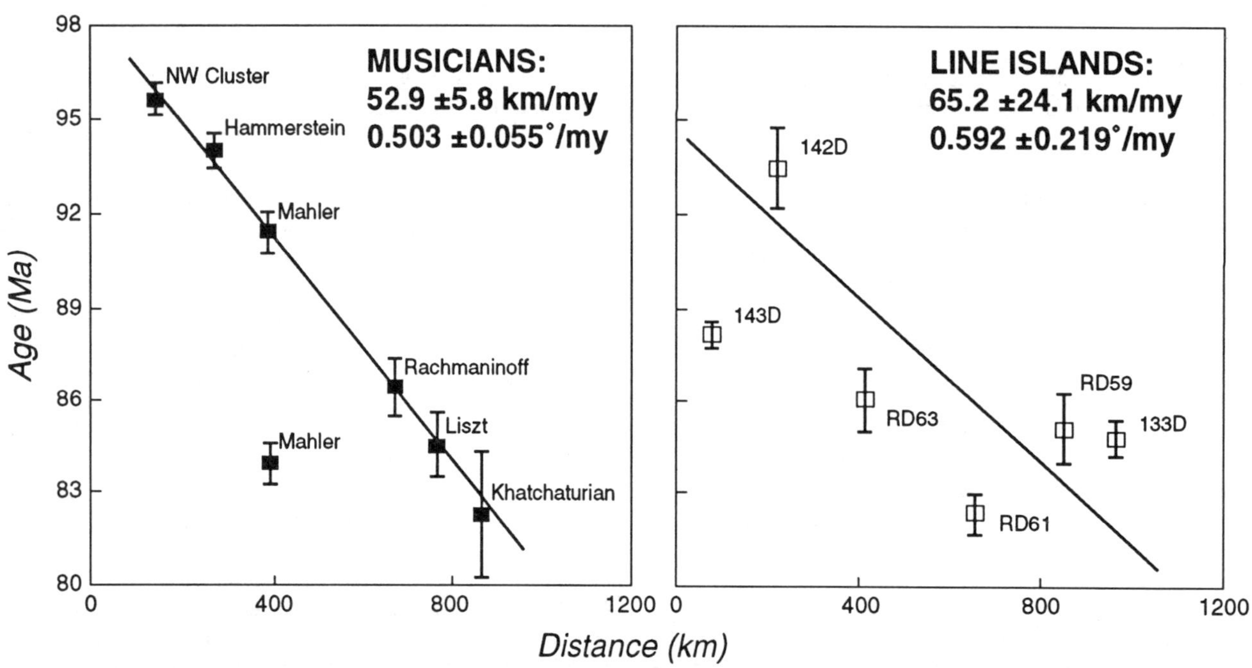

Figure 13. Volcanic age progressions along NW-SE trending, Late Cretaceous Pacific seamount chains. Musicians ages from this study, Line Islands ages from Schlanger et al. [1984]. Error bars drawn ±1 s.d. of reported analytical precision. Note that although linear rather than angular distances are plotted, only a 5% correction between the linear and angular rates is necessary because the Musicians trend and the Line Islands trace are about 71° and 82° away from the pole, respectively. A visual comparison of the two rates is sufficient here.

progression rates, which includes only seamounts older than 80 Ma.

It is also clear that no age progressive, hot spot origin is possible for the N-S seamount lineament. Brahms Seamount in the north is approximately the same age as the seamounts along the same latitude on the NW-SE lineament, and clearly cannot be part of a younger hot spot trace. Mendelssohn Seamount in the south is clearly older than any extension of the NW-SE trend, and also cannot be part of a younger hot spot trace. There is no evidence for volcanism in the Musicians Seamounts after 70 Ma, therefore, none of the volcanism which formed the Musicians Seamounts was coeval with volcanism along the (presently dated) Emperor Seamounts.

On an age versus distance plot of the six dated seamounts along the NW-SE seamount chain (Figure 13), the remarkably good linearity of the age progression is apparent. The calculated progression rate for the NW-SE seamount chain is 52.9 ±5.8 km/m.y., or 0.503 ±0.055°/m.y. around a pole at 36°N, 76°W.

Test of the Fixed Hot Spot Hypothesis

A plot of paleolatitude of formation, determined by modeling sea surface magnetic anomalies [Sager and Pringle, 1987], versus distance along the hot spot trace indicates that all of these seamounts formed at the same latitude, several degrees north of the equator (Figure 14). This is consistent with the fixed hot spot hypothesis for the NW-SE chain. However, for relatively short seamount lineaments, where the total length of the age progression (here approximately 8°) approaches the resolution of the technique (approximately ±5° at the 95% confidence level), the paleolatitude data are not very useful in testing the fixed hot spot hypothesis.

The best test of the fixed hot spot hypothesis is to show identical *angular* rates of volcanic propagation for copolar seamount chains. Seamount lineaments formed by plate motion over a stationary melting anomaly should trace small circles about the rotation pole describing plate motion over the hot spot reference frame. In deriving 36°N, 76°W as the rotation pole for the Late Cretaceous Pacific plate, Epp [1978] used the three NW-SE trending seamount lineaments north of the Hawaiian Ridge (the Musicians lineament studied here, the North Gardner or Liliuokalani Ridge, and the Wentworth Seamount Chain). He then noted that this pole was also a best fit to the trend of the northern Line Islands.

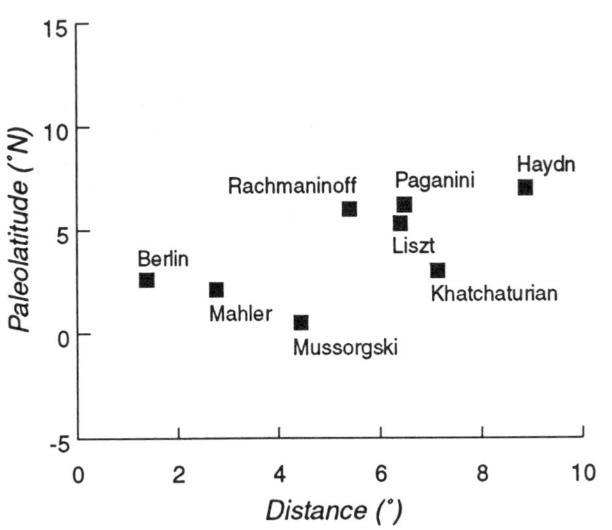

Figure 14. Paleolatitude versus distance for seamounts in NW-SE age progressive lineament of the Musicians Seamounts. Distance is plotted in degrees of arc from a point in the center of the cluster of seamounts in the northwestern corner of the province. Typical paleolatitude error is ±5° at 95% confidence level.

Of these other copolar seamount chains, age data are available only for the Line Islands and Wentworth Seamount Chain. An age progression along four sites of volcanism on the Wentworth Seamount chain, including two well-dated seamounts, is consistent with the Musicians results [Pringle and Dalrymple, this volume]. However, the data set is too limited to be included in a pooled estimate of Pacific plate velocity.

Schlanger et al. [1984] determined generally age progressive volcanism along the entire Line Islands chain, from 90 to 45 Ma at a rate of 96 ±4 km/m.y. However, they did not differentiate any change in the Pacific/hot spot rotation direction during that period. Epp [1984], using the same age data set, recognized at least two periods Pacific plate/hot spot motion, and estimated a rotation rate of about 0.9°/m.y. for the period 90 to 67 Ma.[1]

The six northern Line Islands seamounts which define the Line Islands age progression before 80 Ma are shown in Figure 13. Only those seamounts older than 80 Ma are included, in order to compare the two provinces directly and to avoid including any younger seamounts which might trace a different pole of Pacific plate rotation. Weighting the ages by the inverse of the reported variance [Schlanger et al., 1984] gives a rate of volcanic propogation along these six seamounts of 65.2 ±24.1 km/m.y., or 0.592 ±0.219 °/m.y. This rate is consistent with the Musicians results, but it is clear that the Line Islands age progression is not as well-defined. The imprecision in the Line Islands age progression may be due to the fact that the age data are derived from total-fusion experiments of whole rock samples. We now know that a $^{40}Ar/^{39}Ar$ whole rock total-fusion age determination can depart by more than ±20% of the crystallization age (Figure 11). A more conclusive test awaits higher quality age data from the northern Line Islands.

Assuming that both the Musicians and Line Islands age progressions represent fixed hot spot tracks, a combined estimate for the rotation rate between the Pacific plate and the hot spot frame of reference can be calculated. The best estimate of Pacific/hot spot motion for 82 to 96 Ma, 0.508 ±0.053 °/m.y., is calculated by weighting the two rates by the inverse of the variance and is dominated by the Musicians results. A more unbiased estimate awaits higher quality data from the Line Islands and further confirmation of age progressions along other Pacific plate seamount lineaments of similar age and orientation. The Line Islands also may offer the only sample set which can be used to determine the Pacific rotation rate and pole for the period between 82 and 70 Ma.

Age Progressive Volcanism in a Near-Ridge Environment

Not all age progressive, hot spot volcanism need be the result of long-lived, mantle plumes ascending from deep within the mantle. Of the Pacific hot spot tracks shown in Figure 1, only the Hawaiian-Emperor ridge has had long-lived, stable volcanism over the 70 m.y. represented by its track on the Pacific plate. The Line Islands and Louisville Ridge represent volcanism over 40-60 m.y., but the extension of that volcanism is either difficult to follow (Line Islands) or seems to dissipate (Louisville Ridge). All of the other age progressions shown in Figure 1 are less than 20 Ma, and, obviously, less than 20 m.y. long. In fact, the median length of the known age progressive seamount and island lineaments in the Pacific Basin is on the order of 15 m.y. This does not diminish the use of these shorter hot spot tracks in calculating the current Pacific/hot spot rotation rate (e.g., Duncan and Clague, 1985). It is suggested here that similarly "short" hot spot traces can be just as important in determining older Pacific/hot spot rotation rates, especially if precise, reliable ages are available.

In order to use an age progressive seamount lineament to model plate/hot spot rotation, all that is required is that the source of the melting anomaly (1) be deeper than the asthenosphere and (2) remained relatively fixed with respect to the hot spot reference frame (i.e., lower mantle?) over the life span of the hot spot trace. For example, an intermediate sized plume, say the size of a cylinder 50 km in diameter and 670 km high, undergoing 5% partial melting, could generate 15 seamounts 50 km in diameter and 3 km high. Even for the Iceland-Faroes hot spot trace, a region which most would describe as a long-lived, deep mantle plume, Schilling and Nygaard [1974] suggested that the upwelling of the plume material is cyclic on a timescale of 10-20 m.y., and in the form of "blobs" rather than a continuous column. Sun [1985] noted that the plume and plum-pudding geochemical models of the mantle need not be mutually exclusive. Many isolated seamounts with no obvious connection to long-lived hot spot traces share geochemical characteristics with long-lived hot spots such as Hawaii. All that is really required for both the plume and plum-type hot spot models is that their source region must be deeper than the immediate source region for MORB. Thus, seamount lineaments resulting from both intermediate size "plums" lasting only 10-20 m.y. and long-lived, deep mantle plumes lasting at least 70 m.y. can be useful in determining plate/hot spot motions.

Earlier, it was noted that gravity data indicate that many of the Musicians seamounts formed near the ridge crest and that all apparently formed on sea floor less than 15-20 m.y. old. We have concluded [Pringle, 1992; Pringle et al., 1990] that the trace element and isotopic geochemistry of the Musicians Seamounts, including those seamounts that are part of the age progressive NW-SE lineament, is more similar to that of near-ridge alkalic seamounts than to that of typical intraplate or hot spot islands and seamounts. This is still consistent with a kinematic hot spot origin of the age progressive seamounts of the NW-SE lineament, and may be consistent with a fixed hot spot origin.

Consider that the source of the seamounts in the NW-SE lineament was a hot spot maintained by an intermediate-sized "plum" of upwelling mantle. Melting may have proceeded by a mechanism similar to those described by Sleep [1984] or Zindler et al. [1984]. Whether or not the "plum" was fixed with respect to the hot spot reference frame, or migrated more than a few km/m.y. with convection under the ridge crest, can be tested only by the geophysical and geometric tests described above. For at least 16 m.y., the hot spot produced 15 or so seamounts in an age progressive trend. The melting anomaly probably contributed to the seamount and ridge volcanism between it and the spreading-center ridge crest to the east. Note that the 90 Ma age for Brahms Seamount is the same as that predicted for seamounts formed due west on the hot spot track (Figure 2). The hot spot track ends either in the volcanism of the ridges in the south-eastern quarter of the province or was carried off on the Farallon plate.

A modern analog for the Late Cretaceous Musicians hot spot trace is the Cobb-Eickelberg seamount chain in the northeastern Pacific, and whose hot spot is still active near the Juan de Fuca spreading center. Eaby and Clague [1984] and Desonie and Duncan [1990] found little geochemical distinction between lavas from these seamounts and those from typical mid-ocean ridge basalt found nearby. However, Desonie and Duncan [1990] did find a well-defined age progression along this chain, albeit less than 10 m.y. long, which accurately reflects the current Pacific/hot spot motion. This example emphasizes that relatively short hot spot tracks, although they may not fit the geochemical model of an ocean island basalt (OIB) plume, can represent examples of both kinematic and fixed hot spot models. Considering the success in using similarly-sized hot spot traces to model current plate/hot spot motion in the Pacific, it seems reasonable to assume that the Musicians hot spot was relatively fixed with respect to a Late Cretaceous hot spot reference frame.

Comparisons with the Pacific Apparent Polar Wander Path

The Pacific apparent polar wander path (APWP, Figure 15), which reflects Pacific plate/spin axis motion, diverges significantly from the motion predicted by Pacific plate/hot spot motion, especially for ages older than 40 Ma (Figure 15; Sager and Pringle, 1988; Gordon, 1983). Sager and Pringle [1988] concluded that rapid east-west polar wander, i.e., rapid Pacific/spin axis motion, occurred at a rate of 5.1 ±2.5 °/m.y. between 82 and 85 Ma. Their conclusion seems contrary to a major conclusion of this study, namely that Pacific/hot spot motion occurred at a rate of 0.51 ±0.05 °/m.y. in a NW- SE direction during the same time. At the very least, this discrepancy between Pacific/spin axis and Pacific/hot spot motion suggests that any test of the fixed hot spot hypothesis based on paleomagnetic data, such as the paleolatitude test described above, probably can not be justified for this time period.

Sager and Pringle [1988] concluded that most of the divergence of the Pacific hot spot and paleomagnetic polar wander paths between 39 and 65 Ma can be explained by true polar wander (TPW). They note that most of the studies of TPW suggest that the spin axis has moved about 8°-10° away from the Pacific since the Late Cretaceous [Morgan, 1981; Jurdy, 1981; Gordon and Cape, 1981]. Sager and Bleil [1987] concluded that most of the divergence occurred between 40 and 75 Ma.

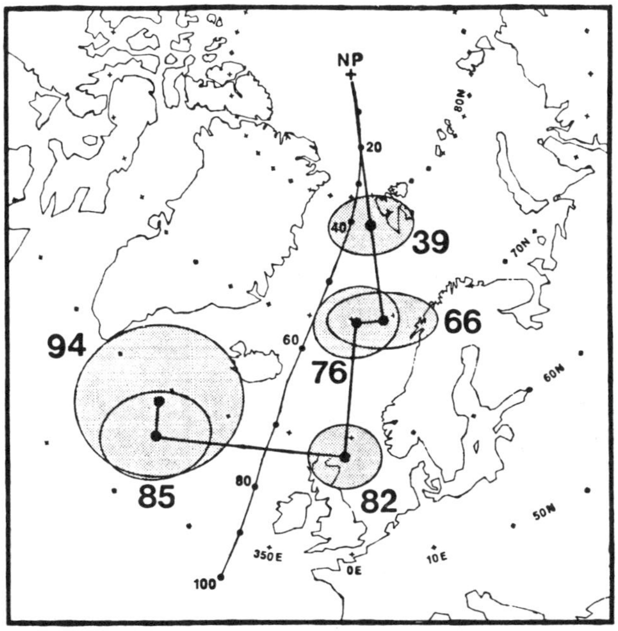

Figure 15. Pacific apparent polar wander path (APWP) for the Late Cretaceous and Early Tertiary (APWP II from Sager and Pringle, 1988). Large solid dots connected by heavyline are average poles on the APWP and shaded ellipses are 95% confidence regions. Large numbers are mean ages in Ma. Thin line represents polar wander path predicted by model of motion of Pacific plate relative to hot spots [Duncan and Hargraves, 1984]; dots are placed along path at 10 m.y. intervals. Map is polar equal area projection with Greenwich meridian vertical; NP stands for the north geographic pole. Note rapid east-west shift in APWP between the 82 and 85 Ma stage poles.

However, the east-west segment of the Pacific APWP has not been as easy to explain. Gordon [1983] preferred TPW to explain the discrepancy, while Jarrard and Sasajima [1980] proposed that the east-west trend of the Pacific APWP was the result of an east-west rotation of the Pacific plate between 82 and 85 Ma.

If the Jarrard and Sasajima [1980] hypothesis is correct, then the Pacific/hot spot models which predict NW-SE motion are in error. The kinematic hot spot hypothesis for the origin of the age progression along the NW-SE seamount lineament would still be valid. However, the location of the melting anomaly could no longer be fixed in the hot spot reference frame, and the volcanic propagation rate could no longer be used to constrain Late Cretaceous Pacific motion. The Late Cretaceous 36°N, 76°W Pacific/hot spot rotation pole determined by Epp [1978] would then be based solely on geometric arguments. Late Cretaceous Pacific hot spot models based on such a pole (for example, Henderson, 1985; Duncan and Hargraves, 1984; Duncan and Clague, 1985) would be similarly unwarranted.

However, if TPW is the primary cause of the rapid east-west shift in the Pacific APWP, then the hot spot APWP would be the correct path to use in describing Pacific plate motion. The 36°N, 76°W rotation pole derived by Epp [1978] is the best description of Pacific/hot spot rotation, and the age progression found for the NW-SE seamount lineament in the Musicians Seamounts provides most of the constraint on the rate of that rotation for the period 96 to 82 Ma. Further work confirming these hypotheses can only aid in trying to understand why such rapid rates and changes in direction in the Pacific APWP occurred at the very end of the Cretaceous Normal Polarity Superchron.

Conclusions

Thirteen ages for 12 seamounts in the Musicians Seamounts are reported in Table 3. These ages are based on information gained from 27 conventional K-Ar ages, 33 $^{40}Ar/^{39}Ar$ total-fusion ages, and 19 $^{40}Ar/^{39}Ar$ incremental-heating experiments. These data highlight the difficulty in obtaining high quality age data that accurately estimate the crystallization age of even relatively fresh whole rock splits and mineral separates from samples that have been altered in the submarine environment since Cretaceous time. Conventional K-Ar ages of apparently fresh looking feldspar separates, $^{40}Ar/^{39}Ar$ total-fusion ages of single feldspar separates, and even some groups of $^{40}Ar/^{39}Ar$ whole rock total-fusion ages all proved unable to provide the high quality age estimates necessary to decipher the history of the Late Cretaceous seamounts examined here. Nine of the 13 ages reported here are based on concordant incremental-heating experiments, whereas the other four ages are averages of at least three $^{40}Ar/^{39}Ar$ total-fusion ages from plagioclase separates.

The dates determined for the seamounts studied here describe volcanism occurring in the Musicians Seamounts for over 20 Ma, from 96 Ma in the northwest corner of the province to 75 Ma on the east-west trending ridges in the southeast quadrant of the province. Any model of the formation of the Musicians Seamounts must explain this distribution and duration of volcanism. Age data are not available for seamounts from the Musicians Horst in the northern part of the province, nor for isolated seamounts found in the eastern half of the province.

Two seamount lineaments were tested for age progression consistent with a hot spot mechanism of formation. A well-defined age progression was found for the NW-SE seamount lineament, from 96 Ma just southeast of the cluster of seamounts in the NW corner of the province, to 82 Ma for Khatchaturian Seamount in the center of the province. The rate of volcanic propagation was determined to be 52.0 ±5.8 km/m.y., or 0.503 ±0.055 °/m.y. about a rotation pole at 36°N, 76°W. The extension of this age progression is not clear south of the Murray fracture zone. No age progression along a N-S trending lineament of seamounts could be found. There is no evidence for volcanism in the Musicians Seamounts coeval with the Emperor Seamounts.

Two tests, paleolatitude of formation and equivalent rates of volcanic propagation along copolar and coeval seamount chains, proved consistent with, but inconclusive proof of, a fixed hot spot mechanism of formation for the NW-SE trending age progression. The age data from Schlanger et al. [1984] for seamounts in the northern Line Islands older than 80 Ma are consistent with a model of age progressive volcanism at a rate of 65.2 ±24.1 km/m.y. or 0.592 ±0.219 °/m.y. Assuming that both chains represent the traces of relatively fixed hot spots, combining the Musicians and Line Islands age progressions yields an estimate of 0.508 ±0.053 °/m.y., which is the present best estimate of Pacific/hot spot motion for 96 to 82 Ma.

A discrepancy exists in the motion of the Pacific plate as described by the Pacific/hot spot and Pacific/spin axis reference frames, especially for the period 85-82 Ma. If the east-west portion of the Pacific APWP proves to be the result of an east-west tectonic rotation of the Pacific plate, then the current hot spot models describing the Late Cretaceous motion of the Pacific plate are in error. If the current hot spot models are correct, then the Pacific APWP may describe a unique period of rapid true polar wander at the end of the Cretaceous normal polarity superchron. To adequately test the current hot spot models as best described by the age progression along the NW-SE trend in the Musicians Seamounts, either higher quality age data must be obtained for the northern Line Islands or well-defined age progressions must be found for coeval seamount lineaments elsewhere in the Pacific. The catalog of long-lived, fixed hot spots still cannot be definitively extended into the Late Cretaceous Pacific.

---

[1] Epp [1984] also presented a second, alternative model for Line Islands hot spot volcanism The model assumed that the northernmost Line Islands are on the order of 130 Ma, based on a single $^{40}Ar/^{39}Ar$ incremental-heating age determination of 127 Ma for sample 142D-1 [Saito and Ozima, 1976; 1977]. However, recalculating the age spectrum and isochron analysis from the data given in Saito and Ozima [1977], we find that the experiment fails 3 out of the 4 criteria used in the present study to test the validity of incremental-heating apparent ages (some of the individual age spectrum steps are significantly different at the 95% confidence level, a SUMS/N-2 of greater than 10 shows that there is significant excess scatter in the isochron, and the isochron $^{40}Ar/^{36}Ar$ intercept of 281 ±6 is significantly different from the atmospheric value). For samples which display similar age spectra (decreasing apparent ages with increasing temperature indicative of significant amounts of argon recoil), we have shown that apparent $^{40}Ar/^{39}Ar$ ages can easily be 20-30% *older* than the actual crystallization age. Thus, we must conclude that the 127 Ma apparent age reported by Saito and Ozima has little geological meaning. Note that Schlanger et al. [1984] determined a total-fusion age of 93.4 ±1.3 Ma on sample 142D-11 (shown in Figure 13), which is from the same dredge haul as 142D-1. There seems to be no reliable geochronological evidence for any Line Islands volcanism older than about 100 Ma, and we can only empathize with Epp's valiant efforts to reconcile significantly different $^{40}Ar/^{39}Ar$ apparent ages that had been reported in the refereed literature.

Acknowledgements. This work represents a significant portion of the author's Ph.D. dissertation at the University of Hawaii at Manoa [Pringle, 1992]. Special thanks go to J.M. Sinton and G. B. Dalrymple, who provided guidance, support, and ample opportunity for independence throughout the course of this study. W.J. Morgan first inspired the author to study Cretaceous hot spot tracks. J. Saburomaru, S. Makita, and E. Kollman generously shared their knowldege of mineral separation and argon extraction techniques. The staff of the U.S.G.S. TRIGA reactor provided all of the required fast neutrons. Valuable comments on various versions of this manuscript were provided by J.M. Sinton, G.B. Dalrymple, D.A. Clague, F. Duennebier, R.A. Duncan, D. Epp, M Garcia, M.A. Lanphere, J. Mammerickx, R. Moberly, D. Muenow, W.W. Sager, and S. Stein.

REFERENCES

Baksi, A.K., Isotopic fractionation of a loosely held atmospheric argon component in the Picture Gorge basalts, *Earth Planet. Sci. Lett.*, *21*, 431-438, 1974.

Baksi, A.K., $^{40}Ar/^{39}Ar$ incremental heating studies on terrestrial whole-rock basalts: an attempt to unravel the effects of recoil loss of ^{39}Ar and ^{37}Ar [abstract], *Trans. Am. Geophys. Un.*, *71*, 653, 1990.

Brooks, C., S.R. Hart, and I. Wendt, Realistic use of two-error regression treatments as applied to rubidium-strontium data, *Rev. Geophys. Space Phys.*, *10*, 551-557, 1972.

Campbell, J.F., N. Frazer, R. Moberly, W. Sager, J.M. Sinton, and B.H. Wallin, Musicians Seamount Province in the North Central Pacific [abstract], *Trans. Am. Geophys. Un.*, *61*, 989, 1980.

Christofferson, E., The relationship of sea-floor spreading in the Pacific to the origin of the Emperor Seamounts and the Hawaiian chain [abstact], *Trans. Amer. Geophys. Un.*, 65, 1099, 1968.

Clague, D.A., and G.B. Dalrymple, Cretaceous K-Ar ages of volcanic rocks from the Musicians Seamounts and the Hawaiian Ridge, *Geophys. Res. Lett.*, *2*, 305-309, 1975.

Clague, D.A., G.B. Dalrymple, and R. Moberly, Petrography and K-Ar ages of dredged volcanic rocks from the western Hawaiian ridge and southern Emperor seamount chain, *Geol. Soc. Amer. Bull.*, *86*, 991-998, 1975.

Clague, D.A., and R.D. Jarrard, Tertiary plate motion deduced from the Hawaiian-Emporer Chain, *Geol. Soc. Amer. Bull.*, *84*, 1135-1154, 1973.

Cox, A. and G.B. Dalrymple, Statistical analysis of geomagnetic reversal data and the precision of potassium-argon dating, *J. Geophys. Res.*, *72*, 2603-2614, 1968.

Dalrymple, G.B, E.C. Alexander, Jr., M.A. Lanphere, and G.P. Kraker, Irradiation of samples for $^{40}Ar/^{39}Ar$ dating using the Geological Survey TRIGA reactor, *U.S. Geol. Surv. Prof. Pap. 1176*, 55 pp., 1981.

Dalrymple, G.B., and D.A. Clague, Age of the Hawaiian-Emperor bend, *Earth Planet. Sci. Lett.*, *31*, 313-329, 1976.

Dalrymple, G.B., and J.M. Moore, ^{40}Ar excess in submarine pillow basalts from Kilauea Volcano, Hawaii, *Science*, *161*, 1132-1135, 1968.

Dalrymple, G.B., and M.A. Lanphere, *Potassium-Argon Dating*, San Francisco, W.H. Freeman Co., 258 pp., 1969.

Dalrymple, G.B., M.A. Lanphere and D.A. Clague, Conventional and $^{40}Ar/^{39}Ar$ ages of volcanic rocks from Ojin (Site 430), Nintoku (Site 432), and Suiko (Site 433) Seamounts and the chronology of volcanic propagation along the Hawaiian-Emperor chain, *Initial Reports of the Deep Sea Drilling Project, 55*: Washington (U.S. Government Printing Office), 659-676, 1980.

Dalrymple, G.B., M.A. Lanphere, and M.S. Pringle, Correlation diagrams in $^{40}Ar/^{39}Ar$ dating: Is there a correct choice?, *Geophys. Res. Lett.*, *15*, 589-591, 1988.

Dalrymple, G.B., M.A. Lanphere and D.A. Clague, Conventional and $^{40}Ar/^{39}Ar$ ages of volcanic rocks from Ojin (Site 430), Nintoku (Site 432), and Suiko (Site 433) Seamounts and the chronology of volcanic propagation along the Hawaiian-Emperor chain, *Initial Reports of the Deep Sea Drilling Project, 55*: Washington (U.S. Government Printing Office), 659-676, 1980.

Dalrymple, G.B., M.A. Lanphere, and M.S. Pringle, Correlation diagrams in $^{40}Ar/^{39}Ar$ dating: Is there a correct choice?, *Geophys. Res. Lett.*, *15*, 589-591, 1988.

Desonie, D.L., and Duncan, R.A., The Cobb-Eickelberg Seamount chain: Hot spot volcanism with mid-ocean ridge basalt affinity, *J. Geophys. Res.*, *95*, 12,697-12,712.

Duncan, R.A., $^{40}Ar/^{39}Ar$ geochronology of basalts from ocean basins: some successes with samples from aseismic ridges, in R.E. Zartman, ed., Short Papers on the Fourth International Conference on Geochronology, Cosmochronology, and Isotope Geology, *U.S. Geol. Surv. Open File Rep. 78-701*, 100-103, 1978.

Duncan, R.A., and D.A. Clague, Pacific plate motions recorded by linear volcanic chains, *in* A.E.M. Nairn, F.G. Sthehli, and S. Uyeda, eds., *Ocean Basins and Margins*, *7A*, Plenum, N.Y., 89-121, 1985.

Duncan, R.A., and R.B. Hargraves, Plate tectonic evolution of the Caribbean region in the mantle reference frame, *in* W.E. Bonini, R.B. Hargraves, and R. Shagam, eds., The Caribbean-South American Plate Boundary and Regional Tectonics, *Mem. Geol. Soc. Am.*, *162*, 81-93, 1984.

Eaby, J., D.A. Clague, and J. Delaney, Sr isotopic variations along the Juan de Fuca Ridge, *J. Geophys. Res.*, *89*, 7883-7890, 1984.

Epp, D., Age and tectonic relationships among volcanic chains on the Pacific plate, Ph.D.thesis, University of Hawaii, Honolulu, HI, 199 pp., 1978.

Epp, D., Possible perturbations to hotspot traces and implications for the origin and structure of the Line Islands, *J. Geophys. Res.*, *89*, 11273-11286, 1984.

Fleck, R.J., J.F. Sutter, and D.H. Elliot, Interpretation of discordant $^{40}Ar/^{39}Ar$ age spectra of Mesozoic tholeiites from Antartica, *Geochim. Cosmochim. Acta*, 41, 15-32, 1977.

Foland, K.A., J.S. Linder, T.E. Laskowski, and N.K. Grant, $^{40}Ar/^{39}Ar$ dating of glauconites: measured ^{39}Ar recoil loss from well-crystallized specimens, *Isot. Geosci.*, *2*, 241-264, 1984.

Freedman, A.P., and B. Parsons, Seasat-derived gravity over the Musicians Seamounts, *J. Geophys. Res.*, *91*, 8325-8340, 1986.

Gordon, R.G., Late Cretaceous apparent polar wander of the Pacific plate: evidence for a rapid shift of the Pacific hotspots with respect to spin axis, *Geophys. Res. Lett.*, *10*, 709-712, 1983.

Gordon, R.G., and C.D. Cape, Cenozoic latitudinal shift of the Hawaiian hotspot and its implications for true polar wander, *Earth Planet. Sci. Lett.*, *55*, 37-47, 1981.

Halliday, A.N., $^{40}Ar/^{39}Ar$ stepheating of clay concentrates from Irish ore bodies, *Geochim. Cosm. Acta*, v. *42*, p. 1851-1858, 1978.

Harrison, T.M., Some observations on the interpretation of $^{40}Ar/^{39}Ar$ age spectra, *Isot. Geosci.*, *1*, 319-338, 1983.

Henderson, L.J., Motion of the Pacific plate relative to the hotspots since the Jurassic and model of oceanic plateaus of the Farallon plate, Ph.D. thesis, Northwestern University, Evanston, IL, 312 pp., 1985.

Hess, J.C., and H.J. Lippolt, Kinetics of Ar isotopes during neutron irradiation: ^{39}Ar loss from minerals as a source of error in $^{40}Ar/^{39}Ar$ dating, *Isot. Geosci.*, *4*, 223-236, 1986.

Huneke, J.C., and Smith, S.P., The realities of recoil: ^{39}Ar recoil out of small grains and anomalous patterns in $^{40}Ar/^{39}Ar$ dating, *Geochim. Cosm. Acta Suppl. 7* (Proceedings of the Seventh Lunar Science Conference), 1987-2008, 1976.

Ingamells, C.O., Lithium metaborate flux in silicate analysis, *Anal. Chim. Acta*, 52, 332-334, 1970.

Jarrard, R.D., and S. Sasajima, Paleomagnetic synthesis for southeast Asia: constraints on plate motions, *in* D.E. Hayes, ed., *The Tectonic and Geologic Evolution of Southeast Asian Seas and Islands, Geophys. Monogr. Ser.*, *23*, Am. Geophys. Un., Washington, D.C., 293-317, 1980.

Jurdy, D.M., True polar wander, *Tectonophysics*, *74*, 1-16, 1981.

Lanphere, M.A., and G.D. Dalrymple, A test of the $^{40}Ar/^{39}Ar$ age spectrum technique on some terrestrial material, *Earth Planet. Sci. Lett.*, *12*, 359-372, 1971.

Lanphere, M.A., and G.D. Dalrymple, The use of $^{40}Ar/^{39}Ar$ data in evaluation of disturbed K-Ar systems, in R.E. Zartman, ed., Short Papers on the Fourth International Conference on Geochronology, Cosmochronology, and Isotope Geology, *U.S. Geol. Surv. Open File Rep. 78-701*, 241-243, 1978.

Mankinen, E.A., and G.B. Dalrymple, Electron microprobe evaluation for whole rock K-Ar dating, *Earth Planet. Sci. Lett.*, *17*, 89-94, 1972.

McIntyre, G.A., C. Brooks, W. Compston, and A. Turek, The statistical assessment of Rb-Sr isochrons, *J. Geophys. Res.*, *71*, 5459-5468, 1966.

Morgan, W.J., Hotspot tracks and the opening of the Atlantic and Pacific oceans, *in* C. Emilliani, ed., *The Sea*, *7*, John Wiley, New York, 443-487, 1981.

Morgan, W.J., Plate motions and deep mantle convections, *Geol. Soc. Amer. Mem. 132 (Hess Volume)*, 7-22, 1972a.

Morgan, W.J., Deep mantle convection plumes and plate motions, *Amer. Assoc. Petrol. Geol. Bull.*, *56*, 203-213, 1972b.

Morgan, W.J., Convection plumes in the in the lower mantle, *Nature*, *230*, 42-43, 1971.

Nobel, C.S., and J.J. Naughton, Deep-ocean basalt: Inert gas content and uncertainties in age dating, *Science*, *162*, 265-267, 1968.

Ozima, M., M. Honda, and K. Saito, $^{40}Ar/^{39}Ar$ ages of guyots in the western Pacific and discussion of their evolution, *Geophys. J. R. Astr. Soc.*, *51*, 475-485, 1977.

Pringle, M.S., Geochronolgy and petrology of the Musicians Seamounts, and the search for hot spot volcanism in the Cretaceous Pacific, Ph.D. disseration, University of Hawaii, Honolulu, HI, 235 pp., 1992.

Pringle, M.S., J.M. Sinton, and H. Staudigel, Petrology of the Musicians and S. Hawaiian seamounts: Anomalous near-ridge seamount volcanism [abstract], *Trans. Amer. Geophys. Un.*, *71*, 1667, 1990.

Rea, D.K., and F.P. Naugler, Musicians Seamounts province and related crustal structures north of the Hawaiian ridge, *Mar. Geol.*, *10*, 89-111, 1971.

Sager, W.W., and U. Bleil, Latitudinal drift of Pacific hotspots during the Late Cretaceous and early Tertiary, *Nature*, *326*, 488-490, 1987.

Sager, W.W., and M.S. Pringle, Paleomagnetic constraints on the origin and evolution of the Musicians and South Hawaiian Seamounts, Central Pacific Ocean, in B.H. Keating, P. Fryer, R. Batiza, and G.W. Boehlert, eds., *Seamounts, Atolls, and Islands,* Geophysical Monograph Series 43 (Menard Volume), Am. Geophys. Un., Washington, D.C., 133-162, 1987.

Sager, W.W., and M.S. Pringle, Mid-Cretaceous to Early Tertiary apparent polar wander path of the Pacific plate, *J. Geophys. Res.*, *93*, 11,753-11,771, 1988.

Saito, K., and M. Ozima, $^{40}Ar/^{39}Ar$ ages of submarine rocks from the Line Islands: implications on the origin of the Line Islands, in G.H. Sutton, M.H. Manghnani, and R. Moberly, eds., *The Geophysics of the Pacific Ocean Basin and its margin,* Geophysical Monograph Series 19, Am. Geophys. Un., Washington, D.C., 205-226, 1976.

Saito, K., and M. Ozima, $^{40}Ar/^{39}Ar$ geochronological studies of submarine rocks from the western Pacific area, *Earth Planet. Sci. Lett.*, *33*, 353-369, 1977.

Schilling, J-G., and A. Noe-Nygaard, Faeroe-Iceland plume: rare earth evidence, *Earth Planet. Sci. Lett.*, *24*, 1-14, 1974.

Schlanger, S.O., M.O. Garcia, B.H. Keating, J.J. Naughton, W.W. Sager, J.A. Haggerty, J.A. Philpotts, and R.A. Duncan, Geology and geochronology of the Line Islands, *J. Geophys. Res.*, *89*, 11201-11272, 1984.

Schwank, D.C. and A.R. Lazarewicz, Estimation of seamount compensation using satellite altimetry, *Geophys. Res. Lett.*, *9*, 907-910, 1982.

Seideman, D.A., $^{40}Ar/^{39}Ar$ studies of deep-sea igneous rocks, *Geochim. Cosmochim. Acta*, *42*, 1721-1734, 1978.

Sleep, N.H., Tapping of magmas from ubiquitous mantle heterogeneities, *J. Geophys. Res.*, *89*, 10029-10041, 1984.

Stacey, J.S., N.D. Sherrill, G.D. Dalrymple, M.A. Lanphere, and N.V. Carpenter, A five-collector system for the simultaneous measurement of argon isotope ratios in a static mass spectrometer, *Int. J. Mass Spectrom. Ion Phys.*, *39*, 167-180, 1981.

Steiger, R.H., and E. Jäger, Subcommission on geochronology: convention on the use of decay constaints in geochronology and cosmochronology, *Earth Planet. Sci. Lett.*, *36*, 359-362, 1977.

Sun, S., Ocean islands: plums or plumes?, *Nature*, *316*, 103-104, 1985.

Taylor, J.R., *An Introduction to Error Analysis*, Mill Valley, California, University Science Books, 270 pp., 1982.Turner, G., and P.H. Cadogan, Possible effects of ^{39}Ar recoil in $^{40}Ar/^{39}Ar$ dating, *Geochim. Cosm. Acta Suppl. 5* (Proceedings of the Fifth Lunar Science Conference), 1601-1615, 1974.

Walker, D.A., and I. McDougall, $^{40}Ar/^{39}Ar$ and K-Ar dating of altered glassy volcanic rocks: the Dabi volcanics, P.N.G., *Geochim. Cosmochim. Acta*, *46*, 2181-2190, 1982.

Watts, A.B., J.H. Bodine, and N.R. Ribe, Observations of flexure and the geologic evolution of the Pacific Ocean basin, *Nature*, *238*, 532-537, 1980.

Watts, A.B., J.K. Weissel, R.A. Duncan, and R.L. Larson, Origin of the Louisville ridge and its relationship to the Eltanin fracture zone, *J. Geophys. Res.*, *93*, 3051-3077, 1987.

Wilson, J.T., A possible origin of the Hawaiian Islands, *Can. J. Phys.*, *41*, 863-870, 1963a.

Wilson, J.T., Contiental Drift, *Scientific American*, *208*, 86-100, 1963b.

York, D., Least squares fitting of a straight line with correlated errors, *Earth Planet. Sci. Lett.*, *5*, 320-324, 1969.

Zindler, A., H. Staudigel, and R. Batiza, Isotope and trace element geochemistry of young Pacific seamounts: implications for the scale of upper mantle heterogeneity, *Earth Planet. Sci. Lett.*, *70*, 175-195, 1984.

M.S. Pringle, U.S. Geological Survey, Menlo Park, CA 94025.

Petrology and Isotope Geochemistry of Lavas from the Line Islands Chain, Central Pacific Basin

MICHAEL O. GARCIA[1], KYE-HUN PARK[2,3], GEOFFREY T. DAVIS[1,4],

HUBERT STAUDIGEL[2,5] AND DAVID P. MATTEY[1,6]

The Line Islands chain is a product of multiple hotspots. The main portion of the chain formed between 42 and 93 Ma. Along some parts of the chain, this phase of volcanism was preceded and succeeded by hotspot volcanism. There is no chemical distinction among the lavas from the different episodes of hotspot volcanism. Ratios of Sr and Pb isotopes and trace elements are similar even for tholeiitic and alkalic lavas. These ratios are similar to those for lavas from Easter and Sala y Gomez islands. The range of Sr and Pb isotopes and the Nb/Zr ratios for Line Islands lavas require three source components. The most likely candidates are a high $^{206}Pb/^{204}Pb$ and Nb/Zr, low $^{87}Sr/^{86}Sr$ source (HIMU); an enriched source (EMI); and a depleted mantle source. Backtracking the positions of the age-dated Line Islands volcanoes to where they probably formed indicates that the main group is unrelated to any known hotspot. The pre-hotspot volcanoes may be related to the Easter and/or the Sala y Gomez hotspots. The post-hotspot volcanoes do not backtrack to any known hotspot but are closest to the Marquesas. The compositions, positions and ages of the backtracked Line Islands volcanoes indicate that the eastern portion of the South Pacific isotopic and thermal anomaly (SOPITA) has persisted for at least 90 m.y. and that the Easter and Sala y Gomez hotspots may have produced geochemically similar lavas for the last 70 m.y.

INTRODUCTION

The Line Islands chain is part of one of the three major, linear volcanic chains in the Pacific basin. It forms the northern segment of the Line-Tuamotu chain, a nearly continuous chain of elongate ridges, seamounts, atolls and a few islands, which extends over 6,000 km from Pitcairn Island northwest to Horizon Guyot (Figure 1). The Line Islands segment of this chain is about 4,200 km long. Wilson [1963] and Morgan [1971] considered the Line-Tuamotu chain to have formed over a relatively fixed, deep-seated melting anomaly or hotspot, in the same manner and at the same time as the Emperor-Hawaii chain. Clague and Jarrard [1973], however, pointed out that the northern part of the Line Islands chain has a different orientation than the Emperor Seamounts and therefore is probably older.

The $^{40}Ar/^{39}Ar$ total fusion ages for Line Islands lavas confirmed that many of the seamounts in the chain were formed by hotspot volcanism and that at least the northern half of the Line Islands formed before the Emperor Seamounts [Schlanger et al., 1984]. However, these ages plus other radiometric ages [Saito and Ozima, 1976, 1977; Lanphere and Dalrymple, 1976; Davis, 1986] and paleontological evidence [Jackson and Schlanger, 1976; Haggerty et al., 1982] indicate that the Line Islands chain area has undergone multiple periods of volcanism [Winterer, 1976; Crough and Jarrard, 1981; Henderson and Gordon, 1982; Epp, 1984; Schlanger et al., 1984], both before and after the main phase of hotspot volcanism. The parallel chains of ridges along the entire length of the chain and the cross-trending ridges between 2°S and 15°N may be topographic expressions of the multiple periods of volcanism in the Line Islands (Schlanger et al., 1984).

Other mechanisms have been proposed to explain Line Islands volcanism, including (1) a leaky transform fault [Orwig and Kroenke, 1980; Farrar and Dixon, 1981] and (2) a slow-spreading rift similar to east Africa [Natland, 1976; Winterer, 1976]. The radiometric ages make these models untenable [Schlanger et al., 1984].

We collected samples from 19 dredge hauls spanning the entire length of the Line Islands including the northern end of the Tuamotus and Horizon Guyot (Figure 1). These samples

[1]Department of Geology and Geophysics, University of Hawaii, Honolulu.
[2]Lamont-Doherty Geological Observatory, Columbia University, Palisades, New York.
[3]Now at Korea Basic Center, Seoul, Korea.
[4]Now at Maersk Energy Inc., Houston, Texas.
[5]Scripps Institution of Oceanography, University of California, La Jolla, California.
[6]Now at Geology Department, Royal Holloway and Bedford New College, Surrey, United Kingdom.

The Mesozoic Pacific: Geology, Tectonics, and Volcanism
Geophysical Monograph 77

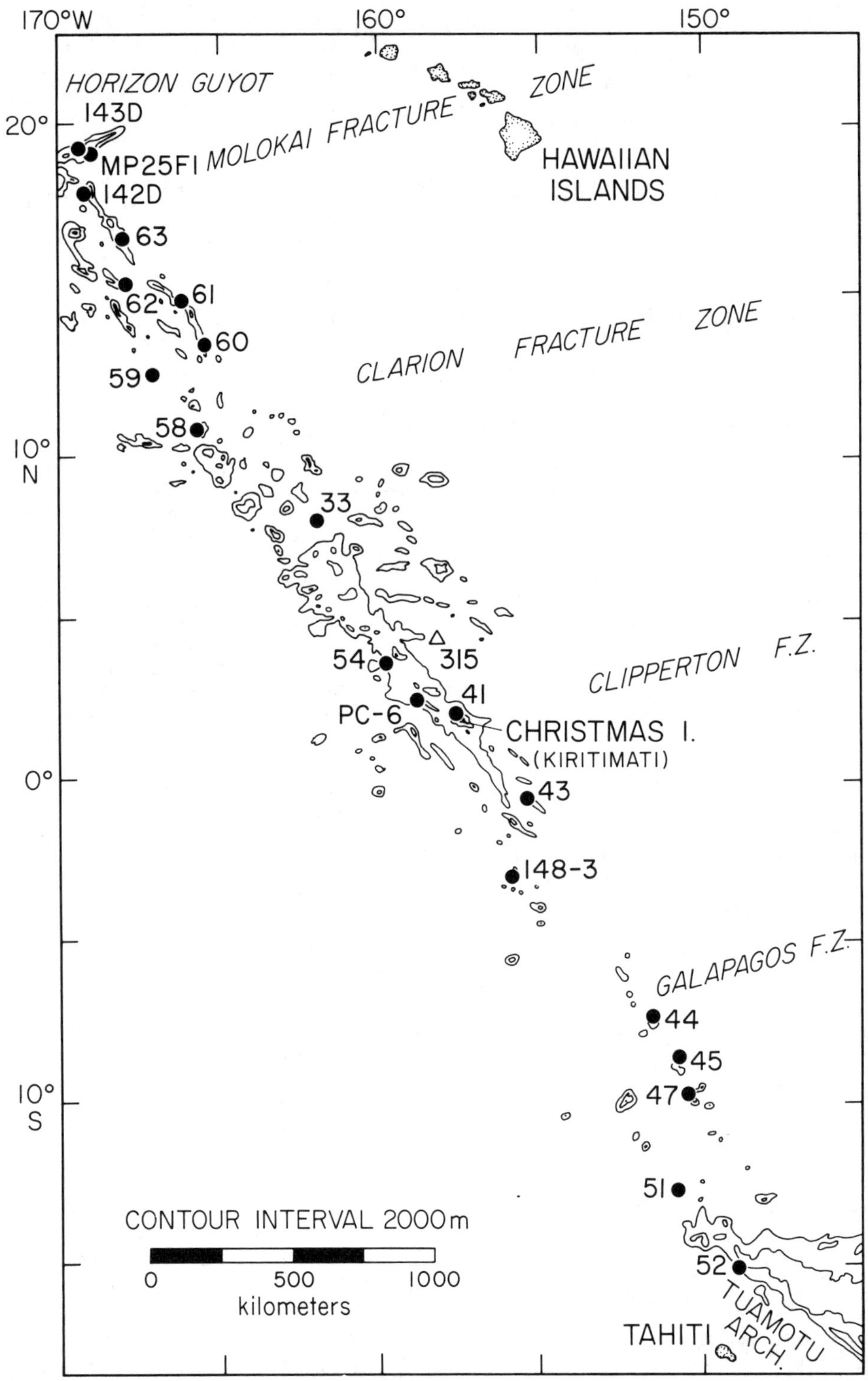

Fig. 1. Bathymetric map of the Line Island chain, central Pacific basin. Circles show location of dredge hauls for samples discussed in this paper. Triangle shows the location of DSDP Site 315.

(see Davis, 1986 for more details on samples) plus those collected at DSDP Site 315 [Jackson et al., 1976] and from previous dredging [Natland, 1976] provide a solid basis for interpreting the petrologic evolution of the Line Islands.

We present here petrologic and geochemical studies of selected lavas from the Line Islands to assess the nature of compositional variation among the lavas generated during the multiple periods of hotspot volcanism to better understand the spatial and temporal variation in the composition of Pacific oceanic island lavas. We focus on those samples that have age control. Our results help constrain the location and magnitude of the South Pacific isotopic anomaly [Hedge, 1978; Hart, 1984] during the late Cretaceous and early Tertiary when the Line Islands chain was formed.

PETROGRAPHY AND MINERAL CHEMISTRY

A wide range of rock types has been recovered from the Line Islands, including tholeiites, alkalic basalts, trachytes, phonolites and basanitoids [Natland, 1976; Jackson et al., 1976; Schlanger et al., 1984]. Many of these rocks are pillow lava fragments with preserved pillow rims. Rock textures in the basalts range from hyalopilitic to subophitic and in the evolved rocks from interstitial to trachytic. Most of the lavas are porphyritic with olivine (usually replaced by clay and iron oxides) as the most common phenocryst, followed in abundance by plagioclase and clinopyroxene (Table 1), which are unaltered. Amphibole occurs rarely and only in some strongly evolved and basanitoid lavas [Natland, 1976], which is also true for Hawaiian lavas (see Garcia et al., 1987 for a summary). Vesicularity in Line Islands rocks ranges from <1% to almost 50%. The most vesicular rocks are also the most altered. Most samples show at least some signs of alteration including devitrification of glass, replacement of olivine by clay and iron oxides, zeolites in some vugs, calcite in veins, and patches of secondary phosphorite.

Groundmass clinopyroxenes (cpx) were analyzed using an electron microprobe to assist in assessing the magmatic affinity of these somewhat altered volcanic rocks. Previous studies have shown that cpx is resistant to alteration and is diagnostic of rock type [Hashimoto, 1972; Vallance, 1974]. High Al and Ti content is generally considered to be indicative of alkaline magmas. However, Coish and Taylor [1979] have shown that cpx in pillow rims commonly has higher Al and Ti contents than cpx in pillow interiors due to rapid quenching. In this study only cpx from pillow interiors were used to avoid this quenching problem.

The Line Islands cpx have a wide range in TiO_2 (0.7-4.2 wt.%), Mg# (63-83), and Ca (38-50% wollastonite component; see Table 2 and Figure 2). Magmatic affinities also range widely: from tholeiitic (low wollastonite component, Ti and Al, high Si; Figure 2) to peralkaline. Lavas from the Hawaiian and Louisville chains show a similar range in composition [Fodor et al., 1975; Clague et al., 1980; Hawkins et al., 1987]. Most of the Line Islands lavas are alkalic based on their groundmass cpx compositions. Assuming a KD of 0.24 for (Fe/Mg) clinopyroxene/(Fe/Mg) liquid [Mahood and Baker, 1986], then the groundmass grains in these Line Islands lavas grew in liquids with Mg numbers of 30 to 56. Thus, most of the Line Island lavas were evolved liquids at the time of their eruption.

TABLE 1. Modes of Lavas From the Line Islands

	63-7	62-1	61-1	61-3	60-6	59-1	59-2	59-10	59-12	59-13	58-1	33-1	33-4	54-2	PC6-1	41-2	43-1	44-3	45-1	47-1	51-1	52-1	52-2
Phenocrysts																							
Olivine	5	0	0	0	3	5	7	4	0	0	0	13	18	0	0	12	15	0	7	2	9	14	17
Plagioclase	0	14	0	0	1	1	3	5	1	2	0	5	3	0	1	0	1	1	4	<1	0	5	8
Clinopyroxene	2	0	0	0	0	4	1	18	0	0	0	0	0	0	0	9	0	0	0	1	0	1	3
Amphibole	0	0	0	0	0	0	0	0	2	16	6	0	0	0	0	0	0	0	0	0	0	0	0
Opaque	0	0	0	0	1	0	0	1	0	0	0	0	0	0	0	0	0	0	0	0	0	0	0
Microphenocrysts																							
Plagioclase	18	42	36	38	29	27	30	27	33	28	38	39	29	29	41	32	46	29	33	16	2	26	29
Olivine	19	2	23	23	24	19	16	8	4	0	10	8	4	8	15	6	20	20	17	2	4	4	3
Clinopyroxene	23	1	13	0	9	17	1	3	2	1	0	2	0	5	1	13	0	1	0	0	0	16	10
Amphibole	0	0	0	0	1	0	0	9	37	11	12	0	0	0	0	0	0	0	0	0	0	0	0
Opaque	8	14	5	13	2	12	12	8	10	18	17	14	12	14	13	16	9	11	6	<1	<1	16	11
Matrix	25	27	23	26	30	15	30	17	11	24	17	19	34	44	29	12	9	38	33	78	86	18	19
Vesicularity	7	29	25	23	7	22	32	21	18	7	0	10	14	4	0	18	18	0	18	34	6	23	22

Phenocrysts are >0.5 mm; microphenocryst minerals are 0.5 to 0.01 mm; matrix includes fine-grained minerals (>0.01 mm) and altered glass. Olivine is completely replaced by iddingsite. Values are volume %, vesicle-free and based on 500 counts per sample. Samples are listed according to the latitude they were collected with north on the left.

Analyst: G. Davis, M. Garcia.

TABLE 2. Microprobe Analyses of Groundmass Clinopyroxene and Phenocrystic Amphibole in Lavas From the Line Islands

	Clinopyroxene									Amphiboles	
	63-7	59-1	59-9	59-10	54-2	41-2	41-2	52-1	52-1	59-7	59-13
SiO_2	48.83	48.29	47.27	47.96	50.99	49.55	50.78	49.42	52.32	39.24	39.80
TiO_2	2.45	1.71	1.87	2.17	1.46	1.66	1.18	1.75	0.81	4.96	4.03
Al_2O_3	5.84	6.35	7.31	6.88	2.48	4.60	3.57	5.07	2.10	13.70	13.59
Cr_2O_3	0.25	0.01	0.21	0.60	0.00	0.17	0.45	0.78	0.35	0.02	0.01
FeO*	6.47	9.31	8.64	6.24	9.58	6.86	5.71	6.76	7.69	12.40	11.93
MnO	0.14	0.23	0.23	0.09	0.22	0.12	0.12	0.11	0.19	0.15	0.12
MgO	13.40	13.22	13.02	14.26	15.77	15.26	15.82	15.08	18.95	12.11	13.17
CaO	22.54	19.81	20.51	21.30	19.02	21.32	21.75	20.59	16.61	11.75	11.28
Na_2O	0.39	0.65	0.88	0.64	0.23	0.22	0.23	0.30	0.14	2.78	2.72
K_2O	--	--	--	--	--	--	--	--	--	1.02	1.49
Total	100.31	99.58	99.94	100.14	99.75	99.76	99.61	99.86	99.16	98.13	98.14
Mg#	78.7	71.7	7.52	81.4	83.1	79.8	83.1	79.6	73.4	63.5	66.3

See Garcia et al. [1986] for methods used. Analyst: G. Davis.

Amphibole in the lavas from dredge haul 59 are Si-poor ranging from ferroan pargasites to kaersutites (Table 2; Leake, 1978, classification scheme) in contrast to edenite and richterite phenocrysts that are observed in some Hawaiian trachytes [e.g., Velde, 1978].

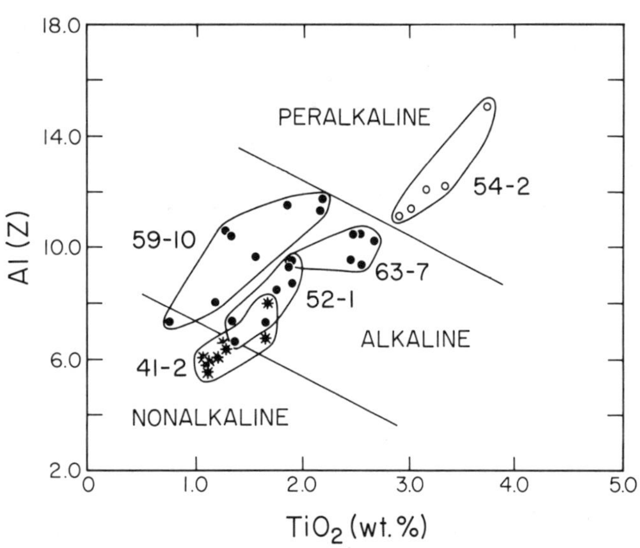

Fig. 2. TiO_2 wt. % vs. Al in the Z site (tetrahedral position) of clinopyroxene groundmass grains from the holocrystalline portion of five representative Line Island lavas. Field classification is from LeBas [1962]. Symbols: * - post-hotspot lavas; ● - main hotspot lavas; and o - undated samples.

WHOLE ROCK MAJOR AND TRACE ELEMENT GEOCHEMISTRY

The freshest samples (based on petrographic criteria; e.g., low vesicularity, least amount of secondary minerals and most holocrystalline) from each dredge haul were analyzed by XRF for major elements. In Table 3, we present 14 new major element analyses; 8 additional analyses from Schlanger et al. [1984] are given to complement their new trace element analyses (Table 4). Most of the samples have moderate to high H_2O, Fe_2O_3 and, in rare cases, P_2O_5 contents. The high P_2O_5 contents could be due to secondary phosphorite [Natland, 1976]. However, our samples with high P_2O_5 contents (e.g., 61-3 with 3.7 wt.% P_2O_5) do not have high CaO contents (Table 3). High P_2O_5 contents (up to 2.3 wt.%) have been noted in some fresh Hawaiian, mildly alkalic lavas and been attributed to their mantle source [Spengler and Garcia, 1988]. Thus, the high P_2O_5 contents of some Line Island lavas may, in part, be a primary feature. Nevertheless, all of the least-altered samples are similar in composition to unaltered, oceanic-island alkalic lavas. Our Line Islands alkalic lavas range from basalt to mugearite (Figure 3). Rare transitional tholeiites were also recovered from the Line Islands [Jackson et al., 1976; Table 3].

Trace element analyses were made on the least altered Line Islands samples by XRF for Cr, Ni, Rb, Y, Zr, Nb, Pb and Th (see Mattey, 1982 for methods used) and by isotope dilution for REE, Li, K, Rb and Sr (Tables 4 and 5). Most of the trace elements define relatively coherent trends (Figure 4) except Rb and Y, which are mobile during alteration [Fodor et al., 1987].

TABLE 3. Major Element, Whole-Rock Analyses and Selected Components of CIPW Normative Mineralogy

	MP25F-1*	63-7	62-1	61-1	61-3	60-6	59-1	59-2	59-10	59-12	59-13	58-1	33-1	33-4	54-2	PC6-1	41-2	43-1	148-3*	44-3	52-1	52-2
SiO_2	43.36	41.70	45.90	45.20	40.50	46.45	47.30	43.60	45.45	51.60	42.90	48.85	45.60	45.20	40.05	47.60	45.20	42.75	54.04	41.75	47.05	44.55
TiO_2	3.84	3.48	3.32	3.88	4.32	2.02	2.85	2.80	3.69	1.44	2.78	3.26	2.64	3.54	5.42	2.89	2.80	2.60	2.39	3.22	3.09	2.93
Al_2O_3	14.74	13.53	18.19	17.02	15.82	16.20	16.57	16.62	16.94	17.71	17.64	17.04	15.37	15.19	13.27	18.69	13.74	17.97	18.15	15.50	16.01	15.87
Fe_2O_3	13.39	9.19	8.14	9.48	10.78	8.44	11.63	12.70	9.91	4.68	5.81	7.29	8.88	11.41	11.12	5.69	11.12	13.27	9.27	13.68	9.55	9.70
FeO	--	3.62	1.46	2.60	2.60	2.56	1.52	1.12	2.96	2.44	5.56	2.76	3.60	1.44	5.80	2.60	1.96	0.79	--	0.92	3.24	3.04
MnO	0.18	0.14	0.16	0.09	0.09	0.09	0.11	0.19	0.11	0.11	0.18	0.15	0.18	0.16	0.15	0.06	0.18	0.13	0.11	0.12	0.13	0.15
MgO	6.48	5.48	1.19	2.20	1.16	4.57	1.59	1.48	2.76	1.49	4.57	3.64	4.58	1.82	4.69	1.06	6.85	0.97	0.66	1.92	4.85	4.05
CaO	9.74	11.85	8.45	8.66	9.90	10.30	5.55	6.46	6.96	7.40	8.11	8.45	11.23	6.26	9.20	6.92	12.55	8.58	5.05	10.19	10.61	11.53
Na_2O	2.57	1.97	3.84	3.22	2.57	2.95	3.16	3.13	2.82	5.02	2.86	3.91	2.82	3.20	2.46	3.98	2.58	3.34	4.75	3.02	2.59	2.60
K_2O	1.89	1.21	2.34	2.46	2.77	1.03	3.19	2.51	2.74	3.74	1.94	1.92	0.86	2.85	0.64	1.56	0.91	1.75	3.17	1.29	0.72	0.92
P_2O_5	0.75	0.93	2.68	1.98	3.67	0.58	1.15	1.81	1.48	2.14	1.88	0.80	0.56	2.12	0.63	1.45	0.37	2.12	1.31	2.81	0.56	1.48
H_2O^+	2.50	3.29	1.99	1.95	3.11	2.43	2.67	3.46	2.25	0.93	3.60	0.86	2.38	3.43	2.86	3.23	0.62	3.39	0.60	2.54	0.31	1.50
H_2O^-	--	2.48	1.78	1.26	2.14	1.92	2.74	3.67	1.65	0.67	1.93	0.66	1.23	3.09	3.64	3.82	0.74	1.90	--	2.16	1.12	1.27
CO_2	--	1.05	0.28	0.02	0.45	0.08	0.09	0.14	0.12	0.35	0.17	0.28	0.01	0.26	0.12	0.14	0.16	0.29	--	0.37	0.00	0.24
Total	99.38	99.92	99.72	100.02	99.88	99.62	100.12	99.69	99.84	99.72	99.93	99.87	99.94	99.97	100.02	99.68	99.78	99.85	99.48	99.79	99.83	99.83
										Normative Mineralogy												
Q	--	--	--	--	--	--	--	--	--	--	--	--	--	--	--	--	--	3.1	2.5	--	--	--
Or	11.6	7.7	14.5	15.0	17.4	6.4	19.9	16.1	16.9	22.6	12.2	11.5	5.3	18.1	4.1	10.0	5.5	11.0	19.1	8.0	4.3	5.6
Ab	15.4	15.2	33.9	28.2	23.1	26.2	28.3	28.7	24.9	34.9	25.6	32.8	23.8	29.0	22.3	36.4	14.2	29.5	41.2	27.0	22.3	22.8
An	23.8	26.3	25.6	25.5	24.9	29.3	21.2	21.9	25.9	15.0	29.7	23.8	27.8	18.5	24.9	26.9	23.6	30.4	16.8	27.2	30.4	29.8
Ne	3.8	1.5	--	--	--	--	--	--	--	4.7	--	0.5	0.5	--	--	--	4.3	0.2	--	--	--	--
Hy	--	--	5.1	2.0	2.7	2.9	15.7	10.6	12.1	--	2.5	--	--	17.3	1.0	10.7	--	--	10.0	4.4	13.5	4.9
D.I.[a]	30.8	24.4	48.4	43.2	40.5	32.6	48.2	44.8	41.8	62.2	37.8	44.8	29.6	47.1	26.3	49.5	24.1	40.8	62.8	35.0	26.6	28.4

See Garcia et al., 1986 for methods used. The CIPW normative mineralogy was calculated using dry, iron adjusted ($Fe_2O_3/FeO = 0.15$), analyses normalized to 100%.
[a]D.I. - differentiation index.
Analyst: K. Ramlal, Univ. of Manitoba except samples with * are by T. Hulsebosch, Univ. of Hawaii.

TABLE 4. XRF Trace Element Analyses of Representative Lavas From the Line Islands

Sample	Cr	Ni	Sr	Y	Rb	Zr	Nb	Pb	Th
63-7	493	171	790	36	30	364	72	3	9
62-1	44	145	1103	95	40	295	45	7	7
61-3	149	73	792	72	31	366	40	3	3
61-1	176	103	836	71	41	532	58	4	7
60-6	401	208	449	32	23	139	29	1	2
59-1	327	100	559	73	50	251	43	4	4
59-2	364	106	626	116	42	249	43	2	1
59-12	0	30	705	101	63	430	93	8	14
59-10	528	242	423	83	58	204	32	--	4
59-13	23	58	1387	51	40	501	134	6	18
58-1	6	24	851	37	51	429	79	5	5
33-1	327	92	716	50	50	348	61	1	1
33-4	311	114	623	42	82	340	60	3	6
54-2	13	43	531	47	10	317	53	2	7
PC6-1	6	97	919	56	23	603	76	<1	1
41-2	540	188	502	29	12	249	25	2	1
43-1	449	94	326	104	30	217	11	7	3
44-3	128	60	482	43	37	232	23	<1	<1
45-1	255	122	493	48	29	215	20	3	4
52-2	521	215	416	41	15	223	24	--	2
52-1	527	139	346	40	18	210	23	--	--

Values are in ppm.
Analyst: G. Hendry, Univ. Birmingham, UK.

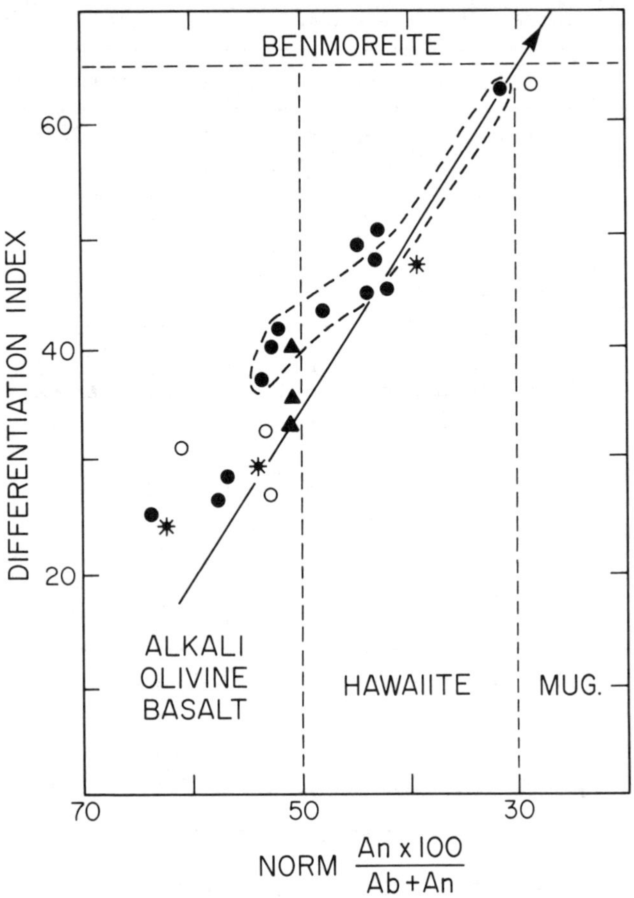

Fig. 3. Normative plagioclase composition (An-anorthite; Ab-albite) vs. differentiation index (normative quartz, albite, orthoclase and nepheline). Five samples from dredge haul 59 are enclosed by a dashed line. Arrow shows the trend of typical Hawaiian alkalic lavas. The rock classification fields and arrow are from Coombs and Wilkinson [1969]. Symbols as in Figure 3 and triangle - prehotspot lavas.

Significant lithological variations are present among the lavas recovered within single dredge hauls (33, 41, 52 and 59-63; see Tables 1 and 3). However, coherent trends for trace elements are present for samples from individual dredge hauls, especially for haul 59. Overall, Ni and Cr decrease and Sr increases with increasing Nb (Figure 4) indicating that olivine and probably Cr-spinel and/or clinopyroxene were the dominant fractionating phases and that plagioclase fractionation was minor or nonexistent. The samples with high Cr and Ni contents generally have common to abundant olivine and clinopyroxene (7 to 22 vol.%).

The abundances of immobile highly incompatible elements (e.g., Zr and Nb) in the Line Islands basalts are high for some evolved samples from dredge hauls 59 and 61 (Figure 5). However, alkalic lavas from other oceanic islands such as Hawaii [West et al., 1988], the Carolines [Mattey, 1982], the Australs and Marquesas [Dupuy et al., 1988], and Pitcairn [Woodhead and McCulloch, 1989] also have high incompatible element contents. Abundances of Sr and Th versus Nb define positive slopes with limited scatter despite the wide age, lithological, and spatial variation of the samples (Figure 4). However, there is a substantial variation in the Nb/Zr ratio of the Line Islands lavas (Figure 5). Other studies have correlated the variation of this ratio to variable degrees of partial melting of the mantle [Mattey, 1982] or to mixing of magmas from different sources [le Roex, 1985].

There is no systematic correlation of rock type with Nb/Zr ratio for Line Islands lavas. The tholeiites from dredge haul 60 and DSDP Site 315 have high ratios (0.16-0.22) like many of the alkalic lavas from the central and northern parts of the Line Islands chain. Furthermore, some of the alkalic lavas have lower ratios than the tholeiites, and one has a ratio like normal mid-ocean ridge basalts (N-MORB; Figure 5). There is no correlation of Nb/Zr ratio with period of eruption. The DSDP Site 315 and dredge haul 45 samples are both pre-hotspot, but they have quite different ratios (0.20 vs. 0.10). However, Site 315 lavas have similar Nb/Zr ratios to the hotspot (dredge hauls 59, 60, 63) and post-hotspot (dredge haul 33) samples (Figure 5) from the same section of the chain. These results indicate that the rock type (tholeiitic vs. alkalic) and period of volcanism (relative to the main hotspot) have little influence on the Nb/Zr ratio of the Line Islands lavas.

The REE patterns for most of the Line Islands lavas analyzed are similar to other oceanic island alkalic lavas (Figure 6). Sample 60-6 is distinct; it has moderate LREE abundances (Table 5) and has a REE pattern similar to tholeiites from Mauna Kea Volcano, Hawaii (Figure 6). The patterns of the other samples (except sample 59-2) are similar to Hawaiian mildly alkalic basalts, hawaiites, and mugearites, although the dredge haul 59 samples tend to be more concave upward (Figure 6). A few of the evolved lavas have negative Eu anomalies (Figure 6). These samples contain plagioclase phenocrysts (1-3 vol. %). However, some other evolved lavas with greater plagioclase contents lack negative Eu anomalies (e.g., sample 62-1 with 14 vol. % plag). Negative Eu anomalies are rare in Hawaiian lavas, even among evolved alkalic lavas that have experienced extensive plagioclase fractionation [e.g., Spengler and Garcia, 1988; West et al., 1988]. Although two of the samples analyzed for REE have high Y contents (59-7, 101 ppm; 62-1, 95 ppm), their REE patterns are similar to the other samples with low Y contents. Thus, there is no indication that the REE concentrations were affected by the same alteration processes that caused high Y contents in some samples. Therefore, the crossing REE patterns may be indicative of garnet in the source for the lavas.

SR AND PB ISOTOPE RATIOS

A previous Sr isotope study of Line Islands lavas included analyses of unleached and HCl-leached samples [Matsuda, 1980]. Most of the leached samples have lower $^{87}Sr/^{86}Sr$ ratios than the unleached samples. However, the standard deviations for individual leached analyses are large (±0.0005-9), so they provide only a first-order characterization of the Line Islands lavas. Most of the leached lavas have low ratios (0.7027 to 0.7036), except for a trachyte (two analyses: 0.7041, 0.7044 ± 0.0005, 8) from the northern portion of the chain. Such large standard deviations indicate incomplete leaching of the samples

TABLE 5. Rare Earth Element, Alkali Metal and Sr and Ba Abundances in Lavas From the Line Islands Chain Determined by Isotope Dilution

	62-1	61-3	60-6	59-1	59-2	59-12	59-13	PC6-1	44-3	BCR-1 (Standard)	Blank
Li	17.7	12.6	28.5	17.3	23.5	17.7	60.5	32.2	21.1	12.6	0.004
K	19800	22200	7640	27500	18100	31100	17000	13200	10500	13100	1.369
Rb	37.1	35.6	13.3	57.9	39.5	97.8	41.7	23.7	29.7	45.4	0.002
Sr	635	727	414	583	611	633	1350	866	428	318	0.049
Ba	761	267	188	406	354	1090	981	311	153	628	0.035
Ce	70.9	61.0	35.4	--	--	121	175	80.4	89.9	50.2	0.006
Nd	53.1	43.5	21.3	45.3	87.7	46.1	72.0	71.6	34.5	27.3	0.004
Sm	11.6	8.93	4.89	9.00	15.8	7.52	11.2	14.1	7.83	6.40	0.001
Eu	3.58	2.84	1.68	2.76	4.52	2.08	3.19	4.32	2.55	1.86	<0.001
Gd	12.4	--	--	10.1	18.5	6.50	9.17	11.8	8.95	--	<0.001
Dy	10.7	5.63	4.59	8.52	17.0	5.25	7.41	10.8	8.75	6.18	<0.001
Er	5.42	2.73	2.19	4.45	10.3	2.88	3.81	4.97	5.32	3.50	<0.001
Yb	4.47	2.31	1.77	3.75	8.41	2.82	3.05	4.66	4.46	3.15	<0.001
Lu	0.81	0.36	0.26	0.49	1.22	0.41	0.52	0.54	0.69	0.48	<0.001

Values are in ppm.
Analysts: G. Davis, R. Harma and D. Mattey.

(see Staudigel et al. [1991] for a discussion on problems of leaching lavas). There are no previous analyses of Pb isotopes in Line Island lavas.

Seventeen new Sr and Pb isotope ratio analyses were made on lavas sampled from along the entire length of the Line Islands, including the northern end of the Tuamotu chain and from different periods of volcanism. To minimize the effects of alteration, samples were leached in 6N HCl over a 24-72 hour period. The acid bath was repeatedly replaced until the solutions remained clear. Then the sample was leached for 1 hr at 60° to 80°C (see Staudigel et al. [1991] for details on leaching techniques).

We have used initial $^{87}Sr/^{86}Sr$ ratios in all diagrams to compare our data with present-day Sr isotopic data from other oceanic volcanoes (Figures 7-10). Ideally, a comparison of ancient and modern volcanoes requires both calculation of the initial ratio at the time of eruption and the subsequent hypothetical radiogenic growth. For our 42 to 93 Ma samples, radiogenic growth in the source region is probably small when compared to the cumulative errors involved in age estimates, and to parent/daughter and isotopic ratio measurements on leached samples (see Staudigel et al. [1991] for more detail). The Pb isotope ratios have not been corrected because Pb and U abundances were not measured. However, our experience has shown that the U/Pb ratios of leached oceanic basalts are typically too low to generate significant radiogenic growth in 100 million years.

The Line Islands lavas have a relatively limited range of initial Sr and Pb isotopic ratios ($^{87}Sr/^{86}Sr_i$, 0.70268-0.70421; $^{206}Pb/^{204}Pb$, 19.15-20.17) given their length (4200 km), age span (~50 m.y.), and multiple periods of volcanism. There is no apparent correlation of any isotopic ratio with period of volcanism (e.g., hotspot vs. post-hotspot), location along the chain (Figure 7) or age (Table 6).

Most of the Line Islands lavas have relatively low Sri ratios (<0.7033), except near the northern end of the chain where, over a 500-km section, values range from 0.70298 to 0.70421 (Figure 7). Low $^{87}Sr/^{86}Sr_i$ ratios are not uncommon for oceanic island basalts; among South Pacific Islands, such low ratios are found in Easter and Sala y Gomez islands (0.7030-0.7033; White and Hofmann, 1982), a few Austral islands (Mangaia and Rurutu, 0.7027 to 0.7032; Palacz and Saunders, 1986) and a few tholeiites from Ua Pou in the Marquesas (0.7029-0.7032; Vidal et al., 1984).

The Pb isotope data for the Line Islands data define a positive correlation for $^{206}Pb/^{204}Pb$ vs. $^{207}Pb/^{204}Pb$ (Figure 8). A slightly broader trend caused by one outlier is present for the $^{206}Pb/^{204}Pb$ vs. $^{208}Pb/^{204}Pb$ data (Figure 8). Like lavas from many south Pacific oceanic island chains (e.g., Australs, Louisville Ridge, Marquesas, Easter and Sala y Gomez), Line Islands lavas have moderate to high Pb ratios that parallel the 1.77 b.y. secondary isochron (Figure 8). Unlike the linear trends for the Pb isotope ratio plots, the Line Islands data form a broad field on a Sr vs. $^{206}Pb/^{204}Pb$ isotope ratio plot (Figure 8). At least three source components are required to explain this broad field.

SPATIAL VARIATIONS IN TRACE ELEMENT AND ISOTOPE RATIOS

Lanphere et al. [1980] noted a general northward decrease in the ratio of $^{87}Sr/^{86}Sr$ in lavas along the Emperor portion of the Hawaiian-Emperor chain. They related this change to the decrease in the distance of a mid-ocean ridge (Kula) from the

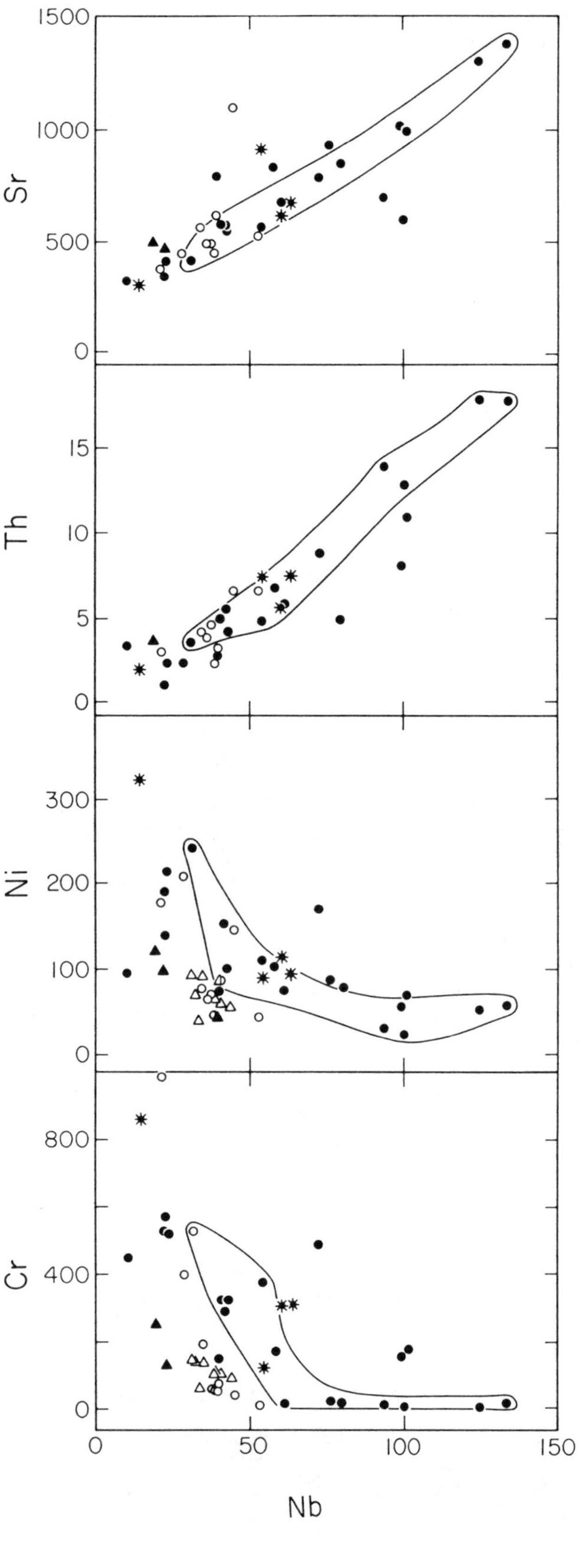

Fig. 4. Nb vs. Sr, Th, Ni and Cr (in ppm) for Line Island lavas. Data from Table 4 and Clague et al. [1980]. Symbols as in Figures 3 and 4; open triangle DSDP Site 315 lavas [Jackson et al., 1976]. Field encloses the five dredge haul 59 samples.

hotspot that produced these lavas. This distance can be related to the difference in age between the oceanic crust and the overlying oceanic island lava. A positive correlation of age difference with Sr isotope and/or Nb/Zr ratio might be indicative of a greater contribution from a N-MORB type source (i.e., depleted) in the oceanic island lavas formed on young crust compared to those formed on older crust.

Using the $^{40}Ar/^{39}Ar$ total fusion ages for the Line Islands lavas [Schlanger et al., 1984] and the age of the oceanic crust inferred from magnetic anomalies [Epp, 1984], age difference was compared with $^{87}Sr/^{86}Sr_i$ ratio for eight dated Line Islands samples (Figure 9). There is no apparent correlation of $^{87}Sr/^{86}Sr_i$ ratio with differences in age for the Line Islands, if the younger age is used for dredge haul 52 (which is probably more likely given that the seamount is located south of the bend in the Line Islands-Tuamotu chain, Figure 9). Likewise, there is no correlation of Nb/Zr ratio (Figure 9) or Pb isotopes with age difference. In part, this may be a consequence of large intra-volcano variations in Nb/Zr. For example, samples in one dredge haul (59) have ratios that vary from 0.15 to 0.26 and the least-evolved sample (13) has the highest ratio. Since these samples are all mildly alkalic, the variation in the Nb/Zr ratios are not related to partial melting variation. Instead, they probably reflect source heterogeneity.

There is a positive correlation for $^{87}Sr/^{86}Sr_i$ and Nb/Zr ratio for the Line Islands lavas (excluding sample 62-1) unlike lavas from other South Pacific Island groups (Figure 10). The Austral and Pitcairn lavas have large ranges in $^{87}Sr/^{86}Sr$ ratios and little or no variation in their Nb/Zr ratio; the Louisville Ridge, Easter, and Sala y Gomez lavas have restricted ranges for both ratios. In contrast, the Line Islands lavas vary substantially for both ratios. The low Sr isotope and Nb/Zr ratios endmember for the Line Islands lavas are similar in those ratios to N-MORB (Figure 10). Lavas from the southern part of the chain (except sample 61-1) have a greater N-MORB component than the northern lavas. However, more analyses are needed to confirm this trend.

The Line Islands lavas have relatively high $^{206}Pb/^{204}Pb$ ratios like many other south Pacific oceanic island basalts (Figure 8). The combination of relatively high $^{206}Pb/^{204}Pb$ and moderate to low $^{87}Sr/^{86}Sr_i$ is distinctive of the HIMU component [c.f., Zindler and Hart, 1986]. The third component in the source for Line Islands lavas is probably an enriched source like the EMI component (Figure 8).

ORIGIN OF THE LINE ISLANDS CHAIN

Previous studies of the Line Islands chain place its hotspot in the south Pacific [e.g., Duncan and Clague, 1985]. The mantle under the south Pacific is thermally anomalous because of its shallow depth [McNutt and Fisher, 1987], thin elastic lithosphere [Calmant and Cazenave, 1987], and numerous hotspots (7, Duncan and Clague [1985]; 8, Staudigel et al. [1991]). Most of the lavas from these south Pacific hotspots

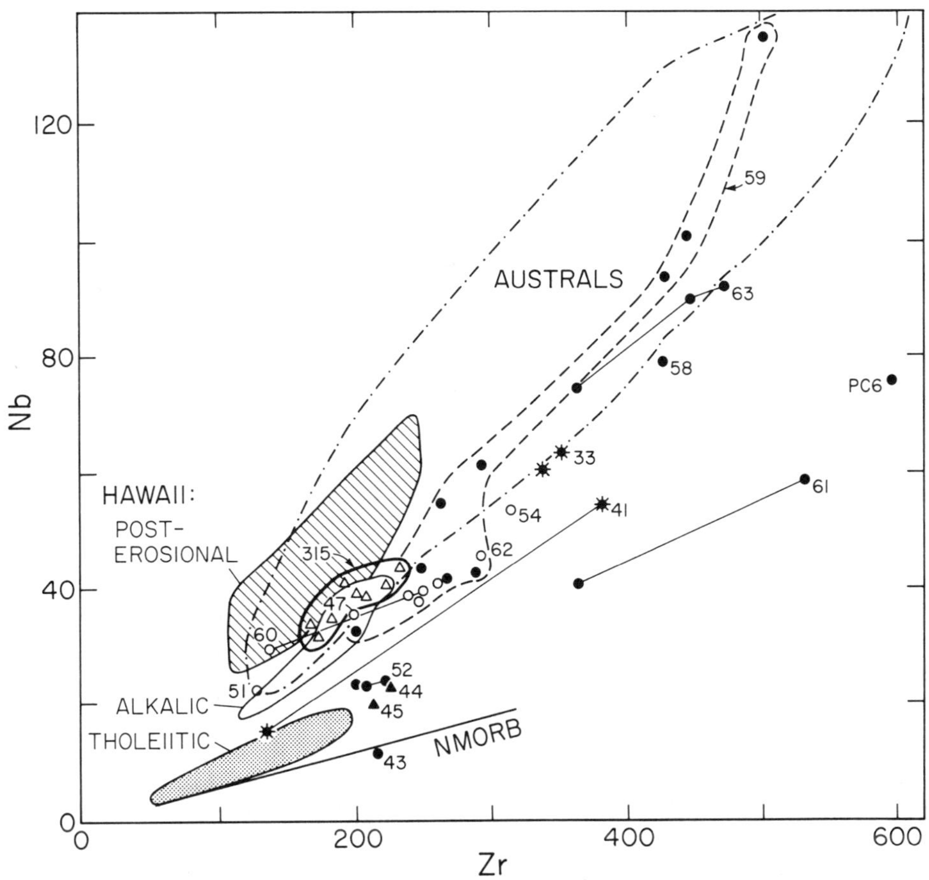

Fig. 5. Nb vs. Zr in ppm for Line Islands lavas. Data from Table 4 except for DSDP Site 315, which is from Jackson et al. [1976]. Tie lines connect samples from the same dredge haul except for dredge haul 59 which is enclosed by a dashed line. Fields for Hawaii post-erosional, alkalic and tholeiitic lavas from Mattey [1982]; trend for N-MORB from Richardson et al. [1984]. Australs field from Dupuy et al. [1988]. Note the wide range in Nb/Zr ratio compared to the other groups of lavas. There is no correlation of rock type with Nb/Zr ratio. Symbols as in Figures 3, 4 and 5.

have enriched radiogenic isotopes including a HIMU component [Staudigel et al., 1991]. This region includes part of the DUPAL-anomaly [Hart, 1984] and is roughly located between 10 to 30°S and 170°E to 110°W. Smith et al. [1989] and Staudigel et al. [1991] have called this region the south Pacific isotopic and thermal anomaly (SOPITA) and have documented that its central portion has existed for at least 120 Ma.

The Line Islands probably formed near the eastern margin of this region [Duncan and Clague, 1985] about 42 to 93 Ma [Schlanger et al., 1984]. Thus, the lavas from the Line Islands are useful in evaluating the longevity and the regional limitations of the SOPITA. To facilitate this analysis, the locations of the well-dated Line Islands volcanoes were backtracked to their probable points of origin using the poles and plate rotations of Duncan and Clague [1985]. There are many sources of errors that are inherent in backtracking positions: for example, unknown duration of volcanism of each volcano, uncertainties in the radiogenic age determination for each volcano, and ambiguities in plate motion vectors, especially for the Cretaceous. The change in the backtracked position for two volcanoes due to age errors of ± 5 m.y. is shown on Figure 11 for a pre- and a hotspot volcano.

The main hotspot-produced Line Island volcanoes that formed between 59 and 90 Ma define a region centered at about 123°W and 18°S. This region is somewhat larger in size than the ± 5 m.y. range for sample 43-1 (Figure 11) but it is well away from any known hotspot. For example, it is 9 to 10°NNE of the Pitcairn hotspot and about 13°NW of the Easter hotspot. In contrast, the northern Tuamotu volcano backtracks to about 114°W and 27°S, which is close to Easter hotspot (Figure 11).

The pre-hotspot volcanoes backtrack to near the Easter and Sala y Gomez hotspots (Figure 11). Their location on the east side of the chain and their ages fit nicely with the predicted path of the Sala y Gomez hotspot [see Duncan and Clague, 1985]. The post-hotspot volcanoes backtrack to about 131°W and 3 to 8°S. The closest hotspot is the Marquesas at about 138°W and 10°S (Figure 11). However, the post-hotspot Line Islands lavas are distinct in trace elements and isotopes from lavas from the Marquesas (see Tables 4 and 6 and Dupuy et al. [1988]).

The misfit between the backtracked positions of the Line

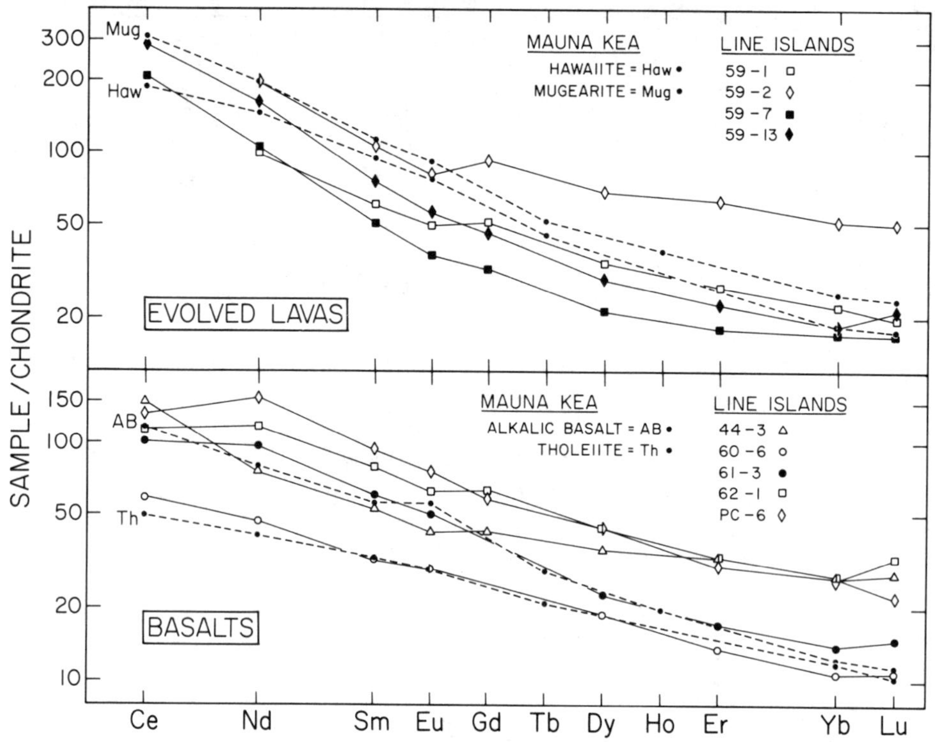

Fig. 6. REE abundances normalized to C1 chondrite values of Anders and Grevesse [1989] for representative Line Island lavas (data from Table 5). For comparison, a typical REE pattern for a tholeiite, a mildly alkalic basalt, a hawaiite and a mugearite from Mauna Kea, Hawaii [Frey et al., 1990; Frey et al., 1991) are plotted with similar composition Line Islands lavas.

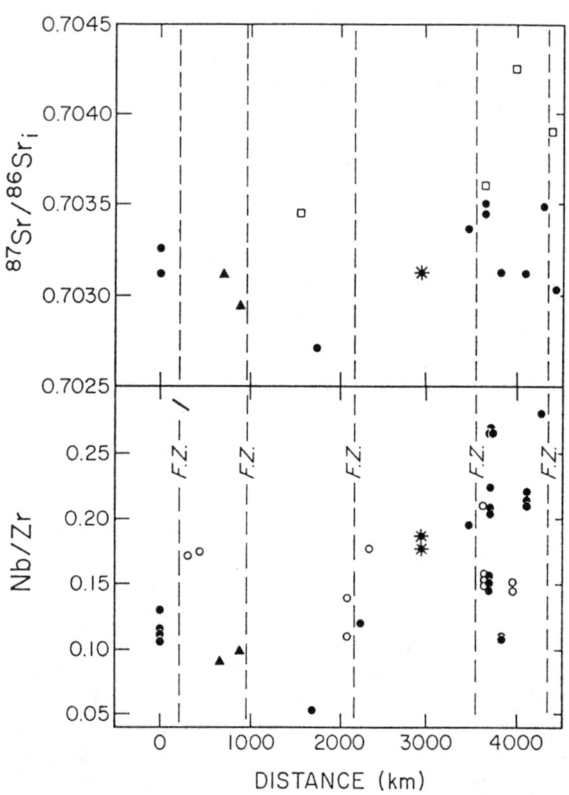

Islands main hotspot volcanoes and the location of any known hotspot is problematic. Two possible explanations are that all of their ages are much too young (~20 m.y.) or they are related to a hotspot that is no longer active or has not been discovered. It seems unlikely that the all of the $^{40}Ar/^{39}Ar$ ages are systematically low. Furthermore, the bathymetry of the south Pacific is not well known (e.g., a 300-km ridge was discovered in the southern Line Islands on our 1979 cruise, Schlanger et al. [1984], and Macdonald Seamount was not discovered as an active hotspot until 1967 following an earthquake swarm, Johnson [1970]). Even in Hawaiian waters, which are relatively well-surveyed, the active and youngest part of the Hawaiian chain wasn't discovered until 1978 when fresh lavas were dredged from Loihi Seamont [Moore et al., 1979]. Thus, it seems possible that an "active" seamount might exist somewhere near the inferred source of the main Line Islands trend at 123°W and 18°S. Alternatively, the hotspot that produced the Line Islands main hotspot trend may have died 40-50 m.y. ago. The south Pacific has several hotspot trends that have had abrupt starts or stops in activity (e.g. Marquesas) or been rejuvenated (Australs; Turner and Jarrard [1982]).

Fig. 7. Ratios of Nb/Zr and initial $^{87}Sr/^{86}Sr$ vs. distance in km from the Line Islands-Tuamoto bend. Data are from Tables 4 and 6 except for boxes [Matsuda,1980]. Symbols as in Figures 3, 4 and 5. Location of fracture zones (F.Z.) shown by dashed lines.

TABLE 6. Sr and Pb Isotope Ratios for Lavas From the Line Islands

Sample	Distance from bend (km)	Age	Hotspot	$^{87}Sr/^{86}Sr$	$(^{87}Sr/^{86}Sr)_i$	$^{206}Pb/^{204}Pb$	$^{207}Pb/^{204}Pb$	$^{208}Pb/^{204}Pb$
143D	4440	88.1	Y	0.703045±18	0.702977	19.198	15.565	38.981
MP25F1	4380	90*	?	0.704026±18	0.703849	19.466	15.602	39.279
142D	4305	93.4	Y	0.703503±49		19.153	15.528	39.344
63-7	4100	86.0	Y	0.703070±42	0.703062	19.630	15.607	39.614
62-1	3960	85*	?	0.704362±32	0.704207	19.855	15.585	39.924
61-1	3820	81.4	Y	0.703193±21	0.703081	20.123	15.618	39.728
60-5	3655	85*	?	0.703939±40	0.703550	19.653	15.609	39.676
59-12	3645	85.0	Y	0.703663±31	0.703442	19.982	15.630	39.985
59-10	3645	85*	?	0.703553±33	0.703392	19.928	15.600	39.734
58-1	3450	75.6	Y	0.703549±33	0.703335	19.667	15.549	39.402
33-1	2940	39.3	Post	0.703099±44	0.703090	19.369	15.580	39.082
43-1	1730	59.0	Y	0.702852±38	0.702683	19.214	15.532	38.703
148-3	1550	65*	?	0.703572±43	0.703413	20.169	15.664	39.967
44-3	875	71.9	Pre	0.703081±31	0.702862	19.609	15.591	39.051
45-20	705	70.5	Pre	0.703212±21		19.417	15.586	38.808
52-1	0	47.4	Y	0.703139±39	0.703132	19.294	15.543	38.888
52-2	0	41.8	Y	0.703127±35	0.703090	19.366	15.542	38.879

Ratios are listed in order of decreasing distance from the bend in the Line-Tuamoto Chain. Ages are ^{40}Ar-^{39}Ar total fusion ages from Schlanger et al. [1984], except sample 58-1 which is based on two K/Ar analyses of acid-treated rock [Davis, 1986]; samples with * are estimates based on the location where the sample was collected and assuming it was formed during the main hotspot period of volcanism. Initial Sr isotope ratios (i) were calculated using the age of the sample and its Rb and Sr contents (see Park [1990], for element abundances). Elemental abundance data for samples 143D and 142D are given in Natland [1976].

Analysts: K.-H. Park and H. Staudigel.

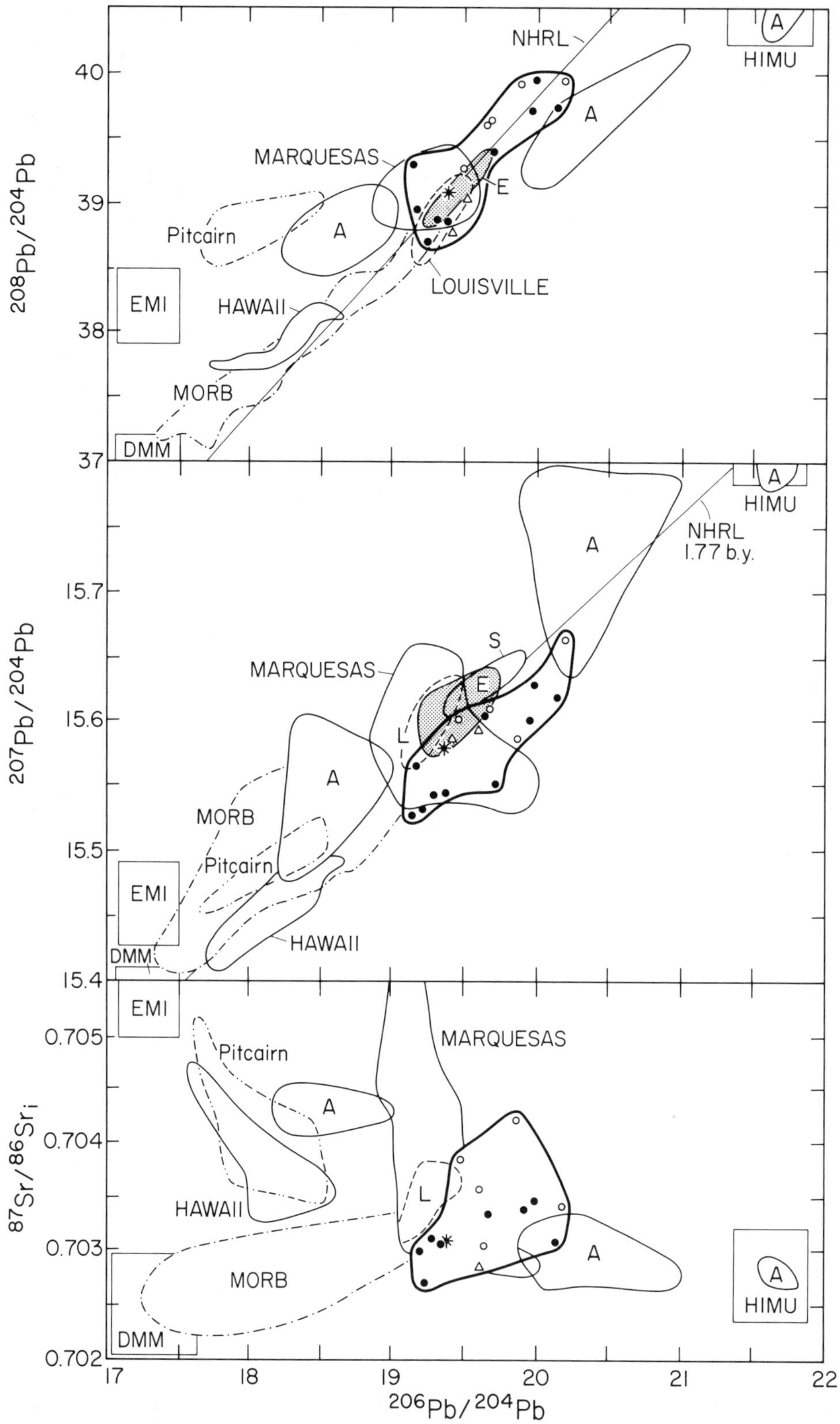

Fig. 8. Isotope variation diagrams: upper, $^{206}Pb/^{204}Pb$ vs. $^{208}Pb/^{204}Pb$; middle, $^{206}Pb/^{204}Pb$ vs. $^{207}Pb/^{204}Pb$; and lower, $^{206}Pb/^{204}Pb$ vs. $^{87}Sr/^{86}Sr_i$. Symbols as in Figures 3, 4 and 5. Fields: Heavy line - Line Islands, E - Easter, S - Sala y Gomez (both from Hanan and Schilling [1989]; for Pitcairn Island, data are from Woodhead and McCulloch [1989]; A - Australs; see Staudigel et al. [1991] for references to other data. The theoretical mantle source components are: DMM, depleted mantle; HIMU, high $^{206}Pb/^{204}Pb$, low $^{87}Sr/^{86}Sr$ mantle; EMI, enriched mantle one [Zindler and Hart, 1986]. The EMII field is not shown; it would plot in the Marquesas field on the Pb-Pb isotope plots and offscale on the Pb-Sr plot. The Northern Hemisphere Reference Line (NHRL) at 1.77 b.y. is shown for comparison [from Hart, 1984].

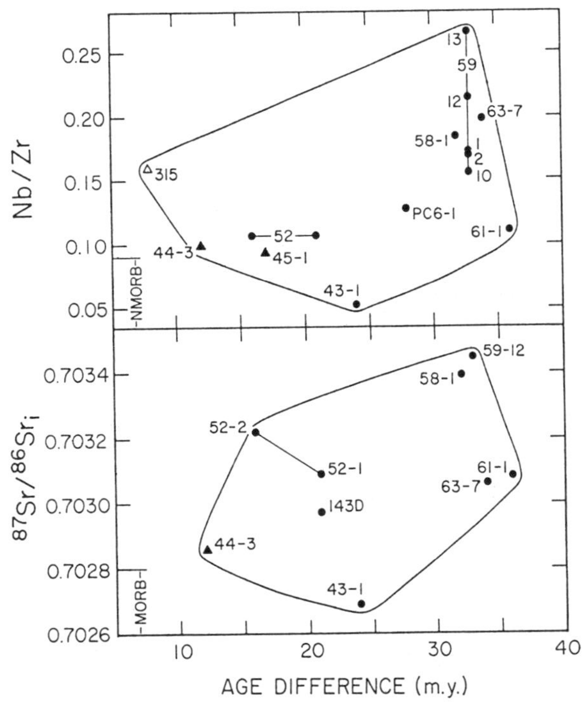

Fig. 9. Ratios of Nb/Zr and initial $^{87}Sr/^{86}Sr$ vs. age difference in million of years for well-dated Line Islands lavas. Age difference is calculated using the age of the oceanic crust based on the magnetic anomaly interpretations of Epp [1984] minus the age of the lava [from Schlanger et al., 1984]. There is no apparent correlation of these ratios with distance. Normal MORB values are from le Roex [1985 and personal communication]. Vertical line connects samples from dredge haul 59. Symbols as in Figures 3, 4 and 5.

Relationship of the Line Islands to the SOPITA

Our present-day location estimates of the Line Islands hotspots fall within a region of the south Pacific that is part of the globe-encircling DUPAL anomaly [Hart, 1984] and the SOPITA [Staudigel et al., 1991; see Figure 11]. The SOPITA may be seen as part of the DUPAL anomaly, but it is not necessarily globe-encircling. The pre-hotspot and post-hotspot volcanoes backtrack into groups near the southeastern and northern margins, respectively, of the SOPITA. The samples that backtrack to near the Easter hotspot (Figure 11) tend to have relatively low Nb/Zr and $^{87}Sr/^{86}Sr_i$ ratios (especially sample 44-3) like Easter Island lavas (Figure 10). The post-hotspot volcano that produced sample 33-1 has an Austral-like source component (Figure 10), although it was formed north of the assumed SOPITA region but within the DUPAL band. However, sample 33-1 does not have a DUPAL-type component (e.g., high $^{87}Sr/^{86}Sr$). Thus, if the backtracked positions for the Line Islands volcanoes are reliable, then they indicate that the SOPITA was present in the southeast Pacific from 93 to 42 Ma and extended north to 3°S, but it may have not have extended as far east as the Easter hotspot. Staudigel et al. [1991] showed that lavas with a HIMU signature have been erupting from the central SOPITA for at least the last 120 m.y. The Line Islands data presented here document that similar lavas have erupted from the eastern SOPITA for at least the last 90 m.y.

Summary

The Line Islands chain is a major bathymetric feature in the central Pacific basin that owes its origin to multiple periods of volcanism. The main hotspot portion of the chain is not related

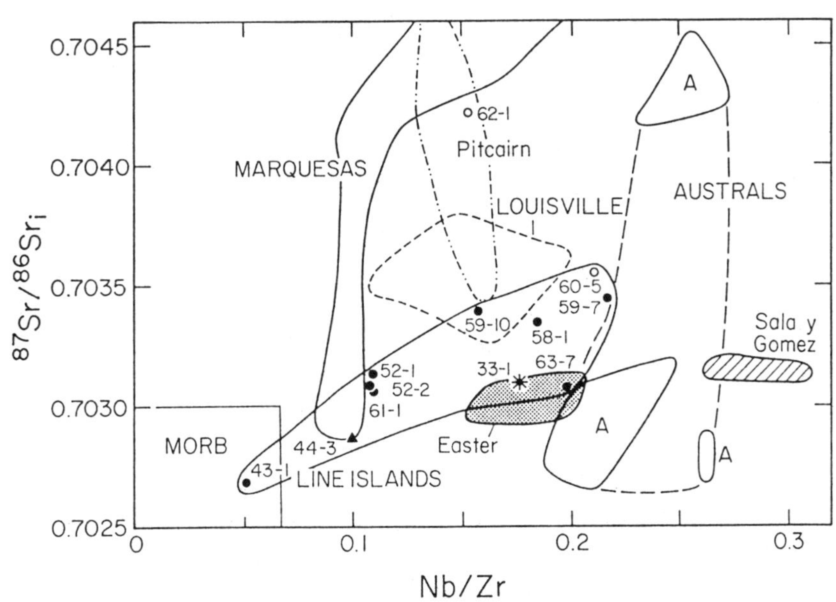

Fig. 10. Nb/Zr vs. initial $^{87}Sr/^{86}Sr$ ratios for Line Islands (see Table 6), Louisville Ridge [Cheng et al., 1987], Pitcairn [Woodhead and McCulloch, 1989], Easter and Sala y Gomez [Cheng, 1989], Marquesas [Vidal et al., 1984; Liotard et al., 1986] and Austral Islands [A; Dupuy et al., 1988] lavas. Note the good mixing correlation for Line Islands lavas between MORB and an average Austral Islands lavas. Symbols as in Figures 3, 4 and 5.

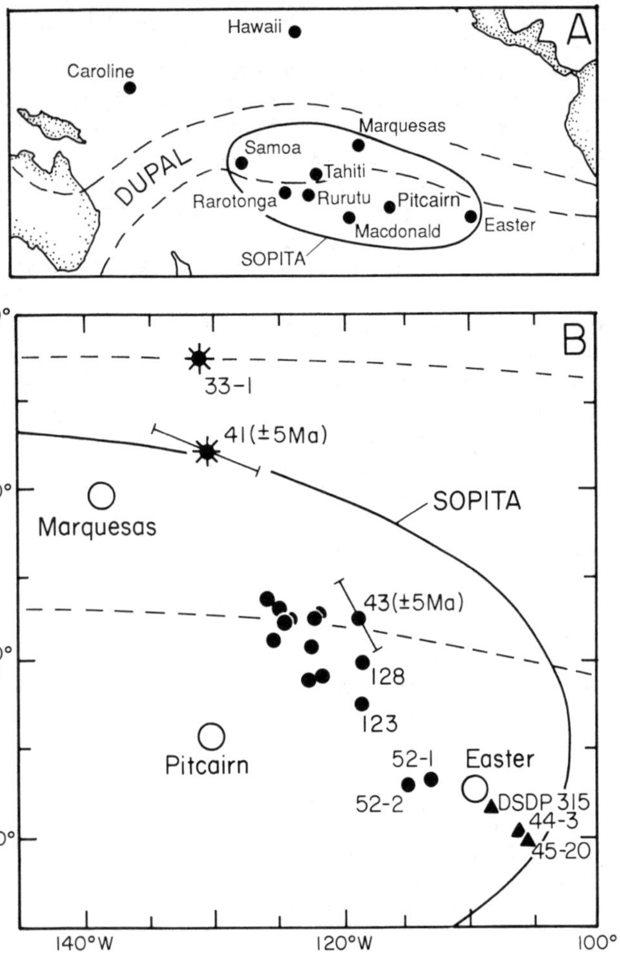

Fig. 11. A. Map of the south Pacific ocean with locations of the major hotspots, and the DUPAL [Hart, 1984] and SOPITA [Staudigel et al., 1991] regions [after Staudigel et al., 1991]. B. Backtracked position for Line Islands volcanoes using the poles and plate rotations of Duncan and Clague [1985]. Note that the field for Line Island hotspot volcanoes plots away from the location of any known hotspot (large open circles). Pre-hotspot volcanoes backtrack well to the south near the location of the Easter and Sala y Gomez hotspots. The post-hotspot volcanoes plot to the north of these fields. The closest potential hotspot is the Marquesas but most of its lavas are geochemically distinct from Line Islands lavas. Location of hotspots from Duncan and Clague [1985].

to any known hotspot. The pre-hotspot volcanism may be related to the Easter or Sala y Gomez hotspot. This would indicate that this hotspot has produced geochemically similar lavas for the last 70 m.y. The post-hotspot volcanism may be from the precursor to the Marquesas hotspot but its mantle source was different. There is no geochemical distinction among the lavas formed during these different periods of volcanism. The range in Pb and Sr isotope and Nb/Zr ratios for the Line Islands requires at least three source components. These components include depleted mantle, and two enriched sources, one with a high $^{206}Pb/^{204}Pb$ ratio. The backtracked positions of the Line Islands volcanoes extend the size of the SOPITA and indicate that it has been a major South Pacific feature for at least the last 90 m.y.

Acknowledgments. This study was part of a multidisciplinary study of the Line Islands, which was led by S. Schlanger. We thank R. Duncan, W. Sager, and B. Keating for their cooperation in this study, the captain and crew of the R/V KANA KEOKI for their assistance in surveying and dredging the Line Islands chain, the Republic of Kiribati for allowing us to work in their waters, and D. Macdougall for sharing unpublished isotope and trace element data for lavas from Easter and Sala y Gomez islands. Isotopic analyses were performed at the Lamont isotope laboratory, which is operated and maintained by A. Zindler and J. L. Rubenstone. We appreciate their support. We thank J. Natland and M. Pringle for their comments on a draft of this paper. This research was supported by the Office of Naval Research and the National Science Foundation (OCE90-12030, M.G.; OCE89-11603, H.S. and OCE87-11798 to A. Zindler in support of K.-H. P.). SOEST contribution no. 3035.

REFERENCES

Anders, E., and N. Grevesse, Abundances of the elements: meteoritic and solar, *Geochim. Cosmochim. Acta, 53*, 197-214, 1989.

Calmant, S., and A. Cazenave, Anomalous elastic thickness of the oceanic lithosphere in the south-central Pacific, *Nature, 328*, 236-238, 1987.

Cheng, Q., Geochemical studies of hotspot volcanism in the southern Pacific and its implications for mantle structure and dynamics, unpubl. Ph.D. thesis, Univ. Calif. San Diego, 237 pp., 1989.

Cheng, Q., K.-H. Park, J. D. MacDougall, Z. Zindler, G. W. Lugmair, J. Hawkins, P. Lonsdale, and H. Staudigel, Isotopic evidence for a hotspot origin of the Louisville seamount chain, *Amer. Geophys. Union Monogr., 43*, 283-296, 1987.

Clague, D. A., Jackson, E. D., and T. L. Wright, Petrology of Hualalai Volcano, Hawaii: implication for mantle composition, *Bull. Volcanol., 43,* 641-656 1980.

Clague, D. A., and R. D. Jarrard, Tertiary Pacific Plate motion deduced from the Hawaiian-Emperor Chain, *Bull. Geol. Soc. Amer., 84,* 1135-1154, 1973.

Coish, R. A., and L. A. Taylor, The effects of cooling rate on texture and pyroxene chemistry in D.S.D.P. Leg 34 basalt: a microprobe study, *Earth Planet. Sci. Lett., 42,* 389-398, 1979.

Coombs, D. S., and J. F. G. Wilkinson, Lineages and fractionation trends in undersaturated volcanic rocks from the East Otago Volcanic Province (New Zealand) and related rocks, *J. Petrol., 10,* 440-501, 1969.

Crough, S. T., and J. D. Jarrard, The Line-Marquesas Swell, *J. Geophys. Res., 86,* 11,763-11,771, 1981.

Davis, G. T., Petrology, geochemistry, and ages of volcanic rocks from the Line Islands, Central Pacific Ocean, M.S. thesis, Univ. Hawaii, 171 pp., 1986.

Duncan, R. A., and D. A. Clague, Pacific plate motion recorded by linear volcanic chains, in *The Ocean Basins and Margins,* edited by A. E. Nairn, F. G. Stehli, and S. Uyeda, 7A, pp. 89-121, Plenum Publishing Corp., 1985.

Dupuy, C., H. G. Barsczus, J. M. Liotard, and J. Dostal, Trace element evidence for the origin of oceanic island basalts: an example from the Austral Islands (French Polynesia), *Contrib. Mineral. Petrol., 98,* 293-302, 1988.

Epp, D., Possible perturbations to hotspots trace and implications for the origin and structure of the Line Islands, *J. Geophys. Res., 89,* 11,273-11,286, 1984.

Farrar, E., and J. M. Dixon, Early Tertiary rupture of the Pacific Plate: 1700 km of dextral offset along the Emperor Trough-Line Islands lineament, *Earth Planet. Sci. Lett., 53,* 307-322, 1981.

Fodor, R. V., G. R. Bauer, R. S. Jacobs, and T. J. Bornhorst, Kahoolawe Island, Hawaii: Tholeiitic, alkalic, and usual hydrothermal (?) "Enrichment" characteristics, *J. Volcanol. Geotherm. Res., 31,* 171-176, 1987.

Fodor, R. V., K. Keil, and T. E. Bunch, Contributions to the mineral chemistry of Hawaiian rocks - IV. Pyroxenes in rocks from Haleakala and West Maui volcanoes, Maui, Hawaii, *Contrib. Mineral. Petrol., 50,* 173-195, 1975.

Frey, F. A., M. O. Garcia, W. S. Wise, A. Kennedy, P. Gurriet, and F.

Albarede, The evolution of Mauna Kea Volcano, Hawaii: petrogenesis of tholeiitic and alkalic basalts, *J. Geophys. Res., 96,* 14,347-14,375, 1991.

Frey, F. A., W. S. Wise, M. O. Garcia, H. West, S.-T. Kwon, and A. Kennedy, Evolution of Mauna Kea Volcano, Hawaii: petrologic and geochemical constraints on postshield volcanism, *J. Geophys. Res., 95,* 1271-1300, 1990.

Garcia, M. O., F. A. Frey, and D. G. Grooms, Petrology of volcanic rocks from Kaula Island, Hawaii: implications for the origin of Hawaiian phonolites, *Contrib. Mineral. Petrol., 94,* 461-471, 1986.

Garcia, M. O., D. G. Grooms, and J. J. Naughton, Petrology and geochronology of volcanic rocks from seamounts along and near the Hawaiian Ridge: implications for propagation rate of the ridge, *Lithos., 20,* 323-336, 1987.

Haggerty, J. A., S. O. Schlanger, and I. Premoli-Silva, Late Cretaceous and Eocene volcanism in the southern Line Islands and implications for hotspot theory, *Geology, 10,* 433-437, 1982.

Hanan, B. B., and J.-G. Schilling, Easter microplate evolution: Pb isotope evidence, *J. Geophys. Res., 94,* 7432-7448, 1989.

Hart, S. R., A large scale isotope anomaly in the Southern Hemisphere mantle, *Nature, 309,* 753-757, 1984.

Hashimoto, M., Relic clinopyroxenes of Paleozoic greenstones of Tamba and Mikabu terraces, southwest Japan, *J. Japan Assoc. Mineral. Petrol. Econ. Geol., 67,* 323-331, 1972.

Hawkins, J. W., P. F. Lonsdale, and R. Batiza, Petrologic evolution of the Louisville seamount chain, in *Seamounts, Islands and Atolls, 43,* edited by B. H. Keating, P. Fryer, R. Batiza, and G. Boehlert, pp. 235-254, AGU, Washington, 1987.

Hedge, C. E., Strontium isotopes in basalts from the Pacific Ocean Basin, *Earth Planet. Sci. Lett., 38,* 88-94, 1978.

Henderson, L. J., and R. G. Gordon, Fixed hot spots and recurrent volcanism along the Line Islands chain, *Geol. Soc. Amer. Abstr. Programs, 14,* 513, 1982.

Jackson, E. D., K. E. Bargar, B. P. Fabbi, and C. Heropoulos, Petrology of basaltic rocks drilled on Leg 33 of the D.S.D.P., *Initial Rep. Deep Sea Drill. Proj., 33,* 571-613, 1976.

Jackson, E. D., and S. O. Schlanger, Regional syntheses: Line Islands Chain, Tuamotu Island Chain, and Manihiki Plateau, Central Pacific Ocean, *Initial Rep. Deep Sea Drill. Proj., 33,* 915-927, 1976.

Johnson, R. H., Active submarine volcanism in the Austral Islands, *Science, 167,* 977-979, 1970.

Lanphere, M. A., and G. B. Dalrymple, K-Ar ages of basalts from DSDP Leg 33, *Initial Rep. Deep Sea Drill. Proj., 33,* 649-653, 1976.

Lanphere, M. A., G. B. Dalrymple, and D. A. Clague, Rb-Sr systematics of basalts from the Hawaiian-Emperor volcanic chain, *Initial Rep. Deep Sea Drill. Proj., 55,* 695-706, 1980.

Leake, B. E., Nomenclature of amphiboles, *Can. Mineral., 16,* 501-520, 1978.

LeBas, M. J., The role of aluminum in igneous clinopyroxenes with relation to their parentage, *Am. J. Sci., 260,* 267-288, 1962.

le Roex, A. P., Geochemistry, mineralogy and magmatic evolution of the basaltic and trachytic lavas from Gough Island, South Atlantic, *J. Petrol., 26,* 149-186, 1985.

Liotard, J.-M., H. G. Barsczus, C. Dupuy, and J. Dostal, Geochemistry and origin of basaltic lavas from Marquesas archipelago, French Polynesia, *Contrib. Mineral. Petrol., 92,* 260-268, 1986.

Mahood, G. C., and D. R. Baker, Experimental constraints on depths of fractionation of mildly alkalic basalts and associated rocks: Pantelleria, Strait of Sicily, *Contrib. Mineral. Petrol., 93,* 251-264, 1986.

Matsuda, J.-I., Low Sr isotopic ratios of alkali basalts from the Line Islands in the Pacific Ocean, *Geochem. Jour., 14,* 41-46, 1980.

Mattey, D. P., Minor and trace element geochemistry of volcanic rocks from Truk, Ponape, and Kusaie, eastern Caroline Islands: evolution of a young hot spot trace across old oceanic crust, *Contrib. Mineral. Petrol., 80,* 1-13, 1982.

McNutt, M. K., and K. Fisher, The Southern Pacific Superswell, *Am. Geophys. Union Monogr., 43,* 25-34, 1987.

Morgan, W. J., Convective plumes in the lower mantle, *Nature, 230,* 42-43, 1971.

Moore, J. G., W. R. Normark, and P. W. Lipman, Loihi Seamount: a young submarine Hawaiian volcano, Hawaii Symposium on Intraplate Volcanism and Submarine Volcanism, Abst. Vol., 127, 1979.

Natland, J. H., Petrology of volcanic rocks dredged from seamounts in the Line Islands, *Initial Rep. Deep Sea Drill. Proj., 33,* 749-777, 1976.

Orwig, T. L., and L. W. Kroenke, Tectonics of the eastern Central Pacific Basin, *Mar. Geol., 34,* 29-43, 1980.

Palacz, Z., and A. D. Saunders, Coupled trace element and isotope enrichment in the Cook-Austral-Samoa Islands, Southwest Pacific, *Earth Planet. Sci. Lett., 79,* 270-280, 1986.

Richardson, S. H., A. J. Erlank, D. L. Reid, and A. R. Duncan, Major and trace elements and Nd and Sr isotope geochemistry of basalts from the Deep Sea Drilling Project Leg 74 Walvis ridge transect, *Initial Rep. Deep Sea Drill. Proj., 74,* 739-754, 1984.

Saito, K., and M. Ozima, $^{40}Ar/^{39}Ar$ ages of submarine rocks from the Line Islands; implications on the origin of the Line Islands, *Am. Geophys. Union Mono., 19,* 369-374, 1976.

Saito, K., and M. Ozima, $^{40}Ar/^{39}Ar$ geochronological studies on submarine rocks from the western Pacific area, *Earth Planet. Sci. Lett., 33,* 353-369, 1977.

Schlanger, S. O., M. O. Garcia, B. H. Keating, J. J. Naughton, W. W. Sager, J. A. Haggerty, J. A. Philpotts, and R. A. Duncan, Geology and geochronology of the Line Islands, *J. Geophys. Res., 89,* 11,261-11,272, 1984.

Smith, W.H.F., H. Staudigel, A. B. Watts, and M. S. Aingle, The Magellan Seamounts: Early Cretaceous record of the south Pacific isotopic and thermal anomaly, *J. Geophys. Res., 94,* 10,501-10,523, 1989.

Spengler, S. R., and M. O. Garcia, Geochemistry of the Hawi lavas, Kohala Volcano, Hawaii, *Contrib. Mineral. Petrol., 99,* 90-104, 1988.

Staudigel, H., K.-H. Park, M. Pringle, J. L. Rubenstone, W. H. F. Smith, and A. Zindler, The longevity of the South Pacific isotopic and thermal anomaly, *Earth Planet. Sci. Lett., 102,* 24-44, 1991.

Turner, D. L., and R. D. Jarrard, K-Ar dating of the Cook-Austral island chain, *J. Volcanol. Geotherm. Res., 12,* 187-220, 1982.

Vallance, T. J., Pyroxenes and the basalt-spillite relation, in *Spillites and Spillitic Rocks,* edited by G. C. Amstutz, Springer-Verlag, New York, 1974.

Velde, D., An aenigmatite-richerite-olivine trachyte from Puu Koae, West Maui, Hawaii, *Am. Mineral., 63,* 771-778, 1978.

Vidal, P., C. Chauvel, and R. Brousse, Large mantle heterogeneities beneath French Polynesia, *Nature, 307,* 536-538, 1984.

West, H. B., M. O. Garcia, F. A. Frey, and A. Kennedy, Nature and cause of compositional variation among the alkalic cap lavas of Mauna Kea volcano, Hawaii, *Contrib. Mineral. Petrol., 100,* 383-397, 1988.

White, W. M., and A. W. Hofmann, Sr and Nd isotope geochemistry of oceanic basalts and mantle evolution, *Nature, 296,* 821-825, 1982.

Wilson, J. T., A possible origin of the Hawaiian Islands, *Can. J. Physics, 41,* 863-870, 1963.

Winterer, E. L., Bathymetry and regional setting of the Line Islands Chain, *Initial Rep. Deep Sea Drill. Proj., 33,* 731-747, 1976.

Woodhead, J. D., and M. T. McCulloch, Ancient seafloor signals in Pitcairn Island lavas and evidence for large amplitude, small length-scale mantle heterogeneities, *Earth Planet. Sci. Lett., 94,* 257-273, 1989.

Zindler, A., and S. Hart, Chemical geodynamics, *Ann. Rev. Earth Planet. Sci., 14,* 493-571, 1986.

M. Garcia, Department of Geology and Geophysics, University of Hawaii, Honolulu, HI 96822.

K.-H. Park, Korea Basic Science Center, P. O. Box 1922, Youngdong Station, Seoul 135-280, Republic of Korea.

G. T. Davis, Maersk Energy Inc., Houston, TX 77042.

H. Staudigel, Scripps Institution of Oceanography, University of California, La Jolla, CA 92093.

D. P. Mattey, Geology Department, Royal Holloway and Bedford New College, Surrey, TW20 OEX, U. K.

Geochemistry and Age of the Ontong Java Plateau

J.J. Mahoney[1], M. Storey[2], R.A. Duncan[3], K.J. Spencer[1], and M. Pringle[4]

Basement rocks from the Ontong Java Plateau are tholeiitic basalts which appear to record very high degrees of partial melting, much like those found today in the vicinity of Iceland. They display a limited range of incompatible element and isotopic variation, but small differences are apparent between sampled sites and between upper and lower groups of flows at Ocean Drilling Program Site 807. ^{40}Ar-^{39}Ar ages of lavas from Site 807, Deep Sea Drilling Project Site 289, and basement from the island of Malaita at the southern edge of the plateau are indistinguishable about a weighted mean of 122.4 ± 0.8 Ma, indicating that plateau-building eruptions ended more or less simultaneously at widely separated locations. These data also provide the best minimum age available for magnetic reversal M0 and for the Aptian-Barremian boundary. Pb-Nd-Sr isotopes for lavas from Sites 289, 803 and 807, as well as southern Malaita reflect a hotspot-like source with $\epsilon_{Nd}(T)$ = +4.0 to +6.3, $(^{87}Sr/^{86}Sr)_T$ = 0.70423-0.70339, $^{206}Pb/^{204}Pb$ = 18.245-18.709, and possessing consistently greater $^{208}Pb/^{204}Pb$ for a given $^{206}Pb/^{204}Pb$ than Pacific MORB. The combination of hotspot-like mantle source, very high degrees of melting, and lack of a discernible age progression is best explained if the bulk of the plateau was constructed rapidly above a surfacing plume head, possibly that of the infant Louisville hotspot. Basalt and feldspar separates indicate a substantially younger age of ~90 Ma for basement at Site 803; in addition, volcaniclastic layers of mid-Cenomanian through Coniacian age occur at DSDP Site 288, and beds of late Aptian-Albian age are found at Site 289. Therefore, at least some volcanism continued on the plateau for 30 m.y. or more. The basalts at Site 803 are chemically and isotopically very similar to those at the ~122 Ma sites, suggesting that the same hot plume-type mantle was present beneath the plateau for an extended period or at two different times. Surviving seamounts of the Louisville Ridge formed between 70 and 0 Ma have much higher $^{206}Pb/^{204}Pb$ than any of the plateau basalts. Thus, assuming the Louisville hotspot was the source of the plateau lavas, a change in the hotspot's isotopic compositions may have occurred between roughly 70 and 90 Ma; such a change may have accompanied the plume-head to plume-tail transition. Similar shifts from early, lower $^{206}Pb/^{204}Pb$ to subsequent higher $^{206}Pb/^{204}Pb$ values are found in several other oceanic plateau-hotspot and continental flood basalt-hotspot systems, and could reflect either a reduction in the supply of low $^{206}Pb/^{204}Pb$ mantle or an inability of some off-ridge plume-tails to melt refractory low $^{206}Pb/^{204}Pb$ material.

Introduction

Oceanic plateaus are extensive volcanic constructional features with crustal thicknesses of ~18-40 km [Hussong et al., 1979] that occur in all three major ocean basins. Despite their often immense size, their origins are understood poorly, in large part because of a dearth of basement sampling and, hence, of information on age and mantle sources. In the Pacific, by far the best-sampled of these features at present is the Ontong Java Plateau (Figure 1), where recent drilling during ODP (Ocean Drilling Program) Leg 130 penetrated 149 m of basement at Site 807 and 26 m at Site 803 [Kroenke, Berger et al., 1991]. Prior to Leg 130, the submarine crust of the plateau was known only from a single, altered basalt flow drilled at DSDP (Deep Sea Drilling Project) Site 289 [e.g. Stoeser, 1975]. Some probable plateau basement also is exposed subaerially at the southwestern edge of the plateau on the islands of Malaita and Small Malaita; although we have completed preliminary geochemical work on a collection of Malaitan rocks, comprehensive studies are still underway. In this paper we present chemical, Nd-Sr-Pb isotopic, and ^{40}Ar-^{39}Ar

[1]School of Ocean and Earth Science and Technology, University of Hawaii, Honolulu, HI 96822.
[2]Department of Geology, University of Keele, Staffordshire STF 5BH, United Kingdom.
[3]College of Oceanography, Oregon State University, Corvallis, OR 97331.
[4]U.S. Geological Survey, Menlo Park, CA 94025.

The Mesozoic Pacific: Geology, Tectonics, and Volcanism
Geophysical Monograph 77

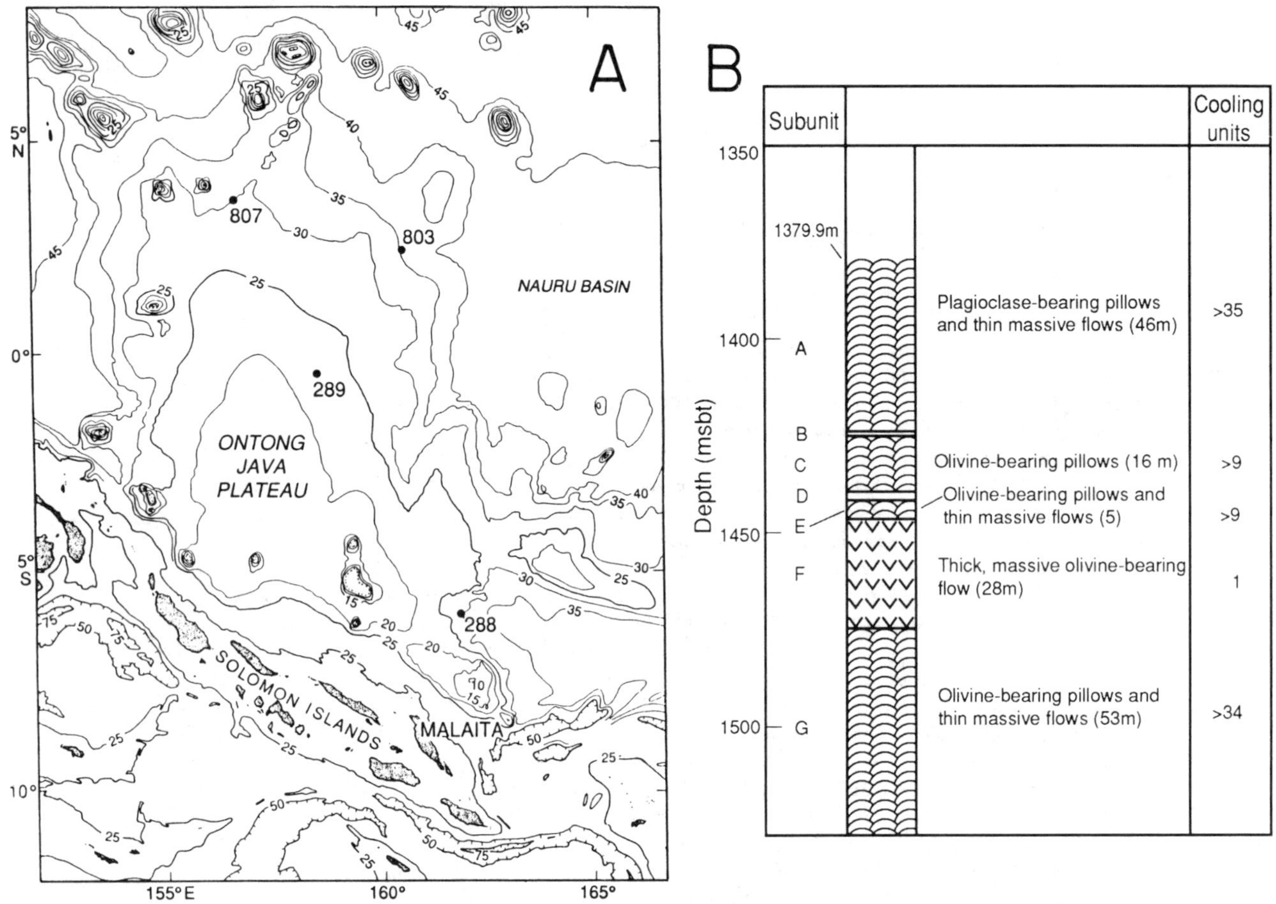

Fig. 1. (a) Map of the main part of the Ontong Java Plateau, showing locations of Leg 130 drill sites 803 and 807, DSDP sites 288 (which did not reach basement) and 289, and the island of Malaita. Bathymetric contour interval is 2500 m; additional contours have been added (at 500 m intervals) on and near the plateau to illustrate its morphology. (b) Lithologic column of basement at Site 807 illustrating unit thickness and estimated number of cooling units (individual massive or pillowed flows) in each. Units B and D are sedimentary interbeds.

geochronological data for the Leg 130 lavas, as well as ^{40}Ar-^{39}Ar determinations on two Malaitan basement basalts; the results are used to address the petrogenesis, timing, and environment of the plateau's formation.

Combined geochemical, gravity, and plate-tectonic evidence indicates that most of the Pacific plateaus may have been formed above near-ridge or ridge-centered Cretaceous hotspots [Mahoney, 1987]. At least the largest ones may reflect the arrival at the base of the lithosphere (not necessarily near ridges) of the inflated "heads" of new mantle plumes [Richards et al., 1989]; this plume-initiation hypothesis predicts plateau-forming volcanism to be cataclysmic in nature, beginning suddenly and lasting only a few million years. In contrast, plateaus created at robust but more or less steady-state near-ridge hotspots—that is, plume "tails"—could take tens of millions of years to form, their size depending upon both the size of the hotspot and the rate of plate movement relative to the hotspot [Mahoney and Spencer, 1991]. The medium-sized Icelandic Plateau (~250,000 km^2) in the North Atlantic is an example of the latter type, having been created near the slowly spreading Mid-Atlantic Ridge over a period of at least 25 m.y. by the already mature Icelandic hotspot [e.g. Oskarsson et al., 1985]. At present, the best evidence for the plume-initiation mechanism is found in the probable continental analogues of many oceanic plateaus; namely, large continental flood basalt provinces. Recent high-precision ^{40}Ar-^{39}Ar dating in the Siberian [Renne and Basu, 1991], Deccan [Duncan and Pyle, 1988; Courtillot et al., 1988], and Columbia River [e.g. Baksi, 1989] provinces, for example, indicates that enormous portions of each were formed in only 1-2 m.y., or even less. Because large, initial plume-heads should exert considerable influence on rift propagation, plume heads in oceanic areas are likely to become associated with spreading centers even if they first appear in a location distant from a ridge [Mahoney and Spencer, 1991]; thus, the plume-initiation and near-ridge hotspot concepts need not be mutually exclusive. A third hypothesis proposed recently for the Pacific plateaus views them all as manifestations of a single gigantic "superplume" that originated at the core-mantle boundary, surfaced in the Early Cretaceous, and largely spent itself by 100 Ma [Larson, 1991]. In this case, the Pacific plateaus may not be associated with specific present-day hotspots.

The two largest oceanic plateaus in the world are the Ontong Java ($>1.5 \times 10^6$ km^2) and the Kerguelen Plateau-Broken Ridge pair (with nearly as great an extent) in the southeastern Indian Ocean. Apart from their great size, these two plateaus are significantly different. For example, the Kerguelen Plateau was formed in a new ocean basin, in close proximity to rifting continental lithosphere, just after the breakup of Indo-Madagascar from Antarctica-Australia; this factor makes it difficult to distinguish geochemical characteristics inherited by plateau lavas from sublithospheric mantle sources from those caused by the influence of mobilized Gondwanan lithosphere [e.g. Storey et al., 1989, 1992]. In contrast, available evidence indicates that the Ontong Java Plateau originated in an intra-oceanic setting, far away from any sizeable continental mass [e.g. Yan and Kroenke, 1992]; thus, interpretation of Ontong Java geochemical signatures can be made with much less concern about direct continental lithospheric contributions. On the other hand, whereas nearly all workers consider the Kerguelen Plateau to be the early product of the present Kerguelen hotspot [e.g. Duncan, 1991], the identity of any hotspot responsible for the formation of the Ontong Java Plateau is much less certain. Plate reconstructions of the paleo-Pacific suggest that the Louisville hotspot (50°S, 139°W) may have played a key role [e.g. Richards et al., 1989; Mahoney and Spencer, 1991; Tarduno et al., 1991], but the connection has remained speculative owing to a lack of data on Ontong Java basement age, large uncertainties in pre-90 Ma Pacific reconstructions, as well as pronounced geochemical differences between the few basement lavas analyzed previously and volcanic rocks of the Louisville Ridge, the 0-70 Ma trace of the hotspot.

GEOCHRONOLOGY

Samples and Methods

Shipboard descriptions of the Site 803 and 807 basement rocks have been given by Kroenke, Berger et al. [1991]. Briefly, basement recovered at both sites consists of sparsely porphyritic to aphyric tholeiitic basalts. Those at Site 807 have been subdivided into five stratigraphic units or packets of flows, labeled A, C, E, F, G (Units B and D are sedimentary interbeds); Unit F is a single 28-m-thick massive flow (Figure 1b). Several thin, pillowed lavas, all classified as one stratigraphic unit, were drilled at Site 803. At Site 807, flows comprising igneous Units C-G contain sparse, ~0.5-mm-diameter olivine phenocrysts, extensively altered to dark green clay. These are lacking in Unit A lavas, as well as in the Site 803 basalts, which instead possess rare, relatively fresh plagioclase phenocrysts (two examples of fresh, subhedral grains ~5 mm to 1 cm across were observed). The Site 807 basalts vary from quite fresh to moderately altered in hand specimen and thin section, displaying medium-gray groundmasses and <10 % to ~30% total alteration. Groundmass plagioclase and clinopyroxene are relatively fresh but mesostasis is variably altered to clays, and in places veins containing celadonite, calcite and possibly zeolites are present. Basalts from Site 803 range from moderately to highly altered, with gray to brownish-gray groundmasses and roughly 25-50% alteration of mesostasis and groundmass phases. The Site 289 flow has been described by Stoeser [1976] and the Malaitan samples by Hughes and Turner [1977]; the former is altered to a level comparable to the Site 803 lavas, whereas the latter are more comparable to some of the Site 807 rocks.

In an attempt to establish the space-time relationships of volcanism across the plateau, ^{40}Ar-^{39}Ar incremental heating methods were employed. Because seawater alteration of primary igneous phases (olivine, pyroxene, feldspar, oxides, glass) to clays, zeolites, celadonite, and calcite causes loss of radiogenic ^{40}Ar and addition of K, standard K-Ar radiometric dates usually seriously underestimate the crystallization age of altered samples. The ^{40}Ar-^{39}Ar method is a variation of K-Ar chronometry that relies on neutron irradiation of samples to produce ^{39}Ar from ^{39}K. Subsequent incremental heating of the sample under vacuum allows extraction of parent (^{39}Ar, representing K) and daughter (^{40}Ar) isotopes from discrete temperature steps. Alteration phases are less retentive of Ar because of their open structures, so Ar diffuses out of them at low temperatures. In fact, much of the K added from seawater probably resides on grain boundaries, in vesicles and cracks, and ^{39}Ar formed from it will be lost from the sample during irradiation. Conversely, the remaining igneous phases outgas at higher temperatures, and their Ar isotopic compositions reflect the desired crystallization age of the sample. Hence, it is often possible to separate the contributions of alteration and igneous phases to the sample's total Ar-isotopic composition.

For the samples studied here, pieces of basalt were trimmed to obtain the freshest and best-crystallized material. These samples were crushed to 0.5-1 mm size, hand-picked to remove vein alteration, and ultrasonically washed in distilled water. Approximately 0.5-g splits of the prepared whole-rock chips were sealed in evacuated quartz tubes and irradiated for 6-8 hr in the core of the Oregon State University TRIGA reactor, where they received a neutron dose of $\sim 0.7 \times 10^{18}$ nvt. The efficiency of conversion of ^{39}K to ^{39}Ar by neutron capture was monitored with samples of hornblende standard Mmhb-1 (520.4 ± 1.7 Ma [Samson and Alexander, 1987]). Further details of the flux characteristics, monitor minerals, and corrections for interfering K- and Ca-derived Ar isotopes are reported by Dalrymple et al. [1981].

Argon extractions were performed at Oregon State University in a conventional high-vacuum glass line using radio-frequency induction heating. Heating steps were set from power levels on the generator and were determined from previous experience to divide the total whole-rock Ar into five or six roughly equal portions. Samples were held at each temperature-step setting for 30 min, and the Ar composition of each gas increment was measured mass spectrometrically with an AEI MS-10S instrument with computer-controlled peak switching and data acquisition. The system background (blank) was 3×10^{-14} moles ^{40}Ar.

In addition, megacrystic (0.5-1 cm) plagioclase feldspars were separated from two samples of Site 803 basalts. These were acid-treated (in HCl and HF), washed and dried, and irradiated in the U.S. Geological Survey's TRIGA reactor under similar conditions as the whole-rock chips. Argon extractions were performed at the U.S. Geological Survey (Menlo Park) using a 5W Ar-ion laser-probe connected directly with a MAP 216 mass spectrometer. Pre-heating of the samples (each ~1 mg in size) using an unfocused laser beam (~600°C) removed most surficial atmospheric ^{40}Ar. Step-heating was achieved in six increments by increasing power to the focused laser beam until sample fusion. Typical step-gas compositions were more than 50% radiogenic ^{40}Ar. The system background was $<1 \times 10^{-15}$ moles ^{40}Ar. The reactor flux monitor (J) was measured with the 27.92 Ma sanidine standard 85G003 from the Taylor Creek rhyolite; this is equivalent to an age of 513.9 Ma for the Mmhb-1 hornblende standard. To enable a direct comparison between the Oregon State and U.S.G.S. data, the results for the Site 803 plagioclases have been recalculated to be equivalent to a 520.4 Ma age for Mmhb-1.

Results

Results from our ^{40}Ar-^{39}Ar incremental heating experiments are reported in Table 1, and representative examples are illustrated in Figures 2-6. Data are displayed in age-spectra diagrams, in which the ages calculated from the composition of gas in each heating step for a given sample are plotted against cumulative gas released (%^{39}Ar). In most samples (e.g. Figure 2a), concordant ages for middle- to high-temperature steps define a "plateau" age, which provides the best estimate of the age of rock crystallization (note that such a plateau is a type of ^{40}Ar-^{39}Ar age, not the age of an oceanic plateau sample). Gas compositions for the various steps also can be correlated in isotope-ratio diagrams (e.g. Figure 2b) where colinear points are fitted by least squares to yield a slope, which determines an isochron age, and an intercept, which defines the composition of Ar in the sample when it crystallized (expected to be atmospheric, ^{40}Ar/^{36}Ar = 295.5). Unlike the age-spectra plots, which assume an initial ^{40}Ar/^{36}Ar composition equal to the atmospheric ratio, isotope correlation diagrams allow independent determination of sample age and initial ^{40}Ar/^{36}Ar. The goodness-of-fit parameter SUMS [York, 1969] has a χ^2 distribution with N-2 degrees of freedom, N being the number of steps included in the linear regression. In Table 1, the F-variate statistic SUMS/N-2 (i.e., the ratio of total scatter about the isochron to the scatter attributable to analytical error alone) is given. For reference, a value of SUMS/(N-2) = 2.6 for a regression fitted with five plateau steps would indicate that an isochron relationship could not be dismissed at the 95% confidence limit. Experiments that yield such an acceptable measure of goodness of fit, an atmospheric ^{40}Ar/^{36}Ar intercept, and concordant isochron and plateau ages define reliable crystallization ages [Lanphere and Dalrymple, 1978].

In addition to age determination, the Ca/K of individual steps can be calculated from Ar isotopic compositions using $1.852 \times {}^{37}$Ar/^{39}Ar [Walker and McDougall, 1982]; the phases outgassing at each step then can be identified from known Ca/K ratios for different minerals: 500-2000 for clinopyroxene, 20-500 for plagioclase, 1-20 for basaltic groundmasses, and <1 for clays. Plots of Ca/K vs. %^{39}Ar for our experiments (e.g. Figure 2c) show that Ca/K increased with temperature from values of ~10 to ~100. The temperature steps that define the plateau ages in Table 1 therefore are associated primarily with Ca/K appropriate for groundmass and plagioclase feldspars.

Discussion and interpretation of age determinations. The incremental heating experiments produced variably disturbed age spectra; most developed nearly ideal plateaus comprising over 90% of the total Ar in the samples. Spurious step ages generally resulted from small amounts of gas released at low temperatures, which we attribute to nonconcordant loss of radiogenic ^{40}Ar (during alteration) and ^{39}Ar (during irradiation) from alteration minerals. Plateau ages, defined by three or more concordant step ages making up more than 50% of the total Ar, were calculated as the mean of the step ages (weighted inversely by their variances). An integrated sample age was calculated by adding together all the step compositions; this age should be equivalent to an ^{40}Ar-^{39}Ar total-fusion age. Isochron ages were determined from the slope of the best-fit line through the isotopic values of the plateau-defining steps. For all of our samples, ^{40}Ar/^{36}Ar intercepts were near the atmospheric value; for the Site 807 basalts, at least, this is not surprising in view of the evidence from Unit B (a limestone interbed) that they were erupted above the calcite compensation depth [Kroenke, Berger et al., 1991].

Six whole-rock samples from Site 807 yield crystallization ages indistinguishable from one another, with a weighted mean plateau value of 122.3 ± 1.0 Ma, corresponding to an early Aptian age [e.g. Harland et al., 1990]. The presence of Aptian microfossils in sediments overlying and interbedded with these basalts is compatible with this age [Kroenke, Berger et al., 1991]. Paleomagnetic data [Tarduno et al., 1991] show that all of the basalts and overlying limestones are normally

TABLE 1. ^{40}Ar-^{39}Ar Plateau and Isochron Ages for Basaltic Rocks from the Ontong Java Plateau

Sample	Plateau age ±1 σ (Ma)	Integrated age (Ma)	^{39}Ar (% of total)	Isochron age ±1 σ (Ma)	SUMS (N-2)	$^{40}Ar/^{36}Ar$ intercept
Oregon State University:						
130-807C-						
75R-2, 129-131	121.0 ± 4.5	118.8	80	121.7 ± 3.6	0.08	297.3 ± 11.0
78R-1, 67-69	121.4 ± 1.9	124.1	75	124.0 ± 2.5	0.40	294.0 ± 3.3
79R-1, 17-18	None developed	98.7	--	None	--	--
80R-1, 52-55	119.9 ± 2.6	118.2	100	122.4 ± 4.0	1.22	290.5 ± 4.9
84R-6, 0-3	123.4 ± 2.5	124.5	97	122.6 ± 1.7	0.05	292.6 ± 17.5
90R-1, 38-41	124.7 ± 2.2	126.3	100	119.5 ± 9.9	0.28	307.6 ± 29.9
93R-3, 15-18	122.1 ± 2.2	143.9	68	126.0 ± 4.8	0.30	294.0 ± 3.0
130-803D-						
71R-1, 114-115	None developed	55.5	--	None	--	--
71R-2, 87-88	88.2 ± 1.1	88.3	99	85.7 ± 1.3	0.49	294.7 ± 2.6
30-289-						
132-4, 79-81	121.7 ± 2.7	121.0	100	128.5 ± 2.6	0.53	288.8 ± 4.1
132-4, 122-125	122.8 ± 2.4	121.4	100	122.7 ± 2.2	0.22	293.0 ± 3.1
Malaita 43	123.3 ± 2.9	127.0	79	118.9 ± 2.4	0.15	297.5 ± 3.6
Malaita 8374	122.3 ± 2.4	125.4	67	114.7 ± 4.9	1.50	300.6 ± 2.1
U.S. Geological Survey:						
130-803D-						
69R-1, 87-89	86.1 ± 4.6	86.1	100	83.1 ± 2.5	1.03	483 ± 117
71R-1, 14-16	83.7 ± 3.1	84.2	100	82.7 ± 1.5	0.16	304.9 ± 14.6
	93.9 ± 1.4*	93.6	100	93.8 ± 2.8	1.06	296.2 ± 9.8

*Weighted mean of six total-fusion analyses of ~1 mg crystals; corresponding isochron age from colinear total-fusion gas compositions.

magnetized, and it is likely that they were magnetized during the earliest part of the long Cretaceous interval of normal polarity. In fact, our new age determinations provide the best minimum estimate available for the age of M0, the last reversal before the long normal interval, and for the Aptian-Barremian boundary.

Two samples from the basement flow at DSDP Site 289 (located some 500 km from Site 807) and two basalts from the southern part of the island of Malaita (some 1500 km south of Site 807) were analyzed at the same time as the Leg 130 lavas and also produced good results. Significantly, their plateau ages are identical to those at Site 807 (Table 1, Figures 3 and 4). Sample 30-289-132-4 (79-81) is the single instance of discordant plateau and isochron ages; its $^{40}Ar/^{36}Ar$ intercept (288.8) is slightly lower than the atmospheric value. We see no evidence to conclude that the initial Ar composition in this sample was non-atmospheric; rather, the low intercept is an artifact of the grouping of step gas compositions. Hence the isochron age for this sample is somewhat too old and we prefer the plateau age (121.7 ± 2.7 Ma).

Two whole-rock and two feldspar samples from Site 803 were analyzed (Table 1 and Figures 5, 6); good plateau ages were obtained for three of them. The feldspar sample from 130-803D-71R-1 (14-16) was run in two laser-heating modes: in the first, incremental heating produced a plateau age of 83.7 ± 3.1 Ma (six steps), whereas in the second a series of total-

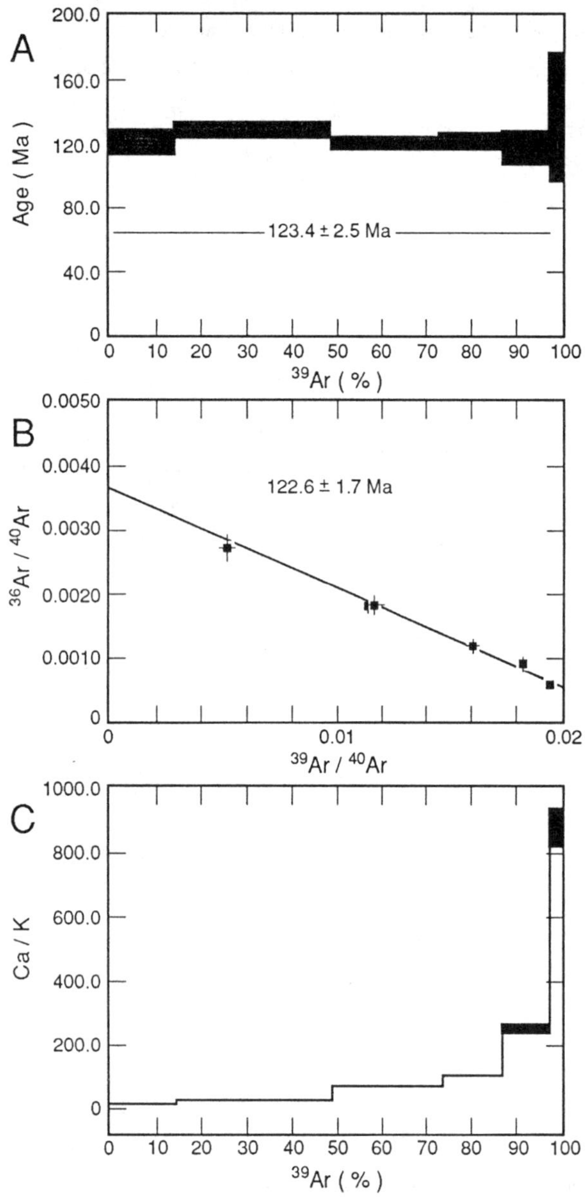

Fig. 2. (a) Apparent age-release diagram for ^{40}Ar-^{39}Ar incremental heating experiment on basalt sample 130-807C-84R-6 (0-3). Horizontal boxes indicate estimated analytical error ($\pm 1\sigma$) about each step age. A plateau age (indicated) has been determined from the weighted mean of contiguous, concordant step-ages. (b) $^{36}Ar/^{40}Ar$ vs. $^{39}Ar/^{40}Ar$ isotope correlation diagram for the step Ar-compositions measured. The isochron age (indicated) is calculated from the best-fitting line through colinear step compositions, after York [1969]; analytical uncertainties are shown. (c) Ca/K values inferred from step Ar-compositions reflect the phases (pyroxene, feldspar, groundmass, clays) contributing to gas released at each temperature.

fusion measurements on six ~1-mg crystals gave a weighted mean age of 93.9 ± 1.4 Ma. Corresponding isochron ages are similar. Petrographically, the second mineral fraction, which gave the greater age, was fresher (i.e., translucent, euhedral

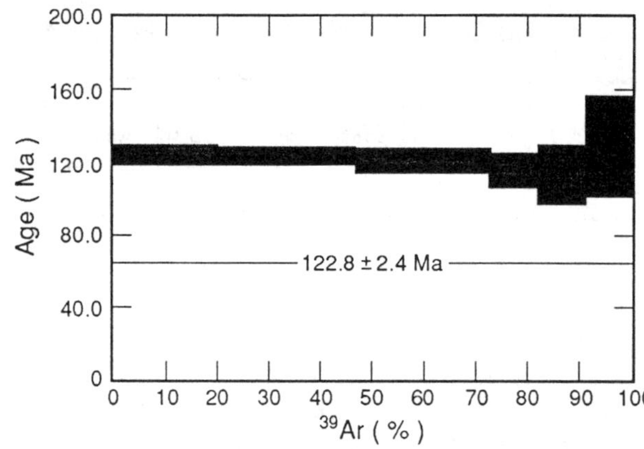

Fig. 3. Apparent age-release diagram for ^{40}Ar-^{39}Ar incremental heating experiment of DSDP basalt sample 30-289-132-4 (122-125). Details as in Figure 2a.

crystals vs. more opaque ones in the first case). Although it is conceivable that the more altered grains of the first fraction record a post-magmatic recrystallization event some 8-10 m.y. younger than the volcanism at Site 803, there is no evidence for this; the Ca/K values of feldspars in the two fractions are virtually identical (in the 100-1000 range), with no hint of K-feldspar overgrowths. Argon loss in the more altered crystals could explain the age difference, but we see no evidence of an older igneous age in the higher temperature steps. The whole-rock age (88.2 ± 1.1 Ma) falls within the range of the feldspar ages. A possible mid-Cenomanian age for the lowermost sediments at this site [Kroenke, Berger et al., 1991] is most compatible with the older of the feldspar ages. Regardless of the precise crystallization age of the Site 803 basalts, it is clear that volcanism at this location occurred about 30 m.y. after the formation of basement (strictly, the termination of volcanism) at Sites 807 and 289, and in southern Malaita.

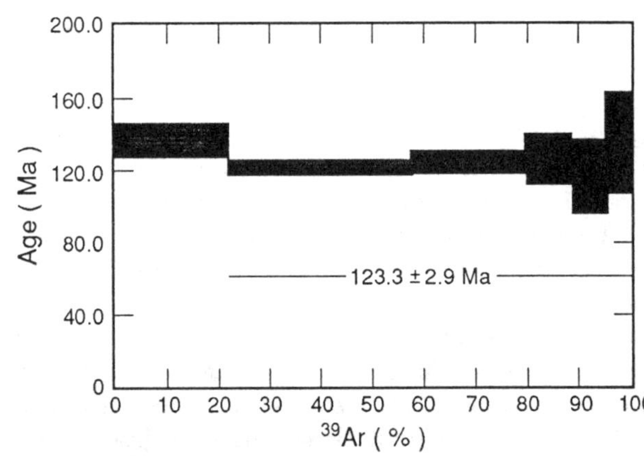

Fig. 4. Apparent age-release diagram for ^{40}Ar-^{39}Ar incremental heating experiment of Malaita basalt 43. Details as in Figure 2a.

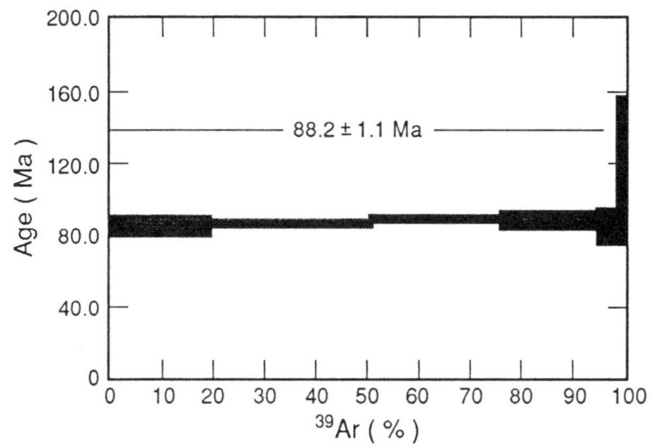

Fig. 5. Apparent age-release diagram for ^{40}Ar-^{39}Ar incremental heating experiment of basalt sample 130-803D-71R-2 (87-88). Details as in Figure 2a.

GEOCHEMISTRY

Samples and Methods

The Leg 130 samples analyzed here were chosen aboard ship by the two senior authors on the basis of relative freshness and representative coverage of section–not always compatible goals, particularly near the more altered top of basement at both sites. Major elements and Rb, Zr, Sr, Y, V, Cr, and Ni were measured on agate- or alumina-ground bulk-sample (free of obvious veins) powders by standard X-ray fluorescence methods at the University of Keele. Trace elements were analyzed on powder briquets and major elements on fused disks consisting of a mixture (by weight) of one part sample powder to five parts $LiBO_2$ flux. Further information on methodology is given by Floyd [1986]. Our results appear in Table 2, along with estimated analytical uncertainties.

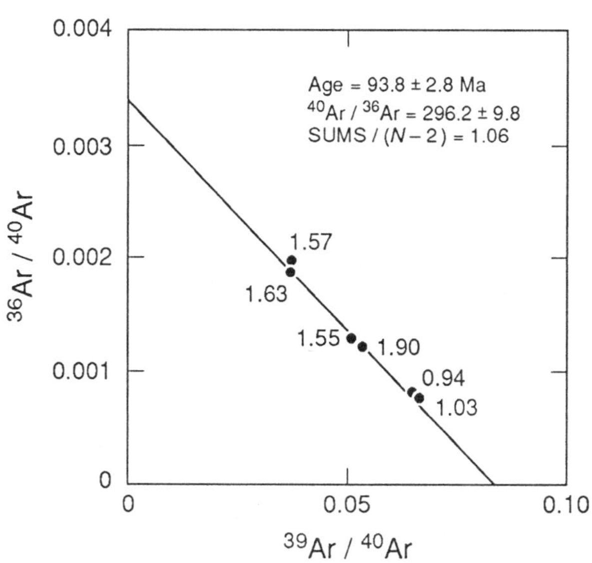

Fig. 6. $^{36}Ar/^{40}Ar$ vs. $^{39}Ar/^{40}Ar$ isotope correlation diagram for feldspar separated from basalt sample 130-803D-71R-1 (14-16). Details as in Figure 2b, except points are total-fusion Ar compositions of individual crystals (weights of which are given in milligrams).

Nine lanthanide rare-earth elements, plus Th, Ta, Hf, and Sc were determined on 0.2-g splits of ten of the powders by instrumental neutron activation analysis at the University of Leicester following Saunders [1986]. In addition, neutron activation data were obtained for a sample from the flow at DSDP Site 289, as well as for two Manihiki Plateau basalts from Site 317. Splits of 21 powders were analyzed for Cs, Rb, Ba, Th, U, Nb, Ta, Zr, Hf, Sc, and rare-earth elements by inductively coupled plasma-mass spectrometry at the University of Durham. For each sample, 0.1 g of powder was dissolved (with HF, $HClO_4$, and HNO_3), spiked with Rh and Re internal standards, and run on a VG Plasmaquad instrument in a dilute HNO_3 matrix (i.e., ~0.1 g total dissolved solids; total dilution factor = 500) with a dwell-time of 250 s for each element analyzed. Correction procedures included drift-monitoring and subtraction of blank contributions, oxide and hydroxide interferences, and overlap of neighboring isotopes on the peaks of interest. Calibration was effected by synthetic standards checked against a set of laboratory and international standards. Table 3 lists both the neutron activation and plasma-mass-spectrometric data, along with estimated analytical uncertainties. Splits of five samples were analyzed by both methods; in general, values for elements determined by both techniques are in good agreement (Sc shows by far the most variability). Also, comparison of Tables 1 and 2 shows that agreement between X-ray fluorescence and plasma-mass-spectrometric data is good for Rb and excellent for Zr and Y, the three elements analyzed in common.

Isotopic ratios of Nd, Sr, and Pb, as well as isotope-dilution abundances of Nd, Sm, Sr, Rb, and Pb (Tables 4, 5) were measured in the Isotope Laboratory at the University of Hawaii, following methods described by Mahoney and Spencer [1991] and Mahoney et al. [1991]. These methods included, before grinding, the ultrasonic cleaning of hand-picked, interior rock chips (~3 mm across) in ultra-pure, weak HNO_3-HF, 6N HCl, and H_2O (in sequence) to remove possible surface Pb contamination. In one case, we picked fresh, crystal-free glass from a flow margin (Unit A sample 807C-77R-2 (108-110)). Note that the selective picking of fresher-appearing chips or crystal-free glass from the bulk sample, and the subsequent cleaning procedure, often mean that the portion of sample analyzed for isotopes does not faithfully represent the whole-rock chemically; therefore, the isotope-dilution abundance measurements in Tables 4 and 5 should not, in general, be compared directly with the whole-rock chemical data in Tables 2 and 3. As an illustration, isotope-dilution values for two bulk samples (~2 cm across, interior pieces) that were not picked and were cleaned only in ultra-pure water are given in Tables 4 and 5 ("UP" samples 803D-71R-1 (124-125) and 807C-79R-5

TABLE 2. Major Element and X-Ray Fluorescence Trace Element Data for Leg 130 Basalts

Sample	SiO_2	TiO_2	Al_2O_3	Fe_2O_3	MnO	MgO	CaO	Na_2O	K_2O	P_2O_5	LOI	Total	Cr	Ni	Rb	Sr	V	Y	Zr
130-807C-																			
Unit A																			
74R-1,46-49	47.12	1.70	15.09	14.05	0.14	4.77	12.24	2.37	0.76	0.16	1.77	100.2	168	94	15	250	365	32	101
74R-1,124-127	48.65	1.66	14.73	12.54	0.18	6.05	12.43	2.45	0.41	0.14	0.66	99.9	170	96	10	218	331	30	102
74R-2,118-122	48.55	1.61	14.23	13.01	0.22	6.82	12.12	2.46	0.14	0.14	0.59	99.9	164	99	<4	203	327	30	98
74R-3,86-90	48.65	1.59	14.09	13.45	0.20	7.28	11.82	2.40	0.09	0.14	0.25	100.0	150	95	<4	177	310	30	100
74R-4,8-12	48.60	1.60	14.11	13.42	0.19	7.10	11.84	2.25	0.08	0.14	0.45	99.8	151	98	<4	183	309	29	98
75R-1,27-30	47.31	1.52	13.45	14.00	0.19	6.51	12.06	2.21	0.84	0.14	1.69	99.9	154	72	28	182	322	29	94
75R-2,61-64	48.17	1.59	14.07	13.51	0.20	7.26	11.77	2.32	0.35	0.14	0.62	100.0	154	115	8	198	327	29	99
75R-2,125-129	48.26	1.63	14.51	13.00	0.19	7.20	12.16	2.34	0.05	0.14	0.50	100.0	149	100	<4	210	314	31	101
75R-3,104-107	48.69	1.59	14.26	13.30	0.19	6.79	11.91	2.33	0.13	0.14	0.35	99.7	160	98	<4	178	314	30	99
75R-4,48-52	48.75	1.62	14.16	13.43	0.21	6.74	12.13	2.38	0.13	0.14	0.45	100.1	162	99	<4	174	313	30	98
75R-5,78-83	48.85	1.60	14.22	13.39	0.19	7.03	11.87	2.38	0.13	0.14	0.18	100.0	159	97	<4	170	315	31	100
76R-1,69-71	49.06	1.61	14.35	13.33	0.18	6.42	11.94	2.43	0.12	0.13	0.20	99.8	156	100	<4	173	325	30	99
77R-1,48-51	48.88	1.61	14.33	13.27	0.19	6.70	11.99	2.49	0.13	0.14	0.27	100.0	153	99	<4	172	316	29	99
78R-1,63-67	49.01	1.60	14.28	13.37	0.19	7.25	11.89	2.43	0.09	0.14	0.20	100.5	154	96	<4	165	299	29	98
78R-2,69-72	48.76	1.59	14.05	13.49	0.27	7.18	11.75	2.34	0.31	0.14	0.40	100.3	143	95	6	166	305	30	97
78R-3,55-60	49.14	1.62	14.28	13.53	0.19	6.72	12.19	2.42	0.11	0.14	0.13	100.5	157	95	6	170	333	29	98
79R-1,13-17	48.83	1.61	14.22	13.29	0.19	6.84	12.17	2.40	0.10	0.14	0.28	100.1	156	94	7	171	324	29	98
79R-3,110-114	48.31	1.61	14.42	13.53	0.23	6.90	12.12	2.40	0.30	0.14	0.07	100.0	162	86	12	173	358	28	98
79R-4,25-28	47.55	1.60	13.89	14.86	0.22	6.97	11.94	2.32	0.38	0.14	0.27	100.1	160	78	15	166	349	29	98
79R-5,28-30	49.46	1.65	14.62	12.75	0.18	6.45	12.30	2.49	0.11	0.14	0.41	100.6	150	97	<4	175	332	31	102
79R-5,34-38	48.67	1.60	14.51	12.67	0.18	6.72	12.40	2.45	0.08	0.14	0.47	99.9	153	105	6	184	327	28	98
80R-1,49-51	48.44	1.66	14.79	12.77	0.40	7.02	12.27	2.41	0.07	0.14	0.10	100.1	146	97	<4	176	327	30	101
Units C-G																			
80R-2,34-37	49.51	1.16	14.25	12.30	0.24	7.39	11.97	2.21	0.41	0.10	0.05	99.6	157	88	12	117	359	25	66
81R-2,74-77	50.21	1.20	15.08	10.53	0.22	8.01	11.42	2.28	0.63	0.11	0.67	100.3	167	103	9	121	347	24	67
82R-2,42-45	50.57	1.21	14.91	10.96	0.23	7.61	11.84	2.29	0.41	0.10	0.38	100.5	140	99	7	123	346	23	68
82R-2,117-120	49.66	1.16	14.24	12.27	0.26	7.41	11.97	2.21	0.39	0.10	0.03	99.7	158	94	12	120	363	24	66
82R-4,102-105	49.51	1.17	14.28	12.93	0.23	7.06	11.69	2.31	0.51	0.10	0.27	100.1	161	79	12	124	366	25	67
83R-1,7-10	49.98	1.21	15.01	10.49	0.19	8.37	11.69	2.39	0.16	0.10	0.50	100.1	152	99	7	129	394	24	67
83R-2,30-33	50.07	1.14	13.91	12.88	0.19	7.41	11.85	2.08	0.06	0.09	-0.05	99.6	120	86	<4	113	331	26	68
84R-1,126-130	50.01	1.13	13.77	13.04	0.19	7.76	12.07	2.11	0.06	0.09	-0.02	100.2	131	86	6	111	315	24	64
84R-2,111-115	50.48	1.14	13.95	12.94	0.19	7.55	11.77	2.05	0.06	0.10	-0.15	100.1	126	88	<4	110	321	25	67

TABLE 2. (continued)

Sample	SiO_2	TiO_2	Al_2O_3	Fe_2O_3	MnO	MgO	CaO	Na_2O	K_2O	P_2O_5	LOI	Total	Cr	Ni	Rb	Sr	V	Y	Zr
85R-3,16-19	49.85	1.06	13.78	12.78	0.19	7.98	11.81	2.07	0.06	0.09	0.19	99.9	173	91	5	107	321	22	60
85R-5,112-115	50.01	1.10	13.85	12.99	0.18	7.89	11.89	2.01	0.06	0.09	0.03	100.1	152	90	<4	107	323	22	62
86R-1,100-103	49.19	1.14	14.25	12.92	0.22	7.20	11.94	2.15	0.35	0.11	0.02	99.5	157	92	7	109	343	25	65
86R-2,122-125	50.04	1.20	14.84	10.78	0.18	7.87	12.28	2.37	0.20	0.11	0.25	100.1	160	101	7	117	375	27	68
87R-2,37-40	50.45	1.20	15.20	10.25	0.19	7.93	12.51	2.40	0.06	0.10	0.07	100.4	169	118	<4	115	347	27	69
87R-2,131-135	48.79	1.15	14.46	12.63	0.21	6.92	12.02	2.25	0.57	0.10	0.32	99.4	166	89	16	113	358	23	65
87R-2,66-70	49.35	1.17	14.21	12.99	0.27	7.12	11.83	2.28	0.54	0.10	0.19	100.1	150	84	16	118	386	23	66
88R-1,18-21	49.28	1.13	13.89	13.27	0.22	7.35	12.19	2.14	0.12	0.10	-0.17	99.5	152	89	8	109	349	23	64
88R-1,70-75	50.22	1.15	14.08	13.24	0.21	7.22	12.14	2.17	0.19	0.10	-0.19	100.5	154	87	5	107	339	24	65
88R-3,18-22	50.07	1.16	14.39	12.01	0.20	7.52	12.41	2.24	0.09	0.10	-0.05	100.1	154	128	5	112	366	24	66
88R-3,78-81	49.74	1.13	14.27	13.44	0.22	7.43	12.00	2.06	0.28	0.09	-0.24	100.4	163	87	10	107	349	24	64
89R-1,38-42	49.44	1.17	14.10	13.05	0.22	7.73	12.28	2.08	0.03	0.10	-0.02	100.2	144	94	<4	110	346	23	67
89R-1,110-113	50.02	1.14	14.15	11.96	0.17	7.49	12.28	2.20	0.04	0.09	-0.05	99.5	151	93	5	112	353	24	67
89R-2,87-91	49.89	1.14	13.96	12.69	0.21	7.56	12.39	2.20	0.08	0.10	-0.29	99.9	150	88	6	108	333	23	65
89R-3,11-15	49.64	1.11	13.79	13.31	0.20	7.38	12.19	2.11	0.18	0.10	-0.17	99.8	148	92	8	107	340	26	63
89R-3,58-62	50.54	1.12	13.91	12.86	0.19	7.07	12.09	2.15	0.25	0.10	-0.17	100.1	155	99	10	108	347	23	64
90R-1,21-25	49.87	1.13	13.70	12.97	0.18	7.71	12.11	2.07	0.05	0.10	0.07	100.0	145	92	5	108	339	24	64
90R-3,72-76	49.37	1.15	14.09	12.64	0.21	7.71	12.30	2.16	0.06	0.09	-0.13	99.7	155	103	<4	107	331	24	64
91R-1,15-18	49.17	1.14	14.06	12.75	0.23	7.75	12.48	2.33	0.07	0.10	-0.28	99.8	159	97	6	108	356	24	65
92R-1,107-110	49.94	1.13	14.00	12.82	0.20	7.55	12.18	2.17	0.05	0.10	-0.14	100.0	163	97	5	108	351	23	64
92R-2,100-103	50.35	1.16	14.46	12.28	0.20	7.40	12.02	2.17	0.05	0.09	-0.07	100.1	156	90	<4	110	337	26	68
93R-1,25-28	50.17	1.15	14.24	12.72	0.21	7.45	12.15	2.13	0.15	0.10	-0.29	100.2	157	96	3	108	348	24	67
93R-1,55-57	50.21	1.15	14.14	12.67	0.21	6.82	12.20	2.31	0.26	0.10	-0.21	99.9	163	96	10	109	343	23	65
93R-2,59-63	50.54	1.12	13.85	12.74	0.21	6.99	11.89	2.19	0.33	0.10	-0.13	99.8	159	89	11	107	352	23	64
93R-3,4-7	49.86	1.12	13.74	12.81	0.20	7.40	12.16	2.07	0.07	0.10	0.11	99.6	145	90	6	109	332	23	65
130-803D-																			
68R-4,27-31	47.68	1.46	16.43	12.80	0.15	4.56	11.35	2.48	0.87	0.16	1.79	99.7	290	185	22	206	357	26	83
69R-1,10-12	49.23	1.41	15.73	11.21	0.15	6.07	12.95	2.44	0.31	0.12	0.70	100.3	266	153	7	174	337	28	83
69R-1,92-94	48.72	1.31	14.51	12.72	0.18	6.74	12.54	2.13	0.42	0.11	0.71	100.1	249	97	9	158	310	26	76
69R-1,138-141	48.51	1.40	15.87	12.15	0.15	5.29	11.55	2.44	0.93	0.20	1.04	99.5	242	87	15	188	322	30	81
69R-2,19-21	48.23	1.34	15.03	12.54	0.17	6.02	13.23	2.20	0.41	0.12	1.16	100.5	255	138	8	167	345	25	77
69R-2,130-133	49.86	1.34	15.01	11.01	0.17	7.36	12.72	2.25	0.31	0.11	0.36	100.5	241	119	6	157	301	26	78
69R-3,41-43	48.09	1.37	15.41	13.33	0.16	5.28	11.37	2.44	0.98	0.13	0.98	99.5	235	83	16	172	333	27	81
69R-3,81-84	48.90	1.36	15.38	10.96	0.15	6.17	13.03	2.31	0.42	0.13	1.02	99.8	236	116	9	171	313	27	78
69R-3,136-139	48.99	1.44	15.98	11.76	0.14	5.79	11.55	2.43	0.84	0.25	0.99	100.2	284	118	12	183	325	30	83
69R-4,29-31	50.35	1.49	16.78	9.75	0.11	5.32	10.83	2.46	1.58	0.36	1.35	100.4	245	83	21	197	267	35	87

TABLE 2. (continued)

Sample	SiO_2	TiO_2	Al_2O_3	Fe_2O_3	MnO	MgO	CaO	Na_2O	K_2O	P_2O_5	LOI	Total	Cr	Ni	Rb	Sr	V	Y	Zr
70R-1,58-61	49.11	1.38	15.53	11.50	0.15	6.56	11.81	2.36	0.71	0.14	0.70	100.0	243	114	12	173	311	27	81
70R-2,45-48	50.92	1.49	17.04	9.88	0.11	4.31	9.58	2.49	2.30	0.55	1.74	100.4	296	83	28	212	217	62	83
70R-3,25-29	49.13	1.33	14.92	11.98	0.18	6.54	12.16	2.29	0.57	0.11	0.49	99.7	246	92	12	161	315	26	79
70R-3,55-59	49.17	1.31	15.07	10.78	0.18	6.88	12.79	2.24	0.38	0.11	0.78	99.7	239	112	5	163	293	27	79
71R-1,125-127	48.64	1.28	14.54	11.94	0.19	7.24	12.80	2.09	0.33	0.11	0.86	100.0	237	109	7	151	311	27	76
71R-2,88-92	48.76	1.30	14.81	11.58	0.23	7.53	12.68	2.13	0.15	0.11	0.47	99.8	236	114	<4	151	314	26	78
71R-2,134-137	48.75	1.31	14.79	12.45	0.18	6.65	12.74	2.24	0.30	0.11	0.75	100.3	250	98	8	155	324	25	77
71R-3,32-35	48.28	1.29	14.54	11.88	0.18	6.60	13.08	2.24	0.42	0.11	1.28	99.9	245	86	10	152	319	27	77
Standards																			
W-1 Meas.	51.89	1.10	14.95	11.21	0.17	6.72	11.07	2.14	0.65	0.14	0.04	100.1	116	75	25	190	271	23	98
W-1 Rec.	52.46	1.07	15.00	11.11	0.17	6.62	11.00	2.16	0.64	0.13	-	100.4	119	75	21	186	257	26	99

Notes: Major elements are all in weight percent; trace elements are in ppm. LOI=loss on ignition. Fe_2O_3=all Fe as Fe_2O_3. Estimated relative errors on major and minor elements are 1%. For trace elements, errors (% standard deviation at levels measured) are estimated at 1% for Cr, 3% for Ni, 3% for Rb at 30 ppm and >20% at <10 ppm, 2% for Sr and V, and 4% for Y and Zr. An indication of accuracy is given by measured values of standard W-1 (average of two measurements, analyzed as an unknown) and recommended values [Govindaraju, 1989].

(38-40)) together with the respective analyses for their picked and cleaned chips. As expected, the bulk-sample results for Nd, Sm, and Sr are within about 10% of the corresponding X-ray fluorescence or inductively coupled plasma-mass-spectrometric values in Tables 2 and 3, but show significant differences from those for the picked and cleaned chips. Tables 4 and 5 also list our isotopic and isotope-dilution analyses of the U.S.G.S. standard rock BCR-1.

In addition to the above measurements, Sr isotopes were determined on severely acid-leached splits of powder. For moderately to rather highly altered tholeiitic samples, the leaching procedure used largely removes the effects of seawater alteration on $^{87}Sr/^{86}Sr$ by producing a residue of relatively fresh material, mainly clinopyroxene, plagioclase, and if present, fresh glass [Mahoney, 1987]. Comparison of the calculated initial $(^{87}Sr/^{86}Sr)_T$ values of the two leached Site 807 Unit A powders with that of the fresh glass, for example, shows no significant differences. In the illustrations and in the discussion that follows, we use the Sr isotopic values for the leached aliquants. Unlike Sr isotopes, Nd isotopic ratios appear not to be affected significantly by levels of alteration greater than those suffered by the Leg 130 samples chosen for isotopes [Mahoney, 1987; Cheng et al., 1987; Staudigel et al., 1991; Castillo et al., 1991]. Likewise, Pb isotopic ratios of unleached and acid-leached pairs previously were found to be very similar for moderately altered tholeiites from the Manihiki Plateau, and for the Ontong Java basalt from Site 289 [Mahoney and Spencer, 1991]. In Table 5, the Pb isotope values for the fresh glass and the two unleached Unit A samples are identical within errors.

Results

All of the samples analyzed are quartz- or olivine-normative tholeiitic basalts, broadly similar in chemical composition to the flow at DSDP Site 289 (see Table 3), to basement lavas on the island of Malaita (K. J. Spencer et al., unpubl. data, 1991), as well as to the Early Cretaceous flows and sills filling the Nauru Basin [e.g. Floyd, 1986; Saunders, 1986; Castillo et al., 1986, 1991], which borders the plateau on the east and probably is genetically related to it [e.g. Tokuyama and Batiza, 1981; Castillo et al., 1991; Pringle, 1992]. As with the Site 289, Malaitan, and Nauru Basin basalts, both the major and many trace element abundances of the Leg 130 lavas overlap the range of values for mid-ocean ridge basalts (MORB). For example, TiO_2 (1.06-1.70 wt.%), P_2O_5 (generally 0.09-0.20 wt.%), Zr (60-102 ppm), and the rare earth elements have values typical of normal MORB. Ba (9.4-36 ppm), Th (0.31-0.56 ppm), Ta (0.19-0.37 ppm), Nb (3.4-6.2 ppm), Sr (107-250 ppm), and, in the fresher samples, K_2O (<0.2 wt.%) also overlap the MORB range but exceed average values for normal MORB (approximately 6.3, 0.12, 0.13, 2.3, and 90 ppm, and 0.07 wt.%, respectively [e.g. Sun and McDonough, 1989]).

TABLE 3. Instrumental Neutron Activation (N) and Inductively-Coupled Plasma-Mass-Spectrometric (I) Data for Leg 130, Site 289, and Manihiki Basalts

Sample	I/N	Sc	Y	Zr	Nb	Cs	Rb	Ba	La	Ce	Pr	Nd	Sm	Eu	Gd	Tb	Dy	Ho	Er	Yb	Lu	Hf	Ta	Th	U
130-807C -																									
Unit A																									
74R-1,46-49	N	48	-	-	-	-	-	-	4.7	10.6	-	7.7	3.2	1.1	3.6	0.7	-	-	-	2.3	0.31	2.18	0.28	0.39	-
74R-2,118-122	I	42	31	101	6.2	0.04	5	24	6.1	15.0	2.3	11.2	-	1.3	4.5	-	4.9	1.1	2.9	2.8	0.42	2.64	0.44	0.45	0.13
75R-1,27-30	I	50	28	93	5.6	0.77	33	24	5.3	13.2	2.0	10.0	3.0	1.2	3.8	-	4.2	0.9	2.6	2.4	0.33	2.25	0.35	0.46	0.16
75R-2,125-129	I	50	32	102	5.8	0.01	3	17	5.8	14.9	2.3	11.2	3.3	1.3	4.7	-	5.3	1.2	3.3	3.0	0.50	2.72	0.45	0.58	0.14
75R-4,48-52	N	46	-	-	-	-	-	-	6.2	14.6	-	11.7	3.4	1.3	4.5	0.8	-	-	-	2.7	0.42	2.64	0.34	0.56	-
77R-1,48-51	N	43	-	-	-	-	-	-	6.0	14.4	-	11.4	3.5	1.4	4.6	0.8	-	-	-	2.7	0.42	2.64	0.35	0.56	-
78R-1,63-67	I	44	31	102	5.8	0.03	4	30	5.8	15.2	2.3	11.1	3.5	1.4	4.5	-	5.3	1.2	3.4	3.0	0.43	2.75	0.41	0.62	0.15
79R-3,110-114	I	46	29	97	6.1	0.21	10	25	6.0	15.4	2.4	11.5	3.5	1.3	4.3	0.8	5.0	1.0	2.8	2.8	0.43	2.54	0.36	0.51	0.12
79R-4,25-28	I	44	29	93	5.6	0.28	12	27	6.3	15.8	2.4	11.6	3.6	1.3	4.3	-	5.0	1.0	3.0	2.7	-	2.50	0.40	0.54	0.13
79R-5,34-38	I	52	31	100	6.1	0.01	3	22	5.7	14.8	2.3	10.5	3.2	1.3	4.3	-	4.9	1.0	2.9	2.8	-	2.49	0.38	0.56	0.14
Units C-G																									
80R-2,34-37	I	47	26	66	3.7	0.15	10	10	3.4	9.4	1.5	7.5	2.5	0.9	3.3	-	4.2	0.9	2.6	2.6	-	1.73	0.22	0.30	0.08
83R-1,7-10	I	61	24	63	3.5	0.01	4	14	3.2	9.1	1.4	7.1	2.4	0.9	3.3	-	3.8	0.9	2.2	2.2	0.34	1.57	0.21	0.26	0.21
	N	51	-	-	-	-	-	-	3.3	9.6	-	8.1	2.6	1.0	3.5	0.7				2.4	0.38	1.81	0.21	0.35	-
86R-2,122-125	I	52	28	68	3.5	0.02	4	11	4.0	10.4	1.7	7.9	2.6	1.0	3.6	-	4.8	1.0	2.8	2.8	-	1.89	0.24	0.27	0.18
88R-3,18-22	I	50	26	65	3.5	0.02	5	10	3.6	9.5	-	7.5	2.4	1.0	3.2	-	4.2	0.9	2.6	2.4	-	1.67	0.24	0.28	0.08
88R-3,78-81	N	52	-	-	-	-	-	-	3.0	8.1	-	6.5	2.3	0.9	3.4	0.7	-	-	-	2.4	0.38	1.74	0.20	0.34	-
89R-2,87-91	I	58	24	66	3.7	0.10	5	13	3.0	7.8	1.3	6.2	-	0.9	2.9	-	3.5	0.8	2.1	2.1	-	1.50	0.21	0.26	0.18
90R-1,21-25	I	47	25	63	3.4	0.01	2	17	3.4	8.7	1.4	7.0	2.5	0.9	3.3	0.6	3.9	0.9	2.4	2.3	0.35	1.66	0.20	0.23	0.09
	I	41	25	63	3.5	0.01	2	17	3.5	9.0	1.4	7.0	2.4	0.9	3.3	0.6	3.9	0.9	2.4	2.5	-	1.79	0.22	0.28	0.10
92R-1,107-110	I	49	24	63	3.5	0.01	4	17	3.6	9.1	1.5	7.4	2.3	0.9	3.2	-	4.1	0.9	2.5	2.4	-	1.79	0.22	0.28	0.08
	N	48	-	-	-	-	-	-	3.6	8.0	-	7.0	2.3	0.9	3.1	0.6	-	-	-	2.3	0.36	1.70	0.19	0.31	-
93R-3,4-7	I	52	24	62	3.3	0.01	4	17	3.2	8.8	1.4	6.8	2.1	0.9	3.1	-	4.0	0.9	2.5	2.4	0.33	1.69	0.19	0.29	0.08
	N	46	-	-	-	-	-	-	3.4	7.6	-	6.8	2.1	0.9	3.0	0.6	-	-	-	2.3	0.36	1.70	0.19	0.31	-
130-803D																									
68R-4,27-31	N	48	-	-	-	-	-	-	4.7	10.6	-	7.7	3.2	1.1	3.6	0.7	-	-	-	2.3	0.31	2.18	0.28	0.39	-
69R-1, 10-12	I	53	26	76	4.5	0.13	8	11	4.0	10.7	-	8.6	2.7	1.1	3.5	-	4.2	0.9	2.5	2.4	-	1.94	0.30	0.33	0.14
69R-2,130-133	I	53	26	79	4.1	0.09	8	13	3.8	10.1	1.7	8.3	2.7	1.1	3.8	0.7	4.7	1.0	2.8	2.7	0.37	2.04	0.31	0.33	0.35
69R-4,29-31	I	54	36	86	4.7	-	23	36	5.7	12.3	2.1	9.9	-	1.2	4.1	-	5.0	1.1	3.1	3.0	-	2.28	0.33	0.37	0.34
	N	49	-	-	-	-	-	-	5.6	11.7	-	10.0	3.0	1.3	4.1	0.8	-	-	-	2.9	0.48	2.33	0.28	0.41	-
70R-3,25-29	I	43	27	82	5.1	0.27	15	20	4.1	10.6	1.7	8.5	-	1.1	3.7	-	4.2	0.9		2.6	0.37	2.01	0.31	0.33	0.22
71R-1,125-127	I	44	26	76	4.3	0.10	10	9	4.1	10.4	1.7	8.2	-	1.1	3.6	-	4.1	0.9	2.5	2.5	-	1.92	0.29	0.36	0.09
71R-2,134-137	I	47	26	74	4.0	0.31	9	10	3.8	10.2	1.7	8.0	2.8	1.1	3.8	-	4.5	1.0	2.8	2.6	0.39	2.14	0.29	0.33	0.10
	N	49	-	-	-	-	-	-	4.3	10.2	-	7.9	2.7	1.0	3.6	0.7	-	-	-	2.3	0.36	1.96	0.25	0.36	-

TABLE 3. (continued).

Sample	I/N	Sc	Y	Zr	Nb	Cs	Rb	Ba	La	Ce	Pr	Nd	Sm	Eu	Gd	Tb	Dy	Ho	Er	Yb	Lu	Hf	Ta	Th	U
30-289-																									
132-3, 52-55	N	46	-	-	-	-	-	-	4.3	11.7	-	9.9	3.3	1.2	4.4	0.8	-	-	-	3.1	0.48	2.46	-	0.45	-
Manihiki Plateau																									
33-317A-																									
31-4-116-118	N	38	-	-	-	-	-	-	2.6	7.5	-	6.3	1.9	0.8	3.0	-	-	-	-	2.1	0.32	1.50	-	0.35	-
34-4,83-85	N	45	-	-	-	-	-	-	4.4	10.8	-	9.1	2.7	1.1	3.7	0.6	-	-	-	2.5	0.38	2.08	-	0.52	-
Standards																									
BOB-1 meas.	N	33	-	-	-	-	-	-	4.8	12.4	-	9.9	3.0	1.2	3.8	0.8	-	-	-	2.6	0.39	2.43	0.43	0.40	-
BOB-1 rec.		-	-	-	-	-	-	-	4.7	13.8	-	10.7	3.3	1.3	-	0.7	-	-	-	2.6	0.44	2.53	0.51	0.45	-
W-2 meas.	I	36	22	94	7.9	0.96	21	171	10.9	23.4	3.1	13.1	3.3	1.1	3.8	0.6	3.7	0.8	2.1	2.0	0.30	2.35	0.53	2.37	0.52
W-2 rec.		35	24	102	8	1.03	21	172	10.6	22.6	3.2	15.1	3.5	1.2	4.2	0.6	4.3	0.8	2.4	2.1	0.36	2.62	0.50	2.57	0.54

Notes: All abundances are in ppm. For NAA data, precision was determined from eight replicate analyses of Standard BOB-1, except for Sc, for which six replicates were analyzed. Precisions, as estimated by standard deviations (in ppm), are: Sc=1.4, La=0.08, Ce=0.32, Nd=0.26, Sm=0.10, Eu=0.03, Tb=0.07, Yb=0.08, Lu=0.01, Hf=0.07, Ta=0.01, and Th=0.02. A measure of NAA accuracy is provided by recommended vs. our average measured values for BOB-1, and of ICPMS accuracy by the average of six recent analyses of W-2 vs. Jenner et al.'s [1990] preferred values. The standard deviations of the W-2 analyses are (in ppm): Sc=1.7, Y=1.3, Zr=4.5, Nb=0.2, Cs=0.08, Rb=0.9, Ba=7.0, La=0.48, Ce=0.76, Pr=0.16, Nd=0.59, Sm=0.17, Eu=0.05, Gd=0.13, Tb=0.05, Dy=0.17, Ho=0.05, Er=0.07, Tm=0.03, Yb=0.06, Lu=0.03, Hf=0.13, Ta=0.06, Th=0.18, and U=0.02.

However, their upper limits still lie well within the range of values found at spreading centers, among so-called "transitional" or "enriched" MORB. Compared to normal MORB, all of the Leg 130 rocks display slight enrichments in highly incompatible elements relative to moderately incompatible elements (e.g. Zr/Nb = 16-19 vs. >30 for normal MORB [e.g. Sun and McDonough, 1989]; also see Figure 7); the same is true of the Site 289, Malaitan, and Nauru Basin lavas and, indeed, is a feature common to the small number of basaltic samples available from the other Pacific plateaus [Floyd, 1986; Saunders, 1986; Mahoney, 1987].

Isotopes. Figures 8 and 9 show that, isotopically, the Leg 130 basalts are very similar to the few samples measured previously from DSDP Site 289 and southern Malaita [Mahoney and Spencer, 1991]. Total variation in initial $\epsilon_{Nd}(T)$ is +4.8 to +6.3 for the Leg 130 lavas; initial $(^{87}Sr/^{86}Sr)_T$ lies between 0.70339-0.70410 for acid-leached powders and glass. As expected, unleached powders have greater (as much as 0.0006 greater) and more variable Sr-isotope ratios, depending on their level of alteration. Present-day $^{206}Pb/^{204}Pb$ is between 18.385 and 18.691, $^{207}Pb/^{204}Pb$ = 15.489 to 15.547, and $^{208}Pb/^{204}Pb$ = 38.368 to 38.535. An estimate of the range of initial Pb isotopic ratios is provided by bulk-sample U/Pb and Th/Pb for 803D-71R-1 (124-127) ($^{238}U/^{204}Pb$ = 16 and $^{232}Th/^{204}Pb$ = 65) and 807C-79R-5 (34-40) ($^{238}U/^{204}Pb$ = 14 and $^{232}Th/^{204}Pb$ = 56). Assuming U has not been elevated appreciably by alteration in these two samples (as implied by their relatively high, oceanic-island-like Th/U values of 4.0), these ratios are 18.13-18.42, 15.51-15.51, and 38.04-38.15.

Although the total range of isotopic variation is rather limited, the Leg 130 data can be seen to cluster into two groups on Figure 9. The Unit A flows have lower $^{206}Pb/^{204}Pb$ (18.385-18.395), higher $(^{87}Sr/^{86}Sr)_T$ (0.70403-0.70410), and lower $\epsilon_{Nd}(T)$ (+4.8 to +5.4) than those of Units C-G (18.609-18.668, 0.70339-0.70345, +5.9 to +6.3, respectively). The Site 803 samples are like those of Units C-G in Pb isotopes (18.626-18.691), with slightly greater $(^{87}Sr/^{86}Sr)_T$ and lower $\epsilon_{Nd}(T)$ (0.70364-0.70373, +5.7 to +5.9, respectively). In comparison, the Site 289 flow (18.708, 0.70346, +6.3) is virtually identical to the Unit C-G lavas, whereas the basalts from southern Malaita (18.245-18.521, 0.70404-0.70423, +4.0 to +4.9) closely resemble the Unit A flows. These results require the existence of subtly different mantle sources among the plateau lavas sampled to date. Note also that a recent suggestion that Malaita may not be part of the plateau but an unrelated piece of the Australian Plate [Musgrave, 1990] is not supported by either the strong isotopic resemblance of the Malaitan samples to the Unit A flows, their very similar chemical characteristics (K. J. Spencer et al., unpubl. data, 1991), or their ^{40}Ar-^{39}Ar dates of 122-123 Ma, which are identical to those for Sites 807 and 289. In the subsequent discussion we therefore include the Malaitan data with results for the drillholes.

TABLE 4. Nd and Sr Isotopic and Isotope-Dilution Data for Leg 130 Basalts

Sample		$(^{143}Nd/^{144}Nd)_0$	$(^{87}Sr/^{86}Sr)_0$	Nd (ppm)	Sm	Sr	Rb	$(^{87}Sr/^{86}Sr)_T$	$(^{143}Nd/^{144}Nd)_T$	$\epsilon Nd(T)$
130-803D-										
69R-2, 54-56	UL	0.512944	0.70472	8.225	2.710	137.9	14.0	0.70434	0.512827	+5.9
	L		0.70391			119.4	5.77	0.70373		
71R-1, 124-126	UP	0.512944	0.70439	8.616	2.800	145.6	3.61	0.70430	0.512828	+5.9
	UL	0.512950	0.70389	4.862	1.823	130.5	2.07	0.70383	0.512817	+5.7
	L		0.70369			109.3	1.33	0.70364		
130-807C-										
75R-4, 46-48	UL	0.512897	0.70438	9.331	3.031	166.2	1.60	0.70433	0.512743	+5.0
Unit A	L		0.70408			165.3	1.59	0.70403		
77R-2, 108-110	GL	0.512878	0.70417	10.98	3.370	148.9	3.53	0.70405	0.512732	+4.8
Unit A										
79R-5, 38-40	UP	0.512913	0.70444	11.53	3.552	170.0	1.12	0.70441	0.512767	+5.5
Unit A	UL	0.512926	0.70426	7.582	2.587	165.8	0.964	0.70423	0.512764	+5.4
	L		0.70413			159.8	0.951	0.70410		
82R-2, 121-123	UL	0.512965	0.70383	5.336	1.932	105.8	4.37	0.70363	0.512793	+6.0
Unit C	L		0.70356			85.66	2.11	0.70343		
82R-4, 105-107	UL	0.512964	0.70407	6.255	2.158	106.2	6.13	0.70378	0.512800	+6.1
Unit E	L		0.70367			80.53	4.08	0.70343		
85R-3, 19-21	UL	0.512983	0.70357	3.594	1.314	97.55	0.743	0.70354	0.512810	+6.3
Unit F	L		0.70347			68.22	0.211	0.70345		

TABLE 4. (continued)

Sample		$(^{143}Nd/^{144}Nd)_0$	$(^{87}Sr/^{86}Sr)_0$	Nd (ppm)	Sm	Sr	Rb	$(^{87}Sr/^{86}Sr)_T$	$(^{143}Nd/^{144}Nd)_T$	$\epsilon Nd(T)$
88R-3, 76-79	UL	0.512965	0.70377	5.142	1.891	93.53	4.97	0.70356	0.512790	+5.9
Unit G	L		0.70358			100.1	3.81	0.70339		
USGS Standard										
BCR-1		0.512633	0.70502	28.51	6.576	329.4	46.2			

Notes: T = 120 Ma for Site 807, 90 Ma for Site 803. UL = unleached, but chips handpicked and acid-cleaned; L = strongly acid-leached powder; UP = bulk sample, not picked, cleaned only in H_2O. GL = handpicked, fresh glass. Isotopic fractionation corrections are $^{148}NdO/^{144}NdO$=0.242436 ($^{148}Nd/^{144}Nd$=0.241572), $^{86}Sr/^{88}Sr$=0.1194. Data are reported relative to University of Hawaii standards values: for La Jolla Nd, $^{143}Nd/^{144}Nd$=0.511855; for BCR-1, see above; for NBS 987 Sr, $^{87}Sr/^{86}Sr$=0.71025; for E and A Sr, $^{87}Sr/^{86}Sr$=0.70803. The total range measured for La Jolla Nd is ± 0.000012 (0.2ε units); for NBS 987 it is ± 0.000022. Within-run errors on the isotopic data above are less than or equal to the external uncertainties on these standards. Total blanks are negligible: <20 picograms for Nd and <120 picograms for Sr. Uncertainties on Nd and Sm abundances are estimated at <0.2%; on Sr and Rb, <0.5% and -1%, respectively.

$\epsilon Nd(T)$ = 0 today corresponds to $^{143}Nd/^{144}Nd$ = 0.512640; $\epsilon Nd(T)$=0 at 120 Ma corresponds to $(^{143}Nd/^{144}Nd)_T$= 0.512486 for $^{147}Sm/^{144}Nd$=0.1967, and to 0.512524 at 90 Ma.

The combined Ontong Java measurements fall within the overall field of oceanic mantle-derived rocks but lie outside the specific fields of Pacific MORB and the majority of present-day Pacific oceanic islands, the latter being characterized by greater $^{206}Pb/^{204}Pb$. Rarotonga, Pitcairn, and several Hawaiian shields (e.g. Koolau) are exceptions, but they are all located very far away from the estimated 120-Ma position of the plateau. The Ontong Java isotopic data also are unlike those of many continental tholeiites, which often fall well outside the oceanic mantle field [e.g. Hawkesworth et al., 1990]. In addition, the elemental ratios Nb/Th (11-16 for our samples) and Nb/U (37-51 for the less altered samples) are sensitive indicators of continental crustal contamination, as crustal values tend to be much lower than oceanic mantle ones [e.g. Hofmann et al., 1986]; the Leg 130 data, however, are firmly within the oceanic mantle range. Thus, there is no geochemical evidence of a continental basement to the plateau, as suggested by some earlier workers on geophysical grounds (see summary and references of Mahoney and Spencer [1991]).

Although the four basement sites now sampled are perhaps unlikely to be truly representative of an edifice as extensive as the Ontong Java Plateau (~1/4 the area of Australia), the entire range of isotopic variation observed at these widely separated locations is quite small. Indeed, their relative uniformity is emphasized by comparison to the Manihiki and Kerguelen plateaus. Basement lavas have been sampled at only two locations on the Manihiki Plateau: (1) DSDP Site 317 (with 33.5 m of basement penetration) and (2) a dredge haul ~300 km to the northwest (data for the latter are not included on Figures 8 or 9); yet they show a significantly greater spread of isotopic variation than the Ontong Java rocks ($^{206}Pb/^{204}Pb$ varies from 18.10 to 19.35, and $\epsilon_{Nd}(T)$ from +0.8 to +6.8 [Mahoney and Spencer, 1991]). Basalts from several drillholes on the Kerguelen Plateau cover an extreme range of $\epsilon_{Nd}(T)$ (+5.2 to -4) and possess variable–and low–$^{206}Pb/^{204}Pb$ (17.47-18.28) as well [Weis et al., 1989; Storey et al., 1992; Salters et al., 1992]. Despite their differences, all three plateau data sets are characterized by a combination of relatively high $^{208}Pb/^{204}Pb$ and low $\epsilon_{Nd}(T)$ for a given $^{206}Pb/^{204}Pb$ value in comparison to Pacific MORB. This feature imparts a distinctive "DUPAL"-like [e.g. Hart, 1988] signature to all.

Trace and major elements. Trace element differences also are evident between Units A and C-G, with Unit A lavas having higher abundances of (alteration-resistant) incompatible elements and small relative enrichments in highly incompatible elements compared with moderately incompatible ones. The chondrite-normalized rare-earth-element patterns (Figure 7) of Unit A lavas, for example, show slight enrichment in the light rare earths, whereas Units C-G have lower overall abundances of rare-earth elements and flat to slightly light-rare-earth-depleted patterns. Patterns of Site 803 samples are similar in shape to those of the Site 807 rocks, with abundances

TABLE 5. Pb Isotopic and Isotope-Dilution Data for Leg 130 Basalts

Sample		$^{206}Pb/^{204}Pb$	$^{207}Pb/^{204}Pb$	$^{208}Pb/^{204}Pb$	Pb (ppm)
130-803D-					
69R-2, 54-56		18.691	15.529	38.448	0.368
71R-1, 124-126		18.626	15.522	38.426	0.255
	UP	18.643	15.523	38.436	0.363
130-807C-					
75R-4, 46-48 Unit A		18.395	15.531	38.383	0.581
77R-2, 108-110 Unit A	GL	18.392	15.533	38.398	0.562
79R-5, 38-40		18.385	15.521	38.368	0.505
Unit A	UP	18.392	15.526	38.387	0.659
82R-2, 121-123 Unit C		18.635	15.523	38.455	0.295
82R-4, 105-107 Unit E		18.659	15.489	38.402	0.227
85R-3, 19-21 Unit F		18.609	15.522	38.391	0.278
88R-3, 76-79 Unit G		18.668	15.547	38.535	0.270
USGS Standard BCR-1		18.805	15.619	38.669	13.53

Notes: UP = bulk sample, not picked, cleaned only in H_2O. Chips of other samples were handpicked and acid-cleaned in HF-HNO_3 and HCl. GL = handpicked, fresh glass. Pb isotopic ratios are present-day values, corrected for fractionation using the NBS 981 standard values of Todt et al. (1984); the total ranges measured for NBS 981 are +0.008 for $^{206}Pb/^{204}Pb$, ±0.008 for $^{207}Pb/^{204}Pb$, and ±0.030 for $^{208}Pb/^{204}Pb$. Within-run uncertainties ($2\sigma_m$) on the isotopic data above are generally less than these values, and in all cases less than ±0.012, ±0.012, and ±0.038. Estimated uncertainty on Pb abundances is <1%. Total procedural blanks are negligible at 5-30 picograms.

intermediate between Units A and C-G. The isotopic differences noted above between units A and C-G suggest that differences in mantle source composition are partly responsible for their respective incompatible element characteristics. Nonetheless, just as with isotopes, the overall range of incompatible element variation for the Leg 130 rocks is small. Cretaceous basement samples from the Kerguelen Plateau show much greater variability and are significantly more enriched in highly incompatible elements such as Th and Ba (i.e., they are much less MORB-like; see Figure 10). Few comprehensive trace element analyses of good quality are available for the Manihiki Plateau, but as Figure 10 illustrates, the two samples measured in this study have slighter higher Th/Ta and Ba/Ta than most of the Ontong Java lavas. Their normalized rare-earth-element patterns are flat, however, much like those of the Ontong Java rocks. Significantly, despite

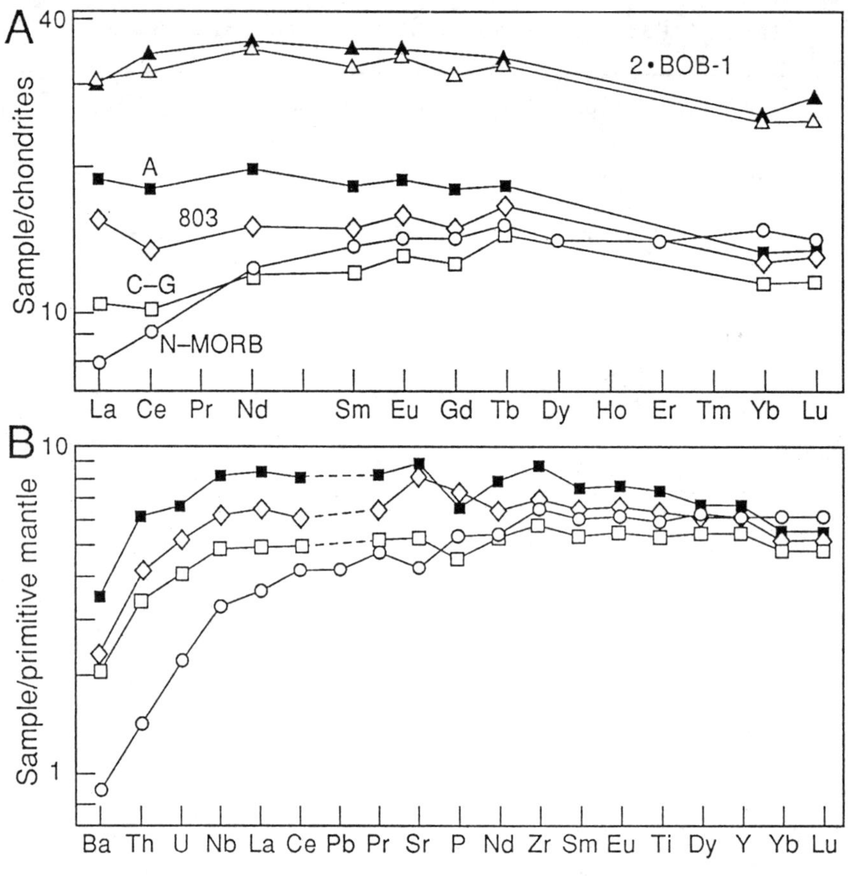

Fig. 7. (a) Average chondrite-normalized rare-earth-element patterns for Site 803 (open diamonds), Site 807 Unit A (filled squares), and Units C-G (open squares). All of the Leg 130 basalts are enriched slightly in the light rare earths relative to normal MORB (represented by open circles). Also shown are measured (open triangles) and recommended (filled triangles) values for standard BOB-1 (plotted as twice the true values for clarity). Normalizing values are those of Hanson [1980]. (b) Average primitive-mantle-normalized element patterns; normalizing values are those of Sun and McDonough [1989]. Uranium values for several Site 803 and Units C-G samples were not included in the average because these samples are comparatively highly altered (e.g., their Th/U <2).

their DUPAL-like isotopic signatures, neither the Ontong Java nor the Manihiki Site 317 basalts have the high Ba/Ta (or Ba/Nb) values so characteristic of South Atlantic and Indian Ocean DUPAL-type plateaus or islands (see Figure 10a).

Major element compositions of the Site 807 basalts reflect variable but rather high amounts of fractionation: MgO ranges from 6.1 to 8.4 wt.% (excluding the uppermost, highly altered sample at the top of basement), whereas total Fe as Fe_2O_3 is elevated, varying from 10.3 to 14.9 wt.%. Site 803 lavas have slightly lower MgO (most samples are 5.3-7.5 wt.%) but also lower Fe_2O_3 (9.8-13.3 wt.%) and higher Cr (236-296 vs. 126-173 ppm for Site 807) and Ni (83-185 vs. 72-126 ppm), suggesting that some are less evolved than the Site 807 flows. Also, they have higher Al_2O_3 and lower CaO/Al_2O_3. Figure 11 displays major element variations for the Leg 130 basalts. Interestingly, the Site 807 data again fall into two clusters corresponding to Units A and C-G, the Unit A lavas being characterized by lower MgO and SiO_2, and greater TiO_2 and Na_2O than Units C-G. Predicted one-atmosphere liquid evolution paths, determined by the method of Weaver and Langmuir [1990], are depicted for 25% fractional crystallization from two high-MgO samples that bracket the range of Na, Ca, Al, and Fe oxide values. These trajectories illustrate that Unit A and Units C-G cannot be related to a common parental liquid composition by low-pressure differentiation (see the diagram for SiO_2, in particular). At pressures greater than 8-10 kb, plagioclase is not stable and the fractionating phase assemblage would consist of clinopyroxene and orthopyroxene [e.g. Bender et al., 1978]. Under such conditions, FeO and Al_2O_3 should both increase, and CaO/Al_2O_3 decrease, with progressive fractionation, but this is contrary to what is observed. Several Site 807 samples from Units C-G do have lower FeO (and higher MgO) than the Unit A basalts, but they also have higher Al_2O_3. Thus, the major element results

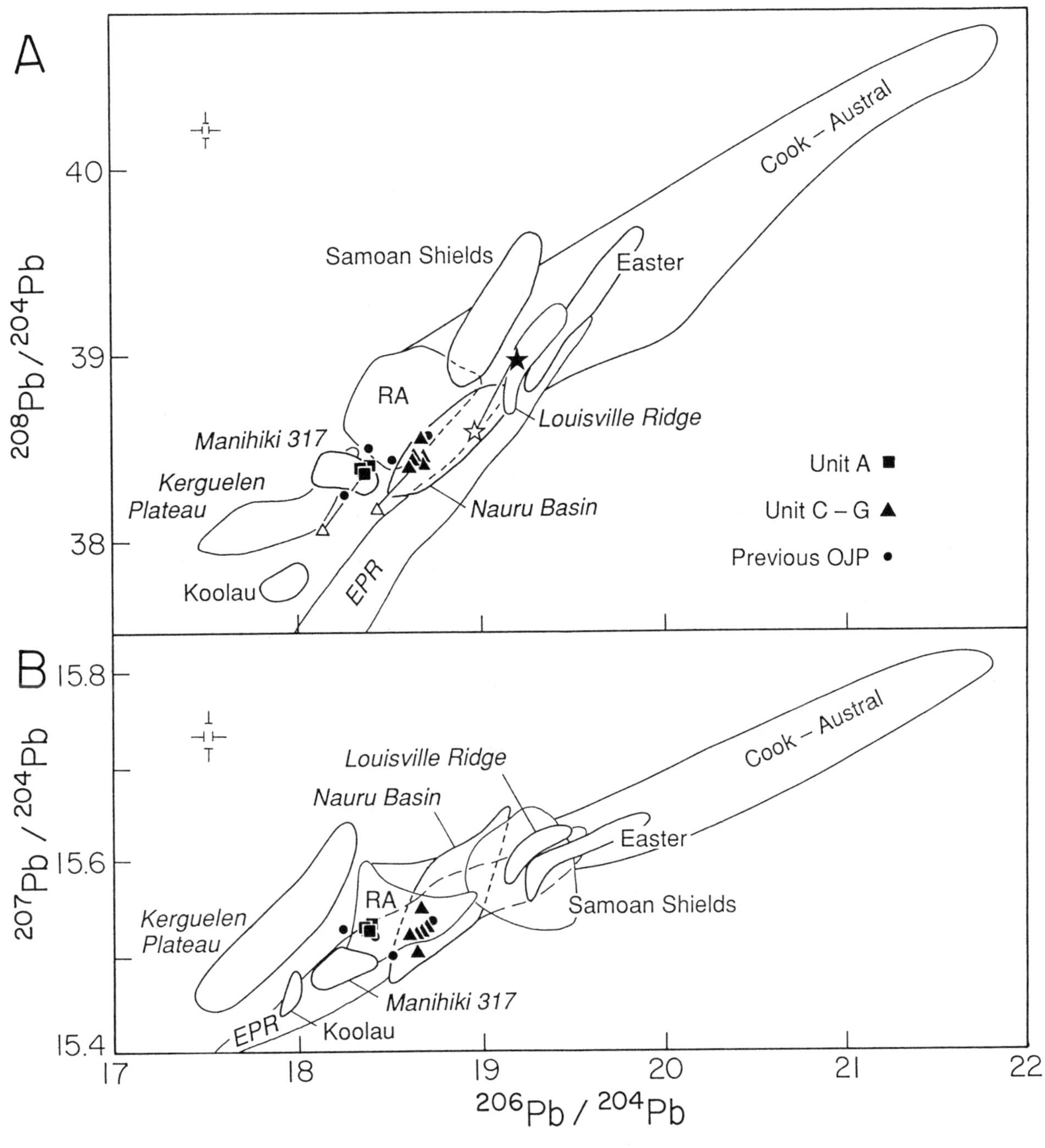

Fig. 8. (a) Present-day $^{208}Pb/^{204}Pb$ vs. $^{206}Pb/^{204}Pb$ and (b) $^{207}Pb/^{204}Pb$ vs. $^{206}Pb/^{204}Pb$ for Leg 130, Site 289, and southern Malaitan basalts. Triangles indicate Site 807 Units C-G and Site 803 data; squares are for Site 807 Unit A; circles are for Site 289 and Malaitan lavas (= Previous OJP [Mahoney and Spencer, 1991]). Open triangles represent initial Pb isotopic ratios estimated from U, Th, and Pb abundances for bulk samples 130-803D-71R-1 (124-127) and 130-807C-79R-5 (34-40). Solid and open stars in panel (a) illustrate the estimated 120-m.y. change in Pb isotopes for the mantle source of a typical Louisville Ridge basalt (calculated using isotopic ratios of Cheng et al. [1987], measured U and Th, and estimated Pb (assuming Ce/Pb = 25) abundances from Hawkins et al. [1987]). Note that this is a probable maximum change as the Th/Pb and U/Pb values for the source are likely to be lower than in these alkalic basalts. Fields for the Kerguelen Plateau [Weis et al., 1989; Storey et al., 1992] and the Manihiki Plateau at DSDP Site 317 [Mahoney and Spencer, 1991] are also shown, along with ones for East Pacific Rise MORB (EPR) [White et al., 1987; Hanan and Schilling, 1989; Prinzhofer et al., 1989; J. Mahoney et al., unpubl. data, 1991], Nauru Basin, Louisville Ridge, and selected Pacific oceanic islands (see references of Mahoney and Spencer [1991]). RA denotes Rarotonga. Error bars are for data reported in this paper.

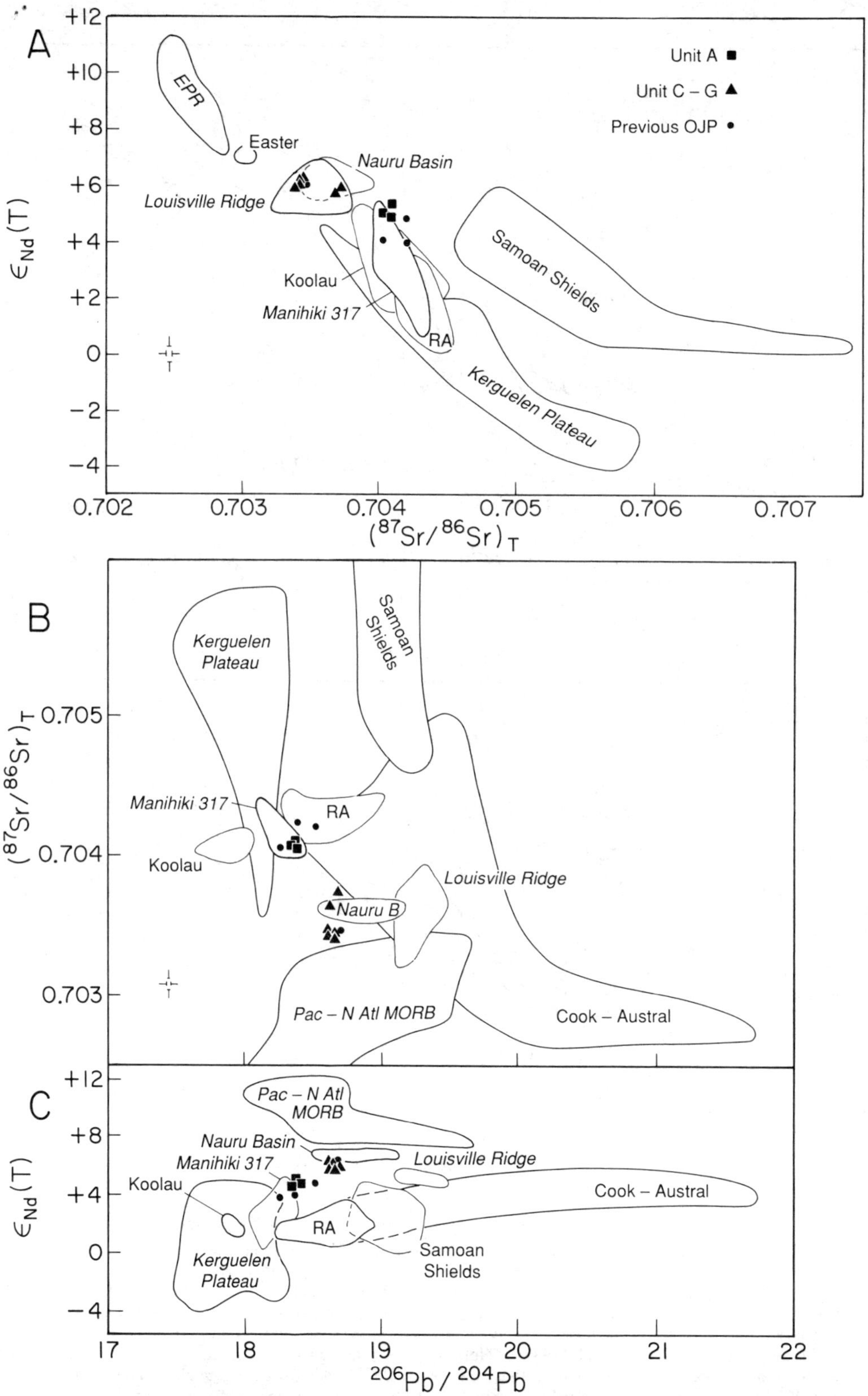

Figure 9. (a) $\epsilon Nd(T)$ vs. $(^{87}Sr/^{86}Sr)_T$, (b) $(^{87}Sr/^{86}Sr)_T$ vs. $^{206}Pb/^{204}Pb$, and (c) $\epsilon Nd(T)$ vs. $^{206}Pb/^{204}Pb$ for Leg 130, Site 289, and Malaitan basalts. Symbols are as in Figure 8; data sources are as in Figure 8 (this study) and in Figure 3 of Mahoney and Spencer [1991]. Error bars are for results reported in this paper.

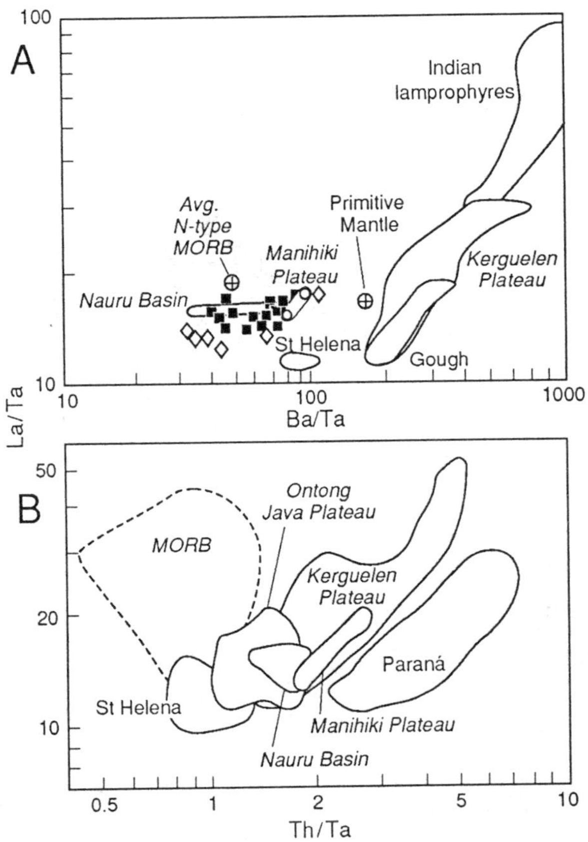

Fig. 10. (a) Plot of Ba/Ta vs. La/Ta, illustrating that the Pacific plateaus (as represented by the Ontong Java and Manihiki Site 317 data) are quite distinct from the Kerguelen Plateau and so-called DUPAL oceanic islands (such as Gough) or high $^{206}Pb/^{204}Pb$ islands (such as St. Helena). Filled squares = Site 807; diamonds = Site 803; circles = Site 317. A field for Cretaceous Indian lamprophyres (thought to have originated in the Gondwanan lithospheric mantle) illustrates the possible role of Gondwanan lithosphere in the formation of the Kerguelen Plateau. Data sources: primitive mantle and average N-type MORB, Sun and McDonough [1989]; St. Helena, Chaffey et al. [1989]; Gough, Weaver et al. [1987]; Nauru Basin, Saunders [1986]; Kerguelen Plateau, Storey et al. [1992] and Mehl et al. [1991]; Indian lamprophyres, Paul and Potts [1981]. (b) Th/Ta vs. La/Ta. Note that the Ontong Java basalts have more MORB-like Th/Ta than those for either the Kerguelen or Manihiki plateaus, whereas the Paraná continental flood basalts of Brazil (ca. 130 Ma) possess even greater Th/Ta values than the Kerguelen Plateau.

appear to require somewhat different parental liquids for Unit A than for Units C-G. Similar reasoning suggests the same is true for Site 803.

DISCUSSION

Melting Conditions of Plateau Magmas

A major unanswered question about oceanic plateaus in general concerns the conditions of melting under which plateau lavas were generated. Figure 12 shows average Na_8 vs. average Fe_8 for Units A and C-G (i.e., the average fractionation-corrected Na_2O and total Fe as FeO backtracked to 8 wt.% MgO [Klein and Langmuir, 1987]), along with regional averages for portions of several present-day oceanic ridges. The Site 807 data lie on the trend of MORB regional averages defined by Klein and Langmuir's [1987, 1989] worldwide compilation for spreading centers; moreover, the values for each group of flows are quite similar to one another. Average CaO/Al_2O_3 also is similar for the two groups and, when plotted versus Na_8 or Fe_8, falls on the regionally averaged, global MORB array as well. Because Na_8, Fe_8, and CaO/Al_2O_3 appear to reflect the integrated degree and depth of melting controlled by regional variations in the potential temperature of upwelling mantle (having homogeneous major element composition), these results indicate generally similar conditions of melting for Units A and C-G.

The position of the Site 807 data on the MORB array in Figure 12 is consistent with the hypothesis that the Ontong Java Plateau was formed near a spreading center; results for off-ridge hotspots and other magmatic settings often fall off the trend for regionally averaged MORBs [e.g. Klein and Langmuir, 1987]. Significantly, the Site 807 data plot at the low-Na_8, high-Fe_8 end of the global MORB correlation; that is, at the end corresponding to the highest observed extents of partial melting. Indeed, their closest present-day analogs are Iceland and the Kolbeinsey and Reykjanes ridges flanking the Icelandic hotspot, a region characterized by notably greater degrees of melting than ordinary spreading centers far from hotspots [Dick et al., 1984; Klein and Langmuir, 1987]. Lavas from the Nauru Basin adjacent to the plateau on the east have similar but slightly greater Na_8 and lesser Fe_8, suggesting slightly lower extents of partial melting at the margins of the plateau. The ~90 Ma Site 803 basalts have low Na_8 but also lower Fe_8 than the Site 807 lavas; assuming the Site 803 samples give a representative average for this site, they could reflect a high degree but shallower overall depth of melting than at Site 807. We also have included an average for the Deccan Traps of India (for the upper-level Ambenali formation, which is essentially uncontaminated by continental crust). This large flood basalt province appears to have been formed by the nascent Réunion hotspot [e.g. Morgan, 1981; Richards et al., 1989; White and McKenzie, 1989] in conjunction with continental rifting and the initiation of seafloor spreading above the hotspot about 66 Ma. Like the Ontong Java, the Ambenali plots relatively near the high-melting end of the MORB array; but neither the Deccan nor Ontong Java points lie at lower Na_8 than data for Iceland or the Reykjanes Ridge. Thus, if the Ontong Java Plateau and the Deccan Traps reflect melting in the heads of new mantle plumes, these results provide no evidence for melting to an extent greater than occurs around a mature near-ridge plume tail such as Iceland.

Klein and Langmuir [1987] showed that, for their worldwide compilation of MORB data, reasonably good correlations exist between average regional Na_8, CaO/Al_2O_3, and Fe_8 versus

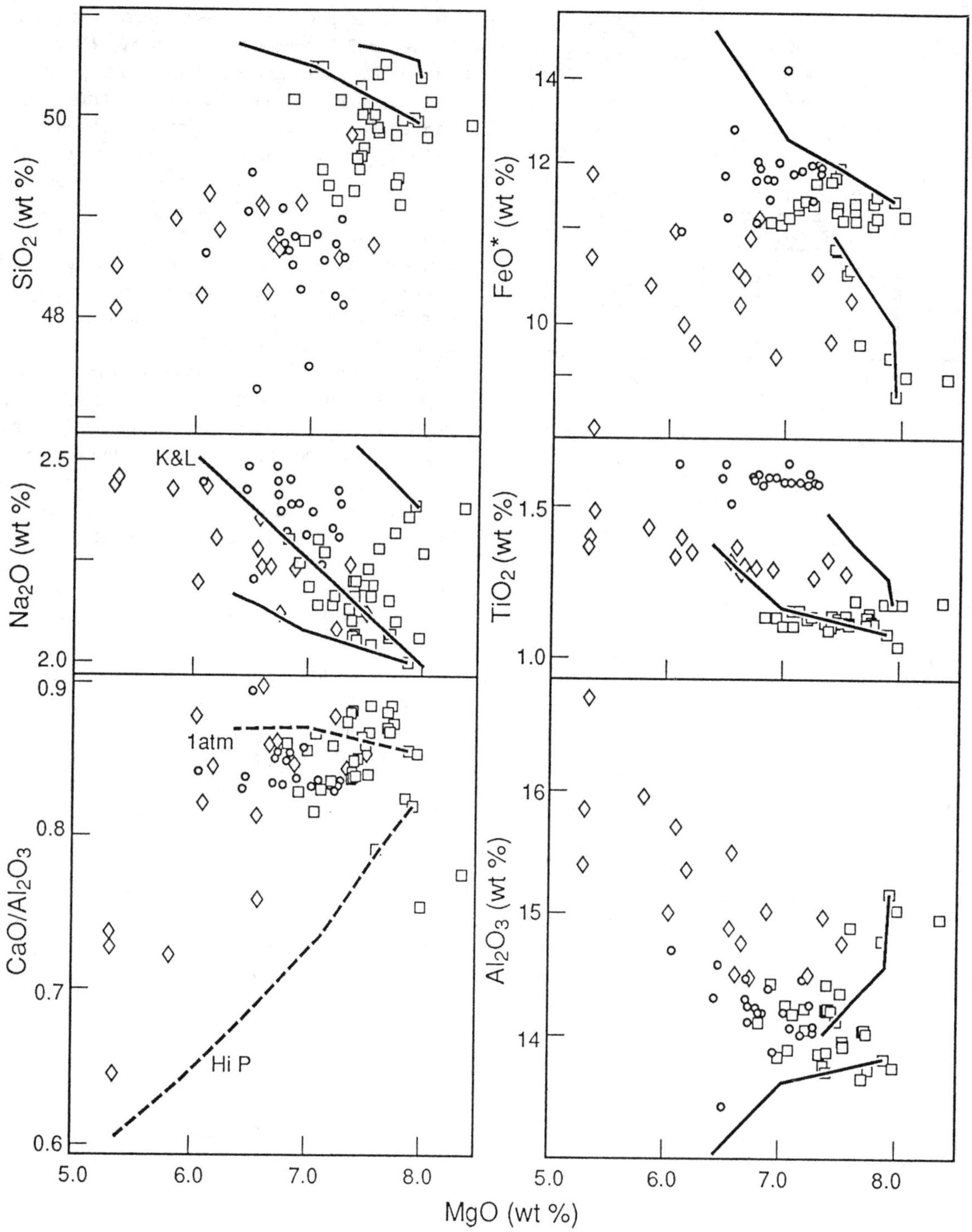

Fig. 11. (a) Major element variations versus MgO for the Site 807 (Unit A = circles, Units C-G = squares) and Site 803 (diamonds) samples. Also shown are two trajectories representing 0-25% low-pressure fractional crystallization calculated using the procedure of Weaver and Langmuir [1990]. The two starting compositions for these paths were chosen because they span the ranges in most oxides (at high-MgO) found in our samples. One low-pressure (dashed line, labeled 1 atm) and one high-pressure (dashed line, labeled Hi P) crystallization path are shown for CaO/Al_2O_3 (high-pressure path also calculated with procedure of Weaver and Langmuir [1990]). The line labeled "K&L" on the Na_2O diagram is for MORB average from Klein and Langmuir [1987]. Note that FeO* values are recalculated from Fe_2O_3* in Table 2.

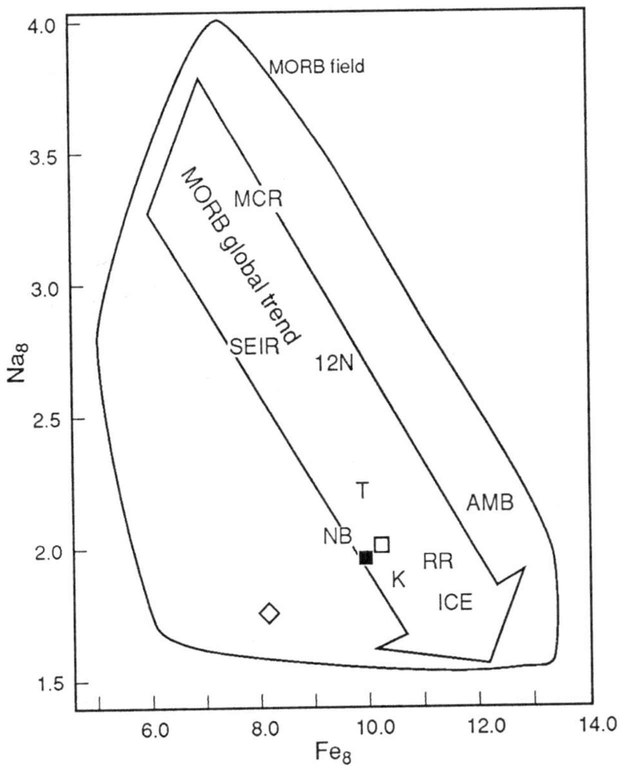

Figure 12. Average Na_8 vs. average Fe_8 for Site 807 Unit A (filled square) and Units C-G (open square), along with an outline of the overall field for MORBs, an arrow representing the trend defined by regionally averaged MORB data, and average values for several specific present-day spreading centers [from Klein and Langmuir, 1987, 1989; Klein et al., 1991] and Nauru Basin (NB; data from Castillo et al. [1986]). ICE = Iceland, RR = Reykjanes Ridge, K = Kolbeinsey Ridge, T = Tamayo Fracture Zone, SEIR = eastern Southeast Indian Ridge, 12N = Mid-Atlantic Ridge 11.4°N-11.97°N, MCR = Mid-Cayman Rise. Note that Units A and C-G have similar average values and that both plot near the high-melting end of the global MORB array, along with ridges flanking the Icelandic hotspot. Site 803 (open diamond) has both low Na_8 and low Fe_8. Also shown is a point for the Ambenali formation (AMB) of the Deccan Traps of India (from Z. Peng et al., unpubl. data, 1991).

axial depth. When the average values of these chemical parameters for the Leg 130 rocks are plotted on the MORB correlations, they fall near the end of the MORB array corresponding to shallow axial depth, in the range for the Kolbeinsey and Reykjanes ridges. This result could be taken to imply a shallow eruption depth (less than about 1000 m) for the plateau lavas. However, although parts of the plateau originally may have been at shallow levels, no evidence for such shallow paleodepths has been found at Sites 803, 807, or 289, or on southern Malaita; instead, the basement rocks and basal sediments sampled to date seem to indicate paleodepths of more than 1000 m [e.g. Kroenke, Berger et al., 1991]. Some Leg 130 flows perhaps could have issued from distant eruptive vents at shallow depths (especially the thick Unit F), but many of the thin pillow lavas appear unlikely to have traveled very far. At present, we prefer not to draw any conclusions regarding paleodepths from the above petrochemical parameters. In particular, the Ontong Java Plateau could have been formed at least partly off-axis, but the correlations between eruption depth and Na_8, Fe_8, and CaO/Al_2O_3 hold only for axial lavas proper [Niu and Batiza, 1992]; they do not obtain even for other near-ridge (but off-axis) basalts such as those on near-axis seamounts or flow fields; nor do they hold, in general, for intraplate basalts.

To provide somewhat more quantitative bounds on melting, the fractional melting inversion of McKenzie and O'Nions [1991] was applied. This inversion uses averaged rare-earth-element concentrations to estimate (1) melt distribution as a function of depth, (2) a total integrated melt fraction, and (3) equivalent crustal thickness corresponding to total melt produced. The rare-earth-element composition of the source, source mineralogy, and distribution coefficients for the rare earths are assumed, and fractional melting, rather than batch melting, is taken to be the operative process (as indeed is indicated by recent work on abyssal peridotites [Johnson et al., 1990]). For this exercise, only data for samples with MgO > 6 wt.% were used [cf. McKenzie and O'Nions, 1991], and the data were averaged together irrespective of site (Leg 130 and Site 289) or stratigraphic location (Units A, C-G); although Units A and C-G, for example, clearly had somewhat different mantle sources, averaging is reasonable because the total range of variation is small (e.g., only 1.4 ϵ_{Nd} units). The inversion was run on this average assuming mantle sources with either primitive or depleted (MORB mantle) rare-earth-element compositions.

Figure 13 illustrates the results (for the neutron activation data only) in the form of the concentration (normalized to assumed source) and melt fraction versus depth diagrams used by McKenzie and O'Nions [1991]. Individual points on Figures 13a and 13c are the normalized average concentrations of the data, with error bars indicating standard deviation for each element; the heavy solid line is the pattern determined by the inversion, with the lightly drawn solid and dashed lines representing estimated upper and lower limits. Both a MORB-type source and one with a primitive-mantle rare-earth pattern yield acceptable fits (i.e., the heavy solid line is close to the average value and well within the error bars for each rare-earth element). The primitive-source model produces a pattern with slightly greater light-rare-earth and slightly lower middle-rare-earth element abundances than the actual data, whereas the MORB-source model yields a pattern that is slightly lower in La and higher in Nd and Sm. In conjunction with the Nd isotopes, which require a source more enriched over time in the light rare earths than that of normal MORB, but less so than for primitive mantle, these results suggest that a mixed source is the most appropriate.

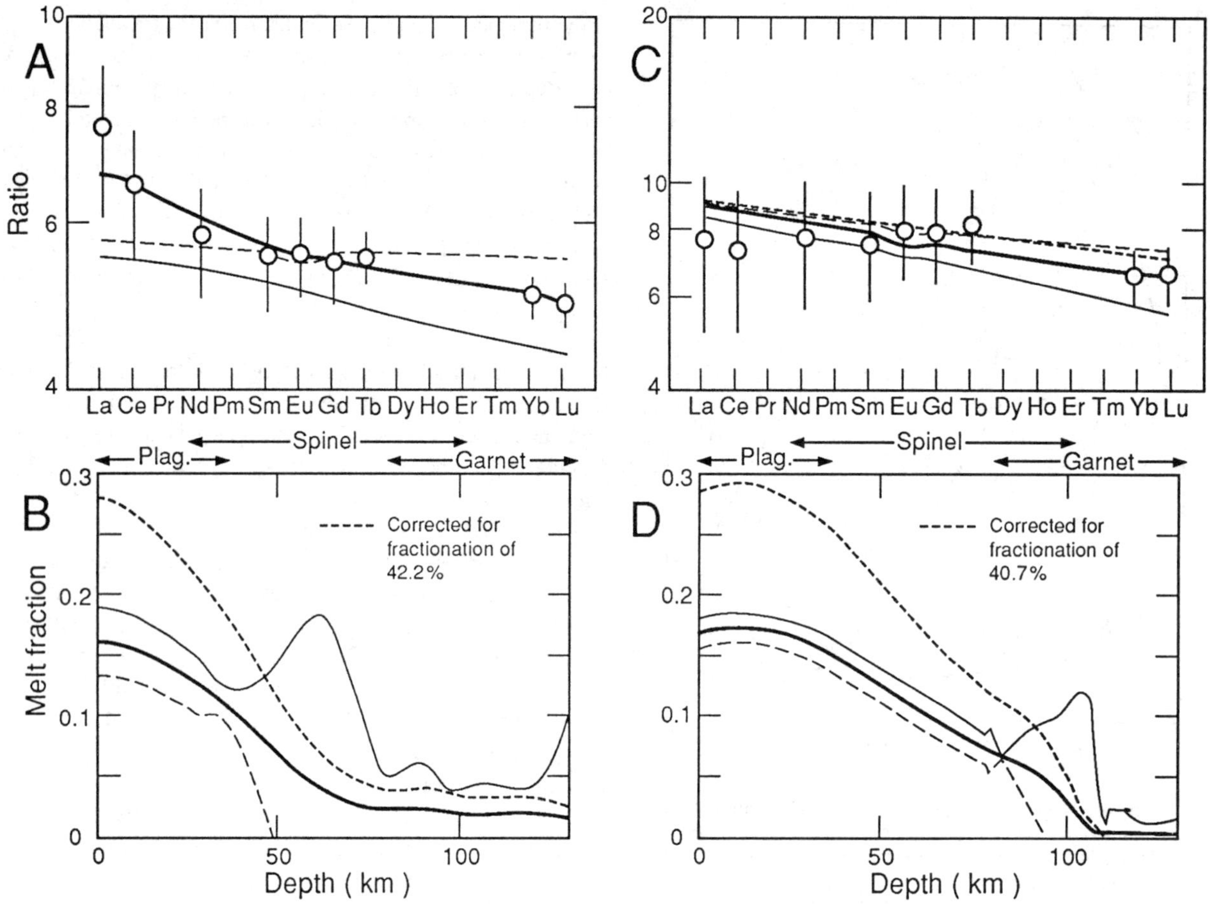

Fig. 13. (a) Element concentration ratio with respect to a normal-MORB mantle source [McKenzie and O'Nions, 1991] for average of Leg 130 and Site 289 basalts with MgO >6 wt.% (average of neutron activation data only). Error bars indicate standard deviation of data for each element. The heavy solid line shows the rare earth pattern determined by fractional-melting inversion; the fine continuous and dashed lines represent patterns calculated from the estimated upper and lower bounds (see McKenzie and O'Nions [1991] for details). Observed and calculated major element values (normalized to 100 wt.% without K_2O and P_2O_5) are given below; fractionation correction is calculated from the difference between observed and calculated MgO and FeO.

	SiO_2	TiO_2	Al_2O_3	FeO	MgO	CaO	Na_2O	K_2O
Obs.	50.37	1.33	14.61	11.63	7.40	12.36	2.3	00.16
±	0.59	0.22	0.44	0.93	0.62	0.32	0.17	0.10
Calc.	50.24	1.24	15.10	8.67	11.33	11.14	2.28	0.26
±	0.81	0.24	0.93	0.97	1.00	1.07	0.29	0.23
(Obs.-Calc.) / S.D.	0.13	0.27	-0.48	2.20	-3.34	1.10	0.07	

(b) Melt fraction (by weight) with depth calculated before (heavy continuous line) and after (heavy dashed line) correction for crystal fractionation. Fine dashed and continuous lines again represent the estimated lower and upper limits, respectively. Total melt thickness calculated from the inversion is 17.2 km after fractionation correction.

(c), (d) Same as Figures 13a and 13b except for a primitive mantle source. Here the estimated total melt thickness after the fractionation correction is 24.3 km. Calculated major element values and difference between observed and calculated values are given below.

	SiO_2	TiO_2	Al_2O_3	FeO	MgO	CaO	Na_2O	K_2O
Calc.	49.82	1.08	14.81	9.00	12.10	11.05	2.15	0.26
±	0.81	0.24	0.93	0.97	1.00	1.07	0.29	0.23
(Obs.-Calc.) / S.D.	0.55	0.75	-0.19	1.96	-3.99	1.17	0.46	

Estimated melt fraction as a function of depth is presented in Figures 13b and 13d. Here, the heavy dashed curve represents the calculated melt distribution after correction for crystal fractionation (the correction is based on the misfit between calculated and measured average MgO and total Fe as FeO, assumed to be caused by olivine fractionation; see McKenzie and O'Nions [1991] for details). The heavy solid curve is the calculated distribution before the fractionation correction, and the lightly drawn lines the lower and upper limits. The fractionation correction is ~40%, owing to the evolved nature of the Leg 130 and Site 289 samples. Significantly, the estimated maximum melt fractions are very high for both the primitive- and MORB-source models, around 28-30%, although melting of the primitive source begins at greater depth. The total thickness of melt produced in the primitive-source model (after the fractionation correction) is 24.3 km; for the MORB-source model, it is 17.2 km. Results obtained by McKenzie and O'Nions [1991] for enriched MORBs from Iceland are remarkably similar: total amount of partial melt ~30%, equivalent crustal thickness = 22.3 km for a primitive-source model after a 39% fractionation correction, and melt distribution with depth much like that for the Ontong Java primitive-source model. The estimated maximum (fractionation-corrected) extent of melting for normal MORBs, on the other hand, is only around 15%, and predicted crustal thickness is 9 km or less [McKenzie and O'Nions, 1991].

As a comparison to the Ontong Java case, the inversion also was applied to Manihiki Plateau data for DSDP Site 317. The Site 317 basalts exhibit a greater range of ϵ_{Nd} than the Leg 130 rocks, but suitably comprehensive chemical analyses are available for only seven of the ten flows drilled, and only six rare-earth elements have been measured for most of them; therefore, the data were averaged and Hf and Ti, for which there are good measurements, were included in the inversion. Figure 14 presents the results. Not surprisingly, the variance associated with most of the rare-earth elements is greater than for the Ontong Java rocks, and Tb appears to be systematically low compared with the other rare earths; however, a primitive-source model produces an acceptable fit (though not shown graphically, the deviation for Ti and Hf is much smaller than for any of the rare-earth elements). Estimated total melt fraction and crustal thickness are again large at ~30% and 23.7 km, repectively, very similar to results for the Ontong Java Plateau and Iceland (the fractionation correction here is only 22%, however, as the Site 317 lavas are relatively undifferentiated).

In short, the Ontong Java and Manihiki Plateau basalts appear to represent very high-degree melts with affinities to present-day lavas from the vicinity of Iceland. The above results support a near-ridge hotspot or plume-head origin for the two Pacific plateaus. A completely off-ridge plume-head origin would appear to be possible only with very substantial lithospheric thinning above the head (i.e., localized rifting that may not have linked up with a spreading center) or if the potential temperature [e.g. McKenzie and Bickle, 1988] in the head was sufficiently above that now beneath Iceland to produce comparably high-degree melts with lesser amounts of thinning. Although there is good evidence that the Manihiki Plateau formed at a ridge crest [Winterer et al., 1974; Sharman and Mammerickx, 1990], tectonic evidence for the original setting of the Ontong Java Plateau is somewhat more equivocal. The plateau is isostatically compensated and its structural features are consistent with a spreading center origin [e.g. Kroenke, 1972], but seafloor magnetic lineations to the east of Sites 807 and 289 [Larson, Schlanger et al., 1981] are >25 m.y. older than the lavas at these drill sites. The lineations decrease in age to the south such that, if east-west continuity of the seafloor is assumed, the southern parts of the plateau (such as Malaita and the eastern lobe) may be the same age as the ocean crust to the east. However, the lineations cannot be traced westward beyond the edge of the plateau; instead, structural trends and a weak magnetic fabric on the plateau are oriented at an angle to those of seafloor on the east, possibly indicating a major offset in the paleo-Pacific ridge system [e.g. Kroenke, 1972].

It is interesting that there is no indication, from either major elements or the inversion results, of melting to an extent exceeding that observed today around Iceland, as has been proposed to exist near the axes of large plume heads [e.g. Campbell and Griffiths, 1990]. (Based on their high-Cr spinels, near-primitive Manihiki Plateau dredge samples studied by Clague [1976] and Dick and Bullen [1984] may be an exception; the near-primary picrites of Curacao and the komatiitic lavas on the island of Gorgona, both probable fragments of the Caribbean Plateau [Storey et al., 1991], may be another.) The Ontong Java lavas sampled so far, of course, simply may not correspond to the axial region of a plume head. Perhaps more serious is that the crustal thicknesses estimated by the inversion model, though considerable and, indeed, in the range of values estimated for the Kerguelen Plateau [e.g. Houtz et al., 1977], are significantly less than those actually measured for the Ontong Java Plateau by seismic refraction (30-40 km, with an average of ~36 km [Furumoto et al., 1976; Hussong et al., 1979]). Assuming that both the model prediction and the refraction measurements are accurate, the difference may be because the available samples are not truly representative of the plateau's crust; in particular, they clearly record only the final episodes of volcanism, at least locally. Earlier, deeper parts of the crustal section might reflect even greater degrees of melting [cf. Storey et al., 1991], corresponding to greater crustal thicknesses. Also, if parts of the plateau were built on pre-existing oceanic crust, an extra 5-8 km can be added to the calculated (melt) thickness. In addition, some crustal thickening could have occurred after eruption of the sampled surface lavas, through underplating and intrusion.

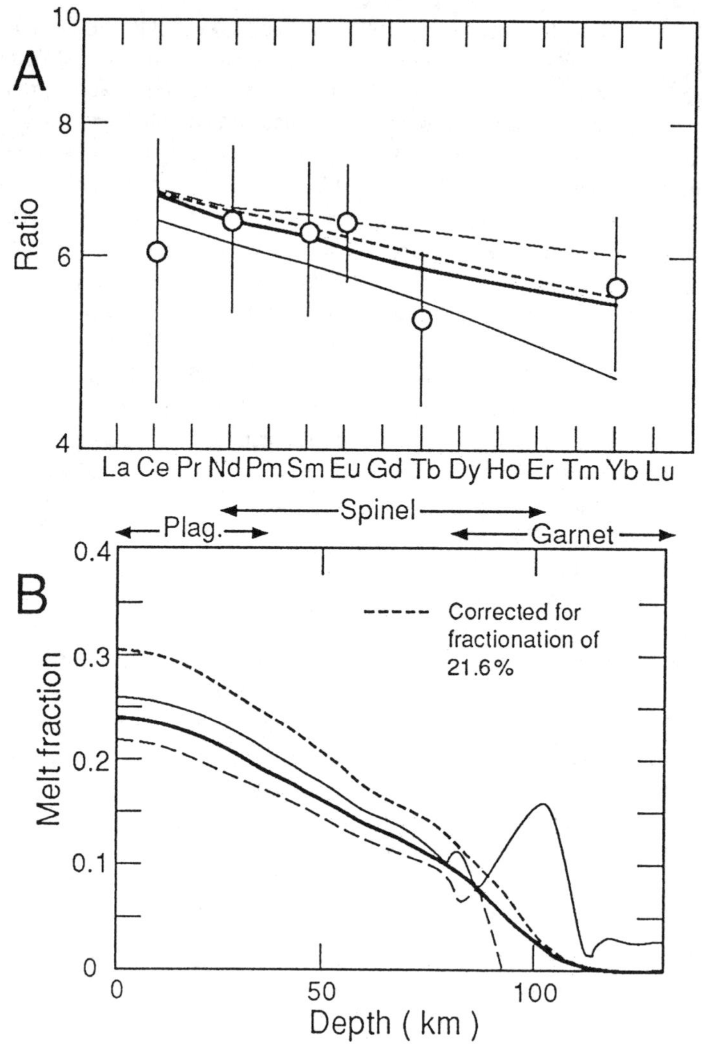

Fig. 14. Results of inversion model for Manihiki Plateau lavas from Site 317, using data from Jackson et al. [1976], Mahoney [1987], and this paper (Table 3). See caption to Figure 13 and text for explanation. Here, Hf and Ti were used in addition to rare earths. Calculated total melt thickness is 23.7 km after fractionation correction of 21.6%. Observed and calculated major element concentrations are given below.

	SiO_2	TiO_2	Al_2O_3	FeO	MgO	CaO	Na_2O	K_2O
Obs.	51.27	1.00	15.61	9.81	8.45	11.65	2.21	0.07
±	0.36	0.06	0.21	0.38	0.64	0.14	0.08	0.04
Calc.	49.84	0.91	14.34	9.14	12.79	11.01	1.97	0.25
±	0.81	0.24	0.93	0.97	1.00	1.07	0.29	0.23
(Obs.-Calc.) / S.D.	1.61	0.33	1.34	0.64	-3.65	0.59	0.82	

Relation of Site 803 Basalts to Other Sites

The plume-initiation model predicts a very short span of basement ages (only a few million years) across the plateau [e.g. Richards et al., 1989]. Our early Aptian ^{40}Ar-^{39}Ar dates for Site 807, Site 289, and southern Malaita are essentially identical and thus strongly support the model. Likewise, although basement was not reached at DSDP Site 288, the oldest sediments recovered were Aptian (112-124.5 Ma [Harland et al., 1990]), indicating that local basement is Aptian or older [Packham and Andrews, 1975]. On the other hand, the three

^{40}Ar-^{39}Ar plateau ages obtained for Site 803 demonstrate that volcanism occurred there some 30 m.y. after it had ceased at Site 807. Whether appreciable amounts of plateau construction took place over a period of more than 30 m.y. or whether, for example, the 26 m of lavas penetrated at Site 803 represent only a brief local reactivation of volcanism long after most of the crust was emplaced around ~122 Ma cannot be determined conclusively at present, given so few basement sites (all with shallow penetration) over such a great area. There is evidence of at least limited activity elsewhere on the plateau around 88-95 Ma, as recorded by volcaniclastic beds of Coniacian through mid-Cenomanian age at DSDP Site 288, which lies nearly 1000 km south of Site 803; also, somewhat older volcaniclastic layers of late Aptian to Albian age occur above basement at DSDP Site 289 [Packham and Andrews, 1975]. What is clear from the present study is that the end of basalt eruption at several widely spaced locations was virtually contemporaneous in the early Aptian. Furthermore, the lack of any post-Aptian, Early Cretaceous ash layers or sills above basement at Site 807 and on Malaita [e.g. Packham and Andrews, 1975; Hughes and Turner, 1977] suggests that volcanism in the 120-90 Ma range, although present in places, was not particularly vigorous or widespread.

For the Manihiki Plateau, only basalts from Site 317 have been dated, yielding a weighted ^{40}Ar-^{39}Ar age of 123 ± 1.5 Ma (R.A. Duncan, unpubl. data, 1991). Basement ages have been obtained for several locations on the Kerguelen Plateau and Broken Ridge; they are remarkably similar to those for the Ontong Java both in absolute value and in total duration encompassed. Ages for the southern and central Kerguelen Plateau are between 110 and 118 Ma [Leclaire et al., 1987; Duncan, 1991; Whitechurch et al., 1992], whereas several lavas from Broken Ridge—originally contiguous with the central part of the Kerguelen Plateau—are 83-88 Ma [Duncan, 1991]. Therefore, despite being separated by considerable distances, construction of all three of these large oceanic plateaus appears to have overlapped in time, as also inferred by Tarduno et al. [1991] on the basis of biostratigraphic dates of basal sediments. Moreover, on the only two plateaus where sampling has been done at several sites, a ca. 30-m.y. record of volcanic activity has emerged. Unfortunately, the number of dated basement sites is as yet too small to construct a credible plot of erupted volume versus age for any plateau. In the case of the Ontong Java, however, the present results suggest that the great bulk of edifice building may have transpired in a very short period around 122 Ma. If true, the plateau marks the largest igneous event in the last 200 m.y.

The ~30-m.y. span of ages for samples from the Kerguelen Plateau and Broken Ridge is consistent with a northward-younging age progression expected from the migration of the Antarctic Plate over the youthful Kerguelen hotspot [Duncan, 1991]. For the Ontong Java Plateau, however, the ages of the widely dispersed Site 807, Site 289 and southern Malaitan locations are identical within errors, whereas Site 803, with a much younger age, sits at a latitude between Sites 807 and 289. Thus, there is no indication of any simple age progression across the plateau, and the roughly 90 Ma event recorded at Site 803 could have been imposed on the plateau after most of it was built. Volcanism of this age is observed throughout much of the western Pacific [Rea and Vallier, 1987; E. L. Winterer et al., unpubl. data, 1991] and may reflect a plate-wide extensional regime that resulted from a major change in the direction of plate motion. That similar ages mark the formation of Broken Ridge on the northeastern side of the Kerguelen Plateau could indicate that this reorganization was global in scope.

The geochemical information for Site 803 is surprising in view of the ^{40}Ar-^{39}Ar ages because, as noted above, the lavas isotopically resemble the lower units at Site 807 and the basalt at Site 289. They also possess very similar incompatible-element characteristics (Zr/Nb, La/Sm, Th/Ta, etc.; see Figure 7), and like the basalts at the other sites, are olivine- or quartz-normative tholeiites. In other words, the mantle source and petrogenesis of lavas at Site 803 appear to have been much like those in the other, older locations. These features are contrary to what would be expected if, for instance, the Site 803 rocks reflected late-stage, small-degree remelting of older plateau lithosphere or non-plume asthenosphere (such as the unusual 34 Ma alnoite intrusions of Malaita appear to represent [e.g. Neal and Davidson, 1989]), or the influence of a second hotspot overridden by the plateau at 90 Ma. As we have no reason to doubt the general validity of the ^{40}Ar-^{39}Ar ages for Site 803, the simplest explanation of the geochemical results is that the plateau was underlain by isotopically similar, hot plume-type mantle for at least 30 m.y. or passed over the same hotspot twice.

Louisville Hotspot Connection

Though not all plate reconstructions place the original location of the Ontong Java Plateau as far south as the Louisville hotspot [Yan and Kroenke, 1992], several reconstructions suggest the plateau was near the hotspot around 120 Ma [Richards et al., 1989; Mahoney and Spencer, 1991; Tarduno et al., 1991; Pringle, 1992]. Because our ^{40}Ar-^{39}Ar age determinations for southern Malaita and Sites 807 and 289 are very close to this value, they offer general support for a Louisville hotspot source for the plateau. Isotopic fields for the 0-70 Ma seamounts of the Louisville Ridge are even more restricted than those of the Ontong Java basalts, and their initial Nd and Sr isotopic ratios are similar to those of Units C-G, Site 289, and Site 803. However, the Louisville Ridge has much higher $^{206}Pb/^{204}Pb$ (19.1-19.5 [Cheng et al. [1987]) than any of the analyzed plateau lavas (Figures 8, 9b and 9c), even after allowance is made for age differences. If the Louisville hotspot was, in fact, the principal source of Ontong Java magmas, could the difference in Pb isotopes be a consequence of the evolution of Pb by radiogenic ingrowth in the plume? Pb abundances have not been determined for Louisville Ridge samples, but rather typical ocean-island-like $^{238}U/^{204}Pb$ values of ~15-35 are indicated by their measured U and Ce abundances [Hawkins et al., 1987], assuming that Ce/Pb is relatively constant [Hofmann et al., 1986]. Yet for the Louisville Ridge mantle source to evolve from the initial $^{206}Pb/^{204}Pb$ of Site 803 at 90 Ma, for example, to the 70-0 Ma values of the ridge lavas would require much higher $^{238}U/^{204}Pb$ in the source, from ~60 to more than 200, for which there is no evidence. Moreover, there is no systematic variation of $^{206}Pb/^{204}Pb$ with age along the ridge itself. Of course, a wider range of isotopic values may eventually be found on the plateau and possibly along the Louisville Ridge as well; but the latter data set covers the entire ~70-m.y. length of the seamount chain. Accepting a Louisville hotspot model for the plateau, the implication (assuming the existing Louisville Ridge isotopic field indeed is representative) is that products of the hotspot changed markedly in isotopic composition sometime between about 70 and 90 Ma, apparently losing an early, relatively low $^{206}Pb/^{204}Pb$ fingerprint. Unfortunately, samples–or seamounts–from the 70-90 Ma period are not available to document such a shift, having been subducted beneath the Australian Plate. Notably, this change cannot be explained by a decrease in the contribution of Pacific MORB-type mantle to the hotspot, in contrast to temporal isotopic variations associated with, for example, the Réunion hotspot [White et al., 1990].

The apparent shift to higher $^{206}Pb/^{204}Pb$ between the plateau and subsequent seamount chain could reflect a loss of low $^{206}Pb/^{204}Pb$ mantle in the hotspot itself, or conceivably only a loss of the ability to melt such material. In a study of Indian Ocean MORB, Mahoney et al. [1989] suggested that low $^{206}Pb/^{204}Pb$ mantle might be widely present in the Indian MORB source but be relatively refractory [see also, O'Connor et al., 1992]. Similarly, refractory low $^{206}Pb/^{204}Pb$ material could have persisted in the Louisville plume but the extent of melting in the last 70 m.y. (after the near-ridge and/or plume-head stage) might not have been great enough for it to be expressed significantly. Much-diminished extents of melting are in fact indicated by the Louisville Ridge lavas, which, in contrast to the high-degree plateau melts, are alkalic basalts, hawaiites, and basanitoids representing degrees of melting estimated in the 4-9% range [Hawkins et al., 1987]. A sustained, substantially lower overall amount of melting also is implied by the total volume of the seamount chain, which is very small (~2 × 10^5 km^3 [Lonsdale, 1988]) compared with that of the plateau (~5 × 10^7 km^3 [Mahoney, 1987]).

Additional evidence exists elsewhere for an early expression of low $^{206}Pb/^{204}Pb$ compositions not found in later products of a hotspot. In the Kerguelen system, the lowest $^{206}Pb/^{204}Pb$ values (to 17.47) occur only among the Early Cretaceous lavas of the southern Kerguelen Plateau and not in the sampled portions of the subsequent hotspot track, which predominantly have $^{206}Pb/^{204}Pb$ > 17.9 (as do many of the early plateau lavas also) [e.g. Storey et al., 1989, 1992; Weis et al., 1989; Barling and Goldstein, 1990; Weis and Frey, 1991; Saunders et al., 1991]. The Manihiki Plateau is not linked with certainty to any hotspot, but paleoceanographic reconstructions indicate that it originally lay in the southeast Pacific; none of the present hotspots in this region, with which it has sometimes been associated (Juan Fernandez, Easter, San Felix), have $^{206}Pb/^{204}Pb$ nearly as low as the Site 317 basalts [Mahoney and Spencer, 1991]. Again, in the Tristan hotspot system (Paraná flood basalts, Walvis Ridge, Rio Grande Rise, Tristan da Cunha and Gough islands), by far the lowest Pb isotopic ratios measured (to 17.1) are in the early products of the hotspot, the ~130-Ma HPT (high phosphorus and titanium) flood basalts [e.g. Hawkesworth et al., 1986]; the same may be true of the Marion hotspot system [Mahoney et al., 1991].

In the Paraná, Kerguelen, and possibly Marion examples, the early low $^{206}Pb/^{204}Pb$ signatures could be caused by plume-induced melting of continental lithosphere, the availability of which decreased as the continents drifted away from the hotspots [e.g. Storey et al., 1992]. Such a mechanism is far less probable for the intra-oceanic Ontong Java and Manihiki plateaus, where low $^{206}Pb/^{204}Pb$ mantle would more likely be indigenous to the plume. In either case, however, the loss of low $^{206}Pb/^{204}Pb$ characteristics could be a result of the decreasing ability of a plume to melt refractory mantle (whether originally continental or not), either as the plume's distance from a rift or spreading center increased (and hence the amount of decompressional melting was reduced [e.g. McKenzie and Bickle, 1988]) or as it went from an expanded plume-head to a restricted plume-tail stage. The Hawaiian example provides a possible argument against this hypothesis because both low $^{206}Pb/^{204}Pb$ (~17.7) and higher $^{206}Pb/^{204}Pb$ (~18.6) components are present in recent Hawaiian shield volcanoes [e.g. West et al., 1987], but the shields possessing the lowest $^{206}Pb/^{204}Pb$ signatures are not the largest ones [Bargar and Jackson, 1976]. For the present, therefore, we suggest the above idea only as a possibility.

CONCLUSIONS

The Ontong Java Plateau probably represents an early stage of a hotspot, possibly the Louisville hotspot, and very likely corresponds to its initial plume-head phase. Supporting this conclusion are the ^{40}Ar-^{39}Ar ages for basement at Site 807, Site 289, and in southern Malaita (all 120-124 Ma), which (1) show that volcanism ceased in a period of only a few million years at several widely spaced locations across the plateau, and (2) are consistent with recent plate reconstructions placing it near the Louisville hotspot at this time. Moreover, Ontong

Java lavas appear to represent very high-degree melts; their closest analogs today are found around Iceland, a vigorous near-ridge hotspot, consistent with a near-ridge origin for the plateau. The plateau could at least partly have formed in an off-axis setting, however, if lithospheric thinning induced by the plume head was substantial enough and/or the plume head was sufficiently hot enough to cause extents of partial melting comparable with those in Iceland. There is no evidence for a monotonic north-south age progression across the plateau; but at least some activity continued to ~90 Ma, as recorded by basement at Site 803 and late Aptian to Coniacian volcaniclastic layers at sites 288 and 289. Because the plume-initiation model predicts that plateaus are constructed in only a few million years, the measured span of more than 30 m.y. suggests that the model should be embraced with some caution—at least in its details. Presumably, the bulk of such an edifice can be built rapidly, but if it remains within the vicinity of hot plume mantle, some volcanism may continue for an extended period or be reactivated by favorable plate-tectonic developments.

Isotopically and chemically, the Leg 130 rocks are very similar to those previously analyzed from Site 289 and southern Malaita, but distinct from 0-70 Ma lavas of the Louisville hotspot's trace, the Louisville Ridge. Accepting a Louisville source for the plateau, the isotopic data suggest a shift of hotspot compositions to significantly higher $^{206}Pb/^{204}Pb$ sometime between 70 and 90 Ma, possibly coincident with the transition from the plume-head to plume-tail stage of the hotspot's development. A similar loss of early low $^{206}Pb/^{204}Pb$ compositions is seen in several other plateau-hotspot or continental flood basalt-hotspot systems; it could reflect a loss of low $^{206}Pb/^{204}Pb$ mantle in (or in contact with) the plume, or an inability of at least some off-ridge plume-tails to melt refractory low $^{206}Pb/^{204}Pb$ material.

Finally, our results neither particularly support nor discredit the recent "superplume" hypothesis [Larson, 1991]. However, because our age data and recent plate reconstructions generally bolster a connection between the plateau and the present Louisville hotspot, we see no need, at present, to appeal to an ocean-wide superplume to explain the Ontong Java Plateau.

Acknowledgments. We are grateful to Dan McKenzie for invaluable help with his inversion procedure. Peter Floyd supported M. Storey during the early part of the study, David Emley assisted with the XRF work, and Andy Saunders provided the NAA data. Nancy Hulbirt illustrated the figures and Carol Koyanagi got the text and tables into respectable shape. We thank official reviewers Bill White and Pat Castillo; Andy Saunders and Clive Neal also provided helpful comments. This research was supported by NSF grant ODP-TAMRF 20477, by the University of Keele, and the NERC. SOEST Contribution No. 2945.

REFERENCES

Baksi, A. K., Reevaluation of the timing and duration of extrusion of the Imnaha, Picture Gorge and Grande Ronde Basalt, *Geol. Soc. Am. Spec. Pap., 239*, 105-111, 1989.

Bargar, K. E., and E. D. Jackson, Calculated volumes of individual shield volcanoes along the Hawaiian-Emperor chain, *J. Res. USGS, 2*, 545-551, 1974.

Barling, J., and S. L. Goldstein, Extreme isotopic variations in Heard Island lavas and the nature of mantle reservoirs, *Nature, 348,* 59-62, 1990.

Bender, J. F., F. N. Hodges, and A. E. Bence, Petrogenesis of basalts from the Project FAMOUS area: experimental study from 0 to 15 Kbars, *Earth Planet. Sci. Lett., 41,* 277-302, 1978.

Campbell, I. H., and R. W. Griffiths, Implications of mantle plume structure for the evolution of flood basalts, *Earth Planet. Sci. Lett., 99,* 79-93, 1990.

Castillo, P., R. Batiza, and R. J. Stern, Petrology and geochemistry of Nauru Basin igneous complex; large volume, off-ridge eruptions of MORB-like basalt during the Cretaceous, *Initial Rep. Deep Sea Drill. Proj., 89,* 555-576, 1986.

Castillo, P. R., R. W. Carlson, and R. Batiza, Origin of Nauru Basin igneous complex: Sr, Nd, and Pb isotopes and REE constraints, *Earth Planet. Sci. Lett., 103,* 200-213, 1991.

Chaffey, D. J., R. A. Cliff, and B. M. Wilson, Characteristics of the St. Helena magma source, in *Magmatism in the Ocean Basins, Geol. Soc. London Spec. Publ., 42,* edited by A. D. Saunders and M. J. Norry, pp. 257-276, 1989.

Cheng, Q., K.-H. Park, J. D. Macdougall, A. Zindler, G. W. Lugmair, J. Hawkins, P. Lonsdale, and H. Staudigel, Isotopic evidence for a hot spot origin of the Louisville seamount chain, in *Seamounts, Islands, and Atolls, Am. Geophys. Union Monogr., 43,* edited by B. Keating, P. Fryer, R. Batiza, and G. Boehlert, pp. 283-296, AGU, Washington, D. C., 1987.

Clague, D. A., Petrology of basaltic and gabbroic rocks dredged from the Danger Island Troughs, Manihiki Plateau, *Initial Rep. Deep Sea Drill. Proj., 33,* 891-911, 1976.

Courtillot, V., G. Feraud, H. Maluski, D. Vandamme, M. G. Moreau, and J. Besse, Deccan flood basalts and the Cretaceous/Tertiary boundary, *Nature, 333,* 843-846, 1988.

Dalrymple, G. B., M. A. Lanphere, and D. A. Clague, Conventional and ^{40}Ar-^{39}Ar K-Ar ages of volcanic rocks from Ojin (Site 430), Nintoku (Site 432) and Suiko (Site 433) seamounts and the chronology of volcanic propagation along the Hawaiian-Emperor chain, *Initial Rep. Deep Sea Drill. Proj., 55,* 659-676, 1981.

Dick, H.J.B., and T. Bullen, Chromium spinel as a petrogenetic indicator in abyssal and alpine-type peridotites and spatially-associated lavas, *Contrib. Mineral. Petrol., 86,* 54-76, 1984.

Dick, H.J.B., R. L. Fisher, and W. B. Bryan, Mineralogic variability of the uppermost mantle along mid-ocean ridges, *Earth Planet. Sci. Lett., 69,* 88-106, 1984.

Duncan, R. A., The age distribution of volcanism along aseismic ridges in the eastern Indian Ocean, *Proc. Ocean Drill. Proj. Results, 121,* 507-518, 1991.

Duncan, R. A., and D. G. Pyle, Rapid eruption of the Deccan flood basalts at the Cretaceous/Tertiary boundary, *Nature, 333,* 841-843, 1988.

Floyd, P. A., Petrology and geochemistry of oceanic intraplate sheet-flow basalts, Nauru Basin, Deep Sea Drilling Project, Leg 89, *Initial Rep. Deep Sea Drill. Proj., 89,* 471-497, 1986.

Furumoto, A. S., J. P. Webb, M. E. Odegard, and D. M. Hussong, Seismic studies on the Ontong Java Plateau, 1970, *Tectonophysics, 34,* 71-90, 1976.

Govindaraju, K., 1989 compilation of working values and sample descriptions for 272 geostandards, *Geostandards Newsletter, 13,* Spec. Issue, 1-113, 1989.

Hanan, B. B., and J.-G. Schilling, Easter microplate evolution: Pb isotope evidence, *J. Geophys. Res., 94,* 7432-7448, 1989.

Hanson, G. N., Rare-earth elements in petrogenetic studies of igneous systems, *Annu. Rev. Earth Planet. Sci., 8,* 371-406, 1980.

Harland, W. B., R. L. Armstrong, A. V. Cox, L. E. Craig, A. G. Smith, and D. G. Smith, *A Geologic Time Scale 1989,* Cambridge Univ. Press, Cambridge, 1990.

Hart, S. R., Heterogeneous mantle domains: signatures, genesis and mixing chronologies, *Earth Planet. Sci. Lett., 90,* 273-296, 1988.

Hawkesworth, C. J., M.S.M. Mantovani, P. N. Taylor, and Z. Palacz, Evidence from the Parana of South Brazil for a continental contribution to Dupal basalts, *Nature, 322,* 356-359, 1986.

Hawkesworth, C. J., P. D. Kempton, N. W. Rogers, R. M. Ellam, and P. W. van Calsteren, Continental mantle lithosphere, and shallow level enrichment processes in the Earth's mantle, *Earth Planet. Sci. Lett., 96,* 256-268, 1990.

Hawkins, J. W., P. F. Lonsdale, and R. Batiza, Petrologic evolution of the Louisville seamount chain, in *Seamounts, Islands, and Atolls, Am. Geophys. Union Monogr., 43,* edited by B. Keating, P. Fryer, R. Batiza, and G. Boehlert, pp. 235-254, AGU, Washington, D. C., 1987.

Hofmann, A. W., K. P. Jochum, M. Seufert, and W. M. White, Nb and Pb in oceanic basalts: new constraints on mantle evolution, *Earth Planet. Sci. Lett., 79,* 33-45, 1986.

Houtz, R. E., D. E. Hayes, and R. G. Markl, Kerguelen Plateau bathymetry, sediment distribution and crustal structure, *Marine Geol., 25,* 95-130, 1977.

Hughes, G. W., and C. C. Turner, Upraised Pacific Ocean floor, southern Malaita, Solomon Islands, *Geol. Soc. Am. Bull., 88,* 412-424, 1977.

Hussong, D. M., L. K. Wipperman, and L. W. Kroenke, The crustal structure of the Ontong Java and Manihiki oceanic plateaus, *J. Geophys. Res., 84,* 6003-6010, 1979.

Jackson, E. D., K. E. Bargar, B. P. Fabbi, and C. Heropoulos, Petrology of basaltic rocks drilled on Leg 33 of the Deep Sea Drilling Project, *Initial Rep. Deep Sea Drill. Proj., 33,* 571-630, 1976.

Jenner, G. A., H. P. Longerich, S. E. Jackson, and B. J. Fryer, ICP-MS–a powerful tool for high-precision trace-element analysis in Earth sciences: evidence from analysis of selected U.S.G.S. reference samples, *Chem. Geol., 83,* 133-148, 1990.

Johnson, K. T., H.J.B. Dick, and N. Shimizu, Melting in the oceanic upper mantle: an ion microprobe study of diopsides in abyssal peridotites, *J. Geophys. Res., 95,* 2661-2678, 1990.

Klein, E. M., and C. H. Langmuir, Global correlations of ocean ridge basalt chemistry with axial depth and crustal thickness, *J. Geophys. Res., 92,* 8089-8115, 1987.

Klein, E. M., and C. H. Langmuir, Local versus global variations in ocean ridge basalt composition: a reply, *J. Geophys. Res., 94,* 4241-4252, 1989.

Klein, E. M., C. H. Langmuir, and H. Staudigel, Geochemistry of basalts from the Southeast Indian Ridge, 115°E-138°E, *J. Geophys. Res., 96,* 2089-2107, 1991.

Kroenke, L. W., Geology of the Ontong Java Plateau, *Haw. Inst. Geophys. Rep., 72-75,* 119 pp, 1972.

Kroenke, L. W., W. H. Berger, T. R. Janecek, et al., *Proc. Ocean Drill. Progr., Init. Rep., 130,* Ocean Drilling Program, College Station, TX, 1991.

Lanphere, M. A., and G. B. Dalrymple, The use of ^{40}Ar-^{39}Ar data in evaluation of disturbed K-Ar systems, in *Short Papers of the Fourth International Conference, Geochronology, Cosmochronology, Isotope Geology, U.S. Geol. Surv. Open-File Rept., 78-701,* edited by R. E. Zartman, pp. 241-243, 1978.

Larson, R. L., S. O. Schlanger et al., *Initial Rep. Deep Sea Drill. Proj., 61,* U.S. Govt. Printing Office, Washington, D.C., 1981.

Larson, R. L., Latest pulse of the Earth: evidence for a mid-Cretaceous super plume, *Geology, 19,* 547-550, 1991.

Leclaire, L., Y. Bassias, M. Denis-Clocchiatti, H. Davies, I. Gautier, B. Gensous, P. J. Giannesini, P. Patriat, J. Segoufin, M. Tesson, and J. Wannesson, Lower Cretaceous basalt and sediments from the Kerguelen Plateau, *Geo-Mar. Lett., 7,* 169-176, 1987.

Lonsdale, P., Geography and history of the Louisville hotspot chain in the Southwest Pacific, *J. Geophys. Res., 93,* 3078-3104, 1988.

McKenzie, D., and M. J. Bickle, The volume and composition of melt generated by extension of the lithosphere, *J. Petrol., 29,* 625-679, 1988.

McKenzie, D., and R. K. O'Nions, Partial melt distributions from inversion of rare earth element concentrations, *J. Petrol.,* 32, 1021-1091, 1991.

Mahoney, J. J., An isotopic survey of Pacific oceanic plateaus: implications for their nature and origin, in *Seamounts, Islands, and Atolls, Am. Geophys. Union Monogr., 43,* edited by B. Keating, P. Fryer, R. Batiza, and G. Boehlert, pp. 207-220, AGU, Washington, D.C., 1987.

Mahoney, J. J., and K. J. Spencer, Isotopic evidence for the origin of the Manihiki and Ontong Java oceanic plateaus, *Earth Planet. Sci. Lett., 104,* 196-210, 1991.

Mahoney, J. J., J. H. Natland, W. M. White, R. Poreda, S. H. Bloomer, R. L. Fisher, and A N. Baxter, Isotopic and geochemical provinces of the western Indian Ocean spreading centers, *J. Geophys. Res., 94,* 4033-4053, 1989.

Mahoney, J., C. Nicollet, and C. Dupuy, Madagascar basalts: tracking oceanic and continental sources, *Earth Planet. Sci. Lett., 104,* 350-363, 1991.

Mehl, K. W., P. R. Bitschene, H.-U. Schminke, and J. Hertogen, Composition, alteration, and origin of the basement lavas and volcaniclastic rocks at Site 738, southern Kerguelen Plateau, *Proc. Ocean Drill. Progr., Sci. Results, 119,* 299-321, 1991.

Morgan, W. J., Hotspot tracks and the opening of the Atlantic and Indian oceans, in *The Sea, Vol. 7,* edited by C. Emiliani, pp. 443-487, Wiley, New York, 1981.

Musgrave, R.J., Palaeomagnetism and tectonics of Malaita, Solomon Islands, *Tectonics, 9,* 735-760, 1990.

Neal, C. R., and J. P. Davidson, An unmetasomatized source for the Malaitan alnoite (Solomon Islands): petrogenesis involving zone refining, megacryst fractionation, and assimilation of oceanic lithosphere, *Geochim. Cosmochim. Acta, 53,* 1975-1990, 1989.

Niu, Y., and R. Batiza, Chemical variations at fast- and slow-spreading mid-ocean ridges: a diapiric melting model, *J. Geophys. Res.,* in press, 1992.

O'Connor, J. M., W. M. White, D. M. Christie, and S. E. Sichel, South Atlantic hotspot volcanism: 2. isotopic variability in response to mixing of three mantle components, *Earth Planet. Sci. Lett.,* in press, 1992.

Oskarsson, N., S. Steinthorsson, and G. E. Sigvaldason, Icelandic geochemical anomaly: origin, volcanotectonics, chemical fractionation, and isotope evolution of the crust, *J. Geophys. Res.,* 90, 10011-10025, 1985.

Packham, G., and J. E. Andrews, Results of Leg 30 and the geological history of the southwest Pacific arc and marginal sea complex, *Initial Rep. Deep Sea Drill. Proj., 30,* 691-706, 1975.

Paul, D. K., and P. J. Potts, Rare-earth abundances and origin of some Indian lamprophyres, *Geol. Mag., 118,* 393-399, 1981.

Pringle, M., Radiometric ages of basaltic basement recovered at Sites 800, 801, and 802, ODP Leg 129, Western Pacific Ocean, *Proc. Ocean Drill. Progr., Sci. Results, 129,* in press, 1992.

Prinzhofer, A., E. Lewin, and C. J. Allègre, Stochastic melting of the marble cake mantle: evidence from local study of the East Pacific Rise at 12°50'N, *Earth Planet. Sci. Lett., 92,* 189-206, 1989.

Rea, D. K., and T. L. Vallier, Cretaceous volcanic episodes in the Pacific, *Geol. Soc. Am. Bull., 94,* 1430-1437, 1987.

Renne, P. R., and A. R. Basu, Rapid eruption of the Siberian Traps flood basalts at the Permo-Triassic boundary, *Science, 253,* 176-179, 1991.

Richards, M. A., R. A. Duncan, and V. Courtillot, Flood basalts and hot-spot tracks: plume heads and tails. *Science, 246,* 103-107, 1989.

Salters, V.J.M., M. Storey, J. H. Sevigny, and H. Whitechurch, Trace element and isotopic characteristics of Kerguelen-Heard Plateau basalts, *Proc. Ocean. Drill. Progr., Sci. Results, 120,* 55-62, 1992.

Samson, S. D., and E. C. Alexander, Calibration of the interlaboratory ^{40}Ar-^{39}Ar dating standard, Mmhb-1, *Chem. Geol., 66,* 27-34, 1987.

Saunders, A. D., Geochemistry of basalts from the Nauru Basin, Deep Sea Drilling Project Legs 61 and 89: implications for the origin of oceanic flood basalts, *Initial Rep. Deep Sea Drill. Proj., 89,* 499-518, 1986.

Saunders, A. D., M. Storey, I. L. Gibson, P. Leat, J. Hergt, and R. N. Thompson, Chemical and isotopic constraints on the origin of basalts from Ninetyeast Ridge, Indian Ocean: results from DSDP

Legs 22 and 26 and ODP Leg 121, *Proc. Ocean. Drill. Progr., Sci. Results, 121,* 559-590, 1991.

Sharman, G. F., and J. Mammerickx, Eastern boundary of the Manihiki Plateau; a propagating rift site, *Eos, 71,* 1668, 1990.

Staudigel, H., K.-H. Park, M. Pringle, J. L. Rubenstone, W.H.F. Smith, and A. Zindler, The longevity of the South Pacific isotope and thermal anomaly, *Earth Planet. Sci. Lett., 102,* 24-44, 1991.

Stoeser, D. B., Igneous rocks from Leg 30 of the Deep Sea Drilling Project, *Initial Rep. Deep Sea Drill. Proj., 30,* 401-444, 1975.

Storey, M., A. D. Saunders, J. Tarney, I. L. Gibson, M. J. Norry, M. F. Thirlwall, P. Leat, R. N. Thompson, and M. A. Menzies, Contamination of Indian Ocean asthenosphere by the Kerguelen-Heard mantle plume, *Nature, 338,* 574-576, 1989.

Storey, M., J. J. Mahoney, L. W. Kroenke, and A. D. Saunders, Are oceanic plateaus sites of komatiite formation? *Geology, 19,* 376-379, 1991.

Storey, M., R. Kent, A. D. Saunders, V. J. Salters, J. Hergt, H. Whitechurch, J. H. Sevigny, M. F. Thirlwall, P. Leat, N. C. Ghose, and M. Gifford, Lower Cretaceous volcanic rocks along continental margins and their relationship to the Kerguelen Plateau, *Proc. Ocean Drill. Progr., Sci. Results, 120,* 33-54, 1992.

Sun, S.-s., and W. F. McDonough, Chemical and isotopic systematics of oceanic basalts: implications for mantle composition and processes, in *Magmatism in the Ocean Basins, Geol. Soc. London Spec. Publ., 42,* edited by A. D. Saunders and M. J. Norry, pp. 313-345, 1989.

Tarduno, J. A., W. V. Sliter, L. W. Kroenke, M. Leckie, H. Mayer, J. J. Mahoney, R. Musgrave, M. Storey, and E. L. Winterer, Rapid formation of Ontong Java Plateau by Aptian mantle plume volcanism, *Science, 254,* 399-403, 1991.

Todt, W., R. A. Cliff, A. Hanser, and A. W. Hofmann, $^{202}Pb+^{205}Pb$ double spike for lead isotopic analyses, *Terra Cognita, 4,* 209, 1984.

Tokuyama, H., and R. Batiza, Chemical composition of igneous rocks and origin of the sill and pillow-basalt complex of Nauru Basin, southwest Pacific, *Initial Rep. Deep Sea Drill. Proj., 61,* 673-687, 1981.

Walker, D. A., and I. McDougall, $^{40}Ar/^{39}Ar$ and K-Ar dating of altered glassy volcanic rocks: the Dabi Volcanics., P.N.G., *Geochim. Cosmochim. Acta, 46,* 2181-2190, 1982.

Weaver, J. S., and C. H. Langmuir, Calculation of phase equilibrium in mineral-melt systems, *Computers and Geosci., 16,* 1-19, 1990.

Weis, D., Y. Bassias, I. Gautier, and J.-P. Mennessier, Dupal anomaly in existence 115 Ma ago: evidence from isotopic study of the Kerguelen Plateau (South Indian Ocean), *Geochim. Cosmochim. Acta, 53,* 2125-2131, 1989.

Weis, D., and F. A. Frey, Isotope geochemistry of Ninetyeast Ridge basement basalts: Sr, Nd, and Pb evidence for the involvement of the Kerguelen hot spot, *Proc. Ocean Drill. Progr., Sci. Results, 121,* 591-610, 1991.

West, H. B., D. C. Gerlach, W. P. Leeman, and M. O. Garcia, Isotope constraints on the origin of Hawaiian lavas from the Maui Volcanic Complex, Hawaii, *Nature, 330,* 216-220, 1987.

White, R., and D. McKenzie, Magmatism at rift zones: the generation of volcanic continental margins and flood basalts, *J. Geophys. Res., 94,* 7685-7729, 1989.

White, W. M., A. W. Hofmann, and H. Puchelt, Isotope geochemistry of Pacific mid-ocean ridge basalt, *J. Geophys. Res., 92,* 4881-4893, 1987.

White, W. M., M. M. Cheatham, and R. A. Duncan, Isotope geochemistry of Leg 115 basalts and inferences on the history of the Réunion mantle plume, *Proc. Ocean Drill. Progr., Sci. Results, 115,* 53-61, 1990.

Whitechurch, H., R. Montigny, J. Sevigny, M. Storey, and V. J. Salters, K-Ar and $^{40}Ar/^{39}Ar$ ages of central Kerguelen Plateau basalts, *Proc. Ocean Drill Progr., Sci. Results, 120,* 71-78, 1992.

Winterer, E. L., P. F. Lonsdale, J. L. Matthews, and B. R. Rosendahl, Structure and acoustic stratigraphy of the Manihiki Plateau, *Deep-Sea Res., 21,* 793-814, 1974.

Yan, C. Y., and L. W. Kroenke, A plate-tectonic reconstruction of the Southwest Pacific, 100-0 Ma, *Proc. Ocean Drill. Progr., Sci. Results, 130,* in press, 1992.

York, D., Least squares fitting of a straight line with correlated errors, *Earth Planet. Sci. Lett., 5,* 320-324, 1969.

R. A. Duncan, College of Oceanography, Oregon State University, Corvallis, OR 97331.

J. J. Mahoney and K. J. Spencer, School of Ocean and Earth Science and Technology, University of Hawaii, Honolulu, HI 96822.

M. Pringle, U.S. Geological Survey, Menlo Park, CA 94025.

M. Storey*, Department of Geology, University of Keele, Staffordshire STF 5BH, United Kingdom.

---

*Presently at: Department of Geology, University of Leicester, Leicester LE1 7RH, United Kingdom.

Geochronological Constraints on a Possible Hot Spot Origin for Hess Rise and the Wentworth Seamount Chain

MALCOLM S. PRINGLE AND G. BRENT DALRYMPLE

U.S. Geological Survey, Menlo Park, CA

New $^{40}Ar/^{39}Ar$ age data are consistent with plume initiation/hot spot models for the origin of the Late Cretaceous Hess Rise and Wentworth Seamount chain, north-central Pacific Ocean. $^{40}Ar/^{39}Ar$ age spectra on two samples of highly altered trachyte from DSDP Site 465A are discordant, revealing only minimum ages of ca. 90-94 Ma. A better minimum age estimate for the formation of SE Hess Rise is the late Albian age of the overlying sedimentary rocks at Site 465A, ca. 98-100 Ma. Two unnamed seamounts on the Wentworth Seamount chain trending SE from Hess Rise have concordant $^{40}Ar/^{39}Ar$ whole rock and laser mineral fusion ages of 93.5 ±0.6 Ma and 86.4 ±0.6 Ma. The age progression along this lineament, ca. 70 km/m.y., is consistent with rates found for other copolar Late Cretaceous hot spot tracks in the Musicians Seamounts and northern Line Islands. Samples from Wentworth Seamount itself are too altered to yield reliable $^{40}Ar/^{39}Ar$ ages, but are consistent with an age of about 90 Ma predicted by the observed age progression. The Hess Rise/Wentworth Seamount province deviates from the typical plume initiation/hot spot model in two important ways. First, no extensions of the Wentworth Seamount chain have been found south of the youngest seamount dated, and the hot spot track is only 14 m.y. long. Second, $^{40}Ar/^{39}Ar$ analysis of two plagioclase separates from alkalic basalts dredged from a seamount on SE Hess Rise have concordant ages of 92 and 87 Ma. Thus, active volcanism on Hess Rise continued at least 12 m.y. after the initial formation of the plateau, after the southern Hess Rise had begun to submerge, and after the site had drifted 800 km to the northwest of the proposed hot spot. Such continuing volcanism also has been observed for Ontong Java Plateau and may be characteristic of oceanic flood basalt provinces, even if they form by a mechanism consistent with the plume initiation model. Alternative hypotheses for the origin of Hess Rise and the Wentworth seamount chain cannot be ruled out.

INTRODUCTION

According to the hot spot hypothesis, age progressive linear island and seamount chains form as the lithosphere moves over deep melting anomalies, or hot spots. If these melting anomalies are essentially fixed with respect to each other, then the tracks of coeval hot spots form copolar small circles and have the same angular rate of volcanic propagation [Morgan, 1972a; 1972b]. Such tracks have become an important tool of in the development of plate tectonic models -- for a specific time period, both the direction and rate of absolute motion of a particular plate can be estimated.

Hess Rise is a prominent, yet complex, oceanic plateau covering 800,000 km^2 in the central North Pacific Ocean (Figure 1). Its origin is uncertain. Vallier et al. [1980] noted that the age difference between the oldest sediments cored at Deep Sea Drilling Project (DSDP) Site 464 in the north (late Aptian to early Albian) and at DSDP Site 465 in the south (late Albian to Cenomanian) is consistent with age progressive volcanism associated with a hot spot, but that such a hypothesis needed to be tested. Subsequently, Vallier et al. [1981] concluded that Hess Rise formed at the Pacific-Farallon spreading center during Early to mid-Cretaceous times; Mammerickx and Sharman [1988] preferred a model where most of Hess Rise formed just west (on the Pacific side) of the ridge crest.

The Wentworth Seamount Chain extends SE from southern Hess Rise (Figure 1). Jarrard and Clague [1977] and Epp [1978] proposed that the Wentworth chain is one of three parallel hot spot tracks north of the Hawaiian ridge and south of the Mendocino Fracture Zone (Figure 2). Epp used the strike of these chains to calculate a Late Cretaceous rotation pole for the Pacific plate, and also noted that the northern Line Islands and the Emperor Seamounts north of 50° were copolar, i.e., presumably coeval, with these three chains. A lack of reliable age data prevented Epp [1978] from determining actual Pacific plate velocities for the Late Cretaceous, but the rotation pole that he determined, 36°N, 76°W, is still used to describe the direction of Pacific plate motion prior to about 70 Ma [e.g., Duncan and Clague, 1985; Henderson, 1985].

The Mesozoic Pacific: Geology, Tectonics, and Volcanism
Geophysical Monograph 77

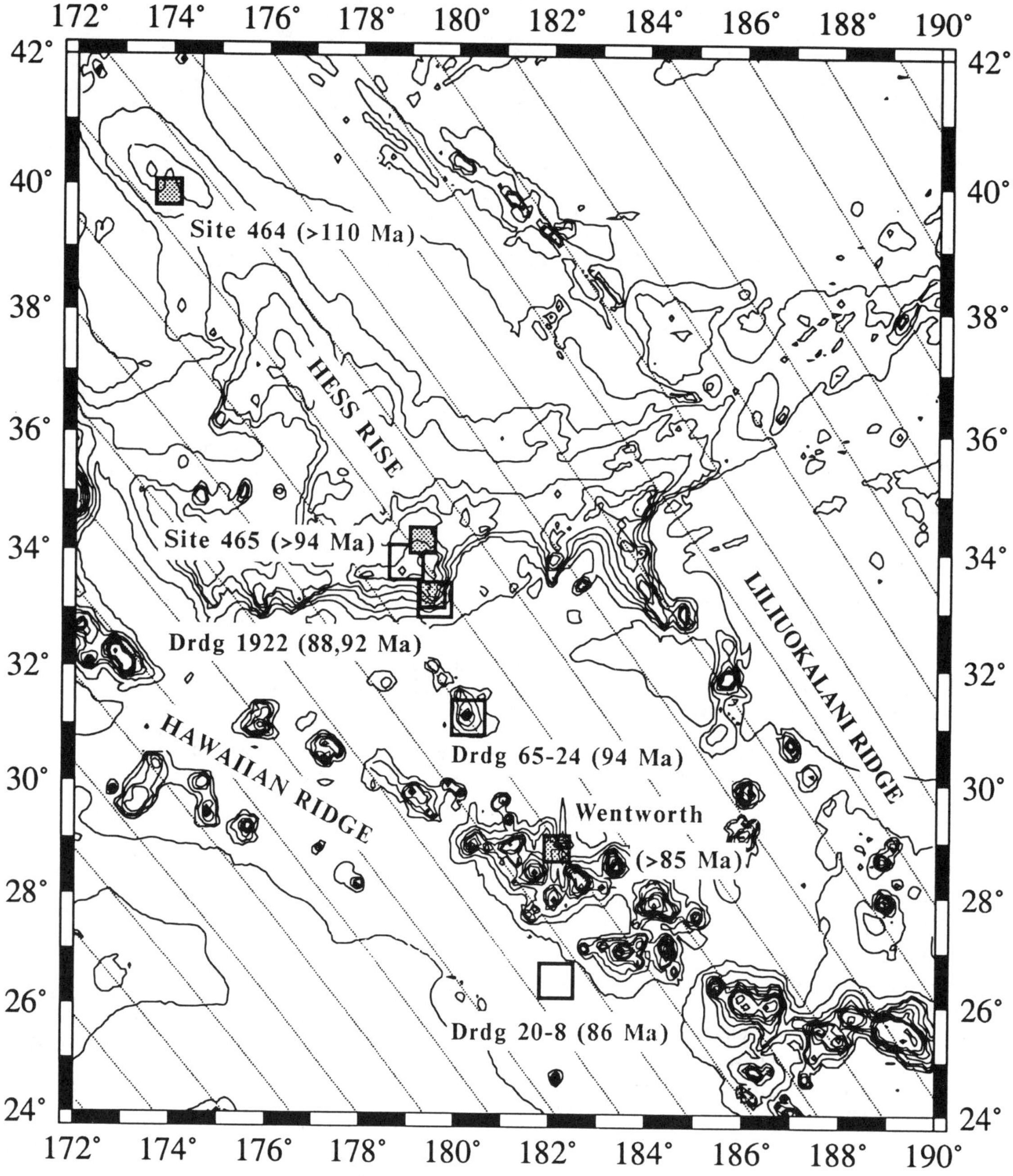

Figure 1. Location and ages of volcanic rocks recovered from Hess Rise and the Wentworth Seamount Chain. Large squares mark those sites for which reliable $^{40}Ar/^{39}Ar$ ages have been determined; small squares mark locations for which only minimum $^{40}Ar/^{39}Ar$ or fossil ages are available. Thin dotted lines are small circles, drawn 1° apart, around the Late Cretaceous Pacific rotation pole of 36°N, 76°W. Figure drawn using GMT mapping utilities of Wessel and Smith [1991].

Pringle [this volume] shows that age progressive volcanism occurred along the NW-SE trending chain in the Musicians Seamounts (Figure 2) from 96 Ma to 82 Ma. Schlanger et al. [1984] and Epp [1984] concluded that at least some of the Line Islands chain formed by age progresive, hot spot volcanism between 100 and 45 Ma. They estimated a Late Cretaceous Pacific plate/hot spot rotation rate of about 0.8-0.9 °/m.y. Pringle [this volume] combined the Musicians results with the older dates for the northern Line Islands from Schlanger et al. [1984] and calculated the best estimate of Pacific plate/hot spot rotation velocity for the period 96 to 82 Ma, 0.508 ±0.053 °/m.y.

The purpose of this study to determine the age of volcanism for samples collected from Hess Rise and the Wentworth Seamount Chain. The limited number and altered state of the

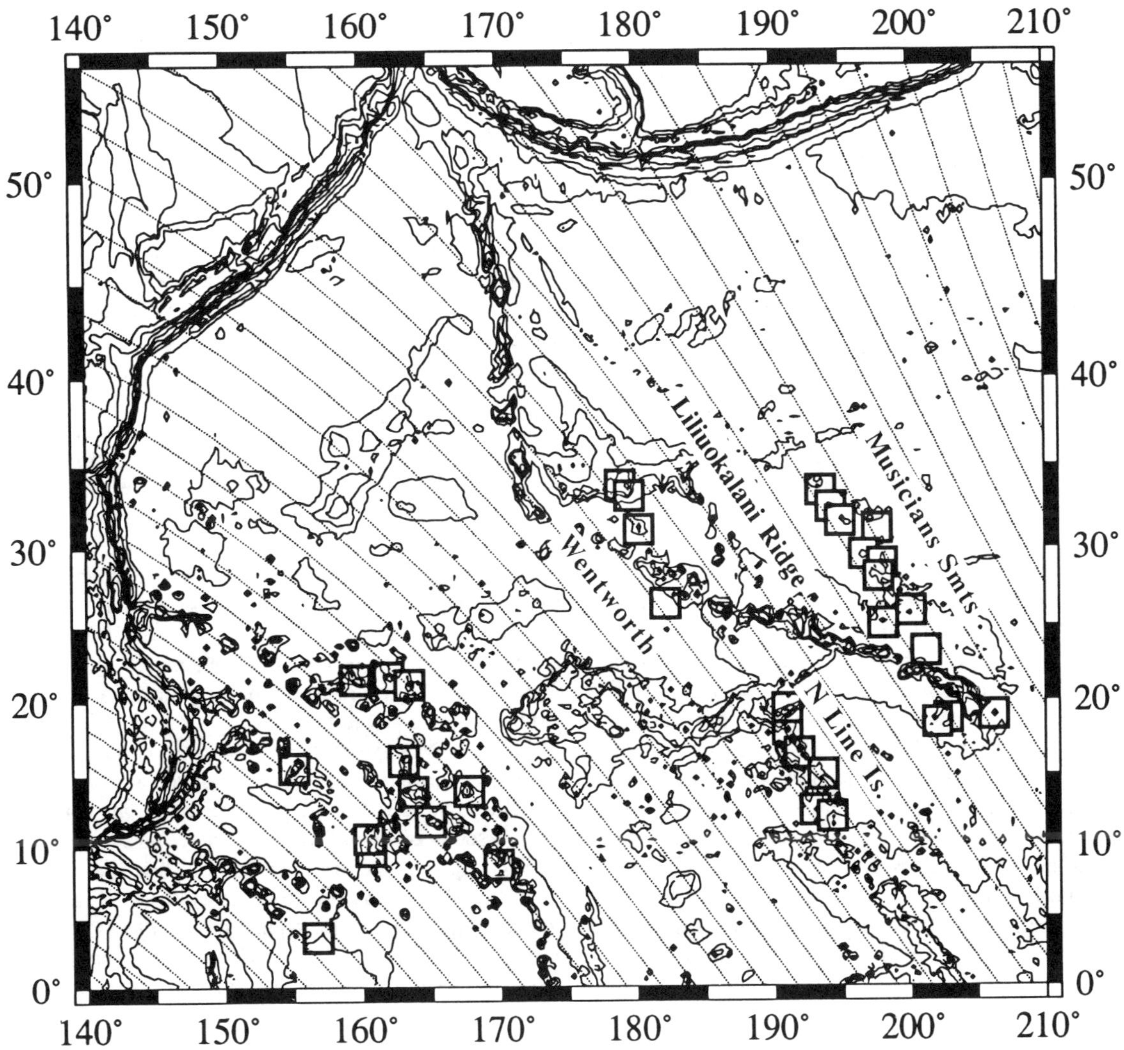

Figure 2. Location of all sites of volcanism dated between 80 Ma and 100 Ma in the north central Pacific Ocean (squares). Thin dotted lines are small circles, drawn 2° apart, around the Late Cretaceous Pacific rotation pole of 36°N, 76°W [Epp, 1978]. Late Cretaceous hot spot tracks should be parallel to these small circles; note that the Wentworth trend deviates more than the Musicians, Liliuokalani, and Line Islands trends. Data sources: this work; Pringle, this volume; Winterer et al. [this volume]; Lincoln et al. [this volume]; Mahoney et al. [this volume]; Schlanger et al. [1984], and Pringle, unpublished data. Figure drawn using GMT mapping utilities of Wessel and Smith [1991].

samples available constrains our ability to fully test the hot spot hypothesis for either feature. However, we will test (1) whether age progressive volcanism is found along these features, (2) if any age progression found is consistent with the Musicians and Line Islands results, and (3) whether the plateau and seamount volcanism on Hess Rise could mark the inception of volcanism along the Wentworth Seamount chain. We also compare our results to models of the origin of oceanic plateaus, and use them to place constraints on the origin of volcanism on Hess Rise.

SAMPLES STUDIED

Wentworth Seamount Chain

Clague and Dalrymple [1975] and Garcia et al. [1987] determined conventional K-Ar ages of 70 to 80 Ma for three seamounts along the Wentworth seamount chain. For reasons discussed below and more fully elsewhere (for example, Pringle, this volume), we regard conventional K-Ar ages of altered whole rock samples as minimum ages.

For Wentworth Seamount itself (Figure 1), Clague and Dalrymple [1975] determined a conventional K-Ar minimum age of 71 Ma for an altered basalt, sample STYX7 25-1. We re-sampled two other rocks from the same dredge haul which appeared more suitable for $^{40}Ar/^{39}Ar$ dating. Sample STYX7 25-2 is a holocrystalline, vesicular, aphyric basalt with less than 5% colorless clays in an intergranular groundmass. Sample STYX7 25-3 is a fine-grained, vesicular mugearite with 10% yellow-green clays after volcanic glass in a trachytic groundmass.

For an uncharted seamount south of the Hawaiian Ridge at 26°27'N, 177°52'W (Figure 1), Garcia et al. [1987] determined a conventional K-Ar minimum age of 74 Ma for two samples which we have re-analyzed, KK71 20-8-J and KK71 20-8-O. Both are sparsely plagioclase- and hornblende-phyric mugearites with less than a few percent clays in a fine-grained, holocrystalline, trachytic groundmass. We also separated the phenocryst phases from KK71 20-8-O and a similar sample KK71 20-8-E, which was too altered for whole rock analysis (groundmass up to 20% clays).

For a seamount north of the Hawaiian Ridge at 31°09'N, 179°45'W (Figure 1), Garcia et al. [1987] determined a conventional K-Ar minimum age of 77 Ma for two samples which we have re-analyzed. Sample KK71 65-24A is a fine-grained, vesicular basalt with 10-15% brown clays after glass in an intersertal groundmass. Sample KK71 65-24D is a fine-grained, vesicular hawaiite with 5-10% brown clays after glass(?) in a trachytic groundmass.

Hess Rise

Volcanic rock has been recovered at from only five sites on Hess Rise: DSDP Sites 464, 465, and 466 [Vallier, et al., 1981], and two dredges by the *R/V D. Mendeleev* from a seamount on the southeastern margin of Hess Rise, 80 km to the southeast of Site 465A (Figure 1). Site 464 on northern Hess Rise recovered only 16 cm of highly altered tholeiitic basalt beneath late Aptian to early Albian sedimentary rocks. Site 465A on southern Hess Rise recovered 24 m of altered trachyte, erupted in a subaerial or very shallow water environment, beneath upper Albian sediments. Site 466 on southern Hess Rise recovered a few alkalic basalt clasts within Campanian to Maastrichtian sedimentary rocks, but their stratigraphic relationship to the Hess Rise volcanism is ambiguous so we did not analyze them. The dredges by the *R/V D. Mendeleev* recovered alkalic basalt, including the two samples discussed below.

Only the trachytes from Site 465A were fresh enough to attempt $^{40}Ar/^{39}Ar$ whole rock incremental-heating. Two samples were selected from Site 465A. Sample 465A-43-2 72-74 cm appeared the least altered of the two, with a nearly holocrystalline, fine-grained trachytic groundmass and only a few percent of brown clays after glass. Sample 465A-45-2 107-109 cm had up to 10% brown clay after glass and some carbonate replacing the cores of plagioclase microphenocryts.

Two samples from *R/V D. Mendeleev* dredge DM23-1922 (33°10.5'N, 179°31.7'E) had plagioclase suitable for $^{40}Ar/^{39}Ar$ dating. Sufficient plagioclase was separated from one sample, DM23-1922-08, to try both incremental-heating with a conventional induction-heating system and laser heating with the new very small sample techniques. From the other sample, DM23-1922, we recovered only enough plagioclase for laser-heating.

K-AR GEOCHRONOLOGY

The K-Ar clock, especially the $^{40}Ar/^{39}Ar$ technique, has been widely used to date oceanic lavas, but care must be taken when applying the method to altered samples. Conventional K-Ar dating of untreated and acid treated whole rock samples and mineral separates, $^{40}Ar/^{39}Ar$ total-fusion analysis of individual mineral separates, and even groups of $^{40}Ar/^{39}Ar$ total-fusion ages of whole rock samples all are unable to consistently provide reliable age determinations [c.f., Pringle, this volume; Davis et al., 1989].

For example, Garcia et al. [1987] tried different acid treatments on whole rock samples in an attempt to remove the effects of alteration, including a light (2% HNO_3) leach of four samples that are re-examined here. They compared ages of acid treated samples to $^{40}Ar/^{39}Ar$ incremental-heating results from earlier studies of two samples dredged from the Hawaiian-Emperor Seamounts, and concluded that conventional K-Ar analysis of some acid-treated, altered whole rock samples gave reliable estimates of crystallization age. However, they also noted that another study of similar experiments on samples dredged for the Line Islands [Schlanger et al., 1984] had given inconsistent and inconclusive results. The principal problem is that, even if some acid treatments work for some rocks, there are no independent criteria to decide which are the valid ages. For the four samples from Garcia et al. [1987] that are re-examined in this study, the conventional K-Ar ages of the acid-treated whole samples underestimate the concordant $^{40}Ar/^{39}Ar$ ages by at least 20%. These earlier results are clearly only minimum age estimates and will not be discussed further.

The real advantage of using $^{40}Ar/^{39}Ar$ geochronology in deciphering the crystallization age of altered rocks is that accurate crystallization ages can be distinguished from unreliable ages using a series of *internal* tests for a set of apparent ages for a given sample. We use the criteria listed in Pringle [this volume], which were modified from those of Lanphere and Dalrymple [1978] and Dalrymple et al. [1980]. It may not be possible to determine precisely *why* one particular age is reliable while another is not, although alteration in the submarine environment clearly plays a role. However, one can test *whether* the K-Ar systematics of a particular sample are too disturbed to reveal a meaningful age.

The best ages for each of the three sites that have reliable crystallization ages are listed in Table 1. In general, the isochron age is preferred as the best estimate of each sample because it (1) combines both an estimate of analytical error and the scatter about the isochron in the final error estimate, and (2) makes no assumption about the composition of the initial non-radiogenic, or trapped, argon component.

Analytical Techniques

Whole rock samples were prepared as 6 mm cores drilled from the freshest sections of 1 cm thick, sawed slabs. Mineral separates were prepared using conventional heavy liquid and magnetic techniques, and further cleaned with appropriate acids and finally with distilled water in an ultrasonic bath. Plagioclase was cleaned with cold 5-10% HF for 5-8 minutes followed by 30 minutes in 2.5N HCl; hornblende was cleaned with the 2.5N HCl only. The two plagioclase separates DM23-1922 and DM23-1922-08 were treated with HF only.

TABLE 1. Summary of $^{40}Ar/^{39}Ar$ Ages of Samples from Hess Rise and the Wentworth Seamount Chain

	Plateau or Weighted TF Age				Isochron Age					
Sample	Material	% of ^{39}Ar	AGE	±1 SD	AGE	±1 SD	$^{40}Ar/^{36}Ar_i$	±1 SD	SUMS/(N-2)	N
UNNAMED SEAMOUUT	(33°10.5'N, 179°31.7'E)		BEST AGE[1]:		87.1 91.8	±0.6 Ma; ±0.6 Ma				
DM23-1922-08	Plag IH	64.8	87.6	±0.7 Ma	89.4	±3.1 Ma	255.1	±64.6	2.85	3
	Plag laser DG	96.5[2]	86.9	±0.5 Ma	87.0	±0.6 Ma	280.1	± 8.7	2.63	9
DM23-1922	Plag laser DG	93.8[2]	91.7	±0.6 Ma	91.7	±0.7 Ma	258.5	±39.0	0.59	5
	Plag laser TF	(100)[3]	89.5	±0.5 Ma	96.3	±5.7 Ma	276.4	±15.7	1.21	4
UNNAMED SEAMOUNT	(31°8.5'N, 179°44.8'W)		BEST AGE[1]:		93.4	±0.6 Ma				
KK71 65-24A	Whole Rock	59.8	93.7	±0.6 Ma	93.4	±0.6 Ma	302.4	± 3.5	2.71	4
		48.3	95.1	±0.5 Ma	94.8	±1.2 Ma	300.6	± 8.8	4.79	3
UNNAMED SEAMOUNT	(26°27.2'N, 177°51.6'W)		BEST AGE[1]:		86.4	±0.6 Ma				
KK71 20-8-J	Whole Rock	61.4	85.3	±0.5 Ma	86.9	±0.8 Ma	131.0	±66.3	2.27	5
KK71 20-8-O	Whole Rock	74.7	85.4	±0.5 Ma	85.8	±0.8 Ma	178.0	±188.1	2.82	6

[1]: See text for determination of "BEST AGE".
[2]: Laser-fusion analyses, including degassing steps (DG).
[3]: Total-fusion (TF) analyses, without any heating or degasing steps.

Whole rock samples, 0.5 to 0.8 g each, and mineral separates, up to 20 mg each wrapped in Al-foil, were sealed in air in quartz vials. They were irradiated for 24-30 hours in the core of the U.S.G.S. TRIGA reactor [Dalrymple et al., 1981] or 60 hours in the H-5 facility of the Ford Nuclear Reactor at the University of Michigan. Whole rock samples were analyzed as incremental-heating experiments using either (1) an all-metal resistance furnace extraction line and on-line Nier-type single-collector mass spectrometer, or (2) a glass-based induction-heater extraction line and off-line multiple-collector mass spectrometer. Mineral separates were analyzed using a 5W Ar-ion continuous laser extraction system and an ultraclean, ultrasensitive mass spectrometer [Dalrymple, 1989]. The reactor flux, J, was measured with the intralaboratory standards biotite SB-3 (162.9 Ma) for the whole rock samples and sanidine TCR 85G003 (27.88 Ma) for the mineral separates.

Errors given for the calculated ages of individual measurements are estimates of the standard deviation of analytical precision. $^{40}Ar/^{39}Ar$ incremental-heating data were reduced both as age spectra and isochrons. Incremental-heating plateau ages and mean mineral laser heating ages were calculated as weighted means, where weighting is by the inverse of the estimated variance [Taylor, 1982]. Isochron ages were calculated using the $^{40}Ar/^{36}Ar$ vs $^{39}Ar/^{36}Ar$ correlation and the York2 least-squares fit with correlated errors [York, 1969]. There are no significant differences between the results of this correlation calculation and the one based on the $^{36}Ar/^{40}Ar$ vs $^{39}Ar/^{40}Ar$ correlation [Dalrymple et al., 1988].

Wentworth Seamount Chain Results

For two samples from the seamount south of the Hawaiian Ridge at 26°27'N, 177°52'W (Dredge KK71-20-8), both incremental-heating experiments were concordant (Tables 1 and 2, Figure 3). The best age estimate for this seamount is the weighted mean of the two isochron ages, 86.4 ±0.6 Ma. Note, however, that both samples are slightly disturbed, i.e., not all of the steps lie on the plateau and the total gas age (Table 2) is slightly lower than the best age. Both age spectra (Figure 3) have a pattern of decreasing ages, with anomalously old ages for the lowest temperature steps and anomalously young ages for the highest temperature steps (especially for the fusion step with the lowest K/Ca ratio). This pattern is often seen in the incremental-heating age spectra of fine-grained, altered whole rock samples. It is usually attributed to argon redistribution and loss, especially the recoil of ^{39}Ar from higher-potassium sites degassed at low temperatures to lower-potassium steps degassed at high temperatures (see Pringle, this volume, for a more complete discussion). However, the mid-temperature gas release for many of these age spectra often meets the criteria for concordant ages used here, and we have previously assumed that

TABLE 2. Analytical Data for Incremental-Heating Experiments on Whole Rock Samples from Hess Rise and the Wentworth Seamount Chain

TEMP (°C)	$^{40}Ar/^{39}Ar$ (b)	$^{37}Ar/^{39}Ar$ (b)	$^{36}Ar/^{39}Ar$ (b)	$^{40}Ar_R$ (c)	$^{40}Ar_K$ (c)	$^{39}Ar_{Ca}$ (c)	$^{36}Ar_{Ca}$ (c)	K/Ca	^{39}Ar (%)	APPARENT AGE (a) (Ma)
465A-43-2 72-74		TRACHYTE CORE		J = 0.004749				Total gas age =		82.66 +- 0.45
500	30.37	0.13714	0.05956	42.1	0.0	0.0	0.1	3.57	1.6	106.23 +- 5.95
560	20.11	0.13596	0.03225	52.6	0.0	0.0	0.1	3.60	0.9	88.52 +- 1.64
620	13.504	0.13278	0.014796	67.7	0.0	0.0	0.2	3.69	5.8	76.63 +- 0.55
680	11.032	0.17559	0.005718	84.8	0.1	0.0	0.8	2.79	9.2	78.39 +- 0.49
740	11.387	0.2170	0.003599	90.8	0.1	0.0	1.6	2.26	9.8	86.44 +- 0.52
800	12.631	0.2306	0.004980	88.4	0.0	0.0	1.2	2.12	14.5	93.27 +- 0.56
860	13.177	0.2037	0.006356	85.8	0.0	0.0	0.8	2.41	11.9	94.38 +- 0.57
950	11.644	0.17851	0.003843	90.3	0.1	0.0	1.2	2.74	12.9	87.93 +- 0.52
1050	10.472	0.2349	0.003846	89.3	0.1	0.0	1.6	2.09	13.5	78.37 +- 0.65
1200	9.162	0.2493	0.002548	91.9	0.1	0.0	2.6	1.97	9.4	70.77 +- 0.43
FUSE	11.315	0.3313	0.012904	66.5	0.1	0.0	0.7	1.48	10.3	63.33 +- 0.45
465A-45-2 107-109		TRACHYTE		J = 0.004659				Total gas age =		83.53 +- 0.42
500	15.909	0.07881	0.016256	69.8	0.0	0.0	0.1	6.22	5.2	91.01 +- 0.49
600	12.013	0.09634	0.006468	84.1	0.0	0.0	0.4	5.09	5.4	82.99 +- 0.42
660	10.497	0.10348	0.003017	91.5	0.1	0.0	0.9	4.73	6.2	79.00 +- 0.36
720	11.007	0.14648	0.002724	92.7	0.1	0.0	1.4	3.34	15.3	83.82 +- 0.28
800	11.880	0.18532	0.003000	92.6	0.0	0.0	1.6	2.64	16.3	90.19 +- 0.29
870	11.970	0.16410	0.003016	92.6	0.0	0.0	1.4	2.99	13.7	90.86 +- 0.30
960	11.202	0.14120	0.003238	91.5	0.1	0.0	1.2	3.47	10.0	84.17 +- 0.31
1060	10.302	0.14707	0.001672	95.3	0.1	0.0	2.3	3.33	12.1	80.66 +- 0.28
1200	9.541	0.17114	0.002038	93.8	0.1	0.0	2.2	2.86	11.0	73.68 +- 0.27
FUSE	10.038	0.3550	0.006949	79.8	0.1	0.0	1.3	1.38	4.9	66.08 +- 0.43
KK71 65-24-D		WR CORE		J = 0.004000				Total gas age =		73.72 +- 0.37
450	25.55	0.6297	0.05002	42.2	0.1	0.0	0.3	0.778	5.7	76.17 +- 0.61
500	12.494	0.7912	0.002614	94.0	0.3	0.1	7.6	0.619	7.2	82.84 +- 0.31
520	12.805	0.8208	0.002033	95.5	0.3	0.1	10.1	0.597	7.1	86.19 +- 0.31
550	12.458	0.7241	0.001593	96.3	0.3	0.0	11.4	0.676	4.2	84.63 +- 0.40
580	12.382	0.6673	0.002812	93.4	0.3	0.0	6.0	0.734	3.9	81.60 +- 0.41
620	12.090	0.5783	0.003422	91.7	0.3	0.0	4.2	0.847	4.4	78.29 +- 0.38
660	11.512	0.4807	0.002528	93.5	0.3	0.0	4.8	1.02	3.9	76.06 +- 0.41
720	11.356	0.4255	0.002425	93.6	0.3	0.0	4.4	1.15	8.1	75.16 +- 0.28
800	11.008	0.4047	0.001553	95.8	0.3	0.0	6.5	1.21	18.4	74.53 +- 0.23
900	9.775	1.1142	0.001677	95.4	0.4	0.1	16.7	0.439	32.3	66.10 +- 0.20
1000	9.450	4.313	0.003733	91.3	0.4	0.3	29.0	0.113	4.7	61.39 +- 0.35
KK71 65-24-D		WR CORE		J = 0.004895				Total gas age =		74.99 +- 0.39
500	51.46	0.3694	0.13555	22.2	0.0	0.0	0.1	1.33	1.0	98.19 +- 2.31
600	26.41	0.4600	0.05895	34.1	0.0	0.0	0.2	1.06	3.8	77.97 +- 0.89
680	11.346	0.5835	0.007583	80.6	0.1	0.0	1.8	0.839	6.7	78.98 +- 0.40
800	10.744	0.5873	0.003195	91.5	0.1	0.0	4.3	0.834	9.8	84.85 +- 0.33
860	10.341	0.4803	0.004177	88.3	0.1	0.0	2.7	1.02	9.0	78.93 +- 0.33
950	10.024	0.3748	0.004372	87.3	0.1	0.0	2.0	1.31	6.2	75.70 +- 0.40
1050	9.392	0.3622	0.002298	93.0	0.1	0.0	3.7	1.35	21.5	75.53 +- 0.25
1200	8.934	0.4384	0.002250	92.8	0.1	0.0	4.6	1.12	29.4	71.81 +- 0.24
FUSE	8.484	3.642	0.003953	89.1	0.1	0.2	21.6	0.134	12.7	65.73 +- 0.26

(a) Lamda E = $0.581 \times 10^{-10} yr^{-1}$, lamda B = $4.962 \times 10^{-10} yr^{-1}$. Errors are s.d. of analytical precision.

(b) Corrected for ^{37}Ar and ^{39}Ar decay, 35.1 days and 259 years, respectively.

(c) Subscripts indicate radiogenic (R), calcium-derived (Ca), and potassium-derived (K) argon (%).

TABLE 2 (continued). Analytical Data for Incremental-Heating Experiments on Whole Rock Samples from Hess Rise and the Wentworth Seamount Chain

TEMP (°C)	$^{40}Ar/^{39}Ar$ (b)	$^{37}Ar/^{39}Ar$ (b)	$^{36}Ar/^{39}Ar$ (b)	$^{40}Ar_R$ (c)	$^{40}Ar_K$ (c)	$^{39}Ar_{Ca}$ (c)	$^{36}Ar_{Ca}$ (c)	K/Ca	^{39}Ar (%)	APPARENT AGE (a) (Ma)
KK71 65-24-A		WR CORE		J = 0.004660				Total gas age =		92.60 +- 0.49
500	54.47	0.5902	0.13211	28.3	0.1	0.0	0.1	0.830	0.9	125.39 +- 4.17
600	55.64	0.5985	0.13453	28.6	0.1	0.0	0.1	0.818	1.2	128.96 +- 3.24
700	60.31	0.5428	0.15691	23.1	0.1	0.0	0.1	0.902	3.2	113.65 +- 1.99
740	33.56	0.4794	0.07309	35.6	0.1	0.0	0.2	1.02	6.2	97.91 +- 1.05
800	12.390	0.5155	0.002916	93.0	0.3	0.0	4.4	0.950	15.3	94.44 +- 0.36
860	12.039	0.9506	0.001440	96.7	0.3	0.1	16.6	0.515	26.7	95.40 +- 0.31
950	11.601	0.6211	0.002069	94.8	0.3	0.0	7.5	0.789	24.4	90.20 +- 0.31
1050	11.319	0.9898	0.003634	90.8	0.3	0.1	6.8	0.495	12.2	84.47 +- 0.40
1200	12.315	3.017	0.008669	80.7	0.3	0.2	8.7	0.162	9.1	81.84 +- 0.51
FUSE	18.076	3.498	0.03769	39.6	0.2	0.2	2.3	0.140	0.7	59.35 +- 5.69
KK72 65-24-A		WR CORE		J = 0.010880				Total gas age =		91.06 +- 0.48
500	33.14	0.8817	0.09125	18.8	0.0	0.1	0.3	0.555	2.9	118.60 +- 2.50
550	10.405	0.4290	0.018372	48.1	0.0	0.0	0.6	1.14	8.8	95.74 +- 0.75
600	5.098	0.4016	0.000710	96.5	0.0	0.0	14.9	1.22	18.9	94.07 +- 0.34
650	5.053	0.7846	0.000772	96.7	0.0	0.1	26.8	0.624	19.6	93.47 +- 0.33
700	5.044	0.8942	0.000879	96.2	0.0	0.1	26.9	0.548	12.5	92.88 +- 0.41
760	4.915	0.7185	0.001097	94.5	0.0	0.0	17.3	0.682	9.7	88.99 +- 0.48
820	4.827	0.7480	0.001215	93.7	0.0	0.1	16.3	0.655	9.3	86.75 +- 0.49
880	4.907	0.8800	0.001818	90.4	0.0	0.1	12.8	0.556	7.9	85.10 +- 0.55
950	4.834	1.0987	0.003029	83.2	0.0	0.1	9.6	0.446	3.9	77.35 +- 1.04
1300	6.188	2.959	0.008188	64.6	0.0	0.2	9.5	0.165	6.4	76.96 +- 0.70
STYX7-25-2		WR CORE		J = 0.004895				Total gas age =		75.23 +- 0.40
500	24.57	1.4144	0.04592	45.2	0.0	0.1	0.7	0.346	11.4	95.50 +- 0.69
600	10.446	2.689	0.003775	91.0	0.1	0.2	16.7	0.182	13.3	82.23 +- 0.35
680	10.533	3.708	0.003012	93.9	0.1	0.2	28.8	0.132	15.1	85.52 +- 0.33
740	10.806	2.694	0.003990	90.8	0.1	0.2	15.8	0.182	8.5	84.74 +- 0.45
800	10.571	1.9208	0.004776	87.9	0.1	0.1	9.4	0.255	6.6	80.29 +- 0.54
860	9.634	1.4839	0.003843	89.2	0.1	0.1	9.0	0.330	5.4	74.42 +- 0.62
950	8.815	1.4388	0.002536	92.6	0.1	0.1	13.3	0.340	7.2	70.72 +- 0.49
1050	7.719	3.183	0.002972	91.4	0.1	0.2	25.1	0.154	23.5	61.37 +- 0.23
1200	6.923	26.19	0.010708	80.4	0.1	1.7	57.2	0.018	8.8	49.31 +- 0.45
FUSE	15.708	36.96	0.04172	37.7	0.0	2.4	20.7	0.013	0.4	52.87 +- 8.92
STYX7 25-3		HAWAIITE		J = 0.004700				Total gas age =		85.88 +- 0.45
500	54.12	0.5470	0.13601	25.7	0.1	0.0	0.1	0.895	1.3	114.48 +- 2.54
600	56.49	0.6056	0.11653	39.1	0.1	0.0	0.1	0.809	2.2	178.1 +- 1.9
700	36.27	0.4237	0.07201	41.3	0.1	0.0	0.1	1.16	4.0	122.78 +- 1.15
740	21.89	0.4465	0.03424	53.8	0.2	0.0	0.3	1.10	6.2	97.13 +- 0.71
800	12.407	0.4503	0.006085	85.5	0.3	0.0	1.9	1.09	10.6	87.77 +- 0.40
860	12.130	0.5463	0.005990	85.4	0.3	0.0	2.3	0.897	11.0	85.82 +- 0.39
950	11.759	0.6790	0.006112	84.7	0.3	0.0	2.8	0.721	24.9	82.61 +- 0.30
1050	10.161	0.6503	0.002762	92.1	0.4	0.0	5.9	0.753	15.9	77.66 +- 0.30
1200	10.188	1.1533	0.003894	89.2	0.4	0.1	7.4	0.425	22.9	75.49 +- 0.27
FUSE	13.338	28.91	0.03167	45.6	0.3	1.9	22.9	0.017	1.0	51.85 +- 2.96

(a) Lamda E = $0.581 \times 10^{-10} yr^{-1}$, lamda B = $4.962 \times 10^{-10} yr^{-1}$. Errors are s.d. of analytical precision.

(b) Corrected for ^{37}Ar and ^{39}Ar decay, 35.1 days and 259 years, respectively.

(c) Subscripts indicate radiogenic (R), calcium-derived (Ca), and potassium-derived (K) argon (%).

TABLE 2 (continued). Analytical Data for Incremental-Heating Experiments on Whole Rock Samples from Hess Rise and the Wentworth Seamount Chain

TEMP (°C)	$^{40}Ar/^{39}Ar$ (b)	$^{37}Ar/^{39}Ar$ (b)	$^{36}Ar/^{39}Ar$ (b)	$^{40}Ar_R$ (c)	$^{40}Ar_K$ (c)	$^{39}Ar_{Ca}$ (c)	$^{36}Ar_{Ca}$ (c)	K/Ca	^{39}Ar (%)	APPARENT AGE (Ma) [(a)]
KK71 20-8-O		WR CORE		J = 0.005397				Total gas age =		84.95 +- 0.43
500	11.330	0.5645	0.004828	87.8	0.1	0.0	3.3	0.868	3.2	94.35 +- 0.74
600	9.021	0.2868	0.000566	98.3	0.1	0.0	14.2	1.71	10.5	84.39 +- 0.51
680	9.060	0.2739	0.000282	99.3	0.1	0.0	27.2	1.79	15.0	85.52 +- 0.50
740	9.116	0.3312	0.000344	99.1	0.1	0.0	27.0	1.48	17.8	85.92 +- 0.50
800	9.143	0.2837	0.000476	98.7	0.1	0.0	16.7	1.73	12.2	85.76 +- 0.51
860	9.163	0.2991	0.000629	98.2	0.1	0.0	13.3	1.64	8.5	85.55 +- 0.63
950	9.130	0.3988	0.000892	97.4	0.1	0.0	12.5	1.23	3.8	84.60 +- 0.65
1050	9.029	0.4371	0.000446	98.9	0.1	0.0	27.5	1.12	17.4	84.91 +- 0.50
1200	8.837	1.9681	0.002060	94.9	0.1	0.1	26.8	0.249	11.1	79.96 +- 0.49
FUSE	10.104	15.663	0.015352	67.9	0.1	1.1	28.6	0.031	0.5	66.29 +- 3.66
KK71 20-8-J		WR CORE		J = 0.005397				Total gas age =		84.52 +- 0.43
450	17.283	0.7228	0.02536	56.9	0.0	0.0	0.8	0.678	0.7	93.41 +- 3.22
500	11.365	0.4718	0.008245	78.9	0.1	0.0	1.6	1.04	1.7	85.23 +- 1.38
600	9.905	0.3890	0.003993	88.4	0.1	0.0	2.7	1.26	1.0	83.28 +- 2.32
660	9.432	0.3214	0.002225	93.2	0.1	0.0	4.1	1.52	4.0	83.68 +- 0.74
720	9.292	0.2760	0.001313	96.0	0.1	0.0	5.9	1.78	9.7	84.85 +- 0.54
780	9.209	0.3214	0.000769	97.8	0.1	0.0	11.7	1.52	16.4	85.60 +- 0.51
850	9.310	0.3625	0.000877	97.5	0.1	0.0	11.6	1.35	18.4	86.28 +- 0.51
930	9.244	0.3596	0.001009	97.0	0.1	0.0	10.0	1.36	12.9	85.30 +- 0.52
1020	9.119	0.4184	0.000892	97.4	0.1	0.0	13.2	1.17	15.3	84.51 +- 0.51
1120	9.027	0.5385	0.001302	96.2	0.1	0.0	11.6	0.910	14.1	82.62 +- 0.50
FUSE	9.443	3.912	0.005506	86.1	0.1	0.3	19.9	0.125	5.7	77.72 +- 0.61

(a) Lamda E = $0.581 \times 10^{-10} yr^{-1}$, lamda B = $4.962 \times 10^{-10} yr^{-1}$. Errors are s.d. of analytical precision.
(b) Corrected for ^{37}Ar and ^{39}Ar decay, 35.1 days and 259 years, respectively.
(c) Subscripts indicate radiogenic (R), calcium-derived (Ca), and potassium-derived (K) argon (%).

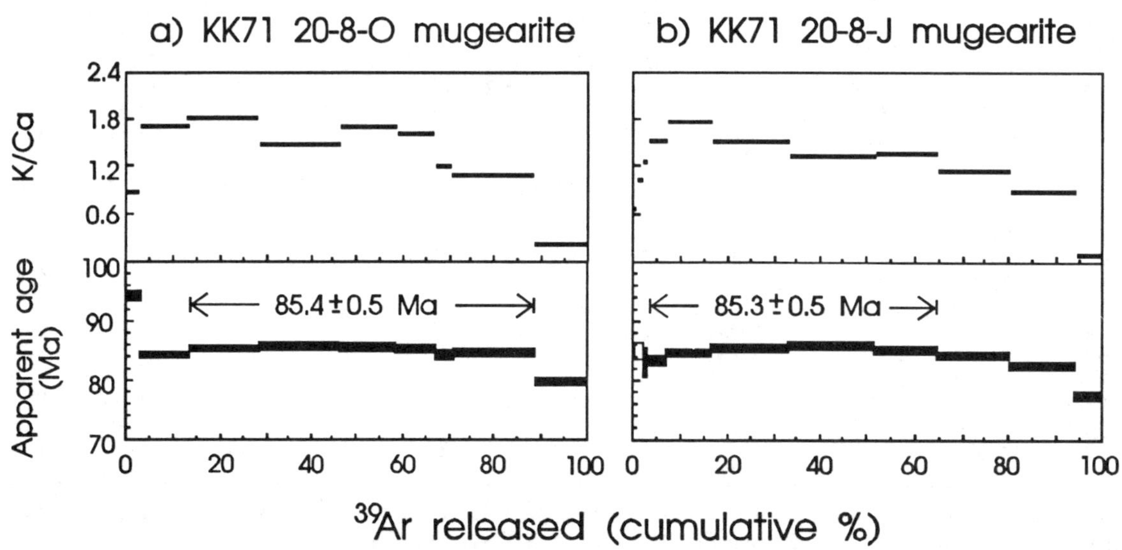

Figure 3. $^{40}Ar/^{39}Ar$ age spectra and K/Ca plots for two mugearite samples from an unnamed seamount at 26°27'N, 177°52'W, south of the Hawaiian Ridge. Individual increment apparent ages are drawn 2 standard deviations high.

TABLE 3. Analytical Data for $^{40}Ar/^{39}Ar$ Laser-Fusion Ages of Mineral Seperates from Samples from a Seamount on the Wentworth Seamount Chain

Sample Material	$^{40}Ar/^{39}Ar$ (b)	$^{37}Ar/^{39}Ar$ (b)	$^{36}Ar/^{39}Ar$ (b)	$^{40}Ar_R$ (c)	$^{40}Ar_K$ (c)	$^{39}Ar_{Ca}$ (c)	$^{36}Ar_{Ca}$ (c)	Age (a) (Ma)	Weighted Age (d)
KK71-20-8-E	11.107	5.240	0.002139	97.6	0.26	0.35	61.5	86.0 ±0.5	86.8 ±0.4
	11.262	5.026	0.001776	98.4	0.25	0.34	71.0	87.9 ±0.6	
Plagioclase	11.141	4.933	0.001670	98.6	0.26	0.33	74.1	87.1 ±0.5	
J=0.00449	11.270	4.780	0.002069	97.5	0.25	0.32	58.0	87.1 ±0.6	
	11.077	5.022	0.001996	97.8	0.26	0.34	63.1	85.9 ±0.7	
	11.490	4.278	0.002443	96.2	0.25	0.29	44.0	86.7 ±0.5	86.7 ±0.4
	11.456	4.275	0.002078	97.2	0.25	0.29	51.6	87.3 ±0.6	
Hornblende	11.385	4.258	0.002170	96.9	0.25	0.29	49.2	86.5 ±0.5	
J=0.00444	11.469	4.546	0.002329	96.7	0.25	0.31	49.0	87.0 ±0.5	
	11.386	4.355	0.002277	96.7	0.25	0.29	48.0	86.3 ±0.5	
	11.449	4.385	0.002436	96.3	0.25	0.29	45.2	86.5 ±0.5	
KK71-20-8-O	10.772	5.663	0.001449	99.7	0.26	0.38	98.1	84.3 ±1.3	86.5 ±0.5
	10.863	6.095	0.002163	98.0	0.26	0.41	70.7	83.7 ±0.9	
Plagioclase	11.707	5.875	0.002037	98.3	0.24	0.39	72.4	90.3 ±0.9	
J=0.00444	10.900	5.927	0.001918	98.6	0.26	0.40	77.6	84.4 ±1.0	
	10.890	6.345	0.002286	97.9	0.26	0.43	69.7	83.7 ±0.9	
	11.973	5.827	0.003201	95.5	0.24	0.39	45.7	89.7 ±0.8	
	11.310	4.106	0.001766	97.8	0.25	0.28	58.4	86.7 ±0.6	85.3 ±0.4
	11.161	4.337	0.001878	97.7	0.26	0.29	58.0	85.5 ±0.5	
Hornblende	11.182	4.183	0.001924	97.4	0.25	0.28	54.6	85.5 ±0.5	
J=0.00444	11.077	4.111	0.001990	97.2	0.26	0.28	51.8	84.5 ±0.6	
	12.227	4.129	0.005374	89.3	0.23	0.28	19.3	85.6 ±0.7	
	11.197	4.731	0.002506	96.3	0.25	0.32	47.4	84.6 ±0.5	

(a) Lamda E = $0.581 \times 10^{-10} yr^{-1}$, lamda B = $4.962 \times 10^{-10} yr^{-1}$. Errors are s.d. of analytical precision.
(b) Corrected for ^{37}Ar and ^{39}Ar decay, 35.1 days and 259 years, respectively.
(c) Subscripts indicate radiogenic (R), calcium-derived (Ca), and potassium-derived (K) argon (%).
(d) Weighted ages are weighted by the inverse of the variance.

such middle temperature plateaus give reliable age estimates (for example, Pringle, this volume).

The recent development of very small sample techniques for $^{40}Ar/^{39}Ar$ dating, and the presence of separable plagioclase and hornblende phenocrysts in these samples, allows us to check the validity of this assumption. Twenty-three laser fusion analyses of both plagioclase and hornblende from two samples from dredge KK71-20-8 are presented in Table 3. The best age estimate is the weighted mean of the total-fusion ages, 86.3 ±0.2 Ma. This is not significantly different from the results obtained from the whole rock incremental-heating analyses, and validates the use of mid-temperature plateaus for samples which have age spectra characteristic of slight to moderate argon redistribution.

For the Wentworth Seamount samples (Dredge STYX7 25), neither of the incremental-heating experiments provided reliable estimates of the crystallization age (Table 2, Figure 4 a and b) Both show disturbed age spectra characteristic of argon redistribution. Both have mid-temperature "pseudo-plateau" steps with ages around 84-85 Ma, which is close to the total gas age of 86 Ma for the hawaiite (Table 2). We interpret these results to indicate a minimum age of about 84-85 Ma for the volcanism at Wentworth Seamount.

For samples from the seamount north of the Hawaiian Ridge at 31°09'N, 179°45'W (Dredge KK71-65-24), only one of four incremental-heating experiments (Figure 4 c) yielded a reliable age according to the criteria used here. The isochron for that experiement gives the best estimate for the age of this seamount, 93.4 ±0.6 Ma (Table 1). A second incremental-heating experiment on the same rock (Figure 4 d) gave a similar isochron age of 94.8 ±1.2, but had only 48.3% of the gas released on the age spectrum plateau and a SUMS/N-2 of 4.79, neither of which meet the criteria for a reliable estimate of crystallization age. The results from the second sample were clearly disturbed (Figure 4 e and f), and suggest only a minimum age of about 75 Ma for that sample.

Hess Rise Results

The two samples of altered trachyte from DSDP Site 465A on Hess Rise yielded highly disturbed age spectra (Table 2, Figure 4 g and h). The high temperature steps exhibit the pattern of

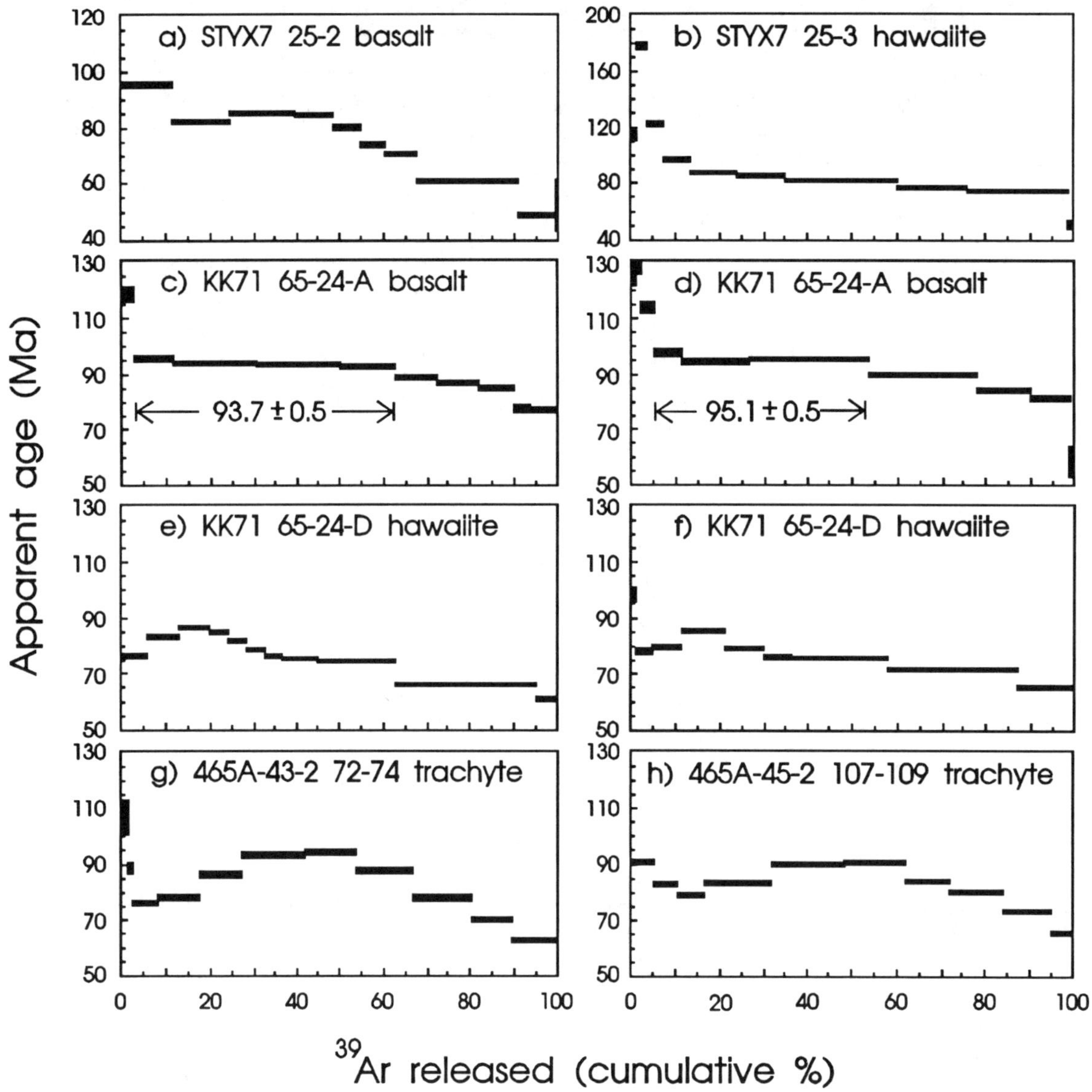

Figure 4. $^{40}Ar/^{39}Ar$ age spectra for whole rock samples from Hess Rise and seamounts along the Wentworth Seamount Chain. Individual increment apparent ages are drawn 2 standard deviations high.

decreasing ages attributed to argon recoil. The low temperature steps show both anomalously old and young ages; we interpret the anomalously young ages to indicate ^{40}Ar loss in excess of any ^{39}Ar loss for those steps. The intermediate temperature steps of the age spectra are about 90-94 Ma, which we interpret as a minimum age.

For the seamount on southern Hess Rise, several experiments were done on each of the two plagioclase separates from dredge DM23-1922 (Table 1 and 3.4). The ages of the two separates are distinct.

For sample DM23-1922-08, an incremental-heating experiment done on 0.34 g of plagioclase yielded an age spectrum plateau at 87.6 ±0.6 Ma and an isochron age of 89.4 ±3.1 Ma, with almost 70% of the gas released on the plateau (Figure 5 a). Because the lower temperature steps had low apparent ages, presumably because of ^{40}Ar loss from a small amount of alteration phases, and because three laser total-fusion analyses also showed anomalously young ages (Table 4), we tried a series of laser degassing experiments to see if we could "clean" the feldspar. Using a defocussed laser beam, the sample was first degassed at 2W and then at 4W; the gas released from these "cleaning" steps had low apparent ages and large amounts of atmospheric ^{40}Ar (Table 4). Figure 5 b shows the results of the rest of the experiment: 2 groups of the degassed plagioclase (about 20-40 crystals each) were fused and analyzed (Fus1 and Fus2), the remaining sample degassed again at 7W and analyzed

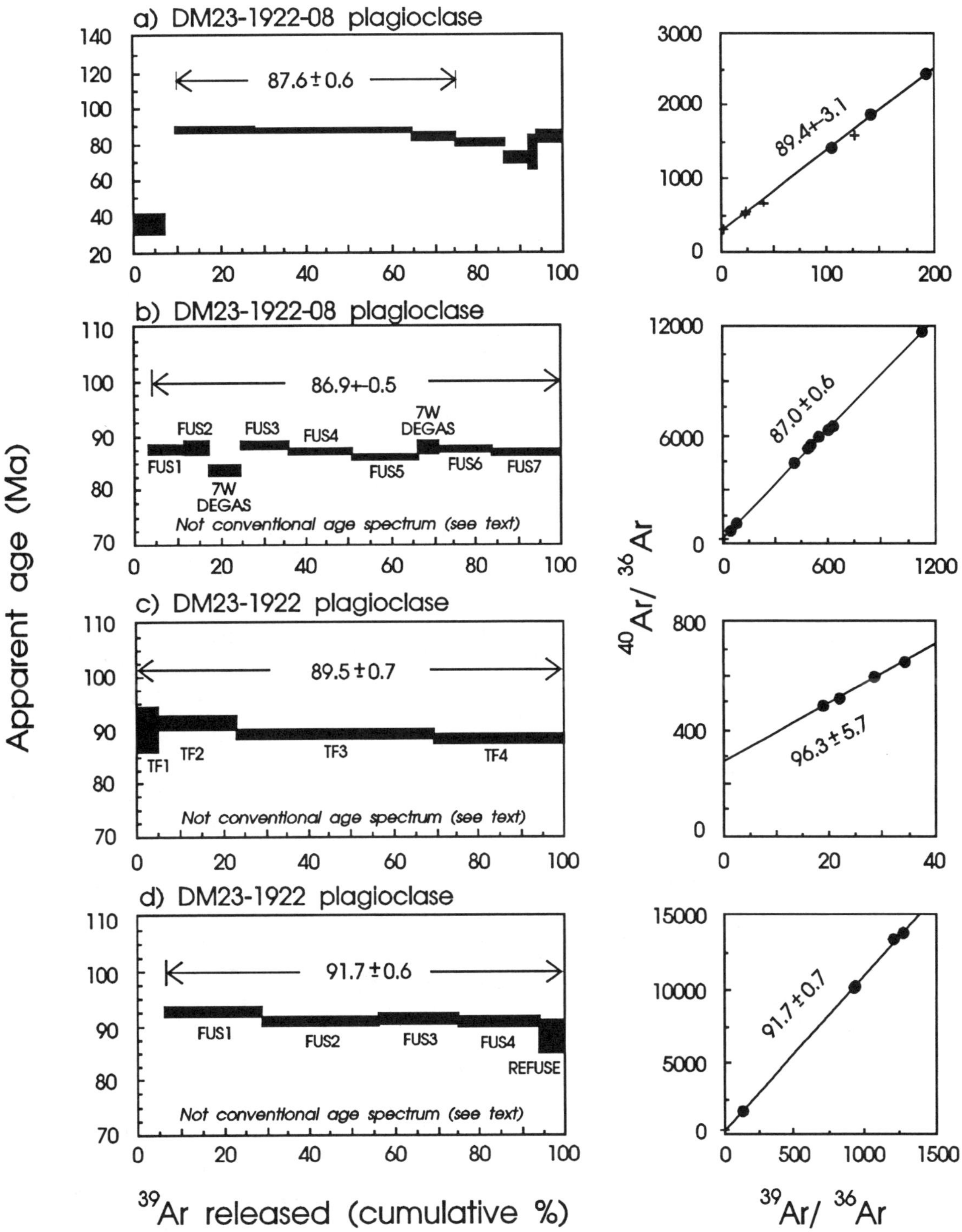

Figure 5. $^{40}Ar/^{39}Ar$ age spectra, laser-heating results, and isochron correlation diagrams for plagioclase separates from a seamount on southern Hess Rise. Note that data shown in b), c), and d) are not from traditional incrementing-heating experiments, but are the results of different laser-heating techniques (see text). On age spectra diagrams, individual apparent ages are drawn 2 standard deviations high. On isochron diagrams, (●) are data included in isochron fit, (+) are not included in fit.

TABLE 4. Analytical Data for Incremental-Heating and Laser-Fusion Experiments on Plagioclase Separates from a Seamount on Southern Hess Rise

STEP	$^{40}Ar/^{39}Ar$ (b)	$^{37}Ar/^{39}Ar$ (b)	$^{36}Ar/^{39}Ar$ (b)	$^{40}Ar_R$ (c)	$^{40}Ar_K$ (c)	$^{39}Ar_{Ca}$ (c)	$^{36}Ar_{Ca}$ (c)	K/Ca	^{39}Ar (%)	APPARENT AGE (a) (Ma)
DM23-1922-08		PLAGIOCLASE		J = 0.004470				Total gas age =		85.63 ± 0.83
550°C	136.49	10.846	0.4496	3.3	0.0	0.7	0.6	0.045	7.4	36.11 ± 5.57
650	(lost)	------	------	----	---	---	----	-----	[2.7]	[220.7 ± 10.0]
720	13.162	17.490	0.011584	84.3	0.0	1.2	39.9	0.028	18.7	88.35 ± 1.07
800	12.539	16.601	0.009495	87.9	0.0	1.1	46.2	0.029	35.9	87.72 ± 0.71
880	13.482	15.334	0.013496	79.2	0.0	1.0	30.0	0.032	10.2	85.02 ± 1.78
960	12.623	14.567	0.011703	81.6	0.0	1.0	32.9	0.033	11.6	81.96 ± 1.59
1040	16.494	11.200	0.02755	55.9	0.0	0.8	10.7	0.043	5.7	73.41 ± 3.16
1200	22.51	10.440	0.04644	42.6	0.0	0.7	5.9	0.047	1.9	76.31 ± 9.44
FUSE	23.42	13.666	0.04665	45.7	0.0	0.9	7.7	0.036	5.9	85.01 ± 3.04
DM23-1922-08 (d)		PLAGIOCLASE		J = 0.004859				Total gas age =		83.80 ± 0.50
Heat @ 2W	308.7	2.680	1.0547	-0.9	0.0	0.2	0.1	0.183	1.6	-24.06 ± 11.88
Heat @ 4W	118.12	6.713	0.3961	1.3	0.0	0.5	0.5	0.073	1.9	13.97 ± 4.98
Fus1	10.839	15.391	0.006588	93.4	0.3	1.1	64.9	0.031	8.2	87.75 ± 0.76
Fus2	13.931	15.547	0.017016	72.8	0.2	1.1	25.4	0.031	5.5	87.76 ± 1.10
7W Degas	15.031	15.016	0.02222	64.3	0.2	1.1	18.8	0.032	7.6	83.67 ± 0.86
Fus3	10.833	16.051	0.006478	94.2	0.3	1.1	68.9	0.030	11.0	88.27 ± 0.61
Fus4	10.363	15.377	0.005149	97.2	0.3	1.1	83.0	0.032	15.1	87.11 ± 0.49
Fus5	10.453	14.304	0.005588	95.1	0.3	1.0	71.2	0.034	15.6	85.96 ± 0.48
7W Degas	10.717	16.476	0.006334	94.9	0.3	1.2	72.3	0.029	4.8	87.97 ± 1.22
Fus6	10.753	15.422	0.006351	94.0	0.3	1.1	67.5	0.031	12.4	87.44 ± 0.56
Fus7	10.560	15.469	0.005961	95.0	0.3	1.1	72.1	0.031	16.3	86.82 ± 0.48
DM23-1922-08		PLAGIOCLASE		J = 0.004859				Total gas age =		74.99 ± 0.65
TF1 (e)	22.32	7.872	0.05032	36.1	0.1	0.6	4.3	0.062	67.9	69.72 ± 0.69
TF2	11.640	14.704	0.010217	84.2	0.3	1.0	40.0	0.033	17.0	84.74 ± 1.08
TF3	16.477	15.227	0.02561	61.5	0.2	1.1	16.5	0.032	15.1	87.56 ± 1.24
DM23-1922 (d)		PLAGIOCLASE		J = 0.004856				Total gas age =		89.93 ± 0.66
4W Degas	85.81	14.211	0.2691	8.7	0.0	1.0	1.5	0.034	6.2	64.57 ± 3.94
Fus1	11.074	14.061	0.004757	97.4	0.3	1.0	82.2	0.035	22.5	93.03 ± 0.85
Fus2	10.856	13.940	0.004690	97.5	0.3	1.0	82.6	0.035	27.3	91.27 ± 0.72
Fus3	10.984	13.366	0.004726	97.0	0.3	0.9	78.6	0.036	18.8	91.83 ± 0.99
Fus4	10.891	15.132	0.005208	97.0	0.3	1.1	80.8	0.032	19.1	91.17 ± 0.98
Refuse	12.862	15.561	0.013095	79.6	0.3	1.1	33.0	0.031	6.0	88.47 ± 2.99
DM23-1922		PLAGIOCLASE		J = 0.004856				Total gas age =		89.53 ± 0.71
TF1 (e)	19.756	14.017	0.03527	52.9	0.2	1.0	11.0	0.035	5.0	90.18 ± 4.12
TF2	21.18	15.200	0.03991	50.1	0.2	1.1	10.6	0.032	18.1	91.54 ± 1.26
TF3	24.00	14.488	0.05015	43.1	0.1	1.0	8.0	0.033	46.2	89.28 ± 0.81
TF4	22.77	14.268	0.04619	45.1	0.1	1.0	8.6	0.034	30.7	88.60 ± 0.92

(a) Lamda E = $0.581 \times 10^{-10} yr^{-1}$, lamda B = $4.962 \times 10^{-10} yr^{-1}$. Errors are s.d. of analytical precision.
(b) Corrected for ^{37}Ar and ^{39}Ar decay, 35.1 days and 259 years, respectively.
(c) Subscripts indicate radiogenic (R), calcium-derived (Ca), and potassium-derived (K) argon (%).
(d) Sample heated with defocused laser beam (see text).
(e) TF: total fusion, without any preheating or degassing of the plagioclase (see text).

(7W Degas), three more groups of sample fused and analyzed (Fus3, Fus4, Fus5), the remaining sample degassed once more at 7W (7W Degas), and finally the remaining feldspar was fused in two groups (Fus6 and Fus7). All but the first degassing steps had concordant ages, and the isochron analysis for this degassing experiment gave an age of 87.0 ±0.6 Ma. Combining the laser-degassing results with the incremental-heating results gives the best age estimate for this sample, 87.1 ±0.6 Ma.

For sample DM23-1922, four laser total-fusion analyses (TF1-TF4) have a weighted mean age of 89.5 ±0.5 Ma and a total-fusion isochron age of 96.3 ±5.7 (Tables 1 and 4; Figure 5 c). The imprecision of the isochron age is mainly the result of the limited range of Ar ratios and the low amounts of radiogenic ^{40}Ar in this sample, 43.1 to 52.9% (Table 4). Because these are relatively high atmospheric ^{40}Ar contents for laser-fusion analyses of plagioclase, we did a laser degassing experiment on this sample also (Figure 5 d). The whole sample was degassed at 4W, then the degassed sample was fused in 4 groups (Fus1 through Fus4, Figure 5 d), and finally the glass beads from all 4 fusions were fused again in one group (Refuse). All of the analyses except the initial degassing step were concordant and had significantly higher radiogenic ^{40}Ar, 79.6 to 97.5%. The weighted "plateau" and isochron age of 91.7 Ma (Table 1) is slightly higher than the total gas age of 89.9 Ma, confirming the presence of a small amount of alteration products which were degassed in the first step. The best age for this sample is the weighted mean of the degassing and total-fusion ages, 91.8 ±0.7 Ma, which is significantly older than the age for the other sample from this dredge.

DISCUSSION AND INTERPRETATION

Possible hot spot origin

The new $^{40}Ar/^{39}Ar$ age data suggest age progressive volcanism from NW to SE for the Wentworth Seamount Chain, from before about 100 Ma at DSDP Site 465A on Hess Rise to 86 Ma at the unnamed seamount 800 km to the SE. The best evidence for an age progression comes from two unnamed seamounts. An age of 86 Ma for the southernmost seamount in the chain at 26°27'N, 177°52'W is indicated by both laser-fusion analysis of mineral separates and concordant incremental-heating experiments. An age of 94 Ma for a seamount at 31°09'N, 177°52'W, 550 km to the NW, is indicated by concordant whole rock incremental-heating experiments. The measured velocity of about 70 km/m.y. (ca. 0.6 °/m.y.) is consistent with, but perhaps slightly faster than, rates found for the other Late Cretaceous hot spot tracks in the Musicians Seamounts and northern Line Islands [Pringle, this volume].

Extrapolating the age progression from the two well-dated seamounts predicts ages of 90 Ma for Wentworth Seamount, 99 Ma for Site 465A on southern Hess Rise, and about 110 Ma for Site 464 on northern Hess Rise. These ages are consistent with our minimum $^{40}Ar/^{39}Ar$ age estimates of 85 Ma for Wentworth Seamount and 90-94 Ma for Site 465A discussed above. The predicted ages also agree with the late Albian (ca. 98-100 Ma) age of the oldest sedimentary rocks at Site 465A, and the early Albian to late Aptian (ca. 110-114 Ma) age of the oldest sedimentary rocks at Site 464 [Theide, et al., 1981]. Thus, much of the volcanism along the Hess Rise/Wentworth Seamount chain is consistent with that expected from an age-progressive hot spot track.

However, the lack of any documented Pacific hot spot track older than 96 Ma, and, thus, the lack of any reliable estimate of Pacific/hot spot motion prior to 96 Ma, severely limits our ability to compare our results with a hypothetical hot spot track across Hess Rise. Most hot spot models of Pacific plate motion before 96 Ma (i.e., Duncan and Clague, 1984; Henderson, 1985) suggest mainly an east-west rotation of the Pacific plate, while paleomagnetic studies (i.e., Larson et al., 1992) find a significant component of northward drift.

We also do not mean to imply that all of the volcanism on Hess Rise can be related to a simple hot spot model. The morphology and geologic evolution of Hess Rise is complex, and marked by tectonic activity and volcanism which continued at least through the Cenomanian [Kroenke and Neomoto, 1982; Vallier et al., 1981; Vallier et al., 1983]. Evidence for normal faulting is common, and the vertical tectonics of Hess Rise also seems complicated. For example, the minimum age for the subaerially-erupted trachytes at Site 465A indicates that there is no lengthy hiatus between the eruption of the trachyte and the deposition of the overlying sediments, and that subsidence for at least southern Hess Rise must have been relatively rapid.

The ages of the alkalic basalts dredged by the *R/V D. Mendeleev* from the seamount on southern Hess Rise indicate that active volcanism continued for at least 12 m.y. after the initial volcanism at nearby Site 465A. This is well after the southern Hess Rise had begun to submerge, after Hess Rise had drifted 800 km to the northwest of the hot spot, and almost as long as volcanism lasted on the proposed hot spot track itself. The later alkalic volcanism may be similar to the late stage alkalic volcanism that has been shown to last at least 5-8 m.y. on seamounts and islands throughout the Pacific [e.g., Pringle, 1991]. Alternatively, it could be related to a period, ca. 80-90 Ma, of major spreading direction changes, ridge jumps, and volcanism along fracture zones undergoing significant tectonic change [Rea and Dixon, 1983; Sager and Pringle, 1987; Mammerickx and Sharman, 1988; Windom et al., 1981].

Constraints on the origin of Hess Rise

The existing models for the origin of Hess Rise concentrate mainly on a description of the tectonic setting and timing of volcanism. A complete model for the origin of Hess Rise must also explain both the irregular morphology and strong structual asymmetyry of Hess Rise [i.e., Kroenke and Neomoto, 1982; Vallier et al., 1981] and the excess basaltic volcanism that oceanic plateaus such as Hess Rise represent. Schubert and Sandwell [1989] estimate that the total volume of Hess Rise is 9.08x10^6 km^3, 75% greater than would be expected for normal ocean crust alone.

Morgan [1981a, 1981b] proposed two different mechanisms for the formation of plateau basalts, which he had noted seem to mark the beginning of at least some hot spot tracks [1972a;

1972b]. The first is a rifting mechanism, where an *existing* hot spot weakens and warms the continental lithosphere and generates flood basalts following rifting and extension related to continental breakup. The second is a plume initiation mechanism, where a *new* mantle plume surfaces with a large, warm "head" capable of producing flood basalt volcanism in one large melting event.

White and McKenzie [1989] have further developed the rifting mechanism model. According to their view, a mantle plume warms a large region of the mantle below continental lithosphere to temperatures at least 100-200°C higher than surrounding mantle, extension by at least a factor of 2 thins the lithosphere, and large volumes of decompression melting produce a melt thickness of about 25 km. Their rifting model describes many of the characteristics of volcanic rift margins such as those in the North Atlantic Tertiary province. Although they mainly explored models for continental rifting, they suggest that such a model might also apply to thinning of older oceanic lithosphere, since such lithosphere has many characteristics similar to Phanerozoic continental crust. However, such a model is incapable of producing plateau basalts erupted on young ocean crust near a spreading ridge, such as those which formed Hess Rise.

Other workers (i.e., Richards et al., 1989; Campbell and Griffiths, 1990; Duncan and Richards, 1991) preferred the model of direct plume initiation for the origin of flood basalts. According to this model, a low viscosity diapir originates deep within the mantle, grows during its ascent, and flattens when it reaches the base of the lithosphere. Voluminous flood basalts are the eruptive products of this plume "head"; hot spot tracks are the eruptive products of the following conduit or plume "tail". This plume-head model works especially well at describing the apparent short duration and characteristic size of many of the world's flood basalts, and the ratio of the eruptive rates of the flood basalts to their respective hot spot tracks. It also does not require any particular tectonic setting at the earth's surface, i.e. it describes equally well the origin of either continental or oceanic flood basalts.

The age data presented here are consistent with a plume initiation-hot spot model for the origin of the Hess Rise and the Wentworth Seamount Chain. However, the Hess Rise example deviates from the typical plume initiation-hot spot model in two important ways. First, no extension of the Wentworth Seamount Chain has been found to the southeast of the youngest seamount dated here, i.e. the hot spot track originating at Hess Rise is only 14 m.y. long. If Hess Rise was formed by a plume head, then the seamount volcanism to the southeast marks the track of a plume tail that never developed into a productive source of sustained hot spot volcanism, and has a much lower hot spot track/flood basalt volume ratio than that suggested by Richards et al. [1989]. Short hot spot tracks have been documented in the Austral-Cook Islands [Turner and Jarrard, 1982], the Gulf of Alaska [Dalrymple et al., 1987], the Musicians Seamounts [Pringle, this volume], etc., but none are associated with large flood basalts or volcanic plateaus.

Second, the age of the alkalic basalts dredged from a seamount on the southeast margin of Hess Rise indicates that active volcanism continued on Hess Rise for at least 12 m.y., almost as long as volcanism lasted on the proposed hot spot track to the south. This is in contrast to the models of plume initiation which require the formation of most of the oceanic flood basalt plateau in 3 m.y. or less [Campbell and Griffiths, 1989; Richards et al., 1989]. This extended period of volcanism may be the result of tectonic controls unique for this period and region of the Pacific plate [i.e., Rea and Dixon, 1983; Mammerickx and Sharman, 1988]. However, on Ontong Java Plateau, another Pacific oceanic plateau which apparently erupted vast amounts of basalt in a relatively short period [Richards et al., 1989; Tarduno et al., 1991], some volcanism continued for at least 30 m.y. after the main eruptive event [Mahoney et al., this volume]. These observations indicate either that (1) the plume initiation model successfully predicts the relatively short period of volcanism during which the bulk of an oceanic plateau is built, but that small amounts of volcanism can erupt for a significant period after the construction of the main edifice, or (2) some other model for the formation of oceanic plateaus is needed.

Further work is necessary to more fully test models for the origin of Hess Rise. This work would require (1) more samples than currently available, (2) further attempts at deciphering reliable ages for the existing sites, (3) isotopic and trace element documentation of the characteristics of the mantle source region of the Hess Rise volcanism, and (4) the development of a more complete tectonic model for this part of the Pacific, as well as quantitative, pre-96 Ma estimates of Pacific/hot spot and Pacific/spin axis relative motion. Until then, models on the origin of the plateau basalts of Hess Rise will remain speculative.

Acknowledgements. We would like to thank M.O. Garcia and G.B. Roudnik for providing samples from their collections on the Wentworth Seamounts and the *R/V Mendeleev* dredges, respectively. J. Saburomaru provided assistance with the $^{40}Ar/^{39}Ar$ analyses. The staffs of the U.S.G.S. TRIGA and University of Michigan Ford Nuclear reactors provided the ~10^{22} fast neutrons necessary for this study. J.M. Sinton, D.A. Clague, D. Epp, M.O. Garcia, J. Mammerickx, D.K. Rea, and T.L. Vallier provided valuable reviews of various versions of the manuscript.

REFERENCES

Campbell, I.H., and R.W. Griffiths, Implications of mantle plume structure for the evolution of flood basalts, *Earth Planet. Sci. Lett.*, *99*, 79-93, 1990.

Clague, D.A., and G.B., Dalrymple, Cretaceous K-Ar ages of volcanic rocks from the Musicians Seamounts and the Hawaiian Ridge, *Geophys. Res. Lett.*, *2*, 305-309, 1975.

Dalrymple, G.B., The GLM continuous laser system for $^{40}Ar/^{39}Ar$ dating: Description and performance characteristics, *U.S. Geol. Survey Bull. 1890*, 89-96, 1989.

Dalrymple, G.B, E.C. Alexander, Jr., M.A. Lanphere, and G.P. Kraker, Irradiation of samples for $^{40}Ar/^{39}Ar$ dating using the Geological Survey TRIGA reactor, *U.S. Geol. Surv. Prof. Pap. 1176*, 55 pp., 1981.

Dalrymple, G.B., M.A. Lanphere and D.A. Clague, Conventional and $^{40}Ar/^{39}Ar$ ages of volcanic rocks from Ojin (Site 430), Nintoku (Site 432), and Suiko (Site 433) Seamounts and the chronology of volcanic propagation along the Hawaiian-Emperor chain, *Initial Reports of the Deep Sea Drilling Project, 55*: Washington (U.S. Government Printing Office), 659-676, 1980.

Dalrymple, G.B., M.A. Lanphere, and M.S. Pringle, Correlation diagrams in $^{40}Ar/^{39}Ar$ dating: Is there a correct choice?, *Geophys. Res. Lett.*, *15*, 589-591, 1988.

Dalrymple, G.B., D.A. Clague, T.L. Vallier, and H.W. Menard, $^{40}Ar/^{39}Ar$ age, petrology, and tectonic significance of some seamounts in the Gulf of Alaska, in B.H. Keating, P. Fryer, R. Batiza, and G.W. Boehlert, eds., *Seamounts, Atolls, and Islands,* Geophysical Monograph Series 43 (Menard Volume), Am. Geophys. Un., Washington, D.C., 297-315, 1987.

Davis, A.S., M.S. Pringle, L.B.G. Pickthorn, D.A. Clague, and W.C. Schwab, Petrology and age of alkalic lava from the Ratak chain of the Marshall Islands, *J. Geophys. Res.*, *94*, 9757-9774, 1989.

Duncan, R.A., and D.A. Clague, Pacific plate motions recorded by linear volcanic chains, *in* A.E.M. Nairn, F.G. Shehli, and S. Uyeda, eds., *Ocean Basins and Margins, 7A*, Plenum, N.Y., 89-121, 1985.

Duncan, R.A., and M.A. Richards, Hotspots, mantle plumes, flood basalts, and true polar wander, *Reviews of Geophysics*, *29*, 31-50, 1991.

Epp, D., Age and tectonic relationships among volcanic chains on the Pacific plate, Ph.D. Thesis, University of Hawaii, Honolulu, 199 pp., 1978.

Epp., D., Possible perterbations to hotspot traces and implications for the origin and structure of the Line Islands, *J. Geophys. Res.*, *89*, 11,273-11,286, 1984.

Garica, M.O., D.G. Grooms, and J.J. Naughton, Petrology and geochronology of volcanic rocks from seamounts along and near the Hawaiian Ridge: Implications for propagation rate of the ridge, *Lithos*, *20*, 323-336, 1987.

Harland, W.B., R.L. Armstrong, A.V. Cox, L.E. Craig, A.G. Smith, and D.G. Smith, *A geologic time scale*, Cambridge, England, Cambridge University Press, 131 pp., 1990.

Henderson, L.J., Motion of the Pacific plate relative to the hotspots since the Jurassic and model of oceanic plateaus of the Farallon plate, Ph.D. thesis, Northwestern University, Evanston, IL, 312 pp., 1985.

Jarrard, R.D., and D.A. Clague, Implications of Pacific island and seamount ages for the origin of volcanic chains, *Rev. Geophys. Space Phys.*, *15*, 57-76, 1977.

Kroenke, L., and K. Neomoto, Marine Geology of the Hess Rise, 2, Basement morphology, sediment thickness, and strutural geology, *J. Geophys. Res.*, *87*, 9259-9278, 1982.

Lanphere, M.A., and G.D. Dalrymple, The use of $^{40}Ar/^{39}Ar$ data in evaluation of disturbed K-Ar systems, in R.E. Zartman, ed., Short Papers on the Fourth International Conference on Geochronology, Cosmochronology, and Isotope Geology, *U.S. Geol. Surv. Open File Rep. 78-701*, 241-243, 1978.

Larson, R.L., M.B. Steiner, E. Erba, and Y. Lancelot, Paleolatitudes and tectonic reconstructions of the oldest portion of the Pacific plate, in Larson, R.L., Lancelot, Y., et al., *Proc. ODP, Sci. Results, 129*, College Station, TX (Ocean Drilling Program), 615-631, 1992.

Mammerickx, J., and G.F. Sharman, Tectonic evolution of the North Pacific during the Cretaceous Quiet Period, *J. Geophys. Res.*, *93*, 3009-3024, 1988.

Mammerickx, J., and S.M. Smith, Bathymetry of the northeast Pacific, *Map Series MC-52*, Geol. Soc. Amer., Boulder, Colorado, 1981.

Morgan, W.J., Hotspot tracks and the early rifting of the Atlantic, *Tectonophysics*, *94*, 123-129, 1981a.

Morgan, W.J., Hotspot tracks and the opening of the Atlantic and Pacific oceans, *in* C. Emilliani, ed., *The Sea*, *7*, John Wiley, New York, 443-487, 1981b.

Morgan, W.J., Plate motions and deep mantle convections, *Geol. Soc. Amer. Mem. 132 (Hess Volume)*, 7-22, 1972a.

Morgan, W.J., Deep mantle convection plumes and plate motions, *Amer. Assoc. Petrol. Geol. Bull.*, *56*, 203-213, 1972b.

Pringle, M.S., Geochronolgy and petrology of the Musicians Seamounts, and the search of hot spot volcanism in the Cretaceous Pacific, Ph.D. thesis, University of Hawaii, Honolulu, 235 pp., 1992.

Pringle, M.S., H. Staudigel, and J. Gee, Jasper Seamount: Seven million years of volcanism, *Geology*, *19*, 364-368, 1991.

Rea, D.K., and J.M. Dixon, Late Cretaceous and Paleogene tectonic evolution of the North Pacific Ocean, *Earth Planet. Sci. Lett.*, 65, 145-166, 1983.

Richards, M.A., R.A. Duncan, and V.E. Courtillot, Flood basalts and hotspot tracks, *Science*, *246*, 103-107, 1989.

Saito, K., and M. Ozima, $^{40}Ar/^{39}Ar$ ages of submarine rocks from the Line Islands: implications on the origin of the Line Islands, in G.H. Sutton, M.H. Manghnani, and R. Moberly, eds., *The Geophysics of the Pacific Ocean Basin and its margin,* Geophysical Monograph Series 19, Am. Geophys. Un., Washington, D.C., 205-226, 1976.

Saito, K., and M. Ozima, $^{40}Ar/^{39}Ar$ geochronological studies of submarine rocks from the western Pacific area, *Earth Planet. Sci. Lett.*, *33*, 353-369, 1977.

Schlanger, S.O., M.O. Garcia, B.H. Keating, J.J. Naughton, W.W. Sager, J.A. Haggerty, J.A. Philpotts, and R.A. Duncan, Geology and geochronology of the Line Islands, *J. Geophys. Res.*, *89*, 11201-11272, 1984.

Schubert, G., and D. Sandwell, Crustal volumes of the continents and of oceanic and continental submarine plateaus, *Earth Planet. Sci. Lett.*, 92, 234-246, 1989.

Tarduno, J., W.V. Sliter, L. Kronke, M. Leckie, H. Mayer, J.J. Mahoney, R. Musgrave, M. Storey, and E.L. Winterer, *Science*, 254, 399-403, 1991.

Taylor, J.R., *An Introduction to Error Analysis*, Mill Valley, California, University Science Books, 270 pp., 1982.

Thiede, J., Vallier, T.L., et al., *Initial Reports of the Deep Sea Drilling Project, 62*: Washington (U.S. Government Printing Office), 1981.

Vallier, T.L., W.E. Dean, D.K. Rea, and J. Theide, Geologic evolution of Hess Rise, central North Pacific Ocean, *Geol. Soc. Amer. Bull.*, *94*, 1289-1307, 1983.

Vallier, T.L., D.K. Rea, W.E. Dean, J. Thiede, and C.G. Adelseck, The geology of Hess Rise, central North Pacific Ocean, in Thiede, J., Vallier, T.L., et al., *Initial Reports of the Deep Sea Drilling Project, 62*: Washington (U.S. Government Printing Office), 1031-1072, 1981.

Vallier, T.L., K.E. Windom, K.E. Siefert, and J. Thiede, Volcanic rocks cored on Hess Rise, western Pacific Ocean, *Nature*, 286, 48-50, 1980.

White, R., and D. McKenzie, Magmatism at rift zones: The generation of volcanic margins and flood basalts, *J. Geophys. Res.*, *94*, 7685-7729, 1989.

Windom, K.E., K.E. Siefert, and T.L. Vallier, Igneous evolution of Hess Rise: Petrologic evidence from DSDP Leg 62, *J. Geophys. Res.*, *86*, 6311-6322, 1981.

York, D., Least squares fitting of a straight line with correlated errors, *Earth Planet. Sci. Lett.*, *5*, 320-324, 1969.

M.S. Pringle and G.B Dalrymple, U.S. Geological Survey, Menlo Park, CA 94025.

Early and Late Cretaceous Volcanism and Reef-Building in the Marshall Islands

JONATHAN M. LINCOLN

Department of Physics and Geoscience, Montclair State College, Upper Montclair, New Jersey

MALCOLM S. PRINGLE

U. S. Geological Survey, Menlo Park, California

ISABELLA PREMOLI SILVA

Dipartmento di Scienze della Terra, Universita Degli Studi di Milano, Milano Italy

Radiometric and biostratigraphic ages for samples recovered from fourteen guyots, seamounts and atolls in the Marshall Islands document construction of carbonate platforms and the volcanos on which the platforms grew during the Early and Late Cretaceous. Previously, the only evidence for reef-building during the Early and Late Cretaceous in the Marshall Islands was shallow-bank fossils deposited in turbidites penetrated at DSDP Site 462 in the Nauru Basin. The distribution of volcano and platform ages reported in this paper suggests a complex history that cannot be explained by passage of the region over a single hotspot.

Assemblages of whole and fragmented rudists recovered from two guyots may be equivalent to the lower to middle Albian of the Gulf of Mexico coast. Limestones dated Cenomanian or older that contain shallow-water debris document the existence of carbonate banks during the Early Cretaceous at three additional guyots. A $^{40}Ar/^{39}Ar$ age of 138.2±0.8 Ma for Look Seamount is the first direct evidence of seamount volcanism in the Marshall Islands during the Early Cretaceous.

Limestones of Late Cretaceous age that contain shallow-water debris were recovered from nine seamounts or guyots in the Marshall Islands; limestones of Eocene age have been recovered at two other locations. Late Cretaceous $^{40}Ar/^{39}Ar$ ages, including a revised age of 75.9±0.6 Ma for the basalt recovered from the bottom of the Anewetak Atoll drill core, have been determined for 6 volcanic edifices in this region.

Four Marshall Islands sites of volcanism and reef-building during the Late Cretaceous are located at, or adjacent to, sites of volcanism or reef-building during the Early Cretaceous. This superposition of ages suggests multiple periods of volcanism, uplift, reef-building, and subsidence that cannot be explained with a single hotspot model. The origin of the Marshall Islands could be explained by the passage of the region over more than one hotspot over a period of at least 70 m.y.

INTRODUCTION

While working for the U. S. Geological Survey between 1951 and 1958, Sy Schlanger participated in the study of limestones recovered during drilling of Anewetak Atoll in the northern Marshall Islands [Schlanger, 1963]. Schlanger returned to the Marshall Islands and the nearby Nauru Basin on several occasions as he investigated the complex interactions among Pacific intraplate volcanism, seafloor uplift and subsidence, reef-building, and sea-level change [Schlanger, 1981; Larson, Schlanger, et al., 1981; Moberly, Schlanger et al., 1986; Schlanger et al., 1987]. In the spring of 1988, Schlanger returned to the Marshall Islands for the last time as co-chief aboard the R/V *Moana Wave*. The geological history of the Marshall Islands discussed in this paper is based primarily on the interpretation of limestones and basalts dredged from guyots during this expedition and a fall of 1989, R/V *Farnella*, cruise.

All reef limestones recovered from drilling atolls and dredging guyots in the Marshall Islands prior to the spring of 1988 are early Eocene or younger [Emery et al., 1954; Schlanger, 1963; Schlanger et al., 1987]. Between 1963 and 1989, the only radiometric ages available from this region were two late

The Mesozoic Pacific: Geology, Tectonics, and Volcanism
Geophysical Monograph 77

Paleocene to early Eocene K/Ar dates for basalt drilled beneath Anewetak Atoll [Kulp, 1963]. These data are not consistent with models that explain the formation of the Marshall Islands by passage of the region over one or more hotspots during the Cretaceous [e.g., Duncan and Clague, 1985]. Five Late Cretaceous edifice ages have recently been reported from this region: three are minimum ages based on planktonic foraminifers from dredged samples [Schlanger et al., 1981a; Hein et al., 1990] and two are $^{40}Ar/^{39}Ar$ ages for dredged basalts [Davis et al., 1989]. However, based only on these five edifice ages, the monotonic age progression predicted by a simple hotspot model cannot be confirmed.

Since the publication of Hamilton's [1953, 1956] papers reporting the recovery of Cretaceous foraminifers and rudist reef limestones from the Mid-Pacific Mountains, a steady flow of publications has proposed that, during one or two intervals in the Cretaceous, the Pacific basin may have been the site of widespread intraplate volcanism [e.g., Heezen et al., 1973; Matthews et al., 1974; Winterer, 1976a, b; Vallier et al., 1980]. Schlanger and Premoli Silva [1981] and Schlanger et al. [1981b] proposed two distinct pulses of Cretaceous volcanic activity in the Central Pacific, one from ~115 to ~90 Ma and another from ~83 to ~65 Ma, based on dredging and Deep Sea Drilling Project (DSDP) results in the Line Islands and Nauru Basin. Based on the stratigraphic distribution of volcanogenic sediments recovered at DSDP sites, Rea and Vallier [1983] reached a similar conclusion: that widespread volcanism in the Pacific was concentrated during the two intervals ~110 to ~95 Ma and ~80 to ~65 Ma. Winterer and Metzler [1984] proposed two phases of volcanic activity in the Mid-Pacific Mountains: one widespread, major phase during the Early Cretaceous followed by volcanic rejuvenation during the Late Cretaceous in the easternmost Mid-Pacific Mountains. Neither the hotspot or episodic, intraplate volcanism models have been tested in the Marshall Islands owing to a paucity of data.

In this paper, we report the recovery of limestones from three guyots in the Marshall Islands that contain shallow-water debris with Albian age, agglutinated benthic and near-shore planktonic foraminifers. Limestones from two additional guyots contain rudist debris that may be Albian in age (see note in proof); the rudist debris is redeposited with a shallow-water fauna that is Campanian to Maastrichtian in age. Mineral separates from basalt dredged from a sixth seamount have been $^{40}Ar/^{39}Ar$ dated at 138±0.4 Ma. These data confirm that intraplate volcanism began in the northern Marshall Islands as early as Aptian to Albian time. Evidence for volcanism, uplift, and reef building during the Late Cretaceous was found at, or near, all locations of volcanism, uplift, and reef building during the Early Cretaceous. We discuss whether these new data support two phases of widespread, coeval volcanism in the Cretaceous Pacific Ocean or whether they can be explained by "fixed" hotspot processes. Since multiple phases of volcanism in the Marshall Islands, separated by as much as 30 m.y., cannot be reconciled with the monotonic, edifice age progression that would result from the movement of the oceanic lithosphere over a single hotspot, we use plate rotation models to test whether passage of this region over more than one hotspot can provide an adequate explanation for its evolution.

REGIONAL SETTING

The atolls, guyots and seamounts of the Marshall Islands lie atop three roughly parallel, en echelon, northwest-southeast trending ridges located between the Central Pacific Basin to the east and the Mariana Basin to the west; the Nauru Basin lies south of the northwestern Marshall Islands (Figure 1). The ridges that comprise the Marshall Islands lie perpendicular to magnetic anomalies M21 through M29, which date the underlying lithospheric plate as ~154 Ma (M23) beneath the southernmost islands to ~165 Ma (M29) beneath the more central Kwajalein Atoll. The northern islands and guyots lie in the Jurassic magnetic quiet zone on a portion of the Pacific plate presumed to be older than 165 Ma [Larson, 1976; Cande et al., 1978].

The atoll and seamount names used in this paper that differ from previously published names are from Hein et al. [1990] (Table 1). The Harland [1990] time scale is used throughout this paper.

RESULTS FROM PREVIOUS INVESTIGATIONS

Operation Crossroads

Pikinni and Anewetak atolls, both located in the northern Marshall Islands (Figure 1) were drilled in 1947 and 1952, respectively, as part of Operation Crossroads. The deeper of the two holes at Pikinni Atoll penetrated 780 m of reef sediments and reached strata of early Miocene and Oligocene(?) age; volcanic basement was not reached in either hole [Emery et al., 1954; Cole, 1954]. A stratigraphic section equivalent to that drilled at Pikinni Atoll was penetrated at the top of two holes drilled at Anewetak Atoll in 1952 [Cole, 1957; Schlanger, 1963]. Basalt, directly overlain by reef, fore-reef and outer-slope sediments of late Eocene age, was reached at depths of 1271 m and 1388 m below Anewetak Atoll. Subsequent K/Ar dating of this basalt yielded ages, in the range of 51.4 to 61.4 Ma, that are compatible with the age of the overlying sediments [Kulp, 1963]. Similar subsidence histories for the two islands were inferred from the correlation of biostratigraphically dated horizons [Cole, 1957]. Seismic refraction studies at Pikinni Atoll revealed a high surface relief layer with a velocity of ~4 km/sec, believed to be the top of the volcanic edifice, at a depth of ~1300 m [Raitt, 1954]. A basement depth of ~1500 to ~1600 m at Pikinni was estimated from aeromagnetic surveys [Alldredge et al., 1954]. Based on similar depths to volcanic basement and similar subsidence histories, the edifice beneath Pikinni Atoll was assumed to be contemporaneous with the edifice beneath Anewetak Atoll.

Dredging efforts at Wodejebato Guyot, which adjoins Pikinni Atoll to the northwest, as a part of the Scripps Institution of Oceanography Mid-Pacific Expedition of 1950 (MP locations on Figure 2), recovered basalts and a tuff breccia, the cracks of which were filled with early Eocene, phosphatized, planktonic foraminifers [Hamilton and Rex, 1959]. Emery et al. [1954]

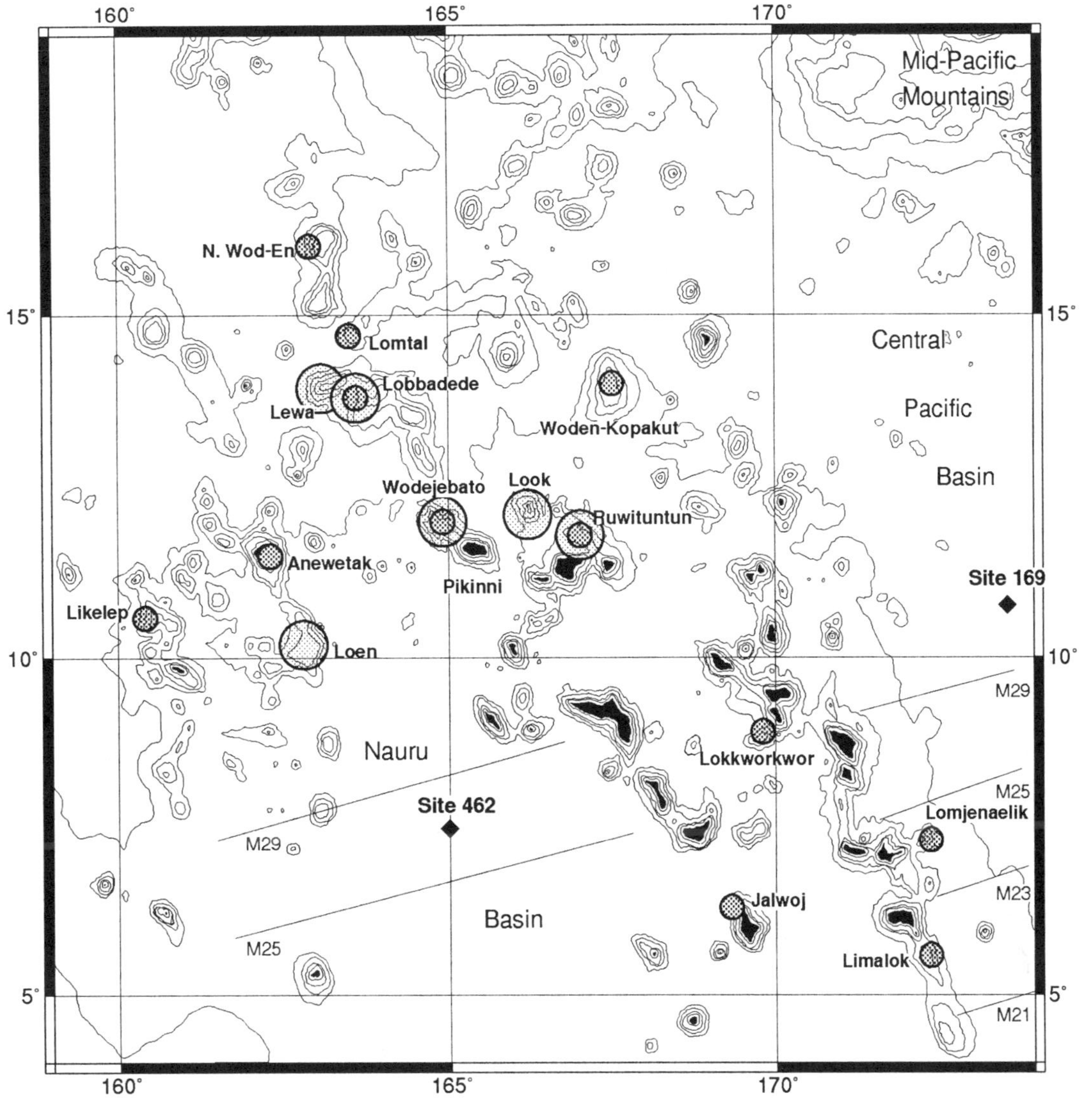

Fig. 1. Map of the Marshall Islands region with locations of seamounts, guyots, atolls, and Deep Sea Drilling Project sites discussed in this paper. Magnetic anomalies are from Larson [1976] and Cande et al., [1978]. The locations of Early Cretaceous seamount volcanism and reef-building discussed in the text are plotted as large circles. The locations of Late Cretaceous volcanism and reef-building discussed in the text are plotted as smaller circles.

suggested that eruption in shallow water was indicated by the highly vesicular and scoriaceous nature of the dredged volcanic rocks; the high proportion of hyaloclastic material in the dredge hauls indicate submarine eruptions.

Subsidence

Detrick and Crough [1978] compared the subsidence histories of Anewetak and Pikinni atolls, derived from the depths to dated strata penetrated during drilling on these atolls, to the subsidence path predicted by assuming that the volcanos formed on ~90 Ma crust that subsequently subsided in accord with Parsons and Sclater's [1977] empirical depth-age relationship for maturing oceanic lithosphere (Figure 3). From this comparison, they concluded that the atolls subsided as if they had formed on much younger crust, ~25 Ma. They inferred from their results that the ~90 Ma oceanic lithosphere had been thermally rejuvenated (thinned and uplifted) at the time the volcanic edifices beneath Anewetak and Pikinni atolls erupted. The fact that the crust underlying Anewetak and Pikinni atolls is now presumed to be ~70 m.y. older than Detrick and Crough [1978] assumed only increases the difference between the observed and predicted subsidence paths of these atolls.

Gravity

Duennebier and Petersen [1982] measured a free-air gravity anomaly of 180 milligals over Wodejebato Guyot; the contours of the gravity anomaly roughly parallel the bathymetric con-

TABLE 1. New Atoll and Guyot Names[1]

Published Name	New Name
Enewetak	Anewetak
Bikini	Pikinni
Sylvania	Wodejebato
Ratak	Woden-kopakut
Erikub	Lokkworkwor
Harrie	Limalok
von Valtier	Lomjenaelik

[1]Compiled by Alfred Capelle and Rev. Kanki Amlej, Alele Museum, Republic of the Marshall Islands [Hein et al., 1990].

tours. A gravity anomaly of this magnitude that parallels bathymetry indicates eruption of the volcano on mature lithosphere that is stronger and less flexible than lithosphere near a ridge crest [Watts et al. , 1985]. A similar free-air gravity anomaly (178 milligals) over Limalok Guyot, in the southern Marshall Islands [Schlanger et al., 1987] indicates that the origin of edifices, i.e. off-ridge volcanism on mature lithosphere, is similar throughout the region.

Results from Nearby DSDP Sites

Drilling at Site 65, east of the southern Marshall Islands, penetrated Paleocene to Oligocene turbidites rich in volcanic material that included semi-consolidated clasts containing Late Cretaceous (*Globotruncana fornicata*) foraminifers and rare, altered radiolarians of probable Cretaceous age [Winterer, Riedel et al., 1971; Heath and Moberly, 1971]. The source of this sediment, most likely a submerged ridge adjacent to the easternmost Marshall Islands, is probably a Late Cretaceous edifice topped by Paleocene to Oligocene calcareous sediments.

Drilling at Site 169 in the Central Pacific Basin, just east of the Marshall Islands, recovered extrusive alkalic basalts beneath

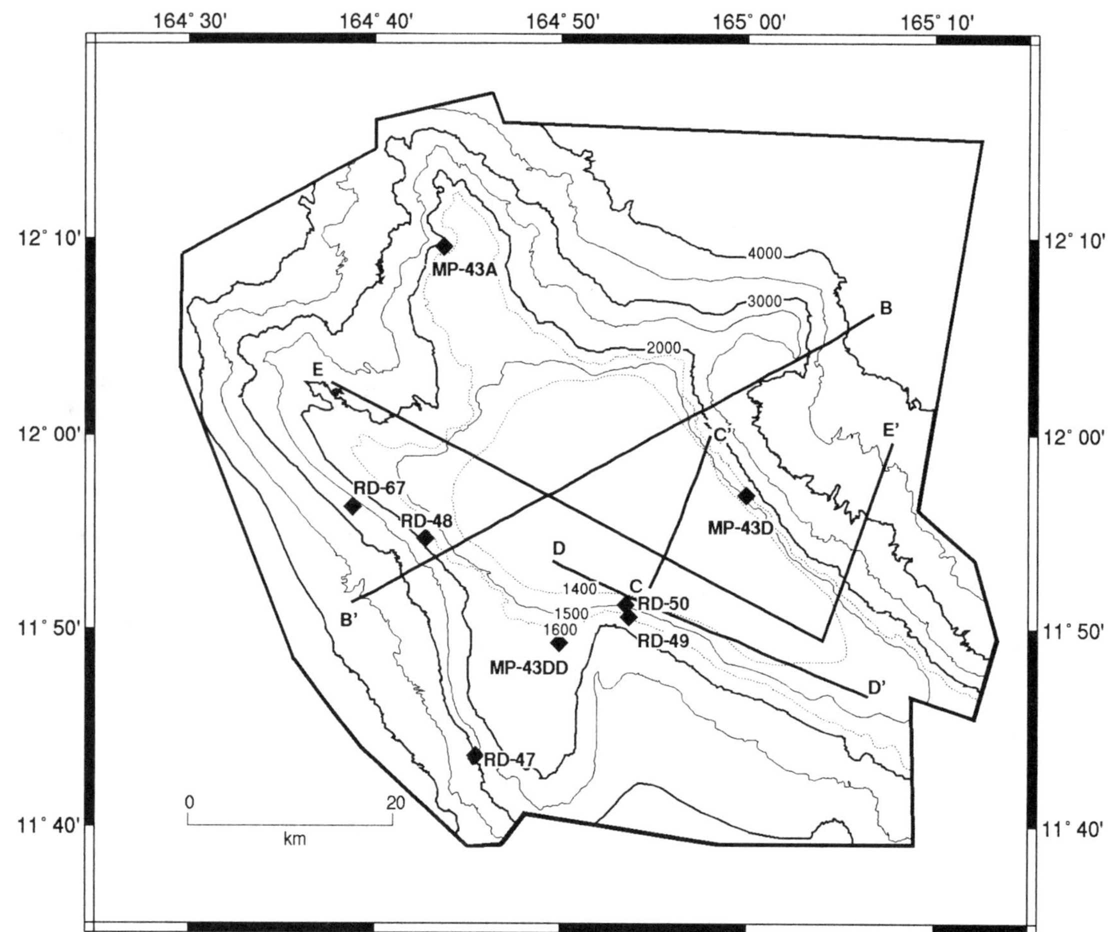

Fig. 2. Bathymetric chart of Wodejebato Guyot based on seismic reflection profiles and SeaMARC II bathymetry acquired in the spring of 1988 during R/V *Moana Wave* cruise MW8805. Dredge locations prefixed with MP were recovered during the 1950 Mid-Pacific Expedition [Hamilton and Rex, 1959], RD-67 was recovered during R/V *Kana Keoki* cruise KK81062604 [Petersen et al., 1986] and RD-47 through RD-50 were recovered during R/V *Moana Wave* cruise MW8805. Lines C-C′ and D-D′ are the locations of the 3.5 kHz profile and the sidescan image, respectively, shown in Figure 6. Lines B-B′ and E-E′ are the locations of seismic profiles shown in Figure 12.

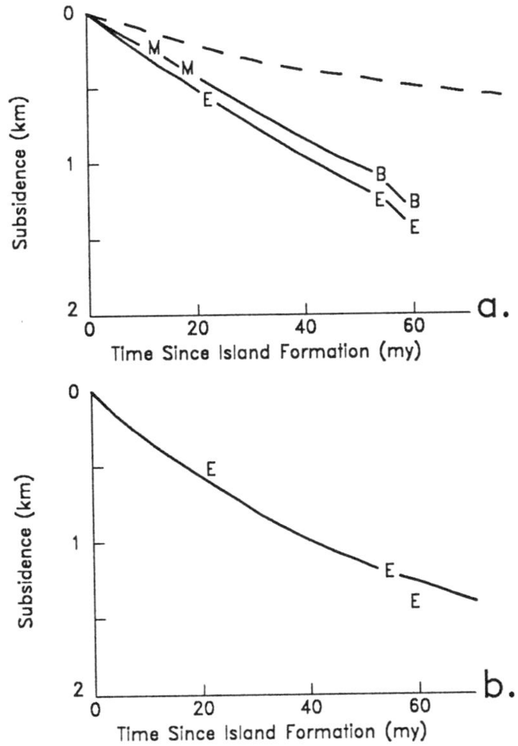

Fig. 3. (*a*) This figure shows the subsidence history of Enewetak (E), Bikini (B) and Midway (M) atolls derived from drilling and biostratigraphic dating. The dashed line shows the predicted subsidence assuming that these islands formed on 90 Ma crust and have subsided according to Parsons and Sclater's [1977] empirical depth-age relation for normal sea floor. (*b*) The predicted subsidence history (solid line) of Enewetak is shown assuming the lithosphere was thinned to the equivalent of normal 25 Ma lithosphere. E's are dated horizons derived from drilling on Enewetak. After Detrick and Crough [1978].

late Albian zeolitic clays [Winterer, Ewing et al., 1973]. A diabase sill was penetrated 20 m beneath Campanian sediments and a few meters above Cenomanian-Turonian sediments providing further evidence of Late Cretaceous volcanism along the easternmost ridge of the Marshall Islands [Winterer, Ewing et al., 1973]. Ozima et al. [1983] calculated an age of 90±4 Ma for the sill based on a well defined $^{40}Ar/^{39}Ar$ isochron and step-heating spectrum.

A thick section of Early Cretaceous submarine flows was penetrated at DSDP Site 462 in the Nauru Basin (Figure 1) [Larson, Schlanger et al., 1981; Moberly, Schlanger et al., 1986]. Takigami et al. [1986] calculated a $^{40}Ar/^{39}Ar$ age of 129.7±4.6 Ma for one of the lower flows. This date is uncertain since it is based on a plateau in the age spectrum for the 800°C and 1000°C ^{39}Ar fractions only. Two other uncertain ages have been reported for these flows: a total fusion $^{40}Ar/^{39}Ar$ date of 130±15 Ma from near the top of the flow sequence [Ozima et al., 1981], and a Rb-Sr isochron age of 127±9 Ma for secondary zeolites and smectites [S. Hart in Castillo et al., 1986]. Pringle [1992] reports a more credible, younger age of 110.8±1 Ma for two of the lower flows at Site 462 based on four whole rock incremental-heating experiments that are concordant according the criteria described in Dalrymple et al. [1981].

The uppermost sills at Site 462 are no older than late Albian since the youngest intruded sediments are late Albian to Cenomanian [Schlanger and Moberly, 1986]. Ozima et al. [1981] report a $^{40}Ar/^{39}Ar$ age, with a reasonably flat step-heating age spectrum, of 110±3 Ma for a basalt from one of these sills. Occasional rudist fragments and one specimen attributable to the mid-Cretaceous orbitolinids (foraminifers) were found in turbidites at Site 462. One or more nearby edifices, possibly Kwajalein Atoll to the northeast, may have supported a mid-Cretaceous reef atop an Early to mid-Cretaceous volcano. Terrestrial plant fragments in the Aptian volcaniclastic sediments recovered at Site 462 [Jenkyns and Schlanger, 1981] suggest the existence of forested volcanic islands in the vicinity of the Nauru Basin during this Aptian to Albian volcanic episode.

Late Campanian to late Maastrichtian volcaniclastic breccias drilled at Site 462 provide evidence for Late Cretaceous volcanism around the Nauru basin. Displaced shallow water faunas within the thick turbidite sequences penetrated at Site 462 are evidence that edifices reached the photic zone adjacent to the Nauru Basin and supported carbonate banks during late Campanian to middle Maastrichtian time [Premoli Silva and Brusa, 1981; Premoli Silva, 1986].

The post-Cretaceous sediments drilled in the Nauru Basin now lie at depths that are shallower than predicted by the Parsons and Sclater [1977] seafloor depth-age curve. Schlanger [1981] proposed that the thermal age of the lithosphere in the Nauru Basin was reset to an age younger than the ~160 Ma inferred from magnetic lineations during a period of regional intraplate volcanism spanning late Aptian time to Campanian-Maastrichtian time.

Results from More Recent Dredging

Rocks similar to those recovered during the Mid-Pacific Expedition were recovered from the south side of Wodejebato Guyot (RD 67 in Figure 2) during a 1981 Hawaii Institute of Geophysics cruise [Petersen et al., 1986]. They included vesicular basalts and phosphatized pelagic sediment of early to middle Eocene age. One piece of hyaloclastite contained shallow-water peloids, fragments of a larger benthic foraminifer, calcareous red algae, and bryozoan fragments.

Eocene reef limestones and early Eocene pelagic limestones containing volcanic debris were recovered from Limalok guyot in the southern Marshall Islands [Schlanger et al., 1987]. Basalts dredged from two guyots during a 1984 U. S. Geological Survey cruise [Davis et al., 1989] yielded a $^{40}Ar/^{39}Ar$ total fusion isochron age of 87.3±0.6 Ma for Lokkworkwor Seamount and a 82.2±1.6 Ma incremental heating age for Woden-kopakut Guyot. Lomjenaelik Seamount in the southern Marshall Islands has been dated as Coniacian-Santonian (~83-87.5 Ma) or older based on planktonic foraminifers from a pelagic limestone containing volcanic debris [Schlanger et al., 1981a]. Dredging in 1989 during USGS-KORDI cruise F10-89-CP [Hein et al., 1990] recovered a breccia with Campanian age planktonic foraminifers from Likelep Guyot and a mudstone with Campanian or Maas-

TABLE 2. Dredge Locations

Dredge	Latitude	Longitude	Seamount
MW8805:[1]			
RD33	10° 1.0′	162° 42.5′	Loen
RD40	11° 46.6′	167° 2.0′	Ruwituntun
RD41	11° 43.4′	167° 1.2′	Ruwituntun
RD43	11° 51.3′	166° 47.0′	Ruwituntun
RD45	12° 0.7′	166° 8.8′	Look
RD47	11° 43.6′	164° 45.6′	Wodejebato
RD48	11° 54.5′	164° 42.6′	Wodejebato
RD49	11° 50.5′	164° 53.6′	Wodejebato
RD50	11° 51.1′	164° 53.5′	Wodejebato
RD56	13° 52.0′	163° 36.5′	Lobbadede
RD58	13° 56.4′	163° 4.7′	Lewa
RD59	14° 42.4′	163° 27.1′	Lomtal
RD62	15° 08.2′	162° 58.8′	North Wod-En
F1089CP:[2]			
D4	12° 11.0′	166° 18.5′	Look
D19	10° 36.0′	160° 22.3′	Likelep

[1]*R. V. Moana Wave*, May, 1988 [Duennebier and Schlanger, 1988]

[2]*R. V. Farnella*, October, 1989 [Hein et al., 1990].

trichtian age planktonic foraminifers from Jalwoj Atoll. Reexamination of the breccia from Likelep Guyot as well as new $^{40}Ar/^{39}Ar$ ages from basalt recovered in the same dredge, both reported in this paper, confirm the age determined in Hein et al. [1990].

NEW $^{40}AR/^{39}AR$ AGES

Samples Studied

The basalts and limestones discussed in this and subsequent sections were recovered by dredging during R/V *Moana Wave* cruise MW8805 in May of 1988 [Duennebier and Schlanger, 1988] and during R/V *Farnella* cruise F1089CP in 1989 [Hein et al., 1990]. The location of dredges discussed in this paper are listed in Table 2.

K-Ar Geochronology

The K-Ar clock has been widely used to date oceanic lavas. However, several fundamental assumptions of the method may be violated by samples that have been altered in the submarine environment. Conventional K-Ar apparent ages of such samples are only minimum age estimates, because of both K-addition and radiogenic ^{40}Ar-loss over geologic time. For example Pringle [this volume] shows that conventional K-Ar ages of the least altered samples dredged from Cretaceous seamounts can be 10 to 30% younger than the original crystallization age.

$^{40}Ar/^{39}Ar$ apparent ages are generally better estimates of the original crystallization age, presumably because ^{39}Ar is lost during the experimental process from the same sites of K-addition and radiogenic ^{40}Ar loss over geologic time [Clague et al., 1975; Dalrymple and Clague, 1976]. However, Pringle [this volume] shows that even the freshest Cretaceous seamount samples can have $^{40}Ar/^{39}Ar$ total-fusion ages as much as 10% older or younger than the original crystallization age; more altered samples can easily have total-fusion ages as much as 30% too old or too young. Ages of individual steps of incremental-heating experiments can deviate even more from the original crystallization age.

The real advantage of using $^{40}Ar/^{39}Ar$ geochronology in deciphering the crystallization age of altered rocks is that one can generate a set of apparent ages for a particular sample, and then use a series of internal tests on this set to differentiate accurate crystallization ages from unreliable apparent ages. It may not be possible to determine why a particular age from a sample is unreliable. However, one can test whether the K-Ar clock of a particular sample is too disturbed to reveal a meaningful age. Here we use the criteria of Pringle [this volume], modified from those suggested by Lanphere and Dalrymple [1978] and Dalrymple et al. [1980].

Analytical Techniques

Whole rock samples were prepared as 6 mm cores drilled from the freshest sections of 1 cm-thick sawed slabs. Mineral separates were prepared using conventional heavy liquid and magnetic techniques, and further cleaned with appropriate acids and distilled water in an ultrasonic bath. Plagioclase was cleaned with cold 5-10% HF for 5-8 minutes followed by 30 minutes in 2.5N HCl; hornblende was cleaned with the 2.5N HCl only.

Whole rock samples, 0.5 to 0.8 g each, and mineral separates, up to 15 mg each wrapped in Al-foil, were sealed in air in quartz vials. They were irradiated for 24-48 hours in the core of the U.S.G.S. TRIGA reactor [Dalrymple et al., 1981]. Whole rock samples were analyzed as incremental-heating experiments using an all-metal resistance furnace extraction line and on-line Nier-type single-collector mass spectrometer. Mineral separates were analyzed using a 5W continuous argon-ion laser extraction system and a Mass Analyzer Products model 216 rare-gas mass-spectrometer with a Bauer-Signer source [Dalrymple, 1989]. The reactor flux, J, was measured with the intralaboratory standards biotite SB-3 (162.9 Ma) and sanadine TCR 85G003 (27.88 Ma).

Errors given for the calculated ages were calculated as weighted means, where weighting is by the inverse of the estimated variance [Taylor, 1982]. Isochron ages were calculated using the $^{40}Ar/^{36}Ar$ vs. $^{39}Ar/^{36}Ar$ correlation and the York2 least-squares fit with correlated errors [York, 1969]. There are no significant differences between the results of this correlation calculation and the one based on the $^{36}Ar/^{40}Ar$ vs. $^{39}Ar/^{40}Ar$ correlation [Dalrymple et al., 1988].

Results

The best ages for the volcanic edifice at each of five seamounts, guyots, or atolls studied here are listed in Table 3. In general, the isochron age is preferred as the best estimate of each sample because it (1) combines both an estimate of analytical error and the scatter about the isochron in the final error estimate, and (2) makes no assumption about the composition of the initial non-radiogenic, or trapped, argon component.

Anewetak Atoll. Two incremental heating experiments were done on each of two samples from Anewetak drill core (Figure 4; Table 3). All four experiments represent reliable age determi-

TABLE 3. Summary of Radiometric $^{40}Ar/^{39}Ar$ age determinations for the northern Marshall Islands.

	Plateau or Weighted Total Fusion Age			Isochron Age			
Sample	Material	% of 39Ar	Age[1] (Ma)	Age[1] (Ma)	(40/36)[1]	sums/n−2	n
Anewetak Atoll (Best Age[2]: 75.9 ± 0.6)							
E-1-4-3A	Whole Rock	84.2	77.1 ± 0.6	76.3 ± 1.2	299.1 ± 3.9	0.97	8
		95.6	77.8 ± 0.5	76.6 ± 0.8	309.1 ± 10.5	1.25	12
E-1-5-6A	Whole Rock	73.1	75.3 ± 0.6	74.3 ± 1.3	310.0 ± 15.6	0.65	10
		95.9	74.8 ± 0.5	73.7 ± 1.9	333.4 ± 76.8	0.76	8
Likelep Seamount (Best Age[2]: 82.1 ± 0.7, 75.5 ± 0.8							
F1089CP D19-18	Whole Rock	62.6	82.4 ± 0.4	82.1 ± 0.7	133.1 ± 92.6	1.94	7
F1089CP D19-20	Whole Rock	95.6	77.2 ± 0.4	75.5 ± 0.8	426.5 ± 56.7	1.36	8
Lobbadede Guyot (Best Age[2]: 82.4 ± 2.4)							
MW8805 RD-56-1	Whole Rock	59.1	82.4 ± 0.7	82.4 ± 2.4	289.1 ± 23.2	3.46	7
North Wod-En Seamount (Best Age[2]: 85.3 ± 0.4)							
MW8805 RD-62-1	Plagioclase	97.0	85.4 ± 0.5	84.7 ± 0.5	307.4 ± 7.8	1.23	9
MW8805 RD-62-4	Whole Rock	90.1	85.6 ± 0.5	85.8 ± 0.6	161.9 ± 47.5	0.51	8
		77.5	86.6 ± 0.5	86.1 ± 1.0	287.5 ± 107.5	1.35	10
Look Seamount (Best Age[2]: 138.2 ± 0.8)							
MW8805 RD-45-1	Hornblende	(All)[3]	140.0 ± 0.8				
F1089CP 4-2	Hornblende	(All)[3]	138.1 ± 0.7	138.2 ± 0.5	343.3 ± 28.5	1.59	13
F1089CP 4-10-1	Hornblende	(All)[3]	138.7 ± 0.7				

[1]± one standard deviation.
[2]"Best Age" is calculated by weighting the individual isochron ages by the inverse of the variance (see text).
[3]Laser total fusion analyses of groups (45-1) or individual (4-2 and 4-10-1) hornblende crystals.

nations according to the criteria listed in Pringle [this volume]. The best age estimate for the basalt recovered from the bottom of the Anewetak Atoll drill core is the weighted mean of the four isochron ages, 75.9±0.6 Ma. This is significantly older than conventional K-Ar ages of ca. 61-51 Ma determined by Kulp [1963] and emphasizes that conventional K-Ar ages of even relatively fresh Cretaceous seamount basalts are minimum age estimates only.

This new age is significantly older than the middle to late Eocene (~36-44 Ma) limestones that were recovered only 10's of meters above the basalt edifice. At Midway Atoll in the Hawaiian Island chain, hawaiite and mugearite pebbles were recovered beneath over 380 m of early Miocene to Recent reef limestone [Ladd et al., 1970; Cole, 1969]. Dalrymple et al., [1977] dated the basalt by the K/Ar method at 27.7±0.6 Ma. Drilling at a second hole recovered basalt beneath ~160 m of Pliocene to Recent sediments. Based on these drilling results from Midway Atoll, a reasonable hiatus between the cessation of volcanic activity and the deposition of reef limestones can range from near zero on the flanks of the edifice to greater than 20 m.y. at the peak, which is the last part of the volcano to subside below sea level. Furthermore, due to the poor recovery during drilling of Anewetak, we cannot exclude the possibilities that either the oldest limestones were not recovered or that there may have been a period of post-Cretaceous and pre-Eocene uplift and erosion. Based on a comparison with Midway Atoll and poor recovery, we conclude that the apparent hiatus between the basalt and the oldest reef recovered beneath Anewetak (~30 m.y.) is not unreasonably long.

Likelep Seamount. Two incremental-heating experiments on samples of different lithologies dredged from Likelep Seamount had significantly different $^{40}Ar/^{39}Ar$ ages (Figure 4, Table 3). The experiment on sample F1089CP D19-18 met all the criteria for a reliable age, and is the best estimate for inception of volcanism at Likelep, 82.1±0.7 Ma. The experiment on sample F1089CP D19-20 yielded a younger age estimate, 75.5±0.8 Ma. Since the $^{40}Ar/^{39}Ar$ of the isochron is significantly higher than 295.5, this age does not meet all the criteria for a reliable age and must be treated with some caution. We interpret these two ages to represent two separate periods of volcanism over at least 5-7 m.y., consistent with the conclusion of Pringle et al. [1991] that volcanism can last at least 7 m.y. at even only moderately-sized seamounts. The radiometric ages for basalt from Likelep are consistent with the Campanian minimum fossil age for a breccia from the same dredge reported by Hein et al. [1990] and discussed later in this paper.

Lobbadede Guyot. Only one sample from Lobbadede Guyot, MW8805 56-1, was suitable for $^{40}Ar/^{39}Ar$ dating (Figure 4, Table 3). The plateau age for the age spectrum is concordant with the isochron age, but some of the plateau steps are significantly different at the 95% confidence level and the F-variate statistic of the isochron, the SUMS/N-2 ratio of 3.46, indicates greater scatter than can be attributed to analytical error alone. The isochron age of 82.4±2.4 Ma is the best estimate for the eruptive age of this basalt because it does include an estimate of the excess scatter about the regression line in the error estimate. However, because this age is based on a single sample from this seamount, and because it does not meet all the

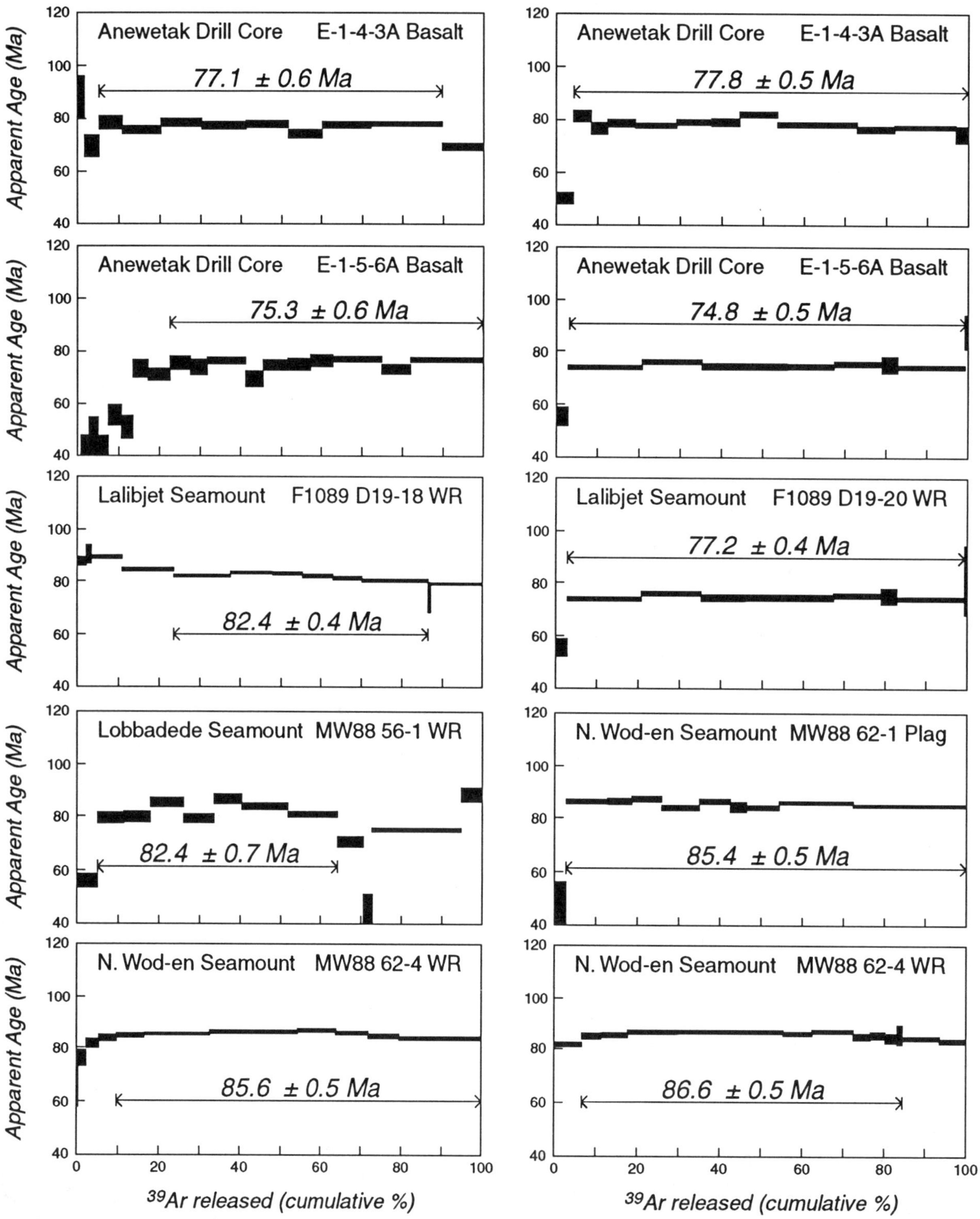

Fig. 4. $^{40}Ar/^{39}Ar$ age spectra for whole rock and mineral separates from Anewetak Atoll and three seamounts in the northern Marshall Islands.

criteria for a reliable age, it should be interpreted with some caution.

North Wod-En Seamount. Two samples from North Wod-En Seamount were suitable for $^{40}Ar/^{39}Ar$ dating (Figure 4, Table 3). An incremental-heating experiment on 15 mg of plagioclase separated from sample MW8805 62-1 was conducted with the laser extraction line by defocussing the laser beam to 2-3 mm wide and varying the laser output to achieve differential temperature steps. For a whole rock sample, MW8805 62-4, two incremental heating experiments were conducted because the $^{40}Ar/^{39}Ar$ intercept of the first was significantly lower than the atmospheric ratio of 295.5, although the experiment met all other criteria for determinating a reliable age. The best age estimate for the seamount is the weighted mean of all isochron ages, 85.3±0.4 Ma.

Look Seamount. Hornblende was separated from three samples from two dredges about 60 km apart on Look Seamount. Thirteen laser fusion analyses (Table 2) were run on either single hornblende crystals weighing 2 to 4 mg each (samples F1089CP 4-2 and 4-10-1) or 8 to 10 smaller crystals, also totaling about 2 to 4 mg (sample MW8805 45-1). The age for Look Seamount is an isochron analysis of all 13 analysis (Figure 5), 138±0.8 Ma. This is the oldest documented seamount volcanism in the Marshall Islands; this is also the oldest age calculated for any seamount basalt in all of the ocean basins.

NEW PALEONTOLOGIC AGES

Wodejebato Guyot

During R/V *Moana Wave* cruise MW8805 in the spring of 1988, manganese coated vesicular basalts, hyaloclastites, breccias and moderately to well lithified pelagic sediments were recovered from the talus slopes that flank Wodejebato Guyot

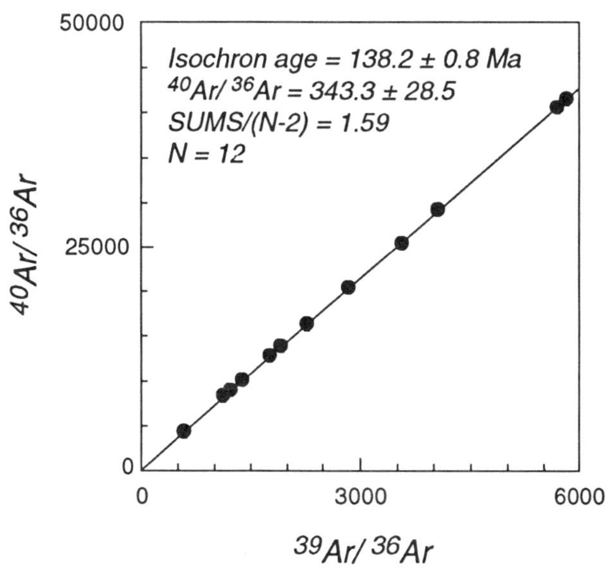

Fig. 5. $^{40}Ar/^{39}Ar$ isochron for Look Seamount based on 13 hornblende analyses from 2 different dredges.

(RD-47, RD-48, RD-49; locations in Figure 2 and Table 2, talus slopes visible in Figure 6).

Fossil Evidence. Planktonic foraminifers in the matrices from breccias recovered in RD-48 and RD-49 only constrain the age of volcanism at Wodejebato to Eocene or older. However, the early Tertiary matrix from one sample recovered in RD-47 contains clasts of pelagic sediment with poorly preserved Late Cretaceous planktonic foraminifers. The only identifiable species is *Globotruncana bulloides* which ranged from late Santonian through the middle Maastrichtian. The lack of *Rugoglobigerina* sp., that became prevalent in mid to Late Campanian time, as well as the presence of elongated species of *Heterohelix*, that were typical of Turonian through Santonian time, suggests these clasts are late Santonian to earliest Campanian in age. The Late Cretaceous and early Tertiary faunas are mixed at the edges of the clasts suggesting erosion and redeposition of poorly consolidated Late Cretaceous pelagic sediment in early Tertiary time. Based on the rocks recovered in RD-47, Wodejebato Guyot is Santonian or older. Rounded basalt pebbles in several conglomerates recovered in RD-49 indicate subaerial erosion of the volcanic edifice.

RD-50 (Figure 2) was successful in recovering ~300 lbs. of manganese encrusted limestones from the slope break (Figure 6) on the south side of Wodejebato Guyot. Although most of the limestone has undergone pervasive dissolution, several larger pieces contain well-preserved cores of coarse grained, poorly sorted, shallow-bank limestone that comprises larger benthic foraminifers, calcareous algae, coral fragments, echinoid debris, occasional bryozoan fragments, and molluscan debris (Figure 7).

The larger foraminifers are dominated by *Sulcoperculina vermuti* with minor numbers of *S. obesa* and *S. globosa.* Species of *Asterorbis*, similar to *A. havanensis* and *A. cubensis*, are also abundant in these rocks. *Sulcoperculina* and *Asterorbis* are common constituents of turbidites penetrated at Site 462 in the Nauru Basin [Premoli Silva and Brusa, 1981; Premoli Silva, 1986]; the discovery of these genera on Wodejebato Guyot confirms a source area for the shallow-water debris recovered in the Nauru Basin. Premoli Silva [1986] assigned an age of Campanian to Maastrichtian to the Nauru Basin shallow-water fossil assemblages that contained these genera.

Occasional specimens of a third large foraminifer are present in these rocks. Based on equatorial layers that are very thick relative to the thin, asymmetric lateral chambers, only one or two stolon in large equatorial chambers, and a heavily armored bilocular to trilocular embryonic apparatus, these foraminifers can be attributed to primitive *Orbitoides tissoti* [van Gorsel, 1978; Butterlin, personal communication, December 1990]. The range of this species of *Orbitoides* is estimated as early to mid Campanian, possibly coeval with or slightly older than the *Sulcoperculina-Asterorbis* assemblage.

Most notable among the molluscan debris are abundant fragmented and rare whole rudists. The fragmentary nature of the rudist debris make the identifications uncertain (see Note in Proof). The most abundant rudist debris consists of shell fragments that have a radiating, finely reticulate microstructure similar to that found in the calcitic outer shell wall of the radiolitid

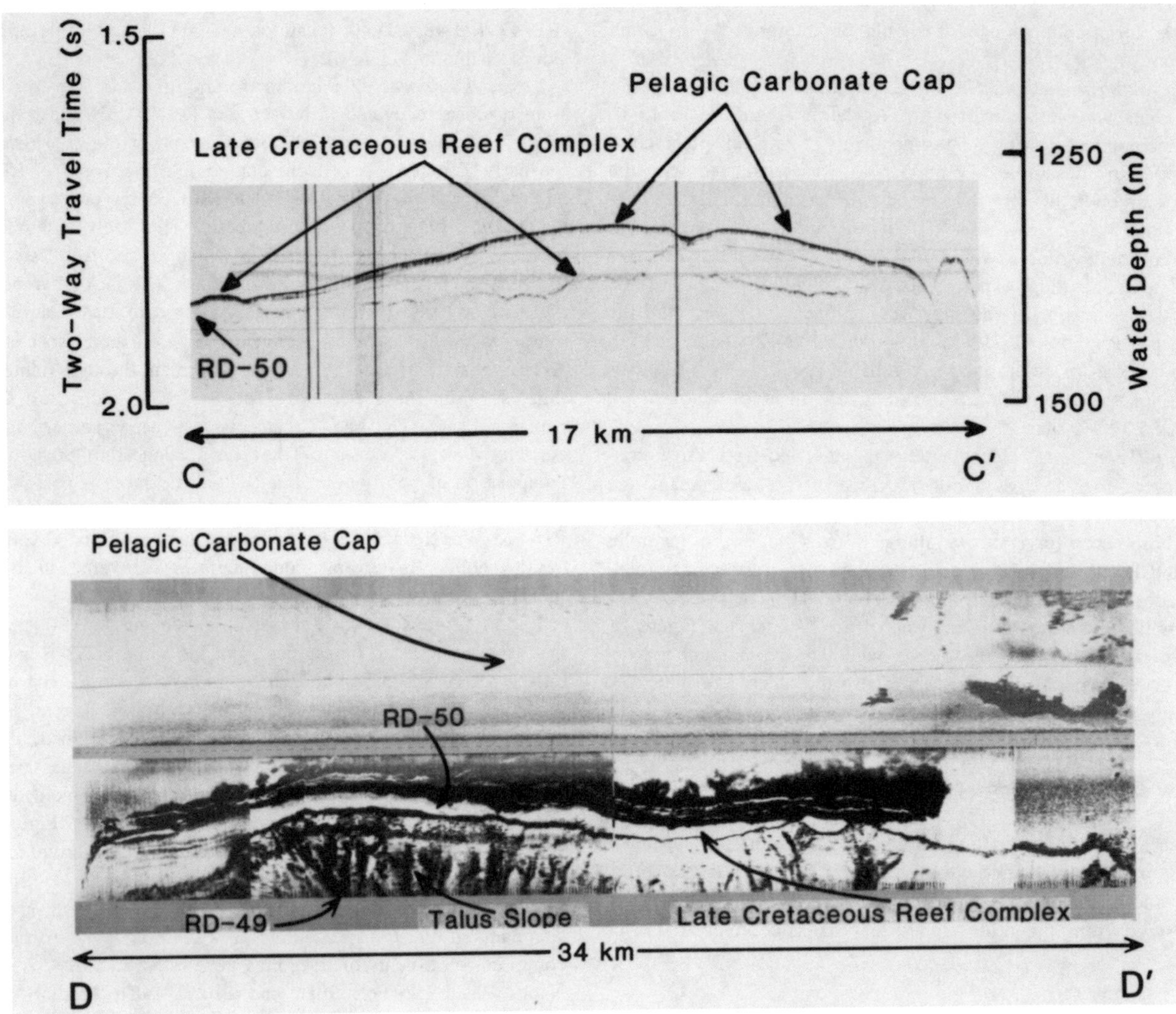

Fig. 6. Line C-C′ is a 3.5 kHz seismic-reflection profile across the southeastern end of Wodejebato Guyot taken while leaving the site of RD-50. Line D-D′ is a side-scan image parallel to the slope break from which RD 50 was recovered. Note the white pelagic sediment cap, highly reflective (black) fringing reef from which rocks were recovered in RD-50 and the gullied talus slopes from which rocks were recovered in RD-47, 48, and 49. The locations of both lines are shown in Figure 2.

rudist *Sphaerulites* sp. from the late-early to early-middle Albian Glen Rose Limestone of central Texas [Coogan, 1969, 1973, 1977]. More evolved radiolitid rudists, characteristic of post-middle Albian time, generally have a coarser reticulate shell structure that is not found in the rudist fragments dredged from Wodejebato Guyot.

Occasional remnants of finely reticulate outer shell wall are found in place within shell molds of radiolitid lower valves (Figure 8). While the outer shell wall of the radiolitid rudists was thick and calcitic, the original composition of the thin inner shell layer is believed to have been aragonitic since this portion of the shell is rarely found well preserved [Majewski, 1969]. The inner shell walls surrounding the body cavities of radiolitids recovered from Wodejebato are pseudomorphs composed of well-cemented planktonic foraminiferal ooze. The small size of these planktonic foraminifers suggest a Paleocene age, although the preservation is too poor for identification. Typically the calcitic outer shell wall of the radiolitid rudists in RD-50 are completely dissolved, leaving behind only the pseudomorphed inner wall and body cavity within a mold of the outer wall (Figure 8c). Since the same Late Cretaceous shallow-bank sediments fill the rudist body cavities and comprise the surrounding matrix, replacement of the inner shell wall and dissolution of the outer shell wall took place following the Late Cretaceous reef building episode.

A second rudist is represented by only a few thick shell fragments that consist of radiating channels and plates (Figure 9). The original composition was probably aragonitic since these shell fragments have been pseudomorphed by the same planktonic ooze as the thin, aragonitic inner wall of the radiolitid

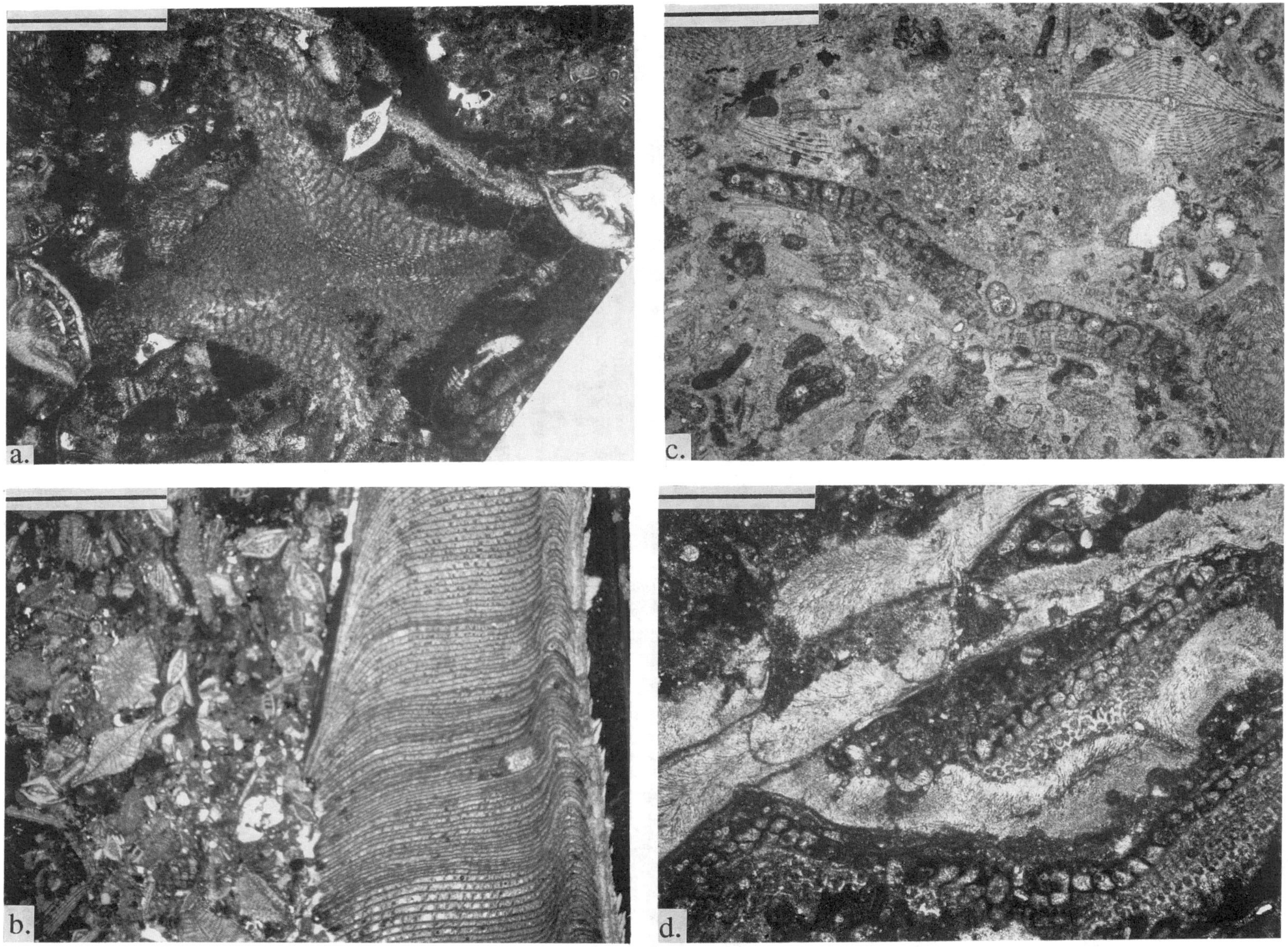

Fig. 7. (*a*) Late Cretaceous large foraminifer species of *Asterorbis* (center) and *Sulcoperculina* (left and right) in limestones recovered in RD-50 from Wodejebato Guyot. Scale bar = 1 mm. (*b*) Photomicrograph of a rudist shell fragment (right half of figure) within Campanian-Maastrichtian foraminiferal matrix (left half of figure) recovered from Wodejebato Guyot in RD-50. The finely reticulate microstructure of the shell fragment is characteristic of the thick, calcitic outer shell wall of the radiolitid rudist *Sphaerulites sp.* Scale bar = 5 mm. (*c*) Photomicrograph of large foraminifers *Orbitoides* (center) and *Asterorbis* (right side) from limestone in RD 50. Thin asymmetric lateral chambers, heavily armored embryonic apparatus and only one to two stolon per large equatorial chamber suggests a primitive form similar to *Orbitoides tissoti*. Scale bar = 1 mm. (*d*) Photomicrograph of encrusting foraminifer *O. tissoti* (dark) and calcareous alga *Ethelia?* (light) from the same sample as in (*c*). Scale bar = 1 mm.

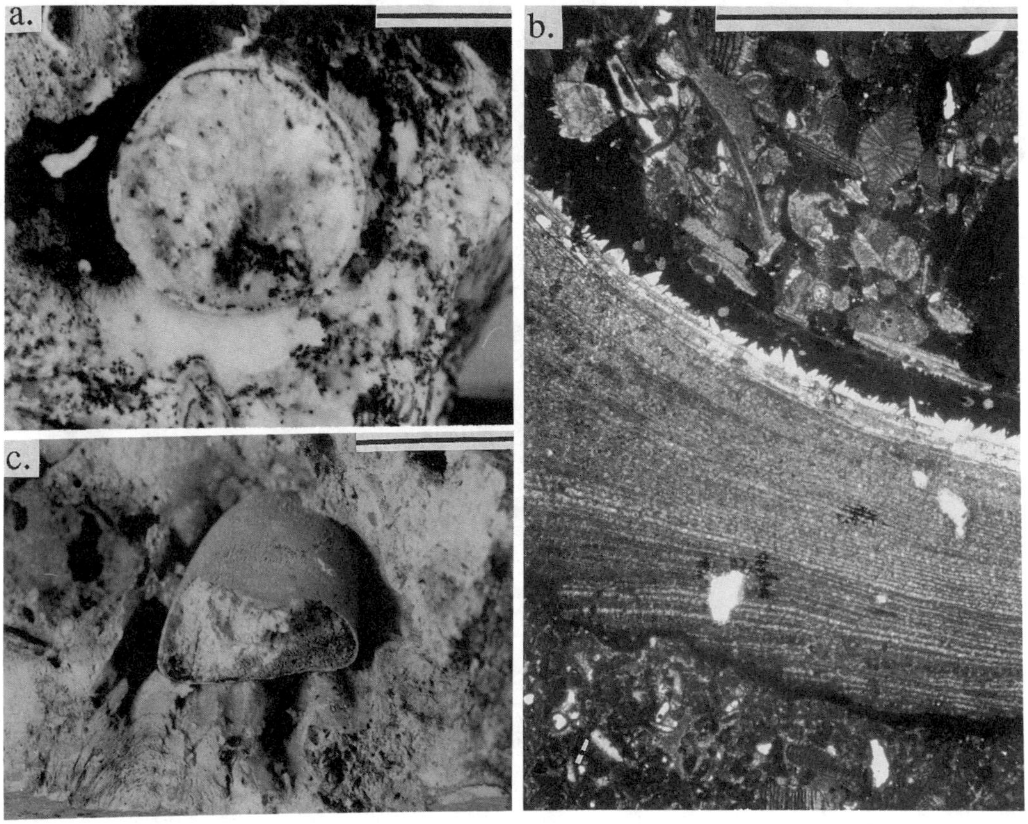

Fig. 8. (*a*) Cross section through radiolitid rudist recovered in RD-50 from Wodejebato Guyot. The circular body cavity in center of the figure is filled by Campanian-Maastrichtian shallow-bank sediments. A thin, inner shell wall, pseudomorphed by planktonic foraminiferal sediments, is visible surrounding the body cavity. The white shell fragment in the lower center of the figure is a small portion of the original outer shell wall. The dark shadow in the upper left corner is a mold of the dissolved outer shell wall. Scale bar = 1 cm. (*b*) Photomicrograph from the slab in (*a*) shows Late Cretaceous benthic foraminifers in the body cavity (top third of the figure) surrounded by the pseudomorphed inner shell wall (dark band in center of figure) and the finely reticulate outer shell wall. Scale bar = 5 mm. (*c*) Radiolitid rudist body cavity with pseudomorphed inner shell wall (center). Only a mold of the dissolved outer shell wall remains. Scale bar = 1 cm.

shell fragments. These are fragments from the thick, porous, aragonitic shell wall of the rudist family Caprinidae. Based on the bifurcating to trifurcating pattern of the radial plates and channels, fragments from Wodejebato Guyot are Caprinidae equivalent in evolution to *Coalcomana* sp. or *Caprinuloidea perfecta* [Coogan, 1969, 1973, 1977]. A few small *Monopleura* cf. *M. marcida* [White, 1884] have also been preserved with both valves intact. By correlation with the Glen Rose Limestone in central Texas, where *Coalcomana* occur together with species of the radiolitid rudist *Sphaerulites* and *Monopleura* [Coogan, 1977], the rudist reef debris recovered from Wodejebato Guyot may be Albian in age.

Pseudomorphed, imbricate shell fragments (Figure 10) were also recovered in RD 50 and are tentatively identified as mytilid pelecypod shells, possibly of the oyster-like genus *Chondrodonta*. These shells are now pseudomorphed by sediments containing Late Cretaceous larger benthic foraminifers. *Chondrodonta* were common accessory organisms that apparently thrived in clear water near the top of Cretaceous shelf margin rudist reef mounds; they commonly occur as densely packed heaps of shell debris [Wilson, 1975]. *Chondrodonta* are associated with caprinid and radiolitid rudists in the middle Cretaceous of central Mexico and Texas [Nelson, 1959; Coogan et al., 1972] and are found in the middle Cretaceous reef-crest facies of the Mishrif Formation, Fateh field, Dubai [Jordan et al., 1985]. The association of a caprinid and radiolitid rudist fauna with *Chondrodonta* at Wodejebato Guyot indicates these redeposited fossils grew in a clear water patch or sand reef environment at or near the reef rim of an Albian reef complex that once capped the guyot's volcanic basement.

Interpretation. Based on fossil evidence, the foundation of Wodejebato Guyot reached the photic zone prior to or during Aptian or early Albian time (Figure 11). Reheating of the underlying Late Jurassic lithosphere during this period of volcanism resulted in crustal thinning and uplift.

As this Aptian-Albian volcanic edifice subsided along with its thermally maturing lithosphere, an Albian rudist reef complex developed on top of the edifice. Based on the thermal sub-

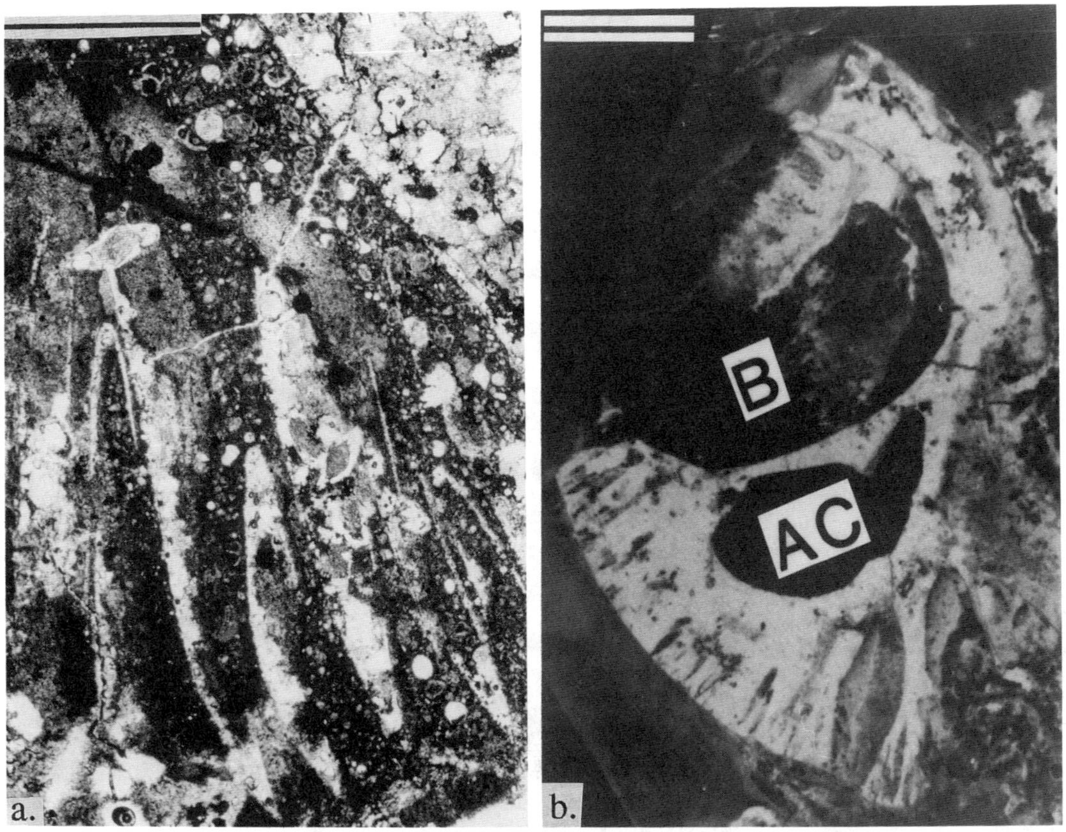

Fig. 9. (*a*) Photomicrograph from caprinid rudist fragment shows that the shell material has been pseudomorphed by planktonic foraminiferal sediments. Scale bar = 2 mm. (*b*) Photograph of a rudist shell fragment (white) showing the complex pattern of trifurcating radial plates typical of the thick inner wall of the Early Albian caprinid genus *Coalcomana*. Body cavity (B) and accessory canal (AC) are labeled. Scale bar = 1 cm.

sidence model proposed by Detrick and Crough [1978], the edifice beneath Wodejebato Guyot should have subsided nearly 600 m between middle Albian time and Campanian time, which is the age for deposition of the large benthic foraminiferal sediments. If reef growth was continuous during this time interval, Wodejebato should now be capped by at least 600 m of reef; however, seismic profiles from Wodejebato (Figure 12) reveal a reef cap less than 200 m thick.

There are three possible explanations. First, the reef may have drowned prior to the end of Albian time when it was no more that 200 m thick and then was later uplifted and eroded into Late Cretaceous shallow bank sediments. Second, reef growth may have been continuous from Albian time to the second period of uplift at which time 300-400 m of Albian and younger reef was eroded into a new Campanian reef. Third, following formation of the Albian reef, Wodejebato did not subside until the end of the Cretaceous and the thickness of Cretaceous reef growth was controlled only by changes in sea level.

Although it is possible the Albian reef was exposed, eroded and redeposited due to a Late Cretaceous global sea-level fall, we eliminate the third explanation based on the lack of a thermal or dynamic mechanism to support the lithosphere and its overlying volcanic edifices at sea level for the 30 m.y. span of reef building at Wodejebato. We propose that, regardless of whether the reefs capping Wodejebato drowned in mid-Cretaceous time or continued to thrive through Late Cretaceous time, they were uplifted in Late Cretaceous time by a second episode of intraplate volcanism at or near Wodejebato Guyot.

One indication that the reef on top of Wodejebato may have drowned prior to uplift in the Late Cretaceous is the recovery of late Santonian pelagic sediments with a few grains of volcanic debris in RD 47. These sediments could have been deposited on top of a drowned Albian reef prior to the second uplift. Furthermore, no reef debris of Cenomanian through Campanian is recorded in the area. The primitive *O. tissoti* may represent debris from the first reef deposited following uplift. The fact that the *O. tissoti* appear to be deposited with the *Sulcoperculina-Asterorbis* assemblage inside the molds of Albian molluscs implies that all the larger foraminifers are similar in age.

Reef growth ceased during or shortly after Late Cretaceous time and Wodejebato guyot was carried to its present depth of ~1500 m by thermal subsidence which has continued since the cessation of volcanism. The abundance of early Tertiary

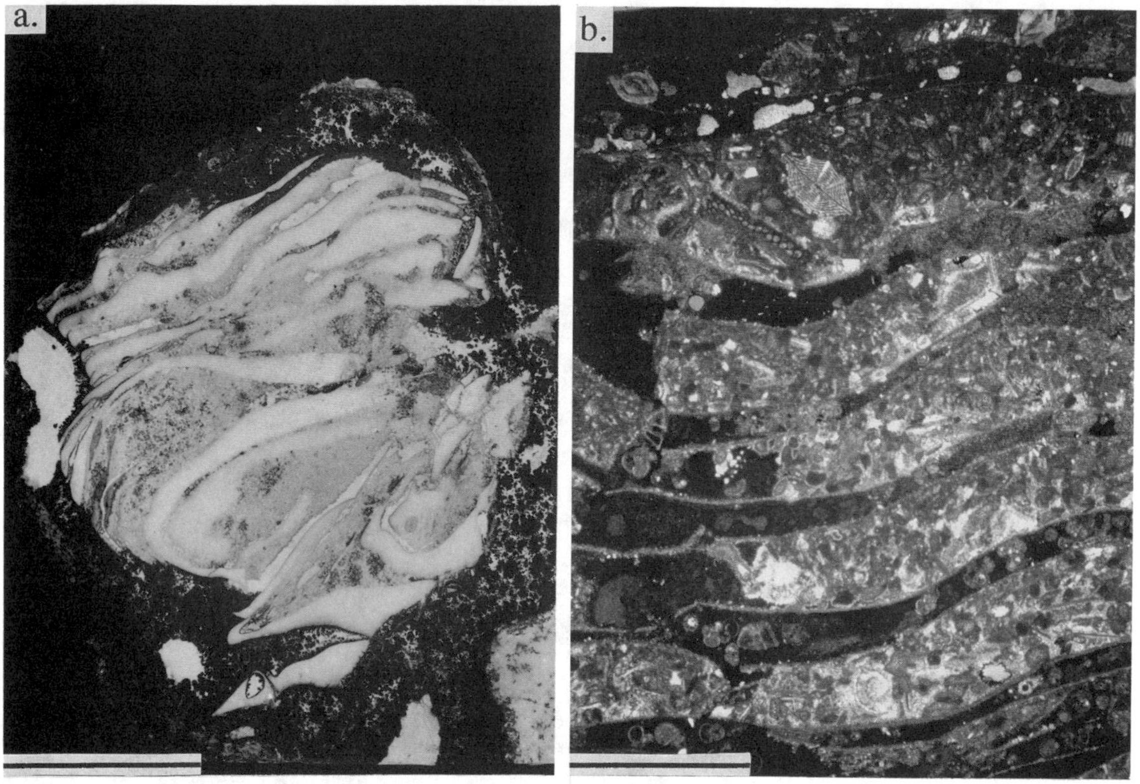

Fig. 10. (*a*) Photograph of imbricate, pseudomorphed, mytilid pelecypod shells (white) similar to the genus *Chondrodonta*. Scale bar = 2 cm. (*b*) Photomicrograph from slab in (*a*) shows that the shell fragments are pseudomorphed by sediments containing Late Cretaceous large foraminifers and other shallow water components. Scale bar = 1 mm.

planktonic foraminifers filling the pore space in the Cretaceous limestones indicates the guyot was well below sea level by early Tertiary time.

Ruwituntun

Ruwituntun Guyot is a flat-topped platform adjacent to Ruwituntun and Rongerick atolls, similar in its setting to Wodejebato Guyot with respect to Pikinni Atoll (Figure 1). Dredge locations, 3.5 kHz profile and side-scan image locations are shown on the bathymetric map in Figure 13. The minimum depth at Ruwituntun Guyot (<1300 m) is shallower than at Wodejebato Guyot although the shelf break occurs at ~1500 on both guyots (compare Figures 2 and 13). Side-scan images collected across Ruwituntun Guyot show numerous highly reflective, "bare rock" exposures, especially at the southeast end of the guyot (Figure 14). The pelagic sediment cover across Ruwituntun guyot, as shown in the 3.5 kHz profile in Figure 14, is thinner than at Wodejebato Guyot; this absence of sediment may be due to local currents. The rock that crops out above the thin pelagic sediment cap may be part of a reef complex that caps the volcanic edifice. Alternatively, based on the cone-shaped morphology apparent in the side-scan images, some of the highly reflective outcrops may be volcanic basement protruding through the sediment cap.

The rocks recovered in RD-41 and RD-43, from the talus slopes flanking Ruwituntun, consist of vesicular basalts and volcaniclastic breccias with planktonic foraminiferal matrices. The matrix in the RD 41 rocks contains foraminifers that only constrain the edifice age to be Eocene or older. Rocks in RD 43 contain *Globotruncana bulloides*, *G. arca*, *G. rosetta*, *Globotruncanita elevata* and *G. stuartiformis*, which date these samples as late Campanian to early Maastrichtian in age. There appear to be some older (Turonian to Santonian), flattened, double keeled forms similar to species of *Marginotruncana* mixed with the younger assemblage, however, poor preservation prohibits identification.

About 200 lbs. of manganese coated foraminiferal-algal limestone similar to that recovered in RD 50 from Wodejebato Guyot were recovered from one of the highly reflective outcrops on Ruwituntun Guyot (RD-40 in Figures 13 and 14). Tests of larger benthic foraminifers, including Late Cretaceous *Sulcoperculina* sp. and *Asterorbis* sp., are abundant along with algal, coral, echinoid and molluscan debris (Figure 15). A few *O. tissoti* were also found in these rocks. Rare specimens of Late Cretaceous planktonic foraminifers (*Globotruncana* sp.) were found in these rocks but are too poorly preserved for identification at the species level. Rudist shell fragments were identified in thin section only. Most of these shell fragments show the same finely reticulate microstructure as the radiolitid rudist outer shell wall fragments from the Wodejebato Guyot limestones (Figure 15). A few poorly preserved caprinid rudist shell fragments were also found. These samples suggest that the

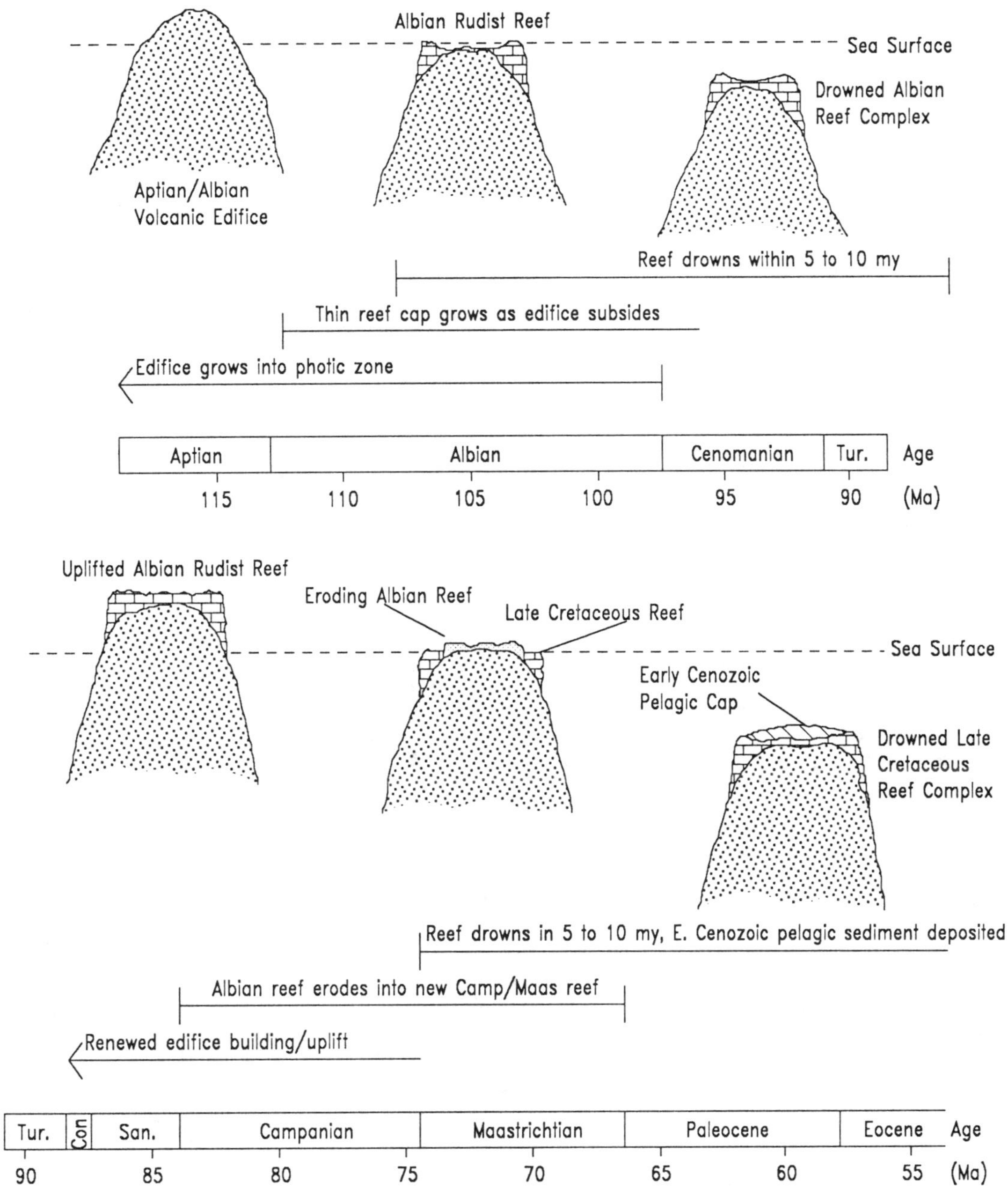

Fig. 11. History of uplift, volcanism, subsidence and reef building at Wodejebato Guyot deduced from dredged rocks and discussed in the text.

geologic history of Ruwituntun Atoll closely parallels that of Wodejebato Guyot. An Early Cretaceous phase of volcanism and uplift in the vicinity of Ruwituntun is further supported by the $^{40}Ar/^{39}Ar$ age of 138± for Look seamount, discussed earlier in this paper. Look is a conical seamount adjoining the northwestern end of Ruwituntun Guyot (Figure 1).

Loen

Loen Guyot is another flat-topped platform that adjoins an atoll, extending to the south of Anewetak (Figure 1). The association of active atolls and adjacent guyots is common in the Marshall Islands; a fourth example is Limalok Guyot-Mili Atoll located in the southern Marshall Islands. The slope break, interpreted as the location of the reef rim, is located at a depth of ~1500 m at Loen Guyot, similar to the slope-break depth at Wodejebato and Ruwituntun.

RD 33 from Loen contained breccia consisting of angular fragments of vesicular basalt floating in a calcareous matrix of shallow-water debris and planktonic foraminifers (Figure 16).

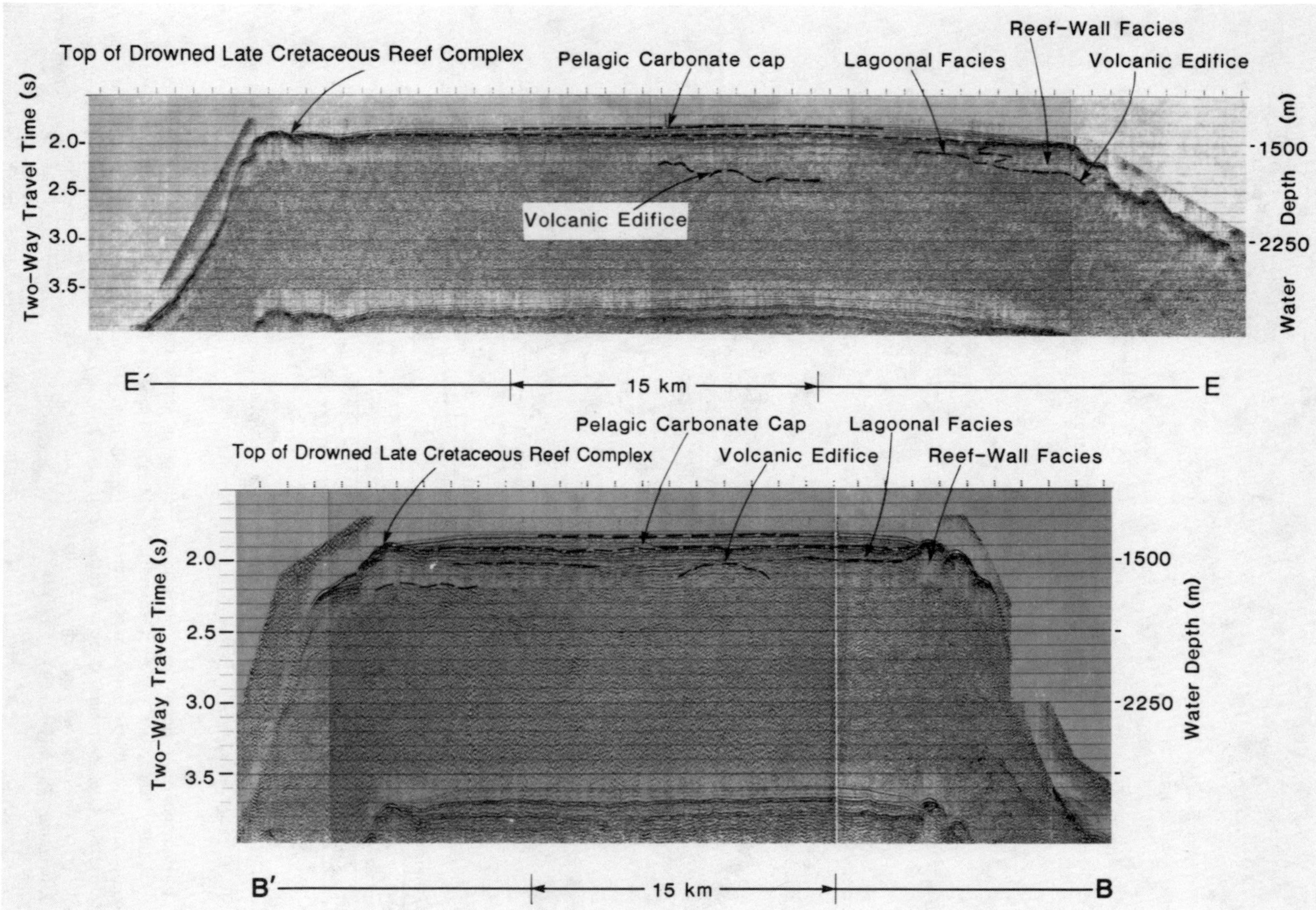

Fig. 12. Seismic profiles across Wodejebato Guyot. The upper profile is along track E-E′ and the lower profile is along track B-B′ shown in Figure 2. These profiles were run from R/V *Kana Keoki* cruise KK810626 Leg 4 by F. Duennebier. The energy source was an 80 in^3 water gun. The digitally recorded data was processed by T. Shipley at the Scripps Institution of Oceanography. Geological interpretations are modified from Schlanger et al. [1987]. The top of the drowned Late Cretaceous reef complex is now at a corrected depth of 1480 m below sea level and the reef complex is apparently no more than 200 m thick at the reef wall.

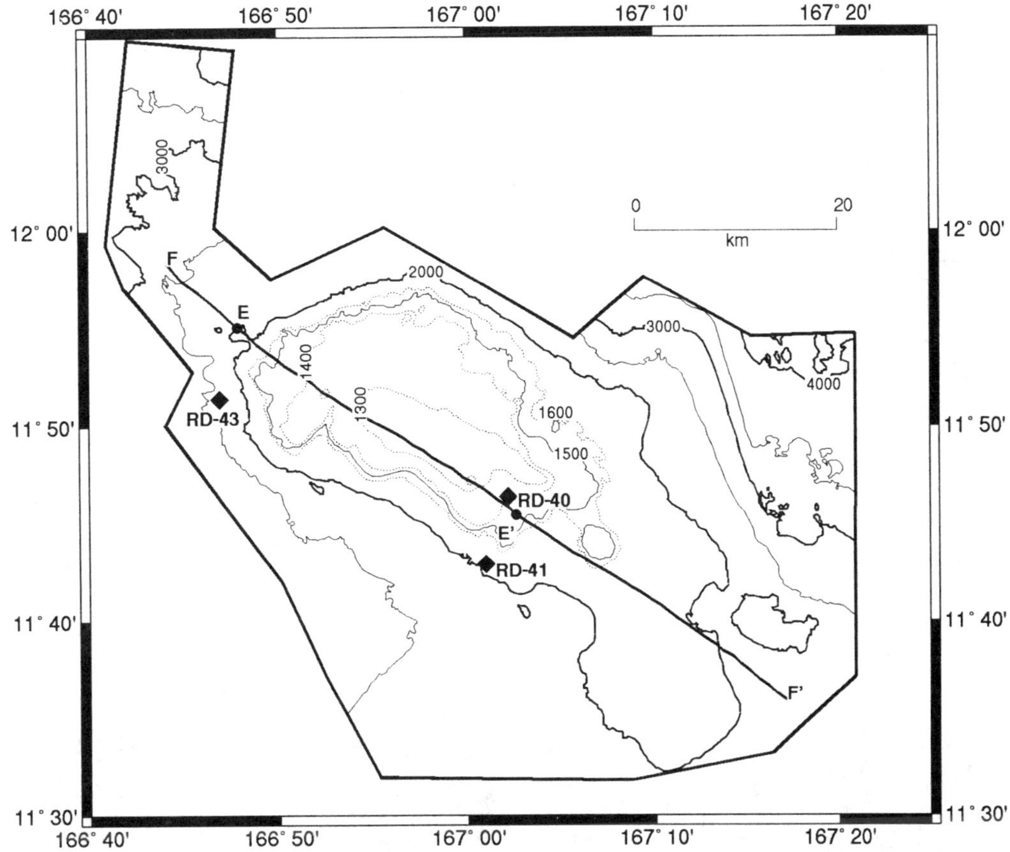

Fig. 13. Bathymetric chart of Ruwituntun Guyot based on seismic reflection profiles and SeaMARC II bathymetry acquired in the spring of 1988 during R/V *Moana Wave* cruise MW8805. Dredge locations (diamonds) are labeled.

The most abundant planktonic foraminifers are specimens belonging to the genus *Favusella*, including *Favusella washitensis* and possibly other species. The genus and *F. washitensis* have been reported from close to the Aptian/Albian boundary to early Cenomanian with maximum abundance in the middle to late Albian [Longoria and Gamper, 1977]. The favusellids generally occur in near-reef environments and are missing in the open ocean deeper sediments. Associated in these rocks are also abundant specimens of the calpionellid species *Colomiella recta*, which ranges from latest Aptian through middle Albian [Longoria and Gamper, 1977; Longoria, personal communication, March 1990]. This taxon also characterized a near-shore environment like the favusellids. The occurrence of these species on Loen indicates that a shallow platform capped this guyot at least in the early to middle Albian. Poorly preserved foraminifers that resemble *Praeglobotruncana stephani* also occur and suggest that Loen was already under a more pelagic regime by latest Albian time.

This rock also contains abundant bryozoan fragments, echinoid debris, sponge spicules, and occasional reticulate patches the may be poorly preserved rudist shell fragments. Agglutinated, partitioned, benthic foraminifers, typical of mid-Cretaceous time, are also common in this rock. These agglutinated foraminifers may include species of *Sabaudia*, although poor preservation prevents a firm identification. This genus occurs mainly in the Aptian/Albian [Schroeder and Neumann, 1985; Sartorio and Venturini, 1988].

The stratigraphic ranges of the shallower fossil components more or less overlap and suggest an Albian age, but not the latest Albian, by which time Loen was already deepening as suggested by the occurrence of *Praeglobotruncana stephani*. These rocks were probably deposited in an outer slope environment as indicated by the mixture of basalt fragments, shallow water debris and open marine planktonic foraminifers.

Another sample collected in RD 33 is a breccia consisting of angular basalt pebbles floating in a planktonic foraminiferal matrix consisting of *Globotruncana bulloides, G. linneiana, G. arca, Contusotruncana fornicata, Globotruncanita elevata* and *G. stuartiformis*. One specimen of *Marginotruncana coronata* was identified; *M. coronata* has been reported into the early Campanian and would be consistent with the *Globotruncana elevata* Zone that is suggested by the rest of the fauna. Species of *Heterohelix* are also relatively abundant, which is consistent with an early Campanian age for these breccias.

RD 33 also recovered a manganese encrusted foraminiferal limestone containing abundant *Marginotruncana renzi, M. marginata*, and *M. pseudolinneiana*. Several *Whitenella* sp. and one *Dicarinella concavata* were identified. Based on the high abun-

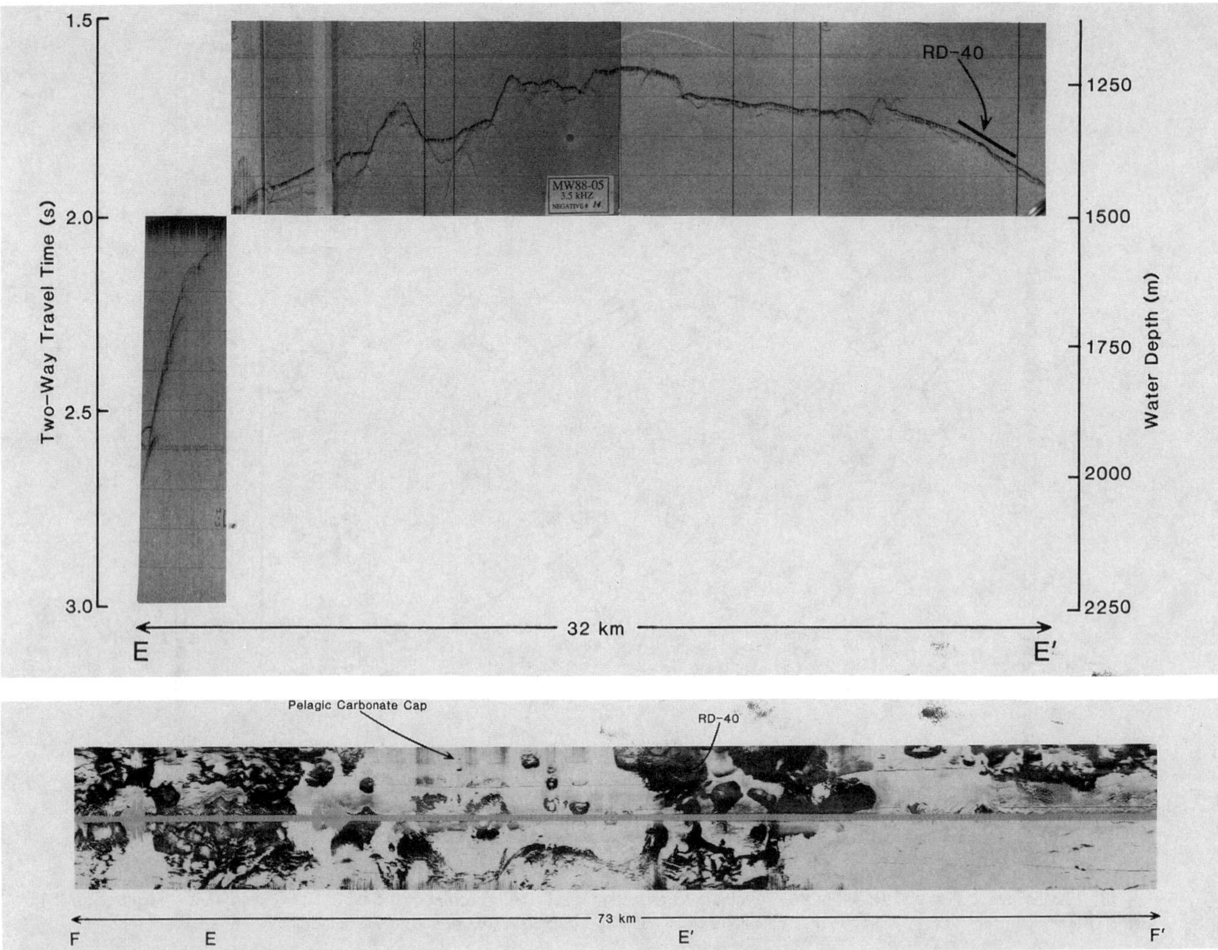

Fig. 14. E-E′ is a 3.5 kHz seismic-reflection profile across Rongelap guyot. Note the thin pelagic cap compared to the cap on Wodejebato Guyot. F-F′ is a side-scan image across Rongelap Guyot along the 3.5 kHz profile shown above. Shallow-bank limestone was recovered in RD-40 from the highly reflective outcrop (black) seen in this image. Locations of both lines are shown in Figure 13.

dance of marginotruncanids and the presence of some whiteinellids and *Dicarinella concavata*, this rock is possibly Coniacian in age (*Dicarinella concavata* Zone).

Loen erupted prior to early Albian time and supported a shallow carbonate bank during most of the Albian except during latest Albian time; this carbonate bank was coeval with the rudist reefs on nearby Wodejebato and Ruwituntun. There is no evidence that this guyot returned to sea level during the Late Cretaceous, although the Coniacian to Campanian age of volcanic breccias recovered from the flanks of this guyot could indicate an erosional event that coincides with eruption of adjacent Anewetak Atoll. The new $^{40}Ar/^{39}Ar$ age of 75.9±0.6 Ma for basalt from Anewetak (Table 3) indicates eruption at this site was coeval with the Late Cretaceous uplift events at Wodejebato and Ruwituntun.

Likelep

We have examined one breccia (sample 16) recovered in U. S. Geological Survey dredge 19 from Likelep Seamount, which is located in a seamount group west of Loen (Figure 1). This seamount was dredged during R/V *Farnella* cruise F10-89-CP in the Fall of 1989 [Hein et al., 1990].

The breccia consists of angular fragments of basalt and lithoclasts floating in a planktonic foraminiferal matrix that contains *Globotruncana bulloides*, *G. arca*, *Globotruncanita elevata*, *G. stuartiformis*, and *Heterohelix globulosa* indicating an early to middle Campanian age. The lithoclasts consist of occasional basalt fragments, bryozoan and echinoid debris, *Inoceramus* shell fragments and poorly preserved algal debris. These results are the same as those reported in Hein et al. [1990]. Likelep erupted and supported a shallow carbonate bank sometime prior to Campanian time.

Lobbadede

One successful dredge was recovered from Lobbadede Guyot which is located in the center of a seamount group northwest of the Pikinni-Wodejebato atoll-guyot pair (Figure 1). Well indurated phosphatized limestones were recovered along with

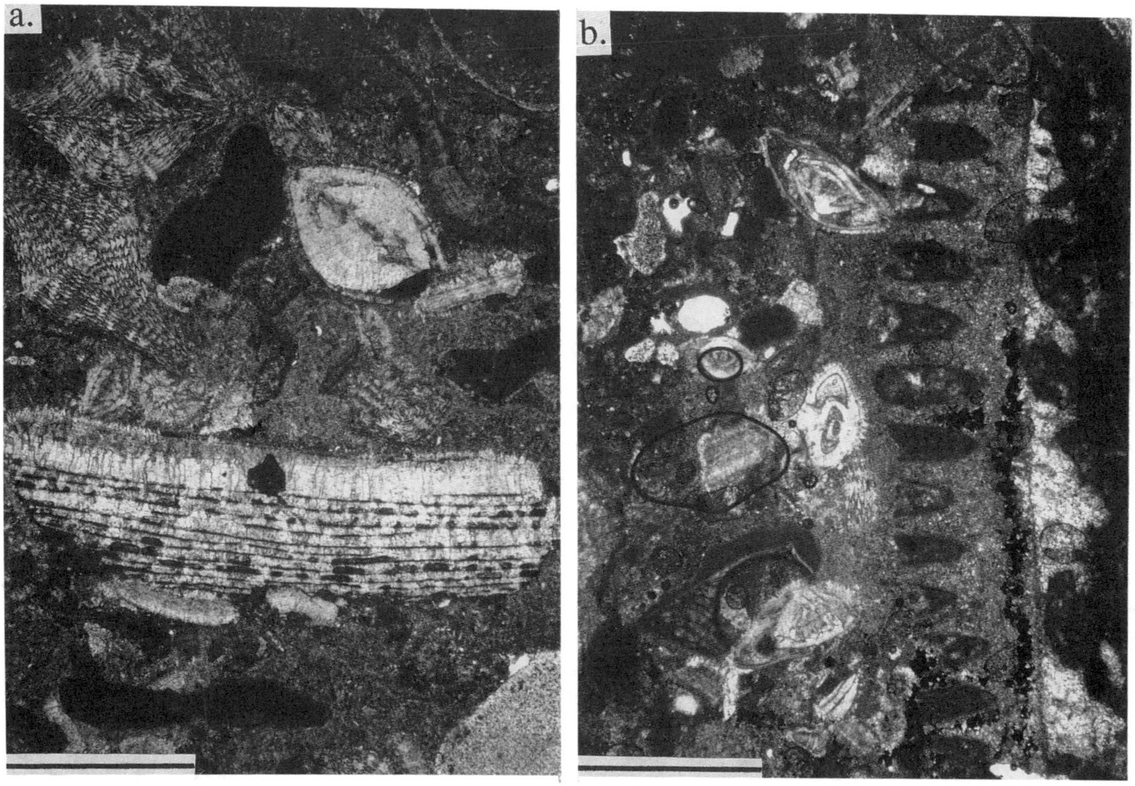

Fig. 15. (*a*) Photomicrograph of limestone recovered in RD-40 from Ruwituntun Guyot shows fragment of radiolitid rudist (center), echinoid debris (lower right corner), and Late Cretaceous large foraminifers including species of *Sulcoperculina* and *Asterorbis*. Algal debris appears as irregular dark shapes. Scale bar = 1 mm. (*b*) Rudist fragment from same sample as (*a*) with radial plates typical of the caprinid rudists. Scale bar = 1 mm.

relatively fresh basalt cobbles. One limestone sample contains fragments of shallow-water agglutinated foraminifers that are similar to those recovered from Loen, but are too poorly preserved to make a firm identification. These shallow-water foraminifers are found in a matrix containing a mixed assemblage of planktonic foraminifers. The oldest among those are Cenomanian in age including *Rotalipora appenninica*, *R. cushmani* and *Praeglobotruncana gibba*. The youngest foraminifers are latest Santonian to possibly early Campanian and include *Dicarinella asymetrica*, *Globotruncana bulloides*, *Globotruncanita elevata*, and *G. stuartiformis*.

A second limestone sample from this dredge also contains the Cenomanian species *Rotalipora cushmani*, *R. appenninica*, and possibly *R. greenhornensis*. As with the other samples from this dredge, the Cenomanian fauna is mixed with Late Cretaceous species of *Marginotruncana* and *Globotruncana*. The youngest limestone recovered in this dredge contains abundant *Globotruncana ventricosa*, *Heterohelix*, and no *G. elevata*, indicating an age of late Campanian.

Lobbadede Guyot was submerged and accumulating planktonic foraminiferal sediments during Cenomanian time, therefore the underlying volcano must have erupted prior to or during the Cenomanian. Based on the poorly preserved agglutinated foraminifers, Lobbadede may have supported a shallow-carbonate bank prior to or during latest Albian-early Cenomanian time that was coeval with the platforms supported at Wodejebato, Ruwituntun, and Loen. The radiometric age of 82.4±2.4 Ma for basalt dredged from Lobbadede is evidence for an eruptive event at this site that may be coeval with Late Cretaceous eruption or uplift at Wodejebato, Ruwituntun, Likelep, North Wod-En and Anewetak.

Lewa

A single successful dredge (RD-58) from Lewa Seamount, just west of Lobbadede Guyot (Figure 1) recovered a piece of conglomerate consisting of pebble to large cobble-sized oolitic limestone clasts encased in a matrix of hyaloclastic and other volcanic debris (Figure 17). The limestone contains abundant agglutinated, biserial, partitioned foraminifers that are similar to the Aptian/Albian genus *Sabaudia*. Other components of this rock include abundant whole and fractured ooids, molluscan debris including some gastropods, echinoid debris, other agglutinated partitioned foraminifers (*Everticyclammina* cf. *hedbergi*), abundant *Favusella washitensis*, and possibly some *Hedbergella libyca*. The *Favusella washitensis*, *Sabaudia* sp., and abundant ooids indicate that Lewa supported an ooid bank in Albian time that was supplying shallow-water debris to the outer slope environment where the oolitic limestone was deposited. The

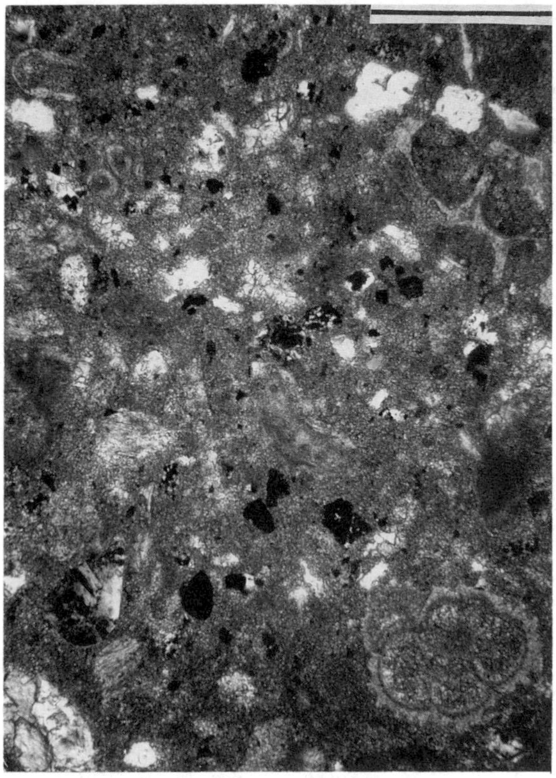

Fig. 16. Photomicrograph of breccia recovered in RD-33 from Loen Guyot shows *Favusella washitensis* (lower right corner) and *Colomiella recta* (upper left corner). Scale bar = 1 mm.

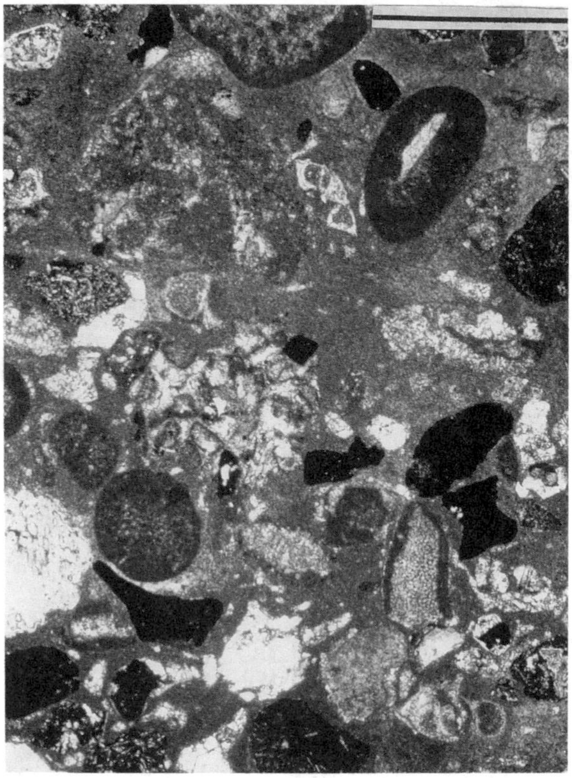

Fig. 17. Photomicrograph of the limestone recovered in RD-58 from Lewa Guyot. Note the ooids, shallow-water debris, and volcaniclastics deposited with planktonic foraminifers. Latest Albian to Cenomanian planktonic foraminifers include *Rotalipora appenninica*. Scale bar = 1 mm.

oolitic limestone was deposited at the same time as the breccia recovered in RD 33 from Loen Guyot and is apparently coeval with the rudist debris from Wodejebato and Ruwituntun.

The volcanic matrix contains latest Albian to Cenomanian planktonic foraminifers including *Rotalipora appenninica*, *Praeglobotruncana stephani*, and *Globigerinelloides bollii*. It is notable that near-shore species *Favusella washitensis* is not found in the volcanic matrix; apparently the conglomerate formed by the erosion, and redeposition in deeper water, of hyaloclastic deposits and outer-slope accumulations of shallow-water, oolitic debris. The rocks from RD 58 corroborate Albian volcanism and reef building in the Lobbadede seamount group that was previously inferred from the Cenomanian planktonic foraminifers and fragments of mid-Cretaceous agglutinated foraminifers recovered in RD 56 from Lobbadede guyot.

Lomtal

A single successful dredge from Lomtal seamount (MW8805 RD-59), just north of the Lobbadede seamount group, recovered more basalt conglomerates with poorly preserved planktonic foraminifers in a phosphatized limestone matrix. The identification of one *Dicarinella concavata* with poorly rounded chambers suggests a probable age of Santonian. It is likely this seamount erupted prior to or during Santonian time. Radiometric ages for seamounts to the south and north (82.4±2.4 Ma and 85.3±0.4 Ma, respectively) are consistent with the minimum fossil age.

SUMMARY OF VOLCANISM, REEF BUILDING AND UPLIFT

Radiometric and fossil minimum ages for basalts and volcanic breccias dredged from guyots or drilled beneath atolls are listed in Tables 4 and 5. Because reef establishment in a mid-oceanic setting such as the Marshall Islands requires eruption of a volcanic foundation or, at the very least, uplift of a pre-existing foundation into the photic zone, we have also included the location and ages of known carbonate platforms from this region in Tables 4 and 5.

Evidence for both Early and Late Cretaceous intraplate volcanism and/or uplift have now been reported for six locations in and near the northern Marshall Islands (Figure 1). These include Wodejebato Guyot, Ruwituntun Guyot, the Lobbadede Seamount group, the Anewetak-Loen atoll-guyot pair, and DSDP Sites 462 and 169. Based on the data summarized in Tables 4 and 5, the two episodes of volcanism and/or uplift were separated at each site by about 15 to 30 m.y. and apparently coincide with the widespread, coeval Cretaceous volcanic events proposed by Winterer [1976a, b, Schlanger and Premoli Silva [1981], Schlanger et al. [1981b], Rea and Vallier [1983] and Winterer and Metzler [1984]. However, another

TABLE 4. Early Cretaceous Volcanism, Uplift and Reef-Building in the Marshall Islands.

Location	Age	Type†	Source‡
Lewa Guyot 13.94N, 163.08E	Albian	Fossil	1
Lobbadede Guyot 13.8N, 163.6E	pre-Cenomanian (Albian)	Fossil	1
Loen Guyot 10.2N, 162.8E	Albian	Fossil	1
Ruwituntun Guyot 11.8N, 167.0E	Albian	Fossil	1
Wodejebato Guyot 12.0N, 164.9E	Albian	Fossil	1
Look Seamount 12.1N, 166.2E	138±0.4 Ma	$^{40}Ar/^{39}Ar$	1
Site 169 10.7N, 173.6E	≥late Albian	Fossil	2
Site 462 7.4N, 165.0E	Albian to Cenomanian	Fossil	3 4
	late Aptian	Fossil	5
	107.0–113.0 Ma	$^{40}Ar/^{39}Ar$	6, 7

†See text for full description.
‡Sources: (1) This paper, (2) Winterer, Ewing et al. [1973], (3) Premoli Silva and Brusa [1981], (4) Premoli Silva [1986], (5) Jenkyns and Schlanger [1981], (6) Ozima et al. [1981], (7) Pringle, [1992].

possible explanation for multiple episodes of volcanism and uplift in the Marshall Islands is passage of the region over more than one hotspot.

TESTS FOR A HOTSPOT ORIGIN

Age Progression

The distribution of edifice ages in the Marshall Islands cannot be reconciled with the edifice age progression predicted by passage of the region over any one hotspot. For example, Davis et al. [1989] have already pointed out that, although their $^{40}Ar/^{39}Ar$ ages for Woden-kopakut Guyot and Lokkworkwor Seamount are consistent with the Late Cretaceous volcanism predicted by the Pacific plate rotation model of Duncan and Clague [1985], their date for Lokkworkwor seamount is older than Woden-kopakut Guyot, rather than younger, as would be predicted for a single hotspot origin. They note that the more alkalic basalt from Woden-kopakut Guyot could represent a post-erosional stage of volcanism that was coeval with volcanism at Lokkworkwor seamount. However, a single hotspot model cannot explain the fossil minimum age for Lomjenaelik Seamount, which is as old, or older, than both Lokkworkwor and Woden-kopakut.

When all available edifice ages are plotted against distance in the Marshall Islands (Figure 18), no age progression can be demonstrated for radiometric ages or for the fossil minimum ages. Two phases of carbonate bank deposition, that appear to be coeval throughout the region, can be inferred from the fossil ages for shallow-marine carbonate banks in the Marshall Islands. However, these fossil ages only place a minimum constraint on the age of edifice building; furthermore, the broad age ranges available for shallow-marine fossils may mask any true age progression in the region. The predicted maximum age difference for the dated edifices in the Marshall Islands (based on ~10 cm/year [Duncan and Clague, 1985]) is not much greater than the uncertainty in the age determinations. This test for age progression may not be meaningful over the short distance for which data are available in the Marshall Islands.

An Alternative Test for A Hotspot Origin

The apparent lack of age progression during Cretaceous edifice building may be the result of passage over more than one hotspot. Pacific plate rotation models proposed by Duncan and Clague [1985] and R. Gordon [personal communication, 1990] predict that present-day active volcanic centers near Macdonald Seamount and the island of Tahiti produced hotspot tracks dur-

TABLE 5. Summary of Late Cretaceous volcanism, uplift and reef-building in the Marshall Islands region.

Location	Age	Type†	Source‡
Anewetak Atoll 11.5N, 162.3E	75.9±0.6 Ma	$^{40}Ar/^{39}Ar$	1
Lokkworkwor Smt. 8.9N, 169.8E	86.7–87.9 Ma	$^{40}Ar/^{39}Ar$	2
Limalok Guyot 5.6N, 172.4E	≥Early Eocene	Fossil	3
Likelep Smt. 10.6N, 160.4E	≥Campanian 82.1±0.7 Ma	Fossil $^{40}Ar/^{39}Ar$	1,4 1
Jalwoj Atoll 6.3N, 169.3	≥Campanian to Maastrichtian	Fossil	4
Lobbadede Guyot 13.8N, 163.6E	82.4±2.4 Ma	$^{40}Ar/^{39}Ar$	1
Lomtal Smt. 14.7N, 163.5E	≥Santonian	Fossil	1
Woden-kopakut Guyot 14.0N, 167.5E	83.8–80.6 Ma	$^{40}Ar/^{39}Ar$	2
Ruwituntun Guyot 11.8N, 167.0E	Campanian	Fossil	1
Wodejebato Guyot 12.0N, 164.9E	Campanian	Fossil	1
Lomjenaelik Smt. 7.3N, 172.4E	≥Coniacian to Santonian	Fossil	3
North Wod-En Smt. 15.0N, 162.9E	85.3±0.4 Ma	$^{40}Ar/^{39}Ar$	1
Site 65 4.4N, 177.0E	≥mid-Campanian to mid-Maastrichtian	Fossil	5
Site 169 10.7N, 173.6E	Cen./Tur. to Campanian	Fossil	6
Site 462 7.4N, 165.0E	≥late Campanian to late Maastrichtian	Fossil	7, 8, 9

†See text for complete description.
†Sources: (1) This paper, (2) Davis et al. [1989], (3) Schlanger et al. [1981a], (4) Hein et al. [1990], (5) Winterer, Riedel et al. [1971], (6) Winterer, Ewing et al. [1973], (7) Moberly, Schlanger et al. [1986], (8) Premoli Silva and Brusa [1981], (9) Premoli Silva [1986].

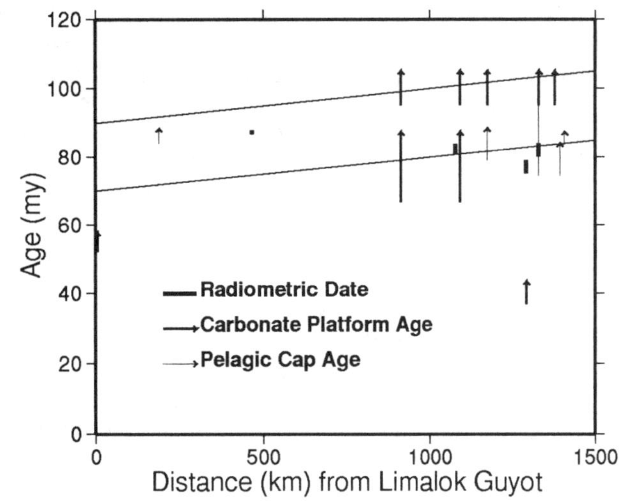

Fig. 18. Plot of radiometric and fossil minimum ages from the Marshall Islands versus distance along the island chains from Limalok Guyot, which is the southernmost dated edifice. The two parallel sloped lines show the predicted edifice age progression if Limalok Guyot had formed over a hotspot at either 70 Ma or 90 Ma and assuming a plate velocity of 10 cm/yr.

ing Cretaceous time that lie within 200 to 500 km of the Marshall Islands. The Aptian to Albian portion of the Macdonald hotspot track lies along the northern Marshall Islands, according to both of these models. For example, the middle and Late Cretaceous portions of the Macdonald track lies near Lokkworkwor and Lomjenaelik Seamounts, both sites of early Late Cretaceous volcanism. The Aptian-Albian Macdonald hotspot track lies near Wodejebato and Ruwituntun guyots, both sites of Aptian-Albian volcanism and reef building. The Paleocene Tahiti track lies near Limalok Guyot, the site of early Eocene reef building, and the Late Cretaceous Tahiti track lies near Wodejebato, Ruwituntun and Woden-kopakut guyots, all sites of Late Cretaceous volcanism, uplift, and reef building. However, even taking into account the possibility of relative motion between hotspots, it is difficult to use published hotspot tracks to explain volcanism in the western Marshall Islands at sites such as Anewetak Atoll, Loen Guyot, or Likelep Seamount.

We restore all dated edifices in the Marshall Islands to their positions at their times of formation using Pacific plate rotation poles, relative to fixed hotspots, published by Duncan and Clague [1985] (dashed lines in Figure 19) and rotation poles provided by R. Gordon [personal communication, 1990] (solid lines in Figure 19). As target hotspots, we use the present-day positions of Tahiti and Macdonald seamounts, both sites of active volcanism. Following Turner and Jarrard [1982], we also include the present-day locations of Rarotonga and Rurutu (Austral-Cook Islands) as possible sites of past and recent hotspot activity based on the occurrence of basalts at these sites that are too young to have erupted over any other presently active hotspot. We use a 300 km radius around each hotspot as a reasonable area of influence for each each hotspot through time. This target size is based on the estimated positions at time of formation for volcanos in the Hawaiian-Emperor chain [Henderson, 1985]. This 600 km circle takes into account uncertainty in dating the volcanos, uncertainty in past Pacific plate rotation, variation in eruptive style through time and uncertainty in the fixity of the hotspot.

The solid and dashed lines in Figures 19a and 19b show the trajectory of past positions for Wodejebato and Ruwituntun guyots, respectively, from 0 to 120 Ma. The line that marks the trajectory for each guyot is thickened for times corresponding to the ages of reef building on each guyot. According to these models, intraplate volcanos erupted and Albian carbonate platforms were established as these two guyots passed over the Macdonald hotspot. As these sites moved off the swell associated with Macdonald hotspot, they would have subsided, either drowning the reef or allowing the reef to thicken (Figure 11). These sites then moved over the Rurutu hotspot during the Late Cretaceous. If Rurutu was an active hotspot at the time, the reefs atop Wodejebato and Ruwituntun would have been uplifted as the lithosphere was reheated and thinned; there may have been nearby volcanism at sites such as Pikinni and Rongelap atolls. The Albian reef was then eroded into a new Campanian carbonate platform during this period of uplift. The Campanian reef then drowned as Wodejebato and Ruwituntun moved off the Cretaceous Rurutu hotspot swell.

The trajectory of past locations is shown for Lobbadede and Lewa guyots in Figures 19c and 19d, respectively. The Albian reef atop these two guyots formed after eruption of the underlying volcanos over the Macdonald hotspot in Aptian-Albian time and then drowned as they moved off the Macdonald hotspot swell. The 82.4± Ma basalt dredged from Lobbadede may have erupted as these guyots passed over the Rurutu hotspot in Late Cretaceous time.

The guyot-atoll pair of Loen and Anewetak are backtracked in Figures 19e and 19f. Loen erupted and the overlying Albian platform was established after passage of the site over Macdonald hotspot. Following subsidence and drowning of the Albian platform atop Loen, it may have been uplifted during the eruption of nearby Anewetak Atoll over either the Rurutu or Rarotonga hotspots in Late Cretaceous time.

Past locations based on the radiometric and fossil minimum ages for Lomtal, Woden-kopakut, Likelep, Lokkworkwor, and Lomjenaelik all coincide reasonably well with the 600 km radius targets surrounding Rarotonga, Rurutu, Tahiti, or Macdonald Seamount.

The past position of Limalok guyot does not fall within any hotspot target area. The minimum age of Limalok guyot is based on early Eocene reef recovered by dredging. If Limalok guyot erupted 10 to 30 m.y. prior to reef building, as did Anewetak and Midway atolls (see earlier discussion) then the past position for Limalok guyot would plot well within the Tahiti target area in Figure 19.

CONCLUSIONS

Based on the new volcano and reef ages reported in this paper, no monotonic age progression can be demonstrated along

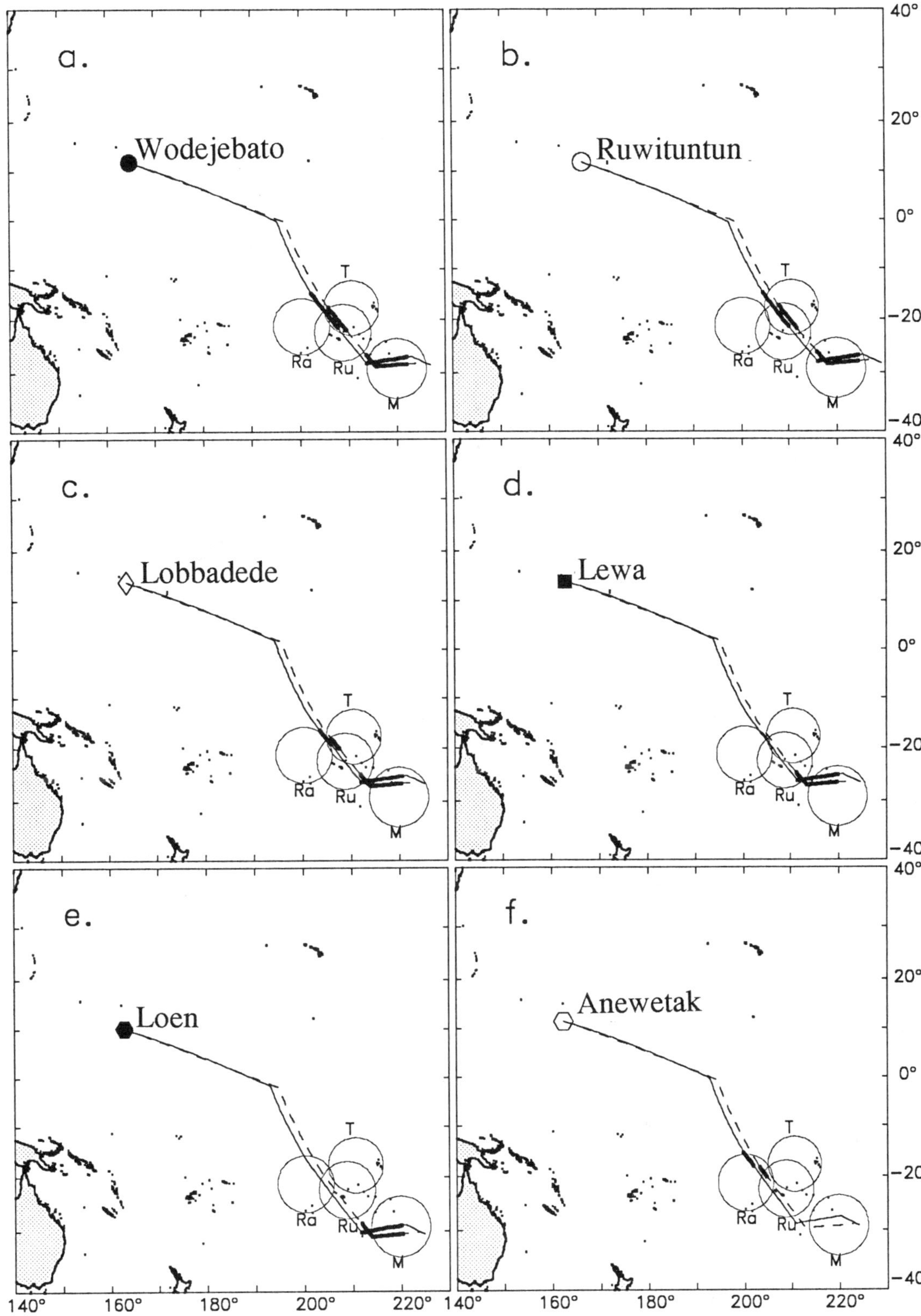

Fig. 19. Each plot shows the present-day location of a dated atoll, guyot or seamount in the Marshall Islands and the locus of past positions from 0 to 120 Ma. The solid line on each plot is calculated using rotation poles for the Pacific plate relative to fixed hotspots from R. Gordon (personal communication, 1990). The dashed line on each plot is calculated using rotation poles for the Pacific plate relative to fixed hotspots from Duncan and Clague [1985]. The locations of each edifice during the times of eruption and/or reef-building are shown by thickened segments of each trajectory. Labeled circles show the locations of active or recently active hotspots proposed by Turner and Jarrard [1982] and Duncan and Clague [1985]. T=Tahiti, M=Macdonald, Ru=Rurutu, Ra=Rarotonga. Each hotspot circle has a 300 km radius which is an estimate of the area of influence for the Hawaiian hotspot through time. [Henderson, 1985].

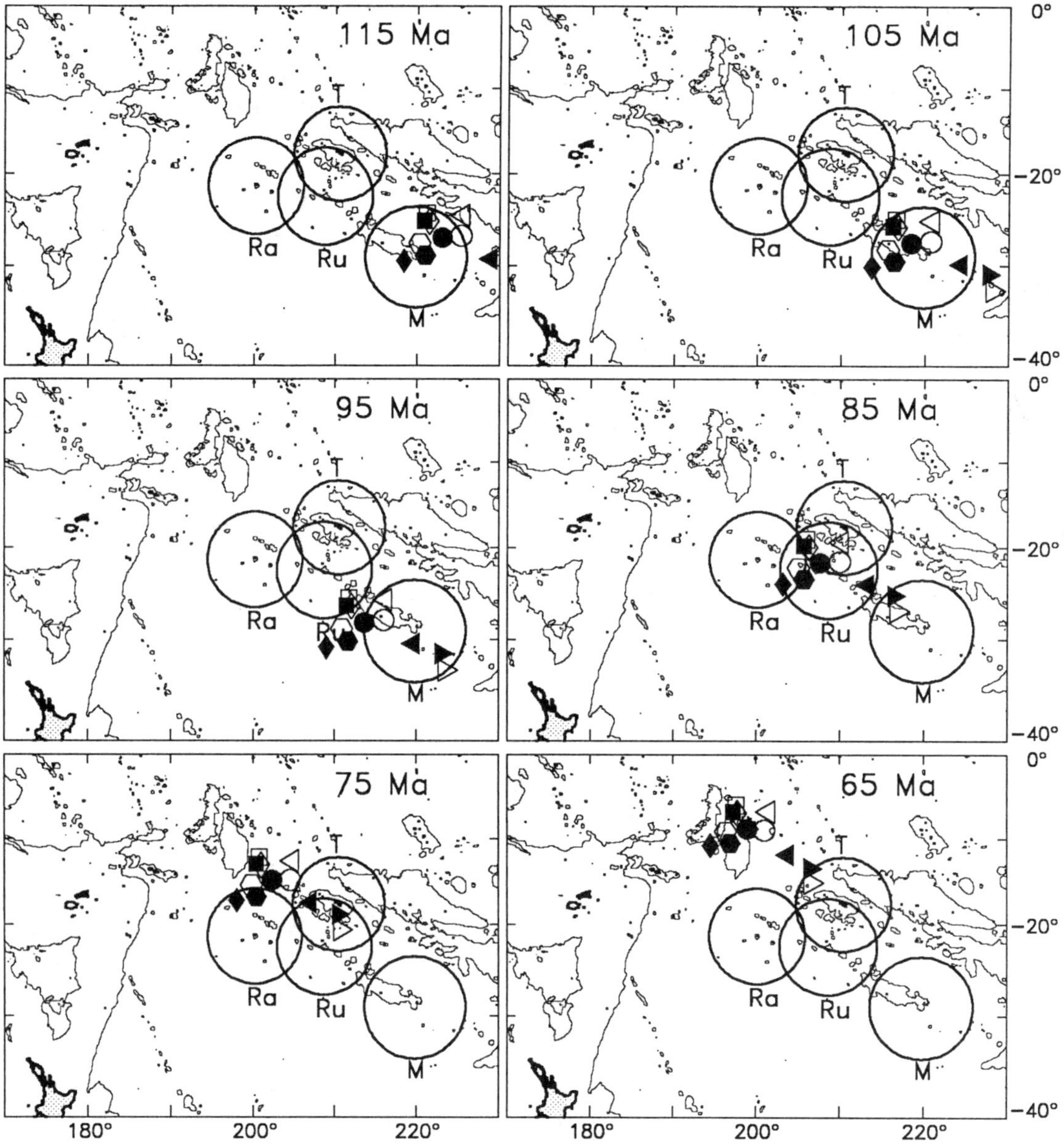

Fig. 20. Each plot shows the past location, at one of six time slices, of all dated atolls, guyots and seamounts in the Marshall Islands based on the rotation poles for the Pacific plate relative to fixed hotspots from R. Gordon (personal communication, 1990). The locations of active or recently active hotspots are shown by 600 km circles as discussed in the text and in the caption to Figure 19. Black right triangle = Lomjenaelik, open right triangle = Limalok, black left triangle = Erikub, open left triangle = Ratak, black diamond = Likelep, open diamond = Lobbadede, black circle = Wodejebato, open circle = Ruwituntun, black square = Lewa, open square = Lomtal, black hexagon = Loen, open hexagon = Anewetak.

the northwest-southeast trend of the Marshall Islands. Although there is evidence for both early and late volcanism, uplift and reef building at several sites within and near the Marshall Islands, we see no need to invoke widespread, episodic Pacific volcanism during the Cretaceous as an explanation for edifice building in the Marshalls. Based on the rotation of dated edifices in the Marshall Islands to their positions at the times of eruption and reef building, the passage of the Marshall Islands over more that one hotspot can account for the history of uplift, volcanism and subsidence inferred from rocks dredged in this region. All dated edifices in the Marshall Islands are plotted in a series of reconstructions shown as Figure 20 for 115, 105, 95, 85, 75, and 65 Ma. Nearly all edifices topped by Cenomanian or older carbonate platforms, such as Wodejebato, Ruwituntun, Loen, Lobbadede and Lewa plot well within 300 km of Macdonald hotspot between ~115 and ~105 Ma. By ~95 Ma, most of the northern edifices in the Marshall Islands had moved off the Macdonald hotspot and had not yet reached the Rurutu hotspot. Many of these platforms may have drowned as the region moved off the swell associated with Macdonald hotspot. Evidence such as the recovery of late Albian reef debris deposited with Cenomanian planktonic foraminifers, indicates this was a time of subsidence and reef drowning in the northern Marshall Islands. Volcanos in the southern Marshall Islands, such as Lomjenaelik may have erupted over Macdonald hotspot during this time. At ~85 Ma, some guyots returned to sea level (Wodejebato and Ruwituntun) and new volcanos erupted (Anewetak, Likelep, North Wod-en, Lobbadede) as the region moved onto the swells associated with three probable hotspots located at present-day Tahiti, Rurutu and Rarotonga, all sites of active or recently active intraplate volcanism. By ~75 to ~65 Ma, the northern Marshall Islands were subsiding as the region moved away from the hotspot swells; the abundance of dredged early Tertiary planktonic foraminiferal limestones suggests the guyots in this region had drowned by early Tertiary time. The Late Cretaceous reefs that formed on uplifted Early Cretaceous volcanos such as Wodejebato, Loen, Lobbadede and Ruwituntun drowned prior to early Tertiary time and have subsided to their present depths of 1500 to 2000 m. Reefs may not have accumulated on some Late Cretaceous volcanos, such as the one beneath Anewetak Atoll, until early Tertiary time; a few of these reefs have remained active since that time.

Acknowledgements. The majority of samples discussed in this paper were collected on a Northwestern University-Hawaii Institute of Geophysics cruise (MW8805) funded by National Science Foundation grant OCE-8709917. Additional funding was provided by grant 19419-AC2-C from the donors of the Petroleum Research Fund, administered by the American Chemical Society. James R. Hein, U. S. Geological Survey (U.S.G.S) and Jung-Keuk Kang, Korea Ocean Research and Development Institute (KORDI) kindly provided the samples from USGS-KORDI cruise F10-89-CP. We greatly appreciate discussions with Alan Coogan, Kent State University, regarding Aptian/Albian rudist assemblages. Doug Bergersen, University of Hawaii, provided assistance with the side scan images. We are especially grateful to J. Haggerty, E. Winterer and J. Hein for their thorough and thoughtful reviews of the manuscript.

This research would not have been completed without the inspiration, dedication and guidance of Sy Schlanger. His absence is felt by all who studied and worked with him.

Note Added in Proof. The data contained in this paper was the basis for planning several sites for ODP Legs 143 and 144. Some data and interpretations presented in this paper are corroborated by the preliminary results of drilling. For example, paleomagnetic inclination data for basalts penetrated at Lo En indicate a paleolattitude of 30° S [Nakanishi and Gee, 1992] that is consistent with eruption in the vicinity of Macdonald hotspot in Aptian/Albian time, as proposed in this paper. The recovery of Campanian clay with redeposited Cenomanian nannofossils overlying basalt at Wodejebato [Nakanishi and Gee, 1992] indicates that the guyot first erupted during or prior to Cenomanian time, was submerged during Cenomanian time, and then was rejuvenated in Late Cretaceous (Campanian?) time, permitting growth of a Campanian-Maastrichtian reef. Except for the initial age of eruption, this scenario is not significantly different than the one illustrated in Figure 11. All basalt recovered during drilling at Wodejebato is reversely magnetized with a paleolattitude of about 10° S [Nakanishi and Gee, 1992]. These basalts may have erupted during the Late Cretaceous period of rejuvenation proposed in this paper. Other data in the paper may be superceded by the results of drilling during Legs 143 and 144. In particular, the results of Leg 144 indicate that the rudists recovered by dredging on Wodejebato may be younger than originally thought [Haggerty et al., 1992]. The basic conclusion of this paper: that the Marshall Islands region has seen more than one episode of uplift, volcanism, reef-building and subsidence, remains unchanged.

REFERENCES

Alldredge, L. R., F. Keller, and W. J. Dichtel, Magnetic structure of Bikini Atoll, *U. S. Geol. Surv. Prof. Paper, 260-M*, 529-536, 1954.

Cande, S. C., R. L. Larson, and J. L. La Brecque, Magnetic Lineations in the Pacific Jurassic quiet zone, *Earth Planet. Sci. Lett., 41*, 434-440, 1978.

Castillo, P., R. Batiza, and R. J. Stern, Petrology and geochemistry of Nauru Basin igneous complex: large-volume, off-ridge eruptions of MORB-like basalt during the Cretaceous, *Initial Rep. Deep Sea Drill. Proj., 89*, 555-576, 1986.

Clague, D. A., G. B. Dalrymple, and R. Moberly, Petrography and K-Ar ages of dredged volcanic rocks from the western Hawaiian ridge and southern Emperor seamount chain, *Geol. Soc. Amer. Bull., 84*, 991-998, 1975.

Cole, W. S., Larger foraminifera and smaller diagnostic foraminifera from Bikini drill holes, *U. S. Geol. Surv. Prof. Paper, 260-O*, 569-608, 1954.

Cole, W. S., Larger foraminifera from the Eniwetok Atoll drill holes, *U. S. Geol. Surv. Prof. Paper, 260-V*, 743-784, 1957.

Cole, W. S., Larger foraminifera from deep drill holes on Midway Atoll, *U. S. Geol. Surv. Prof. Paper, 680-C*, 1969.

Coogan, A. H., Evolutionary trends in rudist hard parts, in *Treatise on Invertebrate Paleontology, N, 2*, ed. Moore, R. C., N766-N776, 1969.

Coogan, A. H., New rudists from the Albian and Cenomanian of Mexico and south Texas, *Revista del Instituto Mexicano del Petroleo, V*, 51-83, 1973.

Coogan, A. H., Early and Middle Cretaceous Hippuritacea (Rudists) of the Gulf Coast, in *Cretaceous carbonates of Texas and Mexico, Applications to Subsurface Exploration, Report of Investigations, vol 89*, ed. Bebout, D., and Loucks, R., Texas Bureau of Economic Geology, 32-70, 1977.

Coogan, A. H., D. G. Bebout, and C. Maggio, Depositional environments and geologic history of Golden Lane and Poza Rica trend, Mexico, an alternative view, *Am. Assoc. Petrol. Geol. Bull., 56*, 1419-1447, 1972.

Dalrymple, G. B., The GLM continuous laser system for $^{40}Ar/^{39}Ar$ dating: Description and performance characteristics, *U. S. Geol. Survey Bull., 1890*, 89-96, 1989.

Dalrymple, G. B., E. C. Alexander, Jr., M. A. Lanphere, and G. P. Kraker, Irradiation of samples for $^{40}Ar/^{39}Ar$ dating using the Geological Survey TRIGA reactor, *U. S. Geol. Surv. Prof. Pap., 1176*, 55 pp., 1981.

Dalrymple, G. B., and D. A. Clague, Age of the Hawaiian-Emperor bend, *Earth Planet. Sci. Lett., 31*, 313-329, 1976.

Dalrymple, G. B., D. A. Clague, and M. A. Lanphere, Revised age for Midway volcano, Hawaiian volcano chain, *Earth Planet. Sci. Lett., 37*, 107-116, 1977.

Dalrymple, G. B., M. A. Lanphere and D. A. Clague, Conventional and $^{40}Ar/^{39}Ar$ ages of volcanic rocks from Ojin (Site 430), Nintoku (Site 432), and Suiko (Site 433) Seamounts and the chronology of volcanic propagation along the Hawaiian-Emperor chain, *Initial Reports of the Deep Sea Drilling Project, 55*, 659-676, 1980.

Dalrymple G. B., M. A. Lanphere, and M. S. Pringle, Correlation diagrams in $^{40}Ar/^{39}Ar$ dating: Is there a correct choice?, *Geophys. Res. Lett., 15*, 589-591, 1988.

Davis, A. S., M. S. Pringle, L.-B. G. Pickthorn, D. A. Clague, and W. C. Schwab, Petrology and age of alkalic lava from the Ratak chain of the Marshall Islands, *J. Geophys. Res., 94(B5)*, 5757-5774, 1989.

Detrick, R. S. and S. T. Crough, Island subsidence, hotspots, and lithospheric thinning, *J. Geophys. Res., 83*, 1236-1244, 1978.

Duennebier, F. K. and L. K. Petersen, *Summary report, IPOD Site Surveys in the western Pacific, R/V Kana Keoki* cruise KK810626, Leg 4, Hawaii Institute of Geophysics, 1982.

Duennebier, F. K. and S. O. Schlanger, *MW8805 Marshall Islands Guyots Cruise Report, April 24-May 30, 1988, Guam to Guam*, Hawaii Institute of Geophysics, 80pp, 1988.

Duncan, R. A. and D. A. Clague, Pacific plate motion recorded by linear volcanic chains, in *The Ocean Basins and Margins, The Pacific Ocean, 7A*, ed. Nairn, A. E., Stehli, F. G., and Uyeda, S., New York, Plenum Press, 89-121, 1985.

Emery, K. O., J. I. Tracey, Jr., and H. S. Ladd, Geology of Bikini and nearby atolls, *U. S. Geol. Surv. Prof. Paper, 260-A*, 265p, 1954.

Haggerty, J. A., I. Premoli Silva, and Shipboard Scientific Party ODP Leg 144, Western Pacific Guyot Drilling (ODP Leg 144) Preliminary Results: Sunken Atolls vs. Carbonate Banks [abstract], *EOS, Trans. Amer. Geophys. Un. 1992 Fall Meeting Supplement*, 587, 1992.

Hamilton, E. L., Upper Cretaceous, Tertiary, and Recent planktonic foraminifera from mid-Pacific flat-topped seamounts, *Journal of Paleontology, 27*, 204-237, 1953.

Hamilton, E. L., Sunken Islands of the Mid-Pacific Mountains, *Geol. Soc. Am. Mem., 64*, 97p., 1956.

Hamilton, E. L. and R. W. Rex, Lower Eocene phosphatized *Globigerina* ooze from Sylvania Guyot, *U. S. Geol. Surv. Prof. Paper, 260-W* 785-797, 1959.

Harland, W. B., R. L. Armstrong, A. V. Cox, L. E. Craig, A. G. Smith, and D. G. Smith, *A Geologic Time Scale*, Cambridge, Cambridge University Press, 1990.

Heath, G. R. and R. Moberly, Jr., Deep-sea turbidites from the western pacific: Leg 7, Deep Sea Drilling Project, *Initial Rep. Deep Sea Drill. Proj., 7*, 1009-1010, 1971

Heezen, B. C., J. L. Matthews, R. Catalano, J. Natland, A. Coogan, M. Tharp, and M. Rawson, Western Pacific guyots, *Initial Rep. Deep Sea Drill. Proj., 20*, 653-723, 1973.

Hein, J. R., J.-K. Kang, M. S. Schulz, B.-K. Park, H. Kirschenbaum, S.-H. Yoon, R. L. Olson, V. K. Smith, D.-W. Park, G. O. Riddle, P. J. Quinterno, Y.-O. Lee, A. S. Davis, S. R. Kim, M. S. Pringle, D.-L. Choi, L. Pickthorne, S. O. Schlanger, F. K. Duennebier, D. D. Bergersen, and J. M. Lincoln, Geological, Geochemical, Geophysical, and Oceanographic Data and Interpretations of Seamounts and Co-rich Ferromanganese Crusts from the Marshall Islands, KORDI-USGS R/V *Farnella* Cruise F10-89-CP, *U. S. Geol. Surv. Open File Rep., 90-407*, 246 pp., 1990.

Henderson, L. J., Motion of the Pacific Plate relative to the hotspots since the Jurassic and model of oceanic plateaus of the Farallon Plate, Ph. D. thesis, Northwestern University, Evanston, Illinois, 311 pp., 1985.

Jenkyns, H. C. and S. O. Schlanger, Significance of plant remains in redeposited Aptian sediments, Hole 462A, Nauru Basin, to Cretaceous oceanic-oxygenation models, *Initial Rep. Deep Sea Drill. Proj., 61*, 557-562, 1981.

Jordan, C. F., T. C. Connally, Jr., and H. A. Vest. Middle Cretaceous carbonates of the Mishrif Formation, Fateh field, offshore Dubai, U.A.E., In *Carbonate petroleum reservoirs*, ed. Roehl, P. O., and Choqquette, P. W., New York, Springer-Verlag, 427-442, 1985.

Kulp, L. J., Potassium-Argon dating of volcanic rocks, *Bull. Volcanol., 26*, 247-258, 1963.

Ladd, H. S., J. I. Tracey, Jr., and M. G. Gross, Deep drilling on Midway Atoll, *U.S. Geol. Surv. Prof. Paper, 680-A* 22 p., 1970.

Lanphere, M. A., and G. D. Dalrymple, The use of $^{40}Ar/^{39}Ar$ data in evaluation of disturbed K-Ar systems, in R. E. Zartman, ed., Short Papers on the Fourth International Conference on Geochronology, Cosmochronology, and Isotope Geology, *U. S. Geol. Surv. Open File Rep., 78-701*, 241-243, 1978.

Larson, R. L., Late Jurassic and Early Cretaceous evolution of the western central Pacific Ocean, *J. Geomag. Geoelec., 28*, 219-236, 1976.

Larson, R. L., S. O. Schlanger, and et al., Site 462: Nauru Basin, western Pacific Ocean, Deep Sea Drilling Project Leg 61, *Initial Rep. Deep Sea Drill. Proj., 61*, 19-394, 1981.

Longoria, J. F., and Gamper, M. A., Albian planktonic foraminifera from the Sabinas basin of northern Mexico, *J. Foraminiferal Research, 7*, 196-215, 1977.

Majewski, O. P., *Recognition of invertebrate fossil fragments in rocks and thinsections*, Leiden, E. J. Brill, 1969.

Matthews, J. L., B. C. Heezen, R. Catalano, A. Coogan, M. Tharp, J. Natland, and M. Rawson, Cretaceous drowning of reefs on mid-Pacific and Japanese Guyots, *Science, 184*, 462-464, 1974.

Moberly, R., S. O. Schlanger, and et al., Site Reports, *Initial Rep. Deep Sea Drill. Proj., 89*, 5-284, 1986.

Nakanishi, M, J. S. Gee, and Scientific Party ODP Leg 144, Paleolatitude of guyots in the northwestern Pacific Ocean: Preliminary results of ODP Leg 144 [abstract], *EOS, Trans. Amer. Geophys. Un., 1992 Fall Meeting Supplement*, 586-587, 1992.

Nelson, H. F., Deposition and alteration of the Edwards Limestone, central Texas, in *Symposium on Edwards Limestone in central Texas*, ed. Lozo, F. E., Nelson, H. F., Young, K., Shelburne, O. B., and Sandidge, J. R., University of Texas Bureau of Economic Geology Publication No. 5905, 21-96, 1959.

Ozima, M., I. Kaneoka, K. Saito, M. Honda, Y. Yanagisawa, and Y. Takigami, Summary of geochronological studies of submarine rocks from the western Pacific Ocean, in *Geodynamics of the Western Pacific-Indonesian Region, Geodynamics Series, 11*,, ed. Hilde, T. W. C., and S Uyeda, Washington, D. C., American Geophysical Union, 137-142, 1983.

Ozima, M., K. Saito, and Y. Takigami, $^{40}Ar-^{39}Ar$ geochronological studies on rocks drilled on Leg 61, Holes 462 and 462A, *Initial Rep. Deep Sea Drill. Proj., 61*, 701-704, 1981.

Parsons, B. and J. G. Sclater, An analysis of the variation of the ocean floor bathymetry and heat flow with age, *J. Geophys. Res., 82*, 803-827, 1977.

Petersen, L. D., F. K. Duennebier, and T. H. Shipley, Site surveys in the western Pacific conducted aboard the *Kana Keoki* cruise KK810626 Leg 4, *Initial Rep. Deep Sea Drill. Proj., 89*, 603-628, 1986.

Premoli Silva, I., A new biostratigraphic interpretation of the sedimentary record recovered at site 462, Leg 61, Nauru Basin, western equatorial Pacific, *Initial Rep. Deep Sea Drill. Proj., 89*, 311-319, 1986.

Premoli Silva, I. and C. Brusa, Shallow-water skeletal debris and larger foraminifers from Deep Sea Drilling Project site 462, Nauru Basin, western equatorial Pacific, *Initial Rep. Deep Sea Drill. Proj., 61*, 439-473, 1981.

Pringle, M. S., Age progressive volcanism in the Musicians Seamounts: A test of the hotspot hypothesis for the Late Cretaceous Pacific, in Pringle, M., A. Sager, W. Sliter, and S. Stein, eds., *The Mesozoic Pacific (Schlanger Volume), Am. Geophys. Un. Monogr.*, this volume.

Pringle, M. S., Geochronology and petrology of the Musicians Seamounts, and the search for hotspot volcanism in the Cretaceous Pacific, Ph. D. dissertation, University of Hawaii, Honolulu, 1992.

Pringle, M. S., H. Staudigel, and J. Gee, Jasper Seamount: Seven million years of volcanism, *Geology, 19*364-368, 1991.

Raitt, R. W., Seismic refraction studies of Bikini and Kwajalein Atolls and Sylvania Guyot, *U. S. Geol. Surv. Prof. Paper, 260-K*, 507-528, 1954.

Rea, D. K. and T. L. Vallier, Two Cretaceous volcanic episodes in the western Pacific Ocean, *Geol. Soc. Am. Bull., 94*, 1430-1437, 1983.

Sartorio, D., and Venturini, S., *Southern Tethys Biofacies*, AGIP S.p.A., San Donata Milanese, 235 pp., 1988.

Schlanger, S. O., Subsurface geology of Eniwetok atoll, *U. S. Geol. Surv. Prof. Paper, 260-BB*, 991-1066, 1963.

Schlanger, S. O., Shallow-water limestones in ocean basins as tectonic and paleoceanographic indicators, in *The Deep Sea Drilling Project: a decade of progress, Soc. Econ. Mineral Petrol. Spec. Pub., 32*, ed. Warme, J. E., Douglas, R. G., and Winterer, E. L., 209-226, 1981.

Schlanger, S. O., J. F. Campbell, J. A. Haggerty, and I. Premoli Silva, Cretaceous volcanism and Eocene failed atolls in the Radak Chain: Implications for the geological history of the Marshall Islands [abs.], *Trans. Am. Geophys. Union, 62*, 1075, 1981a.

Schlanger, S. O., J. F. Campbell, and M. W. Jackson, Post-Eocene subsidence of the Marshall Islands recorded by drowned atolls on Harrie and Sylvania Guyots, in *Seamounts, Islands and Atolls, Am. Geophys. Union Monograph Ser., 43*,, ed. Keating, B. H., Fryer, P., Batiza, R., and Boehlert, G. W., 165-174, 1987.

Schlanger, S. O., H. C. Jenkyns, and I. Premoli-Silva, Volcanism and vertical tectonics in the Pacific Basin related to global Cretaceous transgressions, *Earth Planet. Sci. Lett., 52,* 435-449, 1981b.

Schlanger, S. O. and R. Moberly, Sedimentary and volcanic history: East Mariana Basin and Nauru Basin, *Initial Rep. Deep Sea Drill. Proj., 89,* 653-678, 1986.

Schlanger, S. O. and I. Premoli Silva, Tectonic, volcanic, and paleogeographic implications of redeposited reef faunas of Late Cretaceous and Tertiary age from the Nauru Basin and the Line Islands, *Initial Rep. Deep Sea Drill. Proj., 61,* 817-827, 1981.

Schroeder, R., and Neumann, M. (eds.), Les grands foraminiferes du Cretace moyen de la region Mediterraneenne, *Geobios, Mem. Speciale, 7,* 161 pp., 68 pl., 1985.

Takigami, Y., S. Amari, M. Ozima, and R. Moberly, $^{40}Ar/^{39}Ar$ geochronological studies of basalts from hole 462A, Nauru Basin, Deep Sea Drilling project Leg 89, *Initial Rep. Deep Sea Drill. Proj., 89,* 519-522, 1986.

Taylor, J. R., *An Introduction to Error Analysis,* Mill Valley, California, University Science Books, 270 pp., 1982.

Turner, D. L. and R. D. Jarrard, K-Ar dating of the Cook-Austral Island Chain: A test of the hot-spot hypothesis, *Journal of Volcanology and Geothermal Research, 12,* 187-220, 1982.

Vallier, T. L., K. E. Windom, K. E. Siefert, and J. Thiede, Volcanic rocks cored on Hess Rise, western Pacific Ocean, *Nature, 286,* 48-50, 1980.

van Gorsel, J. T., Late Cretaceous orbitoidal foraminifera, in *Foraminifera, 3,,* ed. Hedley, R. H., and Adams, C. G., Academic Press, 1978.

Watts, A. B., D. P. McKenzie, B. E. Parsons, and M. Roufosse, The relationship between gravity and bathymetry in the Pacific, *Geophys. J. R. astron. Soc., 83,* 263-298, 1985.

White, C. A., On Mesozoic Fossils: Description of certain aberrant forms of the Chamidae from the Cretaceous rocks of Texas, *U. S. Geol. Surv. Bull., 4,* 93-97, 1884.

Wilson, J. L., *Carbonate Facies in Geologic History* New York, Springer-Verlag., 1975.

Winterer, E. L., Bathymetry and regional tectonic setting of the Line Islands Chain, *Initial Rep. Deep Sea Drill. Proj., 33,* 731-748, 1976a.

Winterer, E. L., Anomalies in the tectonic evolution of the Pacific, in *The Geophysics of the Pacific Ocean Basin and its Margins, Geophys. Monogr. ser., vol. 19,* edited by G. H. Sutton, M. H. Manhnani, and R. Moberly, pp. 269-278, AGU, Washington, D. C., 1976b.

Winterer, E. L., and Metzler, C. V., Origin and subsidence of guyots in Mid-Pacific Mountains, *J. Geophys, Res., 89,* 9969-9979, 1984.

Winterer, E. L., J. I. Ewing, and et al., Site Reports, *Initial Rep. Deep Sea Drill. Proj., 17,* 5-334, 1973.

Winterer, E. L., W. R. Riedel, and et al., Site Reports, *Initial Rep. Deep Sea Drill. Proj., 7,* 27-841, 1971.

York, D., Least squares fitting of a straight line with correlated errors, *Earth Planet. Sci. Lett., 5,* 320-324, 1969.

J. Lincoln, Department of Physics and Geoscience, Montclair State College, Upper Montclair, NJ 07043

M. Pringle, U. S. Geological Survey, MS937, Menlo Park CA 94025

I. Premoli Silva, Dipartmento di Scienze della Terra, Universita Degli Studi di Milano, I-20133 Milano Italy

Cretaceous Guyots in the Northwest Pacific: an Overview of Their Geology and Geophysics

EDWARD L. WINTERER, JAMES H. NATLAND[1], AND ROBERT J. VAN WAASBERGEN

Scripps Institution of Oceanography, La Jolla, California

ROBERT A. DUNCAN

College of Oceanography, Oregon State University, Corvallis, Oregon

MARCIA K. MCNUTT AND CECILY J. WOLFE

Dept. Earth, Atmospheric, and Planetary Sci., M.I.T., Cambridge, Massachussetts

ISABELLA PREMOLI SILVA

Dipartimento Scienze della Terra, Università di Milano, Milan, Italy

WILLIAM W. SAGER

Dept. Of Oceanography, Texas A&M Univ., College Station, Texas

WILLIAM V. SLITER

U.S. Geological Survey, Menlo Park, California

Many Early Cretaceous seamounts in the northwest Pacific have summits that were once above sea level, where they were flattened by stream and wave erosion and then capped by rudist reefs. Subsidence during Barremian-early Aptian times allowed reefs in the Mid-Pacific Mountains (MPM) and on scattered seamounts farther west to evolve into atolls, with thicknesses of as much as 800 m of limestone. Available paleontological data suggest that these reefs drowned in early Aptian time and continued subsiding. The Japanese seamounts (JPS), which were emplaced in Albian times, have much thinner capping reefs (100-200 m) and evolved only to the barrier reef stage. During the latest Albian, a time of a major plate reorganization in the Pacific, and roughly coincident with the beginning of volcanism at the northernmost (Wake seamounts, WKS) end of the Marshall seamount chains, all the MPM and JPS reefs were uplifted, became emergent, and developed karstic topography. The WKS, although eroded nearly flat at sea level, did not develop reefs. MPM and JPS reefs drowned and were accumulating pelagic sediments by Turonian time.

Whole-rock $^{40}Ar/^{39}Ar$ incremental heating ages range from Barremian-Aptian (123-128 Ma) for edifice building in the MPM and at MIT Guyot, Albian (109-103 Ma) for the JPS, and Late Albian-Campanian (100-82 Ma) in the WKS. Some Barremian-Aptian seamounts may be overprinted by mid-Cretaceous volcanism.

Petrologically, MPM volcanic rocks resemble near-rise-crest tholeiites such as those in the Galapagos Islands, while JPS and WKS have strong affinities to modern Samoan alkalic basalts and to volcanic rocks in French Polynesia with SOPITA signatures.

Modeling of seamount magnetism yields paleolatitudes of formation of about 2°S-10°S for JPS and WKS, in agreement with previous estimates of the polar wander path of the Pacific plate, but at positions slightly north of the main region of the modern Superswell region of abnormally thin elastic plate thickness of the lithosphere.

Modeling of shipborne and satellite gravity data over MPM, JPS, and WKS guyots yields elastic plate thicknesses of less than 15 km, consistent with the hypothesis that the seamounts formed on a superswell - the Darwin Rise - during mid-Cretaceous times.

INTRODUCTION

A singular and striking feature of the Pacific Plate is the wide band of seamount chains and oceanic plateaus that stretches southeast from near Japan to Easter Island. While the origin of this huge belt, some 12,000 km long and 3000 km wide and morphologically different from the rest of the Pacific, is only beginning to be understood, we can recite some of its characteristics: very large seamounts, many of which are now atolls or guyots; guyot summit depths shallower than expected for the elapsed time since drowning; recurrent episodes of volcanism long after initial edifice building; seamount lavas with radiogenically-enriched isotopic compositions; and widespread mid-plate volcanism in the deep basins between seamount chains and plateaus, predominantly mid-Cretaceous in age. Based on the

[1]Now at Rosenstiel School of Marine and Atmospheric Sciences, University of Miami, Coral Gables, Florida

The Mesozoic Pacific: Geology, Tectonics, and Volcanism
Geophysical Monograph 77

guyot summit-depth data, Menard [1964] envisioned a large region of abnormally shallow sea floor in the northwest Pacific during the Cretaceous, which he termed the Darwin Rise.

Current plate-motion models [e.g., Duncan and Clague, 1985] indicate that the Cretaceous volcanoes of the northwest Pacific formed in the southeast Pacific, where presently active hot-spot volcanism and late Cenozoic seamount chains exhibit the same characteristics as the older Darwin Rise volcanoes. The southeast Pacific region, termed the Superswell by McNutt and Fischer [1987], is characterized by lithosphere that is both shallower and subsides more slowly than normal, a condition ascribed by them to thinned and abnormally hot lithosphere, although abnormal heat flow is not documented in the region today [Stein and Abbott, 1991].

Based on geological and geophysical data collected during November and December, 1988, on the Roundabout, Leg 10, expedition on R/V *Thomas Washington* (RNDB10WT) and previous expeditions, we have argued that the Darwin Rise was a mid-Cretaceous Superswell [McNutt et al., 1990; Wolfe and McNutt, 1991]. In brief, the hypothesis is that Early Cretaceous reef-bearing volcanoes in the northwest Pacific formed in the southeast Pacific on normal lithosphere and subsided at a normal rate during reef build-up. Then, in mid-Cretaceous time, Superswell-type lithosphere formed beneath the region, uplifting the reefs by as much as 200 m, and exposing them to subaerial dissolution and erosion. For the next 20 m.y. or so, until plate motion carried the old volcanoes northwest off the thermally disturbed region, subsidence rates were abnormally slow. Then, as the plate cooled and thickened, subsidence rates gradually returned to normal. We herein summarize those aspects of the geology and geophysics of the northwest Pacific that bear critically on the origin and geologic history of the guyots in the region, pulling together the older data along with the results of RNDB10WT, in an effort to reach a fresh synthesis in advance of the JOIDES drilling scheduled for the spring of 1992. The several coauthors are not in complete agreement about interpretations of the existing data, and these differences remain unresolved in this paper. The drilling will not only form a test of the hypotheses put forward here, but will doubtless also add so massively to the data base as to require formulation of new hypotheses. Each of the sections of this paper is but a summary; for details the reader is referred to individual topical papers by the various coauthors [Van Waasbergen and Winterer; Sager; Duncan and Handschumacher, all in this volume; Wolfe and McNutt, 1991] .

REGIONAL GEOLOGIC SETTING

As shown in the simplified crustal age map of the northwest Pacific (Fig. 1), the Cretaceous guyots stand on ocean crust generated at the Pacific-Izanagi, Pacific-Farallon and Pacific-Phoenix spreading ridges, and ranging in age from mid-Jurassic (*ca.* 160 Ma) to mid-Cretaceous (*ca.* 100 Ma). In many places the seamounts are located on fracture zones. Most of the seamounts in the northwest Pacific are organized in clusters or chains, but some stand alone. The degree to which the chains are the product of Pacific plate motion over hot spots, and the timing and velocities of such motions, are open questions. The main groupings of Cretaceous seamounts are as follows (Fig. 1):

- the Japanese seamount group (JPS), a loosely strung chain that begins east of Tokyo with Takuyo-Daini Seamount and extends ESE about 1000 km to Makarov Guyot. Daiichi-Kashima Seamount, which is being subducted at the Japan Trench (Konishi, 1989], may be the westernmost member of this chain.
- the Marcus-Wake chain (MWC), which begins at the kink joining the Bonin and Mariana Trenches, and extends in an arc ESE for about 2500 km. The eastern part of this chain, near Wake Island, is here termed the Wake seamount group (WKS).
- a loose cluster of seamounts between the JPS and MWC, around 150°E.
- the Magellan chain, which begins at the Trench at 20°N and extends ESE for about 2200 km, but with a large gap in the middle. Both this chain and the MWC are complex in map view, and may be composite, comprising two or more chains, formed at the same or different times.
- the Marshall Islands chains (three distinct chains are discernable), which trend SSE and intersect the eastern parts of the Magellan chain and may extend northward to intersect the eastern end of the MWC. The assignment of many individual seamounts in this region to either SSE-or ESE-trending chains is uncertain.
- the Line Islands chain, which extend SSE for about 3800 km from their intersection with the MPM.
- the Line Islands Crosstrend groups, which trend ESE for about 1600 km and about midway along their length intersect the Line Islands chain at about 10°N.
- the Mid-Pacific Mountains (MPM), which comprise several dozen seamounts surmounting a boxwork set of ENE- and SSE-trending ridges and plateaus about 2000 km long.

We herein report only on guyots in and near the JPS, MWC, and MPM.

PREVIOUS WORK

The modern era of work in the region dates from the 1950 Scripps MIDPAC expedition, which first explored the guyots of the MPM and showed that these are mainly ancient submarine volcanoes, flattened at sea level as Hess [1946] had predicted. Hamilton [1956] showed that instead of their being Precambrian, as Hess had inferred, they were much younger. Hamilton [1956] discovered that in mid-Cretaceous times the seamounts were at sea level in low latitudes, where they were encrusted by rudist-coral reefs before their summits subsided beneath sea level. Later dredging, drilling, and reflection profiling in and near the MPM [Lonsdale et al., 1972; Winterer et al., 1973; Heezen et al., 1973; Matthews et al., 1974; Ladd et al., 1974; Larson et al., 1975; Thiede et al., 1981; Kroenke et al., 1985] showed that reef-building had begun by the late Barremian (*ca.* 120 Ma), and was over by Turonian time (*ca.* 91 Ma). Neither the inception nor the demise of the reefs could be dated more precisely than these broad limits. Reflection profiles of MPM guyots [Heezen et al., 1973] show a thickness of several hundred meters of reefal

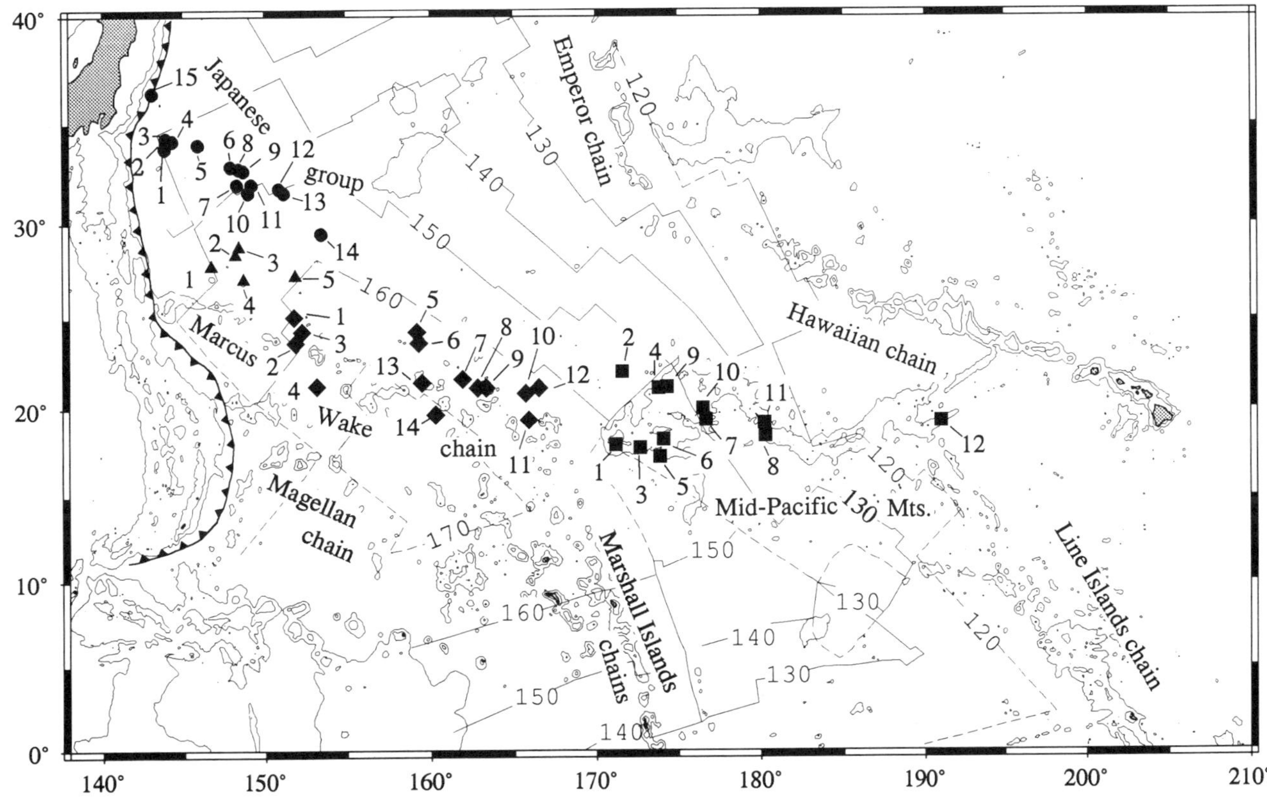

Fig. 1. Crustal age map of the northwest Pacific, showing the location of Cretaceous seamounts. The 10my isochrons are interpolated and extrapolated from the magnetic anomaly map of Nakanishi et al. (1992). Symbols and numbers on the map designate seamounts mentioned in the text and tables or on other figures, as follows:

Circle symbol (Japanese Group): 1, "Jenson"; 3, Takuyo Daini; 4, Takuyo Daisan; 5, "Maiko"; 7, "Charlie Johnson"; 8, "Winterer"; 10, "Stout"; 11, "Thomas Washington"; 13, Isakov; 14, Makarov; 15, Daiichi-Kashima.

Triangle symbol (Seamounts between Japanese and Marcus-Wake Groups): 1, Seamount "Z44"; 2, Seamount "Z42"; 3, Seamount "Z41"; 4, Seamount "Z43"; 5, MIT.

Diamond symbol (Marcus-Wake Group): 1, unnamed seamount; 2 Seamount "948"; 3, unnamed seamount; 4, "Golden Dragon"; 5, "UCSD"; 6, "Scripps"; 7, "Miami"; 8, "Vibelius"; 9, "Wilde"; 10, "Birdseye"; 11, unnamed seamount; 12, "Woods Hole"; 13, "Lamont"; 14, "Pot".

Square symbol (Mid-Pacific Mountains Group): 1, Sio; 2, Darwin; 3, "Harvey"; 4, "Heezen"; 5, "Thomas"; 6, "Allen"; 7, "Jacqueline"; 8, Allison; 9, Resolution; 10, "Caprina"; 11, "Shepard"; 12, Horizon.

sediments. Drilling results [Winterer, Ewing et al., 1973] and profiler data [Heezen et al., 1973] show from 0-200 m of post-reefal pelagic sediments. Ozima et al. [1983] reported a Late Cretaceous $^{40}Ar/^{39}Ar$ age from an altered basalt from a MPM guyot.

Proceeding westward from the MPM, dredges from WKS guyots, in the eastern part of the MWC, recovered volcanic and pelagic but no reefal material [Heezen et al., 1973], and $^{40}Ar/^{39}Ar$ dates reported from dredges of three WKS guyots fall in the range 86-97 Ma [Ozima et al., 1983]. Farther to the northwest, in the JPS group, dredges returned reefal fossils with age ranges of Aptian-Cenomanian (119-91 Ma) [Heezen et al., 1973; Matthews et al., 1974]. A seamount in the Japan Trench yielded shallow-water fossils of Albian age [Konishi, 1989]. $^{40}Ar/^{39}Ar$ ages on two reef-capped seamounts in the JPS, as determined by Ozima et al. [1983], are in the range 94-102 Ma.

Details of the morphology of the JPS and some of the MWC seamounts, as revealed by multibeam mapping, are given by Smoot [1983] and Vogt and Smoot [1984]. The 100-m contour interval of the maps in Vogt and Smoot is too wide to show crucial details of the summit topography, and therefore could not be used to test in these groups the suggestion put forward (for guyots in the MPM) by Winterer and Metzler [1984] that the guyots had been uplifted and developed a karstic erosion surface prior to final drowning. Both "Nelson" and "Smoot" Guyots (seamount names not recognized officially by the U.S. Board of Geographic Names are shown in inverted commas), in the western part of the MWC [Smoot, 1983], show elongate reef-like ridges at the edge of their summit platforms. "Nelson" Guyot, depicted partly at a 10-m contour interval, shows a number of closed depressions on the summit area, interpreted by Smoot as craters and calderas, but which could equally well be interpreted as solution depressions. For neither guyot are reflection profiler data or dredges available.

THE VOLCANIC EDIFICE

Morphology

The fundamental shape of a guyot is the result of the volcanic processes of edifice building and the subsequent modifying erosional processes. Morphologic characteristics such as elongation, volume, number of flank rift zones, volcanic slope angles, and other larger scale features are defined early on the history of a guyot. The major features of edifice morphology of many northwest Pacific guyots, as known from multibeam bathymetric surveys, are well illustrated in Vogt and Smoot [1984] and in Smoot [1991].

As inferred from dredging and seismic data, several guyots in the Wake group ("Lamont", "Miami", "Scripps", and "Wilde" guyots) are virtually sediment-free and are morphologically different from sedimented guyots. Typically, the tops of the volcanic guyots are flat, with a distinctly beveled outer edge. This is most likely the result of having been eroded at sea level, without protection from reefs. This is in contrast to the reef-bearing summits, on which the buried basement relief can measure several hundred meters, generally in the form of a central volcano. Except where the island parts of the volcano were protected by barrier reefs, no stream-valleys are seen on the volcanic summits, which may indicate that erosion at sea level wipes out any landforms created on the island, before submergence.

Samples Studied

We obtained bulk analyses and mineral compositions on a suite of volcanic rocks from 26 locations on 19 guyots and ridges from the JPS, WKS, and MPM seamount groups. Also, we present 11 new $^{40}Ar/^{39}Ar$ whole-rock, incremental-heating experiments from 10 of the seamounts studied here. The seamount samples were obtained on several expeditions of Scripps Institution of Oceanography over nearly 40 years. Since reefs were the usual dredge targets, the rocks were obtained chiefly from breaks in slope, guyot summits, or outcrops just below the bases of large carbonate platforms. These represent approximately comparable high structural levels on most seamounts, probably accounting for the predominance of alkalic lavas in the collections.

Most of the rocks are highly vesicular pillows or pillow fragments from aquagene tuffs. A number are compact, non-vesicular cobbles eroded from once-subaerial outcrops. Almost all are cemented by marine phosphorite and encrusted with hydrogenous ferromanganese oxyhydroxides. In addition, many lavas, particularly the vesicular ones, are largely transformed to smectite, zeolites, and authigenic K-feldspar; such samples may be variably impregnated with marine phosporite in veins and vesicles. Although olivine was once abundant in many samples, most of them now do not retain even a trace of fresh olivine. Interpretation of geochemical data must take into account the effects of alteration. We have used procedures for discounting the effects of partial incorporation of finely disseminated marine phosphorite and Fe-Mn oxyhydroxides in chemical analyses, to reveal the general geochemical characteristics of remaining altered rock.

Rock Types

As expected, the rocks are varieties of ocean-island basalts and differentiated rocks. Nonetheless, even extensive residual alteration in some of them does not conceal that the rocks are mainly of alkalic lineage and in many samples have strongly enriched trace-element compositions. The interpretation follows from comparison of relative abundances of elements that can be mobile during alteration (K, Rb, and Ba) with those that are not (especially Zr and Ti), plus consideration of petrography and mineral compositions.

Once alteration is taken into account, three classes of volcanic rocks can be defined. The classes are those of many Pacific volcanic islands. They are 1) tholeiitic basalts; 2) alkalic basalts plus allied differentiates; and 3) strongly undersaturated nephelinite-series lavas. The question is, what are the most apt contemporary comparisons? For example, all three classes can be found on single islands of modern-day island chains such as Hawaii. For the Mesozoic seamounts, the most precise analogies are not to Hawaiian volcanoes, but to other islands in different parts of the Pacific.

Lavas from the eastern Mid-Pacific Mountains. Rocks dredged from the eastern MPM indicate a mixed tholeiitic and alkalic basaltic province. The dredged tholeiites are aphyric to sparsely plagioclase-phyric, ilmenite-bearing, thus strongly fractionated and TiO_2-rich rocks with similarities to fractionated Icelandic tholeiites or similar lavas from the Galapagos Islands. Abundances of Ba, Sr, and Zr suggest that they had depleted or only moderately enriched geochemical attributes before alteration. The general characteristics of the rocks, and the ages of reef limestones [Hamilton, 1956] and of sediments above basalts at Horizon Guyot [Winterer, Ewing et al., 1972] support the suggestion of Heezen et al. [1974] that the large elongate ridges of this region are composite volcanoes that formed on young ocean crust at or near a spreading ridge.

Three groups of seamounts are represented: the MPM, the WKS, and the JPS. The chemical analyses indicate that from the western MPM to the JPS the volcanoes make up a broadly coherent alkalic and distinctly incompatible-element-enriched province, with petrographic and geochemical similarities to far-off-axis volcanic chains in the southwest Pacific, particularly Samoa and Tahiti. The eastern MPM, on the other hand, resemble near-axis, hot-spot volcanoes such as the Galapagos Islands. The results lend support to the hypothesis that a large, geochemically distinct province has persisted in the Pacific since the Mesozoic [Natland and Wright, 1984]. The hypothesis has more recently been supplemented on the basis of geophysical (mainly gravity) data and subsidence histories. Staudigel et al. [1990] have integrated isotopic data with geophysical inferences that support the persistence or recurrence of a Pacific superswell, now centered on French Polynesia [McNutt and Fischer, 1987; McNutt et al., 1990], and explicitly relate the continuing volcanism to passage of the Pacific plate over an isotopic and thermal anomaly now located in the South Pacific (SOPITA) [Smith et al., 1989].

Western MPM and WKS Lavas. A single boulder of vesicular and highly altered, fractionated tholeiite, similar to MPM tholeiites, was recovered from MIT Guyot, west of the WKS. It is the only tholeiite in the collection dredged outside the MPM. The guyot is a larger volcano than most in the WKS, and differs from these as well in having a thick carbonate platform. These facts suggest that MIT guyot may not be related to the WKS, but is a different type of volcanic structure, or an older one formed on thinner, younger lithosphere.

Apart from this tholeiite, samples from the western MPM, and from the WKS are principally mildly alkalic olivine basalts and related differentiates. More strongly undersaturated basanites and olivine nephelinites, plus altered, amphibole-bearing ankaramites that fractionated from basanite or nephelinite parents, were recovered from "Miami", "Lamont", "Vibelius", and "Scripps" Guyots. No *bona fide* tholeiites have been sampled from these seamounts.

Basalts and hawaiites, probably representing late-stage shield activity, generally resemble basaltic-ankaramitic lava suites of both Samoa and Tahiti in thin section. In detail, total alkali abundances, normative characteristics, ratios of CaO/Al_2O_3, and persistence of plagioclase among phenocryst assemblages more closely resemble Samoa [*cf.* Natland and Turner, 1985] than Tahiti. Geochemically, the lavas are consistently enriched in incompatible elements. Such relationships suggest derivation from enriched (EM-II) mantle sources carrying both recycled terrigenous (EM-II) and ocean-crust (HIMU) components [Saunders, 1986; Weaver, 1991]. One sample from "Wilde" Guyot has initial $^{87}Sr/^{86}Sr$ = 0.7053 [Staudigel et al., 1990], and thus isotopically resembles Samoan and Society basalts with EM-II source characteristics. Four other samples, three of them nephelinitic lavas from "Wilde" and "Scripps" Guyots, isotopically have HIMU characteristics. Based on trace elements alone, Allison and "Jacqueline" Guyots in the western MPM and possibly "Vibelius" Guyot in the WKS have EM-II characteristics; Darwin Guyot in the western MPM and "Woods Hole" Guyot in the WKS have the trace-element attributes of HIMU lavas.

Nephelinite-series lavas from the WKS are petrographically distinctive rocks. Basanites and olivine nephelinites from "Miami" and "Lamont" Guyots have little or no feldspar and abundant groundmass Ti-augite and Ti-magnetite. These are the freshest mafic lavas in the dredge collection; some samples even contain a few grains of fresh olivine. The ankaramites, from "Scripps" and "Vibelius" Guyots, also have abundant clinopyroxene, but they are feldspathic differentiated lavas and carry kaersutitic amphibole as phenocrysts, in the groundmass, or both. Since "Miami", "Lamont", and "Scripps" Guyots do not have carbonate platforms, the rocks are plausibly post-erosional summit eruptives.

The basanites and olivine nephelinites are systematically enriched in incompatible elements when compared with Hawaiian nephelinite-series lavas, but are comparable in these respects to post-erosional lavas from Samoa and valley-filling basanites from Tahiti. The amphibole-bearing ankaramites are potassic, with potash nearly matching or exceeding soda. Three of these rocks, including one ankaramite, analyzed for isotopes have HIMU characteristics [Staudigel et al., 1991], and in this respect do not resemble Samoan or Tahitian rocks. High K_2O contents and Rb concentrations seemingly contradict this result, but are not - in these freshest of mafic lavas in the collection - a consequence of alteration. The cause of the elevated alkalis may have been reaction of comparitively small partial melts with hydrous phases in the mantle or crust. In support of reaction within the crust, one ankaramite from "Scripps" Guyot contains partially resorbed kaersitite phenocrysts with magnetite-rich reaction coronas.

Japanese Seamounts. In the JPS, five of six samples analyzed from Takuyo Daisan and Takuyo-Daini Guyots are fractionated alkalic yet silica-saturated lavas ranging from mafic mugearite to trachyte in composition. Like many such highly evolved lavas, they are silica-saturated despite being derived from undersaturated, alkalic basalt parents. Such rocks are typical of the final stages of waning alkalic volcanism on many shield volcanoes in the South Pacific. The rocks are distinctly potassic, with low soda/potash and high K/Zr and K/Rb ratios. Samples from Takuyo-Daini Guyot show clear petrographic evidence for mixing between hawaiitic and trachytic melts, the compositions of which were obtained on perlitic glassy material by electron probe. The geochemical characteristics suggest comparison to mildly alkalic but potassic lineages from the southwest Pacific (Samoa; Ua Poa in the Marquesas), although excess alkalis in rocks with intermediate compositions may have resulted from the mixing. The lavas were evidently derived from enriched basaltic parents, but precisely which types of mantle sources were involved cannot be said on the basis of trace elements alone.

Strongly undersaturated olivine nephelinites, similar to those from Lamont Guyot in the WKS, were recovered from Isakov Guyot in the JPS. These are quite different from the differentiated mildly alkalic lavas dredged elsewhere in this seamount group. They probably represent a later post-erosional or rejuvenated stage of volcanism.

Radiometric ages of dredged volcanic rocks from guyots

In an attempt to improve models for the time-space relationships of volcanism within the northwest Pacific seamount province, we here present 11 new $^{40}Ar/^{39}Ar$ whole-rock incremental-heating ages from 10 of the seamounts studied here. The rationale for using this method rather than the standard K-Ar dating method for volcanic rocks that have been in contact with seawater over extended periods is set out in Pringle (this volume).

Experimental Methods. Pieces of dredged rock were slabbed and trimmed to obtain the freshest and best crystallized interior material for radiometric dating. These samples were crushed to 0.5-1 mm size, hand-picked to remove vesicle and vein alteration, and ultrasonically washed in distilled water. Approximately 0.5-g splits of the prepared chips were sealed in evacuated quartz tubes and irradiated for 10-15 hours in the core of the Oregon State University TRIGA reactor, where they received a neutron dose of

1.0-1.5 x 10^{18} nvt. The efficiency of conversion of ^{39}K to ^{39}Ar by neutron capture was monitored with samples of hornblende standard Mmhb-1 (520.4 Ma; Samson and Alexander, 1987]. Further details of the flux characteristics, monitor minerals, and corrections for interfering K- and Ca-derived Ar isotopes are reported in Dalrymple et al. [1981].

Argon extractions were performed in a conventional high-vacuum glass line using radio-frequency induction heating. Heating steps were set from power levels on the generator and were determined from previous experience to divide the total whole-rock Ar into six roughly equal portions. Samples were held at each step setting (temperature) for 30 minutes. The Ar composition of each gas increment was measured mass spectrometrically with an AEI-10S instrument with computer-controlled peak switching and data acquisition. The system background was $3x10^{-14}$ moles ^{40}Ar.

Results. Results from $^{40}Ar/^{39}Ar$ incremental-heating experiments are reported in Table 1 and illustrated in Fig. 2. Data are displayed in age-spectra diagrams, in which the ages calculated from heating-step gas compositions for a given sample are plotted against cumulative gas released (%^{39}Ar). In most samples, concordant ages for middle- to high-temperature steps define a "plateau" age which best estimates the age of rock crystallization (e.g., Fig. 2k). Step gas compositions can also be correlated in isotope-ratio diagrams (e.g. Fig. 3a) where colinear points are least-squares fitted to yield a slope, which determines an isochron age, and an intercept, which defines the composition of Ar in the sample when it crystallized (expected to be atmospheric $^{40}Ar/^{36}Ar$=295.5). Finally, the Ca/K of individual-step Ar compositions can be calculated from 1.852 $^{37}Ar/^{39}Ar$ [Walker and McDougall, 1982] and outgassing phases then identified from mineral-composition data: Ca/K for clinopyroxene (500-2000), plagioclase feldspar (20-500), groundmass (1-20), and clays (<1). Plots of Ca/K vs % ^{39}Ar for our experiments (e.g. Fig. 3b) show Ca/K increasing with temperature from values <1 to ~100. The plateau age-defining steps are primarily associated with Ca/K appropriate for groundmass and feldspar phases.

The $^{40}Ar/^{39}Ar$ incremental-heating experiments have produced variably disturbed age spectra for the northwest Pacific seamounts samples. Some, like A5-19-1, A5-22-1, A5-23-1, and WP-49-1, developed nearly ideal plateaus, comprising over 90% of the total gas in the samples. Spurious step ages generally resulted from small amounts of gas released at low temperatures, which we attribute to nonconcordant loss of radiogenic ^{40}Ar (during alteration) and ^{39}Ar (during neutron irradiation) from alteration minerals. Examples are A5-21-3, WP-54-4, WP-64-7, and WP-72-1. A few samples (A5-27-1, WP-55-1a and WP-66-25) showed decreasing ages at higher temperatures, an effect generally attributed to reactor-induced ^{39}Ar and ^{37}Ar recoil [Turner and Cadogan, 1974; Huneke and Smith, 1976], which produces erroneously young ages for (K-poor, Ca-rich) high-temperature steps in fine-grained samples. Sample WP-66-25 is the most extreme example of this effect, and no plateau was obtained; for this sample, the total fusion age, calculated by adding all step gas compositions together under the assumption that Ar isotopes have been redistributed but not lost from the sample, is the best

Table 1. $^{40}Ar/^{39}Ar$ incremental-heating age determinations for northwest Pacific seamounts.

Heating Step	$^{40}Ar/^{36}Ar$	$^{40}Ar/^{39}Ar$	$^{40}Ar/^{37}Ar$*	% ^{39}Ar of total	Step Age ±1σ (Ma)
Sample A5-19-1, "Wilde" Smt mugearite, J=0.002115					
1	1004.5	31.623	192.07	3.0	80.7 ± 1.0
2	20004.	25.257	346.88	30.1	91.8 ± 0.4
3	25931.	24.796	321.30	45.1	90.4 ± 0.3
4	8819.7	25.149	108.58	17.9	89.7 ± 0.4
5	1272.0	30.554	19.36	2.3	87.2 ± 1.5
6	592.6	45.972	23.83	1.7	85.8 ± 0.6
			plateau age		90.6 ± 0.3
			isochron age		91.4 ± 1.0
Sample A5-21-3, "Miami" Smt basalt, J=0.002311					
1	428.59	38.880	32.000	8.6	50.1 ± 1.2
2	672.50	37.117	11.988	11.5	86.9 ± 1.6
3	1972.0	28.002	9.201	15.8	98.7 ± 0.6
4	1368.1	30.263	23.750	23.8	96.9 ± 0.6
5	1941.3	26.254	1.853	34.9	94.8 ± 0.6
6	654.9	35.757	0.766	5.6	89.9 ± 0.8
			plateau age		96.8 ± 0.6
			isochron age		97.0 ± 1.6
Sample A5-22-1, "Lamont" Smt basalt, J=0.003250					
1	873.37	23.451	29.998	23.7	89.2 ± 0.7
2	1969.3	17.934	38.071	10.2	87.4 ± 0.9
3	1455.3	19.358	29.772	9.1	88.6 ± 2.0
4	907.26	22.163	17.445	7.4	86.2 ± 1.1
5	1022.9	20.889	6.3277	23.4	86.8 ± 0.7
6	1164.7	18.023	1.0249	26.3	86.0 ± 0.6
			plateau age		87.2 ± 0.3
			isochron age		86.9 ± 2.2
Sample A5-23-1, "Lamont" Smt basalt, J=0.002115					
1	1075.1	31.293	37.251	16.5	84.1 ± 1.4
2	1413.2	29.270	49.254	17.8	85.7 ± 0.8
3	469.4	58.569	85.173	9.1	80.5 ± 1.7
4	1101.9	28.531	171.669	7.4	77.8 ± 1.5
5	2967.3	23.249	2.528	47.0	80.6 ± 0.4
6	554.7	34.107	1.171	2.3	69.4 ± 2.6
			plateau age		81.5 ± 0.6
			isochron age		81.9 ± 1.8

Table 1 (continued)

Sample A5-27-1, "Scripps" Smt basalt, J=0.003250

1	397.69	70.821	170.44	35.5	103.9 ± 1.2
2	717.46	28.886	12.624	13.8	98.2 ± 1.6
3	2560.00	19.886	6.0729	14.6	102.1 ± 1.4
4	424.52	56.910	7.9128	26.2	99.6 ± 1.9
5	353.52	32.931	3.5351	9.9	36.2 ± 2.6
			plateau age		101.6 ± 0.7
			isochron age		100.1 ± 2.0

Sample WP-49-1, "Winterer" Smt mugearite, J=0.002115

1	887.9	61.706	37.798	2.5	150 ± 1.3
2	8741.8	31.167	118.005	22.2	110.9 ± 1.0
3	23820.	30.428	186.619	43.3	110.7 ± 0.8
4	6928.9	30.203	53.010	14.8	106.8 ± 0.8
5	4859.2	30.106	20.884	16.7	104.8 ± 0.4
6	377.5	54.675	2.471	0.5	49.3 ± 18.1
			plateau age		108.3 ±1.0
			isochron age		108.5 ±1.8

Sample WP-54-4, Isakov Smt nephelinite, J=0.002115

1	659.3	59.022	81.760	7.4	120.1 ± 1.2
2	1626.2	35.014	41.786	26.0	106.1 ± 1.4
3	3470.6	32.257	51.078	20.0	109.1 ± 1.8
4	1062.1	36.699	54.694	24.9	100.9 ± 1.2
5	1131.3	34.779	4.005	19.3	98.5 ± 1.8
6	392.2	71.412	1.251	2.4	89.9 ± 4.8
			plateau age		103.7 ± 1.8
			isochron age		103.4 ± 2.8

Sample WP-55-1A, MIT Smt alkali basalt, J=0.002115

1	496.0	69.319	112.499	12.5	123.6 ± 0.7
2	465.7	91.127	106.372	11.1	122.1 ± 0.8
3	2195.3	37.077	28.793	24.2	117.9 ± 0.5
4	914.0	36.896	24.301	12.8	92.6 ± 0.6
5	1348.6	23.567	24.011	34.0	68.6 ± 0.3
6	744.1	28.266	44.218	5.3	63.6 ± 0.9
			plateau age		120.3 ± 0.8
			isochron age		118.0 ± 8.5

Sample WP-64-7, "Heezen" Smt basalt, J=0.001430

1	381.57	249.34	255.24	24.3	138.7 ± 1.6
2	1790.0	59.634	15.802	29.4	124.3 ± 1.3
3	1659.4	63.705	38.153	26.3	124.9 ± 0.8
4	3580.5	52.594	22.705	14.2	120.3 ± 1.6
5	3118.9	52.631	5.729	2.6	120.3 ± 1.3
6	2660.0	54.010	4.567	3.2	122.6 ± 2.9
			plateau age		123.1 ± 0.6
			isochron age		122.6 ± 1.5

Sample WP-66-25, "Jacqueline" Smt basalt, J=0.001430

1	389.41	237.85	81.582	22.3	142.1 ± 2.1
2	539.92	89.611	21.435	22.2	102.3 ± 1.6
3	1308.4	53.828	10.340	9.9	105.1 ± 2.9
4	948.31	46.684	4.5935	17.0	84.0 ± 1.2
5	1195.1	37.429	3.4352	21.2	73.5 ± 1.9
6	630.32	32.424	0.9980	7.5	52.9 ± 1.2
			no plateau		
			total fusion age		98.5 ± 1.4

Sample WP-72-1, Allison Smt diabase, J=0.002311

1	389.11	62.485	74.052	5.2	62.9 ± 5.3
2	545.75	49.344	18.370	9.3	94.5 ± 3.6
3	1931.3	29.571	6.9562	24.3	104.9 ± 2.2
4	1417.0	30.746	33.970	34.3	100.7 ± 0.8
5	1090.8	29.799	6.6939	22.8	91.7 ± 2.8
6	625.24	33.343	4.0691	4.1	76.2 ± 3.9
			plateau age		101.2 ± 0.8
			isochron age		102.7 ± 2.7

* corrected for decay since neutron irradiation
Neutron interference factors used in age calculations:
$(^{36}Ar/^{37}Ar)_{Ca}$ = 0.000264, $(^{39}Ar/^{37}Ar)_{Ca}$ = 0.000673,
$(^{40}Ar/^{39}Ar)_{K}$ = 0.001.

estimate of the crystallization age. Even so, this age estimate must be treated with caution because, as Pringle has shown (this volume), apparent ages from such discordant experiments may be in error by as much as 30% of the true crystallization age.

Most samples developed clear plateau ages (Fig. 2), defined by three or more concordant step ages comprising more than 50% of the total gas. These ages are matched by isochron ages with atmospheric-Ar intercepts, which excludes any significant excess ^{40}Ar. Exceptions are previously mentioned WP-66-25, sample WP-72-1, which shows a two-step plateau and an age spectra perhaps reflecting ^{39}Ar-recoil plus ^{40}Ar-loss, and WP-55-1a, which shows a three-step, low-temperature, quasi-plateau, and severe high-temperature recoil. Because of the problematic spectra, these three age estimates are not as reliable as the others.

Radiometric age determinations demonstrate a total range in age of volcanic activity of about 40 m.y. (82 to 123 Ma) within this northwest Pacific seamount province. The earliest volcanism began about 120-123 Ma and is recorded at "Heezen" Guyot at

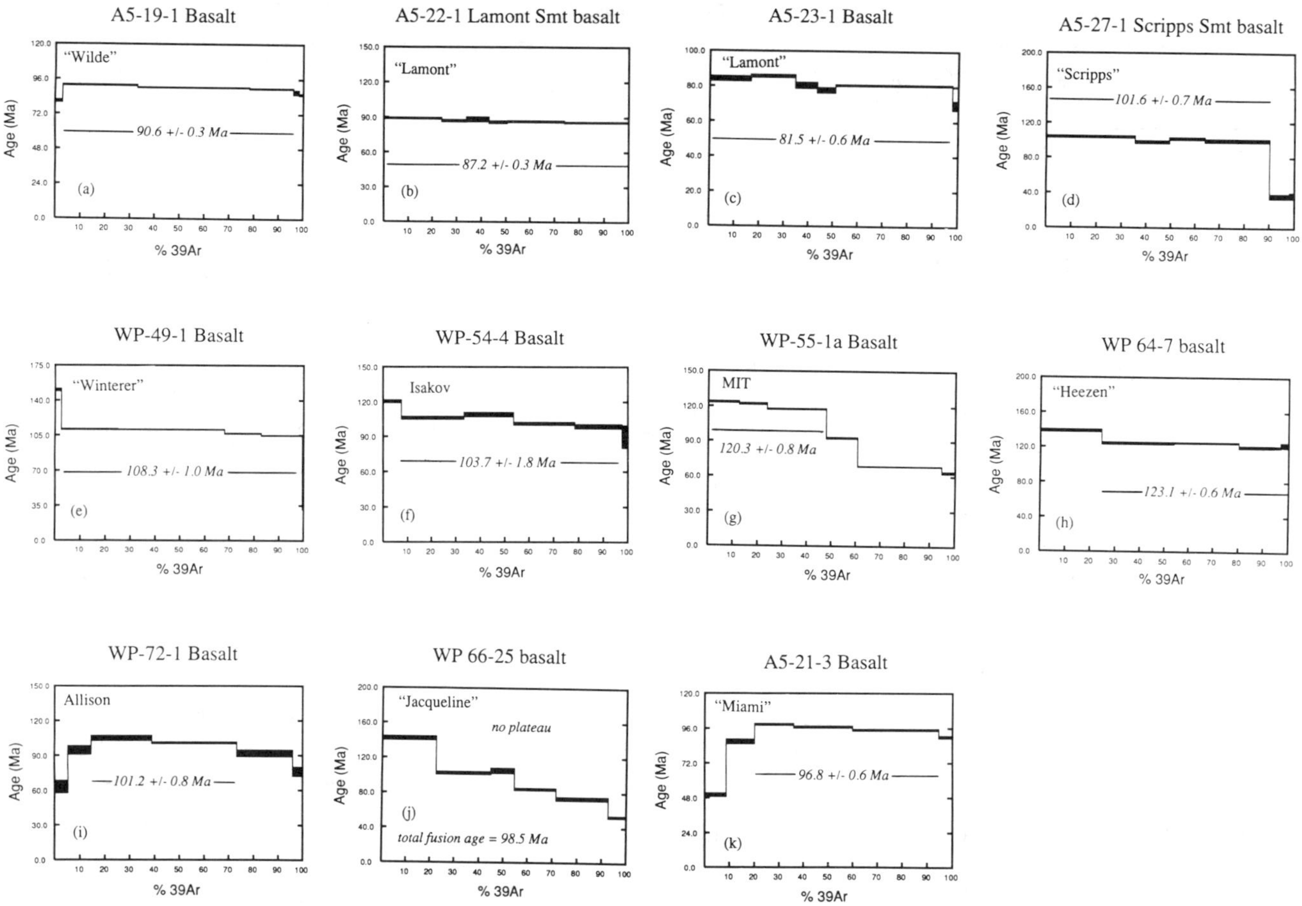

Fig. 2. Apparent age release diagrams for $^{40}Ar/^{39}Ar$ incremental heating experiment of basalt samples. Horizontal boxes indicate estimated analytical error (± 1σ) about each step age. A plateau age (indicated) has been determined from the weighted mean of contiguous, concordant step ages. (a) Sample A5-19-1; (b) Sample A5-22-1; (c) Sample A5-23-1; (d) Sample A5-27-1; (e) Sample WP-49-1; (f) Sample WP-54-4; (g) Sample WP-55-1a; (h) Sample WP-64-7; (i) Sample WP-66-25; (j) Sample WP-72-1; (k) Sample A5-21-3.

the western end of the MPM and at the rather isolated MIT Guyot located to the south of the JPS. Volcanic activity within the range 91 to 108 Ma occurs in the eastern JPS, the eastern Wake Seamounts, and the eastern MPM. The youngest volcanic rocks (82 to 87 Ma) occur at two locations within the Wake Seamounts and MPM.

Origin of NW Pacific Cretaceous Seamounts in South Pacific Hot-Spot Models. It is difficult to draw firm conclusions about the origin of this province from such a small number of age determinations scattered over this large region. We can, however, examine the distribution of ages within the context of the hot-spot model for seamount lineaments. These seamounts date from a period older than the well-constrained period of Pacific plate motion over the hot-spot reference frame (0-74 Ma) [Duncan and Clague, 1985; Henderson, 1985]. Several of the prominent central Pacific hot-spot chains, such as the Line Islands and Marshall-Gilbert seamounts, extend north and west to presumably older related lineaments. Hence, Duncan and Clague [1985] speculated that the MPM recorded the earlier history of the hot spot that built the Line Islands, and that the Wake Seamounts are the precursors to the Marshall-Gilbert track. To these groups we can add the JPS trend which we now know formed at least in part contemporaneously with some of the WKS and MPM seamounts.

Figure 4a shows predicted trails of three hot spots on the Pacific plate for the period 96-120 Ma. These have been generated by 11.5° clockwise rotation about a pole at (67°N, 120°W), obtained by trial-and-error fitting to the seamount trends and new ages. As can be seen by comparing predicted with measured ages, many of the northwest Pacific seamounts could have been generated by stationary hot spots. Alignments of seamounts in the WKS and MPM are poorly expressed, and the actual trends of motions over hotspots may have been more complicated that in the simple model used here.If the trends in Fig. 4a are accepted, the implied northwestward motion of the Pacific plate occurred at *ca.* 0.5°/m.y., or *ca.* 50 km/m.y., - a velocity not especially supportive of suggested increased seafloor spreading rates in middle Cretaceous time [e.g. Larson, 1991]. The trails "backtrack" to

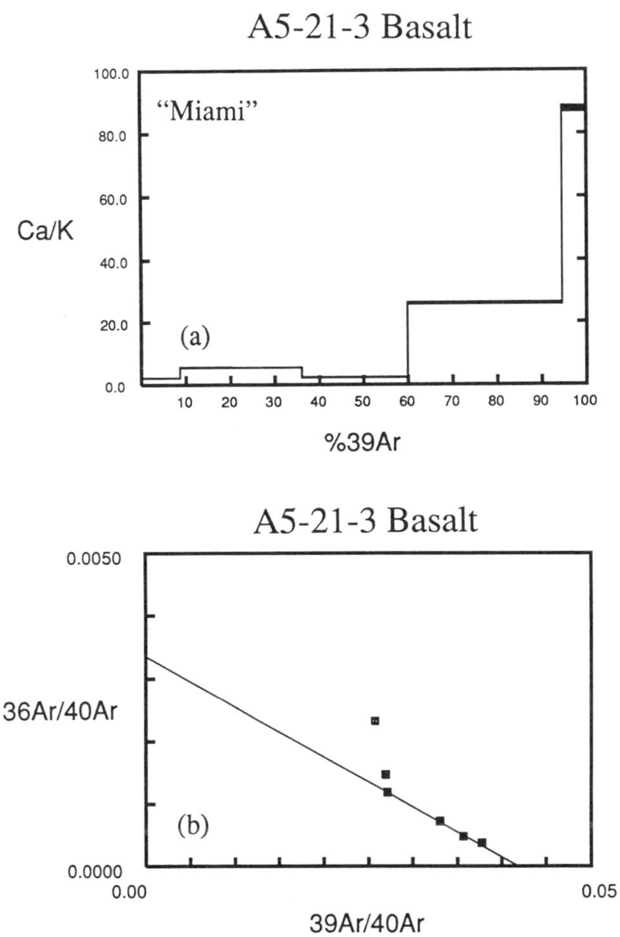

Fig. 3. (a) $^{36}Ar/^{40}Ar$ vs. $^{39}Ar/^{40}Ar$ isotope correlation diagram for the step Ar-compositions measured. The isochron age (indicated) has been calculated from the best-fitting line to colinear step compositions, after York (1969); analytical uncertainties are shown.
(b) Ca/K inferred from the step Ar-compositions reflect the phases (pyroxene, feldspar, groundmass, clays) contributing to gas release at each temperature.

hot-spot positions in the south-central Pacific basin: within a few hundred km of the Society (WKS) and Pitcairn (MPM) hotspots (Fig. 4b). The JPS backtrack to a position northwest of the Society hotspot, not currently the site of hotspot activity. Geochemical characteristics of the younger ends of these volcanic chains might be expected in the associated Cretaceous seamounts.

The 82-87 Ma volcanism detected at "Lamont" and Allison guyots does not fit the simple hot-spot model. This may be due to a rejuvenescent volcanism that reflects some unknown regional extensional regime in Late Cretaceous time, or to overprinting of older WNW-trending chains by younger NNW chains (Marshall and Line Islands trends).

Source-Region Comparisons. Petrologic and geochemical data, coupled with radiometric-age data and with the isotopic data of Smith et al. (1990] from the Magellan seamounts near the Mariana trench, demonstrate that during the Early Cretaceous, many large, reef-capped seamounts were built up from several kinds of enriched-mantle sources beneath the center of the hemispheric Pacific Ocean. The term "enriched" here has the same meaning as that applied to the modern-day province of linear island chains in the southwest Pacific, where both HIMU and potassic EM-II mantle components are especially prominent. As is currently the case, several chains were active, and there is some evidence for overprinting. The petrological data lend considerable substance to the interpretation from isotopes suggesting both within-chain and between-chain diversity of mantle sources, as well as during the histories of individual volcanoes. They also reveal complexities in the processes of partial melting, crystallization, differentiation, and magma mixing, which complicate interpretations of mantle-source, trace-element characteristics. Enriched alkalic basalts and their differentiates predominate in the region, except for the mainly tholeiitic eastern MPM. Nephelinitic - plausibly post-erosional - lavas erupted in the WKS and JPS. The rock associations, and their compositions, call to mind the petrological successions of present-day islands in the southwest Pacific. In this regard, the link to SOPITA is further supported. However, an earlier interpretation, based solely on examination of thin sections [Heezen et al., 1974] was that there are three petrographic provinces spanning this portion of the west-central Pacific, which formed on crust of different ages. This interpretation is still valid in the light of new analytical data and radiomentric ages. From east to west these provinces are: 1) a near-axis tholeiitic province in the eastern MPM; 2) a fairly strongly alkalic province comprising undersaturated basalts and their differentiates as well as nephelinitic lavas in the WKS that formed on crust 40-80 million years old; and 3) a mildly alkalic province in the JPS, which formed on crust no older than 50 million years at the time. To these, based on more recent dredge hauls, we now add a fourth province - the large composite volcanoes of the western MPM, which formed on crust *ca.* 10 million years old.

One contrast to SOPITA is that the breadth of the enriched (EM-II plus HIMU) Mesozoic province is now greater than in the present south Pacific. We note that the eruption of the mid-Cretaceous volcanoes was also broadly contemporaneous with, and continued to follow after, massive eruption of plateau basalts on or near accretionary boundaries of the Pacific Plate at Ontong-Java Plateau and Nauru Basin, Manihiki Plateau, and Magellan Rise to the south [e.g. Larson, 1991; Tarduno et al., 1991]. Perhaps, rather than viewing the province of mid-Cretaceous Guyots as being simply linked to modern linear island chains and superswells in the southwest Pacific, they should be thought of as the northern outliers of a huge province of mid-Cretaceous volcanism in the Pacific, one reflecting fundamental disturbance of a large volume of mantle beneath the central portion of the Mesozoic world ocean. The present-day SOPITA region may be a residual manifestation of that disturbance.

Summit Regions with Shallow-water Carbonate Sediments

The morphology of the summit, margin, and upper flank regions is controlled not only by the processes of volcanic edifice building and erosion of volcanic rock but also by processes of

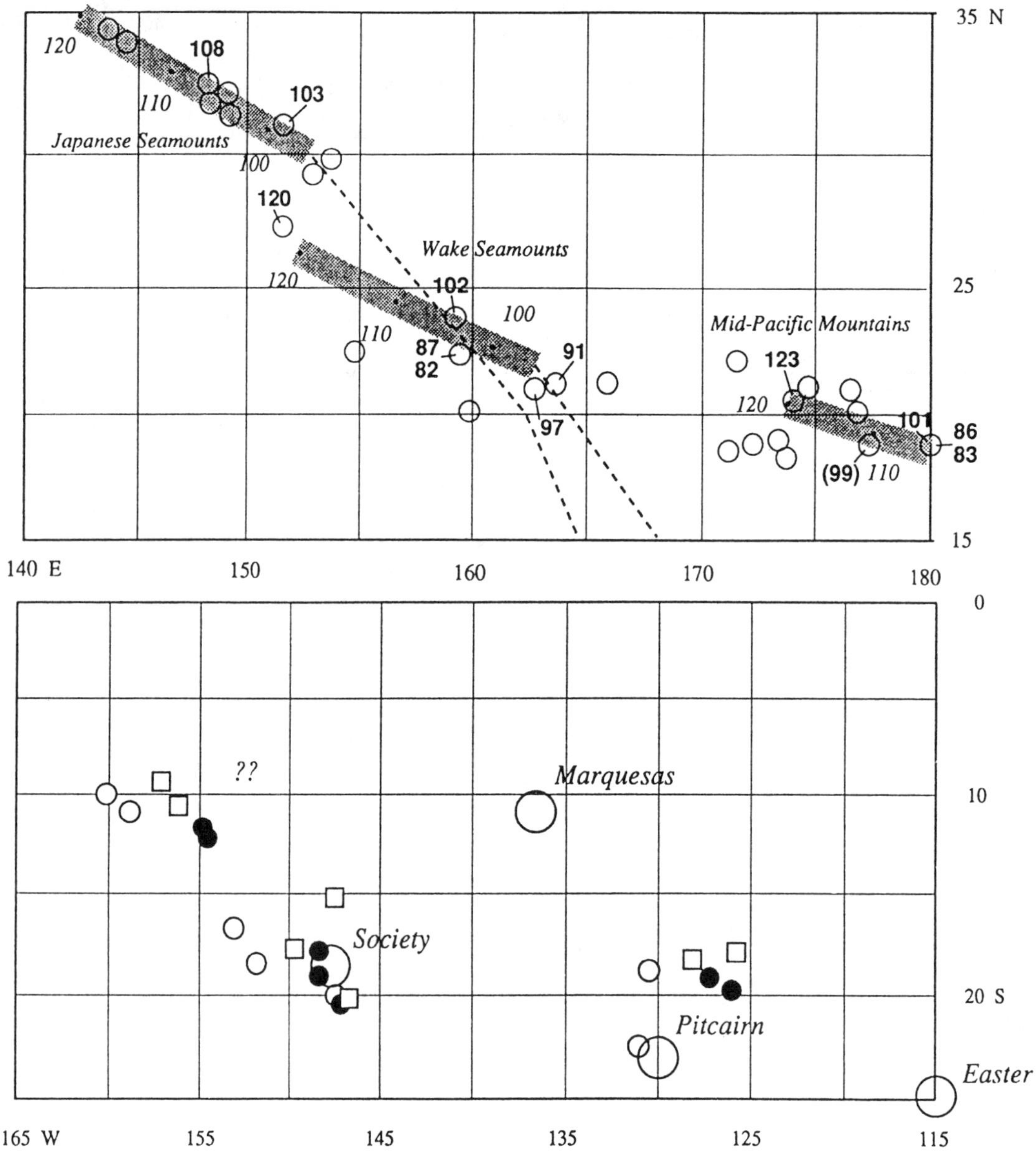

Fig. 4. (a) Computer-modeled hot-spot tracks (stippled band) for the period 96 to 120 Ma in the vicinity of the western Pacific seamounts. The rotation pole is at 67°N, 120°W and the tracks have been generated by 11.5° clockwise rotation. The outheast ends of these segments connect with younger portions of the hotspot tracks (Dashed lines) formed during more northerly motion of the Pacific plate. Italicized numbers are predicted ages (Ma), while bold numbers are measured ages (Ma). Seamount positions are shown by open circles (see Fig. 1).

(b) The dated seamounts backtrack to present hot-spot positions in the south-central Pacific basin, within a few hundred km of the Society (WKS) and Pitcairn (MPM) hotspots The JPS backtrack to a position northwest of the Society hotspot not currently associated with hotspot activity. Symbols relate to different proposed rotationpoles for the Pacific plate motion over hotspots for the period 96 to 120 Ma (open squares, Duncan and Clague, 1985; open circles, Henderson, 1985; closed circles, this paper, as above). Rotations for 0 to 96 Ma follow Duncan and Clague [1985) for all points. The present positions of active hot spots are shown by large open circles of 200 km diameter.

sedimentation, especially the construction of carbonate reefs and platforms, and by subaerial dissolution of the limestone during periods of emergence. There may also have been structural deformation. Some of the guyots appear to be tilted or warped, as evidenced by variations around guyots of the depth of the margins of their summit plateaus. On some guyots, slope break depths vary so as to give the appearance that part of the summit or summit edge has sagged, or "drooped".

In the northwest Pacific, there are three main classes of guyot tops: drowned atolls (thick reef deposits arranged in perimeter and lagoon facies, no volcanic rocks on the summit), barrier-reef islands and "almost-atolls" (thin reef deposits with a barrier reef but with volcanic rocks at the surface), and guyots without limestone cover. On some drowned atolls with thick reef deposits, residual volcanic peaks remain unburied. The summit morphology of guyots lacking a limestone cover was discussed in the section on volcanic edifices; we here take up the morphology of the guyots with shallow-water limestone capping or partly capping the summit region.

Drowned Atolls. All the drowned atolls include a steep outer slope, a rim 20-40 m above the main summit level, and a slightly deeper interior, the lagoon (see Fig. 3, Darwin Guyot, in color, in van Waasbergen and Winterer, this volume). The upper parts of the outer slopes of the atolls are significantly steeper than those of the purely volcanic guyots. Slope angles of as much as 40° occur, and the average is about 25°. Dredges suggest that this forereef slope consists largely of talus from the reef, with fragments of rudists, corals, echinoids, and algae. At the base of the upper slope, where the limestone slope meets the volcanic edifice, the slope angle changes abruptly to less than 20° and the slopes also become much more irregular: ridges and knobs of presumably volcanic origin occur at this depth and deeper. The depth at the break matches well with the depth of the volcanic basement reflectors in the seismic profiles, and the larger flank rift zones commonly terminate against the upper slopes at this depth, suggesting that it corresponds roughly to the base of shallow-water deposits within the drowned atoll.

The rim generally consists of sections which taper off both in height and width toward their ends, like link sausages, and likely represents a chain of islands interrupted by passages through the reef, as on modern atolls. Rocks dredged from the rim are generally rudist and coral boundstones, commonly with abundant algae.

On many guyots, elongate depressions 20-40 m deep occur just behind and parallel to the raised rim; we believe these are part of a karst drainage system developed during a period of emergence. Dredges from these backreef environments are rich in peloidal mud matrix between fragments of reefal rock.

The lagoon has a generally smooth surface, dotted in some places with small elevations (patch reefs?), and in other places with sinkholes and other karst depressions. On many guyots, the lagoon surface is arched: the central parts of the lagoon are shallower than the perimeter reef (we return to this feature in the section on deformation). Dredges from lagoonal facies yield muddy wackestones, commonly with large gastropods. Surprisingly, we have seen virtually no miliolid foraminifers, which dominate lagoonal facies rocks in Barremian-Albian lagoonal rocks in the classic Mediterranean and Mexican sections [Arnaud-Vanneau, 1980; Aguayo-C., 1978].

Seismic reflection profiles show the drowned atolls to consist of an acoustically characterless perimeter reef surrounding parallel-layered lagoonal sediments that commonly conform to the general topography of the underlying basement but show centripetal onlap of basal sediments onto a central basement high. The overall resemblance in seismic stratigraphy among widely separated guyots suggests a common origin: the reflectors may be the result of sea-level fluctuations during the growth of the atolls. In modern atolls such as Anewetak, lagoonal reflectors are the result of Tertiary sea-level low-stands producing indurated solution unconformities [Lincoln and Schlanger, 1991; Schlanger, 1963; Grow et al., 1986]. The reflectors in the Cretaceous guyots mirror the domal surface of the lagoon in sloping away from the guyot center (Fig. 5).

Guyots with thin, shallow-water limestone deposits. Only thin veneers of shallow-water limestone mantle the Japanese guyots and some of the Wake seamounts. Although reefal material shows up in dredges, it can not be distinguished in seismic profiles from volcanic basement. As seen in multibeam (SeaBeam) bathymetric data, several stages of development of carbonate deposition are evident in the summit morphology of these guyots. The earliest stage is represented on Wake guyots, where SeaBeam data show small reef-like rims on some guyot edges. In some places, reef rock was dredged where neither SeaBeam nor seismic data indicated reef.

The JPS guyots represent more advanced stages of shallow-water deposition. Barrier reefs are dramatically revealed by complete summit SeaBeam surveys of "Charlie Johnson" and Takuyo Daisan Guyots where karstic dissolution has resulted in a highly exaggerated topography (see Fig. 11, in color, in van Waasbergen and Winterer, this volume). Their summits have a central mound and a discontinuous but very prominent rim, generally about 100 m high and, judging from the contents of dredge hauls, almost certainly of reefal origin, but with far more relief than one can reasonably expect from reef construction alone. More likely, a barrier reef and shallow lagoon were exposed above sea level and dissolved until only the resistant rim remained. Both "Charlie Johnson" and Takuyo Daisan have fairly steep upper slopes (24°-26°) and volcanic rift zones extend up to the base of the reef rims, so the upper slopes are almost certainly volcanic, with only a thin cover of reef debris.

An entirely different style of summit is displayed by other JPS guyots: The summits of "Winterer" and Isakov Guyots (Fig. 6) consist of unrimmed, flat tops with small, conical, volcanic peaks in the center. A significant aspect of these guyots is the lower terrace that surrounds their summits. The terraces are about 1 km wide and bounded on the inside by 100-m scarps or steep slopes. In many places, the terraces are edged by short, narrow ridges, most likely reefal in origin. Perhaps the main summit plateau is entirely volcanic, and a reef system developed along a base level 100 m below the summit; alternately, the summit is a near-atoll, with reefs eroded away or too low to be resolved by SeaBeam, while the lower plateau is a reef-rimmed wave-cut terrace formed during a relative lowstand of sea level.

Erosional Effects. Subaerial dissolution and both subaerial and submarine erosion account for much of the detailed morphology of the guyot summits. Erosion occurred at several stages: during and after volcanism, affecting the volcanic islands; during sedimentation; during relative lowstands of sea level and temporary exposure of carbonate platforms; and after drowning, in the form of large slumps and debris flows. A significant relative sea level fall occurred prior to final drowning of the carbonate platforms on the guyots, resulting in the creation of

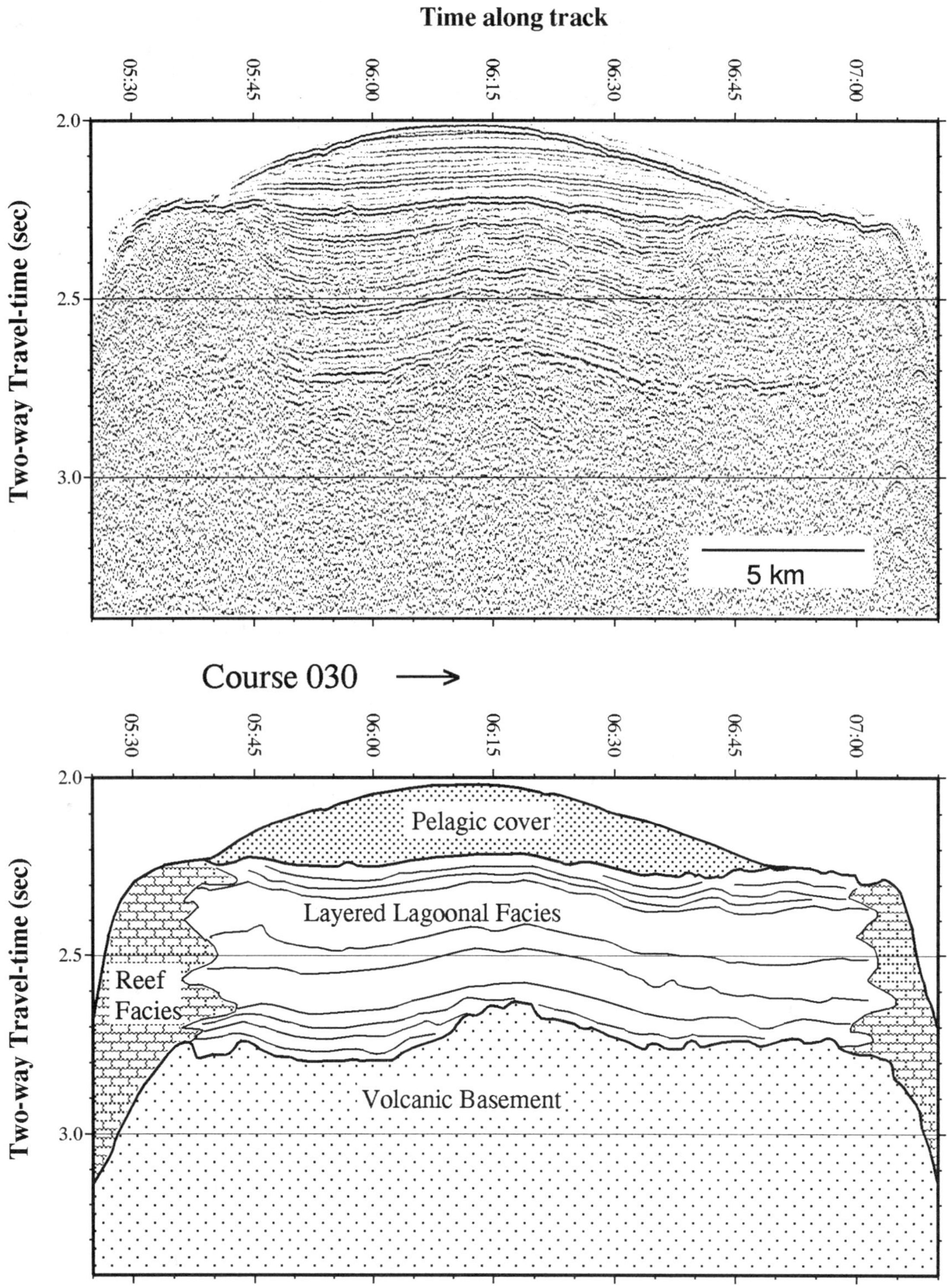

Fig. 5. Seismic reflection profile (upper panel) and line drawing interpretation (lower panel) of Allison Guyot, a drowned atoll in the Mid-Pacific Mountains. There appears to be as much as 0.58 seconds of sediments above volcanic basement. Prominent reflectors are shown in the interpretation, and may correspond to similar reflectors in the profiles of Allison guyot. The location of the section is shown in Fig. 7.

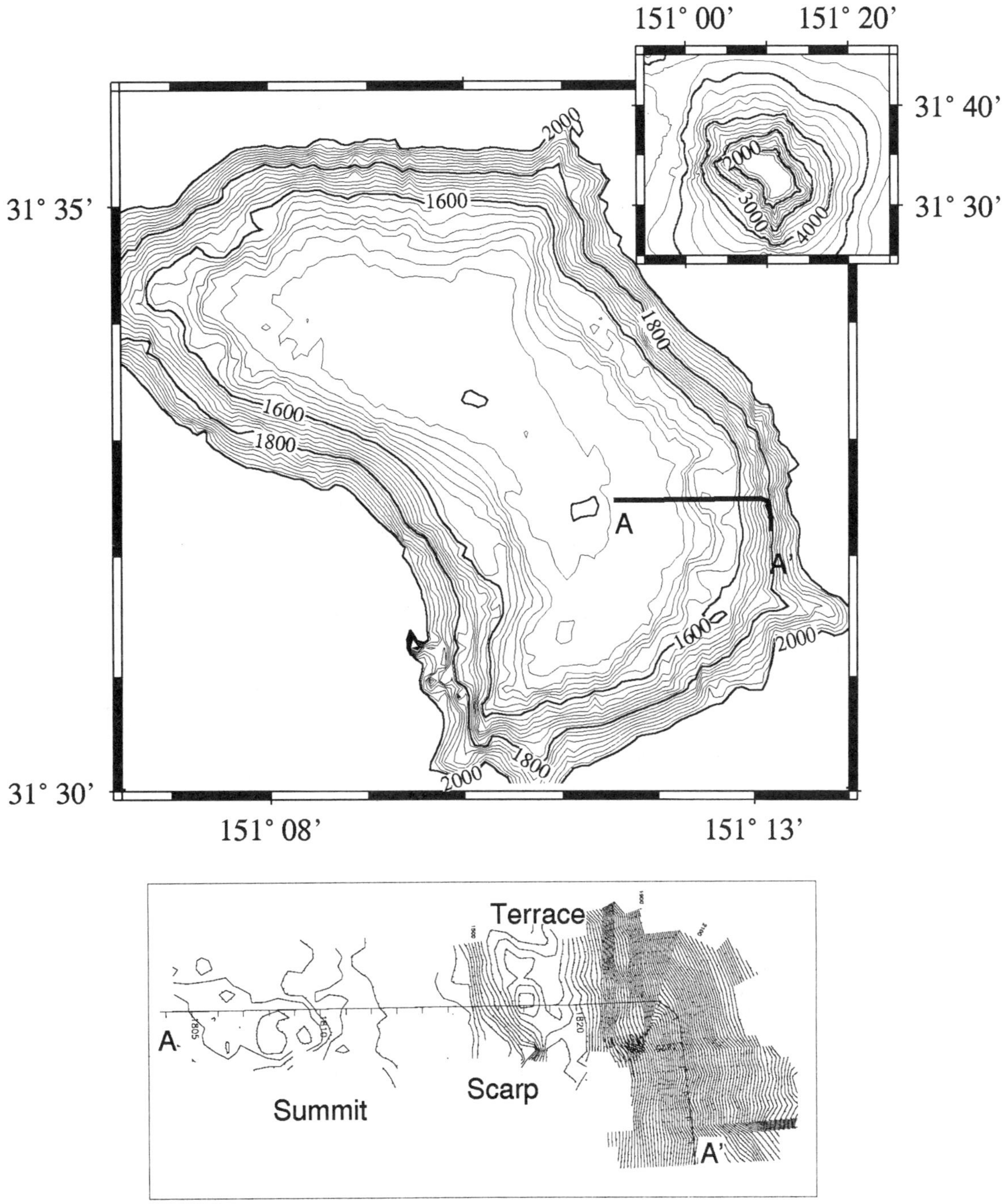

Fig. 6. Bathymetry of Isakov Guyot, in the Japanese seamounts. General shape from Vogt and Smoot [1984]. Detailed summit bathymetry from Roundabout, leg 10 SeaBeam survey. The summit plateau lies at *ca.* 1440 m, and has a few small pinnacles which reach to 1380 m. Along the upper slope, at a depth of *ca.* 1550 m, a 0.5 to 1 km wide terrace occurs. An enlarged single SeaBeam swath shows the summit level, upper scarp, and terrace (contour interval: 10 m). The terrace has a 20 m tall outer rim, just inward from which lies a small depression. This may represent the original lowstand-reef configuration.

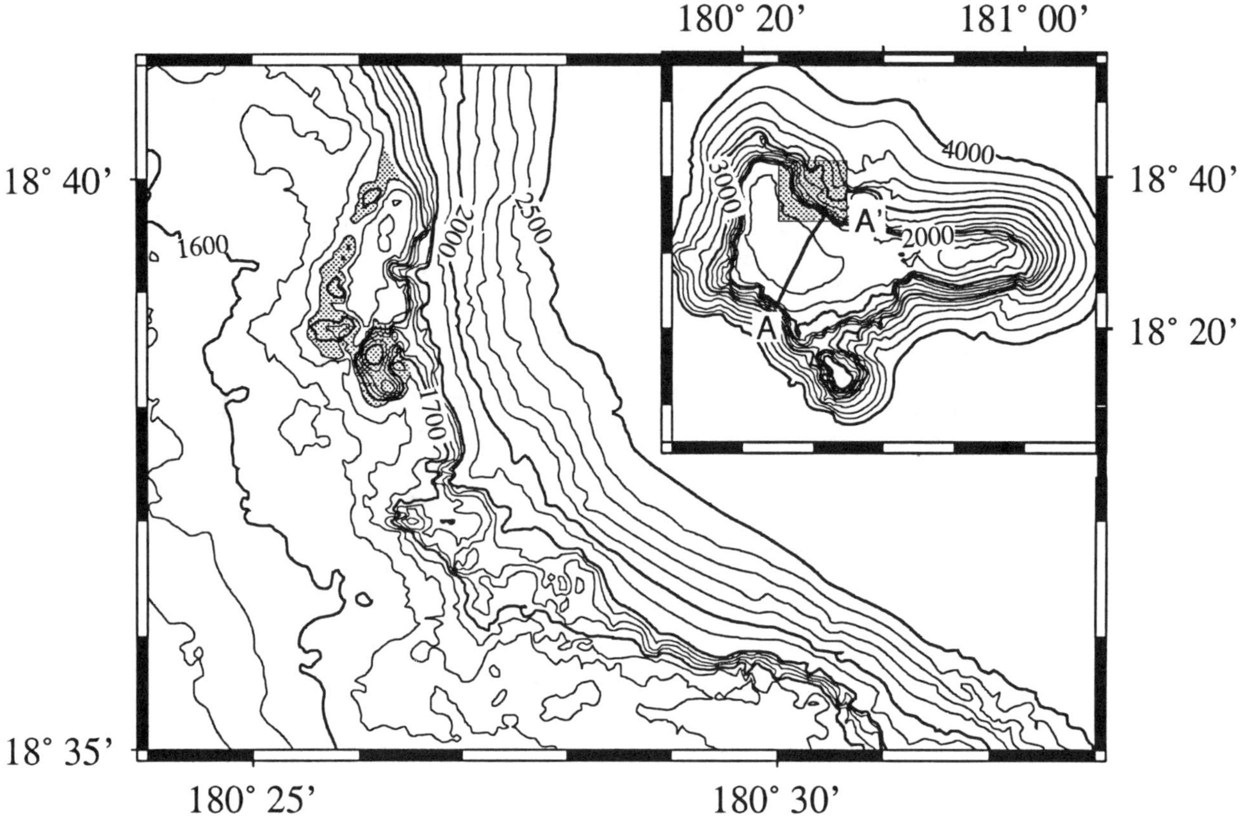

Fig. 7. Bathymetry of summit of Allison guyot, in the Mid Pacific Mountains. A 5-km-wide volcanic ridge extends to the east, which was only very incompletely surveyed. The south-east side of the summit has a sloped surface leading to a poorly defined slope-break at ~2400 m. In seismic reflection profiles across this side of the atoll structure layered reflectors can be seen to crop out along the slope, indicating that the reef wall is missing, possibly the result of a large slump event which affected the upper slopes of the structure along a 14 km long path. The small drop-shaped summit to the south may be related to this event. The enlarged map of the north end of the guyot-summit shows a 180 m deep trough just inside the outer rim of the guyot. the edge of the pelagic sediment cap lies just to the west of the trough. The trough is thought to be the product of erosion during exposure of the summit above sea level.

considerable relief, in the form of karst topography, stream channels and wave-cut terraces.

Our best example of karst topography is MIT Guyot (see Fig. 15, in color, in van Waasbergen and Winterer, this volume). The summit plateau is pocked by doline-like depressions, the deepest of which is a circular hole about 500 m in diameter and more than 180 m deep below the summit level. Seismic profiles show layered lagoonal sediments cut by the summit relief, indicating erosion. The fact that the floors of the depressions have no surface connection to the outer slope indicates they were formed through dissolution.

Other guyots also show karst. For example, just inside the outer rim of Allison Guyot (Fig. 7) there is a 100 m-deep erosional channel indented by numerous deeper sinkhole-like depressions. A similar channel, 80 m deep, scores the surface of "Vibelius" Guyot as a closed depression. Karst topography also shows up on guyots with thin reefal caps. On "Charlie Johnson" and Takuyo Daisan Guyots, the deepest parts of the lagoons are closed depressions 40-60 m deep, which is significantly deeper than the passages between rim sections.

Even where the surface topography of the Cretaceous carbonate platforms is covered by pelagic sediments, 3.5 Khz profiles show buried holes in that reflecting surface, beneath the pelagic sediments. This demonstrates that the holes formed before the deposition of the pelagic cap, rather than by more recent, post-drowning dissolution in deep water.

Stream channels can be seen on the summit plateaus and upper slope regions of several of the guyots. On "Winterer" and Isakov Guyots, distinct "V's" in the contours can be traced from the summit plateau down the upper scarp to the lower terrace level. On "Charlie Johnson" and Takuyo Daisan Guyots, stream channels cut through the gaps between the rim sections, and connect the summit plateau to the outside. The channels terminate on small terraces on the outer slope at the same depth as that of the bottom of the depressions in the summit plateau, indicating that this level represents a new base level during a time of relative lowstand of sea level.

Wave erosion during a relative lowstand of sea level is shown by wave-cut terraces on the upper slopes of many guyots, such as "Winterer" and Isakov. Takuyo Daisan and "Charlie Johnson"

Guyots have small terraces which are better developed below the gaps between the rim sections where stream channels end. On the drowned atolls, small terraces break slopes on "Caprina", MIT and Allison Guyots. Strictly volcanic guyots show no lower terraces.

Structural deformation. Because the flat tops of the reef-capped guyots are the result of processes occurring at sea level, the main slope break, between reefal platform and upper slope, should occur at the same depth all around the guyot. Surprisingly, this is not everywhere the case. The depth of the main slope-break varies by as much as 240 m on any one guyot. This variation occurs in several ways: 1) by consistent deepening toward one side of a guyot (tilting); 2) by gradual local variations along summit margins ("droop"); or 3) by sudden changes, marked by a short, straight slope or scarp on the summit (faulting or slumping). The deformation of the original, pre-drowning surface of the guyot is echoed in the slope of underlying seismic reflectors. Guyots affected by deformation are Allison, "Heezen", MIT, Resolution, Darwin, "Vibelius", "Thomas Washington", and "Scripps". Although most of these are drowned atolls, the other guyot types are also represented, which means that the deformation is not confined to the sedimentary cover.

A number of causes of tilting of seamounts have been suggested: 1) lithospheric flexure by loading of a nearby, younger volcano [McNutt and Menard, 1978]; 2) loading of one side of the seamount flank by a large mass derived from mass-wasting of the upper slopes [Keating, 1987]; and 3) differential growth or erosion on the summit during exposure [Ashmore, 1973]. We suggest that long-term compaction of volcanoclastic layers in the upper levels of the seamount may be important in the tilt and droop phenomena. When a seamount attains a height above the sea floor that brings the summit to within about 1 km of the sea surface, escape of volatiles leads to vesiculation and fragmentation of erupted volcanics, and this porous mass is subject to hydrothermal alteration that can transform the framework into more compactible material. The central vent region, which goes on feeding the later, subaerial parts of the volcano, may become stiffened by dike intrusion and by higher-temperature metamorphism, resulting in a centripetal gradient in compactibility.

Faults cut several guyots. On "Jacqueline" Guyot, the summit plateau has an abrupt step-down of almost 100 m. The western portion of Makarov guyot also appears to be faulted down more than 100 m. The faulting may be related to events of mass-collapse of portions of the summits, as along the southeast margin of Allison Guyot [Van Waasbergen and Winterer, this volume].

Post-reef Pelagic Sediments and Crusts

The Late Cretaceous volcanic guyots and the Early Cretaceous reefal guyots are commonly capped by variable thicknesses of pelagic carbonate sediments. Guyots north of about 20°N are virtually bare of pelagic cover, and the thicknesses increase toward the equator. This is most likely the result of the competing effects of rate of sedimentation, which increases equatorward, and of winnowing by bottom currents, which tend to sweep pelagic sediments off the guyots: the two effects are in rough balance at about 20°N. The pelagic sediments tend to occur as a broad mound, which on some guyots has layers draped progressively over a central nucleus, on others shows flat layers shaped into a mound by erosion (Fig. 5).

At outcrop in the summit regions of the guyots, both the shallow-water carbonate strata and volcanic rocks are virtually everywhere encrusted with phosphorite and manganese oxides. The phosphorite (calcium fluorapatite and calcium hydroxyapatite) replaces shallow-water fossils and the overlying pelagic foraminiferal nannofossil limestone, and coats and fills cavities. The phosphatized rock is generally brecciated in its upper part and the degree of phosphatization generally increases downward.

The phosphorite is overlain by black manganese oxides, which are generally layered and have broccoli-like, botryoidal or mammiform upper surfaces. Mn coatings as much as 7 cm thick are common. Mn generally completely surrounds phosphatic clasts. In some places, the lowest Mn layers are interlaminated with phosphorite, and Mn locally replaces phosphorite. We interpret the superposition of Mn on phosphorite to indicate a sequence of ages, but we have no age determination for either the phosphatization or for the oldest Mn. Because Eocene pelagic sediments are phosphatized on some of the Wake guyots [Heezen et al., 1973], we suspect that most of the phosphatization and formation of manganese coatings is of Neogene age.

Paleontological Ages of Sediments

The subsidence, drowning, and later emergence histories of the guyots is very dependent on reliable estimates of ages of dredged fossils. Dredge hauls from the guyots have yielded an abundance of megafossils and a lesser abundance of microfossils, but most of these are so comminuted or poorly preserved that specific identifications are impossible. The dredges are mainly from the upper parts of the outer slopes and from the reef platform, and sample chiefly the younger strata in the reefal complex.

Megafossils. The most abundant fossils in most hauls are rudists, and according to R.W. Scott [pers. comm., 1991], who kindly examined our collections, these mainly belong to two genera, the caprinid *Praecaprina,* and the radiolitid *Agriopleura.* The rudist *Praecaprina sp.* was recovered from "Heezen", "Jacqueline", and Allison Guyots in the MPM and from Darwin and "Woods Hole" Guyots a little farther west. According to Scott the range of this genus in Europe, Trinidad, and the U.S. Gulf Coast is Barremian to lower Aptian. This is consistent with the late Barremian to early Aptian age of pelagic sediments interbedded with redeposited reefal debris at DSDP Site 463, on a turbidite apron flanking Resolution Guyot [Thiede, Vallier, et al., 1981]. From the drowned atoll farthest west, MIT Guyot, we have no diagnostic fossils to fix the age of the youngest reef material.

Specimens of the rudist *Agriopleura sp. X* were recovered from "Charlie Johnson", Isakov, "Thomas Washington", "Winterer", Takuyo Daisan, and Takuyo-Daini Guyots, all in the JPS Group. According to Scott, the genus ranges from Barremian to Maestrichtian in Europe and North America. It is distinctive and may be a new species. Given that the two genera are mutually exclusive on the Pacific guyots, we suspect that the JPS guyots with *Agriopleura sp. X* may be younger than early Aptian.

Microfossils. From the sunken atolls (Allison, Darwin, "Woods Hole", and MIT Guyots), dredges of carbonate-platform rocks contain (mainly as fragments) dasycladacean, red, *Cayeuxia*, and solenoporacean algae; *Cuneolina cf. C. pavonia* [J. Grötsch, pers. comm., 1991], *Orbitolina*, and rare miliolid foraminifers. *Cuneolina pavonia* ranges from late Albian through the Cenomanian. On Allison Guyot, from a level about 250 m down within the lagoonal facies, exposed in a giant slump scar, came a sample with *Orbitolina parva* [J. Grötsch, pers. comm., 1991], a form ranging from Upper Aptian to earliest Albian. From this same slope the dredge recovered the planktonic foraminifers *Planomalina buxtorfi, P. praebuxtorfi, Ticinella primula,* and *Biticinella breggiensis,* which are of latest Albian (*Rotalipora appenninica* Zone) age, and are mixed into the lagoonal sediments. *Favusella washitensis*, a planktonic foraminifer generally found only near significant land, also occurs associated with the reefal material on this slope.

On Horizon Guyot, at the easternmost end of the MPM, about 1000 km east of Allison Guyot, pelagic sediments of late Cenomanian age overlie the carbonate platform strata across an erosional unconformity [Winterer, Ewing et al., 1973], and Turonian planktonic foraminifers are reported in dredges from "Jacqueline" and "Shepard" Guyots in the central MPM [Heezen et al., 1973].

From the JPS guyots, dredges recovered several types of red and green algae and fragments of *Orbitolina*, a large benthic foraminifer. Konishi [1989] reports *Orbitolina texana*, a species ranging from late Aptian to early Albian, in samples from Daiichi-Kashima Guyot in the Japan Trench. Upper Cretaceous planktonic foraminifers occur in phosphatized foraminiferal limestone encrusting the reefal strata on the summits of many guyots in both the MPM and in the JPS guyots.

Comparison of Paleontologic and Radiometric Ages. The paleontological data on reefal materials indicate two sets of ages: late Barremian to early Aptian (117-121 Ma) for the sunken atolls (Darwin, "Woods Hole" Guyots and the guyots in the MPM), and latest Aptian to Albian (114-97.5 Ma) age for the JPS guyots. The underlying volcanic edifices must thus be at least this old.

Except for Allison Guyot, the radiometric data for the atolls are consistent with the paleontology in indicating ages no younger than late Barremian or early Aptian age for the edifices. The paleomagnetic data, discussed below, suggests edifice formation of most of the atolls prior to 119 Ma, the beginning of the Cretaceous Quiet Zone (KQZ). The thicknesses of reefal sediment in the sunken atolls requires either rapid subsidence rates (i.e., the seamounts were on young oceanic crust) or a long period of subsidence (ie., the seamounts are old). We will return to this question in the section on subsidence. Allison Guyot radiometric dates are much younger than the Barremian-early Aptian date assigned to the rudists dredged there. The dated diabase may be altered. The atoll data are inadequate to test for age progression. The oldest pelagic fossils on the atolls are of late Cenomanian age (92-95 Ma) if we leave aside the latest Albian (98 Ma) pelagic fossils recovered on the side slope of Allison Guyot, which were in water about 250 m deep when the guyot summit was at sea level.

For the JPS guyots, the paleontological and radiometric data are in agreement, and suggest an Albian age for the entire chain, including both edifice construction and reef building, from Daiichi Kashima seamount, in the Japan Trench to Isakov Guyot, a distance of about 1000 km. As discussed below, this is consistent with paleomagnetic evidence indicating formation during the KQZ. Whether there is age progression along the chain cannot be determined from the data at hand, but we note that the most evolved reefal morphology, at Takuyo Daisan and "Charlie Johnson" Guyots, occurs in the western half of the group.

The non-reefal Wake guyots ("Scripps" Guyot is regarded as having a small reef remnant) give edifice ages of 87-97 Ma (Cenomanian to Coniacian). Elsewhere in the north Pacific, seamounts ranging from about 80-100 Ma are abundant, for example in the SSE-trending Musician and Line Islands Seamount chains, where radiometric data suggest progressive volcanism [Schlanger et al., 1984; Pringle, this volume].

PALEODEPTHS AND ELASTIC PLATE THICKNESS OF LITHOSPHERE

In this section, we review recently published results [Wolfe and McNutt, 1991] from an analysis of elastic-plate thickness values beneath seamounts of the JPS, Wake, and MPM groups, and the implications for the thermal structure of the lithosphere in Cretaceous time. The thin-elastic-plate model is just a first approximation to the true rheology of the lithosphere, which includes brittle failure at the surface and ductile creep at depth [Goetze and Evans, 1979]. Nonetheless, for the small deviatoric stresses induced by volcanic loading, the flexure of a purely elastic plate closely resembles the observed deformation of the lithosphere. The effective elastic thickness of the plate T_e is thought to represent the depth to an isotherm marking the elastic/ductile boundary. The isotherm at the base of the elastic plate may be as high as 800°C [Kirby, 1980], but lower temperatures (450°C to 600°C) are implied from a comparison of geotherms from the thermal-plate model [Parsons and Sclater, 1977] with observed values for elastic-plate thickness estimated from the wavelength of bathymetric moats and arches surrounding volcanoes [Walcott, 1970; McNutt and Menard, 1978] or from geoid, gravity, or deflection-of-the-vertical data over the loads [Watts and Cochran, 1974; Watts and Ribe, 1984; Cazenave et al., 1980]. As the oceanic lithosphere continues to cool after emplacement of the load, the flexure of the elastic plate is frozen in, and thus T_e marks the depth to the isotherm at the elastic/ductile transition at the time of volcanism. In accordance with the prediction of this simple model, the observed thickness of the elastic plate beneath midplate volcanoes increases with the square root of the age of the lithosphere at the time of loading [Watts et al., 1980], just as the isotherms in a conductively cooling plate deepen with time [Parsons and Sclater, 1977].

In general, volcanoes that form on lithosphere between 50 and 90 m.y. old show elastic-plate thickness values between 20 and 40 km, except in the area of French Polynesia where T_e values are anomalously low. McNutt and Fischer [1987] designated this region the South Pacific Superswell and have noted that the coincidence of shallow bathymetry, low Love-wave phase velocities, and low elastic-plate thicknesses on the Superswell can all be explained by an unusually warm lithosphere. They speculated that the area now in the North Pacific known as the Darwin Rise [Menard, 1964] is an ancient example of a superswell, based on identical depth/age relationships [Menard, 1984; McNutt et al., 1990] (Fig. 8a), similar isotopic enrichment of hot-spot lavas [Hart, 1984; Natland and Wright, 1984; Smith et al., 1989; Staudigel et al., 1990], and plate-motion models, which place the Darwin Rise over French Polynesia in Cretaceous time when the volcanoes erupted [Duncan and Clague, 1985]. Smith et al. [1989] extended the list of similarities between the Superswell and Darwin Rise by finding T_e values of 10 km and 15 km for two Cretaceous seamounts of the Darwin Rise that formed on lithosphere approximately 42 m.y. and 65 m.y. old, respectively, at the time of volcanism. Wolfe and McNutt [1991] presented widespread evidence for low elastic-plate thicknesses by examining a larger number of Darwin Rise volcanoes using both readily available digital data sets (e.g., dbdb5 bathymetry, Seasat-derived gravity [Haxby et al., 1983] and deflection-of-the-vertical [Sandwell, 1984] and shipborne data from RNDB10WT. Based on ages of the seamounts as estimated from radiometric and paleontological data summarized in this paper, and the age of the underlying lithosphere [Nakanishi et al., 1991], all the constrained T_e values are less than expected for normal oceanic lithosphere, and fall in the superswell region on Figure 8b. The age of the lithosphere at the time of loading for most of the Japanese seamounts is 40 to 60 m.y., and for some seamounts this is just marginally large enough to distinguish the superswell range of T_e values from those of normal lithosphere elsewhere. For most of the Wake seamounts, the age of the lithosphere at the time of loading was greater.

Based on modeling Seasat gravity and deflection of the vertical, the MPM seamounts formed on the weakest plate, consistent with the formation of these volcanoes in a near-ridge environment, as proposed by Winterer and Metzler [1984]. The T_e values for the Japanese and Wake region show more scatter, but nowhere are values definitely greater than 15 km.

Elsewhere in the North Pacific, Atlantic, and Indian Oceans, the elastic-plate thickness values lie between the predicted depths of the 300°C and 600°C isotherms (Fig. 8b). The fact that the elastic-plate thicknesses for the Darwin Rise guyots fall in the field of values characteristic of the Superswell (i.e., above the predicted depth of the 300°C isotherm as predicted by the 125-km-thick-plate model) rather than the field characteristic of hot-spot volcanism elsewhere (i.e., between the 300°C and 600°C isotherms) makes it less likely that the Darwin Rise depth anomalies were localized swells, such as the Hawaiian swell. The exact mechanism for producing the low values of elastic-plate thickness is as yet uncertain, and may not be simply related to thermal anomalies in the plate responsible for the shallow bathymetry.

As inferred from the age-depth relations in Fig. 8a, all of the Darwin Rise volcanoes were on anomalously shallow lithosphere at the time they erupted. For the most part, they agree with the empirical depth-age relationship for seafloor in French Polynesia on the Superswell. However, because we only have depth estimates at the locations of the volcanoes, we cannot rule out the possibility that the depth anomalies were localized swells, such as the Hawaiian swell, as opposed to the broad, regional doming of the Superswell. "Jacqueline" Guyot stands out as being more anomalous than the other Darwin Rise volcanoes or the Superswell analogues. It could be that the radiometric date for "Jacqueline" is too young, and that the volcano actually formed on younger lithosphere.

PALEOMAGNETISM

Using all the sea-floor and multibeam bathymetric and magnetic data available, we now have magnetic models for 38 Cretaceous seamounts in the northwest Pacific: JPS (14), MWS(12), and MPM (9) (Fig. 9 and Table 2). The details of the surveys and the modeling are given in Sager et al. (this volume). Seminorm-minimization, a magnetic-inversion technique developed by Parker et al. [1987] was used for the newly modelled seamounts, and the results are combined with the older least-squares inversion poles for the previously modelled seamounts. The method of Parker et al. is considered to be somewhat better because it models magnetization inhomegeneities and also allows computation of the 95% confidence region of the paleomagnetic pole [Parker, 1988].

In general, the modeling gives exceptionally consistent results for the JPS and MWS seamounts; the MPM show a large amount of scatter, probably because of the effects of complex magnetizations (i.e., complex magnetic anomalies) as well as effects of the magnetic anomaly of the underlying plateau.

Magnetization strength

The JPS seamounts show large, dipolar, magnetic anomalies: Takuyo Daisan Guyot, for example gives an anomaly of 2050 nT. In part, this is due to high magnetizations, all of which are higher than the median value of 5 A/m for Pacific seamounts [Sager et al., this volume]. The highest magnetization of any seamount modelled in the northwest Pacific region is 23.1 A/m, for a small seamount between Jensen and Takuyo-Daini.

By contrast, seamounts in the MPM show only small, complex anomalies. Allison Guyot, for example, has a complex anomaly with numerous highs and lows with a maximum amplitude of only 500 nT, despite the large size (*ca.* 6200 km^3) of the guyot.

Paleomagnetic poles

The three groups of seamounts show differing degrees of consistency in yielding paleomagnetic poles (Fig. 10). The JPS seamount poles display strong consistency, and all but two cluster around the 88-125 Ma end of the Apparent Polar Wander Path (APWP)[Sager, 1992], implying that these seamounts are good paleomagnetic-pole recorders (Fig. 10a). Poles derived from seamounts in the Marcus-Wake group are not as tightly clustered as the JPS seamounts, but still fall mainly on the 82-125 Ma part of the APWP (Fig. 10b). Poles derived for the MPM are highly scattered, with 5 poles near the young end of the APWP ("Heezen", "Jacqueline", "Harvey", "Allen", "Thomas"), one near the old end (Darwin), and two (Sio and Allison) far from the APWP (Fig. 10c). The scattering is probably partly due to the complexity of MPM anomalies (e.g., Allison and "Jacqueline"),

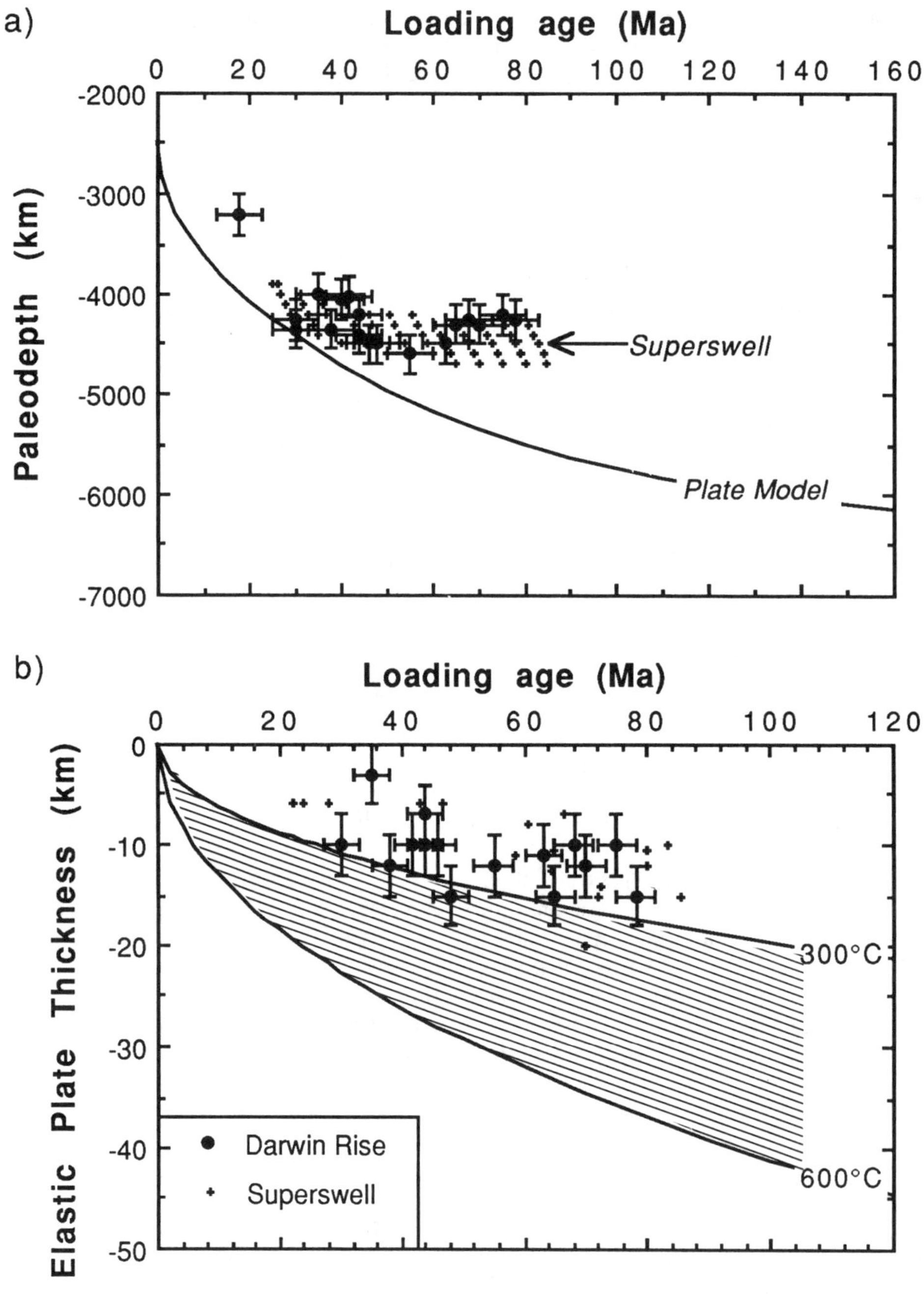

Fig. 8. (a) Paleodepth of Darwin Rise guyots as a function of age of the lithosphere at the time of loading. Paleodepth is measured as the height of the wave-cut basement (beneath any reef cap) above the regional depth of the surrounding seafloor at the base of the volcanic pedestal. Only volcanoes for which the age is reasonably well constrained from radiometric dating are included. Small crosses show the depth-age relationship for the Superswell at present. Normal oceanic lithosphere follows the "Plate Model," which predicts the subsidence of a 125-km-thick thermal plate [Parsons and Sclater, 1977], and is significantly deeper at seafloor ages of 40 to 90 m.y. than inferred for the past for the Darwin Rise or observed over the Superswell.

(b) Summary of elastic plate thickness values for the volcanoes of the Darwin Rise taken from Wolfe and McNutt [1991]. Small crosses indicate the elastic plate thickness values for the Superswell volcanoes [Calmant and Cazenave, 1987]. Diagonally-ruled region corresponds to the emperically-determined relationship between elastic plate thickness and loading age for volcanoes elsewhere in the North Pacific, Atlantic, and Indian Oceans. The ruled area lies roughly between the depths of the 300°C and 600°C isotherms as predicted by the plate models [Watts et al., 1980).

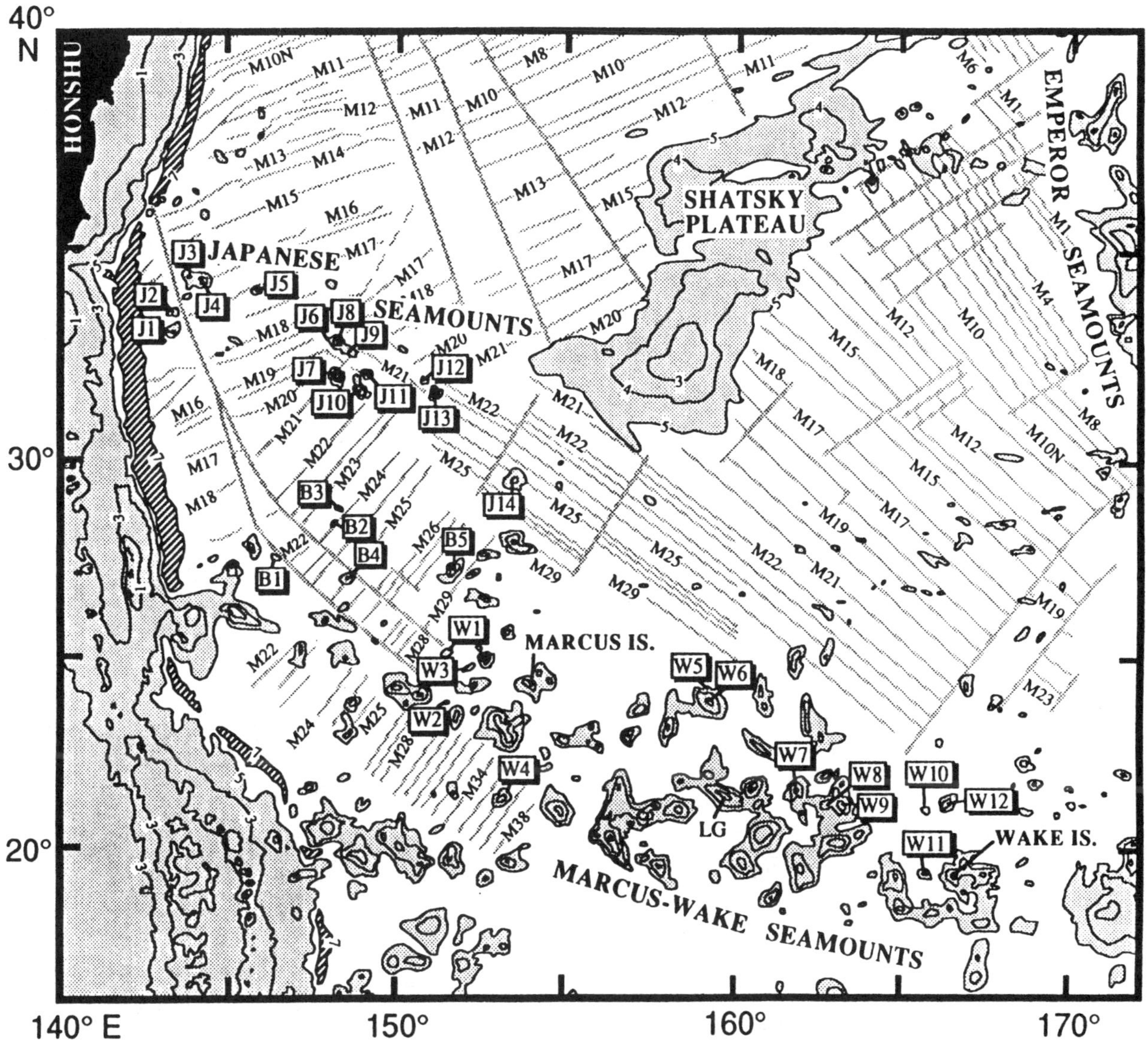

Fig. 9. Map of northwest Pacific, showing locations of seamounts examined in paleomagnetism study. Bathymetry from GEBCO Chart. Depths <5 km in gray; >7 km hachured. Magnetic lineations from Handschumacher et al. [1988] and Sager et al. [1988]. Numbers in boxes label seamounts with paleomagnetic studies. For details see Sager et al., this volume.

partly by the sometimes small anomalies poorly correlated with bathymetry (e.g., Sio), and partly to the confusion introduced in the inversion by the anomaly of the underlying plateau.

Paleolatitudes

As shown in Fig. 11a, calculated paleolatitudes for the JPS seamounts, exclusive of Takuyo-Daini, range from 2°N to 7°S (ave. 2.4°S ±3.5°); and for the Marcus-Wake seamounts (except for "Vibelius" and MIT, which are probably older than the others), from 2°S to 12°S (ave. 8.8°S ±3.3°). MPM paleolatitudes show too much scatter to be considered reliable, and are not plotted. The estimated net northward drift of the JPS and Marcus-Wake seamounts (present latitude minus paleolatitude) is shown in Fig. 11b.

Conclusions from seamount paleomagnetic data

At least two classes of seamounts are present in the region: 1) those with large magnetizations, and 2) those with small magnetizations. In general, the high-magnetization seamounts give poles and paleolatitudes that are consistent and seemingly little affected by viscous overprint or induced magnetization (a notable exception is Takuyo-Daini). On the other hand, many of the seamounts with smaller magnetizations (e.g., "Heezen," "Vibelius," MIT) give results that appear strongly affected by such a bias. We infer that the first group (mainly in the JPS Group, but including several reefless guyots in the WKS) formed during the Cretaceous Quiet Zone (KQZ), while the second group (which includes atolls both in the MPM and farther west in the MWS) formed mainly prior to the KQZ, that is, prior to about 124 Ma

Table 2. Paleomagnetic poles of western Pacific seamounts.

		Location		Paleopole		95% Confidence Ellipse[a]			
ID	Name	Lat (°N)	Lon (°E)	Lat (°N)	Lon (°E)	Maj	Min	Az	Ref
Japanese seamounts									
J1	Jenson	33.8	143.8	55.5	340.5	5.9	2.6	102	1
J2	Jenson N	34.1	143.9	57.1	327.1	20.9	9.7	91	1
J3	Takuyo-Daini	34.3	143.9	75.4	315.7	5.5	3.0	85	1
J4	Takuyo-Daisan	34.2	144.3	58.2	328.2	4.3	1.9	91	1
J5	Maiko	34.0	145.9	51.3	327.0	4.1	1.7	94	1
J6	Winterer-NW	32.9	147.9	51.9	315.7	6.6	2.6	97	1
J7	Charlie Johnson	32.0	148.3	47.4	284.5	5.4	1.7	105	1
J8	Winterer	32.8	148.4	51.6	315.5	4.1	1.5	97	1
J9	Winterer-SE	32.7	148.7	58.7	327.9	15.9	5.0	91	1
J10	Stout	31.6	149.0	50.4	342.5	10.1	4.2	103	1
J11	Thomas Washington	32.0	149.2	53.9	321.5	3.5	1.2	83	1
J12	Isakov-NW	31.8	150.9	58.0	311.9	9.4	3.4	75	1
J13	Isakov	31.6	151.2	52.3	331.5	4.2	1.7	87	1
J14	Makarov	29.5	153.5	59.2	346.2	2.9	0.9	103	1
Marcus-Wake seamounts									
W1	Unnamed	25.1	151.8	46.2	323.7				2
W2	Smt-948	23.7	151.9	57.5	329.5				3
W3	Unnamed	24.3	152.3	74.2	276.9				2
W4	Golden Dragon	21.3	153.2	56.2	337.4	5.6	1.9	98	1
W5	UCSD	24.3	159.2	55.3	338.4	9.2	3.7	85	1
W6	Scripps	23.7	159.3	63.9	350.2	11.1	2.1	87	1
W7	Miami	21.7	161.9	58.9	5.9	6.5	2.1	116	4
W8	Vibelius	21.2	162.8	84.0	310.1	8.9	3.5	65	1
W9	Wilde	21.2	163.3	61.8	333.1	7.1	2.1	82	1
W10	Birdseye	20.9	165.7	58.6	342.3				
W11	Unnamed	19.4	165.9	58.4	5.9				
W12	Woods Hole	21.2	166.5	53.2	319.7	6.8	2.0	61	1
Seamounts between Japanese and Marcus-Wake groups									
B1	Z44	27.8	146.7	60.0	306.0				5
B2	Z42	28.4	148.2	59.9	292.0	7.9	2.7	59	4
B3	Z41	28.8	148.4	55.0	19.0				5
B4	Z43	27.1	148.7	52.5	331.9	6.7	2.5	93	1
B5	MIT	27.3	151.9	29.8	328.6	27.4	9.0	85	1
Mid-Pacific Mountains seamounts									
M1	Sio	18.0	171.2	50.9	277.9				6
M2	Darwin	22.1	171.6	43.4	316.6	15.5	5.9	55	4
M3	Harvey	17.8	172.7	65.6	32.3	9.8	3.1	122	2
M4	Heezen	21.2	173.8	84.4	321.5	11.1	3.3	52	2
M5	Thomas	17.3	173.9	73.3	324.2				2
M6	Allen	18.3	174.1	69.2	328.6				2
M7	Jacqueline	19.4	176.7	79.0	35.4	20.0	4.4	116	6
M8	Allison	18.5	180.3	40.9	348.7	10.5	7.4	147	6

[a]95% Confidence ellipse: Maj = major semi-axis length (degrees), Min = minor semi-axis length (degrees), Az = azimuth of major semi-axis (degrees clockwise from north).

References: (1) Sager et al. (this volume), (2) Sager (1992), (3) Ueda (1988); (4) Hildebrand and Parker (1987), (5) Vacquier and Uyeda (1967), (6) Sager, in preparation.

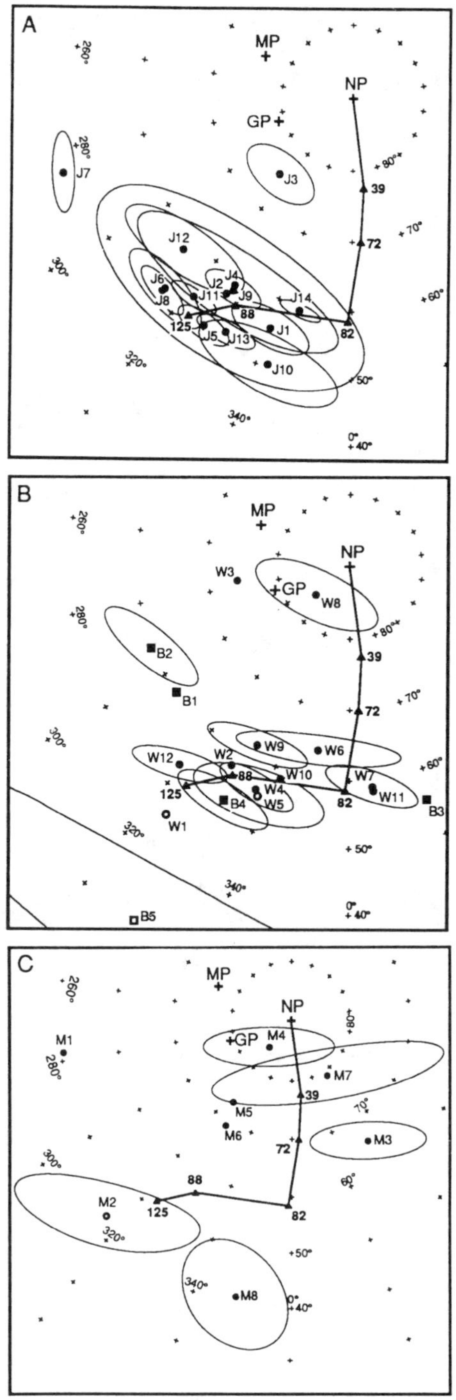

Fig. 10. Paleomagnetic poles for studied northwest Pacific seamounts. Filled circles denote pole positions and surrounding ellipses are their 95% confidence regions. Triangles represent Pacific apparent polar wander path. Numbers correspond to those shown of Fig. 9.

(a) Poles of the Japanese seamounts.

(b) Poles of the Marcus-Wake seamounts (circles) and "between" seamounts (squares). Filled symbols denote poles from normally polarized seamounts; open symbols, reversely polarized seamounts.

(c) Poles of the Mid-Pacific Mountains seamounts.

in the Harland et al. [1990], or about 119 Ma in the Kent and Gradstein [1986] time scale.

The calculated paleolatitudes of formation of the JPS and Wake seamounts were mainly from about 2-10°S, which is significantly north of the present South Pacific Superswell, which centers at about 30°S. Thus, assuming the reliability of the calculated paleolatitudes, if these seamounts indeed formed on the Cretaceous ancestor of the modern Superswell, then either there

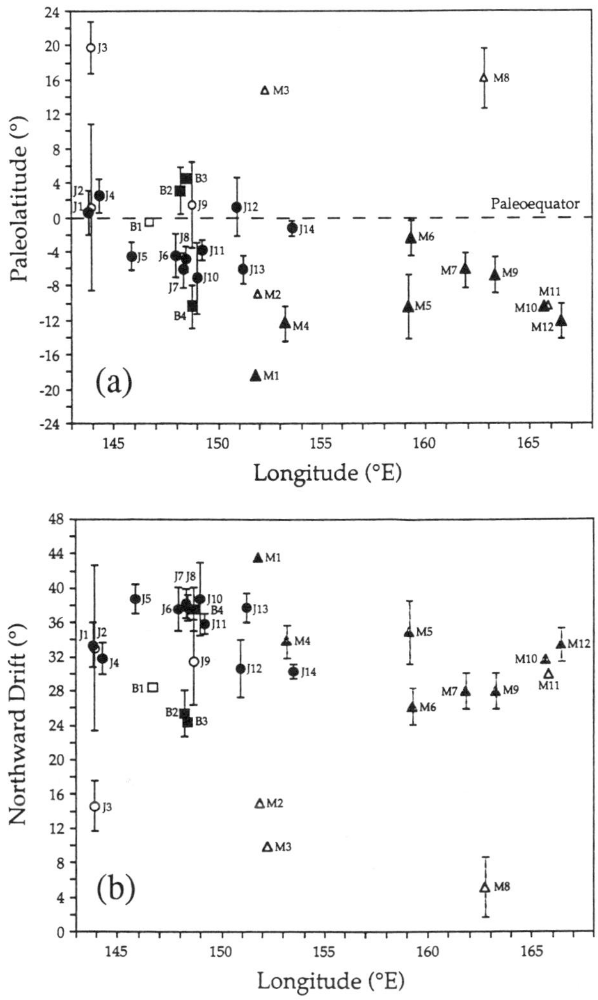

Fig. 11. Paleolatitudes and changes in paleolatitudes for northwest Pacific seamounts.

(a) Paleolatitudes of Japanese (circles), Marcus-Wake (squares), and "between" (triangles) seamounts. Error bars show latitudinal extent of 95% confidence ellipse for seamounts inverted by the seminorm method. Filled symbols denote most reliable data; open symbols show data from seamounts with outlier poles or low-reliability inversions.

(b) Northward drift of seamounts shown in Fig. 11a, calculated as present minus paleolatitude. Symbols as in Fig. 11a.

has been a southward displacement or a southward shrinking of this region of anomalous mantle. A number of studies suggest 10°-12° of southward drift of the Pacific hot spots relative to the spin axis has occurred since Late Cretaceous time [Gordon and Cape, 1981; Sager and Bleil, 1987] and this could account for part of the discrepancy. Additionally, most paleomagnetically studied seamounts are normally polarized, having formed during the KQZ, and thus may have paleolatitudes biased northwards by induced or viscous overprint magnetizations [see Sager et al., this volume]. Indeed, a Pacific APWP derived from the skewness of M-series magnetic lineations implies paleolatitudes that are 5°-15° farther south than the seamount data indicate [Larson et al., 1992]. This factor could account for some of the remaining difference between seamount paleolatitudes and present hot-spot locations, although it is not clear that the skewness data are necessarily more accurate than the seamount data.

The near-equatorial paleolatitudes of several young (90-97 Ma) volcanic guyots in the WKS (e.g., "Miami" and "Wilde"), well within the normal Cretaceous latitudinal belt of rudist/coral reefs, is puzzling because no evidence of reefs could be found on these guyots. Because they give paleolatitudes very near the equator, one is forced to argue either that the seamount paleomagnetic data are greatly in eror (>10°) in order to place these guyots at the edge of the equatorial belt of active reef growth, or that paleogeographic/environmental conditions did not permit establishment of reefs in the equatorial Pacific at that time.

Comparisons of predictions from seamount magnetic data and hot-spot models

Another perspective may be had by comparing predictions derived from the seamount magnetic survey data with plate rotation models. We here use the plate rotation parameters of Henderson [1985], which position the Pacific plate with respect to hot spots (Fig. 12a). If the magnetic data perfectly agreed with the hot-spot backtracking, then the data should fall on the diagonal line. Significantly, almost all data points fall above this line, suggesting a more northerly paleolatitude from the magnetic data than implied by the hot-spot backtracking. Several possible explanations for this discrepancy are discussed below.

Errors in the age of the volcanoes. If the hot-spot model predicts a large N-S component of velocity at the time of volcanism, then small errors in volcanic age translate into large errors in the predicted paleolatitude. If the volcanoes are consistently younger than the K-Ar ages indicate, then less northerly motion would be predicted. This explanation seems unlikely for most of these seamounts; for old rocks, the sense of errors in age is generally to underestimate the age on account of Ar loss in weathered rocks. The one example for which this explanation might be correct is "Woods Hole", whose K-Ar age of 68 Ma is anomalously young.

Errors in the plate rotation parameters. If the ages are correct, the possibility exists that the plate rotation parameters are wrong, and that too much northerly motion of the plate is currently predicted compared to what actually occurred. In both the Henderson [1985] and Duncan and Clague [1985] models, the hot-spot rotation parameters prior to 90 Ma are very uncertain because they are based on connecting alignments of old volcanoes to the present-day position of known hot spots, even where there are no continuous volcanic chains in between. This procedure assumes that the same hot spots have been intermittently active since Early Cretaceous time.

The possibility of error in the rotation parameters in turn implies that the hot spots of French Polynesia have moved south since 120 Ma. This alternative is examined below. Alternately, some other set of hot spots located north of French Polynesia, all presently inactive, produced the Darwin Rise volcanoes. Larson [1991] and Tarduno et al. [1991] discuss the possibility of widespread Early Cretaceous volcanism over a region larger than

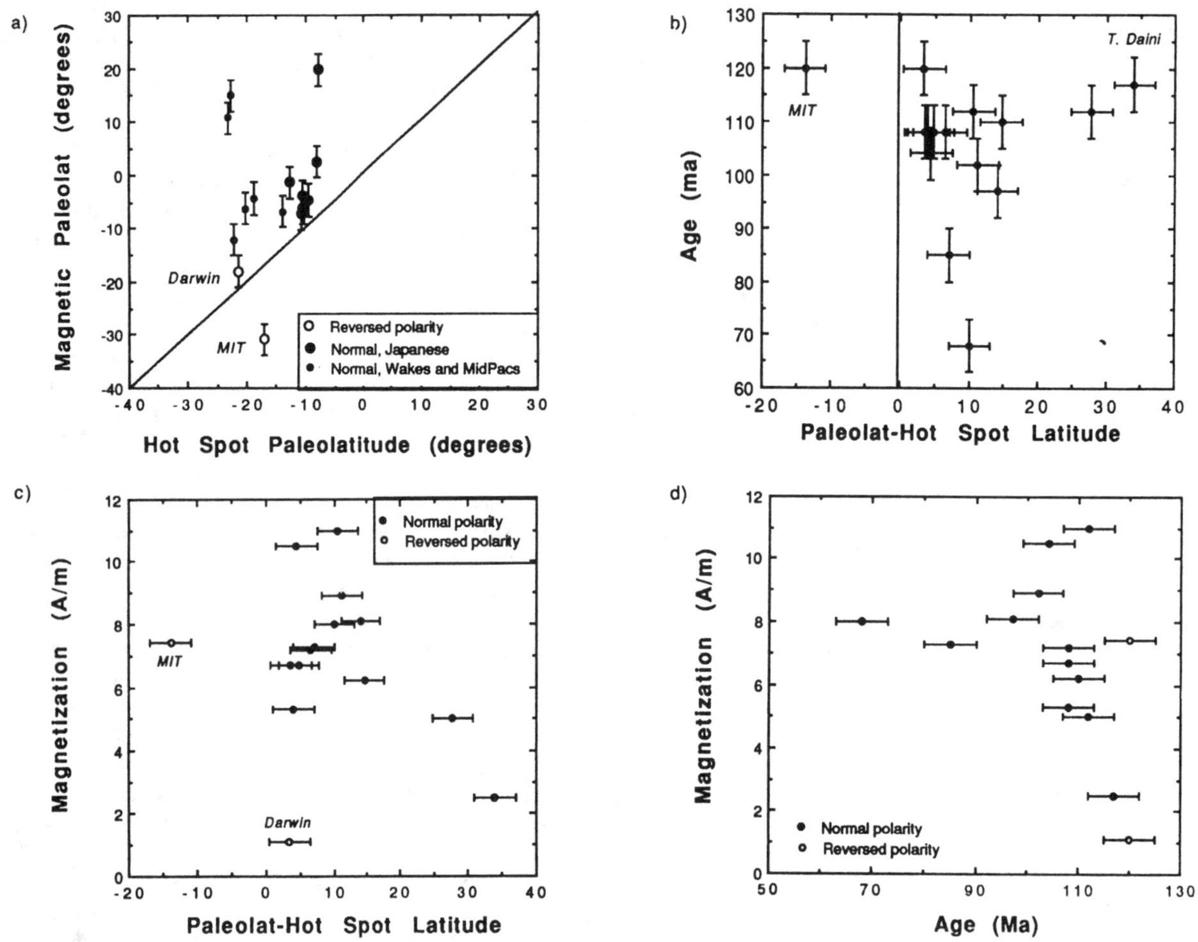

Fig. 12. (a) Comparison of the paleolatitude at time of the formation for the Darwin Rise guyots as computed from their paleomagnetic inclination (vertical axis) and from hot-spot backtracking (horizontal axis) [Henderson, 1985]. All guyots except MIT have magnetic paleolatitudes north of that predicted by the absolute-plate-motion model.

(b) Difference between the magnetic paleolatitude and the hot-spot paleolatitude plotted as a function of the age of the volcano. If the discrepancy in (a) were caused by a steady drift southward of the hot spots with respect to the geomagnetic pole, then the discrepancy should be larger for older seamounts and decrease with time. Rather, the discrepancy is fairly uniform, except for a few outliers like Takuyo Daini and MIT.

(c) Difference between the magnetic paleolatitude and the hot-spot paleolatitude plotted as a function of the strength of the magnetization. More strongly magnetized guyots show less discrepancy, suggesting that viscous overprinting of weakly magnetized seamounts is significant.

(d) Magnetization vs. Age for the Darwin Rise guyots. As expected, the most strongly magnetized volcanoes formed during the Cretaceous Superchron.

the present-day south Pacific area of hot spots and Superswell lithosphere.

Errors in the magnetic paleolatitude. The magnetic paleolatitude might be systematically biased by viscous overprinting of the present-day normal field vector on the original paleomagnetism [see Sager et al., this volume]. The sense of the discrepancy is opposite for normal versus reverse magnetizations. Figure 12a does show some evidence of this effect, as MIT falls on the opposite of the dashed line compared to all of the normally-polarized seamounts. On the other hand, Darwin Guyot, although reversed in polarity, falls quite close to the diagonal line.

The plot in Fig. 12c supports the suggestion that some of the discrepancy in Fig. 12a is caused by viscous overprinting because the highly magnetized guyots have less discrepancy than the more weakly magnetized ones, on average. However, there are still some weakly magnetized guyots that appear to have avoided the overprinting, perhaps on account of more stable magnetic mineralogy. Although there is much scatter, it appears that the Cretaceous Superchron guyots have higher magnetization, on average, compared to older ones and all have normal polarity, as predicted (Fig. 12d). The older guyots have either normal or reversed polarity and are lower on average magnetization. Even within the superchron, however, there is substantial variability in magnetization.

Motion of the Hot Spots with Respect to the Geomagnetic Reference Frame. There is some indication (Fig. 12b) that the maximum discrepancy (as measured by magnetic paleolatitude minus hot-spot paleolatitude) decreased with increasing age. However, a large number of relatively old guyots (100-110 Ma) have hot-spot paleolatitudes that lie within 5°-15° of the magnetic paleolatitude. The bulk of the data suggest that proper motion of hot spots with respect to the geomagnetic reference system is of the order of 10° since the mid-Cretaceous.

Gordon and Cape [1981], and Sager and Bleil [1987] also suggest 10°-12° of True Polar Wander (TPW) that would have moved Pacific hot spots southward since the Late Cretaceous. Any TPW for the Early and mid-Cretaceous is indeterminate because the plate-*vs.*-hot-spot rotation poles for pre-75 Ma is extremely speculative in our present state of knowledge of trends and ages along older seamount chains.

The general impression from this analysis is that absolute drift of the hot spots produces about 10° of southward motion of the hot spots with respect to the geomagnetic pole. Viscous overprinting might contribute another 20° (e.g., Takuyo Daini, Fig. 12b) to the discrepency in paleolatitudes.

DEPTH ANOMALIES

We can examine the history of depth anomalies associated with guyots of the Darwin Rise at various times in the past, using the hot-spot rotation parameters as shown in Fig. 13.

At 120 Ma (Fig. 13a), the depth anomalies in the region are estimated from the paleodepths at the time of volcanism, as calculated from the height above the surrounding sea floor of the volcanic basement beneath the carbonate platforms. Only for MIT and Darwin Guyots do we have adequate data at this time. MIT rotates back to near the present-day Society hot spot, with a depth anomaly of about 600-800 m. The depth anomaly is about 250 m smaller than that for Tahiti at present. Darwin rotates back to approximately the position of Pitcairn, but has a larger depth anomaly than what is consistent with the Superswell at that location today (mostly because the plate there is so young that large depth anomalies relative to the 125-km-thick model are not possible). For 95 Ma (Fig. 13b), about the time of uplift and karstification of the guyot summits, we estimate the depth anomalies for all guyots from the elevation of their flat tops (reefal or volcanic) above the surrounding sea floor. The depth anomalies are largest in the area of the WKS, which are actively forming at this time. The older JPS and MPM seamounts have drifted north from their hot spots and are characterized by smaller depth anomalies. In general, the amplitude, pattern, and latitude of the anomalies is similar to that of the Superswell today, but it appears to be slightly shifted to the west by about 10° to 20° compared to present-day longitude. This discrepancy may be caused by errors in the east-west component of the hot-spot backtracking. Therefore, this result lends support to the theory that the uplift of the guyots may have been caused by elevation of a paleosuperswell.

At the present time Fig. 13c), now that the guyots have drifted far north from the south-central Pacific, the depth anomalies calculated from the depth of the seafloor at the base of the guyots are quite small. The lack of appreciable depth anomalies today supports a thermal origin for those that developed at about 95 Ma.

SUBSIDENCE HISTORY

A tentative geologic history of the northwest Pacific guyots is here cast in terms of the history of their vertical motions (construction, subsidence, re-elevation) through time (Fig. 14). The essential facts about the morphology and age of the guyots are summarized in Van Waasbergen and Winterer (this volume, Tables 1 and 2). We underline the fact that if and when better-preserved fossils from precisely-controlled stratigraphic levels becomes available, and when a larger set of reliable radiometric dates is obtained, the preliminary account of the geologic history given below will doubtless need to be modified.

Initial construction and subsidence

Drowned atolls (Fig. 14, upper panel). Construction of the atoll-bearing seamounts in the MPM was on very young crust. After initial construction to form islands, the seamounts were eroded by streams and waves and subsided at rates moderated by isostatic response to erosional unloading. Reefs grew upward apace with subsidence, forming fringing, barrier, and atoll reefs successively. Atoll seamounts farther to the west, such as MIT, formed on older crust and subsided more slowly than eastern seamounts.

Given the assigned ages of reefal fossils dredged from near the summits of the atolls, we infer that most (MIT Guyot is a possible exception) first drowned in Aptian times, possibly because of southward drift into latitudes where rudist reef growth rates were insufficient to match subsidence rates [Winterer and Metzler, 1984]. They would have subsided further to depths of about 500 m (depending on the age of the underlying oceanic crust) by latest Albian time, some 17 m.y. later.

Fringing and barrier reefs (Fig. 14, middle panel). The construction of the JPS Guyots began at least 108 Ma ("Charlie Johnson" Guyot), and guyots farther west in the group may be older. By latest Albian time the seamounts had been nearly flattened by erosion and capped or encircled by reefs. Reefs in the older part of the chain, toward the northwest, subsided about 200 m while barrier reefs formed. Seamounts to the southeast (Isakov and Makarov), subsided less and formed only fringing reefs or carbonate platforms. The JPS guyots were likely still at sea level in latest Albian time.

At the time of emplacement of the JPS, nearby, older seamounts may have been uplifted (Fig. 14, upper panel), and in some places these seamounts may have been brought back to sea level, where shallow-water carbonates could accumulate.

Late Albian emergence

Near the close of Albian time, a period of emergence for both the atolls and the fringing and barrier reefs (Fig. 14, lower panel) resulted in formation of karstic topography and low-stand wave-cut terraces. Many of the JPS guyots were elevated about 200 m above sea level, and almost all their lagoonal strata were stripped

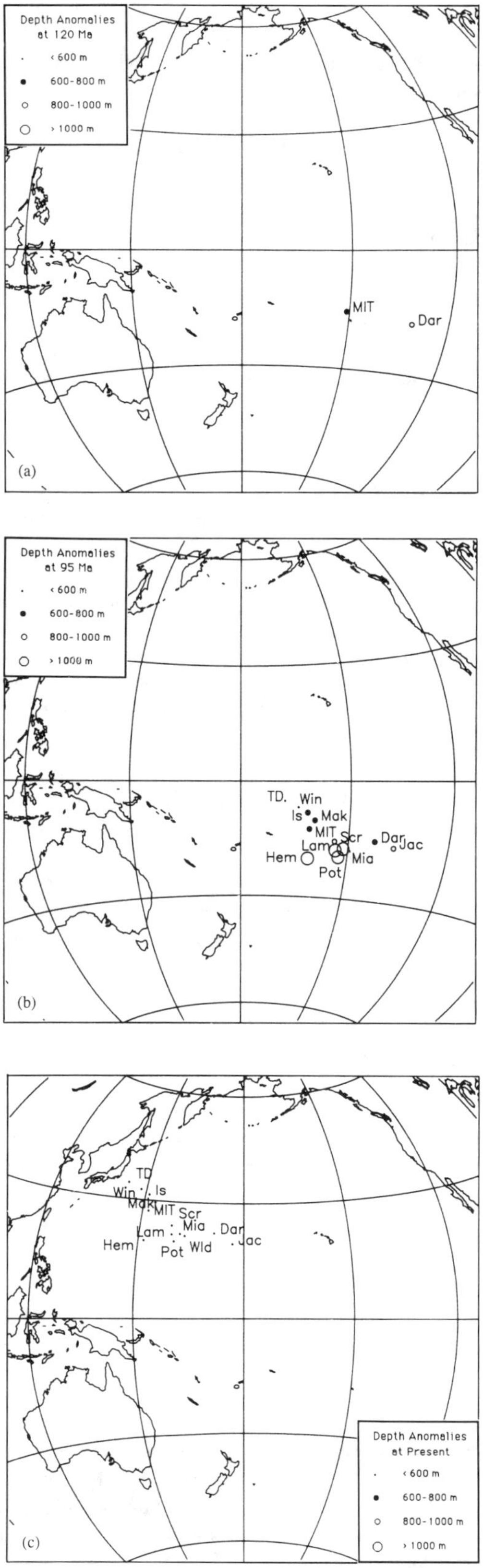

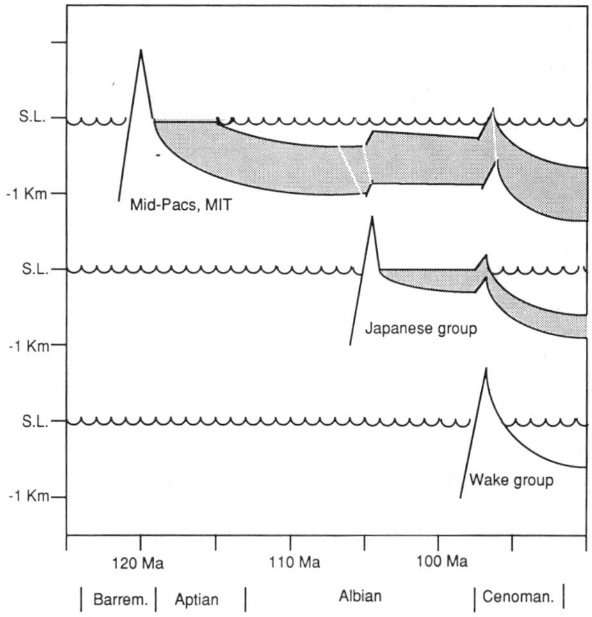

Fig. 14. Schematic subsidence paths for typical Early-mid Cretaceous seamounts in the northwest Pacific. The time of emplacement for each seamount group was longer than for the specimen guyots depicted. Shallow-water carbonate strata shaded.

away, leaving only the perimeter reefs. On the atoll guyots, low-stand terraces and sinkholes on several (MIT, Allison, Resolution) suggest 100-200 m of karst relief and/or low-stand terraces. A few atolls (Darwin, "Heezen", "Jacqueline") show only very slight karstic relief, while two atolls ("Vibelius", "Woods Hole") appear to have been more deeply eroded. On "Vibelius" Guyot, about 500 m of lagoonal sediments remain, except for an erosional mesa where an additional 100 m are preserved. On "Woods Hole" Guyot, virtually all reefal sediment is gone, and it would have been judged a volcanic guyot were it not for the dredge haul of reefal Barremian-Aptian fossils from the west flank. To estimate the total amount of uplift, the height of emergence must be added to the late Albian paleodepths of the drowned Aptian atolls, giving a total relative uplift of about 0.5 km for some guyots.

Except for the latest Albian planktonic foraminifers on the slope of Allison Guyot, the oldest planktonic fossils on the reefs are of Late Cenomanian and Turonian age (about 95-90 Ma). The youngest reef-bearing guyot, Isakov, is dated at 103.4 ± 2.8 Ma ("Scripps" is 100 ± 2 Ma, and probably has a reef). Thus, the time of emergence for the reefs is probably from about 100 to 95 Ma, i.e., late Albian to mid-Cenomanian time.

Fig. 13 Depth anomalies in the Pacific Ocean at (a) 120 Ma, (b) 95 MA, and (c) present, based on the paleodepths and subsidence history of the Darwin Rise guyots. Locations of the guyots at various times in the past are averages of positions predicted using both the Duncan and Clague [1985] and the Henderson [1985] poles of rotation of the Pacific plate with respect to the hot-spot reference frame and allowing for a ±5 m.y. uncertainty in guyot age.

The cause of this regional emergence is unknown. It is tempting to ascribe it to volcanism that created a broad swell, but, except for "Scripps" Guyot, we know of but one seamount of this age in the affected region. "Miami" Guyot, in the Wake group, has a date of 97 ± 1.6 Ma. There are still many undated seamounts along the Marcus-Wake chain.

Regional tectonics may have played a role. There may have been a major change in Pacific plate motion at this time, from WNW to NNW, marked out by the northernmost end of the SSW-trending Musician, Line Islands Chains, and Marshall seamount chains [Duncan and Clague, 1985; Henderson, 1985; Pringle, this volume]. Flow patterns in the upper mantle may have responded to this change in direction by creation of swells in the region of the newly formed bends in the chains.

The alternative to uplift, a short-lived eustatic lowering of sea level by about 200 m, also presents difficulties. The first is that it requires removal of water from the ocean basins on a scale larger than during the Pleistocene glaciations, though no evidence for contemporaneous large-scale continental glaciation has been reported from the continents. It might explain the karstification of the Late Albian JPS guyots, which were still at the surface, but it cannot explain the karstification and deep erosion of the Early Aptian atolls, which available paleontological dates on rudists indicate had been submerged and gradually subsiding into deeper water for about 15 m.y at the time of uplift. Sea level would have to have fallen by additional hundreds of meters just to reach them. The estimates of Haq et al. [1988], based chiefly on coastal onlap relations, place a global sea level fall of only about 50 m at the appropriate time.

Final Drowning

Final drowning of all the guyots in the MPM, JPS, and WKS was accomplished by Turonian time, as evidenced by pelagic sediments. It is very likely that drowning happened much earlier, but that pelagic sediments could not accumulate permanently until subsidence carried the summits to depths of several hundred meters. Today, for example, currents sweep guyot summits now at 1500 m depth virtually clean of loose sediment in latitudes north of about 20°N.

An open question is why new reefs of Cenomanian and Turonian age failed to attach themselves to the seamounts as they subsided through sea level, given that the seamounts were still in very low latitudes. There is a dearth — perhaps a total absence — of post-Albian, pre-Campanian reefs in the whole Pacific Basin. This is not so in the circum-Mediterranean Tethys, where rudist reefs are plentiful in this time period. Vogt [1989] has suggested volcanogenic upwelling of anoxic water as a possible factor in the demise of carbonate banks, but it is difficult to see how this would operate as a reef-killer or reef-preventer over such a large region, extending over at least 15° of latitude and 35° of longitude. Another approach is to ask if colonization during subsidence was hindered or prevented by cutting off of larval migration routes, a mechanism explored by Grigg and Hey [1992]. The paleogeography of Central America during the Cenomanian-Santonian time, when there was little or no reef growth in the central Pacific is poorly understood. Thus, we can only speculate that westward migration routes for current-borne larvae may have been blocked, or that equatorial currents moved so slowly that larvae could not survive the long journey to Pacific islands. Establishment of reefs of this same age in the Caribbean/Gulf of Mexico region, which was wide open to the Central Atlantic and Mediterranean Tethys, may likewise have been hindered by slow currents for migrating larvae.

CONCLUSIONS

Seamounts in the northwest Pacific were constructed during at least three episodes in the Early and mid-Cretaceous. An older group, built mainly between about 123 and 117 Ma, includes seamounts in the ENE-trending western and central MPM as well as scattered seamounts farther west: Darwin, "Woods Hole" (?), "Vibelius", and MIT. During a second episode, from about 108 to 100 Ma, most of the JPS were constructed along a roughly-defined chain trending ESE. Radiometric data are still too sparse to provide a test of age progression along these trends or to define Pacific plate velocity for that time period. During a third episode, beginning about 100 Ma, many seamounts in the WKS group were built, probably from the same hot-spot sources that created the SSE-trending Marshall Islands chains, located immediately adjacent to the WKS.

Volcanic rocks dredged from the older group of seamounts, in the MPM, resemble tholeiitic rocks erupted on seamounts near active spreading ridges, such as the modern Galapagos Islands. Volcanic rocks dredged from the JPS and WKS show strong geochemical affinities to young alkalic rocks on Samoa and the islands of French Polynesia, in the modern SOPITA [Staudigel et al., 1991] and Superswell [McNutt and Fischer, 1987] region.

Modeling of magnetism of 38 Early and mid-Cretaceous seamounts in the northwest Pacific yields paleolatitudes of formation of about 2°-12°S for the JPS and WKS, a position somewhat north of the main part of the Superswell. There are several plausible explanations for this misfit. The volcanoes may have been built over hot spots no longer active, during a time when the supposed ancestor of the Superswell — the Darwin Rise — was larger than its modern descendent. True polar wander in Cenozoic time may have shifted the relative latitude of present-day active Superswell hot spots. Viscous effects overprint the original Cretaceous magnetizations on many seamounts. Current hot-spot models may not accurately describe Pacific-plate *vs.* hot-spot motions for times earlier than 100 Ma.

The radiometrically dated seamounts fit well along previous estimates of the APWP.

Modeling of shipborne and satellite gravity data from the northwest Pacific seamounts yields elastic-plate thicknesses of less than 15 km, a figure much less than normal oceanic lithosphere, but consistent with the hypothesis that the seamounts were located on a superswell — the Darwin Rise — during mid-Cretaceous times.

The summits of the oldest seamounts are crowned by late Barremian-Aptian carbonate platforms, some with thicknesses of sediment of about 800 m, edged by rudist-coral reefs surrounding central lagoons. The JPS seamounts are also surmounted by Albian rudist reefs, but these advanced only to the barrier reef stage of development and are not more than about 200 m thick.

Many of both the Barremian-Aptian and the Albian reefs were uplifted, probably at the close of the Albian. The uplift resulted in emergence and development of a karstic topography with sinkholes and dolines. In the JPS, erosion during emergence resulted in the almost complete removal of lagoonal sediments. The uplift may have been caused by a major change in plate organization and a shift of plate motion in the hot-spot frame of reference at the time of emplacement of the WKS and the inception of volcanism in the Marshall Islands seamount chain.

By Turonian time, about 91 Ma, the seamounts had all subsided once more below sea level and pelagic sediments began to accumulate on the guyot summits. None of the seamounts developed reefs again, even though they were at sea level in low latitudes at the time of drowning, perhaps because of difficulties in seed larvae surviving long transport from the Caribbean region.

NOTE ADDED IN PROOF

Between initial submission and publication of this paper, two drilling legs (ODP Legs 143 and 144) drilled four of the guyots in this study. Since the results of these expeditions have direct bearing on this work, the ODP gave permission to include a brief statement on the impact of the drilling results on our conclusions.

Holes were drilled through the shallow-water limestone caps to basement on Allison, Resolution, MIT and Takuyo Daisan guyots [*van Waasbergen, Sliter, et al.*, in press; *ODP Leg 143 Scientific Party*, in press; *Haggerty, Premoli Silva, et al.*, in press]. The thicknesses of limestone matches our estimates on all guyots except Resolution, where the limestone is 1619 m thick. Ages of the drilled carbonate caps are: Albian (Allison) and Barremian through late Albian (Resolution). No new age data are available at this time for MIT and Takuyo Daisan guyots. On Allison and Resolution guyots, cores show evidence of karstification in the form of speleothem-lined solution cavities, some of which are filled with Turonian and Campanian pelagic material. Recovery rates of the shallow-water limestone material, especially in the upper parts of the sections, were uniformly poor (about 2%).

These findings have the following impact on our conclusions:
1: Guyots in the Mid Pacific Mountains apparently drowned at the same time as those farther west, in latest Albian time.
2: The western MPM started as a series of platforms of moderate relief above a shallow (1000 m waterdepth) plateau, rather than as tall, isolated seamounts.
3: The drowned atolls of the MPM were exposed to an extent about equal to that of the guyots farther west.

Acknowledgements. We are indebted to Captain T. Desjardins, the officers and crew of R/V *Thomas Washington*, and to U. Albright, P. Crampton, J. Goff, G. Ito, M. Lynch, M. Moore, M. Nakanishi, G. Pillard, and D. Stuber for the success of the RNDB10WT Expedition. We thank Robert Scott for his identifications of rudists, and Jürgen Grötsch for sharing paleontological data. The work was partly supported by NSF Grants 8717079-OCE to Winterer, Natland and Duncan, 8717826-OCE to McNutt, and 8715733-OCE to Sager.

REFERENCES

Aguayo-C., J.E., Sedimentary environments and diagenesis of a Cretaceous reef complex, eastern Mexico, *Ann. Centro del Mar y Limnol. Univ. Nat. Autón. México, 1, 83-140, 1978.*

Arnaud-Vanneau, A., Micropaléontologie, paléoécologie et sédimentologie d'une plate-forme carbonatée de la marge passive de la Téthys: l'Urgonien du Vercors septentrional e de la Chartreuse (Alpes Occidentales), *Géologie Alpine*, Grenoble, mém. *10*, 874p. (?).

Ashmore, S.A., The geomorphology of Johnston Atoll, *Naval Oceanographic Off. Tech. Rept., TR-237*, Washington, D.C., 25pp., 1973.

Bass, M.N., R.M. Moberly, J.M. Rhodes, and C. Shih, Volcanic rocks cored in the Central Pacific, Leg 17, Deep Sea Drilling Project. In: Winterer, E.L., J. I. Ewing, et al., *Init. Repts. DSDP, 17*, Washington (U.S. Gov't. Printing Office), 429-504, 1973.

Calmant, S., and A. Cazenave, Anomalous elastic plate thickness of the oceanic lithosphere in the south-central Pacific, *Nature, 328*, 236-238, 1987.

Carlisle, D., 1963. Pillow breccias and their aquagene tuffs, Quadra Island, British Columbia. *J. Geol., 71*, 48-71, 1963.

Cazenave, A., B. Lago, K. Dominh, and K. Lambeck, On the response of the ocean lithosphere to sea-mount loads from Geos 3 satellite radar altimeter observations, *Geophys. J.R. astr. Soc., 63*, 233-252, 1980.

Clague, D.A., and G.B. Dalrymple, G.B., The Hawaiian-Emperor volcanic chain Part I. Geologic Evolution. In Decker, R., T.L. Wright, and P.H. Stauffer, (eds.), *Volcanism in Hawaii*, U.S. Geol. Surv. Prof. Paper *1350*, 5-54, 1987.

Dalrymple, G.B., M.A. Lamphere, and D.A. Clague, Conventional $^{40}Ar/^{39}Ar$ K-Ar ages of volcanic rocks from Ojin (Site 430), Nintoku (Site 433) and Suiko (Site 433) Seamounts and the chronology of volcanic propagation along the Hawaiian-Emperor Chain, in Jackson, E.D., I. Koisumi, et al., *Init. Repts., DSDP, 55*, Washington, D.C., U.S. Gov't Print. Off., 659-676, 1981.

Duncan, R.A., and D.A. Clague, Pacific plate motion recorded by linear volcanic chains, in *The Ocean Basin and Margins, 7A*, edited by A.E.M. Nairn, F.G. Stehli, and S. Uyeda, Plenum, New York, 89-121, 1986.

Goetze, C. and B. Evans, Stress and temperature in the bending lithosphere as constrained by experimental rock mechanics, *Geophys. J. R. astr. Soc., 59*, 463-478, 1979.

Gordon, R.G., and C.D. Cape, Cenozoic latitudinal shift of the Hawaiian hotspot and its implications for true polar wander, *Earth Planet. Sci. Lett., 55*, 37-47, 1981.

Grigg, R.W., and R. Hey, Paleoceanography of the tropical eastern Pacific Ocean, *Nature, 255*, 172-178, 1992.

Grow, J.A., W. Lee, J.J. Miller, J.C. Agena, J.C. Hampson, D.S. Foster, and R.A. Woellner, Multichannel seismic-reflection survey of KOA and OAK craters, *U.S. Geol. Survey Bull., 1678*, D1-D46, 1986.

Haggerty, J.A., I. Premoli Silva, and ODP Leg 144 Shipboard Scientific Party, Western Pacific guyot drilling (ODP Leg 144) Preliminary results: sunken atolls vs carbonate banks, *EOS, 73, No. 43*, 587, 1992.

Haq, B.U., J. Hardenbol, and P.R. Vail, Mesozoic and Cenozoic chronostratigraphy and cycles of sea level change: in: C.K. Wilgus et al.(eds.), *Sea-Level Changes: An Integrated Approach*, Soc. Econ. Paleo. Mineral., Spec. Pub. *42*, 71-108, 1988.

Hamilton, E.L., Sunken islands of the Mid-Pacific Mountains, *Geol. Soc. Amer. Mem., 64*, 1956.

Hart, S.R., A large-scale isotope anomaly in the southern hemisphere mantle, *Nature, 309*, 753-757, 1984.

Harland, W.B., R.L. Armstrong, A.V. Cox, L.E. Craig, A.G. Smith, and D.G. Smith, *A Geologic Time Scale 1989*, Cambridge University Press, New York, 263p, 1990.

Haxby, W.F., G.D. Karner, J.L. LeBrecque, and J.K. Weissel, Digital images of combined oceanic and continental data sets and their use in tectonic studies, *EOS Trans. AGU*, *64*, 995-1004, 1983.

Hedge, C., Z. Peterman, and W. Dickinson, Petrogenesis of lavas from Western Samoa, *Geol. Soc. Amer. Bull.*, *83*, 2709-2714, 1972.

Heezen, B.C., J.L. Matthews, R. Catalano, J. Natland, A. Coogan, M. Tharp, and M. Rawson, Western Pacific Guyots, *Init. Rep. Deep Sea Drill. Proj.*, *20*, 653-723, 1973.

Henderson, L.J., *Motion of the Pacific plate relative to the hot spots since the Jurassic and models of the oceanic plateaus of the Farallon plate*, Ph. D. thesis, Northwestern University, 1985.

Hess, H.H., Drowned ancient islands of the Pacific Basin, *Amer. Jour. Sci.*, *244*, 772-791, 1946.

Hildebrand, J.A., and R.A. Parker, Paleomagnetism of Cretaceous Pacific seamounts revisited, *J. Geophys. Res.*, *92*, 12695-12712, 1987.

Huneke, J.C., and S.P. Smith, The realities of recoil: ^{39}Ar recoil out of small grains and anomalous age patterns in ^{39}Ar-^{40}Ar dating, *Geochim. Cosmochim. Acta, Suppl.*, *7*, 1987-2008, 1976.

Keating, B.H, Structural failure and drowning of Johnston Atoll, Central Pacific Basin, in: Keating, B.H., P. Fryer, R. Batiza, and B. Boehlert, eds., *Seamounts, Islands and Atolls*, Amer. Geophys. Union, Washington, D.C., 49-59, 1987.

Kellogg, J.N., and J. Ogujiofor, Gravity filed analysis of Sio Guyot: and isostatically compensated seamount in the Mid Pacific Mountains, *Geo-Marine Lett.*, *5*, 91-97.

Kent, D.V. and F.M. Gradstein, A Jurassic to recent geochronology, in: P.R. Vogt and B.E. Tucholke, eds., *The Geology of North America, vol. M, The Western North Atlantic Region*, 45-50, Geol. Soc. Amer., Boulder, 1986.

Kirby, S.H., Tectonic stresses in the lithosphere: Constraints provided by experimental deformation of rocks, *J. Geophys. Res.*, *85*, 6353-6363, 1980.

Konishi, K., Limestone of the Daiichi Kashima Seamount and the fate of a subducting guyot: fact and speculation from the Kaiko "Nautile" dives, *Tectonophysics*, *160*, 249-265, 1989.

Kroenke, L.W., J.N. Kellogg, and K. Nemoto, Mid-Pacific Mountains revisited, *Geo-Marine Lett.*, *5*, 77-81, 1985.

Ladd, H.S., W.A. Newman, and N.F. Sohl, Darwin Guyot, the Pacific's oldest atoll, *Proc. Int. Symp. Coral Reef, 2nd*, The Great Barrier Reef Committee, Brisbane, 513-522, 1974.

Larson, R.L., et al., Site 313, *Init. Repts. Deep Sea Drill. Proj.*, *32*, 313-390, 1975.

Larson, R.L., Latest pulse of Earth: evidence for a mid-Cretaceous superplume. *Geology*, *19*, 547-550, 1991.

Larson, R.L. and W.W. Sager, Skewness of Pacific magnetic anomalies M0-M29, *Proc. Ocean Drill. Proj.*, *129*, in press.

Larson, R.L., M.B. Steiner, E. Erba, and Y. Lancelot, Paleolatitudes and tectonic reconstructions of the oldest portion of the Pacific plate, a comparitive study, *Proc. ODP, Sci. Res.*, *129*, in press.

Lincoln, J.M., and S.O. Schlanger, Atoll stratigraphy as a record of sea level change - problems and prospects, *J. Geophys. Res.*, *96*, 6727-6752, 1991.

Lonsdale, P.F., W.R. Normark, and W.A. Newman, Sedimentation and erosion on Horizon Guyot, *Geol. Soc. Amer. Bull.*, *83*, 289-316, 1972.

Matthews, J.L., B.C. Heezen, R. Catalano, A. Coogan, M. Tharp, J. Natland, and M. Rawson, Cretaceous drowning of reefs on mid-Pacific and Japanese guyots, *Science*, *184*, 462-464, 1974.

McNutt, M.K., and K.M. Fischer, The south Pacific superswell, in *Seamounts, islands, and atolls*, edited by B. Keating, P. Fryer, R. Batiza, and G.W. Boehlert, Geophys. Mon. *43*, Amer. Geophys. Union, Washington, D.C., 25-34, 1987.

McNutt, M.K., and H.W. Menard, Lithospheric flexure and uplifted atolls, *J. Geophys. Res.*, *83*, 1206-1212, 1978.

McNutt, M.K., E.L. Winterer, W.W. Sager, J.H. Natland, and G. Ito, The Darwin Rise: a Cretaceous superswell? *Geophys. Res. Lett.* *17*, 1101-1104, 1990.

Menard, H.W., *Marine Geology of the Pacific*, McGraw-Hill, New York, 138-146, 1964.

Menard, H.W., Darwin reprise, *J. Geophys. Res.*, *89*, 9960-9968, 1984.

Nakanishi, M., K. Tamaki, and K. Kobayashi, Magnetic anomaly lineations from Late Jurassic to Early Cretaceous in the west-central Pacific Ocean, *Geophys. J. Int.*, *109*, 701-719, 1992.

Natland, J.H., Petrology of volcanic rocks dredged from seamounts in the Line Islands. In: Schlanger, S.O., E.D. Jackson, et al., *Init. Repts. DSDP*, *33*, Washington (U.S. Govt. Print. Off.), 749-777, 1976.

Natland, J.H., and E. Wright, Magmatic lineages and mantle sources of Cretaceous seamounts of the central Pacific. *Eos*, *65*, 1075-1076, 1984.

Natland, J.H., and D.L. Turner, Age progression and petrological development of Samoan shield volcanoes: Evidence from K-Ar ages, lava compositions, and mineral studies. In: Brocher, T. (ed.), *Geological Investigations of the Northern Melanesian Borderland, Circum-Pacific Council for Energy Resources Earth Sciences Series*, *3*, Houston (Circum-Pacific Council for Energy and Mineral Resources), 139-171, 1985.

ODP Leg 143 Scientific Party, Regional and global implications of drilling of Early Createaceous guyots: ODP Leg 143, Mid Pacific Mountains, *EOS, Trans. Am. Geophys. Union*, in press, 1992.

Ozima, M., I. Kaneoka, K. Saito, M. Honda, M. Yanagisawa, and Y. Takigami, Summary of geochronological studies of submarine rocks from the western Pacific Ocean, in: T.W.C. Hilde and S. Uyeda, eds., *Geodynamics of the western Pacific-Indonesian region*, Geodynamics Series, *11*, edited by Amer. Geophys. Un., Washington, D.C., 137-142, 1983.

Palacz, Z,. and A.D. Saunders, Coupled trace element and isotope enrichment in the Cook-Austral-Samoa islands, southwest Pacific. *Earth Planet. Sci. Lett.*, *79*, 270-280, 1986.

Parker, R.L., A statistical theory of seamount magnetism, *J. Geophys. Res.*, *93*, 3105-3115, 1988.1988.

Parker, R.L., L. Shure, and J. A. Hildebrand, The application of inverse theory to seamount magnetism, *Rev. Geophys.*, *25*, 17-40, 1987.

Parsons, B., and J. G. Sclater, An analysis of the variation of ocean floor bathymetry and heat flow with age, *J. Geophys. Res.*, *82*, 803-827, 1977.

Sager, W.W., Seamount age estimates from paleomagnetism and their implications for the history of volcanism on the Pacific plate, in: B.H. Keating and B. Bolton, eds., *Geology and Offshore Mineral Resources of the Central Pacific Basin*, Circum-Pacific Council for Energy and Mineral Resources, Earth Sci. Ser., *14*, Springer-Verlag, New York, 21-37, 1992.

Sager, W.W., and U. Bliel, Latitudinal shift of Pacific hotspots during the Late Cretaceous and early Tertiary, *Nature*, *326*, 488-490, 1987.

Samson, S.D., and E.C. Alexander, Jr., Calibration of the interlaboratory ^{40}Ar/^{39}Ar dating standard, MMhb-1, *Chem. Geol.*, *66*, 27-34, 1987.

Sandwell, D.T., Along-track deflection of the vertical from Seasat: GEBCO overlays, *NOAA Tech. Memorandum, NOS NGS-40*, 1984.

Schlanger, S.O., Subsurface geology of Eniwetok atoll, *U.S. Geol. Surv., Prof. Paper, 260-BB*, 991-1066, 1963.

Schlanger, S.O., M.O. Garcia, B.H. Keating, J.J. Naughton, W.W. Sager, J.A. Haggerty, J.A Philpotts, and R.A. Duncan, Geology and geochronology of the Line Islands. In: Watts, A.B., (ed.), *The Origin and Evolution of Seamounts*, J. Geophys. Res., *89*, 11,261-11,272, 1984.

Smith, W.H.F., H. Staudigel, A.B. Watts, and M. S. Pringle, The Magellan Seamounts: Early Cretaceous record of the south Pacific isotopic and thermal anomaly, *J. Geophys. Res.*, *94*, 10501-10523, 1989.

Smoot, N.C., North Pacific Guyots, *U.S. Naval Oceanographic Office, Tech. Note TN 01-91*, Stennis Space Center, Mississippi, 93 pp., 1991.

Staudigel, H., K.-H. Park, M. Pringle, J.L. Rubenstone, W.H.F. Smith, and A. Zindler, The longevity of the South Pacific is%wopic and thermal anomaly. *Earth Planet. Sci. Lett.*, *102*, 24-44, 1991.

Stearns, H.T., 1940. Four-phase volcanism in Hawaii (abst.), *Geol. Soc. Amer. Bull.*, *51*, 1947-1948, 1940.

Stearns, H.T., *The Geology of the State of Hawaii*, Palo Alto (Pacific Books), 1966.

Stein, C., and D.H. Abbott, Heat flow constraints on the South Pacific Superswell, *J. Geophys. Res.*. *96*, 16083-16099, 1991.

Tarduno, J. A., W.V. Sliter, L. Kroenke, M. Leckie, H. Mayer, J.J. Mahoney, R. Musgrave, M. Storey, and E.L. Winterer. Rapid formation of Ontong Java Plateau by Aptian mantle plume volcanism, *Science, 254*, 399-403, 1991.

Thiede, J., T.L. Vallier, et al., Site 463, *Init. Rep. Deep Sea Drill. Proj., 62*, 33-156, 1981.

Turner, G., and P.H. Cadogan, Possible effects of ^{39}Ar recoil in $^{40}Ar/^{39}Ar$ dating of lunar samples, *Geochim. Cosmochim. Acta, Suppl., 5*, 1601-1615, 1974.

Ueda, Y, Geophysical study of two seamounts near Minami-Tori Shima (Marcus) Island, western Pacific Ocean, *J. Geomag. Geoelectr., 49*, 1481-1501, 1988.

Vacquier, V., and S. Uyeda, Paleomagnetism of nine seamounts in the western Pacific and of three volcanoes in Japan, *Bull. Earthquake Res. Inst., Univ. Tokyo, 44*, 815-848, 1967.

Van Waasbergen, R.J., W.V. Sliter, and ODP Leg 143 Scientific Party, Results of ODP drilling in the Mid Pacific Mountains: stratigraphy of two Cretaceous drowned atolls, *EOS, 73, No. 43*, 587, 1992.

Vogt, P.R., Volcanogenic upwelling of anoxic, nutrient-rich water: A possible factor in carbonate-bank/reef demise and benthic faunal extinction, *Bull. Geol. Soc. Amer., 101*, 1317-1322, 1989.

Vogt, P.R., and N. C. Smoot, The Geisha guyots: multibeam bathymetry and morphometric interpretation, *J. Geophys. Res., 89*, 11085-11107, 1984.

Walcott, R.I., Flexure of the lithosphere at Hawaii, *Tectonophys.*, *9*, 435-446, 1970.

Walker, D.A., and I. McDougall, $^{40}Ar/^{39}Ar$ and K-Ar dating of altered glassy volcanic rocks: the Dabi Volcanics, P.N.G., *Geochim. Cosmochim. Acta, 46*, 2181-2190, 1982.

Watts, A.B., and J.R. Cochran, Gravity anomalies and flexure of the lithosphere along the Hawaiian-Emperor seamount chain, *Geophys. J. R. ast. Soc., 38*, 119-141, 1974.

Watts, A.B., and N.M. Ribe, On geoid heights and flexure of the lithosphere at seamounts, *J. Geophys. Res., 89*, 11152-11170, 1984.

Watts. A.B., J. H. Bodine, and N. R. Ribe, Observations of flexure and the geological evolution of the Pacific ocean basin, *Nature, 283*, 532-537, 1980.

Weaver, B.L, The origin of ocean island basalt end-member compositions: trace element and isotopic constraints, *Earth Planet. Sci. Lett., 104*, 381-397, 1991.

Winterer, E.L., J. Ewing, et al., Site 171, *Init. Repts. Deep Sea Drill. Proj., 17*, 283-334, 1973.

Winterer, E.L., and Metzler, C.V., Origin and subsidence of guyots in Mid-Pacific Mountains, *Jour. Geophys. Res., 89*, 9969-9979, 1984.

Wolfe, C.J., and M.K. McNutt, Compensation of Cretaceous guyots of the Darwin Rise, northwest Pacific Ocean, *J. Geophys. Res., 86*, 2363-2374, 1991.

Wright, E. and W.M. White, W.M., The origin of Samoa: new evidence from Sr, Nd, and Pb isotopes, *Earth Planet. Sci. Lett., 81*, 151-162, 1986.

York, D., Least squares fitting of a straight line with correlated errors, *Earth Planet. Sci. Lett., 5*, 320-324, 1969.

Zindler, A., and S. Hart, Chemical geodynamics, *Ann. Rev. Earth Planet. Sci., 14*, 493-571, 1986.

E.L. Winterer and R. van Waasbergen, Scripps Institution of Oceanography, University of California San Diego, La Jolla, CA 92093-0212.

R. Duncan, College of Oceanography, Oregon State University, Corvallis, OR 97331.

M.K. McNutt and C.J. Wolfe, Dept. Earth and Planetary Sci., M.I.T, Cambridge, MA 02139

J.H. Natland, R.S.M.A.S, University of Miami, 4600 Rickenbacker Parkway, Coral Gables, FL 33139.

I. Premoli Silva, Dipartimento Scienze della Terra, Università di Milano, 20133 Milan, Italy.

W.W. Sager, Dept. of Oceanography, Texas A&M UNiv., College Station, TX 77843.

W.V. Sliter, U.S. Geological Survey, 345 Middlefield Rd., Menlo Park, CA 94025.

Summit Geomorphology of Western Pacific Guyots

ROBERT J. VAN WAASBERGEN AND EDWARD L. WINTERER

Scripps Institution of Oceanography, University of California San Diego, La Jolla, California

Bathymetric and seismic reflection data for 21 guyots in the Japanese, Wake and Mid Pacific Mountains seamount groups show that many guyots bear thick shallow-water limestone deposits that represent nearly undeformed Cretaceous rudist-reef-bounded carbonate platforms. Three types of guyot summits are distinguished: those with more than 200 m of shallow-water limestone, sufficient to bury the underlying volcanic relief, follow the same basic morphologic patterns as their coral/algal-dominated modern counterparts. Surface morphology is controlled by the original reef/platform configuration, modified by subaerial erosion and by long-term deformation of the sedimentary deposits and the underlying volcanic edifice. These seamounts range in age from Barremian to Aptian, with Barremian or Aptian to late Albian limestone deposits. Guyots with thin limestone deposits represent reef-bearing volcanic islands at various stages of fringing-reef and barrier-reef development. These edifices and their overlying limestone deposits are of Albian age. Reef growth on these was suddenly halted when regional emergence led to cessation of shallow-water-limestone accumulation on all guyots in the western Pacific, probably during the latest Albian. The karstic surfaces were resubmerged by middle Turonian time and, in latitudes south of about 20°N, blanketed by pelagic ooze. Guyots without reef deposits appear to be products of post-Albian volcanism and erosion at sea level. Among the three types of summit-configurations a range of stages of development of Cretaceous carbonate-platforms can be observed. The exposure of the guyot summits indicated by the occurence of karstic relief of 100-200 m on many of the limestone caps suggests that the sea floor of the western Pacific was raised several hundered meters. The age- and platform thickness-relationships among the different types of guyots suggests that this uplift occurred at late Albian time.

INTRODUCTION

After their first description [*Hess*, 1946], flat topped seamounts, or guyots, were recognized to have great potential as indicators of horizontal and vertical tectonic processes of the sea floor. Hess suspected that their flat tops were the result of erosion of volcanic peaks at sea level. *Hamilton* [1956] first discovered the presence of early Cretaceous reefal deposits on guyot summits in the Mid Pacific Mountains. This confirmed the former existence at sea level for those guyots, and gave an indication of their age. Since then it has been discovered that many of the large guyots in the western Pacific region represent former islands and atolls of Cretaceous age [*Karig et al.*, 1970; *Heezen et al.*, 1973; *Ladd et al.*, 1974]. The details of their formation, reef growth history, and the causes of their ultimate demise as islands and atolls remained elusive, however [*Matthews et al.*, 1974; *Winterer and Metzler*, 1984]. The fact that these islands and atolls existed at sea level and tracked sea level through accumulation of carbonate sediments makes them useful in constraining the history of vertical movement of the Pacific plate.

The Mesozoic Pacific: Geology, Tectonics, and Volcanism
Geophysical Monograph 77

Menard [1964] tried to use these characteristics to unravel some part of the Cretaceous vertical tectonic history of the Pacific. The shallow paleo-seafloor-depth indicated by the heights of the guyots above the surrounding seafloor led him to conclude that a significant part of the sea floor had been unusually shallow during the Cretaceous. He named this region of the western Pacific the "Darwin Rise". *Crough* [1978] recognized that hot-spot activity affects the overlying lithosphere, and estimated, using guyot-heights both from the Darwin rise and elsewhere, that hot-spot activity causes older seafloor to rise to the depth normally expected for 25 my old lithosphere.

Guyots and other seamounts have also been used to track the horizontal motion of the Pacific plate. Magnetic paleopoles derived from marine magnetic anomaly data over seamounts and guyots have been used to constrain the latitudinal movement of the volcanos with respect to the magnetic poles [*Francheteau et al.*, 1970; *Sager and Pringle*, 1987]. Age progressions along seamount chains such as the Hawaii-Emperor chain were of critical importance in establishing the concept of "hot-spot chains". Chemical signatures of seamount-basalts are used to track mantle anomalies over great spans of distance and time [*Smith et al.*, 1989; *Staudigel et al.*, 1991].

It has become apparent that the Cretaceous volcanic and tectonic history of the Pacific is far more complex than simple models can explain. A far greater part of the Mesozoic Pacific seafloor is covered with large seamounts and guyots than is Cenozoic seafloor, even when one takes into account the fact that older seafloor has had more time to accumulate seamounts than younger seafloor [*Batiza*, 1982; *Smith*, 1990]. In addition, the western Pacific is the site of large abyssal flood basalts [*Menard*, 1984] which apparently formed during episodes of extensive submarine volcanism late in the Early Cretaceous. The total volume of volcanic material produced in this time-period appears to be very large compared to times before and after [*Larson*, 1991].

The horizontal motion of the Pacific plate in the hot-spot reference frame is not well-known for the Early and early-Late Cretaceous. Part of the problem is that there are no clear hot-spot tracks in the area. Seamount alignments either overlap in complex patterns, or seamounts appear randomly scattered over the seafloor, without neat progressive tracks. On some of the oldest seafloor, in the Marshall Islands area, modern atolls at sea level exist side by side with Cretaceous drowned atolls [*Schlanger et al.*, 1987; *Lincoln et al.*, this volume]. Several of these guyots show evidence of multiple reef-growth and reef-drowning episodes, indicating complicated vertical trajectories.

Guyot tops are very sensitive recorders of processes which affect them, especially at sea level. This paper examines recent detailed information on the morphology and structure of the summits of the guyots and interprets these data in terms of constructive and erosional processes. We examine the structure and evolution of Cretaceous atolls, the remains of which cap many of the guyots, as far as can be done with bathymetric and seismic reflection information. We introduce new radiometric and paleontological age data and intrepret these in context of the timing of events that the morphologic and structural analyses indicate. We hereby further constrain the timing of formation, growth and demise of the volcanoes and their sediment caps, as well as the tectonic processes that affected them. Much more detailed analyses of the sediment caps will surely result from the drilling of several guyot tops which has been planned for 1992. A long-standing problem of why the carbonate platforms failed to keep up with subsidence and drowned [*Schlager*, 1984; *Vogt*, 1989], must also await this drilling.

METHODS

During the Scripps Institution of Oceanography ROUNDABOUT expedition, Leg 10, the R/V *Thomas Washington* collected bathymetric data, seismic reflection profiles, and magnetic and gravity data over 21 seamounts and guyots in the Western Pacific (Figure 1).

The SeaBeam system allows detailed, high-resolution bathymetric mapping of the sea floor [*Renard and Allenou*, 1979]. We used

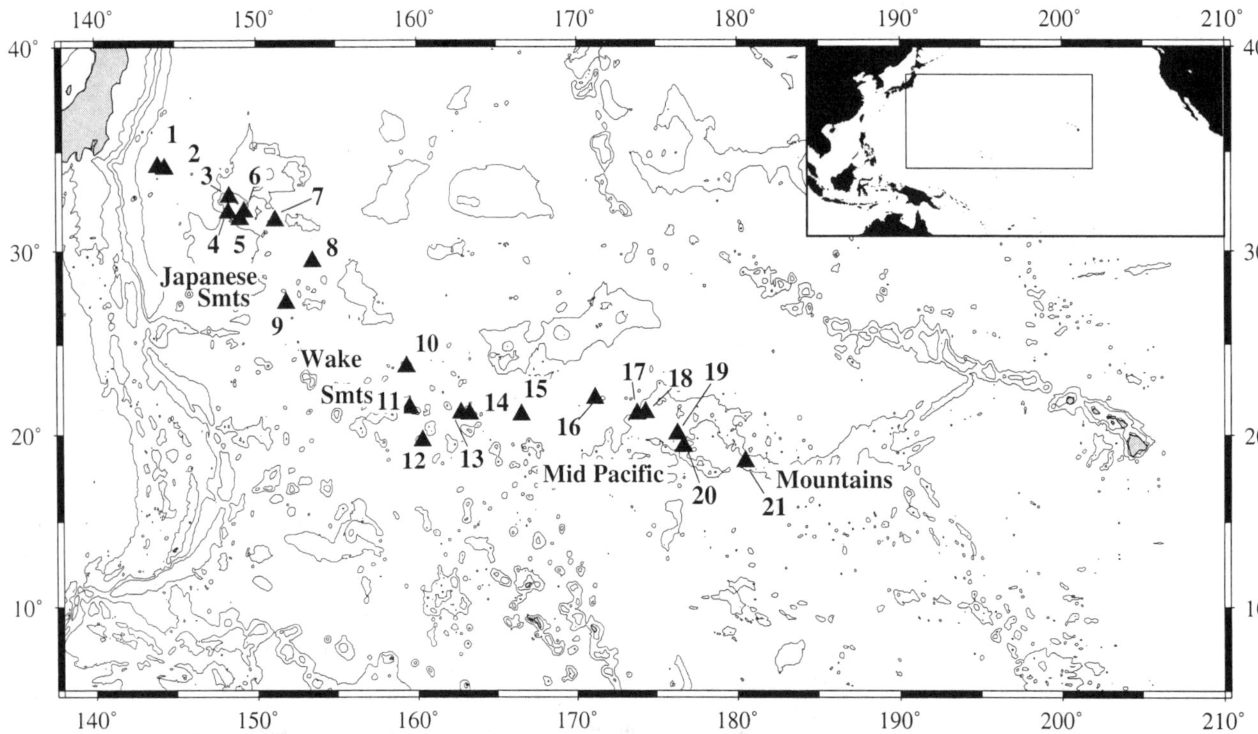

Fig. 1. Bathymetry of the western Pacific Ocean, based on DBDB-5 five-minute gridded bathymetry. The contour interval is 1000 m. Guyots surveyed during Roundabout, leg 10, are indicated as black triangles. 1: Takuyo-Daini. 2: Takuyo Daisan. 3: "Winterer". 4: "Charlie Johnson". 5: "Stout". 6: "Thomas Washington". 7: Isakov. 8: Makarov. 9: MIT. 10: "Scripps". 11: "Lamont". 12: "Pot". 13: "Vibelius". 14: "Wilde". 15: "Woods Hole". 16: Darwin. 17: "Heezen". 18: Resolution. 19: "Caprina". 20: "Jacqueline". 21: Allison.

this tool to construct topographic maps of the tops of 21 guyots and some seamounts between Japan and Hawaii. With the SeaBeam system, at typical guyot-summit depths (1200-2000 m), it is possible to produce reliable maps at 20 m contour intervals, and to resolve structures on the sea floor as small as 80 m in diameter. At this scale the tops of many guyots, though flat at larger scales, appear to have significant morphological features. Seismic reflection and 3.5 kHz echo sounder were used to image structures in the subsurface. The seismic system, using a single-channel hydrophone streamer and an 80 inch^3 watergun source allowed penetration of platform deposits thicker than about 100 m and imaging of the surface of the underlying volcanic basement. The 3.5 kHz system was able to penetrate the relatively thin (<100 m) pelagic sediments that mask the relief of the underlying Cretaceous platform surface on many guyots.

Results of the analyses of magnetic and gravity data are presented in *Sager* [1992], *McNutt et al.*, [1990], *Wolfe and McNutt*, [1991], and summarized in *Winterer et al.*, [this volume].

Our surveys generally covered the summit plateau areas, the plateau margins, and parts of the upper slopes of the seamounts. Data from the flanks and lower slopes were generally obtained only from the approach and departure tracks, and in a few places where we made magnetic surveys or dredged the lower flanks to obtain volcanic material. Coverage of the summit plateaus and upper margins was better than 90% on 5 of the 22 guyots, better than 60% on 5 others, and limited to a few crossings on the remaining 11.

RESULTS

We observed and measured elements of morphology and structure on guyots in the Japanese, Wake and Mid Pacific Mountains seamount groups. These observations and measurements are summarized in Table 1.

Large-scale morphology

The overall dimensions of a guyot at the 10-100 km scale is the result of the volcanic processes of edifice building that produced it. Morphologic characteristics such as elongation, volume, number and orientation of flank rift zones, volcanic slope angles, and other larger scale features are defined early in the history of a guyot [*Vogt and Smoot*, 1984]. Many studies of guyots and seamounts exist which take these aspects into consideration and relate them to hot-spot activity, mantle processes, stresses in the plate, etc. [*Hollister et al.*, 1978; *Vogt and Smoot*, 1984; *Smoot*, 1989; *Lonsdale*, 1988]. Our surveys confirmed the presence of many features pointed out by *Vogt and Smoot* [1984] and by *Smoot* [1989], but for the most part our surveys concentrated on the summit regions, to examine them in detail.

Medium-scale morphology

The main feature which distinguishes guyots from ordinary seamounts at the 1-10 km scale is the flat top. Although there are several possible ways these flat tops may have formed, processes at sea level are the most prominent [*Winterer and Metzler*, 1984]. Two main types of processes flatten the top at or near sea level: erosional and constructional. Erosion produces a flat top when a volcanic island is truncated by streams and by waves. When reefs are present around a volcanic island, they protect the interior from erosion by wave-planation, and allow subsidence to take place unbalanced by isostatic uplift which would result from the removal of the top of the volcano [*Menard*, 1983], except that due to removal by stream erosion and mass wasting. While the volcanic island subsides, the reefs continue to grow up to sea level, eventually burying the volcanic rocks under limestone, forming a carbonate platform, commonly in the form of an atoll [*Darwin*, 1842]. If the platform then drowns, it becomes a guyot with has a purely constructional flat top.

Of the 21 guyots surveyed during Roundabout 10, Cretaceous shallow-water limestone deposits were found on 16. The presence of shallow-water limestone was determined by recovery of this material in dredges, or by the presence of distinct layered reflectors in the seismic profiles. Pelagic sediments tend to mound or drape over their substrate and are generally easy to distinguish from the more nearly horizontal shallow-water carbonates. Cretaceous limestone was dredged from 14 of the 21 guyots, and seen as reflectors in two that were not dredged (Table 1). Many other guyots in the Mid Pacific Mountains and elsewhere in the western Pacific are known to have Cretaceous shallow-water limestone deposits [*Hamilton*, 1956; *Heezen et al.*, 1973; *Winterer et al.*, 1973; *Heezen and MacGregor*, 1973; *Nemoto and Kroenke*, 1985]. In addition, nearby sources of shallow-water limestone is evidenced by debris in turbidites in DSDP/ODP cores of middle Cretaceous age from the Mid Pacific, Magellan, and Marshall seamounts regions [*Larson et al.*, 1981; *Schlanger and Premoli-Silva*, 1981; *Thiede et al.*, 1981; *Lancelot et al.*, 1990]. Guyots that appear to be strictly volcanic are of such large size that it seems likely that they were also at sea level. On some guyots, evidence for subaerial alteration of the basalt dredged from their summits, and the presence of rounded stream and beach-pebbles in dredges from summit areas confirms this.

The guyot-tops with reef sediments can be distinguished from non-reefed tops at the 1-10 km scale: most of the guyots surveyed during Roundabout 10 which had no evidence for the presence of reef-deposits ("Scripps" Guyot, "Lamont" Guyot, "Pot" Guyot, "Wilde" Guyot, all in the Wake seamounts) have summits with a distincly bevelled edge: a broad gentle slope of 1° - 5° between the flat upper summit and the main (much steeper) upper slope. Such bevels were also observed on non-reefed guyots in the Louisville seamount chain [*Lonsdale*, 1988]. This effect is most likely the result of wave erosion at sea level, without protection from reefs. As erosion slowly works its way inland, the island subsides, so that the interior is eroded to a shallower depth than the outer edge. At some point, subsidence is slow enough, and the horizontal progress fast enough to produce a flat top on the remainder of the guyot. In contrast, the effect of reef-buildup is to form a sharp margin between the outer slope and a completely flat platform.

Another important feature which affects the guyot tops at the intermediate scale is the presence of a pelagic cover. Six of the 21 guyots in our study have deposits of pelagic sediments which are clearly visible in the 3.5 kHz profiles. Where pelagic caps are more than a few tens of meters thick, they are commonly dome-

TABLE 1. Morphologic Measurements of Western Pacific Guyots

Guyot	Name	Latitude	Longitude	A	B	C	D	E	F	G
21	Allison	18° 30'N	179° 30'W	70%	1300 m	1700-1800 m	588	North 88 East	3-5	5500 m
19	"Caprina"	20° 03'N	176° 30'E	30%	1600 m	1700 m	105	none	insuff. data	5800 m
4	"Charlie Johnson"	31° 59'N	148° 15'E	90%	1460 m	1720 m	146	none	5	5900 m
16	Darwin	22° 03'N	171° 38'E	90%	1280 m	1320 m	32	North 35 West	1?	5750 m
17	"Heezen"	21° 11'N	173° 48'E	40%	1240 m	1280-1300 m	187	North 70 West	1?	5500 m
7	Isakov	31° 34'N	151° 10'E	90%	1380 m	1440 m	40	North 40 West	4-6	5900 m
20	"Jacqueline"	19° 21'N	176° 40'E	40%	1540 m	1680 m	217	North 20 West	> 1	4800 m
11	"Lamont"	21° 31'N	159° 30'E	10%	1260 m	1300-1360 m	915	North 65 West	6[(smoot)]	5350 m
8	Makarov	29° 28'N	153° 29'E	40%	1340 m	1360-1340 m	234	East	5-6	5800 m
9	MIT	27° 17'N	151° 50'E	70%	1280 m	1320-1380 m	214	North 55 East	> 2	5900 m
12	"Pot"	19° 40'N	160° 18'E	10%	1260 m	1300 m	874[(Smoot)]	North 56 East	4[(Smoot)]	5300 m
18	Resolution	21° 15'N	174° 20'E	20%	1300 m	1320-1340 m	I/D	none		3950 m
10	"Scripps"	23° 47'N	159° 20'E	10%	1260 m	1280-1300 m	588	North 18 West	3-4	5800 m
5	"Stout"	31° 38'N	148° 59'E	15%	1440 m	1500 m	57	North 5 West	5[(smoot)]	5800 m
1	Takuyo Daini	34° 18'N	143° 52'E	10%	1420 m	1450 m	75	none	5[(smoot)]	5600 m
2	Takuyo Daisan	34° 12'N	144° 18'E	70%	1440 m	1540-1560 m	86	none	4[(smoot)]	5700 m
6	"Thomas Washington"	32° 01'N	149° 15'E	85%	1440 m	1480 m	21	none	4[(smoot)]	5800 m
13	"Vibelius"	21° 13'N	162° 46'E	80%	1840 m	2000 m	206	North 82 West	> 1	5500 m
14	"Wilde"	21° 10'N	163° 18'N	40%	1300 m	1340-1360 m	404[(Smoot)]	none	3[(Smoot)]	5250 m
3	"Winterer"	32° 47'N	148° 19'E	90%	1360 m	1460 m	98	North 45 East	5[(smoot)]	5900 m
15	"Woods Hole"	21° 08'N	166° 29'E	90%	1040 m	1040 m	31	North 32 West	3	5600 m

Table 1. Measurements of morphological elements and other aspects of guyots in this study. Some data obtained from *Smoot* [1989] and *Vogt and Smoot* [1984] included for completeness. Numbers in left-hand column refer to numbers in Figure 1. A: SeaBeam coverage of summit area. B: Depth to shallowest point on summit. C: Depth (-range) of main summit plateau area. D: Surface area inside main break-in-slope. E: Main direction of elongation of the summit. F: Number of volcanic rift zones. G: Regional seafloor depth around guyot.

TABLE 1. (Continued) Morphologic Measurements of Western Pacific Guyots

Guyot	Name	H	I	J	K	L	M	N	O
21	Allison	100.0	21°- 29° (24°)	700 m	1680-2000 m	0.4° SW	2600 m	v,a,p	dredge & seismic
19	"Caprina"	14.0	23°- 28° (25°)	100 m	1680-1720 m	no tilt	2100-2200 m	a,p	seismic
4	"Charlie Johnson"	17.4	21°- 28° (24°)	360 m	1780-1820 m	no tilt	not observed	v,a	dredge
16	Darwin	10.2	28°- 32° (29.5°)	140 m	1320-1400 m	no tilt	1620 m	a	dredge & seismic
17	"Heezen"	23.7	20°- 26° (23°)	100 m	1280-1300 m	0.1° N 90° E	2100-2300 m	a	seismic
7	Isakov	10.6	21°- 25° (24°)	240 m	1560-1620 m	no tilt	not observed	v,a	dredge
20	"Jacqueline"	28.0	22°- 30° (24°)	260 m	1700-1800 m	no tilt	2400-2500 m	a,p	dredge & seismic
11	"Lamont"	100.0	10°- 28° (19°)	280 m	1300-1380 m	insuff. data	not observed	v,p	dredge
8	Makarov	22.0	21°- 30° (26.5°)	140 m	1400-1440 m	no tilt	not observed	v,a	dredge
9	MIT	38.5	21°- 32° (26°)	240 m	1360-1500 m	0.13° N 55°E	1900 m	a	dredge & seismic
12	"Pot"	39.4	23°- 26.5° (25°)	260 m	1400-1520 m	insuff. data	not observed	v	seismic
18	Resolution	25.6	23°- 40° (29°)	120 m	1360-1400 m	0.2 north	2100 m	a	dredge & seismic
10	"Scripps"	58.6	19°- 27° (23°)	260 m	1280-1560 m	0.3° N 18°W	2400 m	v (a?)	dredge, (morph.)
5	"Stout"	11.2	20°- 23° (21°)	240 m	1600-1680 m	no tilt	not observed	v (a?)	dredge, (morph.)
1	Takuyo Daini	10.0	24.2°	100 m	1500-1520 m	insuff. data	insuff. data	v,a	dredge
2	Takuyo Daisan	15.3	22°- 33° (26°)	260 m	1540-1660 m	no tilt	not observed	v,a	dredge
6	"Thomas Washington"	6.1	24°- 31° (27°)	160 m	1480-1560 m	0.8° N 60° E	not observed	v,a	dredge
13	"Vibelius"	21.3	15°- 19° (18°)	260 m	1980-2100 m	0.23° N 80° E	2400-2800 m	a,p	dredge & seismic
14	"Wilde"	25.4	21°- 34° (24.8°)	140 m	1420-1460 m	no tilt	not observed	v	dredge
3	"Winterer"	14.5	20°- 25° (22.4°)	260 m	1600-1660 m	no tilt	not observed	v,a	dredge
15	"Woods Hole"	9.0	18°- 25° (22°)	240 m	1140-1400 m	no tilt ?	not observed	v,a	dredge

Table 1 (continued). H: Maximum summit diameter. I: Range and arithmetic mean of slope angles measured over the top 300 m meter below the main break in slope. J: Relief between highest and lowest observed points of summit. K: Range of depth of main break-in-slope. L: Tilt of summit (angle and azimuth). M: Depth to lower slope-break. N: Observeable rock types which crop-out on summit v=volcanic rocks, a=shallow-water carbonates, p=pelagic sediments. O: Data-types from which summit rock types were determined: dredge = material was recovered in dredges, seismic = layered shallow-water carbonate deposits can be seen in seismic profiles, morph. = reef-like structures were observed in SeaBeam data.

TABLE 1. (Continued) Morphologic Measurements of Western Pacific Guyots

Guyot	Name	P	Q	R	S	T	U	V	W	X
21	Allison	650 m	200 m	40	20 (1600)	100 (1720)	0.37	1740-1800 m	> 1000	145
19	"Caprina"	780 m	100 m	20	20 (1680)	20 (1720)	0	1720	atoll	149
4	"Charlie Johnson"	200 m	0	60-100	100 (1620)	180 (1820)	11.2	1820	260	148
16	Darwin	500 m	0	< 20	40 (1280)	20 (1340)	0	none	atoll	155
17	"Heezen"	900 m	20 m	20	30	100	> 0	none	atoll	153
7	Isakov	< 200 m	0	< 20	20	40	0	1560-1600 m	50	153
20	"Jacqueline"	500-600 m	140 m	20-40	40-60	0	> 0 ?	none	atoll	148
11	"Lamont"	0	0	60	none	0	0	1540 m	60	165
8	Makarov	< 200 m	0	20-40	20	40	0	1420-1440 m	20-40	155
9	MIT	650 m	0	80-100	<>20	200	3.16	1460-1500 m	atoll	160
12	"Pot"	0	0	20-40	none	none	N/A	none	80	168
18	Resolution	900 m	23 m	20	20	80	> 0	none	atoll	152
10	"Scripps"	40 m ??	0	20	20 ?	20 ?	N/A	none	80	163
5	"Stout"	< 200 m	0	60	0 (1500)	80	?	1620-1680	60 ?	152
1	Takuyo Daini	< 200 m	0	20	30 (1420)	70 (1520)	?	insuff. data	100	141
2	Takuyo Daisan	240 m	0	20-40	80 (1460)	160 (1640)	3.44	1600 m	> 200	141
6	"Thomas Washington"	< 200 m	0	20-40	20 (1460)	80 (1560)	0.12	1600 m	60	150
13	"Vibelius"	400 m	100 m	20-40	20 (1980)	80 (2080)	> 0	none	atoll	164
14	"Wilde"	0	0	40	0	0	N/A	0	140	162
3	"Winterer"	> 200 m	0	20-40	0	20	> 0	1580-1600	> 200	147
15	"Woods Hole"	> 0 m	0	20-40	0	0	0	none	240	163

Table 1 (continued). P: Thickness of shallow-water limestone deposits on summits. Q: Thickness of pelagic sediments on summits. R: Roughness of summit, depths and heights of small pits and mounds visible in seabeam data. S: Height of outer rims above main summit plateau (absolute depth of rims in brackets (meters)). T: Depths of karstic depressions relative to main summit plateau. U: Volume of material apparently removed by karstic solution. V: Depth of outer terrace level. W: Relief of the volcanic basement (not known where covered by thick atoll deposits). X: Estimate of age of seafloor on which guyot is located.

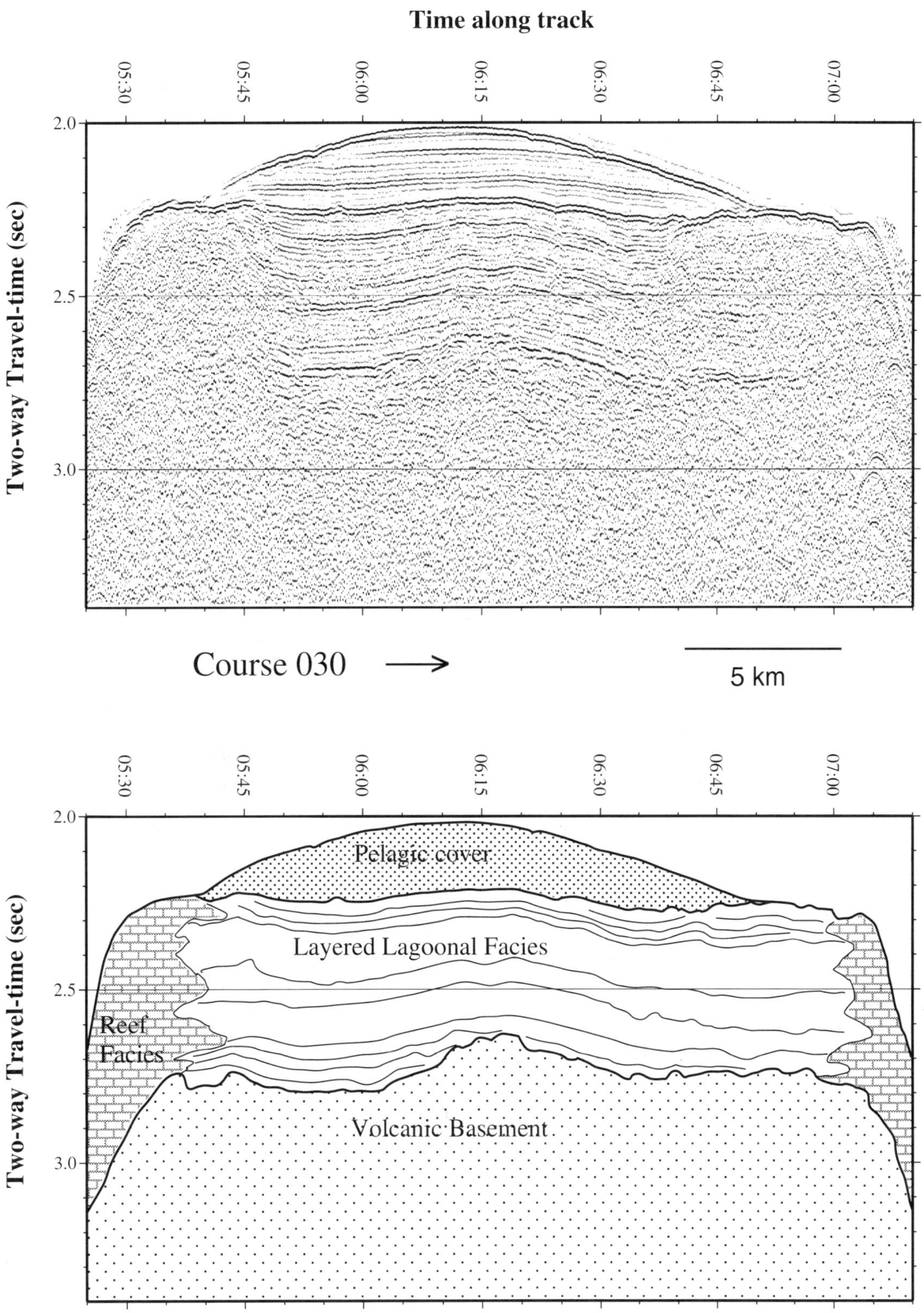

Fig. 2. Seismic reflection profile and line drawing interpretation of Allison Guyot, showing typical atoll structure. The location of the profile is indicate in Figure 18 as line A-A'.

shaped, with internal reflectors that suggest that the deposits have been affected by scouring by currents (Figure 2). Studies on Horizon Guyot [*Lonsdale et al.*, 1972] showed that tidal currents across the guyot-summit sweep the sediments away from the summit margins and onto the center of the summit. Pelagic sediment caps on guyot-tops in the western Pacific are thickest (up to 200 m) in tropical latitudes, become thin (a few tens of meters) around 21° N, and are absent north of 23° N latitude.

Small-scale morphology

Elements of morphology at scales of 20 m to 1 km include the dimensions of small summit structures, holes and scarps, the angle and smoothness of the flanks and upper slopes, of the guyots, and the variability of the slope-break depths along the summit margins. Also included are the topography of the summit plateaus, and the surface of the volcanoes and reef deposits where those were buried by pelagic sediments. Finally, seismic reflection profiles provide images of the internal structures of the sedimented guyot-tops.

SeaBeam, 3.5 kHz and seismic reflection data allow us to look into how deposition and erosion have affected the morphologic development and history of the former islands. There are three main classes of guyot tops: drowned atolls (thick reef deposits, no volcanic rocks on the summit), barrier-reef islands and "almost-atolls" (thin reef deposits, volcanic rocks crop out on the surface), and islands without limestone cover (Table 1).

Atoll structures

Many of the guyots, especially those in the Mid Pacific Mountains, but also MIT Guyot in the Japanese chain and "Vibelius" Guyot in the Wake seamounts, have shallow-water limestone deposits as thick as 700 m on top of their volcanic edifices (Table 1). Seismic reflection records show parallel-layered reflections in the interior of the summits, and acoustically structureless walls at the margins (Figure 2). These are interpreted as layered lagoonal material enclosed by perimeter reefs. These structures represent nearly undeformed Cretaceous atolls.

Atoll summits. The summits of the atolls are relatively flat. The relief relative to the main summit plateau is rarely more than 40 m. Morphologic elements common to all the drowned atolls include a rim 20 to 40 m above the main summit level, and a slightly deeper interior, the "lagoon" (Figure 3). The rim is up to 1 km wide, but is not continuous. Rather, it generally consists of sections up to 3 km long which taper off both in height and width toward their ends, like loops of link sausages. The rim is set 500 to 1000 m inward from the summit edge. On some guyots, e.g. Resolution, the rim structure is not everywhere visible in the SeaBeam data. In these places the 3.5 kHz profiles show that it is buried under a thin blanket of acoustically transparent pelagic sediments (Figure 4). The slope breaks on the summit plateaus lie generally between 40 and 60 m below the tops of the rims.

In the interior of the summit platform, small elevations, generally less than 100 m in diameter, near the limit of resolution of the SeaBeam data, commonly dot the surface. These may be lagoonal patch reefs. On Darwin Guyot (Figure 3), the deepest part of the lagoon is in the center. On other guyots elongate depressions 20 to 40 m deep occur just behind and parallel to the exterior rim.

There does not appear to be a preferred direction of development of reefs along the summit plateaus, as might be expected to occur if the rims were strongly affected by predominant current or wind directions. On many guyots, the rims are better developed on some sides than others but there is no regional pattern that is consistent among guyots. In some places the rim structure is entirely absent along parts of the atoll, and layered lagoonal deposits crop out on the flanks of the atoll (Figure 5).

The segmented raised rims strongly resemble the chains of islets which occur just inside the reef-flats on modern atolls. While development of reef-flats, islets, and inter-islet passages on modern atolls can be influenced markedly by prevailing wind- and current-directions [*Guilcher*, 1988], the lack of consistent directions of development of these features may mean that the Cretaceous atolls were not subject to tradewind-like conditions. It is not at all certain, however, that Cretaceous carbonate platforms were affected by prevailing weather-patterns to the same extend as modern platforms.

Atoll slopes. The upper slopes of the atolls are significantly steeper than those of the purely volcanic guyots. Measured over the top 300 m below the platform-margins, slopes of as much as 40° occur, and the average is about 25° (Table 1). "Vibelius" guyot, a drowned atoll, has unusually gentle upper slope angles of only 18°. At the base of the atoll structure a second slope break commonly occurs. Here, at depths of 500 to 800 m below the summits, the slopes change quite suddenly from about 25° to less than 20°. At the same depth levels, the slopes also become much more irregular: ridges and knobs of presumably volcanic origin occur at this depth and deeper. The depths at which this occurs match well with the those calculated for the volcanic basement reflectors in the seismic profiles, using a velocity of 2.7 km/s. Seismic refraction studies on Anewetok Atoll yielded seismic velocities of 2.5 to 2.7 km/s for the deeper part of the limestone cap [*Grow et al.*, 1986]. In addition, the larger flank rift zones can commonly be seen to terminate against the upper slopes at these depths. These observations suggest that these depths, which are very consistent on any one guyot, represent the base of shallow-water deposition.

Internal structure. The drowned atolls consist of acoustically characterless outer walls surrounding layered interior sediments (Figure 2, 5 and 6). This seismic stratigraphy is generally conformable to the topography of the basement. On Allison guyot (Figure 2), the reflectors appear draped over an arched basement. Occasional small positive bumps within the reflectors may be buried mounds. The fact that the layers are not perfectly horizontal in the profiles is only partly the result of the velocity pull-up associated with the presence of a dome-shaped pelagic cap (Figure 5). On Resolution guyot, where there is no significant pelagic sediment cover, the reflectors appear to sag slightly away from the center, perhaps the result of long-term deformation of the limestone cap over the topography of the volcanic basement (Figure 6). Such deformation might also explain why the larger atoll structures commonly are shallower in the lagoon than at their rims.

The acoustic layering may be the result of sea-level fluctuations during the growth of the atolls. Such fluctuations would produce

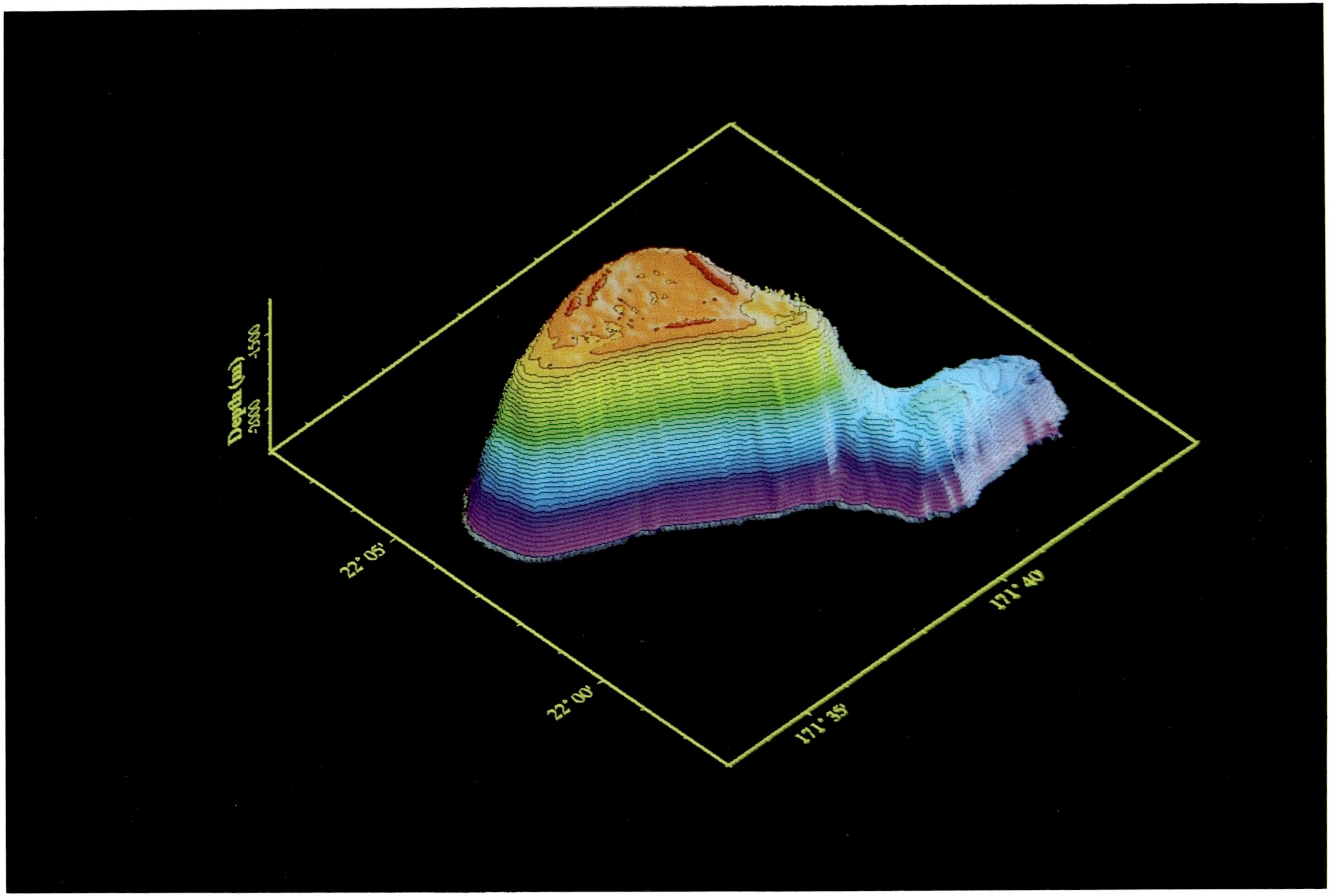

Fig. 3. Shaded relief image of the summit area of Darwin Guyot. Three rim segments at 1300 m depth surround a lagoon which reaches a depth of 1340-1360 m. The contour interval is 25 m. The image was made from regridded seabeam data, with a grid-spacing of 100 m.

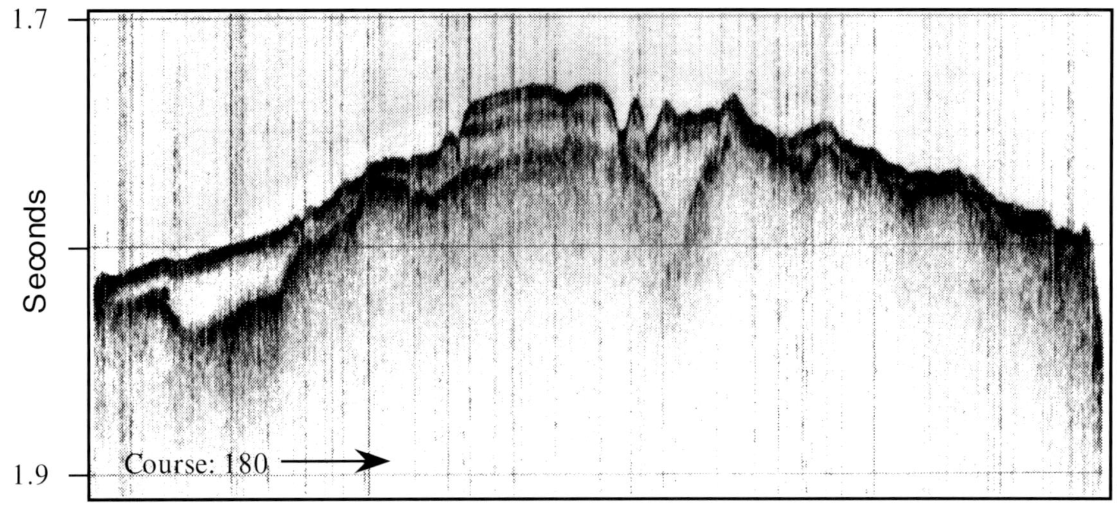

Fig. 4. 3.5 kHz profile of Resolution Guyot. The profile shows a thin blanket of transparent pelagic sediments covering substantial relief in the underlying atoll surface. The section is 31 km long.

highly reflective solution unconformities within the sedimentary sequence similar to those in modern atolls, where they are the result of Tertiary sea level low-stands [*Schlanger,* 1963; *Grow et al.,* 1986; *Lincoln and Schlanger,* 1991]. Alternate possibilities include storm layers, changes from more restricted to more open environments, and selective diagenetic changes allociated with any of the foregoing.

Guyots with thin shallow-water limestone deposits

The Japanese guyots and some of the Wake seamounts appear to have only very thin (less than 200 m) deposits of shallow-water limestone: this material shows up in dredges, but can not be detected in the seismic profiles as in the atolls. The summit configurations do not include bevelled edges, and rims occur along the edges of the summit-plateaus. The volcanic flank rift zones can generally be traced close to and, in a few places, onto the summit-plateaus.

On "Woods Hole" guyot (Figure 7) some shallow-water material was recovered together with volcanic rocks in a dredge haul on the upper slope, but neither seismic profiles nor summit morphology show any sign of sediments. The volcanic radial flank rift zones [*Vogt and Smoot,* 1984] can be traced all the way to the summit, which has the characteristic bevelled edge of a strictly volcanic guyot [*Lonsdale,* 1988]. It is conceivable that "Woods Hole" Guyot at some point had more sediment cover, which was removed by erosion later on, but the "bevelled edge" of the summit plateau is typical of guyots which had no reef-protection when they were being eroded at sea level. The only radiometric age for this guyot indicates it is much younger than the reef-bearing guyots (Table 2) but rudist bivalves of Barremian to early Aptian age [*R. Scott,* pers. comm., 1991] have been dredged from the summit, indicating the presence of the remains of an ancient reef. On "Scripps" Guyot, also in the Wake group, no limestone was dredged, but a partial SeaBeam survey over the north side revealed the presence of a rim-and-lagoon configuration that resembles those of atolls. This flanks a small central high which is probably volcanic (Figure 8). Although reflection profiles show no clear evidence of sediment cover, this configuration is very much like a barrier-reef surrounding a volcano. It is possible that at least an initial reef system had developed on "Scripps" Guyot, and that dredging on the east margin of the north side would recover shallow-water limestone.

The Japanese guyots represent a more advanced stage of shallow-water deposition. All dredge attempts within a 200 m depth range below the main slope break have recovered copious quantities of rudist, corpal and echiniod debris [*Heezen et al.,* 1973], demonstrating that shallow-water limestone deposits exist on the guyot top. They are either not thick enough or do not contrast in acoustic impedance with the volcanic rocks sufficiently to be visible in the seismic reflection data. "Thomas Washington" Guyot (Figure 9), a very small guyot with a summit plateau area of only 14 km^2, has a low rim around a summit plateau with a central high, presumably an eroded volcano. The depth of the depression between the outer rim and the central high is 20 to 30 m, while the central high, a round edifice about 2 km in diameter, is at least 20 m taller than the top of the shallowest rim. The summit morphology is complicated by the fact that the whole guyot appears to be tilted to the east and by the occurrence of holes 60 m deep in the southwest and northeast margins. The shape of the summit fits that of a barrier-reef surrounding a central volcano. Reef development appears to be strongest on the west side, whereas the east side may have a 500 m-wide terrace or lower platform 60 m below the main summit level, separated by a steep scarp from the rest of the summit.

The barrier-reef configuration also characterizes some of the other Japanese guyots, namely "Charlie Johnson", Takuyo Daisan and Takuyo Daini (Figure 10, 11 and 12). The summit areas of "Charlie Johnson" and Takuyo Daisan guyots were completely surveyed with SeaBeam. The summit of "Charlie Johnson" Guyot (Figure 10) consists of a main summit plateau at 1720 m with a

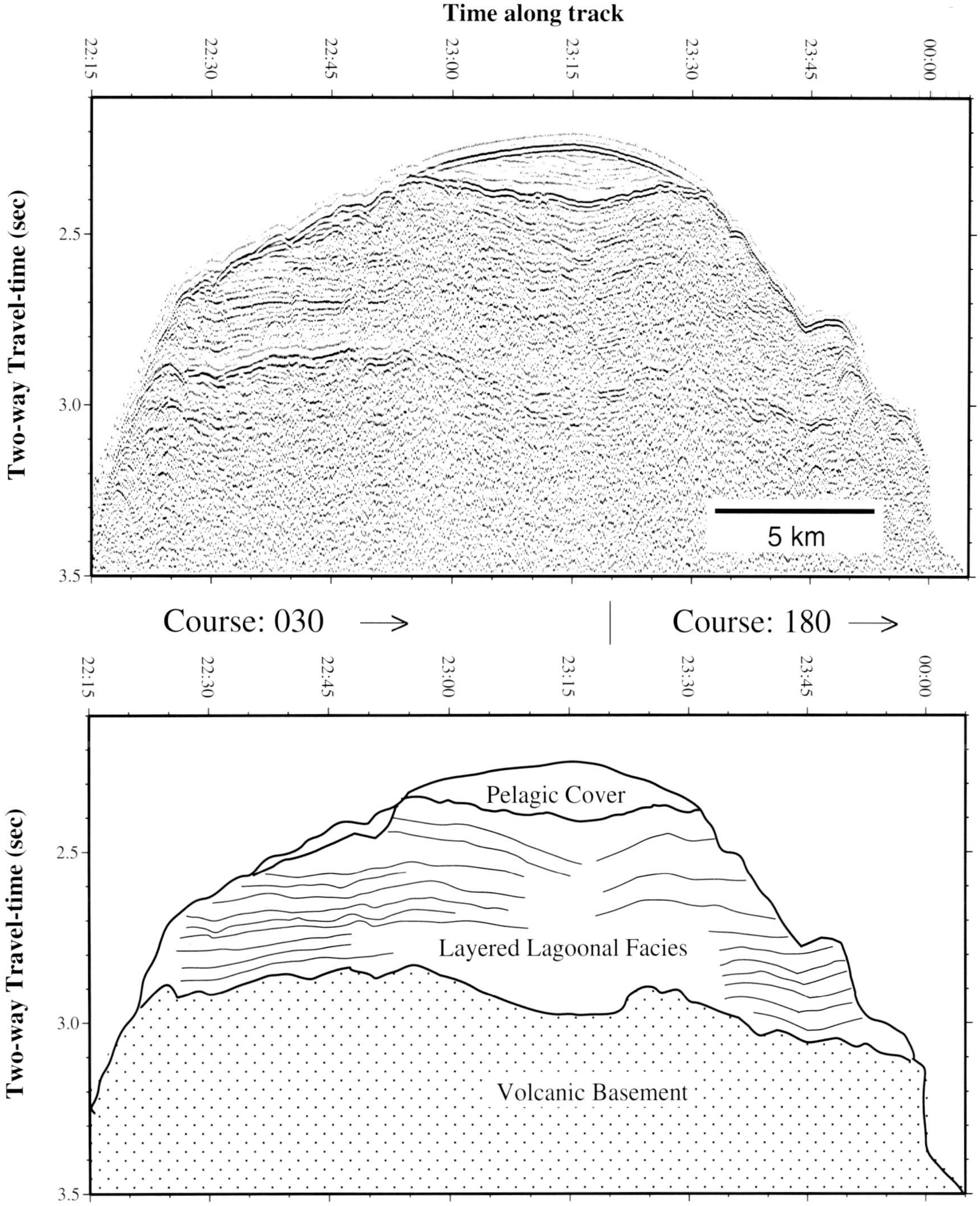

Fig. 5. Seismic reflection profile and line drawing interpretation of Allison Guyot (Line B-B' in Figure 18), showing reflectors at the south-east side of the atoll structure, where no reef wall occurs. This side is believed to have been affected by a large collapse-event, which removed most of the reef-wall along that side of the guyot, exposing the layered lagoonal sediments.

central mound 140 m high, another peak of 240 m on the south end, and a discontinuous but very prominent rim as much as 100 m high around the edge. This rim consists of sections 2 to 4 km long which vary in width between 500 and 1500 m. Between the sections are narrow gaps or passages less than 1 km wide, the bottoms of which reach 40 to 60 m below the main summit level.

Takuyo Daisan Guyot (Figure 11) has very similar features: its summit has a central volcano, which stands 110 m above the main

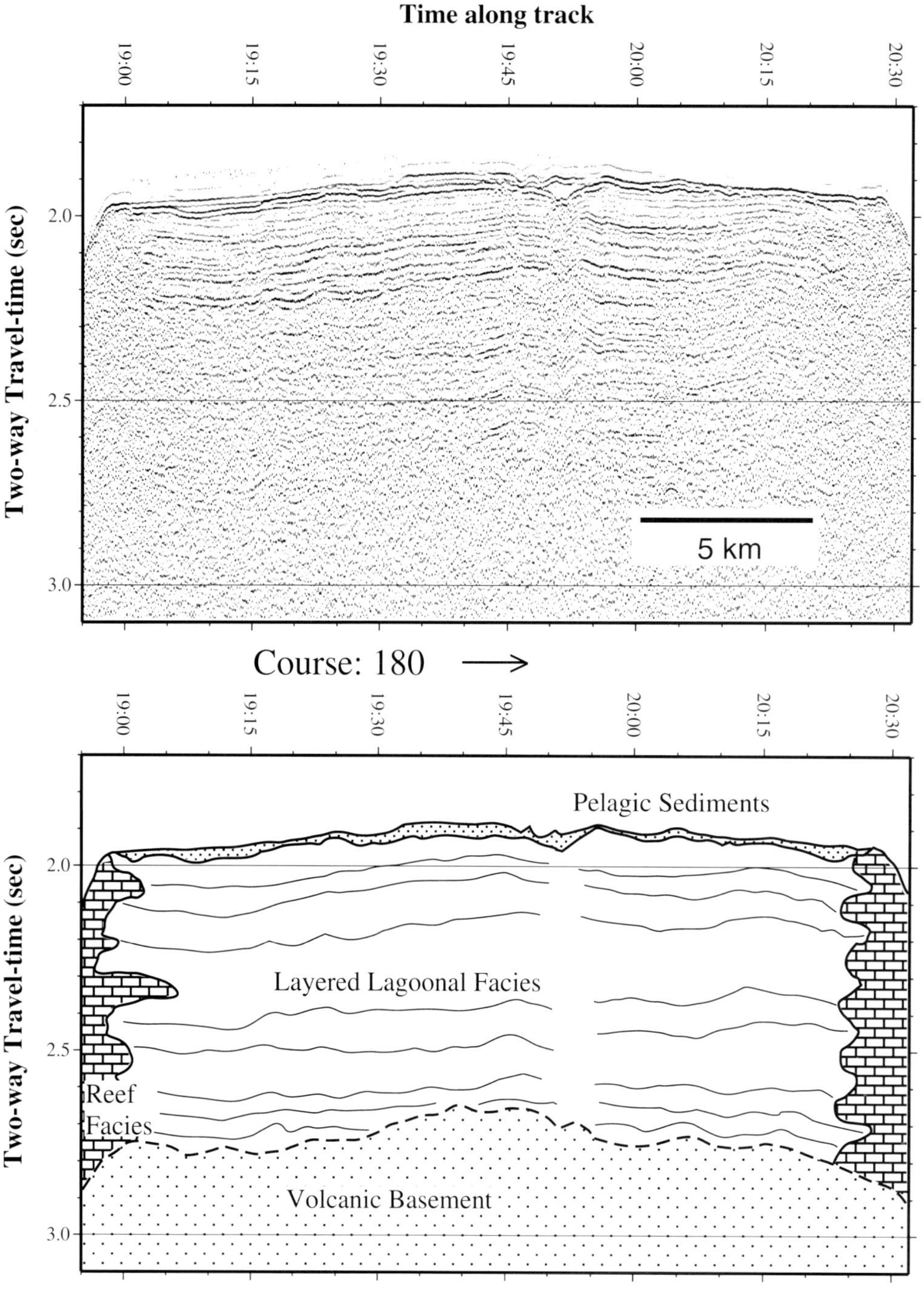

Fig. 6. Seismic reflection profile and line drawing interpretation of Resolution Guyot. There appears to be as much as 0.75 seconds of sediments above the volcanic basement. Prominent reflectors are highlighted in the interpretation, and may correspond to similar reflectors in the profiles of Allison Guyot (Figure 2).

summit level at 1540 m, and a tall rim along the outer edge of the summit platform. This consists of discontinuous segments 600 to 1000 m wide and up to 110 m high, which are best developed along the northeast and south sides. The average height of these rim sections is about 70 m above the main summit level. The gaps between rim sections go down to 40 m below the main summit level and connect to small outer terraces at a depth of 1600 m.

Both "Charlie Johnson" and Takuyo Daisan have fairly steep (24°-26°) upper slopes, which was also noted by *Vogt and Smoot* [1984]. Volcanic flank rift zones can be seen to extend all the

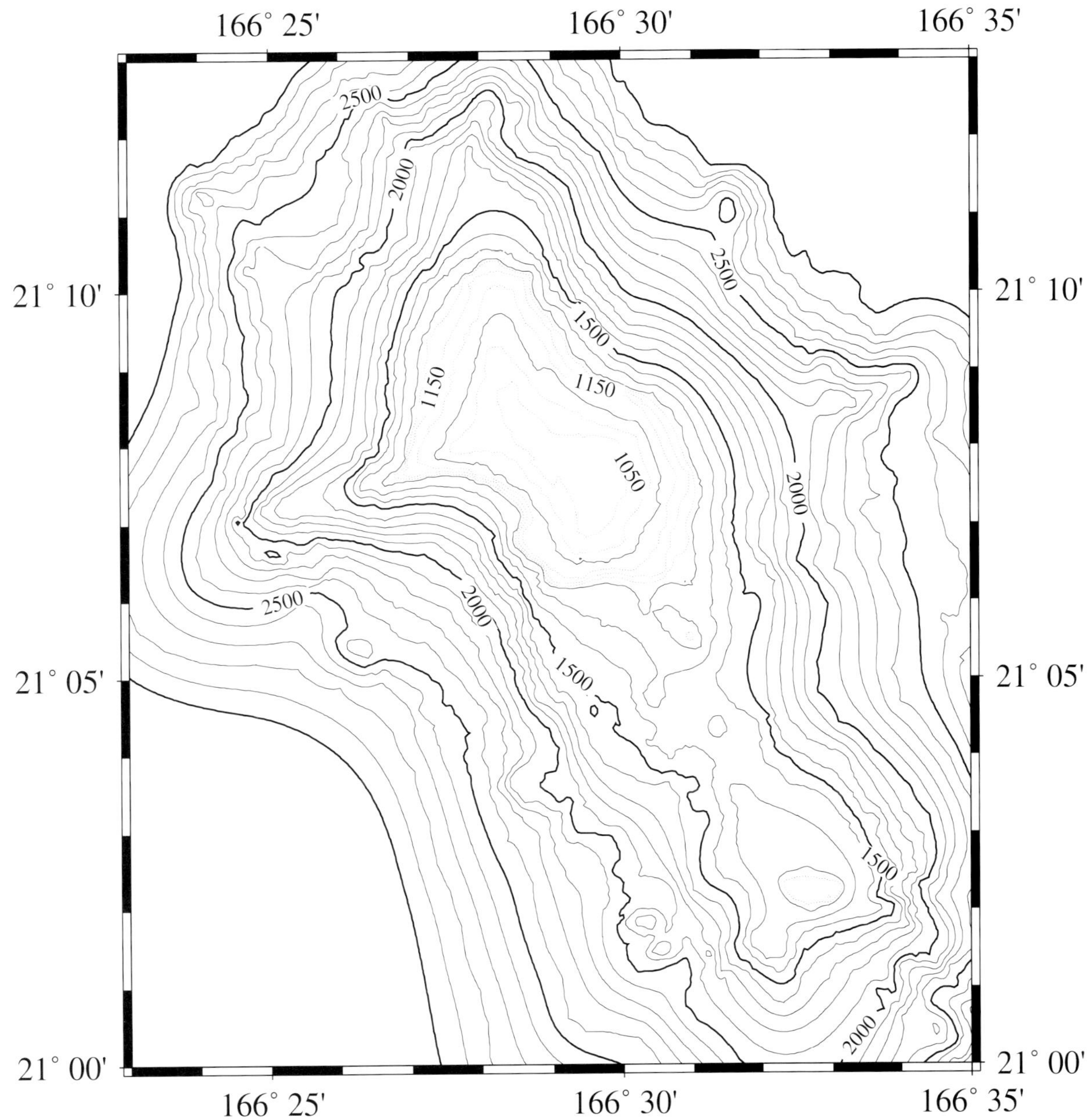

Fig. 7. Bathymetry of the summit of "Woods Hole" Guyot. The main break in slope lies at a depth of 1200 m for most of the summit. Above this depth, a gentle slope leads to the summit plateau at 1050 m. Volcanic rift zones on the west and north sides can be traced onto the summit plateau. A long ridge extends to the southeast, with several more summits, none of which reach shallower than 1200 m. The contour interval is 100 m, while on the summit dotted contours are drawn at an interval of 25 m.

way up to the base of the reef rims, and on "Charlie Johnson" perhaps even on the summit, so the upper slopes are almost certainly volcanic, with only a very thin cover of reef debris. Takuyo Daini Guyot appears to have a very similar summit morphology: our single crossing with the SeaBeam shows tall rims and a central high, a profile similar to those of "Charlie Johnson" and Takuyo Daisan (Figure 12). "Stout" Guyot may have a more complex morphology, with a rugged topography of canyons and rims on the west side, possibly forming a tall rim structure, and an upper platform 100 m above the main summit level on the east side. The limited survey does not provide enough information to say more about this particular guyot.

An entirely different style of summit is displayed by the other Japanese guyots in our surveys: "Winterer", Isakov and Makarov

TABLE 2: Guyot Age Data

Guyot name	Volcanic ages	Reef ages	Pelagic ages
Allison	101.2 Ar/Ar[1] 85.6 Ar/Ar[1] 83.4 Ar/Ar[1]	Barremian-Aptian[5] late Albian[9]	Upper Cretaceous[6]
"Charlie Johnson"	108.3 Ar/Ar[2]	Albian-Cenomanian[4]	
Darwin	118-158 Magnetic[3]	Barremian-Aptian[5]	Minimum age: 109 Ma[3]
Isakov	103.7 Ar/Ar[1]	Albian-Cenomanian[4] Late Cretaceous[5] Early Cretaceous[8]	
"Jacqueline"	117.1 Ar/Ar[1]	Early Cretaceous[8]	
"Lamont"	96.2 Ar/Ar[1] 81.5 Ar/Ar[1] 97.0 Ar/Ar[1] 93 Ar/Ar[2]		
Makarov	93.9 Ar/Ar[2]		Cenomanian-Turonian[4]
MIT	120.3 Ar/Ar[1] 118.0 Ar/Ar[1]	late Albian[9]	
Resolution		Barremian-Aptian[5] reworked Barremian[7]	
"Scripps"	110.8 Ar/Ar[1] 109.5 Ar/Ar[1] 98.0 Ar/Ar[2]		
Takuyo Daini	102.0 Ar/Ar[2]	Albian[8]	
Takuyo Daisan	101.8 Ar/Ar[2]	middle. Albian - Senonian[9]	
"Thomas Washington"		Late Cretaceous[5]	
"Vibelius"	139 Magnetic[3]		
"Wilde"	90.6 Ar/Ar[1] 86.4 Ar/Ar[2]		
"Winterer"		middle Albian - Senonian[10]	Cenomanian[9]
"Woods Hole"	68.3 Ar/Ar[1]	Barremian-Aptian[5] late Albian - early Cenomanian[9]	

Table 2. Available information for the ages of volcanic edifices, reef-limestones, and pelagic sediments of guyots in this study. Sources of data: 1: *R.Duncan*, [pers.comm, 1990]. 2: *Ozima et al.*, [1977]. 3: *Sager*, [1990]. 4: *Matthews et al.*,[1974]. 5: *R.Scott*, [pers.comm., 1991]. 6: *T.Bralower*, [pers.comm., 1990]. 7: *Thiede et al.*, [1981]. 8: *Heezen et al.*, [1973]. 9: *Grotsch et al.* [unpub. ms., 1990]. 10: *A.Coogan*, [pers.comm., 1991].

(Figures 13, 14 and 15). These guyots cannot have more than a thin cap of shallow-water limestone, since volcanic debris shows up in dredges 200 m below the summit, along with reefal limestone [*Heezen et al.*, 1973], and no sediments can be seen in the seismic profiles. Their summits consist of flat tops with small peaks ("Winterer" and Isakov) in the center. On "Winterer" Guyot, these peaks are conical shapes, up to 80 m high above the summit plateau, 2 km wide at the base, 500 m at the top. The summit plateaus are flat, or very gently sloped, with no indication of any rims. On Makarov guyot, the shallowest part of the summit is in the center, where there is a gentle dome, about 5 km in diameter and 40 m high. Small elevated patches about 20 m tall occur on the west half of the summit of Makarov Guyot. These patches are 1.5 km in diameter, and may represent reef mounds of some kind. Lack of 3.5 kHz data over these guyots makes it difficult to determine if any pelagic sediments are present, but none are visible in the seismic reflection profiles.

The most significant aspect of these three guyots is the lower

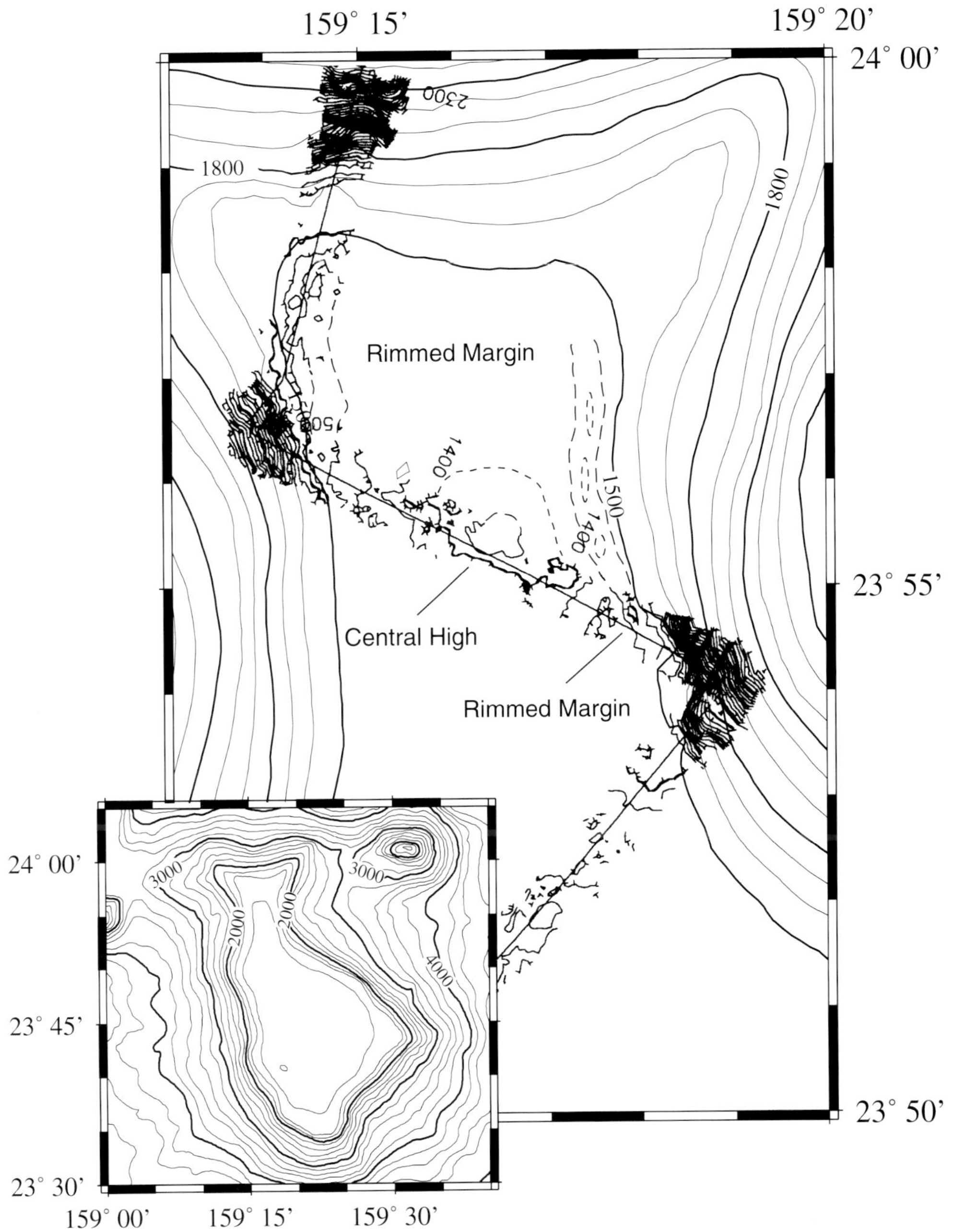

Fig. 8. Bathymetry of "Scripps" Guyot. Regional map redrawn from *Smoot* [1989], with the track of the Roundabout expedition, leg 10 drawn in. The enlarged figure shows the bathymetry of the northern neck of the guyot, with a SeaBeam swath contoured at 20 m. The SeaBeam data indicate the presence of a small rim near the outer slope break of the summit plateau, indicating the possible presence of drowned reefs on this part of the guyot.

terrace or platform that surrounds their summits. These are 500 to 1000 m wide semi-continuous terraces bounded on the inside by 80 to 140 m high scarps or steep slopes, and on the outside by sharp slope breaks to the upper flanks. In many places along the outer margins of the plateaus are short, narrow ridges, 20 to 40 m high, which may be reefal in origin. Just inward from these there commonly are holes in the terraces 20 to 40 m deep and round or slightly elongate parallel to the margin (Figures 13 and 14). There are several ways to explain this morphology: perhaps the main summit plateau is entirely volcanic, and a reef system developed along a base level 100 m below the summit, or the summit is a near-atoll, with the main reefs eroded away or too low to be

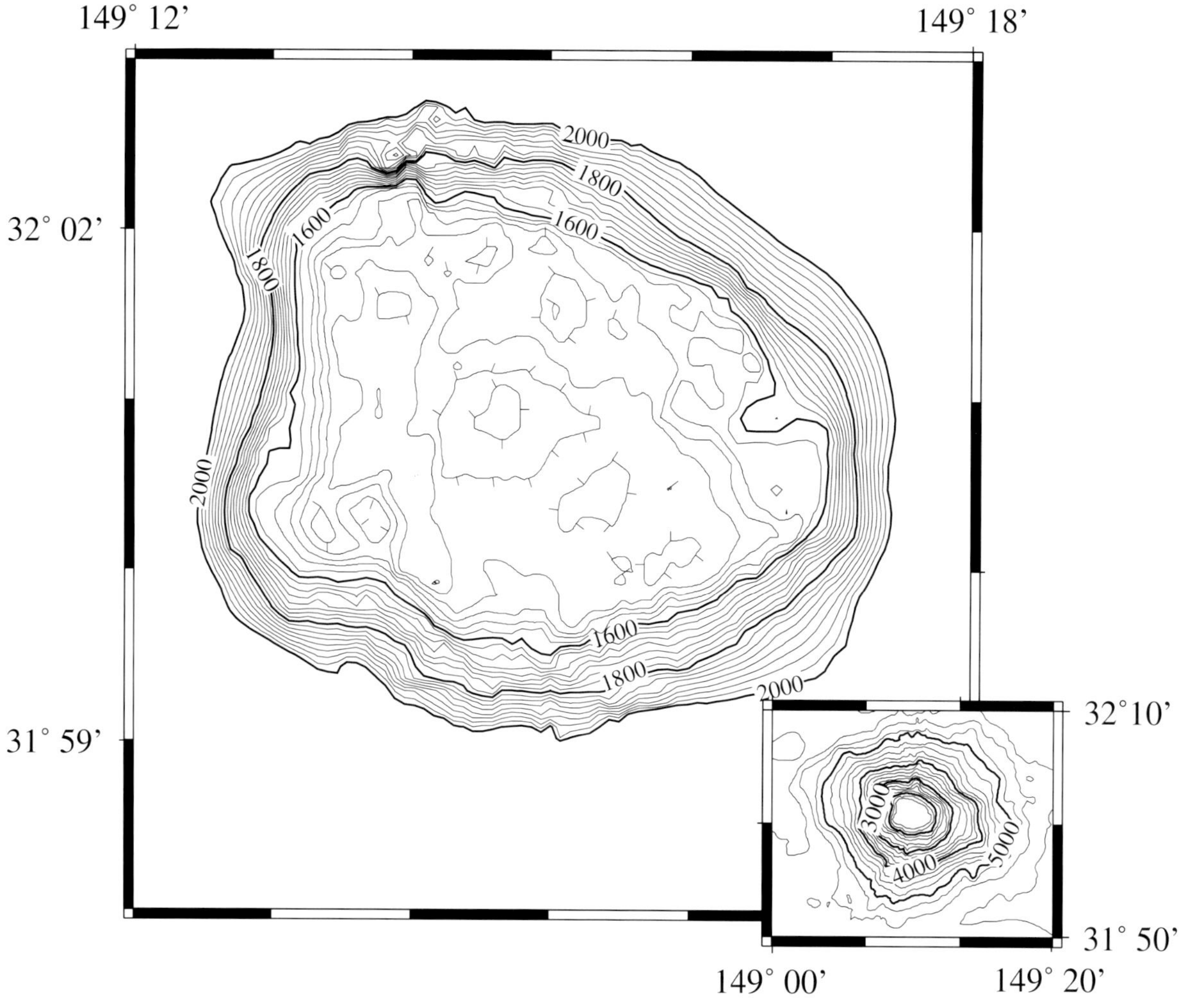

Fig. 9. Bathymetry of "Thomas Washington" Guyot. Regional map redrawn from *Vogt and Smoot* [1984]. Detailed bathymetry of summit from Roundabout, leg 10 SeaBeam survey, with 25 m contour interval. The map shows irregular rims surrounding a central high. The summit slopes to the north-east, which may due to tilting of the edifice, or to erosion during exposure of the guyot summit, which also produced depressions up to 80 m deep in the southwest and northeast sides of the summit (shaded).

resolved by SeaBeam, whereas the lower plateau represents reef growth along a wave-cut terrace during a relative lowstand of sea level. It is possible that there are pelagic sediments on the main summits, obscuring more significant relief. The noted absence of significant pelagic cover on guyot-tops at these latitudes makes this unlikely. Makarov Guyot appears to have the least reef cover. Although it has a lower terrace level, the east side of the summit may be completely bare: shallow, gently dipping reflectors in the seismic data may be interpreted as volcanic flows on the northeast side, indicating complete absence of sediments there. Similar dipping reflectors occur on "Wilde" Guyot in the Wake seamounts, which lacks any sediment cover. The lower terraces appear to be best developed on the shoulders of the volcanic rift zones, which can be traced almost to the terrace level, again indicating that the carbonate cover is at most very thin.

Strictly volcanic guyots

No evidence for the presence of shallow-water limestone deposits has yet been found for several of the guyots in the Wake group: "Lamont", Miami, "Scripps", and "Wilde" guyots have been dredged during our and previous expeditions, and only volcanic material was recovered [*Heezen et al.*, 1973]. Seismic reflection data show no signs of a sedimented summit, and other morphological details indicate these summits are different from the ones with sediments. In addition to the bevelled summits visible in the medium-scale morphology, the summit surfaces appear to

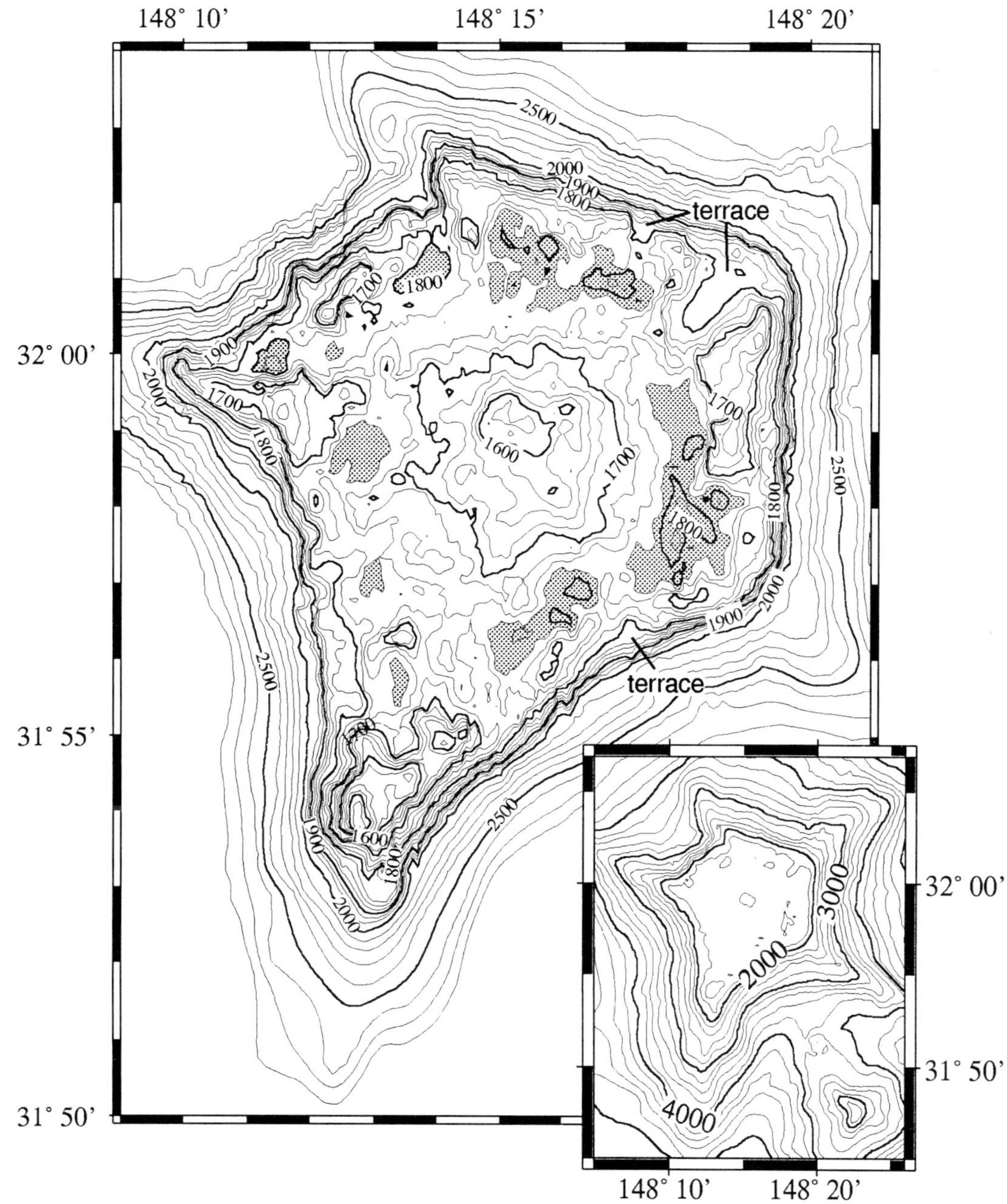

Fig. 10. Bathymetry of "Charlie Johnson" Guyot. Regional map redrawn from *Vogt and Smoot* [1984]. Detailed bathymetry from Roundabout, leg 10 SeaBeam survey. The summit topography includes closed depressions in the interior of the summit platform (shaded), as well as tall rim-segments along the outer edge. Irregular terraces on the flanks of the central high may represent initial fringing reef growth, or may be products of erosion. The shallowest part of the guyot is a steep pinnacle on the southernmost tip, on a volcanic rift zone. Note small terraces on the outer slopes, which occur at the same depths as the bottom of the closed summit depressions, usually below gaps between rim-segments. The contour interval is 25 m above 2000 m, 100 m below 2000 m.

be much rougher than the sedimented guyot-tops. Sharp-crested relief of 40 to 60 m can be seen in 3.5 kHz profiles. Resistant structures such as the summit trace of the flank rift zone on "Woods Hole" Guyot stand out.

EROSION AND WEATHERING OF SEDIMENTED GUYOTS

Since carbonate platforms maintain themselves close to sea level, they are exposed every time sea level is lowered relative to the platform. These periods of emergence result in weathering and the

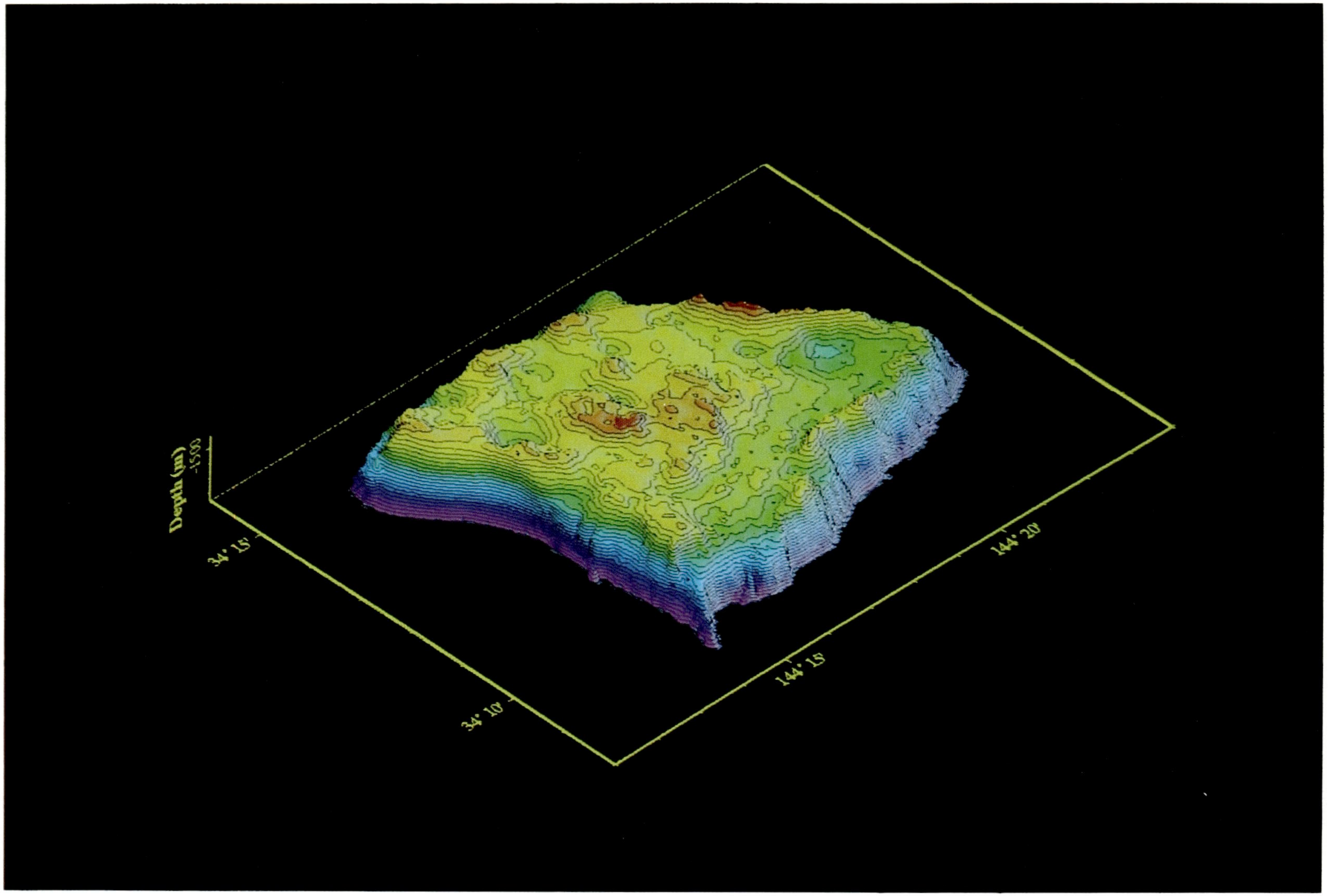

Fig. 11. Shaded relief image of summit region of Takuyo Daisan Guyot. Well-developed rims surround the summit plateau. The irregular-shaped central high is probably only partially volcanic. Seismic reflection profiles show a thin (~100 m) sediment cover between the northeast rim and the south side of the central high. The image was made from regridded seabeam data, with a grid-spacing of 100 m, and a contour interval of 25 m.
The shallowest contour is (red) is 1430 m.

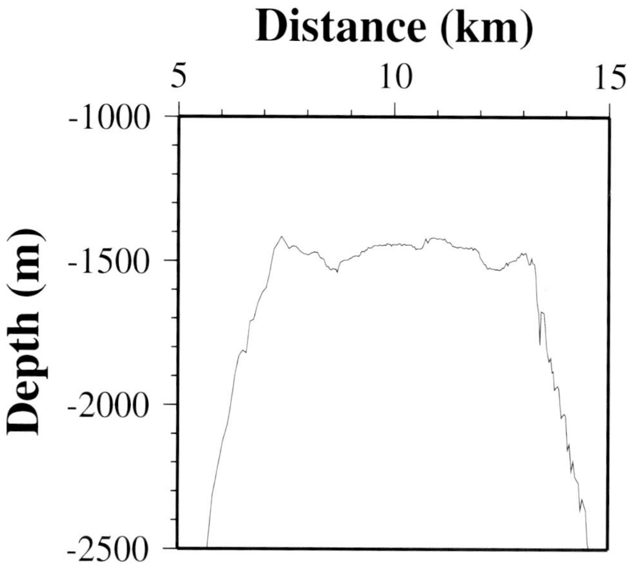

Fig. 12. Profile of Takuyo Daini Guyot, showing the exaggerated rim-lagoon-central high morphology of some of the Japanese guyots. Vertical exaggeration: ~10X The lagoon is at least 100 m deeper than the southwest rim.

formation of ersosional relief on the platform. This relief takes the form of karst topography, stream channels and wave-cut terraces.

The surface of MIT Guyot (Figure 15) is the best example of karst topography in our survey. The summit plateau is pocked by deep depressions, the deepest of which is a circular hole 500 m in diameter which reaches a depth in excess of 1520 m, or more than 200 m below the summit level. In the center of the summit, a series of holes divides the surface into an interconnected but closed drainage pattern. As shown in the seismic profiles across this surface (Figure 16), layering of the sediments is cut by the summit relief, indicating that the relief is the result of erosion. The floors of these depressions, which lie at depths of 80 to 180 m below the summit, do not have any resolvable surface connection to the outer slope. This indicates they were formed through dissolution, using a subsurface drainage system. On other guyots of the drowned-atoll type this pattern can also be found, if not as spectacularly: Allison Guyot (Figure 17) has a deep channel in the surface just inside the eastern rim on the north side, which cuts 100 m into the surface. The channel is pocked by numerous deeper sinkhole-like depressions. Although it has a connection to the outside, the channel is clearly erosional in origin. A similar channel, up to 80 m deep, occurs on the surface of "Vibelius" Guyot, in the Wake group, with a silled connection to the outer slope so that at least part of the channel is a closed depression. On other drowned atolls small holes 20 to 40 m deep and 500 m in diameter can be seen on the summit plateaus. These may also be formed by exposure and dissolution, but they might be part of the normal relief of the carbonate platforms. It is the large scale of the relief on guyots such as MIT and Allison that suggests an erosional origin. In some places, the surface topography of the Cretaceous carbonate platforms is covered by pelagic sediments. In the 3.5 kHz profiles holes can be seen in that surface, beneath the pelagic sediments (e.g. on Resolution (Figure 4) and "Vibelius" guyots).

On the guyots with thin carbonate deposits, karst topography can also be seen: on "Charlie Johnson" and Takuyo Daisan guyots, the deepest parts of the lagoons are closed depressions 40 to 60 m deep, significantly deeper than the deepest part of the passages between the rim sections (Figure 10, gray shaded depressions). These rim sections, doubtless of reefal origin, are 80 to 100 m high. Under normal platform building conditions, the lagoons behind the rims would have filled with sediments, and such a free standing reef wall would not be maintained, which leads us to conclude that the reef rims have been exaggerated by erosion of the lagoonal sediments.

Drowned stream channels can be seen on the summit plateaus and upper slope regions of several of the guyots. On "Winterer" Guyot, a distinct "V" in the contours can be traced from the summit plateau, at a depth of 1480 m, down the upper scarp at the northwest corner to the lower terrace level, and off the main break in slope to a depth of 1660 m (Figure 13). Another such channel cuts into the southwest corner, and runs to the northeast, but can not be traced on the lower terrace. Likewise on Isakov Guyot (Figure 14) small channels can be discerned cutting into the scarp which separated the summit plateau from the lower terrace. None of these extends below the terrace level. On "Charlie Johnson" and Takuyo Daisan Guyots (Figure 10 and 11), stream channels cut through the gaps between the rim sections, and connect the summit plateau to the outside. In some places they can be traced to an origin on the flanks of the central volcano. The gaps between the rim sections appear to be the equivalents of "avas", the deep (more than 50 m) passages from lagoon to ocean that occur on many modern atolls [*Guilcher*, 1988]. Generally, on modern atolls there is only one ava, which occurs on the side of the atoll that is protected from the prevailing winds [*Guilcher*, 1988]. These passages were formed during Pleistocene sea level lowstands, and relict stream valleys in the lagoon floor can be traced through them. More than one channel was cut, but all but the one were filled again when sedimentation resumes during episodes of higher sea level [*Guilcher*, 1988]. On the guyot-tops, the channels which pass through the rim gaps terminate on small terraces on the upper slope. The depth of these terraces is the same as that of the bottom of the depressions in the summit plateau, indicating that this level represents a new base level during the time of relative lowstand of sea level.

Further signs of erosion during a relative lowstand can be found in the occurrence of terraces on the upper slopes of many guyots, generally at depths coinciding with the maximum depth of dissolution-erosion on the summits. The most striking examples of terraces are found on "Winterer" and Isakov guyots. These guyots have platforms around their summits, 100 to 120 m below the summit level. More subtle terraces can be seen on other guyots: on the upper slopes of Takuyo Daisan and "Charlie Johnson" are small terraces, less than 500 m wide, which are slightly wider and better developed below the gaps between the rim sections, where the stream channels end. They occur at the same depth as the

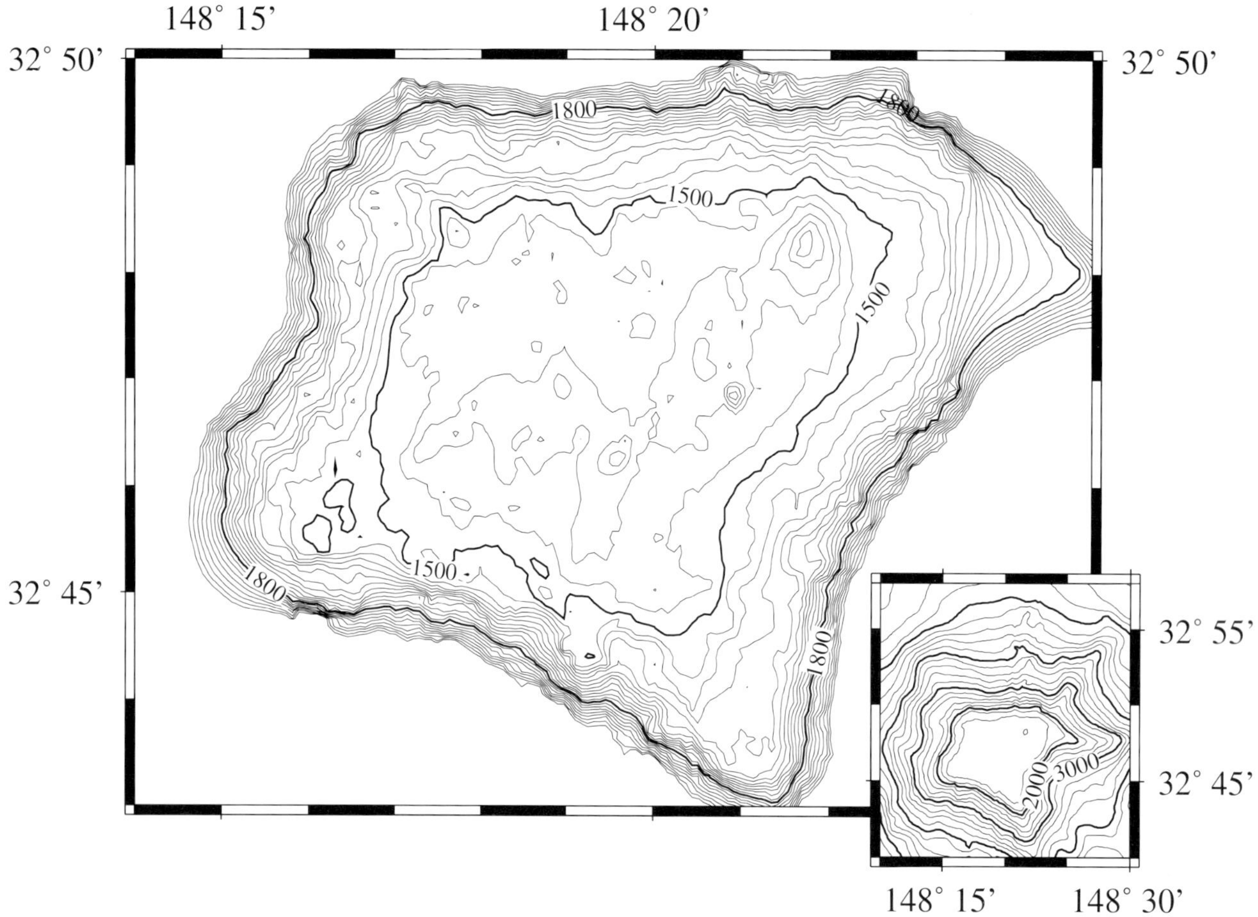

Fig. 13. Bathymetry of "Winterer" Guyot. Regional map redrawn from *Vogt and Smoot* [1984]. Detailed summit bathymetry from a partial SeaBeam survey. The main summit level lies at ~1460 m. A series of small peaks extending up to 1375 m form a small chain over the summit, and may be related to the orientation of the volcanic rift zones. At a depth of 1600-1675 m, a 0.5-to-1-km-wide terrace occurs on the upper slope, which is particularly well-developed along the west, northwest and north sides. The outer edges of this terrace have 20 m tall, narrow raised rims (not resolved on figure). The contour interval is 25 m on the summit map, 200 m on the regional map.

deepest parts of the summit. On the drowned atolls, small terraces can be seen below the break in slope on "Caprina", MIT and Allison guyots. On Allison Guyot a major platform or terrace level occurs at a depth of 1740 to 1800 m, 100 m below the main summit plateau-depth, in the bend of the summit margin on the north side (Figure 17). On the other guyots, the widest terraces tend to occur on extremities of the platforms, such as on the shoulders of the flank rift zones, or at the ends of ridges. There does not appear to be any systematic orientation to these platforms which might be interpreted in terms of predominant wind directions. The biggest terrace visible on MIT Guyot occurs on the northern margin of the southwest end, at a depth of 1460 to 1500 m, close to the depth of the deepest summit depression. Smaller terraces 300 to 500 m wide and up to 700 m long occur sporadically along the southern margin. On the strictly volcanic guyots, no lower terraces were observed.

The observation of karst topography, stream channels and terraces on the summit-plateaus of the reef-bearing guyots shows they were affected by subaerial weathering and erosion well after carbonate-platforms had become estabished. The amplitude of karst-topography and the location of the terraces indicates the height of emergence reached as much as 200 m. The distribution of guyots with erosional topography suggests emergence of islands over a region more than 5000 km long, from the Mid Pacific Mountains to Japan. The presence on the karsted surfaces of pelagic sediments not much younger than the reefs (Table 2) demonstrates that the relief the result of processes taking place before the deposition of the pelagic cap, rather than by later marine dissolution in deep water. A significant relative sea level lowstand appears to have occurred over a wide area prior to final drowning of the carbonate platforms on the guyots, which resulted in the development of this relief.

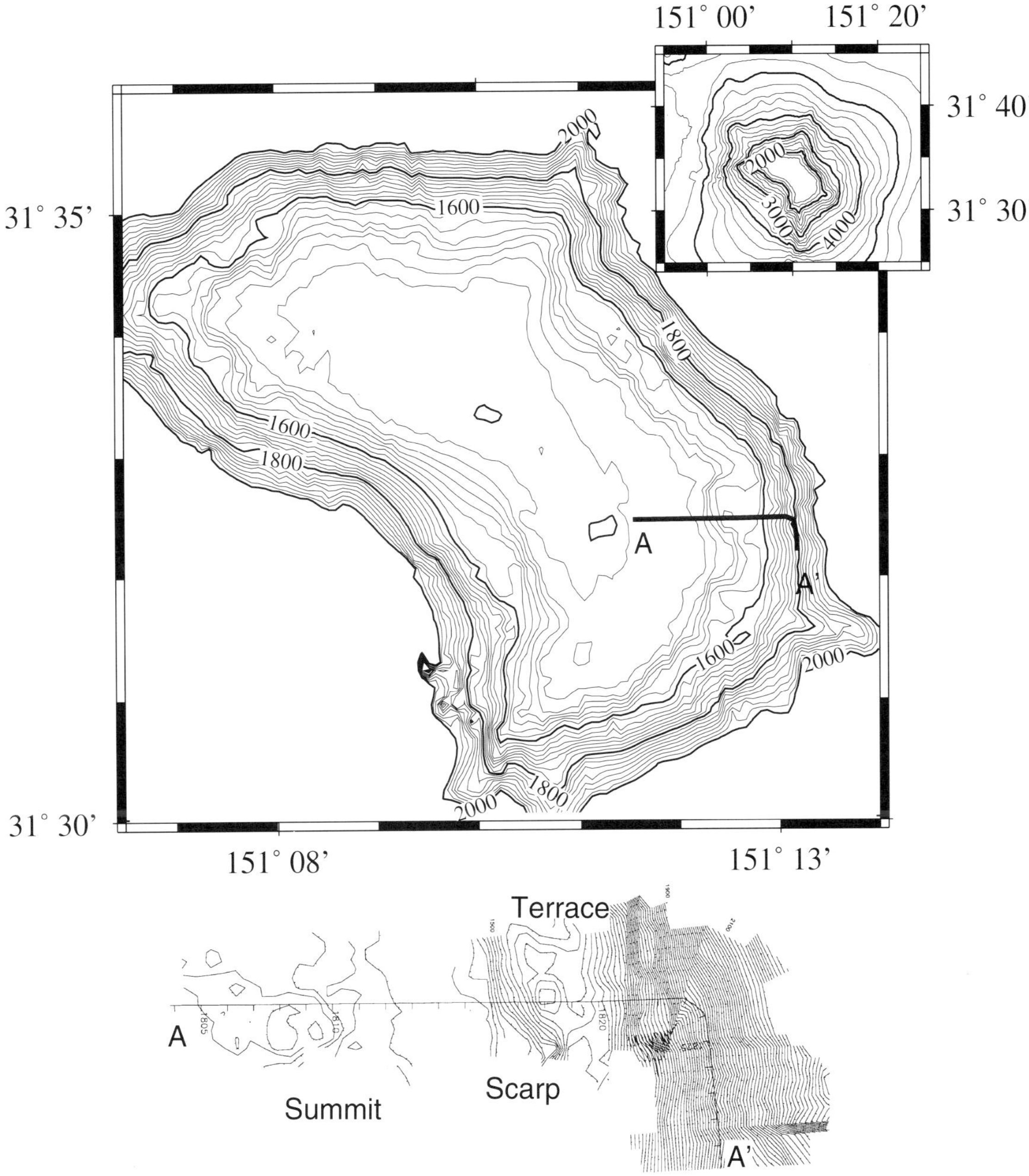

Fig. 14. Bathymetry of Isakov Guyot. Regional map redrawn from *Vogt and Smoot* [1984]. Detailed summit bathymetry from a partial SeaBeam survey. The summit plateau lies at ~1440 m, and has a few small pinnacles which reach to 1380 m. Along the upper slope, at a depth of ~1550 m, a 0.5-to-1-km-wide terrace occurs. An enlarged single SeaBeam swath shows the summit level, upper scarp, and terrace (contour interval: 10 m). The terrace has a 20 m tall outer rim, just inward from which lies a small depression. The contour interval is 25 m on the summit map, 200 m on the regional map.

STRUCTURAL AND TECTONIC DEFORMATION

If the flat tops of the guyots are the result of processes occurring near sea level, then the main slope breaks should occur at the same depths all around any given guyot top, but this is not always the case. The depth of the main slope break varies by as much as 320 m on any one guyot. This slope-break depth variation occurs in several ways: there can be a consistent deepening of the summit depth toward one side of a guyot (tilting), or the summit margins depth can vary locally. This local variation can be gradual, giving a "sagged" or "drooped" appearance to a section of the margin, or

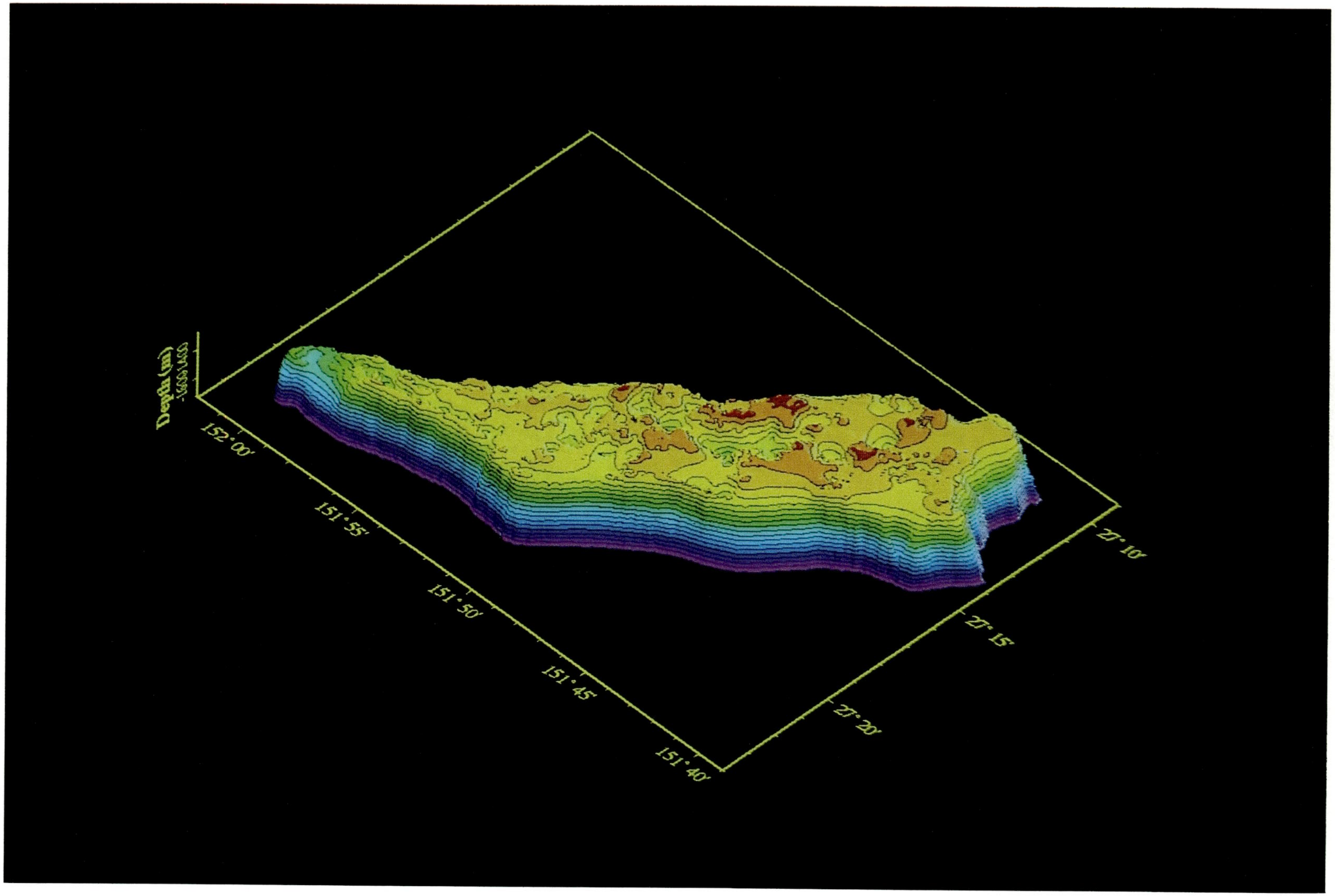

Fig. 15. Shaded relief image of MIT guyot, from regridded SeaBeam data. Most of the summit lies at a depth of ~1280 m (red color). Large holes, some of which are interconnected, but with deep closed depressions, mark the surface. The narrow northeast end deepens to the northeast, possibly a result of compaction of the sediments. The contour interval is 25 m.

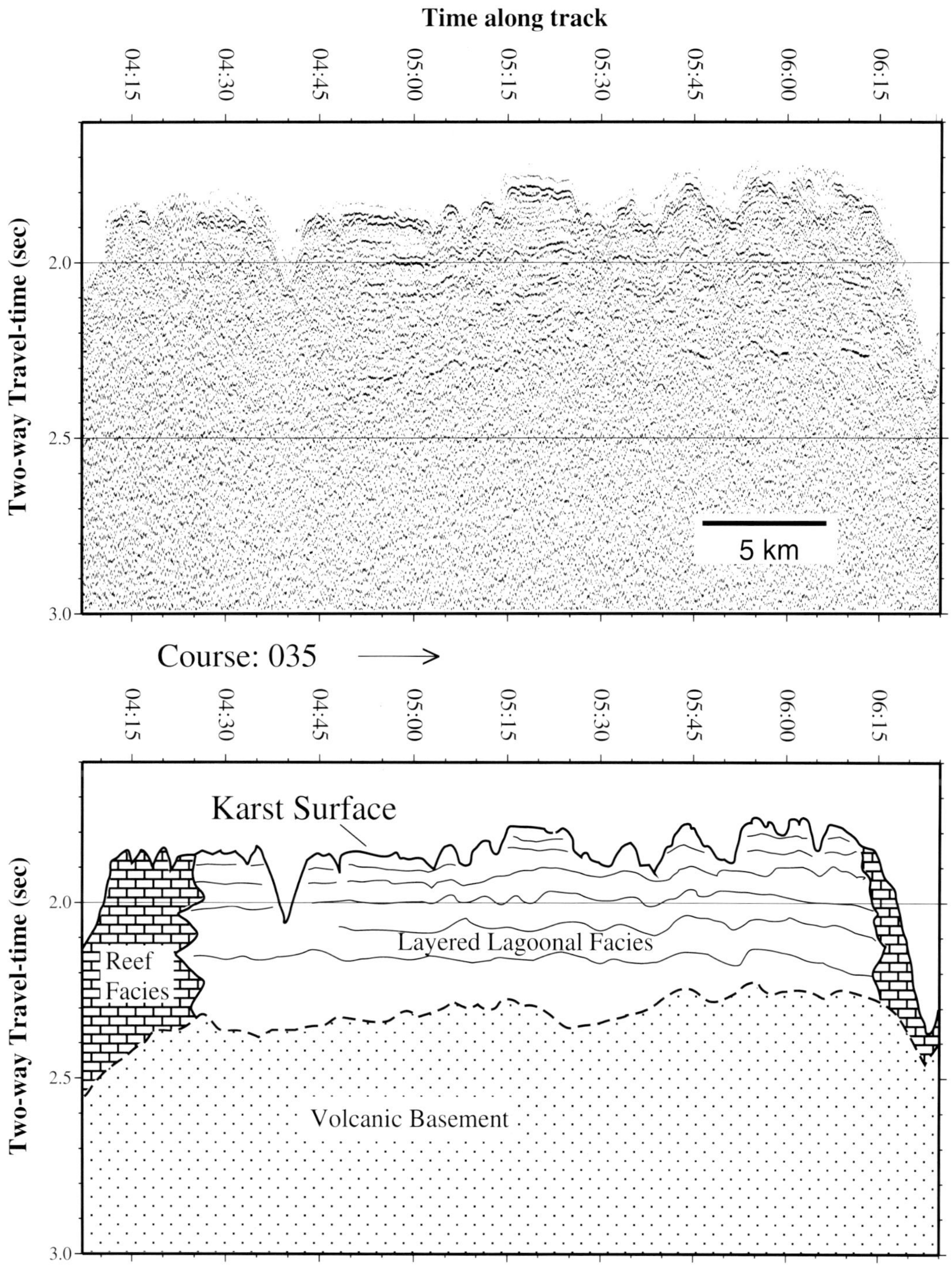

Fig. 16. Seismic reflection profile and line drawing interpretation of MIT Guyot, a drowned atoll in the Japanese Seamounts. Reflectors appear to be cut into by the summit relief, indicating the relief is erosional.

abrupt, with a short slope or scarp on the summit which cuts in a straight line across part of the summit, and can be attributed to faulting or slumping.

Depth variations of as much as 240 m along the break in slope of any one guyot occur on all types of guyots. Averaged over the width of the summits this comes out to tilt angles of 0.1° to 0.8° from horizontal. Guyots affected by tilt include Allison, "Heezen", MIT, Resolution, "Scripps", "Thomas Washington", and "Vibelius" (Table 1). Although most of these are drowned atolls, the other guyot types are also represented, which suggests that the tilt is a

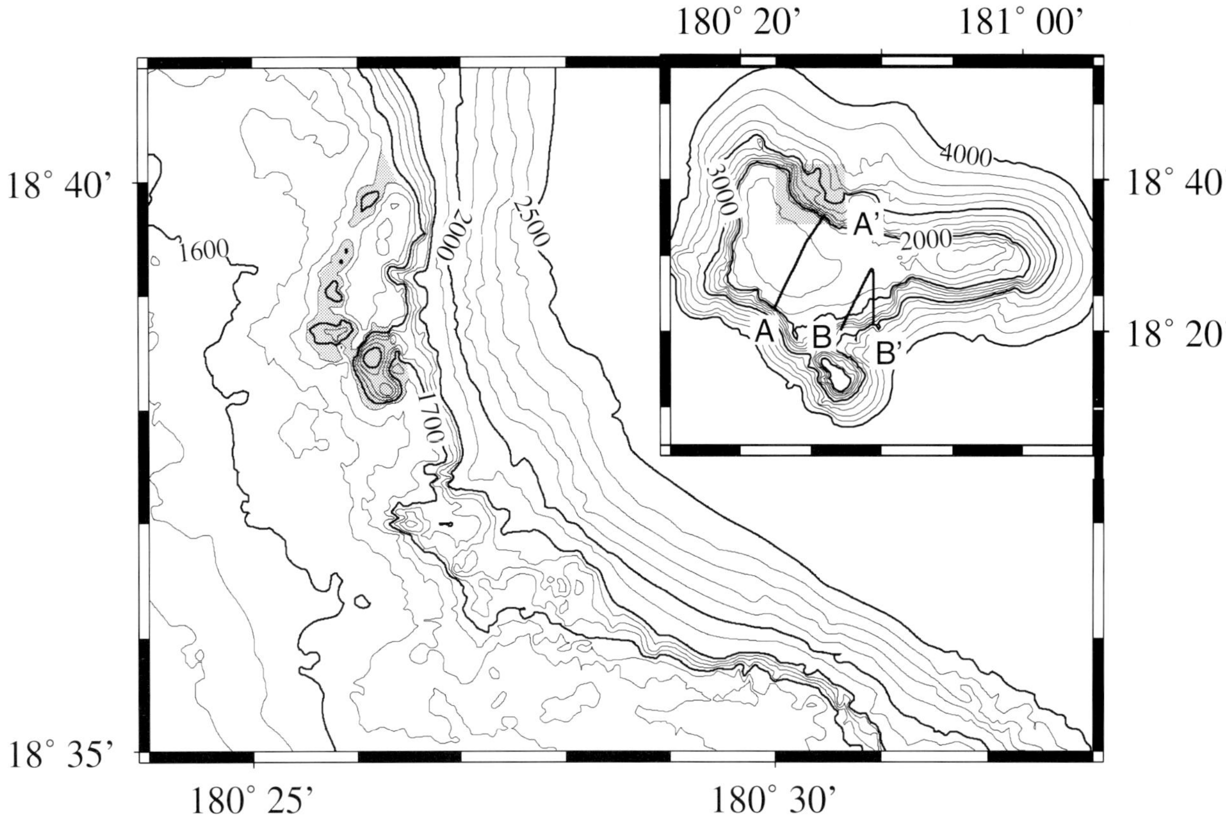

Fig. 17. Bathymetry of summit of Allison guyot. The main body of the guyot includes a 5 km wide volcanic ridge to the east, which was only very incompletely surveyed (inset). The summit of this ridge is shallower than the summit of the atoll structure. The enlarged map of the north end of the guyot-summit shows a 180 m deep trough (shaded) just inside the outer rim of the guyot. The edge of the pelagic sediment cap lies at the 1600 m contour just to the west of the trough. The trough is thought to be the product of solution-weathering during exposure of the summit above sea level.

characteristic of the entire guyot, not just the sediments. The exact cause of tilting of the guyots is beyond the scope of this work. Many causes are possible, including flexure of the seafloor due to the nearby presence of a younger volcano or a trench [*McNutt and Menard,* 1978; *Lonsdale,* 1986]. Other possible causes include uneven summit erosion [*Lonsdale,* 1988], and uneven isostatic adjustment after large-scale mass-wasting on the upper flanks of the guyot [*Keating,* 1987; *Smoot,* 1989].

The slope breaks of some of the guyot summits vary in depth locally along the summit margins. On Darwin guyot, the break in slope of the two ends is 80 m deeper than on the rest of the summit (Figure 3). This gives the summit plateau the appearance of drooping down at the ends. Similarly, the narrow northeast end of MIT Guyot is much deeper than the rest (Figure 15). On Allison Guyot, sections of the summit margin at the southwest end bow down almost 100 m along a section a few km long. Guyots with round summits, lacking extremities (e.g. Resolution Guyot), do not appear to be affected by this, and it is apparent only on summits with thick sections of shallow-water limestone. This suggests that the droop effect may be due to processes affecting sediments on extremities of guyots. Since the sediments are thickest far away from the volcanic core of the atoll, differential compaction may have caused this sagging. One of the effects of droop is that on many of the drowned atolls, the central (lagoonal) part of the section is actually shallower than the (sagged) reef-rims (e.g. on MIT Guyot).

Other processes which have affected summit morphology include faulting and mass wasting/slumping. The effects of faulting can be seen on "Jacqueline" Guyot, a drowned atoll in the Mid Pacific Mountains. On the southern end of the summit platform of this guyot the slope-break becomes abruptly deeper. On the summit plateau a sudden steep slope or scarp occurs, forming a step down of almost 100 m. Such a sudden break in the platform may be due to downfaulting of a section of the guyot. The western portion of Makarov Guyot also appears to be faulted down more than 100 m [*Heezen et al.*, 1973, their Figure 28], which may contribute to the presence of a terrace on that guyot. Similarly downfaulted margins are thought to occur on some of the Emperor guyots [*Greene et al.*, 1980].

The effects of mass-collapse can be seen on the southeast margin of Allison Guyot: in the seismic reflection profiles, layered (lagoonal) sediment can be seen to crop out along the outer slope (Figure 5). The upper flank consists of two gentle slopes separated by a short scarp between 2100 and 2200 m depth. There is no well defined slope break. Rather, the slope increases gradually from the flat top to the steep outer slope over a depth of several hundred meters. It appears that an entire section of reef wall has disappeared from the top of the drowned atoll to the volcanic basement along a 14 km stretch. Dredges on this margin recovered fine grained lagoonal lime-mudstones and a mixed shallow-water-peloidal/pelagic facies. On "Vibelius" Guyot, seismic reflection profiles show that no reef wall exists along several portions of its margin, including the deep embayment on the west side. It is possible that such embayments are created by large slumps. *Vogt and Smoot* [1984] identified slump deposits on the lower flanks of Makarov guyot which originated in the embayment along the northern margin of the summit. MIT Guyot may be similarly affected in some parts.

The results from our surveys suggest that the morphology of the summit, margin, and upper flank regions of Western Pacific Cretaceous guyots is controlled not only by the processes of volcanic edifice building and erosion of volcanic rock but also by processes of sedimentation, especially the construction of carbonate banks, reefs and platforms, and by subaerial dissolution of the limestone during periods of emergence. There may also have been tectonic and/or structural deformation. Some of the guyots appear to be tilted, as evidenced by variations of summit plateau margin depths around the guyot. On some guyots, slope break depths may also vary so as to give the appearance that part of the summit or summit edge has sagged, or drooped.

DISCUSSION

Cretaceous Atolls

The guyots present unique views of the configuration of Cretaceous carbonate platforms. Atolls form a high-energy end-member in the spectrum of rudist-dominated, carbonate platforms. Unlike those along the Tethys margins, in shallower, less open regions, these atolls were located in the middle of the Mesozoic Superocean, surrounded by deep water. After sinking below the surface, they were never buried, and remain essentially undisturbed. From the morphological information from different guyot tops representing different stages of development, we can construct a composite picture of the stages of formation and maturation of a Cretaceous atoll. This picture is very similar to the development of modern atolls, as outlined by *Darwin* [1842].

1: Formation of a volcanic island, fringing reefs. Once a seamount reaches sea level, a platform suitable for reef growth is in place, and, assuming environmental conditions are right, reefs may establish themselves. The initial development of reefs on the Cretaceous volcanic islands is difficult to discern in our data, but may have begun with the growth of fringing reefs, close to the central volcano, forming small platforms such as the ones on the summit of Makarov Guyot, and around the central high on "Charlie Johnson" Guyot. On modern atolls, there is a transitional period of sedimentation from volcanogenic material derived from early rapid erosion of the volcanic island to deposition of pure carbonate sediments [*Bardintzeff et al.*, 1985].

2: Barrier reefs. Initially, erosion on the volcanic islands produces large amounts of silt, making reef growth difficult except as fringing reefs outside the path of discharge [*Bardintzeff et al.*, 1985; *Winterer et al.* ,1973]. Once the island has eroded down sufficiently so that there is not as much silt being produced, reef growth can proceed around the edge of the island, producing a barrier reef with a shallow lagoon behind it. The establishment of such a barrier restricts the access of open marine waters to the initial fringing reefs, which may cease to grow. In addition, the barrier reefs provide protection from wave erosion for the volcanic interior. This stage of atoll formation is represented by "Thomas Washington" Guyot (Figure 9), and possibly by the north end of "Scripps" Guyot (Figure 8).

3: Upward growth during subsidence. As the volcanic pedestal subsides, upward growth of the reefs and back-filling of the lagoon with carbonate sediments gradually buries the volcanic interior. Early stages of this process can be found on "Charlie Johnson" and Takuyo Daisan guyots, and possibly on "Winterer" and Isakov guyots. On the former two guyots, subsequent erosion appears to have removed the lagoonal fill, whereas on the summits of the latter only small (40-60 m) peaks stand above the summit plateau.

The next stage is represented by Darwin Guyot, a fully formed (drowned) atoll without any trace of volcanic rocks on and around the summit. Seismic profiles show that the shallow-water limestone cap of Darwin Guyot is about 0.5 seconds (about 600 m) thick. The atoll structure of Darwin Guyot appears to rise above a broader plateau at a depth of 1800 m, which may be the volcanic basement. This situation is similar to modern atoll-guyot co-occurrences in the Marshall Islands, such as Bikini Atoll with Wodejebato (formerly Sylvania) Guyot. Wodejebato is a Cretaceous edifice on which a drowned Upper Cretaceous carbonate platform occurs [*Lincoln et al.*, this volume]. It is very close to modern Bikini Atoll, and the two share the same volcanic pedestal.

Seismic profiles of the guyot tops indicate that the volcanic basement beneath the thick sedimentary caps have relief consistent with this interpretation. There appears to be a volcanic basement high in the center of Wodejebato [*Schlanger et al.*, 1987]. Likewise, on modern atolls, drilling on Mururoa and Anewetak have shown considerable relief of the volcanic basement, typically with a 300-m-tall central volcano buried in the limestone [*Bardintzeff and Brousse*, 1985]. The Cretaceous atolls have steep upper slopes, 25° to 30°, although not as steep as those of modern atolls, which typically range from 36° to 46° [*Guilcher*, 1988].

Our observations are consistent with a model for Cretaceous atolls which is not much different from modern ones. A few notable differences occur which might be attributed to differences in reef-communities between modern and Cretaceous carbonate platforms. On middle Cretaceous carbonate platforms the dominant organisms were algae and rudist bivalves [*Scott*, 1990]. Material dredged from the guyot-tops contains fossil remains of these in abundance, with only minor contributions of framework-building corals [*Heezen et al.*, 1973]. Rudist-bioherms could not have produced the massive, rigid reef structures which form the margins of modern atolls [*Kaufmann and Sohl,* 1974]. This may explain why the upper slopes of the sedimented guyot-tops are less steep than those of modern atolls, but rather appear to remain at angles which are expected for regular carbonate slopes [*Schlager and Camber,* 1986].

4: Secondary Volcanism. Even after the volcanic island is eroded, reefs have grown up around and over it, and an atoll has formed, renewed volcanic eruptions can modify it. Allison Guyot (Figure 17), which on its west side is a mature drowned atoll with a shallow-water limestone cover at least 650 m thick, has on its east side a volcanic ridge, the top of which is shallower than the shallowest part of the drowned atoll, and which has elements characteristic of an eroded volcano: a flat top, bevelled edge, upper slopes less than 20°. In addition, numerous volcanic peaks connect the ridge to the main atoll body, eventually disappearing below the sediments to the west. These peaks are steep symmetric cones, similar in appearance to those found on some of the Louisville Seamounts [*Lonsdale,* 1988] which were interpreted to be post-erosional. Radiometric ages of around 88 m.y. [*R. Duncan, pers comm,* 1990] from volcanic rocks dredged on the eastern ridge indicate that the ridge is much younger than the fossils from limestone dredged from the top of the atoll, which appear to be Aptian in age [*R. Scott, pers comm,* 1991]. Similar secondary volcanic activity may have affected "Scripps" Guyot, where a possible barrier reef-lagoon configuration can be seen next to and at greater depth than the main volcanic summit, which has been dated as post-Albian as well (Table 2).

ORIGIN OF EROSIONAL TOPOGRAPHY

The occurrence of closed depressions, stream channels and terraces on and around the summits of the guyots is an indication that these summits were exposed above sea level 100 to 200 m prior to their final drowning. Studies of the effects of exposure on modern carbonate islands indicate that such features can be the result of exposure-erosion. Lowstand terraces are found in front of the modern edge of the Great Barrier Reef [*Harris and Davies,* 1989], as well as along the south Florida Bank margin [*Lidz et al.,* 1991]. Modern uplifted islands such as Raratonga, Makatea and Mangaia in the South Pacific and Nauru in the central Pacific typically have a 500- to 1000-m-wide reef flat along their outer edges, separated by a near-vertical cliff from the inland plateau [*Stoddard et al.,* 1985; *Hill and Jacobsen,* 1989]. The inland part of the upper plateau is commonly hollowed out by dissolution, producing landforms similar to those observable on some of the guyots. On Mangaia Island, a tall rim of limestone surrounds the hollow interior of an uplifted atoll [*Marshall,* 1927; *Stoddard et al.,* 1985]. In the center, dissolution has exposed the central volcano. The resulting topography is very similar to that of the summit of Takuyo Daisan Guyot (Figure 13). The high-energy limestone facies that make up the rim are lithified rather than dissolved during exposure, while the low-energy lagoonal facies are dissolved. This effect is enhanced by the flow of runoff-water along the limestone-volcano contact in the lagoon area. Solution caves along the base of the limestone rim of Mangaia Island indicate that much runoff occurs along this boundary.

The depth of the erosional topography is considerable: relief between the deepest depression and the shallowest part of the summit platform of MIT Guyot is more than 200 m, and the lowstand terraces on Isakov and "Winterer" guyots are 150 m below the summits (Table 1). These values form minimum estimates on the height of exposure of the summits, since erosion might have removed more than presently appears, while post-drowning sedimentation may have partially filled some depressions with pelagic sediments. Whereas it is possible that the formation of a freshwater lens during time of exposure may have caused dissolution and the formation of holes deeper than the height of exposure of the summits, the terrace levels could not have formed anywhere except at sea level, indicating that the summits were exposed 150 to 200 m above sea level. It is equally unlikely that the topography is the result of submarine erosion: this process would also not be able to form the terraces, and the summit depressions on at least 4 of the guyots are filled with pelagic sediments not much younger than the reefs themselves (Table 2) [*Matthews et al.,* 1974]. Since submarine dissolution could have formed the depressions only after the guyot had subsided into waters undersaturated for calcium-carbonate, it could not have predated the pelagic sediments. Only through an exposure of the summits 150 to 200 m above sea level could these topographic features have formed. There are two ways to account for this: either sea level fell, or the seafloor rose.

Karst morphology and lowstand reef-tracts on modern carbonate platforms are in many places the result of erosion during Pleistocene glacial sea level lowstands. In others the same morphology occurs after tectonic uplift of the platform. It is possible that sea level went down. If there were an ice age of the scale of the Pleistocene glaciations, a sea level drop of 150 to 200 m could be accounted for. This has been suggested by *Grotsch et al.* [unpub. manus., 1990] on the basis of relative sea level falls recorded in late Albian strata in Europe. We have not yet seen evidence from Cretaceous high-latitude deposits of Pleistocene-size ice sheets. Cool polar temperatures, sea-ice and drop-stones do not necessarily add up to large-scale glaciation. An alternate possibility is that the seafloor rose tectonically. This has been suggested by *McNutt et al.* [1990], who propose that during the middle Cretaceous, an area of seafloor on which the guyots were located was uplifted during the onset of a "Superswell" [*McNutt and Fisher,* 1987], a thermal anomaly in the mantle which is thought to cause anomalously shallow bathymetry over a large area in the South Pacific today.

The time difference between initial uplift (the youngest shallow-water deposits) and drowning (the oldest pelagic sediments on the summits) may tell us which of these two models is the most likely. A large Pleistocene-scale ice age could not have lasted very long, or there would be more world-wide evidence for it. The area of the northwest Pacific over which karsted Cretaceous guyots are known is about 0.5-1 x10^6 km^2. Thermal uplift of such a relatively small region of seafloor, would cause a sea level rise of about 5 m worldwide, and would decay over millions of years [*McNutt et al.,* 1990; *Vogt and Smoot,* 1984]. Another way to estimate the minimum required time is to determine the time required to produce the observed topography. The paleontologic and radiometric age data are summarized in Table 2.

The volcanoes and reefs of the Japanese guyots ("Charlie Johnson", Isakov, Makarov, Takuyo Daisan and Takuyo Daini) all appeared to have formed in the late Albian to Cenomanian ages [*Heezen et al.*, 1973; *Matthews*, 1984; *A. Coogan*, pers. comm., 1989; *R. Scott*, pers.comm., 1991]. MIT Guyot, the westernmost guyot in our study with a thick atoll structure, apparently formed at about 120 Ma, after which shallow-water sedimentation continued until late Albian time. The reefs of the drowned atolls of the Mid Pacific Mountains appear to be Barremian-Aptian in age, although shallow water limestone containing late Albian microfossils were found on Allison guyot (based on identifications of rudist megafossils by *Hamilton* [1956], *A. Coogan* [pers.comm., 1989], and *R. Scott* [pers.comm., 1991], and microfossils by *W.Sliter* and *I. Premoli-Silva* [pers.comm., 1990], as well as *Grotsch et al.*, [unpub. manus., 1990]). Radiometric ages on their volcanic edifices show they are also about 118 m.y. old [*R. Duncan*, pers. comm., 1990]. The oldest pelagic sediments on the Japanese seamounts are Cenomanian-Turonian, those in the Mid Pacific Mountains late Albian or younger [*Heezen at al.*, 1973; *Matthews et al.*, 1984; *W.Sliter, pers. comm.*, 1989]. This suggests, in both areas, a time-span of at least 6 million years between the end of shallow water deposition and pelagic sedimentation after resubmergence.

Modern uplifted atolls in the Pacific such as Mangaia, Makatea, Nauru, and Mare, which have erosional morphologies similar to those observed on the guyot tops, are all tectonically uplifted, and have been above sea level for at least 300,000 (Nauru) to perhaps several million years (Mangaia) [*Montaggioni et al.*, 1985; *Hill and Jacobson*, 1989]. Submerged platforms which were exposed by sea level falls during Pleistocene glaciations tend to have less well-developed karst features, restricted to occasional "blue holes" and small lowstand reef terraces [*Shepard*, 1970; *Guilcher*, 1988]. We can estimate the amount of time necessary to produce the relief observed on the guyots. We measured the volume of material eroded from the guyot summits assuming a pre-erosional level lagoon floor 20 m below the top of the shallowest reef. For example: on "Charlie Johnson" Guyot, there is 11.2 km^3 of empty space between 1640 m (20 m below the top of the tallest reef) and 1820 m depth (the deepest depression in the interior of the summit). Averaged over the surface area of the summit, this is equivalent to a slab 76.7 m thick. Based on a tropical limestone denudation rate of 100 mm/1000 years [*Palmer*, 1991] it would have taken at least 0.77 million years to form the relief of "Charlie Johnson" Guyot. The higher erosion rates occur in regions with thick soil layers, which increases the equilibrium-pCO_2 of the water [*Smith and Atkinson*, 1976]. Since the guyots were far away from any source of siliceous sediments, soil layers were probably rather thin, and the high-end modern tropical erosion rates are thus likely a high estimate for limestone erosion of the Cretaceous atolls. The rate calculation further assumes that all of the erosion is "external", i.e. there are no caves in the subsurface, that the eroded layer was no thicker than the height of the top of the shallowest remaining reef, and that rain collected over the entire summit area contributed to the erosion. The calculation is therefore a minimum estimate of the amount of time necessary to produce the observed topography. Because of limited surveying, estimates on karst volumes could be made for only a few of the guyot summits. Similar calculations yield: Takuyo Daisan Guyot: 0.4 m.y., MIT Guyot: 0.15 m.y.. These are not overwhelmingly longer than could be accounted for by an ice age, and the calculation does not take into account a possible increase in rainfall during an ice age, but since the volumes and rates are probably underestimates, it appears that for the western guyots a longer-lasting event such as tectonic uplift is more likely.

On Allison Guyot, significant erosional topography is visible only on a small part of the summit, whereas the rest is obscured by a thick blanket of pelagic sediments. On most of the drowned atolls of the Mid Pacific Mountains the only observable erosional relief consists of holes on the surface of the lagoons. For these atolls, a very short period of exposure would account for the observed relief. Whereas all of the western guyots in our survey show considerable relief, many of the eastern ones have no sign of erosion.

TIMING OF THE EROSION

Exposure of the summits occurred just prior to final drowning. If significant sedimentation on the summits had occurred after the exposure, the erosional relief would have been covered with shallow-water sediments and reefs. As it is, many of the summits are covered with pelagic sediments, typically as dome-shaped mounds up to 200 m thick. Only the differences in acoustic properties between the pelagic sediments and the underlying shallow water limestone allows the interface between them to be detected in profiler records. It is possible that some of the reefs had already drowned, and became re-exposed during the events that led to the creation of summit relief. At this time, reef-growth could have been temporarily re-initiated, allowing some younger reef deposits to form over the previously drowned platform. This has been demonstrated to have occurred at Wodejebato and Rongelap guyots in the Marshall Islands region, where reworked Lower Cretaceous reef sediments were found in dredge hauls containing Upper Cretaceous reef material [*Lincoln et al.*, this volume]. The presence of late Albian microfossils in shallow water limestone samples dredged from Allison Guyot may be an indication that the summit of that guyot was brought back up to sea level at that time. Radiometric dates on volcanic-rock samples from Allison give ages of ~100 and ~85 m.y., perhaps an indication that this guyot has had a long history of recurrent volcanism. Karstic erosion of the summit of the atoll could have occured following or even contemporaneous with minor sedimentation as the summit was carried up through sea level.

The younger Japanese guyots appear to have formed about 100-108 Ma, and were exposed before they could accumulate a thick section of shallow-water sediments. Exposure and subsequent drowning therefore apparently occurred sometime in the late Albian-early Cenomanian. MIT Guyot, near the Japanese group, formed ~120 Ma, and accumulated a thick section of shallow-water limestone prior to being exposed and drowning, probably at the same time. Of the reef-bearing members of the Wake group, only "Woods Hole" yielded any ages. The presence of both Aptian-Barremian and late Albian shallow-water fauna on this guyot, which has almost no reef cover, may mean that it was an older atoll, which had already drowned and became re-exposed at about late Albian time. The nearly complete absence of sediment cover could, in this instance, be a result of extreme erosion.

Of the drowned atolls of the Mid Pacific Mountains, Darwin Guyot bears no sign of any erosion on its summit. Barremian-Aptian reef fossils were found on its summit, indicating that this atoll formed and drowned in the Early Cretaceous, and was never re-exposed afterward. Other drowned atolls in the Mid Pacific Mountains similarly do not show any sign of erosion [*Heezen et al.*, 1973; *Kroenke,* 1982 (unpub. cruise rep.)], and may have histories similar to Darwin Guyot.

The ages of seamounts and sediments and the distribution of erosional relief are consistent with the following scenario:

1: Volcanic islands formed on what is now the western Pacific seafloor starting in the Early Cretaceous, probably in Barremian time.

2: Reefs formed, and developed the islands into atoll structures over several million years of subsidence. On many of these, including Darwin Guyot, shallow-water sedimentation ceased sometime during Barremian-Aptian time, and the platforms drowned.

3: In the Albian-early Cenomanian a new episode of volcanism occurred and created new islands in the Japanese group, on which reefs began to develop.

4: Not long after, widespread exposure of the summits of the carbonate platforms occurred. Some, but not all, of the previously drowned eastern platforms were returned to sea level and shallow water sedimentation resumed briefly. The platforms in the Japanese group were deeply eroded. Perhaps it was during this period that the reefless, apparently younger guyots of the Wake group formed.

5: Exposure ended, but shallow-water sedimentation did not resume. The platforms drowned, and continued to subside to their present positions. Pelagic sedimentation began on many of the guyot-tops at Cenomanian-Turonian time.

Given the present state of knowledge, the apparent failure of the shallow-water carbonate platforms to re-establish themselves on the low-latitude guyots as they subsided following emergence and subaerial erosion remains an enigma. Only new data can solve this major puzzle, and we will not add to the speculations [*Vogt*, 1989; *Grigg and Hey*, 1992; *Hallock and Schlager*, 1986].

INFERENCES FROM GUYOTS ON THE TECTONIC HISTORY OF THE WESTERN PACIFIC.

Menard [1964] tried to use the height of the guyots above the surrounding sea floor to estimate the depth of the ocean during the Cretaceous in the western Pacific. These paleo-depths are much shallower than one would expect, which led him to conclude that the seamount-dense region of the western Pacific was once the site of anomalously shallow seafloor, which he called the "Darwin Rise". *McNutt and Fisher* [1987], described a region of seafloor in the south Pacific which is also too shallow for its age. This region is thought to be underlain by thermally anomalous lithosphere and mantle, and was dubbed the Superswell [*McNutt and Fisher*, 1987; *Menard,* 1984]. *McNutt et al.* [1990], used the more recent bathymetric data of the Cretaceous western Pacific guyots to determine their paleodepth and concluded that the Darwin Rise may have been a similar superswell, affecting the crust in the later part of the Early Cretaceous, when that region of the Pacific plate was located where the modern superswell exists today. Table 1 includes the main information necessary to determine the height of the guyots above the seafloor at the time the volcanoes formed. Some of the values in this table are a little different from those used by *McNutt et al.* [1990], although they are based on the same sources. The differences are due to the inherent subjectivity of some of the observations.

To determine the paleodepth of the ocean at the time a guyot formed, one needs the height of the guyot above sea-floor, and the age of the volcano. In order to compare this with an expected depth, one needs the age of the seafloor. The height of the guyot above the seafloor is the difference between the depth of the guyot top and the depth of the seafloor, corrected for the thickness of the sediment cap on the guyot, as well as for the thickness of the sediments on the seafloor.

Of these measurements, the depth to the top of the guyot is the most straightforward. One can determine the depth to the shallowest reef-tops prior to drowning of the guyot from the SeaBeam data over the guyot-summits. For the reef-bearing guyots that is the top of the shallowest observed reef. For non-reef-bearing guyots, it is the depth to the main summit plateau. The depth then must be corrected for sediment thickness on the guyot-top, measured from seismic reflection profiles. Differences in interpretation of these data have led to small differences in measurement of the depth of the old sea level on the guyot summits between those listed in Table 1, and those used by *McNutt et al.* [1990]. In addition, some of the guyots, especially in the Japanese set, are reported by *McNutt et al.* [1990] as having no sediments on top, while multiple dredges have indicated that such sediments do exist. On Takuyo Daini, Takuyo Daisan, "Winterer", "Charlie Johnson", "Thomas Washington", Isakov, and Makarov guyots, abundant shallow water limestone has been dredged as much as 200 m below their summits [*Heezen et al.*, 1973]. For this reason we have estimated the sediment thickness there as 200 m. Although "Stout" Guyot was never dredged, the considerable relief on its summit suggests that this guyot also has an eroded limestone cap. Overestimation of the sediment thickness would make the paleodepth too small.

A much more subjective measurement is that of the depth of the regional seafloor. Because of the very gentle lower slopes of the seamounts and the existence of flexural moats, one must measure the regional seafloor depth some distance away from the seamount. Differences of 200 to 300 m can occur in these measurements, depending on where one makes them. In addition to that uncertainty, part of the height of the guyot above regional seafloor has been covered by deep-sea sediments, so the thickness of this layer must be added to the height. No isostatic correction is necessary for the regional sediment load, because the seamount and the surrounding seafloor would be equally affected by such long-wavelength isostasy. We used regional sediment thicknesses from the map of *Ludwig and Houtz* [1971]. For the most part this correction is small, but in some areas as much as 350 m is added to the height. Some of the thickness of the atolls might be

(Barremian-Aptian for both Allison and "Woods Hole", using an estimate of 118 Ma.), since they are minimum ages. For many of the guyots the age remains uncertain.

A significant source of uncertainty is the age of the lithosphere on which the guyots are located. From the magnetic lineation map of *Nakanishi et al.* [1989], we get reasonable estimates of the present age of the lithosphere in the Japanese and Wake seamounts area. But even in regions where the lineations have been identified, the location of the guyots often falls on small offsets in the lineations (fracture zones), causing uncertainties of up to 5 m.y. In the Mid Pacific Mountains, magnetic lineations are obscured by the volcanic plateau on which the guyots rest, making estimates of the age of the lithosphere for these guyots even more uncertain.

The resulting plot of paleodepths vs. age of lithosphere-loading (Figure 18) is somewhat different from the one presented by *McNutt et al.* [1990]. All of the guyots appear to have formed on anomalously shallow seafloor. There are two distinct sets: those guyots located on the Mid Pacific Plateau, and all the others. If isostatic adjustment of the thick sediments caps of the Mid Pacific guyots is taken into account, all the guyots have approximately the same paleodepth level. Regardless of the age of the lithosphere on accounted for by isostatic response of the guyots to the load of the reefs. Simple local compensation of the weight of the limestone would account for about 2/3 of the total thickness, but it seems unlikely that the relatively small mass of the limestone cap on the top of a much larger volcano would be compensated locally. Instead, flexure of the lithosphere would distribute the load over a larger area. We estimated the effects of plate flexure, treating the limestone caps as disk loads [*McNutt and Menard,* 1978; *M.McNutt, pers. comm.,* 1990]. We assumed a density of 2500 kg/m^3 for limestone and 3300 km/m^3 for the asthenosphere. The elastic plate thicknesses were obtained from *Wolfe and McNutt* [1991], who derived them from gravity data. Typical elastic plate thicknesses for the Mid Pacific Mountains were very low (1-5 km). For very low Te values, flexural response to the weight of the limestone caps ranges from 130 m to 230 m for the larger guyots, indicating that a significant part of the thickness of these caps could be accounted for by isostasy. For MIT guyot, no elastic plate thickness estimate exists, but it formed on crust which was older than that of the Mid Pacific Mountains, and may have been much less flexible. For the smaller guyots, such as Darwin and "Caprina", the calculation yields a very small (<50 m) flexural response. These estimates are maximum values for isostatic adjustments, since they use the Te-value of the guyots when they had just formed.

We used the volcanic ages of the seamounts as reported in Table 1, which includes several new radiometric dates. In some places radiometric dates and fossil identifications are apparently contradictory: for example, for Allison Guyot a radiometric age of basalt is reported which is much younger than the age-range of the megafossils dredged from there. The same is true for "Woods

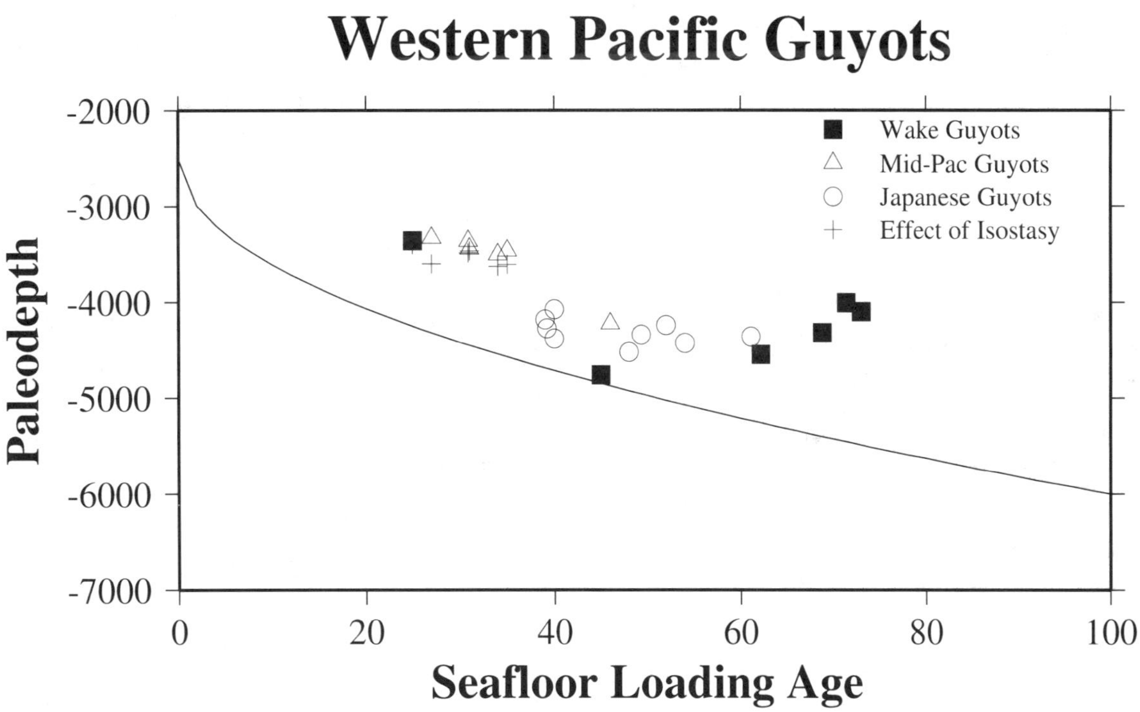

Fig. 18. Plot of height of guyots above seafloor vs. age of seafloor at time of volcanism. The height represents the paleodepth of the seafloor. The "expected" paleodepth based on square-root-of-age subsidence is shown as a solid line. The "+" marks represent the paleodepths of atoll-bearing guyots after isostatic correction for the thick atoll caps, using flexure of the lithosphere under a disk-load.

Hole" Guyot. In these places we have used the fossil dates which they formed, the guyots indicate a paleodepth of 4100-4400 m, equivalent to that expected for seafloor approximately 25 m.y. old. This is in keeping with the observations of *Crough* [1984] as well as *Detrick and Crough* [1978], that hot spot volcanoes tend to "rejuvenate" the seafloor on which they sit to approximately 25 m.y., providing it is already older than that, or to about 40% of the original age [*Epp*, 1984]. If this is the case, the guyots subside at a rate different from the surrounding seafloor, so that their "height" is not an indication of paleodepth of normal seafloor but only of the swells of the volcanoes themselves [*Vogt and Smoot*, 1984].

Does this mean that there was no superswell? The height of the guyots above the seafloor can be adequately accounted for by their having formed each on their own little swell, (as opposed to a regional, contemporaneous superswell), which reset the effective age of the lithosphere to 25 m.y. Pacific seafloor older than 100 m.y. lies on a much flatter curve than the simple square-root-of-age relationship [*Parsons and Sclater*, 1977; *Renkin and Sclater*, 1990]. In addition, a compilation of seafloor depths [*Renkin and Sclater*, 1990] shows that the seafloor with ages between 100 and 120 m.y. forms a distinct bulge of almost 1000 m above a smooth curve. If the seafloor of the Pacific had been affected by a superswell, older lithosphere would be uplifted and the normal subsidence of the seamounts interrupted. This might account for the facts that the guyots have not subsided as far as expected, and that only the guyots which were at sea level at the time of uplift (the Japanese guyots) are significantly eroded. The older platforms of the drowned atolls of the Mid Pacific Mountains were already drowned at the time of uplift and either remained submerged, or only underwent minor exposure.

CONCLUSIONS

Western Pacific Cretaceous guyots formed as a result of a complex interplay of constructional volcanism and sedimentation and marine and subaerial erosion. Over a period spanning Barremian (Mid Pacific Mountains) to Cenomanian (Japanese group and Wake seamounts) volcanic islands formed on the seafloor of what is now the western Pacific. Subsidence and reef development built many of the summits of these islands into the shape of atolls. On the younger islands, reef evolution was less complete, and in some places entirely absent. On islands on which no reefs were established the interplay between erosion at sea level and subsidence of the seafloor produced guyots with characteristic bevelled upper slopes.

During the Albian to Cenomanian ages the tops of many of the carbonate platforms became exposed as much as 200 m above sea level. Those that were still at sea level just prior to this event were most exposed and developed deep erosional topography and terraces in and around their summits. The older platforms, many of which had already drowned and subsided below sea level were either not exposed at all, or only to a minor extent.

The cause of exposure was either a eustatic lowering of sea level or tectonic uplift of the seafloor. The minimum time necessary to produce the observed relief, combined with the observation of renewed volcanism on some of the edifices, suggests that the cause was tectonic rather than eustatic, supporting the idea that a large portion of the Pacific sea floor was uplifted sometime during the late Albian to Cenomanian, at which time a Cretaceous "Superswell" developed, similar to and perhaps the ancestor of the one observed in the central-south Pacific today.

NOTE ADDED IN PROOF

Between initial submission and publication of this paper, two drilling legs (ODP Legs 143 and 144) drilled four of the guyots in this study. Since the results of these expeditions have direct bearing on this work, the ODP gave permission to include a brief statement on the impact of the drilling results on our conclusions.

Holes were drilled through the shallow-water limestone caps to basement on Allison, Resolution, MIT and Takuyo Daisan guyots [*van Waasbergen, Sliter, et al.*, in press; *ODP Leg 143 Scientific Party*, in press; *Haggerty, Premoli Silva, et al.*, in press]. The thicknesses of limestone matches our estimates on all guyots except Resolution, where the limestone is 1619 m thick. Ages of the drilled carbonate caps are: Albian (Allison) and Barremian through late Albian (Resolution). No specific new age-data is available at this time for MIT and Takuyo Daisan guyots. On Allison and Resolution guyots, cores show some evidence of karstification in the form of solution cavities, some of which are filled with Turonian and Campanian pelagic material. Recovery rates of the shallow water limestone material, especially in the upper parts of the sections, were uniformly poor (about 2%).

These findings have the following impact on our conclusions:
1: Guyots in the Mid Pacific Mountains apparently drowned at the same time as those farther west, in latest Albian time.
2: The western Mid Pacific Mountains started as a series of platforms of moderate relief above a shallow (1000 m waterdepth) plateau, rather than as tall, isolated volcanic islands.
3: The drowned atolls of the Mid Pacific Mountains were exposed to an extent about equal to that of the guyots farther west.

Acknowledgements. We are indebted to Captain T. Desjardins, the officers and crew of R/V *Thomas Washington*, and to U. Albright, P. Crampton, J. Goff, G. Ito, M. Lynch, M. Moore, M. Nakanishi, G. Pillard, and D. Stuber for the success of the RNDB10WT Expedition. We thank R. Scott and A. Coogan for their identifications of rudists. We thank B. Sliter and Isabella Premoli Silva for their identification of microfossils, and Jürgen Grötsch for sharing paleontological data. Many of the figures were made using software developed by P. Wessel and W.H.F. Smith. The work was partly supported by NSF Grant 8717079-OCE.

REFERENCES

Bardintzeff, J.M., R. Brousse, and A. Gachon, Conditions of building coral reefs on a volcano: Mururoa in Tuamotu and Rurutu in Australes (French Polynesia), *Proc. 5th Int. Coral Reef Congr.*, *6*,401-405, 1985.

Batiza, R., abundances and distribution of volcanoes in the Pacific Ocean and implications for the origin of non-hotspot volcanoes, *Earth Planet. Sci. Let.*, *60*, 195-206, 1982.

Crough, S.T., Thermal origin of mid-plate hot-spot swells, *Geophys. J. R. Astron. Soc.*, *55*, 451-469, 1978.

Crough, S.T., Seamounts as recorders of hot-spot epeirogeny, *Geol. Soc. Am. Bull.*, *95*, 3-8, 1984.

Darwin, C., *The structure and evolution of coral reefs*, Appleton, New York, 1842.

Detrick, R.S. and S.T. Crough, Island subsidence, hot spots, and lithospheric thinning, *J. Geophys. Res.*, *83*, 1236-1244, 1978.

Epp, D., Implications of volcano and swell-heights for thinning of the lithosphere by hot-spots, *J. Geophys. Res.*, 89, 9991-9996, 1984.

Francheteau, J., C.G.A. Harrison, J.G. Sclater, M.L. Richards, Magnetization of Pacific seamounts: a preliminary polar curve for the northeastern Pacific, *J. Geophys. Res.*, *75*, 2035-2061, 1970.

Greene, H.G., D.A. Clague, G.B. Dalrymple, Seismic stratigraphy and vertical tectonics of the Emperor Seamounts, DSDP leg 55, in *Initial reports of the Deep Sea Drilling Project*, *55*, edited by Jackson, E.D., I. Koisumi et al., Washington, (U.S. Government Printing Office), 759-788, 1980.

Grigg, R.W. and R. Hay, Paleoceanography of the tropical eastern Pacific Ocean, *Science*, *255*, 172-178, 1992.

Grow, J.A., M.W. Lee, J.J. Miller, J.C. Agena, J.C. Hampson, D.S. Foster, and R.A. Woellner, Multichannel seismic-reflection survey of Koa and OAK craters, *U. S. Geological Survey Bulletin*, *1678*, D1-D46, 1986.

Guilcher, A., *Coral reef geomorhology*, Chichester, New York, 1988.

Haggerty, J.A., I Premoli Silva and ODP Leg 144 Scientific Party, Western Pacific Guyot Drilling (ODP Leg 144), preliminary results: sunken atolls vs. carbonate banks, *Eos, Proc. American Geophysical Union*, 1992 fall meeting, in press, 1992, (abstract).

Hallock, P. and W. Schlager, Nutrient excess and the demise of coral reefs and carbonate platforms, *Palaios*, *1*, 389-398, 1986.

Hamilton, E.L., Sunken islands of the Mid-Pacific Mountains, *Mem. Geol. Soc. Am.*, *64*, 1-97, 1956.

Harris, P.T. and P.J. Davies, Submerged reefs and terraces on the shelf edge of the Great Barrier Reef, Australia - Morphology,occurence and implications for reef evolution, *Coral reefs*, *8*, 87-98, 1989.

Heezen, B.C. and I.D. MacGregor, *Initial Reports of the Deep Sea Drilling Project*, *20*, US Govt. Printing Office, Washington, DC, 1973.

Heezen, B.C., J.L. Matthews, R. Catalano, J. Natland, A. Coogan, M. Tharp, and M. Rawson, Western Pacific guyots, in *Initial reports of the Deep Sea Drilling Project*, *20*, edited by Heezen, B.C., I.D. MacGregor, et al., 653-723, U.S. Govt. Printing Office, Washington, D.C., 1973.

Hess, H.H., Drowned ancient islands of the Pacific Basin, *Am. J. Sci.*, *244*, 772-791, 1946.

Hill, P.J. and G. Jacobson, Structure and evolution of Nauru Island, Central Pacific Ocean, *Austr. J. Earth. Sc.*, *36*, 365-381, 1989.

Hollister, C.D., M.F. Glenn, and P.F. Lonsdale, Morphology of seamounts in the western Pacific and Philippine Basin from multibeam sonar data, *Earth Plan. Sci. Let.*, *41*, 405-418, 1978.

Karig, D.E., M.N.A. Peterson, and G.C. Shor, Sediment capped guyots in the Mid Pacific Mountains, *Deep Sea Res.*, *17*, 373-378, 1970.

Kauffman, E.G. and N.F. Sohl, Structure and evolution of Carribean Cretaceous rudist frameworks: Festschrift für Hans Kugler, *Verh. Nat. Gesell. Basel*, *84*, 399-647, 1978.

Keating, B.H., Structural failure and drowning of Johnston Atoll, Central Pacific Basin, in *Seamounts, Islands and Atolls*, edited by Keating, B.H., P. Fryer, R. Batiza, and G. Boehlert, 49-59, American Geophysical Union, Washington D.C., 1987.

Ladd, H.S., W.A. Newman, and N.F. Sohl, Darwin guyot, the Pacifics oldest atoll, Proc. 2nd Int. Coral Reef Symp., The Great Barrier Reef Committee, Brisbane, 513-522, 1974.

Lancelot, Y., R.L. Larson, et al., Site 800, *Proc. O. D. P.*, *129*, 33-89, 1990.

Larson, R.L., S.O. Schlanger, et al., *Initial Reports of the Deep Sea Drilling Project*, *v.61*, US Government Printing Office, Washington, DC, 1981.

Larson, R.L., Latest pulse of the Earth - evidence for a mid-Cretaceous Superplume, *Geology*, *19*, 547-550, 1991.

Lidz, B.H., A.C. Hine, E.A. Shinn, and J.L. Kindinger, Multiple outer-reef tracts along the south Florida bank margin: Outlier reefs, a new windward-margin model, *Geol.*, *19*, 115-118, 1991.

Lincoln, J.M. and S.O Schlanger, Atoll stratigraphy as a record of sea level change - problems and prospects, *J. Geophys. Res.*, *96*, 6727-6752, 1991.

Lincoln, J.M., M.S. Pringle, and I. Premoli Silva, Early and Late Cretaceous volcanism and reef-building in the Marshall Islands, *this volume*.

Lonsdale, P., W.R. Normark and W.A. Newman, Sedimentation and erosion on Horizon Guyot, *Geol. Soc. Am. Bull.*, *83*, 289-315, 1972.

Lonsdale, P., A multibeam reconnaissance of the Tonga Trench axis and its intersection with the Louisville guyot chain, *Mar. Geophys. Res.*, *8*, 295-327, 1986.

Lonsdale, P., Geography and history of the Louisville hotspot chain in the southwest Pacific, *J. Geophys. Res.*, *93*, 3078-3104, 1988.

Ludwig, W.J. and R.E. Houtz, *Isopach map of sediments in the Pacific Ocean basin and marginal sea basins, scale: 1:13,999,369*, American Association of Petroleum Geologists, Tulsa, OK, 1979.

Marshall, P., Geology of Mangaia, *B. P. Bishop Mus. Bull.*, *36*, 1-47, 1927.

Matthews, J.L., B.C. Heezen, R. Catalano, A. Coogan, M. Tharp, J. Natland, and M. Rawson, Cretaceous drowning of reefs on Mid-Pacific and Japanese guyots, *Science*, *184*, 462-463, 1974.

McNutt, M.K. and K.M. Fisher, The South Pacific Superswell, in *Seamounts, islands and atolls*, edited by Keating, B.H., P. Fryer, R. Batiza, and G.W. Boehlen, 25-34, American Geophysical Union, Washington, DC, 1987.

McNutt, M.K. and H.W. Menard, Lithospheric flexure and uplifted atolls, *J. Geophys. Res.*, *83*, 1206-1212, 1978.

McNutt, M.K., E.L. Winterer, W.W. Sager, J.H. Natland, and G. Ito, The Darwin Rise, a Cretaceous Superswell? *Geophys. Res. Let.*, *17*, 1101-1104, 1990.

Menard, H.W., *Marine geology of the Pacific*, McGraw-Hill, New York, 1964.

Menard, H.W., Insular erosion, isostasy and subsidence, *Science*, *220*, 913-918, 1983.

Menard, H.W., Darwin reprise, *J. Geophys. Res.*, *89*, 9960-9968, 1984.

Montaggioni, L.F., G. Richard, F.G. Bourrouilh-Le Jan, C. Gabrie, L. Humbert, M. Monteforte, O. Naim, C. Payri, and B. Salvat, Geology and Marine Biology of Makatea, an uplifted atoll, Tuamotu archipelago, central Pacific ocean, *J. Coast. Res.*, *1*, 165-171, 1985.

Nakanishi, M., K. Tamaki, and K. Kobayashi, Mesozoic magnetic anomaly lineations and seafloor spreading history of the northwest Pacific, *J. Geophys. Res.*, *94*, 15437-15462, 1989.

Nemoto, K. and L.W. Kroenke, Sio guyot: a complex volcanic edifice in the western Mid-Pac Mountains, *Geo-Mar. Let.*, *5*, 83-89, 1985.

ODP Leg 143 Scientific Party, Regional and global implications of drilling of Early Createaceous guyots: ODP Leg 143, Mid Pacific Mountains, *Eos, Trans. Am. Geophys. Union*, in press, 1992.

Ozima, M., M.Honda and K. Saito, ^{40}Ar-^{39}Ar ages of guyots in the western Pacific and discussion of their evolution, *Geophys. J. R. astr. Soc.*, *51*, 475-485, 1977.

Palmer, A.N., Origin and morphology of limestone caves, *Geol. Soc. Am. Bull.*, *103*, 1-21, 1991.

Parsons, B. and J.G. Sclater, an analysis of the variation of ocean floor bathymetry and heat flow with age, *J. Geophys. Res.*, *8*, 803-827, 1977.

Renard, V. and J.P. Allenou, SeaBeam, multi-beam echo-sounding in "Jean Charcot", *Int. Hydrogr. Rev.*, *56*, 35-67, 1979.

Renkin, M.L. and J.G. Sclater, Depth and age in the North Pacific, *J. Geophys. Res.*, *93*, 2919-2935, 1988.

Sager, W.W. and M.S. Pringle, Paleomagnetic constraints on the origin and evolution of the Musicians and south Hawaiian seamounts, central Pacific Ocean, in *Seamounts, islands and atolls*, edited by Keating, B.H., P. Fryer, R. Batiza, and G.W. Boehlen, pp. 133-162, American Geophysical Union, Washington,D.C., 1987.

Sager, W.W., Seamount age estimates from paleomagnetism and their implications, in *Geology and offshore mineral resources of the Central Pacific Basin*, edited by Keating, B.H and B.R. Bolton, 27-37, Springer-Verlag, New York, 1992.

Schlager, W., The paradox of drowned reefs and carbonate platforms, *Geol. Soc. Am. Bull.*, *92*, 197-211, 1981.

Schlager, W. and O. Camber, Submarine slope angles, drowning unconformities and self-erosion of limestone escarpments, *Geology*, *14*, 762-765, 1986.

Schlanger, S.O., Subsurface geology of Eniwetok atoll, *Geol. Surv. Prof. P.*, *260-BB*, 991-1066, 1963.

Schlanger, S.O., J.F. Campbell, and M.W. Jackson, Post-Eocene subsidence of the Marshall Islands recorded by drowned atolls on Harrie and Sylvania guyots, in *Seamounts, Islands and Atolls*, edited by Keating, B.H., P. Fryer, R. Batiza, and G. Boehlert, pp. 165-173, American Geophysical Union, Washington, D.C., 1987.

Schlanger, S.O. and I. Premoli Silva, Tectonic, volcanic, and paleogeographic implications of redeposited reef faunas of Late Cretaceous and Tertiary age from the Nauru Basin and the Line Islands, *Init. Rep. DSDP*, *61*, 817-827, 1981.

Scott, R.W., Models and stratigraphy of mid-Cretaceous reef communities, Gulf of Mexico, *Concepts in Sedimentology and Paleontology*, *2*, S.E.P.M., Tulsa, Oklahoma, 1990.

Shepard, F.P., Lagoonal topography of Caroline and Marshall Islands, *Geol. Soc. Am. Bull.*, *81*, 1905-1914, 1970.

Smith, D.I. and T.C. Atkinson, Process, landforms and climate in limestone regions, in *Geomorphology and Climate*, edited by Derbyshire, E.D., 367-410, Wiley, London, 1976.

Smith, W.H.F., Staudigel, H., Watts, A.B., and Pringle, M.S., the Magellan seamounts - Early Cretaceous record of the South Pacific Isotopic and Thermal Anomaly, *J. Geophys. Res.*, *94*, 10501-10523, 1989.

Smith, W.H.F., Marine Geophysical studies of seamounts in the Pacific Ocean basin, *PhD. Dissertation*, Columbia Univ., New York, NY, 1990.

Smoot, N.C., The Marcus-Wake seamounts and guyots as paleo-fracture indicators and their relation to the Dutton Ridge, *Mar. Geol.*, *88*, 117-131, 1989.

Staudigel, H., K.H. Park, M. Pringle, J.L. Rubenstone, W.H.F. Smith, and A. Zimler, the longevity of the South Pacific isotopic and thermal anomaly, *Earth Plan. Sci. Lett.*, *102*, 24-44, 1991.

Stoddard, D.R., T.P. Scoffin, T. Spencer, R.S. Harmon, and M. Scott, Sea-level change and karst morphology, Mangaia (Cook Islands), *Proc. 5th Int. Coral Reef Congr.*, *3*, 201-205, 1985.

Thiede, J., T.L. Valler, et al., *Initial Reports of the Deep Sea Drilling Project*, *62*, US Government Printing Office, Washington DC, 1981.

Vogt, P.R. and N.C. Smoot, The Geisha Guyots: Multibeam bathymetry and morphometric interpretation, *J. Geophys. Res.*, *89*, 11085-11107, 1984.

Vogt, P.R., Volcanogenic upwelling of anoxic, nutrient-rich water; a possible factor in carbonate bank/reef demise and benthic faunal extinctions?, *Geol. Soc. Am. Bull*, *101*, 1225-1245, 1989.

Van Waasbergen, R.J., W.V. Sliter, ODP Leg 143 Scientific Party, Results of ODP drilling in the Mid Pacific Mountains: stratigraphy of two Cretaceous drowned atolls, *Eos, Proc. American Geophysical Union*, 1992 fall meeting, in press, 1992, (abstract).

Winterer, E.L., J.I. Ewing, and et al., Site 171, in *Initial Reports of the Deep Sea Drilling Project*, edited by Winterer, E.L., J.I. Ewing, and et al., 283-334, US Govt. Printing Office, Washington DC, 1973.

Winterer, E.L. and C.V. Metzler, Origin and subsidence of guyots in the Mid Pacific Mountains, *J. Geophys. Res.*, *89*, 9969-9979, 1984.

Wolfe, C.J. and M.K. McNutt, Compensation of Cretaceous seamounts of the Darwin Rise, northwest Pacific ocean, *J. Geophys. Res.*, *96*, 2363-2373, 1991.

R. van Waasbergen and E.L. Winterer, Scripps Institution of Oceanography, University of California San Diego, La Jolla, CA 92093-0208.

Geology and Geomorphology of Wodejebato (Sylvania) Guyot, Marshall Islands[1]

DOUGLAS D. BERGERSEN

School of Ocean and Earth Science and Technology, University of Hawaii, Honolulu, HI

For much of his life, Sy Schlanger's work centered around understanding the affects of Cretaceous volcanism on such global phenomena as eustatic changes in sea-level and ocean-wide anoxia events. The Marshall Islands, which typify the complex history of volcanism in the western Pacific Ocean, pose a number of vexing tectonic and sedimentologic questions essential to our understanding of Cretaceous events in the Pacific. Within this island group clusters of atolls and guyots lie in roughly linear chains, but the transposition of edifice ages in the direction of plate motion argues against a simple hot spot model of formation. An important step towards deciphering the geologic history of these island chains, and this region as a whole, is to more fully understand the history of their individual components. This paper examines in detail the morphology and internal structure of Wodejebato Guyot (formerly Sylvania) and discusses the constraints these observations place on the history of this drowned atoll.

Wodejebato Guyot lies ~74 km northwest of Pikinni (formerly Bikini) Atoll. A volcanic ridge 8.7 km wide and 1600 m deep attaches the two edifices. Shallow-water, Cretaceous carbonate sediments overlie volcanic basement, and in turn are overlain by 100 m of pelagic sediment. Along the edge of the summit plateau, the carbonate platform crops out from the pelagic sediments. Acoustically massive perimeter ridges, interpreted as reefs, bound the presumed lagoon sediments along the east, north and west flanks. Along the south flank, a distinct 100 m to 1000 m wide terrace lies outside of the carbonate platform. The absence of a perimeter ridge on this flank and the strong correlation between the stratigraphy of the platform sediments and that of the terrace sediments in seismic profiles suggests large-scale block-faulting affects this area on Wodejebato. High backscatter channels, presumably of debris-flow origin, mar the flanks of the guyot outside the carbonate platform.

Given the morphologic and stratigraphic observations presented in this paper, models examining the evolution of Wodejebato must take into account not only the existing radiometric and fossil evidence for two episodes of uplift, but also such information as the depth of volcanic basement along the guyot's flanks and beneath the carbonate cap, the presence of perimeter ridges along the west and north flanks, and the presumed block-faulting along the south flank. Volcanic basement lies at a depth of ~1600 m along most of the guyot flanks and an isopach map of carbonate platform thickness shows that basement shallows to the northeast across the summit plateau. Paired perimeter ridges along the north and northeast flank ridges suggest more than one episode of platform growth. Modification of the guyot flanks through block-faulting is apparently a process which continues to occur well after the edifice has moved away from hot spot swell.

INTRODUCTION

For much of his life, Sy Schlanger's work centered around trying to understand the world-wide distribution of Cretaceous volcanic features and events, including episodes of volcanism and carbonate platform growth in the Marshall Islands. Beginning early in his career with the U.S.G.S. sponsored Operation Crossroads, through a number of ship-based sampling and surveying programs, and culminating with his leadership in securing drilling time with the Ocean Drilling Program, Sy attempted to explain how Marshall Islands volcanic events (and those of a more regional scale) related to such global phenomena as eustatic sea level rises and ocean-wide anoxia events [e.g. Schlanger and Premoli Silva, 1981; Schlanger et al., 1981; Schlanger et al., 1984; Schlanger and Moberly, 1985]. As part of his ever-diligent efforts to expand the data base upon which the regional history of the western Pacific is built, Sy obtained funding to conduct a detailed surveying and sampling program across selected guyots in the Marshall Islands. This program, conducted during a R/V Moana Wave cruise (MW8805), consisted of side-scan sonar and swath bathymetry mapping, 3.5 kHz and seismic profiling, gravity and magnetic measurements, and dredging. A subsequent cruise aboard the R/V Moana Wave (MW9009) collected additional 6-channel seismic data over those guyots selected as potential drill sites. In anticipation of drilling on

The Mesozoic Pacific: Geology, Tectonics, and Volcanism
Geophysical Monograph 77

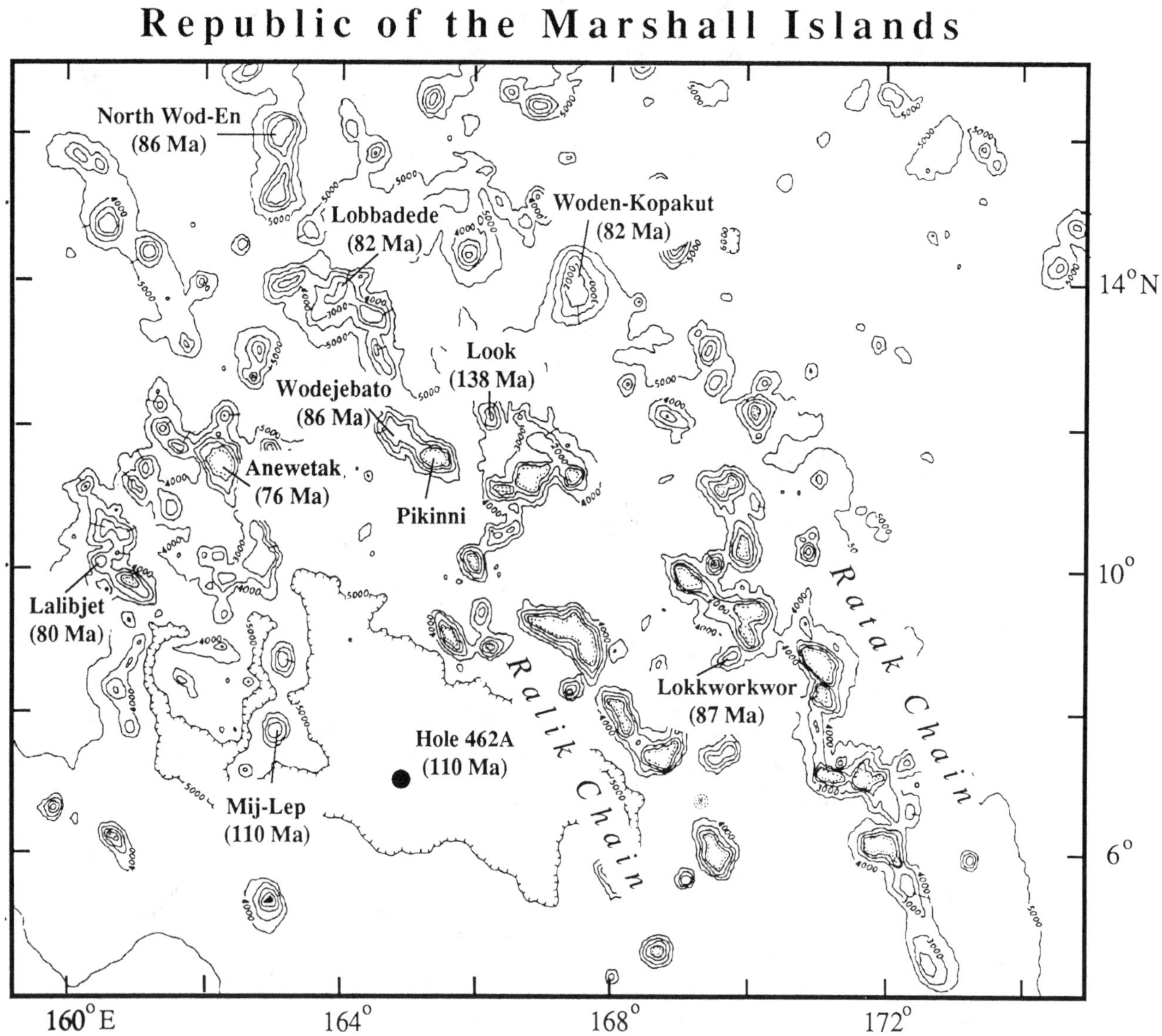

Fig. 1. Regional bathymetry of the Marshall Islands. Guyots are labeled by their Marshallese name, and only those guyots with radiometric ages are listed. Radiometric ages come from Pringle [1992], Davis et al. [1989], and Moberly et al. [1986]. Figure revised from Hein et al.[1990].

ODP Leg 144, this paper presents a portion of the data collected over Wodejebato Guyot (formerly Sylvania), and from these data proposes a model outlining the geologic history of this drowned atoll.

BACKGROUND

The Marshall Islands, located between 5° and 18° N latitude and 160° and 170° E longitude, represent a complex aggregate of mid-plate volcanoes comprising at least 3 separate island chains generally trending in a northwest direction (Figure 1). The multiple episodes of mid-plate volcanism in and around the Marshall Islands indicated by drilling results [e.g. Emery et al., 1954; Winterer, Ewing et al., 1973; Larson and Schlanger, 1981; Moberly, Schlanger et al., 1986] are difficult to delineate by such simple physical traces as hot spot chains. Radiometric ages of basalt samples from various guyots in the Marshall Islands emphasize the complications associated with identifying hot spot chains (Figure 1). Some of this scatter may result from sampling biases or analytical errors, but a simple age-progression of volcanoes in the direction of plate motion is not readily apparent. For example, even within the geographically well-defined Ratak Chain, seamount ages fail to exhibit a linear correlation to seamount position [Davis et al., 1989]. In this chain, Woden-Kopakut Guyot (formerly Ratak), the younger edifice, lies "down-drift" from Lokkworkwor Guyot (formerly Erikub).

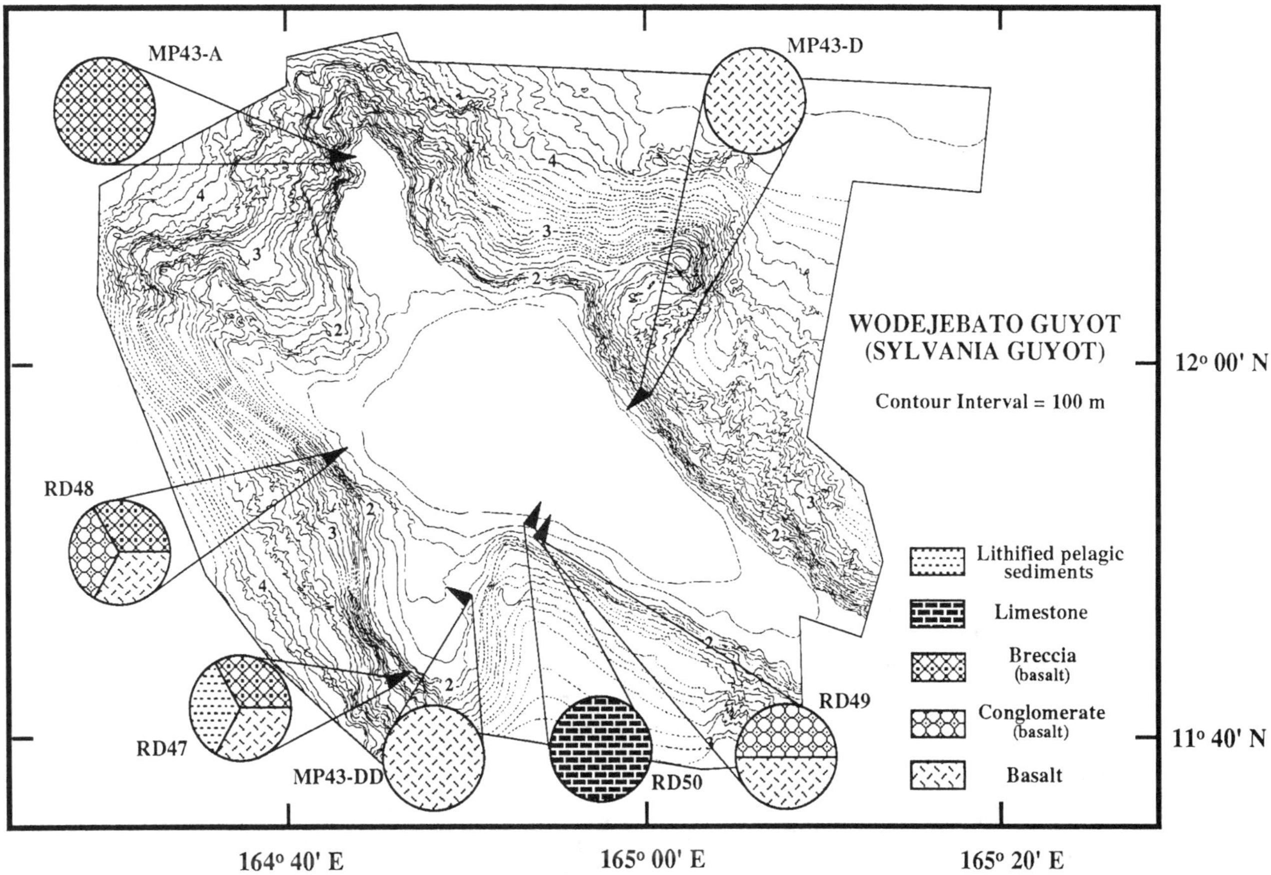

Fig. 2. Contoured SeaMARC II bathymetry over Wodejebato Guyot. Dredge locations are shown by triangles. The dredges represent work done during two cruises: the RD labeled dredges were collected during MW8805 [Schlanger and Duennebier, 1988] and the MP labeled dredges were collected during the Mid-Pacific Expedition [Hamilton and Rex, 1959].

Radiometric ages of basalt samples collected from the various guyots provide, at present, evidence for a single episode of constructional volcanism for each edifice. Additional evidence concerning the history of volcanism and carbonate platform growth on these islands comes in the form of dredged limestone. Early dredges on Wodejebato Guyot (the "MP" labeled dredges shown in Figure 2) recovered only basaltic rocks and tuff breccia [Hamilton and Rex, 1959; Duennebier and Peterson, 1982]. Eocene foraminifers fill the cracks of these rocks. This information, along with the Early Eocene limestone recovered from Limalok Guyot (formerly Harrie) to the southeast and the results from drilling on Anewetak Atoll to the west, led Schlanger et al. [1987] to propose that the latest stage of volcanism in the Marshall Islands was Eocene in age.

During MW8805, four more dredges attempted across Wodejebato recovered basalt, basalt breccia, basalt conglomerate, and limestone (dredges labeled "RD" in Figure 2). The fossil content of the limestone indicates two distinct periods of carbonate platform growth across many of the summits in the Marshall Islands [Lincoln et al., 1989]. On Wodejebato Guyot, RD50 recovered rudist fragments within a Late Cretaceous foraminifer matrix. The rudists, Albian in age, and the foraminifer matrix, Campanian to Maastrichtian in age, suggest two sequences of shallow-water carbonate accumulation separated by approximately 30 m.y. [Lincoln, 1990]. Radiometric dating of a basalt from dredge MP43-D along the north flank of Wodejebato places the latest stage of volcanism at ~86 Ma [Pringle, 1992]. Thus, while the fossil assemblage in the limestone suggests that this edifice existed during the mid-Cretaceous, the basalt date reveals an episode of volcanism at least 20 m.y. later. Evidence for similar episodes of volcanism and carbonate platform growth exists on Lo-En Guyot (formerly Hess) and Lobbadede Guyot west and northwest of Wodejebato, respectively [Lincoln, 1990]. According to Lincoln [1990] the history of volcanism in this area of the Pacific extends back to the late Jurassic/early Cretaceous with secondary phases during Albian/Cenomanian time (105 Ma to 95 Ma), Santonian/Campanian time (85 Ma to 75 Ma), and possibly into the late Paleocene/early Eocene (60 Ma to 50 Ma).

Given that volcanic overprinting of the initial sea floor

spreading event occurred in this area, it's likely that the volcano summits were subject to a complex history of subaerial exposure in response to local plate flexure from volcanic loading, regional plate flexure from mantle upwelling (the hot spot swell), and eustatic rises and falls in sea level. Two models can explain the age data and inferred tectonic and carbonate platform accumulation history for these guyots. The models differ by the degree to which a second episode of uplift affected the existing carbonate cap.

1. In the first model, uplift during the Late Cretaceous resulted in erosion of the mid-Cretaceous carbonate platform and transportation of debris into a younger platform growing primarily outside of the existing complex [Lincoln, 1990]. In this model, the older carbonate platform, lying topographically higher than the younger platform, should presumably show signs of subaerial exposure (e.g., meteoric oxygen isotope values or pronounced karst topography).
2. In the second model, substantial erosion of the older carbonate platform during the Late Cretaceous uplift left a thin veneer of carbonate sediments across the summit. The younger carbonate platform recolonized the existing topographic highs, burying the remnants of the mid-Cretaceous complex.

In light of these models, it is worthwhile to examine how the surface morphology and subsurface structure of a typical drowned atoll in this area constrain the proposed episodes of volcanism, uplift, and carbonate platform growth and erosion. The original subsurface mapping of Wodejebato Guyot was based on a limited number of digital single-channel seismic lines crossing the summit [Figures 8 and 9 of Schlanger et al., 1987]. From Operation Crossroads, estimates of Pikinni's basement depth range between 1300 m and 1600 m [Raitt, 1954; Alldredge et al., 1954]. The single-channel profiles across Wodejebato show several acoustic units typical of a completely developed atoll, specifically "a high massive rim enclosing bedded lagoonal deposits" [Schlanger et al., 1987], but the paucity of both dredge data and geophysical data across this guyot precluded any detailed geologic interpretations. With the wealth of new information collected during MW8805 and MW9009, this paper provides detailed geomorphic descriptions of Wodejebato Guyot using side-scan images, swath bathymetry, 3.5 kHz profiles, selected seismic lines, and dredge data. These new data apply important constraints to the geologic history of this drowned atoll by showing how events after the Late Cretaceous uplift are responsible for most of the existing morphology.

METHODS

Data used in this paper include side-scan images, swath bathymetry, 3.5 kHz profiles, single- and 6-channel digital seismic data, dredge descriptions, and radiometric and fossil ages. Swath mapping of the sea floor was accomplished with SeaMARC II, a shallow-towed (50 to 150 m) side-scan sonar mapping system which collected acoustic backscatter data in swaths up to 10 km wide and bathymetric data with a swath width equal to 3.4 times the water depth. The system operated at frequencies of 11 kHz on the port side and 12 kHz on the starboard side. A detailed description of the system can be found in Blackinton [1986]. At typical survey speeds of about 8 kts, over 3000 km^2 of sea floor could be surveyed per day.

Side-scan resolution is a function of many variables and is both range- and orientation-dependent. The resolution of the SeaMARC II system is on the order of several tens of meters [Johnson and Helferty, 1990]. Corrections applied to the backscatter data compensate for errors in slant range (based on water depth below the tow vehicle), beam pattern, bottom tracking, and gain variations. The data are then displayed as gray-scale images. Individual image swaths for the various tracks are assembled into a mosaic of the survey area. Dark gray to black tones represent areas of high backscatter and shades of light gray to white represent areas of low backscatter and acoustic shadows. During MW8805 the starboard transducer arrays malfunctioned, causing clipping of the data at both the high and low ends of the recording spectrum and resulting in images with few data values represented by mid-range gray tones. Further distortion of the images occurs by the application of a "flat-bottom" assumption (i.e. the sea floor is flat beneath the tow-vehicle) prior to data recording. The result is an incorrect geographical positioning of features where the bottom topography slopes in relation to the tow-vehicle track (e.g., when surveying parallel to the slope of the guyot, features on the up-dip side of the tow-vehicle are positioned anomalously far from the ship track, and vice-versa).

The measurement of phase differences between incoming acoustic signals at a pair of transducer arrays mounted on either side of the SeaMARC II tow vehicle helps determine bathymetry. Comparison of the resulting acoustic angle to a "lookup" table (generated from data collected over flat sea floor) results in an estimate of depth. Bathymetric resolution is about 2 % of the water depth [Blackinton, 1986].

A single-channel streamer using a Masscomp-HIGHRES digitizing system collected digital seismic data during MW8805. When the SeaMARC II tow-vehicle was in the water, the source was a 120 cubic inch air gun; at other times, an 80 cubic inch water gun was used. Processing of the data using the SIOSEIS software package included band-pass filtering from 10 to 100 Hz, automatic gain control, deconvolution, and muting of water noise. During MW9009, a 6-channel streamer was used with the Masscomp system. In addition to the processing routines applied to the MW8805 data, the 6-channel data were gathered, stacked, and migrated.

The preliminary dredge descriptions made during MW8805 were augmented by more detailed fossil descriptions and thin-section analyses [Lincoln, 1990]. Radiometric dating was attempted on basalts recovered during the Mid-Pacific Expedition [Hamilton and Rex, 1959] and MW8805. Fossil ages presented in this paper are from Lincoln [1990], while radiometric ages are from Pringle [1992].

GEOMORPHIC DESCRIPTION

Early measurements on Pikinni Atoll and Wodejebato Guyot come from the late 1940's Operation Crossroads [Emery et al., 1954]. Pikinni Atoll, 42 km long from east to west and 24 km wide from north to south, lies around 12° N and 165° E. Its 26 islands, elongated in a northwest direction, encompass approximately 440 km^2. Subsurface information on Pikinni comes primarily from the four holes bored on the northeast side of the atoll. The deepest of these holes penetrated 780 m of shallow-water carbonate sediments of Early Miocene to Oligocene age [Cole, 1954; Emery et al., 1954]; none of the holes encountered basalt. As the subsidence histories of Pikinni and Anewetak Atoll, lying over 300 km to the west, were considered coeval based on the correlation of their stratigraphy [Schlanger, 1963], the basement depth of Pikinni was inferred to lie between 1300 m and 1600 m.

The center of Wodejebato Guyot lies approximately 74 km to

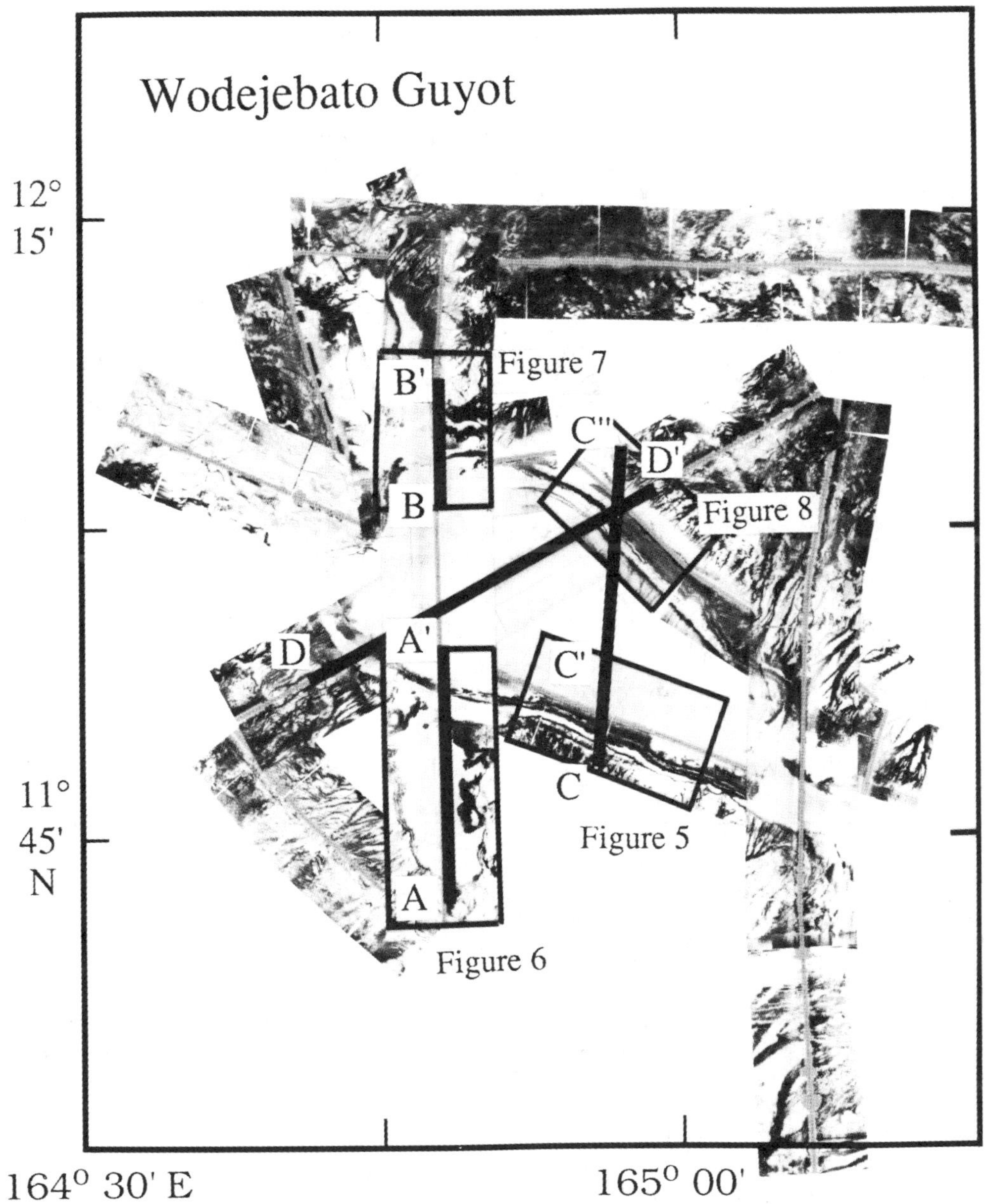

Fig. 3. Side-scan mosaic of Wodejebato Guyot showing the gross morphology of the guyot and the drowned Cretaceous carbonate platform (innermost dark band) capping the volcanic edifice. The boxes with figure labels indicate areas chosen for detailed discussion. Thick black lines show the location of seismic profiles across the south flank ridge (A-A'), the north flank ridge (B-B'), the south flank (C-C'), and two profiles used to interpret the drowned carbonate platform (C-C'' and D-D').

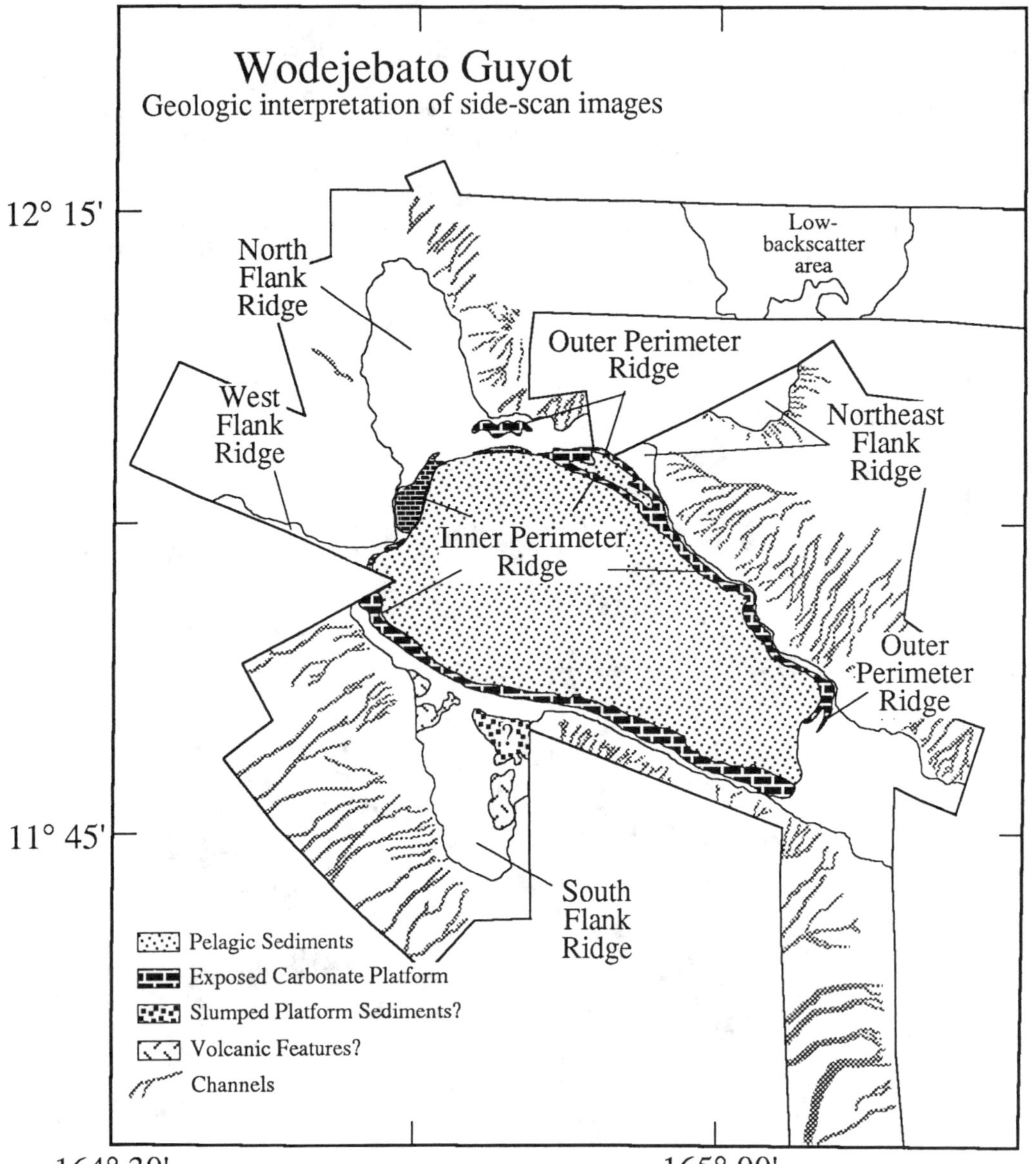

Fig. 4. Geologic interpretation of the side-scan mosaic based on rocks recovered during dredging (compare to Figures 2 and 3) and selected seismic profiles. Although pelagic sediments in this figure are shown only across the summit plateau, they also cover the shelves formed by the flank ridges.

the northwest of Pikinni Atoll, a distance slightly less than that separating Hawaii (Kohala volcano) from East Maui Volcano in the Hawaiian Islands. A volcanic ridge 20 km long, 8.7 km wide and ~1600 m deep connects the two edifices. The summit of Wodejebato is about 43 km long (three-fourths the size of East Maui Volcano) and increases in width from less than 12 km in the southeast to greater than 25 km in the northwest (Figures 2, 3, and 4). Four volcanic ridges project from the northwest half of the edifice, and along with the volcanic spur attaching Wodejebato to Pikinni give the guyot a distinct "starfish" appearance. A drowned carbonate platform, slightly inset from the steep flanks of the guyot, covers the summit plateau (Figures 3 and 4). The regional gradient divides the flanks of the guyot into two provinces: a steep upper slope (~20°-24°) gives way to a more gently inclined lower slope (~7°). In general, the transition depth between upper and lower slopes is around 2500 m. The following subsections describe and discuss the importance of various areas and features across Wodejebato for constraining the evolution of this drowned atoll (see Figure 3 for area locations).

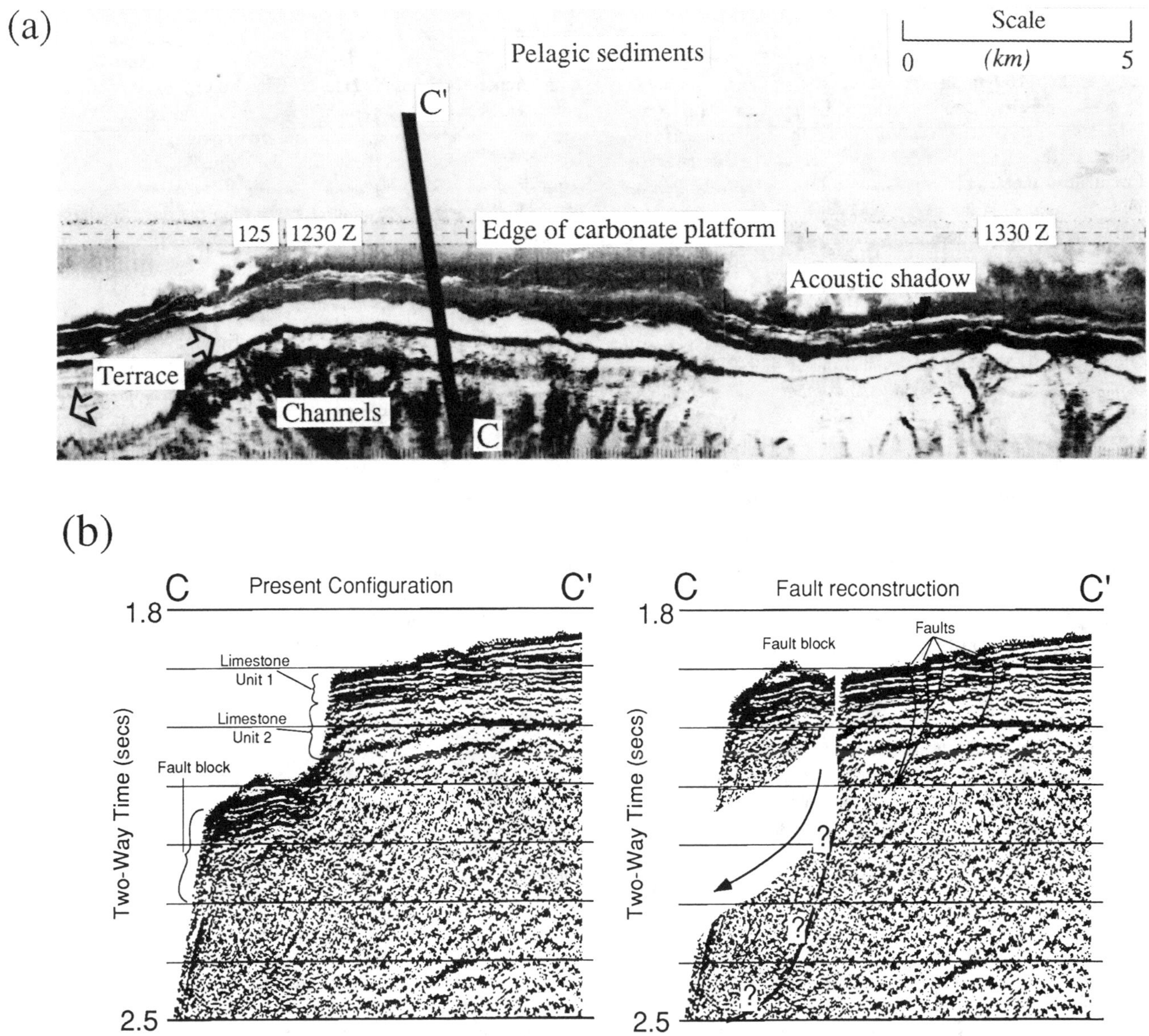

Fig. 5. Side-scan image (a) and seismic profile (b) across the south flank of Wodejebato. Along this flank, the exposed carbonate sediments are interpreted as lagoon or back reef deposits. In the 6-channel seismic profile, these presumed lagoon sediments (represented by the horizontal reflectors) do not truncate against an acoustically-massive perimeter ridge. Dredge RD50 sampled the scarp above the terrace and recovered only carbonate sediment. Dredge RD49 recovered only basalt from the scarp below the terrace in ~1800 m water depth. The upper 0.1 sec of terrace reflectors correlate with those of the topographically higher carbonate platform, suggesting that the terrace is a down-dropped block of carbonate sediments.

Southern Flank

The carbonate platform cropping out in this area appears in the side-scan images as a fairly continuous band of high-backscatter material disrupted along-strike by thin acoustic shadows (Figure 5a). Small-scale changes in topography related to normal faulting or slumping of the carbonate sediments cause these acoustic shadows. The scarp truncating the carbonate platform is about 90 m high. In the side-scan images, the band of exposed carbonate sediments is between 500 m to 2000 m wide along-strike. Seismic data across this flank (Figure 5b) suggest the exposed sediments are lagoonal in origin; horizontal reflectors typical of lagoon sediments across the summit plateau do not appear to truncate against an acoustically-massive perimeter ridge.

Seaward of the carbonate platform lies a terrace, shown in the side-scan images as a region of low-backscatter between the high-backscatter carbonate sediments and the channels marking the upper slope. The terrace lies at a depth of around 1600 m in this area, but shallows to the northwest. It widens from less than 100 m in the southeast to over 1 km in the northwest, where it disappears along the southern flank ridge. In the seismic data, the stratigraphy of the presumed lagoon sediments matches that of the terrace (Figure 5b). Consequently, the terrace represents a down-dropped block of lagoon sediments.

Dredge hauls along the slopes below and above the terrace recovered basalt and limestone, respectively (Figure 2). RD49, sampling the upper slope of the edifice (~1800 m water depth), collected a mixture of basalt conglomerates containing rounded basalt pebbles, hyaloclastites, and pieces of previously formed conglomerates, but no reef debris. Lincoln [1990] noted that Mn coatings 2 to 3 cm thick on earlier-generation breccias and basalt pebbles suggest considerable hiatuses between erosional events. RD50, sampling the carbonate scarp at ~1500 m water depth, recovered about 300 lbs of Mn-encrusted limestone. The limestone contained Albian-age whole and fragmented rudists (along with other shell fragments) within a Campanian- to Maastrichtian-age foraminifer matrix [Lincoln, 1990].

Flank Ridges

As mentioned previously, four volcanic ridges extend from the northwest half of Wodejebato. These volcanic ridges are probably equivalent to the flank rift zones noted by Vogt and Smoot [1977] on seamounts and guyots in the north Pacific. Their relationship to the zones of active volcanism observed on the islands of Hawaii (i.e., loci of vents fed by dikes) is unclear. In this paper I refer to the volcanic ridges extending radial to the edifice as flank ridges; an examination of their origin follows their description.

The southern and northern flank ridges are the most prominent, appearing as broad (~4.5 km wide), relatively flat shelves extending ~15 km outward from the summit plateau (Figure 3). Although less broad than their northern and southern counterparts, the western and northeastern flank ridges extend outward about the same distance.

The southern flank ridge deepens away from the summit by 500 m, at a gradient of about 2°. In the side-scan images it is difficult to discern where the terrace, so prominent along the south flank, merges with the flank ridge (Figure 6a). The high backscatter band marking the areas where the carbonate platform crops out from the overlying pelagic sediments coalesces from two narrow bands in the east (lower half of the swath) to a single band in the west (upper half of the swath). No apparent perimeter ridge bounds the presumed lagoon sediments (Figure 6b). The first break in slope down from the summit is around 1500 m, suggesting that either the fault-block terrace shallows towards the northwest or portions of the carbonate platform slump on to the true terrace. Given the faulting to the east of this flank ridge, it is likely that the two narrow bands seen in the side-scan images are fault-related and that the area to the west of profile A-A' is relatively unaffected by the erosion occurring along the southern flank. Along the edge of the flank ridge, away from the summit plateau, high-backscatter features crop out from the overlying pelagic sediment (Figure 6a). A few of these features resemble cones with diameters around 2 km. The height of the cones is unknown because the swath width of the SeaMARC II bathymetry was too narrow to provide coverage at this depth. A fathogram collected during the Mid-Pacific expedition [Emery et al., 1954, Figure 57, profile 9] possibly crosses one of the high backscatter cones. If so, then the cone in the area of profile 9 is less than 60 m high. Dredge MP43-DD, from about 1600 m water depth in this area, recovered 200 lbs of coarse-grained olivine basalt tuff. In contrast, RD47 sampled the lower slope of this flank ridge and recovered basalt, hyaloclastic conglomerates, and lithified pelagic sediments from a water depth of around 3250 m. Seismic data across this flank ridge show an undulating layer of pelagic sediment covering what may be a thin layer of slumped sediment or volcaniclastics (Figure 6b). As alluded to previously, horizontally-bedded, presumably lagoonal sediments crop out along the edge of the carbonate platform and an acoustically-massive perimeter ridge is absent. As neither dredge recovered rocks directly related to the drowned carbonate platform, the shallow-water carbonate sediments do not appear to extend on to the flank ridge shelf. If indeed the carbonate platform is eroding through block-faulting, some of the high backscatter features seen across the flank ridge near the summit edge may simply be slumped carbonate sediment. On the basis of the basalt tuff recovered in dredge MP43-DD, at least some of the cones visible across the flank ridge shelf must be volcanic in origin.

Moving clockwise around the guyot, side-scan data over the western flank ridge only partially map this feature, limiting the available geomorphic information (Figures 3 and 4). As stated previously, its length is comparable to the other flank ridges but it does not appear to be as planar as its northern and southern counterparts. North of the flank ridge, well-developed channels along the western flank of Wodejebato are noticeably absent in the side-scan images (Figure 3). This absence of channels could result either from a low-gradient upper slope hindering channel formation or from "recent" large-scale erosion along this flank erasing the acoustic record of such channels. Regional bathymetry in this area reveals the gradient to be a fairly uniform ~6°, much less than the 20° to 24° upper slope along the southern and northern flanks. The cause for this lower gradient is unclear at this time.

The northern flank ridge is similar in length and width to its southern counterpart, but it does not appear to dip away from the summit. The 1600 m contour outlines the flank ridge shelf in the bathymetry (Figure 2). A dredge haul (MP43-A) across the upper slope (1900 m to 1600 m) of this ridge recovered tuff breccia [Emery et al., 1954], again suggesting that the carbonate platform does not extend on to the shelves. Unlike the southern flank ridge, the side-scan data do not show high-backscatter cones across the shelf (Figure 3). Instead, two perimeter ridges appear as high-backscatter bands across the area where the flank ridge intersects the volcanic edifice (Figure 7a). In contrast to the flank ridges which extend radial

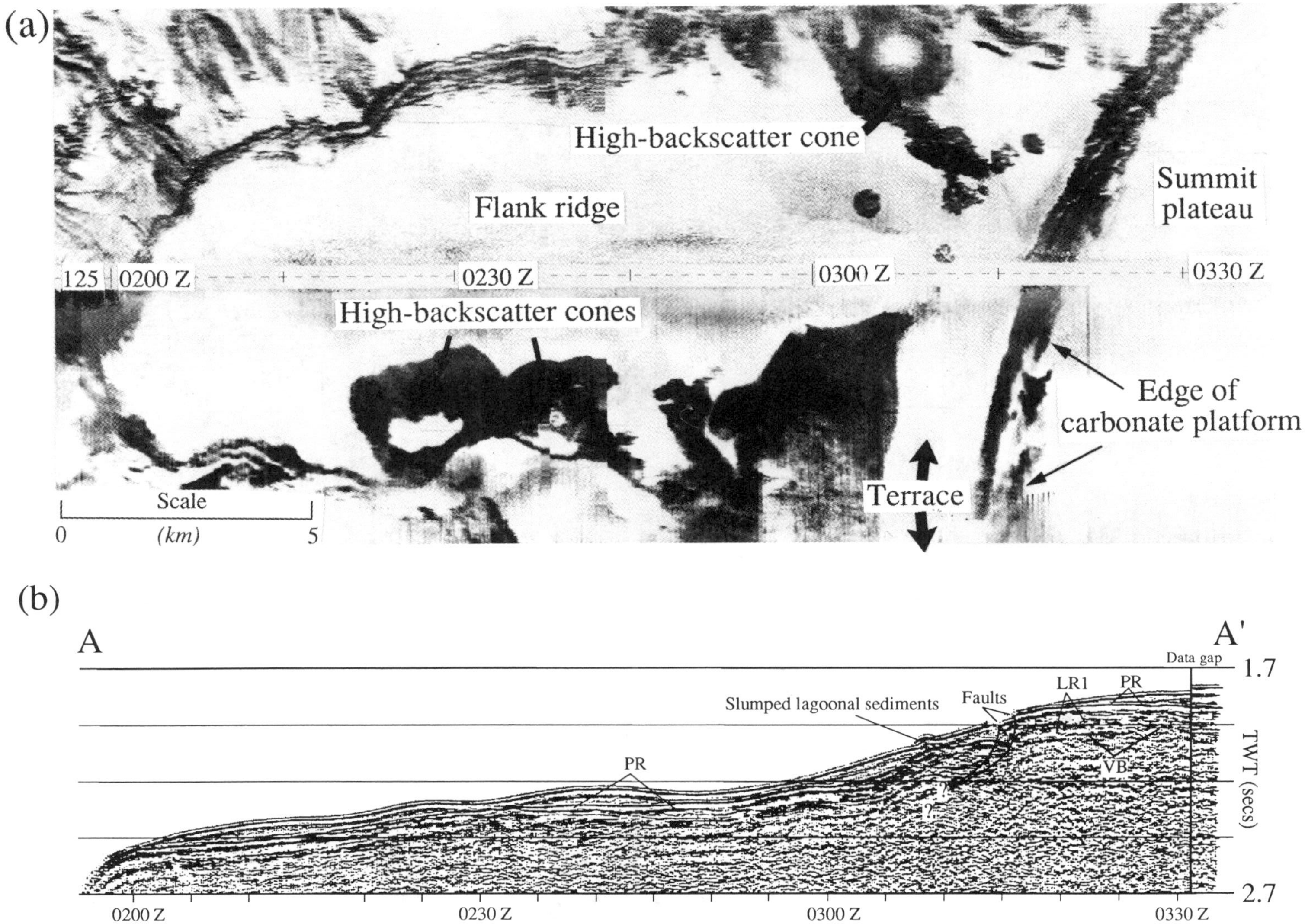

Fig. 6. Side-scan image (a) and seismic profile (b) across the south flank ridge. Profile A-A' shows that the block-faulting noted along the south flank continues to erode the carbonate platform. The reflectors labeled in the seismic profile represent the base of pelagic sediments (PR), the base of Limestone Unit 1 (LR1), and the base of Limestone Unit 2 or the top of volcanic basement (VB). Figures 9 and 10 show how these units vary across the summit of Wodejebato. The vertical exaggeration of profile A-A' = 6.5X.

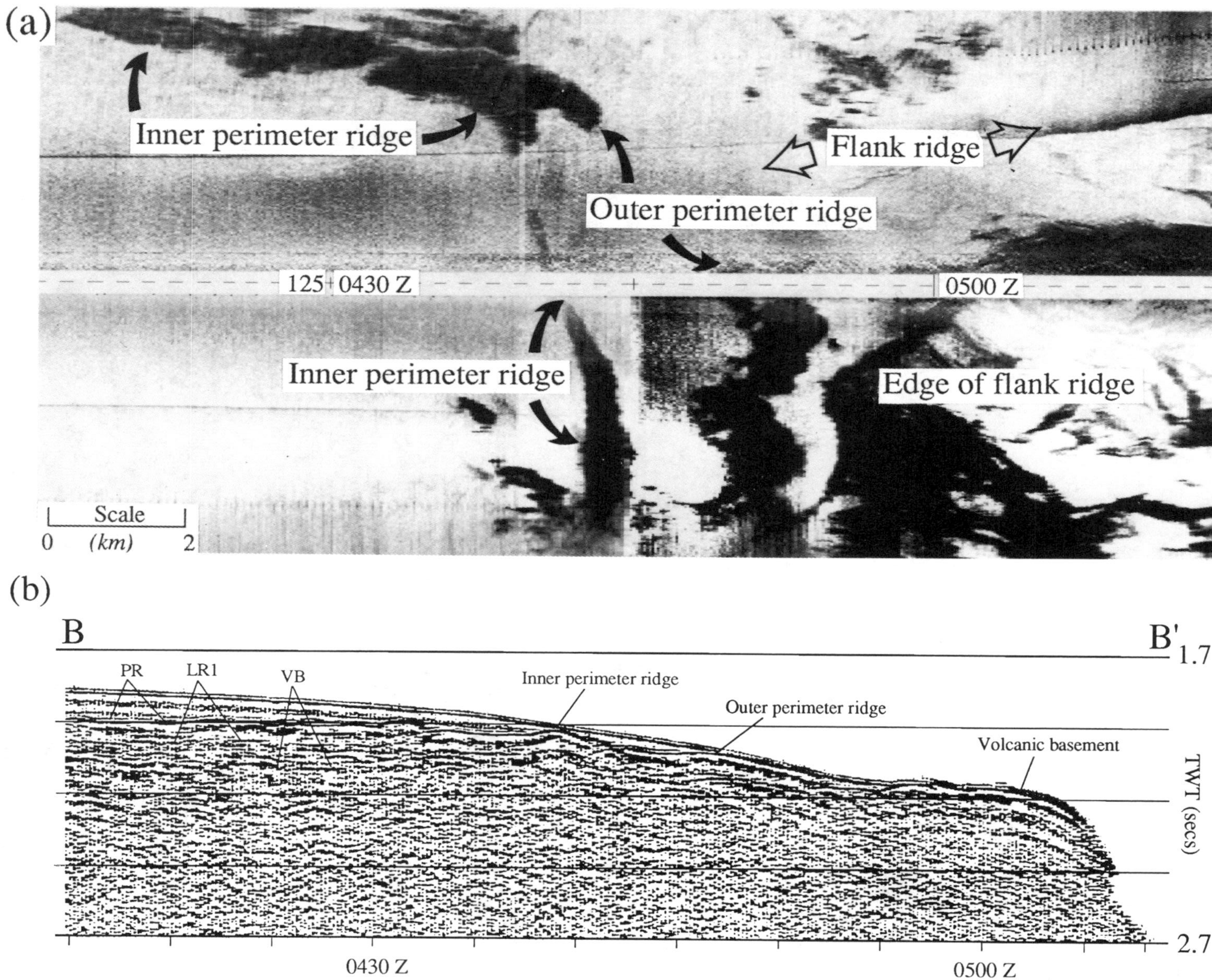

Fig. 7. Side-scan image (a) and seismic profile (b) across the north flank ridge. The side-scan image shows two perimeter ridges (represented by the high backscatter bands) cropping out from the pelagic sediments. These ridges are interpreted to be carbonate buildups. The inner perimeter ridge is relatively continuous around most of the guyot, while the outer ridge only appears on the north and northeast flank ridges, and the volcanic ridge attaching Wodejebato Guyot to Pikinni Atoll. The reflectors labeled in B-B' are the same as those listed in Figure 6. Vertical exaggeration of B-B' = 4.3X.

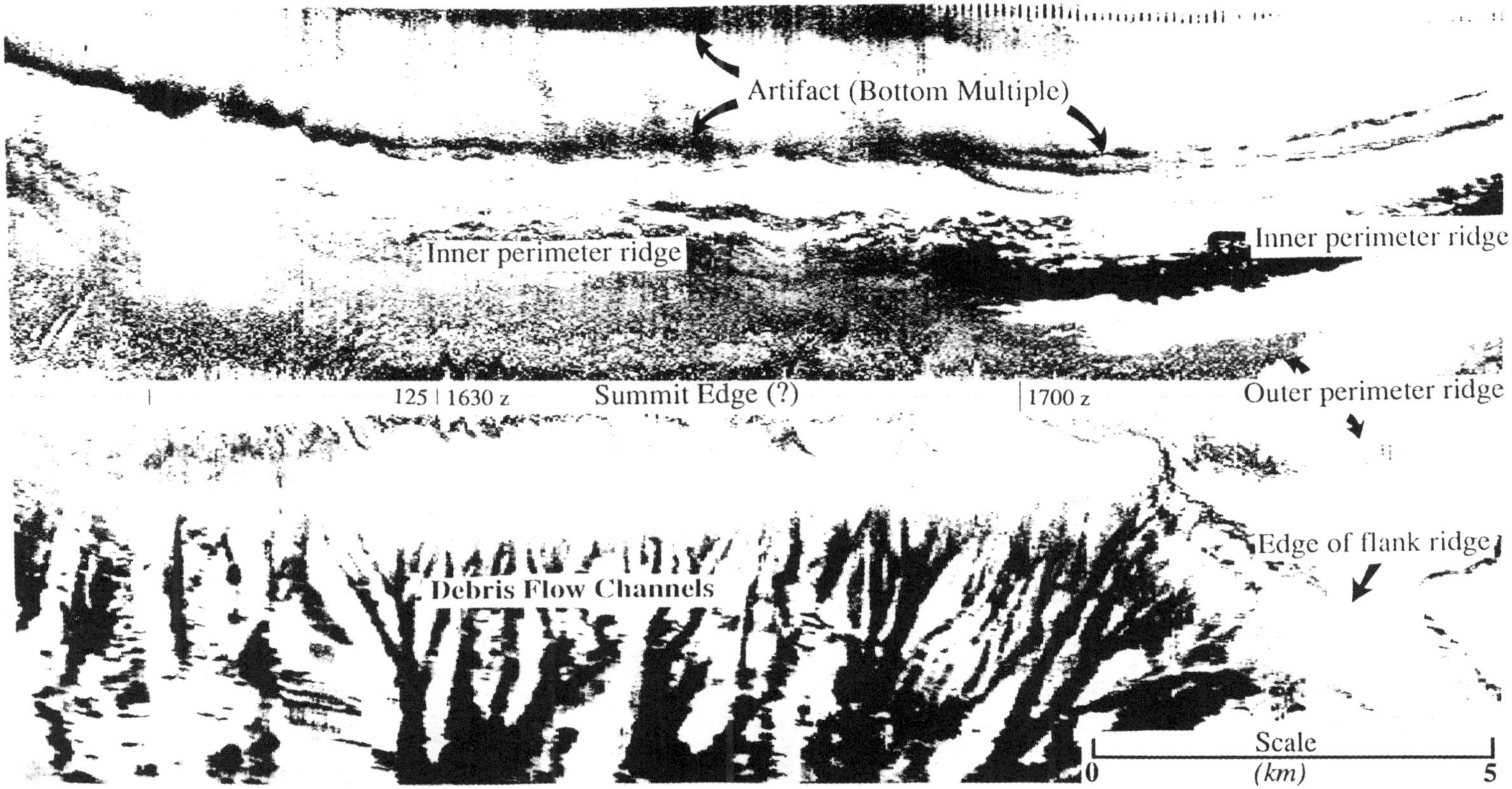

Fig. 8. Side-scan image showing a portion of the northeast flank ridge and northern flank. The two perimeter ridges visible in the side-scan images adjacent to the north and northeast flank ridges are indistinguishable from one another along the north flank. Seismic data across the north flank (not presented) do not show an outer perimeter ridge, suggesting either that this feature is unique to areas where a broad shelf extends away from the guyot flank (i.e., the flank ridges) or that it has eroded.

to the edifice, the perimeter ridges parallel the summit plateau edge. In the seismic data, the perimeter ridges in this area appear as low-relief (~75 m) mounds covered by a thin layer of pelagic sediment (Figure 7b). The inner ridge is slightly shallower (~45 m) and less broad than the seaward ridge. The two ridges extend around the summit along the northern flank, becoming more broad across areas where the shelf margin widens (i.e., adjacent to flank ridges) and finally disappearing on the volcanic ridge attaching Wodejebato to Pikinni. Given their location along the edge of the summit plateau, their relatively continuous distribution around the summit, and their mound-like morphology in seismic profiles, these features probably represent reefs.

The northeast flank ridge exhibits a unique morphology in comparison to the other flank ridges. As opposed to extending from the summit as a broad shelf, a narrow (~1.7 km) volcanic spur connects a 300 m high hill to the main edifice (Figure 2). The hill, elongate in the direction of the summit, is about 2.2 km in diameter. As in the case of the north flank ridge, the perimeter ridges broaden across the region where the flank ridge intersects the edifice (Figure 8).

Vogt and Smoot [1984] discuss factors controlling the number and length of flank rift zones on edifices in the Emperor, Geisha, Michelson, Dutton, and Mid-Pacific chains. They conclude that the height of the edifice correlates with the length of the rift zones and is independent from the number of rift zones. The flank ridges on Wodejebato Guyot are comparable in length and number to those observed in the Geisha guyots, although no direct evidence exists in either the seismic data or the dredge data confirming a rift zone origin for these features. In the Hawaiian chain, volcanoes generally exhibit only two rift zones, although sometimes a third smaller rift exists [J. Sinton, personal communication]. Consequently, the four flank ridges on Wodejebato Guyot could be attributed to: 1) some feature unrelated to Hawaiian-type volcanic rift zones; 2) a magma plumbing system which behaved differently from that of Hawaiian volcanoes, or 3) a multiple-volcano platform. From the side-scan and bathymetry data it appears all the flank ridges are attached to the main volcanic platform of Wodejebato and hence it is unlikely that any of the ridges are remnants of large-scale slumps or debris flows. It is possible that the northeast flank ridge represents a small seamount attached to the main edifice by a narrow volcanic ridge but the absence of age information or geophysical data directly over this feature reduces such a hypothesis to mere speculation. In a similar vein, there is not enough information at present to discern how the magma plumbing systems of Cretaceous volcanoes may have behaved differently from those associated with the two- to three-rift zone volcanoes in the Hawaiian chain. Perhaps the simplest explanation for the four flank ridges on Wodejebato is that they represent the rift zones from a multiple-volcano platform.

Northern Flank

Along the northern edge of the summit plateau, the two distinct perimeter ridges observed across the north and northeast flank ridges merge to become essentially indistinguishable from one another in the side-scan images (Figure 8), and barely identifiable in the seismic data. Near the northeast flank ridge the inner perimeter ridge is ~75 m high and extends ~35 m shallower than the outer perimeter ridge. The outer ridge is ~115 m high. Although not visible in the SeaMARC II data, a narrow basement shelf (< 100 m wide) lies outside of the carbonate platform at ~1600 m depth [H. Staudigel and R. van Wassbergen, personal communication]. Channels, presumably from debris flows, incise the flanks outside the reefs. Upper and lower slope gradients are comparable to those measured along the south flank (>20° and around 7°, respectively).

Drowned Carbonate Platform

A layer of pelagic sediment covers the drowned carbonate platform and hence the summit plateau appears fairly featureless in the side-scan images (Figure 3). Seismic profiles provide information on the internal structure of the carbonate platform and the depth to volcanic basement. Figures 9 and 10 show two representative profiles across the summit. The migrated 6-channel line (Figure 9) was used as a tie-line for the single-channel profiles.

This paper identifies two major lagoon sediment units (Figure 9). The lowermost unit (Limestone 2) exhibits a maximum thickness of 0.11 sec (two-way travel time, TWT) or 165 m assuming a velocity of 3.0 km/sec, and constitutes the majority of carbonate sediment deposited on this guyot. Within this unit, horizontal reflectors define subunits which onlap the basement high. Age information on these subunits awaits drilling during ODP Leg 144. Overlying the lower limestone unit are a series of closely spaced reflectors (Limestone 1). The reflector defining the bottom of this unit could be a diagenetic horizon or a sequence of overbank sediments deposited shortly before the carbonate platform drowned. Both limestone units show relatively little relief (essentially horizontal reflectors). As noted previously, faulting exposes these units along the southern flank of Wodejebato. The overlying pelagic sediments attain a maximum thickness of ~100 m near the center of the summit plateau and thin towards its edges.

The basement reflector identified in this paper may represent a layer of weathered volcaniclastics, as opposed to "true" volcanic flows, but it still marks the lowermost extent of the carbonate sediments. In figures 9 and 10, the basement reflector shallows towards the center of the summit plateau. Profile C-C" (Figures 3 and 9) shows that the basement high extends almost into the pelagic sediments. The two perimeter ridges along the northeast flank ridge are prominent features in this profile. Referred to earlier in the paper as being "acoustically massive", it is apparent from Figure 9 that the ridges exhibit some internal structure. A 0.03 sec (TWT) packet of strong reflectors overlies 0.12 sec of less-defined reflectors. The relatively thin packet of strong reflectors appears to extend ~2 km behind the inner ridge, and may be related to the Limestone 1 unit identified across the interior portions of the platform. The seaward ridge is 125 m high whereas the inner ridge, extending about 50 m shallower than

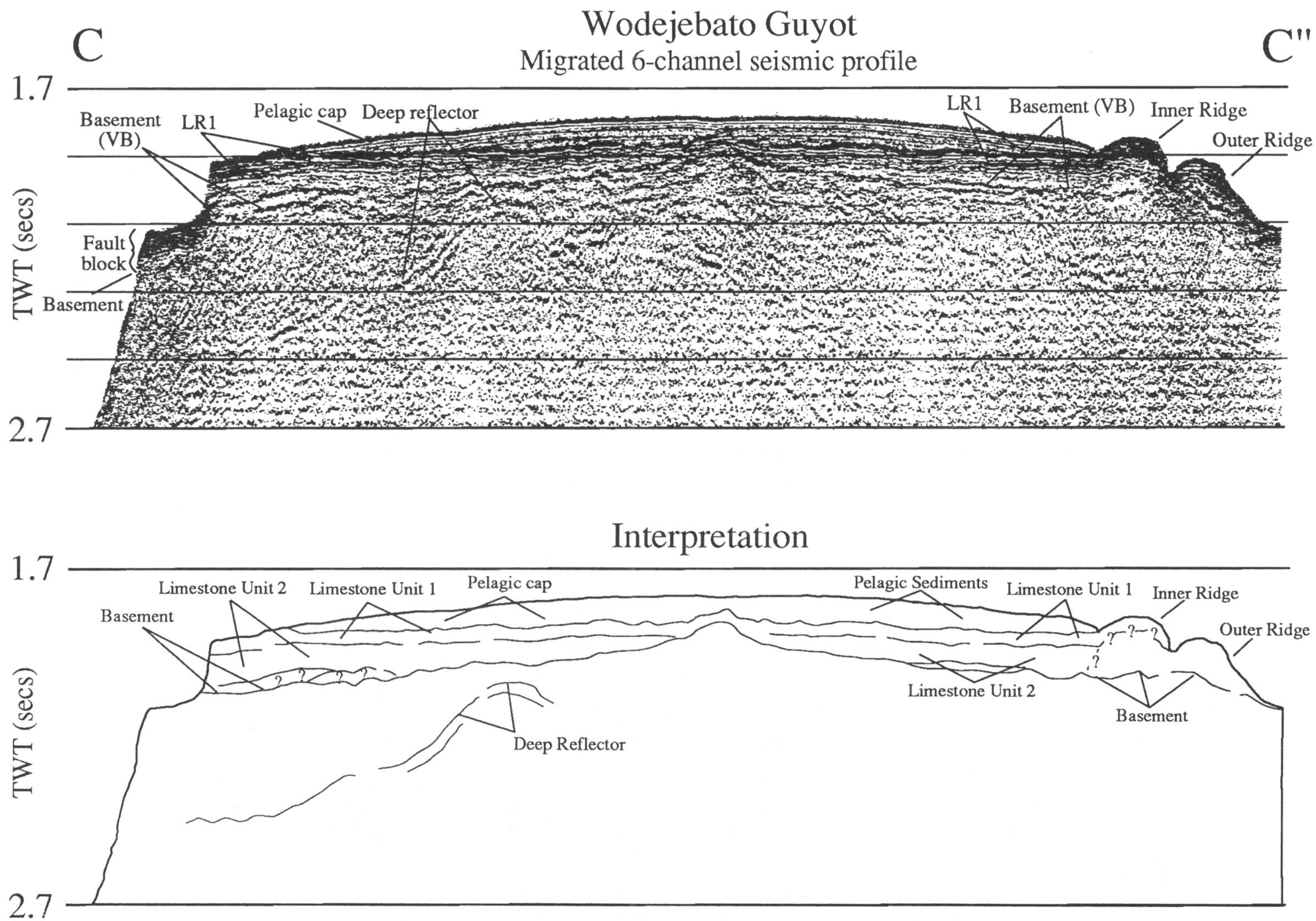

Fig. 9. 6-channel seismic profile extending from the south flank to the northeast flank ridge (profile location shown in Figure 3). The reflectors labeled in C-C" are the same as those listed in Figure 6. The reflector interpreted as volcanic basement shallows towards the center of the guyot, almost extending into the pelagic sediments. Subunits within Limestone Unit 2 onlap the basement high. The packet of strong reflectors forming Limestone Unit 1 lies conformably over the lower carbonate unit and the inner perimeter ridge. The basement reflector beneath the perimeter ridge is located around 2.0 sec. The nature of the deep reflector shown in this figure is unclear; it may represent a flow unit overlain by volcaniclastic sediments. Vertical exaggeration of C-C" = 7.5X.

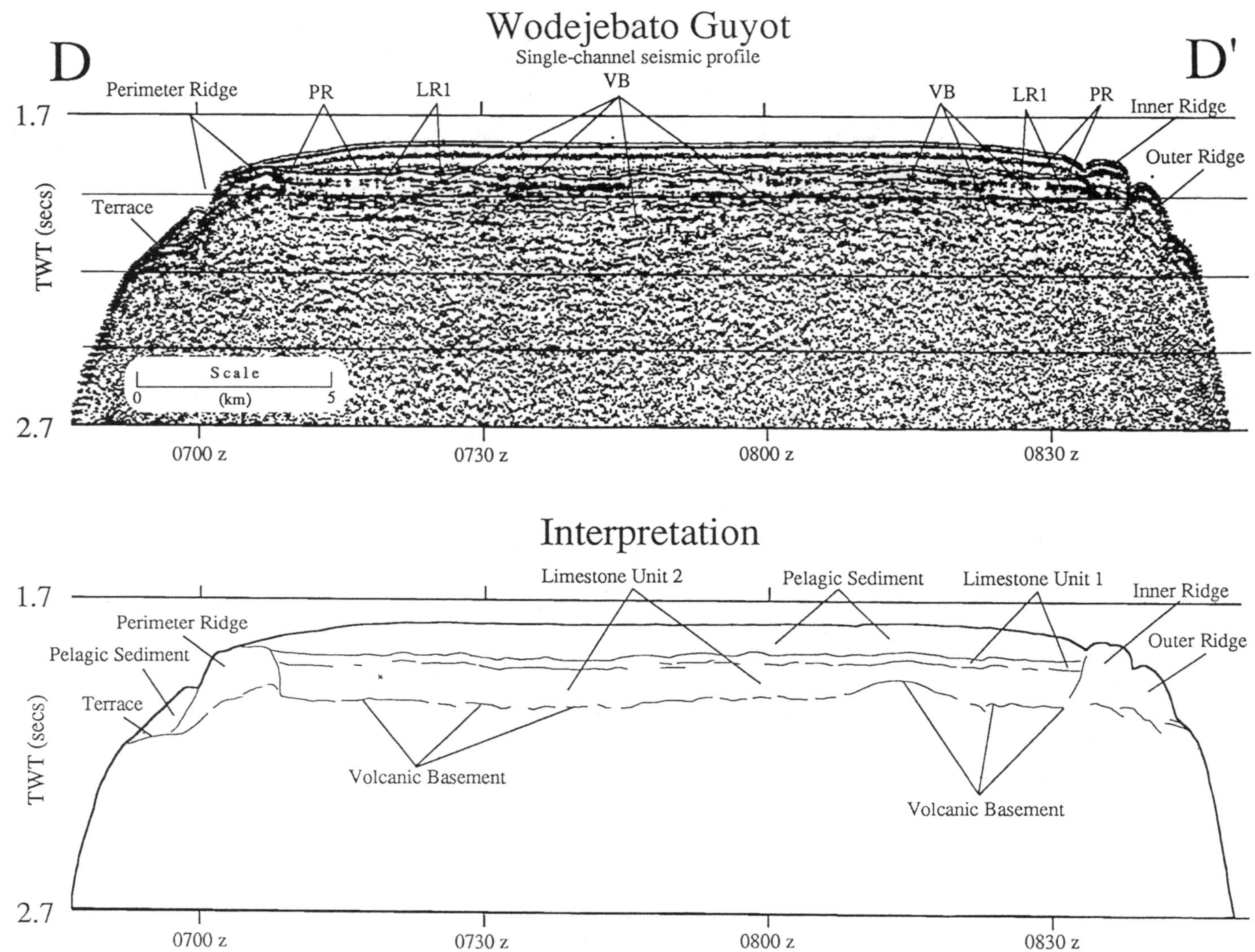

Fig. 10. Seismic profile extending from west of the south flank ridge to east of the northeast flank ridge (profile location shown in Figure 3). The peak in volcanic basement is less pronounced and displaced to the northeast relative to profile C-C''. A perimeter ridge appears to exist along the southwest flank of Wodejebato, and it may be a continuation of the inner ridge noted along the north and northeast flank ridges. It is difficult to determine in the single-channel profiles the depth to which the perimeter ridge extends. Limestone Unit 2 appears to partially onlap the basement high. Vertical exaggeration of D-D' = 11X.

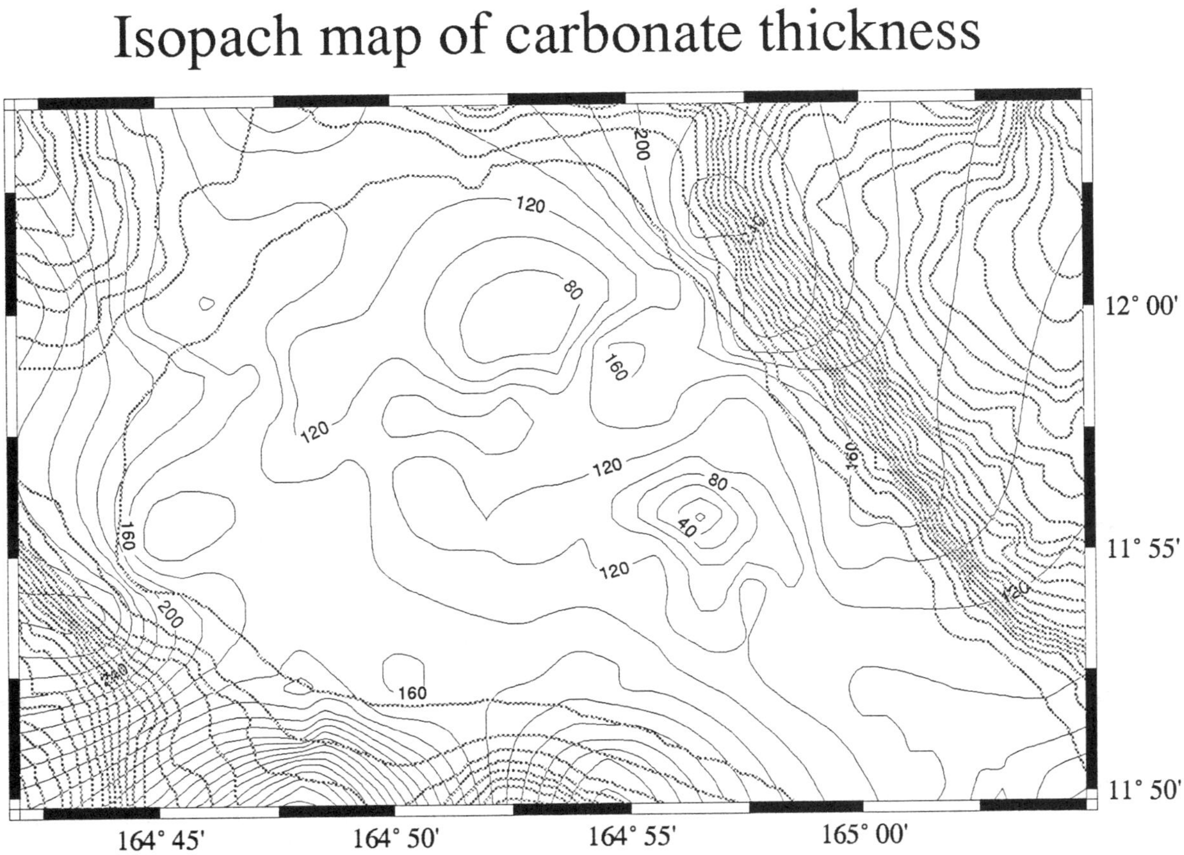

Fig. 11. Isopach map of total carbonate thickness (Limestone Units 1 and 2) over the summit plateau of Wodejebato. The thin black contours represent carbonate thickness in millisec, while the thicker gray contours represent bathymetry (showing the geographic extent of the summit plateau). As the interface between the carbonate and pelagic sediments is a relatively flat surface, the carbonate thickness contours actually map the basement topography. Volcanic basement shallows to the northeast, and apparently has two peaks displaced from one another by ~5 km.

the seaward ridge, is around 90 m high. In profile B-B' (Figures 3 and 10), the basement peak is offset to the northeast. The single, 150 m to 200 m high perimeter ridge shown in this profile suggests that this flank is unaffected by the faulting occurring along the south flank farther to the southeast.

The isopach map of carbonate thickness shown in Figure 11 maps basement topography. Areas where the carbonate platform sediments thin represent basement highs. Volcanic basement shallows to the northeast and appears to have two peaks. Although the apparent peaks in basement provide additional support for the hypothesis that two volcanoes compose the main edifice of Wodejebato, the size of these volcanoes must have been small in comparison to Hawaiian volcanoes as the maximum summit diameter of Wodejebato is only about three-fourths the size of East Maui Volcano. Alternatively, the trough separating the apparent peaks in basement could be an artifact of the wide ship-track spacing.

DISCUSSION

An important step towards deciphering the volcanic and tectonic history of the Marshall Islands as a whole is to more fully understand the history of its individual components. With the wealth of information now available through dredging and geophysical surveying (and soon drilling), it's inevitable that the existing interpretations for this area of the Pacific plate will undergo revisions. An example of such revisions is the reclassification of the carbonate platform capping Wodejebato Guyot. The Eocene age proposed by Schlanger et al. [1987] has been replaced by a Cretaceous age derived from the MW8805 dredge data [Schlanger and Duennebier, 1988]. This preliminary observation has in turn been revised to include two stages of carbonate platform growth over the course of two episodes of tectonic uplift [Lincoln et al., 1989; Lincoln, 1990]. The data presented in this paper provide a more detailed picture of Wodejebato's morphology, and furnish some additional insights to its geologic history. Any model explaining the evolution of Wodejebato must take into account not only the radiometric and fossil evidence for two episodes of uplift and volcanism, but also such geomorphic information as the depth of volcanic basement along the guyot's flanks and beneath the carbonate cap, the presence of the perimeter ridges along the eastern, northern, and western flanks, and the existence of the terrace along the southern flank.

Implications of Guyot Morphology

The depth of volcanic basement along the flanks is about 1600 m. This contour outlines most of the flank ridges on this guyot, and marks the first break in slope down from the drowned carbonate platform along margins where an acoustically massive perimeter ridge exists. Along the southern flank of Wodejebato, volcanic basement can not be deeper than 1800 m as shown by RD49. From the isopach map of carbonate thickness (Figure 11), it's apparent that volcanic basement shallows towards the northeast, although the implications of this shallowing are unclear.

The perimeter ridges extending around the northern flank of Wodejebato may represent the two episodes of carbonate platform accumulation suggested by fossil and radiometric data [Lincoln, 1990], but they could just as well have formed during a single episode of uplift. From the seismic data alone it is unclear to what depth the reefs extend, or upon what type of substrate they grew (i.e., volcanic basement or existing carbonate sediments). The strong reflector at ~2.01 sec (TWT) in Figure 9 may be volcanic basement, but testing of this hypothesis awaits drilling during Leg 144. Without drilling information, it's also impossible to determine the age of lagoon sediments identified in the seismic profiles. In lieu of this additional information, three different growth models may apply to Wodejebato's drowned carbonate platform:

1. The perimeter ridges mark two distinct periods of platform growth (i.e. mid-Cretaceous and Late Cretaceous) with the younger ridge lying outside of and deeper than an older carbonate platform [Lincoln 1990].
2. The perimeter ridges again mark 2 distinct periods of platform growth, but in this case the younger ridge (and carbonate platform) lies inside of and shallower than the older carbonate platform.
3. Both ridges (and the carbonate platform as a whole) grew during the last stage of uplift but are offset from one another because of some mechanism other than regional uplift (e.g., a eustatic sea level change or plate flexure from loading on nearby volcanoes).

If the carbonate platform grew during separate episodes of hot spot induced uplift, then the first or second growth models apply. The absence of a pronounced karstic surface in the seismic profiles weakens support for the first model. It also seems unlikely that subsidence rates after the Late Cretaceous uplift event were so great that a younger carbonate platform was unable to grow on top of the existing platform. The second growth model conforms better to observed reef growth patterns along modern shelves, where younger reefs tend to colonize preexisting highs formed by older reefs [e.g. Purdy, 1974; Hopley, 1982], but the third growth model is equally applicable given the current data. The ultimate test for these models will come with the drilling. If the first model applies, then the lagoon sediments should be primarily mid-Cretaceous in age and should contain evidence for extensive subaerial exposure and diagenesis. If the second or third models apply, Late Cretaceous lagoon sediments should unconformably overlie a thin veneer of mid-Cretaceous sediments. In this case, drilling directly behind the perimeter ridges may determine whether the ridges formed contemporaneously or during two separate uplift events.

For this study, it's important not to focus too much on when the perimeter ridges grew or what mechanism was responsible for their growth (i.e., tectonic or eustatic). The important point here is that a complete atoll existed across the summit at the time of platform drowning. Alterations to the expected atoll morphology (i.e., a massive reef rim enclosing layered lagoon deposits) place constraints on the timing and the extent of selected tectonic events. For example, the 100 to 1000 m wide terrace seaward of the carbonate platform on the southern flank and its proposed fault-block origin suggest that relatively large-scale submarine landslides continue to modify the flanks of volcanic islands well after they have moved away from the hot spot. In the case of Wodejebato, the faulting had to occur after the atoll drowned otherwise a new perimeter ridge would presumably have grown on top of the lagoon sediments. As shown in Figure 5b, smaller faults continue to disrupt the lagoon sediments along this flank.

Model for the Geologic Evolution of Wodejebato Guyot

Although the sequence of events responsible for the formation of Wodejebato Guyot was undoubtedly complex, Figure 12 shows a simple model of guyot evolution favored in this paper. The model combines some of the hypotheses of Lincoln [1990] with the geomorphic observations given in this paper. Information on the early history of Wodejebato comes exclusively from fossil ages; it is impossible to deduce anything except the most recent episode of uplift and island formation from the geomorphic information provided in this paper. As suggested by Lincoln [1990], the edifice first formed in the mid-Cretaceous, possibly in the vicinity of the Macdonald hot spot. Cooling and subsidence of the plate away from the hot spot, along with wave erosion, eventually formed the summit plateau of the volcano. A carbonate platform established itself on top of the wave-planated summit. At some point during its history, the rates of subsidence and sea level rise surpassed the rate of carbonate platform growth, and the developing platform drowned.

Uplift of the plate (and edifice) as it passed over the Rurutu hot spot during the Late Cretaceous [Lincoln, 1990] resulted in restricted volcanism across the edifice and extensive erosion of the mid-Cretaceous carbonate platform. As the plate cooled and subsided again (along with the edifice) a Late Cretaceous carbonate platform recolonized the existing veneer of carbonate sediments. The carbonate platform morphology was similar to many modern atolls: raised perimeter ridges bounded relatively flat-lying lagoon deposits. The two perimeter ridges along the northern and eastern flanks of Wodejebato suggest that more than one episode of carbonate platform growth occurred during this last stage of uplift. After full atoll development a large-scale submarine landslide, possibly comparable to those noted along the Hawaiian Islands [Moore et al., 1989], removed a portion of the southern flank including the perimeter ridge. Smaller-scale faults continue to

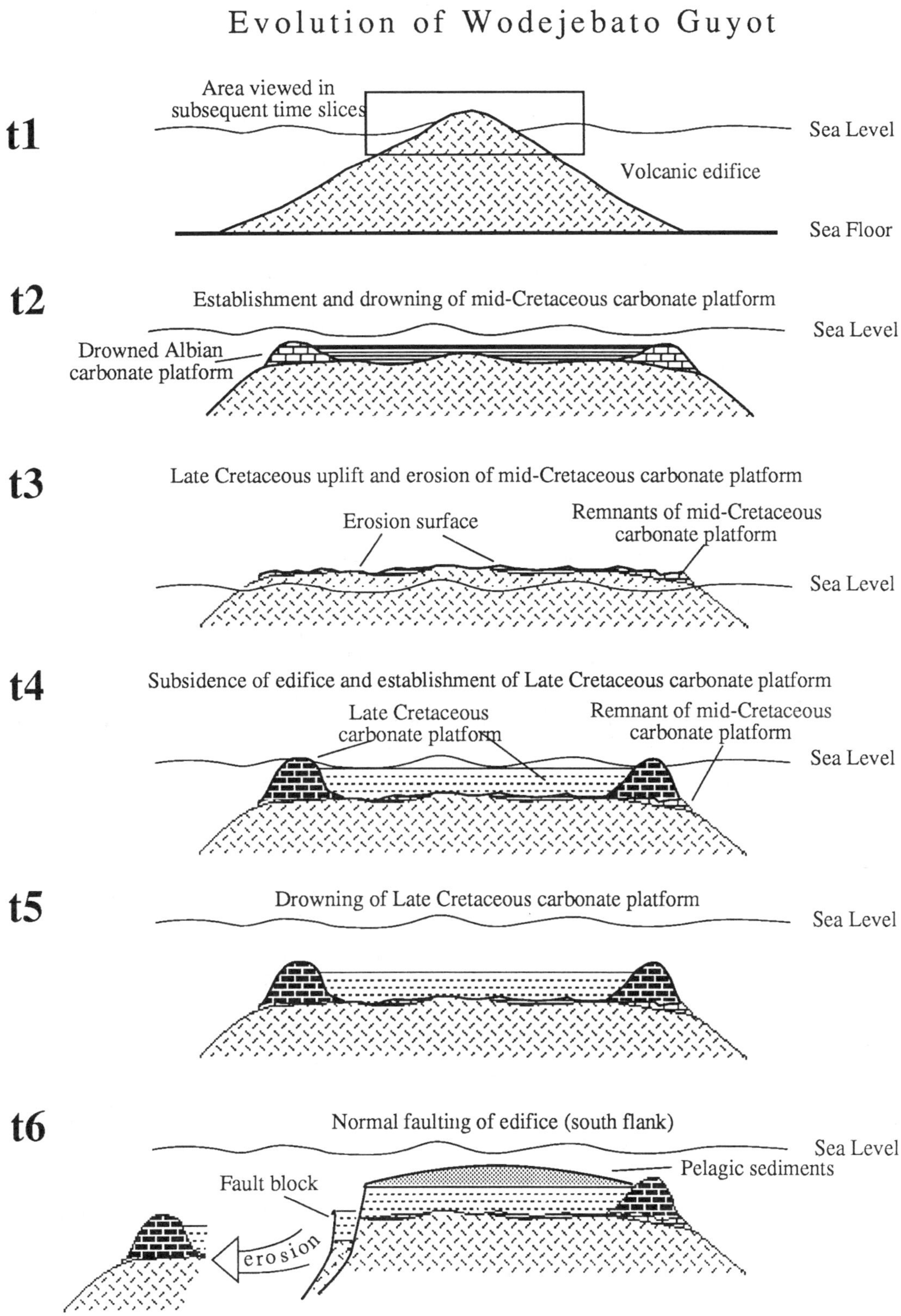

Fig. 12. A simple model depicting the evolution of Wodejebato Guyot. The cross-section shown in this cartoon is comparable to the profile shown in C-C" (excluding the flank ridge). Details of the model are provided in the text.

erode blocks of basement and lagoon sediments, one of which forms the terrace along the southern flank.

CONCLUSIONS

By examining the morphology and subsurface structure of Wodejebato guyot in detail, and through the analysis of basalt and limestone samples collected from its slopes, a relatively complete history for this volcano is emerging. Initial edifice construction occurred during the mid-Cretaceous, with a later stage of volcanic activity around 86 Ma. Carbonate platform accumulation across the summit apparently accompanied each period of plate uplift. In areas where shelves exist along the flanks of the edifice (e.g., next to flank ridges), the perimeter ridges broaden. The southern flank of this guyot exhibits extensive erosion. Lagoon sediments crop out where normal faulting has removed the reef rim. With the planned ODP drilling during Legs 143 and 144, answers to many of the vexing tectonic, sea level, and carbonate platform-growth questions still posed by this complex chain of atolls and guyots should be forthcoming. Future generations of scientists involved in these areas of research will undoubtedly benefit from this new information, thanks in part to Sy Schlanger's perseverance in furthering our understanding of Cretaceous events.

---

[1]*Note Added in Proof.* During ODP Leg 144, nine holes at five sites were drilled across the summit plateau of Wodejebato. Preliminary analysis of the fossil components suggests the carbonate platform was established by the late Campanian and survived until the late Maastrichtian [Haggerty, Premoli-Silva, et al., in press]. At Site 873, located on the south-central portion of the summit plateau, the lagoon sediments record a number of flooding events. These events manifest themselves as the subunits composing the lower limestone unit discussed in this paper, with a major flooding event separating the upper and lower limestone units. The perimeter ridges adjacent to the northeast flank ridge were sampled at Sites 874 through 877. The inner ridge records 2 episodes of bioherm growth separated by a major flooding event, while the outer ridge consists of abraded, well-winnowed, and moderately sorted skeletal sands.

Acknowledgements. This work was supported by NSF grant OCE-8709568. The author is grateful to F. Duennebier and A. Shor for support and advice over the past few years, and to P. Vogt and R. vanWaasbergen for critical reviews.

REFERENCES

Alldredge, L. R., F. Keller, and W. J. Dichtel, Magnetic structure of Bikini atoll, *U. S. Geol. Surv. Prof. Paper, 260-M*, 529-536, 1954.

Bergersen, D. D., and W. H. F. Smith, *MW9009: Marshall Islands Transit Cruise*, 56 pp., 1990.

Blackinton, J. G., *Bathymetric Mapping with SeaMARC II: An Elevation-Angle Measuring Side-Scan Sonar System*, Ph. D. Dissertation, University of Hawaii, Honolulu, 1986.

Cole, W. S., Larger foraminifera and smaller diagnostic foraminifera from Bikini drill holes, *U. S. Geol. Surv. Prof. Paper 260-O*, 569-608, 1954.

Davis, A. S., M. S. Pringle, L. G. Pickthorn, D. A. Clague, and W. C. Schwab, Petrology and age of alkalic lava from the Ratak Chain of the Marshall Islands, *J. Geophys. Res., 94*, 5757-5774, 1989.

Duennebier, F. K. and L. D. Petersen, *Summary report, IPOD Site Surveys in the western Pacific, R/V Kana Keoki cruise KK810626, Leg 4*, HIG, 1982.

Emery, K. O., J. I. Tracey, Jr., and H. S. Ladd, Geology of Bikini and nearby atolls, *U. S. Geol. Surv. Prof. Paper 260-A*, 265 pp., 1954.

Hamilton, E. L., and R. W. Rex, Lower Eocene phosphatized *Globigerina* ooze from Sylvania Guyot, *U. S. Geol. Surv. Prof. Paper 260-W*, 785-797, 1959.

Haggerty, J. A., I. Premoli-Silva, et al., *Proc. Init. Repts. (Pt. A), ODP*, 144, in press.

Hein, J. R., J. K. Kang et al., Geological, geochemical, geophysical, and oceanographic data and interpretations of seamounts and Co-rich ferromanganese crusts from the Marshall Islands, KORDI-USGS R. V. *Farnella* cruise F10-89-CP, *U. S. Geol. Surv. Open File Report 90-407*, 246 pp., 1990.

Hopley, D., *The Geomorphology of the Great Barrier Reef: Quaternary Development of Coral Reefs*, John Wiley and Sons, New York, NY, 453 pp, 1982.

Johnson, H. P., and M. Helferty, The geological interpretation of side-scan sonar, *Reviews of Geophys., 28*, 357-380, 1990.

Larson, R. L., S. O. Schlanger, et al., *Initial Rep. of the Deep Sea Drilling Proj., 61*, 1981.

Lincoln, J. M., *Regional Tectonic History and Cenozoic Sea Levels Deduced from Drowned Carbonate Banks and Atoll Stratigraphy in the Marshall Islands, West Central Pacific Ocean*, Ph. D. Dissertation, 242 pp., Northwestern University, Evanston, 1990.

Lincoln, J. M., S. O. Schlanger, D. D. Bergersen, A. H. Coogan, and I. Premoli Silva, Cretaceous foundations and multiple periods of uplift in the northern Marshall Islands, west central Pacific (abstract), *Geol. Soc. Am. Abstr. Programs, 21*, A195, 1989.

Moberly, R., S. O. Schlanger, et al., *Initial Rep. of the Deep Sea Drilling Proj., 89*, 1986.

Moore, J. G., D. A. Clague, R. T. Holcomb, P. W. Lipman, W. R. Normark, and M. E. Torreson, Prodigious submarine landslides on the Hawaiian Ridge, *J. Geophys. Res., 94*, 17,465-17,484, 1989.

Pringle, M. S., Jr., *Geochronology and Petrology of the Musicians Seamounts, and the Search for Hot Spot Volcanism in the Cretaceous Pacific*, Ph.D. Dissertation, 192 pp., 1992.

Purdy, E. G., Reef configurations, cause and effect., in *Reefs in Time and Space, S. E. P. M. Spec. Pub.. 18*, edited by L. F. LaPorte, pp. 9-76, 1974.

Raitt, R. W. Seismic refraction studies of Bikini and Kwajalein atolls, *U. S. Geol. Surv. Prof. Paper, 260-K*, 507-527, 1954.

Raitt, R. W., Seismic-refraction studies of Eniwetok Atoll, *U. S. Geol. Surv. Prof. Paper 260-S*, 685-698, 1957.

Schlanger, S. O., Subsurface geology of Eniwetok atoll, *U. S. Geol. Surv. Prof. Paper 260-BB*, 991-1066, 1963.

Schlanger, S. O., and I. Premoli Silva, Tectonic, volcanic, and paleogeographic implications of redeposited reef faunas of late Cretaceous and Tertiary age from the Nauru Basin and the Line Islands, *Initial Rep. of the Deep Sea Drilling Proj., 61*, 817-828, 1981.

Schlanger, S. O., H. C. Jenkyns, and I. Premoli Silva, Volcanism and vertical tectonics in the Pacific basin related to global Cretaceous transgressions, *Earth and Planet. Sci. Lett., 52*, 435-449, 1981.

Schlanger, S. O., M. O. Garcia, B. H. Keating, J. J. Naughton, W. W. Sager, J. A. Haggerty, and J. A. Philpotts, Geology and geochronology of the Line Islands, *J. Geophys. Res., 89*, 11261-11272, 1984.

Schlanger, S. O., and R. Moberly, Sedimentary and volcanic history: East Mariana Basin and Nauru Basin, *Initial Rep. of the Deep Sea Drilling Proj., 89*, 653-678, 1986.

Schlanger, S.. O., J. F,. Campbell, and M. W. Jackson, Post-Eocene subsidence of the Marshall Islands recorded by drowned atolls on Harrie and Sylvania Guyots, in *Seamounts, Islands, and Atolls, Geophys. Monogr. Ser.*, v. 43, edited by B. H. Keating, P. G. Fryer, R. Batiza, and G. W. Boehlert, pp. 165-174, AGU, Washington, D. C., 1987.

Schlanger, S. O., and F. K. Duennebier, *MW8805 : Marshall Islands Cruise Report*, 80 pp., 1988.

Winterer, E. L., Ewing, M., et al., *Initial Rep. of the Deep Sea Drilling Proj.*, *17*, 1973.

Wessel, P., and W. H. F. Smith, Free software helps map and display data, *EOS*, *72*, 441, 445-446, 1991.

Vogt, P. R., and N. C. Smoot, The Geisha Guyots: Multibeam bathymetry and morphometric interpretation, *J. Geophys. Res.*, *89*, 11085-11107, 1984.

D. Bergersen, School of Ocean and Earth Science and Technology, University of Hawaii, Honolulu, HI 96822.

A Paleomagnetic Pole and Estimated Age for Lo-En Guyot, Republic of the Marshall Islands

PHILLIP C. BRYAN

Code 7420, Naval Research Laboratory, Washington, DC

TOM SHOBERG, RICHARD G. GORDON, KATERINA E. PETRONOTIS[1]

Department of Geological Sciences, Northwestern University, Evanston, IL

DOUGLAS D. BERGERSEN

School of Ocean and Earth Science and Technology, University of Hawaii, Honolulu, HI

Marine magnetic data collected over Lo-En guyot in the Republic of the Marshall Islands during Moana Wave Cruise MW88-05 are inverted to estimate the best-fitting, least-squares, homogeneous magnetization for the guyot. The magnetization of Lo-En appears to differ from that of the spur ridge north-northwest of Lo-En. Over the shallowest part of the guyot the observed anomaly is of larger amplitude than calculated from the best-fitting model, which suggests that the guyot is more strongly or coherently magnetized in its shallow portions than at depth. Over the western part of the guyot the observed anomaly is of larger amplitude, whereas over the eastern part it is of smaller amplitude, than calculated from the best-fitting model, which also suggests that the guyot is more strongly or coherently magnetized in the western part than in the eastern part. Data importances, which measure how strongly different observations influence the values of the best-fitting parameters, tend to be highest on the portions of the survey over the steepest topography of the guyot. The paleomagnetic pole that we find for Lo-En, 64°N, 5°E, lies between the 65 Ma and 82 Ma reference poles for the Pacific plate, and comes closest to the Pacific apparent polar wander path at 74^{+11}_{-29} Ma, which is much younger than the ~170-m.y.-old seafloor on which Lo-En lies.

INTRODUCTION

S.O. Schlanger began his investigations of carbonate rocks in the western Pacific in the early 1950s. His interest in the extent, timing, and spatial pattern of Pacific basin volcanism grew during his participation on DSDP-ODP legs 17, 33, 61, and 89 in the Pacific and during cruises with colleagues and students at the University of Hawaii. With his broad view of seemingly disparate problems, Schlanger envisioned possible connections between episodes of mid-Cretaceous Pacific volcanism, rapid uplift and subsidence of Pacific volcanoes and adjacent seafloor, global Cretaceous transgressions, the migration of reefal fauna, the origin of many circum-Pacific allochthonous terranes, and episodes of global oceanic anoxia (see, e.g., Schlanger et al. [1981]). Two of us (RGG and DDB) joined Schlanger on the RV Moana Wave during April and May of 1988 on what was to be Schlanger's final cruise; during the cruise we surveyed and dredged western Pacific seamounts in the Marshall Island and the Magellan seamount group. Our goals included the acquisition of new geological and geophysical data that could help constrain the age of volcanism and the amount and timing of vertical motion of these volcanoes.

One approach for inferring the age of an extinct volcano comes from estimates of its average direction of

[1]Now at Department of Earth and Planetary Sciences, University of New Mexico, Albuquerque, NM

The Mesozoic Pacific: Geology, Tectonics, and Volcanism
Geophysical Monograph 77

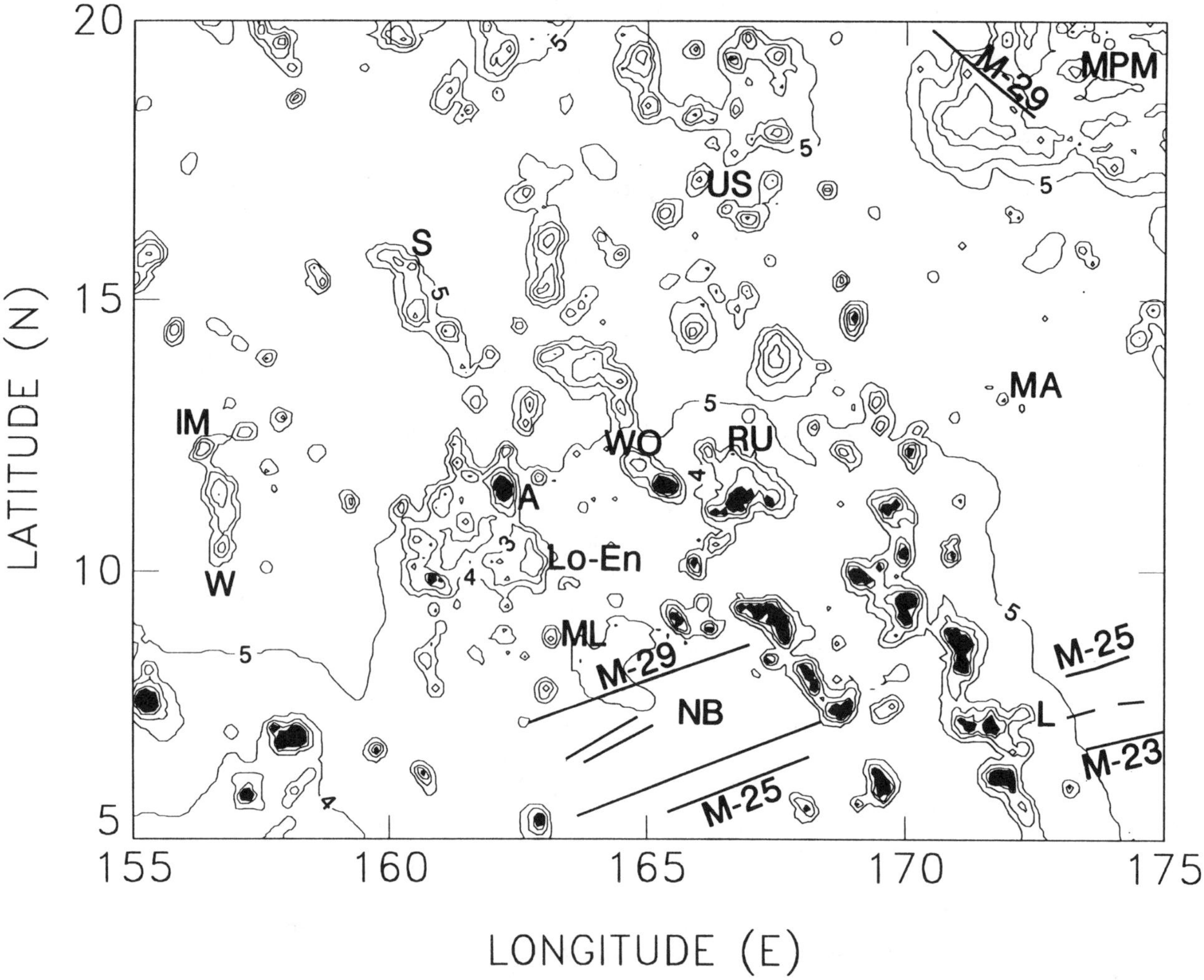

Figure 1. The western Marshall Islands seamount group, showing Lo-En at 10.1°N, 162.8°E, and other guyots. The contours shown were determined from DBDB5, which is a 5 arc-minute by 5 arc-minute digital bathymetric data file created by the U.S. Naval Oceanographic Office. The magnetic lineations are from Cande et al. [1989]. Filled-in contours show sub-aerial exposures. NB = Nauru Basin. Seamounts that have previously been used to calculate paleomagnetic poles [Sager, 1992] are also labeled: A = Anewetak, IM = Ita Mai Tai, L = Lomjenaelik, MA = Magnet, ML = Mij-lep, MPM = Mid-Pacific Mountains, RU = Ruwituntun, S = Seascan, US = Unnamed Seamount at 16.7°N, 166.4°E, W = Winchester, and WO = Wodejebato.

remanent magnetization (see, e.g., Sager [1992]). If the average remanent magnetization is known, a paleomagnetic pole can be inferred by simple calculations and compared with a reference apparent polar wander (APW) path for the Pacific plate to estimate the age of the volcano. Here we use the data that we collected over Lo-En (formerly Hess) guyot, a normally magnetized extinct volcano located at 10.1°N, 162.8°E (Figure 1), to estimate its magnetization and corresponding paleomagnetic pole. Because SeaMARC II data were collected during most of the survey of Lo-En, the available bathymetric data are of higher quality than typically have been available in prior studies of seamount magnetism. The paleomagnetic pole we find for Lo-En, 64°N, 5°E, is northeast of the 82-Ma Pacific plate reference pole and south of the 65-Ma reference pole, suggesting that the age of the guyot lies between 65 and 82 Ma.

WESTERN PACIFIC SEAMOUNT POLES

A useful paleomagnetic pole can be determined from magnetic profiles above a seamount if the seamount acquired its characteristic magnetization over enough time to average secular variation, if the seamount was dominantly or entirely formed during a single geomagnetic chron while little APW occurred, if the magnetization has negligible induced or secondary remanence components, if the shape of the seamount is well known, and if the anomaly due to the seamount can be separated from all other anomalies [Vacquier, 1962; Talwani, 1965; Plouff, 1976; Sager, 1983; Parker et al., 1987]. This paleomagnetic pole can be used to estimate the age of formation of the seamount by comparing the location of the seamount pole with locations of reference poles of known ages from the same lithospheric plate [Sager, 1992]. Ages inferred from seamount poles cannot be as precise as is potentially available from radiometric or biostratigraphic methods. However, the age obtained may be much more representative of the whole volcanic edifice, although weighted more towards shallow than deep parts of the edifice, than is true for most alternative methods, which must rely on very small samples of the volcano.

Paleomagnetic poles have been determined previously for twelve seamounts within the area shown in Figure 1 [Sager, 1992], three of which have independent age dates, two from fossil dates and one from a lithospheric flexure model. When seamount ages are estimated from the locations of the paleomagnetic poles of these seamounts, there is no evident pattern in seamount location or age that would simply explain all volcanism within even this limited area. Additional information is needed if the space-time pattern of volcanism in the western Pacific is to be unraveled.

METHODS AND MODELING PROCEDURES

The observed magnetic anomaly is the total field amplitude measured at sea during the cruise minus the amplitude of the International Geophysical Reference Field (IGRF) and minus the amplitude of a diurnal variation correction. We estimated diurnal variation corrections from the magnetic field recorded at the magnetic observatory on Guam (13.4°N, 144.7°E) during the cruise. The magnetic anomaly data along ship track (and not contours of these data) were inverted using the least-squares algorithm of Plouff [1976] to determine a best-fit uniform magnetization for the guyot. In addition to the assumptions listed in the introduction, this approach requires one further assumption, that the seamount should be uniformly magnetized. (As will be shown below, however, the results suggest the existence of significant non-uniformities of magnetization of this seamount.) Plouff's method requires that the seamount be represented by a vertical stack of polygonal prisms, which we discuss further in a later section.

We used two statistics to quantify the agreement between the field calculated from the best-fit magnetization and the observed anomaly. The first is the correlation coefficient, cc:

$$cc = P_{xy} / \sqrt{P_{xx}P_{yy}} \tag{1}$$

$$P_{xx} = \sum_i (A_i - \bar{A})^2 \tag{2}$$

$$P_{yy} = \sum_i (T_i - \bar{T})^2 \tag{3}$$

$$P_{xy} = \sum_i (A_i - \bar{A})(T_i - \bar{T}) \tag{4}$$

where A_i is the observed magnetic anomaly amplitude measured along ship profiles, $\bar{A}$ is the mean value of the A_i, T_i is the corresponding calculated anomaly, and $\bar{T}$ is the mean of the T_i. The correlation coefficient equals one if the model exactly fits the data.

The second statistic is a goodness-of-fit ratio, GFR, which is the sum of the magnitudes of the observed magnetic anomaly divided by the sum of the magnitudes of the misfit between the magnetic anomaly calculated from the best-fit uniform magnetization and the observed magnetic anomaly [Richards et al., 1967]:

$$GFR = \frac{\sum_i |A_i|}{\sum_i |A_i - T_i|} \tag{5}$$

The GFR, as defined above and typically used in seamount inversions, has at least one weakness: its value can be significantly larger if a magnetic anomaly, as defined above, has a non-zero mean. The GFR may therefore be large for a poor fit to the data if the regional field model, typically the IGRF or its successors, poorly describes the true field. Thus, GFRs of different data sets are incomparable and should not be used to judge relative goodness of fits to different data sets. Herein we define and use a modified goodness-of-fit ratio, MGFR, calculated by,

$$MGFR = \frac{\sum_i |A_i - \bar{A}|}{\sum_i |A_i - T_i|} \tag{6}$$

We calculated the confidence limits on the seamount pole that would apply if the errors were independent and uncorrelated (which is unlikely). These formal confidence limits were calculated by linear propagation of errors from $\sigma^2 W$, where σ^2 is the variance in the magnetic observations and W is the matrix given by

$$W = \left[A^T A\right]^{-1} \tag{7}$$

where A is the least-squares design matrix. From this four by four matrix we extracted the three by three submatrix corresponding to the 3 components of magnetization. We then used the linear transformation tensor described by Parker [1988] to calculate the variance-covariance matrix of the pole from that of the magnetization. This variance-covariance matrix was then transformed into an orthogonal coordinate system with one axis parallel to the best-fitting dipole axis. Deleting rows and columns associated with dipole moment leaves a 2 by 2 variance-covariance matrix that describes the error in the pole position, which can be simply transformed into a 95% confidence limit, as described for example by Gordon and Cox [1980].

To estimate the confidence limits we used two different values for σ. Parker et al. [1987] found in the vicinity of a seamount in the south Pacific that the magnetic field fluctuations had an rms amplitude of about 30 nanoTeslas (nT), which they attributed mainly to local variations in the magnetization intensity of the crust. Insofar as the background fluctuations do have an rms value of 30 nT, the errors so determined reflect how much error would be induced in the paleomagnetic pole by background fluctuations if the seamount were uniformly magnetized. The second estimate we used for σ was the rms misfit between the data and the anomaly calculated from the best-fitting model. This scaling would be appropriate if all the assumptions underlying the model, in particular that of uniform magnetization, were correct. As is discussed below, both these scalings give confidence limits that seem unrealistically small.

We also calculated "data importances" [Minster et al., 1974], which are the diagonal elements of the matrix Q where

$$Q = AWA^T \tag{8}$$

Q is the matrix representative of the orthogonal projection operator that projects the vector of magnetic field measurements onto the column space of the design matrix. The importances are a measure of the information content of each datum and are independent of the value of the datum. Instead, the importances depend on the seamount geometry, the orientation of Earth's present magnetic field, and the survey geometry relative to the seamount. The sum of the importances equals the number of adjustable parameters, which in our analysis is four–three to specify the assumed-uniform magnetization vector and one to estimate a constant-offset regional field. Importances show which data most strongly influence the least-squares solution and how concentrated or dispersed is the information distribution in the data. We hope that importances may prove to be helpful in improving the design of seamount magnetic surveys.

BATHYMETRIC DATA AND MODELS AND MAGNETIC DATA OF LO-EN GUYOT

Lo-En guyot is located in a cluster of seamounts in the western Republic of the Marshall Islands (Figure 1), northwest of the Nauru Basin. The cluster extends to the north and west of Lo-En and most prominently includes Anewetak Atoll, less than 150 km north-northwest of Lo-En. Lo-En's nearest neighbor is a small seamount about 50 km west of Lo-En (Figure 2). Lo-En rises to 1.1 km below sea level from seafloor that is more than 5 km deep, and is elongated in a north-south direction, with a spur ridge on the north extending nearly to the southern flank of Anewetak. The seafloor on which it lies is older than chron M29 and is probably of Middle to earliest Late Jurassic age. There are no recognized magnetic lineations in this region that might affect the analysis of the seamount magnetic anomaly.

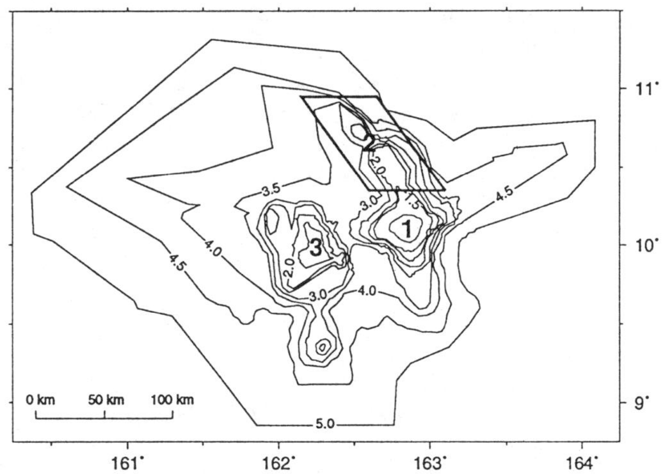

Fig. 2. Bathymetric model of Lo-En Guyot. The thin lines show contours for the complete bathymetric model. Numbers labeling depth contours are in km. Contours are at 0.5 km intervals except for the shallowest contour, which is at 1.2-km depth. The labels 1, 2, and 3 denote elevated regions referred to in the text. The northern spur from Lo-En guyot (region 2) is outlined by the thick lines.

We constructed a bathymetric model of the guyot from contoured SeaMARC II data [Bergersen and Duennebier, 1990] augmented by bathymetric soundings from Moana Wave cruise MW88-05. The SeaMARC II contours define the summit of Lo-En down to 3.0 km below sea level on the southern flank of the guyot and to 4.0 km on the eastern flank. A 5'-wide swath of SeaMARC bathymetric data extends NNW of the edifice from 10° 20'N to 10° 50'N, defining much of the spur ridge that extends northward from Lo-En guyot. SeaMARC II bathymetric data to the west extends from 162° 40'E to 162° 5'E in a swath between 9° 53'N and 10° 10'N, giving details on the highest part of the small seamount west of Lo-En. The rest of our bathymetric model is based on bathymetric contours from a regional map [Chase et al., 1988]. MW88-05 bathymetric measurements gave point locations for the contours outside

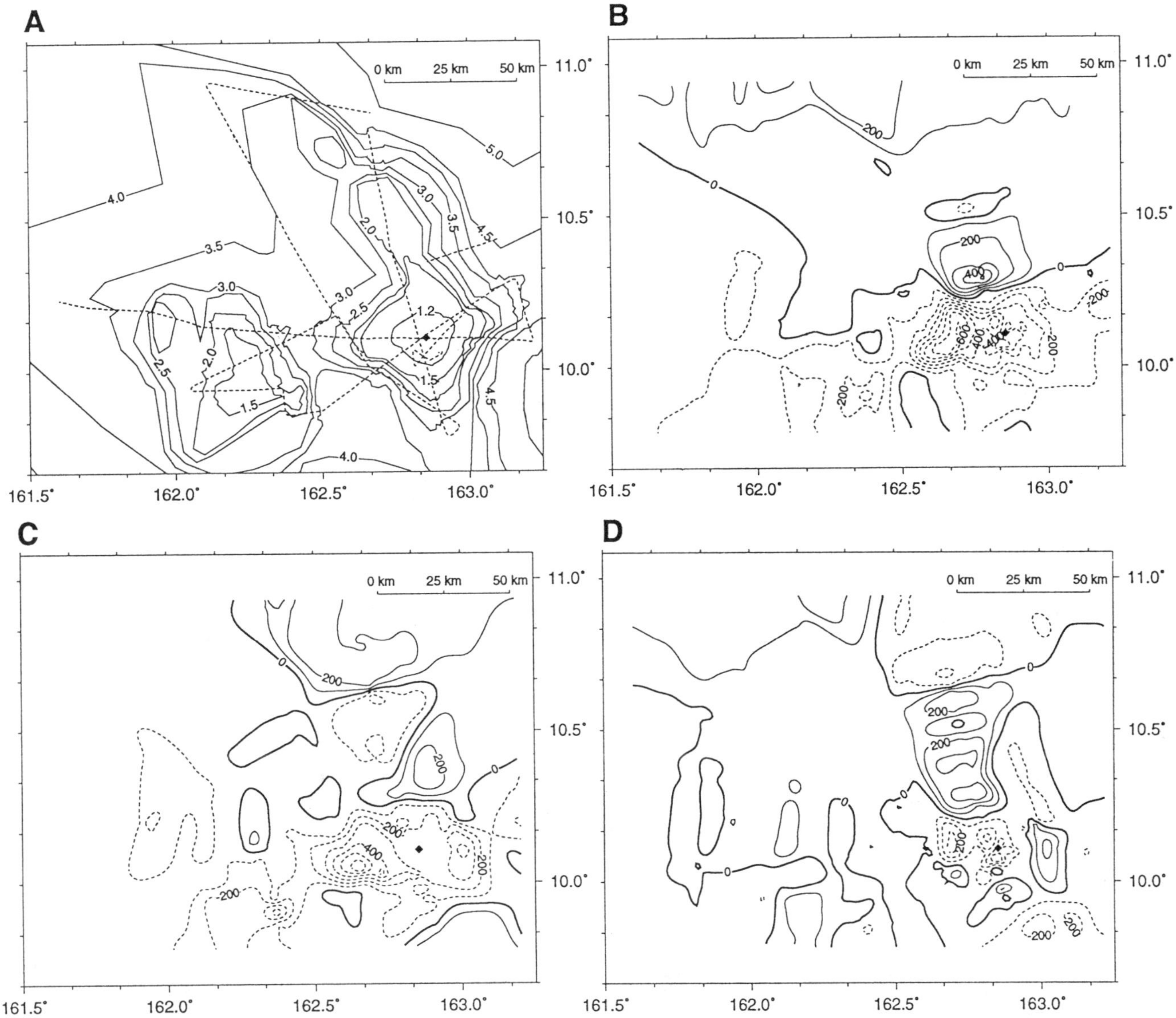

Figure 3. (a) Complete bathymetric model (Bath0) of the regions. Numbers labeling depth contours are in km. Dashed lines show ship tracks from *Moana Wave* cruise MW88-05. (b) Contour of complete observed magnetic anomaly (Data0). Numbers labeling anomaly-contours are in nanoTeslas (nT). Contours are at 100 nT intervals. The zero contour is shown by a bold solid line, contours of positive anomaly by thin solid lines, and contours of negative anomaly by dashed lines. (c) Contour of modeled magnetic anomaly. Contours labeled as in 3b. (d) Contour of residual magnetic anomaly. Contours labeled as in 3b.

of the area defined by SeaMARC II; bathymetric contours from the regional map were shifted or modified to fit these points. This bathymetric model was idealized as a set of layered prisms with each layer centered in depth on a particular bathymetric contour. Lo-En's volcanic platform contains three elevated areas: the main edifice of Lo-En, (area 1 in Figure 2), the spur ridge running north from Lo-En (including the transition, north of 10.2°N latitude, from Lo-En to the northern spur ridge (area 2)), and the small seamount to the west of Lo-En (area 3).

All magnetic data used in this report are total field measurements collected during Moana Wave cruise MW88-05. The only navigational error that we detected occurred where the ship exited the Lo-En guyot area to the east (Figure 3a, 10.4°N), which was navigated after several dredges and appears to include a 0.05° discrepancy in longitude relative to the previous ship track.

TABLE 1. Selected results using various data subsets and bathymetric models

		Bath0	Bath8	Bath11	Bath15
Data0	Pole	52°N, 34°E	53°N, 27°E	52°N, 34°E	54°N, 25°E
	GFR	1.93	2.22	1.98	2.29
	MGFR	1.72	1.98	1.76	2.05
	CC	0.814	0.860	0.820	0.866
	Int.	6.7 A/M	6.8 A/M	7.0 A/M	7.0 A/M
Data1	Pole	55°N, 25°E	54°N, 26°E	57°N, 23°E	56°N, 24°E
	GFR	2.21	2.29	2.31	2.38
	MGFR	1.90	1.97	1.99	2.05
	CC	0.847	0.856	0.856	0.864
	Int.	6.5 A/m	6.6 A/m	6.8 A/m	6.9 A/m
Data8	Pole	57°N, 22°E	56°N, 24°E	57°N, 23°E	56°N, 26°E
	GFR	2.35	2.44	2.33	2.43
	MGFR	2.09	2.16	2.07	2.15
	CC	0.862	0.869	0.861	0.868
	Int.	6.7 A/m	6.8 A/m	6.8 A/m	6.9 A/m
Data14	Pole	65°N, 357°E	64°N, 3°E	65°N, 359°E	**64°N, 5°E**
	GFR	3.01	3.05	2.99	**3.03**
	MGFR	2.29	2.32	2.27	**2.30**
	CC	0.870	0.875	0.868	**0.874**
	Int.	6.2 A/m	5.9 A/m	6.1 A/m	**5.9 A/m**
Data19	Pole	65°N, 0°E	65°N, 3°E	65°N, 3°E	64°N, 6°E
	GFR	3.30	3.34	3.29	3.34
	MGFR	2.06	2.09	2.05	2.08
	CC	0.824	0.831	0.823	0.831
	Int.	5.7 A/m	5.6 A/m	5.7 A/m	5.6 A/m

GFR is the goodness of fit ratio, described in the text. *MGFR* is the de-meaned goodness of fit ratio, described in the text. *CC* is the correlation coefficient, also described in the text. *Int.* is the intensity of magnetization. Bold face denotes the preferred model.

The discrepancy was identified by comparing shipboard precision depth recorder data with the SeaMARC bathymetric data. Because the slope is very steep, the offset was well constrained. We added 0.05° longitude to that part of the ship track to minimize the crossover error between the north-south SeaMARC II bathymetric data and the east-west precision depth recorder soundings. From a search of the Lamont Geobase data base [Menke et al., 1991], we found two cruises that had previously passed within the Lo-En area, but only one of these crossed over part of the main edifice, and this track was nearly coincident with the nearly north-south track from the Moana Wave cruise. Given this redundancy and the potential errors from navigation and diurnal variation correction, we omitted this track.

We analyzed the results from nineteen bathymetric models, from which we present results from four representative models (Table 1). The Bath0 bathymetric model, shown in Figure 2, was created using all the SeaMARC II and shipboard bathymetric data. In Bath8 the northern spur ridge, which has a top surface at 2-km depth, was represented by a 0.5-km thick edifice with its top at 3.0 km below sea level. In Bath11 the spur ridge was represented by the contours from the observed bathymetry while the small seamount west of Lo-En was modeled as if its summit rose only to 2.5-km depth instead of the observed 1.5-km depth. In Bath15 both the spur ridge and the small seamount were modeled as if they rose only to depths of 3 km and 2.5 km, respectively, 1 km deeper than the observed depths. All bathymetric models were contoured at 0.5-km depth intervals, with an extra contour at 1.2-km depth to represent the shallowest part of the main edifice of Lo-En. The bottom of the basal prism of all models was at 5.25 km below sea level. The polygonal prisms in the bathymetric model extended vertically 0.25 km above and 0.25 km below each contour, except for the shallowest two prisms (i.e., those representing the 1.2- and 1.5- km contours), which extend from 1.1- to 1.35-km depth and from 1.35- to 1.75-km depth, respectively.

We initially inverted a magnetic data set that included not only the survey over Lo-En, but also data over the seamount to the west, the spur ridge to the north-northwest, and a west-northwest-striking profile far to the north of Lo-En. The results showed that the magnetization far from Lo-En differed from that of Lo-En, so we subsequently focused only on profiles over and near Lo-En. The various magnetic data sets we analyzed combine nearly all permutations of data collected over five areas: (i) the main edifice of Lo-En (area 1 in Figure 2), which was included in every data subset, (ii) the spur ridge (area 2), (iii) the ship track north of the spur ridge that strikes nearly east-west, (iv) the long ship track west of, and roughly parallel to, the spur ridge, and (v) the small seamount to the west of Lo-En (area 3). In this paper we present results from five representative subsets of the magnetic data (Table 1). The first three data sets include data far from Lo-En: Data0 is the complete magnetic data set, Data1 excludes magnetic data subset (ii), and Data8 excludes magnetic data over subsets (ii) and (v). The final two data sets only include the data likely to be relevant to the magnetization of Lo-En. Data14 excludes magnetic data subsets (ii) and (v) and all data north of 10.6°N. Data19 excludes magnetic data subsets (ii) and (v) and all data north of 10.3°N.

RESULTS

Table 1 shows the inversion results from all combinations of the representative magnetic data subsets and bathymetric models. Correlation coefficients from these representative inversions ranged from 0.814 to 0.875. The corresponding modified goodness-of-fit ratios ranged from 1.72 to 2.32.

When we inverted Data0 (i.e., the complete magnetic data set) over the Bath0 bathymetric model (the most complete bathymetric model), several areas beyond the main edifice of Lo-En guyot were poorly fit (Figures 3b-3d). The rms misfit was 117 nT. Both the GFR (1.93) and the MGFR (1.72) were less than 2, a value used as a minimum reliability criteria by prior workers [e.g., Sager, 1983]. The magnetic anomaly calculated from our best-fit magnetization differed in places by more than 370 nT from the observed anomaly, which had a peak-to-trough amplitude of 1200 nT. Positive residuals exceeding 300 nT occur over the spur ridge and negative residuals less than -350 nT (i.e., of magnitude exceeding 350 nT) occur over the top of Lo-En. The spur ridge extending north-northwest of Lo-En has an associated magnetic anomaly that is large and positive over the transition from Lo-En's main edifice, and small and negative at the center of the spur ridge (Figure 3b). Inversions of the magnetic data from over the spur ridge and over the transition area give the best results when most of the spur ridge is removed from the bathymetric model, suggesting that the magnetic anomaly over the spur ridge was one of the largest, if not the largest, contributor to the misfit between data and model.

When we inverted the whole data set minus the profile over the spur ridge (i.e., the Data1 magnetic data set) using the original bathymetric model (i.e., model Bath0), the correlation coefficient improved from 0.814 to 0.847, also suggesting that the spur ridge magnetic data were a major source of misfit. Inverting Data1 with bathymetric models Bath1 to Bath9, which progressively removed the northern spur ridge from the bathymetry, gave additional modest improvement of the fit to the magnetic anomaly. It seems likely that this spur ridge is only weakly or incoherently magnetized. On the other hand, the poles and correlation coefficients were neither strongly nor consistently influenced by varying the bathymetric model for the small seamount to the west of Lo-En.

The paleomagnetic pole changes little for any bathymetric model used with data subsets Data14 and Data19, both of which exclude magnetic data from the spur ridge and the small seamount west of Lo-En (Table 1). The highest values of correlation coefficient and modified goodness-of-fit ratio occur for the combination of Bath8 and Data14. This bathymetric model is inconsistent, however, with this magnetic data set. Bath8 includes the western seamount whereas the Data14 data exclude the profiles over this seamount. Hence, our preferred inversion is the one combining the Data14 magnetic data with the Bath15 bathymetric model, which excludes this seamount (Figures 4 and 5, Table 1). The statistics for this combination are only slightly worse than for the Bath8-Data14 combination. Figure 4c shows contours of the magnetic anomaly calculated from our best-fit magnetization and Figure 4d shows contours of the difference between the observed and this calculated anomaly. An along-track comparison of the observed anomaly with that calculated from the best-fitting model illustrates the inability of the best-fitting model to fit the high-amplitude, and in some places short-wavelength, anomaly over the top of the guyot (see lines A-B, C-D, F-G, J-K, and K-L in Figure 5).

The SeaMARC II data show that the depth to the summit of Lo-En is about 1.1 km, which is the depth that we used in our model. The 30-kHz record from the cruise show that Lo-En has a pelagic cap [Bergersen and Duennebier, 1990], as has been confirmed by recent drilling [Bergersen et al., 1992]; thus the top of our bathymetric model is slightly too shallow. The strongly negative magnetic anomaly over the summit of Lo-En (see lines A-B, C-D, F-G, and K-L in Figure 5), would be better fit by a hypothetical edifice even shallower than modeled, which may suggest a strongly magnetized layer immediately below the reefal cap. This layer may outcrop in the strongly magnetized mound on the south flank of Lo-En observed during the cruise. The MGFR was even larger when we experimented with hypothetical bathymetric models with summits between 0.9- and 1.0-km depths, which are, of course, shallower than observed. The paleomagnetic pole found using the 0.95-km depth moved 4° to the south and east from the paleomagnetic pole found using the 1.1-km depth. Removing magnetized prisms from the bottom of the model of the guyot also gradually moved the paleomagnetic pole south and east, but the MGFR decreased. (All models in Table 1 retain a 5.25-km depth to the

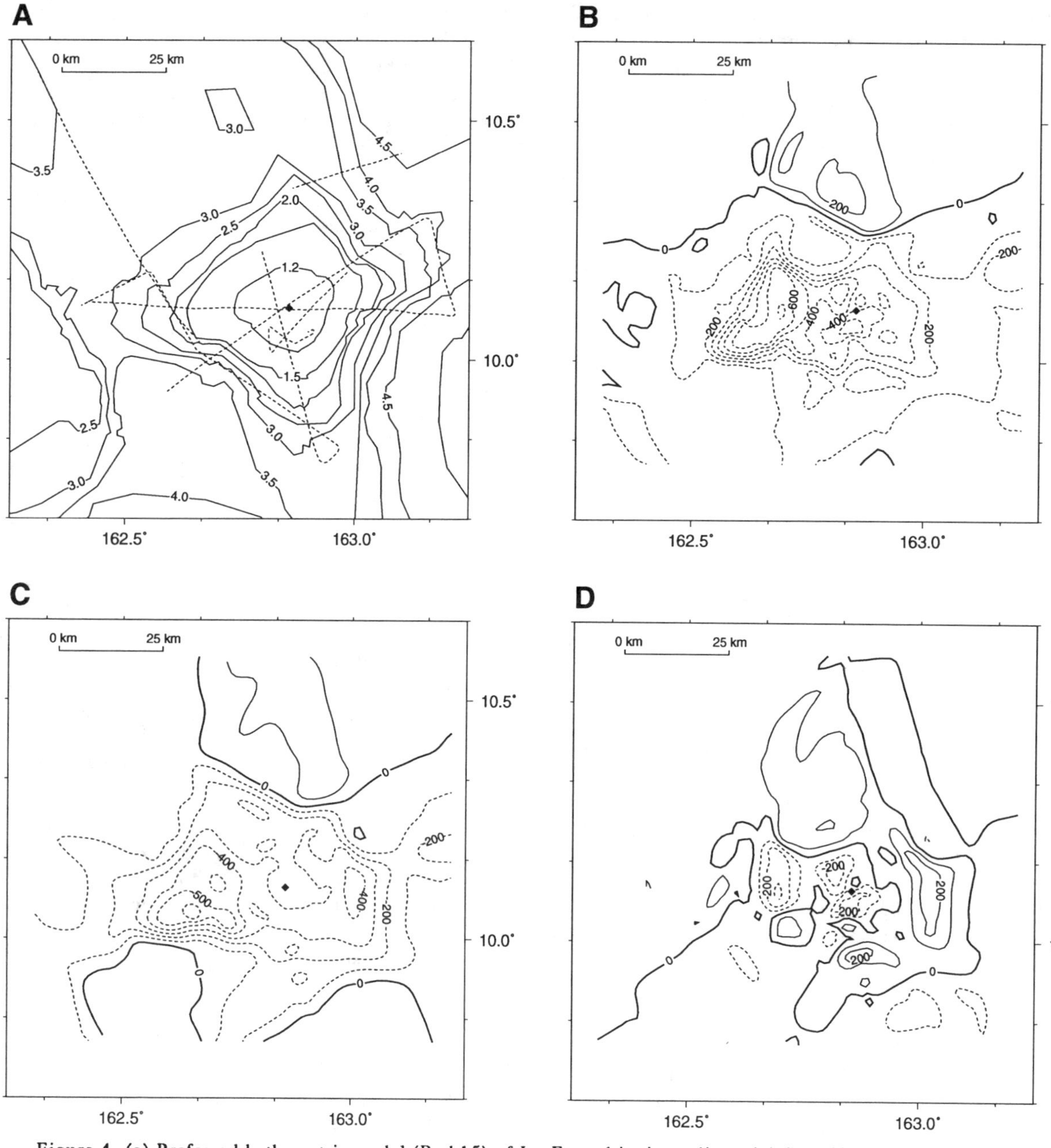

Figure 4. (a) Preferred bathymetric model (Bath15) of Lo-En and its immediate vicinity. Numbers labeling depth contours are in km. (b) Contour of observed magnetic anomaly of the preferred data set (Data14). Numbers labeling anomaly-contours are in nT. Contours are at 100 nT intervals. The zero contour is shown by a bold solid line, contours of positive anomaly by thin solid lines, and contours of negative anomaly by dashed lines. (c) Contour of modeled magnetic anomaly. Contours labeled as in 4b. (d) Contour of residual magnetic anomaly. Contours labeled as in 4b. (e) Data importances of magnetic anomaly observations plotted along ship track. The symbol size is approximately proportional to calculated importance, i.e., the largest symbols mark the data with the greatest importance.

bottom of the deepest prism in the bathymetric model of the guyot.)

Linear propagation of errors indicates very small uncertainties. When we scale the confidence limits by an assumed standard deviation of 30 nT, an elliptical standard error with a major axis of 0.8° oriented N69°W and a minor axis of 0.2° is found. Thus, the error induced in the seamount pole by background fluctuations in the magnetic field is minuscule. If instead of 30 nT, we adopt a value of 93 nT, which is the rms value of

E

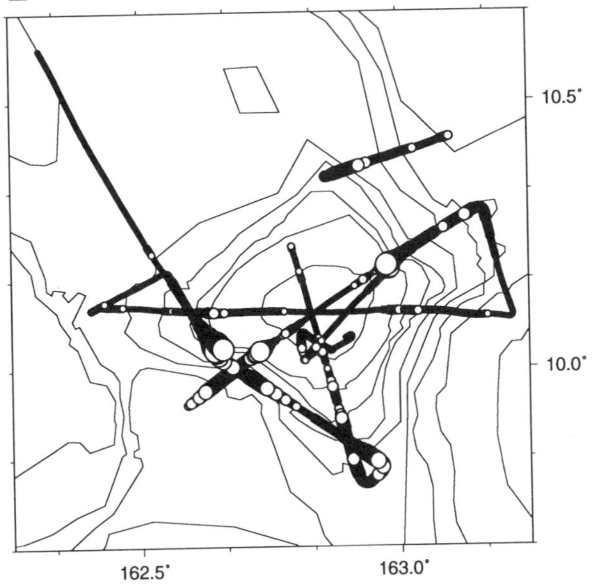

Fig. 4. (continued)

the misfit of our magnetic observations to the best-fitting model, an elliptical standard error with a major axis of 2.4° oriented N69°W and a minor axis of 0.5° is found. As will be discussed further below, these small errors are very likely an underestimate of the true uncertainty in the seamount pole position.

The data importances measure how much each datum influences the parameters describing the best-fit magnetization and regional field correction (Figure 4e). The information appears to be widely distributed throughout the magnetic measurements, but with distinct variations; individual importances vary in size by a factor of 15 (from 8.5×10^{-3} for the most important datum to 5.7×10^{-4} for the least important datum). The highest information density appears to be concentrated along profile segments above the steepest and shallowest bathymetric slopes, which are usually also the regions with the highest gradients in the predicted and observed magnetic anomaly. Thus observations made above these steep and shallow slopes are probably the most sensitive to changes in the direction of magnetization. The high importances of the southernmost magnetic measurements are exceptions to this pattern; these importances suggest that a more optimal survey would have continued making measurements even farther to the south.

DISCUSSION

Evaluation of Some Assumptions

A key assumption in the method of inversion that we have used is that the seamount is uniformly magnetized. An examination of how well the data are fit by the model permits us to examine the veracity of this assumption. Several aspects of the results suggest significant non-uniformities in the magnetization. First, Lo-En has a complex anomaly that is not reflected in its bathymetry. It has three distinct magnetic lows (Figure 4b), which are not reflected in topographic variations (Figure 4a). This lack of correlation is verified in the best-fitting magnetic model (Figure 4c), which has a single low that coincides partly with the largest and deepest of the 3 lows. Second, Parker et al. [1987] found in the vicinity of a seamount in the south Pacific that the magnetic field fluctuations had an rms amplitude of about 30 nT, which they attributed mainly to local variations in the intensity of the crust. If we assume that fluctuations not due to seamount magnetization are of similar size near Lo-En, then the rms misfit we obtain of 93 nT far exceeds that expected from background fluctuations, thus indicating a non-uniform magnetization of Lo-En itself. Third, that the fit of the model to the data improved by adding hypothetical layers to the top of the known magnetic layers also indicates non-uniform magnetization, i.e., that the top is more strongly or more coherently magnetized than are lower layers. Fourth, as emphasized by Parker [1988], if the magnetization is truly uniform, then the residuals of the best-fitting model should be uncorrelated and uniformly distributed in space. Parker [1988] asserts that observed residuals contradict this, with large misfits to uniform-magnetization models always being concentrated near the peak anomaly. A comparison of Figure 4d with Figure 4b shows that the residuals for Lo-En are consistent with this assertion. Each of the separate magnetic lows of Lo-En anomaly (Figure 4b) is matched with a corresponding residual low of less than −200 nT (i.e., of a magnitude exceeding 200 nT) (Figure 4d).

There is also a tendency for negative residuals to occur near the western flank and for positive residuals to occur over the eastern flank of the guyot, suggesting that its average magnetization is stronger or more coherent in the west than in the east. Alternatively, the eastern portion may be magnetized in a somewhat different direction from that of the main body. Figure 5 further shows that the residuals are highly correlated along ship track. Thus we conclude that Lo-En has a significant non-uniform magnetization, part of which likely corresponds to a stronger or more coherent magnetization in its shallow portions than at depth and possibly to stronger or more coherent magnetization in the west than in the east.

Figure 5 also shows that the representation of the seamount as a stack of polygonal prisms introduced discernible error in the model due to discretization. Because the remanent magnetization is shallow and nearly due north, flux from the magnetized body is at its highest density on vertical faces that strike approximately east-west. Because the present field is also

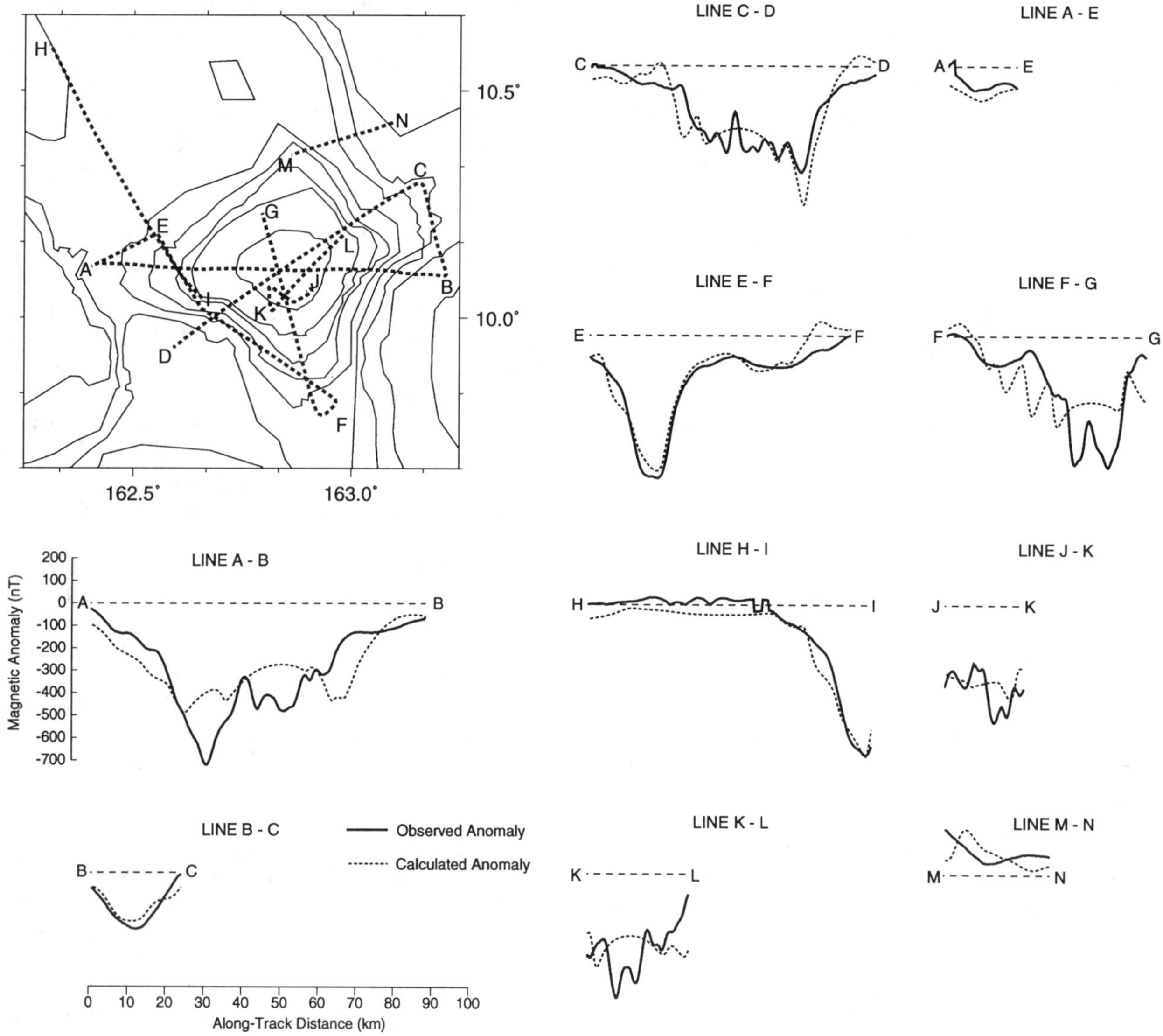

Figure 5. Along track comparison of magnetic anomalies calculated from the best-fitting uniform-magnetization model (short-dashed curves) with observed anomaly (solid curves). The model is our preferred model, i.e., the one bathymetric model Bath15 and data set Data14 (Table 1). All ship track lines are plotted to the same scale with the thin, long-dashed line showing the zero magnetic anomaly value on each. The location of each profile is shown on the inset map.

shallow and nearly due north, it is these concentrated sources of flux that should cause the largest anomalies. The effect of the discretization is to concentrate the flux source on the vertical walls of the prism instead of spreading it more evenly along the steep slopes of the seamount. The effect is strongest for profiles that traverse shallow east-west striking contours used to construct the model. An examination of the index map in Figure 5 suggests this effect should be most pronounced on profile F-G. Indeed, the calculated anomaly along this profile shows short-wavelength anomalies that we infer to be caused by the discretization of the bathymetry. The same effect may cause some of the short-wavelength calculated anomalies shown for line C-D as well, but it seems unimportant on the other profiles. With the advantage of hindsight, it would have been desirable to represent the shallow portions of the bathymetry by more prisms, each of less thickness than we used here, to reduce the size of these artifacts.

It is not easy to predict how much the pole presented here is biased by the assumption of uniform magnetization. Hildebrand and Parker [1987] calculated both a pole assuming a uniform-magnetization and a pole from semi-norm minimization for 14 seamounts. For these 14 pairs of poles, angular distances between pairs of poles determined for the same seamount range from 0.4° to 13.9° with a median difference of 4° and with only one pair differing by more than 7.4°. Uniform-magnetization models that gave poles that differed little from the seminorm poles tended to have smaller rms residuals and larger GFRs than did those that differed by larger amounts. The rms residual (93 nT) and GFR (3.0) for Lo-En are a little below the middle of this group of 14 seamounts. Thus, the difference between the pole we estimate here and a corresponding seminorm pole would most likely be a little larger than 4°, but could be several times larger or smaller.

Age of Lo-En

Figure 6 shows paleomagnetic poles calculated for various combinations of bathymetric models and magnetic data subsets (Table 1). Changes in the bathymetric model had little effect on the calculated paleomagnetic pole. The calculated pole was near our preferred pole as long as the data were restricted to profiles above or near Lo-En (i.e., magnetic data subsets Data14 and Data19). If we include data farther from Lo-En, e.g., data along the ~N80°W-striking profile far to the northwest of Lo-En or data along the spur ridge north-northwest of Lo-En (Figure 3a), the pole moves to the southeast of our preferred pole by as much as ~20°.

The area shown in Figure 1 includes twelve other volcanoes with paleomagnetic poles [Sager, 1992]. Five of the poles fall between the 65 Ma and 82 Ma reference poles (Figure 6): (1) Lomjenaelik (formerly Von Valtier), located at 7.3°N, 172.3°E, with a pole at 63.2°N, 2.2°E from a least-squares inversion [Sager, 1983] and at 54.2°N, 307.2°E (representing an older, mid-Cretaceous age) from a semi-norm minimization inversion of a larger magnetic anomaly data set [Hildebrand and Parker, 1987], (2) Harvey, located at 17.8°N, 172.7°E, with a pole at 68.1°N, 21.9°E, (3) Magnet, located at 12.3°N, 173.2°E, with a pole at 61.0°N, 31.2°E, (4) Seascan, located at 15.1°N, 159.3°E, with a pole at 57.0°N, 4.8°E [Sager, 1983], and (5) an unnamed seamount, located at 19.4°N, 165.9°E, with a pole at 58.4°N, 5.9°E [Harrison et al., 1975].

Figure 6 also shows Pacific plate reference poles including the 39-Ma pole of Sager [1987], the 65-Ma pole of Acton and Gordon [1991], the 82-Ma and 88-Ma poles of Sager and Pringle [1988], the 125-Ma pole of Gordon [1990], and a poorly dated but possibly Early Cretaceous or Late Jurassic pole from Hildebrand and Parker [1987]. The location of the preferred paleomagnetic pole for Lo-En indicates that Lo-En formed near a paleolatitude of 14°S between about 82 Ma and 65 Ma.

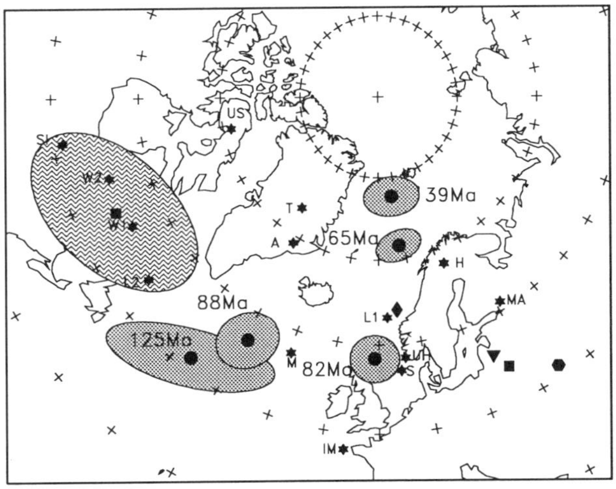

Fig. 6. Paleomagnetic poles selected from our inversions, Pacific plate reference poles, and paleomagnetic poles from seamounts located within the area shown in Figure 1. Diamond: our best estimate of the paleomagnetic pole for Lo-En guyot, using Data14 magnetic anomaly data and Bath15 bathymetric model. Hexagon: the original data, Data0, and the original bathymetric model, Bath0. Square: uses Data1, which deletes data from over the spur ridge extending north from Lo-En, and Bath8 bathymetric model. Triangle: uses Data8, which deletes data from over the small seamount west of Lo-En, and Bath8. Stars: paleomagnetic poles from seamounts located within the area defined by Figure 1 [Sager, 1992]: In the Mid-Pacific Mountains, A = Allen, H = Harvey, SI = Sio, and T = Thomas, in the Marshall Islands, L1 = Lomjenaelik (formerly Von Valtier) from a least-squares inversion, L2 = Lomjenaelik from semi-norm minimization, M = Mij-Lep (formerly Heezen), S = Seascan, and US = Unnamed seamount at 16.7°N, 166.4°E, Other seamounts, MA = Magnet, IM = Ita Mai Tai, W1 = Winchester, from a least-squares inversion, W2 = Winchester, from semi-norm minimization, and UH = Unnamed seamount at 19.4°N, 165.9°E.

A more detailed analysis of the age of the seamount inferred from its paleomagnetic pole is shown in Figure 7. To obtain precise limits on the age of Lo-En we assume that the pole moved at a constant angular rate along a great circle between reference poles. The age at which the position of the Lo-En paleomagnetic pole comes within closest agreement with the Pacific APW path (within 1.4°) is 74 Ma, ~100 m.y. after the formation of the seafloor upon which it rests. The position of the Pacific pole as long ago as 170 Ma is not known, however, and this conclusion should be interpreted cautiously.

We estimated the uncertainty of this age from the uncertainty in the paleomagnetic pole. The confidence limits that we estimated from linear propagation of errors are very likely an underestimate because the nonuniformity of the magnetization of the seamount induces considerable additional uncertainty. Confidence limits that account for the correlation of the errors in the ob-

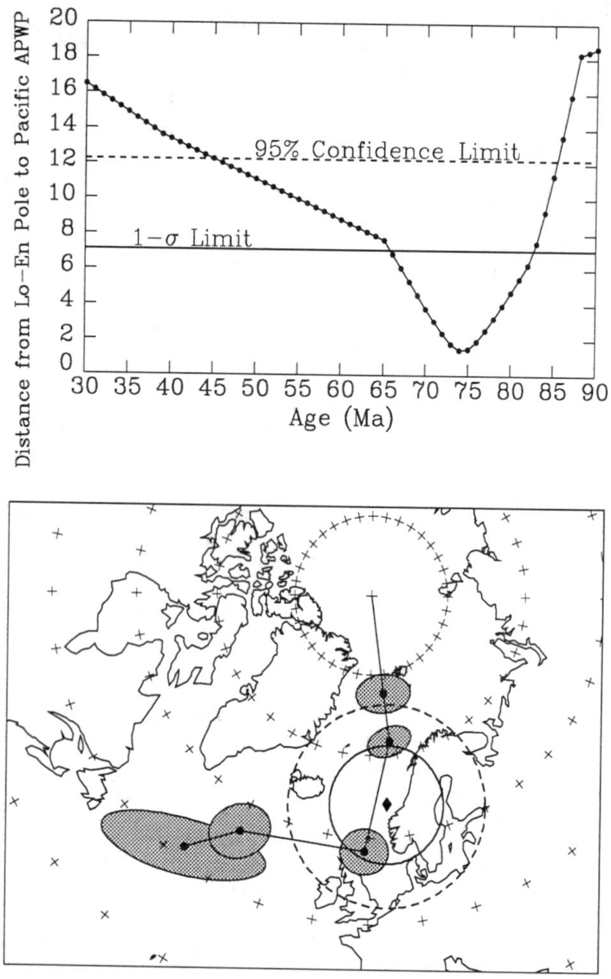

Fig. 7. Age range for Lo-En guyot. The age for Lo-En is estimated by assuming it to be coeval with the age of the position on the Pacific apparent polar wander (APW) path to which it is nearest. For uncertainty calculations a 5° standard error is assumed, and by which is used to determine the two-dimensional 1-σ error (This is the solid line in both plots). The 95% confidence limit for this standard error is shown by the dashed lines. The upper plot shows a graph of the distance of the paleomagnetic pole from Lo-En to the Pacific APW path, and the lower shows this pole (diamond symbol), the 1-σ error (solid small circle about this pole) and the 95% confidence limit (dashed small circle about this pole) plotted with the Pacific APW path.

servations due to non-uniform magnetization [Parker, 1988; Hildebrand and Parker, 1987] are 3 to 14 times larger than those we obtain when we scale the errors by 30 nT, and 1.3 to 6 times larger than those we obtain when we scale the errors by the 93-nT rms misfit of the data to the best-fitting model. Furthermore, prior studies of groups of seamount poles (e.g., Gordon [1983]) suggest that the two-dimensional angular standard deviation of assumed-uniform-magnetization poles of similar ages about their mean pole is much larger, ~5–8°. Thus, hereinafter we adopt a value of 7°, consistent with the dispersion of groups of seamount poles and many times larger than suggested by the error we estimate from linear propagation of errors, and also larger than any of the confidence limits estimated by Hildebrand and Parker [1987]. The solid circle about the Lo-En pole in Figure 7 encloses the region within one two-dimensional standard deviation (i.e., 7°) of the Lo-En pole and the dashed circle encloses the region within the 95% confidence limits. The corresponding age limits are 66–82 Ma for one standard deviation and 45–85 Ma for the 95% confidence limits. The 95% confidence limits are asymmetric about the best estimate because APW was slower between 39 and 65 Ma than it was between 65 and 88 Ma.

These confidence limits may be biased slightly by our assumption of circular confidence limits for the paleomagnetic pole of Lo-En. The results from linear propagation of errors suggest that the errors are about 5 times larger along N69°W than perpendicular to this direction. Moreover, the confidence limits found for non-uniform magnetization models tend to be several times larger to the east and west than to the north and south [cf. Hildebrand and Parker, 1987]. If this applies to Lo-En, then both the young and old age limit should be older than we cite above and a more realistic age limit would thus include ages older, perhaps considerably older, than 88 Ma.

Although we would like to be able to use our age estimate to test the predictions of hotspot models, it is not yet possible to make a useful test of such predictions in this region for several reasons. Age dates for volcanic edifices are too sparse, no published model predicts volcano ages in the region of the Marshall Islands while being consistent with currently available data elsewhere from the Pacific, and no published model for Pacific hotspot traces comes with estimates of the uncertainties in the predicted tracks.

On the other hand, our age estimate for Lo-En is close to the 76 ± 1 Ma age estimate for Anewetak Atoll from an $^{40}Ar/^{39}Ar$ analysis of basalts recovered from beneath the atoll [Lincoln et al., 1993]. The two age dates are consistent with the formation of both Lo-En and Anewetak during a Late Cretaceous volcanic event that affected much of the area around the northern Marshall Islands and Nauru Basin. To decipher the nature and history of such an event requires age estimates for many of the edifices present in this region of the western central Pacific. Ten other edifices in the Marshall Islands have age dates [Davis et al., 1989; Lincoln et al., 1989; Lincoln et al., 1993; Sager, 1983; Schlanger et al., 1981; Schlanger et al., 1987]. Of these, the age estimates for Wodejebato and Ruwituntun [Lincoln et al., 1989] agree well with the age estimates of Lo-En and Anewetak. These four seamounts are within 500 km of one other (Figure 1). Furthermore all these edifices are located less than 550 km from DSDP site 462 in the Nauru Basin. DSDP hole 462 cores show relationships between flows, sills, and intercalated sediments that suggest this region underwent three episodes of volcanism, ridge crest volcanism at ~158-Ma, sill and flow volcanism in Barremian to late Aptian time, and a later stage of volcanism in late Campanian to early Maastrichtian time [Moberly and Jenkyns, 1981]. From these data and from paleontological age estimates from carbonate rocks off Anewetak and Pikinni atolls, Schlanger and Premoli-Silva [1981] postulated a tectonic history for the Central Pacific around the Nauru Basin and Marshall Islands that included a final stage of edifice-building volcanism beginning during Campanian time and ceasing by middle Maastrichtian time (~70 Ma).

Although it is tempting to explain much of the volcanism in this region as a single pulse of activity in and around the Nauru Basin in Campanian to Maastrichtian time, several complications present themselves. First, Mij-Lep, another seamount within 250 km of DSDP hole 462, has a mid-Cretaceous age estimate inferred from its paleomagnetic pole [Sager, 1983]; from Sager's results and the APW path of Figure 6 we estimate its age to be 102 ± 18 Ma. Second, the age dates of seamounts nearest to the Nauru basin are not particularly close to the age dates from the Nauru basin. Third, further analysis of the carbonate rocks dredged from Wodejebato and Ruwituntun suggests that the main edifices formed during Albian time and a later interval of reef building occurred during Campanian time [Lincoln et al., 1993]. This has led Lincoln et al. [1993] and Lincoln [1990] to propose that passage of a hot spot near the area where Wodejebato and Ruwituntun had already been formed could explain the latter episode of reef building. Their explanation is consistent with the ~74-Ma age we infer here for Lo-En guyot.

CONCLUDING REMARKS

A uniform magnetization model for Lo-En guyot fits the observed anomaly, which has a peak-to-trough amplitude exceeding 1200 nT, with an rms misfit of 93 nT and maximum misfit of about 250 nT, corresponding to a peak-to-trough amplitude of the residual anomaly exceeding 400 nT. The spatial pattern of the residuals, along with trial-and-error modeling, show that the magnetization is non-uniform and suggest that the guyot is more strongly or coherently magnetized in its shallow portions than at depth, that the western portion may be more strongly or more coherently magnetized than the eastern portion, and that the spur ridge projecting north from the guyot is only weakly or incoherently magnetized. The best estimate of guyot age from the paleomagnetic pole is ~74 Ma with approximate 95% confidence limits of 45–85 Ma. These confidence limits, although large, nonetheless suggest that the seamount was formed long after the ~170-m.y.-old seafloor underlying it,but before mid-Tertiary time.

Acknowledgments. This work was supported by NSF grant EAR-9005679. The work was completed while PCB held a National Research Council-NRL Research Associateship and while RGG was a visiting scientist at the Laboratoire de Géodynamique sous-marine, Villefranche sur mer, France. We thank Will Sager for providing much-needed technical support, software, advice, and encouragement, Jonathan Lincoln for many useful discussions on Marshall Island seamounts, and Gary Acton, Alice Gripp, and Huaxiao Wang for their contributions to the initial stages of this research. We also thank Fred Duennebier, co-chief scientist on MW88-05, and the crew of MW88-05 for their help. We dedicate this paper to the late S. O. Schlanger, who was co-chief scientist on cruise MW88-05, as well as our teacher and stimulating colleague.

REFERENCES

Acton, G. D., and R. G. Gordon, A 65 Ma palaeomagnetic pole for the Pacific Plate from the skewness of magnetic anomalies 27r-31, *Geoph. J. Int.*, *106*, 407-420, 1991.

Bergersen, D. D., and F. K. Duennebier, Seismic stratigraphy of selected guyots in the Marshall Islands region, *EOS Trans. AGU*, *71*, 1672, 1990.

Bergersen, D. D., R. L. Larson, and Shipboard Science Party, Seismic stratigraphy of guyots drilled during ODP Leg 144–preliminary results, *EOS Trans. AGU*, *73*, 587, 1992.

Cande, S. C., J. L. LaBrecque, R. L. Larson, W. C. Pitman III, X. Golovchenko, and W. F. Haxby, *Magnetic lineations of the world's ocean basins* (map), AAPG, Tulsa, OK, 1989.

Chase, T. E., B. A. Seekins, and J. D. Young, Marine topography of the Republic of the Marshall Islands, *Federated States of Micronesia, Republic of Palau, Commonwealth of the Northern Marianas and Territory of Guam*, USGS Map, 1988.

Davis, A. S., M. S. Pringle, L.-B. G. Pickthorn, D. A. Clague, and, W. C. Schwab, Petrology and Age of Alkalic Lava from the Ratak Chain of the Marshall Islands, *J. Geophys. Res.*, *94*, 5757-5774, 1989.

Gordon, R. G., Late Cretaceous apparent polar wander path of the Pacific plate: evidence for a rapid shift of the Pacific hotspots with respect to the paleomagnetic axis, *Geophys. Res. Letts.*, *10*, 709-712, 1983.

Gordon, R. G., Test for bias in paleomagnetically determined bathymetry from Pacific plate Deep Sea Drilling Project sediments, *J. Geophys. Res.*, *95*, 8397-8404, 1990.

Gordon, R. G., and A. Cox, Calculating palaeomagnetic poles for oceanic plates, *Geophys. J. R. astron. Soc.*, *63*, 619-640, 1980.

Harrison, C. G. A., R. D. Jarrard, V. Vacquier, and R. L. Larson, Palaeomagnetism of Cretaceous Pacific Seamounts, *Geophys. J. R. Astron. Soc.*, *42*, 859-882, 1975.

Hildebrand, J. A., and R. L. Parker, Paleomagnetism of Cretaceous Pacific Seamounts revisited, *J. Geophys. Res.*, *92*, 12695-12712, 1987.

Lincoln, J. M., Regional Tectonic History and Cenozoic Sea Levels Deduced from Drowned Carbonate Banks and Atoll Stratigraphy in the Marshall Islands, West Central Pacific Ocean, Ph. D. thesis, Northwestern University, Evanston, IL, 1990.

Lincoln, J. M., S. O. Schlanger, D. Bergersen, A. H. Coogan, and I. Premoli-Silva, Cretaceous Foundations and Multiple Periods of Uplift in the Northern Marshall Islands, West Central Pacific, *Geol. Soc. Am. Abstr. Programs, 21*, 195, 1989.

Lincoln, J. M., M. S. Pringle, and I. Premoli-Silva, Early and Late Cretaceous Volcanism and Reef Building in the Marshall Islands: New Fossil Evidence and $^{40}Ar/^{39}Ar$ Ages, this volume, 1993.

Menard, H. W., Marine geology of the Pacific, McGraw Hill, New York, 271 pp., 1964.

Menard, H. W., Darwin reprise, *J. Geophys. Res.*, *89*, 9960-9968, 1984.

Menke, W., P. Friberg, A. Lerner-Lam, D. Simpson, R. Bookbinder, and G. Karner, Sharing data over Internet with the Lamont View-Server system, *EOS trans. AGU, 72*, 409-413, 1991.

Minster, J. B., T. H. Jordan, P. Molnar, and E. Haines, Numerical modeling of instantaneous plate tectonics, *Geophys. J. R. astron. Soc.*, *36*, 541-576, 1974.

Moberly, R. and H. C. Jenkyns, Cretaceous Volcanogenic Sediments of the Nauru Basin, Deep Sea Drilling Project Leg 61, *Init. Rep. Deep Sea Drill. Proj.*, *61*, 533-548, 1981.

Parker, R. L., A statistical theory of seamount magnetism, *J. Geophys. Res.*, *93*, 3105-3115, 1988.

Parker, R. L., L. Shure, and J. A. Hildebrand, The Application of inverse theory to seamount magnetism, *Rev. of Geophys.*, *25*, 17-40, 1987.

Plouff, D., Gravity and magnetic fields of polygonal prisms and application to magnetic terrain corrections, *Geophysics*, *41*, 727-741, 1976.

Press, W. H., B. P. Flannery, S. A. Teukolsky, and W. T. Vetterling, *Numerical Recipes: The Art of Scientific Computing*, 818 pp., Cambridge University Press, New York, 1986.

Richards, M. L., V. Vacquier, and G. D. Van Voorhis, Calculation of the magnetization of uplifts from combining topographic and magnetic surveys, *Geophysics, 32*, 678-707, 1967.

Sager, W. W., Seamount Paleomagnetism and Pacific Plate Tectonics, Ph.D. thesis, University of Hawaii, Honolulu, 472 pp., 1983.

Sager, W. W., Late Eocene and Maastrichtian paleomagnetic poles for the Pacific plate: implications for the validity of seamount paleomagnetic data, *Tectonophysics, 144*, 301-314, 1987.

Sager, W. W., Seamount age estimates from paleomagnetism and their implications for the history of volcanism on the Pacific plate, in *Geology and Offshore Mineral Resources of the Central Pacific Basin*, eds. B. H. Keating and B. R. Bolton, Springer-Verlag, New York, 1992.

Sager, W. W., and M. S. Pringle, Mid-Cretaceous to early Tertiary apparent polar wander path of the Pacific plate, *J. Geophys. Res.*, *93*, 11753-11771, 1988.

Schlanger, S. O., H. C. Jenkyns, and I. Premoli Silva, Volcanism and vertical tectonics in the Pacific basin related to global Cretaceous transgressions, *Earth Planet. Science Lett.*, *52*, 435-449, 1981.

Schlanger, S. O., and I. Premoli Silva, Tectonic, Volcanic and Paleogeographic Implications of Redeposited Reef Faunas of Late Cretaceous and Tertiary Age from the Nauru Basin and the Line Islands, *Init. Rep. Deep Sea Drill. Proj.*, *61*, 817-827, 1981.

Talwani, M., Computation with the help of a digital computer of magnetic anomalies caused by bodies of arbitrary shape, *Geophysics*, *30*, 797-817, 1965.

Vacquier, V., A machine method for computing the magnitude and direction of magnetization of a uniformly magnetized body from its shape and a magnetic survey, in *Benedum Earth magnetism symposium*, ed. T. Nagata, University of Pittsburgh Press, Pittsburgh, 123-127, 1962.

P. C. Bryan, Code 7420, Naval Research Laboratory, Washington, DC 20375.

T. Shoberg, R. G. Gordon, K. Petronotis, Department of Geological Sciences, Northwestern University, Evanston, IL 60208.

D. Bergersen, School of Ocean and Earth Science and Technology, University of Hawaii, Honolulu, HI 96822.

Paleomagnetism of the Japanese and Marcus-Wake Seamounts, Western Pacific Ocean

WILLIAM W. SAGER

Departments of Oceanography, Geophysics, and Geodynamics Research Institute, Texas A&M University, College Station, Texas

ROBERT A. DUNCAN

College of Oceanography, Oregon State University, Corvallis, Oregon

DAVID W. HANDSCHUMACHER

Naval Research Laboratory, Stennis Space Center, Mississippi

The western Pacific contains many Mesozoic seamounts of uncertain origin. We examined paleomagnetic and age data from three seamount groups, the Japanese and Marcus-Wake seamount chains and several isolated seamounts located in between, for clues about their formation. The primary data are paleomagnetic poles and parameters determined from seamount magnetic anomalies using the seminorm inversion method. Most poles clustered along the 125 to 88 Ma portion of the Pacific apparent polar wander path, implying the seamounts formed mainly during the mid-Cretaceous. Paleolatitudes indicate the seamounts formed at or just south of the equator. Furthermore, radiometric ages and magnetic polarities indicate that most volcanism occurred during the Cretaceous Long Normal Superchron (K-N), probably between about 90 to 110 Ma. The evidence for post-K-N activity is not compelling; only one radiometric age from any of these seamounts is younger. Several ages, summit depths, and a few paleopoles imply there was volcanism prior to the K-N and of uncertain duration. Neither the Japanese nor the Marcus-Wake seamounts group show a consistant paleolatitude pattern as would be expected from a hot spot origin. Neither was there an obvious progression of poles along the apparent polar wander path to imply age-progressive volcanism. However, apparent polar wander during this period was slow, so a brief period of age-progressive volcanism may not be evident in the paleopole distribution. Additionally, some seamounts displayed evidence that their magnetizations contain induced or normal overprint biases, albeit small for most. Given these uncertainties, with existing data we are unable to distinguish between the hypotheses of a hot spot origin or an amorphous, widespread volcanic pulse for the orgin of these volcanoes.

INTRODUCTION

Perhaps the most notable feature of the western Pacific plate is its profusion of seamounts. Many are guyots, indicating they grew to sea level, and some of these are capped with drowned reefs. What is more, many seem to be aligned in chains, similar to those known to have been created by age-progressive hot spot volcanism in the central and eastern Pacific. Nevertheless, the origin of most western Pacific volcanic edifices remains unclear.

Dating of rocks dredged from western Pacific seamounts implies that they are predominantly of Cretaceous age. Indeed, the abundance of volcanoes in the western Pacific is one factor that has lead investigators to propose that one or more massive volcanic episodes occurred during that period. Such a large, anomalous volcanic event would likely have had a profound effect on the environment [Vogt, 1989]; moreover, it would also be symptomatic of large scale

The Mesozoic Pacific: Geology, Tectonics, and Volcanism
Geophysical Monograph 77

Table 1. Seamount Paleomagnetic Parameters

ID	Name	Location Lat(°N)	Location Lon(°E)	Paleopole Lat(°N)	Paleopole Lon(°E)	Inc.[a]	Dec.[b]	Inten.[c] F_u	Inten.[c] F_n	Paleo-Latitude	Error Ellipse[d] Maj	Error Ellipse[d] Min	Error Ellipse[d] Az	RMS[e] GFR[f]	Type[g]	Ref[h]
Japanese (Geisha) Chain																
J1	Jenson	33.8	143.8	55.5	340.5	0.9	350.7	19.4	8.3	0.5	5.9	2.6	102	55	S	1
J2	Jenson N	34.1	143.9	68.6	341.7	25.7	353.3	23.1	9.6	13.5	18.0	8.5	103	92	S	1
				*57.1	327.1	2.3	358.3	19.7	10.1	1.2	20.9	9.7	91	61	S	1
J3	Takuyo-Daini	34.3	143.9	75.4	315.7	35.8	2.2	6.2	2.6	19.8	5.5	3.0	88	30	S	1
J4	Takuyo-Daisan	34.2	144.3	58.2	328.2	4.9	357.9	11.1	6.8	2.5	4.3	1.9	96	30	S	1
J5	Maiko	34.0	145.9	51.3	327.0	-9.2	359.3	12.6	2.9	-4.6	4.1	1.7	94	30	S	1
J6	Winterer-NW	32.9	147.9	§51.9	315.7	-9.0	7.5	6.1	0.9	-4.6	6.6	2.6	97	30	S	1
J7	Charlie Johnson	32.0	148.3	§47.4	284.5	-2.9	27.9	5.0	2.6	-6.2	5.4	1.8	105	30	S	1
J8	Winterer	32.8	148.4	§51.6	315.5	-9.7	8.0	6.7	2.8	-4.9	4.1	1.5	97	30	S	1
J9	Winterer-SE	32.7	148.7	§42.0	35.5	13.0	-43.5	21.2	22.9	6.6	21.2	7.9	146	32	S	1
				*§58.7	327.9	2.7	0.4	9.1	5.2	1.4	15.9	5.0	91	31	S	1
J10	Stout	31.6	149.0	40.1	339.9	-32.3	-8.7	5.3	7.9	-17.5	9.4	5.5	91	30	S	1
				*50.4	342.5	-14.0	-8.6	6.7	7.1	-7.1	10.1	4.2	103	30	S	1
J11	Thomas Washington	32.0	149.2	53.9	321.5	-7.6	4.5	7.2	1.9	-3.8	3.5	1.2	96	31	S	1
J12	Isakov-NW	31.8	150.9	§58.0	311.9	2.4	10.0	16.4	7.3	1.2	9.4	3.4	75	27	S	1
J13	Isakov	31.6	151.2	§52.3	331.5	-12.1	359.8	10.6	3.5	-6.1	4.2	1.7	87	30	S	1
J14	Makarov	29.5	153.5	59.2	346.2	-1.3	353.5	9.0	2.1	-0.7	2.9	0.9	103	31	S	1
Marcus-Wake Chain																
W1	Unnamed	25.1	151.8	†46.2	323.7	33.7	185.5	3.9		-18.4				5.6	L	3
W2	Smt948	23.7	151.9	*57.3	329.5	-17.5	1.3	5.1		-9.0				2.8	L	4
W3	Unnamed	24.3	152.3	74.2	276.9	27.7	13.4	3.4		14.7				5.5	L	3
W4	Golden Dragon	21.3	153.2	56.2	337.4	-23.8	357.6	3.4	0.8	-12.4	5.6	1.9	98	30	S	1,3
W5	UCSD	24.3	159.2	†§55.3	338.4	20.3	180.5	6.9	0.6	-10.5	9.2	3.7	85	19	S	1,3
W6	Scripps	23.7	159.3	63.9	350.2	-4.6	359.7	5.3	3.1	-2.3	11.1	2.1	87	30	S	1
W7	Miami	21.7	161.9	*59.8	5.9	-12.2	348.1	8.1	4.7	-6.2	6.5	2.1	116	30	S	5
W8	Vibelius	21.2	162.8	84.0	310.1	30.0	3.4	3.2	0.4	16.1	8.7	3.5	65	30	S	1
W9	Wilde	21.2	163.3	61.8	333.1	-13.2	4.9	7.3	3.9	-6.7	7.1	2.1	82	30	S	1,3
W10	Birdseye	20.9	165.7	58.6	342.3	-20.4	1.8	13.0		-10.6				8.1	L	6
W11	Unnamed	19.4	165.9	58.4	5.9	-20.3	349.5	6.2		-10.5				2.2	L	2
W12	Woods Hole	21.2	166.5	53.2	319.7	-23.2	16.0	8.0	4.4	-12.1	6.8	2.0	61	31	S	1
In Between																
B1	Z44	27.9	147.6	60.0	306.0	-1.0	11.0	3.0		-0.5				1.8	L	7
B2	Z42	28.4	148.2	*59.9	292.0	6.1	17.2	4.1	1.6	3.1	7.9	2.7	59	30	S	5
B3	Z41	28.8	148.4	55.0	19.0	9.0	334.0	1.8		4.5				2.3	L	7
B4	Z43	27.1	148.7	52.5	331.9	-20.2	358.0	7.6	5.8	-10.4	6.7	2.5	93	35	S	1,3
B5	MIT	27.3	151.9	†29.8	328.6	52.2	183.3	2.8	6.2	-32.8	27.4	9.0	85	30	S	1

†Antipole (reversed polarity).
*Planar regional included as inversion parameter.
§Crustal background field model subtracted from anomaly prior to inversion.

Table Heading Abbreviations:
[a]*Inc.*, inclination; [b]*Dec.*, declination; [c]*Inten.*, magnetization intensity (F_u, uniform; F_n, nonuniform) units in Amperes/meter; [d]*Error Ellipse*, 95% confidence region; *Maj*, major semi-axis length (in degrees); *Min*, minor semi-axis length (in degrees); *Az*, azimuth of major semi-axis (degrees clockwise from north); [e]*RMS*, root-mean-square error of residual anomaly (allowed misfit for seminorm inversions); [f]*GFR*, goodness-of-fit ratio (quality parameter for least-squares inversions); [g]*Type*, inverse technique; L = least-squares minimization; S = seminorm minimization; [h]*Ref*, reference number (see below).

References:
(1) This study; (2) Harrison et al., 1975; (3) Sager, 1992; (4) Ueda, 1988; (5) Hildebrand and Parker, 1987; (6) Sager, 1983; (7) Vacquier and Uyeda, 1967.

[Larson and Chase, 1972; Sager et al., 1988; Nakanishi et al., 1989]. Most of the Japanese Seamounts trend obliquely across the Japanese lineations from M16 (143 Ma [Harland et al., 1982]) to M21 (154 Ma). Three seamounts from this group, J12, J13, and J14 (see Table 1 for names and locations), are located on M23 (158 Ma) and M25 (161 Ma) in the Hawaiian lineations south of the Shatksy Plateau. The Marcus-Wake Seamounts are all on lithosphere formed prior to M29 (165 Ma). Though correlatable magnetic lineations have not been found in most of this area, seamount W4 sits upon anomaly M36, one of a series of low amplitude pre-M29 lineations mapped by Handschumacher et al. [1988]. Paleomagnetic data is also available for five seamounts located between these two chains (Table 1). Of these, the B1-B4 cluster sit on anomalies M22-M24 (156-160 Ma) and B5 is near M29.

In general, seamounts in the western Pacific appear significantly younger than the seafloor beneath them. Results from lithospheric flexure studies imply that most were formed off-ridge, suggesting that a few tens of millions of years elapsed between the time the seafloor was created and the seamounts formed; however, a small number of the seamounts have flexure signatures that imply near ridge formation [Watts et al., 1980; Watts and Ribe, 1984]. Age data from dredge samples also imply a significant age gap between the seamounts and underlying crust (Table 2). Fossils show that Cretaceous reefs capped many of the guyots. Particularly diagnostic were Albian-Aptian age rudists found on several guyots [Matthews et al., 1974]. Additionally, the few reliable pre-existing radiometric dates are also from the mid-Cretaceous [Ozima et al., 1983].

Accordingly, most models for the origin of western Pacific seamounts call upon intraplate volcanism of some kind. The various hypotheses can be divided into two classes, those that call upon an unusual event and those that do not. In the former division is the "Darwin Rise" hypothesis. Menard [1964; 1984] noted that many of the western Pacific guyots have similar summit depths. He contoured the depths and the results suggested that the volcanoes formed upon a thermal swell 2 km in height, which he named the Darwin Rise. Based upon intrusives and volcanogenic sediments from Deep Sea Drilling Project cores, other investigators have also postulated at least two widespread Cretaceous volcanic events, Aptian to Cenomanian (about 115-95 Ma) and Campanian-Maastrichtian (about 80-65 Ma) [Winterer, 1976; Schlanger et al., 1981; Rea and Vallier, 1983]. The former may have been particularly intense and included the formation of several western Pacific plateaus [Tarduno et al., 1991].

The other type of model conjures up a less unusual mechanism, supposing instead that most western Pacific volcanoes were formed by hot spots. Several investigators used the trends of numerous seamount chains and sparse age data in an attempt to document the motion of the Pacific plate relative to the hot spots back in time to the Early Cretaceous [Duncan and Clague, 1985; Henderson, 1985]. This work implies that the widespread mid-Cretaceous volcanism in the western Pacific is a result of the confluence of many hot spot tracks. Crough

Table 2. Seamount Age Constraints

ID	Name	Age (Ma)	Constraints[1]
Japanese Chain			
J3	Takuyo-Daini	98–113	Albian rudist corals from summit (Matthews et al., 1974).
J4	Takuyo-Daisan (Seiko)	102	^{40}Ar–^{39}Ar age, no error limits given (Ozima et al., 1983).
J8	Winterer	108.3±1.0	^{40}Ar–^{39}Ar plateau age.
		98–113	Albian rudist corals from summit (Matthews et al., 1974).
J13	Isakov	103.7±1.8	^{40}Ar–^{39}Ar plateau age.
		98–119	Cenomanian-Albian rudists from summit (Matthews et al., 1974); Flexure indicates >35 Ma younger than seafloor (Watts et al., 1980).
J14	Makarov	98±1	^{40}Ar–^{39}Ar age (Ozima et al., 1983).
		92–98	Planktonic foraminifera from summit (Matthews et al., 1974).
Marcus-Wake Chain			
W4	Golden Dragon	95	^{40}Ar–^{39}Ar age, no error limits given (Ozima et al., 1983).
W6	Scripps	101.6±0.7	^{40}Ar–^{39}Ar plateau age.
W7	Miami	96.8±0.6	^{40}Ar–^{39}Ar plateau age.
W9	Wilde	90.6±0.3	^{40}Ar–^{39}Ar plateau age.
	Lamont	87.2±0.3	^{40}Ar–^{39}Ar plateau ages.
		81.5±0.6	
Between			
B4	Z43	93.7±4.2	^{40}Ar–^{39}Ar age (Ozima et al., 1983).
B5	MIT	120.3±0.8	^{40}Ar–^{39}Ar plateau age.

[1]Biostratigraphic dates correlated to absolute time using Harland et al. (1982) time scale. ^{40}Ar–^{39}Ar dates not cited were done for this study; further details given in Winterer et al. (this volume).

[1979] suggested that these volcanoes formed as the Pacific plate drifted over hot spots presently located in the region of French Polynesia in the southeast Pacific. This idea has gained additional support from geophysical and petrologic data. The Japanese, Marcus-Wake, and Magellan seamounts, farther south, sit upon lithosphere with anomalously low flexural rigidity, similar to that in the "Superswell" region of French Polynesia [McNutt et al., 1990; Wolfe and McNutt, 1991]. Furthermore, some of the Marcus-Wake and Magellan seamounts also have geochemical signatures similar to patterns documented in French Polynesia [Smith et al., 1989].

The main difference between these two classes of model is an age progression. A volcanic pulse could be synchronous over a large area, but a hot spot track should display an age progression. Although our knowledge of the Cretaceous drift of the Pacific plate relative to the hot spots is poor, current models imply that hot spot seamounts in this area should increase in age from east to west [Duncan and Clague, 1985; Henderson, 1985].

Seamount Paleomagnetism

Because most of the Pacific plate is submerged, it has been difficult to obtain paleomagnetic data from it by traditional means. To fill this need, such data has been often obtained by inversion of magnetic field measurements. Vacquier [1962] first formulated the technique, expressing the magnetic anomaly of a body as a linear combination of its unknown mean magnetization vector components and mathematical expressions of the body geometry. This allowed a linear, least-squares inversion for the magnetization components. Typically, the method has been applied to the magnetic anomalies of seamounts.

Several assumptions are inherent in this model of seamount magnetization and the use of its results to calculate paleomagnetic poles. One of the most important is that the magnetization is assumed to be homogeneous (i.e., with constant magnitude and direction) within the seamount. This is unlikely to be true. If a seamount's volcanism spanned a period with a geomagnetic field reversal, then its magnetization may incorporate normal, reversed, and transitional polarity directions. Furthermore, studies of the magnetizations of samples from seamounts indicate that, even if the polarity is constant, the distribution of magnetization intensity is log-normal [Gee et al., 1989; Parker, 1991], so seamounts probably contain regions with large differences in magnetization.

To account for magnetization inhomogeneities, Parker et al. [1987] developed a seminorm minimization inversion routine in which the magnetization is assumed to consist of uniform and nonuniform parts and the latter is minimized. Because an assumption of this method is that the heterogeneities have small scale length, the authors warned that seamounts with large magnetization nonuniformities were unlikely to give reliable results with this method.

Fortunately, most seamounts with greatly heterogeneous magnetizations give results that are obviously incorrect with magnetization vectors that disagree with other apparently reliable data. It is probably a consequence of this self-selection process that most published seamount paleomagnetic poles are from edifices formed during the Cretaceous Long Normal Superchron (a. k. a., K-N or Cretaceous Quiet Period). Nevertheless, even when a seamount does apparently include a reversal, it is sometimes possible to obtain what appear to reliable results, as long as one polarity predominates [e.g., Sager and Keating, 1984; McNutt, 1986; Sager and Pringle, 1987].

Another important assumption is that the magnetization producing the anomaly is entirely remanent. This is based on observations that ocean basalts often have stable remanent magnetizations that are much larger than their induced counterparts [e.g., Marshall, 1978; Sager, 1983]. For example, the magnetic properties of dredged material from Jasper seamount showed that fine-grained basalts had high Koenigsberger ratios (the ratio of remanent to induced magnetization) and acquired little viscous remanent overprint [Gee et al., 1988]. However, the same study also found coarse-grained rocks with significant induced and viscous magnetizations. Likewise, other studies suggest that an average of 17-20% of seamount magnetization may result from these sources [Hildebrand and Staudigel, 1986; Gee et al., 1989; Bryan and Gordon, 1991]. Consequently, the possibility of a bias by a normal polarity overprint must be considered in the interpretation of seamount paleomagnetic data.

Given the seemingly commonplace violation of these critical assumptions, it is a wonder that seamount paleomagnetism works at all, but comparison of the few well-dated seamount and non-seamount Pacific paleomagnetic data of similar age show excellent agreement [Sager, 1987]. This implies that most seamounts giving apparently reliable inversions probably consist mostly of fine-grained basalts that do a good job of recording the geomagnetic field.

Pacific Apparent Polar Wander Path

Many of the seamounts that gave good paleomagnetic results are not dated. Therefore, the Pacific apparent polar wander path (APWP) is useful because the poles of undated seamounts can be compared with it to estimate their ages [Sager, 1992].

The Pacific APWP is largely constrained by seamount paleomagnetic data. Seamount poles cluster in the north Atlantic Ocean along the Greenwich Meridian from about 80°-50° N (Fig. 2). The cluster trends westward to about 300° E between 50°-60° N. Sager and Pringle [1988] describe the calculation of mean paleomagnetic poles for

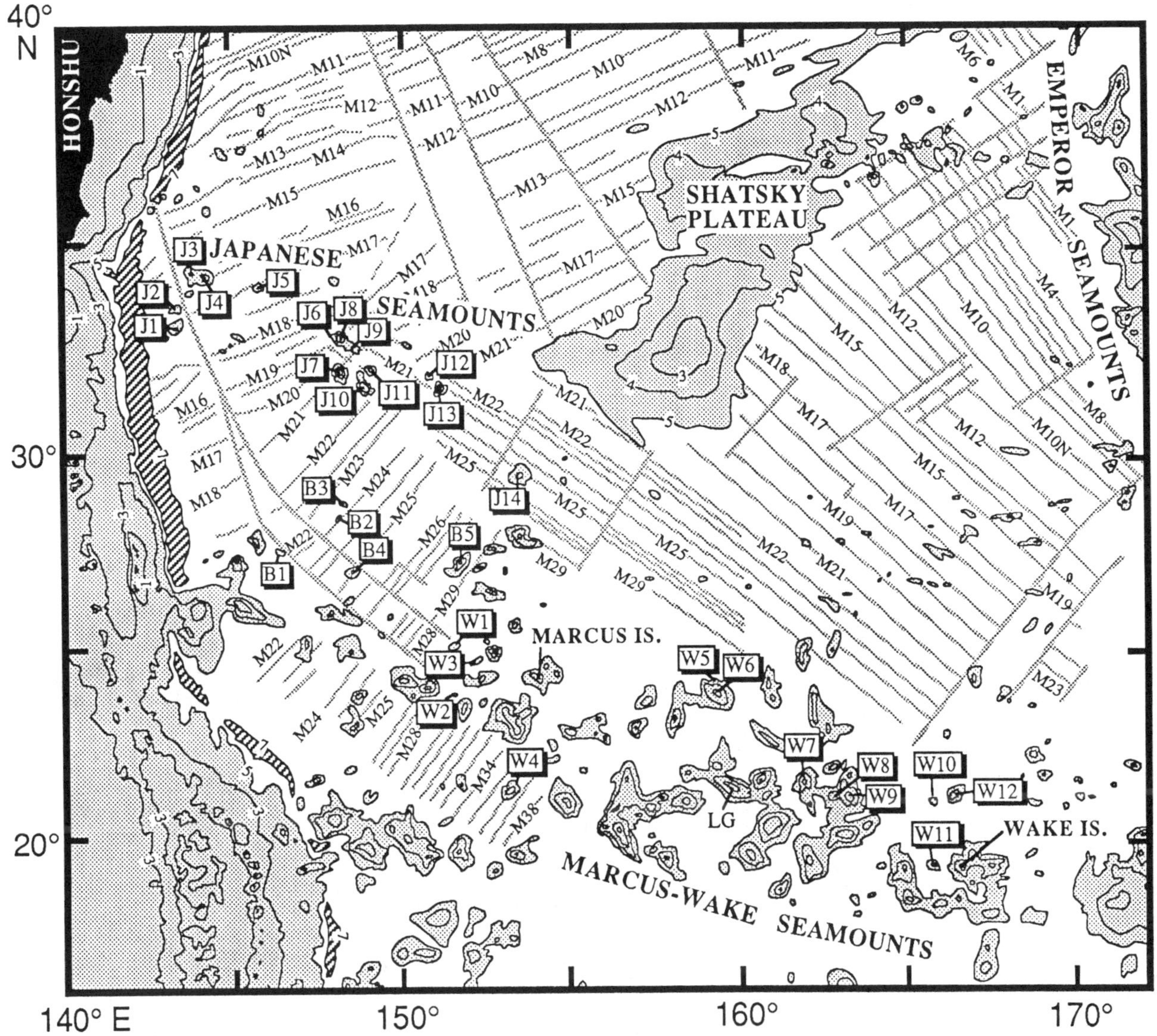

Fig. 1. Map of the northwestern Pacific showing the locations of seamounts examined in this study and major bathymetric features. Bathymetric contours are taken from GEBCO chart #5 and shown at 2 km intervals (except on Shatsky Plateau). Depths shallower than 5 km are shown in gray and depths deeper than 7 km (showing the Japan, Izu-Bonin, and Mariana trenches) are hachured. Magnetic lineations are taken from Handschumacher et al. [1988] and Sager et al. [1988]. Numbers in boxes label the seamounts with paleomagnetic studies; the identification numbers are given in Table 1. "LG" denotes Lamont Guyot, a dated seamount (Table 2) discussed in the text.

convection events within the mantle [Larson, 1991]. Thus, the history of western Pacific volcanism is important for understanding the geology of the Cretaceous.

During 1988, Leg 10 of the Roundabout Expedition, aboard the R/V *Thomas Washington*, collected geophysical data from guyots to the west and south of the Shatsky Plateau as site surveys for cruises of the Ocean Drilling Program. In particular, many seamounts were surveyed in the Japanese and Marcus-Wake seamount groups, two loosely defined chains trending west-northwest (Fig. 1). The abundance of data from these two chains provides the opportunity to study them in greater than usual detail. We examined the paleomagnetic implications of the seamounts' magnetic anomalies with the goal of determining the paleolatitudes at which these volcanoes formed and using the distribution of paleomagnetic poles to constrain the models of their formation.

Geologic Setting

Magnetic lineations in the western Pacific (Fig. 1) indicate that the Japanese and Marcus-Wake seamount groups reside on lithosphere formed at the northern edge of the Pacific plate, near the Pacific-Izanagi-Farallon triple junction, during the Late Jurassic and Early Cretaceous

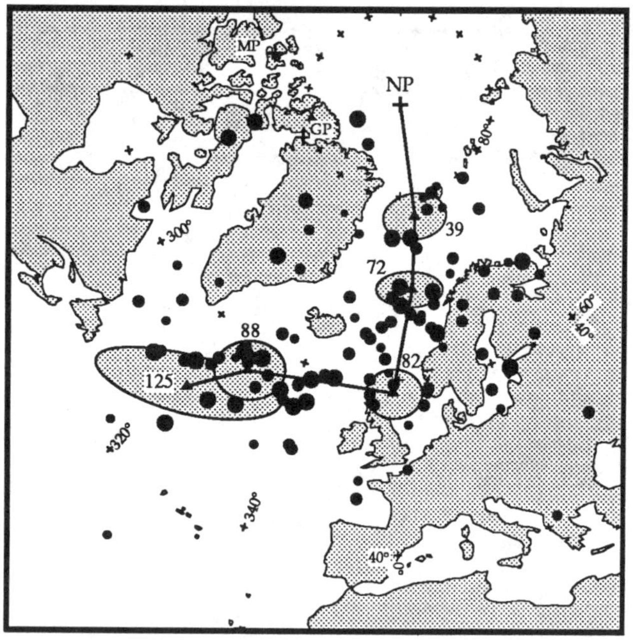

Fig. 2. Pacific seamount paleomagnetic poles and the Pacific apparent polar wander path. Filled circles represent paleomagnetic poles with symbol size indicating quality. The largest symbols denote the most reliable poles, with 95% confidence ellipse major semiaxis lengths <6° or GFR >4.5; medium symbols show poles with major semiaxis lengths 6°-12° or GFR of 4.5-2.5; and smallest symbols represent the least reliable poles, with major semiaxis lengths >12° or GFR <2.5. Triangles show mean paleomagnetic poles of the Pacific apparent polar wander path [Sager and Pringle, 1988; Gordon, 1990]. The stippled ellipses are their 95% confidence limits and large numbers are mean ages in Ma. NP denotes the geographic north pole; MP, the magnetic pole; and GP the geomagnetic pole. Individual poles are identified in Sager [1992].

the APWP. Their four-pole APWP has poles, with ages of 39, 72, and 82 Ma, aligned north-south near the Greenwich Meridian and a pole of 88 Ma age farther to the west. This APWP has the characteristic "hook" shape that has been noted by numerous investigators [Jarrard and Sasajima, 1980; Gordon, 1983; Sager, 1983; Sager and Pringle, 1987; 1988]. To this APWP we add a 125 Ma pole determined by Gordon [1990] from DSDP core data, magnetic lineation skewness, and two seamount paleomagnetic poles. Two more recent studies of marine magnetic lineation skewness have given paleomagnetic poles of about the same age [Petronotis et al., 1992; Larson and Sager, 1992], but these poles are not significantly different in location.

Despite much work by numerous investigators, the older part of the Pacific APWP remains elusive. Many believe that going back in time the APWP turns northward from the vicinity of the 125 Ma pole [Larson and Lowrie, 1975; Jarrard and Sasajima, 1980; Cox and Gordon, 1984; Sager, 1992]. However, Pacific paleomagnetic data older than about 120 Ma are scarce and often of questionable reliability. The distribution of paleopoles from seamounts older than the K-N suggests a possible bias of some of this data by normal overprints [Sager, 1992]. Similarly, paleopoles calculated from the skewness of the M-series magnetic lineations may be biased by anomalous skewness [Larson and Sager, 1992]. Even sediment samples from DSDP cores may have problems with inclination error caused perhaps by compaction [Tarduno, 1989; Gordon, 1990]. Fortunately, we will not rely heavily on the pre-125 Ma part of the APWP in this study. It suffices to note that many apparently older seamounts yield paleopoles to the north and west of the 125 Ma pole.

DATA

Seamount paleomagnetic studies require three types of data: (1) magnetic anomaly values to delineate the magnetic field of the seamount, (2) bathymetry to constrain the shape of the seamount as a magnetic source body, and (3) dates of material from the seamount for interpretation of the paleomagnetic pole. For a seamount magnetization study, a single dense geophysical survey is desirable because this yields better constraints on the seamount's shape and magnetic field which, of course, affect the computed magnetic parameters. Rarely are funds and time sufficient to make a complete seamount survey, so more often than not, seamount surveys are not ideal and it is necessary to combine geophysical data from several sources.

Geophysical Data

For our paleomagnetic inversions, we used data from four main sources (Table 3). Multibeam bathymetric maps were available for most of the seamounts we studied. These maps were made by the U. S. Naval Oceanographic Office using the SASS (Sonar Array Sounding System) and published in Vogt and Smoot [1984] and Smoot [1989, 1991]. Of the seamounts paleomagnetically studied, only four did not have SASS bathymetry maps (W4, W5, W12, and B5), and three had no bathymetry data other than the SASS maps (J1, J2, and J6).

An extraordinarily detailed magnetic anomaly contour map was available for the all of the Japanese seamounts except for J14. Collected by the U. S. Naval Oceanographic Office during the late 1970's, this data set was processed and presented by Handschumacher et al. [1983]. Unfortunately, the ship tracks and original magnetic data are unavailable to the public; nevertheless, the ship tracks for this survey are closely spaced and navigated by satellite methods. Most tracks trend east-west and are spaced about 5-10 km apart. Tie lines run north-south at a spacing of about 55 km. Additionally, diurnal varia-

Table 3. Geophysical Data Used for Paleomagnetic Inversions

J1, J2:	Jenson, Jenson-N	
	Bathymetry:	Vogt and Smoot (1984) Figure 3d.
	Magnetics:	Handschumacher et al. (1983; see text).
J3:	Takuyo-Daini	
	Bathymetry:	Vogt and Smoot (1984) Figure 3d, RNDB10WT.
	Magnetics:	Handschumacher et al. (1983), RNDB10WT, ARES05WT, HUNT01HT, V3214, C2005, SI932009, GECS-CMV.
J4:	Takuyo-Daisan (Seiko)	
	Bathymetry:	Vogt and Smoot (1984) Figure 3d, RNDB10WT.
	Magnetics:	Handschumacher et al. (1983), RNDB10WT, ARES05WT, HUNT01HT, V3214, C2005, SI932009, GECS-CMV, SCAN03AR.
J5:	Maiko	
	Bathymetry:	N. C. Smoot (personal communication, 1987), ARES05WT, SI932009.
	Magnetics:	Handschumacher et al. (1983), ARES05WT, SI932009.
J6:	Winterer-NW	
	Bathymetry:	Vogt and Smoot (1984) Figure 3c.
	Magnetics:	Handschumacher et al. (1983).
J7:	Charlie Johnson	
	Bathymetry:	Vogt and Smoot (1984) Figure 3c, RNDB10WT.
	Magnetics:	Handschumacher et al. (1983), RNDB10WT, SI932009, HUNT03HT.
J8:	Winterer	
	Bathymetry:	Vogt and Smoot (1984) Figure 3c, RNDB10WT, ARES05WT.
	Magnetics:	Handschumacher et al. (1983), RNDB10WT, ARES05WT.
J9:	Winterer-SE	
	Bathymetry:	Vogt and Smoot (1984) Figure 3c.
	Magnetics:	Handschumacher et al. (1983), ARES05WT.
J10:	Stout	
	Bathymetry:	Vogt and Smoot (1984) Figure 3c, RNDB10WT.
	Magnetics:	Handschumacher et al. (1983), RNDB10WT.
J11:	Thomas Washington	
	Bathymetry:	Vogt and Smoot (1984) Figure 3c, RNDB10WT.
	Magnetics:	Handschumacher et al. (1983), RNDB10WT, ARES05WT, KK77031705.
J12:	Isakov-NW	
	Bathymetry:	Vogt and Smoot (1984) Figure 3b, RNDB10WT.
	Magnetics:	RNDB10WT, V3212, HUNT03HT.
J13:	Isakov	
	Bathymetry:	Vogt and Smoot (1984) Figure 3b, RNDB10WT.
	Magnetics:	RNDB10WT, ARES05WT, V3212, LUSI01AR.
J14:	Makarov	
	Bathymetry:	Vogt and Smoot (1984) Figure 3a, RNDB10WT.
	Magnetics:	RNDB10WT, ARES05WT, ARES07WT, HUNT03HT, V2106, KK77031705.
W4:	Golden Dragon	
	Bathymetry:	ZTES04AR, V3214, V3312.
	Magnetics:	ZTES04AR, V3214, V3312.
W5:	UCSD	
	Bathymetry:	RNDB10WT.
	Magnetics:	RNDB10WT.
W6:	Scripps	
	Bathymetry:	Smoot (1989), RNDB10WT.
	Magnetics:	RNDB10WT, ARES05WT.
W8:	Vibelius	
	Bathymetry:	Smoot (1989), RNDB10WT.
	Magnetics:	RNDB10WT, ARES05WT.
W9:	Wilde	
	Bathymetry:	Smoot (1989), RNDB10WT.
	Magnetics:	RNDB10WT, ARES05WT.
W12:	Woods Hole	
	Bathymetry:	RNDB10WT, ARES05WT, CIRC02AR, GH7901.
	Magnetics:	RNDB10WT, ARES05WT, CIRC02AR, GH7901.
B5:	MIT	
	Bathymetry:	RNDB10WT, RAMA04WT, RAMA10WT, UM6503-B, DSDP20GC.
	Magnetics:	RNDB10WT, RAMA04WT, RAMA10WT, UM6503-B, DSDP20GC.

tions and navigation errors were removed during processing.

The key data set was the Roundabout Leg 10 cruise (RNDB10WT in Table 3). This cruise was a site survey for the Ocean Drilling Program and obtained Sea Beam bathymetry, magnetic, gravity, and seismic reflection data from seamounts of the Japanese, Marcus-Wake, and Mid-Pacific Mountains groups. Additionally, most seamounts visited by RNDB10WT were dredged for datable material. Because the cruise visited many seamounts, few were completely surveyed. However, most of the seamounts are guyots whose summits were completely or nearly completely surveyed with the Sea Beam multibeam echo-sounder. Consequently, the typical RNDB10WT guyot survey consisted of an entrance line, summit survey, dredge transit lines, and an exit line. Even though these data did not constitute complete surveys, they were sufficient to allow the other data types to be tied together (see "Methods" section).

Geophysical data for many of the seamounts were also available from the National Geophysical Data Center (NGDC) archives. Mostly these data consist of bathymetry and magnetic data collected along sporadic ship tracks (cruise abbreviations given in Table 3). Notable among these data are those from the 1971 cruise, Aries Leg 5 (ARES05WT), of the R/V *Thomas Washington*, a site survey cruise for the Deep Sea Drilling Project. Many of the seamounts visited by ARES05WT were later surveyed by RNDB10WT, but with better navigation and Sea Beam.

Age Data

Table 2 summarizes age data for the seamounts examined in this study. It includes those fossil dates constrained within one or two stages [Matthews et al., 1974]. It also contains ^{40}Ar-^{39}Ar radiometric dates from two sources, a review by Ozima et al. [1983] and 8 dates, determined during this study, from 7 seamounts. Our new age data are discussed in more detail in a companion paper [Winterer et al., this volume]. Several seamounts that we dated had previous radiometric ages (W6, W7, W9, and Lamont Guyot), but we list only the new data in Table 2. For all of these seamounts the new ages were not significantly different from the old.

Paleomagnetic Poles

Several studies of western Pacific seamounts have previously reported paleomagnetic poles (Table 1) for a few seamounts in the Japanese and Marcus-Wake groups and vicinity [Vacquier and Uyeda, 1967; Harrison et al., 1975; Sager 1983; Hildebrand and Parker, 1987; Ueda, 1988; Sager, 1992]. Some were determined using the least-squares method and a few were calculated with the more sophisticated seminorm minimization method (see next section for discussion). Several of the previously-studied seamounts had poles recomputed in this study. Because we had additional geophysical data and used the seminorm minimization method, we assumed that our results superceed those previously published.

METHODS

Radiometric Age Determination

The freshest and most crystalline material was selected from dredge hauls for radiometric dating using the ^{40}Ar-^{39}Ar incremental-heating technique. This method was used because it has been found to produce the most reliable dates from basalts exposed to low temperature submarine alteration. Ages were calculated using both isochron and plateau methods. Here we use the plateau ages as the most consistent estimate of the crystallization age of the basalts analyzed. Details and discussions of the dating methods are given in Duncan and Hargraves [1990], Winterer et al. [this volume], and Pringle [this volume].

Bathymetry and Magnetic Anomaly Maps

Preparing bathymetry and magnetic anomaly data for analysis was complicated by the diversity of the data sets used (Table 3). RNDB10WT adequately surveyed only a few seamounts, so we combined these data with the U. S. Naval Oceanographic Office magnetic anomaly map and SASS data. Furthermore, because the ship tracks used to collect the Navy data were unavailable, we took steps to mitigate navigation errors and the possibility of bias due to contouring or reproduction.

For bathymetry data, we first plotted the RNDB10WT Sea Beam data in uncorrected meters (i.e., assuming a sonic wave velocity of 1500 ms^{-1}). The SASS contours were enlarged to the same scale and laterally adjusted to fit the Sea Beam contours. This removed most positioning or scaling errors introduced by the reproduction of the SASS maps. We also plotted the bathymetry from previous cruises, obtained from NGDC. This gave additional constraint on the shapes of those seamounts for which no SASS maps were available. Even for seamounts with complete SASS bathymetry, a comparison with the NGDC bathymetry allowed us to correct most navigational errors in these sporadic cruise tracks. This was particularly important for the NGDC lines from which magnetic data were used.

Magnetic data acquired at different times are difficult to combine to make magnetic anomaly maps because of external field variations and because the data are often reduced with different reference fields. External field corrections were not made for most seamounts, because their magnetic anomalies had amplitudes that were much greater than the usual diurnal and short-period magnetic field variations [Onwumechilli, 1967; Matsushita, 1967]. Diurnal corrections were made for the

RNDB10WT magnetic data for two seamounts, J7 and W8, because their magnetic anomalies are smaller (<1000 nT, see Appendix) and they were extensively surveyed on RNDB10WT. We assumed the diurnal variations were small during night hours and compared magnetic values at ship track crossings with daytime measurements to estimate a diurnal variation time series. Seamount W5 also has a small magnetic anomaly; however, it was surveyed completely at night, so the diurnal variations were assumed small.

To compute magnetic anomaly values, the RNDB10WT magnetic data were corrected using the 1985 International Geomagnetic Reference Field. These data were used as a standard and other magnetic data were adjusted into agreement. This process took several forms, described below, depending on the coverage of a seamount by RNDB10WT.

The Japanese seamounts west of 150° E were covered by the Navy magnetic map, so these data played an important role in defining their magnetic anomalies. The Navy magnetic map was treated similar to the Navy multibeam bathymetry; it was plotted at the same scale as the RNDB10WT and NGDC magnetic data and shifted laterally to align anomaly features. If necessary, base level corrections were also applied.

Wherever possible, we used the RNDB10WT and NGDC magnetic data as the primary constraint of the magnetic anomaly because we could not be sure of the placement of ship tracks (hence, contouring bias) in the Navy data set. After making adjustments, the agreement between Navy and non-Navy data was typically good. For input into the magnetic modeling routine, the non-Navy magnetic data were digitized along ship tracks, usually at a spacing of about 1 km (Appendix). The Navy data were gridded in unsurveyed areas, usually with a spacing of about 2-4 km. The spacing difference was used to give the non-Navy data a lesser weight in defining the magnetization.

A few seamounts were not crossed by RNDB10WT or significant numbers of NGDC ship tracks. Therefore, the magnetic inversions of these seamounts relied solely on the Navy magnetic anomaly and multibeam bathymetry maps. Although both Navy data sets are extremely detailed, we could not check the contouring nor could we be certain of the alignment between magnetic anomaly and bathymetry. Registering the two data sets was done by aligning the Navy and non-Navy bathymetry and magnetics data of adjacent seamounts. Because of these uncertainties, such seamounts present the greatest potential for giving erroneous results from their magnetic anomaly inversions. In this category are J1, J2, J6, and J9. Additionally, J5 was not crossed by RNDB10WT; however, it was surveyed by Leg 5 of the Aries expedition (ARES05WT), so we used these magnetic data as primary constraint of its magnetic model. Three other seamounts, J3, J10, and J12 were only crossed by one RNDB10WT track, but the detail and swath coverage of the Sea Beam bathymetry proved adequate for aligning the data sets.

Because the Navy magnetic and SASS bathymetry data are not co-registered, positioning errors from the matching of these data with the RNDB10WT data are independent. The detail of the SASS and Sea Beam systems allowed bathymetric data sets to be closely matched, probably to less than 1 km error for seamounts crossed by RNDB10WT and less than 500 m for those more extensively surveyed during that cruise. The magnetic anomaly data are smoother, so they are somewhat more difficult to register precisely. Nevertheless, for most seamounts it was possible to closely constrain registration by characteristic features or gradients in the magnetic anomaly. We estimate that the better-constrained magnetic anomalies are matched nearly as well as the better bathymetry maps, probably to less than 1 km. Those with less extensive surveys may have errors of 1-2 km. Registration errors of these magnitudes probably do not cause errors in the magnetic anomalies or bathymetry greater than about a few tens of nT or a few hundred meters, respectively. Furthermore, the errors are likely least near the seamount summits, which are generally better surveyed. Such errors should not have a significant effect on the calculated seamount paleomagnetic parameters.

The largest registration errors are potentially for those seamounts not crosed by RNDB10WT, whose bathymetry and magnetic anomalies are known solely from the Navy data and registered by matching the bathymetry and anomaly of adjacent seamounts contained within the same SASS bathymetry map. We have no way of checking the registration accuracy of these seamounts, but believe this error to be less than about 2-4 km. These errors may contribute to some large calculated model magnetizations and rms errors (see below).

Magnetic anomaly maps for the Japanese and other seamounts east of 150° E were constructed entirely from RNDB10WT and NGDC data. We did have Navy contour maps of the magnetic anomalies of J12, J13 and W9, but these were gridded from much less detailed surveys than the other Navy map. Consequently, we could only use these maps as a guide to the general shapes of those anomalies and the background crustal anomalies. Registration errors from matching the NGDC data with the RNDB10WT data can be several kilometers, particularly for older, more poorly navigated NGDC tracks. However, these data make up only a small percentage of the total for all seamounts, so the effect of such errors on magnetic models is thought to be small.

Magnetic Inversions

Magnetic inversions were done using the seminorm minimization method [Parker et al., 1987], mainly because it accounts for magnetic anomaly measurement errors and small inhomogeneities in magnetization and is thus less affected by these complications than the least-squares technique. Futhermore, Parker [1988] developed a statistical model of the magnetization nonuniformities that allows a 95% confidence ellipse to be calculated for seamount paleomagnetic poles. In a later treatment of the problem, he suggests that these error estimates are probably too small for seamounts with signficant nonuniformities [Parker, 1991]. The results from some seamounts support this conclusion, but the difference does not appear to be great for most (see "Discussion" section). Moreover, no similar error estimation scheme is available for the least-squares method, so some error estimate, albeit less than perfect, is better than none at all.

Because the seminorm minimization technique can fit the observed anomaly arbitrarily closely by allowing a larger nonuniform magnetization contribution, a choice must be made as to how close the fit will be and this choice affects pole position. Hildebrand and Parker [1987] typically used a rms misfit of about 30 nT, figuring that this error value was typical of the accuracy of many seamount magnetic surveys. The mean rms-errors for most seamount surveys in this study was difficult to determine because the original Navy data was unavailable, so ship track crossing errors could not compared for that data set. Consequently, for most seamounts, we also used 30 nT (Table 1). The seamounts modeled in this study typically have large, relatively simple anomalies (Appendix), so different, but reasonable choices for rms-misfit (e.g., about 15 to 50 nT) typically did not change the calculated paleopole by more than a few degrees. For two seamounts (J1 and J2), it was impossible to reduce the rms misfit to 30 nT. Neither seamount was crossed by the RNDB10WT cruise, so these high errors probably result because of inaccuracies in registering magnetic and bathymetric data.

Another choice in making a seamount anomaly inversion is how to correct for background crustal fields. It has been found that signficant errors in paleomagnetic pole position can result if no correction is made for large background anomalies, such as magnetic lineations [e.g., Yamazaki, 1988]. Some investigators have allowed the inversion routine to simultaneously calculate the magnetization vector components and a planar regional. However, others have found that the regionals calculated in this manner are often suspiciously large and have a significant affect on the calculated pole location [Sager, 1983; McNutt, 1986]. Inversion routines probably erroneously attribute part of the seamount anomaly to the background field [McNutt, 1986; Parker, 1991]. We made models of background anomalies where they could be defined and appeared to have magnitudes greater than about 10% of the seamount anomaly amplitude. However, for comparison we also routinely calculated the magnetization components with a planar regional. This produced results for a few seamounts (J1, J9, and J10) that are more consistent with the other seamount poles. Although this may be fortuitous, the consistency implies better results. In the case of J2, this may be because the seamount is located between two larger edifices with large magnetic anomalies that probably affect its anomaly. For the other two, it may be a result of inadequate corrections for crustal magnetic anomalies.

The Japanese seamounts were most in need of background corrections because they are on crust recording the Japanese and Hawaiian magnetic lineations (Fig. 1). Corrections were not made for J1-J5 and J14 because the Navy magnetic data showed these seamounts were either in areas of subdued crustal anomalies or not situated near signficant background gradients. Magnetic models of the lineations around the other Japanese seamounts were made using a forward magnetic modeling technique [Plouff, 1976] to represent the anomalies seen in the Navy magnetic data. Background corrections were not made for most of the Marcus-Wake seamounts because they are on crust of the Jurassic Quiet Zone which contains uncorrelatable, low amplitude background magnetic anomalies [Handschumacher et al., 1988].

Several seamounts were close enough to neighbors that it was necessary to make corrections for overlapping anomalies. One is W5, a small volcano on the flank of the much larger W6. W6 was analyzed first to derive its mean magnetization vector. Those results were used to caculate the magnetic anomaly of W6 and these anomaly values were subtracted from the anomaly of W5. In two other instances, it was difficult to accurately calculate the effect of the adjacent volcano. J7 has a smaller volcano on its southeast flank. Because we had no RNDB10WT or NGDC data from the satellite seamount, it was difficult to model its anomaly accurately. Consequently, we only used the anomaly from the other three quadrants of the larger guyot for input to the inversion routine. Similarly, W8 has a small anomaly (Appendix) and is located near W9, whose anomaly is nearly ten times as large. As a result, W9's anomaly seriously perturbs the eastern part of the W8 anomaly. We were unable to make a sufficiently accurate correction for the outer edge of the W9 anomaly, so we only used the unaffected western half of the W8 anomaly in the magnetic inversion.

Finally, in our analysis, some previously published paleomagnetic data, derived using the least-squares method, were considered. This type of data make up about half of the results from the Marcus-Wake and "between" seamounts. Though these inversions gave no confidence limits on the pole positions, it is typical of such data to have a measure of the misfit between the observed and calculated anomalies as an estimate of the quality of the inversion. For this purpose, the "goodness-of-fit" ratio, the ratio of the mean of the absolute values of the observed anomaly to the mean of the absolute values of the residual (observed minus calculated) anomaly, is included in Table 1. The higher the goodness-of-fit ratio, the better the match between the observed and calculated anomalies. A ratio of 2 is often considered a minimum for a reliable inversion, whereas a value of 4 or greater is considered excellent.

RESULTS

Magnetic inversions produced remarkably consistent results for both the Japanese and Marcus-Wake seamount groups. With two exceptions, all of the paleomagnetic poles of the Japanese seamounts cluster along the 82-125 Ma section of the APWP (Fig. 3), with latitudes ranging from 50.4° N to 59.2° N and longitudes from 311.9° E to 346.2° E. Likewise, all but two of the Marcus-Wake poles fall along the same APWP segment, though with a slightly greater range in latitude and longitude, 46.2° N to 63.9° N and 319.7° E to 5.9° E (Fig. 4).

For three of the Japanese seamounts (J1, J9, and J10) the poles from magnetization vectors calculated with a planar regional field were in better agreement with the cluster than those calculated without a background field, so we preferred these data. All three have magnetic anomalies that had to be corrected for the presence of nearby seamounts or crustal magnetic lineations, so this result is not surprising. Moreover, the estimated error ellipses for these poles are the largest of the Japanese seamounts, indicating that they are probably the least reliable.

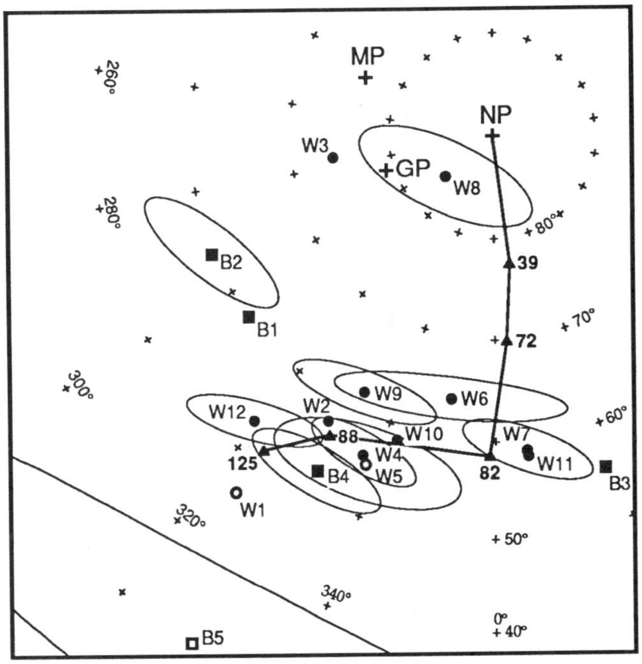

Fig. 4. Paleomagnetic poles of the Marcus-Wake and "between" seamounts. Circles represent the Marcus-Wake seamount poles, whereas squares show "between" poles. Filled symbols denote poles from normally polarized seamounts; open symbols, reversely polarized seamounts. Plot conventions as in Fig. 2.

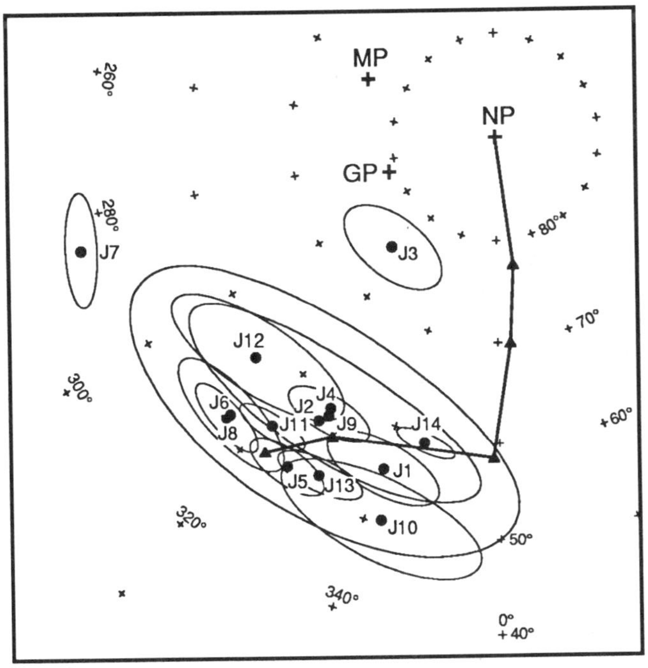

Fig. 3. Paleomagnetic poles of the Japanese seamounts. Filled circles denote pole positions and surrounding ellipses are 95% confidence regions. Triangles represent the Pacific apparent polar wander path. Identification numbers correspond to Table 1. Plot conventions as in Fig. 2.

The seamounts that gave outlier paleopoles are J3, J7, W3, W8, B2, B3, and B5. These poles do display some consistency: all but the B3 and B5 poles are either near the geomagnetic pole or to the northwest of the old end of the APWP (Figs. 3, 4). B5's pole is south of the old end of the APWP, but this may be related to its reversed magnetization, as discussed in the next section. The B3 pole is east of the APWP and was determined from an model with a low goodness-of-fit (Table 1), so this pole may be an outlier owing to an innacurate inversion.

Uniform magnetization intensities of the seamounts in this study range from 1.8 to 19.7 Am^{-1} and average 7.7 Am^{-1}. The Japanese seamounts are remarkable for their large magnetization intensities. They average 10.5 Am^{-1} and all but two have higher values than the median for Pacific seamounts, 5.8 Am^{-1} (Fig. 5). Most of the seamounts also have relatively uniform magnetizations. The average ratio of uniform to nonuniform magnetization intensity is 3.1:1 and this is probably why most seamounts gave consistent paleomagnetic poles.

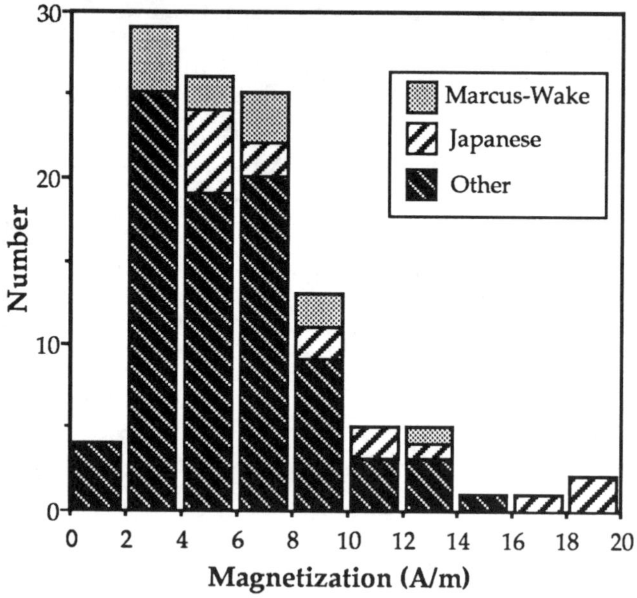

Fig. 5. Histogram comparing mean magnetizations of Japanese and Marcus-Wake seamounts with other Pacific seamounts. Data for other seamounts taken from Sager [1992].

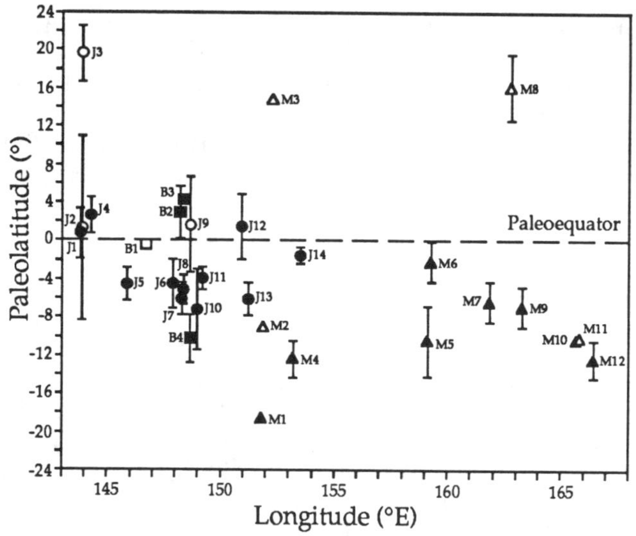

Fig. 6. Paleolatitudes of seamounts examined in this study. Circles, squares, and triangles denote the Japanese, "between", and Marcus-Wake seamounts, respectively. Error bars show latitudinal extent of 95% confidence ellipse for those seamounts whose anomalies were inverted by the seminorm method. Filled symbols denote the most reliable data; open symbols show data from seamounts with outlier poles or low-reliability inversions. Identification numbers correspond to Table 1.

Notably, the seamounts with peculiar paleopoles generally have low magnetizations. J3 and J7 have the lowest and third-from-lowest magnetization intensities of all the Japanese seamounts and W3 and W8 have the lowest intensities of the Marcus-Wake seamounts. Similarly, B2, B3, and B5 have intensities that are low in comparison with the others. These observations imply a link between low magnetization intensity and possibly unreliable paleomagnetic poles. On the other hand, the Japanese seamounts with the highest uniform magnetizations (J1, J2, and J9) are all from the subset of seamounts for which no data were available other than the Navy data. This observation suggests such large apparent magnetizations may be caused by the inversion routine trying to fit inaccurately registered magnetic and bathymetry data. Further evidence for this is that for both J1 and J2, it was impossible to reduce the rms-error to near 30 nT (Table 1). Consequently, the paleomagnetic results of some seamounts with very large apparent magnetizations may be suspect.

The seamounts' magnetizations are mostly of normal polarity; only two seamounts in the Marcus-Wake group and B5 have reversed polarities (Table 1). This bias is not unexpected because the dated seamounts are mostly mid-Cretaceous in age, a period that included the K-N when the geomagnetic field remained in the normal polarity state from about 118-83 Ma [Harland et al., 1982].

There seems to be a small, systematic paleolatitude difference between the Japanese and Marcus-Wake seamounts. The former formed between 2.5° N and 7.1° S and average 2.4° S ±3.5° (this and other error limits are given as standard deviation). The latter have paleolatitudes between 2.3° S and 12.5° S which average 8.8° S ±3.3°. The few "between" seamounts have members consistent with both groups. Outlier paleolatitudes, from J3, W2, W3, W8, and B5, either represent discordant paleopoles or were from the less reliable inversions. A closer look at the Japanese seamount paleolatitudes shows that they cluster at two different paleolatitudes. In the center of the chain, J5-J8, J10-J11, and J12 have paleolatitudes that are significantly below the equator, averaging 5.2° S ±1.2°. The other seamounts, J1, J2, and J3 at the west end of the chain, J9 in the middle, and J12 and J14 at the east end all formed almost precisely on the equator. Their mean paleolatitude is 1.0° N ±1.1°, but two of these seamounts produced paleomagnetic results that have large errors and could fit into the other group as well (Fig. 6).

DISCUSSION

Paleomagnetic data from the Japanese and Marcus-Wake seamounts provide information about the latitudes at which they formed, the progression of volcanism in the seamount groups, and the tectonic drift of the edifices since their formation. Such data are also useful for testing hypotheses about the mechanism that formed these seamounts. In particular, we would like to distinguish between two possible types of formation hypotheses: (1) hot spot volcanism and (2) a large volcanic pulse.

The evidence most often cited for a simple volcanic pulse hypothesis is the observation of widespread volcanic products of similar age [Schlanger et al., 1981; Rea and Vallier, 1983] and of seamounts with similar summit depths [Menard, 1984]. The nature of this phenomenon is not well-defined and so it is not amenable for explaining the formation of a particular seamount chain in detail. However, the volcanic pulse hypothesis implies that the seamounts formed at virtually the same time within a large area. Such a mechanism should not yield a clear age progression, nor would it be likely to form seamounts at consistent latitudes. Instead, seamounts formed by a volcanic pulse would have scattered paleolatitudes related to their relative positions at the time of the pulse.

Conversely, the hot spot hypothesis is sufficiently quantitative that it lends itself to predicting an age progression caused by the drift of the Pacific plate over mantle magma sources. However, no one has yet tried to tie the Japanese or all of the Marcus-Wake seamounts to specific hot spots. The problem is that only a tiny percentage of western Pacific seamounts have been dated, and there is no clearly defined hot spot chain for the mid-Cretaceous and older [Duncan and Clague, 1985]. The hot spot hypothesis implies that the seamounts formed at one or more localized volcanic centers. It also implies that the seamounts in any one chain formed at about the same latitude or at a latitude that varied slowly because of the motion of the hot spot relative to the spin axis. Thus, if formed by a single hot spot, seamounts of a given chain should show consistent paleolatitudes and an age progression defined either by radiometric dates or the progression of paleomagnetic poles along the apparent polar wander path.

Paleomagnetic Data: Errors and Limitations

Before using the data to test hypotheses, it is useful to consider evidence of possible errors, both systematic and random, that may affect the paleomagnetic data. As a whole, the paleomagnetic poles display remarkable consistency. With but a few exceptions, the poles of both the Japanese and Marcus-Wake seamounts, as well as seamounts between them, cluster along the 82 to 125 Ma segment of the Pacific APWP. Although this seems to indicate that these data are accurate, in examining the outlier poles we see some reasons for caution.

Most of the paleopoles that fall outside the cluster are to the north and northwest of the APWP. Poles in this area could indicate one of two possibilities: either the old end of the APWP extends to the north, or these seamount poles are biased by significant amounts of induced or normal polarity overprint magnetization. Several investigators have examined the Jurassic to Early Cretaceous Pacific APWP and concluded that it probably does hook to the northwest [Cox and Gordon, 1984; Sager, 1992; Larson and Sager, 1992]. This implies that the plate moved southward during this period; indeed, the trend of paleomagnetic inclinations in Albian to Aptian age sediments at Deep Sea Drilling Project Site 463 (Mid-Pacific Mountains) is consistent with such a motion [Sayre, 1981].

However, the locations of most of the outlier poles seem to imply that they are biased by induced or normal polarity overprint magnetization. In particular, three normally magnetized seamounts (J3, W3, and W8) have paleopoles that are suspicious because they are located near the geomagnetic north pole (Figs. 3, 4). This is just what would be expected if their magnetizations were almost wholly induced rather than remanent. Furthermore, the poles of two reversely polarized seamounts (W1 and B5) are located to the south of the APWP, also consistent with significant induced magnetization components or normal polarity overprints. The age of W1 is unknown, but its polarity suggests that it is older than the youngest M-series reversal, M0 (118 Ma [Harland et al., 1982]). Its bias appears to be small because its pole is only 7.6° south of the 125 Ma mean pole. Indeed, this difference may not be significant given the scatter of paleopoles. However, the pole of B5 (120.3 ±0.8 Ma) is significantly different. Located 24° south of the 125 Ma pole, its 95% confidence ellipse does not overlap any of those of the other seamounts (Fig. 4).

Assuming that both seamounts should have recorded the 125 Ma pole position, the calculated magnetic directions imply that W1 and B5 have mean Koenigsberger ratios of 4.5 and 1.8, respectively. Both values are within the range of ratios observed in seamount rocks [Gee et al., 1988; 1989], but are lower than average. A compilation of Koenigsberger ratio data from dredge and DSDP samples of seamounts and ocean crustal basalts [Sager, 1983] indicates a modal value of about 16 for the geomagnetic field strength at the latitude of this study. Thus, it appears that some of the seamounts may have given pole positions biased by overprint, but the fact that most poles cluster and the others are so clearly unusual is encouraging and suggests most have small or negligible biases. However, because the seamounts are predominantly of normal polarity, normal overprints may cause a slight overall northward shift of the main cluster of poles, and this should be considered in the interpretation of seamount paleomagnetic data.

Alternate explanations may be more appropriate for several of the outlier poles. Some may just be erroneous. B3 has a pole located well to the east of the APWP (Fig. 4). When its anomaly was inverted [Vacquier and Uyeda, 1967], no correction was made for background fields, even though the magnetic lineations in the vicinity have amplitudes as nearly as large as the seamount anomaly. Similar situations have caused significant errors in magnetization inversions [Yamazaki, 1988]. Other outlier poles may be from older seamounts. J7 and B2 have poles located at about the same latitudes as the cluster poles, but their pole longitudes are farther west (Figs. 3, 4). Many poles from seamounts on Jurassic-age seafloor are found in this region, leading to speculation that such poles represent the Early Cretaceous to Jurassic portion of the APWP [Jarrard and Sasajima, 1980; Hildebrand and Parker, 1987; Sager, 1992]. Thus both J7 and B2 may be older than the ubiquitous mid-Cretaceous seamounts.

Examining paleomagnetic poles that cluster along the APWP, it appears that most poles record a mid-Cretaceous pole position. The short distance between the 88 and 125 Ma mean poles of the APWP implies that polar wander was slow during that period. All of the seamount poles except for J14, W7, and W10 are indistinguishable from these two mean poles. This is consistent with most of the dates for these seamounts. The Japanese seamounts have mid-Cretaceous fossil ages and radiometric dates ranging from 94 to 108.3 Ma, whereas all but one of the dates for the Marcus-Wake seamounts are between 90.6 to 110.8 Ma (Table 2). Of the three poles located near the 82 Ma mean pole, two (J14 and W7) are from seamounts with dates of 94 ±1 and 96.8 ±0.8 Ma, respectively (Table 2). This suggests that these pole positions may have slight errors that cause them to be located too far to the east. Both of these seamounts have small estimated 95% error ellipses, so this observation implies that these error estimates are probably too small. Assuming that all of the paleopoles represent essentially the same pole position and ignoring the more poorly determined poles (those with 95% confidence ellipse major axes longer than 12° or GFR less than 3.0), a Fisher [1953] analysis gave a standard error of 8.5°. Though the distribution of poles may not be strictly Fisherian, (the dispersion in pole longitude appears larger than that in pole latitude), this value gives an idea of the average scatter.

Ages, Poles, and Polarity

Radiometric dates from the Japanese and Marcus-Wake chains are within the mid- to Late Cretaceous, but show no unequivocal trends (Table 2). In the central and eastern end of the Japanese chain there appears to be a younging-east trend from J8 (108.3 ±0.5 Ma) to J13 (103.7 ±0.9 Ma) to J14 (94 ±1 Ma), but J4 (102 Ma) disrupts the trend. The Marcus-Wake seamount dates are even less ordered; from west to east they are: W4 (95 Ma), W6 (110.8 ±1.0 Ma), Lamont Guyot (96.2 ±0.5 and 81.5 ±0.6 Ma), W7 (96.8 ±0.6 Ma), and W11 (90.6 ±0.3 Ma). If we ignore W6, which is north of the main body of the chain and thus may not belong to it, all that can be discerned from these data is that most of the seamounts formed at nearly the same time. The two dated "between" seamounts have dates similar to the other groups, with B4 at 93.7 ±4.2 Ma and B5 at 120.3 ±0.8 Ma.

With the exception of the young date for Lamont Guyot, these dates all fall within the period of slow APW between 88 and 125 Ma, so a significant progression along the APWP is not expected. To check this, we plotted pole longitude versus seamount longitude, expecting to find a correlation if indeed there were an age progression; however, the plot showed random scatter. Thus, we conclude that if there is an age progression within the mid- and Late Cretaceous seamounts, it cannot be discerned with the paleopole data.

There are, however, several observations that suggest the presence of older seamounts. Two guyots (J7 and W8) have summits that are about 180 m and 360 m deeper, respectively, than the others. This difference may indicate that they subsided for a longer period of time. Furthermore, both seamounts have outlier paleopoles. J7's pole may indicate an Early Cretaceous or Jurassic age, as discussed above. W8 has a weak magnetization and pole that probably indicates the magnetization is mainly induced. This may be a result of the seamount having formed during a period of multiple geomagnetic reversals, effectively nullifying the net remanent magnetization.

Similarly, B5 has a magnetization that is relatively weak, partly induced, and complex (the nonuniform part is 2.2 times the uniform part), perhaps for the same reason. It is also one of three seamounts in the study that have reversed polarities. All of these are in the Marcus-Wake and "between" seamount groups. Clearly these seamounts were formed either before or after the K-N. Although one radiometric date, from Lamont Guyot, postdates the K-N, the paleomagnetic poles of the reversed seamounts are located near the 88 to 125 Ma end of the APWP. This implies that these seamounts are older and indeed, B5 is about 10 m.y. older than the next oldest dated seamount in this study.

Magnetization Intensities

Many of the seamounts in this study have strong magnetizations, particularly members of the Japanese group (Fig. 5). Possible explanations include geomagnetic field strength or geochemistry. Some investigators have hypothesized that the geomagnetic field was stronger during the K-N [e.g., Raymond and LaBreque, 1987]. Alternatively, certain seamounts may be built of basalts with greater magnetic mineral concentrations. This hypothesis is somewhat more appealing because it could explain why only certain seamounts, such as the Japanese group, are more magnetic. Furthermore, we note that even within the Japanese group there is a tendency for magnetization intensities to clump. The J6-J11 cluster seamounts have lower magnetizations than the others in the Japanese group (Table 1). Such variations may indicate changes in magma chemistry from a single source or differences in chemistry among different magma sources.

Paleolatitudes

The general pattern of paleolatitudes, displaying an average difference of 6.4° between the Japanese and Marcus-Wake seamounts, is about what is expected considering the present latitude difference between the two groups (6°-7°). It confirms that they formed at different sources in about their same relative locations. Taken as a whole, these paleolatitudes could also be construed to indicate a southward drift of the hot spots if the seamounts were formed by the hot spots of French Polynesia, whose present latitudes are approximately 10° S to 30° S. However, the inferred southward drift could range from 0° to 20°, depending on whether the Japanese and Marcus-Wake seamounts were formed by hot spots on the northern, middle, or southern regions of the superswell.

More detailed interpretations of the paleolatitudes requires that one consider the small differences to be significant. We noted that the Japanese seamount group can be divided into two groups with average paleolatitudes of 1.0 N ±1.1° and 5.2° S ±1.2°. Taken at face value, several implications can be drawn. First, the dichotomy suggests that the chain was not created by a single hot spot, which would be expected to form all of the seamounts at a consistent latitude. Second, it could imply that the seamounts did not form simultaneously; one group might have formed before the other and the paleolatitude difference resulted from the drift of the Pacific plate. However, the radiometric dates indicate that the age difference, if any, is small. Two of the northern paleolatitude group have dates of 94 ±1 Ma and 102 Ma, whereas two of the southern group have ages of 103.7 ±0.9 Ma and 108.3 ±0.5 Ma (Table 2). Consequently, this inference is incorrect. Finally, the different clusters also imply that the Japanese seamounts may not be a chain in the genetic sense.

A similar interpretation could be made of the Marcus-Wake paleolatitudes. Seamounts W6, W7, and W9-W12 display an intriguing linear trend at the eastern end of the chain. Furthermore, W2 and W4 on the western end have paleolatitudes virtually identical with W5. However, this is an arbitrary division and with the small number of data the trend could well be fortuitous. Indeed, if W6 is ignored, the pattern is weakened considerably and the remaining reliable Marcus-Wake paleolatitudes are essentially the same. Futhermore, W6 is the oldest dated member of this group by 4 m.y. (Table 2) and is located with W5 about 150 km north of the loosely-defined main body of the chain, so they may not even be related to the others. Another possible inference is that W6 has the most northern paleolatitude because of Pacific plate drift in the period between its formation and that of the others. However, this trend is exactly the opposite of the age/paleolatitude pattern suggested by the Japanese seamounts.

A different way to examine the paleolatitudes is to plot northward drift, the present seamount latitude minus its paleolatitude (Fig. 7). This quantity is the sum of the latitudinal drift. Most of the seamounts have large components of northward drift because the Pacific plate has drifted about 30° northward since the Late Cretaceous [Sager and Pringle, 1988]. Differences in northward drift are caused by dissimilar ages (i.e., unequal periods of drift), differences in the distance from the site to the Euler pole describing plate motion, and paleomagnetic errors.

The pattern of northward drift values mimics almost exactly the paleolatitude distribution (Figs. 6, 7). Once again, the Japanese seamounts show two clusters, averaging 31.6° ±1.2° and 37.7° ±1.0° and correspond precisely to the groups with paleolatitudes averaging 1.0° N and 5.2° S, respectively. The Marcus-Wake seamounts also display a pattern of northward drift values that closely matches the pattern of paleolatitudes (Figs. 6, 7). W6, W7, and W9-W12 have northward drifts that increase eastward, but once again, considering the small number of points and the values for W4 and W5, this trend may be fortuitous. The most conservative interpretation is to conclude that these seamounts have essentially the same amount of northward drift, 30.6°±3.1°. This is almost the same value as the average for one of the two Japanese groups and implies an identical drift history which in turn suggests equal ages. The two groups of northward drift values could be interpreted as representative of two

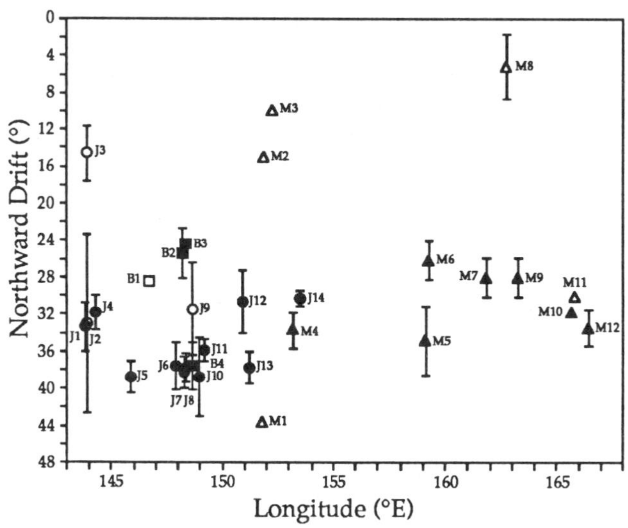

Fig. 7. Northward drift of seamounts examined in this study. Northward drift is calculated as the present latitude minus the paleolatitude. Symbols as in Fig. 6.

periods of volcanism with enough time for about 6° of latitudinal drift in between. However, the group with less northward motion contains seamounts with dates from 110.8 Ma (W6) to 90.6 Ma (W9), which bracket the ages of 93.7 Ma (B4) to 108.3 (J8) in the other group. This makes it necessary to ignore one or more dates or postulate a more complex history of northward motion.

A simpler but less satisfying explanation is to attribute the northward drift pattern to errors in paleolatitude determination. For example, a paleolatitude erroneously 6° too far south would correspond to a northward drift 6° too large, or vice versa. This hypothesis implies that all of the seamounts in the study formed at about the same time and drifted 31° to 38° north and that the northward drift and paleolatitude clusters and trends are actually just data scatter. The disquieting aspect of this explanation is the observation that the paleolatitudes display such tight clusters, particularly in the Japanese seamounts. A random error might be expected to provide a more nearly uniform pattern.

Previously, we noted that some of the seamounts appear to be affected by induced or normal overprint magnetizations, and we suggest that this may be the cause of the patterns of northward drift. A 6° paleolatitude difference in these seamounts could be explained with an induced or overprint magnetization component about 20% as large as the remanent part (i.e., a Koenigsberger ratio of 5). Interestingly, this is nearly the same value of overprint as the mean bias calculated by comparing the magnetization intensities of normal and reversed Pacific seamounts [Hildebrand and Staudigel, 1986; Bryan and Gordon, 1991]. The geometry of the geomagnetic field vector and the seamount magnetization vectors indicates that such an overprint would decrease the amount of apparent northward drift (i.e., bias the paleolatitude northward). Thus, this explanation implies that the tight cluster of Japanese seamounts, with the largest northward drift components, is least affected. A surprising aspect is that these seamounts have most of the lower magnetizations in that group, in contrast to our observation that seamounts with the most obviously unreliable results tend to have lower magnetizations.

CONCLUSIONS

Paleomagnetic data from the Japanese and Marcus-Wake seamounts, as well as seamounts from between these two chains, imply that there were at least two episodes of volcanism in the study area. Most seamounts appear to have formed by intraplate volcanism near the equator during the mid-Cretaceous. These have poles that cluster along the 125 to 88 Ma segment of the Pacific APWP. If the entire span of observed dates is included, the activity may have lasted as long as 39 Ma (81.5 to 120.3 Ma), but most of the ages cluster in about a 20 m.y. period (90.6 to 110.8 Ma). Measured ages and those inferred from paleomagnetic poles are in accord with the observed magnetic polarities which are predominantly normal and imply that most seamounts grew during the K-N superchron. Little evidence exists to suggest younger volcanism (only one 81.5 Ma date from Lamont Guyot); however, several clues, including summit depths, dates, and pole positions, imply older activity. The percentage of these older seamounts is small and their ages are mostly ill-constrained, but one has a radiometric date of 120.3 Ma. For the mid-Cretaceous volcanic episode, the paleomagnetic poles do not show any trend along the APWP that might be construed as evidence of age-progressive volcanism, but polar wander during the mid-Cretaceous was apparently slow, so this does not necessarily mean there was none.

Paleolatitudes of the Japanese and Marcus-Wake seamounts are slightly different. The former erupted at latitudes from 2.5° N to 7.5°, the latter from 2.3° S to 12.5° S. This suggests that the two chains were formed by different magma sources in their current relative positions. The trend of the Pacific APWP implies that these seamount groups remained near the equator until about 82 Ma, before beginning to drift 32° to 38° northward to their present latitudes. Within the seamount groups, some intriguing differences in paleolatitude were noted. The Japanese seamounts can be divided into two groups with paleolatitudes separated by slightly more than 6° and some of the Marcus-Wake seamounts display a linear trend. Both patterns could be used to argue against a hot spot origin for the volcanoes because such an origin should result in consistent paleolatitudes. However, it could also be argued that the Japanese pattern is the result of paleomagnetic errors and that the Marcus-Wake trend is only a fortuitous alignment. Indeed, we noted that several seamounts yielded pole positions apparently biased by a normal polarity overprint or induced magnetization. Although it appears that such bias is small or negligible for most seamounts, this uncertainty suggests that the seamount paleomagnetic data should be interpreted with some caution. Therefore, we cannot rule out either the hot spot or widespread volcanic pulse models for the origins of these seamounts.

APPENDIX

This appendix contains bathymetry and magnetic anomaly maps of seamounts modelled for the first time in this study. Bathymetric contours are shown in the upper (or left) map in each figure; magnetic anomaly contours are shown in the bottom (or right) map. Contours are labelled in hundreds of meters or nanoTeslas; typical contour spacings are 200 or 500 m, 50 or 100 nT. Ship tracks are shown as thin lines in the bathymetry plot. Locations of magnetic readings used to describe the magnetic anomaly in inversion are shown as dots (for RNDB10WT or other academic cruise tracks) and small crosses (for U.S. Navy data). For reference, a representative contour (typically 4000 m) is shown in gray on the magnetic map. Maps are Mercator projections. The scale of each bathymetry and magnetic anomaly pair are the same.

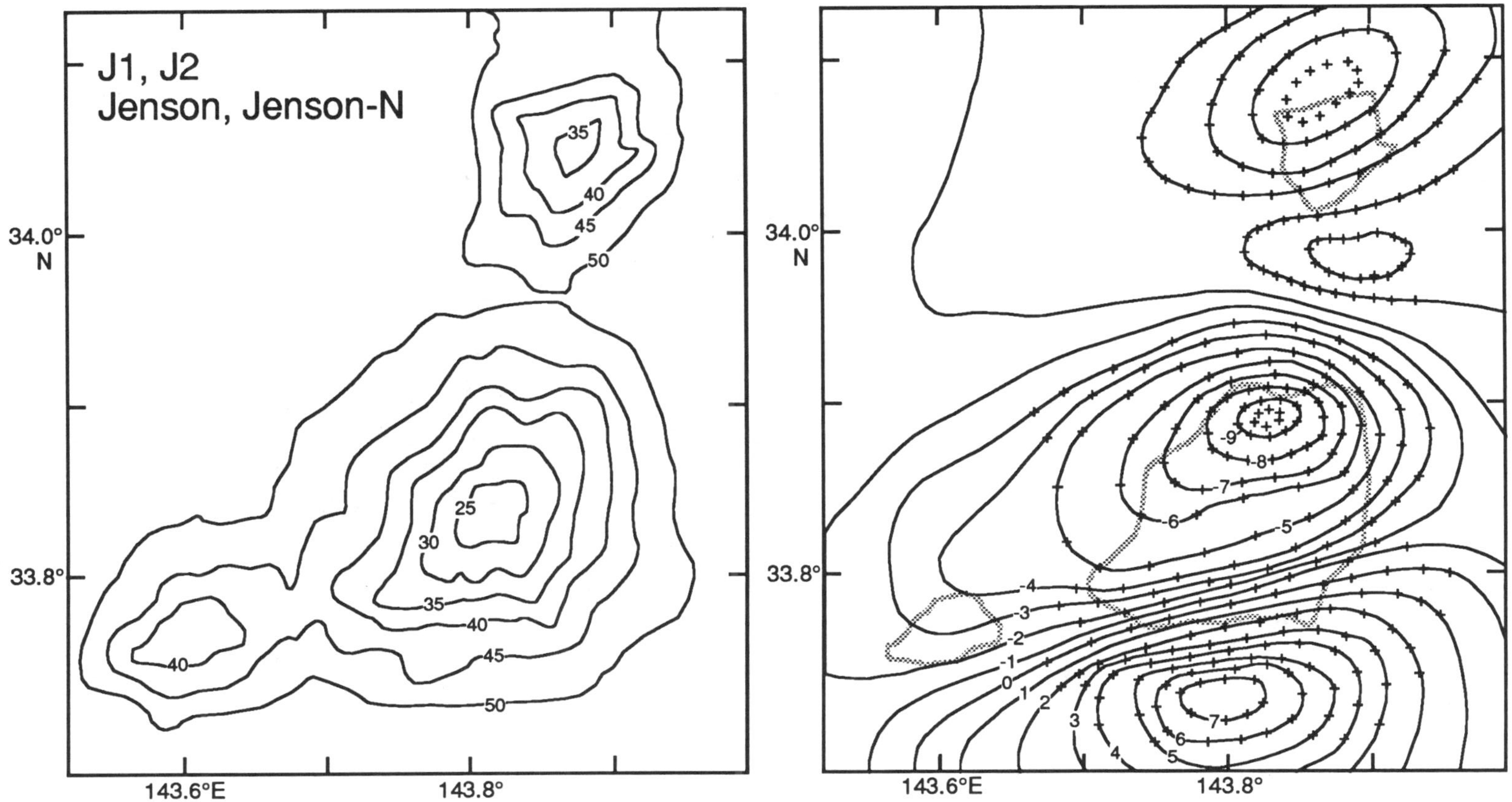

Figure A1. Bathymetry and magnetic anomaly maps of Jenson (J1) and Jenson-N (J2) seamounts.

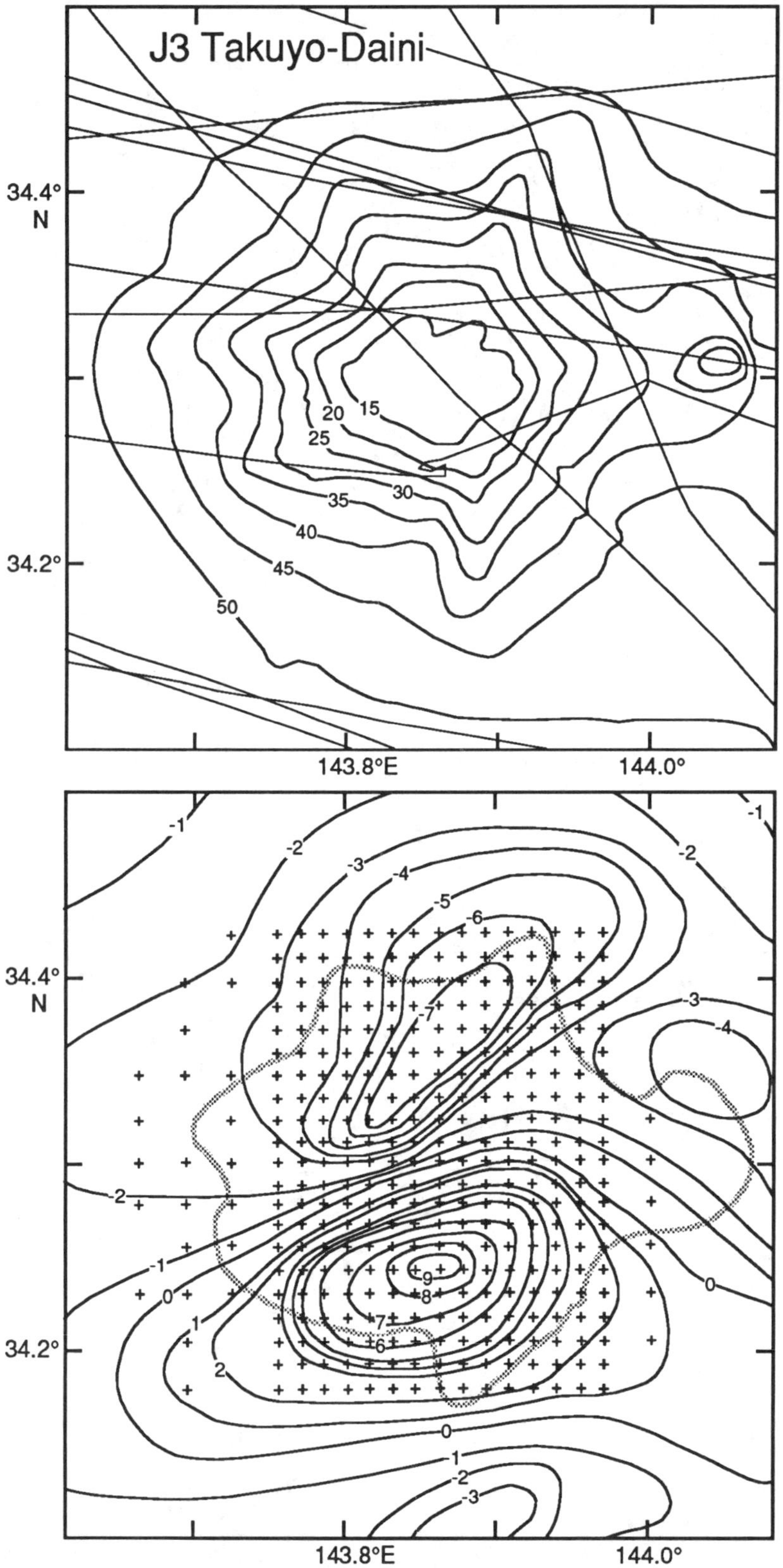

Figure A2. Bathymetry and magnetic anomaly maps of Takukyo-Daini (J3) guyot.

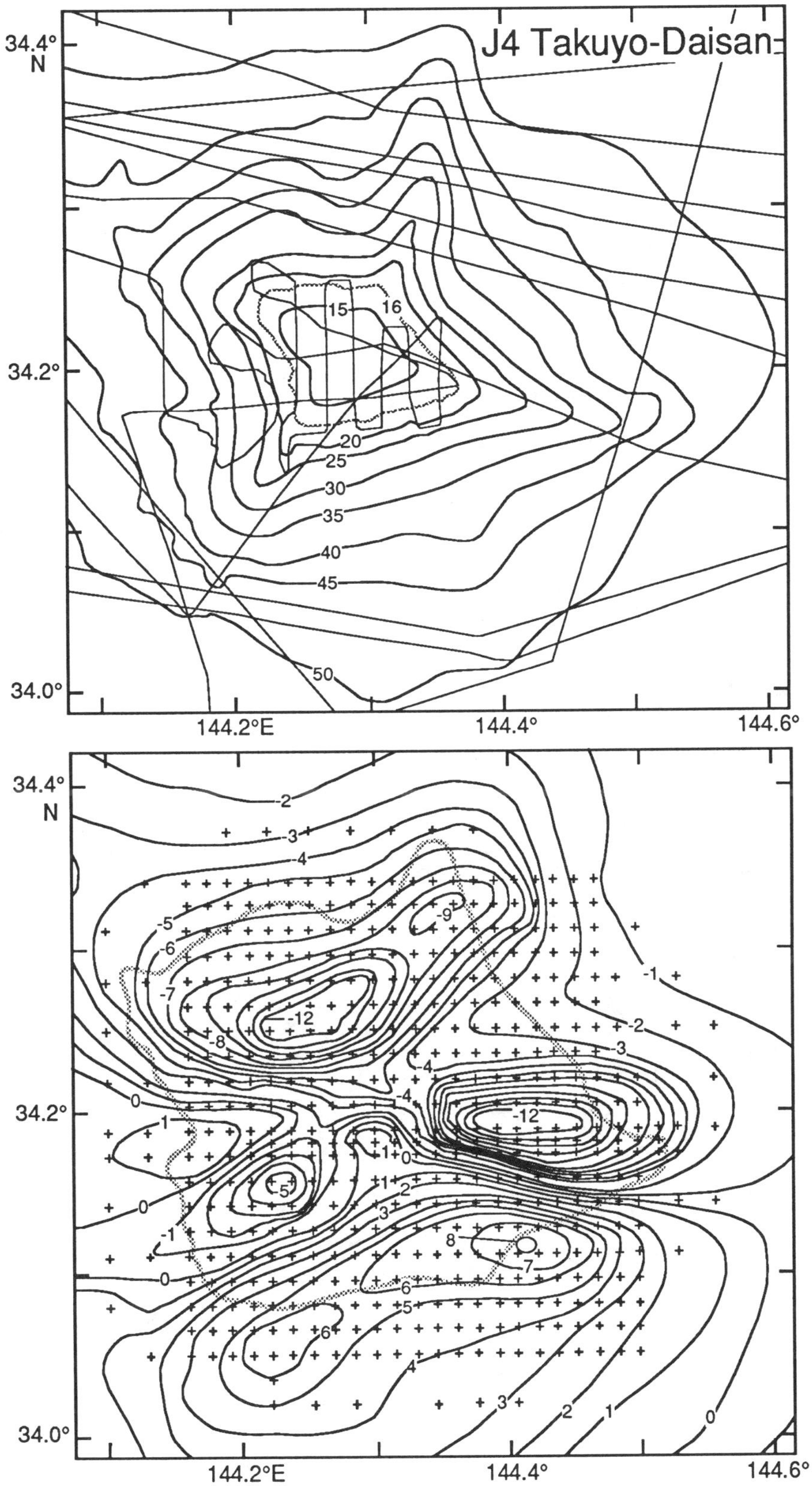

Figure A3. Bathymetry and magnetic anomaly maps of Takuyo-Daisan (J4, Seiko) guyot.

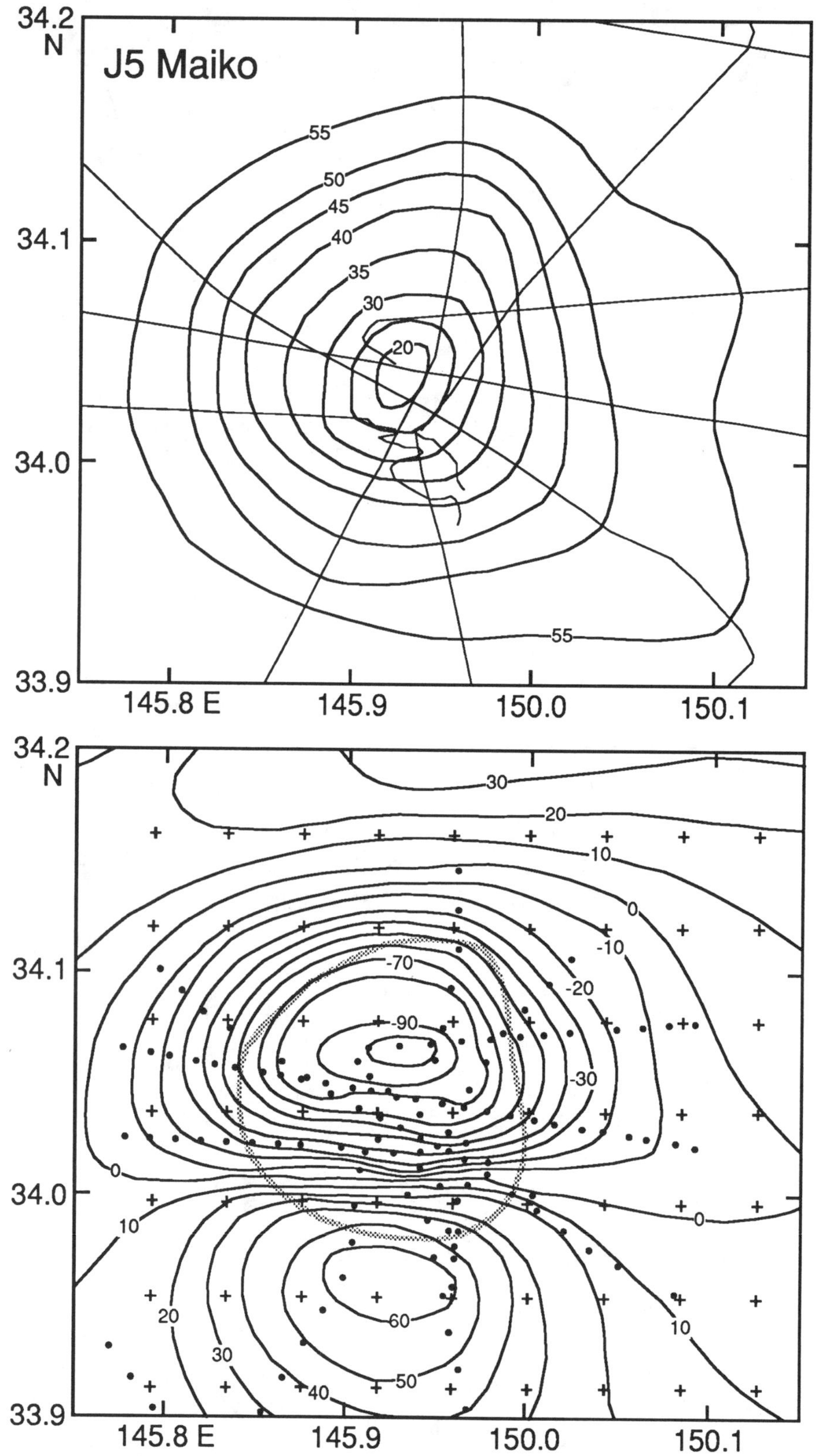

Figure A4. Bathymetry and magnetic anomaly maps of Maiko (J5) seamount.

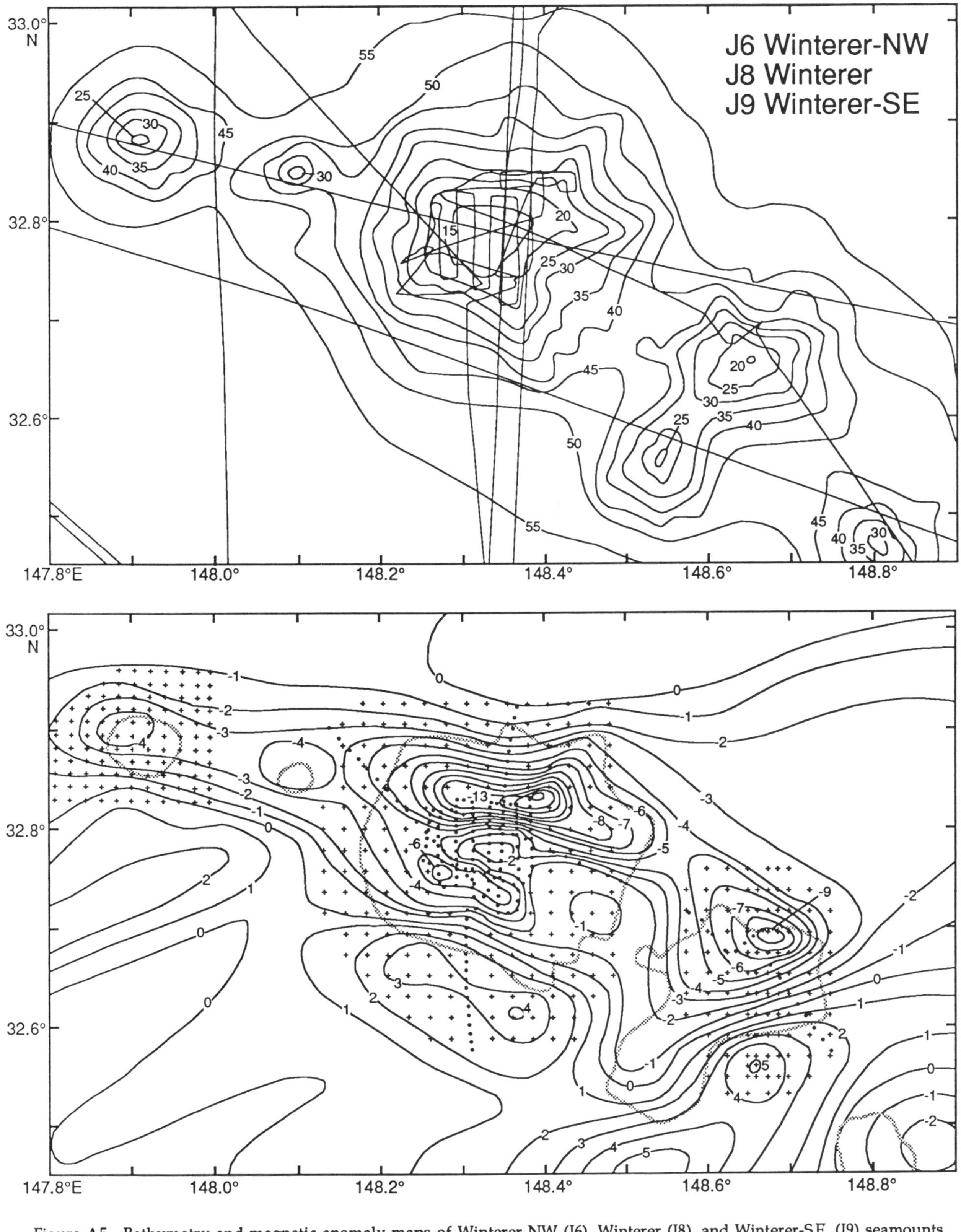

Figure A5. Bathymetry and magnetic anomaly maps of Winterer-NW (J6), Winterer (J8), and Winterer-SE (J9) seamounts.

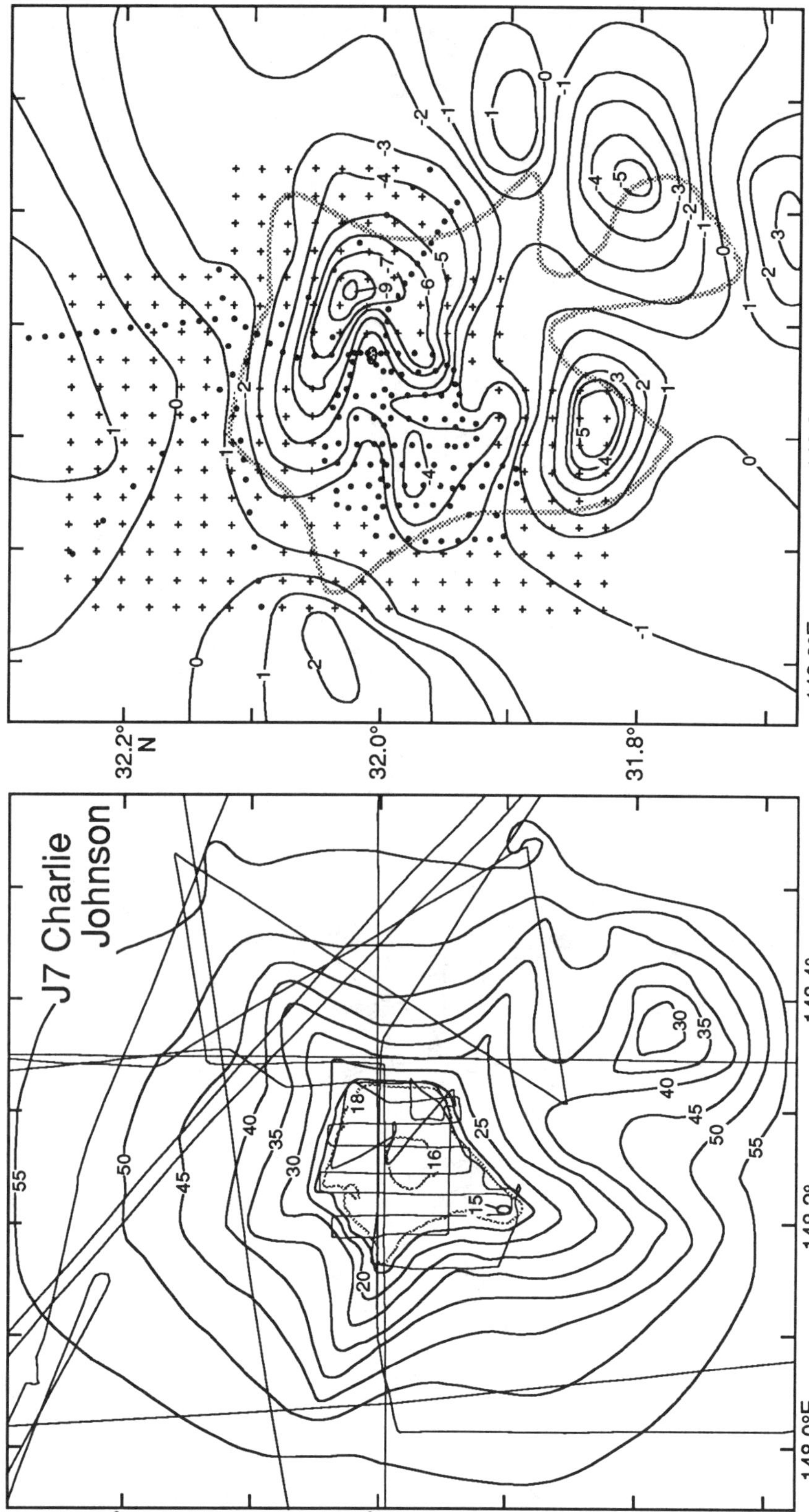

Figure A6. Bathymetry and magnetic anomaly maps of Charlie Johnson (J7) guyot.

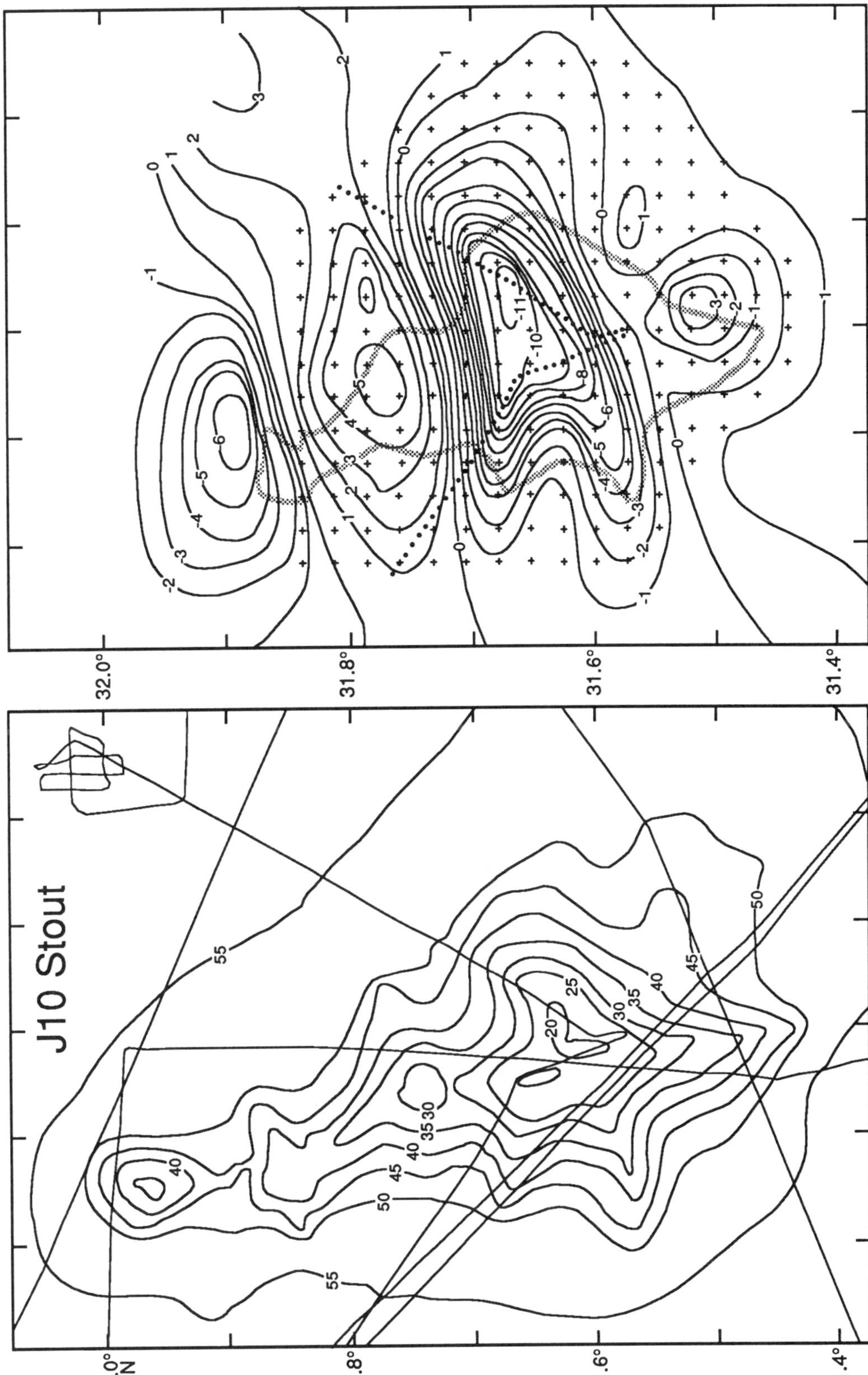

Figure A7. Bathymetry and magnetic anomaly maps of Stout (J10) seamount.

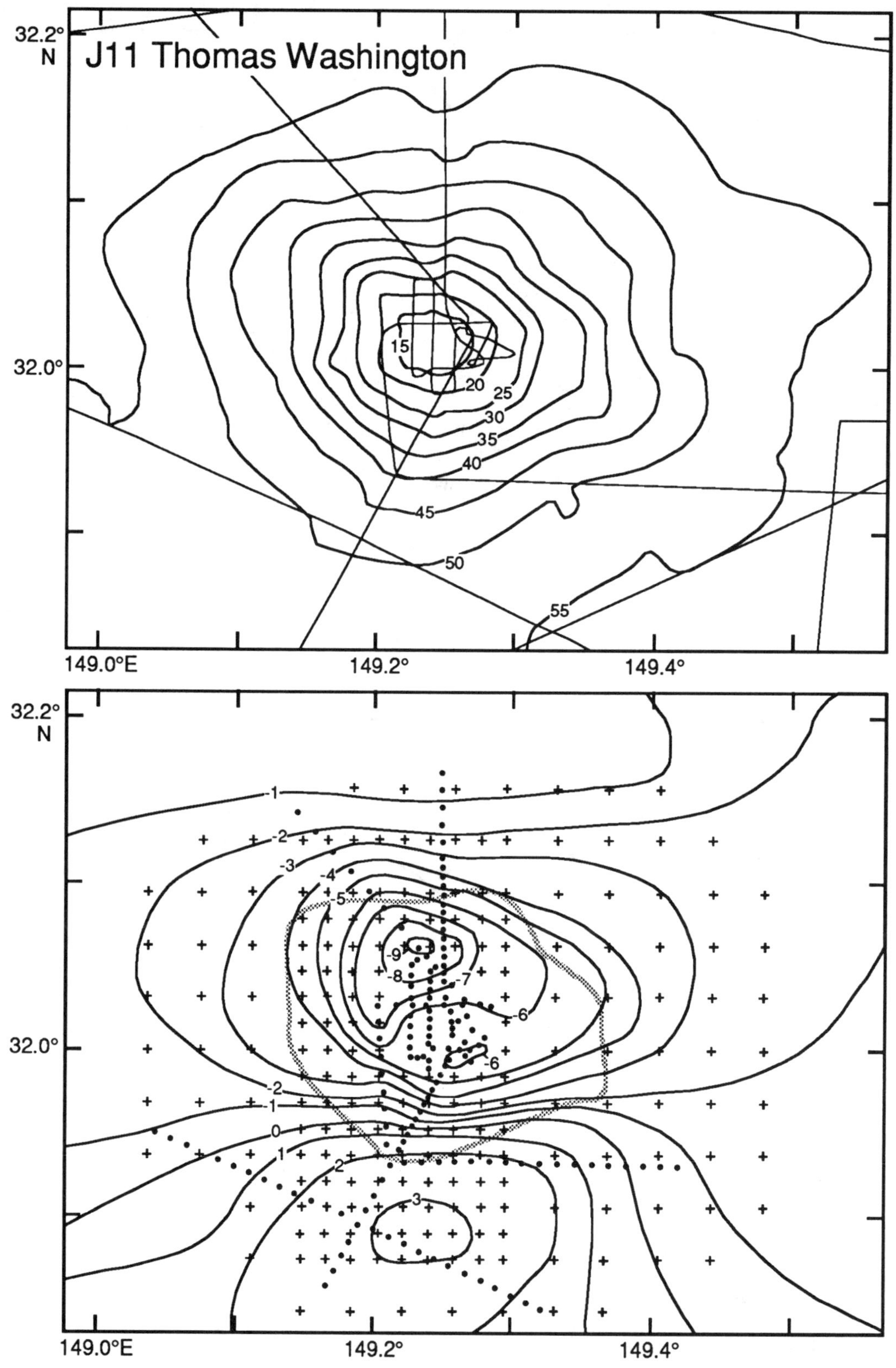

Figure A8. Bathymetry and magnetic anomaly maps of Thomas Washington (J11) guyot.

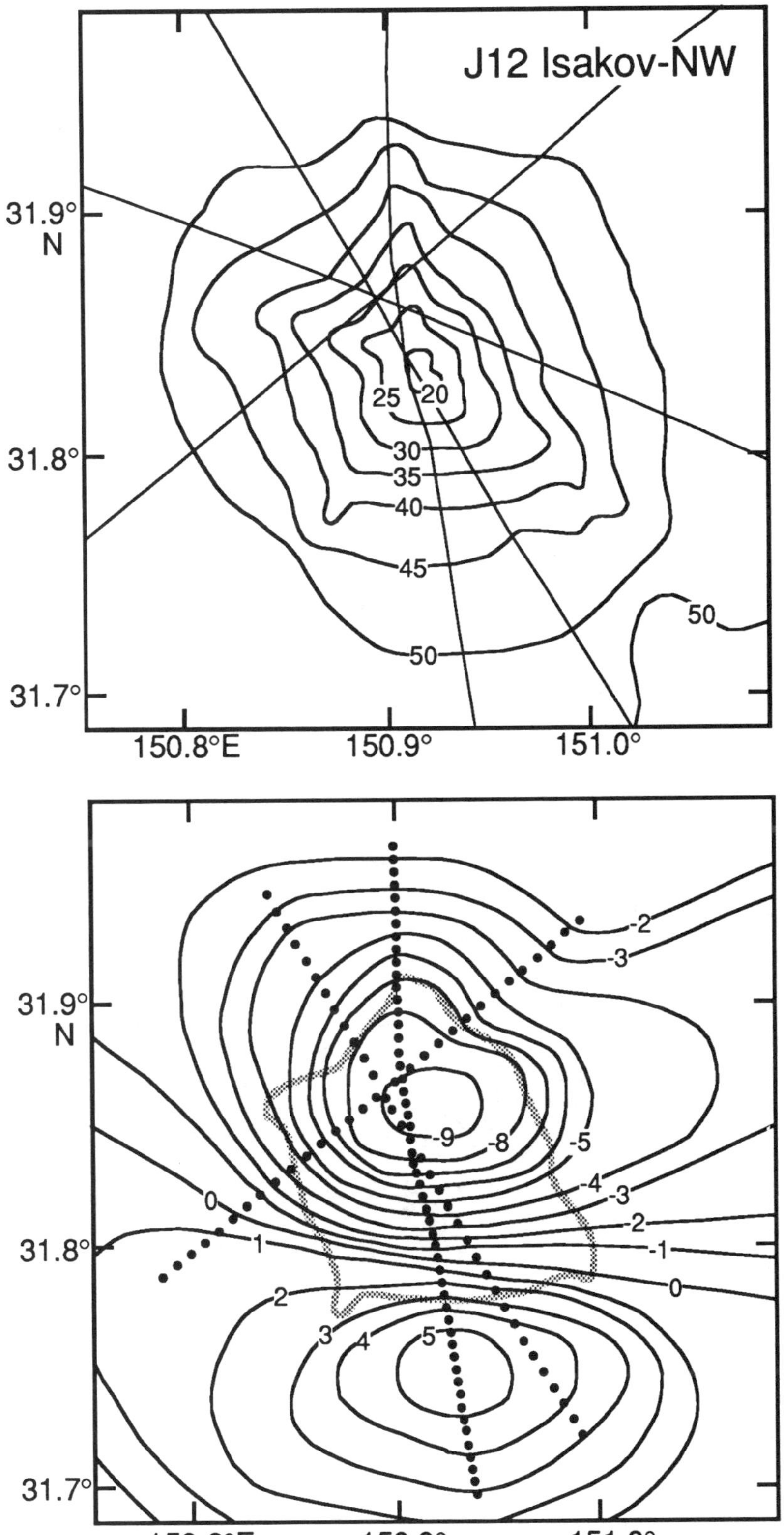

Figure A9. Bathymetry and magnetic anomaly maps of Isakov-NW (J12) seamount.

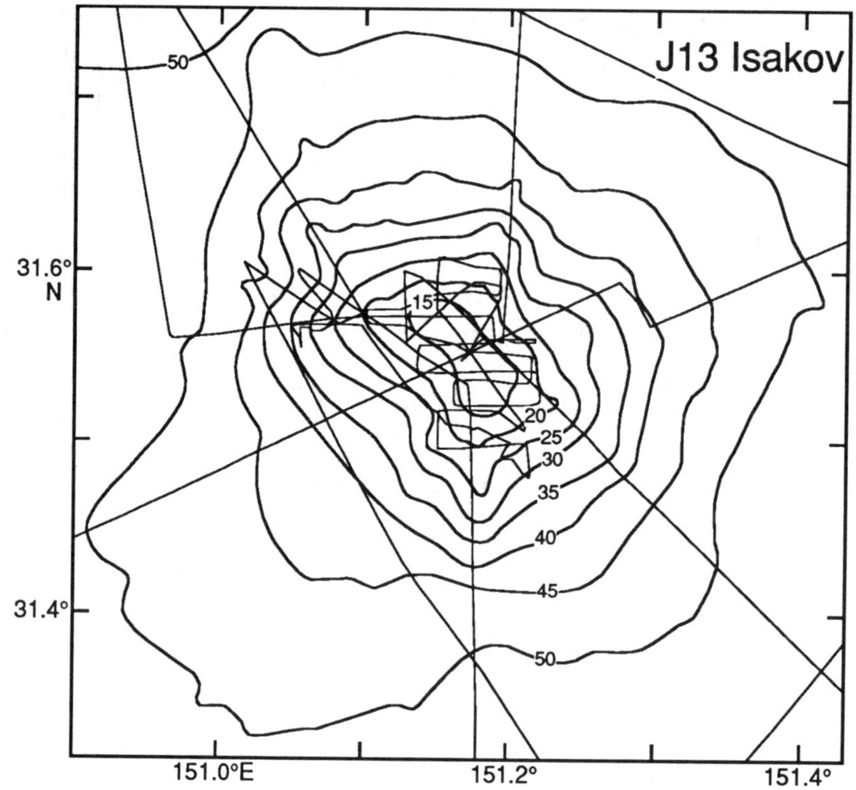

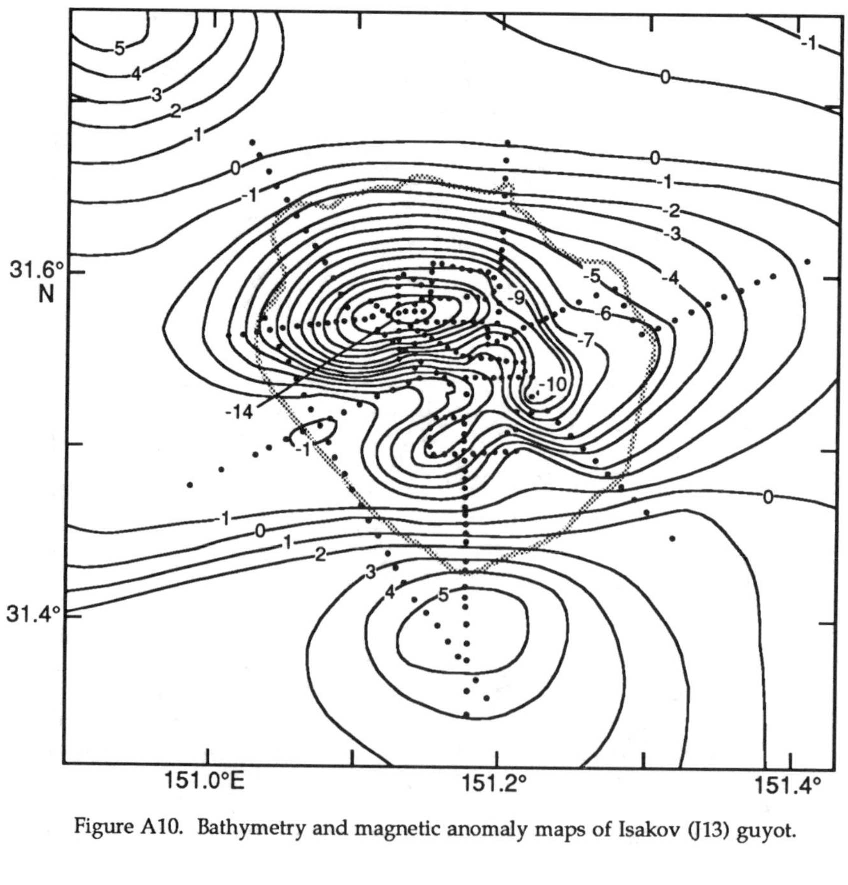

Figure A10. Bathymetry and magnetic anomaly maps of Isakov (J13) guyot.

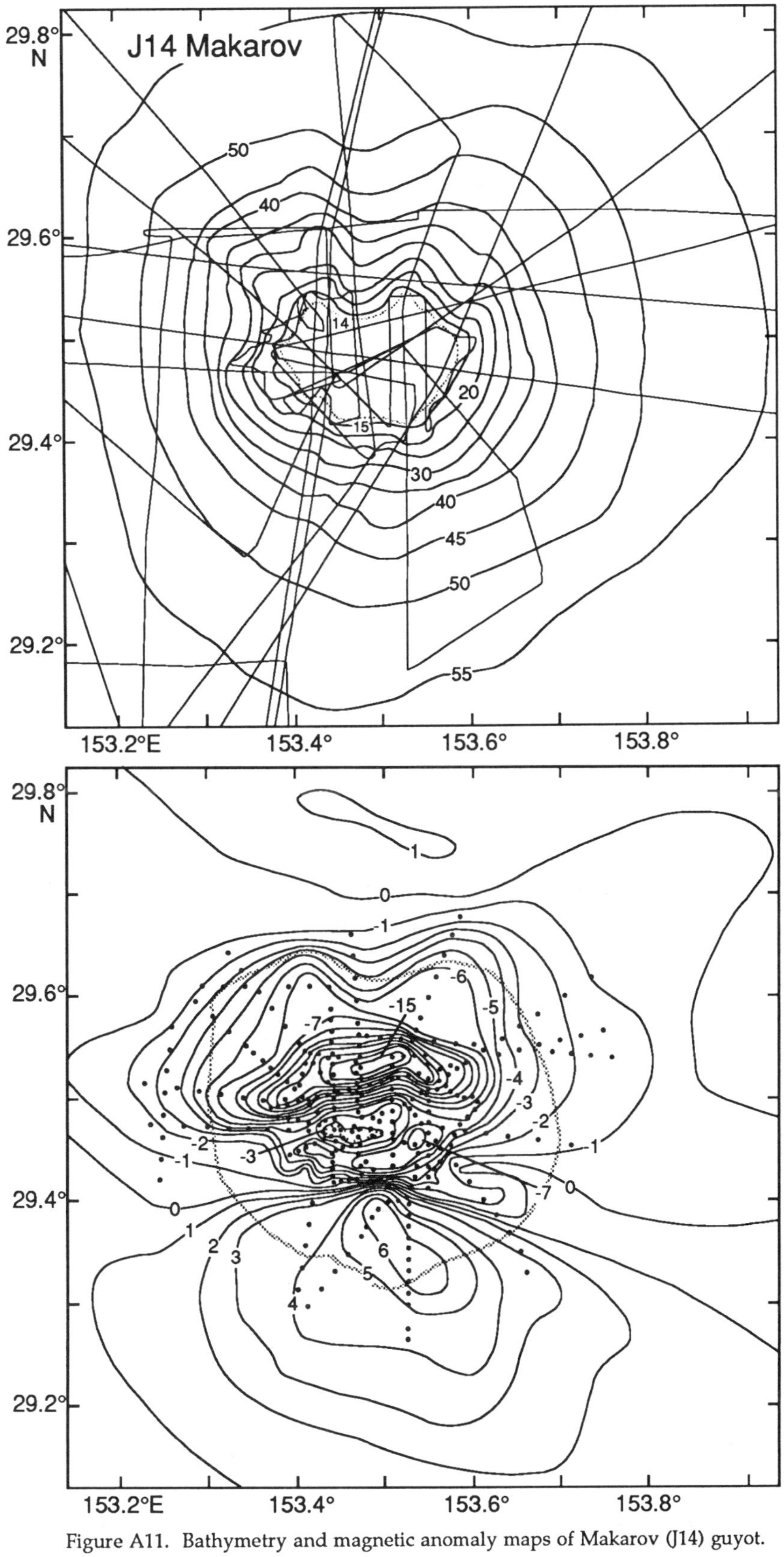

Figure A11. Bathymetry and magnetic anomaly maps of Makarov (J14) guyot.

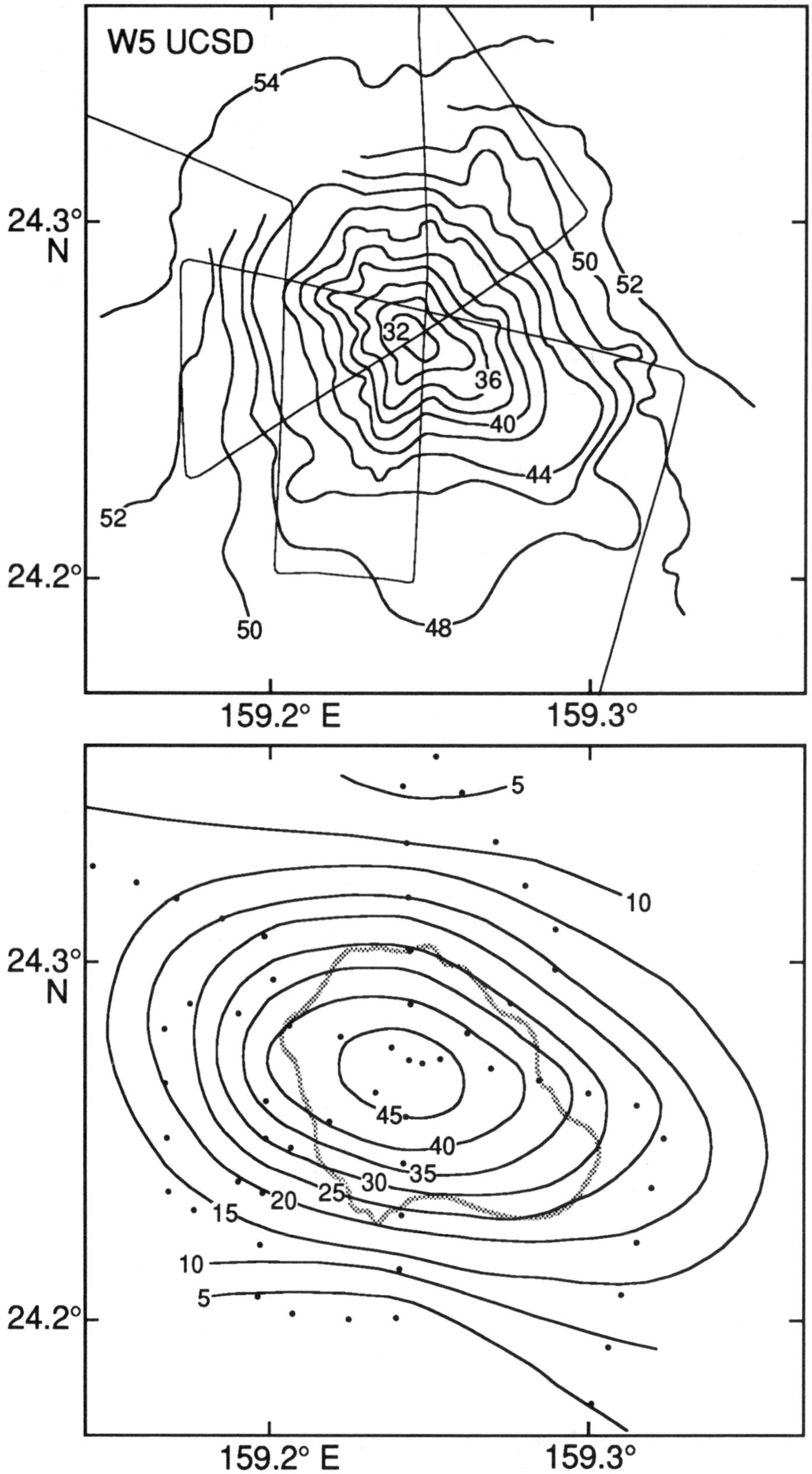

Figure A12. Bathymetry and magnetic anomaly maps of UCSD (W5) seamount.

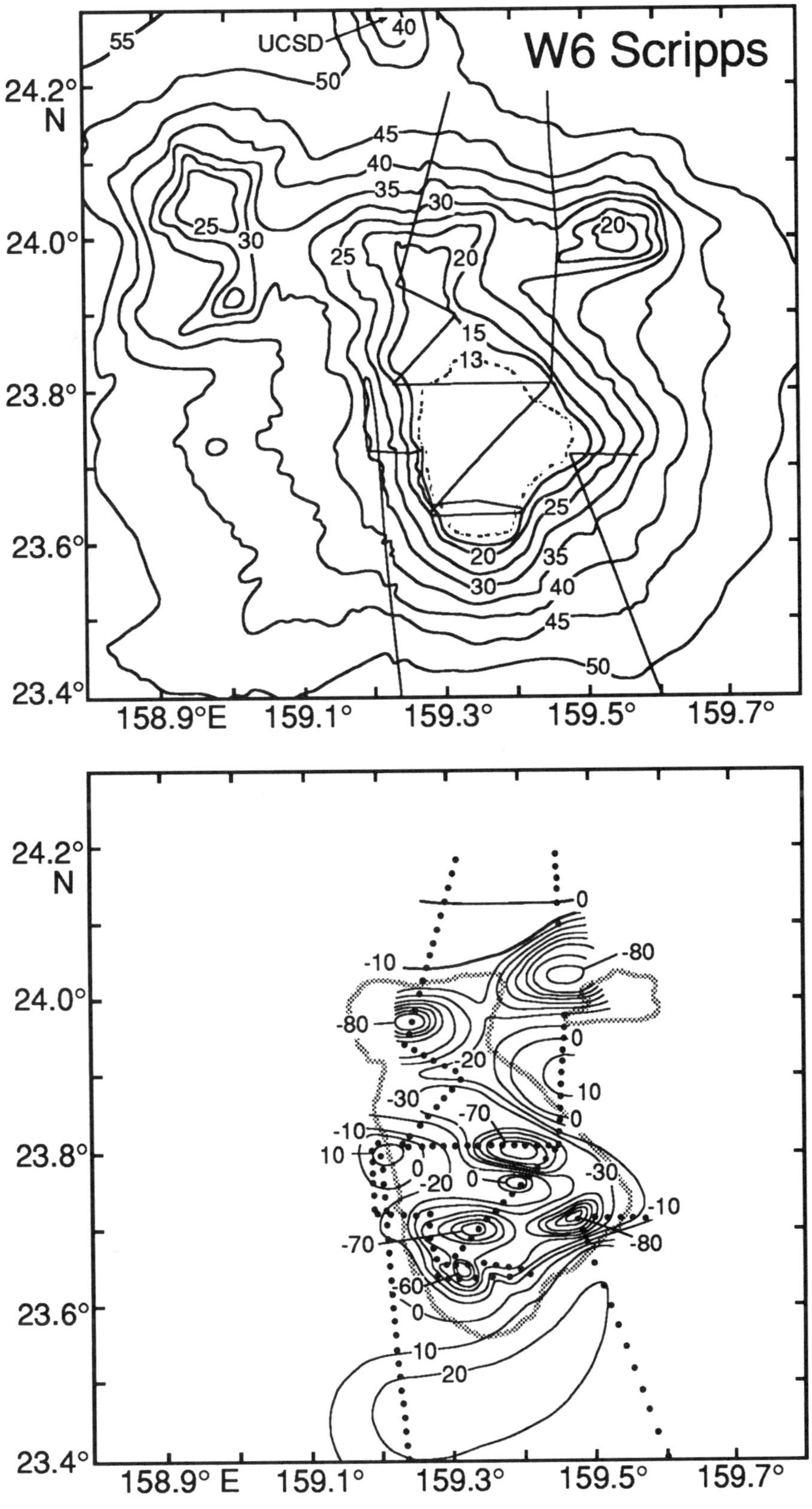

Figure A13. Bathymetry and magnetic anomaly maps of Scripps (W6) guyot.

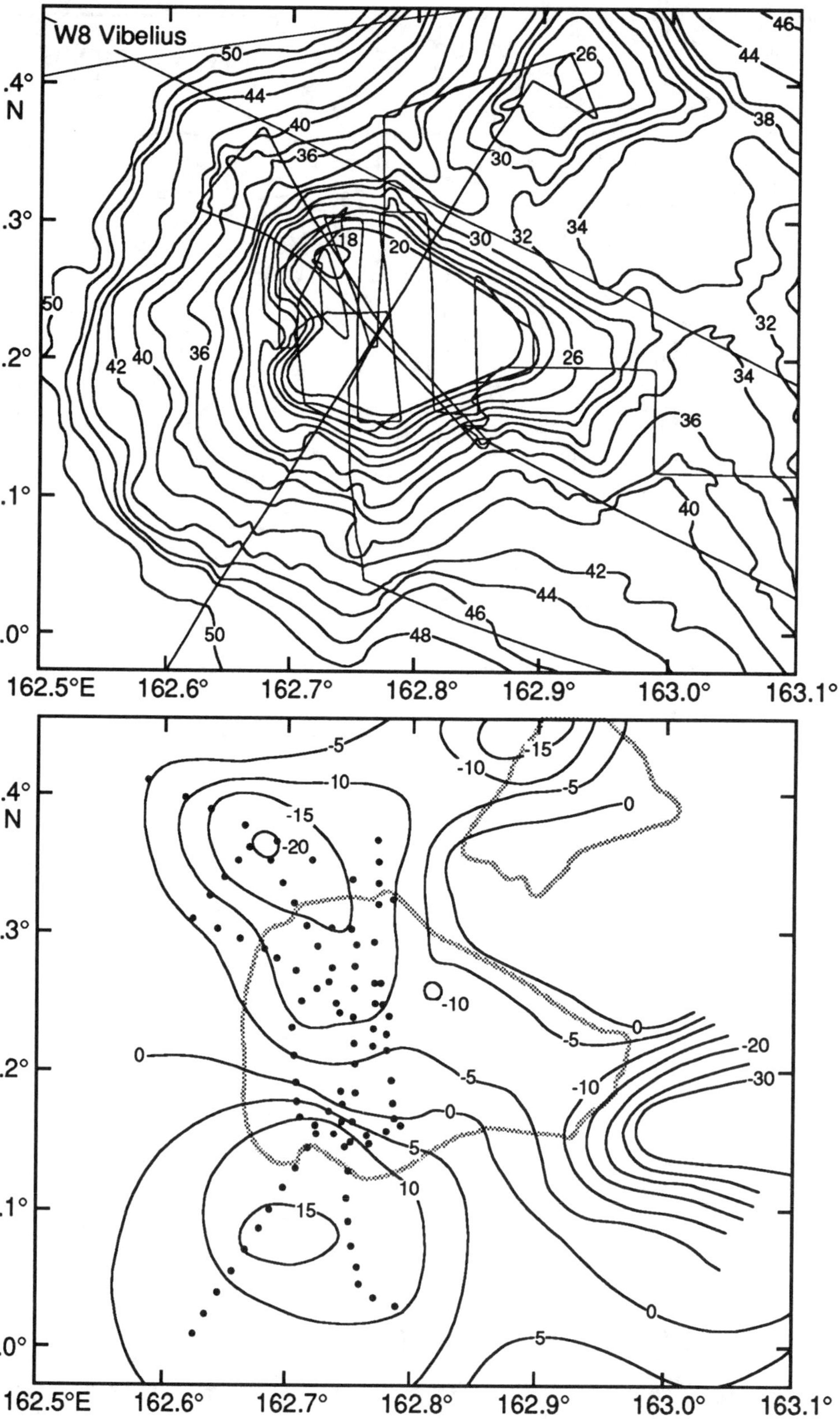

Figure A14. Bathymetry and magnetic anomaly maps of Vibelius (W8) guyot.

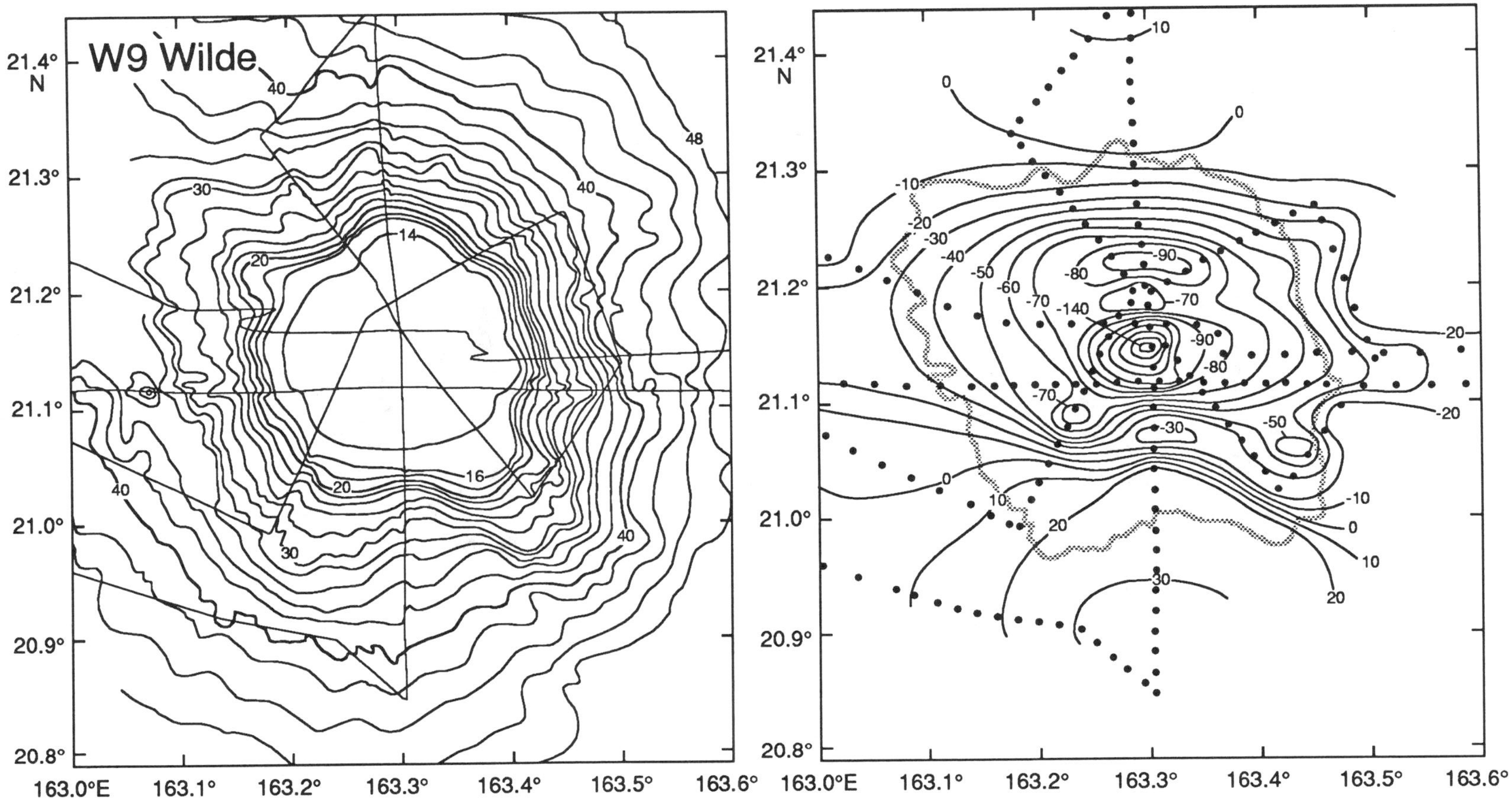

Figure A15. Bathymetry and magnetic anomaly maps of Wilde (W9) guyot.

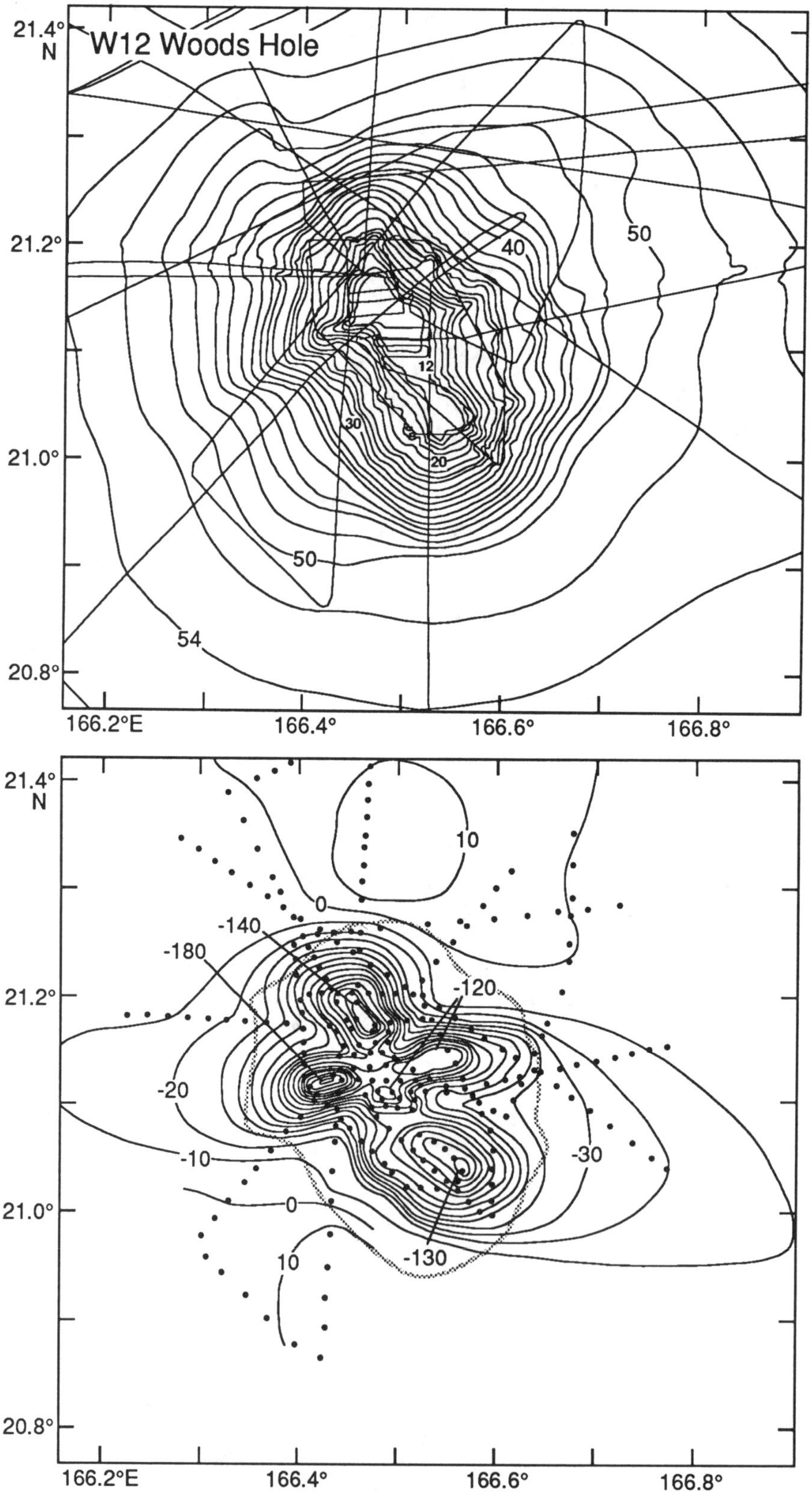

Figure A16. Bathymetry and magnetic anomaly maps of Woods Hole (W12) guyot.

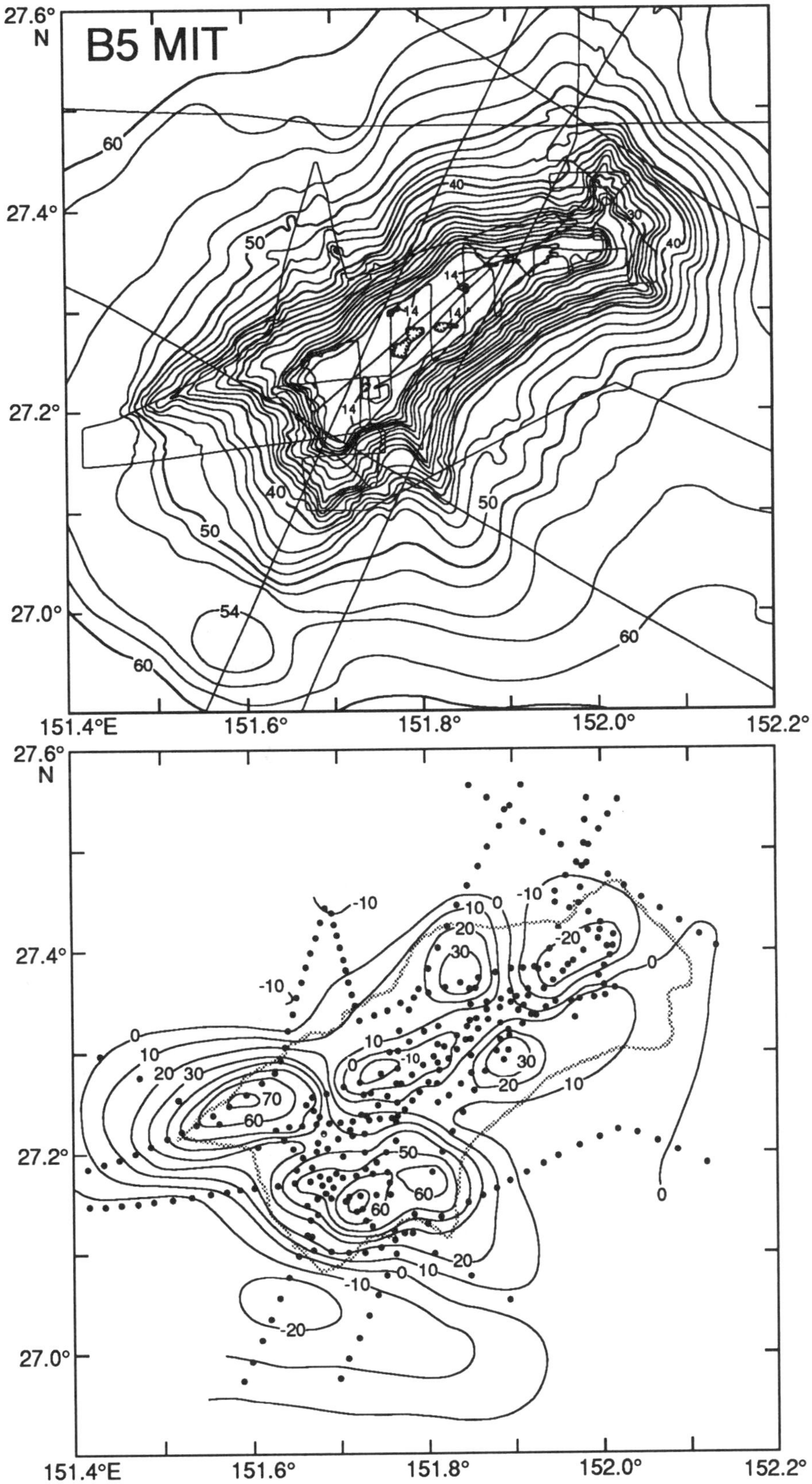

Figure A17. Bathymetry and magnetic anomaly maps of MIT (B5) guyot.

Acknowledgements. We thank Jeff Gee and Ron Merrill for constructive reviews, B. Dayharsh for assisting with paleomagnetic analyses, and John Hildebrand and Robert Parker for providing a copy of and assistance with their seminorm magnetic modeling program. We are indebted to our coproponents, and colleagues on Leg 10 of the Roundabout expedition: Jerry Winterer, Marcia McNutt, Jim Natland, Cecily Wolfe, Garrett Ito, and the captain and crew of the R/V *Thomas Washington*. Support was provided by the National Science Foundation with grant OCE-8715733 and by the Office of Naval Research with grant N0001487K0271. Funding for the U.S. Navy magnetic data collection and analysis was provided by the Office of Naval Research, program element 601153N, under the direction of Herbert C. Eppert, program manager. This is Texas A&M Geodynamics Research Institute contribution No. 88.

REFERENCES

Bryan, P. and R. G. Gordon, Determining the average contribution of viscous overprints to the characteristic magnetization of Pacific seamounts (abstract). *EOS, Trans. AGU, 72*, 125, 1991.

Cox, A. and R. G. Gordon, Paleolatitudes determined from paleomagnetic data from vertical cores, *Rev. Geophys. Space Phys., 22*, 47-72, 1984.

Crough, S. T., Hot spot epeirogeny, *Tectonophysics, 61*, 321-333, 1979.

Duncan, R. A. and R. B. Hargraves, $^{40}Ar/^{39}Ar$ geochronology of basement rocks from the Mascarene Plateau, the Chagos Bank, and the Maldives Ridge, *Proc. ODP, Sci. Res., 115*, 43-51, 1990.

Duncan, R. A. and D. A. Clague, Pacific plate motion recorded by linear volcanic chains, in *The Ocean Basins and Margins*, vol. 7A, edited by A. E. M. Nairn, F. G. Stehli, and S. Uyeda, pp. 89-121, Plenum Publishing, New York, 1985.

Fisher, R. A., Dispersion on a sphere, *Proc. Roy. Soc. Lond., A217*, 295-305, 1953.

Gee, J., L. Tauxe, J. A. Hildebrand, H. Staudigel, and P. Lonsdale, Nonuniform magnetization of Jasper Seamount, *J. Geophys. Res., 93*, 12159-12175, 1988.

Gee, J., H. Staudigel, and L. Tauxe, Contribution of induced magnetization to magnetization of seamounts, *Nature, 342*, 170-173, 1989.

Gordon, R. G., Late Cretaceous apparent polar wander path of the Pacific plate: evidence of a rapid shift of the Pacific hot spots with respect to the paleomagnetic axis, *Geophys. Res., Lett., 10*, 709-712, 1983.

Gordon, R. G., Test for bias in paleomagnetically determined paleolatitudes from Pacific plate Deep Sea Drilling Project Sediments, *J. Geophys. Res., 95*, 8397-8404, 1990.

Handschumacher, D. W., W. W. Sager, T. W. C. Hilde, and D. R. Bracey, Pre-Cretaceous tectonic evolution of the Pacific plate and extension of the geomagnetic polarity reversal time scale with implications for the origin of the Jurassic "Quiet Zone", *Tectonophysics, 155*, 365-380, 1988.

Handschumacher, D. W., P. R. Vogt, and D. R. Bracey, Japanese (M) lineations: systematic anomaly amplitude increase away from fracture zones (abstract), *EOS, Trans. AGU, 64*, 688, 1983.

Harland, W. B., A. V. Cox, P. G. Llewellyn, C. A. G. Pickton, A. G. Smith, and R. Walters, *A Geologic Time Scale*, 132 pp., Cambridge University Press, 1982.

Harrison, C. G. A., R. D. Jarrard, V. Vacquier, and R. L. Larson, Palaeomagnetism of Cretaceous Pacific Seamounts, *Geophys. J. Roy. Astr. Soc., 42*, 859-882, 1975.

Henderson, L. J., *Motion of the Pacific plate relative to the hotspots since the Jurassic and model of oceanic plateaus of the Farallon plate*, Ph. D. dissertation, 321 pp., Northwestern University, Evanston, IL, 1985.

Hildebrand, J. A. and R. L. Parker, Paleomagnetism of Cretaceous Pacific seamounts revisited, *J. Geophys. Res., 92*, 12695-12712, 1987.

Hildebrand, J. A. and H. Staudigel, Seamount magnetic polarity and Cretaceous volcanism of the Pacific Basin, *Geology, 14*, 456-458, 1986.

Jarrard, R. D. and S. Sasajima, Paleomagnetic synthesis for southeast Asia: constraints on plate motions, in *The Tectonic and Geologic Evolution of Southeast Asian Seas and Islands*, Geophys. Mon. Ser. vol. 23, edited by D. E. Hayes, pp. 293-317, AGU, Washington, D. C., 1980.

Larson, R. L., Latest pulse of Earth: evidence for a mid-Cretaceous superplume. *Geology, 19*, 547-550, 1991.

Larson, R. L. and C. G. Chase, Late Mesozoic evolution of the western Pacific Ocean, *Geol. Soc. Amer. Bull., 83*, 3627-3644, 1972.

Larson, R. L. and W. Lowrie, Paleomagnetic evidence for motion of the Pacific plate from Leg 32 basalts and magnetic anomalies, *Init. Rept. DSDP, 32*, 571-577, 1975.

Larson, R. L. and W. W. Sager, Skewness of magnetic anomalies M0 to M29 in the northwestern Pacific, *Proc. ODP, Sci. Res., 129*, 471-481, 1992.

Marshall, M., The magnetic properties of some DSDP basalts from the north Pacific and inferences for plate tectonics, *J. Geophys. Res., 83*, 289-308, 1978.

Matsushita, S., Geomagnetic disturbances and storms, in *Physics of Geomagnetic Phenomena*, v. 2, edited by S. Matsushita and W. H. Campbell, pp. 793-819, Academic Press, New York, 1967.

Matthews, J. L., B. C. Heezen, R. Catalano, A. Coogan, M. Tharp, J. Natland, and M. Rawson, Cretaceous drowing of reefs on Mid-Pacific and Japanese guyots, *Science, 184*, 462-464, 1974.

Menard, H. W., *Marine Geology of the Pacific*, 271 pp., McGraw-Hill, New York, 1964.

Menard, H. W., Darwin Reprise, *J. Geophys. Res., 89*, 9960-9968, 1984.

McNutt, M. K., Nonuniform magnetization of seamounts: a least-squares approach, *J. Geophys. Res., 91*, 3686-3700, 1986.

McNutt, M. K., E. L. Winterer, W. W. Sager, J. H. Natland, and G. Ito, The Darwin Rise: a Cretaceous superswell?, *Geophys. Res. Lett., 17*, 1101-1104, 1990.

Nakanishi, M., K. Tamaki and K. Kobayashi, Mesozoic magnetic anomaly lineations and seafloor spreading history of the northwestern Pacific, *J. Geophys. Res., 94*, 15437-15462, 1989.

Onwumechilli, A., Geomagnetic variations in the equatorial zone, in *Physics of Geomagnetic Phenomena*, v. 1, edited by S. Matsushita and W. H. Campbell, pp. 426-507, Academic Press, New York, 1967.

Ozima, M., I. Kaneoka, K. Saito, M. Honda, M. Yanagisawa, and Y. Takigami, Summary of geochronological studies of submarine rocks from the western Pacific Ocean, in *Geodynamics of the western Pacific-Indonesian region*, Geodynamics Series v. 11, edited by T. W. C. Hilde and S. Uyeda, pp. 137-142, American Geophysical Union, Washington, D. C., 1983.

Parker, R. L., A statistical theory of seamount magnetism, *J. Geophys. Res., 93*, 3105-3115, 1988.

Parker, R. L., A theory of ideal bodies for seamount magnetism, *J. Geophys. Res., 96*, 16101-16112, 1991.

Parker, R. L., L. Shure, and J. A. Hildebrand, The application of inverse theory to seamount magnetism, *Rev. Geophys., 25*, 17-40, 1987.

Petronotis, K. E., R. G. Gordon, and G. D. Acton, Determining palaeomagnetic poles and anomalous skewness from marine magnetic anomaly skewness data from a single plate, *Geophys. J. Int., 109*, 209-224, 1992.

Plouff, D., Gravity and magnetic fields of polygonal prisms and application to magnetic terrain corrections, *Geophysics, 41*, 717-741, 1976.

Raymond, C. A. and J. L. LaBreque, Magnetization of oceanic crust: thermoremanent magnetization or chemical remanent magnetization?, *J. Geophys. Res., 92*, 8077-8088, 1987.

Rea, D. K. and T. L. Vallier, Two Cretaceous volcanic episodes in the western Pacific Ocean, *Geol. Soc. Amer. Bull., 94*, 1430-1437, 1983.

Sager, W. W., *Seamount paleomagnetism and Pacific plate tectonics*, Ph. D. dissertation, 472 pp., University of Hawaii, Honolulu, 1983.

Sager, W. W., Late Eocene and Maastrichtian paleomagnetic poles for the Pacific plate: implications for the validity of seamount paleomagnetic data, *Tectonophysics, 144*, 301-314, 1987.

Sager, W. W., Seamount age estimates from paleomagnetism and their implications for the history of volcanism on the Pacific plate, in *Geology and Offshore Mineral Resources of the Central Pacific Basin*, Circum-Pacific Council for Energy and Mineral Resources, Earth Sci. Ser., vol. 14, edited by B. H. Keating and B. Bolton, pp. 21-37, Springer-Verlag, New York, 1992.

Sager, W. W., D. W. Handschumacher, T. W. C. Hilde, and D. R. Bracey, Tectonic evolution of the northern Pacific plate and Pacific-Farallon-Izanagi triple junction in the Late Jurassic and Early Cretaceous (M21-M10), *Tectonophysics, 155*, 345-364, 1988.

Sager, W. W. and B. H. Keating, Paleomagnetism of Line Islands seamounts: evidence for Late Cretaceous and early Tertiary volcanism, *J. Geophys. Res., 89*, 11135-11151, 1984.

Sager, W. W. and M. S. Pringle, Paleomagnetic constraints on the origin and evolution of the Musicians and South Hawaiian seamounts, central Pacific Ocean, in *Seamounts, Islands, and Atolls*, Geophys. Monogr. Ser. 43, edited by B. Keating, P. Fryer, R. Batiza, and G. Boehlert, pp. 133-162, AGU, Washington, D. C., 1987.

Sager, W. W. and M. S. Pringle, Mid-Cretaceous to early Tertiary apparent polar wander path of the Pacific plate, *J. Geophys. Res., 93*, 11753-11771, 1988.

Sayre, W. O., Preliminary report on the paleomagnetism of Aptian and Albian limestones and trachytes from the Mid-Pacific Mountains and Hess Rise, Deep Sea Drilling Project, Leg 62, *Init. Rept. DSDP, 62*, 983-994, 1981.

Schlanger, S. O., H. C. Jenkyns, and I. Premoli-Silva, Volcanism and vertical tectonics in the Pacific basin related to global Cretaceous transgressions, *Earth Planet. Sci. Lett., 52*, 435-449, 1981.

Smith, W. H. F., H. Staudigel, A. B. Watts, and M. S. Pringle, The Magellan Seamounts: Early Cretaceous record of the south Pacific isotopic and thermal anomaly, *J. Geophys. Res., 94*, 10501-10523, 1989.

Smoot, N. C., The Marcus-Wake seamounts and guyots as paleofracture indicators and their relation to the Dutton Ridge, *Mar. Geol., 88*, 117-131, 1989.

Smoot, N. C., North Pacific guyots, Tech. Note TN 01-91, 93 pp., Naval Oceanographic Office, Stennis Space Center, MS, 1991.

Tarduno, J. A., W. V. Sliter, L. Kroenke, M. Leckie, H. Mayer, J. J. Mahoney, R. Musgrave, M. Storey, and E. L. Winterer, Rapid formation of Ontong Java Plateau by Aptian mantle plume volcanism, *Science, 254*, 399-403, 1991.

Ueda, Y., Geophysical study of two seamounts near Minami-Tori Shima (Marcus) Island, western Pacific Ocean, *J. Geomag. Geolectr., 49*, 1481-1501, 1988.

Vacquier, V., A machine method for computing the magnitude and direction of magnetization of a uniformly magnetized body from its shape and a magnetic survey, in *Benedum Earth Magnetism Symposium*, edited by T. Nagata, pp. 123-127, University of Pittsburgh Press, 1962.

Vacquier, V. and S. Uyeda, Paleomagnetism of nine seamounts in the western Pacific and of three volcanoes in Japan, *Bull. Earthquake Res. Inst., Tokyo, 45*, 815-848, 1967.

Vogt, P. R., Volcanogenic upwelling of anoxic, nutrient-rich water: a possible factor in carbonate-bank/reef demise and benthic faunal extinctions?, *Geol. Soc. Amer. Bull., 101*, 1225-1245, 1989.

Vogt, P. R. and N. C. Smoot, The Geisha guyots: multibeam bathymetry and morphometric interpretation, *J. Geophys. Res., 89*, 11085-11107, 1984.

Watts, A. B. and N. M. Ribe, On geoid heights and the flexure of the lithosphere at seamounts, *J. Geophys. Res., 88*, 11152-11170, 1984.

Watts, A. B., Bodine, J. H., and N. M. Ribe, Observations of flexure and the geological evolution of the Pacific Ocean Basin, *Nature, 283*, 532-537, 1980.

Winterer, E. L., Anomalies in the tectonic evolution of the Pacific, in *The Geophysics of the Pacific Ocean Basin and Its Margin*, Geophys. Monogr. Ser., vol. 19, edited by G. H. Sutton, M. H. Manghnani, and R. Moberly, pp. 269-278, AGU, Washington, D. C., 1976.

Wolfe, C. J. and M. K. McNutt, Compensation of Cretaceous seamounts of the Darwin Rise, northwest Pacific Ocean, *J. Geophys. Res., 96*, 2363-2374, 1991.

Yamazaki, T., Magnetization of Erimo Seamount, *J. Geomag. Geoelectr., 40*, 715-728, 1988.

W. W. Sager, Departments of Oceanography, Geophysics, and Geodynamics Research Institute, Texas A&M University, College Station, TX 77843.

R. A. Duncan, College of Oceanography, Oregon State University, Corvallis, OR 97331.

D. W. Handschumacher, Naval Research Laboratory, Stennis Space Center, MS 39529.